Springer Collected Works in Mathematics

For further volumes:
http://www.springer.com/series/11104

New York, 1951

Harish-Chandra

Collected Papers I

1944 – 1954

Editor
Veeravalli Seshadri Varadarajan

Reprint of the 1984 Edition

 Springer

Author
Harish-Chandra (1923 Kanpur,
    India – 1983 Princeton, USA)

Editor
Veeravalli Seshadri Varadarajan
University of California
Los Angeles, USA

ISSN 2194-9875
ISBN 978-1-4939-2242-0   (Softcover)
      978-0-387-90782-6   (Hardcover)
DOI 10.1007/978-1-4899-7407-5
Springer New York Heidelberg Dordrecht London

Library of Congress Control Number: 2012954381

Printed on acid-free paper

Springer is part of Springer Science+Business Media (www.springer.com)

Harish-Chandra
Collected Papers

Volume I
(1944–1954)

Edited by V. S. Varadarajan

Springer-Verlag Berlin Heidelberg GmbH

Harish-Chandra
1923–1983

Editor

V. S. Varadarajan
Department of Mathematics
University of California
Los Angeles, CA 90024
U.S.A.

AMS Subject Classifications: 01A75, 22-XX

Library of Congress Cataloging in Publication Data
Harish-Chandra.
   Harish-Chandra collected papers.
   Contents: v. 1. 1944–1954—v. 2. 1955–1958—
[etc.]—v. 4. 1970–present.
   1. Mathematics—Collected works. I. Varadarajan,
V. S.   II. Title.
QA3.H294   1983   510   83-4828

9 8 7 6 5 4 3 2 1

ISBN 978-1-4899-7409-9   ISBN 978-1-4899-7407-5 (eBook)
DOI 10.1007/978-1-4899-7407-5

Table of Contents
Volume I

Bibliography of Harish-Chandra

[1944a] (with Bhabha, H. J.) On the theory of point-particles. *Proc. Royal Soc. A.* **183**, 134–141.

[1944b] On the removal of the infinite self-energies of point-particles. *Proc. Royal Soc. A.* **183**, 142–167.

[1945a] On the scattering of scalar mesons. *Proc. Indian Acad. Sci. Sect. A.* **21**, 135–146.

[1945b] Algebra of the Dirac-matrices. *Proc. Indian Acad. Sci. Sect. A.* **22**, 30–41.

[1946a] (with Bhabha, H. J.) On the fields and equations of motion of point particles. *Proc. Royal Soc. A.* **185**, 250–268.

[1946b] On the equations of motion of point particles. *Proc. Royal Soc. A.* **185**, 269–287.

[1946c] A note on the σ-symbols. *Proc. Indian Acad. Sci. Sect. A.* **23**, 152–163.

[1946d] The correspondence between the particle and the wave aspects of the meson and the photon. *Proc. Royal Soc. A.* **186**, 502–525.

[1947a] On the algebra of the meson matrices. *Proc. Camb. Phil. Soc.* **43**, 414–421.

[1947b] On relativistic wave equations. *Phys. Rev.* **71**, 793–805.

[1947c] Equations for particles of higher spin. Report of an International Conference on Fundamental particles and Low temperatures held at the Cavendish Laboratory, Cambridge (1946). Vol. I, Fundamental Particles, 185–188. The Physical Society, London.

[1947d] Infinite irreducible representations of the Lorentz group. *Proc. Royal Soc. A.* **189**, 372–401.

[1948a] Relativistic equations for elementary particles. *Proc. Royal Soc. A.* **192**, 195–218.

[1948b] Motion of an electron in the field of a magnetic pole. *Phys. Rev.* **74**, 883–887.

[1949a] Faithful representations of Lie algebras. *Ann. of Math.* **50**, 68–76.

[1949b] On representations of Lie algebras. *Ann. of Math.* **50**, 900–915.

[1950a] On the radical of a Lie algebra. *Proc. Amer. Math. Soc.* **1**, 14–17.

[1950b] On faithful representations of Lie groups. *Proc. Amer. Math. Soc.* **1**, 205–210.

[1950c] Lie algebras and the Tannaka duality theorem. *Ann. of Math.* **51**, 299–330.

[1951a] On some applications of the universal enveloping algebra of a semisimple Lie algebra. *Trans. Amer. Math. Soc.* **70**, 28–96.

[1951b] Representations of semisimple Lie groups on a Banach space. *Proc. Nat. Acad. Sci. U.S.A.* **37**, 170–173.

[1951c] Representations of semisimple Lie groups. II. *Proc. Nat. Acad. Sci. U.S.A.* **37**, 362–365.

[1951d] Representations of semisimple Lie groups. III. Characters. *Proc. Nat. Acad. Sci. U.S.A.* **37**, 366–369.

[1951e] Representations of semisimple Lie groups. IV. *Proc. Nat. Acad. Sci. U.S.A.* **37**, 691–694.

[1951f] Plancherel formula for complex semisimple Lie groups. *Proc. Nat. Acad. Sci. U.S.A.* **37**, 813–818.

[1952] Plancherel formula for the 2×2 real unimodular group. *Proc. Nat. Acad. Sci. U.S.A.* **38**, 337–342.

[1953] Representations of a semisimple Lie group on a Banach space. I. *Trans. Amer. Math. Soc.* **75**, 185–243.

[1954a] Representations of semisimple Lie groups. II. *Trans. Amer. Math. Soc.* **76**, 26–65.

[1954b] Representations of semisimple Lie groups. III. *Trans. Amer. Math. Soc.* **76**, 234–253.

[1954c] The Plancherel formula for complex semisimple Lie groups. *Trans. Amer. Math. Soc.* **76**, 485–528.

[1954d] On the Plancherel formula for the right K-invariant functions on a semisimple Lie group. *Proc. Nat. Acad. Sci. U.S.A.* **40**, 200–204.

[1954e] Representations of semisimple Lie groups. V. *Proc. Nat. Acad. Sci. U.S.A.* **40**, 1076–1077.

[1954f] Representations of semisimple Lie groups. VI. *Proc. Nat. Acad. Sci. U.S.A.* **40**, 1078–1080.

[1955a] Integrable and square-integrable representations of a semi-simple Lie group. *Proc. Nat. Acad. Sci. U.S.A.* **41**, 314–317.

[1955b] On the characters of a semisimple Lie group. *Bull. Amer. Math. Soc.* **61**, 389–396.

[1955c] Representations of semisimple Lie groups. IV. *Amer. J. of Math.* **77**, 743–777.

[1956a] Representations of semisimple Lie groups. V. *Amer. J. of Math.* **78**, 1–41.

[1956b] Representations of semisimple Lie groups. VI. Integrable and square-integrable representations. *Amer. J. of Math.* **78**, 564–628.

[1956c] The characters of semisimple Lie groups. *Trans. Amer. Math. Soc.* **83**, 98–163.

[1956d] On a lemma of F. Bruhat. *J. Math. Pures. Appl.* (9) **35**, 203–210.

[1956e] Invariant differential operators on a semisimple Lie algebra. *Proc. Nat. Acad. Sci. U.S.A.* **42**, 252–253.

[1956f] A formula for semisimple Lie groups. *Proc. Nat. Acad. Sci. U.S.A.* **42**, 538–540.

[1957a] Representations of semisimple Lie groups. *Proceedings of the International Congress of Mathematicians*, Amsterdam (1954). Vol. 1, 299–304. Erven P. Noordhoff N. V., Groningen, North Holland Publishing Company.

[1957b] Differential operators on a semisimple Lie algebra. *Amer. J. of Math.* **79**, 87–120.

[1957c] Fourier transforms on a semisimple Lie algebra. I. *Amer. J. of Math.* **79**, 193–257.

[1957d] Fourier transforms on a semisimple Lie algebra. II. *Amer. J. of Math.* **79**, 653–686.

[1957e] A formula for semisimple Lie groups. *Amer. J. of Math.* **79**, 733–760.

[1957f] Spherical functions on a semisimple Lie group. *Proc. Nat. Acad. Sci. U.S.A.* **43**, 408–409.

[1958a] Spherical functions on a semisimple Lie group. I. *Amer. J. of Math.* **80**, 241–310.

[1958b] Spherical functions on a semisimple Lie group. II. *Amer. J. of Math.* **80**, 553–613.

[1959a] Automorphic forms on a semisimple Lie group. *Proc. Nat. Acad. Sci. U.S.A.* **45**, 570–573.

[1959b] Some results on differential equations and their applications. *Proc. Nat. Acad. Sci. U.S.A.* **45**, 1763–1764.

BIBLIOGRAPHY OF HARISH-CHANDRA

[1960a] Some results on differential equations.

[1960a'] Supplement to "Some results on differential equations"

[1960b] Differential equations and semisimple Lie groups.

[1961] (with Borel, A.) Arithmetic subgroups of algebraic groups. *Bull. Amer. Math. Soc.* **67**, 579–583.

[1962] (with Borel, A.) Arithmetic subgroups of algebraic groups. *Ann. of Math.* **75**, 485–535.

[1963] Invariant eigendistributions on semisimple Lie groups. *Bull. Amer. Math. Soc.* **69**, 117–123.

[1964a] Invariant distributions on Lie algebras. *Amer. J. of Math.* **86**, 271–309.

[1964b] Invariant differential operators and distributions on a semi-simple Lie algebra. *Amer. J. of Math.* **86**, 534–564.

[1964c] Some results on an invariant integral on a semisimple Lie algebra. *Ann. of Math.* **80**, 551–593.

[1965a] Invariant eigendistributions on a semisimple Lie algebra. *Publ. Math. IHES* No. **27**, 5–54.

[1965b] Invariant eigendistributions on a semisimple Lie group. *Trans. Amer. Math. Soc.* **119**, 457–508.

[1965c] Discrete series for semisimple Lie groups. I. Construction of invariant eigendistributions. *Acta Math.* **113**, 241–318.

[1966a] Two theorems on semisimple Lie groups. *Ann. of Math.* **83**, 74–128.

[1966b] Discrete series for semisimple Lie groups. II. Explicit determination of the characters. *Acta. Math.* **116**, 1–111.

[1966c] Harmonic analysis on semisimple Lie groups. Some recent advances in the basic sciences. Vol. 1 (1962, 1963, 1964), 35–40. *Belfer Graduate School of Science Annual Science Conference Proceedings.* Edited by A. Gelbart. Belfer Graduate School of Science, Yeshiva University, New York, New York.

[1967] Characters of semisimple Lie groups. *Symposia on Theoretical Physics*, 4, 137–142. (Lectures presented at the 1965 Third Anniversary Symposium of the Institute of Mathematical Sciences, Madras, India. Edited by A. Ramakrishnan.) Plenum Press, New York.

[1968a] Harmonic analysis on semisimple Lie groups. *Proceedings of the International Congress of Mathematicians*, Moscow (1966), 89–94. "Mir", Moscow.

[1968b] Automorphic forms on semisimple Lie groups. Notes by J. G. M. Mars. *Lecture Notes in Mathematics*, No. 62, Springer-Verlag, Berlin–Heidelberg–New York.

[1970a] Some applications of the Schwartz space of a semisimple Lie group. *Lectures in Modern Analysis and Applications. II*, 1–7. Edited by C. T. Tamm. *Lecture Notes in Mathematics*. No. **140**, Springer-Verlag, Berlin–Heidelberg–New York.

[1970b] Eisenstein series over finite fields. Functional analysis and related fields, 76–88. *Proceedings of a Conference in honor of Professor Marshall Stone held at the University of Chicago, May 1968.* Edited by F. E. Browder. Springer-Verlag, Berlin–Heidelberg–New York.

[1970c] Harmonic analysis on semisimple Lie groups. *Bull. Amer. Math. Soc.* **76**, 529–551.

[1970d] Harmonic analysis on reductive p-adic groups. Notes by G. van Dijk. *Lecture Notes in Mathematics*, No. **162**, Springer-Verlag, Berlin–Heidelberg–New York.

[1972] On the theory of the Eisenstein integral. Conference on Harmonic Analysis, College Park, Maryland (1971), 123–149. Edited by D. Gulick and R. L. Lipsman. *Lecture Notes in Mathematics*, No. **266**, Springer-Verlag, Berlin–Heidelberg–New York.

[1973] Harmonic analysis on reductive p-adic groups. Harmonic analysis on homogeneous spaces, 167–192. Edited by C. C. Moore. *Proceedings of Symposia in Pure Mathe-*

matics, Vol. **XXVI**, Amer. Math. Soc., Providence, R. I., U.S.A.

[1975] Harmonic analysis on real reductive groups. I. The theory of the constant term. *J. of Functional Analysis* **19**, 104–204.

[1976a] Harmonic analysis on real reductive groups. II. Wave packets in the Schwartz space. *Inventiones Math.* **36**, 1–55.

[1976b] Harmonic analysis on real reductive groups. III. The Maass–Selberg relations and the Plancherel formula. *Ann. of Math.* **104**, 117–201.

[1977a] The characters of reductive p-adic groups. *Contributions to Algebra*, 175–182. A collection of papers dedicated to Ellis Kolchin. Edited by H. Bass, P. J. Cassidy, and J. Kovacic. Academic Press, New York–San Francisco–London.

[1977b] The Plancherel formula for reductive p-adic groups.

[1977b'] Corrections to "The Plancherel formula for reductive p-adic groups"

[1978] Admissible invariant distributions on reductive p-adic groups. Lie theories and their applications, 281–347. *Proceedings of the 1977 annual seminar of the Canadian Mathematical Congress.* Edited by W. Rossmann. Queen's papers in Pure and Applied Math. No. **48**. (Editors: A. J. Coleman and P. Ribenboim). Kingston, Ontario.

[1980] A submersion principle and its applications. Geometry and Analysis, 95–102. Papers dedicated to the memory of V. K. Patodi, Indian Academy of Sciences, Bangalore, and the Tate Institute of Fundamental Research, Bombay.

[1983] Supertempered distributions on real reductive groups. *Studies in Applied Mathematics, Advances in Mathematics, Supplementary Studies Series*, Vol. 8, Academic Press Inc., pp. 139–153, edited by Victor Guillemin.

Abbreviations Used in the Bibliography

Proc. Royal Soc. A. ...	Proceedings of the Royal Society, A...
Proc. Indian Acad. Sci. Sect. A	Proceedings of the Indian Academy of Sciences, Section A
Proc. Camb. Phil. Soc.	Proceedings of the Cambridge Philosophical Society
Phys. Rev.	The Physical Review
Ann. of Math.	Annals of Mathematics
Proc. Amer. Math. Soc.	Proceedings of the American Mathematical Society
Trans. Amer. Math. Soc.	Transactions of the American Mathematical Society
Proc. Nat. Acad. Sci. U.S.A.	Proceedings of the National Academy of Sciences of the United States of America
Bull. Amer. Math. Soc.	Bulletin of the American Mathematical Society
Amer. J. of Math.	American Journal of Mathematics
J. Math. Pures Appl.	Journal de Mathématiques Pures et Appliquées
Publ. Math. IHES	Publications Mathématiques, Institut des Hautes Études Scientifiques
Acta Math.	Acta Mathematica
J. Functional Analysis	Journal of Functional Analysis
Inventiones Math.	Inventiones Mathematicae

Preface

These volumes of the Collected Papers of Harish-Chandra are being brought out in response to a widespread feeling in the mathematical community that they would immensely benefit scholars and researchworkers, especially those in analysis, representation theory, arithmetic, mathematical physics, and other related areas. It is hoped that in addition to making his contributions more accessible by collecting them in one place, these volumes would help focus renewed attention on his ideas and methods as well as lend additional perspective to them.

The papers are arranged chronologically. Harish-Chandra is still an extraordinarily productive mathematician, and so I feel that there is no need to impose any structure on his output (past and current) other than that coming from the evolution of his own thinking. It was also decided to divide the papers into four volumes so that each volume would have a reasonable size. However, I have attempted to make the lines of division in such a way that each volume has a certain unity within itself.

It is impossible to mention all the individuals and organisations that helped me with this project. Nevertheless, I do want to record my special thanks to Phyllis Parris, for typing the two handwritten manuscripts on differential equations; to Walter Kaufmann-Bühler, for his constant enthusiasm and encouragement; and to Elise Oranges for her tireless work in producing the final copy. In addition I am extremely grateful to Roger Howe and to Nolan Wallach for agreeing to write commentaries on selected parts of Harish-Chandra's work, and doing it within the time frame that I gave them. Finally, I want to thank Mrs. Lily Harish-Chandra for providing me with the photographs of Harish-Chandra that appear in these volumes.

Pacific Palisades, California
May, 1983

V. S. VARADARAJAN

xv

Biographical Note

Harish-Chandra was born on October 11, 1923 in Kanpur, India. He received the M.Sc degree from Allahabad University in 1943 and spent the years 1943–1945 at the Indian Institute of Science, Bangalore, India. He went to Cambridge University in England as a graduate student of Dirac in 1945 and received his Ph.D from there in 1947. He was at Columbia University, New York, from 1950 to 1963. In 1963 he became a permanent member of the Institute for Advanced Study at Princeton, New Jersey, where he has been the IBM von Neumann Professor from 1970 onwards.

He was a Guggenheim Fellow from 1957 to 1958 and a Sloan Fellow from 1961 to 1963. He was elected a Fellow of the Royal Society in 1973 and a member of the National Academy of Sciences of the U.S.A. in 1981. He was awarded honorary doctorates by Delhi University in 1973 and by Yale University in 1981.

He and his wife live in Princeton. They have two daughters. Although he has not returned to India except for brief visits, he has always been profoundly concerned with India and Indian science. He remains a figure of great inspiration in his native land.

Postscript

Harish-Chandra is no more. He suffered a fatal heart attack while out for a walk on the evening of Sunday, October 16, 1983. He survives in his work, which is a faithful reflection of his personality—lofty, intense, uncompromising.

Princeton
October 19, 1983

V. S. VARADARAJAN

Introduction

V. S. VARADARAJAN

> *Bottom*: Masters, I am to discourse wonders:
> but ask me not what; for if I tell you, I am
> no true Athenian. I will tell you every-
> thing, right as it fell out.
>
> —Shakespeare, *A Midsummer Night's
> Dream*, Act Four, Scene II.

§0. The first scientific papers of Harish-Chandra were in theoretical physics. However, his interests shifted to mathematics rather early in his career. His prolonged and intense preoccupation with group representations began in the late 1940's when, at the suggestion of Dirac, he started investigating the infinite dimensional unitary representations of the Lorentz group [1947d]. Since then he has erected, almost singlehandedly, a monumental theory of harmonic analysis on reductive groups and their homogeneous spaces. His work is a profound synthesis of algebra, geometry, and analysis. The great force and continued resonance of his ideas have inspired a generation of mathematicians. The singlemindedness and courage with which he has pursued his goals, the beauty of his results, the power and originality of his methods, and the ultimate simplicity of his philosophy, as well as the conviction that sustains it, compel our admiration. There can be no doubt that his achievement is one of the greatest in mathematics in our time.

This introduction attempts to give an overview of this work. I have sketched, in an impressionistic way, the main lines of it in §2; the remaining sections contain additional detail on selected parts of his work so that one of his great achievements —the harmonic analysis of $L^2(G)$ for a real reductive group G—may come into clearer focus. I hope that this account, although incomplete and imperfect, may still be of value as a rough guide to the prospective reader.

§1. The early sources of motivation for representation theory and harmonic analysis were the classical theory of Fourier series and integrals and the theory of linear representations of finite groups. After the rise of topology and functional analysis in the early decades of this century, far-reaching generalizations of these classical themes began to emerge.

xvii

For a locally compact abelian group G let $\hat{G}$ be the group of characters of G. $\hat{G}$ is also locally compact abelian and the Pontryagin–van Kampen theory pairs G and $\hat{G}$ in a canonical duality. Given any function $f \in L^1(G)$ its Fourier transform is the function $\hat{f}$ on $\hat{G}$ given by

$$\hat{f}(\hat{x}) = \int_G f(x)\langle x, \hat{x}\rangle\, dx \qquad (\hat{x} \in \hat{G},\, dx \text{ a Haar measure})$$

and $f \mapsto \hat{f}$ is a homomorphism of the convolution algebra $L^1(G)$ into the algebra of continuous functions on $\hat{G}$ that vanish at infinity. The Fourier transform can also be introduced on $L^2(G)$; it is then a unitary isomorphism of $L^2(G)$ with $L^2(\hat{G})$ where the Haar measure $d\hat{x}$ on $\hat{G}$ is uniquely determined by dx (the measure dual to dx). In particular we have the *Plancherel formula*

$$\int_G |f(x)|^2\, dx = \int_{\hat{G}} |\hat{f}(\hat{x})|^2\, d\hat{x} \qquad (f \in L^2(G))$$

and the *inversion formula*

$$f(x) = \int_{\hat{G}} \hat{f}(\hat{x})\langle x, \hat{x}\rangle^{\mathrm{conj}}\, d\hat{x} \qquad (f \in L^1(G), \hat{f} \in L^1(\hat{G}))$$

which generalize the corresponding formulae from classical Fourier analysis on $\mathbb{R}^n$ ([W 1] [Lo] [Bl]).

The classical Poisson Summation formula can also be generalized to this setting. Let $\Gamma \subset G$ be a lattice in G, i.e., a discrete subgroup of G such that G/Γ is compact. Then $\Gamma^\vee$, the annihilator of Γ, is also a lattice in $\hat{G}$. If Fourier transforms are defined using the Haar measure that gives the measure 1 to G/Γ, then the Poisson formula takes the form

$$\sum_{\gamma \in \Gamma} f(\gamma) = \sum_{\gamma^\vee \in \Gamma^\vee} \hat{f}(\gamma^\vee)$$

where f and $\hat{f}$ are suitably restricted ([W 2]).

The Poisson formula has always been a powerful tool in number theory. The importance of Fourier analysis on general locally compact abelian groups became very clear with the work of Tate who, at the suggestion of Artin, used the general Poisson formula in the setting provided by the adele group of a number field to derive the analytic theory of the Hecke L-series attached to a Grössencharakter ([T] [W 2]).

Suppose now that G is compact. Let dx be the normalized Haar measure on G so that $\int_G dx = 1$. Let $\hat{G}$ be the set of equivalence classes of irreducible (finite dimensional) unitary representations of G. The general facts concerning linear representations of finite groups extend essentially unchanged to G. Thus if $\pi \in \omega \in \hat{G}$ and ϕ, ψ are unit vectors in the representation space of π,

$$\int_G |(\pi(x)\phi, \psi)|^2\, dx = \frac{1}{d(\omega)}$$

where $d(\omega)$ is the degree of ω. Moreover, for $\omega \in \hat{G}$ let Θ_ω be the corresponding character; then

$$\int_G \Theta_\omega \Theta_{\omega'}^{\mathrm{conj}}\, dx = \delta_{\omega\omega'} \qquad (\omega, \omega' \in \hat{G}).$$

The fundamental result is the Peter–Weyl theorem. It asserts the completeness of the irreducible representations and can be stated as the completeness formula

$$\int_G |f|^2\, dx = \sum_{\omega \in \hat{G}} d(\omega)\|\pi_\omega(f)\|_2^2$$

where $f \in L^2(G)$, $\pi_\omega \in \omega$, and $\|\cdot\|_2$ is the Hilbert–Schmidt norm. The measure on $\hat{G}$, which assigns the mass $d(\omega)$ to ω, may thus be viewed as the *Plancherel measure* of G.

Let G be a compact Lie group. For $f \in L^2(G)$ let $\hat{f}$ be its Fourier transform, defined as the function on $\hat{G}$ given by

$$\hat{f}(\omega) = \int_G f\Theta_\omega^{\text{conj}}\, dx \qquad (\omega \in \hat{G}).$$

The completeness formula can then be restated as

$$f(e) = \sum_{\omega \in \hat{G}} d(\omega)\hat{f}(\omega) \qquad (f \in C^\infty(G)).$$

In other words, we have the relation

$$\delta = \sum_{\omega \in \hat{G}} d(\omega)\Theta_\omega$$

in the space of *distributions* on the manifold G, δ being the delta function at the identity element of G.

It is natural to ask for an explicit determination of the irreducible representations of special groups such as the simple Lie groups which had been classified in the late 1890's by Killing and Élie Cartan. The work of Cartan and Hermann Weyl in the 1920's gave a complete solution of this problem. For Cartan it was a question of determining the irreducible finite dimensional representations of the (complex) simple *Lie algebras*; he obtained them as highest weight modules. Weyl had a more global view and worked with the *compact simple Lie groups*. He discovered beautiful formulae for the characters and dimensions of the irreducible representations. Moreover his discovery of the existence of a compact form of a complex simple Lie algebra ("unitarian trick") showed that the representation theory of a complex simple Lie algebra is essentially the same as that of the (simply connected) compact form associated to it, thus unifying his theory with that of Cartan ([We 2]; see also [V 1]).

The notion of distributions, which came into widespread use with the work of Laurent Schwartz [S], led to a deeper understanding of the notion of Fourier transform and to a great extension of the domain of its applicability. In the language of distributions the Plancherel formula for $\mathbb{R}^n$ became the "expansion" of the delta function at the origin in terms of the characters:

$$\delta = (2\pi)^{-n/2}\int_{\mathbb{R}^n} e(\lambda)\, d\lambda \qquad \left(e(\lambda)(x) = e^{i\lambda\cdot x}\right).$$

Suppose now that G is a locally compact group, unimodular and second countable, which is neither compact nor abelian. Let $\hat{G}$ be the set of equivalence classes of its irreducible unitary representations (finite or infinite dimensional). There was no systematic theory of representations of such groups in the 1930's. But in Quantum Physics unitary representations of the various symmetry groups had already begun to play a prominent role. This is because the covariance properties of a system with a (connected) group of symmetries are expressed by means of a unitary representation of the group in question or an extension of it. In relativistic quantum physics this meant the inhomogeneous Lorentz group; its physically significant irreducible unitary representations (both single and double valued in the physicists' terminology) were classified by Wigner in a famous memoir in 1939 [Wi].

The situation changed completely in the 1940's. For, by then, the epoch-making papers of von Neumann (by himself and with Murray) on operator algebras in Hilbert spaces had appeared [MvN] [vN]. This theory provided the framework for the study of infinite dimensional unitary representations; in addition it revealed that new phenomena such as type II and type III algebras, direct integral decompositions, and so on, would have to be taken into account.

The systematic beginnings of unitary representation theory of general locally compact groups go back to 1943 when Gel'fand and Raikov proved that every such group had enough irreducible unitary representations to separate the points of the group [GR]. The subject developed rapidly in the late 1940's and early 1950's mainly at the hands of Godement, Mackey, Mautner, and Segal. Their point of view was functional analytic, and their main tool was the theory of operator algebras. Their work led, among other things, to a general decomposition theory of unitary representations, abstract Plancherel theorems, and to the recognition that a reasonable harmonic analysis can be expected only for type I groups, namely groups whose unitary representations were all of type I [Ma 1] [Ma 2] [Ma 3]. It was also in this period that the papers of Mackey on systems of imprimitivity and induced representations appeared, in which he extended the classical Frobenius theory to all separable locally compact groups. Mackey's work, which went far beyond what is suggested in these remarks, has proved to be very influential in the representation theory of nilpotent and solvable groups (cf. [AM] [Mo]).

These developments led to the following formulation of the central problem of harmonic analysis: given a separable locally compact group G acting transitively on a space X with an invariant measure μ, and an irreducible unitary representation τ of the stabilizer in G of a given point of X, decompose into irreducible constituents the representation of G induced by τ. If τ is the trivial representation, the induced representation is just the natural representation of G in $L^2(X, \mu)$; if $G = X$ and the action is by left translations, the problem is the decomposition of the left regular representation of G. The decomposition may not be entirely discrete so that it has to be understood as a direct integral.

In this generality this problem is hopeless because the category of all separable locally compact groups is simply too vast for classical harmonic analysis to be anything more than a very uncertain guide. New examples were clearly needed, and it was natural to look to the simple Lie groups once again for this purpose. The papers of Gel'fand and Neumark [GN 1] and Bargmann [B], dealing with SL(2, $\mathbb{C}$) and SL(2, $\mathbb{R}$), and published independently of each other in 1947, mark the beginning of the subject of infinite dimensional representations of real semisimple Lie groups.

In [GN 1] Gel'fand and Neumark determined all the irreducible unitary representations of SL(2,$\mathbb{C}$) upto equivalence, discovered an explicit Plancherel formula, and proved that any unitary representation of SL(2,$\mathbb{C}$) could be decomposed as a direct integral of irreducible representations in a natural way. They discovered the surprising fact that this group had *additional* irreducible unitary representations which played no role in the harmonic analysis of $L^2(G)$. They called these the representations of the *supplementary series*. They followed this up with a remarkable and seminal series of papers in which they extended their theory to SL(n,$\mathbb{C}$) and, to a lesser extent, to the other *complex classical groups*. This work, which was presented in detail in their beautiful monograph [GN 2], completely changed the perspective of the theory.

The group SL(n,$\mathbb{C}$) acts transitively on a number of projective varieties such as the Grassmannians and the various flag spaces. The spaces of *holomorphic* sections of the complex analytic line bundles on these homogeneous spaces carry the finite dimensional complex analytic representations of SL(n,$\mathbb{C}$). With remarkable insight Gel'fand and Neumark realized that the *smooth* sections of the *real analytic* line bundles on the same homogeneous spaces would transform (roughly speaking) according to the *infinite dimensional* irreducible representations of the group. These representations would be unitary under appropriate conditions. In this way, one had a single scheme for *all* irreducible representations, which contained the finite dimensional as well as the unitary ones as special cases. For example, $G = \mathrm{SL}(n,\mathbb{C})$ acts transitively on the space of all flags (e_m) $(1 \leqslant m \leqslant n,\ e_1 \subset e_2 \subset \cdots \subset e_n,\ e_m$ is an m-dimensional linear subspace of $\mathbb{C}^n$). Then $X \approx G/B$ where B is the subgroup of the upper triangular matrices. Gel'fand and Neumark proved that the unitary representations of G induced by the unitary one dimensional representations of B, which they called the *principal series* of representations, are all irreducible, and that the regular representation of the group is a direct integral of them. They extended to SL(n,$\mathbb{C}$) (as well as to the other complex classical groups) the construction of the supplementary series.

Although these representations are all infinite dimensional, Gel'fand and Neumark succeeded in associating *characters* to them. If ξ is any one dimensional representation of B and π_ξ the representation of G induced by ξ, the operator

$$\pi_\xi(x) = \int_G x(g)\pi_\xi(g)\,dg$$

is of trace class if x is suitably restricted, and there is a (unique) class function θ_ξ on G such that

$$\mathrm{tr}\big(\pi_\xi(x)\big) = \int_G x(g)\theta_\xi(g)\,dg.$$

It is natural to call θ_ξ the *character* of π_ξ. They evaluated θ_ξ on the subgroup D of diagonal matrices and found for it a formula of the same general structure as Weyl's character formula. In terms of these characters they obtained the *Plancherel formula* for $SL(n, \mathbb{C})$ in the form

$$x(e) = \int_D \hat{x}(\xi) \omega(\xi) \, d\xi \qquad (x \in C_c^\infty(G))$$

where

$$\hat{x}(\xi) = \text{trace}(\pi_\xi(x)) \qquad \text{(Fourier transform!)}$$

and $\omega(\xi)$ is an explicitly determined function $\geqslant 0$. If we rewrite this in the form

$$\delta = \int \omega(\xi) \theta_\xi \, d\xi,$$

its analogy with the Plancherel formula for $\mathbb{R}^n$ and compact Lie groups becomes clear; $\omega(\xi) \, d\xi$ is the Plancherel measure.

Let me now turn to the work of Bargmann which treated both $SL(2, \mathbb{C})$ and $SL(2, \mathbb{R})$. So far as $SL(2, \mathbb{C})$ was concerned the results were essentially the same as those in [GN 1]; however $SL(2, \mathbb{R})$ presented new phenomena. In addition to the principal and supplementary series, defined essentially as for the complex unimodular group, there now appeared a discretely parametrized series of irreducible unitary representations characterized by the remarkable fact that their matrix elements were already square integrable on the group manifold. This property meant that these representations would occur as *discrete direct summands* of the regular representation. The property of having square integrable matrix coefficients makes sense for irreducible unitary representations of arbitrary separable locally compact unimodular groups and is equivalent to their occurring as direct summands of the regular representation; they form the so-called *discrete series* of the group, and the set of the corresponding equivalence classes is denoted by $\hat{G}_d$. Their appearance in the case of $SL(2, \mathbb{R})$ meant that the decomposition of its regular representation had both discrete and continuous parts.

A highlight of the paper of Bargmann is the emergence of another new theme in our subject. He viewed the matrix elements of the irreducible representations as eigenfunctions of the Casimir differential operator defined on the group manifold (originally introduced by the physicist Casimir for $SU(2)$). In this way the problem of harmonic analysis in $L^2(G)$ became one of eigenfunction expansions for the Casimir operator. The matrix elements have simple transformation properties relative to the action (from both left and right) of the rotation subgroup of $SL(2, \mathbb{R})$; and so they are essentially determined by their values at the elements $\begin{pmatrix} e^t & 0 \\ 0 & e^{-t} \end{pmatrix} (t > 0)$ of the group. Thus they may be viewed as functions on the half-line of positive reals; moreover the action of the Casimir operator coincides with that of a second order differential operator on this half-line. The spectral theory of such differential operators goes back to Hermann Weyl [We 1]. Bargmann used Weyl's theory to prove the completeness theorem, i.e., the fact that the linear space, spanned by the matrix elements of the discrete series and the "wave packets" of the matrix elements of the principal series, is dense in $L^2(G)$.

For the matrix elements of the discrete series Bargmann discovered remarkable generalizations of the Schur orthogonality relations. For the principal series the corresponding problem is the explicit determination of the Plancherel measure.

Bargmann found a beautiful expression for it in terms of the leading coefficients of the asymptotic expansions of the matrix elements.

The papers of Gel'fand and Neumark and Bargmann form a great watershed in our subject. Their work showed clearly, at least in retrospect, that to do harmonic analysis on real semisimple Lie groups it is absolutely necessary to operate at a level that allowed for a true interaction among the algebraic, geometric, and analytic properties of these groups.

This was the state of affairs in the early 1950's when Harish-Chandra entered the field. His methods combined in a deep and original manner the infinitesimal and global aspects of these problems, and, through an astonishing sequence of papers spanning almost three decades, led to a profound understanding of harmonic analysis of real reductive groups. Sometime in the 1960's he realized that his work contained deep analogies with the spectral theory of automorphic forms that was then being built through the efforts of Selberg, Gel'fand (and his collaborators), and Langlands. His exploration of these analogies has led in recent years to significant results in the harmonic analysis of reductive groups over local, global, and even finite fields, and has revealed the essential unity of the structure of Fourier analysis in these diverse contexts.

§2. Before starting with a description of Harish-Chandra's work we introduce a few definitions and some notation. G is a real semisimple Lie group, connected and with finite center; K is a maximal compact subgroup of G; and θ, the involution that fixes K pointwise; $\mathfrak{g} = \mathrm{Lie}(G)$, $\mathfrak{k} = \mathrm{Lie}(K)$. For any real Lie algebra $\mathfrak{m}$, $U(\mathfrak{m})$ is the universal enveloping algebra of the complexification $\mathfrak{m}_c$ of $\mathfrak{m}$.

Manifolds are smooth and second countable. They need not be connected but all the connected components will have the same dimension. For any manifold M, $C^\infty(M)$ (resp. $C_c^\infty(M)$) is the space of complex valued functions on M that are smooth (resp. smooth with compact support). If ω is a volume element (= exterior differential form of maximal degree which is nowhere zero) on M, then given any smooth differential operator D on M its transpose D^t is the differential operator such that $\int_M Df \cdot g\omega = \int_M f \cdot D^t g\omega$ for all $f, g \in C_c^\infty(M)$. If T is a distribution on M the distribution DT is defined by $\langle DT, f \rangle = \langle T, D^t f \rangle$ ($f \in C_c^\infty(M)$). Any function F on M that is locally integrable with respect to ω will be usually identified with the distribution $f \mapsto \int_M Ff\omega$. The actions of differential operators on smooth functions and distributions are compatible under this identification. If M is a Lie group we shall always choose ω to be a left invariant volume element. In this case $U(\mathfrak{m}) = U(\mathfrak{m})^t$ is canonically isomorphic to the algebra of all left invariant differential operators on M; $X^t = -X$ for $X \in \mathfrak{m}$.

We write $\mathfrak{Z}$ for the center of $U(\mathfrak{g})$. Then $\mathfrak{Z}$ is canonically isomorphic to the algebra of all differential operators on G that are both left and right invariant. $\mathfrak{Z}$ is a polynomial algebra in l indeterminates where l is the rank of G.

The centralizer in G of a Cartan subalgebra of $\mathfrak{g}$ is called a Cartan subgroup. Cartan subgroups are parameter spaces for conjugacy classes, and so are absolutely fundamental objects in character theory which deals with class functions and distributions. They are abelian if G is linear and are maximal tori if G is compact. A compact Cartan subgroup is always connected and is in fact a torus. An element $x \in G$ is regular if its centralizer in $\mathfrak{g}$ is a Cartan subalgebra. If $l = rk(G)$, t is an indeterminate and $D(x)$ is the coefficient of t^l in $\det(\mathrm{Ad}(x) - 1 + t)$, x is regular if and only if $D(x) \neq 0$. The set of regular points is denoted by G'; it is a dense open subset of G. D is often called the discriminant of G.

Parabolic subgroups (psgrps) are normalizers in G of parabolic subalgebras of $\mathfrak{g}$; a subalgebra $\mathfrak{p}$ of $\mathfrak{g}$ is parabolic if $\mathfrak{p}_c$ contains a Borel subalgebra of $\mathfrak{g}_c$. Any psgrp P has a canonical decomposition $P = MAN \simeq M \times A \times N$. Here N is the unipotent radical of P, namely the maximal normal subgroup of P that consists entirely of unipotent elements; $M_1 = MA$ is reductive and is defined as $P \cap \theta(P)$, so that $P = M_1 N$ is essentially the Levi decomposition of P, while A is the maximal $\mathbb{R}$-split* subgroup lying in the center of M_1. The subgroup M is the intersection of the kernels of all the continuous homomorphisms of M_1 into the positive reals. The decomposition $P = MAN$ is called the Langlands decomposition of P. The parabolic subgroups are needed for defining the various series of irreducible unitary representations of G; moreover the formulation of many of the asymptotic questions in harmonic analysis depend in a fundamental way on the geometry of the set of all parabolic subgroups.

Harish-Chandra's early work was largely devoted to the development of algebraic methods for studying infinite dimensional representations of real semisimple Lie groups. These methods led him to discover the fundamental facts about unitary representations of G. Among these are the results that assert that G is a type I group and that in every irreducible unitary representation of G the irreducible representations of K occur only with finite multiplicities.

Let us define a $(\mathfrak{g}, K)$-module to be a module V for $\mathfrak{g}$ which has a compatible structure as a K-module such that V is the algebraic sum of finite dimensional K-modules. V is unitary if there is a positive definite scalar product $(\, , \,)$ for V such that $(Xu, v) = -(u, Xv)$ for all $u, v \in V$, $X \in \mathfrak{g}$. One of Harish-Chandra's major discoveries was that the space $V(\pi)$ of K-finite vectors in an irreducible unitary representation π is an irreducible $(\mathfrak{g}, K)$-module and that the assignment $\pi \mapsto V(\pi)$ induces a bijection of $\hat{G}$ with the set of isomorphism classes of irreducible $(\mathfrak{g}, K)$-modules that are unitary. He also proved that any irreducible $(\mathfrak{g}, K)$-module is of the form $V(\pi)$ for some irreducible representation π acting on a Hilbert space and possessing an infinitesimal character, thus showing that in the notion of a $(\mathfrak{g}, K)$-module which is irreducible we have the correct formulation of the vague idea of an "arbitrary" irreducible representation of G.

In spite of the promise and power of the algebraic method he never pursued it beyond his initial investigations and was content to use his early work as the point of departure for the analytical theory.

The basic step in the analytical theory is the introduction of characters. The algebraic treatment had already shown that the multiplicities of the irreducible representations of K that occur in an irreducible unitary representation of G do not exceed a constant multiple of the degrees of the representations of K. It follows from this that for any function $f \in C_c^\infty(G)$ the operator

$$\pi(f) = \int_G f(x) \pi(x) \, dx$$

---

*This means that $\mathrm{Ad}(a)$ is diagonalizable over $\mathbb{R}$ for all $a \in A$.

is of trace class, and the linear functional

$$\Theta_\pi : f \mapsto \mathrm{tr}(\pi(f))$$

is a distribution on the group G. Θ_π depends only on the equivalence class ω of π; it is denoted by Θ_ω and is called the character of π (or ω). It is an invariant distribution, i.e., a distribution that is invariant under all inner automorphisms of G. It determines ω completely. If $(\phi_i)_{i \geqslant 1}$ is an orthonormal basis of K-finite vectors for π and f_i is the matrix element defined by ϕ_i, i.e., $f_i(x) = (\pi(x)\phi_i, \phi_i)$ $(x \in G)$,

$$\Theta_\omega = \sum_{i \geqslant 1} f_i,$$

this equation being understood in the sense of distributions, i.e., for all $u \in C_c^\infty(G)$,

$$\Theta_\omega(u) = \sum_{i \geqslant 1} \int_G f_i u \, dx.$$

Since any matrix element satisfies the differential equation

$$zf = \chi_\omega(z)f \qquad (z \in \mathfrak{Z})$$

where χ_ω is the infinitesimal character of ω, it follows that

$$z\Theta_\omega = \chi_\omega(z)\Theta_\omega \qquad (z \in \mathfrak{Z}).$$

Here we recall that the infinitesimal character of ω is the homomorphism according to which $\mathfrak{Z}$ acts (by Schur's lemma) on the space $V(\pi)$. In other words, Θ_ω is an *invariant eigendistribution* on G.

Using the notion of characters one may formulate the problem of harmonic analysis on G as follows (cf. the introductory remarks to [1957e] [1970c]): given any invariant distribution T on G, express T as a "linear combination" of the irreducible characters. For $f \in C_c^\infty(G)$ define its *Fourier transform* $\hat{f}$ to be the function on $\hat{G}$ given by

$$\hat{f}(\omega) = \Theta_\omega(f) \qquad (\omega \in \hat{G}).$$

The problem is then to define a "distribution" $\hat{T}$ on $\hat{G}$ such that

$$T(f) = \hat{T}(\hat{f}) \qquad (f \in C_c^\infty(G)).$$

$\hat{T}$ will be the "Fourier transform" of T. If $T = \delta$, the delta function at the identity element of G, the determination of $\hat{\delta}$ is the problem of finding the explicit Plancherel formula for G; $\hat{\delta}$ will have to be a nonnegative measure, the Plancherel measure. The derivation by Gel'fand and Neumark of the Plancherel formula for $SL(n, \mathbb{C})$, Harish-Chandra's generalization of this to all complex semisimple Lie groups [1951f] [1954c], and his treatment of the case of $SL(2, \mathbb{R})$ [1952], and the work of Gel'fand and Graev on $SL(n, \mathbb{R})$ [GG 1] [GG 2], may be regarded as initial evidence for this view.

A second approach to harmonic analysis is through *eigenfunction expansions*; this goes back to Bargmann as we saw in §1. Let $H \subset G$ be a closed unimodular subgroup, $X = G/H$, and let $\mathcal{H} = L^2(X)$ be the Hilbert space of functions square integrable with respect to the invariant measure on X. Let λ be the natural representation of G in $\mathcal{H}$. If $\mathcal{H}^\infty$ is the space of differentiable vectors in $\mathcal{H}$ for λ, we have a representation $a \mapsto \lambda(a)$ of $U(\mathfrak{g})$ in $\mathcal{H}^\infty$. If $z \in \mathfrak{Z}$ is Hermitian, i.e., $(z^{\mathrm{conj}})^t = z$, the operator $\lambda(z)$ is essentially self adjoint on $\mathcal{H}^\infty$ and is invariant under G. So, in the first approximation, one may view the problem of decomposition of λ as an eigenfunction expansion problem for the commuting algebra of differential operators $\lambda(z)$, $z \in \mathfrak{Z}$ (see the introductory remarks to [1968b]). The eigenspaces are parametrized by the points of the spectrum of the commutative algebra $\mathfrak{Z}$, i.e., by the homomorphisms $\chi \colon \mathfrak{Z} \to \mathbb{C}$. For a given χ, the corresponding eigenspace would be a G-module which, typically, would be infinite dimensional. In view of the relationship between the eigenvalue problem and the representation theoretic question, we should expect that only those eigenfunctions will contribute to the spectral expansion in $L^2(X)$ whose χ are the infinitesimal characters of irreducible *unitary* representations of G (not necessarily all of them either).

The basic example in this approach is the case $X = G$, viewed as a homogeneous space for G (left action) or $G \times G$ (two-sided action). A variant of this in which the eigenspaces are finite dimensional is obtained as follows. Let U be a finite dimensional Hilbert space which is a bimodule for K, i.e., which carries a unitary representation τ of $K \times K^{\vee}$, $K^{\vee}$ being the group opposite to K. The Hilbert space is now the space $L^2(G, \tau)$ of functions $f \colon G \to U$ which are square integrable and satisfy

$$f(k_1 x k_2) = k_1 f(x) k_2$$

for all $k_1, k_2 \in K$ and almost all $x \in G$. Such functions are called *τ-spherical*; if $U = \mathbb{C}$ and $K \times K^{\vee}$ acts trivially, one speaks of *spherical* or *strictly spherical* functions. The terminology reminds us that we have a generalization of the classical theory of spherical harmonics. If χ is a homomorphism of $\mathfrak{Z}$ into $\mathbb{C}$ and $f \in L^2(G, \tau)$ is such that $zf = \chi(z)f$ for all $z \in \mathfrak{Z}$, f is C^∞ (even analytic), and the space of all such f is of finite dimension (cf. [1960b]). One of Harish-Chandra's basic results is that the eigenfunctions needed for the L^2-expansion are precisely those that satisfy the so-called weak inequality ([1970c], p. 539).

The theory of automorphic forms gives another example of this point of view. The classical automorphic forms on the Poincaré half-plane may be interpreted as eigenfunctions for the Laplace–Beltrami operator. They can then be generalized to the context of nonanalytic forms and higher dimensional symmetric spaces, as was done by Maass [M] and Selberg [Se 1]. Following the work of Borel and Harish-Chandra [1962] on the structure of arithmetic subgroups of reductive groups and the associated homogeneous spaces, Gel'fand [G], Selberg [Se 1] [Se 2], and Langlands [L 1] constructed a general theory of automorphic forms. They approached the fundamental problem as the spectral theory of the invariant differential operators in $L^2(G/\Gamma)$, G being the group of real points of a reductive group defined over $\mathbb{Q}$, and Γ is the associated arithmetic subgroup of G. The classical Eisenstein series and its generalizations are examples of eigenfunctions of these differential operators. It is interesting to note that the homomorphisms χ in the spectrum of $\mathfrak{Z}$ that correspond to these eigenfunctions do not come from unitary representations, and so one must make an analytic continuation in χ in order to get the eigenfunctions needed for the spectral theory (cf. [1968b]).

Whichever point of view is adopted, it is not possible to go far without a good supply of irreducible unitary representations. The known results for complex groups and $SL(n,\mathbb{R})$ seemed to suggest that each conjugacy class of Cartan subgroups of G gives rise to a "series" of irreducible unitary representations, and that the Plancherel measure is a *disjoint sum* of the measures contributed by each of these series (see [1957a] [1954e]). As a partial justification of this principle Harish-Chandra sketched in [1954e] a method of constructing a series of irreducible unitary representations associated to any Cartan subgroup L. Given L, one can select a parabolic subgroup $P = MAN$ such that A is the maximal $\mathbb{R}$-split* component of L. If η is an irreducible unitary representation of M and $v \in \hat{A}$, the unitary representation $\pi_{\eta,v}$ of G is defined as the representation induced by the representation $m \cdot a \cdot n \mapsto v(a)\eta(m)$ of P. The $\pi_{\eta,v}$ depend only on the conjugacy class of L and are discretely decomposable into irreducible subrepresentations. If L is chosen so that $\dim(A)$ is maximum, P is a minimal parabolic subgroup, M is compact, and what we have is called the *principal series*; for complex G it *is* the principal series of Gel'fand and Neumark. Bruhat's work [Br 1] had shown that for minimal P the $\pi_{\eta,v}$ are irreducible for v in "general position." So one could hope that this would be so in the case of arbitrary L also.

If P is not minimal, M is not compact and so we are still faced with the problem of determining $\hat{M}$. Harish-Chandra's remarkable idea, which is implicit but very clear in [1957a] and [1954e], was that if we restrict η to vary always over the *discrete series* of M, the various series arising from a complete system of mutually nonconjugate Cartan subgroups would determine essentially disjoint parts of $\hat{G}$; and these representations would suffice to obtain the Plancherel formula for G. The discrete series is thus associated to the (unique) conjugacy class of compact Cartan subgroups, and the general discussion makes it clear that these may have to be constructed a priori and by quite different methods.

Discrete series: beginnings. Already in 1947 Godement had proved that the matrix elements of the discrete series satisfy orthogonality relations [Go]. In particular, one can associate to each $\omega \in \hat{G}_d$ a positive number $d(\omega)$, the *formal degree of ω*, such that

$$\int_G |(\pi(x)\phi, \psi)|^2 \, dx = \frac{1}{d(\omega)}$$

where $\pi \in \omega$ and ϕ, ψ are arbitrary unit vectors in the representation space of π. Moreover if $^0L^2(G)$ is the closed linear span of the irreducible subspaces of $L^2(G)$ (= the "discrete part" of $L^2(G)$), and E is the orthogonal projection $L^2(G) \to {}^0L^2(G)$, then, for all $f \in L^1(G) \cap L^2(G)$,

$$\|Ef\|^2 = \sum_{\omega \in \hat{G}_d} d(\omega)\|\pi_\omega(f)\|_2^2$$

where $\pi_\omega \in \omega$ and $\|\cdot\|_2$ is the Hilbert-Schmidt norm; this shows that the Plancherel

---

*This means that A is the maximal subgroup lying in the connected component of the identity of L such that $\mathrm{Ad}(a)$ is diagonalizable over $\mathbb{R}$ for all $a \in A$.

measure assigns the mass $d(\omega)$ to the class ω [1956b]. These results are actually true for all second countable locally compact groups ([Ma 2], p. 640).

In a succession of beautiful papers Harish-Chandra added new evidence to the picture of harmonic analysis of $L^2(G)$ suggested above [1954f], [1955a], [1955c], [1956a], [1956b]. Assuming not only that $rk(G) = rk(K)$ but further that G/K is Hermitian symmetric (this means that G/K has a G-invariant complex structure) he constructed a whole family of discrete series representations. They were realized on certain Hilbert spaces of holomorphic functions on which G acts naturally; they are the *holomorphic discrete series*. When G/K is not Hermitian symmetric this method would give nothing; even in the Hermitian symmetric case it will not give the entire discrete series. Nevertheless the introductory remarks to [1955c] reveal his conviction that in these cases different methods would have to be invented.

These papers revealed profound analogies between the (holomorphic) discrete series and the finite dimensional representations. These representations were *highest weight modules* with respect to positive systems coming from the complex structure of G/K. Their formal degrees were given *exactly by Weyl's dimension formula* applied to their highest weights. Finally, their *characters made sense as point functions* on the set of regular points of a compact Cartan subgroup and *could be described there by precise analogues of the Weyl character formulae.*

These results were the first serious indication that the discrete series would be parametrized by the characters of the compact Cartan subgroup, and hence that the representations of the series associated to an arbitrary Cartan subgroup would be parametrized by the characters of that Cartan subgroup. One would thus have a natural map

$$\coprod_{1 \leqslant i \leqslant r} \hat{A}_i \to \hat{G}$$

where $A_1, A_2, \ldots, A_r$, is a complete system of mutually nonconjugate Cartan subgroups of G (cf. the introductory remarks to [1957e]).

It is thus clear that by 1956 Harish-Chandra had essentially visualized the lines along which the Fourier analysis of $L^2(G)$ would have to be constructed. In retrospect it is really remarkable that he saw so clearly and so far; the clues available to him were certainly very meager.

Discrete series: construction. The determination of the discrete series turned out to be a prodigious undertaking ([1957b]–[1957e], [1964a]–[1966b]). His method, which was entirely novel, was roughly in three parts.

The first part is a direct and explicit construction of the invariant eigendistributions Θ_ξ, parametrized by the (regular) characters ξ of a compact Cartan subgroup $B \subset K$; these would ultimately be the characters of the representations of the discrete series. In the second part the question is that of showing that the Fourier coefficients (with respect to K) of the Θ_ξ, which are clearly eigenfunctions for $\mathfrak{Z}$ that are K-finite, are in $L^2(G)$. The third part is concerned with the completeness of the Θ_ξ; and the main result here is that these eigenfunctions and their translates span a dense subspace of $^0L^2(G)$.

Before the invariant eigendistributions could be constructed it is absolutely essential to understand the nature of their *singularities*. An initial exploration [1956c] had already shown that any invariant eigendistribution Θ on G is an analytic function F_Θ on G', the set of regular points of G. This function is locally integrable around the singular points also and so defines an invariant distribution $f \mapsto \int F_\Theta f \, dx$ ($f \in C_c^\infty(G)$) on G, also denoted by F_Θ. Experience suggested that $\Theta = F_\Theta$ (cf. [1955b], although the statement there is somewhat guarded), and this actually turned out to be true in complete generality. This result, known as the *Regularity theorem*, lies at a great depth, and is the corner stone for the entire structure of harmonic analysis.

The function F_Θ will in general blow up in the neighborhood of a singular point of the group. This behavior however cannot be completely arbitrary because the differential equations satisfied by the characters remain valid in the neighborhood of the singular points also. Harish-Chandra's work revealed the true nature of the limitations on the behaviour of F_Θ, namely that F_Θ should satisfy conditions of the following form: for suitable (explicitly determined) invariant differential operators ∇ on G', ∇F_Θ should extend continuously to all of G.

The starting point of the construction of the Θ_ξ is now quite clear; one defines Θ_ξ at the regular points on the compact Cartan subgroup by the analogue of the Weyl character formula. But the problem of defining Θ_ξ on the other Cartan subgroups is nontrivial because an invariant eigendistribution is not uniquely determined by its values on $B' = B \cap G'$. Nevertheless Harish-Chandra discovered that there is uniqueness *if the distribution is suitably restricted at infinity on the group*. The condition is that the distributions should be tempered; this however is to anticipate later ideas. At this stage the condition had to be formulated in a more direct manner, and he did it by requiring that $|D|^{-1/2}$ be a *global majorant* for Θ_ξ, D being the discriminant of G. The existence problem, i.e., the construction of the Θ_ξ, is however much deeper and depends in a fundamental way on *Fourier transforms on the Lie algebra of G*. Harish-Chandra proved that the Θ_ξ can be obtained by exponentiating from the Lie algebra the Fourier transforms of the invariant measures on appropriately chosen regular semisimple orbits [1965c].

The regularity theorem is the key that opens all the doors. If we look at the smooth, almost magical manner in which the theory flows once the regularity theorem is established, and also remember that there was very little in the known history of the subject to guide him in his prolonged struggle or to reassure him during moments of doubt, we have no choice except to place this among his greatest creative achievements.

The idea now is to use the method of integration over the conjugacy classes (= orbits) to reduce the harmonic analysis of the discrete series to that on the compact Cartan subgroup. This is of course exactly the method used by Weyl in his famous completeness proof for the compact case. However the present situation is much more complicated, for one must take into account the contributions of the other Cartan subgroups. Before explaining Harish-Chandra's method of doing this I shall discuss in some detail the method of orbital integrals.

It was Gel'fand and Neumark who first realized that the theory of orbital integrals can be made the basis for a proof of the Plancherel formula [GN 2]. They treated SL(n, $\mathbf{C}$) by this method. In [1951f], [1954c] Harish-Chandra pushed this method through for all *complex* semisimple Lie groups. To motivate this let us first assume that G is compact and let $T \subset G$ be a maximal torus. Using Weyl's formulae we can rewrite the Plancherel formula as a sum over $\hat{T}$:

$$f(e) = \sum_{\xi \in \hat{T}} \Omega(\xi) \int_G \Theta_\xi f \, dx \qquad (f \in C^\infty(G)).$$

Here $\Omega(\xi)$ is a polynomial function of ξ, namely a constant multiple of the product of the roots of a positive system; Θ_ξ is the Weyl character, given on T by

$$\Delta(t)\Theta_\xi(t) = \sum_{s \in W} \varepsilon(s)(s\xi)(t) \qquad (t \in T)$$

where W is the Weyl group and Δ is the Weyl denominator function. Integrating over the orbits and setting

$$F_f(t) = \Delta(t) \int_G f(xtx^{-1}) \, dx \qquad (t \in T)$$

we have

$$\int_G \Theta_\xi f \, dx = (-1)^p \hat{F}_f(\xi)$$

where p is the number of positive roots and $\hat{F}_f$ is the Fourier transform of F_f (F_f is obviously in $C^\infty(T)$). So the Plancherel formula is equivalent to the assertion

$$f(e) = (-1)^p \sum_{\xi \in \hat{T}} \Omega(\xi)\hat{F}_f(\xi).$$

But, by Fourier transform theory on T, $(-1)^p\Omega(\xi)\hat{F}_f(\xi)$ is $\widehat{\Omega F_f}(\xi)$ where Ω *is now viewed as a constant coefficient differential operator on T.* So the Plancherel formula is equivalent to the relation

(L.F.) $$f(e) = (\Omega F_f)(e) \qquad (f \in C^\infty(G)).$$

We shall refer to (L.F.) as the *limit formula for the orbital integral*.

This derivation suggests a method for proving the Plancherel formula: first prove (L.F.) and then apply Fourier transform to it. It is remarkable that this method continues to work when G is complex. The maximal torus is now replaced by a Cartan subgroup L. For any $f \in C_c^\infty(G)$, $F_{f,L}(h)$ will denote the integral of f over the orbit of h:

$$F_{f,L}(h) = \Delta(h) \int_{G/L} f(xhx^{-1}) \, d\dot{x}$$

where $d\dot{x}$ is the invariant measure on G/L. Initially this makes sense only for h regular, and in the regular subset of L, $F_{f,L}$ will be a smooth function. It can then be proved that $F_{f,L}$ extends to all of L as an element of $C_c^\infty(L)$. The character formulae for the principal series now show that, with appropriate normalization,

$$\langle T_\xi, f \rangle = \hat{F}_{f,L}(\xi) \qquad (\xi \in \hat{L})$$

where T_ξ is the distribution character of the principal series representation corresponding to the character ξ of L. But now the limit formula

(L.F.) $$f(e) = (\Omega F_{f,L})(e) \qquad (f \in C_c^\infty(G))$$

holds, with same Ω as before, and the Fourier transformation of this relation gives the Plancherel formula

$$f(e) = \int_{\hat{L}} \hat{\Omega}(\xi)\langle T_\xi, f\rangle \, d\xi \qquad (f \in C_c^\infty(G))$$

where $\hat{\Omega}$ is the polynomial function which is the Fourier transform of the differential operator Ω.

The set $\{e\}$ is a very complicated singular point in the space of conjugacy classes. In particular it is not the limit of orbits in general position, and so $f(e)$ is not the limit of $F_{f,L}(h)$ as $h \to e$. It is therefore striking that one can still recover $f(e)$ as the limit as $h \to e$ of *suitable derivatives* of $F_{f,L}(h)$.

It is natural to try to extend this method to all cases by establishing the limit formula and relating the Fourier transforms of the orbital integrals to the irreducible characters. However for arbitrary G the orbital integrals and their derivatives have jumps at the singular points. Moreover there is more than one orbital integral to be considered since there is in general more than one conjugacy class of Cartan subgroups. Nevertheless calculations in special cases [1952] seemed to suggest that the Fourier transform of the orbital integral is very closely related to the irreducible characters, and so a limit formula should still represent a primitive version of the Plancherel formula (see the remarks at the beginning of [1957e]). In a series of papers in the 1950's ([1957b]–[1957e] and [1964c]) Harish-Chandra established the limit formula for the orbital integral associated to a *fundamental* Cartan subgroup, namely a Cartan subgroup whose maximal compact subgroup has the largest possible dimension (these form a single conjugacy class, and when $rk(G) = rk(K)$, fundamental means compact). In the limit formula

$$\text{(L.F.)} \qquad\qquad f(e) = (\Omega F_{f,L})(e) \qquad (L \text{ fundamental})$$

$F_{f,L}$ may still have jumps at singular points of L; *nevertheless the differential operator Ω, which is still the product of roots of a positive system, kills all the jumps, i.e., $\Omega F_{f,L}$ extends continuously to all of L, so that the right side makes sense.*

The jumps of $F_{f,L}$ make the Fourier analysis of the relation (L.F.) a very delicate matter (see [1952]). These jumps are in fact the orbital integrals associated to Cartan subgroups which are "less compact" than L, suggesting that the jumps are caused by the contributions to f from the representations associated to these less compact Cartan subgroups. Thus when L is compact, one must work with matrix elements of the discrete series representations in order to get rid of the jumps. This can only be done by working with a space of rapidly decreasing functions instead of $C_c^\infty(G)$.

For this purpose Harish-Chandra introduced in [1966b] the Schwartz space $\mathcal{C}(G)$ of G. Overcoming many technical difficulties he proved in [1966b] that the theory of orbital integrals can be extended to the Schwartz space, that the limit formula (L.F.) is true for all f in this space, and that the distributions Θ_ξ constructed earlier are *tempered*, with

$$\langle \Theta_\xi, f\rangle = \int_G \Theta_\xi f \, dx,$$

the integrals converging absolutely. Moreover he proved that *if f is an eigenfunction* *for β in the Schwartz space, the integrals of f on the regular non-elliptic orbits are all zero*; here we recall that an orbit is said to be *elliptic* if it meets a compact Cartan subgroup. If G has no compact Cartan subgroup this means that all the orbital integrals of f are zero; the limit formula would then imply that $f = 0$. We shall see presently that eigenfunctions of β in $L^2(G)$ are automatically in Schwartz space; so this argument already shows that the condition $rk(G) = rk(K)$ is necessary for the existence of the discrete series. If $rk(G) = rk(K)$, the same argument would still allow us to conclude that the orbital integral $F_{f,B}$ $(B \subset K)$ has no jumps, i.e., it lies in $C^\infty(B)$. The Fourier transformation of the limit formula coupled with the formula for Θ_ξ on B would show that the harmonic analysis of f is entirely controlled by the Θ_ξ. This is substantially Harish-Chandra's argument for the completeness of the Θ_ξ.

To round out this picture it is necessary to know a method of determining when a given eigenfunction of β belongs to the Schwartz space. In [1966b] Harish-Chandra developed a powerful theory to describe the asymptotic behaviour of τ-spherical eigenfunctions for β that satisfy the weak inequality. If f is any τ-spherical eigenfunction satisfying the weak inequality, Harish-Chandra was able to associate to f, at each point at infinity of the group, a so called *constant term*, that was the dominant term in the asymptotic development of f there. These dominant terms are parametrized by the parabolic subgroups of G. If $P = MAN$ is a parabolic subgroup and f_P is the associated constant term, f_P is essentially a function of the same nature as f, but lives on the group MA; the difference between f and f_P has very rapid decay at infinity "in the direction of P." An important consequence of this theory is the result that *f lies in $L^2(G, \tau)$ if and only if all its constant terms are zero; and that in this case, f is in the Schwartz space*.

In order to apply this theory successfully two things have to be done. First, it must be shown that the Fourier coefficients of the Θ_ξ satisfy the weak inequality. This is technically rather subtle and is done in [1966a]. The second point is to show that all the constant terms of these Fourier coefficients are zero. Harish-Chandra's proof of this made use of a beautiful property of these constant terms, namely that they are *transitive* in a natural sense with respect to the partial ordering among the parabolic subgroups. If f is a τ-spherical eigenfunction built out of the Fourier coefficients of the Θ_ξ, and $P \neq G$ is chosen so that $f_P \neq 0$ and $\dim(M)$ is minimal, the transitivity mentioned above would show that the restrictions of f_P to M are eigenfunctions in the Schwartz space of M (actually linear combinations of such); so M has to have a compact Cartan subgroup, implying that P is associated to a Cartan subgroup of G itself. This Cartan subgroup cannot be compact, and hence as the eigenhomomorphism to which f belongs is regular and is associated to the compact Cartan subgroup, we have a contradiction.

These brief remarks cannot give an adequate idea of the pathbreaking nature of the papers [1965c] [1966b]. The audacity of their conceptions as well as the delicacy and craftsmanship that were needed in executing them, compel one to place them among the greatest works in Fourier analysis.

The continuous spectrum: the theory of the Eisenstein integral and the explicit Plancherel formula. With the discrete series completely determined Harish-Chandra turned to the problem of explicitly decomposing the regular representation of G. He went much farther than a mere L^2-theory and obtained a Fourier analysis of the

Schwartz space of G. The results were worked out in three long papers [1975], [1976a], and [1976b]; but the lectures [1970a], [1970c], and [1972] containing the announcements of these and many other results are interesting in themselves, for they give a bird's eye view of his entire approach.

Let L be a Cartan subgroup written (canonically) as $T \cdot A$ where T is compact and A is connected and split over $\mathbb{R}$. Let $\mathcal{P}(A)$ be the finite set of parabolic subgroups $P = MAN$. Since $rk(M) = \dim(T)$, $\hat{M}_d \neq \phi$. For $\eta \in \omega \in \hat{M}_d$ and $\nu \in \mathfrak{a}^*$ ($\mathfrak{a} = \mathrm{Lie}(A)$), $\pi_{\eta, \nu}$ is the unitary representation of G induced by the representation $m \cdot a \cdot n \mapsto e^{i\nu(\log a)} \cdot \eta(m)$ of P. The character $\Theta_{\omega, \nu}$ of $\pi_{\eta, \nu}$ is independent of P (M and A are the same for all $P \in \mathcal{P}(A)$). Then (see [1976b])

(a) $\pi_{\eta, \nu}$ is finitely decomposable; it is irreducible for regular ν.
(b) Let $W(A)$ be the Weyl group of A, i.e., the normalizer of A modulo the centralizer of A; then $W(A)$ acts on $\hat{M}_d \times \mathfrak{a}^*$ and

$$\Theta_{s\omega, s\nu} = \Theta_{\omega, \nu} \qquad (s \in W(A)).$$

Actually results much stronger than (a) were proved by Harish-Chandra. For example, if L is fundamental, *all* the $\pi_{\eta, \nu}$ are irreducible ([1976b]); special cases of this were known earlier, going all the way back to the irreducibility of the principal series of $SL(n, \mathbb{C})$ due to Gel'fand and Neumark.

One can now associate to L the subspace $\mathcal{C}_L$ of $\mathcal{C}$ consisting of all functions that are orthogonal to the matrix elements of the representations $\pi_{\eta, \nu}$ attached to the Cartan subgroups that are not conjugate to L. $\mathcal{C}_L$ is closed in $\mathcal{C}$ and is invariant under translations. Central to Harish-Chandra's theory is the decomposition (Theorem 12, [1970c])

$$\mathcal{C} = \bigoplus_L \mathcal{C}_L \qquad (*)$$

where the sum is over a complete set of mutually nonconjugate Cartan subgroups; the sum is orthogonal and smooth (i.e., the projections are continuous in $\mathcal{C}$). It will turn out that $\mathcal{C}_L$ is exactly the space of "wave packets" of the matrix elements of the $\pi_{\eta, \nu}$ associated to L (this can be formulated more precisely). Thus $(*)$ is a beautiful and definitive formulation of the heuristic principle that "each Cartan subgroup makes a separate contribution to the Fourier analysis of G."

The splitting $(*)$ allows one to work on each $\mathcal{C}_L$ separately. Let $\Omega_L = \hat{M}_d \times \mathfrak{a}^*$ and let us define, for $f \in \mathcal{C}$, its Fourier transform $\hat{f} = (\hat{f}_L)$ by

$$\hat{f}_L(\omega, \nu) = \int_G \Theta_{\omega, \nu}^{\mathrm{conj}} f \, dx \qquad ((\omega, \nu) \in \Omega_L).$$

Then $\hat{f}_L$ is $W(A)$-invariant, and we have an initial formulation of the Plancherel formula ([1972], Theorem 11): there is a unique $W(A)$-invariant continuous function μ_L on Ω_L ($\hat{M}_d$ is given the discrete topology) such that for all $f \in \mathcal{C}_L$,

$$f(e) = \sum_{\omega \in \hat{M}_d} d(\omega) \int_{\mathfrak{a}^*} \hat{f}_L(\omega : \nu) \mu_L(\omega : \nu) \, d\nu$$

the series converging absolutely. The function μ_L is ≥ 0 and has other nice properties. For fixed ω, it extends to a meromorphic function on $\mathfrak{a}_c^*$ that is holomorphic on $\mathfrak{a}^*$ and > 0 at the regular points of $\mathfrak{a}^*$; moreover it is of at most polynomial growth on $\mathfrak{a}^*$.

The basic question now is of course the explicit calculation of μ_L. Harish-Chandra's method for doing this was based on a remarkable *product formula* for μ_L. To describe this, fix a parabolic subgroup $P \in \mathcal{P}(A)$; then one can associate in a canonical manner a set of pairs (M_i, Q_i) $(1 \leqslant i \leqslant m)$ where M_i is a reductive subgroup of G with no split component and $Q_i = MA_iN_i$ is a parabolic subgroup of M_i with $A_i \subset A$ and $\dim(A_i) = 1$. If $\mathfrak{a}_i = \mathrm{Lie}(A_i)$ and $\nu_i = \nu|_{\mathfrak{a}_i}$, then

$$\mu_L = c \prod_{1 \leqslant i \leqslant m} \mu_{L,i}, \qquad \mu_{L,i}(\omega:\nu) = \mu_{L,i}(\omega:\nu_i)$$

where $\mu_{L,i}$ is the μ-function associated to (M_i, Q_i) (see [1972], pp. 144–145 for a description of the (M_i, Q_i)); c is a constant > 0 independent of ω and ν and can be explicitly evaluated. This reduces the problem of explicit determination of μ_L to the case (G, P) where G has no split component, $P = MAN$, and $\dim(A) = 1$. If $rk(G) > rk(K)$, then L is fundamental and the limit formula, by Fourier transformation on L, leads to the determination of μ_L as a *polynomial function* on $\hat{L}$; here we must remember that we are operating not on the whole Schwartz space but only on $\mathcal{C}_L$ so that the contributions of the other Cartan subgroups of G do not enter the limit formula. Thus the analysis proceeds as if G has a single conjugacy class of Cartan subgroups (see §24, [1976b]). If $rk(G) = rk(K)$, the treatment follows the model where $rk(G/K) = 1$ (cf. [1966a] and §36, [1976b]).

It is impossible to describe except in bare outline the architecture of this beautiful theory. I shall however try to compensate by discussing two important sources of motivation behind Harish-Chandra's treatment. The first is the harmonic analysis of *spherical functions* [1958a] [1958b]. Many of the crucial features that are typical of the continuous spectrum were discovered first in this context. The second source is the spectral theory of $L^2(G/\Gamma)$, due to Selberg [Se 2], Gel'fand and Piatetsky-Shapiro [G] [GP], and Langlands [L 1], Γ being an arithmetic subgroup of G. Harish-Chandra's treatment of this in [1968b] is especially useful for this purpose.

The theory of spherical functions is essentially the harmonic analysis of $L^2(G/K)$. The fundamental objects are the elementary spherical functions $\varphi(\nu:x) = \varphi_\nu(x)$ and the transform that is defined by them. In [1958a] and [1958b] Harish-Chandra made the remarkable discovery that the φ_ν, regarded as functions on A (of the Iwasawa decomposition), look like a sum of plane waves at infinity on A; and that all of these have the same amplitude, in terms of which the Plancherel measure can be determined rather simply. More precisely,

$$d(a)\varphi(\nu:a) \sim \sum_{s \in W(A)} c_s(\nu)e^{is\nu(\log a)},$$

d being a certain homomorphism of A into the positive reals; the amplitudes satisfy

$$|c_s(\nu)| = |c(\nu)| \qquad (c = c_1)$$

and

$$|c(\nu)|^{-2} d\nu$$

is the Plancherel measure. Actually we have the relations

$$c_s(\nu) = c(s\nu)$$

which mirror the *functional equations*

$$\varphi_{s\nu} = \varphi_\nu$$

satisfied by the elementary spherical functions.

The computation of the spherical Plancherel measure thus reduces to determining the function c, which is a special case of the famous c-function of Harish-Chandra. It turned out that this function is meromorphic on $\mathfrak{a}_c^*$ and that it can be expressed as a product of the analogous c-functions associated to the real rank one subgroups of G defined by the root system of (G, A) (the Gindikin–Karpelevič product formula). Since the c-function can be explicitly evaluated for groups of real rank one, the story is complete.

For the harmonic analysis of arbitrary (not necessarily spherical) functions in the Schwartz space Harish-Chandra introduced the eigenfunctions that generalized the elementary spherical functions. Let U be a finite dimensional Hilbert space carrying a unitary double representation τ of K. Let $\tau_M = \tau|_M$ and let $\mathfrak{L}$ be the space of τ_M-spherical maps of M into U whose components (with respect to some basis of U) are in $\mathcal{C}(M) \cap {}^0L^2(M)$. Then $\dim(\mathfrak{L})$ is finite and for each $\psi \in \mathfrak{L}$, $\nu \in \mathfrak{a}_c^*$, and a given choice of P, a parabolic subgroup in $\mathcal{P}(A)$, Harish-Chandra defined an eigenfunction for $\mathfrak{Z}$ in $C^\infty(G, \tau)$, denoted by $E_{\psi, \nu}$ or $E_{P, \psi, \nu}$, called the *Eisenstein integral*. If P is minimal, $U = \mathbb{C}$, τ is the trivial representation, and ψ is 1, this is just the elementary spherical function (see [1972], p. 133).

The Eisenstein integrals satisfy the weak inequality, and so one can determine their behaviour at infinity on the group by applying the asymptotic theory of tempered eigenfunctions developed in [1966b] [1975]. If Q is a parabolic subgroup in $\mathcal{P}(A)$ which may be different from P, then one finds that the constant term of the Eisenstein integral $E_{P, \psi, \nu}$ associated to Q is of the form

$$\sum_{s \in W(A)} (\psi_{s, \nu})(m) e^{is\nu(\log a)}$$

where ν is a regular element of $\mathfrak{a}^*$ and $\psi_{s, \nu} \in \mathfrak{L}$. The maps

$$c_{Q|P}(s : \nu) : \psi \mapsto \psi_{s, \nu}$$

are uniquely determined endomorphisms of $\mathfrak{L}$ and the relation

$$E_{P, \psi, \nu} \underset{Q}{\sim} \sum_{s \in W(A)} c_{Q|P}(s : \nu) \psi e^{is\nu}$$

expresses the fact that $E_{P, \psi, \nu}$ looks like a sum of plane waves at infinity in the direction of Q. This is clearly analogous to what happens in the spherical case.

Let us now fix ω and choose τ so that the subspace $\mathfrak{L}(\omega)$ of $\mathfrak{L}$ that "transforms according to ω" is nonzero; this is always possible. It can then be shown that for regular $\nu c_{Q|P}(1 : \nu)$ is a bijection of $\mathfrak{L}(\omega)$ onto itself. Let $c_{Q|P, \omega}(1 : \nu)$ be the restriction of $c_{Q|P}(1 : \nu)$ to $\mathfrak{L}(\omega)$. Then the fundamental result expressing the relationship between the Plancherel formula and the asymptotics of the Eisenstein integrals can be stated as follows (see [1976b], §13):

$$\mu_L(\omega:\nu)c_{Q|P,\omega}(1:\nu)^{\dagger}c_{Q|P,\omega}(1:\nu) = c \cdot id_\omega$$

where id_ω is the identity endomorphism of $\mathcal{L}(\omega)$, $c > 0$ is a constant independent of ω and ν which can be explicitly calculated. One must also note that as $\mathcal{L} \subset {}^0L(M, \tau_M)$, $\mathcal{L}$ has a natural structure as a Hilbert space, and $\dagger$ is the adjoint operation in this structure.

The c-functions are closely related to the intertwining operators between the induced representations. The latter have natural product formulae which then lead to product formulae for the c-functions and thence to the μ-functions. This is the origin of the product formula for the Plancherel measure.

A central role in Harish-Chandra's treatment of these questions is played by the functional equations of the Eisenstein integrals. The analogy with the theory of automorphic forms is most significant here, and so it may be useful to make a few remarks comparing the Eisenstein integral with the Eisenstein series. The starting point of this comparison is the following characterization of the "discrete part" of the Schwartz space of G, obtained by Harish-Chandra: if ${}^0\mathcal{C} = \mathcal{C}(G) \cap {}^0L^2(G)$, then an element f in $\mathcal{C}(G)$ is in ${}^0\mathcal{C}$ if and only if

$$\int_N f(xn)\, dn = 0 \qquad (x \in G)$$

for all parabolic subgroups $P = MAN \neq G$ (Theorems 14 and 15, [1970c]). This is obviously analogous to the condition that defines a cusp form on G/Γ, so that it is natural to speak of the elements of ${}^0\mathcal{C}$ as *cusp forms* on G. Thus $\mathcal{L}$ may be viewed as the space of τ_M-spherical cusp forms on M; and the definition of the Eisenstein integral imitates that of the Eisenstein series with integration over K replacing the summation over Γ (cf. [1968b], p. 29). Both the characterization of ${}^0\mathcal{C}$ and the finite dimensionality of $\mathcal{L}$ are deep lying consequences of the theory of the discrete spectrum.

The functional equations of the Eisenstein integrals involve the comparison of $E_{P,\psi,\nu}$ and $E_{Q,\psi',s\nu}$, and Harish-Chandra obtained them as a consequence of the remarkable principle asserting that *if just one summand is common to the constant terms of two such functions, they must be identical*. This principle follows from what he called the *Maass–Selberg relations* which may be described as follows: if f is a τ-spherical eigenfunction satisfying the weak inequality and suitable additional conditions (which are satisfied by the Eisenstein integrals), the intensities (= norms in ${}^0L^2(M, \tau_M)$) of the plane waves that occur in its asymptotic forms along the parabolic subgroups in $\mathcal{P}(A)$ are all equal (cf. Theorem 14.1 and the results of §17, especially Lemmas 17.2 and 17.3 of [1976b]). The Maass–Selberg relations originally arose in the theory of automorphic forms ([1968b], Chap. IV, §2); the discovery and proof of their analogues in the present context is a highlight of [1976b].

Representation theory of p-adic groups. The Lefschetz principle and the philosophy of cusp forms. The beginnings of the representation theory of reductive p-adic groups go back to the late 1950's when Mautner's paper [Mau 1] on the representations of $GL(2, \Omega)$ and $PGL(2, \Omega)$ appeared, Ω being a local field (also called a p-adic field), i.e., a locally compact non-discrete field. It developed rapidly in the early 1960's at the hands of Bruhat, Gel'fand–Graev, Satake and others. Harish-Chandra became actively interested in this area in the mid 1960's; his ideas and results since then have proved to be of fundamental importance for the subject. I shall first try to motivate these developments.

Let k be an algebraic number field. Then one can attach to k the local fields k_v which are its completions at the various places v, and the adele ring $\mathbf{A}$ which is a "restricted" direct product of the $k_v \cdot \mathbf{A}$ is an abelian ring which is locally compact in a natural topology; k is diagonally imbedded in it as a discrete subring and $\mathbf{A}/k$ is compact. By the 1950's it had become a well understood principle in number theory that many arithmetic questions on k could be studied over each k_v, and then by working over $\mathbf{A}$, the local data coming from the various k_v could often be put together to reach global results. This is so in classfield theory, in the theory of simple algebras and Brauer groups, and in the arithmetic theory of quadratic forms [W 2].

After the development of the theory of linear algebraic groups in the 1950's it was realized that the formalism of adeles could be applied to any algebraic group defined over k. Let $\mathbf{G}$ be such a group, and for any k-algebra R, let $\mathbf{G}(R)$ be the group of R-points of $\mathbf{G}$. We then have the groups $G_v = \mathbf{G}(k_v)$ for each v and the adele group $G_{\mathbf{A}} = \mathbf{G}(\mathbf{A})$. G_v and $G_{\mathbf{A}}$ are locally compact, separable, and $G_k = \mathbf{G}(k)$ is (diagonally) imbedded naturally as a discrete subgroup of $G_{\mathbf{A}}$. With the availability of the Chevalley theory of reductive algebraic groups and the work of Borel and Harish-Chandra [1962] (see also [Bo 1]) on the structure of $G_{\mathbf{A}}/G_k$ it was natural to hope that one could begin studying the arithmetic aspects of the theory of reductive groups. A succession of closely related ideas and papers, especially those of Weil [W 3] [W 4] [W 5] and Langlands [L 1] [JL] (with Jacquet) [L 2] [L 3] led to an explosive development of this point of view in the 1960's. As a consequence, the supreme importance of the harmonic analysis of $L^2(G_{\mathbf{A}}/G_k)$ for arithmetic was recognized. For example, if $\mathbf{G} = GL(1)$, $G_{\mathbf{A}} = \mathbb{I}$ the idele group and $G_{\mathbf{A}}/G_k = \mathbb{I}/k^\times$ is the idele class group; a character of $\mathbb{I}/k^\times$ is essentially a Grössencharakter of Hecke. Tate's work (cf. §1) in 1947 had derived the Hecke theory of the associated L-series from the Poisson formula for $\mathbf{A}/k$. Also one knows from classfield theory that the characters of $\mathbb{I}/k^\times$ of finite order define (through Artin's reciprocity law) cyclic extensions of k. If $\mathbf{G} = GL(2)$, the modular forms on the Poincaré half plane which are eigenfunctions for the Hecke operators define (under suitable spectral conditions) irreducible representations of $G_{\mathbf{A}}$ that occur in $L^2(G_{\mathbf{A}}/G_k)$, showing that in the notion of an irreducible representation occurring in $L^2(G_{\mathbf{A}}/G_k)$ we have a far reaching generalization of the classical automorphic forms and Eisenstein series. The work of [L 3] suggested that for any integer $n \geqslant 1$ and any n-dimensional representation of the Galois group $\mathrm{Gal}(\bar{k}/k)$ ($\bar{k} =$ an algebraic closure of k) there would be associated an irreducible unitary representation occurring in $L^2(G_{\mathbf{A}}/G_k)$.

The irreducible unitary representations of $G_{\mathbf{A}}$ are tensor products of irreducible unitary representations of the local groups G_v. It is therefore natural to try to understand the representation theory of the groups G_v as a first step. Of course these representations would reflect the arithmetic of the local fields k_v. For example, following [L 3], one would expect any n-dimensional representations of $\mathrm{Gal}(\bar{k}_v/k_v)$ to determine an irreducible unitary representation of $GL(n, k_v)$. This is a very rough sketch of some of the ideas that led to the great interest in the 1960's for constructing a theory of representations of groups $G = \mathbf{G}(\Omega)$ where Ω is a $\mathfrak{p}$-adic field and $\mathbf{G}$ is a reductive group defined over Ω.

From the very beginning Harish-Chandra emphasized the striking similarities between the theory for real groups and the newly developing theory for the $\mathfrak{p}$-adic groups [1968a]. In [1970d] he formulated this as a broad heuristic principle, which he called the *Lefschetz principle*, asserting that whatever is true for real groups is also true for the $\mathfrak{p}$-adic ones. Its thrust was to use the rich theory of Fourier analysis on real reductive groups as a guide in the search for discovering the basic facts to be proved in the representation theory of the $\mathfrak{p}$-adic groups. In [1970d], as a first illustration of this principle, Harish-Chandra explored the theory of characters and orbital integrals on $G = \mathbf{G}(\Omega)$ as above. These investigations were completed in [1978] where he showed, assuming $\operatorname{char} \Omega = 0$, that the theory of characters of G resembles the theory for real groups to a remarkably close extent. Thus, characters of irreducible admissible representations are locally summable functions on G which are locally constant on G'; and they can be obtained by exponentiating the Fourier transforms of the invariant measures on the (nilpotent) orbits in the Lie algebra. One should recall that in the real case such theorems came out of the profound study of the differential equations satisfied by the characters. In the present case there are no differential equations, and so it is quite surprising that such results are true. Harish-Chandra's work uses a beautiful theorem of Howe on invariant distributions on the Lie algebra of G as a substitute for the differential equations.

The philosophy of cusp forms, which he had already introduced in [1970b] as the unifying principle for Fourier analysis on reductive groups is an even more striking illustration of the Lefschetz principle. In [1970b] he had illustrated it for reductive groups over finite fields, and it was the driving force in his work on the Fourier analysis of the Schwartz space of a real reductive group. He showed that the harmonic analysis of these groups can be worked out quite explicitly, based on the concept of cusp forms and parabolic induction. In [1973] and [1977a] he succeeded in pushing this method through for reductive $\mathfrak{p}$-adic groups. The cusp forms are once again (essentially) the matrix elements of the discrete series and this method led him to the theory of the Eisenstein integral, in particular, to the Maass–Selberg relations and Plancherel formula, in virtually complete analogy with the real case.

For groups over $\mathfrak{p}$-adic fields the discrete series is still not completely determined. Nevertheless, by taking Fourier analysis of reductive groups to this stage, Harish-Chandra has shown that its inner structure can be understood in terms of a small number of simple and yet general principles. This is a truly extraordinary accomplishment.

§3. *Orbital Integrals and Limit Formula.* I now wish to supplement this general survey with a little more detailed discussion of selected parts of Harish-Chandra's work. I shall begin with the theory of orbital integrals and the limit formula. The bulk of this work is contained in [1957b]–[1957e] and [1964c]; the extension of the theory to the Schwartz space of G is treated in [1965b], [1966a], and [1966b].

INTRODUCTION

We take G to be the real form of a complex simply connected semi simple group. Let $L \subset G$ be a Cartan subgroup with $\mathfrak{h} = \mathrm{Lie}(L)$; P, a positive system of roots of $(\mathfrak{g}_c, \mathfrak{h}_c)$; and Δ, the Weyl denominator on L. Define ϖ and π by $\varpi = \prod_{\alpha \in P} H_\alpha$, $\pi = \prod_{\alpha \in P} \alpha$. The orbital integral associated to L is then defined for $f \in C_c^\infty(G)$, $h \in L' = L \cap G'$ by

$$F_{f,L}(h) = \varepsilon_R(h)\Delta(h)\int_{G/L = G^*} f(xhx^{-1})\, dx^* \tag{1}$$

where $\varepsilon_R(h)$ is a locally constant sign factor (§22, [1965b]). The analogous definition on $\mathfrak{g}$ is (see §5, [1964c])

$$\psi_{f,\mathfrak{h}}(H) = \varepsilon_R(H)\pi(H)\int_{G^*} f(H^x)\, dx^* \qquad (f \in C_c^\infty(\mathfrak{g})). \tag{2}$$

These are well defined because the regular orbits are closed, and it is easy to show that $F_{f,L}$ and $\psi_{f,\mathfrak{h}}$ are C^∞ on L' and $\mathfrak{h}'$, respectively.

The key to the entire theory is the system of differential equations satisfied by F_f and ψ_f:

$$F_{zf,L} = \mu_{\mathfrak{g}/\mathfrak{h}}(z)F_{f,L} \qquad (z \in \mathfrak{Z}) \tag{3G}$$

$$\psi_{\partial(p)f,\mathfrak{h}} = \partial(p_\mathfrak{h})\psi_{f,\mathfrak{h}} \qquad (p \in I(\mathfrak{g})). \tag{3g}$$

Here, $I(\mathfrak{g})$ is the algebra of G-invariant elements in $S(\mathfrak{g}_c)$ and $p_\mathfrak{h}$ is the "restriction" of p to $\mathfrak{h}$ while $\mu_{\mathfrak{g}/\mathfrak{h}}$ is as in §12 of [1965b]. To illustrate this let us examine how they control the growth of the orbital integrals near a singular point. Consider for instance $\psi_{f,\mathfrak{h}}$. The equations (3g) and the formula for integration on $\mathfrak{g}$ give the estimate

$$\int_{\mathfrak{h}'} |\pi||\partial(p_\mathfrak{h})\psi_{f,\mathfrak{h}}|\, d\mathfrak{h} \leqslant \int_{(\mathfrak{h}')^G} |\partial(p)f|\, d\mathfrak{g}. \tag{4}$$

From this one can conclude that all derivatives of $\psi_{f,\mathfrak{h}}$ are locally bounded, and in fact even more. The question is the often encountered one of obtaining point estimates in terms of L^1-norms of a function and its derivatives; but there is a technical complication because of the weight function $|\pi|$ which vanishes on $\mathfrak{h} \setminus \mathfrak{h}'$. In [1965b] (cf. Theorem 3) Harish-Chandra obtained a global estimate of the Sobolev type that can be used in the present situation. This can be formulated as follows (see [V 3], Proposition 7, p. 157). Let V be a real finite dimensional vector space; $w = \prod_{1 \leqslant j \leqslant q} |\lambda_j|^{a_j}$ where $a_j > 0$, $\lambda_j \in V_c^*$, and $V' =$ the set where $w \neq 0$; let S_0 be a subalgebra of $S = S(V_c)$ such that S is a finite module over S_0. Write $\mathfrak{U}^1(S_0 : w)$ (resp. $\mathfrak{U}^\infty$) for the Frechet space of all $g \in C^\infty(V')$ such that $\int |\partial(u)g|w\, dV < \infty$ (resp. $\sup|\partial(u)g| < \infty$) for all $u \in S_0$ (resp. $u \in S$). Then

$$\mathfrak{U}^1(S_0 : w) \subset \mathfrak{U}^\infty$$

and the inclusion is continuous. Since $\psi_{f,\mathfrak{h}} \in \mathfrak{U}^1(I(\mathfrak{h}):|\pi|)$ by (4), we must have $\psi_{f,\mathfrak{h}} \in \mathfrak{U}^\infty$. Actually, replacing f by qf where q is an invariant polynomial on $\mathfrak{g}$, we find the much stronger result that $\psi_{f,\mathfrak{h}}$ lies in the Schwartz space $\mathcal{C}(\mathfrak{h}')$ of $\mathfrak{h}'$ and that the map $f \mapsto \psi_{f,\mathfrak{h}}$ is continuous with respect to the collection of seminorms of the form

$$f \mapsto \int_{(\mathfrak{h}')^G} |\partial(p)(qf)| \, d\mathfrak{g} \qquad (f \in C_c^\infty(\mathfrak{g})).$$

Since *these are all continuous on* $\mathcal{C}(\mathfrak{g})$, the orbital integral extends to a continuous map $\mathcal{C}(\mathfrak{g}) \to \mathcal{C}(\mathfrak{h}')$; in particular, the invariant measures on the orbits are all tempered. The original proof in [1957c] was essentially a variant of this type of argument.

The same argument works (up to a point) on the group also. The crucial case is when $L = B$ is compact and let us make this assumption. Let $\mathfrak{U}^\infty$ now be the space of all $g \in C^\infty(B')$ such that $\sup|ug| < \infty$ for all $u \in S(\mathfrak{b}_c)(\mathfrak{b} = \mathrm{Lie}(B))$, regarded as a Frechet space in the usual way. Then the above argument will show that $F_{f,B} \in \mathfrak{U}^\infty$ and that $f \mapsto F_{f,B}$ will be continuous with respect to the topology induced on $C_c^\infty(G)$ by the seminorms

$$\nu_z : f \mapsto \int_{G_e} |zf| \, dx \qquad (z \in \mathfrak{Z})$$

where $G_e = (B')^G$ is the elliptic set. If one can show that

$$V(f) = \int_{G_e} |f| \, dx < \infty$$

for any $f \in \mathcal{C}(G)$ and that V is a continuous seminorm on $\mathcal{C}(G)$, the continuity of the ν_z will follow, leading to the extension of the theory of orbital integrals to $\mathcal{C}(G)$. Let φ_B be the characteristic function of the set G_e and let

$$b(x) = \int_K \varphi_B(xk) \, dk \qquad (x \in G).$$

Since

$$\int_{G_e} |f| \, dx \leq \int_G |f| b \, dx$$

it is enough to prove that b satisfies the weak inequality.

Harish-Chandra does this in [1966a] (Theorem 5). It is obvious that $b(x)$ is the measure of the set of all k in K for which xk is elliptic. One can use the boundedness of the finite dimensional characters on G_e to conclude that $b(x) \to 0$ when x goes to infinity on G. For the sharper result needed here, one needs class functions with good growth properties on *all* Cartan subgroups. Harish-Chandra's proof depends on the existence of a locally summable class function Θ such that

(i) Θ is a nonzero constant on G_e.

(ii) If $L = T \cdot A$ is any Cartan subgroup, and $\mathfrak{z}$ is the centralizer of A in $\mathfrak{g}$,

$$|\Theta(x)| \leq \mathrm{const.}|\det(1 - \mathrm{Ad}(x^{-1}))_{\mathfrak{g}/\mathfrak{z}}|^{-1/2} \qquad (x \in L').$$

(iii) $\displaystyle\int_K \Theta(xk) \, dk = 0 \qquad (x \in G).$

Using his theory of the distributions $\Theta_\xi(\xi \in \hat{B})$ he shows that $\Theta = \sum_{s \in W} \varepsilon(s)\Theta_{s\rho}$ (W = the Weyl group of $(\mathfrak{g}_c, \mathfrak{b}_c)$ and ρ has the usual meaning), will have the required properties. Θ is essentially the character of the sum of the discrete series representations which have the same infinitesimal character as the trivial representation. It is interesting to observe that the same argument works in the $\mathfrak{p}$-adic case also; Θ may then be taken as the Steinberg character ([1973]).

The analysis of the jumps of the orbital integrals at the singular points comes down, through general arguments, to the case when the singular point is semi regular, and hence, to a calculation in $\mathfrak{sl}(2,\mathbb{R})$. Here there are two Cartan subalgebras $\mathfrak{h} = \mathbb{R}\cdot H$, $\mathfrak{b} = \mathbb{R}\cdot(X-Y)$ where

$$H = \begin{pmatrix} 1 & 0 \\ 0 & -1 \end{pmatrix}, \qquad X = \begin{pmatrix} 0 & 1 \\ 0 & 0 \end{pmatrix}, \qquad Y = \begin{pmatrix} 0 & 0 \\ 1 & 0 \end{pmatrix}.$$

The orbit E_θ of $\theta(X-Y)$ is a two-sheeted elliptic hyperboloid, and the orbit H_t of tH is a single sheeted hyperbolic hyperboloid; these are separated by C, the cone of nilpotents (light cone!), composed of two halves $C_\pm$ and the vertex 0. $\psi_{f,\mathfrak{h}}$ is C^∞ while $\psi_{f,\mathfrak{b}}$ is not. To see why, let $\theta \to 0+$, $t \to 0$. Then the $E_{\pm\theta}$ tend to $C_\pm$ while H_t tends to C (draw a figure). Thus

$$\lim_{\theta \to \pm 0} \psi_{f,\mathfrak{b}}(\pm\theta(X-Y)) = \pm\int_{C_\pm} f\, dC$$

$$\lim_{t \to 0+} \psi_{f,\mathfrak{h}}(tH) = \int_C f\, dC$$

giving the "jump formula"

$$\psi_{f,\mathfrak{b}}(0+) - \psi_{f,\mathfrak{b}}(0-) = \psi_{f,\mathfrak{h}}(0). \tag{5}$$

Generalized to arbitrary $\mathfrak{g}$, this is the basic relation that links the orbital integrals associated to different Cartan subalgebras ([1964c] Theorem 2); there is of course a corresponding version on the group. It shows that jumps arise only when *crossing root hyperplanes corresponding to singular imaginary roots*, and that for derivatives that are skew with respect to these roots, the jumps are 0 (Theorem 1, [1964c]). The continuity of $\partial(\varpi)\psi_{f,\mathfrak{h}}$ and $\varpi F_{f,L}$ follow from this, as well as the vanishing of their values at 0 and e respectively if there are real roots, i.e., if L is not fundamental. The formula (5) or rather its generalization, namely Theorem 2 of [1964c], implies that if $F_{f,\bar{L}} = 0$ for Cartan subgroups $\bar{L}$ "more split" than L, then $F_{f,L}$ is of class C^∞ on L. We have already seen in §2 how this principle was decisive in the determination of the discrete spectrum. In particular, if G is complex, more generally, if L is "Iwasawa", $F_{f,L}$ is of class C^∞; of course one can prove this by a direct evaluation of $F_{f,L}(h)$:

$$F_{f,L}(h) = d_P(h)\iint_{K\times N} f(khnk^{-1})\, dk\, dn$$

$$\left(d_P(h) = |\det(\mathrm{Ad}(h)_\mathfrak{n})|^{1/2}\right).$$

Let us now turn to the limit formula. It is enough to prove it on the Lie algebra. Write

$$T(f) = \left(\partial(\varpi)\psi_{f,\mathfrak{h}}\right)(0) \qquad (f \in \mathcal{C}(\mathfrak{g})). \tag{6}$$

T is a *tempered* invariant distribution on $\mathfrak{g}$ and the limit formula is the assertion that $T = c\delta$ where $c = c(\mathfrak{h})$ is a real constant which, for fundamental $\mathfrak{h}$, is $(-1)^q$ times a positive number, q being the integer $1/2\{\dim(G/K) - rk(G) + rk(K)\}$.

If $\mathfrak{g}$ is complex or if it has a single conjugacy class of Cartan subalgebras, one can show that (cf. [1975], §§35–36) that

$$\psi_{\hat{f},\,\mathfrak{h}} = c\hat{\psi}_{f,\,\mathfrak{h}} \qquad (f \in \mathcal{C}(\mathfrak{g})) \tag{7}$$

where c is a constant $\neq 0$ (recall that $\psi_{f,\,\mathfrak{h}}$ is smooth and so lies in $\mathcal{C}(\mathfrak{h})$). The integration formula

$$\int_{\mathfrak{g}} u \, d\mathfrak{g} = \text{const.} \int_{\mathfrak{h}} \pi \psi_{u,\,\mathfrak{h}} \, d\mathfrak{h} \qquad (u \in \mathcal{C}(\mathfrak{g})),$$

on Fourier transformation, gives the limit formula ([1954c]). For arbitrary $\mathfrak{g}$ the relation (7) has to be suitably modified, and this requires a deeper study of the Fourier transforms $\hat{\mu}_H$ of the orbital measures $\mu_H\colon f \mapsto \int_{G_*} f(H^x)\, dx^*$. Since the invariant polynomials are constant on the orbits, the $\hat{\mu}_H$ are invariant eigendistributions:

$$\partial(p)\hat{\mu}_H = \tilde{p}(iH)\hat{\mu}_H \qquad (p \in I(\mathfrak{g})) \tag{8}$$

where $\tilde{p}$ is the polynomial corresponding to p. From the theory of invariant eigendistributions (see §4 below) it follows that $\hat{\mu}_H$ is an analytic function on $\mathfrak{g}'$, say $\hat{\mu}(H\colon \cdot)$, which can be studied in detail. If $\bar{\mathfrak{h}}$ is a Cartan subalgebra, $\bar{\mathfrak{h}}^+$ a connected component of $\bar{\mathfrak{h}}'$, $\overline{W}$ the Weyl group of $(\mathfrak{g}_c, \bar{\mathfrak{h}}_c)$, and y is an element of the complex adjoint group taking $\mathfrak{h}_c$ to $\bar{\mathfrak{h}}_c$,

$$\pi(H)\bar{\pi}(\overline{H})\hat{\mu}(H\colon \overline{H}) = \sum_{s \in \overline{W}} c_s(H) e^{i\langle sH^y,\,\overline{H}\rangle} \tag{9}$$

for $H \in \mathfrak{h}'$, $\overline{H} \in \bar{\mathfrak{h}}^+$; the equations (3$\mathfrak{g}$) now imply that *the c_s are locally constant on* $\mathfrak{h}'$ (Lemma 24, [1957c]). With this proviso, (9) is a generalization of (7) and leads directly to the conclusion that $\hat{T}$ is locally constant on $\mathfrak{g}'$. But $\text{supp}(T) \subset$ the set of nilpotents of $\mathfrak{g}$ so that $\hat{T}$ is $I(\mathfrak{g})$-finite. Since an invariant $I(\mathfrak{g})$-finite distribution on $\mathfrak{g}$, which is locally constant on $\mathfrak{g}'$, is constant on $\mathfrak{g}$, we have $\hat{T} = $ a constant. This argument (see §10, [1965a]) makes use of the regularity theorem on the Lie algebra; the earlier proof in [1957d] was more direct but was also more involved.

One should think of the $\hat{\mu}_H(H \in \mathfrak{h}')$ as the analogues of the irreducible characters of G associated to L by the inducing process. For example one can prove that if $\bar{\mathfrak{h}}$ is a Cartan subalgebra, $\hat{\mu}_H$ is 0 on $(\bar{\mathfrak{h}}')^G$ if $\bar{\mathfrak{h}}$ is not conjugate to a Cartan subalgebra of the centralizer of the split component $\mathfrak{h}_{\mathbf{R}}$ of $\mathfrak{h}$; in particular this will be the case if $\dim(\bar{\mathfrak{h}}_{\mathbf{R}}) < \dim(\mathfrak{h}_{\mathbf{R}})$, i.e., $\bar{\mathfrak{h}}$ is "more compact" than $\mathfrak{h}$.

It remains to look at the case when $rk(G) = rk(K)$ and $\mathfrak{h} \subset \mathfrak{k}$. Harish-Chandra uses a fundamental solution Ξ, due to de Rham, of $\partial(\Omega)^{[n/2]}$ ($\Omega = $ the Casimir element, $n = \dim(\mathfrak{g})$), whose restriction to $\mathfrak{h}$ is a fundamental solution to $(-1)^q \partial(\Omega_{\mathfrak{h}})^{l/2}$, $l = \dim(\mathfrak{h})$. He now chooses $f \in \mathcal{C}(\mathfrak{g})$ such that

$$f(0) = 1, \qquad \psi_{f,\,\bar{\mathfrak{h}}} = 0 \quad \text{if } \bar{\mathfrak{h}} \text{ is not conjugate to } \mathfrak{h}. \tag{10}$$

Then in view of earlier remarks on the jumps of ψ_f, $\psi_{f,\,\mathfrak{h}}$ will be smooth and so will lie in $\mathcal{C}(\mathfrak{h})$; and the relation

$$ 1 = \int_{\mathfrak{g}} \Xi \cdot \left(\partial(\Omega)^{[n/2]} f \right) d\mathfrak{g} $$

will reduce to an integral on $\mathfrak{h}$ that gives the limit formula ([1964c]).

Thus everything comes down to the choice of f. Harish-Chandra takes $f = \hat{g}$ where $g \in C_c^{\infty}((\mathfrak{h}')^G)$ and $\int_{\mathfrak{g}} g \, d\mathfrak{g} = 1$; by our earlier remarks on $\hat{\mu}_H$, $\psi_{f,\bar{\mathfrak{h}}} = 0$ for $\bar{\mathfrak{h}}$ not conjugate to $\mathfrak{h}$.

The functions such as f in the above proof are truly remarkable because their Fourier analysis is controlled completely by $\mathfrak{h}$. Their analogues on the group are clearly cusp forms. Since we do not have access to these until a substantial amount of Fourier analysis on the group is done, it is now clear why the proof of the limit formula takes place on the Lie algebra. Nevertheless the reader should note how closely this proof resembles the proof, on the group, of the separation of the discrete spectrum. I think the essential originality of Harish-Chandra's treatment of the limit formula was to have conceived of the transition to $\mathfrak{g}$, and to have realized that the theory of Fourier transforms on the Lie algebra could provide him with the "additive" analogues of the cusp forms that held the key to the solution of the problem.

§4. *The Regularity Theorem.* Announced in [1963], this is the central result in the long series of papers [1964a]–[1965b] on the structure of invariant eigendistributions. Harish-Chandra's approach was to prove it first on $\mathfrak{g}$ and then carry the result over to G. For mostly technical reasons he worked on the group with invariant distributions that were $\mathfrak{Z}$-finite (resp. $I(\mathfrak{g})$-finite on $\mathfrak{g}$). The essential problem is to investigate the structure of invariant distributions around semi simple points.

His main technique for doing this was the *method of descent*. Let M be a manifold on which a Lie group L acts, $m_0 \in M$, and let E be a submanifold of M which contains m_0 and is transversal to the orbit of m_0 at m_0. Then the descent mechanism is a method of reducing questions involving invariant distributions and differential operators defined on M around m_0, to similar questions on E around m_0. The classical example is the reduction of rotationally symmetric problems to problems on a half-line.

To set up the descent machinery (see [1964a]) we need volume elements dE, dM on E and M respectively, with dM invariant under L. Assume $\pi: L \times E \to M$ is submersive and let $M' = \pi(L \times E)$. Integrating on the fibers of π with respect to the volume element $dL \, dE/dM$ will then give a "pull-back" $T \mapsto \beta_T$, taking distributions T on M' to distributions β_T on $L \times E$. If T is invariant, β_T will be of the form $1 \otimes \sigma_T$, and σ_T will be the *transfer* of T; the map $T \mapsto \sigma_T$ is injective. If F is any L-invariant function and $F_E = F|_E$, F is locally integrable on M' if and only if F_E is so on E; and $\sigma_F = F_E$. If L is unimodular and L_0 is a closed subgroup of L such that E is L_0-stable and dE is L_0-invariant, σ_T is L_0-invariant. Suppose now that D is an analytic invariant differential operator on M'. If E is sufficiently small, we can choose an analytic differential operator D' on $L \times E$, invariant under the action $x, (x', m) \mapsto (xx', m)$ of L, such that D' and D are π-related. The analytic differential operator

$\Delta(D)$ on E, obtained by retaining only the terms in D' not involving differentiations with respect to L, is called a *radial component of D*. It is in general not unique because D' is not unique; but it always satisfies

$$\sigma_{DT} = \Delta(D)\sigma_T. \tag{1}$$

If, however, E is a local section at m_0 for the action of L, then $\Delta(D)$ is uniquely determined by the requirement

$$(DF)_E = \Delta(D)F_E \qquad \left(F \in (C^\infty(M'))^L\right) \tag{2}$$

and $D \mapsto \Delta(D)$ will be a homomorphism.

In the present context it is necessary to compute explicitly the radial components of $\mathfrak{Z}$ and $\partial(I(\mathfrak{g}))$ for the actions of G on itself and on $\mathfrak{g}$, around semisimple points. Let a (resp. X) be a semisimple element of G (resp. $\mathfrak{g}$); then its centralizer Z(resp. $\mathfrak{z}$) in G (resp. $\mathfrak{g}$) is transversal to the orbit of a (resp. X). The radial components and distributions on Z (resp. $\mathfrak{z}$) are then defined around a (resp. X) and invariant under $\Xi = Z^0$. If a or X is not central, $Z(X)$ will have lower dimension than $G(\mathfrak{g})$ and so induction on dimension is possible.

Harish-Chandra's main calculations may be summarized as follows.

(i) Here G acts on $\mathfrak{g}$, X is regular, and $\mathfrak{z} = \mathfrak{h}$ is a Cartan subalgebra; $M = \mathfrak{g}'$. We take $E = \mathfrak{h}'$ and find ([1957b] Lemma 8)

$$\Delta(\partial(p)) = \pi^{-1} \circ \partial(p_\mathfrak{h}) \circ \pi \qquad (p \in I(\mathfrak{g})). \tag{3}$$

Let $\mathfrak{D}(\mathfrak{g})$ (resp. $\mathfrak{D}(\mathfrak{h})$) be the algebra of differential operators with polynomial coefficients on $\mathfrak{g}$ (resp. $\mathfrak{h}$). In [1964b] Harish-Chandra proved, as a partial extension of (3), that for $D \in \mathfrak{D}(\mathfrak{g})$, $\delta(D) = \pi \circ \Delta(D) \circ \pi^{-1}$ is in $\mathfrak{D}(\mathfrak{h})$, invariant under the Weyl group, and

$$\delta(\hat{D}) = \widehat{\delta(D)} \quad (\wedge = \text{Fourier transform}). \tag{4}$$

(ii) G acts on itself by conjugacy, $E = L'$ where L is a Cartan subgroup, and $M = G'$. Then the radial components are unique, and ([1956c] Theorem 2)

$$\Delta(z) = \Delta^{-1} \circ \mu_{\mathfrak{g}/\mathfrak{h}}(z) \circ \Delta \qquad (z \in \mathfrak{Z}) \tag{5}$$

where $\Delta = \Delta_L$ is the Weyl denominator and $\mathfrak{h} = \mathrm{Lie}(L)$.

(iii) G acts on $\mathfrak{g}$; $X \in \mathfrak{g}$ semisimple. Let $\zeta(Z) = \det(\mathrm{ad}\, Z)_{\mathfrak{g}/\mathfrak{z}}$; $\mathfrak{z}'$ is the subset of $\mathfrak{z}$ where ζ is not zero. The radial components are not unique; nevertheless, if Ω is an invariant open neighbourhood of X in $\mathfrak{g}$ and T an invariant distribution on Ω,

$$\Delta(\partial(p))\sigma_T = |\zeta|^{-1/2} \circ \partial(p_\mathfrak{z}) \circ |\zeta|^{1/2}\sigma_T \qquad (p \in I(\mathfrak{g})) \tag{6}$$

thus generalizing (3) ([1964b] Theorem 2). In particular, T is $I(\mathfrak{g})$-finite on $(\Omega \cap \mathfrak{z}')^G$ if and only if $|\zeta|^{1/2}\sigma_T$ is $I(\mathfrak{z})$-finite on $\Omega \cap \mathfrak{z}'$.

(iv) G acts on itself; $a \in G$ is semisimple. Let $\nu(y) = \det\,(\mathrm{Ad}((ay))^{-1} - 1)_{\mathfrak{g}/\mathfrak{z}})$ where $\mathfrak{z} = \mathrm{Lie}\,(Z)$. Let Ξ' be the subset of Ξ where ν is nonzero. If Ω is a completely invariant (cf. [1963]) open neighbourhood of a in G,

$$\Delta(z)\theta = \left(|\nu|^{-1/2} \circ \mu_{\mathfrak{g}/\mathfrak{z}}(z) \circ |\nu|^{1/2}\right)\theta \tag{7}$$

for all Ξ-invariant distributions θ on $\Omega_\Xi = \Xi' \cap (a^{-1}\Omega)$. In particular, an invariant distribution T on $(a\Omega_{\Xi})^G$ is $\mathfrak{Z}$-finite if and only if $|\nu|^{1/2}\sigma_T$ is $\mathfrak{Z}_a$-finite on Ω_{Ξ}; $\mathfrak{Z}_a$ = center of $U(\mathfrak{z})$ ([1965b], Lemma 22).

(v) Transfer from $\mathfrak{g}$ to G via the exponential map. Let $z \mapsto p_z$ be the isomorphism (§12, [1965b]) of $\mathfrak{Z}$ with $I(\mathfrak{g})$ such that $\mu_{\mathfrak{g}/\mathfrak{h}}(z) = (p_z)_{\mathfrak{h}}$ for all Cartan subalgebras $\mathfrak{h} \subset \mathfrak{g}$. Define the invariant function $\xi = \xi_{\mathfrak{g}}$ on $\mathfrak{g}$ by

$$\xi(X) = |\det\{(e^{\operatorname{ad} X/2} - e^{-\operatorname{ad} X/2})/\operatorname{ad} X\}|^{1/2} \qquad (X \in \mathfrak{g}).$$

Then $dx = \xi^2 dX$ ($x = \exp X$) and $|\pi|\xi = |\Delta \circ \exp|$ on any Cartan subalgebra $\mathfrak{h}$. Let U be a completely invariant open neighbourhood of 0 in $\mathfrak{g}$ such that $U_G = \exp U$ is completely invariant open in G and $\exp: U \to U_G$ is a diffeomorphism. We then have a G-equivariant pull back isomorphism $T \mapsto \tilde{T}$ of distributions from U_G to U such that at the level of locally summable functions

$$\tilde{F} = (F \circ \exp)\xi. \tag{8}$$

The main formula here is

$$(zT)^{\tilde{}} = \partial(p_z)\tilde{T} \tag{9}$$

for all invariant distributions T (§14, [1965b]). Formula (9) implies that T is $\mathfrak{Z}$-finite if and only if $\tilde{T}$ is $I(\mathfrak{g})$-finite.

The formulae (3) and (6) on the Lie algebra are proved essentially by direct calculation. But such a method cannot handle (7) or (9). It is easy in both cases to determine how the transferred differential operators act on smooth invariant functions; but their actions on invariant distributions are quite difficult to determine directly. Harish-Chandra reduced everything to calculations with invariant *smooth* functions by first proving the following remarkably general and beautiful theorem (Theorem 1, [1965b]): if D is an invariant analytic differential operator on Ω such that $Df = 0$ for all invariant $f \in C^\infty(\Omega)$, then $DT = 0$ for all invariant distributions T on Ω. Its proof comes down, via descent, to an analogous result on $\mathfrak{g}$; the proof of this however uses the regularity theorem on $\mathfrak{g}$ ([1965a] Theorems 4 and 5; for an illuminating sketch of the argument see [1963]).

Fix a completely invariant open set Ω of $\mathfrak{g}$ and an invariant $I(\mathfrak{g})$-finite distribution T on Ω. Using (3) one finds that T is an analytic function F on $\Omega' = \Omega \cap \mathfrak{g}'$; F as well as $\partial(p)F(p \in I(\mathfrak{g}))$ are locally summable on Ω. The first step in proving the regularity theorem is to show that T_F, the distribution defined by F on Ω, is $I(\mathfrak{g})$-finite. If this is done, $\mathfrak{U} = T - T_F$ is invariant, $I(\mathfrak{g})$-finite, and supported on the singular set; the second step is to show that 0 is the only such distribution.

For the first step it is a question of proving that $\partial(p)T_F = T_{\partial(p)F}$ for all $p \in I(\mathfrak{g})$. A formal argument (cf. [1957b], p. 99) reduces this to $p = \omega$, the Casimir operator. Let $J_0 = \partial(\omega)T_F - T_{\partial(\omega)F}$. J_0 can be expressed rather explicitly in terms of boundary values of the orbital integrals ([1965a] p. 15). Theorem 4 of [1964c] guarantees that such a distribution is zero on Ω if it is zero around the semi regular points of the noncompact type in Ω. For J_0, this condition is equivalent, by descent, to the regularity theorem on $\mathfrak{sl}(2,\mathbb{R})$, which can be verified by direct calculation.

The second step, by (6) and induction on $\dim(\mathfrak{g})$, reduces to showing that 0 is the only invariant $I(\mathfrak{g})$-finite distribution on $\mathfrak{g}$ supported by the set $\mathfrak{N}$ of nilpotents. The key to Harish-Chandra's proof of this (Theorems 4 and 5, [1964a]) is to view the space $\mathfrak{J}$ of invariant distributions supported by $\mathfrak{N}$ as a module for a Lie subalgebra

$\mathcal{D} \subset \mathcal{D}(\mathfrak{g})$ which is isomorphic to $\mathfrak{sl}(2,\mathbb{C})$ with a standard basis $\{H', X', Y'\}$ where $2Y' = \partial(\omega)$, $2X' = $ multiplication by $\tilde{\omega}$, and $H' = D + \frac{1}{2}\dim(\mathfrak{g})$ where D is the Euler vector field. Using descent theory and the finiteness of $G \setminus \mathfrak{N}$ it is proved that $\mathfrak{I}$ is a weight module for $\mathcal{D}$ none of whose weights ($=$ eigenvalues of H') is an integer $\geqslant 0$. This implies at once the much stronger result that 0 is the only $\partial(\omega)$-finite element of $\mathfrak{I}$. This completes the sketch of the proof of the regularity theorem on $\mathfrak{g}$; the formulae (7) and (9) then lead to the theorem on the group.

The behaviour around the singular points of the analytic functions defined by the invariant $\mathfrak{Z}$-finite (or $I(\mathfrak{g})$-finite) distributions can be studied in detail. Theorems 2, 3 and Lemma 25 of [1965a] do this on the Lie algebra while Lemmas 31, 34–37 of [1965b] treat the group situation. Let L be a Cartan subgroup and F the invariant locally summable function on Ω that defines a $\mathfrak{Z}$-finite distribution. If $\Phi_L = \Delta_L \cdot (F|_{\Omega \cap L'})$, then Φ_L extends analytically to $L'(R)$, the subset of L where no real (global) root takes the value 1. Moreover, $\varpi_L \Phi_L = \Psi_L$ extends continuously on all of $\Omega \cap L$ and the system of functions (Ψ_L) are compatible in the sense that

$$\Psi_{L_1} = \Psi_{L_2} \quad \text{on } L_1 \cap L_2 \cap \Omega. \tag{10}$$

These are generally known as the Harish-Chandra *matching conditions*; They are equivalent to the statement that $\nabla_G F$ extends to a continuous function on Ω (for the definition of ∇_G see Lemma 36 of [1965b]). Entirely analogous results are true on $\mathfrak{g}$.

§5. *Construction of the Invariant Eigendistributions* Θ_ξ. In [1965c] the regularity theorem and the theory of Fourier transforms are used to construct the distributions Θ_ξ that will be the characters of the discrete series. Let G be as in §3 with $rk(G) = rk(K)$; let $B \subset K$ be a Cartan subgroup, with $\mathfrak{b} = \text{Lie}(B)$. We select a compact form U of G_c such that $K = U \cap G$ and $B \subset U$. W and W_c denote the respective Weyl groups of (K, B) and (U, B). W_c operates on B and the $\xi \in \hat{B}$ that have trivial stabilizer in W_c are called regular. Theorem 3 of [1965c] then asserts that for any regular $\xi \in \hat{B}$ there is a unique invariant eigendistribution Θ_ξ on G such that

(a) $|\Theta_\xi| \leqslant \text{const.}|D|^{-1/2}$ on G'
(b) $\Theta_\xi = \Delta^{-1} \Sigma_{s \in W} \varepsilon(s)(s\xi)$ on $B' = B \cap G'$. $\tag{1}$

The Θ_ξ are obtained by exponentiating suitable invariant eigendistributions on the Lie algebra. The condition (a) is essentially equivalent to the *temperedness* of these distributions on the Lie algebra; it thus allows Fourier transforms to be used for constructing them. In fact, if η is the analogue of D on $\mathfrak{g}$ ($\eta = \pi^2$ on any Cartan subalgebra), the formula (8) of §4 shows that the conditions $|F| \leqslant \text{const.}|D|^{-1/2}$ and $|\tilde{F}| \leqslant \text{const.}|\eta|^{-1/2}$ are equivalent; the theory of orbital integrals can be used to show that the latter condition, for invariant eigendistributions with regular eigenvalues, is equivalent to temperedness. The fundamental result concerning the distributions on the Lie algebra is Theorem 2 of [1965c]: let $\lambda \in i\mathfrak{b}^*$ be regular; then there is a unique invariant distribution $T_\lambda = T_{\mathfrak{g},\lambda}$ on $\mathfrak{g}$ such that

(a) T_λ is tempered, $\partial(p)T_\lambda = p_\mathfrak{b}(\lambda)T_\lambda$ $(p \in I(\mathfrak{g}))$
(b) $T_\lambda = \pi^{-1}\Sigma_{s \in W}\varepsilon(s)e^{s\lambda}$ on $\mathfrak{b}' = \mathfrak{b} \cap \mathfrak{g}'$.

Let $\mathfrak{T}_\lambda$ be the space of all tempered invariant distributions T on $\mathfrak{g}$ such that

$$\partial(p)T = p_\mathfrak{h}(\lambda)T \qquad (p \in I(\mathfrak{g})).$$

Let A_λ be the space of all functions on $\mathfrak{h}$ of the form $\sum_{s \in W_c} c_s e^{s\lambda}(c_s \in \mathbb{C})$ which are W-skew. If $T \in \mathfrak{T}_\lambda$, $\pi \cdot (T|_{\mathfrak{h}'})$ belongs to A_λ (Theorem 2, [1965a]), and the uniqueness part of the theorem above is the statement that the restriction map $R_\mathfrak{h}: T \mapsto \pi \cdot (T|_{\mathfrak{h}'})$ is injective. On the other hand, $-iH_\lambda$ is in $\mathfrak{h}$ and we have

$$(-iH_\lambda)^{G_c} \cap \mathfrak{g} = \coprod_{s \in W \setminus W_c} (-isH_\lambda)^G$$

showing that the Fourier transforms $\hat{\mu}_s(s \in W \setminus W_c)$ of the invariant measures on the G-orbits $(-isH_\lambda)^G$ are linearly independent elements of $\mathfrak{T}_\lambda$ (cf. (8) of §3). Since $\dim(A_\lambda) = [W_c : W]$, the injectivity of $R_\mathfrak{h}$ will imply that $\mathfrak{T}_\lambda$ is *spanned* by the $\hat{\mu}_s$ and that $R_\mathfrak{h}$ is an isomorphism of $\mathfrak{T}_\lambda$ with A_λ. This of course will prove Theorem 2 of [1965c].

The key result is thus the injectivity of $R_\mathfrak{h}$. To get a feeling for it consider $\mathfrak{g} = \mathfrak{sl}(2, \mathbb{R})$ with $\lambda \neq 0$ in $i\mathbb{R}$, $T \in \mathfrak{T}_\lambda$, $\partial(\omega)T = \lambda^2 T$ where $\omega = H^2 + 4YX$. Let $i\lambda > 0$. Then

$$i\theta T(\theta(X - Y)) = c_\mathfrak{h}^+ e^{\lambda\theta} - c_\mathfrak{h}^- e^{-\lambda\theta} \qquad (0 \neq \theta \in \mathbb{R})$$

$$tT(tH) = a_\mathfrak{h}^+ e^{-i\lambda t} - a_\mathfrak{h}^- e^{i\lambda t} \qquad (t > 0).$$

The matching conditions give

$$c_\mathfrak{h}^+ + c_\mathfrak{h}^- = a_\mathfrak{h}^+ + a_\mathfrak{h}^- \tag{2}$$

while the temperedness of T implies that only bounded exponentials can appear in $tT(tH)$, so that $a_\mathfrak{h}^- = 0$. But then $a_\mathfrak{h}^+ = c_\mathfrak{h}^+ + c_\mathfrak{h}^-$, showing that T is completely determined by $T|_{\mathfrak{h}'}$.

Such an argument works in general. In Lemma 28 of [1965c] Harish-Chandra proves that if $\mathfrak{h}$ is a Cartan subalgebra and $\mathfrak{h}^+$ is a component of $\mathfrak{h}'(R)$, one can choose Cartan subalgebras $\tilde{\mathfrak{h}}$ "adjacent" to $\mathfrak{h}$ (i.e., $\dim(\tilde{\mathfrak{h}}_\mathbb{R}) = \dim(\mathfrak{h}_\mathbb{R}) - 1$) and components $\tilde{\mathfrak{h}}^+$ of $\tilde{\mathfrak{h}}'(R)$ such that (a) $Cl(\tilde{\mathfrak{h}}^+)$ and $Cl(\mathfrak{h}^+)$ interface along $\tilde{\mathfrak{h}} \cap \mathfrak{h}$ (which is a hyperplane in both $\mathfrak{h}$ and $\tilde{\mathfrak{h}}$); (b) the coefficients of the exponentials occurring in the formula for T on $\mathfrak{h}^+$ and those coming from $\tilde{\mathfrak{h}}^+$ are related in a manner similar to (2); (c) only bounded exponentials occur. The proof that $T = 0$ on $\mathfrak{h}' \Rightarrow T = 0$ then goes by induction on $\dim(\mathfrak{h}_\mathbb{R})$. The regularity of λ is decisive here.

For technical reasons it is necessary to prove the uniqueness even when T is not everywhere defined. The domains Ω however cannot be arbitrary; roughly speaking $\Omega \cap \mathfrak{h}$ must be "unbounded in real directions." To make this precise note first that any semisimple $X \in \mathfrak{g}$ can be uniquely written as $X = X_e + X_h$ where (i) $X_h \in [\mathfrak{g}, \mathfrak{g}]$ (to allow for cases where $\mathfrak{g}$ is only reductive), X_e, X_h are semisimple and $[X_e, X_h] = 0$; (ii) $\mathrm{ad}\, X_e$(resp. $\mathrm{ad}\, X_h$) has only pure imaginary (resp. real) eigenvalues; X_e is called the *elliptic component of* X. The uniqueness theorem is then true for the class $\mathcal{E}(\mathfrak{g})$ of open sets Ω with the following properties:

(a) Ω is completely invariant; if $X \in \Omega$, then $tX \in \Omega$ for $0 \leqslant t \leqslant 1$;

(b)   if $X \in \mathfrak{g}$ is semisimple, then Ω contains all semisimple elements X' of $\mathfrak{g}$ such that $X'_e = X_e$ (in particular $X_e \in \Omega$).

Typical examples of sets in $\mathcal{E}(\mathfrak{g})$ are $\mathfrak{g}(\varepsilon)$ $(\varepsilon > 0)$, the subset of all X in $\mathfrak{g}$ such that $|\mathrm{Im}\,\lambda| < \varepsilon$ for any eigenvalue λ of ad X. If $\mathfrak{z} \subset \mathfrak{g}$ is a reductive subalgebra containing $\mathfrak{b}$, then $\mathfrak{z} \cap \mathfrak{g}(\varepsilon) \in \mathcal{E}(\mathfrak{z})$. These are the domains that occur in [1965c].

In lifting results to G one can use the exponentiation only around *elliptic* elements of G; for, only these have centralizers in $\mathfrak{g}$ that admit Cartan subalgebras of compact type. For any $b \in B$ let $\mathfrak{z}_b$ be the centralizer of b in $\mathfrak{g}$; and for any $\varepsilon > 0$, let

$$G_\varepsilon^{(b)} = \left(b \exp\left(\mathfrak{z}_b \cap \mathfrak{g}(\varepsilon) \right) \right)^G. \tag{4}$$

The sets $G_\varepsilon^{(b)}$, which are used by Harish-Chandra in the lifting process, have remarkable properties. They are of course completely invariant, open if $\varepsilon > 0$ is sufficiently small. Moreover, if $\varepsilon_b > 0$ are arbitrary numbers, $(b \in B)$,

$$G = \bigcup_{b \in B} G_{\varepsilon_b}^{(b)} \tag{5}$$

while for $b', b'' \in B$ and $\varepsilon', \varepsilon'' > 0$ (small enough),

$$G_{\varepsilon'}^{(b')} \cap G_{\varepsilon''}^{(b'')}$$

is the union of sets of the form $G_\varepsilon^{(b)}$. The uniqueness principle on the Lie algebras $\mathfrak{z}_b$ (in their localized version) will then give, in view of these properties, the uniqueness, not only on the whole group, but also for invariant $\mathfrak{Z}$-finite distributions satisfying (a) and (b) of (1), but defined only on $G_\varepsilon^{(b)}$. For the existence, a suitable linear combination of $T_{\mathfrak{z}_b, s\lambda}$ $(s \in W)$ will lift to a distribution $\Theta_\xi^{(b)}$ on $G_{\varepsilon_b}^{(b)}$ satisfying (a) and (b) of (1) on it if $\varepsilon_b > 0$ is sufficiently small; the compatibility of these on the overlaps of the $G_{\varepsilon_b}^{(b)}$ is a consequence of the strengthened form of the uniqueness mentioned above, and so, in view of (5), we would have constructed Θ_ξ on all of G (cf. [V 3], Part II, §§2, 5).

The technique of using the matching conditions on the interfaces of adjacent Cartan subalgebras, which was used in the proof of uniqueness, has other applications. As an illustration, let us consider the distribution

$$\Theta_\xi^* = \sum_{s \in W_c} \varepsilon(s) \Theta_{s\xi}. \tag{6}$$

On B, Θ_ξ^* coincides, up to a constant factor, with the character of an irreducible finite dimensional representation of U. In particular, it is bounded and W_c-invariant. If L is any other Cartan subgroup, and $W_{L, I}$ is the subgroup of the Weyl group of (G_c, L_c) generated by the reflexions associated to the imaginary roots, the technique mentioned above can be used to prove by induction on $\dim(A)$ (where A is the $\mathbb{R}$-split component of L), that $\Theta_\xi^*|_{L'}$ is $W_{L, I}$-invariant.[†] This will imply that

$$\sup_{L'} \left(|D_L|^{1/2} |\Theta_\xi^*| \right) < \infty \tag{7}$$

where

$$D_L(x) = \det\left(\left(Ad(x^{-1}) - 1 \right)_{\mathfrak{g}/\mathfrak{z}} \right)$$

[†] It can be shown that $W_{L, I}$ operates on L and stabilizes each component of $L'(R)$; see p. 307, [1965c].

($\mathfrak{z}$ = centralizer of A in $\mathfrak{g}$) (§24, [1965c]). We have already seen in §3 the decisive role of the distributions Θ_ξ^* and the estimate (7) in the problem of extending the theory of orbital integrals to $\mathcal{C}(G)$.

§6. *Eigenfunction Asymptotics, Weak Inequality, and Analysis in the Schwartz Space.*
I shall conclude this introduction with a few comments on the Schwartz space and Harish-Chandra's treatment of the asymptotic behaviour of the matrix elements of the irreducible unitary representations and their wave packets [1958a] [1958b] [1966b] [1975] [1976a]. Let $G = KA_0 N_0$ be an Iwasawa decomposition with associated (minimal) parabolic subgroup $P_0 = M_0 A_0 N_0$; A_0^+, the positive chamber of A_0; $\mathfrak{a}_0 = \mathrm{Lie}(A_0)$ and $\rho_0 \in \mathfrak{a}_0^*$ defined as usual by $\rho_0(H) = \frac{1}{2}\mathrm{tr}(\mathrm{ad}\, H)_{\mathfrak{n}_0}$ where $\mathfrak{n}_0 = \mathrm{Lie}(N_0)$. We write σ for the spherical function on G such that $\sigma(a) = \|\log a\| (a \in A_0)$; $\sigma(x)$ is then the distance from K to xK in the Riemannian space G/K.

The concern in [1958a] [1958b] was with the elementary spherical functions φ_ν. The basic estimate for them is (Theorem 3, [1958a]) the following

$$|\varphi_\nu(a)| \leqslant C e^{-\rho_0(\log a)} (1 + \sigma(a))^m \qquad (a \in A_0^+, \nu \in \mathfrak{a}_0^*)$$

where $C > 0$, $m \geqslant 0$ are constants. In [1958b] it was proved that wave packets f of the φ_ν satisfy, for each $m \geqslant 0$,

$$|f(a)| \leqslant C_m e^{-\rho_0(\log a)} (1 + \sigma(a))^{-m} \qquad (a \in A_0^+)$$

$C_m > 0$ being a constant. Since the Jacobian for the polar decomposition $G = K\, Cl(A_0^+)K$ behaves roughly as $e^{2\rho_0}$ on A_0^+, these estimates show that the φ_ν are in $L^{2+\varepsilon}(G)$ for every $\varepsilon > 0$ and that the wave packets are actually in $L^2(G)$. They motivate to some extent the following definition ([1970c], p. 539): a continuous function f on G is said to satisfy the *weak inequality* if there are constants $C > 0$, $m \geqslant 0$, such that

$$|f(k_1 a k_2)| \leqslant C e^{-\rho_0(\log a)} (1 + \sigma(a))^m \qquad (k_1, k_2 \in K,\ a \in A_0^+). \tag{1}$$

The Schwartz space $\mathcal{C}(G)$ is then defined as the space of all $f \in C^\infty(G)$ such that for $u, v \in U(\mathfrak{g})$ and any $m \geqslant 0$, the two-sided derivative ufv of f satisfies

$$|(ufv)(k_1 a k_2)| \leqslant C e^{-\rho_0(\log a)} (1 + \sigma(a))^{-m} \qquad (k_1, k_2 \in K,\ a \in A_0^+) \tag{2}$$

for some constant $C > 0$.

Actually, if we write $\varphi_0 = \Xi$, then

$$e^{-\rho_0(\log a)} \leqslant \Xi(a) \leqslant 1 \qquad (a \in A_0^+) \tag{3a}$$

so that in (1) and (2) we can replace $e^{-\rho_0(\log a)}$ by $\Xi(a)$. Since Ξ has good properties from the point of view of harmonic analysis, this change improves the formal aspects of the theory also (cf. the proof that $\mathcal{C}(G)$ is a topological algebra under convolution, in §14, [1975]). In particular

$$\int_G \Xi^2 (1 + \sigma)^{-m}\, dx < \infty \tag{3b}$$

if $m \gg 0$.

The space $\mathcal{C}(G)$ gives rise to a natural notion of tempered distributions on G. The fundamental nature of the weak inequality is then revealed by Theorem 14.1 of [1975]: for a K-finite $\mathfrak{Z}$-finite function on G, weak inequality is equivalent to temperedness. We also mention the corresponding results for class functions. For invariant $\mathfrak{Z}$-finite distributions Θ temperedness is equivalent to the estimate

$$|\Theta| \leqslant \text{const.}|D|^{-1/2}(1+\sigma)^m \qquad (\text{pointwise on } G') \tag{4}$$

for some $m \geqslant 0$, and then

$$\Theta(f) = \int_G \Theta f \, dx \qquad (f \in \mathcal{C}(G)) \tag{5}$$

the integral converging absolutely; the counterpart to (3b) is

$$\int_G |D|^{-1/2}\Xi(1+\sigma)^{-m} dx < \infty \tag{6}$$

for $m \gg 0$ (cf. [1975], §§10–14; see also the introductory remarks to [1970a]). In (4) the factor $(1+\sigma)^m$ can be dropped if Θ is an eigenfunction for a *regular* infinitesimal character, thereby elucidating the meaning of the condition $\sup|D|^{1/2}|\Theta_\xi| < \infty$ in the construction of the distributions Θ_ξ. Temperedness (of the characters as well as the matrix elements) is the characteristic property of the series of representations $(\pi_{\eta,\nu})$ associated to the various Cartan subgroups of G.

In [1958b] Harish-Chandra discovered that the differential equations satisfied by the φ_ν become (roughly speaking), at infinity on A_0, differential equations with *constant coefficients*. This fact led to the result that φ_ν is asymptotically just a sum of plane waves. In [1966b] Harish-Chandra extended this method to handle all tempered K-finite $\mathfrak{Z}$-finite functions. Let τ be a finite dimensional double unitary representation of K and let $\mathcal{Q}(G, \tau)$ be the space of all τ-spherical tempered $\mathfrak{Z}$-finite functions on G. Let $f \in \mathcal{Q}(G, \tau)$. Fix a parabolic subgroup $P = MAN$. Then $G = K(MA)K$, and one can use the τ-sphericity of f to reduce the differential equations satisfied by f on G to a system of differential equations on MA; in the limit on MA, when $a \in A$ and $a \to {}_P\infty$ (this means that a goes to infinity in such a way that for some $\varepsilon > 0$, $\alpha(\log a) \geqslant \varepsilon\sigma(a)$ for all roots α of (P, A)), the latter differential equations become the ones that define elements of $\mathcal{Q}(MA, \tau_M)$. This suggests that f may be approximated by an element of $\mathcal{Q}(MA, \tau_M)$ at infinity on MA. That this is true is the content of Theorem 21.1 of [1975]; $f_P \in \mathcal{Q}(MA, \tau_M)$ defined there is the constant term of f along P. The fact that f_P is a good approximation to f in a suitable regime is contained in the following estimate (Lemma 23.4 [1975]):

$$|e^{\rho_0(\log a)}f(a) - e^{\rho_{00}(\log a)}f_P(a)| \leqslant C(1+\sigma(a))^m e^{-\kappa\rho_0(\log a)}; \tag{7}$$

here ρ_{00} is the "ρ_0" of MA, $\kappa > 0$, $C > 0$, $m \geqslant 0$ are constants, and a is to be restricted to regions of the following form ($\varepsilon > 0$ can be arbitrary)

$$A_0^+(P:\varepsilon) = \{a \in Cl(A_0^+) \,|\, \alpha(\log a) \geqslant \varepsilon\rho_0(\log a) \quad \text{for all roots } \alpha \text{ of } (P, A)\}. \tag{8}$$

We have already seen in §2 how decisive the theory of the constant term is in the treatment of the discrete spectrum. In particular the estimates (7) on sets $A_0^+(P:\varepsilon)$

show that for an $f \in \mathcal{Q}(G, \tau)$,

$$f \in L^2(G, \tau) \Leftrightarrow f \in \mathcal{C}(G, \tau) \Leftrightarrow f_P = 0 \quad \text{for all } P.$$

Let me mention now another illustration of the method of differential equations. Let g be in $\mathcal{C}(G) \cap {}^0L^2(G)$. We can write $g = \Sigma_j f_j$ where the f_j are K-finite eigenfunctions, the sum being convergent in $L^2(G)$. The f_j of course have vanishing constant terms and so decay exponentially in view of (7); by being careful in the analysis that led to (7) one can keep track of the dependence of (7) on the parameters of f_j (the class $\omega \in \hat{G}_d$ to which f_j belongs, and the K-types according to which f_j transforms). The result is that the series $\Sigma_j f_j$ is convergent *in the topology of the Schwartz space*. On the other hand, the f_j, being $\mathfrak{Z}$-finite functions in $\mathcal{C}(G)$, are cusp forms (Theorem 18.1, [1975]) so that g is a cusp form, showing that $\mathcal{C}(G) \cap {}^0L^2(G)$ is precisely the space of cusp forms ([1970c], §8; [V 3], Part II, §16).

In the theory of the continuous spectrum one considers families of eigenfunctions, for instance the Eisenstein integrals. Harish-Chandra introduces the concept of (τ-spherical) eigenfunctions (ϕ_ν) of type II(λ) associated to a Cartan subgroup $L = T \cdot A$ (§8, [1976a]) here $\lambda \in it^*$ ($t = \mathrm{Lie}(T)$) is regular, ν varies in $\mathfrak{a}^*$ and ϕ_ν is smooth in ν, satisfies the weak inequality in a uniform way, and satisfies the differential equation

$$z\phi_\nu = \mu_{\mathfrak{g}/\mathfrak{h}}(z)(\lambda + i\nu)\phi\nu \quad (z \in \mathfrak{Z}).$$

The fundamental idea now is to form wave packets

$$\phi_\alpha = \int_{\mathfrak{a}^*} \alpha(\nu)\phi_\nu \, d\nu \quad (\alpha \in \mathcal{C}(\mathfrak{a}^*)) \tag{9}$$

and to try to prove that $\phi_\alpha \in \mathcal{C}(G, \tau)$. This may not always be true even if $\alpha \in C_c^\infty(\mathfrak{a}^*)$. The point is that the constant terms $(\phi_{P,\nu})(P \in \mathcal{P}(A))$ are not eigenfunctions on MA but only linear combinations of such (this is because the asymptotic form on MA of the differential operator $z \in \mathfrak{Z}$ is $\mu_{\mathfrak{g}/\mathfrak{h}}(z)$, and $\mu_{\mathfrak{g}/\mathfrak{h}}(\mathfrak{Z}) \neq \mathfrak{Z}_{MA}$, the center of $U(\mathfrak{m} \oplus \mathfrak{a})$). For regular $\nu \in \mathfrak{a}^*$ one can write $\phi_{P,\nu}$ as $\Sigma_{s \in W(A)}\phi_{P,s,\nu}$; the $\phi_{P,s,\nu}$ will now be eigenfunctions (these statements have to be formulated with some care; see §§8–9 of [1976a]) but they may become singular in ν if ν approaches root hyperplanes in $\mathfrak{a}^*$ (for instance, in the spherical case, they are $c(s\nu)e^{is\nu}$). To overcome this, Harish-Chandra introduces the concept of eigenfunctions (ϕ_ν) of type II$'(\lambda)$ in §9, loc.cit, and proves that if (ϕ_ν) is of type II$'(\lambda)$, $\phi_\alpha \in \mathcal{C}(G, \tau)$ for any $\alpha \in \mathcal{C}(\mathfrak{a}^*)$, and $\alpha \mapsto \phi_\alpha$ is a continuous map (Theorem 13.1, loc.cit). The usefulness of this concept becomes clear from Theorem 9.1 of loc.cit asserting that if (ϕ_ν) is of type II(λ), $(\varpi(\lambda + i\nu)\phi_\nu)$ is of type II$'(\lambda)$. Theorem 11.1 of loc.cit actually gives a necessary and sufficient condition that (ϕ_ν) which is of type II(λ), is actually of type II$'(\lambda)$.

The wave packet theorem is proved by induction on $\dim(G)$. The basic point of course is to check that if (ϕ_ν) is of type II$'(\lambda)$ on G, the $(\phi_{P,s,\nu})$ lead to eigenfunctions of type II$'(\lambda)$ on MA (Lemma 9.1, [1976a]). If one had simply restricted oneself to Eisenstein integrals, the inductive step would have become much more complicated.

The wave packets of $E_{P,\psi,\nu}$ define *essentially* the inverse of the Fourier transform. Essentially, because we are using only $d\nu$ as the weight and not the Plancherel measure; or (equivalently), because we have not yet determined the true normalization of the $E_{P,\psi,\nu}$. To correct this one has to compute the Fourier transform of the wave packet ϕ_α and relate it to α. Theorem 13.2 of [1976a] does this. It is actually a generalization of the corresponding theorem 4 of [1958b] for spherical functions, and is based on the same circle of ideas. What it says is that if $\overline{P}$ is the parabolic subgroup opposite to P, the integral

$$\phi_\alpha^{\overline{P}}(m) = d_P(m)\int_{\overline{N}}\phi_\alpha(\bar{n}m)\,d\bar{n} \qquad (m \in MA)$$

can be evaluated by substituting for ϕ_α the "truncated wave packets"

$$\phi_{P,\alpha}(m) = \int_{\mathfrak{a}^*}\alpha(\nu)\phi_{P,\nu}(m)\,d\nu \qquad (m \in MA)$$

(which are obtained by replacing ϕ_ν by its constant term in the formation of the wave packet). If (ϕ_ν) is the Eisenstein integral associated to a $Q \in \mathscr{P}(A)$, $\phi_{P,\nu}$ is expressible as a sum of plane waves $c_{P|Q}(s:\nu)e^{is\nu}$, and the integral of the truncated wave packet has an explicit formula in which the c-functions enter directly (Theorem 19.2, [1976b]). The theory of the c-functions and the functional equations of the Eisenstein integrals are now combined to derive the explicit Plancherel formula. I should mention however one technical difficulty in forming wave packets with the properly normalized weight function, namely, the control of the growth of the Plancherel measure at infinity on $\mathfrak{a}^*$; to overcome this one already needs the product formula and the explicit calculations in the case when $\dim(A)$ is equal to 1. I refer the reader to [1976b] for details.

References

[AM]   Auslander, L., and Moore, C. C., Unitary representations of solvable Lie groups. *Memoirs of the Amer. Math. Soc.* **62** (1966), 1–199.

[B]   Bargmann, V., Irreducible unitary representations of the Lorentz group, *Ann. of Math.* **48** (1947), 568–640.

[Bl]   Blattner, R. J., *General Background. Harmonic Analysis and Representations of Semisimple Lie Groups.* Edited by J. A. Wolf, M. Cahen, and M. De wilde. D. Reidel Publishing Company, Holland, 1980, pp. 1–67.

[Bo 1]   Borel, A., *Introduction aux groupes arithmétiques.* Hermann, Paris, 1969.

[Bo 2]   —, *Formes automorphes et séries de Dirichlet.* Springer Lecture Notes **514** (1976), pp. 183–222.

[Br 1]   Bruhat, F., Sur les représentations induites des groupes de Lie. *Bull. Soc. Math. France*, **84** (1956), 97–205.

[Br 2]   —, Sur les représentations des groupes classiques p-adiques, I, II. *Amer. J. of Math.* **83** (1961), 321–338, 343–368.

[G]   Gel'fand, I. M., Automorphic functions and the theory of representations. *Proc. Int. Cong. of Mathematicians, Stockholm, 1962*, pp. 74–85.

[GG 1] Gel'fand, I. M., and Graev, M. I., On a general method of decomposition of the regular representation of a Lie group into irreducible representations. *Dokl. Akad. Nauk. SSSR*, **92** (1953), 221–224.

[GG 2] —, The analogue of Plancherel's theorem for real unimodular groups. *Dokl. Akad. Nauk, SSSR* **92** (1953), 461–464.

[GN 1] Gel'fand, I. M., and Neumark, M. A., Unitary representations of the Lorentz group. *Izvestiya. Akad. Nauk. SSSR* **11** (1947), 411–504.

[GN 2] —, Unitary representations of the classical groups. *Trudy Mat. Inst. Steklova*, **36** (1950), 1–288.

[GP] Gel'fand, I. M., and Piatetsky-Shapiro, I. I., Unitary representations in the G/Γ space, where G is the group of real n^{th} order matrices and Γ is the subgroup of integral matrices. *Dokl. Akad. Nauk. SSSR* **147** (1962), 275–278.

[GR] Gel'fand, I. M., and Raikov, D. A., Irreducible unitary representations of arbitrary locally bicompact groups. *Mat. Sbornik* (N.S) **13** (55) (1943), 301–316.

[Go] Godement, R., Sur les relations d'orthogonalité de V. Bargmann. I.II. I: Résultats préliminaires, *C.R. Acad. Sci. Paris.* **225** (1947), 521–523; II: Démonstration génerale, *C.R. Acad. Sci. Paris*, **225** (1947), 657–659.

[JL] Jacquet, H., and Langlands, R. P., *Automorphic Forms on* GL (2), Springer Lecture Notes **114** (1970).

[L 1] Langlands, R. P., *On the Functional Equations Satisfied by Eisenstein Series*. Springer Lecture Notes **544** (1976).

[L 2] —, *Euler Products*. Yale University Press, 1967.

[L 3] —, *Problems in the Theory of Automorphic Forms. Lectures in Modern Analysis and Applications*, Springer Lecture Notes, **170** (1970), 18–86.

[Lo] Loomis, L. H., *An Introduction to Abstract Harmonic Analysis*. Van Nostrand, New York, 1953.

[M] Maass, H., Über eine neue Art von nichtanalytischen automorphen Funktionen. *Math. Ann.* **121** (1949), 141–183.

[Mau 1] Mautner, F. I., Spherical functions over p-adic fields, I. *Amer. J. of Math.*, **80** (1958), 441–457.

[Mau 2] —, Spherical functions over p-adic fields, II. *Amer. J. of Math.* **86** (1964), 171–200.

[Ma 1] Mackey, G. W., *The Theory of Unitary Group Representations. Chicago Lectures in Mathematics*. The Chicago University Press, Chicago and London, 1976.

[Ma 2] —, Infinite dimensional group representations. *Bull. Amer. Math. Soc.* **69** (1963), 628–686.

[Ma 3] —, *Unitary Group Representations in Physics, Probability, and Number Theory*. Benjamin, 1978.

[Mo] Moore, C. C., Representations of Solvable and nilpotent groups and harmonic analysis on nil and solvmanifolds. Harmonic analysis on homogeneous spaces. *Proceedings of Symposia in Pure Mathematics*, **XXVI**. Edited by C. C. Moore. *Amer. Math. Soc.* 1973, pp. 3–44.

[MvN] Murray, F. J., and von Neumann, J., On rings of operators. I, II, IV. I: *Ann. of Math.* **37** (1936), 116–229; II: *Trans. Amer. Math. Soc.* **41** (1937), 208–248; IV: *Ann. of Math.* **44** (1943), 716–808.

[Se 1] Selberg, A., Harmonic analysis and discontinuous groups in weakly symmetric Riemannian spaces with applications to Dirichlet series. *J. Indian Math. Soc.* **20** (1956), 47–87.

[Se 2] —, Discontinuous groups and harmonic analysis. *Proc. Int. Cong. of Mathematicians, Stockholm* (1962), 177–189.

[S] Schwartz, L., *Théorie des Distributions*, Hermann, Paris, 1973 (Nouvelle édition).

[T]   Tate, J., Fourier analysis in number fields and Hecke's Zeta functions. Thesis (Princeton), 1950. Reproduced in *Algebraic Number Theory*, Edited by J. W. S. Cassels and A. Frölich, Thompson Book Company Inc., Washington D.C., 1967.

[vN]   von Neumann, J., (a) On rings of operators III. *Ann. of Math.* **41** (1940), 94–161. (b) On rings of operators. Reduction theory. *Ann. of Math.* **50** (1949), 401–485.

[V 1]   Varadarajan, V. S., *Lie Groups, Lie Algebras, and Their Representations*. Prentice Hall, Englewood Cliffs, N.J., 1974.

[V 2]   —, The theory of characters and the discrete series for semisimple groups. Harmonic analysis on homogeneous spaces, *Proceedings of Symposia in Pure Mathematics*, **XXVI**. Edited by C. C. Moore. *Amer. Math. Soc.* 1973, pp. 45–99.

[V 3]   —, *Harmonic Analysis on Real Reductive Groups*. Springer Lecture Notes **576** (1976).

[W 1]   Weil, A., *L'integration dans les groupes topologiques et ses applications*. Hermann, Paris, 1940.

[W 2]   —, *Basic Number Theory*. Springer-Verlag, New York, 1967.

[W 3]   —, (a) Sur certains groupes d'opérateurs unitaires *Acta Math.* **111** (1964), 143–211. (b) Sur la formule de Siegel dans la théorie des groupes classiques. *Acta Math.* **113** (1965), 1–87.

[W 4]   —, Über die bestimmung Dirichletscher Reihen durch Funktionalgleichungen. *Math. Ann.* **168** (1967), 149–156.

[W 5]   —, *Automorphic Forms and Dirichlet Series*. Springer Lecture Notes **189** (1971).

[We 1]   Weyl, H., Über gewöhnliche Differentialgleichungen mit singularitäten und die zugehorigen Entwicklungen willkürlicher Funktionen. *Math. Ann.* **68** (1910), 220–269.

[We 2]   —, Theorie der Darstellung kontinuierlicher halbeinfacher Gruppen durch lineare Transformationen. I, II, III, und Nachtrag. I: *Math. Zeist.* **23** (1925), 271–309; II: *Math. Zeist.* **24** (1926), 328–376; III: *Math. Zeist.* **24** (1926), 377–395; Nachtrag: *Math. Zeist.* **24** (1926), 789–791.

[Wi]   Wigner, E. P., On unitary representations of the inhomogeneous Lorentz group. *Ann. Math.* **40** (1939), 149–204.

Some Additional Aspects of Harish-Chandra's Work on Real Reductive Groups

NOLAN R. WALLACH

The main emphasis in the introduction to these volumes is the work of Harish-Chandra which led to the Plancherel theorem for reductive groups over local fields. This work did not historically follow a straight line and it included important results that are not directly related to the main theme. We include here a partial guide to some of these theorems and ideas. We label each paper to be discussed using the scheme of the Bibliography, e.g. [1951a].

[1951a] In this paper Harish-Chandra shows how to simultaneously construct a semisimple Lie algebra and all of its irreducible finite dimensional representations from its Cartan matrix. Recall that a Cartan matrix is an integral $l \times l$ matrix

$$A = (a_{ij})$$

with

(1) $a_{ii} = 2,\ a_{ij} \leqslant 0,\ i \neq j$;
(2) if $a_{ij} = 0$, then $a_{ji} = 0$;
(3) $\det(A) \neq 0$;
(4) the group generated by the transformations

$$s_i \colon x_j \mapsto x_j - a_{ji} x_i$$

is finite.

Starting with such a matrix A he studies the algebra (either associative or Lie) with generators $X_i, Y_i, H_i,\ 1 \leqslant i \leqslant l$, and the relations

(R)
$$
\begin{aligned}
&[H_i, H_j] = 0, \qquad [X_i, Y_j] = \delta_{ij} H_i \\
&[H_i, X_j] = a_{ji} X_j, \qquad [H_i, Y_j] = -a_{ji} Y_j.
\end{aligned}
$$

At first Harish-Chandra studies the associative algebra subject to (R). He later looks at the Lie algebra subject to (R) (the Lie subalgebra generated by the X_i, Y_i, H_i in the associative algebra). Given a linear functional Λ on the span of the H_i with $\Lambda(H_i) = n_i$, $n_i \in \mathbb{Z}$, $n_i \geq 0$, he constructs a module for the algebra by taking a suitable quotient of what has come to be called a Verma module. He then takes its unique irreducible quotient (which we denote), $L(\Lambda)$. If $\mathfrak{g}'$ is the Lie algebra determined by the relations (R) and if $I = \{ X \in \mathfrak{g}' | X$ acts by zero on every $L(\Lambda)\}$, then $\mathfrak{g} = \mathfrak{g}'/I$ defines the semisimple Lie algebra. The $L(\Lambda)$ are the finite dimensional irreducible representations of $\mathfrak{g}$. Parts of this argument are attributed by Harish-Chandra to Chevalley.

It is of some interest that the conditions (3), (4) are used only to guarantee that $\dim \mathfrak{g} < \infty$ and $\dim L(\Lambda) < \infty$. If the conditions (3) and (4) are dropped, the construction of Harish-Chandra gives the Kac-Moody Lie algebras and their standard modules.

[1953] In the first part of this paper Harish-Chandra initiates the algebraic theory of $(\mathfrak{g}, \mathfrak{k})$-modules (or Harish-Chandra modules). Here $\mathfrak{g}$ is a semisimple Lie algebra over the real numbers, θ is a Cartan involution of $\mathfrak{g}$, and $\mathfrak{k}$ is the fixed point algebra of θ. A $(\mathfrak{g}, \mathfrak{k})$-module is a $\mathfrak{g}$-module, M, over $\mathbb{C}$ that splits into a (not necessarily finite) direct sum of irreducible $\mathfrak{k}$-modules. Harish-Chandra calls such modules quasi-semisimple for $\mathfrak{k}$. The basic theorem in the first part is Theorem 1 which may be stated as follows. Let $U(\mathfrak{g}_{\mathbb{C}}) \supset U(\mathfrak{k}_{\mathbb{C}})$ be the universal enveloping algebras of $\mathfrak{g}_{\mathbb{C}}$ and $\mathfrak{k}_{\mathbb{C}}$ respectively. Let $Z(\mathfrak{g}_{\mathbb{C}})$ be the center of $U(\mathfrak{g}_{\mathbb{C}})$. The theorem then says: if W_1, W_2 are finite dimensional semisimple $\mathfrak{k}_{\mathbb{C}}$-modules then

$$\mathrm{Hom}_{\mathfrak{k}}\left(W_1, U(\mathfrak{g}_{\mathbb{C}}) \underset{U(\mathfrak{k}_{\mathbb{C}})}{\otimes} W_2 \right)$$

is finitely generated as a $Z(\mathfrak{g}_{\mathbb{C}})$-module (the action of $Z(\mathfrak{g}_{\mathbb{C}})$ is $(z \cdot T)(v) = z \cdot T(v)$).

This theorem in particular implies that a $(\mathfrak{g}, \mathfrak{k})$-module M that is finitely generated and is $Z(\mathfrak{g}_{\mathbb{C}})$-finite ($\dim Z(\mathfrak{g}_{\mathbb{C}}) \cdot m < \infty$, $m \in M$) has finite $\mathfrak{k}$-multiplicities, i.e., is admissible. This implication is the starting point for the theory of admissible $(\mathfrak{g}, \mathfrak{k})$-modules. For the special case when $\mathfrak{g}$ is a complex Lie algebra looked upon as a real Lie algebra this result (as well as the basic idea of its proof) is in [1951a]. In [1951a] Harish-Chandra indicates that the result in this special case was suggested to him by Mautner.

In the second part of this paper Harish-Chandra initiates the theory of analytic vectors for Banach space representations of Lie groups. These vectors are called well-behaved in this paper. Let (π, H) be a Banach representation of a connected Lie group G. Let H^{ω} be the space of analytic vectors. His main results are that H^{ω} is dense in H and is $\mathfrak{g}$-invariant, and that the closure of a $\mathfrak{g}$-invariant subspace of H^{ω} is G-invariant.

Let G be connected semisimple with finite center and let $K \subset G$ be the analytic subgroup of G corresponding to $\mathfrak{k}$. Let (π, H) be a Banach representation of G such that if $v \in H^{\omega}$ $\dim Z(\mathfrak{g}_{\mathbb{C}})v < \infty$, and suppose that there exist $v_i \in H^{\omega} \cap H_K$, $1 \leq i \leq N$ such that H is the closed linear span of $\pi(x)v_i$, $x \in G$, $1 \leq i \leq N$; here $H_K = \{v \in H | \pi(K) \cdot v$ spans a finite dimensional space$\}$. Then Theorem 1 and the theory of analytic vectors combine to prove

(1) $H_K \subset H^\omega$; all the K-multiplicities in H are finite.
(2) $V \mapsto \mathrm{Cl}(V)$ is a bijection between the lattice of $\mathfrak{g}$-stable subspaces of H_K and the lattice of G-stable closed subspaces of H.

These results are then used by him to prove that a connected semisimple Lie group is of type I in the sense of Murray and von Neumann.

[1954a] This paper contains the proof of his celebrated subquotient theorem. Let G be linear, connected and semisimple. The theorem asserts that any irreducible $(\mathfrak{g}, \mathfrak{k})$-module that integrates to a K-module is isomorphic to a subquotient of the space of K-finite vectors of a representation (in general nonunitarily) induced from a one dimensional representation of an Iwasawa subgroup of G. He was later able to drop the linearity condition by using his results on differential equations [1960a].

[1954c] The main result of this paper is the Plancherel theorem for semisimple Lie groups defined over the field of complex numbers. Theorems 1 and 2 give the formula for the characters of the principal series representations for arbitrary connected semisimple Lie groups over $\mathbb{R}$, with finite center. The Plancherel theorem is based on this character computation and an integro-differential formula (Lemma 14) which generalizes earlier work of Gel'fand and Naimark. He generalized Lemma 14 to arbitrary semisimple Lie algebras in [1957c,d]; this provided one of the main steps in his proof of the Plancherel theorem for real reductive groups (see Varadarajan's introduction). Lemma 10 and its Corollary are important ingredients in the proofs of Lemma 14 and Theorem 2. He later used this result in his proof [1956d] of the Bruhat lemma.

[1956b] Section 7 of this paper is titled "A Digression on a Theorem of Cartan". Harish-Chandra considers the case $\mathfrak{g}_\mathbb{C}$ simple, and $\mathfrak{g}_\mathbb{C} = \mathfrak{k}_\mathbb{C} \oplus \mathfrak{p}_\mathbb{C}$ (complexified Cartan decomposition) with $\mathfrak{p}_\mathbb{C} = \mathfrak{p}^+ \oplus \mathfrak{p}^-$ a direct sum of two Ad K-invariant nonzero subspaces. He had earlier shown [1955c] that there is $J \in \mathfrak{k}$ with ad $J|_{\mathfrak{p}^+} = iI$, ad $J|_{\mathfrak{p}^-} = -iI$, ad $J|_{\mathfrak{k}} = 0$. Thus $[\mathfrak{p}^+, \mathfrak{p}^+] = [\mathfrak{p}^-, \mathfrak{p}^-] = 0$. Let G be linear; then G is contained in $G_\mathbb{C}$, a connected Lie group with Lie algebra $\mathfrak{g}_\mathbb{C}$. By going to a covering one may assume that $G_\mathbb{C}$ is simply connected. In $G_\mathbb{C}$ there is the connected parabolic subgroup

$$K_\mathbb{C} \exp \mathfrak{p}^+.$$

In [1956a] it is shown that

$$G K_\mathbb{C} \exp \mathfrak{p}^+ \subset (\exp \mathfrak{p}^-) K_\mathbb{C} (\exp \mathfrak{p}^+)$$

is an open subset, and that

$$G \cap (K_\mathbb{C} \exp \mathfrak{p}^+) = K.$$

One therefore has

$$G K_\mathbb{C} \exp \mathfrak{p}^+ = \exp(\Omega) K_\mathbb{C} \exp \mathfrak{p}^+$$

with $\Omega \subset \mathfrak{p}^-$ an open subset.

This gives rise to a diffeomorphism

$$\psi: G/K \to \Omega$$

given by

$$(\exp \psi(g))K_{\mathbb{C}}\exp \mathfrak{p}^{+} = gK_{\mathbb{C}}\exp \mathfrak{p}^{+}$$

Using ad J (defined above) one shows that G/K has a G-invariant complex structure and it is easy to see that $\psi: G/K \to \Omega$ is complex analytic. Using his theory of strongly orthogonal roots and Lemma 20 Harish-Chandra shows that Ω is a bounded, symmetric domain in $\mathfrak{p}^{-}$. This gives a proof without going through classification of Cartan's basic theorem that every Hermitian symmetric domain is biholomorphic with a bounded symmetric domain. The map ψ is now called the Harish-Chandra imbedding and is the starting point in the study of Hermitian symmetric domains.

In this paper there is also a very interesting argument which transfers the calculation of an integral on G to the calculation of a corresponding integral on a compact form of $G_{\mathbb{C}}$ (see Lemma 28).

[1960a] This previously unpublished paper extends work in [1958a,b]. In this paper Harish-Chandra generalizes the classical theory of regular singular points and Frobenius's method to several variables. Most notable is his handling of what is now called the asymptotic expansion along the walls. These results were basic to the Langlands classification of irreducible $(\mathfrak{g}, \mathfrak{k})$-modules.

[1962] In this paper with A. Borel a general reduction theory of arithmetic subgroups of semisimple Lie groups over $\mathbb{R}$ is developed. Theorem 9.4 says that the volume of a fundamental domain for such a discrete subgroup is finite. Theorem 11.8 (which gives a positive answer to Godement's conjecture) is now called the Borel–Harish-Chandra criterion for compactness of a fundamental domain. It says that if $\Gamma \subset G$ is arithmetic then G/Γ is compact if Γ has no unipotent elements. This criterion is now the main method of proving compactness for arithmetic quotients.

[1976b] This paper is the culmination of Harish-Chandra's work on the Plancherel formula for real reductive groups. In addition to the completion of the proof of the Plancherel theorem this paper contains the completeness theorem for intertwining operators for cuspidal principal series representations (Theorem 37.1). This theorem is the basic ingredient in the analysis of the irreducibility of the cuspidal principal series representations.

[1983] The purpose of this paper is to study the basic ingredients that will be necessary for the spectral analysis of tempered, invariant distributions (for example the orbital integrals) on a real reductive group G. As in the case of the Plancherel theorem (which is concerned with the spectral analysis of the Dirac delta function at the identity of G) it is necessary to have the correct notion of the constant term. The theory of the constant term of a tempered, $Z(\mathfrak{g})$-finite, central distribution on G is one of the main themes of this article.

Let Θ be a tempered, invariant, $Z(\mathfrak{g})$-finite distribution on G. Let $K \subset G$ be a maximal compact subgroup of G. Let P be a proper parabolic subgroup of G with standard Levi decomposition $P = MN$. The constant term of Θ along P, Θ_P, should be a $Z(\mathfrak{m})$-finite tempered, central distribution on M which is the limit in a suitable sense of Θ along the positive chamber relative to P in A (= the standard split component of M). If (π, H) is an irreducible tempered representation of G and if Θ is the character of π then there is a natural notion of Θ_P which we now describe. If H_K is the space of K-finite vectors of H then

$$H_K / \mathfrak{n} \cdot H_K$$

is an admissible, finitely generated, $(\mathfrak{m}, K \cap M)$-module. As a module for $\mathfrak{a}$, it splits into a direct sum of generalized weight spaces for $\mathfrak{a}$:

$$(H_K / \mathfrak{n} \cdot H_K)_\lambda.$$

Let ρ be as usual the differential of the square root of the modular function, δ, of P. Let $\bar{P} = \theta(P)$. Then $\Theta_{\bar{P}}$ is $\delta^{1/2}$ times the sum of the characters of the $(\mathfrak{m}, K \cap M)$-modules

$$(H_K / \mathfrak{n} \cdot H_K)_\lambda$$

with

$$\mathrm{Re}(\lambda - \rho, \alpha) = 0$$

for α a root of (P, A). Harish-Chandra's definition in the general case is consistent with this definition for characters.

A tempered, invariant, $Z(\mathfrak{g})$-finite distribution Θ is called *super tempered* if $\Theta_P = 0$ for all proper parabolic subgroups P of G. Thus the super tempered distributions are the analogues of the cusp forms of G. Using the results in [1960a] it is not difficult to see that if Θ is the character of an irreducible representation π of G which is tempered and unitary, then Θ is super tempered if and only if π is square integrable. However, there are super tempered, invariant, eigendistributions that are not linear combinations of characters of discrete series representations. For example, if (π, H) is the reducible unitary principle series representation of SL $(2, \mathbb{R})$, if π^+ and π^- are the irreducible constituents of π, and if $\Theta^{\pm}$ are the characters of $\pi^{\pm}$, then $\Theta^+ - \Theta^-$ is super tempered, although neither Θ^+ nor Θ^- is super tempered. One of the main results of this paper is that all super tempered eigendistributions that are not linear combinations of discrete series characters are given in terms of distribu-. tions generalizing $\Theta^+ - \Theta^-$ for SL $(2, \mathbb{R})$.

The Work of Harish-Chandra on Reductive p-adic Groups

Roger Howe

Harish-Chandra's work on p-adic groups is broadly based on his experience with Lie groups. (In this survey, the term *Lie group* means an analytic group over $\mathbf{R}$.) He himself made the connection quite explicit, formulating what he called the Lefschetz Principle and the Philosophy of Cusp Forms. His Lefschetz Principle [1970d] is a paraphrase for harmonic analysis of the rule of thumb of the same name from algebraic geometry. It asserts that the phenomena of representation theory for groups over p-adic fields, finite fields or adele rings are, suitably interpreted, essentially the same as for groups over $\mathbf{R}$, i.e. Lie groups. In applying this principle the "suitable interpretation" is sometimes less than obvious; there are many respects in which p-adic harmonic analysis might seem rather different from real harmonic analysis. Nevertheless, it is a valuable principle, and there have been many examples where one theory will enlighten the other, or an attempt to reconcile apparent differences will lead to insight in both theories. For example, the cleanness of the theory of the constant term in the p-adic case led to a reworking of it for real groups [CM], [Wa]. It is a tribute to the robustness of Harish-Chandra's methods and to his tenacity that, despite serious obstacles including a lack of detailed understanding of the discrete series, characters and orbital integrals, he was able to reach a main goal, the Plancherel Formula [1977b].

The Philosophy of Cusp Forms is much more specific to harmonic analysis than is the Lefschetz Principle; it might be regarded as the primary concrete embodiment of that Principle in representation theory of reductive groups. It is most explicitly stated in [1970b] where it is traced back as far as a talk of Gel'fand at the Stockholm International Congress. It has become a basic feature of reductive harmonic analysis. In particular it is a unifying feature of Harish-Chandra's work.

The gist of the Philosophy of Cusp Forms is that the collection of all representations of a reductive group G should be partitioned into disjoint classes, with each class being attached to a certain parabolic subgroup. (Or more precisely, to a class of associated parabolic subgroups. Recall that two parabolic subgroups are *associated* if their Levi components (the reductive components in their Levi decompositions) are conjugate.) The representations attached to G itself are called *cuspidal* representations and the matrix coefficients of cuspidal representations are *cusp forms*. The representations associated to a proper parabolic subgroup P are further associated to Weyl group orbits of cuspidal representations of M, the Levi component of P, and this further association is finite-to-one. Moreover, the representations of G associated to a given cuspidal representation σ of M may be recovered as components of the induced representation

$$\text{ind}_P^G\left(\sigma \otimes \delta_P^{1/2}\right)$$

where σ is extended from M to P by letting it be trivial on the unipotent radical of P, and δ_P is the modular function of P. Finally, the method of passing from a representation of G to its associated P and representation σ of M is explicit; it is based on what Harish-Chandra calls the "theory of the constant term".

Thus the Philosophy of Cusp Forms provides an inductive strategy for understanding the representations of G. The total problem is divided into two parts: to determine the cusp forms and to analyze induced representations. Both problems are quite difficult and neither is solved in all cases, but the division has provided a fruitful method of attack and substantial progress has been made on each partial problem.

It should perhaps be pointed out that what one means by "cusp form" varies slightly according to context. For purposes of the Plancherel formula for real or p-adic groups, "cusp form" has an analytic meaning: a representation is cuspidal if and only if it is in the discrete series (modulo the center). For purposes of the theory of admissible representations of p-adic groups only supercuspidal representations are cuspidal, while in the Langlands classification [BW], [L2], [Vo] all representations which are tempered on the commutator subgroup play the part of cuspidal representations. The discrepancy between the cusp forms and the discrete spectrum is perhaps most vexed in the case of automorphic forms. All cusp forms in $L^2(G_A/G_k)$ (for notation see [Bo]) belong to the discrete spectrum, but the non-cuspidal or "residual discrete spectrum" is very rich and almost as poorly understood as the cusp forms. The variability of the notion of cusp form of course complicates the theory, but pays off amply in increased flexibility and applicability.

The determination of the discrete series of a semisimple Lie group was one of Harish-Chandra's great achievements. For semisimple p-adic groups this problem is still unsolved, except in some cases, and in fact it still seems quite far from solution. (See [MS] for some discussion of its current status.) An explicit description even of the representations of the compact group of norm units in a p-adic division algebra has not yet been given.

There are several reasons for the greater mystery of the discrete series, in the p-adic case. Recall that if G is a connected semisimple Lie group, then [V]:

1) Up to conjugacy G contains at most one compact Cartan subgroup T.
2) The discrete series of G are parametrized in a natural way by the regular characters of T.

Some groups, including complex groups and $SL_n(\mathbf{R})$, $n \geqslant 3$, have no compact Cartan subgroup and therefore no discrete series. By contrast, all p-adic groups have compact Cartan subgroups and discrete series in abundance. Further, while it is roughly true that characters of the compact Cartan subgroups parametrize discrete series representations, there can be no such clean correspondence as for real groups. Some characters (even "regular" characters, although the definition of that term is itself problematic) will not correspond to representations. Further, some representations would seem to correspond to characters of several tori. And finally, there are some discrete series, analogous to the unipotent cuspidal representations constructed by Lusztig [Lu1] for groups over finite fields, which seem not to correspond to a character of any torus. Langlands [Bo], [L1] has conjectured that the discrete series should be parametrized by certain homomorphisms of the Weil group (of the local field over which the group is defined) into an appropriate group, called the L-group. This has been verified in some cases [K], [M], [He], [KM], [T]. But even the conjectures on how to describe the discrete series of p-adic groups are still undergoing refinement [Ar], [Lu2].

Because of the relative intractability of the determination of cuspidal representations for p-adic groups, Harish-Chandra along with most other workers in this area concentrated mainly on the second part of the cusp form program, namely, the connection between representations of a reductive group G and its parabolic subgroups P. He also established what qualitative results he could about discrete series, and these were sufficient to yield a version of the Plancherel formula [1977b] quite analogous to, though somewhat less precise than, the real case. We will describe below his progress toward this goal in more detail.

Harish-Chandra's papers on p-adic groups are [1970d], [1973], [1977a], [1977b], [1978], [1980]. However, it should be noted that [1973] is only a summary of results, essentially an extended research announcement. The full proofs of the results in [1973] have not been published by Harish-Chandra. Instead they appear in Allan Silberger's book [Si], which is largely based on lectures of Harish-Chandra at the Institute for Advanced Study in the academic years 1971–1972 and 1972–1973. For understanding Harish-Chandra's work on p-adic groups, this book is an essential supplement to the papers appearing under his own name. Neither have the full proofs of some results stated in [1977b], [1978], or [1980] yet appeared.

The main topics treated in these papers are ones familiar from Harish-Chandra's work on real groups: the theory of characters, orbital integrals, the constant term, discrete series and the Schwartz space. All of these topics are mutually related via the Philosophy of Cusp Forms. In the real case all were subservient to the Plancherel Formula, but in the p-adic case neither character theory nor orbital integrals has been made as precise as in the real case, and eventually were bypassed by Harish-Chandra in his proof of the p-adic Plancherel Formula. We will discuss these topics in turn.

As a prelude to the more detailed discussion, a remark on technique is in order. Throughout his work on real groups, Harish-Chandra relies on the differential equations supplied by the universal enveloping algebra, especially its center. This fundamental avenue of approach is of course not available in the study of p-adic groups and replacing it is a major technical problem. For some purposes, the place of the differential equations is taken by certain finiteness theorems. Thus a result of Jacquet [J] forms the basis for the theory of the constant term, a result of Howe [Ho] is useful in character theory, and a finiteness result in [1977b] provides the final step in the Plancherel Theorem. We will discuss these in more detail below. But in fact, the lack of the differential equations has not been completely made good, and as was already noted, complete information about discrete series, character formulas, and orbital integrals is not yet available.

Character theory is the primary topic of [1977a], [1978], and [1980], and occupies the bulk of [1970d]. The paper [1977a] is an announcement of the results in [1978], which contains the most precise results on characters, and largely supersedes [1970d]. However, the paper [1978] is restricted to groups over fields of characteristic zero; in [1980] some of the results are extended to groups over fields of positive characteristic.

Before one can describe characters one must know in what sense they exist. For this, and much of the rest of representation theory for p-adic groups, a basic notion is that of *admissible representation*, first formalized in [JL]. Let G be a locally compact topological group which is totally disconnected as topological space—a $t.d.$ *group* in Harish-Chandra's terminology [1973]. This is the same as to say G has a basis of neighborhoods of the identity consisting of open compact subgroups. All p-adic algebraic groups are t.d. groups. Let ρ be a representation of G on a vector space V. A vector $v \in V$ is called a *smooth vector* for ρ if there is an open subgroup $K \subseteq G$ such that v is invariant under $\rho(K)$, i.e., $\rho(k)v = v$ for $k \in K$. The set of smooth vectors for ρ form a subspace $V^\infty \subseteq V$; this subspace V^∞ is invariant by G. If V is a complete locally convex topological vector space, and ρ is a strongly continuous representation of G, then V^∞ is a dense subspace of V.

For a given compact open group $K \subseteq G$, let V^K denote the subspace of vectors in V fixed by $\rho(K)$. Then $V^K \subseteq V^\infty$, and V^∞ is a union of the V^K as K varies over all compact open subgroups of G. If $V = V^\infty$, then the representation ρ is called *smooth*. If in addition all the spaces V^K are finite dimensional, then ρ is called *admissible*. A continuous representation of ρ on a locally convex space whose associated smooth subrepresentation is admissible is also called admissible.

An admissible representation has a character in the following sense. Let $C_c^\infty(G)$ denote the space of locally constant, compactly supported, complex-valued functions on G. Then $C_c^\infty(G)$ is a convolution algebra under the usual definition of convolution on G. If ρ is a smooth representation of G on V, then we can define in an obvious and standard manner a representation, also denoted ρ, of $C_c^\infty(G)$ on V. If ρ is an admissible representation, then $\rho(f)$ will be a finite rank operator. (If f is invariant under left translation by the open subgroup $K \subseteq G$, then $\rho(f)(V) \subseteq V^K$.) Hence the trace of $\rho(f)$ will be well defined, and will depend linearly on f. Thus

$$\theta_\rho(f) = \operatorname{trace} \rho(f)$$

defines a linear functional on $C_c^\infty(G)$. By definition a linear functional on $C_c^\infty(G)$ is called a *distribution* on G. Thus we may say that an admissible representation has a character defined by a distribution on G—its "distribution character".

Thus the first question to ask concerning character theory of representations of G is, are all reasonable (say unitary, though one can be more general) irreducible representations of G admissible? The analogous question for Lie groups was one of the first issues resolved by Harish-Chandra [1953]. In the case of a reductive p-adic group the answer was much longer in coming. That the answer was "yes" was first established by Bernstein [Be]. A stronger result was proved for groups over fields of characteristic zero by Harish-Chandra in [1978], generalizing and extending [Ho]. Both results were based on [1970d] and in particular were resolutions of conjectures made in [1970d]. We will explain these conjectures.

The conjectures concern the supercuspidal representations, which are the most characteristic and the most mysterious phenomenon of representation theory on reductive p-adic groups.

Let G be a locally compact group. Let f be a function on G. Define the left and right translate $\lambda_g f$ and $\rho_g f$ of f by g in the usual way:

$$\lambda_g(f)(x) = f(g^{-1}x) \qquad \rho_g(f)(x) = f(xg) \qquad x, g \in G. \tag{1}$$

Let $H \subseteq G$ be a subgroup. We say f is left (right) H-*invariant* if $\lambda_h(f) = f$ $(\rho_h(f) = f)$ for $h \in H$. If f is complex- or vector-valued we say f is left (right) H-*finite* if the left translates $\lambda_h(f)$ (right translates $\rho_h(f)$) span a finite dimensional space. We say f is left (right) compactly supported mod H if there is a compact set $C \subseteq G$ such that $f = 0$ outside HC (outside CH). If H is central in G, we may omit the adjectives "left" and "right" in the definitions above.

Let G be a reductive p-adic group—the rational points of a connected reductive algebraic group $\mathbf{G}$ defined over a p-adic field Ω. Let Z be the center of G. Let P be a parabolic subgroup of G and let N be the unipotent radical of P. Let δ_P be the modular function of P. Let f be a smooth complex- or vector-valued function on G, compactly supported mod Z. Then in analogy with the real case we can define a function $f^{(P)}$ on P/N by the formula

$$f^{(P)}(m) = \delta_P(m)^{1/2} \int_N f(mn)\, dn \qquad m \in P.$$

Here dn is Haar measure on N.

We say a function f on G is a *supercusp form* if

   i) f is smooth;
   ii) f is compactly supported mod Z;
   iii) f is Z-finite; and
   iv) $(\lambda_g f)^{(P)} = 0$ for all proper parabolics $P \subseteq G$ and $g \in G$.

A *supercuspidal representation* is a representation whose matrix coefficients are supercusp forms. Let π be a smooth representation of G on the space V. Consider $T \in \operatorname{End} V$. We say T is a *smooth operator* if there is an open subgroup $K \subseteq G$ such that

$$\pi(k)T = T = T_\pi(k) \qquad k \in K.$$

Let $\operatorname{End}^\circ(V)$ denote the space of smooth operators which also have finite rank. (Observe that if π is admissible, then a smooth operator automatically has finite rank.) Then $\operatorname{End}^\circ(V)$ is a subalgebra of V, and is invariant by right and left multiplication by $\pi(g)$, for $g \in G$. Set

$$f_T(g) = \operatorname{trace} T\pi(g).$$

The function f_T is called a *matrix coefficient* of ρ. It is easy to see that $f_T \in C^\infty(G)$, the space of locally constant functions on G. The space spanned by all matrix coefficients of all admissible representations is denoted $\mathcal{C}(G)$. If π is admissible and all matrix coefficients of π are supercusp forms, then π is called supercuspidal. It turns out ([1973] §6) that for π to be supercuspidal, it is sufficient that its matrix coefficients be compactly supported mod Z.

Supercuspidal representations are obviously discrete series representations (mod Z). A general argument ([1970d], p. 6) using the Schur orthogonality relations for discrete series shows that all discrete series representations are admissible. In fact if π is an irreducible square integrable representation (mod Z) of G on a space V, and V^K is the space of $\pi(K)$-invariant vectors in V for an open subgroup $K \subseteq G$, one has

$$\dim V^K \leqslant \left(d(\pi)\mu(K) \right)^{-1}$$

where $d(\pi)$ is the formal degree of π and $\mu(K)$ is the Haar measure of K. If 1_K denotes the trivial representation of K, then $\dim V^K$ is the multiplicity of the trivial representation of K in π. It will be denoted $[\pi:1_K]$.

Harish-Chandra shows in [1970d], Part II, that the question of admissibility hinges on the possibility of bounding $[\pi:1_K]$ independently of π. He considers three possible facts that would do this. Observe that if π is an irreducible admissible representation of G, then $\pi|Z$ is a multiple of some character $\chi_\pi = \chi$, called the *central character* of π. Consider the following statements.

i)   For each compact open subgroup K there is a number $\delta(K)$ such that $[\pi:1_K] \leqslant \delta(K)$ for all supercuspidal irreducible representations π of G.

ii)   For each compact open subgroup K there is a number $m(K)$ such that, for any character χ of Z,

$$\sum_\pi [\pi:1_K] \leqslant m(K)$$

where π ranges over all (equivalence classes of) irreducible supercuspidal representations of G with central character χ.

iii)   There is $\varepsilon > 0$ such that $d(\pi) > \varepsilon$ for all irreducible supercuspidal representations π of G.

Obviously statement ii) implies i). The Schur orthogonality relations in fact imply that iii) implies ii). In [1970d], Part II, Harish-Chandra shows that statement i) implies that all irreducible unitary representations of G are admissible. In [Be] Bernstein shows by a simple argument based on the structure of $C^\infty(G//K)$, the K bi-invariant functions of compact support, that statement i) is true. In [1978] Harish-Chandra, as a consequence of a study of characters of supercuspidal representations, shows that if the ground field Ω has characteristic 0, then statement iii) is true. In fact, he shows that if Haar measure on G is appropriately normalized, then the formal degrees $d(\pi)$ of supercuspidal representations are integers.

Once the issue of existence of characters is settled, one would like to know more precisely what they look like. For supercuspidal characters and when the ground field is of characteristic 0, Harish-Chandra already establishes an important qualitative result in [1970d]. Let dg denote Haar measure on G, and let φ be a complex-valued function on G which is locally L^1, i.e. the integral

$$\int_C |\varphi|(g)\, dg$$

where $|\varphi|$ indicates the absolute value of φ and C is any open compact subset of G, is finite. Then φ defines a distribution D_φ on G by the obvious formula

$$D_\varphi(f) = \int_G \varphi(g)f(g)\, dg \qquad f \in C_c^\infty(G).$$

If a distribution D is equal to D_φ for some φ as above, we say D is *locally L^1*. If the function φ may be taken to be locally constant on some open set $S \subseteq G$, we say D is *locally constant* on S. In [1970d], Part V, Harish-Chandra shows that the character of an irreducible supercuspidal representation is a locally constant function on the set G' of regular elements of G. In Parts VII and VIII of [1970d] he goes on to show that if the ground field Ω has characteristic zero, then in fact the character of an irreducible supercuspidal representation is locally L^1 on G. These results are based on an integral formula ([1970d], Theorem 9) for the character of a supercuspidal representation and on a detailed analysis of the geometry of conjugacy classes, analogous to his work on real groups.

In [1978] Harish-Chandra extends these basic facts to all irreducible admissible representations. Furthermore he gives more precise information about the behavior of characters near singular points. These results make use of the exponential map from the Lie algebra $\mathfrak{g}$ of G to G, and so are valid only when the ground field Ω has characteristic zero. They depend on a finiteness result, proved first for GL_n in [Ho], and extended and generalized by Harish-Chandra in [1978]. This may be stated as follows. Let $L \subseteq \mathfrak{g}$ be a lattice, and let $C_c^\infty(\mathfrak{g}/L)$ be the functions on $\mathfrak{g}$ which are compactly supported and constant on cosets of L. Let $\mathrm{Ad}\, G$ denote the adjoint action of G on $\mathfrak{g}$. Let $\omega \subseteq \mathfrak{g}$ be a compact set, let $\mathrm{Ad}\, G(\omega)$ denote the subset of $\mathfrak{g}$ swept out by the action of $\mathrm{Ad}\, G$ on ω, and let $J(\omega)$ denote the space of distributions which are supported on $\mathrm{Ad}\, G(\omega)$ and which are invariant under $\mathrm{Ad}\, G$. Let $j_L J(\omega)$ denote the space of linear functionals on $C_c^\infty(\mathfrak{g}/L)$ obtained by restricting elements of $J(\omega)$ to $C_c^\infty(\mathfrak{g}/L)$. Then $j_L J(\omega)$ has finite dimension.

This finiteness result has two main consequences for characters. First, it gives a description of the possible singularities of characters near any point. It is in terms of Fourier transforms of invariant measures on nilpotent conjugacy classes. Let $x \in \mathfrak{g}$ be a nilpotent element, and let $\mathcal{O} = \mathcal{O}_x = \mathrm{Ad}\, G(x)$ denote the G-conjugacy class of x. There are only finitely many nilpotent conjugacy classes. On $\mathcal{O}$ there is supported a G-invariant measure ν_x, unique up to multiples. Furthermore, a result of Deligne and Rao [Ra] says that the measure ν_x assigns finite mass to bounded subsets of $\mathcal{O}$; consequently ν_x defines a distribution on $\mathfrak{g}$.

We want to consider the Fourier transforms of the measures ν_x. These may be defined in the usual way. If $f \in C_c^\infty(\mathfrak{g})$, its Fourier transform $\hat{f}$ is defined by

$$\hat{f}(x) = \int \hat{f}(x')\chi(B(x', x))\, dx'$$

where $B(\,,\,)$ is an Ad G-invariant, non-degenerate symmetric bilinear form on $\mathfrak{g}$ (the Killing form if $\mathfrak{g}$ is semisimple), χ is a unitary character of the (additive group of the) base field Ω, and dx' is a Haar measure on $\mathfrak{g}$, normalized to make the map $f \to \hat{f}$ unitary. We can then extend $\hat{}$ to distributions by the recipe

$$\hat{D}(f) = D(\hat{f}) \qquad f \in C_c^\infty(\mathfrak{g}),\ D \in C_c^\infty(\mathfrak{g})^*. \tag{2}$$

Recall that $J(\omega)$ is the space of Ad G-invariant distributions supported on Ad $G(\omega)$ where $\omega \subseteq \mathfrak{g}$ is some compact set. Harish-Chandra in [1978] has described the Fourier transforms of elements of $J(\omega)$. To state his result we need the functions $\eta_\mathfrak{g}$ and D_G, familiar from the real theory, which measure how regular elements of $\mathfrak{g}$ or G are.

Consider the polynomial $\det(t - \operatorname{ad} x)$ on $\mathfrak{g} \times \Omega$. Regard this as a polynomial in t with coefficients depending on x. Then if l is the rank of $\mathfrak{g}$ (i.e., the dimension of any Cartan subalgebra), the coefficient of t^b is identically zero if $b < l$. The function $\eta_\mathfrak{g}$ is the coefficient of t^l. Thus

$$\det(t - \operatorname{ad} x) = t^l \eta_\mathfrak{g}(x) + t^{l+1} R(x, t). \tag{3}$$

The function $\eta_\mathfrak{g}$ may be described in another way, as follows. Let $x \in \mathfrak{g}$ be regular, let $\mathfrak{t} \subseteq \mathfrak{g}$ be the kernel of $\operatorname{ad} x$ and let $\mathfrak{t}^\perp$ be the orthogonal complement of $\mathfrak{t}$ with respect to the bilinear form B. Then

$$\eta_\mathfrak{g}(x) = \det(\operatorname{ad} x | \mathfrak{t}^\perp).$$

The function D_G is the analogue on G of $\eta_\mathfrak{g}$. The polynomial $\det(1 + t - \operatorname{Ad} g)$ also begins with the term t^l and one writes

$$\det(1 + t - \operatorname{Ad} g) = t^l D_G(g) + t^{l+1} R'(g, t).$$

The functions $\eta_\mathfrak{g}$ and D_G take values in Ω, the ground field. Recall [W] that on Ω there is defined a natural absolute value

$$|\ |: \Omega \to \mathbf{R}.$$

We will need the real-valued functions gotten by taking absolute values of $\eta_\mathfrak{g}(x)$ or $D_G(x)$. Thus we write

$$|\eta_\mathfrak{g}|(x) = |\eta_\mathfrak{g}(x)| \qquad |D_G|(x) = |D_G(x)|.$$

Consider a distribution $D \in J(\omega)$. Harish-Chandra ([1978], Theorem 3) shows that $\hat{D}$ is locally L^1, locally constant on $\mathfrak{g}'$, the set of regular elements in $\mathfrak{g}$, and that $|\eta_\mathfrak{g}|^{1/2}\hat{D}$ is locally bounded on $\mathfrak{g}$. These results apply in particular to the Fourier transforms $\hat{\nu}_x$ of the invariant measures on nilpotent orbits.

Finally we can describe characters. Let θ_π be the character of an irreducible admissible representation π of G. Then θ_π is a locally L^1 function, locally constant on G', and the function $|D_G|^{1/2}\theta_\pi$ is locally bounded. Furthermore, if $\gamma \in G$ is a semisimple element, then the behavior of θ_π near γ is as follows. Let M be the centralizer of γ in G, and $\mathfrak{m}$ the Lie algebra of M. Let C be a small neighborhood of 0 in $\mathfrak{m}$, on which the exponential map exp is defined. Then $\gamma \exp C$ is a neighborhood of γ in M, and $\operatorname{Ad} G(\gamma \exp C)$ is an $\operatorname{Ad} G$-invariant neighborhood of γ in G. If C is chosen sufficiently small, then we may write

$$\theta_\pi(\gamma \exp Y) = \sum_\xi c_\xi \hat{\nu}_\xi(Y) \qquad Y \in C$$

where ξ runs over the nilpotent conjugacy classes *in* $\mathfrak{m}$, and the c_ξ are complex numbers depending on π and γ. Roughly, we may say that locally θ_π is a linear combination of Fourier transforms of nilpotent orbits.

This result is analogous to the knowledge that an invariant eigendistribution on a real reductive group is a locally L^1 function, and is given locally on each Cartan subgroup by $D_G^{-1/2}$ times a linear combination of certain exponentials. What is still missing, except in special cases, are results as precise as Harish-Chandra's character formula for discrete series of real groups.

As was already stated, the above results apply when the ground field Ω has characteristic zero. When Ω is of positive characteristic, much less is known. However, in [1980] it is shown that in all characteristics characters are locally constant on the regular set. The techniques of this paper are different from those of [1973] or [1978]. The basic fact used is the submersiveness of a certain map ([1980], Theorem 1). Also, the following interesting fact is established. Let K be a "good" maximal compact subgroup of G in the sense of Bruhat and Tits [BT]. For an admissible representation π of G on a vector space V and a regular element x in G, set

$$T_x = \int_K \pi(kxk^{-1})\, dk.$$

Then $T_x \in \operatorname{End}^\circ(V)$, and in particular T_x has finite rank. Then evidently $\theta_\pi(x) = \operatorname{tr} T_x$.

The study of the map F_f or, in common current parlance, orbital integrals, is entwined with and in some sense dual to the study of characters, and currently is in roughly the same state. One has substantial knowledge of the qualitative behavior of F_f, but precise control analogous to the "jump formula" for real groups is still lacking. This would be provided by an explicit description of the "Shalika germs" which will be defined and discussed below.

First recall the definition of F_f. Fix a Cartan subgroup $A \subseteq G$, and fix an invariant measure dg^* on G/A. Then for $f \in C_c^\infty(G)$ and $a \in A'$, where $A' = G' \cap A$ one sets

$$F_f(a) = |D_G(a)|^{1/2} \int_{G/A} f(gag^{-1})\, dg^*. \tag{4}$$

There is a close analogue of F_f for the Lie algebra. If $\mathfrak{a}$ is the Lie algebra of A and $x \in \mathfrak{a}$, put

$$\phi_f(x) = |\eta_\mathfrak{a}(x)|^{1/2} \int_{G/A} f(\operatorname{Ad} g(x))\, dg^* \qquad f \in C_c^\infty(\mathfrak{g}). \tag{5}$$

In formulas (4) and (5) the functions D_G and $\eta_\mathfrak{a}$ are the ones defined above in (2) and (3).

Formulas (4) and (5) are very near parallels of the definitions of F_f and ϕ_f for real groups. However, we should remark on one difference that has not been important up to now, but which needs clarification before these maps can be thoroughly understood. For real groups the function D_G is a smooth function. The function $|D_G|^{1/2}$ is not smooth. However, on any Cartan subgroup A, there is a function $D_G^{A\ 1/2}$ such that $(D_G^{A\ 1/2})^2 = D_G$ on A and which yields an optimal theory of orbital integrals. In other words, if one multiplies $|D_G|^{1/2}$ by appropriate phase factors, one obtains a better theory, in which the functions F_f are as smooth as possible. It is the function $D_G^{A\ 1/2}$ rather than the positive $|D_G|^{1/2}$ which is used in Harish-Chandra's definition of F_f for real groups. It would seem that some modification of $|D_G|^{1/2}$ by appropriate phases is also preferable in the p-adic case; but precisely how to do this has not yet been made clear.

The parts of Harish-Chandra's work devoted to F_f on p-adic groups are [1970d], Part VI; [1973], §16; [1978] §§3, 4, 8, 9; and [1980], §5. In [1970d] it is already established that when the ground field Ω has characteristic 0, the function F_f is bounded for any $f \in C_c^\infty(G)$. However, it is not shown that F_f is locally constant. This is done in [1973], §16 (for the proofs see [Si]) by the same technique that establishes local constancy of characters. In fact, local constancy of F_f is even established for f belonging to Harish-Chandra's Schwartz space $\mathcal{C}(G)$ (see below for the definition of $\mathcal{C}(G)$); however local boundedness of F_f for $f \in \mathcal{C}(G)$ is not proven. Local constancy of F_f for Ω of positive characteristic, and $f \in \mathcal{C}(G)$ is proved in [1980], but boundedness not.

For real groups the behavior of $F_f(x)$ as x approaches singular points is of much interest. This is also true for p-adic groups. The standard approach to the study of this limiting behavior is provided by an observation of Shalika [Sh]. Harish-Chandra's version of this result is Theorem 14 of [1978], which applies to the Lie algebra. As above, we let ν_x denote the invariant measures on the nilpotent G-conjugacy classes $\mathcal{O}_x$ in $\mathfrak{g}$. Fix a Cartan subalgebra $\mathfrak{a} \subseteq \mathfrak{g}$. Then there are functions $\Gamma_\mathcal{O}^\mathfrak{a} = \Gamma_\mathcal{O}$, for each nilpotent conjugacy class $\mathcal{O} = \mathcal{O}_x$ such that

$$\phi_f(a) - \sum \nu_x(f)\Gamma_\mathcal{O}(a)$$

vanishes in some neighborhood of the origin in $\mathfrak{a}$. The functions $\Gamma_\mathcal{O}$ are strictly speaking, only germs of functions, an equivalence class of functions equal in some neighborhood of 0; they are generally called "Shalika germs". However, they may be defined uniquely on all of $\mathfrak{a}$ by requiring them to satisfy an appropriate condition of homogeneity, namely

$$\Gamma_\mathcal{O}(t^2 a) = |t|^m \Gamma_\mathcal{O}(a) \qquad t \in \Omega^x, \ a \in \mathfrak{a}$$

where m is an integer depending on $\mathcal{O}$. There are analogous germ expansions around non-zero singular points of $\mathfrak{a}$, and for F_f.

Clearly the functions $\Gamma_\mathcal{O}$ control the singularities of ϕ_f. Considerable effort consequently has been devoted to determining them [Vi], [Ko], [Re1], [Re2], [Ro2]. However, they remain mysterious. Harish-Chandra shows in [1978], §9 that the $\Gamma_\mathcal{O}^\mathfrak{a}$ separate the nilpotent orbits $\mathcal{O}$ in the sense that the only linear combination $\sum c_\mathcal{O}\Gamma_\mathcal{O}^\mathfrak{a}$ which vanishes for all Cartan subalgebras $\mathfrak{a}$ is the trivial combination with all $c_\mathcal{O} = 0$. This is implied by his result ([1978], Theorem 10) that all invariant distributions annihilate any function $f \in C_c^\infty(\mathfrak{g})$ such that $\phi_f = 0$ for all Cartan subalgebras $\mathfrak{a}$. He also shows that $\Gamma_0^\mathfrak{a}$, the germ associated to the origin, is zero if $\mathfrak{a}$ is not an elliptic Cartan subalgebra, and on an elliptic Cartan is a constant times $|\eta_\mathfrak{a}|^{1/2}$. The value of the constant was conjectured by Harish-Chandra and determined by Rogawski [Ro1].

The third major topic in Harish-Chandra's work is the Plancherel Formula, and the associated analysis on G: the constant term, induced series of representations, wave packets, Eisenstein integrals, c-functions. For real groups, character theory and orbital integrals also contribute to the Plancherel Formula, but as explained above, for p-adic groups, they have not reached the state necessary for use in the Plancherel Formula. However certain technical aspects of the p-adic situation allow Harish-Chandra to complete his program without knowledge of the discrete series. The resulting Plancherel Formula is not as explicit as the one for real groups. But except for the lack of explicit knowledge of the discrete series, the Plancherel Formula for p-adic groups is quite parallel to that for real groups. Hence our discussion of it will be relatively brief, and will focus on the aspects which are particular to p-adic groups.

Since the theory of the constant term for real groups depends heavily on the differential equations supplied by the center of the enveloping algebra, its transference to p-adic groups requires new techniques. Harish-Chandra made a preliminary essay in [1970d], establishing the existence of the constant term contingent on a conjecture. However a firmer basis was provided shortly after by a result of Jacquet [J].

Let π be an admissible representation of G on the vector space V. Let $P \subseteq G$ be a parabolic subgroup with unipotent radical N. Let $V(N)$ be the subspace of V spanned by vectors of the form $\pi(n)v - v$. Since N is normalized by P, one can easily see that $V(N)$ is stable under P. Hence there is defined a representation of P on $V_N = V/V(N)$. From the definition of $V(N)$ it is clear that N acts trivially on V_N. Hence the action of P on V_N factors to an action π_M of $P/N \simeq M$ on V_N. Jacquet showed (for $G = GL_n(\Omega)$, but with a general proof) that V_N is also an admissible module for M. It turns out that π is supercuspidal if and only if $V_N = \{0\}$.

Recall that $\mathcal{Q}(G)$ is the space of all matrix coefficients of all admissible representations of G. Using Jacquet's theorem, Harish-Chandra establishes the existence of the constant term in a very strong form. Let M now denote a Levi component of P, so that $M \simeq P/N$. Let A denote the center of M. Let $\{\alpha_i\}$ be the simple roots of A acting on N by conjugation. Each α_i is a rational homomorphism

$$\alpha_i \colon A \to \Omega^\times.$$

For a given $t > 0$, set

$$A^+(t) = \{a \in A \colon |\alpha_i(a)| > t \text{ for each } \alpha_i\}.$$

As above let δ_P denote the modular function of P. Then Harish-Chandra shows ([1973], §6; see [Si], Chapter 2 for proofs) that given $f \in \mathcal{Q}(G)$, there is a unique $f_P \in \mathcal{Q}(M)$ such that, for any compact set $\omega \subseteq M$, one has

$$f(ma) = \delta_P(ma)^{1/2} f_P(ma) \qquad m \in \omega, \ a \in A^+(t)$$

for t sufficiently large. Thus in the p-adic case one eventually has actual equality between f and f_P, its *constant term along* P, and not merely an asymptotic relation.

Since f_P is in $\mathcal{C}(M)$, it is in particular A-finite. The finite dimensional space spanned by the A-translates of f_P will be spanned by generalized eigenspaces for A. If Y is such an eigenspace, then there is a quasicharacter ψ of A such that $\lambda_a - \psi(a)$ is nilpotent on Y. (Here λ_a is the left action of A, as in (1).) Harish-Chandra calls these characters ψ the *exponents* of f relative to the pair (P, A). The A-finiteness of f_P takes the place in many situations of the differential equations governing the constant term in the real case, and the exponents take the place of infinitesimal characters. To place limits on the elements of $\mathcal{C}(M)$ which can be constant terms of matrix coefficients of a fixed irreducible representation of G, Harish-Chandra develops a sharp form of the theory of intertwining operators for induced representations, pioneered by Bruhat for real groups [Br1].

With the constant term, Harish-Chandra can proceed toward the Plancherel Formula along very much the same road used in the real case. The theory of induced representations, of the Eisenstein integral, the c-functions, the Maass–Selberg relations, of wave packets and the Schwartz space proceed very much as in the real case. Already in [1973] he is able to state the Plancherel Formula for the wave packets constructed from a given series of induced representations (Theorem 34). In this Plancherel Formula, the Plancherel measure is determined by a product formula, as it is for real groups. However, it is not explicitly known. But luckily, it is not necessary to know it explicitly in the p-adic case. This is because any p-adic torus A, modulo an open compact subgroup, is discrete. Hence $\hat{A}^0$, the identity component of the Pontrjagin dual of A, is compact. It is a real torus—a product of circles. If M is the Levi component of a parabolic subgroup P, and A the center of M, then a series from which one builds wave packets consists of representations of the form $\pi \otimes \chi$ where π is a square integrable (mod A) unitary representation of M and $\chi \in \hat{A}^0$. In the real case, $\hat{A}^0$ is a vector group, and it is necessary to know that Plancherel measure on $\hat{A}^0$ grows only polynomially at ∞ in order to be able to control the Fourier transform. In fact, for real groups, Harish-Chandra gives an explicit expression for the Plancherel measure, and one can see directly that this expression has moderate growth. But since $\hat{A}^0$ is compact for p-adic groups, the growth of Plancherel measure is not an issue, and no explicit knowledge of it is necessary.

We will close this account by describing what is involved in the passage from the "local" Plancherel Formula of [1973] to the full formula announced in [1977b].

First the notion of constant term described above must be modified so that matrix coefficients of representations which are square integrable but not super-cuspidal have zero constant term. This leads to the notion of the weak constant term.

Let K be a good maximal compact subgroup of G. Let Ξ be, as for real groups, the K-spherical matrix coefficient of the representation unitarily induced from the trivial representation of a minimal parabolic subgroup of G. Let σ be the measure of slow growth on G; roughly σ is the logarithm of Ξ (see [1973]; §14). For each compact open $K_1 \subseteq K$, one defines $\mathcal{C}(G//K_1)$ to be the space of K_1 bi-invariant functions f on G such that

$$|f| \leqslant c(f, k)\Xi(1 + \sigma)^{-k}$$

for an appropriate positive number $c(f, k)$ and for each positive number k. Then $\mathcal{C}(G)$, the *Schwartz space* of G is the union over compact open K_1 of the spaces $\mathcal{C}(G//K_1)$.

Consider $\phi \in \mathcal{C}(G)$. One says ϕ satisfies the *weak inequality* if for some constant c and some integer k one has the estimate

$$|\phi| \leqslant c\Xi(1+\sigma)^k.$$

Such a ϕ is called *tempered*, and the subspace of $\mathcal{C}(G)$ consisting of tempered matrix coefficients is denoted $\mathcal{C}^w(G)$.

Let f be in $\mathcal{C}^w(G)$, and let f_P be its constant term along the parabolic subgroup P. Write

$$f_P = \sum f_{P,\psi}$$

where ψ runs over the exponents of f, and $f_{P,\psi}$ is the component of f in the generalized ψ-eigenspace of the A translates of f, A being the center of a Levi component M of P. The sum of the $f_{P,\psi}$ over those ψ which are unitary is called the *weak constant term* of f along P, and is denoted f_P^w. It is in $\mathcal{C}^w(M)$. A *cusp form* on G is an element of $\mathcal{C}^w(G)$ such that the weak constant terms $(\lambda(g)f)_P^w$ of all left translates of f are zero for all proper parabolic subgroups P of G. The space of cusp forms on G is denoted ${}^0\mathcal{C}^w(G)$.

There is a notion of constant term for $\mathcal{C}(G)$ that is dual to that for $\mathcal{C}^w(G)$. For $f \in \mathcal{C}(G)$ and P a parabolic subgroup of G, set

$$f^{(P)}(m) = \delta_P(m)^{1/2} \int_N g(mn)\, dn \qquad m \in P/N$$

where N is the unipotent radical of P. Let ${}^0\mathcal{C}(G)$ denote the subspace of $\mathcal{C}(G)$ consisting of those f such that $(\lambda(g)f)^{(P)} = 0$ for all proper parabolic subgroups.

If G were semisimple rather than reductive, then it is easy to see that ${}^0\mathcal{C}^w(G) \subseteq {}^0\mathcal{C}(G)$. If G possesses a non-compact center, then one must formulate the relation between ${}^0\mathcal{C}^w$ and ${}^0\mathcal{C}$ in a more complicated way, but the situation is in essence the same. So to simplify the discussion we take G semisimple. Then ${}^0\mathcal{C}^w(G)$ consists of matrix coefficients of the discrete series, while ${}^0\mathcal{C}(G)$ is the subspace of $\mathcal{C}(G)$ which cannot be obtained by inducing tempered representations from proper parabolic subgroups. Hence to pass from the Plancherel Theorem of [1973], §17, to the Plancherel Formula for G, one needs to show that in fact ${}^0\mathcal{C}^w(G) = {}^0\mathcal{C}(G)$. This essentially amounts to the statement that (for G semisimple), the representation of G on the span of the left translates of $f \in {}^0\mathcal{C}(G)$ is admissible, i.e., consists of finitely many irreducible summands. This is a consequence of Lemma 4 of [1977b]. A crucial step in the proof of Lemma 4 is Theorem 11, which states that if $\theta \in {}^0\mathcal{C}(G)$, then the map

$$f \to \int_{G/Z} dy^* \int_G f(x)\theta(yxy^{-1})\, dx$$

is well defined and yields a tempered distribution on G. Although the details of this result are not yet published, in flavor it reminds one of the analysis of supercuspidal characters in [1970d].

ROGER HOWE

The form of Harish-Chandra's p-adic Plancherel Theorem is this. The space $\mathcal{C}(G)$ is the orthogonal direct sum of wave packets formed from series of representations induced unitarily from discrete series of (the Levi components of) parabolic subgroups of P. Moreover if two such series of induced representations yield the same subspace of $\mathcal{C}(G)$, then the parabolics from which they are induced are associate, and the representations of the Levi components are conjugate. To complete the analogy with real groups, one needs only to explicitly determine the discrete series; this is the outstanding problem left in p-adic representation theory.

REFERENCES

[Ar] J. Arthur, On some problems suggested by the trace formula, Special Year in Representation Theory, U. of Maryland, Nov. 1982, Springer Lecture Notes, to appear.

[Be] I. N. Bernstein, All reductive p-adic groups are tame, *Fun. Anal. and App.* **8** (1974), 91–93.

[Bo] A. Borel, Automorphic L-functions, in *Automorphic Forms, Representations, and L-functions, Proc. Sym. Pure Math.*, **XXXIII**, Part 2, Amer. Math. Soc. Providence, R.I., 1974, 27–63.

[BW] A. Borel and N. Wallach, Continuous cohomology, discrete subgroups, and representations of reductive groups, *Ann. of Math. Studies* **94**, Princeton University Press, Princeton, 1980.

[Br1] F. Bruhat, Sur les representations induites des groupes de Lie, *Bull. Soc. Math. France*, **84** (1956), 97–205.

[BT] F. Bruhat and J. Tits, Groupes reductifs sur un corps local, I, Donnees radicielles valuees, *Pub. Math. I.H.E.S.* **41** (1972), 5–251.

[CM] W. Casselman and D. Milicic, Asymptotic behavior of matrix coefficients of admissible representations, *Duke Math. J.* **49** (1982), 869–930.

[He] G. Henniart, La conjecture de Langlands locale pour GL(3), *I.H.E.S. Notes* (1982).

[Ho] R. Howe, The Fourier transform and germs of characters, *Math. Ann.* **208** (1974), 305–322.

[H] H. Jacquet, Representations des groupes lineaires p-adiques, Theory of Group Representations and Harmonic Analysis (C.I.M.E., II, Ciclo, Montecatini terme, 1970) Edizione Cremonese, Roma, 1971, 119–220.

[JL] H. Jacquet and R. Langlands, Automorphic forms on GL(2). *Lec. Notes Math.*, **260**, Springer-Verlag, Berlin, New York, 1972.

[Ko] R. E. Kottwitz, Orbital integrals on GL(3), *Am. J. Math*, **102** (1980), 327–384.

[K] P. Kutzko, The Langlands conjecture for GL_2 of a local field, *Ann. Math.* **112** (1980), 381–412.

[KM] P. Kutzko and A. Moy, On the local Langlands Conjecture in prime dimension, preprint.

[L1] R. Langlands, Problems in the theory of automorphic forms, *Lectures in Modern Analysis and Applications, Lecture Notes in Math.*, **170**, Springer-Verlag, New York, 1973, 18–86.

[L2] R. Langlands, On the classification of irreducible representations of real reductive groups, *Notes, I.A.S.*, Princeton, 1973.

[Lu1] G. Lusztig, Irreducible representations of finite classical groups, *Inv. Math.* **43** (1977), 125–175.

[Lu2] G. Lusztig, Some examples of square integrable representations of semisimple p-adic groups, preprint, *I.H.E.S.*, March 1982.

[M] A. Moy, Local constants and the tame Langlands correspondence, Thesis, University of Chicago, 1982.

[MS] A. Moy and P. Sally, Supercuspidal representations of SL_n over a p-adic field: the tame case, preprint.

[Ra] R. Rao, Orbital integrals in reductive groups, *Ann. of Math.* **96** (1972), 505–510.

[Re1] J. Repka, Shalika's germs for p-adic $GL(n)$: the leading term, preprint.

[Re2] J. Repka, Shalika's germs for p-adic $GL(n)$, II: the subregular term, preprint.

[Ro1] J. Rogawski, An application of the building to orbital integrals, *Comp. Math.* **42** (1981), 417–423.

[Ro2] J. Rogawski, Some remarks on Shalika germs, preprint.

[Sh] J. Shalika, A theorem on semisimple p-adic groups, *Ann. Math.* **95** (1972), 226–242.

[Si] A. Silberger, Introduction to Harmonic Analysis on Reductive p-adic Groups, *Math. Notes*, Princeton University Press, Princeton, N.J., 1979.

[T] J. Tunnell, Report on the local Langlands conjecture for GL_2, *Proc. Symp. Pure Math.*, **XXXIII**, Part 2, Amer. Math. Soc., Providence, R.I., 1979, 135–138.

[V] V. Varadarajan, Harmonic Analysis on Real Reductive Groups, *Lecture Notes in Math.*, **576**, Springer-Verlag, New York, 1976.

[Vi] M. F. Vigneras, Caracterisation des integrales orbitales sur un groupe reductif p-adique, *J. Fac. Sci. Tokyo*, Sec. IA, **28**, (1982), 945–961.

[Vo] D. Vogan, Representation of Real Reductive Lie Groups, *Progress in Math.*, **15**, Birkhauser, Boston, Basel, Stuttgart, 1981.

[Wa] N. Wallach, Asymptotic expansions of generalized matrix entries, special year on representation theory, University of Maryland, Vol. I, *Lecture Notes in Mathematics*. Springer-Verlag, Berlin, Heidelberg, New York, Tokyo, 1983.

[W] A. Weil, *Basic Number Theory*, 2nd ed., Grund. Math. Wiss., **144**, Springer-Verlag, Berlin, Heidelberg, New York, 1973.

Permissions

Springer-Verlag would like to thank the original publishers of Harish-Chandra's papers for granting permissions to reprint specific papers in this collection. The following list contains the credit lines for those articles.

[1944a] Reprinted from *Proc. Royal Soc. A.* **183**, ©1944 by The Royal Society of London.
[1944b] Reprinted from *Proc. Royal Soc. A.* **183**, ©1944 by The Royal Society of London.
[1945a] Reprinted from *Proc. Indian Acad. Sci. Sect. A.* **21**, ©1945 by The Indian Academy of Sciences.
[1945b] Reprinted from *Proc. Indian Acad. Sci. Sect. A.* **22**, ©1945 by The Indian Academy of Sciences.
[1946a] Reprinted from *Proc. Royal Soc. A.* **185**, ©1946 by The Royal Society of London.
[1946b] Reprinted from *Proc. Royal Soc. A.* **185**, ©1946 by The Royal Society of London.
[1946c] Reprinted from *Proc. Indian Acad. Sci. Sect. A.* **23**, ©1946 by The Indian Academy of Sciences.
[1946d] Reprinted from *Proc. Royal Soc. A.* **186**, ©1946 by The Royal Society of London.
[1947a] Reprinted from *Proc. Camb. Phil. Soc.* **43**, ©1947 by Cambridge University Press.
[1947b] Reprinted from *Phys. Rev.*, **71**, ©1947 by The American Physical Society.
[1947c] Reprinted from Report of an International Conf. on Fundamental Particles and Low Temperatures held at the Cavendish Laboratory, Cambridge, (1946). Vol I, Fundamental Particles, ©1947 by The Institute of Physics.
[1947d] Reprinted from *Proc. Royal Soc. A.* **189**, ©1947 by The Royal Society of London.
[1948a] Reprinted from *Proc. Royal Soc. A.* **192**, ©1948 by The Royal Society of London.
[1948b] Reprinted from *Phys. Rev.* **74**, ©1948 by The American Physical Society.
[1949a] Reprinted from *Ann. of Math.* **50**, ©1949 by Princeton University Press.
[1949b] Reprinted from *Ann. of Math.* **50**, ©1949 by Princeton University Press.
[1950a] Reprinted from *Proc. Amer. Math. Soc.* **1**, ©1950 by The American Mathematical Society.
[1950b] Reprinted from *Proc. Amer. Math. Soc.* **1**, ©1950 by The American Mathematical Society.
[1950c] Reprinted from *Ann. of Math.* **51**, ©1950 by Princeton University Press.
[1951a] Reprinted from *Trans. Amer. Math. Soc.* **70**, ©1951 by The American Mathematical Society.

On the theory of point-particles

By H. J. Bhabha, F.R.S. and Harish-Chandra

Cosmic Ray Research Unit, Indian Institute of Science, Bangalore

(*Received* 20 *March* 1944)

It is deduced from the conservation of the energy-momentum tensor that if the flow of energy and momentum into a tube surrounding a time-like world-line, on which the field is singular, become singular as the size of the tube is contracted to zero, then the singular terms are necessarily perfect differentials of quantities on the world-line with respect to the proper time along the world-line. The same can be proved of any other tensor, as, for example, the angular-momentum tensor, which is conserved. It is proved from this that for any *point*-particle whatever having charge, spin or other properties, which need not be specified, it is always possible to deduce exact equations of motion which are finite.

It is proved further that if the energy-momentum tensor is altered by the addition of $\partial K^{\mu\nu\sigma}/\partial x^\sigma$, where $K^{\mu\nu\sigma}$ is any tensor antisymmetric in ν and σ, then the equations of motion are unaltered, but it is possible to choose $K^{\mu\nu\sigma}$ in such a way as to make the flow of energy and momentum into a given tube non-singular.

By a point-particle is understood a particle whose field-producing and inertial properties are all located at a point. The particle may have a finite charge and a finite mass, but the charge density and mass density are exactly zero at every point of space other than the point at which the particle is located at that instant of time. The motion of the particle through space-time is therefore described by a time-like world-line. If the particle possesses a dipole or a higher multipole moment, then this is described by a suitable co-ordinate having a given value at each point of the world-line, but it is again assumed that the dipole density is exactly zero at all points of space-time not lying on the world-line. Thus the motion of the particle through space-time is completely described by a time-like world-line with the values of the co-ordinates describing the spin and other properties of the particle given at each point of it.

It has been shown by Dirac (1938) that an exact theory of a point electron moving in an electromagnetic field can be set up free from singularities, and Bhabha (1940) and Bhabha & Corben (1941) have shown that a similar theory free from singularities can be set up for a point dipole. It has been shown further by one of us (Bhabha 1939, 1941) that the theory can be extended to point-particles interacting with meson fields. These cover all the cases of practical interest, but a general demonstration that it is always possible to set up an exact classical theory for point-particles interacting with any field has not so far been given. This will be done in the present paper.

In all the work referred to above, the method used for finding the equations of motion of the point-particle is the same. From the field equations we calculate the field produced by the point-particle, or more exactly, we take that solution of the *homogeneous* field equations which has a singularity of the required type on the world-line. We now surround a finite length of the world-line by a world tube whose radius is ultimately made to tend to zero, and calculate the flow of energy, momentum and angular momentum into the world tube from outside by using the usual energy-momentum and angular-momentum tensors of the field. For brevity we shall refer to the quantities so calculated as the inflow. The equations of motion are now found from the condition that conservation of energy, momentum and angular momentum require that the flow of all these quantities into the tube must only depend on conditions at the two ends of the tube, that is, it must only be a function of the co-ordinates of the particle and their higher derivatives and also possibly of the field quantities at the two ends of the tube. Since the field is singular on the world-line, the usual energy-momentum tensor is also singular, and in consequence the flow of energy and momentum into the tube is likewise singular. That in spite of this it has always been possible to derive finite equations for the motion of the point-particles has always depended on the circumstance that the singular parts of the inflow over an infinitesimal length of the tube are always perfect differentials. We shall prove that this is a general property which is a consequence only of the conservation of the energy-momentum tensor and therefore that finite equations for the motion of a point-particle of any type whatsoever can always be derived.

We shall also show that it is always possible to modify the energy-momentum tensor by the addition of the divergence of a tensor of higher rank in the manner suggested by Pryce (1938) so that the inflow over a tube of given shape becomes finite. But the inflow would not necessarily be finite over a tube of any other shape. That it is always possible to modify the tensor in the same way as to make finite the inflow over tubes of any arbitrary shape, as also the energy and momentum integrals over arbitrary space-like surfaces is shown by one of us (H.-C.) in the paper which immediately follows this.

1. We use the same notation as in the previous papers by one of us (H.J.B.). We take the metric tensor to have the form $g_{00} = -g_{11} = -g_{22} = -g_{33} = 1$ with $g_{\mu\nu} = 0$ for $\mu \neq \nu$. The world-line of the particle is described by four co-ordinates $z^\mu(\tau)$ which are functions of the proper time τ measured from some point along it. The other co-ordinates of the particle, if any, need not be specified for the present work. A dot denotes differentiation with respect to τ, and $V^\mu = \dot{z}^\mu$ is used to denote the velocity of the particle. x^μ denotes a point of space, $s^\mu \equiv x^\mu - z^\mu(\tau_0)$ the distance from any point of space to the retarded point τ_0 on the world-line defined by

$$s_\mu s^\mu = 0. \tag{1}$$

If τ_0 be kept fixed then equation (1) is also the equation of the light cone whose apex is at τ_0. We further introduce a quantity κ as a function of the co-ordinates x^μ defined by

$$\kappa \equiv s_\mu v^\mu(\tau_0). \tag{2}$$

The energy-momentum tensor is denoted by $T^{\mu\nu}$. It satisfies the conservation equation

$$\frac{\partial T^{\mu\nu}}{\partial x^\nu} = 0. \tag{3}$$

It is necessary to find the generalization of Gauss's theorem to four-dimensional hyperbolic space. Let

$$\zeta^0(x^\mu) = \alpha \tag{4}$$

be a closed three-dimensional surface S surrounding a four-dimensional volume V such that the surface $\zeta^0(x^\mu) = \alpha - \Delta\alpha$ for positive $\Delta\alpha$ is contained inside (4) and lies wholly in the volume V. Let ζ^1, ζ^2 and ζ^3 be parameters defining the position of a point on the surface S. Then, if X^μ be any tensor, the generalization of Gauss's theorem reads

$$\int_V \frac{\partial X^\nu}{\partial x^\nu} dx^0 dx^1 dx^2 dx^3 = \int_S X^\nu \frac{\partial \zeta^0}{\partial x^\nu} |D|^{-1} d\zeta^1 d\zeta^2 d\zeta^3, \tag{5}$$

where D is the determinant of the transformation from the x's to the ζ's and the surface is covered in such a way that $d\zeta^1, d\zeta^2, d\zeta^3$ are always positive:

$$D \equiv \begin{vmatrix} \dfrac{\partial \zeta^0}{\partial x^0} & \dfrac{\partial \zeta^1}{\partial x^0} & \dfrac{\partial \zeta^2}{\partial x^0} & \dfrac{\partial \zeta^3}{\partial x^0} \\[2ex] \dfrac{\partial \zeta^0}{\partial x^1} & \dfrac{\partial \zeta^1}{\partial x^1} & \dfrac{\partial \zeta^2}{\partial x^1} & \dfrac{\partial \zeta^3}{\partial x^1} \\[2ex] \dfrac{\partial \zeta^0}{\partial x^2} & \dfrac{\partial \zeta^1}{\partial x^2} & \dfrac{\partial \zeta^2}{\partial x^2} & \dfrac{\partial \zeta^3}{\partial x^2} \\[2ex] \dfrac{\partial \zeta^0}{\partial x^3} & \dfrac{\partial \zeta^1}{\partial x^3} & \dfrac{\partial \zeta^2}{\partial x^3} & \dfrac{\partial \zeta^3}{\partial x^3} \end{vmatrix} \tag{6}$$

$\partial\zeta^0/\partial x^\mu$ is the normal to the surface S. For a displacement $(\partial\zeta^0/\partial x^\mu)\Delta l$ in the direction of this normal with positive Δl the change $\Delta\zeta^0$ of ζ^0 is

$$\Delta\zeta^0 = \frac{\partial\zeta^0}{\partial x^\mu}\frac{\partial\zeta^0}{\partial x_\mu}\,\Delta l.$$

This is positive if $\partial\zeta^0/\partial x^\mu$ is a time-like vector and negative if it is space-like. Thus on the space-like portions of the surface the normal to the surface on the right-hand side of (5) is directed *outwards*, and on the time-like portions it is directed *inwards*. By writing $dS_\nu = \dfrac{\partial\zeta^0}{\partial x^\nu}\,|\,D\,|^{-1}\,d\zeta^1 d\zeta^2 d\zeta^3$ for an element of the surface S with its normal directed in the sense defined above, the right-hand side of (5) can be written

$$\int X^\nu dS_\nu. \tag{7}$$

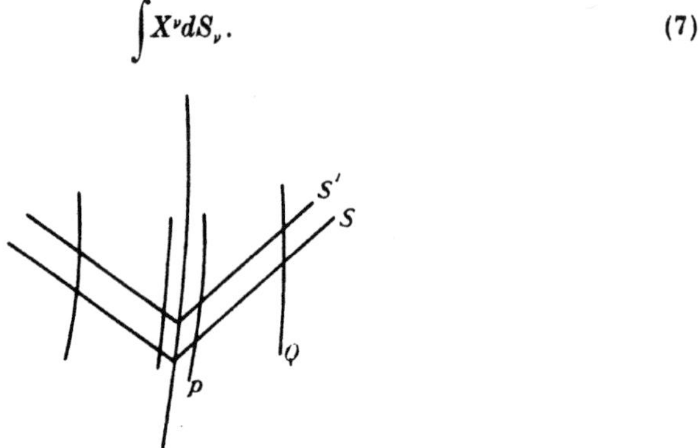

FIGURE 1

Consider a portion of the world-line between the points τ_1 and τ_2 and surround it by a world-tube of any arbitrary shape. We consider the ends of the tube to be given by the two two-dimensional surfaces where it intersects the future light cones from the points τ_1 and τ_2 respectively. The flow of energy and momentum into the tube is then given by

$$\int_{\tau_1}^{\tau_2} T^{\mu\nu} dS_\nu, \tag{8}$$

the normal to the surface being directed outwards. For conservation, (8) must be put equal to a function of the form $A^\mu(\tau_2) - A^\mu(\tau_1)$, where $A^\mu(\tau)$ is a function of v^μ and the other co-ordinates of the particle, possibly also of the field quantities, and their higher derivatives at the point τ only. On differentiating this equation

$$\frac{d}{d\tau}\int_{\tau_1}^{\tau} T^{\mu\nu} dS_\nu = \dot{A}^\mu(\tau). \tag{9}$$

This is not a mathematical identity but furnishes the equations of motion of the point-particle. It should be noted at once that the equations derived in this way are independent of the shape of the tube. For if the given tube is surrounded by another

4

of some other shape, then it follows from (3) and (5) that the difference in the integral (8) over the two tubes is just equal to the integral over the portions of the light cones at τ_1 and τ_2 intercepted between the two tubes. In the limit when the size of both tubes is made to tend to zero, the integrals over the light cones just become functions of the conditions at τ_1 and τ_2 respectively, and hence the difference in (8) calculated over two tubes of different shape is identically of the form $A'^\mu(\tau_2) - A'^\mu(\tau_1)$. This difference can be absorbed into the right-hand side of (9), and hence it does not affect the equation of motion which can be so derived. One may therefore work with whatever tube is most convenient.

Consider now a volume V lying between the two light cones S and S' starting from the points τ' and $\tau' + d\tau'$ respectively, and between the two tubes P and Q defined by $\kappa = \epsilon$ and $\kappa = \eta$, as shown in figure 1. The boundary of V so defined can be characterized by the equation $\zeta^0 = 0$, where ζ^0 is a discontinuous function chosen in the following way:

$$\zeta^0 = \begin{cases} \tau - (\tau' + d\tau') & \text{on } S', \\ \kappa - \eta & \text{on } Q, \\ \epsilon - \kappa & \text{on } P, \\ \tau' - \tau & \text{on } S. \end{cases}$$

Since (3) holds inside this volume V it follows from (5) that

$$\int_{\tau = \tau' + d\tau'} T^{\mu\nu} dS_\nu - \int_{\tau = \tau'} T^{\mu\nu} dS_\nu = \int_{\kappa = \epsilon} T^{\mu\nu} dS_\nu - \int_{\kappa = \eta} T^{\mu\nu} dS_\nu, \tag{10}$$

the normals to the light cones being taken in the direction of increasing τ in both cases, and towards the world-line on the two world-tubes. To evaluate the second integral on the left it is convenient to take $\zeta^1 = \kappa$, $\zeta^2 = s^2$ and $\zeta^3 = s^3$ in addition to $\zeta^0 = \tau$. As shown in the previous papers, it follows from (1) and (2) that

$$\frac{\partial \tau}{\partial x^\mu} = \frac{s_\mu}{\kappa}, \tag{11}$$

$$\frac{\partial \kappa}{\partial x^\mu} \equiv \kappa_\mu = v_\mu - \frac{s_\mu}{\kappa}(1 - \kappa'), \quad \text{where} \quad \kappa' \equiv \dot{v}_\mu s^\mu, \tag{12}$$

$$\frac{\partial s^2}{\partial x^\mu} = \delta_\mu^2 - v^2 \frac{s_\mu}{\kappa}, \tag{13}$$

$$\frac{\partial s^3}{\partial x^\mu} = \delta_\mu^3 - v^3 \frac{s_\mu}{\kappa}. \tag{14}$$

The right-hand sides of (11)–(14) have to be inserted as the columns of the matrix on the right of (6). Since the determinant of a matrix vanishes if it has any two columns identical, it follows that we may omit all except the first terms on the right-hand sides of (11)–(14) and get

$$D = \frac{s_0 v_1 - s_1 v_0}{\kappa}. \tag{15}$$

Therefore

$$\int_{\tau=\text{const.}} T^{\mu\nu}dS_\nu = \iiint_\epsilon^\eta T^{\mu\nu}\frac{s_\nu}{\kappa}\left|\frac{\kappa}{s_0 v_1 - s_1 v_0}\right| ds^2 ds^3 d\kappa = \iint_\epsilon^\eta T^{\mu\nu}\frac{s_\nu}{\kappa}d\Omega d\kappa, \qquad (16)$$

$d\Omega$ being the element of surface of a sphere of radius κ in the 'rest system' in which $v_0 = 1$, $v_1 = v_2 = v_3 = 0$. Similarly,

$$\int_{\kappa=\epsilon} T^{\mu\nu}dS_\nu = \int d\tau \int_{\kappa=\epsilon} T^{\mu\nu}\kappa_\nu d\Omega. \qquad (17)$$

(10) now becomes $\displaystyle \frac{d}{d\tau}\left\{\int_\epsilon^\eta\int_\kappa T^{\mu\nu}\frac{s_\nu}{\kappa}d\Omega d\kappa\right\} = \int_{\kappa=\epsilon} T^{\mu\nu}\kappa_\nu d\Omega - \int_{\kappa=\eta} T^{\mu\nu}\kappa_\nu d\Omega.$ $\qquad (18)$

It is important to note that the left-hand side of (18) is *not* a perfect differential. The integral does not depend in general only on conditions on the world-line at the point τ. Differentiating this equation with respect to ϵ, we get

$$\frac{d}{d\kappa}\int_\kappa T^{\mu\nu}\kappa_\nu d\Omega + \frac{d}{d\tau}\int_\kappa T^{\mu\nu}\frac{s_\nu}{\kappa}d\Omega = 0. \qquad (19)$$

As $\kappa \to 0$ the second term in (19) becomes a perfect differential, and hence if there are any singular terms in the first term of (19) these must be perfect differentials and be compensated identically by those in the second term. (19) is the exact statement that the singular part of the inflow is a perfect differential. In case the energy tensor can be expanded as a series in ascending powers of κ for sufficiently small values of κ starting with some negative power of κ, then (19) shows that all the terms except the term independent of κ in the inflow must be perfect differentials.[*]

The singular terms being perfect differentials can always be compensated by the addition of suitable terms to A^μ in (9) and therefore *play no part in determining the motion of the point-particle*. The term independent of κ is not a perfect differential, and by putting it equal to a suitable perfect differential as in (9) the equation of translational motion of the point-particle is obtained. It is seen that to determine the equation of a point-particle it is only necessary to calculate that part of the inflow which remains finite as $\kappa \to 0$. The rest can be ignored.

It is clear that the same argument holds if two of the boundaries of the volume V are taken as any two surfaces passing through the points τ and $\tau + d\tau$ instead of the light cones. The argument also holds for any other tensor which satisfies a conservation law of the type (3), and therefore for the angular-momentum tensor of the field. In finding the rotational equations of the particle it is therefore only necessary to calculate the finite part of the inflow of angular momentum and to put this equal to a suitable perfect differential.

2. Let $X^{\mu\nu}$ be a tensor antisymmetric in μ and ν. Let us suppose that the co-ordinates ζ^0, ζ^1, ζ^2 and ζ^3 of the last section are so chosen that $\zeta^1 = \beta$ with constant β

---

[*] This result is implicit in Mathisson's paper (1940) and can be derived from his variational equations, but it is much more cumbersome and indirect to prove it by his method.

140   **H. J. Bhabha and Harish-Chandra**

represents a closed two-dimensional surface ω forming the boundary of a portion S' of the three-dimensional surface $\zeta^0 = \alpha$, and assume that $\zeta^1 = \beta - \Delta\beta$ ($\Delta\beta$ positive) is a two-dimensional surface lying entirely within the three-dimensional surface so enclosed. Then it is well known that

$$\int_{S'} \frac{\partial X^{\nu\sigma}}{\partial x^\sigma} dS_\nu \equiv \int_{S'} \frac{\partial X^{\nu\sigma}}{\partial x^\sigma} \frac{\partial \zeta^0}{\partial x^\nu} |D|^{-1} d\zeta^1 d\zeta^2 d\zeta^3 = \int_\omega X^{\nu\sigma} \frac{\partial \zeta^0}{\partial x^\nu} \frac{\partial \zeta^1}{\partial x^\sigma} |D|^{-1} d\zeta^2 d\zeta^3. \quad (20)$$

In this formula let us insert a tensor $K^{\mu\nu\sigma}$ antisymmetric in ν and σ in place of $X^{\nu\sigma}$, and take for the surface ζ^0 the surface of the tube $\kappa = \epsilon$ between the points τ_1 and τ_2. It is convenient in this case to take as the variables ζ, $\zeta^0 = \kappa$, $\zeta^1 = \tau$, $\zeta^2 = s^2$, $\zeta^3 = s^3$, and we obtain, as in the last section, remembering (11)–(15),

$$\int_{\tau_1}^{\tau_2} \int \frac{\partial K^{\mu\nu\sigma}}{\partial x^\sigma} \kappa_\nu d\Omega d\tau = \int_\kappa K^{\mu\nu\sigma} \kappa_\nu \frac{s_\sigma}{\kappa} d\Omega \Big|_{\tau_1}^{\tau_2}$$

$$= \int_\kappa K^{\mu\nu\sigma} v_\nu \frac{s_\sigma}{\kappa} d\Omega \Big|_{\tau_1}^{\tau_2}, \quad (21)$$

on account of the antisymmetry of $K^{\mu\nu\sigma}$ in ν and σ. Differentiating this with respect to τ,

$$\int_\kappa \frac{\partial K^{\mu\nu\sigma}}{\partial x^\sigma} \kappa_\nu d\Omega = \frac{d}{d\tau} \int_\kappa K^{\mu\nu\sigma} v_\nu \frac{s_\sigma}{\kappa} d\Omega. \quad (22)$$

The inflow of $\partial K^{\mu\nu\sigma}/\partial x^\sigma$ over an infinitesimal portion of a world-tube of any arbitrary shape can similarly always be shown to be identically a perfect differential. Hence, if the energy-momentum tensor $T^{\mu\nu}$ is replaced by a new one $T'^{\mu\nu}$ defined by

$$T'^{\mu\nu} = T^{\mu\nu} + \frac{\partial K^{\mu\nu\sigma}}{\partial x^\sigma}, \quad (23)$$

which obviously also satisfies a conservation equation like (3), then the inflow is only altered by the addition of perfect differentials, and these do not alter the equations of motion since they can be compensated by the addition of suitable terms to A^μ in (9).

Denote by $\dot{B}^\mu$ that part of the inflow, calculated by using the original tensor $T^{\mu\nu}$, which is a perfect differential. B^μ depends only on the variables at τ. Then taking

$$K^{\mu\nu\sigma} = \frac{1}{4\pi} \frac{B^\mu}{\kappa^3} (v^\nu s^\sigma - v^\sigma s^\nu) \quad (24)$$

and inserting it into (22), we get, remembering (1), (2),

$$\int_\kappa \frac{\partial K^{\mu\nu\sigma}}{\partial x^\sigma} \kappa_\nu d\Omega = -\dot{B}^\mu. \quad (25)$$

Since it has been shown that the singular parts of the inflow are always perfect differentials they can be included in B^μ, and it has therefore been proved that it is always possible to find a tensor $K^{\mu\nu\sigma}$ such that the inflow calculated from the modified tensor (23) is finite over the particular world-tube chosen.

The tensor (24) is by no means uniquely defined. Thus

$$K^{\mu\nu\sigma} = \frac{1}{4\pi} \frac{B^\mu}{\kappa^3} e^{-b\kappa^n}(v^\nu s^\sigma - v^\sigma s^\nu) \tag{26}$$

could equally well have been taken, with b a positive constant and n a number greater than the highest order of singularity in B^μ. This tensor would in fact have the advantage over (24) that its integral over any surface at infinity would always vanish.

The integral of the modified tensor over any arbitrary three-dimensional surface is not necessarily finite, nor is the inflow finite over a tube of any other shape than the one chosen above. The general properties of the energy tensor $T^{\mu\nu}$ are investigated in more detail in the paper by one of us (H.-C.) which follows this, and it is shown there that it is always possible to find a tensor $K^{\mu\nu\sigma}$ such that the modified tensor has no singularities of order higher than the third.

REFERENCES

Bhabha 1939 *Proc. Roy. Soc. A*, **172**, 384–409.
Bhabha 1940 *Proc. Ind. Acad. Sci. A*, **11**, 247–267, 467.
Bhabha 1941 *Proc. Roy. Soc. A*, **178**, 314–350.
Bhabha & Corben 1941 *Proc. Roy. Soc. A*, **178**, 273–314.
Dirac 1938 *Proc. Roy. Soc. A*, **167**, 148–169.
Mathisson 1940 *Proc. Camb. Phil. Soc.* **36**, 331–350.
Pryce 1938 *Proc. Roy. Soc. A*, **168**, 389–401.

Reprinted from
Proc. Royal Soc. A.
183 (1944), 134–141

On the removal of the infinite self-energies of point-particles

By Harish-Chandra, *J. H. Bhabha Memorial Student*

Cosmic Ray Research Unit, Indian Institute of Science, Bangalore

(*Communicated by H. J. Bhabha, F.R.S.—Received* 20 *March* 1944)

A general method is set up for modifying the energy-momentum tensor so as to remove the singularities in the flow of energy and momentum into the world-line of a particle without affecting the equations of motion of the particle. It is shown how the singularities of different order may be removed one by one.

In the case of the electromagnetic and meson fields it is shown that the modified tensor leads to a finite integral of energy and momentum over any space-like surface. In other cases the corresponding result may be secured by making a further modification in the tensor.

1. It has been shown in the preceding paper* that all singular terms in the expression for the rate of inflow† of energy and momentum are perfect differentials, and therefore the equations of motion of a point-particle calculated from considerations of the conservation of energy and momentum are always finite. One would now expect to be able to alter, without disturbing the equations of motion, the energy-momentum tensor $T^{\mu\nu}$ of the field so as to make the inflow finite, thus avoiding completely the appearance of infinities which are entirely spurious, at any rate, from the point of view of the equations of motion. It would be still better if it could be arranged that the total energy and momentum calculated from the modified tensor $\bar{T}^{\mu\nu}$ were finite. This of course would automatically secure a finite inflow.

It is proved in A that the replacement of the energy-momentum tensor $T^{\mu\nu}$ by

$$\bar{T}^{\mu\nu} = T^{\mu\nu} + \frac{\partial K^{\mu\nu\sigma}}{\partial x^\sigma}, \tag{1}$$

where $K^{\mu\nu\sigma}$ is antisymmetric in (ν,σ), does not alter the equations of motion. Therefore an attempt should be made to alter $T^{\mu\nu}$ in accordance with (1) in trying to get rid of the infinities. This method has already been used by Pryce (1938) to make the energy and momentum of the field finite in the case of the point electron. It will be shown quite generally in this paper that given an energy-momentum tensor $T^{\mu\nu}$ satisfying the conservation equation

$$\frac{\partial T^{\mu\nu}}{\partial x^\nu} = 0, \tag{2}$$

a tensor $K^{\mu\nu\sigma}$ can always be found so as to make the energy-momentum integrals calculated from $\bar{T}^{\mu\nu}$ finite provided $T^{\mu\nu}$ fulfils certain very general restrictions. In general, $K^{\mu\nu\sigma}$ is not symmetric in μ, ν even when $T^{\mu\nu}$ is so, and therefore the modified

---

* Bhabha & Harish-Chandra (1944), referred to in this paper as A.
† This term is used in the same sense as in A.

[142]

tensor $T^{\mu\nu}$ is also not symmetric. But it will be shown that in the case of an electromagnetic charge or a dipole this symmetry can be achieved. However, the symmetry of the energy-momentum tensor, which is needed only for the purpose of constructing an angular-momentum tensor, is not necessary here, because it will be shown that a similar process can be applied to the angular-momentum tensor also, so that the total angular momentum calculated from this modified tensor is finite.

2. Keeping to the notation of A it is now necessary to determine $K^{\mu\nu\sigma}$. Equations (18) and (19) of A are

$$\frac{d}{d\tau}\left\{\int_\epsilon^\eta d\kappa \int_\kappa T^{\mu\nu}\frac{s_\nu}{\kappa}\,d\Omega\right\} = \int_{\kappa=\epsilon} T^{\mu\nu}\kappa_\nu\,d\Omega - \int_{\kappa=\eta} T^{\mu\nu}\kappa_\nu\,d\Omega, \tag{3}$$

$$\frac{d}{d\tau}\int_\kappa T^{\mu\nu}\frac{s_\nu}{\kappa}\,d\Omega = -\frac{d}{d\kappa}\int_\kappa T^{\mu\nu}\kappa_\nu\,d\Omega. \tag{4}$$

Here $\int_\kappa$ denotes integration over the sphere of intersection of the light cone at τ with the tube $\kappa = $ constant. This sphere will be called the 'retarded sphere' of radius κ at τ. In the appendix (equation (98)) it is shown that

$$\kappa\frac{d}{d\kappa}\int_\kappa T^{\mu\nu}\kappa_\nu\frac{s^\alpha}{\kappa}\frac{s^\beta}{\kappa}\ldots\frac{s^\lambda}{\kappa}\,d\Omega = \int_\kappa\left\{\frac{\partial}{\partial x^\sigma}(T^{\mu\nu}s^\sigma - T^{\mu\sigma}s^\nu)\right\}\kappa_\nu\frac{s^\alpha}{\kappa}\frac{s^\beta}{\kappa}\ldots\frac{s^\lambda}{\kappa}\,d\Omega, \tag{5}$$

where $\alpha, \beta, \ldots, \lambda$ are any number of free indices. Now introduce two operators D and θ defined in the following way:

$$D = \kappa\frac{d}{d\kappa}, \tag{6a}$$

$$\theta U'^{\alpha\beta\ldots\lambda\nu} = \frac{\partial}{\partial x^\sigma}(U'^{\alpha\beta\ldots\lambda\nu}s^\sigma - U'^{\alpha\beta\ldots\lambda\sigma}s^\nu)$$

$$= 2U'^{\alpha\beta\ldots\nu} + U'^{\alpha\beta\ldots\sigma}\frac{s_\sigma}{\kappa}v^\nu + s^\sigma\frac{\partial U'^{\alpha\beta\ldots\nu}}{\partial x^\sigma} - s^\nu\frac{\partial U'^{\alpha\beta\ldots\sigma}}{\partial x^\sigma}, \tag{6b}$$

where $U'^{\alpha\beta\ldots\lambda\nu}$ is any arbitrary tensor with any number of indices. It should be noted that

$$\frac{\partial}{\partial x^\nu}(\theta U'^{\alpha\beta\ldots\lambda\nu}) = 0 \tag{7}$$

owing to the antisymmetry in ν, σ of the expression inside the brackets in (6b). Obviously both D and θ are linear operators. Also $D\kappa^n = n\kappa^n$ and therefore by repetition $D^r\kappa^n = n^r\kappa^n$, so that if $f(D)$ is a polynomial expressible in powers of D, $f(D)\kappa^n = f(n)\kappa^n$. This holds whether n is positive or negative.

Using D and θ, (5) can be written in the form

$$D\int_\kappa T^{\mu\nu}\kappa_\nu\frac{s^\alpha}{\kappa}\ldots\frac{s^\lambda}{\kappa}\,d\Omega = \int_\kappa \theta T^{\mu\nu}.\kappa_\nu\frac{s^\alpha}{\kappa}\ldots\frac{s^\lambda}{\kappa}\,d\Omega.$$

Since $\theta T^{\mu\nu}$ is conserved with respect to ν as $T^{\mu\nu}$ was, the same considerations as above can be applied to $\theta T^{\mu\nu}$ instead of $T^{\mu\nu}$ and hence

$$D^2 \int_\kappa T^{\mu\nu}\kappa_\nu \frac{s^\alpha}{\kappa}\cdots\frac{s^\lambda}{\kappa}d\Omega = D\int_\kappa (\theta T^{\mu\nu})\kappa_\nu \frac{s^\alpha}{\kappa}\cdots\frac{s^\lambda}{\kappa}d\Omega = \int_\kappa \theta^2 T^{\mu\nu}.\kappa_\nu\frac{s^\alpha}{\kappa}\cdots\frac{s^\lambda}{\kappa}d\Omega.$$

By repetition it can be proved that

$$D^r\int_\kappa T^{\mu\nu}\kappa_\nu \frac{s^\alpha}{\kappa}\cdots\frac{s^\lambda}{\kappa}d\Omega = \int_\kappa \theta^r T^{\mu\nu}.\kappa_\nu \frac{s^\alpha}{\kappa}\cdots\frac{s^\lambda}{\kappa}d\Omega,$$

and therefore
$$f(D)\int_\kappa T^{\mu\nu}\kappa_\nu \frac{s^\alpha}{\kappa}\cdots\frac{s^\lambda}{\kappa}d\Omega = \int_\kappa f(\theta)\, T^{\mu\nu}.\kappa_\nu \frac{s^\alpha}{\kappa}\cdots\frac{s^\lambda}{\kappa}d\Omega, \tag{8}$$

where $f(D)$ is a polynomial expressible in positive integral powers of D.

It will now be assumed that the part of $T^{\mu\nu}$ which becomes infinite as $\kappa \to 0$ can be expressed for sufficiently small values of κ in the form of a series involving a finite number of negative powers of κ, so that for sufficiently small κ one can write*

$$T^{\mu\nu} = T_0^{\mu\nu} + \sum_{r=1}^{r=m+2}\frac{a_{(-r)}^{\mu\nu}}{\kappa^r}, \tag{9}$$

where $T_0^{\mu\nu}$ remains finite as $\kappa \to 0$ and $a_{(-r)}^{\mu\nu}$ are functions of s^α/κ and τ only. It is assumed that these functions are finite and continuous. That the energy-momentum tensors of the electromagnetic and the meson fields actually satisfy these requirements is shown in the appendix. Therefore

$$\int_\kappa T^{\mu\nu}\kappa_\nu \frac{s^\alpha}{\kappa}\cdots\frac{s^\lambda}{\kappa}\,d\Omega = c_0^{\mu\alpha\ldots\lambda} + \sum_{r=1}^{r=m}\frac{c_{(-r)}^{\mu\alpha\ldots\lambda}}{\kappa^r} \tag{10}$$

can be written, where $c_0^{\mu\alpha\ldots\lambda}$ is a function of κ and τ which remains finite as $\kappa \to 0$ and $c_{(-r)}^{\mu\alpha\ldots\lambda}$ are functions of τ only. It is assumed that for sufficiently small κ, $c_0^{\mu\alpha\ldots\lambda}$ is expressible as a series in positive powers of κ. This also is true for the electromagnetic and meson fields.

Applying the operator $f(D)$ to both the sides of (10).

$$f(D)\int_\kappa T^{\mu\nu}\kappa_\nu \frac{s^\alpha}{\kappa}\cdots\frac{s^\lambda}{\kappa}\,d\Omega = f(D)\,c_0^{\mu\alpha\ldots\lambda} + \sum_{r=1}^{r=m}f(-r)\frac{c_{(-r)}^{\mu\alpha\ldots\lambda}}{\kappa^r}. \tag{11}$$

Taking
$$f(D) = \prod_{k=1}^{k=m}\left(1+\frac{D}{k}\right), \tag{12}$$

we get $f(-r) = 0$ for $1 \leqslant r \leqslant m$.

From (11) and (8) it follows that

$$\prod_{k=1}^{k=m}\left(1+\frac{D}{k}\right)\int_\kappa T^{\mu\nu}\kappa_\nu \frac{s^\alpha}{\kappa}\cdots\frac{s^\lambda}{\kappa}d\Omega = \int_\kappa \left\{\prod_{k=1}^{k=m}\left(1+\frac{\theta}{k}\right)T^{\mu\nu}\right\}\kappa_\nu \frac{s^\alpha}{\kappa}\cdots\frac{s^\lambda}{\kappa}\,d\Omega$$

$$= \prod_{k=1}^{k=m}\left(1+\frac{D}{k}\right)c_0^{\mu\alpha\ldots\lambda}. \tag{13}$$

---

* The highest singularity is written as of the order $m+2$ for convenience. The removal of only those singularities from $T^{\mu\nu}$ is necessary which are of an order higher than the 2nd, since singularities of the 2nd or lesser order give a finite energy integral (Pryce 1938).

Since the right side of this equation is non-singular, the left must be so, i.e. $\prod_{k=1}^{k=m}\left(1+\dfrac{\theta}{k}\right)T^{\mu\nu}$ gives a finite integral.

From this it follows that

$$\left\{\prod_{k=1}^{k=m}\left(1+\frac{\theta}{k}\right)T^{\mu\nu}\right\}\kappa_\nu \tag{14}$$

cannot contain any singularity of an order higher than the 2nd, because if it contained a term of the type b^μ/κ^{r+2} (where $r > 0$), this would give after integration a singular term of the type $c^{\mu\alpha\cdots\lambda}/\kappa^r$, where

$$c^{\mu\alpha\cdots\lambda} = \int b^\mu \frac{s^\alpha}{\kappa}\cdots\frac{s^\lambda}{\kappa}\,d\omega \quad \left[d\omega = \frac{d\Omega}{\kappa^2} = \text{the element of solid angle}\right].$$

Since such a term is absent from the right side of (13) it follows that

$$c^{\mu\alpha\cdots\lambda} = \int b^\mu \frac{s^\alpha}{\kappa}\cdots\frac{s^\lambda}{\kappa}\,d\omega = 0.$$

As in the rest system s^α/κ are nothing else than the direction cosines (except s^0/κ which is 1). it follows from the expansibility of every continuous function defined on the surface of a sphere, in terms of surface harmonics, that $b^\mu = 0$. Thus (14) does not contain any singularities higher than of the 2nd order.

Because

$$\prod_{k=1}^{k=m}\left(1+\frac{\theta}{k}\right) = 1 + a\theta + b\theta^2 + \ldots = 1 + \theta(a + b\theta + \ldots), \tag{15}$$

where $a, b, \ldots$ are numerical coefficients. it is clear that

$$\prod_{k=1}^{k=m}\left(1+\frac{\theta}{k}\right)T^{\mu\nu} = T^{\mu\nu} + \theta(a + b\theta + \ldots)\,T^{\mu\nu} = T^{\mu\nu} + \theta K^{\mu\nu},$$

where

$$K^{\mu\nu} \equiv (a + b\theta + \ldots)\,T^{\mu\nu}.$$

Further. as $\theta K^{\mu\nu} = \partial K^{\mu\nu\sigma}\,\partial x^\sigma$. where $K^{\mu\nu\sigma} = K^{\mu\nu}s^\sigma - K^{\mu\sigma}s^\nu$. the modified tensor

$$\prod_{k=1}^{k=m}\left(1+\frac{\theta}{k}\right)T^{\mu\nu} = T^{\mu\nu} + \frac{\partial K^{\mu\nu\sigma}}{\partial x^\sigma}.$$

which is just of the required form. Put

$$\prod_{k=1}^{k=m}\left(1+\frac{\theta}{k}\right)T^{\mu\nu} = T'^{\mu\nu}. \tag{16}$$

It should be noted in passing that it can be shown directly that $T'^{\mu\nu}$ does not contain singularities of an order higher than the 3rd, because, as shown in the appendix (equation (102)),

$$D\int_\kappa T^{\mu\sigma}(\delta_\sigma^\nu - v_\sigma v^\nu)\frac{s^\alpha}{\kappa}\cdots\frac{s^\lambda}{\kappa}\,d\Omega = \int_\kappa \theta T^{\mu\sigma}\cdot(\delta_\sigma^\nu - v_\sigma v^\nu)\frac{s^\alpha}{\kappa}\cdots\frac{s^\lambda}{\kappa}\,d\Omega.$$

Hence

$$\int_\kappa T'^{\mu\sigma}(\delta_\sigma^\nu - v_\sigma v^\nu)\frac{s^\alpha}{\kappa}\cdots\frac{s^\lambda}{\kappa}\,d\Omega = \prod_{k=1}^{k=m}\left(1+\frac{D}{k}\right)\int_\kappa T^{\mu\sigma}(\delta_\sigma^\nu - v_\sigma v^\nu)\frac{s^\alpha}{\kappa}\cdots\frac{s^\lambda}{\kappa}\,d\Omega. \tag{17}$$

Through the same reasoning as used before it can be shown that, since the right side of (17) is non-singular, so also must be the left. From this it can be inferred exactly as in the case of $T'^{\mu\nu}\kappa_\nu$ that $T'^{\mu\sigma}(\delta^\nu_\sigma - v_\sigma v^\nu)$ does not contain singularities of an order higher than the 2nd.

It is further found (appendix equation (100)) that

$$D\int_\kappa T^{\mu\nu}s_\nu\frac{s^\alpha}{\kappa}\cdots\frac{s^\lambda}{\kappa}\,d\Omega = \int_\kappa \theta T^{\mu\nu}\cdot s_\nu\frac{s^\alpha}{\kappa}\cdots\frac{s^\lambda}{\kappa}\,d\Omega.$$

Hence
$$\int_\kappa T'^{\mu\nu}s_\nu\frac{s^\alpha}{\kappa}\cdots\frac{s^\lambda}{\kappa}\,d\Omega = \prod_{k=1}^{k=m}\left(1+\frac{D}{k}\right)\int_\kappa T^{\mu\nu}s_\nu\frac{s^\alpha}{\kappa}\cdots\frac{s^\lambda}{\kappa}\,d\Omega. \tag{18}$$

From this one infers precisely as before that $T'^{\mu\nu}s_\nu$ does not contain singularities higher than of the 2nd order, so that $T'^{\mu\nu}\frac{s_\nu}{\kappa}$ or $T'^{\mu\nu}\frac{s_\nu}{\kappa}(1-\kappa')$ cannot contain singularities higher than $1/\kappa^3$. Since it has also been proved that

$$T'^{\mu\nu}\kappa_\nu = T'^{\mu\nu}\left\{v_\nu - \frac{s_\nu}{\kappa}(1-\kappa')\right\}$$

does not have singularities higher than of the 2nd order, it follows that $T'^{\mu\nu}v_\nu$ does not possess terms of order higher than $1/\kappa^3$. Now write $T'^{\mu\nu}$ in the form

$$T'^{\mu\nu} = T'^{\mu\sigma}(\delta^\nu_\sigma - v_\sigma v^\nu) + (T'^{\mu\sigma}v_\sigma)v^\nu.$$

The first term does not contain terms of order higher than $1/\kappa^2$ and the second of order higher than $1/\kappa^3$. So the highest singularity in $T'^{\mu\nu}$ can at most be of the order $1/\kappa^3$.

3. We now proceed to examine in detail the process of removal of the singular terms of $T^{\mu\nu}$ by the operator $\prod_{k=1}^{k=m}\left(1+\frac{\theta}{k}\right)$. In addition to θ we define another operator ϕ by the equation

$$\phi U^{\alpha\dots\nu} = 2U^{\alpha\dots\nu} + s^\sigma\frac{\partial U^{\alpha\dots\sigma}}{\partial x^\sigma} + U^{\alpha\dots\sigma}\frac{s_\sigma}{\kappa}v^\nu, \tag{19}$$

so that from (6b) and (19) is obtained

$$\theta U^{\alpha\dots\nu} = \phi U^{\alpha\dots\nu} - s^\nu\frac{\partial U^{\alpha\dots\sigma}}{\partial x^\sigma}. \tag{20}$$

When $U^{\alpha\dots\sigma}$ is conserved with respect to σ,

$$\theta U^{\alpha\dots\nu} = \phi U^{\alpha\dots\nu}. \tag{21}$$

While $\theta U^{\alpha\dots\nu}$ is always conserved with respect to ν, $\phi U'^{\alpha\dots\nu}$ is conserved in general only when $U^{\alpha\dots\nu}$ itself is conserved.

Let $T^{\mu\nu}_{(-n)}$ denote the nth order term in $T^{\mu\nu}$ so that in accordance with (9) $T^{\mu\nu}_{(-n)} = \frac{a^{\mu\nu}_{(-n)}}{\kappa^n}$. Then

$$\phi T^{\mu\nu}_{(-n)} = 2T^{\mu\nu}_{(-n)} + s^\sigma\frac{\partial T^{\mu\nu}_{(-n)}}{\partial x^\sigma} + T^{\mu\nu}_{(-n)}\frac{s_\sigma}{\kappa}v^\nu.$$

As before assume $a^{\mu\nu}_{(-n)}$ to be a function of the arguments s^α/κ and τ only. Now both τ and s^α/κ are to be treated as constants with respect to the operator $s^\sigma \dfrac{\partial}{\partial x^\sigma}$, because

$$s^\sigma \frac{\partial \tau}{\partial x^\sigma} = 0 \quad \text{and} \quad s^\sigma \frac{\partial}{\partial x^\sigma}\left(\frac{s^\alpha}{\kappa}\right) = 0, \quad \text{so that} \quad s^\sigma \frac{\partial a^{\mu\nu}_{(-n)}}{\partial x^\sigma} = 0. \quad \text{Therefore}$$

$$s^\sigma \frac{\partial T^{\mu\nu}_{(-n)}}{\partial x^\sigma} = -n \frac{a^{\mu\nu}_{(-n)}}{\kappa^{n+1}} \kappa_\sigma s^\sigma = -n \frac{a^{\mu\nu}_{(-n)}}{\kappa^n} = -n T^{\mu\nu}_{(-n)}. \tag{22}$$

Thus $\qquad \phi T^{\mu\nu}_{(-n)} = (2-n)\, T^{\mu\nu}_{(-n)} + T^{\mu\sigma}_{(-n)} \dfrac{s_\sigma}{\kappa} v^\nu = \left[(2-n)\,\delta^\nu_\sigma + v^\nu \dfrac{s_\sigma}{\kappa}\right] T^{\mu\sigma}_{(-n)}. \tag{23}$

Now since $s^\rho \dfrac{\partial}{\partial x^\rho}\left(\dfrac{s_\sigma}{\kappa}\, v^\nu\right) = 0$, it follows that

$$\phi^2 T^{\mu\nu}_{(-n)} = \left[(2-n)\,\delta^\nu_\rho + v^\nu \frac{s_\rho}{\kappa}\right]\left[(2-n)\,\delta^\rho_\sigma + v^\rho \frac{s_\sigma}{\kappa}\right] T^{\mu\sigma}_{(-n)}.$$

By repetition $\qquad \phi^l T^{\mu\nu}_{(-n)} = \left(\left[(2-n)\,\delta + v\,\frac{s}{\kappa}\right]^l\right)^\nu_\sigma T^{\mu\sigma}_{(-n)} \tag{24}$

for any positive integer l, where

$$\left(\left[(2-n)\,\delta + v\,\frac{s}{\kappa}\right]^l\right)^\nu_\sigma$$
$$\equiv \left[(2-n)\,\delta^\nu_{\rho_1} + v^\nu \frac{s_{\rho_1}}{\kappa}\right]\left[(2-n)\,\delta^{\rho_1}_{\rho_2} + v^{\rho_1} \frac{s_{\rho_2}}{\kappa}\right]\cdots\left[(2-n)\,\delta^{\rho_{l-1}}_{\sigma} + v^{\rho_{l-1}} \frac{s_\sigma}{\kappa}\right].$$

Put $\qquad a\,\delta^\nu_\sigma + b v^\nu \dfrac{s_\sigma}{\kappa} \equiv \left[A(a,b)\right]^\nu_\sigma \tag{25a}$

and $\qquad \left[\left(a\,\delta + b v\,\dfrac{s}{\kappa}\right)^l\right]^\nu_\sigma \equiv \left[A^l(a,b)\right]^\nu_\sigma. \tag{25b}$

The indices will sometimes be omitted. For example, we shall write

$$\phi^l T_{(-n)} = A^l(2-n,1)\, T_{(-n)}. \tag{26a}$$

meaning thereby $\qquad \phi^l T^{\mu\nu}_{(-n)} = \left[A^l(2-n,1)\right]^\nu_\sigma T^{\mu\sigma}_{(-n)}. \tag{26b}$

It is also seen that $\qquad c A(a,b) = A(ca,cb). \tag{27a}$

$$A(a_1,b_1) + A(a_2,b_2) = A(a_1+a_2, b_1+b_2), \tag{27b}$$

and $\qquad c = A(c,0). \tag{27c}$

where c is an ordinary number. On using (26) and (27)

$$\prod_{k=1}^{k=m}\left(1+\frac{\phi}{k}\right) T_{(-n)} = \prod_{k=1}^{k=m}\left[1+\frac{A(2-n,1)}{k}\right] T_{(-n)} = \prod_{k=1}^{k=m} A\left(1+\frac{2-n}{k}, \frac{1}{k}\right) T_{(-n)}. \tag{28}$$

Consider now $A(a_1, b_1)\, A(a_2, b_2)$. It is found that

$$\left[A(a_1, b_1)\, A(a_2, b_2) \right]_{\sigma}^{\nu} = \left(a_1 \delta_{\rho}^{\nu} + b_1 v^{\nu} \frac{s_{\rho}}{\kappa} \right) \left(a_2 \delta_{\sigma}^{\rho} + b_2 v^{\rho} \frac{s_{\sigma}}{\kappa} \right)$$

$$= a_1 a_2 \delta_{\sigma}^{\nu} + [(a_1 + b_1)(a_2 + b_2) - a_1 a_2]\, v^{\nu} \frac{s_{\sigma}}{\kappa}$$

$$= \left[A(a_1 a_2, [a_1 + b_1][a_2 + b_2] - a_1 a_2) \right]_{\sigma}^{\nu}.$$

It can easily be proved by induction that

$$A(a_1, b_1)\, A(a_2, b_2) \dots A(a_n, b_n)$$
$$= A(a_1 a_2 \dots a_n, [a_1 + b_1][a_2 + b_2] \dots [a_n + b_n] - a_1 a_2 \dots a_n). \qquad (29)$$

Thus we get immediately

$$\prod_{k=1}^{k=m} A\left(1 + \frac{2-n}{k}, \frac{1}{k} \right) = A\left(\prod_{k=1}^{k=m}\left(1 + \frac{2-n}{k} \right), \ \prod_{k=1}^{k=m}\left(1 + \frac{3-n}{k} \right) - \prod_{k=1}^{k=m}\left(1 + \frac{2-n}{k} \right) \right). \qquad (30)$$

Hence if $m \geqslant n - 2$ and $n \geqslant 4$,

$$\prod_{k=1}^{k=m} A\left(1 + \frac{2-n}{k} \cdot \frac{1}{k} \right) = A(0, 0) = 0.$$

Thus $\qquad\qquad \displaystyle\prod_{k=1}^{k=m}\left(1 + \frac{\phi}{k} \right) T_{(-n)} = 0 \quad \text{for} \quad m + 2 \geqslant n \geqslant 4.$

Since $m + 2$ has already been chosen equal to the order of the highest singularity in $T^{\mu\nu}$, it follows that the operator $\displaystyle\prod_{k=1}^{k=m}\left(1 + \frac{\phi}{k} \right)$ when applied to $T^{\mu\nu}$ removes all singular terms higher than the 3rd. For the term of the 3rd order. from (28) and (30),

$$\prod_{k=1}^{k=m}\left(1 + \frac{\phi}{k} \right) T_{(-3)} = A(0, 1)\, T_{(-3)} = v^{\nu} \frac{s_{\sigma}}{\kappa}\, T_{(-3)}^{\mu\sigma}.$$

Thus $\qquad\qquad \displaystyle\prod_{k=1}^{k=m}\left(1 + \frac{\phi}{k} \right) T^{\mu\nu} = \prod_{k=1}^{k=m}\left(1 + \frac{\phi}{k} \right) T_{-2)}^{\mu\nu} + . T_{(-3)}^{\mu\sigma}\frac{s_{\sigma}}{\kappa}\, v^{\nu}.$

where $T_{-2)}^{\mu\nu}$ denotes the part of $T^{\mu\nu}$ remaining after omitting terms of 3rd order or higher.*

---

* $T_{-n)}^{\mu\nu}$ denotes the part of $T^{\mu\nu}$ left over after omitting terms of order higher than the nth, and $T_{(-n}^{\mu\nu}$ the part obtained by retaining only terms of the nth or higher order, so that

$$T_{-n+1)}^{\mu\nu} + T_{(-n}^{\mu\nu} = T^{\mu\nu}.$$

The suffixes $-n)$ and $(-n$ attached to any other quantity have analogous significance. Similarly the suffix $(-n)$ always denotes the term of the nth order.

Since $T^{\mu\nu}$ is conserved with respect to ν, $\phi T^{\mu\nu} = \theta T^{\mu\nu}$ and $\phi' T^{\mu\nu} = \theta' T^{\mu\nu}$, so that

$$T'^{\mu\nu} = \prod_{k=1}^{k=m}\left(1+\frac{\theta}{k}\right) T^{\mu\nu} = \prod_{k=1}^{k=m}\left(1+\frac{\phi}{k}\right) T^{\mu\nu} = \prod_{k=1}^{k=m}\left(1+\frac{\phi}{k}\right) T^{\mu\nu}_{-2)} + T^{\mu\sigma}_{(-3)}\frac{s_\sigma}{\kappa} v^\nu, \quad (31)$$

which is in conformity with the proved fact that $T'^{\mu\nu}$ does not contain singularities higher than of the 3rd order.

In accordance with (15)

$$\prod_{k=1}^{k=m}\left(1+\frac{\phi}{k}\right) = 1+\phi(a+b\phi+\dots),$$

so that

$$T'^{\mu\nu} = T^{\mu\nu}_{-2)} + \phi(a+b\phi+\dots)\, T^{\mu\nu}_{-2)} + T^{\mu\sigma}_{(-3)}\frac{s_\sigma}{\kappa} v^\nu$$

$$= T^{\mu\nu}_{-2)} + \theta(a+b\phi+\dots)\, T^{\mu\nu}_{-2)} + T^{\mu\sigma}_{(-3)}\frac{s_\sigma}{\kappa} v^\nu + s^\nu\frac{\partial}{\partial x^\sigma}[(a+b\phi+\dots)\, T^{\mu\sigma}_{-2)}]. \quad (32)$$

A new tensor $T''^{\mu\nu}$ is now introduced, defined by

$$T''^{\mu\nu} \equiv T'^{\mu\nu} - \theta(a+b\phi+\dots)\, T^{\mu\nu}_{-2)} = T^{\mu\nu}_{-2)} + T^{\mu\sigma}_{(-3)}\frac{s_\sigma}{\kappa} v^\nu + s^\nu\frac{\partial}{\partial x^\sigma}[(a+b\phi+\dots)\, T^{\mu\sigma}_{-2)}]. \quad (33)$$

$T''^{\mu\nu}$ is obviously conserved and is of the form $T^{\mu\nu}+\dfrac{\partial K^{\mu\nu\sigma}}{\partial x^\sigma}$, where $K^{\mu\nu\sigma}$ is antisymmetrical in ν,σ and therefore $T''^{\mu\nu}$ instead of $T'^{\mu\nu}$ may be taken as the modified tensor.

It will now be shown that

$$\frac{\partial}{\partial x^\sigma}[(a+b\phi+\dots)\, T^{\mu\sigma}_{-2)}] = a\left[\frac{\partial T^{\mu\sigma}_{(-2)}}{\partial x^\sigma}\right]_{(-3)} = a\frac{\partial T^{\mu\sigma}_{-2)}}{\partial x^\sigma}. \quad (34)$$

It has been seen above that the order of a term is not changed by application of the operator ϕ. Hence $(a+b\phi+\dots)\, T^{\mu\nu}_{-2)}$ contains no singularities higher than of the 2nd order. Since

$$T^{\mu\nu}_{(-3} = T^{\mu\nu} - T^{\mu\nu}_{-2)}.$$

it follows from the conservation of $T^{\mu\sigma}$ and therefore of $(a+b\phi+\dots)\, T^{\mu\sigma}$ that

$$\frac{\partial}{\partial x^\sigma}[(a+b\phi+\dots)\, T^{\mu\sigma}_{-2)}] = -\frac{\partial}{\partial x^\sigma}[(a+b\phi+\dots)\, T^{\mu\sigma}_{(-3}]. \quad (35)$$

Now

$$\frac{\partial T^{\mu\nu}_{(-n)}}{\partial x^\sigma} = \frac{1}{\kappa^n}\left[\dot{a}^{\mu\nu}_{(-n)}\frac{s_\sigma}{\kappa} + a^{\mu\nu,\rho}_{(-n)}\frac{\partial}{\partial x^\sigma}\left(\frac{s_\rho}{\kappa}\right) - n\frac{a^{\mu\nu}_{(-n)}}{\kappa}\kappa_\sigma\right]$$

$$= \frac{\dot{a}^{\mu\nu}_{(-n)}s_\sigma}{\kappa^n\,\kappa} + \frac{a^{\mu\nu,\rho}_{(-n)}}{\kappa^{n+1}}\left\{g_{\rho\sigma} - \frac{s_\sigma}{\kappa}v_\rho - \frac{s_\rho}{\kappa}v_\sigma + \frac{s_\rho s_\sigma}{\kappa\,\kappa}(1-\kappa')\right\} - n\frac{a^{\mu\nu}_{(-n)}}{\kappa^{n+1}}\left[v_\sigma - (1-\kappa')\frac{s_\sigma}{\kappa}\right]. \quad (36)$$

where

$$\dot{a}^{\mu\nu} = \frac{\partial a^{\mu\nu}_{(-n)}}{\partial\tau} \quad \text{and} \quad a^{\mu\nu,\rho} = \frac{\partial a^{\mu\nu}_{(-n)}}{\partial\left(\frac{s_\rho}{\kappa}\right)}. \quad (37)$$

treating $\tau, s_\rho/\kappa$ as independent variables. It follows therefore that on differentiation $T^{\mu\nu}_{(-n)}$ can give terms only of the nth and $n+1$th order. The same result is obviously applicable to terms of different orders in $(a+b\phi+\ldots)\,T^{\mu\nu}$. Thus the highest singularity in $\dfrac{\partial}{\partial x^\sigma}[(a+b\phi+\ldots)\,T^{\mu\nu}]$ is of the 3rd order, which is also the order of the lowest singularity in $\dfrac{\partial}{\partial x^\sigma}[(a+b\phi+\ldots)\,T^{\mu\sigma}_{(-3)}]$. From equation (35) it follows therefore that $\dfrac{\partial}{\partial x^\sigma}[(a+b\phi+\ldots)\,T^{\mu\sigma}_{(-2)}]$ contains only one term of the 3rd order, because the terms of lower order which it could contain do not exist on the right side of (35). Since only the 2nd order term in $(a+b\phi+\ldots)\,T^{\mu\sigma}_{(-2)}$ can contribute to a 3rd order term in the divergence, it is sufficient to calculate the 2nd order term of $(a+b\phi+\ldots)\,T^{\mu\sigma}_{(-2)}$ and find out the 3rd order term that results from it on differentiation.

Now proceed to determine $(a+b\phi+\ldots)\,T^{\mu\nu}_{(-n)}$. From (26), (27) and (29) it follows immediately that

$$(a+b\phi+\ldots)\,T_{(-n)} = A(p,q)\,T_{(-n)},\tag{38}$$

where p,q are some numerical numbers. However, on using (27)–(30) one gets

$$\prod_{k=1}^{k=m}\left(1+\frac{\phi}{k}\right)T_{(-n)} = [1+\phi(a+b\phi+\ldots)]\,T_{(-n)} = T_{(-n)}+\phi A(p,q)\,T_{(-n)}$$

$$= T_{(-n)}+A(2-n,1)\,A(p,q)\,T_{(-n)}$$

$$= A(1+p(2-n),\,q(3-n)+p)\,T_{(-n)}$$

$$= A\left[\prod_{k=1}^{k=m}\left(1+\frac{2-n}{k}\right),\ \prod_{k=1}^{k=m}\left(1+\frac{3-n}{k}\right)-\prod_{k=1}^{k=m}\left(1+\frac{2-n}{k}\right)\right]T_{(-n)}.$$

Thus

$$1+p(2-n) = \prod_{k=1}^{k=m}\left(1+\frac{2-n}{k}\right),\quad q(3-n)+p = \prod_{k=1}^{k=m}\left(1+\frac{3-n}{k}\right)-\prod_{k=1}^{k=m}\left(1+\frac{2-n}{k}\right),$$

i.e.

$$p = \frac{1}{2-n}\left[\prod_{k=1}^{k=m}\left(1+\frac{2-n}{k}\right)-1\right],\tag{39a}$$

$$q = \frac{1}{3-n}\left[\prod_{k=1}^{k=m}\left(1+\frac{3-n}{k}\right)-\left(1+\frac{1}{2-n}\right)\prod_{k=1}^{k=m}\left(1+\frac{2-n}{k}\right)+\frac{1}{2-n}\right].\tag{39b}$$

By passing to the limit $n\to 2$ it is easy to show that in this case $p=a=1+\frac{1}{2}+\ldots+\frac{1}{m}$ and $q=m-a$. Therefore

$$\frac{\partial}{\partial x^\sigma}[(a+b\phi+\ldots)\,T^{\mu\sigma}_{(-2)}] = \left\{\frac{\partial}{\partial x^\sigma}\left[a\,T^{\mu\sigma}_{(-2)}+(m-a)\,T^{\mu\rho}_{(-2)}\frac{s_\rho}{\kappa}v^\sigma\right]\right\}\bigg|_{(-3)},\tag{40}$$

the suffix denoting that only the 3rd order term resulting after differentiation is to be retained. Now, since

$$\frac{\partial}{\partial x^\sigma}\left(\frac{s_\rho}{\kappa}v^\sigma\right) = 0,\tag{41}$$

it is found that

$$\frac{\partial}{\partial x^\sigma}\left[T^{\mu\rho}_{(-n)}\frac{s_\rho}{\kappa}v^\sigma\right] = \frac{s_\rho}{\kappa}v^\sigma\frac{\partial T^{\mu\nu}_{(-n)}}{\partial x^\sigma} = \frac{s_\rho}{\kappa}\left[\frac{\dot{a}^{\mu\rho}_{(-n)}}{\kappa^n} - \frac{a^{\mu\rho,\lambda}_{(-n)}}{\kappa^n}\frac{s_\lambda}{\kappa}\frac{\kappa'}{\kappa} - n\frac{a^{\mu\rho}_{(-n)}}{\kappa^n}\frac{\kappa'}{\kappa}\right], \tag{42}$$

where the various symbols have the same meaning as in (37). Since $\dot{a}^{\mu\rho}_{(-n)}$, $a^{\mu\rho,\lambda}_{(-n)}$, $a^{\mu\rho}_{(-n)}$ are finite functions of s^α/κ and τ only, it is obvious that $\frac{\partial}{\partial x^\sigma}\left[T^{\mu\rho}_{(-n)}\frac{s_\rho}{\kappa}v^\sigma\right]$ is of the nth order. So that

$$\left\{\frac{\partial}{\partial x^\sigma}\left[T^{\mu\rho}_{(-2)}\frac{s_\rho}{\kappa}v^\sigma\right]\right\}_{(-3)} = 0. \tag{43}$$

By comparing (36) and (42) it is also seen that

$$\frac{s_\rho}{\kappa}v^\sigma\frac{\partial T^{\mu\nu}_{(-n)}}{\partial x^\sigma} = \left[\frac{\partial T^{\mu\nu}_{(-n)}}{\partial x^\rho}\right]_{(-n)}, \tag{44}$$

a result which will be used later. So from (33), (40) and (43)

$$T''^{\mu\nu} = T^{\mu\nu}_{-2)} + T^{\mu\sigma}_{(-3)}\frac{s_\sigma}{\kappa}v^\nu + s^\nu\left\{\frac{\partial T^{\mu\nu}_{(-2)}}{\partial x^\sigma}\right\}_{(-3)}a. \tag{45}$$

It should be observed that

$$\frac{\partial}{\partial x^\nu}\left[s^\nu\left\{\frac{\partial T^{\mu\sigma}_{(-2)}}{\partial x^\sigma}\right\}_{(-3)}\right] = 3\left\{\frac{\partial T^{\mu\sigma}_{(-2)}}{\partial x^\sigma}\right\}_{(-3)} + s^\nu\frac{\partial}{\partial x^\nu}\left\{\frac{\partial T^{\mu\sigma}_{(-2)}}{\partial x^\sigma}\right\}_{(-3)} = 0, \tag{46}$$

since $\left\{\dfrac{\partial T^{\mu\sigma}_{(-2)}}{\partial x^\sigma}\right\}_{(-3)}$ is of the 3rd order, and therefore by a relation analogous to (22)

$$s^\nu\frac{\partial}{\partial x^\nu}\left\{\frac{\partial T^{\mu\sigma}_{(-2)}}{\partial x^\sigma}\right\}_{(-3)} = -3\left\{\frac{\partial T^{\mu\sigma}_{(-2)}}{\partial x^\sigma}\right\}_{(-3)}.$$

Now since $T''^{\mu\nu}$ is conserved with respect to ν, as also $s^\nu\left\{\dfrac{\partial T^{\mu\sigma}_{(-2)}}{\partial x^\sigma}\right\}_{(-3)}$, it follows from (45) that

$$\frac{\partial T^{\mu\nu}_{-2)}}{\partial x^\nu} + \frac{\partial}{\partial x^\nu}\left[T^{\mu\sigma}_{(-3)}\frac{s_\sigma}{\kappa}v^\nu\right] = 0. \tag{47}$$

Further, using again the relation expressed by (22)

$$\theta\left[T^{\mu\sigma}_{(-3)}\frac{s_\sigma}{\kappa}v^\nu\right] = -s^\nu\frac{\partial}{\partial x^\rho}\left[T^{\mu\sigma}_{(-3)}\frac{s_\sigma}{\kappa}v^\rho\right]$$

$$= s^\nu\frac{\partial T^{\mu\rho}_{-2)}}{\partial x^\nu} \quad \text{from (47).} \tag{48}$$

Employing an argument similar to that used in deriving (35) and (40) it can be shown that

$$\frac{\partial T^{\mu\rho}_{-2)}}{\partial x^\rho} = -\frac{\partial T^{\mu\rho}_{(-3}}{\partial x^\rho} = \left[\frac{\partial T^{\mu\rho}_{(-2)}}{\partial x^\rho}\right]_{(-3)} = -\left[\frac{\partial T^{\mu\rho}_{(-3)}}{\partial x^\rho}\right]_{(-3)} \tag{49}$$

152   **Harish-Chandra**

Therefore from (45), (48) and (49)

$$T'''^{\mu\nu} \equiv T''^{\mu\nu} - \theta\left[aT^{\mu\sigma}_{(-3)}\frac{s_\sigma}{\kappa}v^\nu \right] = T^{\mu\nu}_{-2)} + T^{\mu\sigma}_{(-3)}\frac{s_\sigma}{\kappa}v^\nu. \tag{50}$$

It is easy to see from (15), (16), (33) and (50) that

$$T'''^{\mu\nu} = T^{\mu\nu} + \theta K^{\mu\nu},$$

where $\quad \theta K^{\mu\nu} = \theta(a+b\phi+\dots)\,T^{\mu\nu} - \theta(a+b\phi+\dots)\,T^{\mu\nu}_{-2)} - \theta\left[aT^{\mu\sigma}_{(-3)}\frac{s_\sigma}{\kappa}v^\nu \right]$

$$= \theta(a+b\phi+\dots)\,T^{\mu\nu}_{(-3} - \theta\left[aT^{\mu\sigma}_{(-3)}\frac{s_\sigma}{\kappa}v^\nu \right].$$

We can therefore write

$$K^{\mu\nu} = (a+b\phi+\dots)\,T^{\mu\nu}_{(-3} - aT^{\mu\sigma}_{(-3)}\frac{s_\sigma}{\kappa}v^\nu.$$

Using (39) one gets explicitly

$$K^{\mu\nu} = \left[T^{\mu\nu}_{(-3)} - T^{\mu\sigma}_{(-3)}\frac{s_\sigma}{\kappa}v^\nu \right] + \sum_{n=4}^{n=m}\left[\frac{T^{\mu\nu}_{(-n)}}{n-2} + \frac{T^{\mu\sigma}_{(-n)}s_\sigma/\kappa\cdot v^\nu}{(n-2)(n-3)} \right]. \tag{51}$$

Thus $\quad\quad T'''^{\mu\nu} = T^{\mu\nu}_{-2)} + T^{\mu\sigma}_{(-3)}\frac{s_\sigma}{\kappa}v^\nu = T^{\mu\nu} + \frac{\partial}{\partial x^\sigma}[K^{\mu\nu}s^\sigma - K^{\mu\sigma}s^\nu], \tag{52}$

and is therefore of the required form.

Now $T'''^{\mu\nu}\kappa_\nu = T^{\mu\nu}_{-2)}\kappa_\nu + T^{\mu\sigma}_{(-3)}\frac{s_\sigma}{\kappa}\kappa'$ and therefore it possesses terms only up to the 2nd order, thus giving a finite integral over the world tube.

Since $\quad\quad \dfrac{\partial}{\partial x^\rho} = \left(\delta^\sigma_\rho - \frac{s_\rho}{\kappa}v^\sigma \right)\dfrac{\partial}{\partial x^\sigma} + \frac{x_\rho}{\kappa}v^\sigma\dfrac{\partial}{\partial x^\sigma}, \tag{53}$

it follows from (36) and (44) that

$$\left(\delta^\sigma_\rho - \frac{s_\rho}{\kappa}v^\sigma \right)\frac{\partial T^{\mu\nu}_{(-n)}}{\partial x^\sigma} = \left[\frac{\partial T^{\mu\nu}_{(-n)}}{\partial x^\rho} \right]_{(-n-1)}. \tag{54}$$

(44) and (54) hold not only for $T^{\mu\nu}_{(-n)}$ but also for any term of the nth order which is expressed as a/κ^n, where a is dependent on s^α/κ and τ only.

In view of (44) the conservation of $T'''^{\mu\nu}$ is easy to understand. Because, using (49),

$$\frac{\partial T'''^{\mu\nu}}{\partial x^\nu} = \left[\frac{\partial T^{\mu\nu}_{(-2)}}{\partial x^\nu} \right]_{(-3)} + \frac{\partial}{\partial x^\nu}\left[T^{\mu\sigma}_{(-3)}\frac{s_\sigma}{\kappa}v^\nu \right]$$

$$= \left[\frac{\partial T^{\mu\nu}_{(-2)}}{\partial x^\nu} \right]_{(-3)} + \frac{s_\sigma}{\kappa}v^\nu\frac{\partial T^{\mu\sigma}_{(-3)}}{\partial x^\nu} \quad \text{from (41)}$$

$$= \left[\frac{\partial T^{\mu\nu}_{(-2)}}{\partial x^\nu} \right]_{(-3)} + \left[\frac{\partial T^{\mu\nu}_{(-3)}}{\partial x^\nu} \right]_{(-3)} = 0 \quad \text{from (44) and (49).}$$

Precisely in the same way it can be proved that

$$T^{\mu\nu}_{-n)} + T^{\mu\sigma}_{(-n-1)} \frac{s_\sigma}{\kappa} v^\nu \tag{55}$$

is conserved with respect to ν. Further it can be shown that if $n \geqslant 2$

$$T^{\mu\nu}_{-n)} + T^{\mu\sigma}_{(-n-1)} \frac{s_\sigma}{\kappa} v^\nu = T^{\mu\nu} + \frac{\partial K^{\mu\nu\sigma}}{\partial x^\sigma},$$

where $K^{\mu\nu\sigma}$ is a tensor antisymmetrical in ν,σ. This follows from the fact that

$$S^{\mu\nu} \equiv - T'''^{\mu\nu} + T^{\mu\nu}_{-n)} + T^{\mu\sigma}_{(-n-1)} \frac{s_\sigma}{\kappa} v^\nu$$

$$= \left(\delta^\nu_\sigma - \frac{s_\sigma}{\kappa} v^\nu\right) T^{\mu\sigma}_{(-3)} + T^{\mu\nu}_{(-4)} + \ldots + T^{\mu\nu}_{(-n)} + T^{\mu\sigma}_{(-n-1)} \frac{s_\sigma}{\kappa} v^\nu$$

is conserved. It is then obvious that exactly as in the case of $T^{\mu\nu}$ a tensor

$$S'''^{\mu\nu} = S^{\mu\nu}_{-2)} + S^{\mu\sigma}_{(-3)} \frac{s_\sigma}{\kappa} v^\nu = S^{\mu\nu} + \frac{\partial K'^{\mu\nu\sigma}}{\partial x^\sigma}$$

can be formed, where $K'^{\mu\nu\sigma}$ is a suitable tensor antisymmetrical in ν,σ. But since $S^{\mu\nu}$ does not contain terms of the 2nd order or lower, and

$$S^{\mu\rho}_{(-3)} \frac{s_\rho}{\kappa} v^\nu = v^\nu \frac{s_\rho}{\kappa} \left(\delta^\mu_\sigma - \frac{s_\sigma}{\kappa} v^\rho\right) T^{\mu\sigma}_{(-3)} = 0.$$

then $S'''^{\mu\nu} = 0$, i.e. $S^{\mu\nu} = - \dfrac{\partial K'^{\mu\nu\sigma}}{\partial x^\sigma}$, so that

$$T^{\mu\nu}_{(-n)} + T^{\mu\sigma}_{(-n-1)} \frac{s_\sigma}{\kappa} v^\nu = T'''^{\mu\nu} - \frac{\partial K'^{\mu\nu\sigma}}{\partial x^\sigma} = T^{\mu\nu} + \frac{\partial K^{\mu\nu\sigma}}{\partial x^\sigma}.$$

Thus it is found that there exist a large number of tensors formed according to the simple rule (55) and capable of being put in the form (1) in which the order k of the highest singularity can be any number such that $3 \leqslant k \leqslant m$.

4. It will now be proved that the 3rd order term in $T'''^{\mu\nu}$ does not give rise to a singularity in the energy-momentum integral, in the case of the electromagnetic and meson fields.

It has already been shown that $T'''^{\mu\nu}\kappa_\nu$ has terms only up to the order $1/\kappa^2$, even though $T'''^{\mu\nu}$ contains terms up to the order $1/\kappa^3$. If equation (4) is rewritten for $T'''^{\mu\nu}$, then

$$\frac{d}{d\tau}\left(\int_\kappa T'''^{\mu\nu} \frac{s_\nu}{\kappa} d\Omega\right) = - \frac{d}{d\kappa} \int_\kappa T'''^{\mu\nu}\kappa_\nu d\Omega.$$

Since the right side of this equation is finite the left side must also be so. It is, however, evident from (50) that the left side contains a singular term of the order $1/\kappa$ which is

$$\frac{d}{d\tau}\left(\int_\kappa T^{\mu\sigma}_{(-3)}\frac{s_\sigma}{\kappa}d\Omega\right). \tag{56}$$

Therefore this term must be put equal to zero. Therefore

$$\int_\kappa T^{\mu\sigma}_{(-3)}\frac{s_\sigma}{\kappa}d\Omega = \frac{c^\mu}{\kappa}, \tag{57}$$

where c^μ are constants. The physical meaning of this is obvious. Because the rate of outflow of energy and momentum from the particle to the outer space, given by $\int_{\kappa\to 0}T'''^{\mu\nu}\kappa_\nu\,d\Omega$ is finite, any singularity in total field energy or field momentum must remain independent of τ.*

It is now assumed that for values of τ very far in the past the world-line is straight so that the particle (whose motion the world-line represents) was moving a long time back in a straight line. It is further assumed that at that time it had no rotational motion. Consider now the following motion of the particle. In the remote past it was moving uniformly in the absence of an external field. It is then acted upon by an external field which causes acceleration. Finally, the external field dies down and the particle has again in the distant future uniform unaccelerated motion but with a velocity entirely different from the one it had far back in the past. If such a motion is permissible, then since c^μ is independent of τ, it must remain the same in the initial and final stages of motion, i.e. it should be the same whatever the velocity of the particle in the course of its uniform motion. This means that c^μ is independent of the choice of the Lorentz-frame of reference. The only tensors which are so independent are 0 and $g^{\mu\nu}$. The latter alternative is obviously inadmissible. Therefore $c^\mu = 0$.

---

* $\int_{\kappa=\epsilon}^{\kappa=\eta}d\kappa\int_\kappa T^{\mu\sigma}\frac{s_\sigma}{\kappa}\,d\Omega$ is not really the energy and momentum contained inside the volume enclosed between the two spheres $\kappa=\epsilon$ and $\kappa=\eta$. Energy-momentum is in fact defined as the integral of $T^{\mu\nu}$ over a plane three-dimensional space-like surface. However, if in equation (10) of A we take S as the plane defined by $[x^\mu - z^\mu(\tau)]v_\mu(\tau) = \epsilon$ (which is nothing but the space of the rest system taken a time ϵ later than τ), it is easy to see that P intersects S in the retarded sphere of radius ϵ. Remembering that for the plane $dS_\nu = v_\nu(\tau)\,dV$, where dV is the element of three-dimensional volume, relation (10) of A for $T'''^{\mu\nu}$ can be written as

$$\frac{d}{d\tau}\int_S T'''^{\mu\nu}v_\nu\,dV = \int_{\kappa=\epsilon} T'''^{\mu\nu}\kappa_\nu\,d\Omega - T'''^\mu(Q),$$

where $T'''^\mu(Q)$ is the integral over the two-dimensional surface of intersection of S and the tube Q. It can be directly inferred from this that the singularity in $\int T'''^{\mu\nu}v_\nu\,dV$ is independent of τ.

In the case of the electromagnetic and mesic particles it is usually assumed that such a motion is possible.* Therefore

$$c^\mu = 0 \tag{58}$$

for the corresponding fields.† Thus the integral of $T'''^{\mu\nu}$ over the light cone

$$\left(\int_{\kappa_1}^{\kappa_2} d\kappa \int_\kappa T'''^{\mu\nu} \frac{s_\nu}{\kappa} d\Omega \right)_{\kappa_1 \to 0}$$

is finite. From this it will be proved that the integral of $T'''^{\mu\nu}$ over any finite space-like surface is finite, irrespective of the manner in which the point of singularity is approached. One need consider only the 3rd order term $T^{\mu\sigma}_{(-3)} \dfrac{s_\sigma}{\kappa} v^\nu$, since this is the only one which can give rise to a singularity in the integral.

Let the space-like surface be

$$\zeta^0(x_\mu) = \text{constant}, \tag{59}$$

and let it intersect the world-line at c. Using the notation of A, the integral is

$$\int T^{\mu\sigma}_{(-3)} \frac{s_\sigma}{\kappa} v^\nu \frac{\partial \zeta^0}{\partial x^\nu} \, | \, D \, |^{-1} d\zeta^1 d\zeta^2 d\zeta^3. \tag{60}$$

Take $\zeta^1 = x^1$, $\zeta^2 = x^2$, $\zeta^3 = x^3$, where x^1, x^2, x^3 are measured in the rest system at c. Then $D = \partial\zeta^0/\partial x^0$. Now let c be surrounded by a two-dimensional closed surface lying in (59) and given by the equation

$$\zeta^1(x_\mu) = \text{constant} = \epsilon, \tag{61}$$

such that as $\epsilon \to 0$ this surface tends to the point c. It will be assumed that for sufficiently small values of ϵ (61) is always entirely enclosed between the two two-dimensional surfaces resulting from the intersection of the two tubes

$$\kappa = \alpha\epsilon, \tag{62}$$

$$\kappa = \beta\epsilon, \tag{63}$$

with (59). Here α and β are constants $(\alpha < \beta)$ independent of ϵ. Now calculate the integral (60) taken over the portion of (59) lying between (61) and (63). For points lying in this region. it follows from the Taylor expansion of v^ν that for sufficiently small values of ϵ

$$| \, v^\nu - (v^\nu)_0 \, | < \lambda\epsilon,$$

where $(v^\nu)_0$ is the velocity at the point c and λ is a constant independent of ϵ. Therefore without affecting the singular terms v^ν can be replaced by $(v^\nu)_0$ in (60). Remembering that $(v^\nu)_0 = (1, 0, 0, 0)$ and $D = \partial\zeta^0/\partial x^0$ one gets the integral as

$$\int_{\zeta^1 = \epsilon}^{\kappa = \beta\epsilon} T^{\mu\sigma}_{(-3)} \frac{s_\sigma}{\kappa} \, dx^1 dx^2 dx^3. \tag{64}$$

---

* In fact it may even be necessary to demand that only such motions are permissible (Dirac 1938).

† This can also be proved directly, from the form of the energy-momentum tensor of the electomagnetic and the meson fields.

Put
$$T^{\mu\sigma}_{(-3)} \frac{s_\sigma}{\kappa} = \frac{a^{\mu\sigma}_{(-3)}}{\kappa^3} \frac{s_\sigma}{\kappa} \equiv \frac{a^\mu}{\kappa^3},$$

so that a^μ is a finite function of τ and s^α/κ only. Let $|a^\mu|_m$ be the maximum value of $|a^\mu|$ in the region of integration of (64). Then

$$\left| \int_{\zeta^1=\epsilon}^{\kappa=\beta\epsilon} \frac{a^\mu}{\kappa^3} dx^1 dx^2 dx^3 \right| < \int_{\zeta^1=\epsilon}^{\kappa=\beta\epsilon} \left| \frac{a^\mu}{\kappa^3} \right| dx^1 dx^2 dx^3 < \frac{|a^\mu|_m}{(\alpha\epsilon)^3} \int_{\kappa=\alpha\epsilon}^{\kappa=\beta\epsilon} dx^1 dx^2 dx^3. \qquad (65)$$

The spatial volume enclosed between $\kappa = \alpha\epsilon$ and $\kappa = \beta\epsilon$ is obviously of the order ϵ^3. The right side of (65) is thus finite and hence so also must be the left. Now it is seen from figure 1, where AB is the world-line, L the light cone and Q any finite tube, that since the integrals of $T'''^{\mu\nu}$ over P and L are non-singular the integral over the portion of S intercepted between P and Q must also be so. It has just now been shown that the integral taken over the portion of S enclosed between P and $\zeta^1 = \epsilon$ remains finite as $\epsilon \to 0$. Therefore the integral over the portion intercepted between $\zeta^1 = \epsilon$ and Q also remains finite as $\epsilon \to 0$. It has thus been proved that the integral over S is finite, no matter how the point of singularity be approached.*

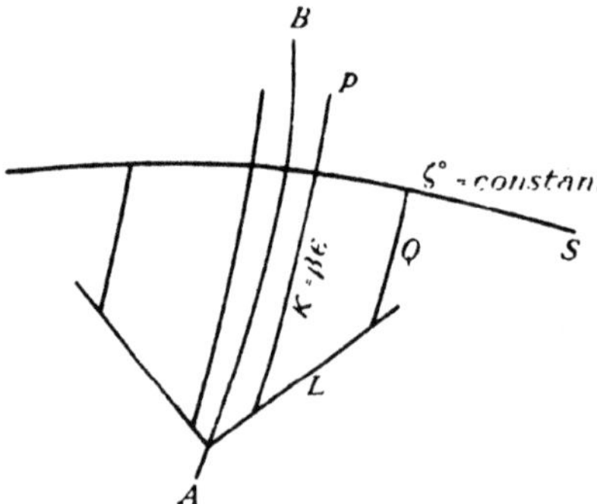

FIGURE 1

If $T'''^{\mu\nu}$ be integrated over the region of S enclosed between $\zeta^1 = \epsilon$ and $\zeta^1 = \eta$ ($\eta > \epsilon$), then from (52) and equation (20) of A it is seen that

$$\int_S T'''^{\mu\nu} dS_\nu = \int_S T^{\mu\nu} dS_\nu + \int_{\zeta^1=\eta} K^{\mu\nu\sigma} dS_{\nu\sigma} - \int_{\zeta^1=\epsilon} K^{\mu\nu\sigma} dS_{\nu\sigma}, \qquad (66)$$

where $dS_{\nu\sigma}$ is written for

$$\frac{\partial \zeta^0}{\partial x^\nu} \frac{\partial \zeta^1}{\partial x^\sigma} |D|^{-1} d\zeta^2 d\zeta^3$$

* It is noteworthy that the energy-momentum integrals though finite are not unique. Their value may depend on the way the point of singularity is approached. The energy is therefore indeterminate to this extent.

and $K^{\mu\nu\sigma}$ for $K^{\mu\nu}s^\sigma - K^{\mu\sigma}s^\nu$. It has been shown that as $\epsilon \to 0$ the left side remains finite. Now $K^{\mu\nu\sigma}$ contains certain terms of the 2nd order, namely, those arising from the 3rd order term $T^{\mu\nu}_{(-3)} - T^{\mu\sigma}_{(-3)}\dfrac{s_\sigma}{\kappa}v^\nu$ in K^μ. These terms can be removed and form a tensor $\hat{T}^{\mu\nu}$ given by

$$\hat{T}^{\mu\nu} = T^{\mu\nu} + \frac{\partial}{\partial x^\sigma}[K'^{\mu\nu}s^\sigma - K'^{\mu\sigma}s^\nu]. \tag{67}$$

where

$$K'^{\mu\nu} = \sum_{n=4}^{n=m}\left[\frac{T^{\mu\nu}_{(-n)}}{n-2} + \frac{T^{\mu\sigma}_{(-n)}\dfrac{s_\sigma}{\kappa}v^\nu}{(n-2)(n-3)}\right]. \tag{68}$$

Since only 2nd order terms have been removed from $K^{\mu\nu\sigma}$ it is easily seen from (66) that $\hat{T}^{\mu\nu}$ also gives a finite integral as $\epsilon \to 0$. But as $K'^{\mu\nu\sigma}$ now contains no term of an order less than the 3rd its integral over the surface $\zeta^1 = \eta$ tends to zero as $\eta \to \infty$,[*] so that

$$\int_S \hat{T}^{\mu\nu}dS_\nu = \int_S T^{\mu\nu}dS_\nu - \int_{\zeta^1 = \epsilon} K'^{\mu\nu\sigma}dS_{\nu\sigma}, \tag{69}$$

where the three-dimensional integrals are to be taken over the entire space outside $\zeta^1 = \epsilon$. Since the total energy and momentum obtained by making ϵ tend to zero remain finite, the inflow must also be finite.

For small values of κ when $T^{\mu\nu}$ can be expanded into terms of different orders with respect to κ, it can be proved without difficulty that

$$\hat{T}^{\mu\nu} = T^{\mu\nu}_{-3)} + s^\nu\frac{\partial T^{\mu\sigma}_{-3)}}{\partial x^\sigma}, \tag{70}$$

where $T^{\mu\nu}_{-3)}$ is the part of $T^{\mu\nu}$ left over after omitting terms of order higher than $1/\kappa^3$.

5. All the above results except (58) can be applied to any arbitrary tensor $U^{\alpha\beta\ldots\nu}$ having any number of indices, provided it satisfies a conservation equation with respect to ν, and can be expanded for small values of κ in a series involving different powers of κ. The tensor $U^{m\alpha\ldots\nu}$, and therefore $\hat{U}^{\alpha\ldots\nu}$ derived from $U^{\alpha\ldots\nu}$ by the above process, may, in general, still contain a singularity of the order $1/\kappa$ in the integral

$$\int_\kappa \hat{U}^{\alpha\ldots\nu}\frac{s_\nu}{\kappa}d\Omega. \tag{71}$$

This singularity, however, must be independent of τ. It is now possible to show that this singularity can be removed.

It is easily seen by actual calculation that

$$\left\{\frac{\partial}{\partial x^\sigma}\left[\frac{(v^\nu s^\sigma - v^\sigma s^\nu)}{\kappa^3}e^{-b\kappa}\log h\kappa\right]\right\}\frac{s_\nu}{\kappa} = \frac{1}{\kappa^3}(1 - b\kappa\log h\kappa)e^{-b\kappa}. \tag{72}$$

[*] Here it is assumed that for sufficiently large values of η, $\xi^1 = \eta$ is completely contained in the space between two spheres of radius $\alpha\eta$ and $\beta\eta$ described about c in the rest system. (α and β are constants independent of η.) The total surface of $\xi^1 = \eta$ is assumed to tend to infinity as η^2.

where b and h are positive constants. Therefore

$$\int_\kappa \frac{\partial}{\partial x^\sigma}\left[\frac{(v^\nu s^\sigma - v^\sigma s^\nu)}{\kappa^3}\, e^{-b\kappa}\log h\kappa\right]\frac{s_\nu}{\kappa}\, d\Omega = \frac{4\pi}{\kappa}\,(1 - b\kappa\log h\kappa)\, e^{-b\kappa}. \tag{73}$$

The singularity in (71) is of the form $c^{\alpha\cdots}/\kappa$, where $c^{\alpha\cdots}$ are certain constants. Therefore

$$\int_\kappa \underline{U}^{\alpha\cdots\nu}\frac{s_\nu}{\kappa}\, d\Omega = bc^{\alpha\cdots}\log h\kappa\, e^{-b\kappa}, \tag{74a}$$

where

$$\underline{U}^{\alpha\cdots\nu} = \tilde{U}^{\alpha\cdots\nu} - \frac{\partial}{\partial x^\sigma}\left[\frac{c^{\alpha\cdots}}{4\pi}\frac{(v^\nu s^\sigma - v^\sigma s^\nu)}{\kappa^3}\, e^{-b\kappa}\log h\kappa\right]. \tag{74b}$$

Even though (74a) contains a singularity it can easily be seen that

$$\int_{\kappa=\epsilon}^{\kappa=\eta} d\kappa \int \underline{U}^{\alpha\cdots\nu}\frac{s_\nu}{\kappa}\, d\Omega$$

remains finite as $\epsilon \to 0$. Very much in the same manner as for $T'''^{\mu\nu}$ it can be shown that the integral of $\underline{U}^{\alpha\cdots\nu}$ over any space-like surface is finite no matter how the point of singularity be approached.

Further on account of $e^{-b\kappa}$ the surface integral of the term inside brackets in (74b) vanishes on the infinite sphere, so that in an equation like (69) the integral may still be omitted on the surface at infinity. It is to be noted that b is entirely arbitrary except for the fact that it is greater than zero.

Now apply these considerations to the case of the angular momentum tensor $M^{\mu\nu,\sigma}$ defined by

$$M^{\mu\nu,\sigma} = x^\mu T^{\nu\sigma} - x^\nu T^{\mu\sigma},$$

which is conserved with respect to σ, due to symmetry of the original tensor $T^{\mu\nu}$. One can therefore build $\tilde{M}^{\mu\nu,\sigma}$, which for small values of κ can be written as

$$\tilde{M}^{\mu\nu,\sigma} = M^{\mu\nu,\sigma}_{-3)} + s^\sigma\frac{\partial}{\partial x^\rho}(M^{\mu\nu,\rho}_{-3)}). \tag{75}$$

Since $x^\mu = z^\mu(\tau) + \dfrac{s^\mu}{\kappa}\kappa$ it is found that

$$M^{\mu\nu,\sigma}_{-3)} = x^\mu T^{\nu\sigma}_{-3)} - x^\nu T^{\mu\sigma}_{-3)} + s^\mu T^{\nu\sigma}_{(-4)} - s^\nu T^{\mu\sigma}_{(-4)}. \tag{76}$$

Therefore, using (57),

$$\int_\kappa \tilde{M}^{\mu\nu,\sigma}_{(-3)}\frac{s_\sigma}{\kappa}\, d\Omega = \frac{z^\mu c^\nu - z^\nu c^\mu}{\kappa} + \int (s^\mu T^{\nu\sigma}_{(-4)} - s^\nu T^{\mu\sigma}_{(-4)})\frac{s_\sigma}{\kappa}\, d\Omega \equiv \frac{c^{\mu\nu}}{\kappa}.$$

Thus a tensor $\underline{M}^{\mu\nu,\sigma}$ can be constructed defined by

$$\underline{M}^{\mu\nu,\sigma} = M^{\mu\nu,\sigma} + \frac{\partial}{\partial x^\rho}[\Theta^{\mu\nu,\sigma}s^\rho - \Theta^{\mu\nu,\rho}s^\sigma], \tag{77a}$$

where
$$\Theta^{\mu\nu,\sigma} = \sum_{n=4}^{n=m} \left[\frac{M^{\mu\nu,\sigma}_{(-n)}}{n-2} + \frac{M^{\mu\nu,\rho}_{(-n)} \frac{s_\rho}{\kappa} v^\sigma}{(n-2)(n-3)} \right] - \frac{c^{\mu\nu}}{4\pi} v^\sigma \frac{e^{-b\kappa} \log h\kappa}{\kappa^3} \tag{77b}$$

and
$$c^{\mu\nu} = (z^\mu c^\nu - z^\nu c^\mu) + \int (s^\mu T^{\nu\sigma}_{(-4)} s_\sigma - s^\nu T^{\mu\sigma}_{(-4)} s_\sigma)\, d\Omega, \tag{77c}$$

such that the total angular momentum calculated from it is always finite.

Now return to the special case of the meson and the electromagnetic fields. Precisely as in the case of c^μ (equation (58)) it can be shown that since $c^{\mu\nu}$ is antisymmetric in μ, ν while $g^{\mu\nu}$ is symmetric

$$c^{\mu\nu} = 0. \tag{78}$$

However, it is important to note that owing to the fact that the meson fields contain a factor $e^{-\chi\kappa}$ at infinity (Appendix, equation (116a)),* the angular-momentum integral as calculated from $M^{\mu\nu,\sigma}$ does not contain a singularity at infinity. In the electromagnetic case, however, a logarithmic singularity would appear at infinity if $c^{\mu\nu}$ did not vanish. However, as it is, it is found that the total angular momentum calculated from $\tilde{M}^{\mu\nu,\sigma}$ is free from singularity both at zero and infinity.†

All the above considerations are immediately applicable to an assembly of more than one particle, because near any particle the field due to all others is regular and can be included in the ingoing field. We have only to take instead of (67) and (68)

$$\tilde{T}^{\mu\nu} = T^{\mu\nu} + \sum_p \frac{\partial}{\partial x^\sigma} K_p^{\mu\nu\sigma}, \tag{78a}$$

where p refers to a particular particle,

$$K_p'^{\mu\nu\sigma} = K_p'^{\mu\nu} s_p^\sigma - K_p'^{\mu\sigma} s_p^\nu, \tag{78b}$$

$$K_p'^{\mu\nu} = \sum_{n=4}^{n=m} \left\{ \frac{T^{\mu\nu}_{p(-n)}}{n-2} + \frac{T^{\mu\sigma}_{p(-n)}}{(n-2)(n-3)} \frac{s_{\sigma,p}}{\kappa_p} v_p^\nu \right\}. \tag{78c}$$

---

* For large spatial distances from the particle (measured in the present space of any Lorentz-frame) the corresponding retarded point goes far back in the past of the particle, so that for sufficiently large distances the retarded point lies in the region of the world-line corresponding to uniform motion of the particle. The value of the fields at infinity may therefore be taken to be the same as in the static case. This is also evident from the consideration that since potentials and therefore fields are propagated with a velocity at most equal to that of light, the value of these quantities at sufficiently large distances must still be unaffected by radiation from the particle. The external field of course is supposed to vanish sufficiently fast at infinity.

† In this connexion it may be mentioned that a misprint has occurred in the expression for $F^{(2)}_{\mu\nu}$ given in the paper by Bhabha & Corben (1941). In equation (114) of their paper, instead of the term

$$\left[-2v_\mu s^\rho \frac{S_{\rho\nu}}{\kappa^3} \right]_-$$

it should read

$$\left[-2v_\mu s^\rho \frac{\dot{S}_{\rho\nu}}{\kappa^3} \right]_-.$$

11-2

Here s_p^σ and v_p^ν refer to the particular particle. $T_{p(-n)}^{\mu\nu}$ is the nth order term in $T^{\mu\nu}$ when it is expanded in the neighbourhood of the particle p (i.e. for small κ_p). Similarly, corresponding to (77)

$$\tilde{M}^{\mu\nu,\sigma} = M^{\mu\nu,\sigma} + \sum_p \frac{\partial}{\partial x^\rho}[\Theta_p^{\mu\nu,\sigma} s_p^\rho - \Theta_p^{\mu\nu,\rho} s_p^\sigma], \tag{79a}$$

$$\Theta_p^{\mu\nu,\sigma} = \sum_{n=4}^{n=m_p}\left[\frac{M_{p(-n)}^{\mu\nu,\sigma}}{n-2} + \frac{M_{p(-n)}^{\mu\nu,\sigma}}{(n-2)(n-3)}\frac{s_{\rho,p}}{\kappa_p}v_p^\sigma\right] \tag{79b}$$

can be written. Here $c_p^{\mu\nu} = 0$.

6. Some special remarks may now be made regarding the symmetrization of the modified tensor for the electromagnetic dipole. As already observed in §1 this symmetrization is, however, of no particular importance.

As proved in the appendix (equation (124)) it is found that for the electromagnetic case

$$T_{(-3)}^{\mu\nu\,\text{ret.}} s_\nu = 0, \tag{80}$$

where $T^{\mu\nu\,\text{ret.}}$ has the same meaning as that in the paper of Bhabha & Corben (1941). Therefore

$$T_{,}^{\prime\prime\prime\mu\nu\,\text{ret.}} = T_{-2)}^{\mu\nu\,\text{ret.}} \quad \text{from (50),} \tag{81}$$

which is obviously symmetrical. Since in the static case $T^{\mu\nu}$ does not contain terms of order lower than $1/\kappa^4$ (appendix, equation (116b))* then

$$T_{-2)\,\text{static}}^{\mu\nu} = 0. \tag{82}$$

Therefore the integral of $T_{-2)}^{\mu\nu\,\text{ret.}}$ does not contain any singularities at infinity.† Using the symbols used by Bhabha & Corben (1941), we write

$$T^{\mu\nu} = T^{\mu\nu\,\text{ret.}} + T^{\mu\nu\,\text{mix.}} + T^{\mu\nu\,\text{in.}}. \tag{83}$$

Because both $T^{\mu\nu\,\text{ret.}}$ and $T^{\mu\nu\,\text{in.}}$ are separately conserved $T^{\mu\nu\,\text{mix.}}$ must also be conserved, and the procedure of the preceding pages can be applied to $T^{\mu\nu\,\text{mix.}}$. We construct

$$T^{\prime\prime\prime\mu\nu\,\text{mix.}} = T_{-2)}^{\mu\nu\,\text{mix.}} + T_{(-3)}^{\mu\sigma\,\text{mix.}}\frac{s_\sigma}{\kappa}v^\nu. \tag{84}$$

Further

$$M^{\prime\prime\prime\mu\nu,\sigma\,\text{mix.}} = (x^\mu T_{-2)}^{\nu\sigma\,\text{mix.}} - x^\nu T_{-2)}^{\mu\sigma\,\text{mix.}}) + (s^\mu T_{(-3)}^{\nu\sigma\,\text{mix.}} - s^\nu T_{(-3)}^{\mu\sigma\,\text{mix.}})$$
$$+ (z^\mu T_{(-3)}^{\nu\rho\,\text{mix.}} - z^\nu T_{(-3)}^{\mu\rho\,\text{mix.}})\frac{s_\rho}{\kappa}v^\sigma + (s^\mu T_{(-4)}^{\nu\rho\,\text{mix.}} - s^\nu T_{(-4)}^{\mu\rho\,\text{mix.}})\frac{s_\rho}{\kappa}v^\sigma. \tag{85}$$

From (84) and (85)

$$M^{\prime\prime\prime\mu\nu,\sigma\,\text{mix.}} = x^\mu T^{\prime\prime\prime\nu\sigma\,\text{mix.}} - x^\nu T^{\prime\prime\prime\mu\sigma\,\text{mix.}} + \left[s^\mu\left\{T_{(-3)}^{\nu\rho\,\text{mix.}}\left(\delta_\rho^\sigma - \frac{s_\rho}{\kappa}v^\sigma\right) + T_{(-4)}^{\nu\rho\,\text{mix.}}\frac{s_\rho}{\kappa}v^\sigma\right\}\right]_- \tag{86}$$

---

* The potentials contain terms of order $1/\kappa$ or higher. The fields therefore contain terms of order $1/\kappa^2$ or higher. Hence $T^{\mu\nu}$ does not contain terms of order lower than $1/\kappa^4$.

† Cf. footnote * on p. 159.

the minus sign at the end denoting the subtraction of terms got by exchange of μ, ν in the expression within the square brackets.

Since the highest singularity in the mixed tensor in the case of an electromagnetic dipole is of order $1/\kappa^3$, $T^{\nu\rho}_{(-4)}{}^{\text{mix.}} = 0$. Using this fact and the conservation of $M'''^{\mu\nu,\sigma\,\text{mix.}}$ and $T'''^{\mu\sigma\,\text{mix.}}$, it is found from (86) that

$$0 = \frac{\partial M'''^{\mu\nu,\sigma\,\text{mix.}}}{\partial x^\sigma} = T'''^{\nu\mu\,\text{mix.}} - T'''^{\mu\nu\,\text{mix.}} + \frac{\partial}{\partial x^\sigma}\left[s^\mu T^{\nu\rho}_{(-3)}{}^{\text{mix.}}\cdot\left(\delta^\sigma_\rho - \frac{s_\rho}{\kappa} v^\sigma\right)\right]_- . \tag{87}$$

Putting
$$L^{\mu\nu,\sigma} \equiv \left[s^\mu T^{\nu\rho}_{(-3)}{}^{\text{mix.}}\cdot\left(\delta^\sigma_\rho - \frac{s_\rho}{\kappa} v^\sigma\right)\right]_- \tag{88}$$

it is found that
$$T'''^{\mu\nu\,\text{mix.}} - \frac{1}{2}\frac{\partial}{\partial x^\sigma}\left[L^{\mu\nu,\sigma} + L^{\sigma\mu,\nu} - L^{\nu\sigma,\mu}\right] \tag{89}$$

is symmetrical and is conserved (Pauli 1941). It is also in the form (1) as required. Further, since $L^{\mu\nu,\sigma}$ is of the order $1/\kappa^2$ the additional term does not introduce any singularity in the energy integral as is obvious from (66). Also, since $T^{\mu\nu}_{(-3)}{}^{\text{mix.}}$ vanishes at infinity (due to the absence of the external field in the remote past),* the integral in (66) over the surface at infinity is zero.

Thus
$$\bar{T}^{\mu\nu} = T^{\mu\nu}_{-2}{}^{\text{ret.}} + T'''^{\mu\nu\,\text{mix.}} - \frac{1}{2}\frac{\partial}{\partial x^\sigma}\left[L^{\mu\nu,\sigma} + L^{\sigma\mu,\nu} - L^{\nu\sigma,\mu}\right] + T^{\mu\nu\,\text{in.}}$$

can be taken as the modified tensor. This tensor is symmetrical. That in the case of a point charge this symmetry is possible has already been shown by Pryce (1938). In this case

$$T^{\mu\nu\,\text{ret.}} = \frac{e^2}{4\pi}\left[-\frac{(1-\kappa')^2}{\kappa^6}s^\mu s^\nu + (1-\kappa')\frac{(v^\mu s^\nu + v^\nu s^\mu)}{\kappa^5} + \frac{(\dot{v}^\mu s^\nu + \dot{v}^\nu s^\mu) - (\dot{v})^2 s^\mu s^\nu - \frac{1}{2}g^{\mu\nu}}{\kappa^4}\right]. \tag{90}$$

Thus according to (51)

$$K^{\mu\nu} = \left(T^{\mu\nu}_{(-3)} - T^{\mu\sigma}_{(-3)}\frac{s_\sigma}{\kappa}v^\nu\right) + \frac{T^{\mu\nu}_{(-4)}}{2} + \frac{T^{\mu\sigma}_{(-4)}}{2\,.\,1}\frac{s_\sigma}{\kappa}v^\nu.$$

* For small κ the retarded potentials U_μ and the retarded field quantities $G_{\mu\nu}$ can be expanded in a series of positive and negative powers of κ, and the ingoing field in a Taylor's series with positive powers of κ

$$G^{\text{in.}}_{\mu\nu} = (G^{\text{in.}}_{\mu\nu})_{z_\rho(\tau_0)} + \left(\frac{\partial G^{\text{in.}}_{\mu\nu}}{\partial x^\sigma}\right)_{z_\rho(\tau_0)}\frac{s^\sigma}{\kappa}\kappa + \frac{1}{2}\left(\frac{\partial^2 G^{\text{in.}}_{\mu\nu}}{\partial x^\sigma \partial x^\lambda}\right)\frac{s^\sigma}{\kappa}\frac{s^\lambda}{\kappa}\kappa^2 + \dots,$$

where the suffix $z_\rho(\tau_0)$ means that the corresponding quantities are to be evaluated at the retarded point. Therefore $T^{\mu\nu}_{(-3)}{}^{\text{mix.}}$ consists of terms of the form $\frac{u.v}{\kappa^3}$, where u and v are both of order zero in κ and u comes from $G^{\text{ret.}}_{\mu\nu}$, while v comes from $G^{\text{in.}}_{\mu\nu}$. The factor in v having the suffix $z_\rho(\tau_0)$ is to be evaluated at the retarded point. At infinity $v = 0$ because the corresponding retarded point goes to the remote past and there the field quantities and their differentials vanish.

(In spite of the fact that $K^{\mu\nu}$ is taken instead of $K'^{\mu\nu}$ given by (68), the third term in (66) vanishes as $\eta \to \infty$ due to (82).*) So that

$$K^{\mu\nu\sigma} = K^{\mu\nu}s^{\sigma} - K^{\mu\sigma}s^{\nu}$$

$$= \frac{e^2}{4\pi}\left[(\tfrac{3}{4}-\kappa')\frac{s^{\mu}}{\kappa^5}(v^{\nu}s^{\sigma}-v^{\sigma}s^{\nu}) + s^{\mu}\frac{(\dot{v}^{\nu}s^{\sigma}-\dot{v}^{\sigma}s^{\nu})}{\kappa^4} - \frac{1}{4}\frac{(g^{\mu\nu}s^{\sigma}-g^{\mu\sigma}s^{\nu})}{\kappa^4}\right]. \qquad (91)$$

It is interesting to see that this is different from Pryce's tensor which is more complicated. It is (Pryce 1938)

$$K^{\mu\nu,\sigma} = \frac{e^2}{4\pi}\left[-\tfrac{9}{4}\kappa'\frac{(v^{\nu}s^{\sigma}-v^{\sigma}s^{\nu})}{\kappa^5}s^{\mu} + \frac{(g^{\mu\nu}v^{\sigma}-g^{\mu\sigma}v^{\nu})}{4\kappa^3} - (1+2\kappa')\frac{(g^{\mu\nu}s^{\sigma}-g^{\mu\sigma}s^{\nu})}{4\kappa^4}\right.$$

$$\left. + \frac{3(v^{\nu}s^{\sigma}-v^{\sigma}s^{\nu})}{4\kappa^4}v^{\mu} + \frac{3(\dot{v}^{\nu}s^{\sigma}-\dot{v}^{\sigma}s^{\nu})}{4\kappa^4}s^{\mu}\right].$$

The modified tensor obtained from both these expressions is symmetric. Our modified tensor is just equal to $T^{\mu\nu}_{-2)}$ret. as is proved quite generally in (81).

In conclusion I wish to express my thanks to Professor H. J. Bhabha for valuable discussions, general guidance and suggested improvements in the writing of this paper.

APPENDIX

Consider now $$\kappa\frac{d}{d\kappa}\int T^{\mu\nu}\kappa_{\nu}\cdot\kappa^2 d\omega, \qquad (92)$$

τ being kept constant during differentiation. Here $d\omega$ is an element of solid angle in the rest system and $d\Omega = \kappa^2 d\omega$. Since

$$\kappa = s_{\mu}v^{\mu} \equiv [x_{\mu} - z_{\mu}(\tau)]\,v^{\mu}(\tau).$$

on differentiation keeping τ constant

$$d\kappa = dx_{\mu}v^{\mu}. \qquad (93a)$$

For constant τ $$s^{\mu}dx_{\mu} = 0. \qquad (93b)$$

These are the only two conditions which the infinitesimal vector which connects a point on the sphere κ to a corresponding point on the slightly larger sphere $\kappa + d\kappa$, need satisfy. The way in which the points on these two spheres may be put in correspondence is immaterial. Now take dx_{μ} along s_{μ} thus satisfying (93b) automatically. (93a) then gives

$$ds_{\mu} = dx_{\mu} = d\kappa\frac{s_{\mu}}{\kappa},$$

i.e. $$\frac{ds_{\mu}}{d\kappa} = \frac{s_{\mu}}{\kappa}. \qquad (94)$$

* Cf. footnote * on p. 159.

From (94) it is easily found that

$$\frac{d}{d\kappa}\left(\frac{s_\mu}{\kappa}\right) = 0,\tag{95}$$

from which it may be concluded that

$$\frac{d}{d\kappa}(d\omega) = 0\tag{96}$$

as $d\omega$ is dependent on s_μ/κ only. Thus

$$\frac{d}{d\kappa}\int T^{\mu\nu}\kappa_\nu\,.\,\kappa^2 d\omega = \int\frac{\partial}{\partial x^\sigma}(T^{\mu\nu}\kappa_\nu)\frac{dx^\sigma}{d\kappa}\kappa^2 d\omega + 2\int T^{\mu\nu}\kappa_\nu\kappa\,d\omega.$$

Since $\qquad\dfrac{\partial\kappa_\nu}{\partial x^\sigma}\dfrac{dx^\sigma}{d\kappa} = \dfrac{\kappa'}{\kappa}\dfrac{s_\nu}{\kappa}$ (from equations (12) of A, (94) and (95))

then

$$\kappa\frac{d}{d\kappa}\int T^{\mu\nu}\kappa_\nu d\Omega = \int\left[2T^{\mu\nu} + s^\sigma\frac{\partial T^{\mu\nu}}{\partial x^\sigma}\right]\kappa_\nu d\Omega + \int T^{\mu\nu}\frac{s_\nu}{\kappa}\kappa'\,d\Omega$$

$$= \int\frac{\partial}{\partial x^\sigma}[T^{\mu\nu}s^\sigma - T^{\mu\sigma}s^\nu]\,.\,\kappa_\nu d\Omega,\tag{97}$$

as $\qquad\dfrac{\partial}{\partial x^\sigma}[T^{\mu\nu}s^\sigma - T^{\mu\sigma}s^\nu] = 2T^{\mu\nu} + s^\sigma\dfrac{\partial T^{\mu\nu}}{\partial x^\sigma} + T^{\mu\sigma}\dfrac{s_\sigma}{\kappa}v^\nu,$ (cf. equation (6b))

if $\dfrac{\partial T^{\mu\nu}}{\partial x^\nu} = 0$. Since $\dfrac{d}{d\kappa}\left(\dfrac{s_\mu}{\kappa}\right) = 0$ it is easy to prove further that

$$\kappa\frac{d}{d\kappa}\int T^{\mu\nu}\kappa_\nu\frac{s^\alpha}{\kappa}\dots\frac{s^\lambda}{\kappa}d\Omega = \int\frac{\partial}{\partial x^\sigma}[T^{\mu\nu}s^\sigma - T^{\mu\sigma}s^\nu]\,.\,\kappa_\nu\frac{s^\alpha}{\kappa}\dots\frac{s^\lambda}{\kappa}d\Omega,\tag{98}$$

as quoted in the text. Remembering (94) and (95) it can similarly be proved that

$$\frac{d}{d\kappa}\int T^{\mu\nu}s_\nu\frac{s^\alpha}{\kappa}\dots\frac{s^\lambda}{\kappa}d\Omega = \frac{1}{\kappa}\int\left[3T^{\mu\nu} + s^\sigma\frac{\partial T^{\mu\nu}}{\partial x^\sigma}\right]s_\nu\frac{s^\alpha}{\kappa}\dots\frac{s^\lambda}{\kappa}d\Omega,\tag{99}$$

so that

$$\kappa\frac{d}{d\kappa}\int T^{\mu\nu}s_\nu\frac{s^\alpha}{\kappa}\dots\frac{s^\lambda}{\kappa}d\Omega = \int\left[2T^{\mu\nu} + s^\sigma\frac{\partial T^{\mu\nu}}{\partial x^\sigma} + T^{\mu\sigma}\frac{s_\sigma}{\kappa}v^\nu\right]s_\nu\frac{s^\alpha}{\kappa}\dots\frac{s^\lambda}{\kappa}d\Omega$$

$$= \int\theta T^{\mu\nu}\,.\,s_\nu\frac{s^\alpha}{\kappa}\dots\frac{s^\lambda}{\kappa}d\Omega.\tag{100}$$

In the same way it is found that

$$\frac{d}{d\kappa}\int T^{\mu\sigma}(\delta^\nu_\sigma - v_\sigma v^\nu)\frac{s^\alpha}{\kappa}\dots\frac{s^\lambda}{\kappa}d\Omega = \frac{1}{\kappa}\int\left[2T^{\mu\sigma} + s^\rho\frac{\partial T^{\mu\sigma}}{\partial x^\rho}\right](\delta^\nu_\sigma - v_\sigma v^\nu)\frac{s^\alpha}{\kappa}\dots\frac{s^\lambda}{\kappa}d\Omega,\tag{101}$$

because $\delta^\nu_\sigma - v_\sigma v^\nu$, which depends on τ alone, is to be treated as a constant here. Since $v^\sigma(\delta^\nu_\sigma - v_\sigma v^\nu) = 0$ then

$$\kappa\frac{d}{d\kappa}\int T^{\mu\sigma}(\delta^\nu_\sigma - v_\sigma v^\nu)\frac{s^\alpha}{\kappa}\dots\frac{s^\lambda}{\kappa}d\Omega = \int\theta T^{\mu\sigma}\,.\,(\delta^\nu_\sigma - v_\sigma v^\nu)\frac{s^\alpha}{\kappa}\dots\frac{s^\lambda}{\kappa}d\Omega.\tag{102}$$

The most general equation which the potentials U_ν of a meson field may have to satisfy is

$$\frac{\partial}{\partial x^\rho}\frac{\partial}{\partial x_\rho} U^\nu + \chi^2 U^\nu = 4\pi \frac{\partial}{\partial x^\alpha} \cdots \frac{\partial}{\partial x^\lambda} \Sigma^{\alpha\ldots\lambda,\nu}, \tag{103}$$

where $\Sigma^{\alpha\ldots\lambda,\nu}$ is a tensor describing the effect of a multipole on the field. In fact, in the most general case the right side is a sum of terms containing different numbers of differentiations, but it will be sufficient to consider one such term since the solution of this general case is just the sum of the solutions of each of the equations involving one term only. Since interest arises only in the case of point-particles we can write (cf. Bhabha 1941, p. 321)

$$\Sigma^{\alpha\ldots\lambda,\nu}_{(x_\rho)} = \int_{-\infty}^{\infty} d\tau\, S^{\alpha\ldots\lambda,\nu}_{(\tau)}\, \delta(x^0 - z^0)\, \delta(x^1 - z^1)\, \delta(x^2 - z^2)\, \delta(x^3 - z^3). \tag{104}$$

As is well known the solution for the retarded potentials is

$$U^{\nu\,\text{ret.}}_{(x_\rho)} = \frac{\partial}{\partial x^\alpha} \cdots \frac{\partial}{\partial x^\lambda}\left(\frac{S^{\alpha\ldots\lambda,\nu}}{\kappa}\right) - \chi\frac{\partial}{\partial x^\alpha} \cdots \frac{\partial}{\partial x^\lambda}\int_{-\infty}^{\tau_0} S^{\alpha\ldots\lambda,\nu}\frac{J_1(\chi u)}{u}\, d\tau, \tag{105}$$

where $u_\rho(\tau) = x_\rho - z_\rho(\tau)$, $u = (u_\rho u^\rho)^{\frac{1}{2}}$ and τ_0 is the retarded time given by $u(\tau_0) = 0$. Here the first term is just the one which alone appears for the electromagnetic field ($\chi = 0$). It is obviously expressible in a finite series involving terms of different orders in κ. The differentiations in the 2nd term act in two ways. First, on the limit of integration τ_0, and secondly, on the integrand. Bearing in mind that

$$\frac{\partial}{\partial x^\lambda}\left(\frac{J_n(\chi u)}{u^n}\right) = -\chi u_\lambda \frac{J_{n+1}(u)}{u^{n+1}} \tag{106}$$

and

$$\frac{\partial u_\lambda}{\partial x^\alpha} = g_{\lambda\alpha}, \tag{107}$$

then

$$\frac{\partial}{\partial x^\nu} \cdots \frac{\partial}{\partial x^\lambda}\left(\frac{J_1(\chi u)}{u}\right) \tag{108}$$

can be evaluated in terms of u^x and $\dfrac{J_n(\chi u)}{u^n}$. Also, since

$$u^\alpha(\tau_0) = s^\alpha \tag{109}$$

and

$$\left(\frac{J_n(\chi u)}{u^n}\right)_{\tau_0} = \left(\frac{J_n(\chi u)}{u^n}\right)_{u\to 0} = \frac{\chi^n}{2^n \cdot n!}, \tag{110}$$

then the value of (108) at the retarded point can be expressed as a finite series in s^α involving terms of different orders in κ. It should be noted that

$$\frac{\partial}{\partial x^\mu}\int_{-\infty}^{\tau_0} S^{\alpha\ldots\gamma}\frac{\partial}{\partial x^\nu} \cdots \frac{\partial}{\partial x^\lambda}\left(\frac{J_1(\chi u)}{u}\right) d\tau$$

$$= \frac{s_\mu}{\kappa}\left[S^{\alpha\ldots\gamma}\frac{\partial}{\partial x^\nu} \cdots \frac{\partial}{\partial x^\lambda}\left(\frac{J_1(\chi u)}{u}\right)\right]_{\tau_0} + \int_{-\infty}^{\tau_0} S^{\alpha\ldots\gamma}\frac{\partial}{\partial x^\mu}\frac{\partial}{\partial x^\nu} \cdots \frac{\partial}{\partial x^\lambda}\left(\frac{J_1(\chi u)}{u}\right) d\tau. \tag{111}$$

Again the first term on the right and its differentials can be written as finite series in s^α containing terms of different orders in κ. Therefore the only term in $U^{\nu\,\text{ret.}}$ which has to be considered is

$$\int_{-\infty}^{\tau_0} S^{\alpha\ldots\lambda,\nu}\, \frac{\partial}{\partial x^\alpha} \cdots \frac{\partial}{\partial x^\lambda}\left(\frac{J_1(\chi u)}{u}\right) d\tau. \tag{112}$$

As shown above this will consist of terms of the type

$$\int_{-\infty}^{\tau_0} S^{\alpha\ldots\lambda,\nu}\, u_\gamma \ldots u_\mu \frac{J_n(\chi u)}{u^n}\, d\tau. \tag{113}$$

It is noteworthy that all these terms are non-singular. They can be expanded in a series containing only terms of positive order in κ in the following way. Write $u_\rho \equiv x_\rho - z_\rho(\tau) = l_\rho - s_\rho$, where $l_\rho \equiv z_\rho(\tau) - z_\rho(\tau_0)$, so that $u^2 = l_\mu l^\rho - 2l^\rho s_\rho = l^2 - 2(ls)$, where $l = (l_\mu l^\mu)^{\frac{1}{2}}$ and $(ls) = (l_\mu s^\mu)$. Then

$$\frac{J_n(\chi u)}{u^n} = \frac{J_n(\chi\sqrt{[l^2 - 2(ls)]})}{(\sqrt{[l^2 - 2(ls)]})^n}. \tag{114}$$

Since $\dfrac{J_n(\chi l)}{l^n}$ is a continuous and differentiable function of l^2, (114) can be written in the form of a Taylor's series for sufficiently small values of s_ρ

$$\begin{aligned}
\frac{J_n(\chi u)}{u^n} &= \frac{J_n(\chi l)}{l^n} + \sum_{p=1}^{\infty} \frac{[-2(ls)]^p}{p!}\left(\frac{d}{dl^2}\right)^p \frac{J_n(\chi l)}{l^n}\\
&= \frac{J_n(\chi l)}{l^n} + \sum_{p=1}^{\infty} \chi^p\, \frac{(ls)^p}{p!}\, \frac{J_{n+p}(\chi l)}{l^{n+p}}.
\end{aligned} \tag{115}$$

It follows from (113) and (115) that the expansion of (105) in a series of s^α is possible. s^α are independent of τ and so can be put outside the integral sign. Since the integrands involve only $S^{\alpha\ldots\lambda,\nu}$ and l_λ which are functions of τ and τ_0 alone, the integral is a function of τ_0 alone. It is now possible to separate terms of different orders by writing s^α/κ instead of s^α and multiplying by a suitable power of κ to compensate it. Thus the expansion of $U^{\nu\,\text{ret.}}$ in the required form is obtained. $G^{\text{ret.}}_{\mu\nu}$, which involves one more differentiation, can be expanded in an entirely similar manner. Since the ingoing field and potentials are always expressible as such a series by Taylor's expansion, $T^{\mu\nu}$ can be written in the required form with different order terms separated. This result is used in (9) and (10) in the text.

As is well known, in the static case when $S^{\alpha\ldots\lambda,\nu}$ is constant we get

$$U^\nu = S^{ik\ldots m,\nu}\, \frac{\partial}{\partial x^i}\frac{\partial}{\partial x^k} \cdots \frac{\partial}{\partial x^m}\left(\frac{e^{-\chi\kappa}}{\kappa}\right). \tag{116a}$$

(Here the latin indices run from 1 to 3 only.) In the electromagnetic case, by putting $\chi = 0$, we get

$$U^\nu = S^{ik\ldots m,\nu}\, \frac{\partial}{\partial x^i}\frac{\partial}{\partial x^k} \cdots \frac{\partial}{\partial x^m}\left(\frac{1}{\kappa}\right). \tag{116b}$$

From (105) it is found that in the electromagnetic case the retarded potentials*
ϕ^ν do not contain terms of order lower than $1/\kappa$. These satisfy the equation

$$\frac{\partial \phi^\nu}{\partial x^\nu} = 0. \tag{117}$$

Using the relations (44) and (54), it is found from (117) that

$$\frac{s_\nu}{\kappa} \iota^\sigma \frac{\partial \phi^\nu_{(-n)}}{\partial x^\sigma} + \left(\delta^\sigma_\nu - \frac{s_\nu}{\kappa} \iota^\sigma\right) \frac{\partial \phi^\nu_{(-n+1)}}{\partial x^\sigma} = 0 \quad \text{for} \quad n > 1 \tag{118a}$$

and

$$\frac{s_\nu}{\kappa} \iota^\sigma \frac{\partial \phi^\nu_{(-1)}}{\partial x^\sigma} = 0. \tag{118b}$$

Here $(-n)$ denotes the nth order term. Now calculate

$$[F_{\mu\nu(-n)} F^{\nu\rho}_{(-1)} + F_{\mu\nu(-1)} F^{\nu\rho}_{(-n)} + \tfrac{1}{2}\delta^\rho_\mu F_{\alpha\beta(-1)} F^{\alpha\beta}_{(-1)}] s_\rho. \tag{119}$$

where $F_{\mu\nu}$ are the retarded electromagnetic field strengths. From the relation

$$F_{\mu\nu} = \frac{\partial \phi_\nu}{\partial x^\mu} - \frac{\partial \phi_\mu}{\partial x^\nu} \tag{120}$$

it is found that

$$F_{\mu\nu(-n)} = \left[\frac{s_\mu}{\kappa} \iota^\sigma \frac{\partial \phi_{\nu(-n)}}{\partial x^\sigma} + \left(\delta^\sigma_\mu - \frac{s_\mu}{\kappa} \iota^\sigma\right) \frac{\partial \phi_{\nu(-n+1)}}{\partial x^\sigma}\right]_- \quad \text{for} \quad n > 1 \tag{121a}$$

and

$$F_{\mu\nu(-1)} = \left[\frac{s_\mu}{\kappa} \iota^\sigma \frac{\partial \phi_{\nu(-1)}}{\partial x^\sigma}\right]_-. \tag{121b}$$

Using (121b) and (118b) it is found that $F_{\mu\nu(-1)} s^\nu = 0$. so that the first term in
(119) does not contribute anything. Now

$$F_{\mu\nu(-n)} s^\nu = \left[\frac{s_\mu}{\kappa} s^\nu \iota^\sigma \frac{\partial \phi_{\nu(-n)}}{\partial x^\sigma} + s^\nu\left(\delta^\sigma_\mu - \frac{s_\mu}{\kappa} \iota^\sigma\right) \frac{\partial \phi_{\nu(-n-1)}}{\partial x^\sigma} - s^\sigma \frac{\partial \phi_{\mu(-n+1)}}{\partial x^\sigma}\right]$$

$$= \left[(s^\nu\delta^\sigma_\mu - s_\mu g^{\sigma\nu}) \frac{\partial \phi_{\nu(-n+1)}}{\partial x^\sigma} - s^\sigma \frac{\partial \phi_{\mu(-n-1)}}{\partial x^\sigma}\right] \quad \text{from (118a).} \tag{122}$$

so that

$$F_{\mu\nu(-1)} F^{\nu\rho}_{(-n)} s_\rho = F_{\mu\nu(-1)}\left| s^\nu \frac{\partial \phi_{\mu(-n-1)}}{\partial x^\nu} - s^\rho \frac{\partial \phi^\nu_{(-n+1)}}{\partial x^\rho}\right| \tag{123}$$

as $F_{\mu\nu(-1)} s^\nu = 0$. Also

$$F_{\alpha\beta(-n)} F^{\alpha\beta}_{(-1)} = 2 F^{\alpha\beta}_{(-1)}\left[\frac{s_\alpha}{\kappa} \iota^\sigma \frac{\partial \phi_{\beta(-n)}}{\partial x^\sigma} + \left(\delta^\sigma_\alpha - \frac{s_\alpha}{\kappa} \iota^\sigma\right) \frac{\partial \phi_{\beta(-n+1)}}{\partial x^\sigma}\right]$$

$$= 2 F^{\alpha\beta}_{(-1)} \frac{\partial \phi_{\beta(-n+1)}}{\partial x^\alpha}.$$

---

* ϕ^ν is used to denote U^{ret} for the electromagnetic field.

Thus (119) becomes

$$\left[\frac{s_\mu}{\kappa}\,v^\sigma - \frac{\partial\phi_{\nu(-1)}}{\partial x^\sigma}\right]\left[s^\rho\,\frac{\partial\phi_{\mu(-n+1)}}{\partial x_\nu} - s^\rho\,\frac{\partial\phi^\nu_{(-n+1)}}{\partial x_\mu}\right] + s_\mu\left[\frac{s^\alpha}{\kappa}\,v^\sigma - \frac{\partial\phi^\beta_{(-1)}}{\partial x^\sigma}\right]_-\frac{\partial\phi_{\beta(-n+1)}}{\partial x^\alpha} = 0.$$

Thus the expression (119) vanishes. Since $4\pi T'^{\mu\rho}_{(-3)}{}^{\text{ret.}}\cdot s_\rho$ is just this expression with $n = 2$, then

$$T^{\mu\nu}_{(-3)}{}^{\text{ret.}}\cdot s_\nu = 0. \tag{124}$$

References

Bhabha 1939 *Proc. Roy. Soc.* A, **172**, 384–409.
Bhabha 1941 *Proc. Roy. Soc.* A, **178**, 315–350.
Bhabha & Corben 1941 *Proc. Roy. Soc.* A, **178**, 273–314.
Bhabha & Harish-Chandra 1944 *Proc. Roy. Soc.* A, **183**, 134.
Dirac 1938 *Proc. Roy. Soc.* A, **167**, 148–169.
Pauli 1941 *Rev. Mod. Phys.* **13**, 203–232.
Pryce 1938 *Proc. Roy. Soc.* A, **168**, 389–401.

Reprinted from
Proc. Royal Soc. A.
183 (1944), 142–167

On the Scattering of Scalar Mesons

Harish-Chandra

(*J. H. Bhabha Student, Cosmic Ray Research Unit, Indian Institute of Science, Bangalore*)

Received January 15, 1945
(Communicated by Prof. H. J. Bhabha, F.R.S.)

In a recent paper (Harish-Chandra, 1944, referred to as **A** in this paper) the equations of motion of a point particle interacting with a scalar meson field have been derived. The object of this note is to use these equations to calculate the scattering of scalar mesons by neutrons (or protons) on the classical theory, taking into account the radiation damping. This calculation is entirely similar to the corresponding one, done by Bhabha (1939, 1941) for the case of the vector-mesons. On account of the neglect of the quantum effects and the charge of the meson these calculations are subject to the same limitations as those of Bhabha. Since it is as yet not at all certain whether the actual meson has a spin of 1 or 0 unit, the scattering formulæ obtained here are to be looked upon as possible alternatives to those given by Bhabha.

§1. We shall keep to the notation of the previous papers (Bhabha and Harish-Chandra, 1944, Harish-Chandra, 1944). τ is the proper time at any point on the world line of the neutron measured from some fixed point on it. $z_\mu(\tau)$ are the co-ordinates of this point. x_μ denote the co-ordinates of any field-point. The fundamental metric tensor is taken to be

$$g_{\mu\nu} = 0, \qquad \mu \neq \nu, \qquad g_{00} = -g_{11} = -g_{22} = -g_{33} = 1.$$

A dot denotes differentiation with respect to $\tau \cdot v_\mu = \dot{z}_\mu(\tau)$ and $u_\mu = x_\mu - z_\mu(\tau)$.

We shall assume that the neutron has a 'charge' and a 'dipole'. Following Bhabha we consider the scattering due to each of these separately. First we calculate the scattering due to the charge alone.

In the notation of **A** the equation of motion of the neutron is

$$m\dot{v}_\mu - g_1 \frac{d}{d\tau}\left(\mathrm{U}'^{\mathrm{mean}}v_\mu\right) = -g_1 \mathrm{U}_\mu'^{\mathrm{mean}} \tag{1}$$

where m is the mass and g_1 the 'charge' of the neutron. $\mathrm{U}'^{\mathrm{mean}}$ and $\mathrm{U}_\mu'^{\mathrm{mean}}$ are the modified mean-potential and field respectively. It has been shown in **A** [Eqs. (3·17) and $(3 \cdot 18)]$ that

$$\mathbf{U}'^{\text{mean}} = \mathbf{U}^{\text{in}} + \tilde{\mathbf{U}},$$

$$\mathbf{U}_{\mu}'^{\text{mean}} = \mathbf{U}_{\mu}^{\text{in}} - g_1\left(\tfrac{1}{3}\ddot{v}_{\mu} + \tfrac{1}{3}v_{\mu}(\dot{v})^2 + \tfrac{1}{2}\chi^2 v_{\mu}\right) + \tilde{\mathbf{U}}_{\mu} \tag{2a}$$

where

$$\tilde{\mathbf{U}} = - g_1\chi \int_{-\infty}^{\tau} \frac{\mathbf{J}_1(\chi u)}{u} d\tau', \qquad \tilde{\mathbf{U}}_{\mu} = g_1\chi^2 \int_{-\infty}^{\tau} u_{\mu} \frac{\mathbf{J}_2(\chi u)}{u^2} d\tau' \tag{2b}$$

$\mathbf{U}^{\text{in}}$ is the ingoing potential and $\mathbf{U}_{\mu}^{\text{in}} = \partial_{\mu} \mathbf{U}^{\text{in}}$ is the ingoing field. We assume the ingoing field to be a plane wave travelling in the x_1-direction. Assuming that the amplitude of oscillation of the neutron is small compared to the incident wave length we can write

$$\mathbf{U}_1^{\text{in}} = \gamma \sin \omega_0 t, \qquad \mathbf{U}_2^{\text{in}} = 0, \qquad \mathbf{U}_3^{\text{in}} = 0 \tag{3}$$

where $t = z_0(\tau)$. We assume that the velocity of the neutron is small so that

$$v_k \approx \frac{dz_k}{dt}$$

for $k = 1, 2, 3$. In conformity with (3) we put (*cf.* Bhabha, 1939)

$$\left.\begin{aligned}
z_1 &= \frac{\beta}{\omega_0} \sin(\omega_0 t + \delta), & z_2 &= 0, z_3 = 0 \\
v_1 &= \beta \cos(\omega_0 t + \delta), & \dot{v}_1 &= -\beta\omega_0 \sin(\omega_0 t + \delta) \\
v_0 &\approx 1, & v_2 &= v_3 = 0.
\end{aligned}\right\} \tag{4}$$

We assume the amplitudes β and γ to be small so that quantities quadratic in them may be neglected. Putting $x_{\mu} = z_{\mu}(\tau)$ we get for (2)

$$u_1 = z_1(\tau) - z_1(\tau') = \frac{\beta}{\omega_0}\left[\sin(\omega_0 t + \delta) - \sin(\omega_0 t' + \delta)\right],$$

$$u_2 = u_3 = 0 \tag{5a}$$

$$u \approx t - t' = u_0. \tag{5b}$$

From (2b), (4) and (5) we get

$$\tilde{\mathbf{U}} = - g_1\chi \int_{-\infty}^{\tau} \frac{\mathbf{J}_1(\chi u)}{u} d\tau'$$

$$= - g_1\chi \int_0^{\infty} \frac{\mathbf{J}_1(\chi u)}{u} du = - g_1\chi \tag{6}$$

$$\tilde{\mathbf{U}}_0 = g_1\chi^2 \int_{-\infty}^{\tau} u_0 \frac{\mathbf{J}_2(\chi u)}{u^2} d\tau'$$

$$= g_1\chi^2 \int_0^{\infty} \frac{\mathbf{J}_2(\chi u)}{u} du = \tfrac{1}{2}g_1\chi^2 \tag{7}$$

$$\tilde{U}_1 = g_1 \chi^2 \int_{-\infty}^{\tau} u_1 \frac{J_2(\chi u)}{u^2} d\tau'$$

$$= g_1 \chi^2 \int_0^{\infty} \frac{J_2(\chi u)}{u} u_1 \, du$$

$$= \frac{\beta g_1 \chi^3}{\omega_0} \int_0^{\infty} \frac{J_2(s)}{s} \left[\sin \alpha - \sin(\alpha - \nu s) \right] ds \tag{8}$$

where $s = \chi u$, $\nu = \omega_0/\chi$ and $\alpha = \omega_0 t + \delta$. Using the well-known result that

$$\int_0^{\infty} ds \, \frac{J_2(s)}{s^2} e^{-i\nu s} = \begin{cases} \frac{1}{3}\left[(1-\nu^2)^{3/2} + i\nu^3 - \frac{3}{2}i\nu \right] & 0 < \nu < 1 \\ \frac{1}{3}\left[-i(\nu^2-1)^{3/2} + i\nu^3 - \frac{3}{2}i\nu \right] & \nu > 1 \end{cases} \tag{9}$$

we get

$$\tilde{U}_1 = -\beta g_1 \omega_0^2 (\mathbf{P}\cos\alpha + \mathbf{Q}\sin\alpha) \tag{10a}$$

where

$$\mathbf{P} = \begin{cases} \frac{1}{3} - \frac{1}{2}\nu^2 & 0 < \nu < 1 \\ -\frac{1}{3}\frac{(\nu^2-1)^{3/2}}{\nu^3} + \frac{1}{3} - \frac{1}{2\nu^2} & \nu > 1 \end{cases}$$

$$\mathbf{Q} = \begin{cases} \frac{1}{3}\frac{(1-\nu^2)^{3/2}}{\nu^3} - \frac{1}{3\nu^3} & 0 < \nu < 1 \\ -\frac{1}{3\nu^3} & \nu > 1. \end{cases} \tag{10b}$$

Substituting the value of the various quantities in (1) and putting $m + g_1^2\chi = \mathbf{M}$ we get up to terms of the first order in β and γ

$$-\beta\omega_0\mathbf{M}\sin\alpha = -g_1\gamma\cos(\alpha-\delta)$$
$$+ \beta g_1^2\omega_0^2[\mathbf{P}\cos\alpha + \mathbf{Q}\sin\alpha]$$
$$- \frac{1}{3}g_1^2\beta\omega_0^2\cos\alpha + \frac{1}{2}\chi^2 g_1^2\beta\cos\alpha. \tag{11}$$

Equating coefficients of $\sin\alpha$ and $\cos\alpha$ on both sides of (11) we get

$$\beta = \frac{\gamma}{g_1\omega_0\left\{ \mathbf{P}'^2\omega_0^2 + \left[\mathbf{Q}\omega_0 + \dfrac{\mathbf{M}}{g_1^2} \right]^2 \right\}^{1/2}} \tag{12a}$$

$$\cos\delta = \frac{\omega_0\mathbf{P}'}{\left[\mathbf{P}'^2\omega_0^2 + \left(\mathbf{Q}\omega_0 + \dfrac{\mathbf{M}}{g_1^2} \right)^2 \right]^{1/2}} \tag{12b}$$

where the positive value of the square root is to be taken. $\mathbf{P}' = \mathbf{P} - \frac{1}{3} + \frac{1}{2}\nu^2$ so that

$$\mathbf{P}' = \begin{cases} 0 & \nu < 1 \\ -\dfrac{1}{3}\dfrac{(\nu^2-1)^{3/2}}{\nu^3} & \nu > 1. \end{cases} \tag{12c}$$

It is clear from (12) that the quantity $\mathbf{M} + g_1^2\mathbf{Q}\omega_0$ behaves as the effective mass. For very slow oscillations $\omega_0 \ll \chi$ ($\nu \ll 1$) $\mathbf{Q} \approx -1/2\nu$

$$\mathbf{M} + g_1^2\mathbf{Q}\omega_0 = \mathbf{M} - \tfrac{1}{2}g_1^2\chi = m + \tfrac{1}{2}g_1^2\chi.$$

Therefore in this case the field contributes a *positive mass* $\frac{1}{2}g_1^2\chi$. On the other hand in the vector meson case the field adds a *negative mass* $-\frac{1}{2}g_1^2\chi$ for slow oscillations (*cf*. Bhabha, 1939).

To calculate the scattering we have to calculate the retarded field at a very distant point $x_\mu = (\mathbf{X},\mathbf{Y},\mathbf{Z},\mathbf{T})$ lying on the future light cone from the point τ on the world line.

$$\begin{aligned}
\mathbf{U}_{(x_\mu)}^{ret} &= \frac{g_1}{\kappa(\tau)} - \chi g_1 \int_{-\infty}^{\tau} \frac{\mathbf{J}_1(\chi u)}{u}\,d\tau' \\
&= g_1 \int_{-\infty}^{\tau} \frac{\mathbf{J}_0(\chi u)}{u}\frac{d}{d\tau}\left(\frac{1}{\kappa}\right)d\tau' \\
&= g_1 \int_{-\infty}^{\tau} \mathbf{J}_0(\chi u)\left\{\frac{1}{\kappa^2} - \frac{\kappa'}{\kappa^2}\right\}d\tau'
\end{aligned} \tag{13}$$

where $\kappa = u^\mu(\tau')v_\mu(\tau')$ and $\kappa' = u^\mu(\tau')\dot{v}_\mu(\tau')$. For a very distant point we can neglect the first term $1/\kappa^2$ and write

$$\mathbf{U}^{ret} \approx -g_1 \int_{-\infty}^{\tau} \mathbf{J}_0(\chi u)\frac{\kappa'}{\kappa^2}\,d\tau' = -g_1 \int_0^{\infty} \mathbf{J}_0(\chi u)\frac{\kappa'}{\kappa^3}\,du. \tag{14}$$

Writing $\mathbf{R} = \sqrt{\mathbf{X}^2 + \mathbf{Y}^2 + \mathbf{Z}^2}$ we get up to terms of the first order in β

$$\left.\begin{aligned}
\kappa' &= -\mathbf{X}\dot{v}_1 = \beta\omega_0\mathbf{X}\sin(\omega_0 t' + \delta) \\
u^2 &= (\mathbf{T}-t')^2 - \mathbf{R}^2 + \frac{2\mathbf{X}\beta}{\omega_0}\sin(\omega_0 t' + \delta) \\
\kappa &= (\mathbf{T}-t') - \mathbf{X}v_1 = (\mathbf{T}-t') - \beta\mathbf{X}\cos(\omega_0 t + \delta)
\end{aligned}\right\} \tag{15}$$

so that in the same approximation

$$\begin{aligned}
\frac{\kappa'}{\kappa^3} &= \frac{\beta\omega_0\mathbf{X}\sin(\omega_0 t' + \delta)}{(u^2 + \mathbf{R}^2)^{3/2}} \\
&= \frac{\beta\omega_0\mathbf{X}\sin\left(\omega_0\mathbf{T} + \delta - \omega_0\sqrt{u^2 + \mathbf{R}^2}\right)}{(u^2 + \mathbf{R}^2)^{3/2}} \\
&= \frac{\beta\omega_0\mathbf{X}\sin\left(\alpha' - \omega_0\sqrt{u^2 + \mathbf{R}^2}\right)}{(u^2 + \mathbf{R}^2)^{3/2}}
\end{aligned} \tag{16}$$

if we write α' for $\omega_0 T + \delta$. Thus from (14) and (16)

$$\mathbf{U}^{ret} = - g_1 \beta \omega_0 \mathbf{X} \int_0^\infty \mathbf{J}_0(\chi u) \frac{\sin\left(\alpha' - \omega_0\sqrt{u^2 + \mathbf{R}^2}\right)}{(u^2 + \mathbf{R}^2)^{3/2}} u\, du$$

$$= - \frac{g_1 \beta \omega_0 \mathbf{X}}{\chi} \int_0^\infty \mathbf{J}_0(s) \frac{\sin\left(\alpha' - \nu\sqrt{s^2 + r^2}\right)}{(s^2 + r^2)^{3/2}} s\, ds \qquad (17)$$

where $s = \chi u$, $r = \mathbf{R}\chi$ and $\nu = \omega_0/\chi$. On evaluating (17) and retaining only the terms of the lowest order in $1/\mathbf{R}$ we find

$$\mathbf{U}^{ret} = \begin{cases} g_1 \beta \dfrac{\mathbf{X}}{\mathbf{R}} \dfrac{\sqrt{\chi^2 - \omega_0^2}}{\mathbf{R}\omega_0} e^{-\mathbf{R}\sqrt{\chi^2 - \omega_0^2}} \sin(\omega_0 \mathbf{T} + \delta), & \omega_0 < \chi \\[3mm] g_1 \beta \dfrac{\mathbf{X}}{\mathbf{R}} \dfrac{\sqrt{\omega_0^2 - \chi^2}}{\mathbf{R}\omega_0} \cos\left(\omega_0 \mathbf{T} + \delta - \mathbf{R}\sqrt{\omega_0^2 - \chi^2}\right), & \omega_0 > \chi. \end{cases} \qquad (18)$$

For the case $\omega_0 < \chi$ the potential falls off exponentially with distance and obviously there is no scattering. We therefore consider the case $\omega_0 > \chi$. In this case the incident wave is consistently with (3)

$$\mathbf{U}^{in} = \frac{-\gamma}{\sqrt{\omega_0^2 - \chi^2}} \sin\left(\omega_0 t - \sqrt{\omega_0^2 - \chi^2}\, x_1\right). \qquad (19)$$

The flow of energy per unit area per unit time in the x_1 direction as calculated from the energy-momentum tensor

$$4\pi \mathbf{T}_{\mu\nu} = \mathbf{U}_\mu \mathbf{U}_\nu - \tfrac{1}{2} g_{\mu\nu}\left(\mathbf{U}_\rho \mathbf{U}^\rho - \chi^2 \mathbf{U}^2\right) \qquad (20)$$

is

$$\frac{\overline{\mathbf{U}_0 \mathbf{U}_1}}{4\pi} = \frac{\gamma^2 \omega_0}{8\pi\sqrt{\omega_0^2 - \chi^2}} \qquad (21)$$

where the bar denotes the average over time. Similarly the flow of energy due to (18) in the direction of the radius $\mathbf{R}$ across an element of surface subtending a solid angle $d\Omega$ is

$$\frac{g_1^2 \beta^2}{8\pi} \cos^2\theta \frac{\left(\omega_0^2 - \chi^2\right)^{3/2}}{\omega_0}\, d\Omega \qquad (22)$$

where θ is the angle between the incident and the scattered wave i.e., $\mathbf{X}/\mathbf{R} = \cos\theta$. The differential-scattering cross-section in the direction θ is therefore

$$g_1^2 \frac{\beta^2}{\gamma^2} \cos^2\theta \frac{\left(\omega_0^2 - \chi^2\right)}{\omega_0^2}\, d\Omega = \frac{\cos^2\theta}{\dfrac{1}{9}\left(\omega_0^2 - \chi^2\right) + \dfrac{\omega_0^4}{\left(\omega_0^2 - \chi^2\right)^2}\left[a + \chi - \dfrac{\chi^3}{3\omega_0^2}\right]^2} \qquad (23)$$

from (12a) if we put $m/g_1^2 = a$. Integrating over all directions θ we find that the total cross-section is

$$12\pi \frac{1}{\left(\omega_0^2 - \chi^2\right) + \dfrac{\omega_0^4}{\left(\omega_0^2 - \chi^2\right)^2}\left[3a + 3\chi - \dfrac{\chi^3}{\omega_0^2}\right]^2}\, d\Omega. \tag{24}$$

For the spin 1 case the scattering of the transverse mesons is given by (Bhabha, 1939)

$$6\pi\left(1 + \frac{1}{2}\frac{\chi^2}{\omega_0^2}\right)\frac{1}{\left(\omega_0^2 - \dfrac{3\chi^4}{4\omega_0^2} - \dfrac{1}{4}\dfrac{\chi^6}{\omega_0^4}\right) + \left(\dfrac{3}{2}a + \dfrac{1}{2}\dfrac{\chi^3}{\omega_0^2}\right)^2}.$$

For high energy this is half of (24) independently of the value of g_1. For $\chi = 0$ (24) becomes

$$\frac{12\pi}{9a^2 + \omega_0^2}$$

which is of the same form as the corresponding formula of Dirac (1938), *viz.*, $24\pi/9a^2 + 4\omega_0^2$ for the scattering of light by an electron. As found in other cases the cross-section (24) decreases with increasing ω_0 as $1/\omega_0^2$ for high frequencies.

§2. We shall now treat the scattering by the neutron due to the dipole-moment alone. We denote the dipole-vector by $g_2 \mathbf{S}^\alpha$. As already pointed out in **A** we have to assume

$$\mathbf{S}^\alpha v_\alpha = 0. \tag{25}$$

The equation of rotational motion is [Eqn. $(5\cdot36b)$ of **A**]

$$\mathbf{I}\epsilon_{\mu\nu\rho\sigma}v^\rho\dot{\mathbf{S}}^\sigma = g_2\left[\mathbf{S}_\mu\overline{\mathbf{U}}_\nu - \mathbf{S}_\nu\overline{\mathbf{U}}_\mu\right] \tag{26}$$

where $\overline{\mathbf{U}}_\nu = \mathbf{U}_\nu'^{\text{mean}} - v_\nu\,(v^\mu\mathbf{U}_\mu'^{\text{mean}})$. $\epsilon_{\mu\nu\rho\sigma}$ is the tensor which is antisymmetric in each pair of indices and $\epsilon_{0123} = -1$. Following Bhabha we put m equal to infinity to simplify the problem. In this case we find from the translational equation that $\dot{v}_\mu = \ddot{v}_\mu = \cdots = 0$, so that we can consider the dipole in the rest system. In the usual three-dimensional vector notation (26) can be written as

$$\mathbf{I}\dot{\mathbf{S}} = g_2[\mathbf{S}\cdot\mathbf{U}'^{\text{mean}}] \tag{27}$$

where the components of $\mathbf{S}$ and $\mathbf{U}'^{\text{mean}}$ are $\mathbf{S}_k$ and $\mathbf{U}_k'^{\text{mean}}$ $(k = 1, 2, 3)$ respectively. The bracket denotes the vector product.

Using (25) we find from $(3\cdot19)$ of **A** that

$$\mathbf{U}_\mu'^{\text{mean}} = \mathbf{U}_\mu^{\text{in}} + g_2\left(\frac{1}{3}\dddot{\mathbf{S}}_\mu + \frac{\chi^2}{2}\dot{\mathbf{S}}_\mu\right) + \tilde{\mathbf{U}}_\mu \tag{28}$$

where

$$\tilde{\mathbf{U}}_\mu = g_2 \chi^2 \int_{-\infty}^{t} \mathbf{S}_\mu \frac{\mathbf{J}_2(\chi u)}{u^2} dt'. \tag{29}$$

Here we have used the fact that $\tau = t$ and $u_1 = u_2 = u_3 = 0$ so that $\mathbf{S}^\sigma u_\sigma = 0$. Also

$$u = u_0 = t - t'$$

so that

$$\tilde{\mathbf{U}} = g_2 \chi^2 \int_0^\infty \mathbf{S}(t - u) \frac{\mathbf{J}_2(\chi u)}{u^2}. \tag{30}$$

It is to be noted that the right side of (30) is the same as $\frac{1}{2}\tilde{\mathbf{G}}$ of Bhabha [1941, Eqn. (68)] if we replace $\mathbf{M}$ by $\mathbf{S}$ there. (27) can now be written as

$$\mathbf{I}\dot{\mathbf{S}} = g_2[\mathbf{S}\cdot\mathbf{U}^{\text{in}}] + \tfrac{1}{3}g_2^2\left[\mathbf{S}\cdot\dddot{\mathbf{S}}\right] + \tfrac{1}{2}g_2^2\chi^2[\mathbf{S}\cdot\dot{\mathbf{S}}] + g_2[\mathbf{S}\cdot\tilde{\mathbf{U}}] \tag{31}$$

This equation is the same as Eqn. (66) of Bhabha (1941) if we replace $\mathbf{H}$ by $2\mathbf{U}^{\text{in}}$, $\mathbf{M}$ by $\mathbf{S}$ and $\mathbf{I}$ by $2\mathbf{I}$ in the latter and put $\mathbf{K} = 0$. Thus corresponding to equations (69) and (70) of Bhabha we put

$$\mathbf{U}^{\text{in}} = \tfrac{1}{2}\mathbf{H}_0\cos\omega_0 t \equiv \tfrac{1}{2}\mathbf{H} \tag{32}$$

$$\mathbf{S}(t) = \mathbf{S}_0 + \mathbf{S}_1\sin\omega_0 t + \mathbf{S}_2\sin(\omega_0 t + \delta) \tag{33}$$

where $\mathbf{S}_0$ is the initial direction of the dipole and $\mathbf{S}_0$, $\mathbf{S}_1$ and $\mathbf{S}_2$ are mutually perpendicular and such that $\mathbf{S}_2$ is along $[\mathbf{S}_0\cdot\mathbf{S}_1]$. We assume $\mathbf{S}_0^2 = 1$. Substituting (32), (33) in (30) and (31) we get corresponding to Eqn. (74) of Bhabha

$$\omega_0\mathbf{S}_1\left[\alpha\cos\omega_0 t + \frac{|\mathbf{S}_2|}{|\mathbf{S}_1|}\{\xi\sin(\omega_0 t + \delta) - \zeta\cos(\omega_0 t + \delta)\}\right]$$

$$+ \omega_0\mathbf{S}_2\left[\alpha\cos(\omega_0 t + \delta) - \frac{|\mathbf{S}_1|}{|\mathbf{S}_2|}\{\xi\sin\omega_0 t - \zeta\cos\omega_0 t\}\right]$$

$$= \frac{3}{2g_2^2}[\mathbf{S}_0\cdot\mathbf{H}_0]\cos\omega_0 t$$

$$\tag{34}$$

where $\alpha = 3\mathbf{I}/g_2^2$ and

$$\xi \equiv \begin{cases} -\dfrac{\chi^3}{\omega_0} + \dfrac{\left(\chi^2 - \omega_0^2\right)^{3/2}}{\omega_0} & \omega_0 < \chi \\[4mm] -\dfrac{\chi^3}{\omega_0} & \omega_0 > \chi \end{cases} \tag{35}$$

$$\zeta \equiv \begin{cases} 0 & \omega_0 < \chi \\[4mm] \dfrac{\left(\omega_0^2 - \chi^2\right)^{3/2}}{\omega_0} & \omega_0 > \chi. \end{cases} \tag{36}$$

Here the value of α is twice that used by Bhabha [Eqn. (37) l.c.], while ξ and ζ are the same as in his case. Since (34) is identical with Eqn. (74) of Bhabha, all the further results derivable from it in terms of α, ξ, ζ remain the same. Thus we get

$$\tan\delta = -\frac{\xi}{\zeta} \tag{37a}$$

$$\frac{|S_1|}{|S_2|} = \frac{\alpha}{\sqrt{\xi^2 + \zeta^2}} \tag{37b}$$

$$|S_1| = \frac{3\alpha|H_0|\sin\theta}{2g_2\omega_0\left[\left(\alpha^2 - \xi^2 - \zeta^2\right)^2 + 4\alpha^2\zeta^2\right]^{1/2}} \tag{37c}$$

θ being the angle between S_0 and H_0. The work done by the external force on the dipole is on the average

$$-\tfrac{1}{2}g_2\left(\overline{H\cdot S}\right) = \frac{3}{8}\frac{H_0^2\sin^2\theta\,\zeta\left(\alpha^2 + \xi^2 + \zeta^2\right)}{\left(\alpha^2 - \xi^2 - \zeta^2\right)^2 + 4\alpha^2\zeta^2}. \tag{38}$$

To obtain the scattering we have to calculate as before the retarded potential U^{ret} at a very distant point X, Y, Z, T lying on the future light cone from the point $z_\mu(\tau) = (0, 0, 0, t)$.

$$U^{ret}_{(x_\mu)} = \partial_\alpha\left\{ g_2\int_{-\infty}^{\tau} J_0(\chi u)\left(\frac{S^\alpha}{\kappa}\right) d\tau \right\} \tag{39}$$

$$u^2 = (T - t')^2 - R^2 \tag{40a}$$

$$\kappa = T - t' = \sqrt{u^2 + R^2}\,. \tag{40b}$$

Retaining only terms of the lowest order in $1/R$ we get

$$U^{ret} = g_2\partial_\alpha\int_0^\infty J_0(\chi u)\frac{\dot{S}^\alpha}{(u^2 + R^2)}\,du$$

$$= -g_2\,div\left\{\int_0^\infty \frac{\omega_0 J_0(\chi u)}{u^2 + R^2}\left\{ S_1\cos\left(\omega_0 T - \omega_0\sqrt{u^2 + R^2}\right)\right.\right.$$

$$\left.\left. + S_2\cos\left(\omega_0 T + \delta - \omega_0\sqrt{u^2 + R^2}\right)\right\}\cdot u\,du \right. \tag{41}$$

where the components of divergence are $\partial_k = \left(\dfrac{\partial}{\partial X}, \dfrac{\partial}{\partial Y}, \dfrac{\partial}{\partial Z}\right)$. The term inside the divergence can be evaluated and up to the terms of lowest order in $1/R$ it is equal to

$$\frac{e^{-R\sqrt{\chi^2 - \omega_0^2}}}{R}\left[S_1\sin\omega_0 T + S_2\sin(\omega_0 T + \delta)\right] \quad \text{for } \omega_0 < \chi$$

$$\frac{1}{R}\left\{ S_1\sin\left(\omega_0 T - R\sqrt{\omega_0^2 - \chi^2}\right) + S_2\sin\left(\omega_0 T + \delta - R\sqrt{\omega_0^2 - \chi^2}\right)\right\} \quad \text{for } \omega_0 > \chi$$

so that to the same approximation

$$
\mathbf{U}^{ret} = \begin{cases}
g_2 \dfrac{\sqrt{\chi^2 - \omega_0^2}}{\mathbf{R}} e^{-\mathbf{R}\sqrt{\chi^2 - \omega_0^2}} \left[\dfrac{(\mathbf{RS}_1)}{\mathbf{R}} \sin \omega_0 \mathbf{T} \right. \\
\quad \left. + \dfrac{(\mathbf{RS}_2)}{\mathbf{R}} \sin(\omega_0 \mathbf{T} + \delta) \right] \quad \text{for } \omega_0 < \chi \\[2ex]
g_2 \dfrac{\sqrt{\omega_0^2 - \chi^2}}{\mathbf{R}} \left[\dfrac{(\mathbf{RS}_1)}{\mathbf{R}} \cos\left(\omega_0 \mathbf{T} - \mathbf{R}\sqrt{\omega_0^2 - \chi^2} \right) \right. \\
\quad \left. + \dfrac{(\mathbf{RS}_2)}{\mathbf{R}} \cos\left(\omega_0 \mathbf{T} + \delta - \mathbf{R}\sqrt{\omega_0^2 - \chi^2} \right) \right] \quad \text{for } \omega_0 > \chi
\end{cases}
\tag{42}
$$

where $\mathbf{R} = (\mathbf{X}, \mathbf{Y}, \mathbf{Z})$. (42) would agree completely with the expression (84) of Bhabha for $\mathbf{U}_k^{ret}$ if we replace the scalar products with $\mathbf{R}$ by vector products. When $\omega_0 < \chi$ there is no radiation. For $\omega_0 > \chi$ the average rate of radiation in the direction $\mathbf{R}$ inside a solid angle $d\Omega$ is from (20)

$$
\frac{g_2^2}{8\pi} \left(\omega_0^2 - \chi^2 \right)^{3/2} \omega_0 \left[\frac{(\mathbf{RS}_1)^2}{\mathbf{R}^2} + \frac{(\mathbf{RS}_2)^2}{\mathbf{R}^2} + 2 \frac{(\mathbf{RS}_1)(\mathbf{RS}_2)}{\mathbf{R}^2} \cos \delta \right] d\Omega
\tag{43}
$$

corresponding to (84) of Bhabha. The total radiation obtained by integrating (43) over all directions is

$$
\tfrac{1}{6} g_2^2 \left(\omega_0^2 - \chi^2 \right)^{3/2} \omega_0 \left(|\mathbf{S}_1|^2 + |\mathbf{S}_2|^2 \right)
\tag{44}
$$

which is the same as (38) due to (37) and (36). The energy flow due to the incident wave (32) is

$$
\frac{|\mathbf{H}_0|^2}{32\pi} \frac{\omega_0}{\sqrt{\omega_0^2 - \chi^2}} .
\tag{45}
$$

The total effective cross-section for the scattering of the scalar meson wave is therefore from (45), (44) and (38) and (36)

$$
12\pi \sin^2\theta \frac{\left(\omega_0^2 - \chi^2 \right)^2}{\omega_0^2} \frac{\alpha^2 + \xi^2 + \zeta^2}{\left(\alpha^2 - \xi^2 - \zeta^2 \right)^2 + 4\alpha^2 \zeta^2} .
\tag{46}
$$

In terms of α, ξ, ζ (46) is the same as Eqn. (82) of Bhabha except for a factor 2. Substituting the value of ξ and ζ the scattering cross-section becomes

$$
12\pi \sin^2\theta \left(\omega_0^2 - \chi^2 \right)^2 \frac{\alpha^2 \omega_0^2 + \left(\omega_0^2 - \chi^2 \right)^3 + \chi^6}{\left[\alpha^2 \omega_0^2 + \left(\omega_0^2 - \chi^2 \right)^3 + \chi^6 \right]^2 - 4\alpha^2 \omega_0^2 \chi^6} .
\tag{47}
$$

Except for a factor 2, (45) agrees entirely with Eqn. (88) of Bhabha in which $\mathbf{K}$ has been put equal to zero. For small g_2 and not very high frequencies we can expand

(47) as a series in ascending powers of g_2. The first term is

$$\frac{4\pi}{3}\sin^2\theta \frac{g_2^4}{I^2}\frac{\left(\omega_0^2-\chi^2\right)^2}{\omega_0^2} \tag{48}$$

which is exactly half of the corresponding expression (86) of Bhabha. Thus for large ω_0 the scattering is double (independently of the value g_2) and for small ω_0 half that of transverse vector-mesons. The formulæ (47) and (48) have already been discussed by Bhabha in detail.

§3. The above theory will now be compared with the quantum theory of neutral mesons. For simplicity we put $c=1$, $\hbar=1$. The total Lagrangian for the neutron and the meson fields together is taken to be

$$\begin{aligned}
\mathbf{L} = \frac{i}{2}\left(\psi^+\beta\gamma^\mu\partial_\mu\psi - \partial_\mu\psi^+\beta\gamma^\mu\psi\right) - \mu\psi^+\beta\psi + g_1\psi^+\beta\psi U + g_2\psi^+\beta\gamma^\mu\psi\,\partial_\mu U \\[1mm]
+ i\frac{g_2'}{3!}\psi^+\beta\gamma^\mu\gamma^\nu\gamma^\sigma\psi\epsilon_{\mu\nu\sigma\rho}\partial^\rho U + \frac{g_1'}{4!}\psi^+\beta\gamma^\mu\gamma^\nu\gamma^\sigma\gamma^\rho\psi\epsilon_{\mu\nu\sigma\rho}\cdot U \\[1mm]
+ \tfrac{1}{2}\left(\partial_\mu U\cdot\partial^\mu U - \chi^2 U^2\right).
\end{aligned} \tag{49}$$

Here γ^μ are the 4-rowed square matrices satisfying

$$\gamma^\mu\gamma^\nu + \gamma^\nu\gamma^\mu = 2g^{\mu\nu}$$

and $\beta=\gamma^0$. ψ refers to the neutron field and ψ^+ is its Hermitian conjugate. γ's are related to the usual Dirac-matrices $\underline{\alpha}$ and β by

$$\underline{\alpha} = \left(\alpha^1, \alpha^2, \alpha^3\right) = \beta\left(\gamma^1, \gamma^2, \gamma^3\right)$$

We assume $\underline{\alpha}$ and β to be Hermitian. μ is the mass of the neutron. The terms containing g_1 and g_2 in (49) represent the usual 'scalar' interactions for the charge and the dipole respectively while those containing g_1' and g_2' represent the corresponding 'pseudo-scalar' interactions. Evidently in the classical theory there is no distinction between the scalar and pseudo-scalar interactions since this distinction arises only from the γ-matrices. To simplify (49) we use the usual representation of $\underline{\alpha}$ and β through two sets of mutually independent Pauli matrices,

$$\underline{\alpha} = \rho_1(\sigma^1, \sigma^2, \sigma^3),\ \beta = \rho_3.$$

With this substitution (49) simplifies to

$$\begin{aligned}
\mathbf{L} = \frac{i}{2}\left(\psi^+\beta\gamma^\mu\partial_\mu\psi - \partial_\mu\psi^+\beta\gamma^\mu\psi\right) - \mu\psi^+\beta\psi + g_1\psi^+\beta\psi\cdot U \\[1mm]
+ g_2\psi^+\left[(\underline{\alpha}\mathbf{U})+U_0\right]\psi + g_2'\psi^+\left[(\underline{\alpha}\mathbf{U})+\rho_1 U_0\right]\psi - g_1'\psi^+\rho_2\psi\cdot U \\[1mm]
+ \tfrac{1}{2}\left(\partial_\mu U\cdot\partial^\mu U - \chi^2 U^2\right)
\end{aligned} \tag{50}$$

where $\mathbf{U} = (\partial_1 U, \partial_2 U, \partial_3 U)$, $U_0 = \partial_0 U$. The Hamiltonian of the system for only one neutron present is found as usual and on ignoring the infinite self-energy of the neutron terms out to be

$$\mathbf{H} = (\underline{\alpha}p) + \mu\beta - \frac{i}{\sqrt{\mathbf{V}}}\Sigma(2k_0)^{-1/2}(g_1\rho_3 - g_1'\rho_2)\{a_k e^{i(kx)} - a_k^\times e^{-i(kx)}\}$$

$$+ \frac{1}{\sqrt{\mathbf{V}}}\Sigma(2k_0)^{-1/2}(g_2\rho_1 + g_2')(\underline{\sigma}k) - \rho_1 k_0\{a_k e^{i(kx)} + a_k^\times e^{i(kx)}\}$$

$$+ \Sigma\left(\mathbf{N}_k + \tfrac{1}{2}\right) \tag{51}$$

where $k_0 = +\sqrt{\chi^2 + |k|^2}$, $\mathbf{N}_k$ is the number of mesons in the momentum state k and a_k and $a_k^\times$ are the corresponding absorption and emission operators respectively.

$$\left[a_k, a_k^\times\right]_- = 1.$$

As usual $\mathbf{V}$ is a large volume in which the field quantities are periodic.

Corresponding to the treatment of §2 we put $g_1 = g_1' = 0$ and regard the neutron as an infinitely heavy particle ($\mu \to \infty$). A straightforward calculation shows that the scattering due to a pure 'scalar' dipole-interaction ($g_2' = 0$) is zero in the usual second order (g_2^4) approximation. This result holds even when the calculation is performed taking into account the finite mass of the neutron. Also the terms containing both g_2 and g_2' vanish for $\mu \to \infty$, so that in this limit the contribution to the scattering comes purely from the pseudo-scalar interaction. The differential scattering cross-section is thus found to be for $\mu \to \infty$

$$d\sigma = 4\,d\Omega\left(\frac{g_2'}{\sqrt{4\pi}}\right)^4 \frac{p^4}{\mathbf{E}^2}\sin^2\phi \tag{52}$$

where p is the momentum, $\mathbf{E}$ the energy and ϕ the angle of scattering of the meson. The total cross-section is therefore

$$\frac{32\pi}{3}\left(\frac{g_2'}{\sqrt{4\pi}}\right)^4 \frac{p^2}{\mathbf{E}^2}. \tag{53}$$

$g_2'/\sqrt{4\pi}$ is the value of the dipole-moment in the usual units as against its value g_2' in the 'Heaviside-units' which were employed till now. On putting $\mathbf{I} = \hbar/2$ ($= \tfrac{1}{2}$ since $\hbar$ is put equal to 1) and averaging over all θ, (48) becomes identical with (53) except for a factor 3. This discrepancy is due to the well-known (*cf.* Bhabha, 1941, footnote on page 340; also see Bhabha and Madhava Rao, 1941) difference in the classical and the quantum average over the direction of the spin of the neutron. In fact (52) also agrees to the same extent with the corresponding differential cross-section obtained by dividing (43) by (45), expanding in powers of g_2 and retaining only the lowest term.

I am thankful to Prof. H. J. Bhabha for his criticism and advice.

Summary

The classical formulæ for the scattering of scalar mesons by a neutron are obtained taking account of the radiation damping. The neutron is assumed to possess a 'charge' and a 'dipole moment'. The scattering due to each of these is treated separately. It is found that the formulæ for the scattering due to the dipole has exactly the same form as the one obtained by Bhabha for the transverse mesons. Due to numerical factors the scattering for large energies of the incident mesons is double, and for small energies half, that of transverse vector-mesons.

The scalar and pseudo-scalar charge and dipole interactions are considered in the quantum theory. The scalar dipole interaction does not give rise to any scattering at all, the whole of the scattering being due to the pseudo-scalar interaction. In this case the quantum-theoretical formulæ agree with the corresponding classical ones if the effect of radiation reaction is neglected in the latter.

REFERENCES

Bhabha, *Proc. Roy. Soc.* A, 1939, **172**, 384–409.
—, *Ibid.*, 1941, **178**, 314–50.
— and Harish-Chandra, *Ibid.*, 1944, in course of publication.
— and Madhava Rao, *Proc. Ind. Acad. Sci.* A, 1941, **13**, 9–24.
Dirac, *Proc. Roy. Soc.* A, 1938, **167**, 148–69.
Harish-Chandra, *Ibid.*, 1944, in course of publication, referred to as **A** in this paper.

Reprinted from
Proceedings of the Indian Academy of Sciences
21 (1945), 135–146

Algebra of the Dirac-Matrices

Harish-Chandra

(*J. H. Bhabha Student, Cosmic Ray Research Unit, Indian Institute of Science, Bangalore*)

Received April 30, 1945
(Communicated by Prof. H. J. Bhabha, F.A.SC., F.R.S.)

§1. The Dirac matrices $\gamma_a(a=1,2,3,4)$ are characterised by the commutation rules

$$\gamma_a\gamma_b + \gamma_b\gamma_a = 2\delta_{ab}. \tag{1}$$

These four matrices give rise to a set of 16 quantities

$$\left.\begin{array}{ll} 1 & \\ \gamma_a & \\ i\gamma_a\gamma_b & (a \neq b) \\ i\gamma_a\gamma_b\gamma_c & (a,b,c \text{ all different}) \\ \gamma_1\gamma_2\gamma_3\gamma_4 & \end{array}\right\} \tag{2}$$

which is closed under multiplication if we regard two quantities which differ by a numerical factor -1, i or $-i$ as essentially the same. As is well known these sixteen matrices are linearly independent and apart from equivalence possess only one four-dimensional representation which is irreducible. Several important identities concerning these matrices, which are independent of any particular representation have been established by Pauli (1936) by making use of the well-known results following from Schur's theorem. The object of the present paper is to point out that the commutation rules of the above 15 quantities (all excluding 1) can be expressed quite elegantly by one single formula and that the above-mentioned identities can therefore be derived directly from the commutation rules. It seems that perhaps the present method is more general and powerful than that of Pauli. The identities are obtained in a form such that the five elements of any pentad (see Eddington, 1936) can be regarded as basal elements. The use of the matrix B of Pauli is avoided so that the identities (34_1) and (34_5) of his paper can now be generalised to the case $\phi^+ \neq \psi^+$, $\phi \neq \psi$. This was not possible previously. Some new tensor identities are also obtained.

Further the present method yields quite easily the matrix determinant of a quantity composed linearly from the above sixteen matrices. As is well known this determinant is independent of the representation used. Eddington (1936) has already calculated it by using the rather cumbersome method of employing a particular representation. As a physical application of the evaluation of this determinant we shall consider the case of a charged particle of spin $\frac{1}{2}$ having an *explicit* spin interaction with the electro-magnetic field. It will be shown that apart from quantum effects and upto the first approximation the particle behaves as if it possessed a pure magnetic dipole moment, which points either along or opposite to the magnetic field in the rest system. This magnetic moment which arises from the explicit spin interaction is to be distinguished from the usual magnetic moment of the electron which is due purely to quantum effects, and which would therefore disappear if the non-commutability of the different operators were ignored.

§2. For the purpose of this paper it is more convenient to multiply the $\dot{\gamma}_s$ by i and use

$$\mathbf{E}_a \equiv i\gamma_a \tag{3}$$

instead of γ_a. Therefore

$$\mathbf{E}_a\mathbf{E}_b + \mathbf{E}_b\mathbf{E}_a = -2\delta_{ab}. \tag{4}$$

The matrices $\mathbf{E}_1, \mathbf{E}_2, \mathbf{E}_3, \mathbf{E}_4$ anticommute and their squares are equal to -1. Put*

$$\mathbf{E}_5 \equiv i\mathbf{E}_1\mathbf{E}_2\mathbf{E}_3\mathbf{E}_4 \tag{5}$$

so that

$$\mathbf{E}_5^2 = -1, \mathbf{E}_a\mathbf{E}_5 = -\mathbf{E}_5\mathbf{E}_a \qquad (a=1,2,3,4). \tag{6}$$

Following Eddington we define

$$\mathbf{E}_{\mu\nu} \equiv \mathbf{E}_\mu\mathbf{E}_\nu \qquad\qquad (\mu,\nu=1,2,3,4,5, \mu \neq \nu)$$

$$\left.\begin{array}{l} \mathbf{E}_{0\nu} \equiv \mathbf{E}_\nu \\ \mathbf{E}_{\nu 0} \equiv -\mathbf{E}_{0\nu} = -\mathbf{E}_\nu \end{array}\right\} \qquad (\nu=1,2,\cdots,5) \tag{7}$$

$$\mathbf{E}_{\nu\nu} \equiv 0 \qquad\qquad (\nu=0,1,\cdots,5).$$

Then the following equations hold (*cf.* Eddington *l.c.*) for $\mu,\nu=0,1,\cdots,5$

$$\mathbf{E}_{\mu\nu} = -\mathbf{E}_{\nu\mu} \tag{8a}$$

$$\mathbf{E}_{\mu\nu}^2 \equiv \mathbf{E}_{\mu\nu}\mathbf{E}_{\mu\nu} = -1 \qquad (\mu \neq \nu) \tag{8b}$$

$$\mathbf{E}_{\mu\nu}\mathbf{E}_{\mu\rho} = \mathbf{E}_{\nu\rho} \qquad (\mu,\nu,\rho \text{ all different}) \tag{8c}$$

$$\mathbf{E}_{\mu\nu}\mathbf{E}_{\sigma\rho} = \mathbf{E}_{\sigma\rho}\mathbf{E}_{\mu\nu} = \mp i\mathbf{E}_{\lambda\tau} \qquad (\mu,\nu,\sigma,\rho,\lambda,\tau \text{ all different}). \tag{8d}$$

In (8d) the positive or the negative sign is to be chosen according as $(\mu,\nu,\sigma,\rho,\lambda,\tau)$ is an odd or even permutation of $(0,1,2,3,4,5)$. It is easy to see that (8) is equivalent

---

*The present definition of $\mathbf{E}_5$ differs in sign from that of Eddington. The advantage is that now

$$\mathbf{E}_5 = i\mathbf{E}_1\mathbf{E}_2\mathbf{E}_3\mathbf{E}_4 = i\gamma_1\gamma_2\gamma_3\gamma_4 = i\gamma_5$$

corresponding to (3). γ_5 is the same as in Pauli's paper.

to the single equation

$$\mathbf{E}_{\lambda\mu}\mathbf{E}_{\nu\rho} = -\delta_{\lambda\nu}\delta_{\mu\rho} + \delta_{\mu\nu}\delta_{\lambda\rho} + \mathbf{E}_{\lambda\nu}\delta_{\mu\rho} - \mathbf{E}_{\mu\nu}\delta_{\lambda\rho}$$
$$-\mathbf{E}_{\lambda\rho}\delta_{\mu\nu} + \mathbf{E}_{\mu\rho}\delta_{\lambda\nu} - \frac{i}{2}\epsilon_{\lambda\mu\nu\rho\sigma\tau}\mathbf{E}^{\sigma\tau}. \tag{9}$$

Here $\delta_{\mu\nu}$ is the usual Kronecker's symbol

$$\delta_{\mu\nu} = \begin{cases} 1 & \mu = \nu \\ 0 & \mu \neq \nu \end{cases}$$

and $\epsilon_{\lambda\mu\nu\rho\sigma\tau}$ is antisymmetric in all 6 indices and $\epsilon_{012345} = 1$. It is convenient to make the convention that the same index appearing once below and once above in the same term implies a summation. Thus for example

$$\mathbf{E}_\rho^\nu \mathbf{E}_{\nu\sigma} = \sum_{\nu=0}^{5} \mathbf{E}_{\rho\nu}\mathbf{E}_{\nu\sigma}$$

while no summation is intended in the expression

$$\mathbf{E}_{\rho\nu}\mathbf{E}_{\nu\sigma}.$$

In fact (9) can be looked upon as a tensor equation in a six-dimensional space whose metric tensor is $\delta_{\mu\nu}$. Equation (9) is invariant to any orthogonal transformation of this six-dimensional space, if we regard $\mathbf{E}_{\lambda\mu}$ as an antisymmetric tensor. Also if $\mathbf{F}_{\lambda\mu}$ is any set of fifteen E-numbers[†], antisymmetric in λ, μ and satisfying the same commutation rules as (9), then it is not difficult to prove that

$$\mathbf{F}_{\lambda\mu} = a_\lambda^\nu a_\mu^\rho \mathbf{E}_{\nu\rho}$$

where a_λ^ν are the coefficients of an orthogonal transformation of the six-dimensional space, *i.e.*,

$$\sum_\nu a_\lambda^\nu a_\mu^\nu = \delta_{\lambda\mu}$$

$$|a_\lambda^\nu| = 1$$

where $|a_\lambda^\nu|$ denotes the six-dimensional determinant of the transformation. Thus every matrix transformation

$$\mathbf{F}_{\lambda\mu} = \Lambda \mathbf{E}_{\lambda\mu} \Lambda^{-1}$$

is equivalent to an orthogonal transformation of the six-dimensional space. The converse is also true due to the equivalence of all four-dimensional representations.

For future use we note the following relations which follow directly from (9):

$$\mathbf{E}_{\lambda\mu}\mathbf{E}_{\nu\rho} + \mathbf{E}_{\lambda\nu}\mathbf{E}_{\mu\rho} = -\delta_{\lambda\nu}\delta_{\mu\rho} - \delta_{\lambda\mu}\delta_{\nu\rho} + 2\delta_{\mu\nu}\delta_{\lambda\rho} + \mathbf{E}_{\lambda\nu}\delta_{\mu\rho}$$
$$+ \mathbf{E}_{\lambda\mu}\delta_{\nu\rho} - 2\mathbf{E}_{\lambda\rho}\delta_{\mu\nu} + \mathbf{E}_{\mu\rho}\delta_{\lambda\nu} + \mathbf{E}_{\nu\rho}\delta_{\lambda\mu} \tag{10a}$$

$$\mathbf{E}_\sigma^\nu \mathbf{E}_{\nu\rho} = 5\delta_{\sigma\rho} - 4\mathbf{E}_{\sigma\rho}. \tag{10b}$$

[†]Any linear combination of the E's and 1 is called an E-number (*cf.* Eddington, *l.c*). Every matrix with 4 rows and 4 columns is an E-number due to the linear independence of the E's and 1.

Also if

$$\mathbf{S} = s + s_{\lambda\mu}\mathbf{E}^{\lambda\mu} \qquad (s_{\lambda\mu} = -s_{\mu\lambda}) \tag{11a}$$

$$\mathbf{T} = t + t_{\lambda\mu}\mathbf{E}^{\lambda\mu} \qquad (t_{\lambda\mu} = -t_{\mu\lambda}) \tag{11b}$$

where s, $s_{\lambda\mu}$, t, $t_{\lambda\mu}$ are ordinary numbers it follows from (9) that

$$\mathbf{ST} = st - 2s_{\lambda\mu}t^{\lambda\mu} + \left(st_{\lambda\mu} + ts_{\lambda\mu} - 4s_{\lambda\nu}t_{\mu}^{\nu} - \frac{i}{2}s^{\alpha\beta}t^{\gamma\delta}\epsilon_{\alpha\beta\gamma\delta\lambda\mu}\right)\mathbf{E}^{\lambda\mu} \tag{12}$$

so that

$$\mathbf{ST} - \mathbf{TS} = -8s_{\lambda\nu}t_{\mu}^{\nu}\mathbf{E}^{\lambda\mu} \tag{13a}$$

$$\mathbf{ST} + \mathbf{TS} = 2st - 4s_{\lambda\mu}t^{\lambda\mu} + 2\left(st_{\lambda\mu} + ts_{\lambda\mu} - \frac{i}{2}s^{\alpha\beta}t^{\gamma\delta}\epsilon_{\alpha\beta\gamma\delta\lambda\mu}\right)\mathbf{E}^{\lambda\mu}. \tag{13b}$$

In particular on choosing $\mathbf{S} = \mathbf{E}_{\alpha\beta}$, (13) gives

$$\mathbf{TE}_{\alpha\beta} - \mathbf{E}_{\alpha\beta}\mathbf{T} = 4\left(t_{\alpha\lambda}\mathbf{E}_{\beta}^{\lambda} - t_{\beta\lambda}\mathbf{E}_{\alpha}^{\lambda}\right) \tag{14a}$$

$$\mathbf{TE}_{\alpha\beta} + \mathbf{E}_{\alpha\beta}\mathbf{T} = -4t_{\alpha\beta} + 2t\mathbf{E}_{\alpha\beta} - it^{\gamma\delta}\epsilon_{\alpha\beta\gamma\delta\lambda\rho}\mathbf{E}^{\lambda\rho}. \tag{14b}$$

On contracting with $t^{\lambda\mu}$ (10a) yields

$$\mathbf{TE}_{\nu\rho} + t^{\lambda\mu}\mathbf{E}_{\lambda\nu}\mathbf{E}_{\mu\rho} = -3t_{\nu\rho} + 3t_{\nu}^{\lambda}\mathbf{E}_{\lambda\rho} - t_{\rho}^{\lambda}\mathbf{E}_{\lambda\nu} + \mathbf{T}\delta_{\nu\rho} + \left(\mathbf{E}_{\nu\rho} - \delta_{\nu\rho}\right)t. \tag{15}$$

Multiplying (14a) by $\mathbf{E}\beta_{\gamma}$ on the right and using (15) we obtain

$$\mathbf{E}_{\alpha\beta}\mathbf{TE}_{\gamma}^{\beta} = \mathbf{T}\delta_{\alpha\gamma} + 4(\delta_{\alpha\gamma} - \mathbf{E}_{\alpha\gamma})t + 4\left(t_{\alpha}^{\beta}\mathbf{E}_{\beta\gamma} + t_{\gamma}^{\beta}\mathbf{E}_{\beta\alpha}\right) - 8t_{\alpha\gamma}. \tag{16}$$

On contracting α, γ (16) gives the well-known result

$$2\mathbf{T} - \mathbf{E}^{\alpha\beta}\mathbf{TE}_{\alpha\beta} = 32t.$$

Multiplying (14a) by $\mathbf{E}_{\alpha\beta}$ on the right we get

$$\mathbf{E}_{\alpha\beta}\mathbf{TE}_{\alpha\beta} = -\mathbf{T} + 4t_{\lambda\alpha}\mathbf{E}_{\alpha}^{\lambda} + 4t_{\lambda\beta}\mathbf{E}_{\beta}^{\lambda} - 8t_{\alpha\rho}\mathbf{E}_{\alpha\beta} \qquad (\alpha \neq \beta). \tag{17}$$

As is well known (*cf.* Pauli, 1936) it follows from the commutation rules (9) that the spur of $\mathbf{E}_{\lambda\mu}$ is zero. Therefore for any four-rowed representation

$$sp(\mathbf{T}) = 4t$$

$$sp(\mathbf{E}_{\lambda\mu}\mathbf{T}) = -8t_{\lambda\mu}.$$

Also since 1, $\mathbf{E}_{\lambda\mu}$ are 16 linearly independent matrices $\mathbf{T}$ in (11b) can be any arbitrary matrix of 4 rows and 4 columns. Let ψ and ϕ be any two matrices with 4 rows and 1 column and ψ^{+} and ϕ^{+} with 1 row and 4 columns. Then $\phi\phi^{+}$, $\psi\psi^{+}$, $\phi\psi^{+}$ and $\psi\phi^{+}$ are square matrices with 4 rows and 4 columns and we can choose $\mathbf{T}$ equal to any of them. We notice that for $\mathbf{T} = \phi\phi^{+}$

$$t = \tfrac{1}{4}sp(\mathbf{T}) = \tfrac{1}{4}\phi^{+}\phi \tag{18a}$$

$$t_{\lambda\mu} = -\tfrac{1}{8}sp(\mathbf{E}_{\lambda\mu}\mathbf{T}) = -\tfrac{1}{8}\phi^{+}\mathbf{E}_{\lambda\mu}\phi. \tag{18b}$$

Substituting these values in (11b) and (14b) and multiplying by ψ^{+} on the left and ψ

on the right we obtain

$$\psi^+\phi \cdot \phi^+\psi = \tfrac{1}{4}\psi^+\psi \cdot \phi^+\phi - \tfrac{1}{8}\psi^+ \mathbf{E}^{\lambda\mu}\psi \cdot \phi^+ \mathbf{E}_{\lambda\mu}\phi \qquad (19)$$

$$\psi^+\phi \cdot \phi^+ \mathbf{E}_{\alpha\beta}\psi + \psi^+ \mathbf{E}_{\alpha\beta}\phi \cdot \phi^+\psi = \tfrac{1}{2}\psi^+\psi \cdot \phi^+ \mathbf{E}_{\alpha\beta}\phi + \tfrac{1}{2}\psi^+ \mathbf{E}_{\alpha\beta}\psi \cdot \phi^+\phi$$

$$+ \frac{i}{8}\phi^+ \mathbf{E}^{\gamma\delta}\phi \cdot \psi^+ \mathbf{E}^{\lambda\rho}\psi \cdot \epsilon_{\alpha\beta\gamma\delta\lambda\rho}. \qquad (20)$$

On choosing $\mathbf{T} = \psi\phi^+$ (16) gives in the same way

$$\psi^+ \mathbf{E}_{\alpha\beta}\psi \cdot \phi^+ \mathbf{E}_\alpha^\beta\phi = \psi^+\psi \cdot \phi^+\phi + \psi^+\phi \cdot \phi^+\psi - \psi^+ \mathbf{E}_{\alpha\beta}\phi \cdot \phi^+ \mathbf{E}_\alpha^\beta\psi \qquad (21)$$

$$\psi^+ \mathbf{E}_{\alpha\beta}\psi \cdot \phi^+ \mathbf{E}_\gamma^\beta\phi = - \psi^+ \mathbf{E}_{\alpha\gamma}\phi \cdot \phi^+\psi + \psi^+\phi \cdot \phi^+ \mathbf{E}_{\alpha\gamma}\psi$$

$$- \tfrac{1}{2}\psi^+ \mathbf{E}_{\beta\gamma}\phi \cdot \phi^+ \mathbf{E}_\alpha^\beta\psi + \tfrac{1}{2}\psi^+ \mathbf{E}_{\beta\alpha}\phi \cdot \phi^+ \mathbf{E}_\gamma^\beta\psi, \qquad (\alpha \neq \gamma). \qquad (22)$$

Similarly on putting $\mathbf{T} = \phi\phi^+$ (17) gives

$$\psi^+ \mathbf{E}_{\alpha\beta}\phi \cdot \phi^+ \mathbf{E}_{\alpha\beta}\psi = - \psi^+\phi \cdot \phi^+\psi - \tfrac{1}{2}\phi^+ \mathbf{E}_{\lambda\alpha}\phi \cdot \psi^+ \mathbf{E}_\alpha^\lambda\psi - \tfrac{1}{2}\phi^+ \mathbf{E}_{\lambda\beta}\phi \cdot \psi^+ \mathbf{E}_\beta^\lambda\psi$$

$$+ \psi^+ \mathbf{E}_{\alpha\beta}\psi \cdot \phi^+ \mathbf{E}_{\alpha\beta}\phi, \qquad (\alpha \neq \beta). \qquad (23)$$

Now let a, b, c, d be indices which run from 1 to 4 only. Then following Pauli (1936) we put

$$\left.\begin{aligned}
\psi^+\psi &\equiv i\Omega_0 \\
\psi^+ \mathbf{E}_{0a}\psi = \psi^+ \mathbf{E}_a\psi &= i\psi^+\gamma_a\psi \equiv iS_a \\
\psi^+ \mathbf{E}_{a5}\psi = - \psi^+\gamma_a\gamma_5\psi &= i\psi^+\hat{\gamma}_a\psi \equiv i\hat{S}_a \\
\psi^+ \mathbf{E}_{ab}\psi = - \psi^+\gamma_a\gamma_b\psi &\equiv \mathbf{M}_{ab} \qquad (a \neq b) \\
\mathbf{M}_{aa} &= 0 \\
\psi^+ \mathbf{E}_{05}\psi = i\psi^+\gamma_5\psi &\equiv i\Omega_5 \\
\hat{\mathbf{M}}_{ab} &= \tfrac{1}{2}\epsilon_{abcd}\mathbf{M}^{cd}.
\end{aligned}\right\} \qquad (24)$$

Here

$$\gamma_5 \equiv \gamma_1\gamma_2\gamma_3\gamma_4 = \mathbf{E}_1\mathbf{E}_2\mathbf{E}_3\mathbf{E}_4 = - i\mathbf{E}_5$$

and

$$\hat{\gamma}_a \equiv i\gamma_a\gamma_5 = \frac{i}{3!}\epsilon_{abcd}\gamma^b\gamma^c\gamma^d$$

where the tensor ϵ_{abcd} is antisymmetric in all the four indices and $\epsilon_{1234} = 1$. If the corresponding quantities constructed from ϕ^+, ϕ be distinguished by a dash, (19) can be written in the following form:

$$\psi^+\phi \cdot \phi^+\psi = - \tfrac{1}{4}\Omega_0\Omega_0' + \tfrac{1}{4}\mathbf{S}_a\mathbf{S}'^a + \tfrac{1}{4}\hat{\mathbf{S}}_a\hat{\mathbf{S}}'^a - \tfrac{1}{8}\mathbf{M}_{ab}\mathbf{M}'^{ab} + \tfrac{1}{4}\Omega_5\Omega_5'. \qquad (25)$$

Choosing $\alpha = 0$, $\beta = 5$ in (20) we get

$$i\psi^+\phi \cdot \phi^+\gamma_5\psi + i\psi^+\gamma_5\phi \cdot \phi^+\psi = - \tfrac{1}{2}\Omega_0\Omega_5' - \tfrac{1}{2}\Omega_5\Omega_0' + \frac{i}{4}\mathbf{M}^{ab}\hat{\mathbf{M}}'_{ab}. \qquad (26)$$

Similarly putting $\alpha = 0$ in (21) we obtain

$$\mathbf{S}_a\mathbf{S}'^a + \Omega_5\Omega_5' = -\Omega_0\Omega_0' + \psi^+\phi\cdot\phi^+\psi - \psi^+\gamma_a\phi\cdot\phi^+\gamma_a\psi - \psi^+\gamma_5\phi\cdot\phi^+\gamma_5\psi. \quad (27)$$

On the other hand we get on taking $\alpha = 5$

$$\hat{\mathbf{S}}_a\hat{\mathbf{S}}'^a + \Omega_5\Omega_5' = -\Omega_0\Omega_0' + \psi^+\phi\cdot\phi^+\psi - \psi^+\hat{\gamma}^a\phi\cdot\phi^+\hat{\gamma}_a\psi - \psi^+\gamma_5\phi\cdot\phi^+\gamma_5\psi. \quad (28)$$

On putting $\alpha = 0$, $\gamma = 5$ (22) gives

$$-\mathbf{S}_a\hat{\mathbf{S}}'^a = -i\psi^+\gamma_5\phi\cdot\phi^+\psi + i\phi^+\gamma_5\psi\cdot\psi^+\phi$$
$$+\tfrac{1}{2}\phi^+\gamma_a\psi\cdot\psi^+\hat{\gamma}^a\phi + \tfrac{1}{2}\psi^+\gamma_a\phi\cdot\phi^+\hat{\gamma}^a\psi. \quad (29)$$

Also if we put $\alpha = 0$, $\beta = 5$ in (23) we get

$$-\psi^+\gamma_5\phi\cdot\phi^+\gamma_5\psi = -\phi^+\psi\cdot\psi^+\phi + \tfrac{1}{2}\mathbf{S}_a\mathbf{S}'^a + \tfrac{1}{2}\hat{\mathbf{S}}_a\hat{\mathbf{S}}'^a \quad (30)$$

which is the same as equation (44P).* From (30) and (25) we get

$$-\psi^+\gamma_5\phi\cdot\phi^+\gamma_5\psi - \psi^+\phi\cdot\phi^+\psi = \tfrac{1}{2}\Omega_0\Omega_0' + \tfrac{1}{4}\mathbf{M}_{ab}\mathbf{M}'^{ab} - \tfrac{1}{2}\Omega_5\Omega_5' \quad (31)$$

which corresponds to equation (43P). Equations (26), (27), (28), (29) and (31) can be written in the following form:

$$\tfrac{1}{4}\mathbf{M}^{ab}\hat{\mathbf{M}}'_{ab} - \frac{i}{2}(\Omega_0\Omega_5' + \Omega_5\Omega_0') = (\psi^+\phi\cdot\phi^+\gamma_5\psi - \phi^+\phi\cdot\psi^+\gamma_5\psi)$$
$$+ (\phi^+\psi\cdot\psi^+\gamma_5\phi - \psi^+\psi\cdot\phi^+\gamma_5\phi) \quad (32)$$

$$2\mathbf{S}_a\mathbf{S}^{a'} = -2\Omega_0\Omega_0' - 2\Omega_5\Omega_5' - (\psi^+\gamma_a\phi\cdot\phi^+\gamma^a\psi - \psi^+\gamma_a\psi\cdot\phi^+\gamma^a\phi)$$
$$- (\psi^+\gamma_5\phi\cdot\phi^+\gamma_5\psi - \psi^+\gamma_5\psi\cdot\phi^+\gamma_5\phi) + (\psi^+\phi\phi^+\psi - \psi^+\psi\phi^+\phi) \quad (33)$$

$$2\hat{\mathbf{S}}_a\hat{\mathbf{S}}'^a = -2\Omega_0\Omega_0' - 2\Omega_5\Omega_5' - (\psi^+\hat{\gamma}_a\phi\cdot\phi^+\hat{\gamma}^a\psi - \psi^+\hat{\gamma}_a\psi\cdot\phi^+\hat{\gamma}^a\phi)$$
$$- (\psi^+\gamma_5\phi\cdot\phi^+\gamma_5\psi - \psi^+\gamma_5\psi\cdot\phi^+\gamma_5\phi) + (\psi^+\phi\cdot\phi^+\psi - \psi^+\psi\cdot\phi^+\phi) \quad (34)$$

$$-2\mathbf{S}_a\hat{\mathbf{S}}'^a = i\{\phi^+\gamma_5\psi\cdot\psi^+\phi - \psi^+\gamma_5\phi\cdot\phi^+\psi\}$$
$$+ \tfrac{1}{2}\{\phi^+\gamma_a\psi\cdot\psi^+\hat{\gamma}^a\phi + \phi^+\hat{\gamma}_a\psi\cdot\psi^+\gamma^a\phi - 2\phi^+\hat{\gamma}_a\phi\cdot\psi^+\gamma^a\psi\} \quad (35)$$

$$\tfrac{1}{2}\mathbf{M}_{ab}\mathbf{M}'^{ab} = \Omega_0\Omega_0' - \Omega_5\Omega_5' - 2(\psi^+\phi\cdot\phi^+\psi - \psi^+\psi\cdot\phi^+\phi)$$
$$- 2(\psi^+\gamma_5\phi\cdot\phi^+\gamma_5\psi - \phi^+\gamma_5\phi\psi^+\gamma_5\psi). \quad (36)$$

Equations (32) to (36) are the generalisation of equations (34_4P), (34_1P), (34_3P), (34_5P) and (34_2P) respectively. Equations (32) and (36) have already been given by Pauli as equations (47P) and (43P). The others were not obtained by him. It is noteworthy that we have derived the identities directly without employing the matrix B of Pauli. One more interesting identity can be derived by interchanging ϕ^+ and ψ^+ in (23) and putting $\alpha = 0$, $\beta = 5$

$$\tfrac{1}{2}\phi^+\gamma_a\psi\cdot\psi^+\gamma^a\phi + \tfrac{1}{2}\phi^+\hat{\gamma}_a\psi\cdot\psi^+\hat{\gamma}^a\phi = -\Omega_5\Omega_5' - \Omega_0\Omega_0'.$$

However on putting $\phi^+ = \psi^+$ and $\phi = \psi$ it degenerates merely into the sum of (34_1P) and (34_3P).

---

*P refers to Pauli's paper.

It may be mentioned here that it is possible to derive *tensor* identities in addition to the *invariant* identities given above, by choosing other suitable sets of values for α, β and γ in the equations (20) to (22). On putting $\phi = \psi$ and $\phi^+ = \psi^+$ the following identities are obtained:

$$\hat{\mathbf{M}}_{ab}\hat{\mathbf{S}}^b - \Omega_0 \mathbf{S}_a = 0 \qquad [\alpha = 0, \beta = a \text{ in } (20)] \tag{37a}$$

$$\hat{\mathbf{M}}_{ab}\mathbf{S}^b + \Omega_0 \hat{\mathbf{S}}_a = 0 \qquad [\alpha = a, \beta = 5 \text{ in } (20)] \tag{37b}$$

$$\mathbf{M}_{ab}\Omega_5 + i\Omega_0 \hat{\mathbf{M}}_{ab} + i\left(\mathbf{S}_a\hat{\mathbf{S}}_b - \mathbf{S}_b\hat{\mathbf{S}}_a\right) = 0 \qquad [\alpha = a, \beta = b \text{ in } (20)] \tag{37c}$$

$$\mathbf{M}_{ab}\mathbf{S}^b + i\Omega_5\hat{\mathbf{S}}_a = 0 \qquad [\alpha = 0, \gamma = a \text{ in } (22)] \tag{*37d}$$

$$\mathbf{M}_{ab}\hat{\mathbf{S}}^b - i\Omega_5\mathbf{S}_a = 0 \qquad [\alpha = a, \gamma = 5 \text{ in } (22)] \tag{37e}$$

$$\mathbf{S}_a\mathbf{S}_b + \hat{\mathbf{S}}_a\hat{\mathbf{S}}_b + \mathbf{M}_{ac}\mathbf{M}_b^c = -\delta_{ab}\Omega_0^2 \qquad [\alpha = a, \beta = b \text{ in } (21) \text{ or } (22)]. \tag{37f}$$

The generalised identities for the case $\phi \neq \psi$ and $\phi^+ \neq \psi^+$ can be obtained by similar substitutions and they need not be given here explicitly.

§3. Now we shall calculate the matrix determinant of $\mathbf{T}$. As is well known this determinant, which we denote by det. $\mathbf{T}$ is the same for all four-dimensional representations of $\mathbf{E}_{\lambda\mu}$. In fact it is equal to the independent term in the characteristic equation of $\mathbf{T}$. It is therefore sufficient to determine the characteristic equation. For this purpose we make use of (12) and find that

$$\mathbf{T} - t = t_{\lambda\mu}\mathbf{E}^{\lambda\mu}$$

$$(\mathbf{T} - t)^2 = -2t_{\mu\nu}t^{\mu\nu} - \frac{i}{2}t^{\alpha\beta}t^{\gamma\delta}\epsilon_{\alpha\beta\gamma\delta\lambda\rho}\mathbf{E}^{\lambda\rho}$$

$$\left\{(\mathbf{T} - t)^2 + 2t_{\mu\nu}t^{\mu\nu}\right\}^2 - \tfrac{1}{2}t^{\alpha\beta}t^{\gamma\delta}\epsilon_{\alpha\beta\gamma\delta\lambda\rho}\epsilon^{\lambda\rho\alpha'\beta'\gamma'\delta'}t_{\alpha'\beta'}t_{\gamma'\delta'}$$

$$= \frac{i}{8}\epsilon^{\alpha\beta\mu\nu\sigma\tau}t_{\mu\nu}t_{\sigma\tau}\epsilon^{\gamma\delta\mu'\nu'\sigma'\tau'}t_{\mu'\nu'}t_{\sigma'\tau'}\epsilon_{\alpha\beta\gamma\delta\lambda\rho}\mathbf{E}^{\lambda\rho}$$

$$= \frac{i}{8}\epsilon^{\alpha\beta\mu\nu\sigma\tau}t_{\mu\nu}t_{\sigma\tau}16\left(t_{\alpha\beta}t_{\lambda\rho} - t_{\lambda\beta}t_{\alpha\rho} - t_{\alpha\lambda}t_{\beta\rho}\right)\mathbf{E}^{\lambda\rho}$$

$$= 2i\epsilon^{\alpha\beta\mu\nu\sigma\tau}t_{\mu\nu}t_{\sigma\tau}\left(t_{\alpha\beta}t_{\lambda\rho}\mathbf{E}^{\lambda\rho} - 2t_{\alpha\lambda}t_{\beta\rho}\mathbf{E}^{\lambda\rho}\right)$$

$$= 2i\epsilon^{\alpha\beta\mu\nu\sigma\tau}t_{\alpha\beta}t_{\mu\nu}t_{\sigma\tau}(\mathbf{T} - t) - 4it_{\alpha\lambda}t_{\beta\rho}\epsilon^{\alpha\beta\mu\nu\sigma\tau}t_{\mu\nu}t_{\sigma\tau}\mathbf{E}^{\lambda\rho}. \tag{38}$$

Now

$$t_{\alpha\lambda}t_{\beta\rho}\epsilon^{\alpha\beta\mu\nu\sigma\tau}t_{\mu\nu}t_{\sigma\tau} = \tfrac{1}{6}t_{\lambda\rho}\epsilon^{\alpha\beta\mu\nu\sigma\tau}t_{\alpha\beta}t_{\mu\nu}t_{\sigma\tau}. \tag{39}$$

(39) is easily verified for a 'tensor' $t_{\mu\nu}$ whose only non-vanishing components are t_{01}, t_{23}, t_{45}. Since by a suitable orthogonal transformation every antisymmetrical tensor $t_{\mu\nu}$ can be brought into this form, it follows that the invariant equation (39) holds for every $t_{\mu\nu}$. Also

$$\epsilon_{\alpha\beta\gamma\delta\lambda\rho}\epsilon^{\lambda\rho\alpha'\beta'\gamma'\delta'}t_{\alpha'\beta'}t_{\gamma'\delta'} = 16\left(t_{\alpha\beta}t_{\gamma\delta} - t_{\alpha\gamma}t_{\beta\delta} - t_{\alpha\delta}t_{\gamma\beta}\right). \tag{40}$$

---

*This identity was mentioned by Prof. Bhabha in a lecture.

Substituting (39) and (40) in (38) we get

$$\left\{ (T-t)^2 + 2t_{\mu\nu}t^{\mu\nu} \right\}^2 - 8\left\{ \left(t^{\alpha\beta}t_{\alpha\beta}\right)^2 - 2t^{\alpha\beta}t_{\beta\gamma}t^{\gamma\delta}t_{\delta\alpha} \right\}$$
$$- \tfrac{4}{8}it_{\alpha\beta}t_{\gamma\delta}t_{\lambda\rho}\epsilon^{\alpha\beta\gamma\delta\lambda\rho}(T-t) = 0. \tag{41}$$

Equation (41) is the characteristic equation of T. Putting $T = 0$ on the left side of (41) we obtain the term independent of T so that

$$\det T = \left(2t_{\mu\nu}t^{\mu\nu} + t^2\right)^2 - 8\left\{ \left(t^{\alpha\beta}t_{\alpha\beta}\right)^2 - 2\left(t^{\alpha\beta}t_{\beta\gamma}t^{\gamma\delta}t_{\delta\alpha}\right)\right\}$$
$$+ \tfrac{4}{3}it_{\alpha\beta}t_{\gamma\delta}t_{\lambda\rho}\epsilon^{\alpha\beta\gamma\delta\lambda\rho}t. \tag{42}$$

(42) agrees with the result given by Eddington* (1936). For the purpose of the following discussion it is convenient to replace T by T' where

$$T' = t + \tfrac{1}{2}t_{\lambda\mu}E^{\lambda\mu} \tag{43}$$

so that

$$\det T' = \left(t^2 + \tfrac{1}{2}t_{\mu\nu}t^{\mu\nu}\right)^2 - \tfrac{1}{2}\left\{ \left(t^{\alpha\beta}t_{\alpha\beta}\right)^2 - 2t^{\alpha\beta}t_{\beta\gamma}t^{\gamma\delta}t_{\delta\alpha} \right\}$$
$$+ \frac{i}{6}t_{\alpha\beta}t_{\gamma\delta}t_{\lambda\rho}\epsilon^{\alpha\beta\gamma\delta\lambda\rho}t. \tag{44}$$

Let us now revert from $E^{\lambda\mu}$ to the original matrices E_a and their products. Obviously T' can be put in the form

$$T' = t + t_a E^a + \tfrac{1}{2}t_{ab}E^{ab} + \frac{i}{6}\epsilon_{abcd}E^a E^b E^c s^d$$
$$+ \frac{1}{24}s\epsilon_{abcd}E^a E^b E^c E^d. \tag{45}$$

Here a, b, c, d run from 1 to 4 only and t_a and s_a are four-dimensional vectors while s is a scalar. From (5) we have

$$\frac{1}{24}\epsilon_{abcd}E^a E^b E^c E^d = -iE_5 = -iE_{05} \tag{46a}$$

$$\tfrac{1}{6}\epsilon_{abcd}E^a E^b E^c = iE_5 E_d = -iE_{d5}. \tag{46b}$$

On comparing (43) and (45) and using (46) we find

$$\left.\begin{aligned} t_{0a} &= t_a \\ t_{a5} &= s_a \\ t_{05} &= s. \end{aligned}\right\} \tag{47}$$

For brevity the following notation for any vector $\mathbf{A}_a$ or a tensor $\mathbf{B}_{ab}$ is introduced

$$|\mathbf{A}_a|^2 \equiv \mathbf{A}_a\mathbf{A}^a \tag{48a}$$

$$|\mathbf{B}_{ab}|^2 \equiv \tfrac{1}{2}\mathbf{B}_{ab}\mathbf{B}^{ab}. \tag{48b}$$

---

*Eddington has chosen $\mathbf{E}_5 = -i\mathbf{E}_1\mathbf{E}_2\mathbf{E}_3\mathbf{E}_4$ and therefore in his case the sign of i in (42) is reversed.

Then

$$\tfrac{1}{2} t^{\mu\nu} t_{\mu\nu} = t_{0a} t^{0a} + |t_{ab}|^2 + t_{05} t^{05} + t_{a5} t^{a5}$$

$$= |t_a|^2 + |t_{ab}|^2 + |s_a|^2 + s^2. \tag{49}$$

Also put

$$(t^2)_{\alpha\gamma} = (t^2)_{\gamma\alpha} \equiv t_\alpha^\beta t_{\beta\gamma} = t_\alpha^0 t_{0\gamma} + t_\alpha^a t_{a\gamma} + t_\alpha^5 t_{5\gamma}$$

so that

$$(t^2)_{00} = -s^2 - |t_a|^2 \tag{50a}$$

$$(t^2)_{0b} = t^a t_{ab} - ss_b \tag{50b}$$

$$(t^2)_{ab} = -t_a t_b + t_a^c t_{cb} - s_a s_b \tag{50c}$$

$$(t^2)_{a5} = -st_a + t_a^b s_b \tag{50d}$$

$$(t^2)_{05} = t^a s_a \tag{50e}$$

$$(t^2)_{55} = -s^2 - |s_a|^2. \tag{50f}$$

Thus we have

$$t_{\alpha\beta} t^{\beta\gamma} t_{\gamma\delta} t^{\delta\alpha} = (t^2)_{\alpha\gamma} (t^2)^{\alpha\gamma}$$

$$= (t^2)_{00}(t^2)^{00} + (t^2)_{55}(t^2)^{55} + 2(t^2)_{05}(t^2)^{05} + 2(t^2)_{0a}(t^2)^{0a}$$

$$+ 2(t^2)_{a5}(t^2)^{a5} + (t^2)_{ab}(t^2)^{ab}$$

$$= \left(s^2 + |t_a|^2\right)^2 + \left(s^2 + |s_a|^2\right)^2 + 2(t^a s_a)^2 + 2|t^a t_{ab} - ss_b|^2$$

$$+ 2|t_a^b s_b - st_a|^2 + 2|t_a t_b + s_a s_b - t_a^c t_{cb}|^2. \tag{51}$$

From (44), (47), (49) and (51) we have

$$\det. \mathbf{T}' = \left(t^2 + |t_a|^2 + |t_{ab}|^2 + |s_a|^2 + s^2\right)^2 - 2\left(|t_a|^2 + |t_{ab}|^2 + |s_a|^2 + s^2\right)^2$$

$$+ \left(s^2 + |t_a|^2\right)^2 + \left(s^2 + |s_a|^2\right)^2 + 2(t^a s_a)^2 + 2|t^a t_{ab} - ss_b|^2$$

$$+ 2|t_a^b s_b - st_a|^2 + 2|t_a t_b + s_a s_b - t_a^c t_{cb}|^2$$

$$+ ist_{ab} t_{cd} \epsilon^{abcd} t + 4it_a t_{bc} s_d \epsilon^{abcd} t \tag{52}$$

since

$$t_{\alpha\beta} t_{\gamma\delta} t_{\lambda\rho} \epsilon^{\alpha\beta\gamma\delta\lambda\rho} = 6t_{05} t_{ab} t_{cd} \epsilon^{05abcd} + 24t_{0a} t_{bc} t_{d5} \epsilon^{0abcd5}.$$

Instead of $\mathbf{E}_a$ or γ_a which satisfy (4) or (1) respectively it is convenient to introduce the matrices $\alpha_\mu (\mu = 0, 1, 2, 3)$ whose commutation rules are

$$\alpha_\mu \alpha_\nu + \alpha_\nu \alpha_\mu = 2g_{\mu\nu} \tag{53}$$

where $g_{\mu\nu}$ is the usual metric tensor of flat space-time ($g_{\mu\nu} = 0$, $\mu \neq \nu$, $g_{00} = -g_{11} = -g_{22} = -g_{33} = 1$). For this purpose it is sufficient to choose $(\alpha_1, \alpha_2, \alpha_3) = (\mathbf{E}_1, \mathbf{E}_2, \mathbf{E}_3)$ and $\alpha_0 = -i\mathbf{E}_4 = \gamma_4$. From now onwards the Greek indices run from

0 to 3. We make the convention that for any tensor $\mathbf{A}_{ab\cdots}$

$$i^n \mathbf{A}_{(0)n} = \mathbf{A}_{(4)n} \tag{54}$$

where $\mathbf{A}_{(4)n}$ denotes that among $a, b, \cdots$ there are n indices equal to 4. The new quantity $\mathbf{A}_{(0)n}$ defined by (54) is obtained by replacing each index 4 by 0. Since for raising the Greek indices we use the tensor $g^{\mu\nu}$ the following relation holds

$$\mathbf{A}_{\mu\nu}\cdots\mathbf{B}^{\mu\nu}\cdots = (-1)^m \mathbf{A}_{ab}\cdots\mathbf{B}^{ab}\cdots$$

where m is the total number of indices $a, b, \cdots$ in each of the tensors $\mathbf{A}_{ab}\cdots$ and $\mathbf{B}^{ab}\cdots$ which are contracted together. Also from (54)

$$\epsilon_{0123} = -\epsilon_{1230} = i\epsilon_{1234} = i.$$

Therefore we put

$$\epsilon_{\mu\nu\sigma\tau} \equiv i\eta_{\mu\nu\sigma\tau}$$

where $\eta_{0123} = 1$. Thus (45) and (52) can be written as

$$\mathbf{T}' = t - t_\mu\alpha^\mu + \tfrac{1}{2}t_{\mu\nu}\alpha^\mu\alpha^\nu - \tfrac{1}{6}\eta_{\mu\nu\sigma\tau}\alpha^\mu\alpha^\nu\alpha^\sigma s^\tau - \frac{1}{24}s\eta_{\mu\nu\sigma\tau}\alpha^\mu\sigma^\nu\alpha^\sigma\alpha^\tau \tag{55}$$

$$\det\mathbf{T}' = \left(t^2 - |t_\mu|^2 + |t_{\mu\nu}|^2 - |s_\mu|^2 + s^2\right)^2 - 2\left(-|t_\mu|^2 + |t_{\mu\nu}|^2 - |s_\mu|^2 + s^2\right)^2$$

$$+ \left(s^2 - |t_\mu|^2\right)^2 + \left(s^2 - |s_\mu|^2\right)^2 + 2\left(t^\mu s_\mu\right)^2 - 2|ss_\mu + t^\nu t_{\nu\mu}|^2$$

$$- 2|st_\mu + t_\mu^\nu s_\nu|^2 + 2|t_\mu t_\nu + s_\mu s_\nu + t_\mu^\sigma t_{\sigma\nu}|^2$$

$$- st_{\mu\nu}t_{\sigma\tau}\eta^{\mu\nu\sigma\tau}t - 4t_\mu t_{\nu\sigma}s_\tau\eta^{\mu\nu\sigma\tau}t. \tag{56}$$

A notation similar to (48) has been employed in (56) for Greek indices also.

The above result can immediately be applied to discuss the case of a particle of spin $\tfrac{1}{2}$ having a charge- and dipole-interaction with an electro-magnetic (or meson) field. If p_μ be the energy-momentum vector of the particle, ϕ_μ the electromagnetic potentials and $\mathbf{F}_{\mu\nu}$ the field-strengths the wave equation for the particle can be written as

$$\left\{\alpha^\mu\left(p_\mu - g_1\phi_\mu\right) + \frac{i}{2}g_2\alpha^\mu\alpha^\nu\mathbf{F}_{\mu\nu} + m\right\}\psi = 0. \tag{57}$$

Here g_1 is the charge and g_2 the dipole-strength and m the mass of the particle. Our object is to determine the classical analogue of this particle. Hence we treat p_μ, ϕ_μ, $\mathbf{F}_{\mu\nu}$ as numerical quantities commuting with each other. This corresponds to the neglect of quantum effects. As is well known the condition for the existence of a solution of (57) is that

$$\det\left\{\alpha^\mu\pi_\mu + \frac{i}{2}g\alpha^\mu\alpha^\nu\mathbf{F}_{\mu\nu} + m\right\} = 0. \tag{58}$$

Here $\pi_\mu = p_\mu - g_1\phi_\mu$ and we have written g instead of g_2 for simplicity. (58) corresponds to the classical equation of motion of the particle. Comparing (58) and

(55) we find

$$t = m$$
$$t_\mu = -\pi_\mu$$
$$t_{\mu\nu} = ig\mathbf{F}_{\mu\nu}$$
$$s_\mu = 0, \quad s = 0$$

so that on account of (56), (58) reduces to

$$\left(\pi_\mu\pi^\mu - m^2 + \tfrac{1}{2}g^2\mathbf{F}_{\mu\nu}\mathbf{F}^{\mu\nu}\right)^2 - 4\pi_\mu\pi^\mu\left(\tfrac{1}{2}g^2\mathbf{F}_{\nu\sigma}\mathbf{F}^{\nu\sigma}\right) - 2\left(\tfrac{1}{2}g^2\mathbf{F}_{\mu\nu}\mathbf{F}^{\mu\nu}\right)^2$$
$$-4g^2\pi_\mu\mathbf{F}^{\mu\nu}\mathbf{F}_{\nu\sigma}\pi^\sigma + g^4\mathbf{F}^\nu_\mu\mathbf{F}_{\nu\sigma}\mathbf{F}^{\sigma\tau}\mathbf{F}^\mu_\tau = 0. \tag{59}$$

Now

$$\mathbf{F}_{\mu\nu}\mathbf{F}^{\nu\sigma}\mathbf{F}_{\sigma\tau}\mathbf{F}^{\tau\mu} = 2\left(\tfrac{1}{2}\mathbf{F}_{\mu\nu}\mathbf{F}^{\mu\nu}\right)^2 + \left(\tfrac{1}{2}\mathbf{F}_{\mu\nu}\mathbf{F}^{*\mu\nu}\right)^2 \tag{60}$$

where $\mathbf{F}^{*\mu\nu}$ is the tensor dual to $\mathbf{F}^{\mu\nu}$ and is defined by

$$\mathbf{F}^{*\mu\nu} = \tfrac{1}{2}\eta^{\mu\nu\sigma\tau}\mathbf{F}_{\sigma\tau}. \tag{61}$$

Equation (60) is easily verified for the particular Lorentz-frame in which only $\mathbf{F}_{01}$ and $\mathbf{F}_{23}$ are different from zero. Since it is a tensor equation it must hold for every other frame also. Thus we obtain from (59) and (60)

$$\left[\pi_\mu\pi^\mu - m^2 + \tfrac{1}{2}g^2\mathbf{F}_{\mu\nu}\mathbf{F}^{\mu\nu}\right]^2 + 4\left\{\left(\tfrac{1}{4}g^2\mathbf{F}_{\mu\nu}\mathbf{F}^{*\mu\nu}\right)^2 - \pi_\mu\pi^\mu\left(\tfrac{1}{2}g^2\mathbf{F}_{\nu\sigma}\mathbf{F}^{\nu\sigma}\right)\right.$$
$$\left. - g^2\pi_\mu\mathbf{F}^{\mu\nu}\mathbf{F}_{\nu\sigma}\pi^\sigma\right\} = 0. \tag{62}$$

When the explicit spin interaction is absent $g = 0$ and (62) reduces to the usual classical equations of a point-charge

$$\pi_\mu\pi^\mu - m^2 = 0. \tag{63}$$

In this case

$$\pi_\mu = mv_\mu \tag{64}$$

where v_μ is the classical four-velocity of the particle. Since we wish to retain only terms of the lowest order in g in (62) we can substitute (64) in the terms of (62) containing g, so that on ignoring terms of the order g^4, (62) becomes

$$\left(\pi_\mu\pi^\mu - m^2 + \tfrac{1}{2}g^2\mathbf{F}_{\mu\nu}\mathbf{F}^{\mu\nu}\right)^2 = 4m^2g^2\left(\tfrac{1}{2}\mathbf{F}_{\mu\nu}\mathbf{F}^{\mu\nu} - \mathbf{F}_\mu\mathbf{F}^\mu\right) \tag{65}$$

where $\mathbf{F}_\mu = \mathbf{F}_{\mu\nu}v^\nu$. Therefore

$$\pi_\mu\pi^\mu - m^2 = \pm 2mg\sqrt{\tfrac{1}{2}\mathbf{F}_{\mu\nu}\mathbf{F}^{\mu\nu} - \mathbf{F}_\mu\mathbf{F}^\mu} \tag{66}$$

if the second and higher powers of g are neglected. Now

$$\tfrac{1}{2}\mathbf{F}_{\mu\nu}\mathbf{F}^{\mu\nu} - \mathbf{F}_\mu\mathbf{F}^\mu = \mathbf{H}^2$$

where $\mathbf{H}$ is the magnetic field in the rest system of the particle. Therefore

$$\pi_\mu \pi^\mu - m^2 = \pm 2mg|\mathbf{H}|. \tag{67}$$

For the non-relativistic case we find on putting $\pi_0 = m + \mathbf{W}$ that

$$\mathbf{W} = \frac{\pi^2}{2m} \pm g|\mathbf{H}| \tag{68}$$

where $\pi = (\pi_1, \pi_2, \pi_3)$. (68) shows that in the rest system and in a weak electromagnetic field the particle manifests only a *magnetic moment g* such that the direction of the moment is either along or opposite to the magnetic component of the field. This is precisely what is to be expected of a particle of spin $\frac{1}{2}$.

I am thankful to Prof. H. J. Bhabha for helpful comments and advice.

Summary

The Dirac-matrices generate an algebra consisting of sixteen linearly independent elements. A formula is given for expressing the product of any two elements as a linear combination of these sixteen. This determines the structure of the algebra completely. It is shown that certain known identities concerning these matrices can be obtained comparatively easily by the present method. Some new identities are also deduced.

The characteristic equation of a general element of the algebra is derived and from it an expression is obtained for the determinant of any four-dimensional matrix representing the element. This expression is used to discuss the case of a particle of spin $\frac{1}{2}$ having an explicit spin interaction with the electromagnetic field. It is shown that in the classical limit $\hbar \to 0$ and upto the first approximation in the interaction constant g the particle manifests only a *magnetic-moment g* in the rest system, the direction of the moment being either along or opposite to the magnetic field in the same system.

REFERENCES

1. Eddington, *Relativity theory of electrons and protons*, Cambridge University Press, 1936.
2. Pauli, *Ann. Inst. Poincaré*, 1936, **6** 109–36.

Reprinted from
Proceedings of the Indian Academy of Sciences
22 (1945), 30–41

On the fields and equations of motion of point particles

By H. J. Bhabha, F.R.S. and Harish-Chandra

Cosmic Ray Research Unit, Indian Institute of Science, Bangalore

(*Received* 1 *November* 1944)

The investigation covers point particles possessing a charge, dipole and higher multipole moments interacting with fields of any spin satisfying the generalized wave equation (8). It is shown that the radiation field defined as the retarded minus the advanced field and all its derivatives is always finite at all points including those on the world line of the point particles. The symmetric field, defined as half the sum of the retarded and advanced fields, is shown to contain a part expressible as an integral along the world line from minus to plus infinity, which is continuous and finite everywhere. This integral vanishes if $\chi = 0$. The modified symmetric field is defined as the symmetric field minus this integral. The actual field is expressed as a sum of the modified symmetric field plus the modified mean field defined as half the sum of the ingoing and outgoing fields plus the integral just mentioned. It is proved that the part of the stress tensor of the field quadratic in the modified symmetric field plays no part in determining the equations of motion of the point particle. Being conserved by itself, it can always be subtracted away, thus defining a new stress tensor which is free from all the highest singularities in the usual stress tensor. The equations of motion of the particle are shown to depend only on the usual 'mixed terms' in the inflow with the modified mean field substituted for the ingoing field. The formulation for several particles is given.

The different fields associated with point particles are investigated and certain general properties established from which results about the general form of the equations of motion of the point particles can be deduced. The investigation covers generalized wave fields* of all spin, the usual scalar and vector meson or generalized Maxwell fields being only particular cases. Similarly, the field producing properties of the point particle can also be treated with great generality, and the particle may possess a charge, dipole or higher multipole moment or any combination of these.

The retarded field is defined by the boundary condition that it vanishes at all points on and before an infinitely extended space-like surface. Usually this surface is taken to be in the infinitely distant past. The advanced field is correspondingly defined by the boundary condition that it vanishes everywhere on and after an infinitely extended space-like surface which is usually taken to be in the infinitely distant future. Both these fields are unsymmetrical with respect to the past and the future by definition. They are both singular on the world line of the particle. In general theory it is more convenient to use two other fields derived from these, namely, the radiation field, defined as the retarded minus the advanced field, and the symmetric field, defined as half the sum of the retarded and advanced fields.

---

* The expression generalized wave field is used to cover any field the components of which satisfy the generalized wave equation (8) in free space, and covers fields of all integral spin. The expression meson field, which has been used in the previous literature, is inaccurate, since the meson having a spin of 0 or 1 unit is but a particular case. The fields with $\chi = 0$ are included in the above as limiting cases, but when it is necessary to distinguish these from the more general fields for which $\chi \neq 0$, we shall refer to them as specialized wave fields.

[250]

The radiation field changes its sign when the direction of the time axis is reversed. while the symmetric field is left unchanged.

The symmetric and radiation fields have important properties near the world line. The symmetric field is made up of a part consisting of terms depending on conditions at the retarded and advanced points on the world line only. and a part composed of two integrals along the world line from minus infinity to the retarded point and from the advanced point to infinity respectively. The latter vanish for a specialized wave field ($\chi = 0$). It is convenient to combine the latter into one integral from minus infinity to plus infinity, and to add a compensating integral from the retarded to the advanced point to the first part of the symmetric field. The integral from minus infinity to plus infinity is now a finite, continuous and differentiable function of the field point even on the world line of the particle, and satisfies the homogeneous generalized wave equation at all points of space. It is important for the theory to be developed here to introduce the *modified symmetric field* defined as the symmetric field minus the integral from minus to plus infinity just mentioned. The modified symmetric field contains all the singularities of the symmetric field. and its value at any point is independent of portions of the world line lying in the past or future light cones of the point. This is an important feature of the modified symmetric field which differentiates it from the symmetric field. Moreover. if from any point in space near the world line we drop a perpendicular to the world line, and call its length ϵ, then it will be shown that the modified symmetric field can be expanded in an ascending series containing only *odd* powers of ϵ, the highest singularity depending on the highest multipole possessed by the particle. This result has an important consequence in the form of the equations of motion of the point particle.

Dirac (1938) has already shown that the radiation field satisfying the Maxwell equations produced by a point charge is finite on the world line. It will be proved in this paper quite generally that the radiation field and all its derivatives are always finite on the world line for every type of point particle and every spin of the field. The radiation field and each of its derivatives can always be written as the sum of two parts, one containing terms depending on conditions at the retarded and advanced points only, and the other containing two integrals along the world line. The former can always be expressed near the world line as an ascending series in *even* powers of ϵ only, starting with a term independent of ϵ. For specialized wave fields ($\chi = 0$) the two integrals vanish and the whole radiation field then becomes expressible as a series in even powers of ϵ.

The actual field at a point can be expressed. following Dirac, as follows:

$$\text{actual field} = \text{retarded field} + \text{ingoing field}$$
$$= \text{advanced field} + \text{outgoing field}$$
$$= \text{symmetric field} + \text{mean field},$$

where the mean field is defined as half the sum of the ingoing and outgoing fields. In view of the simple property of the modified symmetric field mentioned above, which is not possessed by the symmetric field, it is important to split up the actual field as follows: actual field = modified symmetric field + modified mean field, where the modified mean field is defined as the mean field plus the integral from minus to plus infinity which was subtracted from the symmetric field. Since this integral is finite and continuous everywhere and satisfies the homogeneous generalized wave equation at all points, so also does the modified mean field. The stress tensor, which is a homogeneous quadratic expression in the field quantities, now falls into three parts, the first containing only the modified symmetric field, the second containing the modified symmetric and mean fields, and the third part only the modified mean field. It follows from the property of the symmetric field stated above that the first part of the energy tensor must be expressible near the world line as a series in even powers of ϵ only. It will be shown to follow immediately from this that the contribution of this part to the flow of energy and momentum into a thin tube of radius ϵ surrounding the world line must be a series containing odd powers of ϵ only. Now the theorem established in a previous paper (Bhabha & Harish-Chandra 1944) states that if the rate of inflow* be expressed in the vicinity of the world line in a series in powers of the radius of the tube then all the terms except the one independent of the radius of the tube must be identically perfect differentials. For brevity we shall refer to it as the inflow theorem. It follows that the part of the stress tensor containing only the symmetric field contributes only perfect differentials to the rate of the inflow of energy and momentum, which therefore have no effect on the equations of motion of the point particle. The same can be proved of the part containing only the modified mean field, which can in any case have no effect on the equations since it is non-singular. The only part of the energy tensor which determines the equation of motion of the particle is the mixed part. Since the mixed terms in the inflow are formally the same as those that one would have if radiation reaction were neglected, with the only difference that the modified mean field is written in place of the ingoing field, and can in general be derived from a Lagrangian, the process of finding the general equations for a point particle becomes very simple. One has merely to obtain the usual mixed terms in the inflow, and then substitute the mean field in place of the ingoing field. It will be shown by one of us (H.C.) in the paper which follows this that the form of the mixed terms in the inflow is completely determined by certain general considerations, and they can be written down in any given case without any calculation.

Each of the three parts of the stress tensor mentioned above are conserved everywhere except on the world line, and therefore it is possible to take as the new stress tensor for the field the original tensor minus the first part containing only the modified symmetric field. This removes all the worst singularities in the stress tensor.

---

* This word is used in the same sense as in the paper mentioned.

1. The retarded, advanced, radiation and symmetric fields

We use as far as possible the same notation as in the previous papers. The co-ordinates of a point in space time are denoted by x^μ where a Greek index takes on all values from 0 to 3, and the metric tensor is defined by

$$g_{00} = -g_{11} = -g_{22} = -g_{33} = 1.$$

A point on the world line is denoted by $z^\mu(\tau)$, τ being the proper time on the world line measured from some point on it. A dot denotes differentiation with respect to τ. The velocity of the particle is $v^\mu \equiv \dot{z}^\mu$. The following symbols are used with the same meaning as in the previous papers:

$$u^\mu \equiv x^\mu - z^\mu(\tau),$$

$$\kappa = u_\mu v^\mu, \quad \kappa' = u_\mu \dot{v}^\mu, \quad \kappa'' = u_\mu \ddot{v}^\mu, \text{ etc.} \tag{1}$$

For every point x^μ near the world line a point called the 'contemporary' point with the proper time τ_c can be defined on it by the equation

$$(\kappa)_c = (x^\mu - z^\mu(\tau_c)) v_\mu(\tau_c) = 0. \tag{2}$$

The suffix c attached to a symbol will be used to denote that it refers to the contemporary point. Write

$$l^\mu \equiv x^\mu - z^\mu(\tau_c), \quad \epsilon^2 = -l_\mu l^\mu, \tag{3}$$

where ϵ is real and positive. Then

$$\left(\frac{\partial u^2}{\partial \tau}\right)_c = -2(u^\mu v_\mu)_c = 0, \tag{4}$$

so that the distance from the point x^μ to a point on the world line is stationary at the contemporary point. ∂_μ will be used to denote $\partial/\partial x^\mu$. We note the following relations which will be required later:

$$\partial_\mu \tau_c = \left(\frac{l_\mu}{1 - \kappa'}\right)_c,$$

$$\partial_\nu l^\mu = \delta_\nu^\mu - \left(\frac{l^\mu v_\nu}{1 - \kappa'}\right)_c,$$

$$\partial_\nu \epsilon^2 = -2l_\mu \partial_\nu l^\mu = -2l_\nu. \tag{5}$$

The right-hand sides of all three equations are finite and continuous on the world line, and all higher derivatives of them also remain finite and continuous. The same is true of any positive *even* power of ϵ, which can be differentiated an unlimited number of times without the higher derivatives becoming singular as $\epsilon \to 0$. On the contrary, the differentiation of an *odd* power of ϵ a sufficient number of times can certainly lead to a term singular as $\epsilon \to 0$. For example,

$$\partial^\mu \partial_\nu \epsilon = -\delta_\nu^\mu \epsilon^{-1} - l^\mu l_\nu \epsilon^{-3} + \left(\frac{l^\mu v_\nu}{1 - \kappa'}\right)_c \epsilon^{-1}.$$

The retarded point on the world line associated with the point x^μ is defined as the point for which

$$(u_\mu u^\mu)_r \equiv (x_\mu - z_\mu(\tau_r))(x^\mu - z^\mu(\tau_r)) = 0 \tag{6}$$

with $x^0 - z^0(\tau_r) > 0$. The suffix r will be used to denote that the quantities refer to the retarded point. The advanced point is determined by the solution of the equation (6) for which $x^0 - z^0(\tau_a) < 0$. The suffix a will be used to denote quantities referring to this point.

Let $S_{...}$, where the dots stand for any arbitrary number of tensor indices, be a tensor defined at all points of the world line and finite and continuous on it. It may be considered as a continuous function of τ. Let τ_A be a fixed point on the world line. Then the field function defined by

$$O_{...}^{\text{ret.}}(x^\mu) \equiv \left(\frac{S_{...}}{\kappa}\right)_r - \chi \int_{\tau_A}^{\tau_r} S_{...} \frac{J_1(\chi u)}{u} d\tau \tag{7}$$

for $\tau_r \geqslant \tau_A$ and identically zero for $\tau_r < \tau_A$ satisfies the generalized wave equation

$$(\partial_\mu \partial^\mu + \chi^2) U = 0 \tag{8}$$

at all points of space not on the world line. χ is a constant characteristic of the field and J_1 is the first order Bessel function. For later use we note that the nth order Bessel function is defined by

$$J_n(u) = \left(\frac{u}{2}\right)^n \sum_{s=0}^{\infty} (-1)^s \frac{1}{s!(n+s)!} \left(\frac{u}{2}\right)^{2s}. \tag{9}$$

The dots affixed to O stand for the same tensor indices as appear in $S_{...}$.

Take a fixed point $\tau_c > \tau_A$ on the world line and draw the two-dimensional surface around it generated by all points satisfying (2) and (3) with constant ϵ. This surface forms a sphere around the point τ_c in the rest system of the point, that is, in the Lorentz frame in which the velocity v^μ at the point has the components 1, 0, 0, 0. If we denote by $d\Omega$ an element of solid angle subtended by an element of the two-dimensional surface of the sphere at its centre at τ_c, then an element of the surface with its normal directed outwards and perpendicular to the world line is $l^\nu \epsilon\, d\Omega$. Given an arbitrary continuous function ϕ of position inside the sphere expansible in a Taylor series about the point τ_c, then it can be shown easily that

$$\underset{\epsilon \to 0}{\text{Lt}} \int \phi (\partial_\nu O_{...}^{\text{ret.}}) l^\nu \epsilon\, d\Omega = \underset{\epsilon \to 0}{\text{Lt}} \int \phi \left(\frac{\partial}{\partial \epsilon} O_{...}^{\text{ret.}}\right) \epsilon^2\, d\Omega = -4\pi\phi_c(S_{...})_c. \tag{10}$$

The expression (7) is completely and uniquely defined by the requirement that it satisfies (a) the equation (8) at all points not on the world line, (b) the limiting condition (10) at the world line, and (c) the condition that it vanishes at all points on and prior to an infinite space-like surface passing through the point τ_A. We shall call it the fundamental retarded solution. Usually the condition (c) is imposed on a surface in the infinitely distant past, i.e. $\tau_A \to -\infty$, and we shall henceforth consider only this case.

All retarded solutions can be derived from (7) by inserting the appropriate tensors for $S_{...}$, by differentiating with respect to the co-ordinates x^μ a certain number of times and by contraction of tensor indices with those of the differentiations. The tensors $S_{...}$ may have to satisfy different symmetry conditions in their indices in different cases, but this does not concern us here. For example, if the field under consideration is the scalar field, then the retarded potential for a point charge is simply given by

$$U^{\text{ret.}} = g_1 O^{\text{ret.}},\tag{11a}$$

with S a scalar function on the world line and g_1 a constant. The retarded potential for a dipole is

$$U^{\text{ret.}} = g_2 \partial^\mu O^{\text{ret.}}_\mu,\tag{11b}$$

where g_2 is a constant and we have to write S_μ for $S_{...}$ in (7). S_μ gives the direction of the dipole moment.

If the field under consideration is a vector meson field then the retarded potentials for a point charge are given by

$$U^{\text{ret.}}_\mu = g_1 O^{\text{ret.}}_\mu,\tag{11c}$$

where we now have to write v_μ for $S_{...}$. The corresponding retarded potentials for a dipole are

$$U^{\text{ret.}}_\mu = g_2 \partial^\nu O^{\text{ret.}}_{\mu\nu},\tag{11d}$$

and $S_{\mu\nu}$ has to be written for $S_{...}$. Every other tensor field and higher multipole moment for the particle can be treated correspondingly. The general retarded potentials for a point particle can therefore be written in the form

$$U^{\text{ret.}}_{...} = D[O^{\text{ret.}}_{...}],\tag{12}$$

where the operator D denotes a sum of terms each one of which consists of a different number of differentiations of the fundamental solution (7) with appropriate tensors written for $S_{...}$ and contraction of indices. The dots written as suffixes to U denote a number of tensor indices and determine the spin of the field. The number of un-contracted indices in all the terms must be the same as in $U_{...}$.

The function which is uniquely and completely defined by the three conditions that it (a) satisfies the equation (8) at all points of space not on the world line, (b) fulfils the condition (10) at the world line, and (c) vanishes at all points on and after an infinite space-like surface cutting the world line at the point τ_B is given by

$$O^{\text{adv.}}_{...} \equiv -\left(\frac{S_{...}}{\kappa}\right)_a - \chi \int_{\tau_a}^{\tau_B} S_{...}\,\frac{J_1(\chi u)}{u}\,d\tau\tag{13}$$

for $\tau_B \geqslant \tau_a$ and zero for $\tau_B < \tau_a$. This is the fundamental advanced solution and all advanced potentials can be expressed in terms of it by

$$U^{\text{adv.}}_{...} = D[O^{\text{adv.}}_{...}],\tag{14}$$

H. J. Bhabha and Harish-Chandra

where D is the same operator as in (12). The radiation field introduced by Dirac (1938) is defined by

$$U^{\text{rad.}}_{..} = U^{\text{ret.}}_{..} - U^{\text{adv.}}_{..}$$
$$= D[O^{\text{ret.}}_{...} - O^{\text{adv.}}_{...}]. \tag{15}$$

We therefore investigate first the fundamental radiation solution defined by

$$O^{\text{rad.}}_{...} = O^{\text{ret.}}_{...} - O^{\text{adv.}}_{...}$$
$$= \left(\frac{S_{...}}{\kappa}\right)_r + \left(\frac{S_{...}}{\kappa}\right)_a - \chi\int_{-\infty}^{\tau_r} S_{...}\frac{J_1(\chi u)}{u}\,d\tau + \chi\int_{\tau_a}^{\infty} S_{...}\frac{J_1(\chi u)}{u}\,d\tau. \tag{16a}$$

It is convenient to introduce the modified fundamental radiation solution $O'^{\text{rad.}}_{...}$ defined by

$$O'^{\text{rad.}}_{...} = O^{\text{rad.}}_{...} - \chi\int_{-\infty}^{\infty} S_{...}\frac{J_1(\chi u)}{u}\,d\tau$$
$$= \left(\frac{S_{...}}{\kappa}\right)_r + \left(\frac{S_{...}}{\kappa}\right)_a - \chi\int_{-\infty}^{\tau_r} S_{...}\frac{J_1(\chi u)}{u}\,d\tau - \int_{-\infty}^{\tau_a} S_{...}\frac{J_1(\chi u)}{u}\,d\tau. \tag{16b}$$

For later use we introduce explicitly the radiation solution for the specialized wave equation ($\chi = 0$)

$$\bar{O}^{\text{rad.}}_{...} = \left(\frac{S_{...}}{\kappa}\right)_r + \left(\frac{S_{...}}{\kappa}\right)_a. \tag{16c}$$

Let the field point x^μ be kept fixed for the moment. The value of u^2 at a point τ on the world line can be expressed by a Taylor series in powers of $\Delta\tau \equiv \tau - \tau_c$ for sufficiently small values of $\Delta\tau$,

$$u^2 = (u^2)_c + \left(\frac{\partial u^2}{\partial\tau}\right)_c \Delta\tau + \frac{1}{2}\left(\frac{\partial^2 u^2}{\partial\tau^2}\right)_c (\Delta\tau)^2 + \dots. \tag{17a}$$

It can be seen at once that none of the coefficients $(\partial^n u^2/\partial\tau^n)_c$ of this series is singular. Using (1)–(4) we get in particular

$$(u^2)_c = -\epsilon^2, \quad \left(\frac{\partial u^2}{\partial\tau}\right)_c = -2\kappa_c = 0, \quad \left(\frac{\partial^2 u^2}{\partial\tau^2}\right)_c = 2(1-\kappa')_c,$$
$$\left(\frac{\partial^3 u^2}{\partial\tau^3}\right)_c = -2\kappa''_c, \quad \left(\frac{\partial^4 u^2}{\partial\tau^4}\right)_c = -2(\dot{v}^2 + \kappa''')_c. \tag{17b}$$

Now at the advanced and retarded points $u^2 \equiv 0$ by definition, and $\Delta\tau$ takes on one of the two values $\Delta\tau_a \equiv \tau_a - \tau_c$ or $\Delta\tau_r \equiv \tau_r - \tau_c$ respectively which satisfy the equation obtained by putting (17b) into (17a)

$$\epsilon^2 = (1-\kappa')_c(\Delta\tau)^2 - \tfrac{1}{3}\kappa''_c(\Delta\tau)^3 - \tfrac{1}{12}(\dot{v}^2 + \kappa''')_c(\Delta\tau)^4 + \dots. \tag{18a}$$

The essential feature of equation (18a) is that the coefficient of $\Delta\tau$ on the right is zero since the distance u from the point x^μ to the world line is stationary at τ_c, and the coefficient of $(\Delta\tau)^2$ is rational and tends to 1 as x^μ approaches the world line. Taking the root of both sides of (18a) we get

$$\pm\frac{\epsilon}{\sqrt{(1-\kappa')}} = \Delta\tau - \frac{1}{6}\frac{\kappa''}{1-\kappa'}(\Delta\tau)^2 - \frac{1}{24}\left\{\frac{\ddot{v}^2+\kappa'''}{1-\kappa'} + \frac{1}{3}\frac{\kappa''^2}{(1-\kappa')^2}\right\}(\Delta\tau)^3, \qquad (18b)$$

where we have omitted the suffix c for brevity. This clearly shows that for the advanced point $\Delta\tau_a$ is obtained by taking the positive sign on the left, while $\Delta\tau_r$ is got by taking the negative sign. Reverting equation (18b) by successive approximation we obtain

$$\begin{aligned}
\Delta\tau_a &= \frac{\epsilon}{\sqrt{(1-\kappa')}} + \frac{1}{6}\frac{\kappa''\epsilon^2}{(1-\kappa')^2} + \frac{1}{24}\frac{\ddot{v}^2\epsilon^3}{(1-\kappa')^{\frac{5}{2}}} + O(\epsilon^4) \\
&= \epsilon + \tfrac{1}{2}\epsilon\kappa' + \tfrac{3}{8}\epsilon\kappa'^2 + \tfrac{1}{6}\kappa''\epsilon^2 + \tfrac{1}{24}\ddot{v}^2\epsilon^3 + O(\epsilon^4).
\end{aligned} \qquad (19)$$

We get $\Delta\tau_r$ by reversing the sign of ϵ in (19), as follows directly from (18b). Hence $(\Delta\tau_r)^n + (\Delta\tau_a)^n$ is expressible as a series in even powers of ϵ only, while $(\Delta\tau_r)^n - (\Delta\tau_a)^n$ is expressible as a series in odd powers of ϵ.

Similarly, expressing the value of κ at the point τ by a Taylor series in powers of $\Delta\tau$ we get, using the last three of the equations (17b),

$$\kappa = -(1-\kappa')_c\Delta\tau + \tfrac{1}{2}\kappa_c''(\Delta\tau)^2 + \tfrac{1}{6}(\ddot{v}^2+\kappa''')(\Delta\tau)^3 + \dots. \qquad (20)$$

The essential feature of this series is again that the coefficient of $\Delta\tau$ is rational and tends to 1 as $\epsilon \to 0$. It follows that κ^{-n}, where n is an integer, can be expanded as a series in ascending powers of $\Delta\tau$ starting with $(\Delta\tau)^{-n}$, the coefficients of the series being rational and non-singular functions of l^μ and the particle variables at the contemporary point.

Any function of position on the world line expressible as a sum of terms containing only positive integral powers of l^μ and the particle variables at τ, and positive or negative integral powers of κ can therefore be expressed as a series in ascending powers of $\Delta\tau$ the coefficients of which are rational and non-singular functions of l^μ and the particle variables at τ_c. If the highest negative power of κ be κ^{-m} then the series commences with $(\Delta\tau)^{-m}$, thus

$$f(u^\mu,\tau) = \sum_{n=-m}^{\infty} f_n(l^\mu,\tau_c)(\Delta\tau)^n. \qquad (21)$$

where f_n is a rational function of l^μ and the particle variables at τ_c. f_n cannot therefore contain an odd power of $\epsilon \equiv \sqrt{(-l_\mu l^\mu)}$, but may contain an even power of ϵ. Denoting by $(f)_r$ and $(f)_a$ the values of f at the retarded and advanced points respectively, we get

$$(f)_r + (f)_a = \sum_{n=-m}^{\infty} f_n(l^\mu,\tau_c)\{(\Delta\tau_r)^n + (\Delta\tau_a)^n\}$$
$$= \text{ascending series in even powers of } \epsilon, \qquad (22)$$

and

$$(f)_r - (f)_a = \sum_{n=-m}^{\infty} f_n(l^\mu,\tau_c)\{(\Delta\tau_r)^n - (\Delta\tau_a)^n\}$$
$$= \text{ascending series in odd powers of } \epsilon. \qquad (23)$$

All functions with which we have to deal in this paper satisfy the conditions laid down for f.

The result (23) can be applied at once to (16c) to show it must be expressible as a series in even powers of ϵ only. For the Maxwell field of a point charge this fact has already been noted by Dirac (1938). The series can have no negative powers of ϵ since the greatest singularity in either of the terms on the right of (16c) is of the order ϵ, and this cannot appear in the series since it is an odd power. Substituting the value of τ_a given by (19) into (20) we get

$$\kappa_u = -\epsilon \sqrt{(1-\kappa')} + \frac{1}{3}\frac{\kappa''\epsilon^2}{1-\kappa'} - \frac{1}{24}\frac{\dot{v}^2\epsilon^3}{(1-\kappa')^{\frac{1}{2}}} + O(\epsilon^4). \tag{24}$$

Similarly, the Taylor expansion of $S_{...}$ gives in conjunction with (19)

$$(S_{...})_a = (S_{...})_c + (\dot{S}_{...})_c \frac{\epsilon}{\sqrt{(1-\kappa'_c)}} + \tfrac{1}{2}(\ddot{S}_{...})_c \frac{\epsilon^2}{1-\kappa'_c} + O(\epsilon^3). \tag{25}$$

Whence

$$\left(\frac{S_{...}}{\kappa}\right)_a = -\frac{S_{...}}{\epsilon}, \; -\dot{S}_{...} - \tfrac{1}{2}S_{...}\frac{\kappa'}{\epsilon}, \; -S_{...}\left(\frac{3}{8}\frac{\kappa'^2}{\epsilon} + \tfrac{1}{3}\kappa'' - \tfrac{1}{24}\dot{v}^2\epsilon\right) - \dot{S}_{...}\kappa' - \tfrac{1}{2}\ddot{S}_{...}\epsilon, \; + O(\epsilon^2), \tag{26}$$

where we have separated terms of different orders in ϵ by a comma. Reversing the sign of ϵ to obtain $(S_{...}/\kappa)_r$ we get

$$\left(\frac{S_{...}}{\kappa}\right)_r + \left(\frac{S_{...}}{\kappa}\right)_a = -2\dot{S}_{...}, \; -2S_{...}\kappa'' - 2\dot{S}\kappa', \; + O(\epsilon^2). \tag{27}$$

The specialized radiation solution (16c) is therefore finite as $\epsilon \to 0$.

The two integrals on the right of (16a) remain finite as the point x^μ approaches the world line since the integrands always remain finite on account of the well-known property of the Bessel function

$$\operatorname*{Lt}_{u\to 0} \frac{J_n(u)}{u^n} = \frac{1}{2^n n!} \tag{28}$$

which follows from (9). Hence the fundamental radiation solution given by (16a) of the generalized wave equation (8) is always finite on the world line. The same is also obviously true of the modified radiation solution (16b).

Since the right-hand side of (27) only contains even powers of ϵ starting from zero with coefficients which are rational integral functions of l'' and the particle variables, it follows from the equations (5) that all successive derivatives of this series with respect to x^μ can be series containing only positive even powers of ϵ none of the coefficients of which are singular. All successive derivatives of the left-hand side of (27) therefore remain finite and unambiguous as the field point approaches the world line.

Further, by a well-known property of the Bessel function which can be deduced immediately from (9), we have, for fixed τ,

$$\partial_\mu \frac{J_n(\chi u)}{u^n} = \frac{u_\mu}{u}\frac{\partial}{\partial u}\left(\frac{J_n(\chi u)}{u^n}\right) = -\chi u_\mu \frac{J_{n+1}(\chi u)}{u^{n+1}}. \tag{29}$$

Hence
$$\partial_\nu \partial_\rho \ldots \frac{J_1(\chi u)}{u} \tag{30}$$

is always finite even on the world line, whatever the number of differentiations. Writing $\hat{\partial}_{\nu\rho\ldots}$ for this expression for brevity and remembering that

$$\partial_\mu \tau_a = \left(\frac{u_\mu}{\kappa}\right)_a, \quad \partial_\mu \tau_r = \left(\frac{u_\mu}{\kappa}\right)_r, \tag{31}$$

we get

$$\partial_\mu \left\{ -\int_{-\infty}^{\tau_r} S_{\ldots}\hat{\partial}_{\nu\rho\ldots}\,d\tau + \int_{\tau_a}^{\infty} S_{\ldots}\hat{\partial}_{\nu\rho\ldots}\,d\tau \right\}$$

$$= -\left(\frac{u_\mu}{\kappa} S_{\ldots}\hat{\partial}_{\nu\rho\ldots}\right)_r - \left(\frac{u_\mu}{\kappa} S_{\ldots}\hat{\partial}_{\nu\rho\ldots}\right)_a - \int_{-\infty}^{\tau} S_{\ldots}\hat{\partial}_{\mu\nu\rho\ldots}\,d\tau + \int_{\tau_a}^{\infty} S_{\ldots}\hat{\partial}_{\mu\nu\rho\ldots}\,d\tau. \tag{32}$$

Both of the first two terms on the right are finite on the world line, and being of the general form of the right-hand side of (22), all their successive derivatives are also finite. The two integrals on the right are also finite by (30), so that it can be deduced by induction that all derivatives of the two integrals on the right of (16a) are finite on the world line. We have therefore proved that $O^{\text{rad.}}$ and all its derivatives are finite on the world line, and hence the same must be true for $U^{\text{rad.}}$ and all its derivatives. *The radiation field with all its derivatives is always finite on the world line.* The same is true of the modified radiation field.

From (15) it follows that we can write

$$U^{\text{rad.}} = \bar{U}^{\text{rad.}} - \chi \int_{-\infty}^{\tau_r} D\left[S_{\ldots}\frac{J_1(\chi u)}{u}\right]d\tau + \chi \int_{\tau_a}^{\infty} D\left[S_{\ldots}\frac{J_1(\chi u)}{u}\right]d\tau, \tag{33}$$

where $\bar{U}^{\text{rad.}}$ contains only the sum of terms taken at the retarded and the advanced points and is expressible as a series in even powers of ϵ. It should be noted that $U^{\text{rad.}} \neq D[\bar{O}^{\text{rad.}}]$ but contains additional terms like the first two on the right of (32).

Now consider the fundamental symmetric solution $O^{\text{sym.}}$ defined by

$$O^{\text{sym.}} \equiv \tfrac{1}{2}(O^{\text{ret.}} + O^{\text{adv.}})$$

$$= \frac{1}{2}\left(\frac{S_{\ldots}}{\kappa}\right)_r - \frac{1}{2}\left(\frac{S_{\ldots}}{\kappa}\right)_a - \frac{\chi}{2}\int_{-\infty}^{\tau} S_{\ldots}\frac{J_1(\chi u)}{u}\,d\tau - \frac{\chi}{2}\int_{\tau_a}^{\infty} S_{\ldots}\frac{J_1(\chi u)}{u}\,d\tau. \tag{34}$$

Writing
$$O'^{\text{sym.}} \equiv \frac{1}{2}\left(\frac{S_{\ldots}}{\kappa}\right)_r - \frac{1}{2}\left(\frac{S_{\ldots}}{\kappa}\right)_a + \frac{\chi}{2}\int_{\tau_r}^{\tau_a} S_{\ldots}\frac{J_1(\chi u)}{u}\,d\tau, \tag{35}$$

$$O^{\text{sym.}} = O'^{\text{sym.}} - \frac{\chi}{2}\int_{-\infty}^{\infty} S_{\ldots}\frac{J_1(\chi u)}{u}\,d\tau. \tag{36}$$

Although u^2 is negative and u purely imaginary for points on the world line between τ_r and τ_a, since $J_n(\chi u)/u^n$ is a series in ascending powers of u^2 as shown by (9), the integrand of all the integrals is real at every point of the world line and varies continuously along it.

17-2

The first two terms on the right of (35) are of the general form (23) and must be expressible as a series in odd powers of ϵ with rational integral coefficients. All derivatives of this part must have the same form, though they may be singular to a higher degree.

Similarly the integral in (35) can be expressed as a series in odd powers of ϵ. For the value of the integrand at the point τ can be expanded in a Taylor series. Denoting the integrand by $i(\tau)$ for brevity, and writing $\Delta\tau = \tau - \tau_c$ as before,

$$i(\tau) = \sum_{n=0}^{\infty} \frac{1}{n!}\left(\frac{\partial^n i}{\partial \tau^n}\right)_c (\Delta\tau)^n, \tag{37}$$

(9) shows that i is an ascending series in positive powers of u^2. All derivatives of it with respect to τ remain series in positive powers of u^2, and hence every coefficient $(\partial^n i, \partial \tau^n)_c$ is a series in positive powers of ϵ^2. On integration we get

$$\int_{\tau_r}^{\tau_a} i(\tau)\,d\tau = \sum_{n=0}^{\infty} \frac{1}{(n+1)!}\left(\frac{\partial^n i}{\partial \tau^n}\right)_c \{(\Delta\tau_a)^{n+1} - (\Delta\tau_r)^{n+1}\}.$$

As proved earlier, each term in curly brackets is a series in odd powers of ϵ only. The series on the right of (37) must therefore be expressible as an ascending series in odd powers of ϵ the coefficients of which are non-singular and integral functions of the variables at the contemporary point. Hence, in the neighbourhood of the world line $O'^{\text{sym.}}$ can be expressed as an ascending series in *odd* powers of ϵ. (For the Maxwell field of a point charge this fact has also been noted by Dirac (1938).) It is singular on the world line. It follows by using (5) that $O'^{\text{sym.}}$ *and all its successive derivatives are also and always expressible as ascending series in odd powers of ϵ*, each having a higher singularity than the previous one.

The integrand of the integral in (36) is continuous along the world line. Hence it follows from (30) that the integral and all its derivatives are finite even on the world line. Further

$$(\partial_\mu \partial^\mu + \chi^2)\int_{-\infty}^{\infty} S \ldots \frac{J_1(\chi u)}{u}\,d\tau = \int_{-\infty}^{\infty} S \ldots (\partial_\mu \partial^\mu + \chi^2)\frac{J_1(\chi u)}{u}\,d\tau$$

$$= \int_{-\infty}^{\infty} S \ldots \left(\frac{\partial^2}{\partial u^2} + 3\cdot\frac{\partial}{u\,\partial u} + \chi^2\right)\frac{J_1(\chi u)}{u}\,d\tau = 0, \tag{38}$$

since the Bessel function satisfies just the equation obtained by putting the integrand equal to zero. Hence it follows from (36) that $O'^{\text{sym.}}$ satisfies equation (8) at all points of space not on the world line, and condition (10) on it. All its derivatives therefore also satisfy equation (8) at any point not on the world line.

Following Dirac we write for the actual potential $U'^{\text{act.}}$ at a point

$$U'^{\text{act.}} = U'^{\text{ret.}} + U'^{\text{in.}} = U'^{\text{adv.}} + U'^{\text{out.}}. \tag{39}$$

Introducing the mean field defined by

$$U'^{\text{mean}} \equiv \tfrac{1}{2}(U'^{\text{in.}} + U'^{\text{out.}}) \tag{40}$$

we can also write it

$$U^{\text{act.}}_{\dots} = U^{\text{sym.}}_{\dots} + U'^{\text{mean}}_{\dots}$$

$$= U'^{\text{sym.}}_{\dots} + U''^{\text{mean}}_{\dots}, \tag{41}$$

where

$$U'^{\text{mean}}_{\dots} = U^{\text{mean}}_{\dots} - \frac{\chi}{2} \int_{-\infty}^{\infty} D\left[S_{\dots} \frac{J_1(\chi u)}{u} \right] d\tau. \tag{42}$$

the operator D being the same as in (12), (14) and (15). It is clear from (38) that $U'^{\text{mean}}_{\dots}$ satisfies the equation (8) at all points of space and is continuous and finite everywhere. We call it the modified mean field.

2. The equations of motion

Certain general results about the form of the equations of motion of the point particles can be deduced at once from the properties of the radiation and symmetric fields established above. The equations of motion of the point particle are obtained by using the stress tensor of the field to calculate the flow of energy and momentum into a thin tube surrounding the world line. The inflow along an infinitesimal length of the tube must then be put equal to a perfect differential for conservation of energy and momentum, and this provides the equation of motion of the point particle.

Let the tube be defined by $l_\mu l^\mu = -\epsilon^2$ with constant ϵ. The element of three dimensional surface of this tube with its normal directed outwards is given by

$$l^\nu \epsilon (1 - \kappa'_c)\, d\Omega d\tau, \tag{43}$$

$d\Omega$ being the element of solid angle about the contemporary point in its rest system introduced in the previous section. The flow of energy and momentum into a portion of the tube of length $d\tau$ is

$$\left\{ \int T_{\mu\nu} l^\nu \epsilon (1 - \kappa'_c)\, d\Omega \right\} d\tau \equiv T_\mu d\tau. \tag{44}$$

The equations of translational motion are then given by putting

$$T_\mu = \dot{A}_\mu, \tag{45}$$

where A_μ is some function of the variables describing the state of the particle and the field at the point τ and their derivatives.

The stress tensor $T_{\mu\nu}$ is a homogeneous quadratic expression in the potentials and the field strengths which are linear derivatives of them. We denote by $T_{\mu\nu}(U^A_{\dots}, U^B_{\dots})$ the more general expression that is obtained from it containing two independent fields $U^A_{\dots}$ and $U^B_{\dots}$ by replacing one $U_{\dots}$ in each term by $U^A_{\dots}$ and the other by $U^B_{\dots}$ and then making the expression symmetric in $U^A_{\dots}$ and $U^B_{\dots}$. The resulting expression is thus linear in $U^A_{\dots}$ and $U^B_{\dots}$ separately and quadratic in the two. The original stress is just $T_{\mu\nu}(U_{\dots}, U_{\dots})$ in this notation.

For example, the stress tensor for the generalized Maxwell field is

$$4\pi T_{\mu\nu} = G_{\mu\rho} G^\rho_\nu + \tfrac{1}{4} g_{\mu\nu} G_{\rho\sigma} G^{\rho\sigma} + \chi^2 (U_\mu U_\nu - \tfrac{1}{2} g_{\mu\nu} U_\rho U^\rho),$$

the field strengths $G_{\mu\nu}$ being derivatives of the potentials U_μ. Then,

$$4\pi T_{\mu\nu}(U^A_{...}, U^B_{...}) \equiv \tfrac{1}{2}G^A_{\mu\rho}G^{B\rho}_\nu + \tfrac{1}{2}G^B_{\mu\rho}G^{A\rho}_\nu + \tfrac{1}{4}g_{\mu\nu}G^A_{\rho\sigma}G^{B\rho\sigma}$$
$$+ \tfrac{1}{2}\chi^2(U^A_\mu U^B_\nu + U^B_\mu U^A_\nu - g_{\mu\nu}U^A_\rho U^{B\rho}),$$

$G^A_{\mu\nu}$ being the field strengths corresponding to the potentials U^A_μ.

Returning to the general tensor for any field, we have, using (41),

$$T_{\mu\nu} \equiv T_{\mu\nu}(U^{act.}_{...}, U^{act.}_{...}) = T_{\mu\nu}(U'^{sym.}_{...} + U'^{mean}_{...}, U'^{sym.}_{...} + U'^{mean}_{...})$$
$$= T_{\mu\nu}(U'^{sym.}_{...}, U'^{sym.}_{...}) + 2T_{\mu\nu}(U'^{sym.}_{...}, U'^{mean}_{...}) + T_{\mu\nu}(U'^{mean}_{...}, U'^{mean}_{...}). \tag{46}$$

Corresponding to this splitting up we get for the inflow

$$T_\mu = T'^{sym.}_\mu + T'^{mix.}_\mu + T'^{mean}_\mu, \tag{47}$$

where

$$T'^{sym.}_\mu \equiv \int T_{\mu\nu}(U'^{sym.}_{...}, U'^{sym.}_{...})\, l^\nu \epsilon (1 - \kappa'_c)\, d\Omega, \tag{48}$$

$$T'^{mix.}_\mu = 2\int T_{\mu\nu}(U'^{sym.}_{...}, U'^{mean}_{...})\, l^\nu \epsilon (1 - \kappa'_c)\, d\Omega \tag{49}$$

and

$$T'^{mean}_\mu = \int T_{\mu\nu}(U'^{mean}_{...}, U'^{mean}_{...})\, l^\nu \epsilon (1 - \kappa'_c)\, d\Omega. \tag{50}$$

Since $U'^{sym.}$ and its derivatives can be expressed near the world line as series in odd powers of ϵ only, it follows that $T_{\mu\nu}(U'^{sym.}_{...}, U'^{sym.}_{...})$ is a series in even powers of ϵ only. As far as the integration with respect to $d\Omega$ is concerned, ϵ is a constant, and the only quantities which vary are l^ν and terms containing it like $\kappa'_c \equiv l^\nu \dot{\imath}_\nu$, $\kappa''_c = l^\nu \ddot{\imath}_\nu$, etc. The integral of a product of say s factors $l^\nu l^\rho \ldots$ vanishes from symmetry if s is odd, and gives ϵ^s multiplied by a constant if s is even. Hence, as a result of the factor ϵ which appears explicitly in (43) and (48), $T'^{sym.}_\mu$ is a series in *odd* powers of ϵ only, the coefficients of which are just functions of the particle variables on the world line at the point τ_r. The inflow theorem (Bhabha & Harish-Chandra 1944) states that if the rate of inflow be expanded in a series in powers of ϵ, the coefficients of all the terms except the one independent of ϵ must be perfect differentials. $T'^{sym.}_\mu$ being a series in odd powers of ϵ has no such term independent of ϵ, and hence must be identically a perfect differential. It has also been proved that the inflow through two tubes of different shape can only differ by a perfect differential. $T'^{sym.}_\mu$ therefore plays no part in the equations of motion since it can always be eliminated by adding an identical term in A_μ. *Thus the part of the stress tensor containing only the modified symmetric field contributes nothing to the equations of motion.*

Since U'^{mean} is a continuous and non-singular function at all points, T'^{mean}_μ must tend to zero as $\epsilon \to 0$ and hence it cannot contain any term independent of ϵ. Obviously, it can play no part in the equations of motion. It also follows from the inflow theorem that T'^{mean}_μ is identically a perfect differential. This can be easily verified by direct calculation.

The only significant contribution to the right-hand side of (45) comes from the terms in $T_\mu'^{\text{mix.}}$ independent of ϵ, which we denote by $[T_\mu'^{\text{mix.}}]_0$. The equation of motion must therefore be of the form

$$A_\mu = [T_\mu'^{\text{mix.}}]_0. \tag{51}$$

We note that the right-hand side of (51) can only contain $U_{...}'^{\text{mean}}$ and its derivatives multiplied by functions of the particle variables at the point τ, and is linear in them. Another proof of this result which does not depend on an expansion of the symmetric field in powers of ϵ is given below.

For a point charge and a dipole the highest singularity in $U_{...}'^{\text{sym.}}$ is of the order ϵ^{-2}. Since the integration over $d\Omega$ of a function multiplied by the factor (43) gives at least a factor ϵ^3, no term containing $U_{...}'^{\text{mean}}$ can appear in the equations. The equations only contain the field strength $G_{...}'^{\text{mean}}$ and its derivatives. This result is quite general and true also for higher multipoles, as is proved by one of us (H.C.) in the paper which follows this.

Since $U_{...}'^{\text{mean}}$ contains an integral from $-\infty$ to ∞ it might appear as if the motion of the particle at τ_c depended not only on the motion of the particle in the past, but also in the future. Using (42), (40), (39) and (33) we see, however, that on the world line

$$
\begin{aligned}
(U_{...}'^{\text{mean}})_c &= \tfrac{1}{2}(U_{...}'^{\text{in.}} + U_{...}'^{\text{out.}})_c - \frac{\chi}{2}\int_{-\infty}^{\infty} D\left[S_{....}\,\frac{J_1(\chi u)}{u}\right]d\tau \\
&= (U_{...}'^{\text{in.}})_c + \tfrac{1}{2}(U_{...}'^{\text{ret.}} - U_{...}'^{\text{adv.}})_c - \frac{\chi}{2}\int_{-\infty}^{\infty} D\left[S_{....}\,\frac{J_1(\chi u)}{u}\right]d\tau \\
&= (U_{...}'^{\text{in.}})_c + \tfrac{1}{2}(\bar{U}_{...}'^{\text{rad.}})_c - \chi\int_{-\infty}^{\tau_e} D\left[S_{....}\,\frac{J_1(\chi u)}{u}\right]d\tau \\
&= (U_{...}'^{\text{in.}})_c + \tfrac{1}{2}(U_{...}''^{\text{rad.}})_c,
\end{aligned} \tag{52}
$$

which clearly shows that the motion only depends on the actual motion of the particle at τ_c and in the past. It is reasonable to interpret $\tfrac{1}{2}(\bar{U}_{...}^{\text{rad.}})_c$ as the field giving the effects of radiation reaction, and the last integral as an addition to the ingoing field due to the particle's own motion in the past. It has already been mentioned that $\bar{U}^{\text{rad.}}$ is not just $D[\bar{O}^{\text{rad.}}]$ but contains in addition the terms which result from the differentiation of τ_r and τ_a in (16a) with respect to x^μ, as, for example, the first two terms on the right of (32).

We also note that

$$
\begin{aligned}
U_{...}'^{\text{sym.}} &= U_{...}^{\text{sym.}} + \frac{\chi}{2}\int_{-\infty}^{\infty} D\left[S_{....}\,\frac{J_1(\chi u)}{u}\right]d\tau \\
&= U_{...}^{\text{ret.}} - \tfrac{1}{2}U_{...}^{\text{rad.}} + \frac{\chi}{2}\int_{-\infty}^{\infty} D\left[S_{....}\,\frac{J_1(\chi u)}{u}\right]d\tau \\
&= U_{...}^{\text{ret.}} - \tfrac{1}{2}U_{...}'^{\text{rad.}}.
\end{aligned} \tag{53}
$$

The last term is continuous everywhere. Hence

$$[T'^{\text{mix.}}]_0 = \left[2\int T_{\mu\nu}(U_{...}^{\text{ret.}}, U_{...}'^{\text{mean}})\,l^\nu\epsilon(1-\kappa_c')\,d\Omega\right]_0. \tag{54}$$

This gives precisely the mixed terms calculated in the previous papers with $U'^{\text{mean}}_{\ldots}$ substituted in place of the $U^{\text{in.}}_{\ldots}$ which appeared in them. The process of finding the actual equations of motion therefore becomes very simple. We merely calculate the term independent of ϵ in the inflow using the mixed part of the stress tensor, and then substitute $U'^{\text{mean}}_{\ldots}$ given by (52), in place of $U^{\text{in.}}_{\ldots}$. The only part of the calculation which is at all cumbersome is the evaluation of $\bar{U}^{\text{rad.}}_{\ldots}$ required in (52). A convenient way of calculating this part is given by Harish-Chandra in the paper which follows this.

The general results established above hold also for the rotational equations. To calculate these use has to be made of the angular momentum tensor $M_{\lambda\mu\nu}$ instead of the stress tensor. Since

$$M_{\lambda\mu\nu} = x_\lambda T_{\mu\nu} - x_\mu T_{\lambda\nu} = (l_\lambda T_{\mu\nu} - l_\mu T_{\lambda\nu}) + (z_\lambda T_{\mu\nu} - z_\mu T_{\lambda\nu}), \tag{55}$$

we see at once that precisely the same arguments hold as before and the same result follows. Using (51) we have

$$\left[2\int M_{\lambda\mu\nu}(U'^{\text{sym.}}_{\ldots}, U'^{\text{mean}}_{\ldots}) l^\nu \epsilon(1 - \kappa'_c)\, d\Omega \right]_0 = [M'^{\text{mix.}}_{\lambda\mu}]_0 + \{z_\lambda[T'^{\text{mix.}}_\mu]_0 - z_\mu[T'^{\text{mix.}}_\lambda]_0\}$$

$$= [M'^{\text{mix.}}_{\lambda\mu}]_0 + \frac{d}{d\tau}(z_\lambda A_\mu - z_\mu A_\lambda) - (v_\lambda A_\mu - v_\mu A_\lambda), \tag{56}$$

where

$$[M'^{\text{mix.}}_{\lambda\mu}]_0 = \left[2\int \{l_\lambda T_{\mu\nu}(U'^{\text{sym.}}_{\ldots}, U'^{\text{mean}}_{\ldots}) - l_\mu T_{\lambda\nu}(U'^{\text{sym.}}_{\ldots}, U'^{\text{mean}}_{\ldots})\} l^\nu \epsilon(1 - \kappa'_c)\, d\Omega. \tag{57}$$

The rotational equation must therefore have the form

$$\dot{B}_{\lambda\mu} + v_\lambda A_\mu - v_\mu A_\lambda = [M^{\text{mix.}}_{\lambda\mu}]_0. \tag{58}$$

We give an alternative proof of (51) which depends upon the symmetry existing between the retarded and the advanced points. This method of proof avoids the necessity of expanding the various quantities involved in series and therefore has the advantage of compactness. The retarded and advanced points are defined by (6) and indeed τ_r is that solution of (6) for which $(U_0)_r > 0$ and τ_a that solution for which $(u_0)_a < 0$. It is clear therefore that in every equation which does not explicitly utilize the condition $(u_0)_r > 0$ or $(u_0)_a < 0$ we can always replace r by a and vice versa, because whatever holds for τ_r would also hold for τ_a and conversely. In particular this symmetry is exhibited in differentiation:

$$\partial_\mu \tau_r = \left(\frac{u_\mu}{\kappa}\right)_r, \quad \partial_\mu \tau_a = \left(\frac{u_\mu}{\kappa}\right)_a,$$

$$\partial_\mu(u^\nu)_r = \delta^\nu_\mu - \left(\frac{u_\mu v^\nu}{\kappa}\right)_r,$$

$$\partial_\mu(u^\nu)_a = \delta^\nu_\mu - \left(\frac{u_\mu v^\nu}{\kappa}\right)_a. \tag{59}$$

Now if we interchange a and r in (35) $O'^{\text{sym.}}_{...}$ changes sign. $O'^{\text{sym.}}_{...}$ is therefore anti-symmetric in the indices a and r. From the symmetry shown by (59) in a and r it follows that

$$U'^{\text{sym.}}_{..} = D[O'^{\text{sym.}}_{...}]$$

is also antisymmetric in a and r. On the other hand, the modified radiation field (16b) remains unaltered by an interchange of a and r. Therefore $O'^{\text{rad.}}$ and consequently $U'^{\text{rad.}}_{...}$ is symmetric in a and r. The same holds for $U'^{\text{mean}}_{...}$ as is obvious from (42).

We calculate the inflow into two tubes surrounding the world line, called the retarded and the advanced tubes respectively, defined by the equations

$$\begin{aligned} \kappa_r &= \epsilon \equiv \epsilon_r \quad \text{(retarded tube),} \\ \kappa_a &= -\epsilon \equiv \epsilon_a \quad \text{(advanced tube),} \end{aligned} \tag{60}$$

where ϵ is a positive constant. The intersection of these two tubes with the future and the past light cones starting from $z^\mu(\tau)$ respectively are called similarly the 'retarded' and the 'advanced' spheres of radius ϵ at τ. The three dimensional surface elements of the two tubes with the normals outwards can be written as

$$\begin{aligned} dS^\nu_r &= \partial^\nu \kappa_r \, d\Omega_r d\tau \quad \text{(for retarded tube),} \\ dS^\nu_a &= -\partial^\nu \kappa_a \, d\Omega_a d\tau \quad \text{(for advanced tube),} \end{aligned} \tag{61}$$

where $d\Omega_r$ is an element of surface of the two dimensional retarded sphere of radius ϵ at τ and similarly $d\Omega_a$ is an element of the surface of the corresponding advanced sphere. The difference in sign in the two equations in (61) is important and arises from a similar difference in (60), where increasing ϵ corresponds to *increasing* κ_r and *decreasing* κ_a.

For brevity put

$$\int T^{\mu\nu}(U'^A_{...}, U'^B_{...}) \partial_\nu \kappa_r d\Omega_r \equiv T^\mu_r(U^A_{...}, U^B_{...}), \tag{62a}$$

$$\int T^{\mu\nu}(U'^A_{...}, U'^B_{...}) \partial_\nu \kappa_a d\Omega_a \equiv T^\mu_a(U^A_{...}, U^B_{...}). \tag{62b}$$

Let I^μ_r denote the rate of inflow calculated on the retarded tube and I^μ_a that on the advanced tube. For the equations of motion the significant part of the inflow is the part of these independent of ϵ, which we denote by $[I^\mu_r]_0$ and $[I^\mu_a]_0$ respectively. Then by (46) and (62) we get

$$[I^\mu_r]_0 = [T^\mu_r(U'^{\text{sym.}}_{...}, U'^{\text{sym.}}_{...})]_0 + [2T^\mu_r(U'^{\text{sym.}}_{...}, U'^{\text{mean}}_{...})]_0 \tag{63}$$

and

$$[I^\mu_a]_0 = -[T^\mu_a(U'^{\text{sym.}}_{...}, U'^{\text{sym.}}_{...})]_0 - 2T^\mu_a(U'^{\text{sym.}}_{...}, U'^{\text{mean}}_{...})_0. \tag{64}$$

For calculating $T^\mu_r(U'^{\text{sym.}}_{...}, U'^{\text{sym.}}_{...})$ and $T^\mu_a(U'^{\text{sym.}}_{...}, U'^{\text{sym.}}_{...})$ we have to put $U^A_{...} = U^B_{...} = U'^{\text{sym.}}_{...}$ in (62). Since $U'^{\text{sym.}}_{...}$ is antisymmetric in a and r it follows that $T_{\mu\nu}(U'^{\text{sym.}}_{...}, U'^{\text{sym.}}_{...})$ is symmetric in a and r. Thus it is easy to see from (62) that if we interchange* a and r in the calculation of $T^\mu_r(U'^{\text{sym.}}_{...}, U'^{\text{sym.}}_{...})$ we get $T^\mu_a(U'^{\text{sym.}}_{...}, U'^{\text{sym.}}_{...})$.

---

* This interchange also implies interchange of ϵ_r and ϵ_a.

But $[T_r^\mu(U'^{\text{sym.}}_{\cdots}, U'^{\text{sym.}}_{\cdots})]_0$ is independent of ϵ and is a function only of the particle variables at r so that the interchange of a and r does not affect it. Therefore we must have

$$[T_a^\mu(U'^{\text{sym.}}_{\cdots}, U'^{\text{sym.}}_{\cdots})]_0 = [T_r^\mu(U'^{\text{sym.}}_{\cdots}, U'^{\text{sym.}}_{\cdots})]_0. \tag{65}$$

On the other hand, since $U'^{\text{mean}}_{\cdots}$ is symmetric in a and r while $U'^{\text{sym.}}_{\cdots}$ is antisymmetric, it follows that $T^{\mu\nu}(U'^{\text{sym.}}_{\cdots}, U'^{\text{mean}}_{\cdots})$ must be antisymmetric in a and r. Therefore by an analogous reasoning

$$[T_a^\mu(U'^{\text{sym.}}_{\cdots}, U'^{\text{mean}}_{\cdots})]_0 = -[T_r^\mu(U'^{\text{sym.}}_{\cdots}, U'^{\text{mean}}_{\cdots})]_0. \tag{66}$$

Therefore on subtracting (64) from (63) we get

$$[I_r^\mu]_0 - [I_a^\mu]_0 = 2[T_r^\mu(U'^{\text{sym.}}_{\cdots}, U'^{\text{sym.}}_{\cdots})]. \tag{67}$$

The left side being the difference between the rates of inflow on two tubes is a perfect differential and therefore so also is the right side. From this (51) follows immediately.

Since $U'^{\text{sym.}}_{\cdots}$ satisfies the wave equation (8) at all points not on the world line, it follows that

$$\partial^\nu T_{\mu\nu}(U'^{\text{sym.}}_{\cdots}, U'^{\text{sym.}}_{\cdots}) = 0,$$

except on the world line. We can therefore take for the modified stress tensor of the field,

$$\begin{aligned}
T'_{\mu\nu} &= T_{\mu\nu} - T_{\mu\nu}(U'^{\text{sym.}}_{\cdots}, U'^{\text{sym.}}_{\cdots}) \\
&= 2T_{\mu\nu}(U'^{\text{sym.}}_{\cdots}, U'^{\text{mean}}_{\cdots}) + T_{\mu\nu}(U'^{\text{mean}}_{\cdots}, U'^{\text{mean}}_{\cdots}).
\end{aligned} \tag{68}$$

This tensor satisfies the conservation equation except on the world line and leads to the same equations of motion as the original tensor. It has the advantage that the worst singularities of $T_{\mu\nu}$ which are contained in the part $T_{\mu\nu}(U'^{\text{sym.}}_{\cdots}, U'^{\text{sym.}}_{\cdots})$ are absent from it. It does not contain in particular, the infinite static field energy of the point particle. The new angular momentum tensor is then defined by (55) with $T'_{\mu\nu}$ in place of $T_{\mu\nu}$.

The theory given above can be generalized immediately to the case of several point particles. The modified symmetric field $U'^{\text{sym.}\,(s)}_{\cdots}$ of the sth particle is defined precisely as in (35). The actual field can then be split into

$$U^{\text{act.}}_{\cdots} = U'^{\text{mean}}_{\cdots} + \sum_{s=1}^{n} U'^{\text{sym.}\,(s)}_{\cdots}, \tag{69}$$

where

$$\begin{aligned}
U'^{\text{mean}}_{\cdots} &= \tfrac{1}{2}(U'^{\text{in.}}_{\cdots} + U'^{\text{out.}}_{\cdots}) - \sum_{s=1}^{n} \frac{\chi}{2} \int_{-\infty}^{\infty} D\left[S^{(s)}_{\cdots} \frac{J_1(\chi u^{(s)})}{u^{(s)}} \right] d\tau^{(s)} \\
&= U'^{\text{in.}}_{\cdots} + \sum_{1}^{n} \tfrac{1}{2} U'^{\text{rad.}\,(s)}_{\cdots}.
\end{aligned} \tag{70}$$

$u^{(s)}$ denotes the distance from the point x^μ to the point $z^{\mu(s)}$ on the world line of the sth particle. The equation of the sth particle is then

$$A_\mu^{(s)} = [T_\mu(U'^{\text{sym.}\,(s)}_{\cdots}, U'^{\text{mean}(s)}_{\cdots})]_0, \tag{71}$$

where
$$U'^{\text{mean}(s)}_{\cdots} \equiv U'^{\text{mean}}_{\cdots} + \sum_{t \neq s} U'^{\text{sym.}(t)}_{\cdots}$$

$$= U^{\text{act.}}_{\cdots} - U'^{\text{sym.}(s)}_{\cdots}$$

$$= U^{\text{in.}}_{\cdots} + \sum_{t \neq s} U^{\text{ret.}(t)}_{\cdots} + \tfrac{1}{2} U'^{\text{rad.}(s)}_{\cdots}$$

$$= U^{\text{in.}}_{\cdots} + \sum_{t \neq s} U'^{\text{ret.}(t)}_{\cdots} + \tfrac{1}{2} \overline{U}^{\text{rad.}(s)}_{\cdots} - \chi \int_{-\infty}^{\tau} D\left[S^{(s)}_{\cdots} \frac{J_1(\chi u)^{(s)}}{u^{(s)}} \right] d\tau^{(s)}. \tag{72}$$

It has already been shown that $T_{\mu\nu}(U'^{\text{sym.}(s)}_{\cdots}, U'^{\text{sym.}(s)}_{\cdots})$ has no effect in determining the motion of the sth particle. Since $U'^{\text{sym.}(s)}$ is certainly finite and continuous on the world line of the other particles, it is clear that this part of the stress tensor also has no effect in determining the motion of the other particles. Hence we can take as the modified stress tensor

$$T'_{\mu\nu} \equiv T_{\mu\nu} - \sum_{s=1}^{n} T_{\mu\nu}(U'^{\text{sym.}(s)}_{\cdots}, U'^{\text{sym.}(s)}_{\cdots})$$

$$= \sum_{s=1}^{n} T_{\mu\nu}(U'^{\text{sym.}(s)}_{\cdots}, U'^{\text{mean}}_{\cdots}) + \sum_{s>t=1}^{n} T_{\mu\nu}(U'^{\text{sym.}(s)}_{\cdots}, U'^{\text{sym.}(t)}_{\cdots}). \tag{73}$$

This tensor is conserved and gives the same equations of motion for each particle as $T_{\mu\nu}$. It also has the advantage that it does not contain the worst singularities contributed by the static field energies of the individual particles.

Finally, it may be mentioned that a 'Wenzel' field can be introduced in the general theory treated here just as in the case of a point charge moving in a Maxwell field (Dirac 1939). We take the world line of the particle to extend from $-\infty$ to a point τ_B. Now define the fundamental Wenzel solution for the particle by

$$O^W_{\cdots}(x^\mu) = \begin{cases} -\chi \displaystyle\int_{-\infty}^{\tau_B} S_{\cdots} \frac{J_1(\chi u)}{u} d\tau & \text{for } \tau_B < \tau_r. \\[2ex] \left(\dfrac{S_{\cdots}}{\kappa}\right)_r - \chi \displaystyle\int_{-\infty}^{\tau_r} S_{\cdots} \frac{J_1(\chi u)}{u} d\tau & \text{for } \tau_r < \tau_B < \tau_a, \\[2ex] \left(\dfrac{S_{\cdots}}{\kappa}\right)_r + \left(\dfrac{S_{\cdots}}{\kappa}\right)_a - \chi \displaystyle\int_{-\infty}^{\tau_r} S_{\cdots} \frac{J_1(\chi u)}{u} d\tau + \chi \displaystyle\int_{\tau_a}^{\tau_B} S_{\cdots} \frac{J_1(\chi u)}{u} d\tau & \text{for } \tau_a < \tau_B. \end{cases} \tag{74}$$

The first condition is satisfied when the point x^μ lies in the future light cone of the point τ_B, the second when it lies outside the light cone, and the third when it lies in the past light cone. This field is clearly just $O^{\text{ret.}}_{\cdots} - O^{\text{adv.}}_{\cdots}$ as given by (7) and (13) for a world line that extends from $-\infty$ to τ_B only. With the help of the Green's function given by one of us in an earlier paper (Bhabha 1939), which is a generalization of the relativistic delta-function of Jordan and Pauli, $O^W_{\cdots}$ can be written in the form

$$O^W_{\cdots} = \int_{-\infty}^{\tau_B} S_{\cdots} G^{\text{rad.}}(u^2) d\tau, \tag{75}$$

where

$$G^{\mathrm{rad.}}(u^2) = \begin{cases} 2\delta(u^2) - \chi'\dfrac{J_1(\chi u)}{u} & \text{for } u^2 > 0, \ u_0 > 0, \\[2mm] 0 & \text{for } u^2 < 0. \\[2mm] -2\delta(u^2) + \chi\dfrac{J_1(\chi u)}{u} & \text{for } u^2 > 0. \end{cases} \tag{76}$$

$G^{\mathrm{rad.}}$ satisfies the generalized wave equation (8) at all points not excepting $u^\mu = 0$, and hence $O^W_{...}$ satisfies (8) at all points including those on the world line. Define the generalized Wenzel potential by

$$U^W_{...} \equiv U^{\mathrm{in.}}_{...} + D[O^W_{....}]. \tag{77}$$

Now if λ^μ be a small time-like vector, then it is easily seen that

$$\operatorname*{Lt}_{\lambda \to 0} \tfrac{1}{2}\{U^W_{...}(z^\mu(\tau_B) + \lambda^\mu) + U^W_{...}(z^\mu(\tau_B) - \lambda^\mu)\} = U^{\mathrm{in.}}_{...}(z^\mu) + \tfrac{1}{2}\bar{U}^{\mathrm{rad.}}_{...}(z^\mu)$$

$$ - \chi \int_{-\infty}^{\tau_B} D\left[S_{....}\frac{J_1(\chi u)}{u}\right]d\tau = U'^{\mathrm{mean}}_{...}(z^\mu). \tag{78}$$

This is precisely the field (52) which enters into the equation of motion (51) of the point particle.

If n particles are present we have simply to define the generalized Wenzel potential by

$$U^W_{...} = U^{\mathrm{in.}}_{...} + \sum_{s=1}^{n} U^{W(s)}_{...}, \tag{79}$$

where

$$U^{W(s)}_{...} = D[O^{W(s)}_{....}],$$

and it follows at once that

$$\operatorname*{Lt}_{\lambda \to 0} \tfrac{1}{2}\{U^W_{...}(z^{\mu(s)} + \lambda^\mu) + U^W_{...}(z^{\mu(s)} - \lambda^\mu)\} = U'^{\mathrm{mean}(s)}_{...}, \tag{80}$$

which is the field that determines the motion of the sth particle.

REFERENCES

Bhabha 1939 *Proc. Roy. Soc.* A, **172**, 384–409.
Bhabha & Harish-Chandra 1944 *Proc. Roy. Soc.* A, **183**, 134–141.
Dirac 1938 *Proc. Roy. Soc.* A, **167**, 148–169.
Dirac 1939 *Ann. Inst. Poincaré*, **9**, 13–49.
Harish-Chandra 1944 *Proc. Roy. Soc.* A, **185**, 269–287.

Reprinted from
Proc. Royal Soc. A.
185 (1946), 250–268

Reprinted without change of pagination from the
Proceedings of the Royal Society, A, *volume* 185, 1946

On the equations of motion of point particles

By Harish-Chandra, *J. H. Bhabha Student, Cosmic Ray Research Unit,*
Indian Institute of Science, Bangalore

(*Communicated by H. J. Bhabha, F.R.S.—Received 1 November* 1944)

The equations of motion of a point particle, possessing a charge, dipole or higher multipole
moments and having an interaction with a generalized wave field of an arbitrary integral
spin, are obtained explicitly. These equations are independent of the particular choice of the
energy-momentum tensor of the field from among the many alternatives given by Fierz.
A convenient method for the calculation of the radiation field is given.

The definition of the spin of the particle is given, and it is postulated that the magnitude
of this spin is constant. It is shown that this assumption effects a great simplification in the
equations of a charged dipole in a vector-meson field. The equations are determined completely in terms of two arbitrary constants which are to be interpreted as the mass and the
spin. Further, it is shown that only those dipoles for which the 'electric' and the 'magnetic'
moments are parallel in the rest system are consistent with this assumption. The equations
of motion of a charged dipole in a scalar-meson field are derived. These equations also contain
only two arbitrary constants which again are to be interpreted as the mass and the spin.

1. In the preceding paper (Bhabha & Harish-Chandra 1944*a*, referred to in this
paper as A) it has been shown that the equations of motion of a point particle, having
an interaction with a generalized wave field,† are determined only by a few terms in
the inflow, denoted in A by $[T_\mu'^{\text{mix}}.]_0$. It is gratifying to note that the reaction of
radiation is completely taken into account if instead of the ingoing field we consider
the modified mean field as the effective field acting on the particle. In the present
paper it will be shown that the actual calculation of even the mixed terms of the
inflow is not necessary, and it is possible to write down the equations of motion
straight away. After giving the general theory for fields of any integral spin we shall
illustrate the method by considering the vector and scalar-meson fields in particular.

As pointed out in A, the equations of motion contain explicitly only the modified
mean field. However, for these equations to be practically useful it is necessary to
calculate the radiation field or rather the modified radiation field in terms of the
particle variables. In fact, the radiation field is the only quantity which now need
be computed for the determination of the equations of motion. A method, which
makes this calculation easy even for particles possessing multipole moments, will
be given in this paper.

Further, a definition of spin angular momentum of a point particle is given. For
brevity it is called the spin of the particle. It is postulated that the magnitude of
the spin is constant, and it is shown that this postulate has important consequences
in reducing the arbitrariness in the character of the dipole for the vector-meson field
and in determining the value of the spin in terms of the quantities describing the
particle. For instance, it will be shown that only those dipoles for which the 'electric'

† We shall use the various terms, e.g. generalized wave field, mean field, modified mean
field, etc., in the same sense as in A.

[269]

and the 'magnetic' moments are parallel are compatible with this assumption.†
Besides, this constancy of the spin will be found to be of great assistance in determining the actual form of the equations of motion in the case of the vector and
scalar-meson fields.

2. As far as possible we shall keep to the notation of A. Unless it is explicitly
mentioned otherwise, all symbols are to be understood in the same sense as in A.
The fundamental metric tensor is taken to be

$$g_{\mu\nu} = 0, \quad \mu \neq \nu, \quad g_{00} = -g_{11} = -g_{22} = -g_{33} = 1.$$

We shall now derive the equations of motion of a point particle interacting with
a generalized wave field of an arbitrary integral spin f.

A field of spin f is characterized by a symmetric tensor $U_{\alpha\beta...\gamma}$ of fth rank satisfying
the following two conditions (cf. Fierz 1939):

$$\partial^\alpha U_{\alpha\beta...\gamma} = 0, \quad U^\beta_{\beta...\gamma} = 0. \tag{2.1‡}$$

Following Fierz (1939) we define $U^{(q)}_{...}$ by induction for all values of $q \leqslant f$ in the
following way:

$$U^{(1)}_{[\alpha\beta]...\gamma} = \partial_\alpha U_{\beta...\gamma} - \partial_\beta U_{\alpha...\gamma}, \quad U^{(q+1)}_{...[\alpha\beta]...} = \partial_\alpha U^{(q)}_{...\beta...} - \partial_\beta U^{(q)}_{...\alpha...}. \tag{2.2}$$

In these equations the dots indicate that the various indices, their order and the
way they are paired inside square brackets is exactly the same on the two sides of
the equation except for the indices which are explicitly written. An explicitly written
index is not to be taken as paired unless the pairing is explicitly shown. As shown
by Fierz (1939) the energy-momentum tensor can be written as

$$4\pi T^{(q)\nu}_\mu = \left(\frac{1}{2\chi^2}\right)^{q-1}\{ - U^{(q)}_{...[\sigma\mu]...} U^{(q)...[\sigma\nu]...} + \tfrac{1}{4}\delta^\nu_\mu U^{(q)}_{...} U^{(q)...} \\ + \chi^2(U^{(q-1)}_{...\mu} U^{(q-1)...\nu} - \tfrac{1}{2}\delta^\nu_\mu U^{(q-1)}_{...} U^{(q-1)...})\}. \tag{2.3}$$

In (2.3) dots denote the indices which are to be contracted. They appear in the same
order and with the same pairing on the two factors which are contracted with each
other. Utilizing (2.1), (2.2) and the relation

$$\partial_\alpha U^{(q)}_{...[\beta\gamma]...} + \partial_\gamma U^{(q)}_{...[\alpha\beta]...} + \partial_\beta U^{(q)}_{...[\gamma\alpha]...} = 0$$

which follows from (2.2), we get

$$4\pi\partial^\mu T^{(q)\nu}_\mu = \frac{1}{(2\chi^2)^{q-1}}\{\partial^\mu\partial_\mu U^{(q-1)}_{...\sigma...} + \chi^2 U^{(q-1)}_{...\sigma...}\} U^{(q)...[\sigma\nu]...}.$$

We put
$$\partial^\mu\partial_\mu U^{(q-1)}_{...} + \chi^2 U^{(q-1)}_{...} = 4\pi\rho^{(q-1)}_{...}, \tag{2.4}$$

so that
$$\partial^\mu T^{(q)\nu}_\mu = \frac{1}{(2\chi^2)^{q-1}}\rho^{(q-1)}_{\sigma...} U^{(q)[\sigma\nu]...}, \tag{2.5}$$

---

† A similar result has been obtained by Myron Mathisson (1942).

‡ $\partial_\alpha \equiv \dfrac{\partial}{\partial x^\alpha}$.

due to the symmetry of $U_{...}$. If we put

$$\partial^\mu \partial_\mu U_{...} + \chi^2 U_{...} = 4\pi \rho_{...}, \qquad (2\cdot6)$$

it follows that $\rho^{(q-1)}_{...}$ is derived from $\rho_{...}$ by a process of differentiation analogous to $(2\cdot2)$. Except for conditions similar to $(2\cdot1)$ which it has to satisfy, $\rho_{...}$ is arbitrary, and for the present we take it as a finite function which vanishes outside a narrow time-like tube P. We call $\rho_{...}$ the charge density. One of the solutions of $(2\cdot4)$ for a given $\rho_{...}$ is

$$U_{...}(x_\mu) = \int^{x_\bullet} D(x_\mu - x'_\mu)\, \rho_{...}(x'_\mu)\, d_4 x', \qquad (2\cdot7)$$

where $D(x_\mu - x'_\mu)$ is the well-known generalization of the delta function of Jordan and Pauli for the case $\chi \neq 0$, and $\int^{x_\bullet}$ denotes that we have to integrate over the whole four-dimensional volume lying in the past of the point x_μ, i.e. such that $x_0 - x'_0 \geqslant 0$. We call this solution the retarded solution and denote it by $U^{\text{ret.}}_{...}$. The general solution of $(2\cdot6)$ is then $U^{\text{ret.}}_{...} + U^{\text{in.}}_{...}$, where $U^{\text{in.}}_{...}$ is any solution of the homogeneous wave equation satisfying $(2\cdot1)$. As usual we assume $U^{\text{in.}}_{...}$ to be finite and differentiable. We construct $U^{(q)\text{ret.}}_{...}$ and $U^{(q)\text{in.}}_{...}$ from $U^{\text{ret.}}_{...}$ and $U^{\text{in.}}_{...}$ in a manner similar to $(2\cdot2)$. Now, if we use the notation of A and write

$$2T^{(q)}_{\mu\nu}(U^{\text{ret.}}_{...},\, U^{\text{in.}}_{...}) = T^{(q)\text{mix.}}_{\mu\nu},$$

then it is not difficult to show that

$$\partial^\mu T^{(q)\text{mix.}}_{\mu\nu} = \frac{1}{(2\chi^2)^{q-1}}\, \rho^{(q-1)\sigma...} U^{(q)\text{in.}}_{...[\sigma\nu]...}. \qquad (2\cdot8)$$

Let us now surround P by a wider time-like tube Q. Also let A and B be two space-like surfaces intersecting Q such that A lies in the future of B. Integrating $(2\cdot8)$ over the four-dimensional volume V enclosed by Q, A and B we get

$$\int_V \partial^\mu T^{(q)\text{mix.}}_{\mu\nu}\, d_4 x = \frac{1}{(2\chi^2)^{q-1}} \int_V \rho^{(q-1)\sigma...} U^{(q)\text{in.}}_{...[\sigma\nu]...}\, d_4 x. \qquad (2\cdot9)$$

By Gauss's theorem we can write the left side as a surface integral over the boundary of V. Denoting an element of the surface of this boundary by dS_ν, we get

$$\int_V \partial^\mu T^{(q)\text{mix.}}_{\mu\nu}\, d_4 x = -\int_Q T^{(q)\text{mix.}}_{\mu\nu}\, dS^\mu + \int_A T^{(q)\text{mix.}}_{\mu\nu}\, dS^\mu - \int_B T^{(q)\text{mix.}}_{\mu\nu}\, dS^\mu. \qquad (2\cdot10)$$

Here we have to take the future normal for A and B and the outward normal for Q. From $(2\cdot9)$ and $(2\cdot10)$ it follows that the inflow due to $T^{(q)\text{mix.}}_{\mu\nu}$ is

$$\int_Q T^{(q)\text{mix.}}_{\mu\nu}\, dS^\mu = -\frac{1}{(2\chi^2)^{q-1}} \int_V \rho^{(q-1)\sigma...} U^{(q)\text{in.}}_{[\sigma\nu]...}\, d_4 x + \mathscr{J}^{(q)}_\nu(A) - \mathscr{J}^{(q)}_\nu(B), \qquad (2\cdot11)$$

where we have written $\mathscr{J}^{(q)}_\nu(A)$ and $\mathscr{J}^{(q)}_\nu(B)$ for the corresponding integrals over A and B in $(2\cdot10)$. Also by partial integration we prove that

$$-\frac{1}{(2\chi^2)^{q-1}} \int_V \rho^{(q-1)\sigma...} U^{(q)\text{in.}}_{[\sigma\nu]...}\, d_4 x = \mathscr{J}'^{(q)}_\nu(A) - \mathscr{J}'^{(q)}_\nu(B) - \frac{1}{(2\chi^2)^{q-2}} \int_V \rho^{(q-2)\sigma...} U^{(q-1)\text{in.}}_{[\sigma\nu]...}\, d_4 x,$$

where the $\mathscr{I}''$s are surface integrals over A and B respectively. Therefore it follows by successive reduction that

$$-\frac{1}{(2\chi^2)^{q-1}}\int_V \rho^{(q-1)\sigma\cdots}U^{(q)\text{in.}}_{[\sigma\nu]\cdots}\, d_4x = \mathscr{I}'_\nu(A) - \mathscr{I}'_\nu(B) - \int_V \rho^{\sigma\cdots}U^{(1)\text{in.}}_{[\sigma\nu]\cdots}\, d_4x. \qquad (2\cdot12)$$

Now if the charge density $\rho_{\cdots}$ can be written in the form

$$\rho_{\mu\nu\cdots} = \partial^{\alpha\beta\cdots}\Sigma_{\alpha\beta\cdots,\,\mu\nu\cdots}, \qquad (2\cdot13)$$

where $\partial^{\alpha\beta\cdots} \equiv \partial^\alpha\partial^\beta\cdots$, then we call $\Sigma_{\alpha\beta\cdots,\,\mu\nu\cdots}$ the multipole density. Obviously we can always choose $\Sigma_{\alpha\beta\cdots,\,\mu\nu\cdots}$ symmetric in $\alpha\beta\cdots$. Let the number of indices α, β, $\cdots$ corresponding to the number of differentiations in $(2\cdot13)$ be n. Then by further partial integration we can prove from $(2\cdot12)$ that

$$-\frac{1}{(2\chi^2)^{q-1}}\int_V \rho^{(q-1)\sigma\cdots}U^{(q)\text{in.}}_{[\sigma\nu]\cdots}\, d_4x$$

$$= \mathscr{I}''_\nu(A) - \mathscr{I}''_\nu(B) + (-1)^{n+1}\int_V \Sigma^{\alpha\beta\cdots,\,\sigma\cdots}\partial_{\alpha\beta\cdots}\, U^{(1)\text{in.}}_{[\sigma\nu]\cdots}\, d_4x,. \qquad (2\cdot14)$$

where $\mathscr{I}'''_\nu$'s are again surface integrals over A and B. From $(2\cdot14)$ and $(2\cdot11)$ we get, if we write $\mathscr{I}^{(q)}_\nu + \mathscr{I}''_\nu = \mathscr{I}'''_\nu$,

$$\int_Q T^{(q)\text{mix.}}_{\mu\nu}\, dS^\mu = \mathscr{I}'''_\nu(A) - \mathscr{I}'''_\nu(B) + (-1)^{n+1}\int_V \Sigma^{\alpha\beta\cdots,\,\sigma\cdots}\partial_{\alpha\beta\cdots}\, U^{(1)\text{in.}}_{[\sigma\nu]\cdots}\, d_4x. \qquad (2\cdot15)$$

Till now we have calculated with arbitrary functions $\rho_{\cdots}$ and $\Sigma_{\alpha\beta\cdots,\,\cdots}$ which vanish outside a narrow tube P. Let P be characterized by a parameter η such that as $\eta \to 0$, P narrows down and ultimately tends to a world line inside it. We now apply a limiting process in which η is made to tend to zero while the value of $\Sigma_{\alpha\beta\cdots,\,\cdots}$ inside P is made to tend to infinity. Denoting the value of $\Sigma^{\alpha\beta\cdots,\,\cdots}$ for a particular value of η by $\Sigma^{\alpha\beta\cdots,\,\cdots}_{(\eta)}$, we impose the condition that

$$\lim_{\eta\to0}\int_V \Sigma^{\alpha\beta\cdots,\,\cdots}_{(\eta)}\, d_4x = \int_{\tau_1}^{\tau_2} S^{\alpha\beta\cdots,\,\cdots}_{(\tau)}\, d\tau, \qquad (2\cdot16a)$$

where V is any four-dimensional volume and $S^{\alpha\beta\cdots,\,\cdots}_{(\tau)}$ is a given function of τ on the world line to which P tends in the limit $\eta \to 0$; the portion between τ_1 and τ_2 $(\tau_2 > \tau_1)$ of this world line being contained inside V. Using Dirac's δ-function $(2\cdot16a)$ can be conveniently expressed as

$$\lim_{\eta\to0}\Sigma^{\alpha\beta\cdots,\,\cdots}_{(\eta)} = \int_{-\infty}^{\infty} S^{\alpha\beta\cdots,\,\cdots}_{(\tau)}\, \delta(x-z(\tau))\, d\tau, \qquad (2\cdot16b)$$

where $\qquad \delta(x-z(\tau)) \equiv \delta(x_0-z_0(\tau))\,\delta(x_1-z_1(\tau))\,\delta(x_2-z_2(\tau))\,\delta(x_3-z_3(\tau)),$

$z_\mu(\tau)$ being the co-ordinate of a point τ of the world line. From $(2\cdot13)$, $(2\cdot16)$ and $(2\cdot7)$ it is not difficult to prove that in the limit $\eta \to 0$ $U^{\text{ret.}}_{\cdots}$ tends to the value

$$U^{\text{ret.}}_{\cdots} = \partial^{\alpha\beta\cdots}(O^{\text{ret.}}_{\alpha\beta\cdots,\,\cdots}) \qquad (2\cdot17)$$

in the notation of A. Our field $U_{\cdots}$ therefore corresponds to that of a multipole of order n.

From (2·15) and (2·16) we get

$$\lim_{\eta \to 0} \int_Q T^{(q)\,\mathrm{mix.}}_{\mu\nu} \, dS^\mu = (\mathscr{J}_\nu'''(A) - \mathscr{J}_\nu'''(B))_{\lim \eta \to 0} + (-1)^{n+1} \int_{\tau_1}^{\tau_2} S^{\alpha\beta\dots,\,\sigma\dots} \partial_{\alpha\beta\dots} U^{(1)\,\mathrm{in.}}_{[\sigma\nu]\dots} \, d\tau. \tag{2.18}$$

The second term on the right is finite. Therefore the first term on the right must also tend to a finite limit since the left side is finite ($U_{\dots}$ being finite on Q). Now if Q also is characterized by a parameter ϵ such that Q tends to the same world line as P as $\epsilon \to 0$, then we have

$$\lim_{\epsilon \to 0} \lim_{\eta \to 0} \int_Q T^{(q)\,\mathrm{mix.}}_{\mu\nu} \, dS^\mu$$

$$= \lim_{\epsilon \to 0} \lim_{\eta \to 0} (\mathscr{J}_\nu'''(A) - \mathscr{J}_\nu'''(B)) + (-1)^{n+1} \int_{\tau_1}^{\tau_2} S^{\alpha\beta\dots,\,\sigma\dots} \partial_{\alpha\beta\dots} U^{(1)\,\mathrm{in.}}_{[\sigma\nu]\dots} \, d\tau. \tag{2.19}$$

$\mathscr{J}_\nu'''(A)$ and $\mathscr{J}_\nu'''(B)$ now become in the limit $\epsilon \to 0$ functions—though infinite—of the condition at the points τ_2 and τ_1 respectively (cf. Bhabha & Harish-Chandra 1944b). Therefore they can contribute only perfect differentials to the rate of inflow. Ignoring these perfect differentials which are not significant from the point of view of the equations of motion we get the rate of inflow $I_\nu^{(q)\,\mathrm{mix.}}$ due to $T^{(q)\,\mathrm{mix.}}_{\mu\nu}$ in the form

$$I_\nu^{(q)\,\mathrm{mix.}} = (-1)^{n+1} S^{\alpha\beta\dots,\,\sigma\dots} \partial_{\alpha\beta\dots} U^{(1)\,\mathrm{in.}}_{[\sigma\nu]\dots}. \tag{2.20}$$

Therefore the equation of translational motion is in accordance with equation (51) of A

$$A_\nu = (-1)^{n+1} S^{\alpha\beta\dots.\,\sigma\dots} \partial_{\alpha\beta\dots} U^{(1)'\,\mathrm{mean}}_{[\sigma\nu]\dots}. \tag{2.21}$$

Here we have replaced $U^{\mathrm{in.}}_{\dots}$ in (2·20) by the modified mean field $U'^{\mathrm{mean}}_{\dots}$ in (2·21) in conformity with the results of A. It is noteworthy that (2·21) is independent of the particular choice of q.

For obtaining the rotational equation we have to deal with the angular momentum tensor

$$M^{(q)}_{\mu\nu,\,\lambda} = x_\mu T^{(q)}_{\nu\lambda} - x_\nu T^{(q)}_{\mu\lambda}.$$

Putting, as before,

$$M^{(q)\,\mathrm{mix.}}_{\mu\nu,\,\lambda} \equiv 2 M^{(q)}_{\mu\nu,\,\lambda}(U^{\mathrm{ret.}}_{\dots}, U^{\mathrm{in.}}_{\dots}) = x_\mu T^{(q)\,\mathrm{mix.}}_{\nu\lambda} - x_\nu T^{(q)\,\mathrm{mix.}}_{\mu\lambda},$$

we find that

$$\partial^\lambda M^{(q)\,\mathrm{mix.}}_{\mu\nu,\,\lambda} = x_\mu \partial^\lambda T^{(q)\,\mathrm{mix.}}_{\nu\lambda} - x_\nu \partial^\lambda T^{(q)\,\mathrm{mix.}}_{\mu\lambda}$$

$$= \frac{1}{(2\chi^2)^{q-1}} [x_\mu \rho^{(q-1)\sigma\dots} U^{(q)}_{[\sigma\nu]\dots}]_-,$$

where the minus sign at the end of the bracket denotes that the same terms are to be subtracted with the interchange of μ and ν. In the same way as before we get corresponding to (2·18)

$$\lim_{\eta \to 0} \int_Q M^{(q)\,\mathrm{mix.}}_{\mu\nu,\,\lambda} \, dS^\lambda = (\mathscr{M}_{\mu\nu}'''(A) - \mathscr{M}_{\mu\nu}'''(B))_{\lim \eta \to 0}$$

$$+ (-1)^{n+1} \int_{\tau_1}^{\tau_2} S^{\alpha\beta\dots,\,\sigma\dots}_{(\tau)} \{\partial_{\alpha\beta\dots} [x_\mu U^{(1)\,\mathrm{in.}}_{[\sigma\nu]\dots}]_-\}_{x_\mu = z_\mu(\tau)} \, d\tau. \tag{2.22}$$

Ignoring again the perfect differentials the rate of inflow $I^{(q)\,\text{mix.}}_{\mu\nu}$ of angular momentum due to $M^{(q)\,\text{mix.}}_{\mu\nu,\,\lambda}$ becomes

$$\begin{aligned}
I^{(q)\,\text{mix.}}_{\mu\nu} &= (-1)^{n+1}\, S^{\alpha\beta\dots,\,\sigma\dots}\{\partial_{\alpha\beta}\dots[x_\mu\, U^{(1)\,\text{in.}}_{[\sigma\nu]\dots}]_-\}_{x_\mu = z_\mu(\tau)} \\
&= (-1)^{n+1}\, S^{\alpha\beta\dots,\,\sigma\dots}\{z_\mu\,\partial_{\alpha\beta}\dots\, U^{(1)\,\text{in.}}_{[\sigma\nu]\dots} + n g_{\mu\alpha}\,\partial_\beta\dots\, U^{(1)\,\text{in.}}_{[\sigma\nu]\dots}\}_-.
\end{aligned} \tag{2.23}$$

The equation of rotational motion, therefore, is

$$\dot{\mathscr{B}}_{\mu\nu} = (-1)^{n+1}\, S^{\alpha\beta\dots,\,\sigma\dots}\{z_\mu\,\partial_{\alpha\beta}\dots\, U^{(1)'\,\text{mean}}_{[\sigma\nu]\dots} + n g_{\mu\alpha}\,\partial_\beta\dots\, U^{(1)'\,\text{mean}}_{[\sigma\nu]\dots}\}_-. \tag{2.24}$$

From (2.21) and (2.24) we get

$$\begin{aligned}
\dot{B}_{\mu\nu} &\equiv \dot{\mathscr{B}}_{\mu\nu} - \frac{d}{d\tau}(z_\mu A_\nu - z_\nu A_\mu) \\
&= n(-1)^{n+1}[S_{\mu\beta\dots,}{}^{\sigma\dots}\,\partial^{\beta}\dots U^{(1)'\,\text{mean}}_{[\sigma\nu]\dots}]_- - (v_\mu A_\nu - v_\nu A_\mu).
\end{aligned} \tag{2.25}$$

It is interesting to note that only $U^{(1)'\,\text{mean}}_{\dots}$ and not $U'^{\,\text{mean}}_{\dots}$ appears in the equations of motion (2.21) and (2.25).

The above calculation applies only to spin $f > 0$. For $f = 0$ it fails, since in this case $U^{(1)}_{[\alpha\beta]}$ cannot be constructed. For $f = 0$ we have to start from the tensor

$$4\pi T_{\mu\nu} = U_\mu U_\nu - \tfrac{1}{2} g_{\mu\nu}(U_\sigma U^\sigma - \chi^2 U^2), \tag{2.26}$$

where $U_\mu \equiv \partial_\mu U$. Therefore we have

$$4\pi\partial^\mu T_{\mu\nu} = \{\partial^\mu \partial_\mu U + \chi^2 U\}\, U_\nu = 4\pi\rho U_\nu$$

and

$$\partial^\mu T^{\text{mix.}}_{\mu\nu} \equiv 2\partial^\mu T_{\mu\nu}(U^{\text{ret.}}_{\dots}, U^{\text{in.}}_{\dots}) = \rho U^{\text{in.}}_\nu.$$

The rest of the procedure is exactly the same as before and we get the equations of motion in the following form:

$$\dot{A}_\nu = (-1)^{n+1}\, S^{\alpha\beta\dots\gamma}\partial_{\alpha\beta\dots\gamma\nu}\, U'^{\,\text{mean}}, \tag{2.27a}$$

$$\dot{B}_{\mu\nu} = n(-1)^{n+1}[S^{\beta\dots\gamma}_\mu\,\partial_{\nu\beta\dots\gamma}\, U'^{\,\text{mean}}]_- - (v_\mu A_\nu - v_\nu A_\mu). \tag{2.27b}$$

To illustrate our result we apply it first to the vector-meson case. In this case $U^{(1)}_{[\sigma\nu]} \equiv G_{\sigma\nu}$. So we get

$$\dot{A}_\nu = (-1)^{n+1}\, S^{\alpha\beta\dots,\sigma}\partial_{\alpha\beta}\dots G'^{\,\text{mean}}_{\sigma\nu},$$

$$\dot{B}_{\mu\nu} = n(-1)^{n+1}[S_{\mu\beta\dots,}{}^{\sigma}\,\partial^{\beta}\dots G'^{\,\text{mean}}_{\sigma\nu}]_- - (v_\mu A_\nu - v_\nu A_\mu).$$

For the point charge we have to put $n = 0$ and $S^\sigma = gv^\sigma$, where g is a constant. Then we get

$$\dot{A}_\nu = -gv^\sigma G'^{\,\text{mean}}_{\sigma\nu} = g G'^{\,\text{mean}}_{\nu\sigma}\, v^\sigma, \tag{2.28}$$

as found by Dirac (1938) for the case of the electron. For a dipole ($n = 1$) we find

$$\begin{aligned}
\dot{A}_\nu &= S^{\alpha\sigma}\partial_\alpha G'^{\,\text{mean}}_{\sigma\nu} = \tfrac{1}{2}S^{\alpha\sigma}(\partial_\alpha G'^{\,\text{mean}}_{\sigma\nu} - \partial_\sigma G'^{\,\text{mean}}_{\alpha\nu}) \\
&= -\tfrac{1}{2}S^{\alpha\sigma}\partial_\nu G'^{\,\text{mean}}_{\alpha\sigma},
\end{aligned} \tag{2.29a}$$

$$\dot{B}_{\mu\nu} = [S_\mu{}^\sigma G'^{\,\text{mean}}_{\sigma\nu}]_- - (v_\mu A_\nu - v_\nu A_\mu). \tag{2.29b}$$

This is precisely the form of the mixed terms found by Bhabha & Corben (1941).

For the scalar meson case we get for a charge g

$$\dot{A}_\nu = -g U'^{\,\mathrm{mean}}_\nu, \tag{2.30}$$

and for a dipole S^α

$$\dot{A}_\nu = S^\alpha \partial_{\alpha\nu} U'^{\,\mathrm{mean}}, \tag{2.31a}$$

$$\dot{B}_{\mu\nu} = [S_\mu U'^{\,\mathrm{mean}}_\nu - S_\nu U'^{\,\mathrm{mean}}_\mu] - (v_\mu A_\nu - v_\nu A_\mu). \tag{2.31b}$$

3. We shall now consider the calculation of the radiation field. This is important because if we wish to put (2.21), (2.25) or (2.27) to any practical use we must express $U'^{\,\mathrm{mean}}_{\ldots}(z_\mu)$ in terms of $U^{\mathrm{in.}}_{\ldots}(z_\mu)$ and the particle variables at τ and in the past.

In the following we shall use the suffices r and a in the same sense as in A. We put

$$(u_\mu)_r \equiv x_\mu - z_\mu(\tau_r) = r_\mu, \quad (u_\mu)_a \equiv x_\mu - z_\mu(\tau_a) = a_\mu,$$

and try to express $a^2 \equiv a_\mu a^\mu$ in terms of $\tau_a - \tau_r \equiv \epsilon$ and the particle variables at τ_r. We have

$$0 = a^2 = a_\mu a^\mu = r^2 + \left(\frac{\partial u^2}{\partial \tau}\right)_r \epsilon + \frac{1}{2!}\left(\frac{\partial^2 u^2}{\partial \tau^2}\right)_r \epsilon^2 + \ldots$$

$$= \left(\frac{\partial u^2}{\partial \tau}\right)_r \epsilon + \frac{1}{2!}\left(\frac{\partial^2 u^2}{\partial \tau^2}\right)_r \epsilon^2 + \ldots. \tag{3.1}$$

We note that

$$\left.\begin{aligned}
\frac{\partial u^2}{\partial \tau} &= -2\kappa, \\[1ex]
\frac{\partial^2 u^2}{\partial \tau^2} &= -2\dot{\kappa} = -2(\kappa'-1), \\[1ex]
\frac{\partial^3 u^2}{\partial \tau^3} &= -2\ddot{\kappa} = -2\kappa'', \\[1ex]
\frac{\partial^4 u^2}{\partial \tau^4} &= -2\kappa^{\mathrm{III}} = -2\kappa''' + 2(v\ddot{v}), \\[1ex]
\frac{\partial^5 u^2}{\partial \tau^5} &= -2\kappa^{\mathrm{IV}} = -2\kappa'''' - 10(\dot{v}\ddot{v}),
\end{aligned}\right\} \tag{3.2}\dagger$$

where dots denote differentiation with respect to τ keeping x_μ constant and $\kappa^{(n\ \mathrm{dashes})} = u_\mu v^{\mu(n\ \mathrm{dots})}$. Substituting (3.2) in (3.1) and dividing by ϵ we get to the fourth order

$$\left\{-2\kappa + \epsilon, -\kappa'\epsilon, -\frac{\epsilon^2}{3}\kappa'' + \frac{\epsilon^3}{12}(v\ddot{v}), -\frac{\epsilon^3}{12}\kappa''' - \frac{\epsilon^4}{12}(\dot{v}\ddot{v})\right\}_r = 0. \tag{3.3}$$

Reversing (3.3) by the method of successive approximations we find that to the requisite order

$$\epsilon = \{2\kappa, + 2\kappa\kappa', + 2\kappa\kappa'^2 + \tfrac{4}{3}\kappa^2\kappa'' - \tfrac{2}{3}\kappa^3(v\ddot{v}),$$

$$+ 2\kappa\kappa'^3 + 4\kappa^2\kappa''\kappa' - \tfrac{8}{3}\kappa^3\kappa'(v\ddot{v}) + \tfrac{2}{3}\kappa^3\kappa''' + \tfrac{4}{3}\kappa^4(\dot{v}\ddot{v})\}_r. \tag{3.4}$$

$\dagger$ We write κ^{III} and κ^{IV} for the third and fourth derivatives, and so in similar cases.

Expressing κ as a Taylor series in powers of ϵ and using (3·2) and (3·4) we find

$$\kappa_a = \{-\kappa,, + \tfrac{2}{3}\kappa^2\kappa'' - \tfrac{2}{3}\kappa^3(v\ddot{v}), + \tfrac{4}{3}\kappa^2\kappa''\kappa' - 2\kappa^3\kappa'(v\ddot{v}) + 2\kappa^4(\dot{v}\ddot{v}) + \tfrac{2}{3}\kappa^3\kappa'''\}_r, \qquad (3·5)$$

i.e. $$\frac{1}{\kappa_a} = -\frac{1}{\kappa_r}\{1,, + \tfrac{2}{3}\kappa\kappa'' - \tfrac{2}{3}\kappa^2(v\ddot{v}), + \tfrac{4}{3}\kappa\kappa''\kappa' - 2\kappa^2\kappa'(v\ddot{v}) + 2\kappa^3(\dot{v}\ddot{v}) + \tfrac{2}{3}\kappa^2\kappa'''\}_r. \qquad (3·6)$$

If S is any function of τ, which is non-singular on the world line, we can expand $S(\tau)$ as a Taylor's series in powers of $\tau - \tau_r$. Using (3·4) we obtain in this way

$$S_a = \{S, + 2\kappa\dot{S}, + 2\kappa\kappa'\dot{S} + 2\kappa^2\ddot{S}, + \dot{S}(2\kappa\kappa'^2 + \tfrac{4}{3}\kappa^2\kappa' - \tfrac{2}{3}\kappa^3(\ddot{v}v)) + 4\kappa^2\kappa'\ddot{S} + \tfrac{4}{3}\kappa^3 S^{111}\}_r. \qquad (3·7)$$

Using (3·6) and (3·7) we find

$$\begin{aligned}
\left(\frac{S}{\kappa}\right)_a + \left(\frac{S}{\kappa}\right)_r = -\{ & 2\dot{S}, + S(\tfrac{2}{3}\kappa'' - \tfrac{2}{3}\kappa(\ddot{v}v)) + 2\kappa'\dot{S} + 2\kappa\ddot{S}, \\
& + S(\tfrac{4}{3}\kappa'\kappa'' - 2\kappa\kappa'(\ddot{v}v) + \tfrac{2}{3}\kappa\kappa''' + 2\kappa^2(\ddot{v}\dot{v})) \\
& + \dot{S}(\tfrac{8}{3}\kappa\kappa'' - 2\kappa^2(\ddot{v}v) + 2\kappa'^2) + 4\kappa\kappa'\ddot{S} + \tfrac{4}{3}\kappa^2 S^{111}\}_r. \qquad (3·8)
\end{aligned}$$

The right side of (3·8) is in fact a series in powers of r_ρ. Since it has been proved in A that the left side and all its differentials are finite and non-singular on the world line, we can expand it as a Taylor's series in powers of r_ρ. Writing

$$\left(\frac{S}{\kappa}\right)_a + \left(\frac{S}{\kappa}\right)_r \equiv \left(\frac{S}{\kappa}\right)_{ar},$$

and using the suffix 0 for denoting the value of any quantity at $x_\rho = z_\rho(\tau_r)$, we get

$$\left(\frac{S}{\kappa}\right)_{ar} = \left(\left(\frac{S}{\kappa}\right)_{ar}\right)_0 + r^\rho\left(\partial_\rho\left(\frac{S}{\kappa}\right)_{ar}\right)_0 + \frac{1}{2!}r^\rho r^\sigma\left(\partial_{\rho\sigma}\left(\frac{S}{\kappa}\right)_{ar}\right)_0 + \dots. \qquad (3·9)$$

This series must be the same as (3·8). Therefore the coefficients of r^ρ and $r^\sigma r^\rho$ (symmetrized with respect to σ and ρ) in (3·8) are $\left(\partial_\rho\left(\frac{S}{\kappa}\right)_{ar}\right)_0$ and $\frac{1}{2!}\left(\partial_{\sigma\rho}\left(\frac{S}{\kappa}\right)_{ar}\right)_0$ respectively. However, the coefficient of $r^\sigma r^\rho$ is indeterminate to the extent of any terms proportional to $g_{\sigma\rho}$ as $r^\sigma r^\rho g_{\sigma\rho} = r^\sigma r_\sigma = 0$. These terms in $\left(\partial_\sigma\partial_\rho\left(\frac{S}{\kappa}\right)_{ar}\right)_0$, which must be finite and unambiguous on the world line, are, however, easily determined by direct differentiation as follows. We notice that if $R(\tau)$ is a function of τ alone which does not depend on x_μ, then only terms of the type† $(\kappa R)_r$, $(\kappa\kappa^{(n)}R)_r$ and $(\kappa^2 R)_r$ in (3·8) can give rise to *finite and unambiguous* terms proportional to $g_{\sigma\rho}$ on differentiation. It is easy to see that

$$\left.\begin{aligned}
\partial_{\sigma\rho}(\kappa R)_r &\to g_{\sigma\rho}(\dot{R})_r, \\
\partial_{\sigma\rho}(\kappa\kappa^{(n)}R)_r &\to -g_{\sigma\rho}(vv^{(n)})_r R_r, \\
\partial_{\sigma\rho}(\kappa^2 R)_r &\to -2g_{\sigma\rho}R_r,
\end{aligned}\right\} \qquad (3·10)$$

where only the finite and unambiguous terms arising from the differentiations on the left are written on the right of the arrow.

<hr>

† $\kappa^{(n)} \equiv \kappa^{(n\ \text{dashes})}$, while $v^{(n)} \equiv v^{(n\ \text{dots})}$.

In this way we easily find from (3·8) that if S is independent of x_μ then

$$-\left(\partial_\sigma\left(\frac{S}{\kappa}\right)_{ar}\right)_0 = \{S(\tfrac{2}{3}\ddot{v}_\sigma - \tfrac{2}{3}v_\sigma(\ddot{v}v)) + 2\dot{v}_\sigma\dot{S} + 2v_\sigma\ddot{S}\}_0 \tag{3·11}$$

$$-\left(\partial_{\rho\sigma}\left(\frac{S}{\kappa}\right)_{ar}\right)_0 = [S\{\tfrac{4}{3}\dot{v}_\rho\ddot{v}_\sigma + \tfrac{4}{3}\ddot{v}_\rho\dot{v}_\sigma - 2v_\rho\dot{v}_\sigma(\ddot{v}v) - 2\dot{v}_\rho v_\sigma(\ddot{v}v) + \tfrac{2}{3}v_\rho v_\sigma^{\mathrm{iii}} + \tfrac{2}{3}v_\rho^{\mathrm{iii}}v_\sigma + 4v_\rho v_\sigma(\ddot{v}\dot{v})\}$$
$$+ \dot{S}\{\tfrac{8}{3}v_\rho\ddot{v}_\sigma + \tfrac{8}{3}\ddot{v}_\rho v_\sigma - 4v_\rho v_\sigma(\ddot{v}v) + 4\dot{v}_\rho\dot{v}_\sigma\} + 4(v_\rho\dot{v}_\sigma + \dot{v}_\rho v_\sigma)\ddot{S}$$
$$+ \tfrac{8}{3}v_\rho v_\sigma\dddot{S} - \tfrac{2}{3}g_{\rho\sigma}\{(\ddot{v}v)S + (\dot{v})^2\dot{S} + S^{\mathrm{iii}}\}]_0. \tag{3·12}$$

Even if S depends upon x_μ, e.g. $S = u_\mu$, (3·8) holds provided S is non-singular on the world line. Only the separation of terms of different orders by commas in (3·8) would not be correct because S, $\dot{S}$, $\ddot{S}$, ... may not all be of the same order as assumed there. The rule for equating the coefficients of (3·8) and (3·9) as well as (3·10) still holds. As an example we consider the evaluation of $\left(\partial_\sigma\left(\dfrac{Su^\rho}{\kappa}\right)_{ar}\right)_0$. Here S is supposed to be independent of x_μ. From (3·8) we find that up to the terms of the first order

$$-\left(\frac{Su^\rho}{\kappa}\right)_{ar} = \{-2Sv^\rho, +2\dot{S}u^\rho - 2\kappa'Sv^\rho - 2\kappa S\dot{v}^\rho - 4\kappa\dot{S}v^\rho\}_r.$$

From this we get immediately

$$-\left(\partial_\sigma\left(\frac{Su^\rho}{\kappa}\right)_{ar}\right)_0 = [2\dot{S}\delta_\sigma^\rho - 2\dot{v}_\sigma Sv^\rho - 2v_\sigma S\dot{v}^\rho - 4v_\sigma\dot{S}v^\rho]_0. \tag{3·13}$$

It is obvious that for higher differentials of $\left(\dfrac{S}{\kappa}\right)_{ar}$ and $\left(\dfrac{Su^\rho}{\kappa}\right)_{ar}$ we shall have to calculate (3·8) to terms of higher order, and equate the symmetrized coefficient of $r^\alpha r^\beta \ldots r^\sigma$ in the corresponding series to $\dfrac{1}{n!}\left(\partial^{\alpha\beta\ldots\sigma}\left(\dfrac{S}{\kappa}\right)_{ar}\right)_0$ except for terms which contain $g^{\mu\nu}$, where μ, ν are any two of the n indices α, β, ..., σ. These terms are to be obtained by a method similar to (3·10). Clearly such a method would enable us to calculate $(\bar{U}^{\mathrm{rad.}})_0$ of equation (33) of A in terms of the particle variables at τ_0. The modified radiation potential is then easily obtained since it is equal to

$$(U'^{\mathrm{rad.}}_{\ldots})_0 = (\bar{U}^{\mathrm{rad.}}_{\ldots})_0 - 2\chi\int_{-\infty}^{\tau_0}\left\{D\left[S\ldots\frac{J_1(\chi u)}{u}\right]\right\}_{x_\mu=z_\mu(\tau_0)}d\tau. \tag{3·14}$$

Now we shall illustrate our method by considering a charged dipole in the vector-meson field. Following Bhabha & Corben (1941) we use g_1 to denote the charge and $g_2 S_{\mu\nu}$ to denote the dipole. $S_{\mu\nu}$ is an antisymmetric tensor. The modified radiation potential is given by

$$U'^{\mathrm{rad.}}_\nu = U^{\mathrm{rad.}}_\nu - \chi\int_{-\infty}^{\infty}(g_1 v_\nu + g_2 S_{\sigma\nu}\partial^\sigma)\frac{J_1(\chi u)}{u}d\tau$$
$$= g_1\left\{\left(\frac{v_\nu}{\kappa}\right)_{ar} - \chi\left[\int_{-\infty}^{\tau} + \int_{-\infty}^{\tau_a}\right]v_\nu\frac{J_1(\chi u)}{u}d\tau\right\}$$
$$+ g_2\left\{\partial^\sigma\left(\frac{S_{\sigma\nu}}{\kappa}\right)_{ar} - \frac{\chi^2}{2}\left(\frac{S_{\sigma\nu}u^\sigma}{\kappa}\right)_{ar} + \chi^2\left[\int_{-\infty}^{\tau_r} + \int_{-\infty}^{\tau_a}\right]S_{\sigma\nu}u^\sigma\frac{J_2(\chi u)}{u^2}d\tau\right\}. \tag{3·15}$$

The modified radiation field is thus given by

$$G'^{\text{rad.}}_{\mu\nu} = \partial_\mu U'^{\text{rad.}}_\nu - \partial_\nu U'^{\text{rad.}}_\mu.$$

On using (3·12) and (3·13) we find that its value at the point $z_\mu(\tau_0)$ on the world line is given by

$$
\begin{aligned}
(G'^{\text{rad.}}_{\mu\nu})_0 = {} & g_1\Big\{\tfrac{4}{3}(\ddot{v}_\mu v_\nu - v_\mu \ddot{v}_\nu) + 2\chi^2 \int_{-\infty}^{\tau_0} (v_\nu u_\mu - v_\mu u_\nu)\frac{J_2(\chi u)}{u^2}\,d\tau\Big\}_0 \\
& + g_2\Big\{\tfrac{4}{3}(\ddot{v}\dot{v})\,S_{\mu\nu} + \tfrac{4}{3}(\dot{v})^2\,\dot{S}_{\mu\nu} + \tfrac{4}{3}S^{\text{iii}}_{\mu\nu} + 2[\tfrac{4}{3}v_\mu\,S'''_\nu + 2v_\mu(\dot{v})^2\,S'_\nu + \tfrac{1}{3}v^{\text{iii}}_\mu\,\mathcal{S}_\nu \\
& + \tfrac{2}{3}\ddot{v}_\mu S_{\nu\sigma}\dot{v}^\sigma + \tfrac{4}{3}\ddot{v}_\mu S'_\nu + \tfrac{4}{3}v_\mu\dot{S}_{\nu\sigma}\ddot{v}^\sigma + \tfrac{2}{3}\dot{v}_\mu S_{\nu\sigma}\ddot{v}^\sigma + \tfrac{1}{3}v_\mu S_{\nu\sigma}v^{\text{iii}\sigma} \\
& + 2v_\mu\ddot{S}_{\nu\sigma}\dot{v}^\sigma + 2\dot{v}_\mu S''_\nu + 2\dot{v}_\mu\dot{S}_{\nu\sigma}\dot{v}^\sigma + 2v_\mu\mathcal{S}_\nu(\ddot{v}\dot{v}) + (\dot{v})^2 v_\mu S_{\nu\sigma}\dot{v}^\sigma \\
& + (\dot{v})^2\dot{v}_\mu\mathcal{S}_\nu]_- + 2\chi^2[\tfrac{1}{2}\dot{S}_{\mu\nu} - \tfrac{1}{2}S_{\mu\rho}\dot{v}^\rho v_\nu - \tfrac{1}{2}\mathcal{S}_\mu\dot{v}_\nu - S'_\mu v_\nu]_- \\
& + 4\chi^2\int_{-\infty}^{\tau_0} S_{\mu\nu}\frac{J_2(\chi u)}{u^2}\,d\tau - 2\chi^3\int_{-\infty}^{\tau_0}[S_{\sigma\nu}u^\sigma u_\mu]_-\frac{J_3(\chi u)}{u^3}\,d\tau\Big\}_0 . \qquad (3\cdot16)\dagger
\end{aligned}
$$

In (3·16) we have to take $u_\mu = z_\mu(\tau_0) - z_\mu(\tau)$ inside the integrals.

In the scalar-meson field where the dipole is now characterized by $g_2 S^\alpha$ we have

$$
\begin{aligned}
U'^{\text{rad.}} = {} & g_1\Big\{\Big(\frac{1}{\kappa}\Big)_{ar} - \chi\Big[\int_{-\infty}^{\tau_r} + \int_{-\infty}^{\tau_a}\Big]\frac{J_1(\chi u)}{u}\,d\tau\Big\} \\
& + g_2\Big\{\partial_\sigma\Big(\frac{S^\sigma}{\kappa}\Big)_{ar} - \frac{\chi^2}{2}\Big(\frac{S_\sigma u^\sigma}{\kappa}\Big)_{ar} + \chi^2\Big[\int_{-\infty}^{\tau_r} + \int_{-\infty}^{\tau_a}\Big]S_\sigma u^\sigma\frac{J_2(\chi u)}{u^2}\,d\tau\Big\}. \qquad (3\cdot17)
\end{aligned}
$$

The radiation field is given by $U'^{\text{rad.}}_\mu = \partial_\mu U'^{\text{rad.}}$. Using (3·8) we find that the modified radiation field for the charge alone is

$$(U'^{\text{rad.}}_\mu)_0 = -2g_1\Big\{\tfrac{1}{3}\ddot{v}_\mu + \tfrac{1}{3}v_\mu(\dot{v})^2 + \chi^2 v_\mu - \chi^2\int_{-\infty}^{\tau_0} u_\mu\frac{J_2(\chi u)}{u^2}\,d\tau\Big\}_0 , \qquad (3\cdot18)$$

while the radiation field for a dipole at rest (i.e. $\dot{v}_\mu = \ddot{v}_\mu = \ldots = 0$) is

$$
(U'^{\text{rad.}}_\mu)_0 = g_2\Big\{-\tfrac{8}{3}S^{\text{iii}\sigma}v_\sigma v_\mu + \tfrac{2}{3}S^{\text{iii}}_\mu + \chi^2 S_\mu - 2\chi^2 v_\mu\dot{S}_\sigma v^\sigma \\
+ 2\chi^2\int_{-\infty}^{\tau_0} S_\mu\frac{J_2(\chi u)}{u^2}\,d\tau - 2\chi^3\int_{-\infty}^{\tau_0} S_\sigma u^\sigma u_\mu\frac{J_3(\chi u)}{u^3}\,d\tau\Big\}_0 . \qquad (3\cdot19)
$$

Equations (3·18) and (3·19) will be useful in calculating the scattering of scalar mesons by a neutron (or proton), which will be done in a subsequent paper.

As an illustration we now compare our results for the vector-meson field with those of Bhabha & Corben (1941). From equations (72) and (73) of their paper and

† We have used here the notation of Bhabha & Corben (1941)

$$\mathcal{S}_\nu = S_{\nu\sigma}v^\sigma, \qquad S^{(n\ \text{dashes})}_\nu = S^{(n\ \text{dots})}_{\mu\nu}v^\nu.$$

The minus sign at the end denotes that the terms inside the brackets are to be subtracted with interchange of μ, ν.

the results of A it follows that a possible set of values of the radiation reaction terms $T_\mu^{\text{react.}}$ and $C_{\lambda\mu}^{\text{react.}}$ of their paper can be written as[†]

$$T_\mu^{\text{react.}} = \tfrac{1}{2} g_1 F_{\mu\sigma}^{\text{rad.}} v^\sigma - \tfrac{1}{4} g_2 S^{\rho\sigma} \partial_\mu F_{\rho\sigma}^{\text{rad.}}, \qquad (3\cdot20a)$$

$$C_{\lambda\mu}^{\text{react.}} = \tfrac{1}{2} g_2 F_{\lambda\mu}^{\text{rad.}}. \qquad (3\cdot20b)$$

Of these the $C_{\lambda\mu}^{\text{react.}}$ terms are given by them in equations ($137b$) and (140) of their paper, and we find that if we put $h \doteq \tfrac{2}{3}$, $k_1 = -\tfrac{1}{3}$, $k_2 = -\tfrac{7}{15}$, $k_3 = \tfrac{1}{3}$, $k_4 = \tfrac{4}{3}$ and $k_6 = 0$ in these equations[‡] we get the same value as obtained from ($3\cdot20b$) by putting $\chi = 0$ in ($3\cdot16$), apart from terms proportional to $S_{\lambda\mu}$, which, however, cancel out from the equations of motion.

4. In this section the values of A_ν and $B_{\mu\nu}$ of the equations of motion ($2\cdot21$) and ($2\cdot25$) will be considered in detail.

In quantum mechanics every physical entity is to be treated as a field whose equations of motion can be derived by variation from a Lagrangian. We should therefore try to connect our scheme with the Lagrangian mechanics. In fact there already exists such a connexion in so far as we derive the equations of motion of the wave field from the Lagrangian $L^{W(q)}$

$$4\pi L^{W(q)} = \frac{1}{(2\chi^2)^{q-1}} [\tfrac{1}{4} U^{(q)}_{\ldots} U^{(q)\cdots} - \tfrac{1}{2} \chi^2 U^{(q-1)}_{\ldots} U^{(q-1)\cdots}].$$

When this wave field is in interaction with a particle (which itself is a field in the quantum sense), the total Lagrangian L is

$$L = L^P + L^I + L^{W(q)},$$

where L^P is the Lagrangian for the particle alone and L^I contains the interaction terms. In fact even L^I is completely fixed by the equations of motion of the wave field, and we have

$$L^I = -\frac{1}{(2\chi^2)^{q-1}} U^{(q-1)}_{\ldots} \rho^{(q-1)\cdots}.$$

However, L^P is entirely arbitrary. Now if $\rho^{\cdots}$ is given by ($2\cdot13$) where $\Sigma_{\ldots,\ldots}$ is a function only of the particle variables, then it follows by partial integration that L^I can be replaced by an equivalent form

$$L^{I'} = -\frac{(-1)^n}{(\chi^2)^{q-1}} (\partial_{\alpha\beta\ldots} \partial^{\mu'\nu'\cdots} U^{(q-1)}_{[\mu\mu'][\nu\nu']\ldots}) \Sigma^{\alpha\beta\ldots,\mu\nu\ldots},$$

where n is the number of the indices α, β, ... and μ', ν', ... are the $(q-1)$ indices paired to μ, ν, ... Thus L can be written as

$$L = L^P + L^{I'} + L^{W(q)}. \qquad (4\cdot1)$$

† We use $F_{\mu\nu}$ instead of $G_{\mu\nu}$ to denote the field strengths of the electromagnetic field ($\chi = 0$). In this case the modified and the unmodified fields are the same.

‡ The disagreement of at least one of the two coefficients which involve k_5 in equation (140) of their paper seems to be due to some mistake in the calculation of $C_{\rho\mu\,(2)}^{\text{react.}}$ by them.

The equations of motion of the particle should follow by variation of the particle variables occurring in L^P and $L^{I'}$. For the usual expressions for L^P, this process, as is well known, leads to singular equations. However, our object should be to find out a new expression for L^P such that the singularities arising from L^P and $L^{I'}$ in the variation mutually compensate each other and lead to non-singular equations. A somewhat similar compensation is used to cancel off these infinities in the method of Dirac (1938) which has been followed in A and here. Though we do not yet know the correct expression for L^P we can derive valuable information from (4·1). On splitting up $U^{\cdots}$ into $U^{\mathrm{ret.}\,\cdots}$ and $U^{\mathrm{in.}\,\cdots}$, we see from (4·1) that the equations of motion of the particle must be linear in the ingoing field. Secondly, the ingoing field can enter into the equations of motion only in the form

$$\frac{1}{(\chi^2)^{q-1}}\,\partial_{\alpha\beta\ldots}\,\partial^{\mu'\nu'\ldots}U^{(q-1)\,\mathrm{in.}}_{[\mu\mu'][\nu\nu']\ldots}$$

or with more differentiations. As the ingoing field satisfies the homogeneous wave equation we have

$$\frac{1}{(\chi^2)^{q-1}}\,\partial_{\alpha\beta\ldots}\,\partial^{\mu'\nu'\ldots}U^{(q-1)\,\mathrm{in.}}_{[\mu\mu'][\nu\nu']\ldots} = \dot{\partial}_{\alpha\beta\ldots}\,U^{\mathrm{in.}}_{\mu\nu\ldots}. \qquad (4\cdot2)$$

This shows that $U^{\mathrm{in.}}$ must occur in the equations of motion of the particle with at least n differentiations. Of course the explicit number of differentiations can be reduced by contracting two or more of the indices α, β, … and replacing $\partial^\alpha\partial_\alpha$ by $-\chi^2$. However, our rule implies that every term in the equations of motion containing $U^{\mathrm{in.}}$ should be capable of being put in a form such that $U^{\mathrm{in.}}$ appears with at least n differentiations and χ does not appear explicitly. The terms containing $U^{\mathrm{in.}}$ can arise only from the variation of $L^{I'}$, and if, as we have assumed, $\Sigma^{\alpha\beta\cdots,\,\mu\nu\cdots}$ is given by (2·16), it follows that these terms must contain besides $U^{\mathrm{in.}}$, $S^{\alpha\beta\cdots,\,\mu\nu\cdots}$ with or without dots and possibly v_μ and its differentials with respect to τ and only these. Further, these terms must be linear in $S^{\alpha\beta\cdots,\,\mu\nu\cdots}$. This condition is important since it shows that no other dimensional quantity can occur in these terms—not even χ if they are put in a proper form (*vide supra*). We note that the right sides of (2·21) and (2·25) already satisfy these conditions.

In the scalar-meson case we have to put

$$4\pi L^W = \tfrac{1}{2}U_\mu U^\mu - \tfrac{1}{2}\chi^2 U^{\,2},$$

$$L^{I'} = -(-1)^n\,(\partial_{\alpha\beta\ldots}\,U)\,\Sigma^{\alpha\beta\cdots}.$$

The same conditions as those given above follow for this case also.

Let us now consider (2·21). Let the part of A_μ which contains $U^{\mathrm{in.}}$ be called $A_\mu^{\mathrm{mix.}}$. Now each term of $A_\mu^{\mathrm{mix.}}$ must contain $U^{\mathrm{in.}}$ with at least n differentiations. This shows that $A_\mu^{\mathrm{mix.}}$ itself must satisfy the same condition. Otherwise some terms of $A_\mu^{\mathrm{mix.}}$ would contain $U^{\mathrm{in.}}$ with less than n differentiations. From (2·21) it follows that the dimensions of $A_\mu^{\mathrm{mix.}}$ are the same as those of $S^{\alpha\beta\cdots,\,\sigma\cdots}\partial_{\alpha\beta\ldots}\,U^{\mathrm{in.}}_{\nu\ldots}$. Therefore from dimensional considerations we can prove that $A_\mu^{\mathrm{mix.}}$ must be a function of

$\partial_{\alpha\beta...}\, U^{\text{in}}_{\nu...}$, $S^{\alpha\beta...,\sigma...}$ and v_μ. It cannot contain any dots on $S^{......}$ or v_μ or more than n differentiations of $U^{\text{in}}_{...}$. It is to be noted that every dot on $U^{\text{in}}_{...}$ corresponds to one differentiation since $\dfrac{\partial}{\partial\tau} = v^\sigma\partial_\sigma$.

We similarly define $B^{\text{mix}}_{\mu\nu}$. It must consist of terms of the form $Q_{\mu\nu,\,\alpha\beta...,\sigma...}\,\partial^{\alpha\beta...}U^{\text{in}.\sigma...}$, where the number of indices α, β, ... is at least n. From (2·25) the dimensions of $\dot{B}^{\text{mix}}_{\mu\nu}$ are those of $S_{\mu\beta...}\,\partial^{\beta...}_{(n-1)}U^{\text{in}}_{\nu...}$, where $\partial^{\beta...}_{(n-1)}$ has $n-1$ indices. It therefore follows that the dimensions of $Q_{\mu\nu,\,\alpha\beta...,...}$ must be $S_{\alpha\beta...,...} \times [\text{length}]^k$, where $k \geqslant 1$. However, no such quantity can be constructed from $S_{\alpha\beta,...}$, v_μ and differentiations with respect to τ, because v_μ is dimensionless and every dot is equivalent to $(\text{length})^{-1}$ in dimensions. Therefore

$$B^{\text{mix}}_{\mu\nu} = 0. \tag{4·3}$$

This fact, which had already been noticed by Bhabha & Corben (1941) for the special case of the electromagnetic field, will be of great help to us. It holds also for the scalar-meson case.

Now the equation (2·25) or (2·29b) expresses the conservation of angular momentum. In fact in (2·24) we can look upon $\mathscr{B}_{\mu\nu}$ as the total angular momentum of the particle so that $\dot{\mathscr{B}}_{\mu\nu}$ is balanced against the rate of decrease of the angular momentum of the field. In the same way, A_μ can be regarded as the 'mechanical' energy and momentum of the particle (cf. Pryce 1938). Now

$$B_{\mu\nu} = \mathscr{B}_{\mu\nu} - (z_\mu A_\nu - z_\nu A_\mu),$$

which corresponds to the subtraction of the 'orbital angular momentum' from the total. Therefore $B_{\mu\nu}$ can be looked upon as the spin-angular momentum, especially so since it is independent of the ingoing field. In case of an extended body the spin is the angular momentum about the centre of mass. It has only three spatial components in the system in which the centre of mass is at rest. Therefore in accordance with this interpretation we assume

$$B_{\mu\nu} v^\nu = 0. \tag{4·4a}$$

Besides, we postulate that the magnitude of the spin should remain constant, i.e.

$$\Omega^2 \equiv \tfrac{1}{2} B_{\mu\nu} B^{\mu\nu} = \text{constant}. \tag{4·4b}$$

We call Ω the spin of the particle. It will be seen presently that (4·4a) and (4·4b) are of great value in determining completely the equations of motion. Of course they are not essential from the point of view of logical consistency. In fact, Bhabha & Corben (1941) have given the equations of motion of a point dipole without assuming these conditions. However, their equations are too general and contain a large number of arbitrary constants. We shall see in § 5 that the conditions (4·4) reduce this arbitrariness very effectively.

We add here a few remarks about A_ν in (2·21) or (2·27). As pointed out in a previous paper (Bhabha & Harish-Chandra 1944b) A_ν must be a function of the conditions at the point $z_\mu(\tau)$ *alone*. Now we have *defined* $U^{\text{in.}}$ in terms of $U^{\text{ret.}}$ and the actual field $U^{\text{act.}}$ by the equation

$$U^{\text{in.}} = U^{\text{act.}} - U^{\text{ret.}}. \tag{4.5}$$

Since $U^{\text{ret.}}$ contains an integral over the past motion of the particle it is obvious that $U^{\text{in.}}$ at $z_\mu(\tau)$ is *not* a function of the conditions at the point τ alone, one of these conditions being of course $U^{\text{act.}}$ itself. However, we can express $U^{\text{ret.}}$ as (cf. equation (33) of A)

$$U^{\text{ret.}} = \overline{U}^{\text{ret.}} - \chi \int_{-\infty}^{\tau_r} D\left[S_{\dots} \frac{J_1(\chi u)}{u} \right] d\tau,$$

where $\overline{U}^{\text{ret.}}$ depends only on r_μ and the particle variables at τ_r. We can now define

$$\overline{U}^{\text{in.}} \equiv U^{\text{in.}} - \chi \int_{-\infty}^{\tau} D\left[S_{\dots} \frac{J_1(\chi u)}{u} \right] d\tau = U^{\text{act.}} - \overline{U}^{\text{ret.}}. \tag{4.6}$$

The value of $\overline{U}^{\text{in.}}$ at $z_\mu(\tau)$ is the limit of the right side of (4.6) as $r_\mu \to 0$. Therefore $\overline{U}^{\text{in.}}$ is a function of the conditions at τ alone. Thus A_ν cannot contain $U^{\text{in.}}$ alone but must do so only in the combination $\overline{U}^{\text{in.}}$. Since U'^{mean} differs from $\overline{U}^{\text{in.}}$ in the limit $r_\mu \to 0$ only by $\frac{1}{2}\overline{U}^{\text{rad.}}$ which is a function of the particle variables at τ, it follows that $U^{\text{in.}}$ can as well appear in the combination U'^{mean}. We shall see that this is in fact the case for a charged dipole in the scalar and vector-meson fields.

5. We shall now derive the equations of motion of a charged dipole in a vector-meson field on the basis of (4.4).

We denote the charge by g_1 and the dipole tensor by $g_2 S^{\mu\nu}$. $S^{\mu\nu}$ is an antisymmetric tensor and g_1 and g_2 are constants. Following Bhabha & Corben (1941) we assume that during the course of the motion $S_{\mu\nu}$ undergoes only a four-dimensional rotation so that if at the time τ_0 its components in a certain Lorentz frame are $(S_{\mu\nu})_0$, then at every later time τ there always exists a Lorentz frame in which the components have again the same value $(S_{\mu\nu})_0$. From this assumption we get immediately the two invariant equations

$$S_{\mu\nu} S^{\mu\nu} = \text{const.}, \tag{5.1a}$$

$$S_{\mu\nu} S^{*\mu\nu} = \text{const.} \tag{5.1b}$$

Here $S^{*\mu\nu}$ is the tensor dual to $S^{\mu\nu}$ and is given by

$$S^{*\mu\nu} = \tfrac{1}{2}\epsilon^{\mu\nu\mu\sigma} S_{\rho\sigma}, \tag{5.2a}$$

so that

$$S^{\mu\nu} = -\tfrac{1}{2}\epsilon^{\mu\nu\rho\sigma} S^{*}_{\rho\sigma}, \tag{5.2b}$$

where $\epsilon^{\mu\nu\rho\sigma}$ is antisymmetric in each pair of indices and $\epsilon^{0123} = 1$. For later use we note the following relations

$$\epsilon_{\mu\nu\rho\sigma}\epsilon^{\mu\alpha\beta\gamma} = -\Sigma \pm \delta_\nu^\alpha \delta_\rho^\beta \delta_\sigma^\gamma, \tag{5.3a}$$

$$\epsilon_{\mu\nu\rho\sigma}\epsilon^{\mu\nu\alpha\beta} = -2(\delta_\rho^\alpha \delta_\sigma^\beta - \delta_\sigma^\alpha \delta_\rho^\beta), \tag{5.3b}$$

where Σ denotes summation over all permutations of α, β, γ and the positive or the negative sign is to be taken according as the permutation is even or odd.

We put

$$S_{\mu\nu}v^\nu = \mathcal{S}_\mu,$$

$$S^*_{\mu\nu}v^\nu = \mathcal{S}^*_\mu,$$

$$\bar{S}_{\mu\nu} = S_{\mu\nu} - \mathcal{S}_\mu v_\nu + \mathcal{S}_\nu v_\mu,$$

$$\bar{S}^*_{\mu\nu} = S^*_{\mu\nu} - \mathcal{S}^*_\mu v_\nu + \mathcal{S}^*_\nu v_\mu,$$

so that $\bar{S}_{\mu\nu}v^\nu = \bar{S}^*_{\mu\nu}v^\nu = 0$, and in the rest system $\mathcal{S}^*_0 = 0$, $\mathcal{S}^*_1 = S^{23}$, $\mathcal{S}^*_2 = S^{31}$, $\mathcal{S}^*_3 = S^{12}$. We note the following relations:

$$\bar{S}^*_{\mu\nu} = \epsilon_{\mu\nu\rho\sigma}\mathcal{S}^\rho v^\sigma, \tag{5.4a}$$

$$\mathcal{S}^\mu = -\tfrac{1}{2}\epsilon^{\mu\nu\rho\sigma}v_\nu \bar{S}^*_{\rho\sigma}. \tag{5.4b}$$

(5.1) can now be written as

$$\tfrac{1}{2}\bar{S}_{\mu\nu}\bar{S}^{\mu\nu} + \mathcal{S}_\mu\mathcal{S}^\mu = \mathcal{S}_\mu\mathcal{S}^\mu - \mathcal{S}^*_\mu\mathcal{S}^{*\mu} = \text{const.}, \tag{5.5a}$$

$$\mathcal{S}_\mu\mathcal{S}^{*\mu} = \text{const.} \tag{5.5b}$$

From (2.28) and (2.29) we get the following equations of motion:

$$A_\nu = g_1 G'^{\,\text{mean}}_{\nu\sigma}v^\sigma - \tfrac{1}{2}g_2 S^{\rho\sigma}\partial_\nu G'^{\,\text{mean}}_{\rho\sigma}, \tag{5.6a}$$

$$\dot{B}_{\mu\nu} = g_2[S_\mu^{\;\sigma}G'^{\,\text{mean}}_{\sigma\nu}]_- - (v_\mu A_\nu - v_\nu A_\mu). \tag{5.6b}$$

In view of (4.4) we demand that

$$\tfrac{1}{2}\dot{B}_{\mu\nu}B^{\mu\nu} = g_2 S_\mu^{\;\sigma}G'^{\,\text{mean}}_{\sigma\nu}B^{\mu\nu} \equiv 0. \tag{5.7}$$

Since $G^{\text{in}}_{\sigma\nu}$ is entirely arbitrary (5.7) can hold identically only if

$$B^{\mu\nu}S_\nu^{\;\sigma} - B^{\sigma\nu}S_\nu^{\;\mu} = 0. \tag{5.8}$$

From (4.4a) and (5.8) we get in the rest system the following equations:

$$B^{12}S_2^{\;3} - B^{32}S_2^{\;1} = 0, \tag{5.9a}$$

$$B^{12}S_2^{\;0} + B^{13}S_3^{\;0} = B^{12}\mathcal{S}_2 + B^{13}\mathcal{S}_3 = 0, \tag{5.9b}$$

and those obtained from (5.9) by cyclic permutation of the indices 1, 2, 3. From (5.9) and (4.4a) we get

$$B_{\mu\nu} = I\bar{S}_{\mu\nu}, \tag{5.10a}$$

$$B_{\mu\nu} = I'\bar{S}^*_{\mu\nu} = I'\epsilon_{\mu\nu\rho\sigma}\mathcal{S}^\rho v^\sigma. \tag{5.10b}$$

Both these equations are valid when neither $\bar{S}_{\mu\nu}$ nor $\mathcal{S}^\rho$ is zero. In case one of them is zero we have to select that equation from the two for which the right side does not vanish. From (5.8) and (5.10) we derive straight away

$$\mathcal{S}^\mu = J\mathcal{S}^{*\mu} = \tfrac{1}{2}J\epsilon^{\mu\nu\rho\sigma}v_\nu\bar{S}_{\rho\sigma}, \qquad \bar{S}^{*\rho\sigma} = -J\bar{S}^{\rho\sigma}, \tag{5.11a}$$

$$\mathcal{S}^{*\mu} = J'\mathcal{S}^\mu = -\tfrac{1}{2}J'\epsilon^{\mu\nu\rho\sigma}v_\nu\bar{S}^*_{\rho\sigma}, \qquad \bar{S}^{\rho\sigma} = -J'\bar{S}^{*\rho\sigma}, \tag{5.11b}$$

according as we use (5.10a) or (5.10b). Of course the scalar quantities I, I', J and J' are related to each other. (5.11) shows that the 'electric' and the 'magnetic' moments in the rest system are parallel to each other. Therefore it follows from (5·5) that

$$\mathcal{S}_\mu \mathcal{S}^\mu = -\tfrac{1}{2}\bar{S}^*_{\mu\nu}\bar{S}^{*\mu\nu} = \text{const.,} \qquad (5\cdot12a)$$

$$\mathcal{S}^*_\mu \mathcal{S}^{*\mu} = -\tfrac{1}{2}\bar{S}_{\mu\nu}\bar{S}^{\mu\nu} = \text{const.} \qquad (5\cdot12b)$$

From (5·12) and (4·4b) it follows that I, I', J and J' are all constants.

We now put
$$G'^{\,\text{mean}}_{\mu\nu} v^\nu = \mathcal{G}_\mu, \qquad (5\cdot13a)$$

$$\bar{G}_{\mu\nu} = G'^{\,\text{mean}}_{\mu\nu} + v_\mu \mathcal{G}_\nu - v_\nu \mathcal{G}_\mu, \qquad (5\cdot13b)$$

so that $\bar{G}_{\mu\nu} v^\nu = 0$ and
$$A_\mu = \bar{A}_\mu + M v_\mu, \qquad (5\cdot14)$$

where $M = A_\mu v^\mu$ and $\bar{A}_\mu v^\mu = 0$. Thus by contracting (5·6b) with v^ν we get

$$\bar{A}_\mu = \dot{B}_{\mu\nu} v^\nu - g_2 \bar{S}_{\mu\sigma} \mathcal{G}^\sigma - g_2 \mathcal{S}_\sigma \bar{G}^\sigma{}_\mu. \qquad (5\cdot15)$$

Therefore, bearing in mind that $\dot{v}^\mu \dot{B}_{\mu\nu} v^\nu = -\dot{v}^\mu B_{\mu\nu}\dot{v}^\nu = 0$, we get

$$
\begin{aligned}
v^\mu \dot{\bar{A}}_\mu = -\dot{v}^\mu \bar{A}_\mu &= g_2 \dot{v}^\mu \bar{S}_{\mu\sigma}\mathcal{G}^\sigma + g_2 \mathcal{S}_\sigma \bar{G}^\sigma{}_\mu v^\mu \\
&= -g_2 v^\mu \dot{\bar{S}}_{\mu\sigma}\mathcal{G}^\sigma + g_2 \mathcal{S}_\sigma \bar{G}^\sigma{}_\mu \dot{v}^\mu.
\end{aligned}
\qquad (5\cdot16)
$$

On the other hand, from (5·6a) we get

$$v^\mu \dot{A}_\mu = -\tfrac{1}{2}g_2 S^{\rho\sigma} v^\nu \partial_\nu G'^{\,\text{mean}}_{\rho\sigma}$$

$$= -\tfrac{1}{2}g_2 \frac{d}{d\tau}(S^{\rho\sigma} G'^{\,\text{mean}}_{\rho\sigma}) + \tfrac{1}{2}g_2(\dot{S}^{\rho\sigma} G'^{\,\text{mean}}_{\rho\sigma})$$

$$= -\tfrac{1}{2}g_2 \frac{d}{d\tau}(S^{\rho\sigma} G'^{\,\text{mean}}_{\rho\alpha}) + \tfrac{1}{2}g_2(\dot{\bar{S}}^{\rho\sigma} \bar{G}^{\,\text{mean}}_{\rho\sigma}) + g_2(\dot{\mathcal{S}}^\rho \mathcal{G}_\rho) + g_2(\mathcal{S}^\rho \bar{G}_{\rho\sigma}\dot{v}^\sigma) + g_2(\dot{\bar{S}}^{\rho\sigma}\mathcal{G}_\rho v_\sigma). \qquad (5\cdot17)$$

Subtracting (5·16) from (5·17) and using (5·14) we get

$$\frac{dM}{d\tau} = -\tfrac{1}{2}g_2 \frac{d}{d\tau}(S^{\rho\sigma} G'^{\,\text{mean}}_{\rho\sigma}) + \tfrac{1}{2}g_2(\dot{\bar{S}}^{\rho\sigma}\bar{G}_{\rho\sigma}) + g_2(\dot{\mathcal{S}}^\rho \mathcal{G}_\rho). \qquad (5\cdot18)$$

Now (5·6b) can be written in the following two forms according as we use (5·10a) or (5·10b):

$$I\dot{\bar{S}}_{\mu\nu} = g_2[S_{\mu\sigma} G^{\sigma'\,\text{mean}}_\nu]_- - (v_\mu A_\nu - v_\nu A_\mu), \qquad (5\cdot19a)$$

$$I'\dot{\bar{S}}^*_{\mu\nu} = g_2[S_{\mu\sigma} G^{\sigma'\,\text{mean}}_\nu]_- - (v_\mu A_\nu - v_\nu A_\mu). \qquad (5\cdot19b)$$

From (5·11a) and (5·19a) we find

$$\tfrac{1}{2}I\dot{\bar{S}}_{\mu\nu}\bar{G}^{\mu\nu} + I\dot{\mathcal{S}}_\mu \mathcal{G}^\mu = -\tfrac{1}{2}g_2 J \epsilon_{\mu\rho\sigma\alpha} v^\rho \bar{S}^{\sigma\alpha}\bar{G}^{\mu\nu}\mathcal{G}_\nu + g_2 J \epsilon_{\mu\nu\rho\sigma}\mathcal{G}^\mu v^\nu \bar{S}^\rho{}_\alpha \bar{G}^{\alpha\sigma}. \qquad (5\cdot20a)$$

Similarly from (5·11b), (5·19b) and (5·4b) we find

$$
\begin{aligned}
\tfrac{1}{2}I'\dot{\bar{S}}_{\mu\nu}\bar{G}^{\mu\nu} + I'\dot{\mathcal{S}}_\mu \mathcal{G}^\mu &= g_2 J'\mathcal{S}_\mu \mathcal{G}_\nu \bar{G}^{\mu\nu} - \tfrac{1}{2}I'\epsilon^{\mu\nu\rho\sigma}\mathcal{G}_\mu v_\nu \dot{\bar{S}}^*_{\rho\sigma} \\
&= g_2 J'\mathcal{S}_\mu \mathcal{G}_\nu \bar{G}^{\mu\nu} - g_2 \epsilon^{\mu\nu\rho\sigma}\mathcal{G}_\mu v_\nu \bar{S}_{\rho\alpha}\bar{G}^\alpha{}_\sigma \\
&= -\tfrac{1}{2}g_2 J'\epsilon_{\mu\rho\sigma\alpha} v^\rho \bar{S}^{*\sigma\alpha}\bar{G}^{\mu\nu}\mathcal{G}_\nu + g_2 J'\epsilon^{\mu\nu\rho\sigma}\mathcal{G}_\mu v_\nu \bar{S}^*_{\rho\alpha}\bar{G}^\alpha{}_\sigma.
\end{aligned}
\qquad (5\cdot20b)
$$

Now $\bar{S}_{\mu\nu}$ and $\bar{S}^*_{\mu\nu}$ have each only three non-vanishing components in the rest system. By a suitable orientation of the spatial axes in the rest system it may be arranged that all components of $\bar{S}_{\mu\nu}$ except $S_{12} = -S_{21}$ vanish (cf. Bhabha 1940). In this system it is not difficult to verify that the right side of (5·20a) is zero. Similarly in the system in which all components of $\bar{S}^*_{\mu\nu}$ vanish except $S^*_{12} = -S^*_{21} = S^{30}$ we verify that the right side of (5·20b) vanishes. Therefore we have

$$\tfrac{1}{2}I\dot{\bar{S}}_{\mu\nu}\bar{G}^{\mu\nu} + I\dot{\mathcal{S}}_\mu\mathcal{G}^\mu = 0, \tag{5·21a}$$

$$\tfrac{1}{2}I'\dot{\bar{S}}_{\mu\nu}\bar{G}^{\mu\nu} + I'\dot{\mathcal{S}}_\mu\mathcal{G}^\mu = 0. \tag{5·21b}$$

Since (5·21) are tensor equations they hold in all systems. So for I and I' not zero we find from (5·21) and (5·18)

$$M = m - \tfrac{1}{2}g_2 S^{\rho\sigma}G'^{\,\mathrm{mean}}_{\rho\sigma}, \tag{5·22}$$

where m is a constant. This equation holds even when $\bar{S}_{\mu\nu}$ or $\mathcal{S}_\mu$ is zero since it can be derived both from (5·10a) and (5·10b). Therefore we obtain from (5·15), (5·22) and (5·14)

$$A_\mu = mv_\mu - \tfrac{1}{2}g_2(S^{\rho\sigma}G'^{\,\mathrm{mean}}_{\rho\sigma})v_\mu + I\dot{\bar{S}}_{\mu\nu}v^\nu - g_2\bar{S}_{\mu\sigma}\mathcal{G}^\sigma - g_2\mathcal{S}_\sigma\bar{G}^\sigma_{\,\mu}, \tag{5·23a}$$

$$A_\mu = mv_\mu - \tfrac{1}{2}g_2(S^{\rho\sigma}G'^{\,\mathrm{mean}}_{\rho\sigma})v_\mu + I'\dot{\bar{S}}^*_{\mu\nu}v^\nu - g_2\bar{S}_{\mu\sigma}\mathcal{G}^\sigma - g_2\mathcal{S}_\sigma\bar{G}^\sigma_{\,\mu}, \tag{5·23b}$$

according as we use (5·10a) and (5·10b). The equations of motion can therefore be written as follows, if we write $\bar{S}'_\mu = \dot{\bar{S}}_{\mu\nu}v^\nu$ and $\bar{S}^{*\prime}_\mu = \dot{\bar{S}}^*_{\mu\nu}v^\nu$ for brevity:

$$\frac{d}{d\tau}\left(mv_\nu - \tfrac{1}{2}g_2 S^{\rho\sigma}G'^{\,\mathrm{mean}}_{\rho\sigma}v_\nu + I\bar{S}'_\nu - g_2\bar{S}_{\nu\sigma}\mathcal{G}^\sigma - g_2\mathcal{S}_\sigma\bar{G}^\sigma_{\,\nu}\right)$$
$$= g_1 G'^{\,\mathrm{mean}}_{\nu\sigma}v^\sigma - \tfrac{1}{2}g_2 S^{\rho\sigma}\partial_\nu G'^{\,\mathrm{mean}}_{\rho\sigma}, \tag{5·24a}$$

$$I\{\dot{\bar{S}}_{\mu\nu} + v_\mu\bar{S}'_\nu - v_\nu\bar{S}'_\mu\} = g_2[\bar{S}_{\mu\sigma}\bar{G}^\sigma_{\,\nu}]_- - g_2[\mathcal{S}_\mu\mathcal{G}_\nu - \mathcal{S}_\nu\mathcal{G}_\mu], \tag{5·25a}$$

$$\frac{d}{d\tau}\left(mv_\nu - \tfrac{1}{2}g_2 S^{\rho\sigma}G'^{\,\mathrm{mean}}_{\rho\sigma}v_\nu + I'\bar{S}^{*\prime}_\nu - g_2\bar{S}_{\nu\sigma}\mathcal{G}^\sigma - g_2\mathcal{S}_\sigma\bar{G}^\sigma_{\,\nu}\right)$$
$$= g_1 G'^{\,\mathrm{mean}}_{\nu\sigma}v^\sigma - \tfrac{1}{2}g_2 S^{\rho\sigma}\partial_\nu G'^{\,\mathrm{mean}}_{\rho\sigma}, \tag{5·24b}$$

$$I'\{\dot{\bar{S}}^*_{\mu\nu} + v_\mu\bar{S}^{*\prime}_\nu - v_\nu\bar{S}^{*\prime}_\mu\} = g_2[\bar{S}_{\mu\nu}\bar{G}^\sigma_{\,\nu}]_- - g_2[\mathcal{S}_\mu\mathcal{G}_\nu - \mathcal{S}_\nu\mathcal{G}_\mu]. \tag{5·25b}$$

Equations (5·24a) and (5·25a) are equivalent to (5·24b) and (5·25b), except when $\mathcal{S}_\mu$ or $\bar{S}_{\mu\nu}$ is zero. We have to choose the (a) or (b) equations according as we choose (5·10a) or (5·10b). The equations are to be supplemented with the additional conditions $\mathcal{S}^\mu\bar{S}_{\mu\sigma} = 0$, $\mathcal{S}^\mu\mathcal{S}_\mu = $ constant and $\bar{S}^{\mu\nu}\bar{S}_{\mu\nu} = $ constant, which follow from (5·11) and (5·12) and hold in both cases.

Except for the values of the two constants m and I (or I') the equations (5·24) and (5·25) are entirely free from arbitrariness. In the corresponding equations obtained by Bhabha & Corben (1941, equations (72) and (73)) three more *independent* 'mechanical' constants I' (or I), K and K' appear besides I (or I') and m, and, in addition, there is a large number of other constants entering into the radiation reaction terms. In our case we automatically get $K = K' = 0$. Also, on account of the fact that the 'electric' and the 'magnetic' moments are parallel I and I' are no longer independent and reduce to one arbitrary constant I (or I'). Besides, the radiation reaction terms are completely determined. In fact, if we put $\mathscr{S}_\mu = 0$ in (5·24a) and (5·25a), they become identical with the equations (101) and (102) of Bhabha & Corben for $K = K' = 0$, apart from the ambiguity of the radiation reaction terms in the latter. Similarly, on putting $\bar{S}_{\mu\nu} = 0$ in (5·24b), (5·25b) we find that they agree with the corresponding equations for the pure 'electric' dipole given by Bhabha & Corben (p. 299), except that in our equations $K = K' = 0$ and the radiation reaction terms are definite.

Finally, we consider a charged dipole in the scalar-meson field. Let g_1 be the charge and $g_2 S^\alpha$ the dipole moment. We may assume that

$$\mathscr{S} \equiv S^\alpha v_\alpha = 0, \tag{5·26}$$

for otherwise we can put $\qquad S^\alpha \equiv \bar{S}^\alpha + \mathscr{S} v^\alpha,$

where $\bar{S}^\alpha v_\alpha = 0$. The retarded potential due to the part $v^\alpha \mathscr{S}$ of S^α is

$$U = g_2 \left\{ \partial_\alpha \left(\frac{v^\alpha \mathscr{S}}{\kappa} \right)_r - \chi \partial_\alpha \int_{-\infty}^{\tau_r} \mathscr{S} v^\alpha \frac{J_1(\chi u)}{u} d\tau \right\}$$

$$= g_2 \left\{ \left(\frac{\dot{\mathscr{S}}}{\kappa} \right)_r - \chi \int_{-\infty}^{\tau_r} \dot{\mathscr{S}} \frac{J_1(\chi u)}{u} d\tau \right\},$$

if we notice that $v^\alpha \partial_\alpha \dfrac{J_1(\chi u)}{u} = -\dfrac{d}{d\tau} \dfrac{J_1(\chi u)}{u}$. Thus $g_2 \mathscr{S}$ behaves exactly as a charge and can therefore be included in g_1. The field-producing properties of the charged dipole are then completely determined by $g_2 \bar{S}^\alpha$ and g_1. We can therefore assume from the outset that $S^\alpha v_\alpha = 0$. Besides, we assume that g_1 and g_2 are constants and also

$$S^2 \equiv S_\alpha S^\alpha = \text{constant}. \tag{5·27}$$

From (2·30) and (2·31) we find that the equations of motion are

$$\dot{A}_\nu = -g_1 U_\nu'^{\text{mean}} + S^\alpha \partial_\nu U_\alpha'^{\text{mean}}, \tag{5·28a}$$

$$\dot{B}_{\mu\nu} = g_2 [S_\mu U_\nu'^{\text{mean}} - S_\nu U_\mu'^{\text{mean}}] - (v_\mu A_\nu - v_\nu A_\mu). \tag{5·28b}$$

From (4·4) and (5·28b) we get

$$\tfrac{1}{2} B_{\mu\nu} \dot{B}^{\mu\nu} = B_{\mu\nu} S^\mu U'^{\nu\,\text{mean}} \equiv 0,$$

so that $\qquad\qquad\qquad B_{\mu\nu} S^\mu = 0.$

We can therefore write $B^{\mu\nu}$ in the form

$$B^{\mu\nu} = I \epsilon^{\mu\nu\rho\sigma} v_\rho S_\sigma. \tag{5·29}$$

On account of (5·27) and (4·4b) I must be a constant. Substituting (5·29) in (5·28b) we get

$$I \epsilon_{\mu\nu\rho\sigma}(\dot{v}^\rho S^\sigma + v^\rho \dot{S}^\sigma) = g_2 [S_\mu U_\nu'^{\text{mean}} - S_\nu U_\mu'^{\text{mean}}] - (v_\mu A_\nu - v_\nu A_\mu), \tag{5·30a}$$

i.e. $\quad I\{(\dot{S}^\alpha v^\beta - \dot{S}^\beta v^\alpha) + (S^\alpha \dot{v}^\beta - S^\beta \dot{v}^\alpha)\} = g_2 S_\mu U_\nu'^{\text{mean}} \epsilon^{\mu\nu\alpha\beta} - v_\mu A_\nu \epsilon^{\mu\nu\alpha\beta}. \tag{5·30b}$

By contracting (5·30a) with v^ν and using (5·14) we obtain

$$\bar{A}_\mu = I\epsilon_{\mu\nu\rho\sigma} v^\nu \dot{v}^\rho S^\sigma - g_2 S_\mu \dot{U}'^{\text{mean}}, \tag{5·31}$$

so that
$$v^\mu \dot{\bar{A}}_\mu = -\dot{v}^\mu A_\mu = g_2 \dot{v}^\mu S_\mu \dot{U}'^{\text{mean}} = -g_2 v^\mu \dot{S}_\mu \dot{U}'^{\text{mean}}. \tag{5·32}$$

On the other hand, we obtain from (5·28a)

$$v^\mu \dot{A}_\mu = -g_1 \dot{U}'^{\text{mean}} + g_2 \frac{d}{d\tau}(S^\alpha U'^{\text{mean}}_\alpha) - g_2(\dot{S}^\alpha U'^{\text{mean}}_\alpha). \tag{5·33}$$

Subtracting (5·32) from (5·33) we obtain

$$\frac{dM}{d\tau} = \frac{d}{d\tau}\{-g_1 U'^{\text{mean}} + g_2 S^\alpha U'^{\text{mean}}_\alpha\} - g_2 \dot{S}^\alpha(U'^{\text{mean}}_\alpha - v_\alpha U'^{\text{mean}}). \tag{5·34}$$

By contracting (5·30b) with $v_\beta(U'^{\text{mean}}_\alpha - v_\alpha \dot{U}'^{\text{mean}})$, we obtain

$$I\dot{S}^\alpha(U'^{\text{mean}}_\alpha - v_\alpha \dot{U}'^{\text{mean}}) = 0.$$

Therefore we find from (5·34) that

$$M = m - g_1 U'^{\text{mean}} + g_2 S^\alpha U'^{\text{mean}}_\alpha, \tag{5·35}$$

where m is a constant. The equations of motion therefore become, if we put $\bar{U}_\nu = U'^{\text{mean}}_\nu - v_\nu \dot{U}'^{\text{mean}}$,

$$\frac{d}{d\tau}\{mv_\mu - g_1 U'^{\text{mean}} v_\mu + g_2(S^\alpha U'^{\text{mean}}_\alpha) v_\mu + I\epsilon_{\mu\nu\rho\sigma} v^\nu \dot{v}^\rho S^\sigma - g_2 S_\mu \dot{U}'^{\text{mean}}\}$$
$$= -g_1 U'^{\text{mean}}_\mu + g_2 S^\alpha \partial_\mu U'^{\text{mean}}_\alpha, \tag{5·36a}$$

$$I\epsilon_{\mu\nu\rho\sigma} v^\rho \dot{S}^\sigma = g_2[S_\mu \bar{U}_\nu - S_\nu \bar{U}_\mu], \tag{5·36b}$$

since
$$\epsilon_{\mu\nu\rho\sigma} \dot{v}^\rho S^\sigma + [v_\mu \epsilon_{\nu\rho\sigma\alpha} v^\rho \dot{v}^\sigma S^\alpha]_- = 0,$$

as is easily seen by constructing the tensor dual to it and using (5·3). These equations also contain only two arbitrary constants m and I. They will be used to calculate the scattering of scalar mesons by a neutron (or a proton) in a subsequent paper.

This work was done under the supervision of Professor H. J. Bhabha, and I am thankful to him for his help and advice.

REFERENCES

Bhabha 1940 *Proc. Ind. Acad. Sci.* A, **11**, 247–267, 467.
Bhabha & Corben 1941 *Proc. Roy. Soc.* A, **178**, 273–314.
Bhabha & Harish-Chandra 1944a *Proc. Roy. Soc.* A, **185**, 250–268.
Bhabha & Harish-Chandra 1944b *Proc. Roy. Soc.* A, **183**, 134–141.
Dirac 1938 *Proc. Roy. Soc.* A, **167**, 148–169.
Fierz 1939 *Helv. Phys. Acta*, **12**, 3–37.
Myron Mathisson 1942 *Proc. Camb. Phil. Soc.* **38**, 40–60.
Pryce 1938 *Proc. Roy. Soc.* A, **168**, 389–409.

A Note on the σ-Symbols

HARISH-CHANDRA

(*J. H. Bhabha Student, Cosmic Ray Research Unit, Indian Institute of Science, Bangalore*)

Received June 15, 1945
(Communicated by Prof. H. J. Bhabha, F.A.SC., F.R.S.)

As is well known (see Van der Waerden, 1932) the symbols $\sigma^{k}_{\lambda\dot{\mu}}$ are used to set up a connection between tensors and spinors for transformations of the Lorentz group. k is a tensor index running from 0 to 3 while λ and μ are spinor indices which can take the values 1 and 2 only.* Hitherto it has been usual to prescribe the numbers $\sigma^{k}_{\lambda\dot{\mu}}$ explicitly and to show that they remain unaltered when subjected to a Lorentz transformation and the associated spinor transformation simultaneously. In fact the spinor transformation associated to a given Lorentz transformation is, in effect, defined by this condition. In the present paper no use will be made of an explicit representation of the σ's. All their properties will be deduced from the defining equations (1) and (2). Besides compactness, this procedure has the advantage that the same equations and all their consequences remain valid even when the most general transformations not included in the Lorentz group are admitted. They can therefore be directly taken over to the general theory of relativity (*cf*. Infeld and Vander Waerden, 1933).

For the present the space-time is assumed flat and the metric tensor is taken to be $g_{kl} = 0$ $k \neq l$, $g_{00} = - g_{11} = - g_{22} = - g_{33} = 1$. Similarly the antisymmetric spinors $\epsilon^{\lambda\mu}$, $\epsilon^{\dot{\lambda}\dot{\mu}}$, $\epsilon_{\lambda\mu}$, $\epsilon_{\dot{\lambda}\dot{\mu}}$ used for raising and lowering the spinor indices are given by $\epsilon^{12} = \epsilon_{12} = 1$, $\epsilon^{\dot{1}\dot{2}} = \epsilon_{\dot{1}\dot{2}} = 1$. For any spinor a_{μ}

$$a^{\lambda} = \epsilon^{\lambda\mu}a_{\mu}, \qquad a_{\lambda} = a^{\mu}\epsilon_{\mu\lambda}$$

with similar relations for the dotted spinors. The σ's are *defined* by the two following conditions:

$$\overline{\sigma^{k}_{\mu\dot{\lambda}}} = \sigma^{k}_{\lambda\dot{\mu}} \tag{1}$$

$$\sigma^{\lambda\dot{\mu}}_{k}\sigma_{l\dot{\mu}\nu} = \delta^{\lambda}_{\nu}g_{kl} - \frac{i}{2}\epsilon_{klmn}\sigma^{m\lambda\dot{\mu}}\sigma^{n}_{\dot{\mu}\nu}. \tag{2a}$$

---

*In this paper Latin alphabets shall always denote tensor indices, the Greek alphabets being reserved for spinor indices.

Here the bar denotes conjugate-complex and ϵ_{klmn} is a tensor antisymmetric in all the four indices with $\epsilon_{0123} = -1$. From (1) and (2a) it follows that

$$\sigma_k^{\dot{\lambda}\mu}\sigma_{l,\mu\dot{\nu}} = \delta_{\dot{\nu}}^{\dot{\lambda}}g_{kl} + \frac{i}{2}\epsilon_{klmn}\sigma^{m\dot{\lambda}\mu}\sigma_{\mu\dot{\nu}}^n. \tag{2b}$$

The fact that the usual representation of σ's satisfies (2) becomes obvious when, in conformity with the usual method, one regards σ^0 as the two-rowed unit matrix and σ^1, σ^2, σ^3 as the three Pauli matrices and compares their commutation rules with (2). Some results, which are already well known (Infeld and Van der Waerden, 1933; Fierz and Pauli, 1939) follow immediately from (2)

$$\sigma_{\lambda\dot{\mu}}^k\sigma^{l\dot{\mu}\nu} + \sigma_{\lambda\dot{\mu}}^l\sigma^{k\dot{\mu}\nu} = 2\delta_\lambda^\nu g^{kl} \tag{3a}$$

$$\sigma_{\lambda\mu}^k\sigma^{l\mu\dot{\nu}} + \sigma_{\lambda\mu}^l\sigma^{k\mu\dot{\nu}} = 2\delta_{\dot{\lambda}}^{\dot{\nu}} g^{kl} \tag{3b}$$

$$\sigma_k^{\lambda\dot{\mu}}\sigma_{\dot{\mu}\nu}^l = 2\delta_k^l. \tag{3c}$$

Also

$$\sigma_{\lambda\dot{\mu}}^k\sigma^{l\dot{\mu}\nu} - \sigma_{\lambda\dot{\mu}}^l\sigma^{k\dot{\mu}\nu} = -i\epsilon^{klmn}\sigma_{m\lambda\dot{\mu}}\sigma_n^{\dot{\mu}\nu} \tag{4a}$$

$$\sigma_{\lambda\mu}^k\sigma^{l\mu\dot{\nu}} - \sigma_{\lambda\mu}^l\sigma^{k\mu\dot{\nu}} = i\epsilon^{klmn}\sigma_{m\lambda\mu}\sigma_n^{\mu\nu}. \tag{4b}$$

From the irreducibility of the Pauli matrices it follows that (Fierz and Pauli, 1939)

$$\sigma_k^{\dot{\lambda}\mu}\sigma_{\nu\dot{\rho}}^k = 2\delta_{\dot{\rho}}^{\dot{\lambda}}\delta_\nu^\mu. \tag{5}$$

However a direct proof based on (2) can be given as follows. Consider $\sigma_{\lambda\dot{\mu}}^k\sigma^{l\dot{\mu}\nu}\sigma_{\nu\dot{\rho}}^m$. Notice that from (2)

$$\epsilon_{klmn}\sigma_{\mu\dot{\lambda}}^k\sigma^{l\dot{\lambda}\rho}\sigma_{\rho\dot{\nu}}^m = -\frac{i}{2}\epsilon_{klmn}\epsilon^{klpq}\sigma_{p,\mu\dot{\lambda}}\sigma_q^{\dot{\lambda}\rho}\sigma_{\rho\dot{\nu}}^m$$

$$= i(\delta_m^p\delta_n^q - \delta_n^p\delta_m^q)\sigma_{p,\mu\dot{\lambda}}\sigma_q^{\dot{\lambda}\rho}\sigma_{\rho\dot{\nu}}^m$$

$$= 2i\left(g_{mn}\delta_\mu^\rho - \sigma_{n,\mu\dot{\lambda}}\sigma_m^{\dot{\lambda}\rho}\right)\sigma_{\rho\dot{\nu}}^m \quad \text{from (3)}$$

$$= -6i\sigma_{n,\mu\dot{\nu}} \quad \text{from (2).} \tag{6}$$

Also from (2)

$$\sigma_k^{\dot{\lambda}\mu}\left(\sigma_{l,\mu\dot{\nu}}\sigma_{\alpha\dot{\beta}}^l + \sigma_{l,\mu\dot{\beta}}\sigma_{\alpha\dot{\nu}}^l\right) = \delta_{\dot{\nu}}^{\dot{\lambda}}\sigma_{k,\alpha\dot{\beta}} + \delta_{\dot{\beta}}^{\dot{\lambda}}\sigma_{k,\alpha\dot{\nu}}$$

$$+ \frac{i}{2}\epsilon_{klmn}\sigma^{m,\dot{\lambda}\mu}\left(\sigma_{\mu\dot{\nu}}^n\sigma_{\alpha\dot{\beta}}^l + \sigma_{\mu\dot{\beta}}^n\sigma_{\alpha\dot{\nu}}^l\right). \tag{7}$$

Now notice that

$$\sigma_{\lambda\dot{\mu}}^m\sigma_{\alpha\dot{\beta}}^n - \sigma_{\alpha\dot{\beta}}^m\sigma_{\lambda\dot{\mu}}^n = \epsilon_{\lambda\alpha}\sigma_{\gamma\dot{\mu}}^m\sigma_{\dot{\beta}}^{n,\gamma} + \epsilon_{\dot{\mu}\dot{\beta}}\sigma_{\dot{\alpha}\dot{\rho}}^m\sigma_\lambda^{n,\dot{\rho}}. \tag{8}$$

Therefore

$$\epsilon_{klmn}\sigma^{m\dot\lambda\mu}\left(\sigma^n_{\mu\dot\nu}\sigma^l_{\alpha\dot\beta}+\sigma^n_{\mu\dot\beta}\sigma^l_{\alpha\dot\nu}\right)$$

$$=\tfrac{1}{2}\epsilon_{klmn}\sigma^{m,\dot\lambda}_\alpha\left(\sigma^n_{\dot\nu\mu}\sigma^{l,\mu}_{\dot\beta}+\sigma^n_{\dot\beta\mu}\sigma^{l,\mu}_{\dot\nu}\right)$$

$$=\tfrac{1}{4}\epsilon_{klmn}\left[\left(\delta^{\dot\lambda}_{\dot\nu}\sigma^m_{\alpha\dot\rho}\sigma^{n,\dot\rho\mu}\sigma^l_{\mu\dot\beta}-\sigma^m_{\dot\nu\gamma}\sigma^{n,\dot\lambda\gamma}\sigma^l_{\alpha\dot\beta}\right)+\left(\delta^{\dot\lambda}_{\dot\beta}\sigma^m_{\alpha\dot\rho}\sigma^{n,\dot\rho\mu}\sigma^l_{\mu\dot\nu}-\sigma^m_{\dot\beta\gamma}\sigma^{n,\dot\lambda\gamma}\sigma^l_{\alpha\dot\nu}\right)\right]$$

$$=\tfrac{1}{2}i\left(\delta^{\dot\lambda}_{\dot\nu}\sigma_{k,\alpha\dot\beta}+\delta^{\dot\lambda}_{\dot\beta}\sigma_{k,\alpha\dot\nu}\right)+\tfrac{1}{4}\epsilon_{klmn}\sigma^{m,\dot\lambda\mu}\left(\sigma^n_{\mu\dot\nu}\sigma^l_{\alpha\dot\beta}+\sigma^n_{\mu\dot\beta}\sigma^l_{\alpha\dot\nu}\right)\quad\text{from (6).}$$

Therefore

$$\epsilon_{klmn}\sigma^{m,\dot\lambda\mu}\left(\sigma^n_{\mu\dot\nu}\sigma^l_{\alpha\dot\beta}+\sigma^n_{\mu\dot\beta}\sigma^l_{\alpha\dot\nu}\right)=2i\left(\delta^{\dot\lambda}_{\dot\nu}\sigma_{k,\alpha\dot\beta}+\delta^{\dot\lambda}_{\dot\beta}\sigma_{k,\alpha\dot\nu}\right).\tag{9}$$

From (7) and (9) it follows that

$$\sigma^{\dot\lambda\mu}_k\left(\sigma_{l,\mu\dot\nu}\sigma^l_{\alpha\dot\beta}+\sigma_{l,\mu\dot\beta}\sigma^l_{\alpha\dot\nu}\right)=0.$$

Multiplying by $\sigma^k_{\rho\dot\lambda}$ and using (2) one gets

$$\sigma_{l,\mu\dot\nu}\sigma^l_{\alpha\dot\beta}+\sigma_{l,\mu\dot\beta}\sigma^l_{\alpha\dot\nu}=0.\tag{10a}$$

Now

$$\sigma_{l,\mu\dot\nu}\sigma^l_{\alpha\dot\beta}-\sigma_{l,\mu\dot\beta}\sigma^l_{\alpha\dot\nu}=\epsilon_{\dot\nu\dot\beta}\sigma_{l,\mu\dot\rho}\sigma^{l\dot\rho}_\alpha$$

$$=4\epsilon_{\dot\nu\dot\beta}\epsilon_{\mu\alpha}\quad\text{from (2).}\tag{10b}$$

(5) follows immediately from (10).

    A few other useful relations can be derived from (2). Put

$$\sigma^k_{\mu\dot\lambda}\sigma^{l,\dot\lambda\rho}\sigma^m_{\rho\dot\nu}=A^{klm}_{\mu\dot\nu}.$$

Then from (3)

$$A^{klm}_{\mu\dot\nu}=A^{klm}_{\mu\dot\nu}$$

$$=-A^{lkm}_{\mu\dot\nu}+2g^{kl}\sigma^m_{\mu\dot\nu}$$

$$=A^{lmk}_{\mu\dot\nu}+2g^{kl}\sigma^m_{\mu\dot\nu}-2g^{km}\sigma^l_{\mu\dot\nu}$$

$$=-A^{mlk}_{\mu\dot\nu}+2g^{kl}\sigma^m_{\mu\dot\nu}-2g^{km}\sigma^l_{\mu\dot\nu}+2g^{lm}\sigma^k_{\mu\dot\nu}$$

$$=A^{mkl}_{\mu\dot\nu}\qquad\qquad-2g^{km}\sigma^l_{\mu\dot\nu}+2g^{lm}\sigma^k_{\mu\dot\nu}$$

$$=-A^{kml}_{\mu\dot\nu}\qquad\qquad\qquad+2g^{lm}\sigma^k_{\mu\dot\nu}.$$

Adding up the various expressions on the right one gets

$$6A^{klm}_{\mu\dot\nu}=-\epsilon^{klmn}\epsilon_{npqr}A^{pqr}_{\mu\dot\nu}+6\left(g^{kl}\sigma^m_{\mu\dot\nu}-g^{km}\sigma^l_{\mu\dot\nu}+g^{lm}\sigma^k_{\mu\dot\nu}\right)$$

so that from (6)

$$\sigma^k_{\mu\dot\lambda}\sigma^{l,\dot\lambda\rho}\sigma^m_{\rho\dot\nu}=g^{kl}\sigma^m_{\mu\dot\nu}-g^{km}\sigma^l_{\mu\dot\nu}+g^{lm}\sigma^k_{\mu\dot\nu}+i\epsilon^{klmn}\sigma_{n,\mu\dot\nu}.\tag{11a}$$

The conjugate-complex equation is

$$\sigma^{k}_{\dot{\mu}\lambda}\sigma^{l,\lambda\dot{\rho}}\sigma^{m}_{\dot{\rho}\nu} = g^{kl}\sigma^{m}_{\dot{\mu}\nu} - g^{km}\sigma^{l}_{\dot{\mu}\nu} + g^{lm}\sigma^{k}_{\dot{\mu}\nu} - i\epsilon^{klmn}\sigma_{n,\dot{\mu}\nu}. \tag{11b}$$

From (3c) and (11)

$$\sigma^{k}_{\mu\lambda}\sigma^{l,\dot{\lambda}\rho}\sigma^{m}_{\rho\dot{\nu}}\sigma^{n,\dot{\nu}\mu} = 2\big(g^{kl}g^{mn} - g^{km}g^{ln} + g^{kn}g^{lm} + i\epsilon^{klmn}\big) \tag{12a}$$

$$\sigma^{k}_{\dot{\mu}\lambda}\sigma^{l,\lambda\dot{\rho}}\sigma^{m}_{\dot{\rho}\nu}\sigma^{n,\nu\dot{\mu}} = 2\big(g^{kl}g^{nm} - g^{km}g^{lm} + g^{kn}g^{lm} - i\epsilon^{klmn}\big). \tag{12b}$$

From (4) and (12) the following well-known relations for any anti-symmetric tensor $\mathbf{F}_{kl}$ (Laporte and Uhlenbeck, 1931; Fierz and Pauli, 1939) are obtained immediately.

$$\mathbf{F}_{mn} = \tfrac{1}{4}\sigma^{\alpha\dot{\lambda}}\sigma^{\beta\dot{\mu}}_{m}\big(\epsilon_{\alpha\beta}f_{\dot{\lambda}\dot{\mu}} + \epsilon_{\dot{\lambda}\dot{\mu}} + f_{\alpha\beta}\big) \tag{13a}$$

$$f^{\rho}_{\mu}\sigma^{m,\mu\dot{\nu}}\sigma^{n}_{\dot{\nu}\rho} = 4\mathbf{F}^{+\,mn} \tag{13b}$$

$$f^{\dot{\rho}}_{\dot{\mu}}\sigma^{m\mu\dot{\nu}}\sigma^{n}_{\nu\dot{\rho}} = 4^{\mathbf{F}^{-\,mn}} \tag{13c}$$

where

$$f^{\rho}_{\mu} = \tfrac{1}{2}\mathbf{F}_{kl}\sigma^{k}_{\mu\dot{\lambda}}\sigma^{l,\dot{\lambda}\rho} = \tfrac{1}{2}\mathbf{F}^{+}_{kl}\sigma^{k}_{\mu\dot{\lambda}}\sigma^{l,\dot{\lambda}\rho} \tag{14a}$$

$$f^{\dot{\rho}}_{\dot{\mu}} = \tfrac{1}{2}\mathbf{F}_{kl}\sigma^{k}_{\dot{\mu}\lambda}\sigma^{l,\lambda\dot{\rho}} = \tfrac{1}{2}\mathbf{F}^{-}_{kl}\sigma^{k}_{\dot{\mu}\lambda}\sigma^{l,\lambda\dot{\rho}} \tag{14b}$$

and

$$\mathbf{F}^{+}_{kl} = \tfrac{1}{2}\big(\mathbf{F}_{kl} - i\mathbf{F}^{\times}_{kl}\big)$$

$$\mathbf{F}^{-}_{kl} = \tfrac{1}{2}\big(\mathbf{F}_{kl} + i\mathbf{F}^{\times}_{kl}\big)$$

$$\mathbf{F}^{\times}_{kl} = \tfrac{1}{2}\epsilon_{klmn}\mathbf{F}^{mn}.$$

It is yet to be proved from (2) that for every proper Lorentz transformation there exists a spinor transformation such that the two applied together leave $\sigma^{k}_{\lambda\dot{\mu}}$ unchanged. For this purpose it is sufficient to consider infinitesimal transformations. An infinitesimal Lorentz transformation is given by

$$x'_{k} = x_{k} + \epsilon^{l}_{k}x_{l} \qquad \big(\epsilon_{kl} = -\epsilon_{lk}\big) \tag{15}$$

where ϵ_{kl} is a real infinitesimal quantity. Similarly an infinitesimal spinor transformation

$$a'_{\mu} = a_{\mu} + \eta^{\nu}_{\mu}a_{\nu} \qquad \big(\eta_{\mu\nu} = \eta_{\mu\nu}\big) \tag{16a}$$

$$a'_{\dot{\mu}} = a_{\dot{\mu}} + \eta^{\dot{\nu}}_{\dot{\mu}}a_{\dot{\nu}} \qquad \big(\eta_{\dot{\mu}\dot{\nu}} = \overline{\eta_{\mu\nu}}\big) \tag{16b}$$

is characterised by an infinitesimal spinor $\eta_{\mu\nu}$. The symmetry of $\eta_{\mu\nu}$ in μ, ν follows from the invariance of $\epsilon_{\mu\nu}$. It is therefore sufficient to show that for every ϵ^{kl} there exists a $\eta_{\mu\nu}$ such that

$$\epsilon^{kl}\sigma_{l,\alpha\dot{\lambda}} + \eta^{\beta}_{\alpha}\sigma^{k}_{\beta\dot{\lambda}} + \eta^{\dot{\mu}}_{\dot{\lambda}}\sigma^{k}_{\alpha\dot{\mu}} = 0. \tag{17}$$

From (5) and (17) the solution is easily obtained.

$$\eta_{\alpha\beta} = \tfrac{1}{4}\epsilon^{kl}\sigma_{k,\alpha\dot{\lambda}}\sigma^{\dot{\lambda}}_{l,\beta}\eta_{\dot{\alpha}\dot{\beta}} = \overline{\eta_{\alpha\beta}} = \tfrac{1}{4}\epsilon^{kl}\sigma_{k,\alpha\dot{\lambda}}\sigma^{\dot{\lambda}}_{l,\dot{\beta}}. \tag{18}$$

The transformation matrix of (16) is therefore

$$\delta_\alpha^\beta + \tfrac{1}{4}\epsilon^{kl}\sigma_{k,\alpha\dot\lambda}\sigma_l^{\dot\lambda\beta} \tag{19a}$$

$$\delta_{\dot\alpha}^{\dot\beta} + \tfrac{1}{4}\epsilon^{kl}\sigma_{k,\dot\alpha\lambda}\sigma_l^{\lambda\dot\beta}. \tag{19b}$$

It must now be shown that the transformation (19) considered as a representation of the Lorentz group satisfies the integrability conditions (see Van der Waerden, 1932). In any representation of the Lorentz group the transformation (15) is represented by

$$1 + \tfrac{1}{2}\epsilon^{kl}\mathbf{I}_{kl}$$

where $\mathbf{I}^{kl} = -\mathbf{I}^{lk}$ are the representative matrices for the infinitesimal transformations (see Van der Waerden $l.c.$). Therefore in our case, from (19)

$$(\mathbf{I}_{kl})_\alpha^\beta = \tfrac{1}{4}\Big(\sigma_{k,\alpha\dot\lambda}\sigma_l^{\dot\lambda\beta} - \sigma_{l,\alpha\dot\lambda}\sigma_k^{\dot\lambda\beta}\Big) \tag{20}$$

where α and β on the left side are to be looked upon as matrix indices. The integrability conditions for the Lorentz group are well known and are

$$\mathbf{I}^{kl}\mathbf{I}^{mn} - \mathbf{I}^{mn}\mathbf{I}^{kl} = -g^{km}\mathbf{I}^{ln} + g^{lm}\mathbf{I}^{km} + g^{kn}\mathbf{I}^{lm} - g^{ln}\mathbf{I}^{km}. \tag{21}$$

Now from (4) and (20)

$$(\mathbf{I}_{kl})_\alpha^\beta = -\frac{i}{2}\epsilon_{klmn}(\mathbf{I}^{mn})_\alpha^\beta = -\frac{i}{4}\epsilon_{klmn}\sigma_{\alpha\dot\lambda}^m\sigma^{n,\dot\lambda\beta}.$$

Therefore

$$(\mathbf{I}_{kl}\mathbf{I}_{mn} - \mathbf{I}_{mn}\mathbf{I}_{kl})_\alpha^\gamma = -\frac{1}{16}\big(\epsilon_{klpq}\epsilon_{mnrs} - \epsilon_{mnpq}\epsilon_{klrs}\big)\sigma_{\alpha\dot\lambda}^{p}\sigma^{q\dot\lambda\beta}\sigma_{\beta\dot\mu}^{r}\sigma^{s,\dot\mu\gamma}$$

$$= -\frac{1}{16}\epsilon_{klpq}\epsilon_{mnrs}\Big\{ \sigma_{\alpha\dot\lambda}^{k}\sigma^{q,\dot\lambda\beta}\sigma_{\beta\dot\mu}^{r}\sigma^{s,\dot\mu\gamma} - \sigma_{\alpha\dot\lambda}^{s}\sigma^{r,\dot\lambda\beta}\sigma_{\beta\dot\mu}^{q}\sigma^{p,\dot\mu\dot\gamma}\Big\}$$

$$= -\frac{1}{16}\epsilon_{klpq}\epsilon_{mnrs}\Big\{ -2g^{pr}\sigma_{\alpha\dot\mu}^{q}\sigma^{s,\dot\mu\gamma} + i\epsilon^{pqrt}\sigma_{t,\alpha\dot\mu}\sigma^{s,\dot\mu\gamma}$$

$$+ 2\sigma_{\alpha\dot\lambda}^{s}g^{rp}\sigma^{q,\dot\lambda\gamma} + i\sigma_{\alpha\dot\lambda}^{s}\epsilon^{rqpt}\sigma_t^{\dot\lambda\gamma}\Big\} \quad \text{from (11)}$$

$$= -\frac{1}{16}\epsilon_{klpq}\epsilon_{mnrs}\Big\{ -8g^{pr}(\mathbf{I}^{qs})_\alpha^\gamma + 4i\epsilon^{pqrt}(\mathbf{I}_t^s)_\alpha^\gamma\Big\}.$$

Therefore

$$\mathbf{I}_{kl}\mathbf{I}_{mn} - \mathbf{I}_{mn}\mathbf{I}_{kl} = \tfrac{1}{2}\epsilon_{klq}^{r}\epsilon_{mnrs}\mathbf{I}^{qs} - \frac{i}{4}\epsilon_{klpq}\epsilon_{mnrs}\epsilon^{pqrt}\mathbf{I}_t^s$$

$$= -\tfrac{1}{2}\Big(\sum_{klq} \pm g_{klm}g_{ln}g_{qs}\Big)\mathbf{I}^{qs} + \frac{i}{2}\epsilon_{mnrs}\big(\delta_k^r\delta_l^t - \delta_l^r\delta_k^t\big)\mathbf{I}_s^t,$$

where the sum is to be taken over all permutations of $k,\ l,\ q$ with $+$ or $-$ sign according as the permutation is even or odd. So

$$\mathbf{I}_{kl}\mathbf{I}_{mn} - \mathbf{I}_{mn}\mathbf{I}_{kl} = -\tfrac{1}{2}\Big(\sum_{klq} \pm g_{km}g_{ln}g_{qs}\Big)\mathbf{I}^{qs} + \frac{i}{2}\big(\epsilon_{mnks}\mathbf{I}_l^s - \epsilon_{mnls}\mathbf{I}_k^s\big)$$

$$= \tfrac{1}{2}\big[-g_{km}\mathbf{I}_{ln} + g_{lm}\mathbf{I}_{kn} + g_{kn}\mathbf{I}_{lm} - g_{lm}\mathbf{I}_{kn}\big]$$

$$+ \tfrac{1}{4}\big(\epsilon_{mnks}\epsilon_{lpq}^s\mathbf{I}^{pq} - \epsilon_{mnls}\epsilon_{spq}^k\mathbf{I}^{pq}\big)$$

$$= \big[-g_{km}\mathbf{I}_{ln} + g_{lm}\mathbf{I}_{kn} + g_{kn}\mathbf{I}_{lm} - g_{ln}\mathbf{I}_{km}\big].$$

Therefore (21) is fulfilled.

It is now possible to consider finite Lorentz transformations. Let the transformations (15) be denoted by $1 + \tfrac{1}{2}\epsilon_{kl}\mathbf{J}^{kl}$. The transformation matrix of (15) is

$$\delta_k^l + \epsilon_k^l.$$

Therefore

$$\tfrac{1}{2}\big(\epsilon_{kl}\mathbf{J}^{kl}\big)_m^n = \epsilon_m^n \tag{22}$$

where m and n are the matrix indices of the transformation $\mathbf{J}^{kl}$. A finite Lorentz transformation $\mathbf{L}$ can be generated from the infinitesimal transformation by the following common device.

$$\mathbf{L} = \Big(1 + \tfrac{1}{2}\frac{\theta^{kl}}{n}\mathbf{J}_{kl}\Big)_{\text{Lim } n \to \infty}^n = e^{1/2\theta^{kl}\mathbf{J}_{kl}}$$

$$= 1 + \tfrac{1}{2}\theta^{kl}\mathbf{J}_{kl} + \frac{\big(\tfrac{1}{2}\theta^{kl}\mathbf{J}_{kl}\big)^2}{2!} + \frac{\big(\tfrac{1}{2}\theta^{kl}\mathbf{J}_{kl}\big)^3}{3!} + \cdots.$$

On account of (22) one can write symbolically

$$\mathbf{L} = e^\theta = 1 + \theta + \frac{\theta^2}{2!} + \frac{\theta^3}{3!} + \cdots$$

where θ^n is a matrix defined by induction as follows:

$$\big(\theta^{n+1}\big)_p^q = \big(\theta^n\big)_p^r\theta_r^q.$$

It is easy to verify that

$$\theta_{kl}\theta^{lm}\theta_{mn}\theta^{np} + \tfrac{1}{2}\big(\theta_{mn}\theta^{mn}\big)\theta_{kl}\theta^{lp} - \big(\tfrac{1}{8}\theta^{lm}\theta^{nr}\epsilon_{lmn}\big)^2\delta_k^p = 0$$

for any tensor θ_{kl} whose only non-vanishing components are $\theta_{03} = -\theta_{30}$ and $\theta_{12} = -\theta_{21}$. However since every antisymmetrical tensor can be brought to this form by a Lorentz transformation it follows that the above identity is valid for every θ_{kl}. Written in matrix form it becomes

$$\theta^4 + \tfrac{1}{2}\theta_{kl}\theta^{kl}\theta^2 - \big(\tfrac{1}{8}\theta^{kl}\theta^{mn}\epsilon_{klmn}\big)^2 = 0.$$

Making use of this characteristic equation of θ, it can be proved that

$$\mathbf{L} = \tfrac{1}{2}\left[\cosh\sqrt{\Phi_+} + \cosh\sqrt{\Phi_-}\right] + \frac{\cosh\sqrt{\Phi_+} - \cosh\sqrt{\Phi_-}}{\Phi_+ - \Phi_-}\left(\theta^2 + \tfrac{1}{4}\theta_{kl}\theta^{kl}\right)$$

$$+ \tfrac{1}{2}\left[\frac{\sinh\sqrt{\Phi_+}}{\sqrt{\Phi_+}} + \frac{\sinh\sqrt{\Phi_-}}{\sqrt{\Phi_-}}\right]\theta$$

$$+ \left(\frac{\sinh\sqrt{\Phi_+}}{\sqrt{\Phi_+}} - \frac{\sinh\sqrt{\Phi_-}}{\sqrt{\Phi_-}}\right)\frac{1}{\Phi_+ - \Phi_-}\left(\theta^3 + \tfrac{1}{4}\theta_{kl}\theta^{kl}\cdot\theta\right) \qquad (23a)$$

where

$$\Phi_{\pm} = -\tfrac{1}{4}\theta_{kl}\theta^{kl} \pm \tfrac{1}{4}\sqrt{\left(\tfrac{1}{2}\theta^{kl}\theta^{mn}\epsilon_{klmn}\right)^2 + \left(\theta^{kl}\theta_{kl}\right)^2}. \qquad (23b)$$

This is the expression for the most general proper Lorentz transformation.

The spinor transformation Λ associated to (23) is given by

$$\Lambda = \left(1 + \tfrac{1}{2}\frac{\theta^{kl}\mathbf{I}_{kl}}{n}\right)^n_{\text{Lim } n \to \infty} = e^{1/2\theta^{kl}\mathbf{I}_{kl}}$$

$$= 1 + \left(\tfrac{1}{2}\theta^{kl}\mathbf{I}_{kl}\right) + \frac{\left(\tfrac{1}{2}\theta^{kl}\mathbf{I}_{kl}\right)^2}{2!} + \cdots. \qquad (24)$$

Now

$$\left[\left(\tfrac{1}{2}\theta^{kl}\mathbf{I}_{kl}\right)^2\right]^\gamma_\alpha = \left(\tfrac{1}{2}\theta^{kl}\mathbf{I}_{kl}\right)^\beta_\alpha\left(\tfrac{1}{2}\theta^{mn}\mathbf{I}_{mn}\right)^\gamma_\beta$$

$$= \frac{1}{16}\theta^{kl}\theta^{mn}\sigma_{k,\alpha\dot\lambda}\sigma_l^{\dot\lambda\beta}\sigma_{m,\beta\dot\mu}\sigma_n^{\dot\mu\gamma}$$

$$= \frac{1}{16}\theta_{kl}\theta_{mn}\left(-2g^{km}\sigma^l_{\alpha\dot\mu}\sigma^{n,\dot\mu\gamma} + i\epsilon^{klmp}\sigma_{p,\alpha\dot\mu}\sigma^{n,\dot\mu\gamma}\right) \quad \text{from (11)}$$

$$= -\tfrac{1}{8}\theta_{kl}\theta^{kl}\delta^\gamma_\alpha + \frac{l}{16}\theta_{kl}\theta_{mn}\epsilon^{klmp}\left\{\delta^n_p\delta^\gamma_\alpha - i\epsilon^n_{prs}(\mathbf{I}^{rs})^\gamma_\alpha\right\} \quad \text{from (2)}$$

$$= \left(-\tfrac{1}{8}\theta_{kl}\theta^{kl} + \frac{i}{16}\theta_{kl}\theta_{mn}\epsilon^{klmn}\right)\delta^\gamma_\alpha.$$

Put

$$\Theta = -\tfrac{1}{4}\theta_{kl}\theta^{kl} + \frac{i}{16}\theta_{kl}\theta_{mn}\epsilon^{klmn} \qquad (25)$$

Then from (24),

$$\Lambda = \left(1 + \frac{\Theta}{2!} + \frac{\Theta^2}{4!} + \cdots\right) + \left(1 + \frac{\Theta}{3!} + \frac{\Theta^2}{5!} + \cdots\right)\left(\tfrac{1}{2}\theta^{kl}\mathbf{I}_{kl}\right)$$

$$= \cosh\sqrt{\Theta} + \frac{\sinh\sqrt{\Theta}}{\sqrt{\Theta}}\left(\tfrac{1}{2}\theta^{kl}\mathbf{I}_{kl}\right) \qquad (26a)$$

or

$$\Lambda_\alpha^\beta = \cosh\sqrt{\Theta}\,\delta_\alpha^\beta + \frac{\sinh\sqrt{\Theta}}{\sqrt{\Theta}}\,\tfrac{1}{4}\theta^{kl}\sigma_{k,\alpha\dot{\lambda}}\sigma_l^{\dot{\lambda}\beta}. \tag{26b}$$

Therefore the spinor transformation associated to any given Lorentz transformation is completely determined. As an example consider the case of a spatial rotation about the x_3 axis. In this all components of θ_{kl} are zero except $\theta_{12} = -\theta_{21} = 2\phi$ (say). Therefore

$$\Theta = -\phi^2$$

and

$$\Lambda_\beta^\alpha = \cos\phi\,\delta_\alpha^\beta + i\frac{\sin\phi}{i\phi}\frac{\phi}{2}\left(\sigma_{1,\alpha\dot{\lambda}}\sigma_2^{\dot{\lambda}\beta} - \sigma_{2,\alpha\dot{\lambda}}\sigma_1^{\dot{\lambda}\beta}\right)$$

$$= \cos\phi\,\delta_\alpha^\beta + \frac{\sin\phi}{i}\tfrac{1}{2}\epsilon_{1203}\left(\sigma_{\alpha\dot{\lambda}}^0\sigma^{3\dot{\lambda}\beta} - \sigma_{\alpha\dot{\lambda}}^3\sigma^{0,\dot{\lambda}\beta}\right)$$

$$= \cos\phi\,\delta_\alpha^\beta + i\sin\phi\tfrac{1}{2}\left(\sigma_{\alpha\dot{\lambda}}^0\sigma^{3,\dot{\lambda}\beta} - \sigma_{\alpha\dot{\lambda}}^3\sigma^{0,\dot{\lambda}\beta}\right). \tag{27a}$$

Similarly for a Lorentz transformation along the x_3 axis the only non-vanishing component of θ_{kl} is $\theta_{03} = -\theta_{30} = 2\phi$ (say), so that $\Theta = \phi^2$ and

$$\Lambda_\alpha^\beta = \cosh\phi\cdot\delta_\alpha^\beta - \sinh\phi\cdot\tfrac{1}{2}\left(\sigma_{0,\alpha\dot{\lambda}}\sigma_3^{\dot{\lambda}\beta} - \sigma_{3,\alpha\dot{\lambda}}\sigma_0^{\dot{\lambda}\beta}\right). \tag{27b}$$

In the usual representation σ^0 is the unit matrix and $\sigma^{3,\dot{\alpha}\beta} = -\sigma_{\alpha\beta}^3$. Therefore in matrix notation (27a) and (27b) can be written as

$$\Lambda = \cos\phi - i\sin\phi\,\sigma^3$$

$$\Lambda = \cosh\phi - \sinh\phi\,\sigma^3$$

where the matrix elements of σ^3 are $\sigma_{\alpha\beta}^3$.

Till now only the proper Lorentz transformations have been discussed. Reflection can now be included in the following way. From (11)

$$\sigma_\mu^{0,\dot{\lambda}}\sigma_{\dot{\lambda}\rho}^k\sigma_\nu^{0,\rho} = 2g^{0k}\sigma_{\mu\nu}^0 - \sigma_{\mu\nu}^k = g^{kk}\sigma_{\mu\nu}^k. \tag{28}$$

Therefore $\sigma_\mu^{0\lambda}$ is the reflection matrix, and by reflection the spinors $a_{\dot{\mu}}$ and b_μ go over into a_μ and $b_{\dot{\mu}}$ given by

$$a_\mu = \sigma_\mu^{0,\dot{\lambda}}a_{\dot{\lambda}}, \qquad b_{\dot{\mu}} = \sigma_{\dot{\mu}}^{0,\lambda}b_\lambda. \tag{29}$$

The quantities $\sigma_{\mu\nu}^k$ remain unchanged for the simultaneous application of reflection to the tensor index k as well as to the spinor indices μ, ν. In the usual representation $\sigma_{\mu\nu}^0$ is the unit matrix and therefore the only non-vanishing components of $\sigma_\mu^{0,\lambda}$ are $\sigma_1^{0,2} = -\sigma_2^{0,1} = -1$. In this case therefore (29) coincides with the usual rules for reflection.

If $\sigma^k_{\lambda\dot\mu}$ and $\sigma'^k_{\lambda\dot\mu}$ be two different sets of σ's satisfying (1) and (2) then it follows from (5) that

$$\sigma'^k_{\lambda\dot\mu} = a^k_l \sigma^l_{\lambda\dot\mu}$$

where

$$a^k_l = \tfrac{1}{2}\sigma'^k_{\lambda\dot\mu}\sigma_l^{\lambda\dot\mu}$$

so that a^k_l are real and

$$a^k_m a^l_n g^{mn} = g^{kl}.$$

Therefore a^k_l must be the coefficients of a Lorentz transformation apart from the fact that they may reverse the direction of time.

Since the equations (1) and (2) are in a proper covariant form they remain valid for all real transformations of the tensor space and any arbitrary transformations of the spin-space (*cf.* Infeld and Van der Waerden, 1933). However it must be borne in mind that for this general case

$$\epsilon_{0123} = -\sqrt{-g}$$

$$\epsilon^{0123} = \frac{1}{\sqrt{-g}}$$

where g is the determinant of the general g_{ik} matrix. Also ϵ_{12} and ϵ^{12} are no longer 1 but are equal to γ and $1/\gamma$ respectively where γ is a spinor density of weight 1, *i.e.*, on transformation it gets multiplied by the determinant of the transformation in the spin-space (Infeld and Van der Waerden, 1933). All the results [*e.g.*, equations (5), (11) and (12)] therefore hold also for the general case which is of importance in the general theory of relativity.

In conclusion let us consider an interesting application of (11) to the Dirac equation of a particle of spin $\tfrac{1}{2}$. Expressed in terms of spinors it splits up into the following two equations:

$$i\partial^{\alpha\dot\lambda}a_{\dot\lambda} = \chi b^a \tag{30a}$$

$$i\partial_{\alpha\dot\lambda}b^\alpha = \chi a_{\dot\lambda}. \tag{30b}$$

To these are to be added the corresponding conjugate-complex equations. Apart from numerical factors the charge-current-density spinor is given by

$$S_{\alpha\dot\beta} = a_\alpha a_{\dot\beta} + b_\alpha b_{\dot\beta} \tag{31}$$

where a_α and $b_{\dot\beta}$ are the complex-conjugates of $a_{\dot\alpha}$ and b_β respectively. It is obvious that the charge-density given by (21) is positive definite if the usual representation of σ's is used since in this case σ^0 is the unit matrix. For every other representation of σ's the charge-density is therefore either positive or negative definite according as this representation is obtained from the usual one by a Lorentz transformation without or with the reversal of the direction of time. However only the definite character of the charge-density is of importance, the sign being immaterial.

Notice that the equations (30) are completely equivalent to the second order equation

$$\partial_k \partial^k a_{\dot\lambda} + \chi^2 a_{\dot\lambda} = 0 \tag{32}$$

which follows from them. This becomes obvious if one looks upon (30a) as the *definition* of b^α. Therefore (30) can be replaced by

$$\partial^k a_{\dot\lambda} = \chi a_{\dot\lambda}^k \tag{33a}$$

$$\partial^k a_{\dot\lambda}^k = -\chi a^{\dot\lambda} \tag{33b}$$

so that

$$b^\alpha = i\sigma_k^{\alpha\dot\lambda} a_{\dot\lambda}^k.$$

(31) now becomes

$$S_{\alpha\dot\beta} = a_\alpha a_{\dot\beta} + \sigma_{\alpha\dot\lambda}^l \sigma_{\dot\beta\mu}^m a_l^{\dot\lambda} a_m^\mu, \qquad \left(a_m^\mu = \overline{a_m^{\dot\mu}}\right)$$

or

$$\mathbf{S}^k \equiv \tfrac{1}{2}\sigma^{k\alpha\dot\beta} S_{\alpha\dot\beta} = \tfrac{1}{2}\sigma^{k,\alpha\dot\beta} a_\alpha a_{\dot\beta} + \tfrac{1}{2}\sigma_{\dot\lambda\alpha}^l \sigma^{k,\alpha\dot\beta}\sigma_{\dot\beta\mu}^m a_l^{\dot\lambda} a_m^\mu$$

$$= \tfrac{1}{2}\sigma^{k,\alpha\dot\beta} a_\alpha a_{\dot\beta} + \tfrac{1}{2}\left[\sigma_{\dot\lambda\mu}^m\left(a_m^\mu a^{k\dot\lambda} + a_m^{\dot\lambda} a^{k\mu}\right) - \sigma_{\dot\lambda\mu}^k a^{m\dot\lambda} a_m^\mu \right]$$

$$+ i\epsilon^{klmn} a_l^{\dot\lambda} a_m^\mu \sigma_{n,\dot\lambda\mu} \quad \text{from (11)}. \tag{34}$$

The equations (33) and (34) are completely equivalent to the usual formulation of the Dirac-equation in the force-free case. The definite character of the charge-density is not quite obvious from (34). Equations (33) resemble very much the corresponding equations for a particle of spin 0.

However the equivalence of (30) and (33) holds only for the force-free case. In case of interaction with an electromagnetic field (30) go over into

$$i\pi^{\alpha\dot\lambda} a_{\dot\lambda} = \chi b^a \tag{35a}$$

$$i\pi_{\alpha\dot\lambda} b^a = \chi a_{\dot\lambda} \tag{35b}$$

where $\pi^{\alpha\dot\lambda}$ is the spinor corresponding to $\pi_k \equiv \partial_k + ie\phi_k$, ϕ_k being the electromagnetic potentials and e the charge of the particle. The second order equation derived from (35) is

$$\pi_{\alpha\dot\lambda}\pi^{\alpha\dot\mu} a_{\dot\mu} + \chi^2 a_{\dot\lambda} = 0.$$

Or from (3)

$$\pi_k \pi^k a_{\dot\lambda} + \tfrac{1}{2}\left(\pi_k \pi_l - \pi_l \pi_k \right)\sigma_{\alpha\dot\lambda}^k \sigma^{l,\alpha\dot\mu} a_{\dot\mu} + \chi^2 a_{\dot\lambda} = 0$$

which is not the same as that obtained by replacing ∂_k by π_k in (32). Therefore to take electromagnetic interaction into account it is not sufficient to replace ∂^k by π^k

in (33). The correct generalisation of (33) in this case is

$$\pi^k a_{\dot\lambda} = \chi a_{\dot\lambda}^k \tag{36a}$$

$$\pi_k a_{\dot\lambda}^k + \frac{ie}{\chi} f_{\dot\lambda}^{\dot\mu} a_{\dot\mu} = -\chi a_{\dot\lambda} \tag{36b}$$

where

$$f_{\dot\lambda}^{\dot\mu} = \tfrac{1}{2} f_{kl} \sigma_{\dot\lambda\alpha}^k \sigma^{l\alpha\dot\mu} \quad \text{and} \quad f_{kl} = \partial_k \phi_l - \partial_l \phi_k.$$

The extra term in (36b) corresponds precisely to the magnetic moment e/χ of the electron in Dirac's theory. The expressions (31) and (34) for the current vector remain unchanged. From (36) it follows that the Dirac equation is *completely equivalent* to the second order equation

$$\pi_k \pi^k a_{\dot\lambda} + ie f_{\dot\lambda}^{\dot\mu} a_{\dot\mu} + \chi^2 a_{\dot\lambda} = 0$$

provided the current vector is defined by (34). Equations (36) emphasise the fact that even in the simple case of spin $\tfrac{1}{2}$ correct electromagnetic interaction cannot be introduced simply by replacing ∂_k by π_k in any arbitrary formulation which is valid for the force-free case.

Summary

The σ-symbols are defined by means of the equations (1) and (2). All their properties are deduced from their definition without making use of any explicit representation. Certain interesting relations concerning the product of three or more σ's are obtained. They are shown to be useful in transforming tensors into spinors and *vice versa*.

Directly from (1) and (2) it is deduced that corresponding to every proper Lorentz transformation there exists a spinor-transformation such that the two applied together leave the σ's unchanged. The spinor transformation corresponding to the most general proper Lorentz transformation is explicitly given. Also the spinor-transformation corresponding to reflection is obtained. It is pointed out that since the defining equations (1) and (2) and the relations deduced from them are already in a proper covariant form they can be taken over as such to the general theory of relativity.

Finally the Dirac equation for a particle of spin $\tfrac{1}{2}$ is discussed from a new angle. Here only one spinor together with its space-time derivatives (and *not* two spinors) is used to describe the particle. It is shown that the Dirac equation is completely equivalent to a second order equation for this single spinor. The expression for the charge-current density in terms of this single spinor and its derivatives is obtained. In the present formulation correct electromagnetic interaction can be introduced only by the addition of an extra term depending explicitly on the field. This additional term is the one which corresponds to the magnetic moment of the electron.

A NOTE ON THE σ-SYMBOLS

REFERENCES

Fierz and Pauli, *Proc. Roy. Soc.*, A, 1939, **173**, 211–32.
Infeld and Van der Waerden, *Preuss. Akad. Berl. Ber.*, 1933, **9**, 380–474.
Laporte and Uhlenbeck, *Phys. Rev.*, 1931, **37**, 1380–97.
Van der Waerden, "Die gruppen theoretische methode in der Quantummechanik", Berlin, Springer.

Reprinted from
Proceedings of the Indian Academy of Sciences
23(1946), 152–163

The correspondence between the particle and the wave aspects of the meson and the photon

By Harish-Chandra

J. H. Bhabha Student, Tata Institute of Fundamental Research, Bombay

(*Communicated by H. J. Bhabha, F.R.S.—Received* 10 *September* 1945)

In the Kemmer formulation of the meson equation certain Γ_k^* matrices are introduced. They are determined completely by the condition that for a Lorentz transformation $\Gamma_k^*\psi$ transforms as a vector. The Γ^*'s, together with another matrix β (defined in terms of them), turn out to be very convenient for discussing the properties of the meson matrices β_k. With their help, a formula for the spur of a product of any number of β_k's is derived both for the 10- and the 5-row representations. Also they enable one to bring out, without making use of an explicit representation, the complete equivalence of the Kemmer formulation with the vector and scalar meson equations in the usual tensor form. Besides, the use of the Γ^*-matrices now permits the introduction of nuclear interaction in the Kemmer formulation in an elegant way. The extension of the 'particle' formulation to the case of a particle of zero rest-mass is discussed. Finally, a specially convenient representation of the β_k's is given both for the case of spin 1 and of spin 0. The Γ^*'s and the β are explicitly given for this particular representation.

1. Kemmer (1939) has given a formulation of the theory of the meson‡ based on the following equation:

$$i\beta^k\partial_k\psi + \chi\psi = 0. \tag{1}$$

‡ The term 'meson' is used in this paper to denote any particle with spin 1 or 0, having a non-vanishing rest-mass.

Here β^k ($k = 0, 1, 2, 3$) are matrices acting upon ψ which itself is a one-column matrix. The β^k's satisfy the following commutation rules:

$$\beta^k\beta^l\beta^m + \beta^m\beta^l\beta^k = g^{kl}\beta^m + g^{ml}\beta^k, \tag{2}$$

g^{kl} being the usual metric tensor for flat space-time,

$$g_{kl} = 0, \quad k \neq l; \quad g_{00} = -g_{11} = -g_{22} = -g_{33} = 1.$$

It is well known that (1) is completely equivalent to the usual 'wave' formulation of the meson equations. However, until now this fact could be established only by making use of an explicit representation of the β^k's. A method of proving it directly will be given in this paper. It will be shown that the correspondence between the two formulations is indeed so close and complete that one can go over with ease from one formulation to the other at any stage of the calculation.

Further, it was not possible until now to introduce in (1) the interaction between the meson and the heavy particles in an elegant way. In his attempt to take into account this nuclear interaction Wilson (1940) had to make explicit use of the cumbersome representation of the β-matrices. This completely destroyed the beauty of the 'particle' formulation. However, as a result of the recognition of the close correspondence mentioned above the nuclear interaction can now be included in the matrix scheme without affecting its elegance.

In using the matrix method for the quantum-mechanical treatment of the physical processes involving mesons it is necessary to evaluate the spurs of the multiple products of the β_k's. The usual methods for their evaluation are extremely cumbersome (cf. Booth & Wilson 1940). In this paper a general formula for the spur of a product of any number of β_k's will be derived quite easily. It is hoped that it will be of use in future calculations.

It is well known that the Kemmer formulation can be applied only to a particle of non-vanishing rest-mass, i.e. $\chi \neq 0$. The derivation of the second order equation

$$\partial_k \partial^k \psi + \chi^2 \psi = 0 \tag{3}$$

from (1) depends directly on the explicit assumption $\chi \neq 0$. It will be shown that by a slight modification the same formalism can be adapted to the case of zero rest-mass so that it can now be applied with equal elegance to the case of the photon or the electromagnetic field.

2. The Hermitian conjugate of (2) is

$$\beta^{\dagger k}\beta^{\dagger l}\beta^{\dagger m} + \beta^{\dagger m}\beta^{\dagger l}\beta^{\dagger k} = g^{kl}\beta^{\dagger m} + g^{ml}\beta^{\dagger k},$$

where $\beta^{\dagger k}$ is the Hermitian conjugate of β^k. Thus the $\beta_k^\dagger$'s satisfy the same commutation rules as the β_k's. Therefore every irreducible representation of the β_k's is equivalent to the corresponding one of the $\beta_k^\dagger$'s, because it has been shown by Kemmer that two inequivalent irreducible representations of the β_k's having the same number of rows and columns do not exist. Therefore there exists a matrix Λ such that

$$\Lambda\beta^k\Lambda^{-1} = \beta^{\dagger k}. \tag{4}$$

It follows from (4) (see Pauli 1936) that Λ can always be chosen to be Hermitian, i.e.

$$\Lambda^\dagger = \Lambda. \tag{5}$$

The Hermitian conjugate of (1) is

$$-i\partial_k \psi^\dagger \beta^{\dagger k} + \chi \psi^\dagger = 0,$$

so that from (4) $\qquad\qquad -i\partial_k \psi^\dagger \Lambda \beta^k + \chi \psi^\dagger \Lambda = 0.$

Put $\qquad\qquad\qquad\qquad\qquad \psi^\dagger \Lambda = \psi^*, \tag{6}$

then ψ^* satisfies the e   on

$$-i\partial_k \psi^* \beta^k + \chi \psi^* = 0. \tag{7}$$

Let I_{mn} be the matrices representing the infinitesimal transformations of the Lorentz group, so that by an infinitesimal transformation

$$x'_k = x_k + \epsilon_{kl} x^l \quad (\epsilon_{kl} = -\epsilon_{lk}) \tag{8a}$$

ψ is transformed to $\qquad\qquad \psi' = (1 + \tfrac{1}{2}\epsilon_{kl} I^{kl})\,\psi. \tag{8b}$

From the relativistic invariance of (7) it follows that

$$\psi^{*\prime} = \psi^*(1 - \tfrac{1}{2}\epsilon_{kl} I^{kl}), \tag{8c}$$

and from (8) and (6) one obtains

$$I^{\dagger kl}\Lambda + \Lambda I^{kl} = 0. \tag{9}$$

Besides, I^{kl} satisfy the following well-known commutation relations

$$[I^{kl}, I^{mn}] = -g^{km}I^{ln} + g^{lm}I^{kn} + g^{kn}I^{lm} - g^{ln}I^{km}, \tag{10a}$$

$$[\beta^k, I^{lm}] = g^{kl}\beta^m - g^{km}\beta^l, \tag{10b}$$

where $[A, B] = AB - BA$. It is well known that in the present case

$$I^{kl} = \beta^k \beta^l - \beta^l \beta^k, \tag{11}$$

and (9) and (10) are satisfied.

Introduce a one-column vector-matrix T_k such that $\psi^* T_k$ transforms as a vector, so that from (8c)

$$-I_{kl}T_m = g_{km}T_l - g_{lm}T_k. \tag{12}$$

Put $\qquad\qquad\qquad\qquad\qquad \Gamma_k^* = T_k^\dagger \Lambda; \tag{13}$

therefore from (12) and (9)

$$\Gamma_k^* I_{lm} = g_{kl}\Gamma_m^* - g_{km}\Gamma_l^*. \tag{14}$$

From (12) and (14) it follows that

$$I_k{}^l T_l = 3T_k, \tag{15a}$$

$$\Gamma^{*k} I_k{}^l = 3\Gamma^{*l}; \tag{15b}$$

thus from (15a) and (14)

$$3\Gamma_k^* T_l = \Gamma_k^* I_l{}^m T_m = g_{kl}\Gamma_m^* T^m - \Gamma_l^* T_k,$$

which immediately gives $\qquad \Gamma_k^* T_l - \Gamma_l^* T_k = 0,$

and therefore $\qquad \Gamma_k^* T_l = \tfrac{1}{4} g_{kl} \Gamma_m^* T^m. \qquad (16)$

Also from (12) and (14) $\qquad \tfrac{1}{2} I_{mn} I^{mn} T_k = -3 T_k, \qquad (17a)$

$$\tfrac{1}{2} \Gamma_k^* I_{mn} I^{mn} = -3 \Gamma_k^*, \qquad (17b)$$

so that $\qquad \Gamma_k^* [\beta_l, \tfrac{1}{2} I_{rs} I^{rs}] T_m = 0.$

But from (10b) $\qquad [\beta_l, \tfrac{1}{2} I_{rs} I^{rs}] = I_{ls} \beta^s + \beta^s I_{ls}$

$$= 3\beta_l + 2\beta^s I_{ls}; \qquad (18)$$

therefore $\qquad 3\Gamma_k^* \beta_l T_m - 2g_{lm} \Gamma_k^* \beta^s T_s + 2\Gamma_k^* \beta_m T_l = 0.$

On contracting l, m one gets $\qquad \Gamma_k^* \beta_l T^l = 0,$

so that $\qquad 3\Gamma_k^* \beta_l T_m + 2\Gamma_k^* \beta_m T_l = 0.$

On interchanging l, m and subtracting it is found that

$$\Gamma_k^* \beta_l T_m = \Gamma_k^* \beta_m T_l,$$

so that $\qquad \Gamma_k^* \beta_l T_m = 0. \qquad (19)$

It will now be shown that for a given representation of the β_k's the choice of T_k is unique apart from a numerical factor. For this purpose transform (12) into the corresponding spinor equations

$$-I_{\mu\nu} T_{\alpha\beta} = \tfrac{1}{2}(\epsilon_{\mu\alpha} T_{\nu\beta} + \epsilon_{\nu\alpha} T_{\mu\beta}), \qquad (20a)$$

$$-I_{\dot\mu\dot\nu} T_{\alpha\beta} = \tfrac{1}{2}(\epsilon_{\dot\mu\beta} T_{\alpha\dot\nu} + \epsilon_{\dot\nu\beta} T_{\alpha\dot\mu}). \qquad (20b)$$

The Greek alphabets will be used to denote spinor indices, the Latin aphabets being reserved for tensor indices. Here

$$I_{\mu\nu} = \tfrac{1}{4} I_{kl} \sigma_{\mu\lambda}^k \sigma^{l,\lambda}{}_\nu, \qquad (21a)$$

$$I_{\dot\mu\dot\nu} = \tfrac{1}{4} I_{kl} \sigma_{\dot\mu\lambda}^k \sigma^{l,\lambda}{}_{\dot\nu}, \qquad (21b)$$

$$T_{\alpha\beta} = T_k \sigma_{\alpha\beta}^k. \qquad (21c)$$

The σ-symbols have the usual meaning. $\epsilon^{\mu\nu}$ and $\epsilon_{\mu\nu}$ are the fundamental antisymmetric spinors for raising and lowering the indices, $\epsilon^{12} = \epsilon_{12} = 1$. Similarly, $\epsilon^{\dot1\dot2} = \epsilon_{\dot1\dot2} = 1$. Every finite irreducible representation of $I_{\mu\nu}$ is characterized by a positive number k which can assume only integral or half-integral values. The following relation holds for a particular irreducible representation

$$-\tfrac{1}{2} I^{\mu\nu} I_{\mu\nu} = k(k+1). \qquad (22a)$$

Similarly, a finite irreducible representation of $I^{\dot\mu\dot\nu}$ is characterized by a positive number (integral or half-integral) given by

$$-\tfrac{1}{2} I^{\dot\mu\dot\nu} I_{\dot\mu\dot\nu} = l(l+1). \qquad (22b)$$

As is well known (Dirac 1936; Fierz 1939) for any given values of k and l the following relations hold:

$$\left.\begin{aligned}
&I_{\mu\nu}(k) = u_{\mu}(k)\,v_{\nu}(k) - k\epsilon_{\mu\nu}, \\
&I_{\dot\mu\dot\nu}(l) = u_{\dot\mu}(l)\,v_{\dot\nu}(l) - l\epsilon_{\dot\mu\dot\nu}, \\
&u_{\mu}(k)\,u^{\mu}(k-\tfrac{1}{2}) = v_{\mu}(k-\tfrac{1}{2})\,v^{\mu}(k) = 0, \\
&u_{\mu}(k)\,v^{\mu}(k) = 2k, \quad v^{\mu}(k)\,u_{\mu}(k) = 2k+1, \\
&v^{\mu}(k+\tfrac{1}{2})\,u^{\nu}(k+\tfrac{1}{2}) - u^{\nu}(k)\,v^{\mu}(k) = e^{\nu\mu}, \\
&u_{\dot\mu}(l)\,u^{\dot\mu}(l-\tfrac{1}{2}) = v_{\dot\mu}(l-\tfrac{1}{2})\,v^{\dot\mu}(l) = 0, \\
&u_{\dot\mu}(l)\,v^{\dot\mu}(l) = 2l, \quad v^{\dot\mu}(l)\,u_{\dot\mu}(l) = 2l+1, \\
&v^{\dot\mu}(l+\tfrac{1}{2})\,u^{\dot\nu}(l+\tfrac{1}{2}) - u^{\dot\nu}(l)\,v^{\dot\mu}(l) = e^{\dot\nu\dot\mu},
\end{aligned}\right\} \tag{23}$$

where $u_{\mu}(k)$ is a spinor matrix with $2k+1$ rows and $2k$ columns, while $v_{\mu}(k)$ has $2k$ rows and $2k+1$ columns. Similarly for $u_{\dot\mu}(l)$ and $v_{\dot\mu}(l)$. Corresponding to definite values of k and l split up $T_{\alpha\beta}$ in $T_{\alpha\beta}(k,l)$ (cf. Bhabha 1945). From (20) it is found that

$$-\tfrac{1}{2}I^{\mu\nu}I_{\mu\alpha}T_{\alpha\beta} = \tfrac{3}{4}T_{\alpha\beta}, \tag{24a}$$

$$-\tfrac{1}{2}I^{\dot\mu\dot\nu}I_{\dot\mu\dot\nu}T_{\alpha\beta} = \tfrac{3}{4}T_{\alpha\beta}, \tag{24b}$$

from which it follows that $T_{\alpha\beta}(k,l) = 0$ unless

$$k(k+1) = \tfrac{3}{4}, \quad l(l+1) = \tfrac{3}{4},$$

i.e. unless $k = \tfrac{1}{2}, l = \tfrac{1}{2}$. Therefore (20) reduces to

$$-I_{\mu\nu}(\tfrac{1}{2})\,T_{\alpha\beta} = \tfrac{1}{2}\epsilon_{\mu\alpha}\,T_{\nu\beta} + \tfrac{1}{2}\epsilon_{\nu\alpha}\,T_{\mu\beta}, \tag{25a}$$

$$-I_{\dot\mu\dot\nu}(\tfrac{1}{2})\,T_{\alpha\beta} = \tfrac{1}{2}\epsilon_{\dot\mu\beta}\,T_{\alpha\dot\nu} + \tfrac{1}{2}\epsilon_{\dot\nu\beta}\,T_{\alpha\dot\mu}, \tag{25b}$$

where $T_{\alpha\beta}$ is written instead of $T_{\alpha\beta}(\tfrac{1}{2},\tfrac{1}{2})$ for brevity. Multiply (25a) by $v^{\mu}(\tfrac{1}{2})$ so that from (23)

$$-\tfrac{3}{2}v_{\nu}(\tfrac{1}{2})\,T_{\alpha\beta} = \tfrac{1}{2}v_{\alpha}(\tfrac{1}{2})\,T_{\nu\beta} + \tfrac{1}{2}\epsilon_{\nu\alpha}v^{\mu}(\tfrac{1}{2})\,T_{\mu\beta}.$$

On interchanging ν, α and adding it is found that

$$v_{\nu}(\tfrac{1}{2})\,T_{\alpha\beta} = -v_{\nu}(\tfrac{1}{2})\,T_{\nu\beta},$$

so that

$$-v_{\nu}(\tfrac{1}{2})\,T_{\alpha\beta} = \tfrac{1}{2}\epsilon_{\nu\alpha}v^{\mu}(\tfrac{1}{2})\,T_{\mu\beta},$$

or, on multiplying by $u^{\nu}(\tfrac{1}{2})$,

$$T_{\alpha\beta} = u_{\alpha}(\tfrac{1}{2})\,T_{\beta}, \tag{26}$$

where $T_{\beta} = \tfrac{1}{2}v^{\mu}(\tfrac{1}{2})\,T_{\mu\beta}$. Similarly, by substituting (26) in (25b) one obtains

$$T_{\alpha\beta}(\tfrac{1}{2},\tfrac{1}{2}) = u_{\alpha}(\tfrac{1}{2})\,u_{\beta}(\tfrac{1}{2})\,T, \tag{27a}$$

where

$$T = \tfrac{1}{2}v^{\beta}(\tfrac{1}{2})\,T_{\beta} = \tfrac{1}{4}v^{\alpha}(\tfrac{1}{2})\,v^{\beta}(\tfrac{1}{2})\,T_{\alpha\beta}(\tfrac{1}{2},\tfrac{1}{2}). \tag{27b}$$

Similarly from (14)

$$\Gamma^{*}_{\alpha\beta}(\tfrac{1}{2},\tfrac{1}{2}) = v_{\alpha}(\tfrac{1}{2})\,v_{\beta}(\tfrac{1}{2})\,\Gamma^{*}, \tag{27c}$$

where

$$\Gamma^{*} = \tfrac{1}{4}\Gamma^{*}_{\alpha\beta}(\tfrac{1}{2},\tfrac{1}{2})\,u^{\alpha}(\tfrac{1}{2})\,u^{\beta}(\tfrac{1}{2}). \tag{27d}$$

In case‡ k, l, m, n form a complete system of commuting variables, T and Γ^* are ordinary numbers. That k, l, m, n do actually form a complete system of commuting variables for the various irreducible representations of the β_k's follows from the dimensionality of the 5- and 10-row representations (cf. § 6) which are the only non-trivial irreducible representations as shown by Kemmer. Therefore $T_{\alpha\beta}$ and $\Gamma^*_{\alpha\beta}$ are determined completely except for a numerical constant. Incidentally (27) proves that non-vanishing T_k and Γ^*_k satisfying (12) and (14) always exist whenever the irreducible representation $k = \tfrac{1}{2}$, $l = \tfrac{1}{2}$ is contained in the representation of the β_k's This condition is of course satisfied for the case of spin 1 and 0 (cf. § 6).

Notice that from (27)

$$\Gamma^*_k T^k = \tfrac{1}{2}\Gamma^*_{\alpha\beta} T^{\alpha\beta} = \tfrac{1}{2}\Gamma^* T,$$

so that unless Γ^* or T is zero $\Gamma^*_k T^k \neq 0$. However, from (13) and the non-singularity of Λ it follows that if $T = 0$ then $\Gamma^* = 0$ and conversely. Therefore

$$\Gamma^*_k T^k \neq 0,$$

unless $\Gamma^*_k = T_k = 0$. Thus if T_k exists at all it can always be so normalized§ that

$$\Gamma^*_k T^k = 4.$$

This determines T_k completely except for a phase-factor of modulus unity. On introducing this normalization (16) becomes

$$\Gamma^*_k T_l = g_{kl}. \tag{28}$$

In view of (27), (19) is easy to understand. It has been shown by Bhabha (1945) that the only non-vanishing matrix elements of β_r are of the type

$$(k, l \,|\, \beta_r \,|\, k \pm \tfrac{1}{2}, l \pm \tfrac{1}{2}).$$

Since the only non-vanishing elements of Γ^*_k and T_l are $\Gamma^*_k(\tfrac{1}{2}, \tfrac{1}{2})$ and $T_l(\tfrac{1}{2}, \tfrac{1}{2})$, it follows that a product of any odd number of β_r's multiplied by Γ^*_k on the left and T_l on the right would give zero. Also notice that if θ be any operator commuting with I_{kl}, then θT_k and $\Gamma^*_k \theta$ satisfy equations entirely similar to (12) and (14). It therefore follows from the above result that θT_k and $\Gamma^*_k \theta$ can differ from T_k and Γ^*_k by a numerical factor only. One such operator θ is $\beta_m \beta^m$. Thus $\beta_m \beta^m T_k$ and $\Gamma^*_k \beta_m \beta^m$ are numerical multiples of T_k and Γ^*_k. Due to (13) the multiplying factors are conjugate complex of each other.

It may be mentioned here that all the results obtained so far depend only on the existence of I^{kl} and Λ satisfying (9) and (10) and are completely independent of the commutation rules (2). They hold therefore for the matrices β^l occurring in any relativistic wave equation of the type (1) (see Bhabha 1945). From now onwards (2) will be used explicitly.

---

‡ m and n are the eigenvalues of I^{12} and $I^{\dot{1}\dot{2}}$ respectively.

§ If necessary the sign of Λ may be reversed without disturbing (4) and (5) to make $\Gamma_k{}^* T^k$ positive.

508   Harish-Chandra

Notice that from (2)
$$\beta_k\beta^k\beta_l+\beta_l\beta_k\beta^k = 5\beta_l, \tag{29a}$$

and
$$\beta^k\beta_l\beta_k = \beta_l, \tag{29b}$$

so that from (17) and (11)

$$\beta_m\beta_n(\beta^m\beta^n-\beta^n\beta^m)\,T_k = \beta_n\beta^n(\beta_m\beta^m-4)\,T_k = B(B-4)\,T_k$$
$$= -3T_k,$$

i.e.
$$(B-1)(B-3)\,T_k = 0, \tag{30}$$

where B is a number given by

$$\beta^n\beta_n\,T_k = BT_k;$$

therefore $B = 1$ or $B = 3$. For the 10-row representation of the β_k's $\beta_m\beta^m$ has only 2 and 3 as eigenvalues, while for the 5-row representation the eigenvalues are 1 and 4.‡ Therefore for the 10-row representation $B = 3$ while for the 5-row one $B = 1$. Now consider $I_l^m I_{mn}\,T_r$. From (12)

$$-I_l^m I_{mn}\,T_r = I_{ln}\,T_r - g_{nr}I_l^m\,T_{\dot{m}}$$
$$= -g_{ln}\,T_r - 2g_{nr}\,T_l.$$

On the other hand, from (11) and (29)

$$I_l^m I_{mn} = (\beta^l\beta_m - \beta_m\beta_l)(\beta^m\beta_n - \beta_n\beta^m)$$
$$= \beta_l\beta_m\beta^m\beta_n - 2\beta_l\beta_n + \beta_m\beta_l\beta_n\beta^m$$
$$= \beta_l\beta_n(3-\beta_m\beta^m) + \beta_n\beta_l(1-\beta_m\beta^m) + g_{ln}\beta_m\beta^m, \tag{31}$$

therefore
$$\{\beta_l\beta_n(3-B) + \beta_n\beta_l(1-B) + g_{ln}B\}\,T_r = g_{ln}\,T_r + 2g_{nr}\,T_l. \tag{32}$$

For the 10-row representation $B = 3$ and so

$$\beta_n\beta_l\,T_r = g_{nl}\,T_r - g_{nr}\,T_l. \tag{33a}$$

Multiplying (33a) by β^n

$$\beta^n\beta_n\beta_l\,T_r = \beta_l(5-\beta^n\beta_n)\,T_r = 2\beta_l\,T_r = \beta_l\,T_r - \beta_r\,T_l,$$

i.e.
$$\beta_l\,T_r = -\beta_r\,T_l. \tag{33b}$$

---

‡ The easiest way to verify it is to calculate $\tfrac{1}{2}\beta_{\alpha\lambda}\beta^{\alpha\lambda}$ from the equation (100) of § 6. Notice that from (29a) it follows that $(\beta_m\beta^m - 5/2)^2$ commutes with the β_k's and therefore for any irreducible representation of the β_k's it is a multiple of the unit matrix. Thus for any irreducible representation $\mathscr{B} = \beta_m\beta^m$ satisfies the equation

$$(\mathscr{B} - B_1)(\mathscr{B} - B_2) = 0,$$

where B_1, B_2 are numbers with $B_1 + B_2 = 5$. Also it is easy to prove from (2) that the characteristic equation for $\mathscr{B}$ is

$$\mathscr{B}(\mathscr{B} - 1)(\mathscr{B} - 2)(\mathscr{B} - 3)(\mathscr{B} - 4) = 0;$$

this, together with (29a), permits only two possibilities

$$B_1 = 1, \quad B_2 = 4, \quad B_1 = 2, \quad B_2 = 3,$$

besides the trivial one $B = 0$ corresponding to $\beta_l = 0$.

Thus for spin 1, $\beta_l T_r$ is antisymmetric in l, r.

On the other hand, for the 5-row representation $B = 1$, so that from (32)

$$\beta_l \beta_n T_r = T_l g_{nr}. \tag{34a}$$

Multiplying (34a) by β^l

$$\beta^l \beta_l \beta_n T_r = \beta_n (5 - \beta^l \beta_l) T_r = 4\beta_n T_r = \beta^l T_l g_{nr},$$

therefore

$$\beta_n T_r = \tfrac{1}{4} g_{nr} \beta^l T_l = \beta_r T_n. \tag{34b}$$

The corresponding equations for Γ_k^* are

$$\left. \begin{array}{l} \Gamma_k^* \beta_l \beta_m = \Gamma_k^* g_{lm} - \Gamma_l^* g_{km} \\[4pt] \Gamma_k^* \beta_l = - \Gamma_l^* \beta_k \end{array} \right\} \quad \text{for spin 1,} \tag{35}$$

$$\left. \begin{array}{l} \Gamma_k^* \beta_l = \Gamma_l^* \beta_k = \tfrac{1}{4} g_{kl} \Gamma_m^* \beta^m \\[4pt] \Gamma_k^* \beta_l \beta_m = g_{kl} \Gamma_m^* \end{array} \right\} \quad \text{for spin 0.} \tag{36}$$

Now introduce a matrix β defined as follows:

$$\beta = T_k \Gamma^{*k}. \tag{37}$$

From (28) it follows that

$$\beta^2 = \beta, \tag{38}$$

$$\Gamma_k^* \beta = \Gamma_k^*, \tag{39a}$$

$$\beta T_k = T_k; \tag{39b}$$

also from (33) and (35)

$$\beta_n \beta_l \beta = g_{nl} \beta - T_l \Gamma_n^* = \beta \beta_n \beta_l \quad \text{for spin 1,}$$

and similarly from (34) and (36)

$$\beta_l \beta_n \beta = T_l \Gamma_n^* = \beta \beta_l \beta_n \quad \text{for spin 0.}$$

Therefore in both cases

$$\beta_n \beta_l \beta - \beta \beta_n \beta_l = 0. \tag{40}$$

Now put

$$\beta \beta_k + \beta_k \beta = \theta_k; \tag{41}$$

then

$$\theta_k \theta_l = \beta_k \beta \beta_l + \beta \beta_k \beta_l \beta,$$

since $\beta \beta_k \beta = 0$ from (19). Also from (40) and (38)

$$\theta_k \theta_l \theta_m = \beta_k \beta \beta_l \beta_m \beta + \beta \beta_k \beta_l \beta \beta_m$$
$$= \beta_k \beta_l \beta_m \beta + \beta \beta_k \beta_l \beta_m,$$

therefore

$$\theta_k \theta_l \theta_m + \theta_m \theta_l \theta_k = (g_{kl} \beta_m + g_{ml} \beta_k) \beta + \beta (g_{kl} \beta_m + g_{ml} \beta_k)$$
$$= g_{kl} \theta_m + g_{ml} \theta_k; \tag{42}$$

thus the θ_k's satisfy the same commutation rules as the β_k's. Also from (19), (37) and (39)

$$\left. \begin{array}{l} \theta_k T_l = (\beta \beta_k + \beta_k \beta) T_l = \beta_k T_l, \\[4pt] \Gamma_k^* \theta_l = \Gamma_k^* \beta_l, \end{array} \right\} \tag{43a}$$

and

$$\left.\begin{array}{l}\theta_k\theta_l T_m = \theta_k\beta_l T_m = \beta\beta_k\beta_l T_m = \beta_k\beta_l\beta' T_m = \beta_k\beta_l T_m,\\[4pt]\Gamma_k^*\theta_l\theta_m = \Gamma_k^*\beta_l\beta_m.\end{array}\right\} \tag{43b}$$

The representation of the θ_k's must be equivalent to that of the β_k's since from (43)

$$\beta_k\beta^k T_l = B T_l = \theta_k\theta^k T_l,$$

and the B values for two inequivalent representations are different, therefore

$$\beta_k = S\theta_k S^{-1}, \tag{44}$$

where S is a non-singular matrix. Notice that from (41)

$$\theta_k\beta + \beta\theta_k = \theta_k, \tag{45}$$

so that from (44) and (45) $\qquad \beta_k\beta' + \beta'\beta_k = \beta_k, \tag{46}$

where $\qquad\qquad \beta' = S\beta S^{-1} = ST_k\Gamma^{*k}S^{-1}.$

Put $\theta_k\theta_l - \theta_l\theta_k = J_{kl}$. Then from (43)

$$I_{kl}T_m = J_{kl}T_m = -g_{km}T_l + g_{lm}T_k.$$

But $J_{kl} = S^{-1}I_{kl}S$, and so

$$I_{kl}ST_m = -g_{km}ST_l + g_{lm}ST_k.$$

Thus ST_k satisfies an equation similar to (12) and therefore must be equal to cT_k, where c is an ordinary number. Similarly, $\Gamma_k^* S^{-1}$ is equal to $c^*\Gamma_k^*$, where c^* is another number. However,

$$cc^*\Gamma_k^* T_l = \Gamma_k^* S^{-1}ST_l = \Gamma_k^* T_l = g_{kl},$$

therefore $cc^* = 1$ and

$$\beta' = ST_k\Gamma^{*k}S^{-1} = cc^*T_k\Gamma^{*k} = T_k\Gamma^{*k} = \beta,$$

so that $\qquad\qquad\qquad \beta_k\beta + \beta\beta_k = \beta_k. \tag{47}$

Collecting all the equations together

$$\beta = T_k\Gamma^{*k}, \tag{48a}$$

$$\beta^2 = \beta, \tag{48b}$$

$$\beta\beta_k + \beta_k\beta = \beta_k, \tag{48c}$$

$$\Gamma_k^* T_l = g_{kl}, \tag{48d}$$

$$\Gamma_k^* \beta_l T_m = 0, \tag{48e}$$

$$\left.\begin{array}{l}\beta_k\beta_l T_m = g_{kl}T_m - g_{km}T_l\\[4pt]\Gamma_k^*\beta_l\beta_m = \Gamma_k^* g_{lm} - \Gamma_l^* g_{km}\\[4pt]\beta_k T_l = -\beta_l T_k\\[4pt]\Gamma_k^*\beta_l = -\Gamma_l^*\beta_k\end{array}\right\} \text{ for spin 1,} \tag{49a}$$

$$\left.\begin{array}{l}\beta_k\beta_l T_m = T_k g_{lm}\\[4pt]\Gamma_k^*\beta_l\beta_m = g_{kl}\Gamma_m^*\\[4pt]\beta_l T_m = \tfrac{1}{4}g_{lm}\beta_k T^k\\[4pt]\Gamma_k^*\beta_l = \tfrac{1}{4}g_{kl}\Gamma_m^*\beta^m\end{array}\right\} \text{ for spin 0.} \tag{49b}$$

It will now be shown that the conditions (48b) and (48c) are so restrictive that for any irreducible set of matrices β_k (which need not in general satisfy (2)) there exist either none or just two matrices β satisfying them. Notice that if β is a solution then $1-\beta$ is also one. Further $\beta \neq 1-\beta$, since $\beta \neq \frac{1}{2}$ on account of (48b). Now let β' be any other solution. Then

$$\beta_k(\beta-\beta')+(\beta-\beta')\beta_k = 0,$$

so that $\qquad [\beta_k,(\beta-\beta')^2] = [\beta_k,(\beta-\beta')]_+(\beta-\beta')-(\beta-\beta')[\beta_k,(\beta-\beta')]_+,$

where $[A,B]_+ = AB+BA$. Since the β_k's are irreducible it follows that $(\beta-\beta')^2$ is a multiple of the unit matrix, i.e.

$$\beta^2+\beta'^2-\beta\beta'-\beta'\beta = \beta+\beta'-\beta\beta'-\beta'\beta = c, \tag{50}$$

where c is a number. On multiplying (50) by β or β'

$$\beta\beta'\beta = (1-c)\beta, \tag{51a}$$

$$\beta'\beta\beta' = (1-c)\beta'. \tag{51b}$$

From (48c) and (51) $\qquad \beta\beta'\beta\beta_k+\beta_k\beta\beta'\beta = (1-c)\beta_k.$

Now $\qquad [\beta\beta'\beta,\beta_k]_+ = \beta\beta'[\beta,\beta_k]_+-\beta[\beta',\beta_k]_+\beta+[\beta,\beta_k]_+\beta'\beta$

$$= \beta\beta'\beta_k-\beta\beta_k\beta+\beta_k\beta'\beta$$
$$= \beta\beta'\beta_k+\beta_k\beta'\beta = (1-c)\beta_k, \tag{52a}$$

since $\beta\beta_k\beta = 0$ as follows from (48c). Similarly

$$[\beta'\beta\beta',\beta_k] = \beta'\beta\beta_k+\beta_k\beta\beta' = (1-c)\beta_k. \tag{52b}$$

Subtracting (52b) from (52a) one obtains

$$[(\beta'\beta-\beta\beta'),\beta_k] = 0,$$

therefore $\beta'\beta-\beta\beta'$ is a multiple of the unit matrix, i.e.

$$\beta'\beta-\beta\beta' = d, \tag{53}$$

where d is a number. Multiplying (53) by β from both sides

$$d\beta = 0,$$

therefore $d = 0$, since $\beta \neq 0$ from (48c). Thus

$$\beta'\beta = \beta\beta'. \tag{54}$$

Multiplying by β on the right gives

$$\beta'\beta = \beta\beta'\beta = (1-c)\beta,$$

while multiplication by β' on the left gives

$$\beta'\beta = \beta'\beta\beta' = (1-c)\beta',$$

therefore $\qquad\qquad (1-c)\beta = (1-c)\beta'.$

Thus either $\beta = \beta'$ or $c = 1$, in which case $\beta\beta' = \beta'\beta = 0$ and therefore

$$\beta' = 1-\beta$$

from (50).

Now consider the matrix $\qquad \gamma = \tfrac{1}{2}\beta_k\beta\beta^k$ (55)

for the case of spin 1:
$$\begin{aligned}
\gamma^2 &= \tfrac{1}{4}\beta_k\beta\beta^k\beta^l\beta\beta_l = \tfrac{1}{4}\beta_k\beta^k\beta^l\beta\beta_l \\
&= \tfrac{1}{4}\beta^l(5-\beta_k\beta^k)\beta\beta_l \\
&= \tfrac{1}{4}\beta^l(5-B)\beta\beta_l \\
&= \tfrac{1}{2}\beta^l\beta\beta_l = \gamma.
\end{aligned}$$

Further $\qquad \gamma = \tfrac{1}{2}\beta_k\beta\beta^k = \tfrac{1}{2}(1-\beta)\beta_k\beta^k \qquad \tfrac{1}{2}(\beta_k\beta^k-3\beta)$

therefore from (29a) and (48c)
$$\gamma\beta_k + \beta_k\gamma = \beta_k.$$

From the result just proved above it follows that either $\gamma = \beta$ or $\gamma = 1-\beta$. But
$$\begin{aligned}
\beta^k\beta_k\gamma &= \beta^k\beta_k \cdot \tfrac{1}{2}\beta_l\beta\beta^l \\
&= \tfrac{1}{2}\beta_l(5-\beta^k\beta_k)\beta\beta^l \\
&= 2\gamma,
\end{aligned}$$

while $\qquad\qquad\qquad \beta^k\beta_k\beta = 3\beta,$

therefore $\qquad\qquad\qquad \gamma = 1-\beta$

and $\qquad\qquad\qquad \beta + \tfrac{1}{2}\beta^k\beta\beta_k = 1 \quad$ for spin 1. (56)

For spin 0 take $\qquad\qquad \gamma = \tfrac{1}{4}\beta_k\beta\beta^k.$

Then $\qquad\qquad \gamma^2 = \tfrac{1}{16}\beta^l(5-B)\beta\beta_l = \tfrac{1}{4}\beta^l\beta\beta_l = \gamma.$

Also $\qquad \gamma = \tfrac{1}{4}\beta_k\beta\beta^k = \tfrac{1}{4}(1-\beta)\beta_k\beta^k = \tfrac{1}{4}(\beta_k\beta^k-\beta),$

therefore $\qquad\qquad\qquad \gamma\beta_k + \beta_k\gamma = \beta_k$

and $\gamma = \beta$ or $1-\beta$. But
$$\beta^l\beta_l\gamma = \tfrac{1}{4}\beta_k(5-\beta_l\beta^l)\beta\beta^l = 4\gamma,$$

while $\qquad\qquad\qquad \beta^l\beta_l\beta = \beta,$

therefore $\qquad\qquad\qquad \gamma = 1-\beta$

and $\qquad\qquad\qquad \beta + \tfrac{1}{4}\beta^k\beta\beta_k = 1 \quad$ for spin 0. (57a)

Also from (49b) and (57a) it follows that
$$\beta^k\beta\beta^l = g^{kl}(1-\beta) \quad \text{for spin 0.} \tag{57b}$$

It is interesting to observe that for spin 0
$$\begin{aligned}
\beta^k\beta^l\beta^m &= \beta^k\beta\beta^l\beta^m + \beta^k\beta^l\beta\beta^m \\
&= g^{kl}(1-\beta)\beta^m + \beta^k(1-\beta)g^{lm} \\
&= g^{kl}\beta^m\beta + \beta\beta^k g^{lm}.
\end{aligned} \tag{58}$$

The condition (58) is stronger than (2) and may be used to separate the 5-row representation from the 10-row one.

A few other interesting results can also be derived. For example, consider‡
$$\omega = \frac{i}{4}\epsilon^{klmn}\beta_k\beta_l\beta_m\beta_n. \tag{59a}$$

‡ ϵ^{klmn} is a tensor antisymmetric in k, l, m, n and $\epsilon^{0123} = 1$.

It follows from (58) that $\omega = 0$ for spin 0. Therefore only the 10-row representation need be considered. In this case it is easy to prove from (48) and (49) that

$$\omega = \frac{i}{4}\epsilon^{klmn}\beta_k T_l \Gamma_m^* \beta_n, \tag{59b}$$

so that

$$\begin{aligned}
\omega^2 &= -\tfrac{1}{16}\epsilon^{klmn}\epsilon^{k'l'm'n'}\beta_k T_l \Gamma_m^* \beta_n \beta_{k'} T_{l'} \Gamma_{m'}^* \beta_{n'} \\
&= \tfrac{1}{8}\epsilon^{klmn}\epsilon_{mn}{}^{m'n'}\beta_k T_l \Gamma_{m'}^* \beta_{n'} \\
&= -\tfrac{1}{2}\beta_k T_l \Gamma^{*k}\beta^l = \tfrac{1}{2}\beta_k T_l \Gamma^{*l}\beta^k \\
&= \tfrac{1}{2}\beta_k \beta \beta^k = 1-\beta. \tag{59c}
\end{aligned}$$

It follows immediately from (59b) and (59c) that

$$\begin{aligned}
\omega^3 &= \omega, \\
\omega^2\beta_k + \beta_k\omega &= \beta_k, \\
\omega\beta_r\omega &= 0, \\
\beta_r\omega\beta_s + \beta_s\omega\beta_r &= 0.
\end{aligned}$$

Also from (59b)

$$\begin{aligned}
\beta_r\beta_s\omega + \omega\beta_s\beta_r &= \frac{i}{2}\epsilon_s{}^{lmn}\beta_l(T_m\Gamma_r^* - T_r\Gamma_m^*)\beta_n \\
&= \frac{i}{2}\epsilon_s{}^{lmn}\beta_l\beta(\beta_m\beta_r - \beta_r\beta_m)\beta_n \\
&= \frac{i}{2}(1-\beta)\epsilon_s{}^{lmn}(\beta_l\beta_m\beta_r\beta_n - \beta_l\beta_r\beta_m\beta_n) \\
&= \frac{i}{4}(1-\beta)\epsilon_s{}^{lmn}(\beta_r\beta_l\beta_m\beta_n - \beta_l\beta_r\beta_m\beta_n + \beta_l\beta_m\beta_r\beta_n - \beta_l\beta_m\beta_n\beta_r) \quad \text{from (2)} \\
&= g_{sr}\omega^3 = g_{sr}\omega.
\end{aligned}$$

Therefore, as has already been pointed out by Schrödinger (1943) and Kemmer (1943), ω satisfies the same commutation rules as any of the β_k's. Thus if one puts $\omega = \beta_4$ and $g_{44} = 1$, $g_{4k} = 0$ $(k \neq 4)$, then (2) would hold for $k, l, m = (0, 1, 2, 3, 4)$.

3. The above results can now be applied to evaluate the spurs of any matrix composed of the β_k's and their products. Notice that

$$sp(\beta) = \Gamma_m^* T^m = 4, \tag{60a}$$

$$sp(\beta_k) = sp(\beta\beta_k + \beta_k\beta) = 2\Gamma_m^*\beta_k T^m = 0, \tag{60b}$$

and

$$sp(\beta_k\beta_l) = sp(\beta_k\beta\beta_l + \beta_k\beta_l\beta) = \Gamma_m^*\beta_l\beta_k T^m + \Gamma_m^*\beta_k\beta_l T^m,$$

so that from (49)

$$\left.\begin{aligned}
sp(\beta_k\beta_l) &= 6g_{kl} \quad \text{for spin 1,} \\
sp(\beta_k\beta_l) &= 2g_{kl} \quad \text{for spin 0.}
\end{aligned}\right\} \tag{60c}$$

Consider now the spin 1 case, so that

$$\Gamma_k^*\beta_l\beta_m T_n = g_{kn}g_{lm} - g_{km}g_{ln} \equiv \eta_{[kl][mn]}. \tag{61a}$$

Due to (39)

$$\Gamma_{k_0}^*\beta_{k_1}\beta_{k_2}\cdots\beta_{k_n}T_{k_{n+1}} = \Gamma_{k_0}^*\beta\beta_{k_1}\cdots\beta_{k_n}\beta T_{k_{n+1}}.$$

In case n the number of factors $\beta_{k_1}, \ldots, \beta_{kn}$ is odd one finds on using (40) that

$$\Gamma_{k_0}^* \beta_{k_1} \cdots \beta_{k_{2n-1}} T_{k_{2n}} = \Gamma_{k_0}^* \beta_{k_1} \cdots \beta\beta_{k_{2n-1}} \beta T_{k_{2n}} = 0. \tag{61b}$$

On the other hand, for an even number of factors one gets from (38), (39), (40), (37) and (61a)

$$\Gamma_{k_0}^* \beta_{k_1} \cdots \beta_{k_{2n}} T_{k_{2n+1}} = \Gamma_{k_0}^* \beta_{k_1} \beta_{k_2} \beta\beta_{k_3} \beta_{k_4} \beta \cdots \beta\beta_{k_{2n-1}} \beta_{k_{2n}} T_{k_{2n+1}}$$
$$= \eta_{[k_0 k_1][k_2}{}^{t_1]} \eta_{[t_1 k_3][k_4}{}^{t_2]} \cdots \eta_{[t_{n-i} k_{2n-1}][k_{2n} k_{2n+1}]}. \tag{61c}$$

Now
$$sp(\beta_{k_1} \cdots \beta_{k_n}) = sp(\beta\beta_{k_1} \cdots \beta_{k_n} + \beta_{k_1}\beta\beta_{k_2} \cdots \beta_{k_n})$$
$$= \Gamma_m^* \beta_{k_1} \cdots \beta_{k_n} T^m + \Gamma_m^* \beta_{k_2} \cdots \beta_{k_n} \beta_{k_1} T^m,$$

so that
$$sp(\beta_{k_1} \cdots \beta_{k_{2n}}) = \eta_{[t_1 k_1][k_2}{}^{t_2]} \eta_{[t_2 k_3][k_4}{}^{t_3]} \cdots \eta_{[t_n k_{2n-1}][k_{2n}}{}^{t_1]}$$
$$+ \eta_{[t_1 k_2][k_3}{}^{t_2]} \eta_{[t_2 k_4][k_5}{}^{t_3]} \cdots \eta_{[t_n k_{2n}][k_1}{}^{t_1]}, \tag{62a}$$
$$sp(\beta_1 \cdots \beta_{k_{2n+1}}) = 0. \tag{62b}$$

For spin 0
$$\Gamma_k^* \beta_l \beta_m T_n = g_{kl} g_{mn},$$

therefore
$$\Gamma_{k_0}^* \beta_{k_1} \cdots \beta_{k_{2n-1}} T_{k_{2n}} = 0, \tag{63a}$$
$$\Gamma_{k_0}^* \beta_{k_1} \cdots \beta_{k_{2n}} T_{k_{2n+1}} = g_{k_0 k_1} g_{k_2 k_3} \cdots g_{k_{2n} k_{2n+1}}. \tag{63b}$$

Thus
$$sp(\beta_{k_1} \cdots \beta_{k_{2n}}) = g_{k_1 k_2} g_{k_3 k_4} \cdots g_{k_{2n-1} k_{2n}} + g_{k_2 k_3} g_{k_4 k_5} \cdots g_{k_{2n} k_1}, \tag{64a}$$
$$sp(\beta_{k_1} \cdots \beta_{k_{2n+1}}) = 0. \tag{64b}$$

Multiplying (62b) and (64b) by arbitrary vectors $A^{(1)k_1}, A^{(2)k_2}, \ldots, A^{(2n)k_{2n}}$, one finds that

$$sp\{(A^{(1)}\beta)(A^{(2)}\beta)\ldots(A^{(2n)}\beta)\}$$
$$= A^{(1,2)t_1}_{t_1} A^{(3,4)t_2}_{t_2} \ldots A^{(2n-1,2n)t_n}_{t_n} + A^{(2,3)t_1}_{t_1} A^{(4,5)t_2}_{t_2} \ldots A^{(2n,1)t_n}_{t_n} \quad \text{for spin 1,} \tag{65a}$$

while for spin 0

$$sp\{(A^{(1)}\beta)(A^{(2)}\beta)\ldots(A^{(2n)}\beta)\}$$
$$= (A^{(2)}A^{(3)})(A^{(4)}A^{(5)})\ldots(A^{(2n)}A^{(1)}) + (A^{(1)}A^{(2)})\ldots(A^{(2n-1)}A^{(2n)}), \tag{65b}$$

where
$$(A^{(r)}\beta) = A^{(r)k}\beta_k, \quad (A^{(r)}A^{(s)}) = A^{(r)k}A^{(s)}_k,$$
$$A^{(r,s)l}_k = (A^{(r)}A^{(s)})\delta_k^l - A^{(r)l}A^{(s)}_k.$$

4. Let us now return to equation (1). First consider the spin 1 case. Multiplying (1) by Γ_l^* from the left

$$i\Gamma_l^* \beta_k \partial^k \psi + \chi \Gamma_l^* \psi = 0.$$

Similarly, multiplying by $\Gamma_k^* \beta_l$,

$$i\Gamma_k^* \beta_l \beta_m \partial^m \psi + \chi \Gamma_k^* \beta_l \psi = 0,$$

or from (49a)
$$i(\partial_l \Gamma_k^* \psi - \partial_k \Gamma_l^* \psi) + \chi \Gamma_k^* \beta_l \psi = 0.$$

Put
$$\Gamma_k^* \psi = U_k, \tag{66a}$$
$$\frac{\chi}{i} \Gamma_k^* \beta_l \psi = G_{kl}. \tag{66b}$$

On account of the antisymmetry of $\Gamma_k^* \beta_l$ in k, l, G_{kl} is also antisymmetric. The above equations can now be written as

$$\partial_k U_l - \partial_l U_k = G_{kl}, \tag{67a}$$

$$\partial^k G_{kl} + \chi^2 U_l = 0, \tag{67b}$$

which are precisely the usual equations of the wave formulation. To understand the equation

$$\partial_l \psi = \partial_k \beta^k \beta_l \psi \tag{68}$$

which follows from (1), replace β^k by $\beta\beta^k + \beta^k\beta$. Then

$$\beta^k \beta_l \psi = \beta\beta^k \beta_l \psi + \beta^k \beta\beta_l \psi = \beta^k \beta_l \beta\psi + \beta^k \beta\beta_l \psi$$

$$= \beta^k \beta_l T_m \Gamma^{*m}\psi + \beta^k T_m \Gamma^{*m}\beta_l \psi$$

$$= (\delta_l^k T_m - \delta_m^k T_l)\, U^m + \frac{i}{\chi}\beta^k T_m G^{ml}.$$

Similarly

$$\psi = (\beta + \tfrac{1}{2}\beta^k\beta\beta_k)\,\psi = T_m U^m + \tfrac{1}{2}\beta^k T_m \frac{i}{\chi} G^{mk},$$

therefore from (68)

$$T_k \partial_l U^k + \frac{i}{2\chi}\beta^k T^m \partial_l G_{mk} = T_k \partial_l U^k - T_l \partial^k U_k + \frac{i}{\chi}\beta^k T^m \partial_k G_{ml}. \tag{69}$$

Multiplying by Γ^{*r} one obtains

$$\partial^k U_k = 0, \tag{70a}$$

while on multiplying by $\Gamma^{*r}\beta^s$ it is found that

$$\partial_l G_{rs} + \partial_s G_{lr} + \partial_r G_{sl} = 0. \tag{70b}$$

The equations (70) are completely equivalent to (69) or (68) because the latter follow from the former.

In the same way the energy-momentum tensor

$$T_{kl} = \chi\psi^*(g_{kl} - \beta_k\beta_l - \beta_l\beta_k)\,\psi \tag{71}$$

can also be transformed

$$\psi^*\beta_k\beta_l\psi = \psi^*(\beta\beta_k\beta_l + \beta_k\beta\beta_l)\,\psi$$

$$= \psi^* T^m(\Gamma_m^* g_{kl} - \Gamma_k^* g_{ml})\,\psi + \psi^*\beta_k T_m \Gamma^{*m}\beta_l \psi$$

$$= g_{kl} U^{\dagger m} U_m - U_l^\dagger U_k - \frac{1}{\chi^2} G_{km}^\dagger G^m{}_l$$

and

$$\psi^*\psi = \psi^*(\beta + \tfrac{1}{2}\beta^k\beta\beta_k)\,\psi$$

$$= U^{\dagger m} U_m - \frac{1}{2\chi^2} G^{\dagger km} G_{mk} = U^{\dagger m} U_m + \frac{1}{2\chi^2} G^{\dagger mn} G_{mn},$$

therefore

$$T_{kl} = \frac{1}{\chi}\{G_{km}^\dagger G^{ml} + G_{lm}^\dagger G^{mk} + \tfrac{1}{2}g_{kl} G^{\dagger mn} G_{mn} + \chi^2(U_k^\dagger U_l + U_l^\dagger U_k - g_{kl} U^{\dagger m} U_m).\} \tag{72}$$

The angular momentum tensor $M_{kl,m}$ is given by

$$M_{kl,m} = \frac{1}{2i}\left\{(\partial_l \psi^* x_k - \partial_k \psi^* x_l)\,\beta_m \psi - \psi^*(x_k \partial_l - x_l \partial_k)\,\beta_m \psi - \frac{1}{2i}\,\psi^*(I_{kl}\beta_m + \beta_m I_{kl})\,\psi\right\}.$$

$$(73)$$

It would be interesting to transform the spin term to the usual wave formulation. It is found that

$$\psi^*\beta_k\beta_l\beta_m\psi = \frac{i}{\chi}[G_k^{\dagger n}U_n g_{lm} - G_{km}^{\dagger}U_l - U_l^{\dagger}G_{km} + U^{\dagger n}G_{mn}g_{kl}],$$

therefore

$$\psi^*I_{kl}\beta_m\psi = \frac{i}{\chi}[G_k^{\dagger n}U_n g_{lm} - G_l^{\dagger n}U_n g_{km} - G_{km}^{\dagger}U_l + G_{lm}^{\dagger}U_k - U_l^{\dagger}G_{km} + U_k^{\dagger}G_{lm}].$$

$\psi^*\beta_m I_{kl}\psi$ is minus the complex conjugate of $\psi^*I_{kl}\beta_m\psi$. Therefore the spin term $S_{kl,m}$ is

$$S_{kl,m} = -\frac{1}{2i}\psi^*(I_{kl}\beta_m + \beta_m I_{kl})\,\psi$$

$$= -\frac{1}{2\chi}\{G_k^{\dagger n}U_n g_{lm} - G_l^{\dagger n}U_n g_{km} - 2G_{km}^{\dagger}U_l + 2G_{lm}^{\dagger}U_k + \text{complex conjugate}\}, \qquad (74)$$

so that for $\underline{k}$ and $\underline{l}$ not zero

$$S_{\underline{kl},0} = \frac{1}{\chi}[(G_{\underline{k}_0}^{\dagger}U_{\underline{l}} - G_{\underline{l}_0}^{\dagger}U_{\underline{k}}) + \text{complex conjugate}].$$

The commutation rules for ψ and ψ^*, when the second quantization is introduced, have been given by Kemmer. They are‡

$$[(\psi^*\beta_0^2)_\mu, (\beta_0\psi')_\nu] = [(\psi^*\beta_0)_\mu, (\beta_0^2\psi')_\nu] = -(\beta_0^2)_{\nu\mu}\,\delta(x - x'), \qquad (75)$$

where μ and ν refer to the matrix elements of the 1-row and 1-column matrices $\psi^*\beta_0^2, \psi^*\beta_0$ and $\beta_0\psi', \beta_0^2\psi'$ respectively. Now

$$\psi^*\beta_0 = \psi^*(\beta\beta_1 + \beta_0\beta) = U^{\dagger m}\Gamma_m^*\beta_0 + \frac{i}{\chi}G_{0m}^{\dagger}\Gamma^{*m},$$

$$\beta_0\psi = \beta_0 T_m U^m + \frac{i}{\chi}T^m G_{m0},$$

$$\psi^*\beta_0^2 = \psi^*(\beta\beta_0^2 + \beta_0\beta\beta_0) = U^{\dagger m}(\Gamma_m g_{00} - \Gamma_0 g_{m0}) + \frac{i}{\chi}G_{0m}^{\dagger}\Gamma^{*m}\beta_0$$

$$= U^{\dagger \underline{k}}\Gamma_{\underline{k}}^* + \frac{i}{\chi}G_{0\underline{k}}^{\dagger}\Gamma^{*\underline{k}}\beta_0,$$

$$\beta_0^2\psi = T_{\underline{k}}U^{\underline{k}} + \frac{i}{\chi}\beta_0 T^{\underline{k}}G_{0\underline{k}},$$

‡ In Kemmer's paper $[A, B]$ denotes a Poisson bracket, so that in his case

$$[A, B] = i(AB - BA).$$

Also $\delta_{\mu\nu}$ on the right is replaced by $(\beta_0^2)_{\nu\mu}$ here so that the right side vanishes automatically for the matrix elements corresponding to the eigenvalue 0 of β_0. This artifice of replacing $\delta_{\mu\nu}$ by $(\beta_0^2)_{\nu\mu}$ was pointed out to me by Professor Bhabha.

where the underlined index $\underline{k}$ runs from 1 to 3 only. From (75)

$$[(\psi^*\beta_0^2 T_{\underline{k}}),(\Gamma_{\underline{l}}^*\beta_0\psi')] = [(\psi^*\beta_0 T_{\underline{k}}),(\Gamma_{\underline{l}}^*\beta_0^2\psi')]$$
$$= -g_{\underline{kl}}\delta(x-x'),$$

i.e.
$$\left[U_{\underline{k}}^\dagger,\frac{i}{\chi}G_{\underline{l}0}'\right] = \left[\frac{i}{\chi}G_{0\underline{k}}^\dagger,U_{\underline{l}}'\right] = -g_{\underline{kl}}\delta(x-x'),$$

which are the correct commutation rules of the usual wave theory.

It is now extremely easy to include nuclear interaction in the above scheme. The total Lagrangian can be written as

$$L = L^H + L^M + L^E + L^I,$$

where L^H is the Lagrangian of the field of the heavy particle, L^M of the meson field and L^E of the electromagnetic field. L^H and L^M include the interaction of the electromagnetic field with the heavy particle and the meson respectively. L^I is the interaction term between the meson field and the heavy particles, and is given by

$$L^I = g_1\Psi^*\alpha^k(\tau_{PN}\Gamma_k^*\psi+\tau_{NP}\psi^*T_k)\Psi + g_2\Psi^*\alpha^k\alpha^l(\tau_{PN}\Gamma_k^*\beta_l\psi+\tau_{NP}\psi^*\beta_l T_k)\Psi. \quad (76)$$

Ψ and Ψ^* refer to the heavy particle. α^k's are the Dirac matrices acting on Ψ. $\Psi^* = \Psi^\dagger L$, where L is a matrix such that

$$\alpha^{\dagger k} = L\alpha^k L^{-1}.$$

τ_{PN} and τ_{NP} are the operators corresponding to the conversion of a neutron into a proton and vice versa. In fact (76) is exactly the same as the usual interaction term in the wave formulation of the meson theory. Obviously (76) can be extended to include pseudo-vector interaction by adding terms of the type‡

$$i\frac{g_1}{3!}\Psi^*\alpha^k\alpha^l\alpha^m\epsilon_{klmn}(\tau_{PN}\Gamma^{*n}\psi+\tau_{NP}\psi^*T^n)\Psi$$

and
$$\frac{g_2'}{2}\Psi^*\alpha^k\alpha^l\epsilon_{klmn}(\tau_{PN}\Gamma^{*m}\beta^n\psi+\tau_{NP}\psi^*\beta^n T^m)\Psi,$$

where ϵ_{klmn} is a tensor antisymmetric in all the four indices and $\epsilon_{0123} = -1$.

Let us now consider the case of scalar mesons. Multiplying (1) by Γ^{*l} on the left

$$i\Gamma^{*l}\beta^k\partial_k\psi + \chi\Gamma^{*l}\psi = 0.$$

‡ If $\eta = 1-2\beta_0^2$ is taken as the reflexion matrix then ψ^*T_k is a vector and not a pseudo-vector, since

$$\eta T_k = (1-2\beta_0^2)T_k = T_k - 2g_{00}T_k + 2g_{0k}T_0 = 2g_{0k}T_0 - T_k.$$

η satisfies the relations
$$\eta^2 = 1, \quad \eta\beta_k = -\beta_k\eta \quad (k\neq 0), \quad \eta\beta_0 = \beta_0\eta.$$

For the 5-row representation the reflexion matrix must be taken to be $-\eta$ so that ψ^*T_k may transform as a vector since in this case

$$\eta T_k = T_k - 2T_0 g_{0k}.$$

Similarly, multiplying by $\Gamma^{*m}\beta_m$ and using (49b),

$$4i\partial_k\Gamma^{*k}\psi + \chi\Gamma^{*m}\beta_m\psi = 0.$$

Put
$$\Gamma_k^*\psi = U_k \tag{77a}$$

and
$$\Gamma_k^*\beta_l\psi = \tfrac{1}{4}g_{kl}\Gamma_m^*\beta^m\psi = -g_{kl}\frac{\chi}{i}U, \tag{77b}$$

then (1) is equivalent to

$$\partial_l U = U_l, \quad \partial_k U^k + \chi^2 U = 0,$$

which are again the usual equations for the scalar field. Also it is found that

$$\psi = (\beta + \tfrac{1}{4}\beta_k\beta\beta^k)\psi = T_m U^m - \tfrac{1}{4}\beta_k T^k\frac{\chi}{i}U$$

and
$$\beta^k\beta_l\psi = (\beta^k\beta\beta_l + \beta\beta^k\beta_l)\psi = -\frac{\chi}{4i}\delta_l^k\beta_m T^m U + T^k U_l,$$

so that (68) can be written as

$$T_k\partial_l U^k - \tfrac{1}{4}\beta_k T^k\frac{\chi}{i}\partial_l U = -\frac{\chi}{4i}\beta_k T^k\partial_l U + T^k\partial_k U_l$$

or
$$\partial_k U_l - \partial_l U_k = 0.$$

The expression for the current vector and the energy-momentum tensor can be transformed exactly as above. They agree with the usual expression of the wave formulation. Further, from (58) it is found that

$$\psi^*\beta_k\beta_l\beta_m\psi \doteq \frac{\chi}{i}\{U^\dagger g_{kl}U_m - U_k^\dagger g_{lm}U\},$$

so that

$$S_{kl,m} = -\frac{1}{2i}\psi^*(I_{kl}\beta_m + \beta_m I_{kl})\psi = -\frac{\chi}{2}\{(U_k^\dagger g_{lm}U - U_l^\dagger g_{km}U) + \text{complex conjugate}\},$$

therefore $S_{kl,0} = 0$ in this case.

Also
$$\psi^*\beta_0 = \tfrac{1}{4}U_0^\dagger\Gamma^{*m}\beta_m + \frac{\chi}{i}U^\dagger\Gamma_0^*,$$

$$\psi^*\beta_0^2 = U_0^\dagger\Gamma_0^* + \frac{\chi}{4i}U^\dagger\Gamma^{*m}\beta_m,$$

$$\beta_0\psi = \tfrac{1}{4}\beta_m T^m U_0 - \frac{\chi}{i}T_0 U,$$

$$\beta_0^2\psi = T_0 U_0 - \frac{\chi}{4i}\beta_m T^m U,$$

so that
$$\psi^*\beta_0 T_0 = \frac{\chi}{i}U^\dagger, \quad \psi^*\beta_0^2 T_0 = U_0^\dagger$$

and
$$\Gamma_0\beta_0\psi = -\frac{\chi}{i}U, \quad \Gamma_0\beta_0^2\psi = U_0,$$

therefore from (75)

$$\left[U_0^\dagger, \, -\frac{\chi}{i}U'\right] = \left[\frac{\chi}{i}U^\dagger, \, U_0'\right] = -\delta(x-x'),$$

which are the correct commutation rules of the usual scalar theory.

The nuclear interaction can be introduced in the scalar theory by exactly the same expression (76) as in the vector theory. However, on account of (49b) and the fact that $\alpha^k\alpha_k = 4$ the second term in (76) can be written as

$$g_2\Psi^*(\tau_{PN}\Gamma_k^*\beta^k\psi + \tau_{NP}\psi^*\beta_k T^k)\,\Psi.$$

The g_2 term gives rise to the 'charge' interaction while the g_1 term brings about the 'dipole' interaction. This is precisely the reverse of what happens in the vector theory. The pseudo-scalar interaction can be introduced through the terms

$$\frac{ig_1'}{3!}\Psi^*\alpha^k\alpha^l\alpha^m\epsilon_{klmn}(\tau_{PN}\Gamma^{*n}\psi + \tau_{NP}\psi^* T^n)\,\Psi$$

and

$$\frac{ig_2'}{4!}\Psi^*\alpha^k\alpha^l\alpha^m\alpha^n\epsilon_{klmn}(\tau_{PN}\Gamma^{*r}\beta_r\psi + \tau_{NP}\psi^*\beta_r T^r)\,\Psi.$$

In the case of neutral mesons the additional condition

$$\psi^* T_k = \Gamma_k^*\psi \tag{78a}$$

has to be imposed in both the scalar and the vector case. It then follows from the equations of motion that

$$\psi^*\beta_k T_l = -\Gamma_l^*\beta_k. \tag{78b}$$

Also for this case τ_{NP} and τ_{PN} are both to be replaced by $\frac{1}{2}$.

The above method of treating neutral mesons is completely equivalent to the one employed by Majorana (1937) and Belinfante (1939). Let a bar denote the complex conjugate and a curl the transposed of a matrix. On taking the transposed of (2) and reversing the sign it is obvious that the matrices $-\tilde{\beta}_k$ satisfy the same commutation relations as β_k, so that for an irreducible representation of β_k there exists a matrix θ having the property that

$$-\tilde{\beta}_k = \theta\beta_k\theta^{-1}. \tag{79a}$$

From (79a) and the irreducibility of β_k it is easy to prove that (cf. Pauli 1936)

$$\tilde{\theta} = \pm\theta. \tag{79b}$$

To decide between the two possibilities notice that from (11) and (79a)

$$\tilde{I}^{kl}\theta + \theta I^{kl} = 0.$$

Put

$$\Gamma_k = \tilde{T}_k.$$

Then $\Gamma_k\theta$ satisfies an equation similar to (14) and therefore

$$\Gamma_k\theta = a\Gamma_k^*,$$

where a is a number. Since θ is a non-singular matrix $a \neq 0$, and therefore without disturbing (79) θ can be replaced by θa^{-1}. In other words, θ can be so chosen consistently with (79) that

$$\Gamma_k \theta = \Gamma_k^*. \tag{79c}$$

On taking the transposed of (79c) and using (79b) it is found that

$$T_k = \pm \theta^{-1} \tilde{\Gamma}_k^*,$$

therefore
$$\Gamma_k^* T_l = \pm \Gamma_k \theta . \theta^{-1} \tilde{\Gamma}_l^* = \pm \Gamma_k \tilde{\Gamma}_l^* = \pm \Gamma_l^* T_k = \pm g_{kl}.$$

It therefore follows on account of (28) that in (79b) only the positive sign is permissible. (78) can now be written as

$$\psi^* T_k = \Gamma_k^* \psi = \Gamma_k \theta \psi = \tilde{\psi} \theta T_k, \tag{80a}$$

$$\psi^* \beta_k T_l = -\Gamma_l^* \beta_k \psi = -\Gamma_l \theta \beta_k \psi = \Gamma_l \tilde{\beta}_k \theta \psi$$
$$= \tilde{\psi} \theta \beta_k T_l, \tag{80b}$$

so that from (80) and (56), (57)

$$\psi^* = \tilde{\psi} \theta. \tag{81}$$

As a consequence of (81) the current density $\psi^* \beta_k \psi$ vanishes since

$$\psi^* \beta_k \psi = \tilde{\psi} \theta \beta_k \psi = \tilde{\psi} \tilde{\beta}_k \theta \psi = -\tilde{\psi} \theta \beta_k \psi.$$

Also the usual Lagrangian

$$L = \frac{1}{2i} (\partial_k \psi^* \beta^k \psi - \psi^* \beta^k \partial_k \psi) + \chi \psi^* \psi \tag{82a}$$

can be written in the form

$$L = \frac{1}{i} \psi \theta \beta^k \partial_k \psi + \chi \psi \theta \psi, \tag{82b}$$

where ψ is written instead of $\tilde{\psi}$ to emphasize the fact that now there is only one canonical variable ψ in the Lagrangian.

The connexion between θ and Λ is obtained from the equation

$$-\bar{\beta}_k = \theta^{\dagger-1} \Lambda \beta_k \Lambda^{-1} \theta^{\dagger} = \tilde{\Lambda}^{-1} \theta \beta_k \theta^{-1} \tilde{\Lambda},$$

so that
$$\theta^{\dagger-1} \Lambda = c \tilde{\Lambda}^{-1} \theta,$$

where c is a number. Due to $\theta = \tilde{\theta}$ and $\Lambda = \Lambda^{\dagger}$ the above equation can be written as

$$\bar{\theta}^{-1} \Lambda = c \bar{\tilde{\Lambda}}^{-1} \theta.$$

Multiplying by $\bar{\Gamma}_k^*$ on the left

$$\bar{\Gamma}_k^* \bar{\theta}^{-1} \Lambda = c \bar{\Gamma}_k^* \bar{\Lambda}^{-1} \theta = c \Gamma_k \theta.$$

But
$$\bar{\Gamma}_k^* \bar{\theta}^{-1} \Lambda = \bar{\Gamma}_k \Lambda = \Gamma_k^*,$$

therefore from (79c), $c = 1$ and

$$\bar{\theta}^{-1} \Lambda = \bar{\Lambda}^{-1} \theta. \tag{83a}$$

The matrix $\mathscr{L}$ of Belinfante (1939) corresponds to $\Lambda^{-1}\bar{\theta}$.

Let it also be remarked that
$$\theta\beta = \bar{\beta}\theta, \tag{83b}$$

since
$$\theta\beta = \theta T_k \Gamma^k \theta = \bar{\beta}\theta.$$

5. Equation (1) can now easily be adapted for the case of zero rest-mass. Instead of (1) write
$$i\beta^k \partial_k \psi + \gamma\psi = 0, \tag{84}$$

where γ is a matrix satisfying the following conditions:
$$\gamma^2 = \gamma, \tag{85a}$$
$$\gamma\beta^k + \beta^k\gamma = \beta^k. \tag{85b}$$

In (84) γ appears in place of the mass χ of the particle. Multiplying (84) by $(1-\gamma)$ on the left one obtains
$$i\beta^k \partial_k(\gamma\psi) = 0. \tag{86}$$

Thus $\gamma\psi$ satisfies (1) with $\chi = 0$. Also on multiplying (84) by $\partial_m \beta^m \beta^l$ on the left it is found that
$$\partial_m \beta^m \beta^l \gamma\psi = \partial^l \gamma\psi, \tag{87}$$

therefore $\gamma\psi$ also satisfies (68). The second order wave equation
$$\partial_k \partial^k (\gamma\psi) = 0$$

follows from (86) and (87). Further, if ϕ is any solution of the equation
$$\beta^k \partial_k \phi = 0, \tag{88}$$

then
$$\gamma\beta^k \partial_k \phi = \beta^k \partial_k (1-\gamma)\phi = 0.$$

Also since
$$\gamma(1-\gamma)\phi = 0$$

$(1-\gamma)\phi$ satisfies (84). Thus $\psi + (1-\gamma)\phi$ is also a solution of (84). This corresponds to a gauge transformation of the usual theory.

Equation (84) can be derived from the Lagrangian
$$L = \frac{1}{i}[\partial_k \psi^* \beta^k \gamma\psi - \psi^* \gamma\beta^k \partial_k \psi] + \psi^* \gamma\psi. \tag{89}$$

γ must satisfy the condition
$$\gamma^\dagger = \Lambda\gamma\Lambda^{-1}. \tag{90}$$

In fact from the result that for any irreducible representation of the β^k's there can exist only two γ's it follows that either (90) is fulfilled or
$$\Lambda\gamma\Lambda^{-1} = 1 - \gamma^\dagger.$$

In our case γ can be either β or $1-\beta$, and since
$$\Lambda\beta = \Lambda T_k \Gamma^{*k} = \Lambda T_k T^{\dagger k}\Lambda = \beta^\dagger \Lambda$$

(90) is satisfied. The Lagrangian (89) is gauge-invariant, i.e. it remains unchanged when ψ is replaced by $\psi + (1-\gamma)\phi$ and ϕ satisfies (88). This is why the expression (89) has been preferred to the following more obvious and simpler one,

$$L = \frac{1}{2i}\{\partial_k\psi^*\beta^k\psi - \psi^*\beta^k\partial_k\psi\} + \psi^*\gamma\psi,$$

which, though it is not gauge-invariant, yields the same equations of motion.

The expression for the current vector derived from (89) is the same as in the meson theory

$$S_k = \psi^*(\beta_k\gamma + \gamma\beta_k)\,\psi = \psi^*\beta_k\psi, \tag{91}$$

it is not gauge-invariant.

The energy-momentum tensor derived from (89) is

$$T_{kl} = \frac{1}{i}\{\partial_k\psi^*\beta_l\gamma\psi - \psi^*\gamma\beta_l\partial_k\psi\} + g_{kl}\psi^*\gamma\psi.$$

However, it can be verified that

$$T_{kl} = T'_{kl} - \frac{1}{2i}\partial^m\{\psi^*[(g_{kl}\beta_m - g_{km}\beta_l)\gamma - \gamma(g_{kl}\beta_m - g_{km}\beta_l)]\,\psi\}, \tag{92}$$

where
$$T'_{kl} = \frac{1}{2i}[\partial_k\psi^*\beta_l\psi - \psi^*\beta_l\partial_k\psi].$$

T'_{kl} has the same form as in the usual meson theory. Also it is found that due to (84)

$$T'_{kl} = \theta_{kl} - \frac{1}{2i}\partial^m[\psi^*(\beta_m\beta_k\beta_l - \beta_l\beta_k\beta_m)\,\psi],$$

where
$$\theta_{kl} = \psi^*(g_{kl} - \beta_k\beta_l - \beta_l\beta_k)\,\gamma\psi. \tag{93}$$

Therefore θ_{kl} which is symmetric in k and l may be taken to be the energy-momentum tensor instead of T_{kl}. (93) differs from (71) only in having γ instead of χ. Thus in the present theory γ formally plays a part somewhat similar to that of the mass in the meson theory. On account of the factor γ, θ_{kl} is gauge-invariant.

The expression (73) of the angular-momentum tensor remains valid also in the present case. However, the spin is not gauge-invariant.

The Hamiltonian formulation of (84) may be obtained in the same way as for the meson theory (Kemmer 1939). Multiplying (84) by $\beta_0(1-\gamma)$

$$\beta_0^2\partial_0(\gamma\psi) + \beta_0\beta_{\underline{k}}\partial^{\underline{k}}(\gamma\psi) = 0,$$

while from (87) with $l = 0$

$$(1 - \beta_0^2)\,\partial_0(\gamma\psi) - \beta_{\underline{k}}\beta_0\partial^{\underline{k}}(\gamma\psi) = 0,$$

therefore
$$\partial_0(\gamma\psi) + (\beta_0\beta_{\underline{k}} - \beta_{\underline{k}}\beta_0)\,\partial^{\underline{k}}(\gamma\psi) = 0 \tag{94}$$

This gives the equation of motion of $\gamma\psi$. No such equation exists for $(1-\gamma)\psi$. This corresponds to the well-known fact that the Hamiltonian formulation gives the equations of motion only of the gauge-invariant quantities. Obviously $\gamma\psi$ is gauge-invariant while $(1-\gamma)\psi$ is not.

In order to eliminate the 'longitudinal waves' ψ is to be chosen in such a way that it satisfies in addition to (87) the equation

$$\partial_m \beta^m \beta^l \psi = \partial^l \psi. \tag{95}$$

This can be done by a suitable gauge transformation.

There are now two possibilities. Either $\gamma = 1 - \beta$ or $\gamma = \beta$. Consider first the case of the 10-row representation of the β_k's so that now

$$\gamma = 1 - \beta = \tfrac{1}{2}\beta_k \beta \beta^k.$$

It can be verified that in this case (84) corresponds to the usual Maxwell equations. The Lagrangian (89) becomes

$$\tfrac{1}{2} F^{\dagger kl} F_{kl} - (\partial_k U_l^\dagger F^{kl} + F^{\dagger kl}\partial_k U_l),$$

where
$$F_{kl} = \frac{1}{i}\,\Gamma_k^* \beta_l \psi \tag{96a}$$

and
$$U_k = \Gamma_k^* \psi. \tag{96b}$$

From (69) it follows that the elimination of the longitudinal waves in this case corresponds to imposing the condition $\partial^k U_k = 0$. With this choice of γ (84) may be used to describe the photon or the electromagnetic field. In this case the reality condition (78) or (81) has to be imposed. The interaction of the photon with the charged particles can be introduced in a way entirely similar to (76). For example, to introduce interaction with an electron of charge e the interaction term

$$e \Psi^* \alpha^k \Psi \Gamma_k \theta \psi$$

has to be added to the sum of the independent Lagrangians of the electromagnetic and electron fields. On account of (81) and (83), (89) can now be written in the form

$$L' \equiv \tfrac{1}{2} L = i \psi \theta \gamma \beta^k \partial_k \psi + \tfrac{1}{2}\psi \theta \gamma \psi. \tag{97}$$

For the other choice $\gamma = \beta$ (84) is equivalent to

$$\partial_k F^{lk} = U^l, \tag{98a}$$

$$\partial_k U_l - \partial_l U_k = 0, \tag{98b}$$

F^{kl} and U_l being still defined by (96). It follows from (98) that there exists a U such that
$$U_l = \partial_l U \quad \text{and} \quad \partial^l \partial_l U = \partial^l U_l = 0.$$

Equations (98) are thus formally equivalent to the equations of the usual scalar theory. The energy-momentum tensor turns out to be

$$U_k^\dagger U_l + U_l^\dagger U_k - g_{kl} U^{\dagger m} U_m,$$

and therefore agrees with the usual expression. The current vector and the commutation rules for the second quantization are, however, different. A gauge transformation consists in adding to F^{kl} another antisymmetrical tensor F'^{kl} satisfying

$$\partial^k F'_{kl} = 0.$$

The elimination of the longitudinal waves corresponds to imposing the condition

$$\partial_k F_{lm} + \partial_m F_{kl} + \partial_l F_{mk} = 0.$$

Coming to the 5-row representation of the β_k's and taking $\gamma = 1 - \beta = \frac{1}{4}\beta_k\beta\beta^k$, it is found that (78) is equivalent to

$$\partial_m U^m + U = 0, \tag{99a}$$

$$\partial_k U = 0, \tag{99b}$$

where U and U_m are defined by (77) with $\chi = 1$. The energy-momentum tensor comes out to be

$$-\theta_{kl} = g_{kl} U^\dagger U.$$

Gauge transformation consists in adding to U_m a tensor U'_m such that

$$\partial^m U'_m = 0.$$

(88) corresponds to imposing the condition $\partial_k U_l = \partial_l U_k$. The current vector has the same form as in the usual scalar theory

$$\frac{1}{i}\{U^\dagger U_k - U_k^\dagger U\}.$$

Since from (99) U is a constant the energy-momentum density is also constant and therefore the field is of no physical interest.

The case $\gamma = \beta$ corresponds to the usual scalar theory, the equations being

$$\partial^k U = U^k, \quad \partial_k U^k = 0.$$

The Lagrangian is
$$U_k^\dagger U^k - (\partial^k U^\dagger . U_k + U_k^\dagger \partial^k U).$$

Gauge transformation consists in adding to U^k a vector U'^k satisfying

$$\partial_k U'^k = 0.$$

In this case (95) follows automatically from (87) since

$$\partial_k \beta^k \beta^l \beta \psi - \partial^l \beta \psi = \partial_k \beta^k (1 - \beta) \beta^l \psi - \partial^l \beta \psi$$
$$= \partial_k \beta^k \beta^l \psi - \partial_k g^{kl}(1 - \beta)\psi - \partial^l \beta \psi \quad \text{from } (57b)$$
$$= \partial_k \beta^k \beta^l \psi - \partial^l \psi.$$

6. Till now the theory has been developed in an abstract way. It is useful to know what the matrices T_k, Γ^{*k} and β actually are for a particular representation. For this purpose it is best to choose the following representation. For the 10-row representation, ψ and $\beta_{\alpha\lambda} = \beta_k \sigma^k_{\alpha\lambda}$ are

$$\psi = \begin{pmatrix} \psi(0,1) \\ \psi(\tfrac{1}{2},\tfrac{1}{2}) \\ \psi(1,0) \end{pmatrix}, \quad \beta_{\alpha\lambda} = \sqrt{2}\begin{pmatrix} 0 & v_\alpha u^+ & 0 \\ -\tfrac{1}{2}u_\alpha v_\lambda^+ & 0 & -\tfrac{1}{2}v_\alpha^+ u_\lambda \\ 0 & u_\alpha^+ v_\lambda & 0 \end{pmatrix}, \tag{100}$$

where $v_\alpha = v_\alpha(\tfrac{1}{2})$ and $v_\alpha^+ = v_\alpha(1)$ and similarly for others. The splitting up of ψ corresponds to the decomposition of the whole representation with respect to the k and l values so that the components of ψ are $\psi(k,l)$ (cf. Bhabha 1945). For this representation

$$\frac{1}{\sqrt{2}} T_{\alpha\lambda} = \begin{pmatrix} 0 \\ u_\alpha u_\lambda \\ 0 \end{pmatrix}, \quad \frac{1}{\sqrt{2}} \Gamma^*_{\alpha\lambda} = \overbrace{0 \quad v_\alpha v_\lambda \quad 0}, \quad \beta = \begin{pmatrix} 0 & 0 & 0 \\ 0 & 1 & 0 \\ 0 & 0 & 0 \end{pmatrix}. \tag{101a}$$

Also
$$\omega = \begin{pmatrix} -1 & 0 & 0 \\ 0 & 0 & 0 \\ 0 & 0 & 1 \end{pmatrix}. \tag{101b}$$

$(101\,b)$ is easily obtained from $(59\,b)$, (100) and $(101\,a)$ if one notices that the spinor corresponding to e^{klmn} is given by (cf. Harish-Chandra 1945)

$$\tfrac{1}{2} e^{klmn} \sigma_{k,\alpha\lambda} \sigma_{l,\beta\mu} \sigma_{m,\gamma\nu} \sigma_{n,\delta\rho} = \epsilon_{\alpha\beta} \epsilon_{\gamma\delta} \dot{\Sigma}_{(\lambda\mu)(\nu\rho)} + \epsilon_{\lambda\mu} \epsilon_{\nu\rho} \Sigma_{(\alpha\beta)(\gamma\delta)}, \tag{102a}$$

where
$$\Sigma_{(\alpha\beta)(\gamma\delta)} = i(\epsilon_{\alpha\gamma} \epsilon_{\beta\delta} + \epsilon_{\beta\gamma} \epsilon_{\alpha\delta})$$
and
$$\dot{\Sigma}_{(\lambda\mu)(\nu\rho)} = -i(\epsilon_{\lambda\nu} \epsilon_{\mu\rho} + \epsilon_{\mu\nu} \epsilon_{\lambda\rho}). \tag{102b}$$

On the other hand, for the 5-row representation

$$\psi = \begin{pmatrix} \psi(\tfrac{1}{2}, \tfrac{1}{2}) \\ \psi(0,0) \end{pmatrix}, \quad \beta_{\alpha\lambda} = \sqrt{2} \begin{pmatrix} 0 & u_\alpha u_\lambda \\ v_\alpha v_\lambda & 0 \end{pmatrix},$$

$$\frac{1}{\sqrt{2}} T_{\alpha\lambda} = \begin{pmatrix} u_\alpha u_\lambda \\ 0 \end{pmatrix}, \quad \frac{1}{\sqrt{2}} \Gamma^*_{\alpha\lambda} = \overbrace{v_\alpha v_\lambda \quad 0}, \quad \beta = \begin{pmatrix} 1 & 0 \\ 0 & 0 \end{pmatrix}.$$

REFERENCES

Belinfante 1939 *Physica*, **6**, 849–887.
Bhabha 1945 *Rev. Mod. Phys.* **17**, 200–216.
Booth & Wilson 1940 *Proc. Roy. Soc.* A, **175**, 483–518.
Dirac 1936 *Proc. Roy. Soc.* A, **155**, 447–459.
Fierz 1939 *Helv. Phys. Acta*, **12**, 3–37.
Harish-Chandra 1946 *Proc. Ind. Acad. Sci.* **23**, 152–163.
Kemmer 1939 *Proc. Roy. Soc.* A, **173**, 91–116.
Kemmer 1943 *Proc. Camb. Phil. Soc.* **39**, 189–196.
Majorana 1937 *Nuovo Cim.* **14**, 171.
Pauli 1936 *Ann. Inst. Poincaré*, **6**, 109–136.
Schrödinger 1943 *Proc. Roy. Irish Acad.* A, **48**, 135.
Wilson 1940 *Proc. Camb. Phil. Soc.* **36**, 363–380.

Reprinted from
Proc. Royal Soc. A
186 (1946), 502–525

ON THE ALGEBRA OF THE MESON MATRICES

By HARISH-CHANDRA

Received 27 April 1946

Communicated by P. A. M. DIRAC

1. In a recent paper (3) it has been shown that the properties of the β-matrices, which were introduced by Duffin (1) and Kemmer (5) for describing the meson, can be brought out rather vividly by using certain 1-row and 1-column matrices denoted in (3) by Γ_k^* and T_k respectively. It was, however, not possible until now to treat the algebra of the β-matrices with the same elegance and generality as that obtainable in the case of Dirac's matrices (cf. (4)). It is shown in the present paper that this can be achieved by using the Γ-formalism and extending the dimension number from four to five. This is somewhat analogous to the procedure adopted in the case of the Dirac Algebra (cf. Eddington (2, 4)).

The results of (3) can briefly be summarized in the following equations:

$$\beta_k\beta_l\beta_m + \beta_m\beta_l\beta_k = g_{kl}\beta_m + g_{ml}\beta_k, \tag{1}$$

$$\mathsf{T}_k\Gamma^{*k} = \beta, \tag{2}$$

$$\Gamma_k^*\mathsf{T}_l = g_{kl}, \quad (3a) \qquad \Gamma_k^*\beta_l\mathsf{T}_m = 0, \quad (3b)$$

$$\beta_k = \beta\beta_k + \beta_k\beta, \tag{4}$$

$$\beta_k\mathsf{T}_l = -\beta_l\mathsf{T}_k, \quad \Gamma_k^*\beta_l = -\Gamma_l^*\beta_k, \qquad (5a)$$

$$\beta_k\beta_l\mathsf{T}_m = g_{kl}\mathsf{T}_m - g_{km}\mathsf{T}_l, \quad \Gamma_k^*\beta_l\beta_m = \Gamma_k^*g_{lm} - \Gamma_l^*g_{km}, \qquad (5b)$$

$$\beta + \tfrac{1}{2}\beta_k\beta\beta^k = 1, \qquad (5c)$$

for the 10-row representation

$$\Gamma_k^*\beta_l = \tfrac{1}{4}g_{kl}\Gamma_m^*\beta^m = \Gamma_l^*\beta_k, \quad \beta_k\mathsf{T}_l = \tfrac{1}{4}g_{kl}\beta^m\mathsf{T}_m = \beta_l\mathsf{T}_k, \qquad (6a)$$

$$\Gamma_k^*\beta_l\beta_m = g_{kl}\Gamma_m^*, \quad \beta_k\beta_l\mathsf{T}_m = \mathsf{T}_k g_{lm}, \qquad (6b)$$

$$\beta + \tfrac{1}{4}\beta_k\beta\beta^k = 1. \qquad (6c)$$

for the 5-row representation

Here g_{kl} is the usual metric tensor for which

$$g_{kl} = 0 \quad (k \neq l) \quad \text{and} \quad g_{00} = -g_{11} = -g_{22} = -g_{33} = 1.$$

Consider first the 10-row representation. Put

$$\Gamma_{kl}^* = \Gamma_k^*\beta_l, \quad \Gamma_{k4}^* = -\Gamma_{4k}^* = \Gamma_k^*, \quad \Gamma_{44}^* = 0,$$
$$\mathsf{T}_{kl} = \beta_k\mathsf{T}_l, \quad \mathsf{T}_{4k} = -\mathsf{T}_{k4} = \mathsf{T}_k, \quad \mathsf{T}_{44} = 0, \tag{7}$$

and

$$g_{4k} = 0, \quad g_{44} = 1.$$

Then the following equations hold:

$$\Gamma_{\lambda\mu}^* = -\Gamma_{\mu\lambda}^*, \quad \mathsf{T}_{\lambda\mu} = -\mathsf{T}_{\mu\lambda}, \tag{8a}$$

$$\Gamma_{\lambda\mu}^*\mathsf{T}_{\nu\rho} = g_{\lambda\rho}g_{\mu\nu} - g_{\mu\rho}g_{\lambda\nu} \equiv \eta_{[\lambda\mu][\nu\rho]}, \tag{8b}$$

$$\tfrac{1}{2}\mathsf{T}_{\lambda\mu}\Gamma^{*\mu\lambda} = 1, \tag{8c}$$

where the greek indices run from 0 to 4, the tensor $g_{\mu\nu}$ being now used for lowering and raising the indices. The quantities $\Gamma^*_{\lambda\mu}$ and $\mathrm{T}_{\nu\rho}$ are each 10 $(=\tfrac{1}{2}\times 5\times 4)$ in number. Therefore there are 100 quantities of the type $\mathrm{T}_{\lambda\mu}\Gamma^*_{\nu\rho}$. They are all linearly independent since, if for a certain set of coefficients $a^{[\lambda\mu][\nu\rho]}$ (antisymmetrical in λ, μ and ν, ρ)

$$a^{[\lambda\mu][\nu\rho]}\mathrm{T}_{\lambda\mu}\Gamma^*_{\nu\rho} = 0,$$

then $\qquad a^{[\lambda\mu][\nu\rho]}\Gamma^*_{\alpha\beta}\mathrm{T}_{\lambda\mu}\Gamma^*_{\nu\rho}\mathrm{T}_{\gamma\delta} = a^{[\lambda\mu][\nu\rho]}\eta_{[\alpha\beta][\lambda\mu]}\eta_{[\nu\rho][\gamma\delta]} = 4a_{[\alpha\beta][\gamma\delta]} = 0.$

Also from (4) $\qquad\qquad \beta_k = \mathrm{T}_{4\lambda}\Gamma^{*\lambda}{}_k + \mathrm{T}_{k\lambda}\Gamma^{*\lambda}{}_4, \qquad\qquad\qquad (9)$

so that on account of $(8c)$ and (9) every element of the β-algebra can be expressed linearly in terms of $\mathrm{T}_{\lambda\mu}\Gamma^*_{\nu\rho}$. The linear independence of these 100 quantities thus corresponds to the irreducibility of the 10-row representation of the β_k's.

Let S and T be any two quantities of our algebra given by

$$S = \tfrac{1}{4}s^{[\lambda\mu][\nu\rho]}\mathrm{T}_{\lambda\mu}\Gamma^*_{\nu\rho}, \quad T = \tfrac{1}{4}t^{[\lambda\mu][\nu\rho]}\mathrm{T}_{\lambda\mu}\Gamma^*_{\nu\rho}, \qquad\qquad (10)$$

where $s^{[\lambda\mu][\nu\rho]}$ and $t^{[\lambda\mu][\nu\rho]}$ are numerical coefficients antisymmetrical in λ, μ and ν, ρ. For the purpose of the manipulation of indices they are to be regarded as tensors of the 5-dimensional space. From (8) and (10) we find that

$$ST = \tfrac{1}{16}s^{[\lambda\mu][\alpha\beta]}\mathrm{T}_{\lambda\mu}\eta_{[\alpha\beta][\gamma\delta]}t^{[\gamma\delta][\nu\rho]}\Gamma^*_{\nu\rho} = -\tfrac{1}{8}s^{[\lambda\mu][\alpha\beta]}t_{[\alpha\beta]}{}^{[\nu\rho]}\mathrm{T}_{\lambda\mu}\Gamma^*_{\nu\rho}. \qquad (11)$$

The relation (11) determines the structure of our algebra completely. Notice that from (10) the spur of T is given by

$$\mathrm{sp}\,T = \tfrac{1}{4}t^{[\lambda\mu][\nu\rho]}\Gamma^*_{\nu\rho}\mathrm{T}_{\lambda\mu} = \tfrac{1}{4}t^{[\lambda\mu][\nu\rho]}\eta_{[\nu\rho][\lambda\mu]} = -\tfrac{1}{2}t^{[\lambda\mu]}{}_{[\lambda\mu]}. \qquad (12)$$

Also $\qquad \mathrm{sp}(T\mathrm{T}_{\nu\rho}\Gamma^*_{\lambda\mu}) = \mathrm{sp}(\mathrm{T}_{\nu\rho}\Gamma^*_{\lambda\mu}T) = \Gamma^*_{\lambda\mu}T\mathrm{T}_{\nu\rho} = t_{[\lambda\mu][\nu\rho]}, \qquad (13)$

so that $\qquad\qquad T = \tfrac{1}{4}(\Gamma^*_{\lambda\mu}T\mathrm{T}_{\nu\rho})\mathrm{T}^{\lambda\mu}\Gamma^{*\nu\rho}, \qquad\qquad\qquad (14)$

which is also directly evident from $(8c)$. Equation (14) shows that in order to express an arbitrary matrix T in the form (10) one has merely to calculate $\Gamma^*_{\lambda\mu}T\mathrm{T}_{\nu\rho}$. This can be done from formula (61) of (3) if T is given in terms of β_k's and their products.

It is now possible to obtain identities similar to those given by Pauli (6) for the Dirac matrices by a procedure analogous to that used in (4). Let ψ and ϕ be two arbitrary 1-column matrices and ψ^* and ϕ^* two arbitrary 1-row matrices. Put $T = \phi\phi^*$. Then $\qquad\qquad t_{[\lambda\mu][\nu\rho]} = \mathrm{sp}(\phi\phi^*\mathrm{T}_{\nu\rho}\Gamma^*_{\lambda\mu}) = \phi^*\mathrm{T}_{\nu\rho}\Gamma^*_{\lambda\mu}\phi. \qquad (15)$

Substituting (15) in (10) and multiplying by ψ^* on the left and by ψ on the right we obtain $\qquad\qquad \psi^*\phi\cdot\phi^*\psi = \tfrac{1}{4}\phi^*\mathrm{T}_{\nu\rho}\Gamma^*_{\lambda\mu}\phi\cdot\psi^*\mathrm{T}^{\lambda\mu}\Gamma^{*\nu\rho}\psi. \qquad (16)$

Other identities can similarly be derived from the following relations which follow immediately from (10) and (8):

$$\mathrm{T}_{\alpha\beta}\Gamma^*_{\gamma\delta}T = -\tfrac{1}{2}\mathrm{T}_{\alpha\beta}t_{[\gamma\delta]}{}^{[\nu\rho]}\Gamma^*_{\nu\rho}, \qquad\qquad (17a)$$

$$\mathrm{T}_{\alpha\beta}\Gamma^*_{\gamma\delta}T\mathrm{T}_{\alpha'\beta'}\Gamma^*_{\gamma'\delta'} = \mathrm{T}_{\alpha\beta}t_{[\gamma\delta][\alpha'\beta']}\Gamma^*_{\gamma'\delta'}. \qquad\qquad (17b)$$

They are $\qquad \psi^*\mathrm{T}_{\alpha\beta}\Gamma^*_{\gamma\delta}\phi\cdot\phi^*\psi = -\tfrac{1}{2}\psi^*\mathrm{T}_{\alpha\beta}\Gamma^*_{\nu\rho}\psi\cdot\phi^*\mathrm{T}^{\nu\rho}\Gamma^*_{\gamma\delta}\phi, \qquad (18)$

$$\psi^*\mathrm{T}_{\alpha\beta}\Gamma^*_{\gamma\delta}\phi\cdot\phi^*\mathrm{T}_{\alpha'\beta'}\Gamma^*_{\gamma'\delta'}\phi = \psi^*\mathrm{T}_{\alpha\beta}\Gamma^*_{\gamma'\delta'}\psi\cdot\phi^*\mathrm{T}_{\alpha'\beta'}\Gamma^*_{\gamma\delta}\phi. \qquad (19)$$

In order to express the Γ's in terms of the β_k's we observe that

$$\mathrm{T}_k\Gamma^*_l = (g_{kl} - \beta_k\beta_l)\beta.$$

But, from (5c), $\qquad\qquad \beta + \tfrac{1}{2}\beta_k\beta^k(1-\beta) = 1,$

i.e. $\qquad\qquad\qquad\qquad\;\; \beta + \tfrac{1}{2}(\beta_k\beta^k - 3\beta) = 1,$

since $\beta^k\beta_k.\beta = 3\beta$ (cf. (3)). Therefore $\beta = \beta_k\beta^k - 2$, and

$$\mathsf{T}_k\Gamma_l^* = (g_{kl} - \beta_l\beta_k)(\beta_m\beta^m - 2). \tag{20}$$

The remaining quantities $\mathsf{T}_k\Gamma_{lm}^*$, $\mathsf{T}_{kl}\Gamma_m^*$, $\mathsf{T}_{kl}\Gamma_{mn}^*$ can be derived from (20) by multiplication by β_k's from the left or the right. One can therefore now express all quantities $\mathsf{T}_{\lambda\mu}\Gamma_{\nu\rho}^*$ in terms of β_k's so that the identities (16), (18) and (19) can be written in terms of β_k's. One interesting point may be noticed. From (13) and (8),

$$\mathsf{T}^{\gamma\delta}\Gamma_{\lambda\mu}^*\,\mathsf{T}\mathsf{T}_{\nu\rho}\,\Gamma_{\gamma\delta}^* = t_{[\lambda\mu][\nu\rho]}.$$

The right side of this equation is to be regarded as a multiple of the unit matrix. The left side can now be transformed to the β_k-basis. This equation shows that whatever the matrix T may be, on multiplication by suitable β-quantities on both sides and summing, it is always transformed into a multiple of the unit matrix.

We observe that, for $r \geqslant 1$,

$$\mathrm{sp}(T^r) = \frac{(-1)^r}{2^r}\, t^{[\lambda_1\mu_1]}{}_{[\lambda_2\mu_2]}\,t^{[\lambda_2\mu_2]}{}_{[\lambda_3\mu_3]} \ldots t^{[\lambda_r\mu_r]}{}_{[\lambda_1\mu_1]}. \tag{21}$$

Since the characteristic equation of any n-dimensional matrix can always be written in terms of $\mathrm{sp}(T)$, $\mathrm{sp}(T^2)$, ..., $\mathrm{sp}(T^n)$, (21) determines the characteristic equation of T.

The 5-row representation of the β's can be treated similarly. In this case put

$$\Gamma_4^* = \tfrac{1}{4}\Gamma_m^*\beta^m, \quad \mathsf{T}_4 = \tfrac{1}{4}\beta^m\mathsf{T}_m. \tag{22}$$

Then, from (3) and (6),

$$\Gamma_\lambda^*\mathsf{T}_\mu = g_{\lambda\mu} \quad (\lambda,\mu = 0,1,2,3,4), \quad (23a) \qquad \mathsf{T}^\lambda\Gamma_\lambda^* = 1. \quad (23b)$$

The twenty-five quantities $\mathsf{T}_\lambda\Gamma_\mu^*$ are all linearly independent since, if

$$a^{\lambda\mu}\mathsf{T}_\lambda\Gamma_\mu^* = 0,$$

we obtain on multiplying by Γ_ν^* on the left and by T_ρ on the right

$$a_{\nu\rho} = 0.$$

Therefore the quantities $\mathsf{T}_\lambda\Gamma_\mu^*$ form a complete linearly independent basis for the 5-row representation. From (4), $\qquad \beta_k = \mathsf{T}_k\Gamma_4^* + \mathsf{T}_4\Gamma_k^*. \tag{24}$

By means of (23b) and (24) we can pass from the β_k-basis to the Γ-basis.

Corresponding to (10) let

$$S = s^{\lambda\mu}\mathsf{T}_\lambda\Gamma_\mu^*, \quad T = t^{\lambda\mu}\mathsf{T}_\lambda\Gamma_\mu^*. \tag{25}$$

Then, from (23), $\qquad\qquad\qquad ST = s^{\lambda\nu}t_\nu{}^\mu\mathsf{T}_\lambda\Gamma_\mu^*. \tag{26}$

Also $\qquad\qquad\qquad\qquad \mathrm{sp}(T) = t^{\lambda\mu}\Gamma_\mu^*\mathsf{T}_\lambda = t^\lambda{}_\lambda, \tag{27a}$

and $\qquad\qquad \Gamma_\lambda^*T\mathsf{T}_\mu = \mathrm{sp}(T\mathsf{T}_\mu\Gamma_\lambda^*) = \mathrm{sp}(\mathsf{T}_\mu\Gamma_\lambda^*T) = t_{\lambda\mu}, \tag{27b}$

so that $\qquad\qquad\qquad T = (\Gamma_\lambda^*T\mathsf{T}_\mu)\mathsf{T}^\lambda\Gamma^{*\mu}, \tag{28}$

as is also obvious from (23b). Further,

$$\mathsf{T}_\alpha\Gamma_\beta^*T = \mathsf{T}_\alpha t_\beta{}^\mu\Gamma_\mu^*, \quad (29a) \qquad \mathsf{T}_\alpha\Gamma_\beta^*T\mathsf{T}_\gamma\Gamma_\delta^* = \mathsf{T}_\alpha t_{\beta\gamma}\Gamma_\delta^*. \quad (29b)$$

The following identities may be derived immediately from (28) and (29):

$$\psi^*\phi.\phi^*\psi = \phi^*\Gamma_\mu\Gamma_\lambda^*\phi.\psi^*\Gamma^\lambda\Gamma^{*\mu}\psi, \tag{30}$$

$$\psi^*\Gamma_\alpha\Gamma_\beta^*\phi.\phi^*\psi = \psi^*\Gamma_\alpha\Gamma_\mu^*\psi.\phi^*\Gamma^\mu\Gamma_\beta^*\phi, \tag{31}$$

$$\psi^*\Gamma_\alpha\Gamma_\beta^*\phi.\phi^*\Gamma_\gamma\Gamma_\delta^*\psi = \psi^*\Gamma_\alpha\Gamma_\delta^*\psi.\phi^*\Gamma_\gamma\Gamma_\beta^*\phi. \tag{32}$$

In order to return from the Γ-basis to the β_k-basis, we observe that

$$\beta_k\beta_l\beta = \Gamma_k\Gamma_l^*.$$

From (57b) of (3) this can be written as

$$\beta_k\beta_l - g_{kl}(1-\beta) = \Gamma_k\Gamma_l^*.$$

Now from (6c) we can deduce that

$$\beta = \tfrac{1}{3}(4 - \beta_k\beta^k),$$

so that

$$\beta_k\beta_l - \tfrac{1}{3}g_{kl}(\beta_m\beta^m - 1) = \Gamma_k\Gamma_l^*,$$

$$\Gamma_4\Gamma_k^* = \beta_k\beta = \beta_k\frac{(4 - \beta_k\beta^k)}{3}, \quad \Gamma_k\Gamma_4^* = \beta\beta_k = \frac{(4 - \beta_l\beta^l)}{3}\beta_k,$$

$$\Gamma_4\Gamma_4^* = \tfrac{1}{4}\beta_k\beta\beta^k = 1 - \beta = \tfrac{1}{3}(\beta_k\beta^k - 1).$$

It follows from (26) and (27a) that $\mathrm{sp}(T^2) = t^{\lambda\mu}t_{\mu\lambda}$, and that

$$\mathrm{sp}(T^r) = t_{\lambda_1}{}^{\lambda_2}t_{\lambda_2}{}^{\lambda_3}\dots t_{\lambda_r}{}^{\lambda_1}. \tag{33}$$

2. The results of (3) and of the present paper show that the introduction of the Γ's has proved useful in understanding the structure of the β-algebra. One would therefore wish to know if a similar formalism can be introduced in the other well-known case of the Dirac matrices. However, since the Dirac algebra is already sufficiently simple no further practical improvement can be expected from the Γ-formalism.

The Dirac matrices α_k are defined by

$$\alpha_k\alpha_l + \alpha_l\alpha_k = 2g_{kl}, \tag{34a}$$

or, in the spinor form,

$$A_{\lambda\dot\mu}A_{\nu\dot\rho} + A_{\nu\dot\rho}A_{\lambda\dot\mu} = 4\epsilon_{\lambda\nu}\epsilon_{\dot\mu\dot\rho}, \tag{34b}$$

where $A_{\lambda\dot\mu} = \alpha_k\sigma^k_{\lambda\dot\mu}$. From now on all greek suffixes denote spinor indices, roman letters being used for tensor indices. The σ-symbols and $\epsilon_{\lambda\nu}$, $\epsilon_{\dot\mu\dot\rho}$ have their usual meaning ($\epsilon_{12} = \epsilon^{12} = \epsilon_{\dot1\dot2} = \epsilon^{\dot1\dot2} = 1$). Λ is a non-singular matrix such that

$$\Lambda^\dagger = \Lambda, \quad \alpha_k^\dagger = \Lambda\alpha_k\Lambda^{-1}, \tag{34c}$$

where † denotes Hermitian conjugate. $I_{kl} = -I_{lk}$ are the spin matrices such that the wave function ψ in the wave equation

$$i\partial_k\alpha^k\psi + \chi\psi = 0 \tag{35}$$

transforms to

$$\psi' = (1 + \tfrac{1}{2}\epsilon_{kl}I^{kl})\psi,$$

by the Lorentz transformation

$$x_k' = x_k + \epsilon_{kl}x^l \quad (\epsilon_{kl} = -\epsilon_{lk}),$$

where the ϵ_{kl} are infinitesimal quantities. Put $\psi^* = \psi^\dagger\Lambda$. Define a 1-column matrix Γ_α so that $\psi^*\Gamma_\alpha$ transforms as an undotted spinor for a Lorentz transformation. The condition for this is easily found and is

$$-I_{\mu\nu}\Gamma_\alpha = \tfrac{1}{2}(\epsilon_{\mu\alpha}\Gamma_\nu + \epsilon_{\nu\alpha}\Gamma_\mu), \quad (36a) \qquad I_{\dot\mu\dot\nu}\Gamma_\alpha = 0, \quad (36b)$$

where

$$I_{\mu\nu} = \tfrac{1}{4}I_{kl}\sigma^k_{\mu\lambda}\sigma^{l\lambda}{}_\nu = I_{\nu\mu}, \quad I_{\dot\mu\dot\nu} = \tfrac{1}{4}I_{kl}\sigma^k_{\dot\mu\lambda}\sigma^{l\lambda}{}_{\dot\nu} = I_{\dot\nu\dot\mu}. \tag{37}$$

In addition to (34c), Λ satisfies the relation

$$I_{kl}^{\dagger}\Lambda + \Lambda I_{kl} = 0. \tag{38}$$

In the present case

$$I_{kl} = \tfrac{1}{2}(\alpha_k \alpha_l - \alpha_l \alpha_k). \tag{39}$$

Remembering that

$$\overline{\sigma_{\mu\lambda}^{k}} = \sigma_{\dot\mu\lambda}^{k},$$

where the bar denotes the complex conjugate, we deduce from (34c) and (38) that

$$(A_{\lambda\mu})^{\dagger}\Lambda = \Lambda A_{\mu\lambda}, \tag{40a}$$

$$(I_{\lambda\dot\mu})^{\dagger}\Lambda + \Lambda I_{\lambda\mu} = 0, \tag{40b}$$

$$(I_{\lambda\mu})^{\dagger}\Lambda + \Lambda I_{\lambda\dot\mu} = 0. \tag{40c}$$

The Hermitian conjugate of (36) is therefore

$$-\Gamma_{\dot\alpha}^{*} I_{\dot\mu\dot\nu} = \tfrac{1}{2}(\epsilon_{\dot\alpha\dot\mu}\Gamma_{\dot\nu}^{*} + \epsilon_{\dot\alpha\dot\nu}\Gamma_{\dot\mu}^{*}), \quad (41a) \qquad \Gamma_{\dot\alpha}^{*} I_{\mu\nu} = 0, \quad (41b)$$

where $\Gamma_{\dot\alpha}^{*} = (T_\alpha)^{\dagger}\Lambda$. Define a 1-column matrix $T_{\dot\alpha}$ similarly, such that $\psi^{*}T_{\dot\alpha}$ transforms as a dotted spinor for Lorentz transformations. For this case, we find

$$I_{\mu\nu}T_{\dot\alpha} = 0, \quad (42a) \qquad -I_{\dot\mu\dot\nu}T_{\dot\alpha} = \tfrac{1}{2}(\epsilon_{\dot\mu\dot\alpha}T_{\dot\nu} + \epsilon_{\dot\nu\dot\alpha}T_{\dot\mu}), \quad (42b)$$

and therefore

$$\Gamma_{\alpha}^{*} I_{\dot\mu\dot\nu} = 0, \quad (43a) \qquad -\Gamma_{\alpha}^{*} I_{\mu\nu} = \tfrac{1}{2}(\epsilon_{\alpha\mu}\Gamma_{\nu}^{*} + \epsilon_{\alpha\nu}\Gamma_{\mu}^{*}), \quad (43b)$$

where $\Gamma_{\alpha}^{*} = (T_{\dot\alpha})^{\dagger}\Lambda$. From (41) we find that

$$-\tfrac{1}{2}\Gamma_{\dot\alpha}^{*} I_{\dot\mu\dot\nu} I^{\dot\mu\dot\nu} = \tfrac{3}{4}\Gamma_{\dot\alpha}^{*}.$$

Therefore, by (36),

$$\Gamma_{\dot\alpha}^{*}T_\beta = 0. \tag{44a}$$

Similarly,

$$\Gamma_{\alpha}^{*}T_\beta = 0. \tag{44b}$$

Also $\tfrac{3}{4}\Gamma_{\dot\alpha}^{*}T_\beta = -\tfrac{1}{2}\Gamma_{\dot\alpha}^{*} I_{\dot\mu\dot\nu} I^{\dot\mu\dot\nu}T_\beta = -\tfrac{1}{4}\{\epsilon_{\dot\alpha\dot\mu}\Gamma_{\dot\nu}^{*} + \epsilon_{\dot\alpha\dot\nu}\Gamma_{\dot\mu}^{*}\}(-\delta_{\beta}^{\dot\mu}T^{\dot\nu}) = \tfrac{1}{4}\{\epsilon_{\dot\alpha\dot\beta}\Gamma_{\dot\nu}^{*}T^{\dot\nu} - \Gamma_{\dot\beta}^{*}T_{\dot\alpha}\},$

so that

$$\Gamma_{\dot\alpha}^{*}T_\beta = \tfrac{1}{2}\epsilon_{\dot\alpha\dot\beta}\Gamma_{\dot\lambda}^{*}T^{\dot\lambda}. \tag{45a}$$

Similarly

$$\Gamma_{\alpha}^{*}T_\beta = \tfrac{1}{2}\epsilon_{\alpha\beta}\Gamma_{\lambda}^{*}T^{\lambda}. \tag{45b}$$

Also, from (37) and (39),

$$I_{\mu\nu} = \tfrac{1}{8}A_{\mu\rho}A^{\rho}_{\nu} - \tfrac{1}{2}\epsilon_{\mu\nu}, \quad (46a) \qquad I_{\dot\mu\dot\nu} = \tfrac{1}{8}A_{\dot\mu\rho}A^{\rho}_{\dot\nu} - \tfrac{1}{2}\epsilon_{\dot\mu\dot\nu}. \quad (46b)$$

Now

$$\epsilon_{\lambda\mu}I_{\nu\rho} + \epsilon_{\nu\rho}I_{\mu\lambda} = \tfrac{1}{8}\{A_{\lambda\nu}A_{\mu\rho} - A_{\mu\rho}A_{\lambda\nu}\}.$$

We therefore obtain, on multiplying by $A^{\lambda\nu}$ on the left,

$$A_{\mu}^{\nu}I_{\nu\rho} + A^{\lambda}_{\rho}I_{\mu\lambda} = \tfrac{3}{2}A_{\mu\rho}, \tag{46}$$

and, multiplying by T_α on the right and using (36),

$$A_{\mu}^{\nu}I_{\nu\rho}T_\alpha = -\tfrac{1}{2}(A_{\mu\alpha}T_\rho + \epsilon_{\rho\alpha}A_{\mu}^{\nu}T_\nu) = \tfrac{3}{2}A_{\mu\rho}T_\alpha.$$

By interchanging ρ and α, and adding, it follows that

$$A_{\mu\alpha}T_\rho = -A_{\mu\rho}T_\alpha = \tfrac{1}{2}\epsilon_{\alpha\rho}A_{\mu\beta}T^{\beta}. \tag{47a}$$

Similarly

$$A_{\mu\dot\alpha}T_\beta = -A_{\mu\dot\beta}T_{\dot\alpha} = \tfrac{1}{2}\epsilon_{\dot\alpha\dot\beta}A_{\mu\dot\beta}T^{\dot\beta}. \tag{47b}$$

In a way similar to that used in (3) for $T_{\alpha\lambda}$, it can be shown that, apart from a numerical factor, T_α and $T_{\dot\alpha}$ as defined by (36) and (42) are unique. Now consider $A_{\mu\beta}T^{\beta} \equiv \gamma_\mu$. Since

$$A_{\alpha\beta}I_{\mu\nu} - I_{\mu\nu}A_{\alpha\beta} = \tfrac{1}{2}\{\epsilon_{\mu\alpha}A_{\nu\beta} + \epsilon_{\nu\alpha}A_{\mu\beta}\}, \tag{48a}$$

$$A_{\alpha\beta}I_{\dot\mu\dot\nu} - I_{\dot\mu\dot\nu}A_{\alpha\beta} = \tfrac{1}{2}\{\epsilon_{\dot\mu\dot\beta}A_{\alpha\dot\nu} + \epsilon_{\dot\nu\dot\beta}A_{\alpha\dot\mu}\}, \tag{48b}$$

it follows from (48) and (36) that $\gamma_{\dot\alpha}$ satisfies the same equation as $T_{\dot\alpha}$ in (42). Therefore

$$A_{\dot\alpha\beta}T^\beta = cT_{\dot\alpha}, \tag{49a}$$

where c is some complex number. Similarly,

$$A_{\alpha\beta}T^\beta = \dot cT_{\alpha}, \tag{49b}$$

where $\dot c$ is another complex number. Therefore, from (47) and (49),

$$A_{\lambda\mu}A_{\nu\rho}T_\alpha = \tfrac{1}{2}A_{\lambda\mu}\epsilon_{\nu\alpha}cT_\beta = \tfrac{1}{4}\epsilon_{\dot\mu\rho}\epsilon_{\nu\alpha}T_\lambda c\dot c.$$

Multiplying by $\epsilon^{\lambda\nu}$ and $\epsilon^{\dot\mu\rho}$ we find that

$$8T_\alpha = -\tfrac{1}{2}T_\alpha c\dot c,$$

and therefore

$$c\dot c = -16.$$

Since $T_{\dot\alpha}$ can always be multiplied by a numerical factor, we can choose $c = 4i$ in (49a), so that $c = \dot c = 4i$, and

$$A_{\lambda\mu}T_\alpha = 2i\epsilon_{\lambda\alpha}T_\mu, \quad (50a) \qquad A_{\lambda\mu}T_{\dot\alpha} = 2i\epsilon_{\dot\mu\dot\alpha}T_\lambda. \quad (50b)$$

The Hermitian conjugate of (50) is

$$\Gamma^*_{\dot\alpha}A_{\lambda\mu} = 2i\epsilon_{\dot\alpha\lambda}\Gamma^*_{\dot\mu}, \quad (51a) \qquad \Gamma^*_{\alpha}A_{\lambda\mu} = 2i\epsilon_{\alpha\mu}\Gamma^*_{\dot\lambda}. \quad (51b)$$

Therefore, from (50) and (51) and (44),

$$\Gamma^*_{\dot\alpha}A_{\lambda\mu}T_\beta = i\epsilon_{\dot\alpha\lambda}\epsilon_{\mu\beta}\Gamma^{*}_{\dot\rho}T^\rho = i\epsilon_{\dot\alpha\lambda}\epsilon_{\mu\beta}\Gamma^{*}_{\dot\beta}T^\beta,$$

so that

$$\Gamma^{*}_{\dot\rho}T^\rho = \Gamma^{*}_{\dot\beta}T^\beta. \tag{52}$$

The equation (52) shows that $\Gamma^{*}_{\dot\rho}T^\rho$ is a pure imaginary number. Now by the same method as employed in (3) for $\Gamma^{*}_{\dot\alpha\lambda}T^{\alpha\lambda}$ it can be shown that $\Gamma^{*}_{\dot\rho}T^\rho \neq 0$, and therefore by multiplying T_ρ by a suitable constant it can be arranged that $\Gamma^{*}_{\dot\rho}T^\rho = i$. If necessary the sign of Λ may be reversed without affecting (35) and (38) to make $\Gamma^{*}_{\dot\rho}T^\rho$ a positive multiple of i. Then (44) becomes

$$\Gamma^{*}_{\alpha}T_\beta = i\epsilon_{\alpha\beta}, \quad (53a) \qquad \Gamma^{*}_{\dot\alpha}T_\beta = i\epsilon_{\dot\alpha\dot\beta}. \quad (53b)$$

Now consider the matrix J defined by

$$J = iT_\alpha\Gamma^{*\alpha} + iT_{\dot\alpha}\Gamma^{*\dot\alpha}.$$

From (53) we find that

$$J^2 = J. \tag{54}$$

Also, from (50) and (51), $\quad JA_{\lambda\mu} = 2(T_\lambda\Gamma^*_{\dot\mu} + T_{\dot\mu}\Gamma^*_\lambda) = A_{\lambda\mu}J.$

Therefore, since J commutes with the irreducible set of matrices α_k, it must be a multiple of the unit matrix, so that, from (54), either $J = 0$ or $J = 1$. But $J \neq 0$ since

$$\Gamma^{*}_{\alpha}JA_{\lambda\mu} = 2i\epsilon_{\alpha\lambda}\Gamma^*_{\dot\mu} \neq 0.$$

Hence

$$iT_\alpha\Gamma^{*\alpha} + iT_{\dot\alpha}\Gamma^{*\dot\alpha} = 1, \tag{55}$$

and

$$A_{\lambda\mu} = 2(T_\lambda\Gamma^*_{\dot\mu} + T_{\dot\mu}\Gamma^*_\lambda). \tag{56}$$

Further, it is easy to see that

$$\omega = iT_\alpha\Gamma^{*\alpha} - iT_{\dot\alpha}\Gamma^{*\dot\alpha} \tag{57}$$

anticommutes with $A_{\lambda\mu}$, and $\omega^2 = 1$. The only quantities whose squares are equal to unity and which anticommute with α_k are

$$\pm\frac{i}{4!}\epsilon^{klmn}\alpha_k\alpha_l\alpha_m\alpha_n,$$

28-2

where e^{klmn} is antisymmetric in all the four indices, and $e^{0123} = 1$. Starting from the relation

$$A_{\alpha\lambda}A_{\beta\mu}A_{\gamma\nu}A_{\delta\rho} = 16i^3\{\epsilon_{\alpha\beta}\epsilon_{\gamma\delta}\mathsf{T}_\lambda\epsilon_{\mu\nu}\Gamma_\rho^* + \mathsf{T}_\alpha\epsilon_{\beta\gamma}\Gamma_\delta^*\epsilon_{\lambda\mu}\epsilon_{\nu\rho}\} \tag{58}$$

and using equation (102) of (3), we can easily show that

$$\omega = \frac{i}{4!}e^{klmn}\alpha_k\alpha_l\alpha_m\alpha_n. \tag{59}$$

The quantities $\mathsf{T}_\lambda\Gamma_{\dot\mu}^*$, $\mathsf{T}_\lambda\Gamma_\mu^*$, $\mathsf{T}_\lambda\Gamma_\mu^*$, $\mathsf{T}_\lambda\Gamma_{\dot\mu}^*$ are all linearly independent, as follows immediately from (53) and (44). Since there are four quantities in each set corresponding to different values of λ and μ, there are sixteen quantities in all. This corresponds to the order of the algebra of the α-matrices. Any 4-rowed square matrix can be expressed linearly in terms of them. The following relations are easily verified:

$$4\mathsf{T}_\lambda\Gamma_{\dot\mu}^* = A_{\lambda\dot\mu} + i\hat{A}_{\lambda\dot\mu}, \quad (60a) \qquad 4\mathsf{T}_{\dot\mu}\Gamma_\lambda^* = A_{\lambda\dot\mu} - i\hat{A}_{\lambda\dot\mu}, \quad (60b)$$

$$i\mathsf{T}_\lambda\Gamma_\mu^* = I_{\lambda\mu} + \frac{1+\omega}{4}\epsilon_{\lambda\mu}, \quad (60c) \qquad i\mathsf{T}_{\dot\lambda}\Gamma_{\dot\mu}^* = I_{\dot\lambda\dot\mu} + \frac{1-\omega}{4}\epsilon_{\dot\lambda\dot\mu}, \quad (60d)$$

where

$$\hat{A}_{\lambda\dot\mu} = \hat{a}_k\sigma_{\lambda\dot\mu}^k,$$

and

$$\hat{a}_k = -i\alpha_k\omega = \frac{1}{3!}\epsilon_k^{lmn}\alpha_l\alpha_m\alpha_n.$$

The wave equation of the electron may be written as

$$\frac{i}{2}\partial_{\lambda\dot\mu}A^{\lambda\dot\mu}\psi + \chi\psi = 0,$$

or

$$i\partial_{\lambda\dot\mu}(\mathsf{T}^\lambda\Gamma^{*\dot\mu} + \mathsf{T}^{\dot\mu}\Gamma^{*\lambda})\psi + \chi\psi = 0.$$

Multiplying by Γ_α^* and $\Gamma_{\dot\alpha}^*$ on the left we obtain

$$-\partial_{\alpha\dot\mu}\Gamma^{*\dot\mu}\psi + \chi\Gamma_\alpha^*\psi = 0 \quad \text{and} \quad -\partial_{\lambda\dot\alpha}\Gamma^{*\lambda}\psi + \chi\Gamma_{\dot\alpha}^*\psi = 0$$

respectively. These are the usual equations for the electron in the spinor form if we put

$$\Gamma_\alpha^*\psi = a_\alpha, \quad \Gamma_{\dot\alpha}^*\psi = b_{\dot\alpha}.$$

The expression for the charge-current density spinor transforms to

$$s_{\lambda\dot\mu} = \psi^* A_{\lambda\dot\mu}\psi = 2\psi^*(\mathsf{T}_\lambda\Gamma_{\dot\mu}^* + \mathsf{T}_{\dot\mu}\Gamma_\lambda^*)\psi = 2(b_\lambda^\dagger b_{\dot\mu} + a_{\dot\mu}^\dagger a_\lambda).$$

Similarly, the formula for the spin becomes

$$\frac{1}{i}\psi^* I_{\lambda\mu}\psi = \tfrac{1}{2}\psi^*(\mathsf{T}_\lambda\Gamma_\mu^* + \mathsf{T}_\mu\Gamma_\lambda^*)\psi = \tfrac{1}{2}(b_\lambda^\dagger a_\mu + b_\mu^\dagger a_\lambda),$$

$$\frac{1}{i}\psi^* I_{\dot\lambda\dot\mu}\psi = \tfrac{1}{2}\psi^*(\mathsf{T}_{\dot\lambda}\Gamma_{\dot\mu}^* + \mathsf{T}_{\dot\mu}\Gamma_{\dot\lambda}^*)\psi = \tfrac{1}{2}(a_{\dot\lambda}^\dagger b_{\dot\mu} + a_{\dot\mu}^\dagger b_{\dot\lambda}).$$

Corresponding to equation (100) of (3) a particular representation is obtained by putting

$$\psi = \begin{pmatrix} \psi(0, \tfrac{1}{2}) \\ \psi(\tfrac{1}{2}, 0) \end{pmatrix}, \quad A_{\alpha\lambda} = \begin{pmatrix} 0 & -2iu_\lambda v_\alpha \\ -2iu_\alpha v_\lambda & 0 \end{pmatrix}, \tag{61a}$$

$$\Gamma_{\dot\alpha}^* = (v_{\dot\alpha} \quad 0), \quad \Gamma_\alpha^* = (0 \quad v_\alpha), \quad \mathsf{T}_{\dot\alpha} = -i\begin{pmatrix} u_{\dot\alpha} \\ 0 \end{pmatrix}, \quad \mathsf{T}_\alpha = -i\begin{pmatrix} 0 \\ u_\alpha \end{pmatrix}. \tag{61b}$$

Summary

It is shown that the algebra of the Duffin-Kemmer matrices can be handled quite neatly by introducing an extra dimension and by using the Γ-formalism of an earlier paper(3). By this means it is possible to derive identities similar to the De Broglie-Pauli identities for the Dirac matrices. To complete the analogy a Γ-formalism is also set up for the Dirac matrices.

REFERENCES

(1) DUFFIN, R. J. *Phys. Rev.* 54 (1938), 1114.
(2) EDDINGTON, A. S. *Relativity theory of electrons and protons* (Cambridge, 1936).
(3) HARISH-CHANDRA. *Proc. Roy. Soc.* A, 186 (1946), 502–25.
(4) HARISH-CHANDRA. *Proc. Ind. Acad. Sci.* A, 32 (1945), 30–41.
(5) KEMMER, N. *Proc. Roy. Soc.* A, 173 (1939), 91–116.
(6) PAULI, W. *Ann. Inst. Poincaré*, 6 (1936), 109–36.

GONVILLE AND CAIUS COLLEGE
    CAMBRIDGE

Reprinted from
Proc. Camb. Phil. Soc.
43 (1947), 414–421

Reprinted from THE PHYSICAL REVIEW, Vol. 71, No. 11, 793–805, June 1, 1947
Printed in U. S. A.

On Relativistic Wave Equations

HARISH-CHANDRA
Gonville and Caius College, Cambridge, England
(Received May 14, 1946)

The problem of the relativistic invariance of a first-order wave equation with matrix coefficients β_k is examined. It is found that it is intimately connected with the structure of the enveloping algebra of the β-matrices. In particular the center of this algebra can contain only elements of a very restricted type. This fact provides a powerful means for the investigation of the abstract β-algebra, as is illustrated in the general s-dimensional case of the Dirac and the Duffin-Kemmer matrices. Relativistic invariance imposes severe restrictions on the spurs of the β-matrices and their multiple products. Conversely, these restrictions insure relativistic invariance. The theory is applied to the case of particles of higher spin.

1. INTRODUCTION

THE wave equation for a particle of arbitrary spin can be written in the form

$$i\beta^k\partial_k\psi+\chi\psi=0, \quad (\partial_k=\partial/\partial x^k) \tag{1}$$

where χ is a constant determining the mass of the particle, and $\beta^k(k=0, 1, 2, 3)$ are square matrices operating on ψ which itself is a one-column matrix. The object of the present paper is to investigate the conditions which the β's must satisfy in order that Eq. (1) may be relativistically invariant. The conditions so obtained are also found to be largely sufficient. They insure the invariance of Eq. (1) under *proper* Lorentz transformations.

The method of this paper seems to provide a powerful means for the investigation of the algebra generated by the β's, and therefore for the determination of their irreducible representations. In order to find all the irreducible representations of a semi-simple algebra one has to obtain a linearly independent basis for its center.[1] It will be shown, however, that only quantities of a very restricted type can belong to the center, and therefore the task of obtaining a basis becomes much easier. Moreover, once such a basis has been obtained, one can immediately decide whether or not a particular representation is equivalent to its complex conjugate, transposed or Hermitian conjugate. The question of the invariance of the representation under reflection can also be answered at once. It will be shown also that considerable information concerning the

spurs of the β's and their multiple products can be deduced from the relativistic invariance of Eq. (1). Conversely, if the spurs of the β's and their multiple products satisfy certain conditions, the invariance of Eq. (1) for the full Lorentz group is ensured.

As an illustration of these general results, the algebras of the Dirac and the Duffin-Kemmer matrices in the general case of s dimensions are discussed briefly. Although most of our results have already been obtained by Schrödinger[2] and Kemmer,[3] they serve to illustrate the power of the new method.

In the scheme of equations proposed by Dirac[4] one must impose certain other subsidiary conditions on ψ in addition to Eq. (1). However, as is well known, these subsidiary conditions in general become inconsistent with Eq. (1) when the electromagnetic interaction is introduced in the usual manner. A way out of this difficulty has been indicated by Fierz and Pauli,[5] but their method, though logically consistent, is rather cumbersome and consequently lacking in elegance. It therefore seems worth while to investigate alternative methods of approach to the problem. Bhabha[6] has recently proposed a formalism in which the subsidiary conditions are completely eliminated, so that the interaction can be introduced in the usual manner without difficulty. But the particle in general no longer

---

[1] The center of an algebra consists of the set of elements which commute with all elements of the algebra.

[2] E. Schrödinger, Proc. Roy. Irish Acad. [A] **48**, 135 (1943).

[3] N. Kemmer, Proc. Camb. Phil. Soc. **39**, 189 (1943).

[4] P. A. M. Dirac, Proc. Roy. Soc. [A] **155**, 447 (1936).

[5] M. Fierz and W. Pauli, Proc. Roy. Soc. [A] **173**, 211 (1939).

[6] H. J. Bhabha, Rev. Mod. Phys. **17**, 200 (1945).

obeys the second-order wave equation

$$\partial_k \partial^k \psi + \chi^2 \psi = 0 \qquad (2)$$

in the force-free case. The physical result of this circumstance is that the particle described by Bhabha's equation would possess several different mass values. However, on grounds of simplicity, it seems desirable to try to discover a scheme in which the difficulty concerning the subsidiary conditions may be avoided without abandoning the second-order equation. Such an attempt is made in this paper.

As shown by Pauli,[7] the connection between spin and statistics is really based on the relativistic invariance of the wave Eq. (1). It therefore remains valid in the present case, as indeed in every other where the wave equation is relativistically invariant (e.g., for the equations proposed by Bhabha).[6, 8]

2. ALGEBRAIC THEORY

For the relativistic invariance of Eq. (1) it is necessary that for every Lorentz transformation $(t_k{}^l)$ there should exist a non-singular matrix T satisfying the condition

$$\beta_k' = t_k{}^l \beta^l = T\beta_k T^{-1}.$$

However, the problem considered in this paper may be stated somewhat more generally. *If β_k $(k=1, 2, \cdots, s)$ be an irreducible[9] set of s matrices, what are the conditions which the β's must satisfy in order that, for every transformation of the real orthogonal[10] group, there should exist a non-singular matrix T such that*

$$\beta_k' = t_k{}^l \beta_l = T\beta_k T^{-1}, \qquad (3a)$$

$$t_k{}^m t_l{}^n g_{mn} = g_{kl}, \qquad (3b)$$

$$g_{kl} = \begin{cases} 0 & k \neq l, \\ 1 & k = l; \end{cases} \qquad (4)$$

and are these conditions sufficient to insure the

invariance[11] of the β-representation under the orthogonal group?

Let $\mathfrak{A}(\beta)$ denote the enveloping algebra of the β-matrices,[12] i.e., the algebra consisting of all quantities of the type

$$c = c^k \beta_k + c^{kl} \beta_k \beta_l + c^{klm} \beta_k \beta_l \beta_m + \cdots, \qquad (5)$$

where $c^k \cdots$ are numerical coefficients belonging to the field $\mathfrak{F}$ of complex numbers.[13] Since the β_k are irreducible, $\mathfrak{A}(\beta)$ is simple. Let u_p $(p=1, \cdots, h)$ be a linearly independent basis for $\mathfrak{A}$. As elements of $\mathfrak{A}(\beta)$, u_p, must be expressible in the form (5) and therefore one may write $u_p = u_p(\beta)$. On replacing each β_k by β_k' in $u_p(\beta)$ one obtains $u_p(\beta') = u_p(t\beta)$. Now $u_p(t\beta)$ is also an element of $\mathfrak{A}$ and therefore is expressible linearly in terms of the basis $u_p(\beta)$; that is,

$$u_p(t\beta) = \sum_q T_p{}^q u_q(\beta), \qquad (6)$$

where[14] $T_p{}^q \varepsilon \mathfrak{F}$, and may be regarded as coefficients of a transformation T in the u-space. Thus every transformation t in the β-space induces a transformation T in the u-space. It is easy to verify that the correspondence $t \rightarrow T$ is a homomorphism with respect to addition and multiplication and therefore the set of transformations T is a representation of the orthogonal group $O(s)$. Now, it is well known that every representation of $O(s)$ is fully reducible and therefore the u-space can be decomposed into a direct sum of irreducible invariant sub-spaces. Let U denote the entire u-space and let the irreducible invariant sub-spaces be denoted by $V(1)$, $V(2)$, $\cdots$, $V(n)$ such that[15] $(V(m):\mathfrak{F}) = h_m$. Now all irreducible representations of $O(s)$ are known.[16] Any such *single valued* representation of the

---

[7] W. Pauli, Phys. Rev. **58**, 716 (1940).

[8] H. J. Bhabha, Proc. Ind. Acad. Sci. **21**, 291 (1945).

[9] In order to insure that Eq. (1) describes an "elementary" particle, which should be an entity of the simplest possible kind, we assume that the set of matrices β_k is irreducible.

[10] For the moment only the real orthogonal group is considered, but it will be shown in section 3 that the corresponding problem concerning other real groups (for example, the Lorentz group) associated with indefinite ground forms, is intimately connected with the present one, and has an almost identical solution.

[11] The property of the existence of a non-singular matrix satisfying Eq. (3) corresponding to every transformation $t_k{}^l$ of a particular group G, can be briefly denoted as the property of the "invariance of the β-representation under G." The full (real) orthogonal group in s dimensions will be denoted by $O(s)$ while $O^+(s)$ will denote the corresponding *proper* orthogonal group.

[12] The whole set of matrices $\beta_k (k=1, 2, \cdots, s)$ is conveniently denoted by β. In the same way it is also convenient to abbreviate Eq. (3) to $\beta' = t\beta$, where t is now to be looked upon as a transformation operator acting on the set β and transforming it to β'.

[13] The summation convention for contracted indices is used throughout in this paper.

[14] This symbolism indicates that the coefficients $T_p{}^q$ belong to the field $\mathfrak{F}$.

[15] If V is a vector space over $\mathfrak{F}$, then $(V:\mathfrak{F})$ denotes the dimension of V over $\mathfrak{F}$.

[16] H. Weyl, *The Classical Groups* (Princeton University Press, Princeton, New Jersey, 1939).

entire group $O(s)$ is equivalent to the set of transformations of a tensor of suitable rank and symmetry properties. It is therefore possible to find a linearly independent basis $^{(m)}v_p$ ($p=1, \cdots, h_m$) for $V(m)$ such that T transforms $^{(m)}v_p$ as components of a tensor of rank f_m with suitable symmetry properties. Thus, one can define a tensor $^{(m)}v_{k_1 \cdots k_{f_m}}$ (of suitable symmetry) such that its h_m independent components are equal to the $^{(m)}v_p$'s and

$$v_{[k]}(t\beta) = t_{[k]}^{[l]} \cdot v_{[l]}(\beta). \tag{7}$$

Here the superscript (m) has been omitted for simplicity, and the symbol $[k]$ is used to denote the entire set of tensor indices $k_1, \cdots, k_f$. $t_{[k]}^{[l]}$ is merely a convenient abbreviation for $t_{k_1}^{l_1} t_{k_2}^{l_2} \cdots t_{k_f}^{l_f}$ At this point we need the following lemma.

Lemma 1. If $D_{[k]}(\beta)$ is a set of quantities belonging to $\mathfrak{A}(\beta)$ which transform as a tensor of rank f for all orthogonal transformations of the β-space so that

$$D_{[k]}(t\beta) = t_{[k]}^{[l]} \cdot D_{[l]}(\beta), \tag{8a}$$

then

$$D_{[k]}(\beta) = d_{[k]}(g) + d_{[k]l_1}(g) \cdot \beta^{l_1} + d_{[k]l_1l_2}(g) \cdot \beta^{l_1}\beta^{l_2} + \cdots, \tag{8b}$$

where $d_{[k]}...(g)$ are tensors constructed out of g_{kl} alone i.e.

$$d_{[k]}...(g) = \sum a(\alpha, \beta, \gamma, \delta, \cdots)g_{\alpha\beta}g_{\gamma\delta}\cdots \tag{9}$$

where $a(\alpha, \beta, \gamma, \delta \cdots)\varepsilon\mathfrak{F}$ and the sum is to be taken over all permutations $\alpha, \beta, \gamma, \delta \cdots$ of the indices $k_1 \cdots k_f, \cdots$ occurring on the left hand side of Eq. (9). The sum in Eq. (8b) is, of course, finite. When the total number of indices $k_1, \cdots k_f, \cdots$ is odd, a relation such as Eq. (9) is impossible and so in this case $d_{[k]}...(g) = 0$.

Proof: Put

$$D = \xi^{[k]} \cdot D_{[k]}(\beta),$$

where

$$\xi^{[k]} = \xi^{(1)k_1} \ldots \xi^{(f)k_f},$$

and $\xi^{(r)k}$ are indeterminates transforming contra-variantly[17] to β_k. Clearly D is invariant under the transformations of $O(s)$. Since $O(s)$ is a topologically compact group, it is possible to perform the process of averaging over the entire group manifold.[16] Let $\langle D \rangle$ be the average of D over $O(s)$.

From Eq. (5), D must be of the form

$$D = \xi^{[k]} \cdot \{d_{[k]l_1}\beta^{l_1} + d_{[k]l_1l_2}\beta^{l_1}\beta^{l_2} + \cdots\}, \tag{10a}$$

where $d_{[k]}...$ are suitable numerical coefficients. Therefore

$$\langle D \rangle = \xi^{[k]} \cdot \{\langle d \rangle_{[k]l_1}\beta^{l_1} + \langle d \rangle_{[k]l_1l_2}\beta^{l_1}\beta^{l_2} + \cdots\}, \tag{10b}$$

where $\langle d \rangle_{[k]}...$ are the averaged coefficients. Consider the forms

$$D(\eta) = \xi^{[k]} \cdot \{d_{[k]l_1}\eta^{(1)l_1} + d_{[k]l_1l_2}\eta^{(1)l_1}\eta^{(2)l_2} + \cdots\}, \tag{11a}$$

and

$$\langle D \rangle(\eta) = \xi^{[k]} \cdot \{\langle d \rangle_{[k]l_1}\eta^{(1)l_1} + \langle d \rangle_{[k]l_1l_2}\eta^{(1)l_1}\eta^{(2)l_2} + \cdots\}, \tag{11b}$$

where $\eta^{(r)k}$ are also indeterminates transforming contravariantly to β_k. Notice that Eqs. (11a) and (11b) are derived from Eqs. (10a) and (10b) by replacing the first β in each multiple product of the β's by $\eta^{(1)k}$, the second β by $\eta^{(2)k}$, and so on. Also $\langle D \rangle(\eta)$ is the average of $D(\eta)$, and, therefore, clearly $\langle D \rangle(\eta)$ is an absolute invariant of the orthogonal group $O(s)$. Since $x^k y_k = x^k y^l g_{kl}$ is the only typical basic invariant of $O(s)$,[16] $\langle D \rangle(\eta)$ must be expressible as a rational integral function of the scalar products of the ξ's and η's. This, in turn, means that the coefficients $\langle d \rangle_{[k]}...$ are of the form of Eq. (9). Now, because of the invariance of D, $D = \langle D \rangle$ and therefore Eq. (8b) holds.

Lemma 2. If Eq. (8a) is true at least for all transformations $t\varepsilon O^+(s)$, then

$$D_{[k]}(\beta) = d_{[k]}(g, \epsilon) + d_{[k]l_1}(g, \epsilon) \cdot \beta^{l_1} + \cdots, \tag{12}$$

where $d_{[k]}...(g, \epsilon)$ are tensors constructed out of g_{kl} and $\epsilon_{k_1 \cdots k_s}$[18] only.

The proof is similar to that of Lemma 1 with the difference that $O^+(s)$ now has two typical basic invariants $x^k y_k$ and $\epsilon_{k_1 \cdots k_s} x^{(1)k_1} x^{(2)k_2} \ldots x^{(s)k_s}$. Since the product of two ϵ's can be expressed in terms of the g's, we need consider only terms linear in ϵ.

From Eq. (7) and Lemma 1 it follows directly that

$$v_{[k]}(\beta) = v_{[k]}(\beta, g) \tag{13}$$

i.e., as in Eq. (8b) $v_{[k]}(\beta, g)$ is a tensor composed of the β's and g's only. Now $^{(m)}v_{[k]}(\beta, g)$ taken to-

---

[17] Although the difference between co-variant and contra-variant vectors disappears for orthogonal transformations, it is convenient to recognize it formally.

[18] $\epsilon_{k_1...k_s}$ is anti-symmetrical in every pair of indices, and $\epsilon_{12...s} = 1$.

gether for all values of m form a complete basis for $\mathfrak{A}(\beta)$. Hence the product of any two v's is again expressible as a linear sum of the v's so that

$$^{(m)}v_{[k]}(\beta) \cdot {}^{(m')}v_{[l]}(\beta)$$
$$+ \sum_{m''} {}^{(m,\,m')}{}_{(m'')}\Gamma_{[k][l]}{}^{[j]} \cdot {}^{(m'')}v_{[j]}(\beta) \quad (14)$$

where $^{(m,\,m')}{}_{(m'')}\Gamma_{[k][l]}{}^{[j]}$ are certain numerical coefficients. (Here the summation convention is used for the j's.) On imposing suitable symmetry properties on the Γ's corresponding to those of the v's in respect to permutations of the j's, they are uniquely determined by Eq. (14). Now from the similarity of the matrices β and $t\beta$ it follows that the correspondence $\beta \leftrightarrow t\beta$ is an automorphism of $\mathfrak{A}(\beta)$. Therefore

$$^{(m)}v_{[k]}(t\beta) \cdot {}^{(m')}v_{[l]}(t\beta)$$
$$= \sum_{m''} {}^{(m,\,m')}{}_{(m'')}\Gamma_{[k][l]}{}^{[j]} \cdot {}^{(m'')}v_{[j]}(t\beta). \quad (15)$$

So that from Eqs. (7), (14), and (15) one finds on omitting the m's that

$$t_{[k]}{}^{[i]} \cdot t_{[l]}{}^{[j]} \cdot \Gamma_{[i][j]}{}^{[a]} = t_{[b]}{}^{[a]} \cdot \Gamma_{[k][l]}{}^{[b]}.$$

Put
$$\Gamma_{[k][l][a]} = g_{[a][b]} \cdot \Gamma_{[k][b]}{}^{[b]}, \quad (16)$$
where
$$g_{[a][b]} = g_{a_1 b_1} g_{a_2 b_2} \cdots g_{a_f b_f}.$$
Now put
$$\Gamma = \xi^{[k]} \eta^{[l]} \zeta^{[a]} \cdot \Gamma_{[k][l][a]},$$

where $\zeta^{(r)k}$ is a third set of indeterminates transforming contravariantly to β_k. Then Γ is an absolute invariant of $O(s)$ and, therefore,

$$\Gamma_{[k][l][a]} = \Gamma_{[k][l][a]}(g) \quad (17)$$

where the right side denotes a tensor constructed out of the g's only. Equation (14) can therefore be written as

$$^{(m)}v_{[k]}(\beta,\,g) \cdot {}^{(m')}v_{[l]}(\beta,\,g)$$
$$= \sum_{m''} {}^{(m,\,m')}{}_{(m'')}\Gamma_{[k][l][j]}(g) \cdot {}^{(m'')}v^{[j]}(\beta,\,g). \quad (18)$$

Equation (18) has the form

$$c_{k_1 k_2 \cdots}(\beta,\,g) = 0. \quad (19)$$

Since $^{(m)}v_{[k]}(\beta,\,g)$ form a complete basis for $\mathfrak{A}(\beta)$, Eq. (18) describes the structure of $\mathfrak{A}(\beta)$ completely. Therefore *the relations determining completely the structure of an algebra can always be put in the tensor form of Eq. (19), provided the correspondence $\beta \rightarrow t\beta$ is an automorphism of $\mathfrak{A}(\beta)$ for every orthogonal transformation t.*

Conversely, let one or more equations of the type (19) be given to *define* β_k. From Eqs. (19) and (3b) it is clear that

$$c_{k_1 k_2 \cdots}(t\beta,\,g) = t_{k_1}{}^{l_1} t_{k_2}{}^{l_2} \cdots c_{l_1 l_2 \cdots}(\beta,\,g) = 0, \quad (20)$$

so that $t\beta$ satisfy the same relations as Eq. (19). Hence $\beta \rightarrow t\beta$ is an endomorphism of $\mathfrak{A}(\beta)$. Similarly, $\beta \rightarrow t^{-1}\beta$ is also an endomorphism. The product of these two endomorphisms is the identical automorphism $\beta \rightarrow \beta$. Hence each of the two endomorphisms must be an automorphism. Next, assume that the defining relations of Eq. (19) are such that the abstract algebra $\mathfrak{A}(\beta)$ is finite and simple. Now every automorphism of a simple algebra is an inner automorphism. Hence there exists a unit (i.e., a regular element) $T(\beta)\varepsilon\mathfrak{A}(\beta)$ such that

$$t\beta = T(\beta) \cdot \beta \cdot T^{-1}(\beta). \quad (21)$$

This shows that in any representation of $\mathfrak{A}(\beta)$ the matrices β and $t\beta$ are similar. *Thus in order that an irreducible β-representation be invariant under the orthogonal group O, it is necessary and sufficient that the algebra $\mathfrak{A}(\beta)$ be finite and simple and its structure be completely described by relations of the type (19).*

Notice that if $e(\beta)$ is an element belonging to the center of $\mathfrak{A}(\beta)$, then from Eq. (21)

$$e(t\beta) = T(\beta) \cdot e(\beta) \cdot T^{-1}(\beta) = e(\beta).$$

Hence, $e(\beta)$ is invariant under transformations of the orthogonal group, i.e., it transforms as a scalar. Hence, from Lemma 1,

$$e(\beta) = e(g,\,\beta),$$

that is, $e(\beta)$ is a scalar constructed by contracting all the indices of β^k in the multiple products among themselves. As is well known, the center of a simple algebra over the field $\mathfrak{F}$ of complex numbers consists of $\mathfrak{F}$. 1 where 1 is the unity quantity of $\mathfrak{A}(\beta)$. Hence $1 = e(\beta)$ where $e(\beta)$ is a 'scalar.' One can therefore allow a multiple of 1 to appear in Eq. (19) without spoiling the tensor character of the equations.

Since it is not usually easy to see from a set of equations such as (19) whether the algebra $\mathfrak{A}(\beta)$ is simple or not, we drop the restriction about the simplicity of $\mathfrak{A}$. Assume, therefore, that a set of equations such as (19) is given, defining an algebra $\mathfrak{A}(\beta)$ of a finite order. We wish to investigate the invariance of an irreducible representation of $\mathfrak{A}(\beta)$ under $O(s)$. Let $\beta_k \rightarrow {}'\beta_k$ be an irreducible representation of $\mathfrak{A}(\beta)$ by the matrices $`\beta_k$. Since $\beta \rightarrow t^{-1}\beta$ is an automorphism of $\mathfrak{A}$, $t^{-1}\beta \rightarrow `\beta$ is also an irreducible representation of $\mathfrak{A}$. Hence $\beta \rightarrow `\beta$ and $\beta \rightarrow t`\beta$ are both irreducible representations of $\mathfrak{A}$. Choose

$$t_k{}^l = \delta_k{}^l + \epsilon_k{}^l, \quad (\epsilon_{kl} = -\epsilon_{lk}), \qquad (22a)$$

where ϵ_{kl} is an infinitesimal quantity ($t_k{}^l$ corresponds to an infinitesimal rotation). Then

$$\beta_k \rightarrow `\beta_k, \qquad (22b)$$

$$\beta_k \rightarrow `\beta_k + \epsilon_k{}^l \, `\beta_l, \qquad (22c)$$

are both irreducible representations of $\mathfrak{A}$. Now every finite algebra has only a finite number of irreducible representations apart from equivalence. Therefore from continuity, Eqs. (22b) and (22c) must be equivalent representations. Since all *proper* orthogonal transformations can be generated from the infinitesimal transformations, the representations $\beta \rightarrow `\beta$ and $\beta \rightarrow t^+`\beta$ where $t^+ \varepsilon O^+(s)$ are equivalent. *Hence every irreducible representation of $\mathfrak{A}(\beta)$ is invariant under the proper orthogonal group O^+.*

Let $\mathfrak{N}(\beta)$ be the radical of $\mathfrak{A}$. Clearly the radical is invariant under any automorphism of $\mathfrak{A}$. Hence exactly as before, one can find a tensor basis ${}^{(m)}\eta_{[k]}(\beta, g)$ for $\mathfrak{N}$. In any irreducible representation of $\mathfrak{A}$ all elements of $\mathfrak{N}$ are represented by zero. Hence all the irreducible representations of $\mathfrak{A}$ are the same as those of $\mathfrak{S}(\beta)$, where $\mathfrak{S}(\beta)$ is defined by Eq. (19) and the further equations ${}^{(m)}\eta_{[k]}(\beta, g) = 0$. Clearly the radical of $\mathfrak{S}(\beta)$ is zero and therefore $\mathfrak{S}$ is semi-simple. Also $\mathfrak{S}(\beta)$ is defined by proper tensor equations, so that $\beta \rightarrow t\beta$ is an automorphism of $\mathfrak{S}$. Now every semi-simple algebra is expressible as a direct sum of simple algebras.[19] Therefore,

$$\mathfrak{S}(\beta) = \mathfrak{A}_1(\beta) \oplus \mathfrak{A}_2(\beta) \oplus \cdots \oplus \mathfrak{A}_r(\beta) \qquad (23)$$

where $\mathfrak{A}_u(\beta)$ ($u = 1, \cdots, r$) are all simple. This decomposition of $\mathfrak{S}$ is unique apart from the order of $\mathfrak{A}_u$ in the decomposition. Hence under any automorphism θ of $\mathfrak{S}$, the simple algebra goes over into some $\mathfrak{A}_u{}'$ or in symbols

$$\mathfrak{A}_u \theta = \mathfrak{A}_u{}'. \qquad (24)$$

In particular let θ be the automorphism $\beta \rightarrow t\beta$, where t is given by Eq. (21a). Then as before it

follows from continuity considerations that $u' = u$ in Eq. (24), and so the transformation $\beta \rightarrow t^+\beta$ for $t^+ \varepsilon O^+(s)$ is an automorphism of $\mathfrak{A}_u$. Corresponding to Eq. (23) one has

$$\beta = \beta^{(1)} + \beta^{(2)} + \cdots + \beta^{(r)}, \quad \beta^{(u)} \varepsilon \mathfrak{A}_u. \qquad (25)$$

Hence $\mathfrak{A}_u$ is the algebra generated by $\beta^{(u)}$. The automorphism $\beta \rightarrow t^+\beta$ of $\mathfrak{S}$ corresponds to the automorphism $\beta^{(u)} \rightarrow t^+\beta^{(u)}$ of $\mathfrak{A}_u$. Since the latter must be an inner automorphism of $\mathfrak{A}_u$ there exists a unit $T^{+(u)} \varepsilon \mathfrak{A}_u$ such that

$$t^+\beta^{(u)} = T^{+(u)}\beta^{(u)}(T^{+(u)})^{-1}. \qquad (26)$$

Put

$$T^+ = T^{+(1)} + T^{+(2)} + \cdots + T^{+(r)},$$

then

$$(T^+)^{-1} = (T^{+(1)})^{-1} + (T^{+(2)})^{-1} + \cdots + (T^{+(r)})^{-1},$$

since if $a_u \varepsilon \mathfrak{A}_u$ and $b_v \varepsilon \mathfrak{A}_v$, then $a_u b_v = 0$ if $u \neq v$, and

$$1 = e_1 + \cdots + e_r, \qquad (27)$$

where 1 is the unity quantity of $\mathfrak{S}$ and $e_u \varepsilon \mathfrak{A}_u$ is the unity quantity of $\mathfrak{A}_u$. Thus

$$t^+\beta = T^+\beta(T^+)^{-1}. \qquad (28)$$

Now if $c(\beta)$ is an element belonging to the center of $\mathfrak{S}$, then $c(t^+\beta) = T^+c(\beta)(T^+)^{-1} = c(\beta)$. Hence $c(\beta)$ is invariant under proper orthogonal transformations. Therefore, from Lemma 2,

$$c(\beta) = c(\beta, g, \epsilon). \qquad (29a)$$

By the improper orthogonal transformation J, which reverses the direction of the x_1-axis, $c(\beta)$ goes over into

$$c(J\beta) = c(\beta, g, -\epsilon). \qquad (29b)$$

The center of $\mathfrak{S}$ is obviously invariant under all automorphisms of $\mathfrak{S}$. Hence $c(J\beta)$ also belongs to the center of $\mathfrak{S}$. Therefore both

$$c_{sc}(\beta) = \tfrac{1}{2}c(\beta, g, \epsilon) + \tfrac{1}{2}c(\beta, g, -\epsilon),$$

and

$$c_{ps}(\beta) = \tfrac{1}{2}c(\beta, g, \epsilon) - \tfrac{1}{2}c(\beta, g, -\epsilon),$$

belong to the center of $\mathfrak{S}$. c_{sc} is the "scalar" part of c, while c_{ps} is the "pseudo-scalar" part. It is therefore sufficient to consider only those quantities of the center which are either pure scalars or pure pseudo-scalars.

The number of inequivalent irreducible representations of $\mathfrak{S}$ is r, which is the number of simple sub-algebras in the decomposition Eq. (23). Also,

<hr>

[19] A. A. Albert, *Structure of Algebras* (Am. Math. Soc. Coll. Pub., New York, 1939).

$(e_1, \cdots, e_r)$ form a linearly independent basis for the center of $\mathfrak{S}$. Hence the number of irreducible representations is the same as the dimension of the center over $\mathfrak{F}$. One has, therefore, merely to obtain all scalar or pseudo-scalar quantities in $\mathfrak{S}$ which commute with β_k and choose from them a linearly independent set. Let $(c_1(\beta), \cdots, c_r(\beta))$ be such a set. In an irreducible representation of $\mathfrak{S}$ these quantities are represented by multiples of the unit matrix, i.e., by numbers in $\mathfrak{F}$. In general, two irreducible representations of $\mathfrak{S}$ are equivalent only if $c_u(\beta)(u=1, \cdots, r)$ are represented by the same numbers in the two representations. In particular, suppose the c's are so numbered that $c_1(\beta), c_2(\beta), \cdots, c_\mu(\beta)$ are scalars, while $c_{\mu+1}(\beta), \cdots, c_r(\beta)$ are pseudo-scalars. Then an irreducible β-representation is reflection-invariant only if $c_{\mu+1} = \cdots = c_r = 0$ for it.

From the point of view of the physical application it is desirable that the matrices β_k should be equivalent to their transposed $\beta_k{}^\ddagger$ and Hermitian conjugate $\beta_k{}^\dagger$. Hence $\beta_k{}^\ddagger$ also satisfy Eq. (19) i.e.,

$$c_{kl}...(\beta^\ddagger) = 0. \tag{30a}$$

Taking the transposed of (30a) one obtains

$$c_{kl}...{}^T(\beta) = 0, \tag{30b}$$

where $c_{kl}...{}^T(\beta)$ is derived from $c_{kl}...(\beta)$ by reversing the order of the factors in the multiple products. Hence from Eqs. (19) and (30b),

$$c_{kl}...(\beta) + c_{kl}...{}^T(\beta) = 0, \tag{31a}$$

$$c_{kl}...(\beta) - c_{kl}...{}^T(\beta) = 0. \tag{31b}$$

Since Eq. (19) follows from Eqs. (31a and b), the latter define $\mathfrak{S}(\beta)$ completely. Hence the defining equations of $\mathfrak{S}$ can always be put in the form

$$c_{kl}...(\beta, g) = 0 \tag{32}$$

where $c_{kl}...{}^T = \pm c_{kl}...$. From the equivalence of β_k, $\beta_k{}^\ddagger$ and $\beta_k{}^\dagger$ it follows that β_k is equivalent to its complex conjugate $\beta_k{}^*$. Hence from Eq. (32) $c_{kl}...(\beta^*, g) = 0$. Taking the complex conjugate one finds

$$c_{kl}...{}^*(\beta, g) = 0 \tag{33}$$

where $c_{kl}...{}^*$ is derived from $c_{kl}...$ by applying the automorphism $i \to -i$ of $\mathfrak{F}$ to the numerical coefficients in $c_{kl}...$. Hence from Eqs. (32) and (33)

$$c_{kl}...(\beta, g) + c_{kl}...{}^*(\beta, g) = 0, \tag{34a}$$

$$ic_{kl}...(\beta, g) - ic_{kl}...{}^*(\beta, g) = 0. \tag{34b}$$

Equation (32) follows from Eq. (34), and so (34) defines $\mathfrak{S}(\beta)$ completely. Therefore the defining equations of $\mathfrak{S}$ can be written in the form

$$c_{kl}...(\beta, g) = 0 \tag{35}$$

where $c_{kl}...{}^T = \pm c_{kl}...$ and $c_{kl}...{}^* = c_{kl}...$. Thus (35) is an equation with real coefficients. If the algebra $\mathfrak{S}(\beta)$ so defined by Eq. (35) be simple, then apart from equivalence it is capable of only one irreducible representation, say $\beta_k \to \mathfrak{b}_k$ where $\mathfrak{b}_k$ are irreducible matrices. Clearly $\mathfrak{b}_k{}^\dagger$ and $\mathfrak{b}_k{}^\ddagger$ are also irreducible and satisfy Eq. (35), so that they also constitute a representation of $\mathfrak{S}$. All these representations must therefore be equivalent. *Hence if (35) defines a finite simple algebra and β_k be an irreducible set of matrices satisfying (35), then β, $\beta^\ddagger$, β^*, $\beta^\dagger$, and $t\beta$ (for any $t\epsilon O(s)$) are all equivalent.*

In the general case when $\mathfrak{S}(\beta)$ is semi-simple, Eq. (35) is not sufficient to ensure the equivalence of β, $\beta^\dagger$, and $\beta^\ddagger$. Of course, as shown above, it is still necessary; i.e., if for every irreducible representation of $\mathfrak{S}$, β, $\beta^\dagger$ and $\beta^\ddagger$ are equivalent, then the defining equation's of $\mathfrak{S}$ can always be put in the form Eq. (35). To find under what conditions the irreducible sets β, $\beta^\dagger$, and $\beta^\ddagger$ are equivalent we go back to the general criterion for deciding the equivalence of two representations. Take $(c_1(\beta), \cdots, c_r(\beta))$ to be a linearly independent basis for the center of $\mathfrak{S}$. Clearly $c_u{}^T(\beta)$ and $c_u{}^*(\beta)$ also belong to the center. Therefore, for a particular irreducible representation, β and $\beta^\ddagger$ are equivalent only if $c_u(\beta) = c_u{}^T(\beta)$ for this representation. Similarly, the equivalence of β and $\beta^\dagger$ requires that $c_u(\beta) = c_u{}^{T*}(\beta)$.

From Eq. (26) it follows in the usual way that there exist elements $I_{kl}{}^{(u)}$ in $\mathfrak{A}_u$ such that

$$I_{kl}{}^{(u)} = -I_{lk}{}^{(u)} \text{ and }$$
$$[\beta_k{}^{(u)}, I_{lm}{}^{(u)}] = g_{kl}\beta_m{}^{(u)} - g_{km}\beta_l{}^{(u)}, \tag{36a}$$

$$[I_{kl}{}^{(u)}, I_{mn}{}^{(u)}] = -g_{km}I_{ln}{}^{(u)} + g_{lm}I_{kn}{}^{(u)}$$
$$+ g_{kn}I_{lm}{}^{(u)} - g_{ln}I_{km}{}^{(u)}, \tag{36b}$$

where $[A, B] = AB - BA$. It is easy to prove that the choice of $I_{kl}{}^{(u)}$ is unique since if $J_{kl}{}^{(u)}$ were another set satisfying the same equations then

$$C_{kl}{}^{(u)} = J_{kl}{}^{(u)} - I_{kl}{}^{(u)}$$

would commute with $\beta_k{}^{(u)}$ and would therefore

belong to the center of $\mathfrak{A}_u$. Hence

$$0 = [I_{kl}{}^{(u)}, I_{mn}{}^{(u)}] - [J_{kl}{}^{(u)}, J_{mn}{}^{(u)}]$$
$$= g_{km}C_{ln}{}^{(u)} - g_{lm}C_{kn}{}^{(u)} - g_{kn}C_{lm}{}^{(u)} + g_{ln}C_{km}{}^{(u)}$$

from which it follows that $C_{kl}{}^{(u)} = 0$ unless $s = 2$. So assuming that $s \neq 2$, $I_{kl}{}^{(u)}$ is uniquely determined. Put

$$I_{kl} = I_{kl}{}^{(1)} + \cdots + I_{kl}{}^{(r)}. \tag{37}$$

Then $I_{kl} = -I_{lk}$ and

$$[\beta_k, I_{lm}] = g_{kl}\beta_m - g_{km}\beta_l, \tag{38a}$$

$$[I_{kl}, I_{mn}] = -g_{km}I_{ln} + g_{lm}I_{kn} + g_{kn}I_{lm} - g_{ln}I_{km}. \tag{38b}$$

I_{kl} is also uniquely defined by Eq. (37), since from Eq. (23) every I_{kl} is decomposable uniquely in the form (37), and $I_{kl}{}^{(u)}$, in consequence of Eqs. (37) and (38), must satisfy Eq. (36) and therefore are uniquely determined. I_{kl} is an element of $\mathfrak{S}(\beta)$ and hence is expressible in the form of Eq. (5). Write it as $I_{kl}(\beta)$. Consider the automorphism $\beta \rightarrow t^{-1}\beta$ of $\mathfrak{S}$. It is then easy to see from Eqs. (38) that

$$I'_{kl} = t_k{}^m t_l{}^n I_{mn}(t^{-1}\beta)$$

satisfy the same equations as (38). Hence

$$I_{kl}(t\beta) = t_k{}^m t_l{}^n I_{mn}(\beta)$$

so that $I_{kl}(\beta)$ transforms as a tensor. Applying Lemma 1, we immediately get

$$I_{kl}(\beta) = I(\beta_k\beta_l - \beta_l\beta_k)$$
$$+ [\sum P^{\cdots}(\beta)\beta_k Q^{\cdots}_{\cdots}(\beta)\beta_l R\ldots(\beta)]_-, \tag{39}$$

where I is a numerical constant, $P^{\cdots}$, $Q^{\cdots}_{\cdots}$, $R\ldots$ are tensors constructed from β and g only, and the dots denote contracted indices. $\sum$ denotes a sum over similar terms and the minus sign at the end means that terms with l and k interchanged are to be subtracted. The first term $I(\beta_k\beta_l - \beta_l\beta_k)$ is of the simplest possible kind and has been purposely separated out. Bhabha's theory[6, 8] is based on this simplest substitution

$$I_{kl} = I(\beta_k\beta_l - \beta_l\beta_k).$$

By an entirely similar consideration of the other two automorphisms of $\mathfrak{S}$ one concludes that $I_{kl}(\beta) = -I_{kl}{}^T(\beta)$ and $I_{kl}{}^*(\beta) = I_{kl}(\beta)$. Hence

$$I_{kl}(\beta) = I(\beta_k\beta_l - \beta_l\beta_k)$$
$$+ \tfrac{1}{2}[\sum P^{\cdots}(\beta)\beta_k Q^{\cdots}_{\cdots}(\beta)\beta_l R\ldots(\beta)$$
$$- P^{T\cdots}(\beta)\beta_k Q^{T\cdots}_{\cdots}(\beta)\beta_l R^T\ldots(\beta)]_- \tag{40}$$

where the numerical coefficients in Eq. (40) are all real.

3. EXTENSION TO INDEFINITE GROUND FORMS

Let us now consider instead of O, other groups G of linear transformations with real coefficients which leave the indefinite ground form

$$g_{kl}x^k x^l = 0, \tag{41a}$$

$$g_{kl} = 0, \quad k \neq l, \quad g_{kk} = +1 \text{ or } -1 \tag{41b}$$

unchanged. As usual, we use g_{kl} to raise or lower the tensor indices. For every co-variant tensor $A_{k_1 \cdots k_n}$ define

$$\mathbf{A}_{k_1 \cdots k_n} = (i)^r A_{k_1 \cdots k_n}$$

where r is the number of indices k_m among $(k_1 \cdots, k_n)$ such that $g_{k_m k_m} = -1$. In this way for every co-variant tensor in the x-space one constructs a co-variant tensor in the $\mathbf{x}$-space. In particular, $\mathbf{g}_{kl}$ is now the same as defined by Eq. (4). We regard $\mathbf{g}_{kl}$ as the metric tensor of the $\mathbf{x}$-space and use it (instead of g_{kl}) to raise tensor indices in the $\mathbf{x}$-space. It is then easy to see that the tensor corresponding to $A_{k_1 k_2 \cdots} B^{\cdots l_1 l_2}$ is $\mathbf{A}_{k_1 k_2 \cdots} \mathbf{B}^{\cdots l_1 l_2}$ where the dots denote contracted indices. Thus every tensor equation can be taken over from the x-space to the $\mathbf{x}$-space by just using boldface type. However, a real transformation t of the x-space corresponds in general to a complex transformation of the $\mathbf{x}$-space. But it is well known that the necessary and sufficient condition for the invariance of a set of matrices β_k under G is that there should exist matrices $I_{kl} = -I_{lk}$ satisfying Eqs. (38) (now with the altered meaning of g_{kl}), and further that there be a non-singular matrix R such that

$$R\beta R^{-1} = J\beta, \tag{42}$$

where J is the transformation in G corresponding to the reversal of the direction of the x_1-axis. But these conditions are equivalent to

$$[\boldsymbol{\beta}_k, \mathbf{I}_{lm}] = \mathbf{g}_{kl}\boldsymbol{\beta}_m - \mathbf{g}_{km}\boldsymbol{\beta}_l, \tag{43a}$$

$$[\mathbf{I}_{kl}, \mathbf{I}_{mn}] = -\mathbf{g}_{km}\mathbf{I}_{ln} + \mathbf{g}_{lm}\mathbf{I}_{kn} + \mathbf{g}_{kn}\mathbf{I}_{lm} - \mathbf{g}_{ln}\mathbf{I}_{km} \tag{43b}$$

and

$$R\boldsymbol{\beta}R^{-1} = J\boldsymbol{\beta}, \tag{43c}$$

since $\mathbf{J} = J$. Equations (43), however, are just the conditions that the $\boldsymbol{\beta}$-matrices be invariant under

O. Thus it is sufficient to examine the invariance of β under O. This we have already done. Hence all our results in section 2 are still valid, provided that each tensor $A_{k_1}\cdots$ is replaced by $\mathbf{A}_{k_1}\cdots$. On going back from the $\mathbf{x}$-space to the x-space we have to revert to the normal type for the symbols. Therefore, all our conclusions which are expressed as tensor equations remain unaltered in form also when the new meaning is attributed to g_{kl}.

4. THE SPURS

The invariance of the β-representation under G leads to important restrictions on the spurs of the β-matrices and their products. Put

$$Sp(\beta_{k_1}\cdots\beta_{k_n}) = s_{k_1\cdots k_n}, \tag{44}$$

where clearly $s_{k_1\cdots k_n}$ are numerical quantities. Under the transformation $\beta \to \beta' = t\beta$, Eq. (44) goes over to

$$Sp(\beta_{k_1}'\cdots\beta_{k_n}') = t_{k_1}{}^{l_1}\cdots t_{k_n}{}^{l_n} s_{l_1\cdots l_n}.$$

But from Eq. (3a)

$$Sp(\beta_{k_1}'\cdots\beta_{k_n}')$$
$$= Sp(T\beta_{k_1}\cdots\beta_{k_n}T^{-1}) = Sp(\beta_{k_1}\cdots\beta_{k_n}),$$

so that

$$t_{k_1}{}^{l_1}\cdots t_{k_n}{}^{l_n} s_{l_1\cdots l_n} = s_{k_1\cdots k_n}.$$

From this we deduce as before that

$$s_{k_1\cdots k_n} = s(g)_{k_1\cdots k_n}. \tag{45}$$

Since $s_{k_1\cdots k_n}$ cannot be constructed from g_{kl} when n is odd, it follows that

$$Sp(\beta_{k_1}\cdots\beta_{k_{2n+1}}) = 0, \tag{46}$$

and in particular

$$Sp(\beta_k) = 0.$$

Now if β_k and $\beta_k{}^{\ddagger}$ are equivalent,

$$Sp(\beta_{k_1}\cdots\beta_{k_n}) = Sp(\beta_{k_n}{}^{\ddagger}\cdots\beta_{k_1}{}^{\ddagger}) = Sp(\beta_{k_n}\cdots\beta_{k_1})$$

i.e.,

$$s(g)_{k_1\cdots k_n} = s(g)_{k_n\cdots k_1}. \tag{47}$$

Also further, if β_k and $\beta_k{}^{\dagger}$ and, therefore, $\beta_k{}^*$ are equivalent, the numerical coefficients in Eq. (45) are all real. Conversely if β_k are an irreducible set of matrices satisfying Eqs. (45) and (47) with $s(g)_{k_1\cdots k_n} = s^*(g)_{k_1\cdots k_n}$, then the β-representation is invariant under G and β_k, $\beta_k{}^{\ddagger}$, $\beta_k{}^{\dagger}$, and $\beta_k{}^*$ are all equivalent. To prove this it is sufficient to show that if there exists a correspondence $\beta_k \to \beta_k'$

between two sets of *irreducible* matrices such that

$$Sp(\beta_{k_1}\cdots\beta_{k_n}) = Sp(\beta_{k_1}'\cdots\beta_{k_n}'),$$

then β_k and β_k' are equivalent. Consider the algebra $\mathfrak{A}(B)$ generated by the matrices B_k which are the direct sum of β_k and β_k' i.e.,

$$B_k = \begin{array}{|c|c|} \hline \beta_k & 0 \\ \hline 0 & \beta_k' \\ \hline \end{array}. \tag{48}$$

Clearly $\mathfrak{A}(B)$ is semi-simple, and therefore either it is simple or a direct sum of two or more simple algebras. In the first case it is capable of only one irreducible representation apart from equivalence. Since $B_k \to \beta_k$, and $B_k \to \beta_k'$ are both representations of $\mathfrak{A}(B)$ they must be equivalent. On the other hand, in the second case

$$\mathfrak{A}(B) = \mathfrak{A}_1(B) \oplus \mathfrak{A}_2(B) \oplus \cdots,$$

where $\mathfrak{A}_1, \mathfrak{A}_2, \cdots$ are all simple. In case β_k and β_k' are not equivalent, they are equivalent to the regular representations of two different simple sub-algebras, say $\mathfrak{A}_1$ and $\mathfrak{A}_2$, respectively. Consider the unity quantities $e_1(B)$ and $e_2(B)$ of $\mathfrak{A}_1$ and $\mathfrak{A}_2$, respectively. We have, $e_1 e_2 = 0$. In the β-representation e_1 is represented by the unit matrix and e_2 by the zero matrix, while in the β'-representation $e_1 = 0$ and $e_2 = 1$. Hence

$$Sp(e_1(\beta)) \neq Sp(e_1(\beta')), \tag{49}$$

where $e_1(\beta)$ and $e_1(\beta')$ are both obtained from $e_1(B)$ by replacing B_k by β_k and β_k', respectively. Equation (49) therefore contradicts the hypothesis. Hence β_k and β_k' are equivalent.

5. THE DIRAC AND THE DUFFIN-KEMMER ALGEBRAS

As an illustration of the above general theory, let us consider the algebra generated by the Dirac and the Duffin-Kemmer matrices in the general case of s-dimensions. It is well known that in both of these cases

$$I_{kl} = I(\beta_k\beta_l - \beta_l\beta_k)$$

where I is a constant. Therefore, as pointed out by Bhabha,[6] any representation of these algebras is a representation of the orthogonal group $O(s+1)$. Since every representation of the orthogonal group is fully reducible, it follows that every representation of the Dirac (or the Duffin-

Kemmer) algebra is also fully reducible. In particular, the regular representations of these algebras are fully reducible, and so the two algebras are semi-simple.

The Dirac matrices α_k are defined by the commutation rules[20]

$$\alpha_k\alpha_l+\alpha_l\alpha_k=2g_{kl} \tag{50}$$

so that

$$\alpha_k\alpha^k=s. \tag{51}$$

In any scalar quantity all the indices are contracted, and on using Eq. (50) the contracted indices can always be brought adjacent to each other. Therefore, from Eq. (51) every scalar quantity is a multiple of 1. In the same way it is easy to see that every pseudo-scalar quantity must be a multiple of

$$\omega=\frac{1}{s!}\epsilon^{k_1\cdots k_s}\alpha_{k_1}\cdots\alpha_{k_s}. \tag{52}$$

Now, ω commutes or anti-commutes with α_k according as s is odd or even. Clearly 1 and ω are linearly independent, since a pseudo-scalar cannot be linearly dependent on a scalar. Thus a basis for the center of the algebra consists of $(1,\omega)$ or 1 according as s is odd or even. Thus the algebra has one or two inequivalent irreducible representations according as s is even or odd. Hence, for s even, α, $\alpha^{\ddagger}$, $\alpha^{\dagger}$, α^* are all equivalent. For s odd, α and $J\alpha$ are inequivalent since $\omega^2=(-1)^{\frac{1}{2}s(s-1)}\neq0$. Also $\omega^T=(-1)^{\frac{1}{2}s(s-1)}\omega$. Hence α and $\alpha^{\ddagger}$ are equivalent only if $\frac{1}{2}s(s-1)$ is even, i.e., if s is of the form $4n+1$. In a particular representation ω is represented by $\pm\omega_0$ where $\omega_0=i^{\frac{1}{2}s(s-1)}$. Therefore, α and α^* are equivalent only if $\omega_0=\omega_0^*$, i.e., if $s=4n+1$. α and $\alpha^{\dagger}$ are always equivalent.

The Duffin-Kemmer matrices are defined by

$$\beta_k\beta_l\beta_m+\beta_m\beta_l\beta_k=g_{kl}\beta_m+g_{ml}\beta_k. \tag{53}$$

Put $\beta_k\beta^k=B$, so that

$$\beta_kB+B\beta_k=(s+1)\beta_k, \tag{54a}$$

that is,

$$\beta_kB=(s+1-B)\beta_k, \tag{54b}$$

$$\beta^k\beta_l\beta_k=\beta_l. \tag{55}$$

Two contracted indices in a scalar expression can, from Eq. (53), be brought either adjacent to each other, or immediately to the left and right of a

β_k, so that from Eqs. (54) and (55) every scalar is expressible as a polynomial $f(B)$ in B. Further, if $f(B)$ belongs to the center,

$$f(B)\beta_k=\beta_kf(s+1-B)=\beta_kf(B). \tag{56}$$

Since the algebra, being semi-simple, has a 1 which must be expressible as a scalar $e(\beta)$, it follows from Eq. (56) on multiplying on the left by $e^k(\beta)$, where $e^k(\beta)\beta_k\equiv e(\beta)$, that

$$f(s+1-B)=f(B)=\tfrac{1}{2}\{f(s+1-B)+f(B)\}.$$

Put

$$\tfrac{1}{2}(s+1)-B=\Delta \quad\text{and}\quad f\left(\frac{s+1}{2}-\Delta\right)\equiv\varphi(\Delta).$$

Then every scalar quantity of the center is of the form $\frac{1}{2}\varphi(\Delta)+\frac{1}{2}\varphi(-\Delta)\equiv\psi(\Delta^2)$ where $\psi(\Delta^2)$ is a polynomial in Δ^2. Now $\Delta^2=\frac{1}{4}(s+1)^2-B(s+1-B)$. Put

$$\theta=B(s+1-B). \tag{57}$$

Then every scalar quantity in the center is expressible as a polynomial in θ.

In a similar way, it follows that every pseudo-scalar quantity is of the form $\omega f(B)$, where f is a polynomial in B and

$$\omega=\epsilon^{k_1\cdots k_s}\beta_{k_1}\cdots\beta_{k_s}. \tag{58}$$

Now

$$\beta_{k_1}\omega=\epsilon^{\hat{k}_1k_2\cdots k_s}\{\beta\hat{k}_1^2\beta_{k_2}\cdots\beta_{k_s}-\beta\hat{k}_1\beta_{k_2}\beta_{k_1}\beta_{k_3}\cdots\beta_{k_s}$$
$$+\beta\hat{k}_1\beta_{k_2}\beta_{k_3}\beta\hat{k}_1\beta_{k_4}\cdots\beta_{k_s}$$
$$+\cdots+(-1)^s\beta\hat{k}_1\beta_{k_2}\cdots\beta_{k_s}\beta_{k_1}\}$$

where there is no summation over the index $\hat{k}_1$. Using Eq. (53) and remembering that $k_1, k_2, \cdots, k_s$ are all different, we get

$$\beta_{k_1}\omega=\left[\frac{s+1}{2}\right]\epsilon^{\hat{k}_1k_2\cdots k_s}\beta\hat{k}_1^2\beta_{k_2}\cdots\beta_{k_s}, \tag{59a}$$

where $[n]$ denotes the integral part of n. Similarly,

$$\omega\beta_{k_1}=\left[\frac{s+1}{2}\right]\epsilon^{k_2\cdots k_s\hat{k}_1}\beta_{k_2}\cdots\beta_{k_s}\beta_{k_1}^2. \tag{59b}$$

Remembering that $\beta_k^2\beta_l=\beta_l(1-\beta_k^2)$ for $k\neq l$, one finds from Eq. (59) that

$$\beta_k\omega-\omega\beta_k$$
$$=\begin{cases}0 & \text{if } s \text{ is odd,} \tag{60a}\\ \left[\dfrac{s+1}{2}\right]\epsilon^{k_2\cdots k_s}\beta_{k_2}\cdots\beta_{k_s}, & \text{if } s \text{ is even.} \tag{60b}\end{cases}$$

---

[20] For convenience we take g_{kl} as defined in Eq. (4).

Also, from Eq. (59b),

$$\omega\beta k_1{}^2 = \left[\frac{s+1}{2}\right]\epsilon^{k_2\cdots k_s\hat{k}_1}\beta k_2\cdots\beta k_s\beta\hat{k}_1.$$

Hence

$$\omega B = \sum_{k_1}\omega\beta k_1{}^2 = \left[\frac{s+1}{2}\right]\omega. \tag{61}$$

Therefore any pseudo-scalar quantity $\omega f(B)$ is really a numerical multiple of ω. Now it is clear from Eqs. (60) that ω belongs to the center only if s is odd. Hence the center contains pseudo-scalar elements only if s is odd and then they are numerical multiples of ω.

Consider the expression

$$B_r = \sum_{(k_1\cdots k_r)}\beta k_1{}^2\cdots\beta k_r{}^2 \tag{62}$$

where the summation is to be taken over all possible values of $k_1, \cdots, k_r$ such that no two of the k's are the same. Notice that

$$B_r B = \sum_{(k_1\cdots k_r)}\beta k_1{}^2\cdots\beta k_r{}^2\sum_{k_{r+1}}\beta k_{r+1}{}^2.$$

Now $\beta_k{}^2$ and $\beta_l{}^2$ always commute, and $\beta_k{}^4 = \beta_k{}^2$. Hence the right side is

$$\sum_{(k_1\cdots k_{r+1})}\beta k_1{}^2\cdots\beta k_r{}^2\beta k_{r+1}{}^2$$
$$+r\sum_{(k_1\cdots k_r)}\beta k_1{}^2\cdots\beta k_r{}^2,$$

so that

$$B_{r+1} = B_r(B-r).$$

But $B_1 = B$ and, therefore,

$$B_r = B(B-1)\cdots(B-r+1). \tag{63}$$

Clearly, $B_{s+1} = 0$ since $k_1, \cdots, k_{s+1}$ cannot all be different. Hence

$$B(B-1)\cdots(B-s) = 0. \tag{64a}$$

Notice that $B = B_1, B_2, \cdots, B_s$ are all linearly independent and so $B, B^2, \cdots, B^s$ are also linearly independent over $\mathfrak{F}$. Put $f(B) = B(B-1)\cdots(B-s)$. Then $f(B) = 0$ and so belongs to the center. Therefore, $f(B) = f(s+1-B) = 0$; i.e.,

$$(s+1-B)(s-B)\cdots(1-B)$$
$$= (-1)^{s+1}(B-1)\cdots(B-s)$$
$$\times(B-(s+1)) = 0. \tag{64b}$$

Hence from Eqs. (64a) and (64b),

$$(B-1)(B-2)\cdots(B-s) = 0. \tag{65}$$

Equation (65) is the minimum equation for B, since otherwise $B, B^2, \cdots, B^s$ would not be linearly independent. From Eqs. (65) and (57) it follows that the minimum equation for θ is

$$(\theta-1\cdot s)(\theta-2(s-1))\cdots$$
$$\times(\theta-[\tfrac{1}{2}(s+1)][\tfrac{1}{2}(s+2)]) = 0. \tag{66}$$

Hence $1, \theta, \theta^2, \cdots, \theta[\tfrac{1}{2}(s+1)]-1$ form a linearly independent complete scalar set in the center. When s is even, this set is a basis for the center. On the other hand, when s is odd, ω must be added to it to obtain a complete basis. Therefore the dimension of the center over $\mathfrak{F}$ is $[\tfrac{1}{2}(s+1)]$ or $[\tfrac{1}{2}(s+1)]+1$, according as s is even or odd. Hence the algebra has $\tfrac{1}{2}s$ or $\tfrac{1}{2}(s+3)$ irreducible inequivalent representation, according as s is even or odd.

From Eq. (61) it follows that ω is zero in all representations except those in which B can have the eigenvalue $[\tfrac{1}{2}(s+1)]$, i.e., in which θ has the eigenvalue $[\tfrac{1}{2}(s+1)][\tfrac{1}{2}(s+2)]$. But θ is in the center and so

$$\theta = \left[\frac{s+1}{2}\right]\left[\frac{s+2}{2}\right] \equiv \theta_0$$

for this representation.

Proceeding as in Eq. (59a), one easily finds that

$$\beta k_2\beta k_1\omega = \left[\frac{s+1}{2}\right]\left[\frac{s}{2}\right]\epsilon^{k_1k_2k_3\cdots k_s}$$
$$\times(1-\beta k_1{}^2)\beta k_2{}^2\beta k_3\cdots\beta k_s, \quad (k_1\neq k_2),$$

and, similarly,

$$\beta k_s\beta k_{s-1}\cdots\beta k_1\omega = \left[\frac{s+1}{2}\right]\left[\frac{s}{2}\right]\cdots$$
$$\times\left[\frac{2}{2}\right]\epsilon^{k_1\cdots k_s}\beta k_s{}^2(1-\beta k_{s-1}{}^2)\beta k_{s-2}{}^2(1-\beta k_{s-3}{}^2)\cdots,$$

where $k_1, \cdots, k_s$ are all different. Therefore

$$\omega^2 = (-1)^{[\frac{1}{2}s(s-1)]}\left[\frac{s+1}{2}\right]\cdots\left[\frac{2}{2}\right]$$
$$\times P_{(k_1\cdots k_s)}\beta k_s{}^2\beta k_{s-2}{}^2\cdots(1-\beta k_{s-1}{}^2)(1-\beta k_{s-3}{}^2)\cdots,$$
$$\tag{67}$$

where $P_{(k_1\cdots k_s)}$ denotes a sum over all permutations of $(1, 2, \cdots s)$. It is easy to see from deduce that Eqs. (67) and (63) that

$$\omega^2 = (-1)^{[\frac{1}{2}s(s-1)]}\left[\frac{s+1}{2}\right]\left[\frac{s}{2}\right]\cdots\left[\frac{2}{2}\right]$$

$$\times\left\{B_{[\frac{1}{2}(s+1)]} - \left[\frac{s}{2}\right]B_{[\frac{1}{2}(s+3)]} + \frac{\left[\frac{s}{2}\right]\left[\frac{s-2}{2}\right]}{2!}B_{[\frac{1}{2}(s+5)]}\right.$$

$$- \frac{\left[\frac{s}{2}\right]\left[\frac{s-2}{2}\right]\left[\frac{s-4}{2}\right]}{3!}B_{[\frac{1}{2}(s+7)]}$$

$$\left. + \cdots + (-1)^{[\frac{1}{2}s]}B_s\right\}. \quad (68)$$

Using Eqs. (63) and (61), it follows that

$$\omega^3 = (-1)^{[\frac{1}{2}s(s-1)]}\left\{\left[\frac{s+1}{2}\right]\left[\frac{s}{2}\right]\cdots\left[\frac{2}{2}\right]\right\}^2 \omega. \quad (69)$$

Now ω^2 is a scalar and hence a polynomial in B, as is also clear from Eq. (68). Hence ω^2 can be written in the form

$$\omega^2 = P(\theta) + Q(\theta)\cdot B$$

where P and Q are polynomials in θ. Since ω^2 vanishes for all representations except those for which $\theta = \theta_0$, it follows that

$$\omega^2 = \phi(\theta)\{p + qB\}, \quad (70)$$

where

$$\phi(x) \equiv (x - 1\cdot s)\cdots\left(x - \left[\frac{s-1}{2}\right]\left[\frac{s+4}{2}\right]\right).$$

In view of Eq. (66), p and q can be taken to be constants. From Eqs. (61), (70), and (66) one finds that

$$\omega^4 = \phi(\theta_0)\left(p + q\left[\frac{s+1}{2}\right]\right)\omega^2, \quad (71)$$

while on multiplying Eq. (70) by $(s+1-B)$ one gets

$$\omega^2\left[\frac{s+2}{2}\right] = \phi(\theta)\{p(s+1) + q\theta_0 - pB\}. \quad (72)$$

From Eqs. (69), (70), (71), and (72) it is easy to

$$\omega^2 = \begin{cases} (-1)^{[\frac{1}{2}s(s-1)]}\left(\frac{1}{2}s!\right)^4 \dfrac{\phi(\theta)}{\phi(\theta_0)} \\ \qquad \times\left\{\dfrac{s}{2} + 1 - B\right\}, \quad (s\text{ even}), \quad (73a) \\[2em] (-1)^{[\frac{1}{2}s(s-1)]}\left(\dfrac{s+1}{2}\right)^2 \\ \qquad \times\left(\tfrac{1}{2}(s-1)!\right)^4 \dfrac{\phi(\theta)}{\phi(\theta_0)}, \quad (s\text{ odd}). \quad (73b) \end{cases}$$

Now notice that it follows from Eq. (59a) that

$$\beta_{k_1}\omega\beta_{k_1} = \left[\frac{s+1}{2}\right]\epsilon^{k_1k_2\cdots k_s}\beta_{k_1}{}^2\beta_{k_2}\cdots\beta_{k_s}\beta_{k_1}$$

$$= \left[\frac{s+1}{2}\right]\epsilon^{k_1k_2\cdots k_s}\beta_{k_2}\cdots\beta_{k_s}(1 - \beta_{k_1}{}^2)\beta_{k_1}$$

$$= 0, \quad (s\text{ even}). \quad (74)$$

It will presently be shown in Section 6 that Eq. (74) necessarily implies

$$\beta_k\omega\beta_l + \beta_l\omega\beta_k = 0, \quad (s\text{ even}). \quad (75)$$

Now consider $\omega\beta_k\omega$. It is a vector and hence must be expressible in the form $\beta_k f(B)$ where $f(B)$ is a polynomial in B. Then

$$\beta_k{}^2\omega\beta_k\omega = \beta_k{}^3 f(B) = \beta_k f(B) = \omega\beta_k\omega.$$

But the left side is zero from Eq. (74), when s is even. Hence

$$\omega\beta_k\omega = 0, \quad (s\text{ even}). \quad (76)$$

Further it is easy to see from Eq. (61) that

$$\beta_{k_1}{}^2\omega + \omega\beta_{k_1}{}^2 = \omega, \quad (s\text{ even}).$$

As shown in Section 6, this implies that

$$(\beta_k\beta_l + \beta_l\beta_k)\omega + \omega(\beta_k\beta_l + \beta_l\beta_k) = 2\omega g_{kl}, \quad (s\text{ even}).$$

But ω is a pseudo-scalar and therefore commutes with I_{kl} i.e.,

$$(\beta_k\beta_l - \beta_l\beta_k)\omega - \omega(\beta_k\beta_l - \beta_l\beta_k) = 0.$$

Adding the two equations, one gets

$$\beta_k\beta_l\omega + \omega\beta_l\beta_k = \omega g_{kl}, \quad (s\text{ even}). \quad (77)$$

Moreover, Eq. (73a) gives

$$\beta_k\omega^2 + \omega^2\beta_k = (-1)^{[\frac{1}{2}s(s-1)]}\left(\tfrac{1}{2}s!\right)^4\frac{\phi(\theta)}{\phi(\theta_0)}\beta_k. \quad (78)$$

Put

$$\omega_0 = \frac{i^{[\frac{1}{2}s(s-1)]}}{\left[\dfrac{s+1}{2}\right]\cdots\left[\dfrac{2}{2}\right]}\cdot\omega.$$

Then $\omega_0{}^3=\omega_0$. The only irreducible representations in which $\omega_0\neq0$ are those in which $\theta=\theta_0$. In these representations

$$\omega_0{}^2=\begin{cases}\tfrac{1}{2}s+1-B & (s\text{ even}), & (79a)\\[4pt] 1 & (s\text{ odd}). & (79b)\end{cases}$$

In view of Eqs. (75), (76), (77), and (78), one can put $\beta_{s+1}=\omega_0$, (s even) for the representation $\theta=\theta_0$, and allow the indices to run from 1 to $s+1$ in Eq. (53). Also notice that for this representation the matrices γ_k defined by

$$\gamma_k=\beta_k\omega_0+\omega_0\beta_k \tag{80}$$

satisfy the same commutation rules as Eq. (53). In fact

$$\gamma_k=(1+\omega_0-\omega_0{}^2)\beta_k(1+\omega_0-\omega_0{}^2); \tag{81}$$

and since $(1+\omega_0-\omega_0{}^2)^2=1$, γ and β are equivalent. Since $1+\omega_0-\omega_0{}^2$ is a pseudo-scalar quantity it commutes with I_{kl}. Also clearly it commutes with B and ω_0; therefore

$$\gamma_k\gamma_l-\gamma_l\gamma_k=\beta_k\beta_l-\beta_l\beta_k,\quad \gamma^k\gamma_k=\beta^k\beta_k,$$
$$\epsilon^{k_1\cdots k_s}\gamma_{k_1}\cdots\gamma_{k_s}=\epsilon^{k_1\cdots k_s}\beta_{k_1}\cdots\beta_{k_s}.$$

Since $\theta=\theta^*=\theta^T$ it follows that for even s, any irreducible β-representation is equivalent to $J\beta$, β^*, $\beta^\ddagger$, $\beta^\dagger$. For s odd, β, $J\beta$, β^*, $\beta^\ddagger$, $\beta^\dagger$ are equivalent for the representations for which $\theta\neq\theta_0$, $\omega=0$. In the other two $\theta=\theta_0$,

$$\omega=\pm i^{[\frac{1}{2}s(s-1)]}\tfrac{1}{2}(s+1)(\tfrac{1}{2}(s-1)!)^2.$$

Clearly here β and $J\beta$ are inequivalent. β and β^* are equivalent only when s is of the form $4n+1$. The same holds for β and $\beta^\ddagger$, while β and $\beta^\dagger$ are always equivalent.

6. PHYSICAL APPLICATION

We now come to the physical application of the above abstract theory. The tensor indices now run from 0 to 3 only and $g_{00}=-g_{11}=-g_{22}=-g_{33}=1$. Consider the wave equation (1). Let the minimum equation for β_0 be

$$(\beta_0)^n+a_1(\beta_0)^{n-1}+a_2(\beta_0)^{n-2}+\cdots+a_n=0, \tag{82}$$

with a's in $\mathfrak{F}$. Write it in the form

$$(\beta_0)^n+g_{00}a_2(\beta_0)^{n-2}+(g_{00})^2a_4(\beta_0)^{n-4}+\cdots$$
$$+\{a_1(\beta_0)^{n-1}+g_{00}a_3(\beta_0)^{n-3}+\cdots\}=0.$$

Since β and $t\beta$ are equivalent, t now being a Lorentz transformation, it follows that $\beta_0'=t_0{}^k\beta_k$ has the same minimum equation. Put $t_0{}^k=y^k$. Then

$$y^ky_k=t_0{}^kt_0{}^lg_{kl}=g_{00}=1 \tag{83}$$

and $\beta_0'=y^k\beta_k$. Therefore

$$\{(y^l\beta_l)^n+a_2y^{k_1}y^{k_2}g_{k_1k_2}(y^l\beta_l)^{n-2}$$
$$+a_4(y^{k_1}y^{k_2}g_{k_1k_2})^2(y^l\beta_l)^{n-4}+\cdots\}$$
$$+\{a_1(y^l\beta_l)^{n-1}+a_3y^{k_1}y^{k_2}g_{k_1k_2}(y^l\beta_l)^{n-3}+\cdots\}=0.$$

Now the y's are arbitrary except for Eq. (83). Hence put

$$y^l=\frac{z^l}{(g_{mn}z^mz^n)^{\frac{1}{2}}}\equiv\frac{z^l}{Z},$$

where the z's are independent indeterminates. Then on multiplying by Z^n one finds

$$\{(z^l\beta_l)^n+a_2z^{k_1}z^{k_2}g_{k_1k_2}(z^l\beta_l)^{n-2}+\cdots\}$$
$$+Z\{a_1(z^l\beta_l)^{n-1}+\cdots\}=0. \tag{84}$$

This must be an identity in z. Therefore from the irrationality of Z in z it follows that

$$a_1(z^l\beta_l)^{n-1}+\cdots=0.$$

Put $z^0=1$, $z^l=0$, $l\neq0$. Then

$$a_1(\beta_0)^{n-1}+\cdots=0.$$

If $a_1,\ a_3,\ \cdots$ are not all zero, this is an equation, of a degree lower than the minimum equation. Since this contradicts the definition of the minimum equation, it follows that $a_1=a_3=a_5=\cdots=0$. Hence only the terms in the first bracket in Eq. (84) remain. Equating coefficients one gets

$$P_{(k_1\cdots k_n)}(\beta_{k_1}\cdots\beta_{k_n}+a_2g_{k_1k_2}\beta_{k_3}\cdots\beta_{k_n}$$
$$+a_4g_{k_1k_2}g_{k_3k_4}\beta_{k_5}\cdots\beta_{k_n}+\cdots)=0, \tag{85}$$

where $P_{(k_1\cdots k_n)}$ denotes a sum over all permutations of $k_1\cdots k_n$. From Eq. (85) it follows that

$$\{(\partial^l\beta_l)^n+a_2\partial^k\partial_k(\partial^l\beta_l)^{n-2}$$
$$+a_4(\partial^k\partial_k)^2(\partial^l\beta_l)^{n-4}+\cdots\}\psi=0.$$

On making use of Eq. (1), this can be written as

$$\{(i\chi)^n+a_2(i\chi)^{n-2}\partial^k\partial_k$$
$$+a_4(i\chi)^{n-4}(\partial^k\partial_k)^2+\cdots\}\psi=0. \tag{86}$$

Equation (86) is the differential equation (with numerical coefficients) of the lowest order which the most general solution ψ of Eq. (1) would satisfy if no subsidiary conditions were imposed. Now, if we demand that (86) be the usual second-order equation, we must have[21] $a_4 = a_6 = \cdots = 0$ and $a_2 = -1$. Equation (85) then becomes

$$P_{(k_1 \cdots k_n)}(\beta_{k_1} \cdots \beta_{k_n} - g_{k_1 k_2}\beta_{k_3} \cdots \beta_{k_n}) = 0. \quad (87)$$

Equation (87) is completely equivalent to

$$(z^l \beta_l)^n = z^k z_k (z^l \beta_l)^{n-2}, \quad (88)$$

where z's are indeterminates. Equation (88) is the natural generalization of the well-known cases of the Dirac and the Duffin-Kemmer matrices, for which

$$(z^l \beta_l)^2 = z^k z_k$$

and

$$(z^l \beta_l)^3 = z^k z_k (z^l \beta_l),$$

respectively.

Since Eq. (87) is already in a proper tensor form, one might at first sight think that the invariance of an irreducible representation under the proper Lorentz group is insured. This, however, is in general not the case, since for $n \geqslant 3$ Eq. (87) does *not* generate a *finite* algebra. Some other stronger tensor condition compatible with Eq. (87) is needed to make the algebra finite. An example of such a condition for $n = 3$ is provided by the Duffin commutation rules

$$\beta_k \beta_l \beta_m + \beta_m \beta_l \beta_k - g_{kl}\beta_m - g_{ml}\beta_k = 0.$$

In the Dirac and the Duffin-Kemmer case it is possible to choose a representation in which β's are Hermitian (i.e., $\beta_0 = \beta_0^\dagger$ and $\beta_k = -\beta_k^\dagger$, $k = 1, 2, 3$). But this is clearly impossible when $n > 3$, since if β_0 is Hermitian it can be brought to the diagonal from. Now, its minimum equation is

$$\beta_0^{n-2}(\beta_0^2 - 1) = 0, \quad (89)$$

and hence its eigenvalues are ± 1, 0. But then,

since β_0 is diagonal, it clearly satisfies the equation

$$\beta_0(\beta_0^2 - 1) = 0.$$

Hence $n \leqslant 3$.

I have not been able to establish the existence of an irreducible relativistically invariant set β_k satisfying Eq. (87) for $n > 3$, but not for $n \leqslant 3$.[22] All the obvious commutation rules which are compatible with Eq. (87) and with the general restrictions laid down in Section 3, and which make the resulting algebra finite, either turn out to be inconsistent or are found, on closer investigation, to degenerate into the already known cases of the Dirac or the Duffin-Kemmer rules. On the other hand, there seems to be no obvious way of proving that such a representation is impossible for $n > 3$. If it be true that no such β-representation could exist for $n > 3$, our results should be interpreted to mean that no higher spins are possible.

It is comparatively easy to find an explicit expression for β_k with $n = 4$; e.g.,

$$\beta_k = \frac{1}{\sqrt{2}}\{\alpha_k(\tfrac{3}{2})X\omega + i \cdot 1X\alpha_k(\tfrac{1}{2})\}$$

where $\alpha_k(\tfrac{3}{2})$ and $\alpha_k(\tfrac{1}{2})$ are matrices constructed in the way proposed by Bhabha,[8] which satisfy the minimum equations

$$(\alpha_0^2 - \tfrac{1}{4})(\alpha_0^2 - 9/4) = 0, \quad \text{and} \quad (\alpha_0^2 - \tfrac{1}{4}) = 0,$$

respectively. ω is uniquely[23] defined by $\omega^2 = 1$ and

$$\alpha_k(\tfrac{1}{2})\omega + \omega\alpha_k(\tfrac{1}{2}) = 0.$$

$\alpha_k(\tfrac{1}{2})$ are really $\tfrac{1}{2}$ times the usual Dirac matrices. X denotes Kronecker product. It is easy to see that this representation is relativistically invariant (including reflection) and is equivalent to its Hermitian conjugate. But since it is not certain that it is irreducible, one cannot rule out the possibility that the irreducible part may satisfy an equation such as (89) with $n \leqslant 3$.

---

[21] It is assumed here that $\chi \neq 0$, so that one can divide out $(i\chi)^{n-2}$.

[22] Since the manuscript was sent to the press, one such representation has been found corresponding to $n = 4$.

[23] Apart from ambiguity in sign, which is immaterial.

Equations for Particles of Higher Spin

Harish-Chandra

Abstract. The theory of elementary particles based on a linear wave equation with matrix coefficients is developed. The physical assumptions regarding such a particle are translated into algebraic restrictions on these coefficient matrices. The theory is then illustrated by considering a special case, which corresponds to a particle with spin values $\frac{3}{2}$ and $\frac{1}{2}$.

Consider the equation

$$i\gamma^k \partial_k \psi + \chi\beta\psi = 0, \tag{1}$$

where $\gamma^k (k = 0, 1, 2, 3)$ and β are square matrices operating on the 1-column matrix ψ, and χ is a constant $\neq 0$. Without loss of generality β may be assumed to be an idempotent matrix so that $\beta^2 = \beta$. If the set of matrices* γ_k and β is decomposable,[†] then (1) is equivalent to two independent equations of the same form, each of which is simpler than (1), since the corresponding representation space is of a lower dimension. Hence the simplest possible equations of the above type are those for which the set (γ_k, β) is indecomposable. Among these, two essentially different cases arise according as $\beta = 1$ or is singular.

Now it is well known that the wave equations of all the elementary particles discovered so far can be written in the above form, and in all cases except that of the photon (Harish-Chandra, 1946a) $\beta = 1$. It therefore seems worth while to try to build a general theory of elementary particles based on (1). Clearly if the particle is to be elementary the set (γ_k, β) must be indecomposable. We shall, however, make the stronger assumption that it is then irreducible. Conversely the irreducibility of the set (γ_k, β) may be regarded as the mathematical criterion for the "elementary" character of the corresponding particle.

---

*The indices are raised and lowered in the usual way by means of the metric tensor $g_{kl}(g_{kl} = 0, k \neq l;$ $g_{00} = -g_{11} = -g_{22} = -g_{33} = 1)$.

[†]A vector space is said to be decomposable under a given set of matrices if it can be expressed as a direct sum of two sub-spaces (neither of them 0), each of which is invariant under the given set. In the same circumstances the set of matrices itself is said to be decomposable. Decomposability implies reducibility, but not conversely.

In this paper only the case when $\beta = 1$ will be discussed. Further, we make the following three physical assumptions concerning the particle:

(i)   that the wave equation (1) is relativistically invariant;

(ii)   that the particle satisfies the second order equation

$$\left(\partial_k \partial^k + \chi^2 \right) \psi = 0; \tag{2}$$

(iii)   that either the total charge or the total energy of the field, associated with the particle, is positive definite.

(iii) is necessary if the field is to be quantized. The consequences of (i) and (ii) have been discussed in an earlier paper (Harish-Chandra, 1946b); as shown there, it follows from (2) that the minimum equation of γ_0 must be of the form

$$\gamma_0^{n-2}\left(\gamma_0^2 - 1 \right) = 0. \tag{3}$$

Also, it seems essential for constructing an energy-momentum tensor and a current vector in accordance with the usual principles of quantum mechanics, that the γ_k's should be equivalent to their Hermitian conjugates, i.e. there should exist a non-singular matrix Λ such that

$$\gamma_k^\dagger = \Lambda \gamma_k \Lambda^{-1} \tag{4a}$$

$$\Lambda^\dagger = \Lambda. \tag{4b}$$

(Here $\dagger$ denotes Hermitian conjugate.) (4) enables one to derive (1) by variation of the Lagrangian

$$L = \frac{1}{2i}\left(\partial_k \psi^* \gamma^k \psi - \psi^* \gamma^k \partial_k \psi \right) + \chi \psi^* \psi, \tag{5}$$

where $\psi^* = \psi^\dagger \Lambda$. The current vector s_k and the energy-momentum tensor T_{ki}, derived from (5) in the usual way, are

$$s_k = e \psi^* \gamma_k \psi, \tag{6a}$$

$$T_{kl} = \frac{i}{2}\left(\partial_k \psi^* \gamma_l \psi - \psi^* \gamma_l \partial_k \psi \right). \tag{6b}$$

Using (6), it can be shown that (iii) is equivalent to the condition that either $\Lambda \gamma_0^{m+1}$ or $\Lambda \gamma_0^m$ (where m is the greatest even integer $\leqslant n-1$) be a non-negative[†] matrix. Assuming, therefore, that it is so, quantization can be introduced through the following commutation rules:

$$\left[\psi_\rho^*(x), \psi_\sigma(x') \right]_\pm = \frac{1}{2}\left[\left(\frac{\gamma^k \partial_k}{i\chi} \right)^m \left(\frac{\gamma^k \partial_k}{i\chi} + 1 \right) D(x - x') \right]_{\sigma\rho} \tag{7}$$

Here $[A, B]_\pm = AB \pm BA$ and $+$ or $-$ sign is to be taken according as $\Lambda \gamma_0^{m+1}$ or $\Lambda \gamma_0^m$ is non-negative. σ and ρ are matrix indices and $D(x - x')$ is the well-known generalization of Pauli and Jordans' delta function for the case $\chi \neq 0$ (see Pauli, 1940).

[†]A Hermitian matrix is called non-negative if all its eigenvalues, which are necessarily real, are $\geqslant 0$.

Electromagnetic interaction can be introduced by replacing ∂_k by $\partial_k = \partial_k - ie\phi_k$ in (5) and (1). Here ϕ_k are the electromagnetic potentials. It is then found that in the non-relativistic approximation the particle manifests only a magnetic moment given by

$$\frac{e}{\chi}\left[P_+ \gamma \cdot \left(\frac{1-P_+}{1-\gamma_0} \right) \gamma P_+ \right], \tag{8}$$

where γ is the three-dimensional vector $(\gamma_1, \gamma_2, \gamma_3)$ and the brackets denote the usual three-dimensional vector product;

$$P_+ = \frac{\gamma_0^m(1+\gamma_0)}{2} \quad \text{and} \quad \frac{1-P_+}{1-\gamma_0} \equiv 1 + \gamma_0 + \cdots + \gamma_0^{m-1} + \frac{\gamma_0^m}{2}.$$

As an application of the above theory, consider the γ-representation given by

$$\gamma_k = \alpha_k + i\omega\beta_k, \tag{9}$$

where α_k are the Dirac matrices and β_k the Duffin-Kemmer matrices defined respectively by the commutation rules

$$\alpha_k\alpha_l + \alpha_l\alpha_k = 2g_{kl}, \tag{10a}$$

$$\beta_k\beta_l\beta_m + \beta_m\beta_l\beta_k = g_{kl}\beta_m + g_{ml}\beta_k, \tag{10b}$$

and $\omega = (1/i)\alpha_0\alpha_1\alpha_2\alpha_3$. The Kronecker product of the α- and β-representations is implied in (9). It can be shown that the γ-representation so defined is irreducible provided the corresponding β-representation is chosen to be so. Clearly it is relativistically invariant. Also, the minimum equation for γ_0 is

$$\gamma_0^2(\gamma_0^2 - 1) = 0, \tag{11}$$

and $\Lambda = \Lambda_\alpha \cdot \Lambda_\beta$ where Λ_α and Λ_β are Hermitian matrices in the α- and β-spaces respectively, such that

$$\alpha_k^\dagger = \Lambda_\alpha \alpha_k \Lambda_\alpha^{-1}, \tag{12a}$$

$$\beta_k^\dagger = \Lambda_\beta \beta_k \Lambda_\beta^{-1}. \tag{12b}$$

Choosing a representation in which α_0 and β_0 are Hermitian and α_k, $\beta_k(k = 1, 2, 3)$ are anti-Hermitian, so that $\Lambda_\alpha = \alpha_0$ and $\Lambda_\beta = 1 - 2\beta_0^2$, one finds that

$$\Lambda\gamma_0^3 = \alpha_0(1 - 2\beta_0^2)(1 - \beta_0^2)\alpha_0 = 1 - \beta_0^2$$

is a non-negative matrix. All the necessary requirements of our theory are therefore satisfied, and the corresponding equation describes an elementary particle.

Putting $\pi_k = (1/i)\partial_k - e\phi_k$, the equation of motion in the presence of an electromagnetic field can be written as

$$\gamma^k\pi_k\psi = \chi\psi. \tag{13}$$

From (13) one can deduce that

$$\pi_m \beta^m \psi = \frac{ie}{2\chi^2} F_{kl} \{ 3\beta^k g^{lm} - \beta^k \beta^m \beta^l - 2i\omega \alpha^k \beta^m \beta^l \} \pi_m \psi$$

$$+ \frac{ie}{2\chi^2} F_{kl} (\alpha^k \alpha^l - 2\beta^k \beta^l) \pi_m \beta^m \psi$$

$$+ \frac{e}{2\chi^2} \partial_k F_{lm} \cdot (\beta^k \alpha^l \alpha^m - 2i\omega \beta^k \alpha^l \beta^m - \beta^m g^{kl} + \beta^m \beta^k \beta^l) \psi, \qquad (14)$$

where $F_{kl} = \partial_k \phi_l - \partial_l \phi_k$. (14) shows that $\pi_m \beta^m \psi$ vanishes in the absence of interaction and may therefore be treated as a small quantity. The current vector s_k is given by

$$s^k = e\psi^* \gamma^k \psi = \frac{e}{2\chi i} (\psi^* \partial^{-k} \psi - \partial^{+k} \psi^* \cdot \psi)$$

$$+ \frac{e}{4\chi i} \cdot \partial_l \{ \psi^* (\alpha^k \alpha^l - \alpha^l \alpha^k - 2i\omega \alpha^k \beta^l + 2i\omega \alpha^l \beta^k) \psi \}$$

$$- \frac{e}{2\chi} \{ \psi^* \beta^k \beta^l \pi_l \psi + \psi^* \pi_l^\dagger \beta^l \beta^k \psi \}. \qquad (15)$$

Here $\partial_k^+ = \partial_k + ie\phi_k$ and $\psi^* \pi_l^\dagger = -(1/i) \partial_l^+ \psi^*$. The first term in (15) corresponds to the usual orbital current. The second term may be interpreted as the current due to a dipole density M^{kl} given by

$$M^{kl} = \frac{e}{\chi} \psi^* \left\{ \frac{1}{4i} (\alpha^k \alpha^l - \alpha^l \alpha^k) - \frac{\omega}{2} (\alpha^k \beta^l - \alpha^l \beta^k) \right\} \psi.$$

On account of (14) the third term is proportional to e^2 and must be interpreted as the polarization current (cf. Kemmer, 1939). It vanishes in the absence of the external field.

Corresponding to the α- and β-spaces the particle has got two independent spins. Of these, one has the value $\frac{1}{2}$, while the possible values of the other are 0 and 1 or 1 according as the 10-row or the 5-row representation of β_k is used. In either case the total spin of the particle may have the values $\frac{3}{2}$ and $\frac{1}{2}$. In order to describe the state of the free particle completely in the rest system, one must specify not only the sign of the energy and the magnitude and direction of the total spin, but also the angle between the two component spins, i.e. whether they are pointed along or opposite to each other. In this sense, therefore, the particle possesses an inner structure beyond the spin. In the non-relativistic approximation the equation of motion of the particle in an electromagnetic field is the same as that of the Dirac electron. This is due to the fact that in this approximation the electromagnetic field acts only on one of the spins, namely the α-spin. However, on account of the existence of the β-spin, each nonrelativistic state with a given direction of the α-spin (i.e. of the non-relativistic magnetic moment) is degenerate, and, therefore, the Pauli exclusion principle would be apparently inoperative for such particles in the non-relativistic approximation until all the degenerate levels are filled up. This difference can, therefore, be detected in the case of systems consisting of more than one particle.

One can go over from the above "particle formulation" to the "wave formulation" (cf. Kemmer, 1939) by using the Γ-formalism developed in two previous papers (Harish-Chandra, 1946b,c). Using Greek letters to denote spinor indices, the field equations may be written as

$$\partial_k^- G_{kl,\lambda} + \chi^2 U_{l,\lambda} = \chi \partial_{\lambda\mu}^- U_l^\mu, \tag{16a}$$

$$\partial_k^- U_{l,\lambda} - \partial_l^- U_{k,\lambda} + G_{kl,\lambda} = \frac{1}{\chi} \partial_\lambda^{-\mu} G_{kl,\mu}, \tag{16b}$$

$$\partial_k^- G_{kl,\mu} + \chi^2 U_{l,\mu} = \chi \partial_{\mu\lambda}^- U_l^\lambda, \tag{16c}$$

$$\partial_k^- U_{l,\mu} - \partial_l^- U_{k,\mu} + G_{kl,\mu} = \frac{1}{\chi} \partial_\mu^{-\lambda} G_{kl,\lambda}, \tag{16d}$$

in the case of the 10-row representation of $\beta_k \cdot \partial_{\lambda\mu}^- = \sigma_{\lambda\mu}^k \hat{\partial}_k$, where the σ-symbols have their usual meaning. For the 5-row representation of β_k the field equations similarly turn out to be

$$\partial_k^- U_\lambda + U_{k,\lambda} = \frac{1}{\chi} \partial_{\lambda\mu}^- U_k^\mu, \tag{17a}$$

$$\partial_k^- U_\lambda^k + \chi^2 U_\lambda = \frac{1}{\chi} \partial_\lambda^{-\mu} U_\mu, \tag{17b}$$

$$\partial_k^- U_\mu + U_{k,\mu} = \frac{1}{\chi} \partial_{\lambda\mu}^- U_k^\lambda, \tag{17c}$$

$$\partial_k^- U_\mu^k + \chi^2 U_\mu = \chi \partial_\mu^{-\lambda} U_\lambda. \tag{17d}$$

The full details of this investigation will be published later elsewhere.

REFERENCES

Harish-Chandra, 1946a. *Proc. Roy. Soc.*, A, **186**, 502.
Harish-Chandra, 1946b. *Phys. Rev.*, **71**, 793.
Harish-Chandra, 1946c. *Proc. Camb. Phil. Soc.*, **43**, 415.
Kemmer, 1939, *Proc. Roy. Soc.*, A, **173**, 91.
Pauli, 1940. *Phys. Rev.*, **58**, 716.

Reprinted from
Report of an International Conference on
Fundamental Particles and Low Temperatures
held at the Cavendish Laboratory, Cambridge,
on 22–27 July 1946. Volume 1, Fundamental Particles.
The Physical Society, London. pp. 185–188

Infinite irreducible representations of the Lorentz group

By Harish-Chandra, *Gonville and Caius College, Cambridge*

(*Communicated by P. A. M. Dirac, F.R.S.—Received* 8 *August* 1946)

It is shown that corresponding to every pair of complex numbers κ, κ^* for which $2(\kappa-\kappa^*)$ is real and integral, there exists, in general, one irreducible representation $\mathfrak{D}_{\kappa,\kappa^*}$ of the Lorentz group. However, if 4κ, $4\kappa^*$ are both real and integral there are two representations $\mathfrak{D}^{+}_{\kappa,\kappa^*}$ and $\mathfrak{D}^{-}_{\kappa,\kappa^*}$ associated to the pair (κ, κ^*). All these representations are infinite except $\mathfrak{D}^{-}_{\kappa,\kappa^*}$ which is finite if 2κ, $2\kappa^*$ are both integral. For suitable values of (κ, κ^*), $\mathfrak{D}_{\kappa,\kappa^*}$ or $\mathfrak{D}^{+}_{\kappa,\kappa^*}$ is unitary. $\mathfrak{U}$ and $\mathfrak{B}$ matrices similar to those given by Dirac (1936) and Fierz (1939) are introduced for these infinite representations. The extension of Dirac's expansor formalism to cover half-integral spins is given. These new quantities, which are called expinors, bear the same relation to spinors as Dirac's expansors to tensors. It is shown that they can be used to describe the spin properties of a particle in accordance with the principles of quantum mechanics.

1. Introduction

All the finite irreducible representations of the Lorentz group are well known. Every such representation is characterized by an ordered pair of numbers p, q such that $2p$ and $2q$ are integral and $\geqslant 0$. None of these representations, however, is unitary. Dirac (1945) has recently drawn attention to the existence of some unitary, though infinite, representations with a view to their possible physical applications. The present paper is concerned with the investigation of the general irreducible representations of the proper Lorentz group. It is found that such a representation can, in general, be characterized by an ordered pair (κ, κ^*) of complex numbers such that $2(\kappa-\kappa^*)$ is an integer† and may therefore be denoted by $\mathfrak{D}_{\kappa,\kappa^*}$. However, in case both κ and κ^* are of the form $\tfrac{1}{4}n(n \doteq -2)$ there exist two irreducible representations $\mathfrak{D}^{+}_{\kappa,\kappa^*}$ and $\mathfrak{D}^{-}_{\kappa,\kappa^*}$ corresponding to the pair (κ, κ^*). The finite representations correspond to $\mathfrak{D}^{-}_{\kappa,\kappa^*}$ with κ, κ^* both of the form $\tfrac{1}{2}n(n \doteq -1)$, the connexion between (p,q) and (κ, κ^*) being given by $p = |\kappa+\tfrac{1}{2}| - \tfrac{1}{2}$, $q = |\kappa^*+\tfrac{1}{2}| - \tfrac{1}{2}$. In general $\mathfrak{D}_{\kappa,\kappa^*}$ is unitary only if either

$$\kappa = -\tfrac{1}{2}+i\nu+\tfrac{1}{2}n, \quad \kappa^* = -\tfrac{1}{2}+i\nu-\tfrac{1}{2}n, \tag{1a}$$

or
$$\kappa = \kappa^* = -\tfrac{1}{2}+\nu, \quad |\nu| \leqslant \tfrac{1}{2}. \tag{1b}$$

Here ν is an arbitrary real number. However, in the special case when κ, κ^* are both of the form $\tfrac{1}{4}n(n \doteq -2)$, $\mathfrak{D}^{-}_{\kappa,\kappa^*}$ is unitary only if $\tfrac{1}{2} \geqslant |\kappa+\tfrac{1}{2}| = |\kappa^*+\tfrac{1}{2}|$, while $\mathfrak{D}^{+}_{\kappa,\kappa^*}$ is unitary whenever $|\kappa+\tfrac{1}{2}| = |\kappa^*+\tfrac{1}{2}|$.

It is possible to construct a series in the complex variables u_1, u_2 and their complex conjugates $u_{\dot{1}}$, $u_{\dot{2}}$ such that the coefficients of this series transform according to the irreducible representation $\mathfrak{D}_{\kappa,\kappa^*}$ when a spinor transformation corresponding to a

---

† The term 'integer' is used throughout to mean a real integer whether positive, negative or zero. The symbol n always denotes an integer, while s denotes an integer or half odd integer $\geqslant 0$.

[372]

given Lorentz transformation is applied to the variables u_1, u_2 and $u_{\dot{1}}, u_{\dot{2}}$. These coefficients therefore form the components of a vector of the mathematical vector space $\Re_{\kappa, \kappa^*}$ which transforms according to the representation $\mathfrak{D}_{\kappa, \kappa^*}$. It is clear that these vectors of $\Re_{\kappa, \kappa^*}$ bear the same relation to ordinary spinors as the expansors of Dirac to ordinary tensors. They may therefore be called *expinors*. Thus to every series in $u_1, u_2, u_{\dot{1}}, u_{\dot{2}}$, which is convergent in a certain sense, there corresponds an expinor. This expinor transforms according to the representation $\mathfrak{D}_{\kappa, \kappa^*}$ if the series is homogeneous in each of the two sets of variables (u_1, u_2) and $(u_{\dot{1}}, u_{\dot{2}})$ and has the degrees 2κ and $2\kappa^*$ respectively. Such an expinor may therefore be called a homogeneous expinor. In case both 2κ and $2\kappa^*$ are integral and $\geqslant 0$, the corresponding homogeneous expinor is the same as the symmetric spinor with 2κ undotted and $2\kappa^*$ dotted indices. Homogeneous expinors are thus an extension of symmetric spinors.

Further, as pointed out by Dirac (1945), it is possible to use expinors to describe the transformation properties of the wave function of a spinning particle. In a theory based on these expinors it is possible to make the charge density positive definite for particles of integral spin or the energy density positive definite for particles of half-integral spin, in contradistinction with the results of the existing theory (see Pauli 1940). This is made possible by the circumstance that infinite unitary representations of the Lorentz group exist for both integral and half-integral spins. By imposing subsidiary conditions on the wave function, it can be arranged that in the rest system the particle has a 'pure' spin s, i.e. all the non-vanishing components of the wave function in the rest system transform according to the irreducible representation $\mathfrak{D}_s$ of the rotation group.

2. The general representation

The finite representations of the Lorentz group are in fact obtained from those of the four-dimensional real orthogonal group by a device which is called the 'unitarian trick' by Weyl (1939). It consists essentially in replacing in a representation three of the real parameters of the orthogonal group by pure imaginary ones. However, this device does not yield all the possible representations of the Lorentz group. This arises from the fact that the topological structure of the orthogonal group is not the same as that of the Lorentz group, the difference consisting in the aperiodic nature of the 'temporal rotations'. In consequence, unlike the orthogonal group, the Lorentz group is not compact.

We begin with the consideration of the infinitesimal transformations, which, as is well known, play a very important part in the theory of Lie groups. Such a Lorentz transformation is given by

$$x'_k = x_k + \epsilon_k{}^l x_l \quad (\epsilon_{kl} = -\epsilon_{lk}), \tag{2}$$

ϵ_{kl}'s being infinitesimal quantities. (The usual metric tensor g_{kl} is used for raising and lowering the tensor indices $g_{00} = -g_{11} = -g_{22} = -g_{33} = 1$, the other components being zero.) In any representation this transformation will be represented by

$$1 + \tfrac{1}{2}\epsilon_{kl} I^{kl}, \tag{3}$$

where I^{kl} are operators of the representation which may always be chosen to be antisymmetric in k, l. The commutation relations of I^{kl} are completely fixed by the structure of the Lorentz group (see Van der Waerden 1932) and are

$$[I^{kl}, I^{mn}] = -g^{km} I^{ln} + g^{lm} I^{kn} + g^{kn} I^{lm} - g^{ln} I^{km}, \tag{4}$$

where $[A, B] = AB - BA$. Put

$$K_1 = iI^{23}, \quad K_2 = iI^{31}, \quad K_3 = iI^{12}, \quad L_1 = iI^{10}, \quad L_2 = iI^{20}, \quad L_3 = iI^{30}. \tag{5}$$

From the definition (2) it is clear that I^{kl} transform as components of an antisymmetrical tensor, under Lorentz transformations. Therefore it follows from (5) that for spatial rotations both (K_1, K_2, K_3) and (L_1, L_2, L_3) transform as components of a three-dimensional vector. These two vectors will be denoted by $\mathbf{K}$ and $\mathbf{L}$ respectively. In terms of $\mathbf{K}$ and $\mathbf{L}$ (4) can be written as

$$[K_1, K_2] = iK_3, \tag{6a}$$

$$[K_1, L_2] = iL_3 = [L_1, K_2], \tag{6b}$$

$$[L_1, L_2] = -iK_3, \tag{6c}$$

the remaining relations being obtained by cyclic permutations of $(1, 2, 3)$.

Consider now an irreducible representation of the proper Lorentz group. The matrices I^{kl} then form an irreducible set. Denote the representation space by $\mathfrak{R}$. Reduce this space with respect to the subgroup δ_3 consisting of spatial rotations only. Now δ_3 is a compact group, and it is well known that every representation of a compact group is completely reducible into a direct sum of irreducible representations each of which is unitary and of a finite degree (see Pontrjagin 1939, p. 205). Now every irreducible representation of δ_3 is characterized by a number $k \geqslant 0$ which is integral or half-integral. Thus $\mathfrak{R}$ is decomposed into a direct sum of subspaces $\mathfrak{R}_k$ which are all irreducible with respect to δ_3. It will be assumed† that in this decomposition there occurs at most only one subspace $\mathfrak{R}_k$ for any particular value of k. By choosing a suitable coordinate system in $\mathfrak{R}_k$, K_3 can be brought to the diagonal form with its eigenvalues m running through the numbers $k, k-1, \ldots, -k+1, -k$. Further, this coordinate system in $\mathfrak{R}_k$ can always be so chosen that the matrix elements of K_1 and K_2 are given by

$$(m \mid K^{\cdot} \mid m-1) = (m-1 \mid K^{\cdot} \mid m) = \sqrt{\{(k+m)(k-m+1)\}}, \tag{7}$$

where
$$K^{\cdot} = K_1 + iK_2, \quad K^{\cdot} = K_1 - iK_2. \tag{8}$$

Thus K_1, K_2, K_3 are represented by Hermitian matrices and

$$\mathbf{K}^2 = k(k+1). \tag{9}$$

Now since $\mathbf{L}$ transforms as a vector under spatial rotations it follows (see, for example, Weyl 1931, p. 200) that $(k \mid \mathbf{L} \mid k') = 0$ unless $k' = k$ or $k \pm 1$. In fact, the

---

† *Note added in proof.* It is possible to avoid this assumption. The proof is then somewhat more complicated but the final result is the same.

matrix elements of **L** are completely fixed by its vector character, apart from a factor depending on k (but not on m). They are given by Weyl (1931),

$$(k, m \mid L^+ \mid k-1, m-1) = -p^-(k)\sqrt{\{(k+m)(k+m-1)\}},$$
$$(k, m-1 \mid L^- \mid k-1, m) = p^-(k)\sqrt{\{(k-m+1)(k-m)\}}, \qquad (10a)$$
$$(k, m \mid L_3 \mid k-1, m) = p^-(k)\sqrt{\{(k+m)(k-m)\}},$$

$$(k, m \mid L^+ \mid k, m-1) = p(k)\sqrt{\{(k+m)(k-m+1)\}},$$
$$(k, m-1 \mid L^- \mid k, m) = p(k)\sqrt{\{(k-m+1)(k+m)\}}, \qquad (10b)$$
$$(k, m \mid L_3 \mid k, m) = p(k)m,$$

$$(k, m \mid L^+ \mid k+1, m-1) = p^+(k)\sqrt{\{(k-m+1)(k-m+2)\}},$$
$$(k, m-1 \mid L^- \mid k+1, m) = -p^+(k)\sqrt{\{(k+m)(k+m+1)\}}, \qquad (10c)$$
$$(k, m \mid L_3 \mid k+1, m) = p^+(k)\sqrt{\{(k+m+1)(k-m+1)\}},$$

the remaining elements vanishing. Here $p^-(k)$, $p(k)$ and $p^+(k)$ are numerical coefficients depending on k, and $L^+ = L_1 + iL_2$, $L^- = L_1 - iL_2$. Now notice that it is possible to construct two invariants of the proper Lorentz group from I^{kl}, viz. $J^2 = -\tfrac{1}{2}I^{kl}I_{kl}$ and $I = -\tfrac{1}{2}\epsilon^{klmn}I_{kl}I_{mn}$. ($\epsilon^{klmn}$ is a tensor antisymmetric in all the indices and $\epsilon^{0123} = 1$). Since they are left unaltered by a proper Lorentz transformation, J^2 and I commute with I^{kl} and therefore in an irreducible representation they are represented by a numerical constant. Now

$$-\tfrac{1}{2}\epsilon^{klmn}I_{kl}I_{mn} = I = K_1 L_1 + K_2 L_2 + K_3 L_3 = (\mathbf{KL}), \qquad (11a)$$

$$-\tfrac{1}{2}I^{kl}I_{kl} = J^2 = \mathbf{K}^2 - \mathbf{L}^2 = k(k+1) - (L^+ L^- + L_3^2 + K_3). \qquad (11b)$$

Substituting from (7) and (10) in (11) one gets

$$k(k+1)p(k) = I, \qquad (12a)$$

$$J^2 = k(k+1)(1 - p^2(k)) - m(1 + p^2(k)) - p^+(k)p^-(k+1)(k-m+1)(2k+3)$$
$$- p^-(k)p^+(k-1)(k+m)(2k-1), \qquad (12b)$$

it being understood that $p^-(k) \equiv 0$ if $k = 0, \tfrac{1}{2}$. For $k \neq 0$ there are at least two distinct values of m for which (12b) holds. Hence the coefficient of m must vanish for $k \neq 0$, i.e.

$$(1 - p^2(k))k = p^+(k)p^-(k+1)k(2k+3) - p^-(k)p^+(k-1)k(2k-1), \qquad (13a)$$

$$J^2 = k(k+1)\{1 - p^2(k)\} - p^+(k)p^-(k+1)(k+1)(2k+3)$$
$$- p^-(k)p^+(k-1)k(2k-1). \qquad (13b)$$

The factor k has been introduced deliberately in (13a) so as to include the exceptional case $k = 0$. From (13) it follows that

$$p^+(k)p^-(k+1) = \{k(k+2) - k^2 p^2(k) - J^2\}\frac{1}{(2k+1)(2k+3)}. \qquad (14)$$

Now let j be the lowest value of k appearing in the decomposition of $\Re$ into $\Re_k$. First consider the case when $j \neq 0$. Bearing in mind that $p^-(j) = 0$ one obtains from (13) and (12a) on putting $k = j$

$$J^2 = j^2 - \frac{I^2}{j^2} - 1. \tag{15}$$

Hence from (14)

$$p^-(k)\,p^-(k+1) = \frac{(k-j+1)(k+j+1)}{(2k+1)(2k+3)}\left\{1 + \frac{I^2}{(k+1)^2 j^2}\right\} \quad (k \geqslant j). \tag{16}$$

On the other hand, if $j = 0$ then from (12a) $I = 0$ and so $p(k) = 0$, since one can always put $p(0) = 0$ without affecting the representation at all. Therefore from (14)

$$p^+(k)\,p^-(k+1) = \frac{(k+1)^2}{(2k+1)(2k+3)}\left\{1 - \frac{J^2+1}{(k+1)^2}\right\}. \tag{17}$$

(17) may be regarded as a special case of (15) and (16) for $j = 0$ provided one agrees to look upon I as $j\mu$ with $\mu = I/j$ as a definitely given number. This makes I vanish automatically when $j = 0$. Also from (12a)

$$p(k) = \frac{j\mu}{k(k+1)} \quad (k \geqslant j), \tag{18}$$

with the understanding that $p(0) \equiv 0$. By multiplying all the eigenvectors in each of the subspaces $\Re_k$ by a suitable normalization factor $q(k)$ it can always be arranged that

$$p^-(k) = p^-(k+1) = \sqrt{\frac{(k-j+1)(k+j+1)}{(2k+1)(2k+3)}}\sqrt{\left(1 + \frac{\mu^2}{(k+1)^2}\right)}. \tag{19}$$

For convenience the first square root is supposed to be always positive. The second one may then be chosen at will for each k. In particular, if the representation of the Lorentz group is unitary, **K** and **L** are Hermitian so that $p(k)$ is real and

$$p^-(k) = \bar{p}^-(k+1),$$

where the bar denotes complex conjugate. The normalization factors $q(k)$ are then unimodular, and the corresponding transformation of $\Re$ is unitary. Therefore every unitary representation of the Lorentz group can be brought into a form for which (19) holds, by means of a unitary transformation. Since unitary transformations preserve the unitary character it follows that the transformed representation is also unitary. Now (16) and (18) and (19) show that, apart from equivalence, an irreducible representation is completely fixed by the two numbers j and μ. Also as shown above a unitary representation can be transformed to the form (19) without destroying its unitary character. Hence all irreducible representations which are equivalent to a unitary representation are those for which the right-hand sides of (18) and (19) are real. *Thus for a representation to be unitary it is necessary and sufficient that $I = \mu j$ be real and*

$$\mu^2 > -1. \tag{20}$$

The condition (20) is actually effective only when $j = 0$.

3. The spinor matrices $\mathfrak{U}_\varkappa$, $\mathfrak{V}_\varkappa$

To proceed further we must now introduce spinors. For the moment consider only the rotation group δ_3. Then, it is well known that with every vector of the three-dimensional space there can be associated a symmetrical spinor with two indices. Thus to the vector $\mathbf{K} = (K_1, K_2, K_3)$ corresponds a spinor $K_{\varkappa\beta}$ defined† by

$$K_{12} = K_{21} = -K_3, \quad K_{11} = K^-, \quad K_{22} = -K^+. \tag{21}$$

As usual the antisymmetrical spinors $\epsilon^{\varkappa\beta}$ and $\epsilon_{\varkappa\beta}$ ($\epsilon^{12} = \epsilon_{12} = 1$) will be used for raising and lowering the indices according to the rule

$$a^\varkappa = \epsilon^{\varkappa\beta}a_\beta, \quad a_\beta = a^\varkappa\epsilon_{\varkappa\beta}. \tag{22}$$

Now it is well known that $K_{\varkappa\beta}(k)$ (which is the matrix representing $K_{\varkappa\beta}$ in $\mathfrak{R}_k$) can be expressed in terms of the spinor matrices $u_\varkappa(k)$, $v_\beta(k)$ introduced by Dirac (1936) and Fierz (1939). $u_1(k)$, $u_2(k)$ have $2k+1$ rows and $2k$ columns and $v_1(k)$, $v_2(k)$ have $2k$ rows and $2k+1$ columns. Their properties may briefly be summarized as follows:

$$u_\varkappa(k)\,v^\varkappa(k) = 2k, \quad v^\varkappa(k)\,u_\varkappa(k) = 2k+1, \tag{23a}$$

$$u_\varkappa(k+\tfrac{1}{2})\,u^\varkappa(k) = v^\varkappa(k)\,v_\varkappa(k+\tfrac{1}{2}) = 0, \tag{23b}$$

$$v^\varkappa(k) = \overline{u_\varkappa(k)}, \quad u_\varkappa(k) = \overline{v^\varkappa(k)}, \tag{23c}$$

$$v_\varkappa(k+\tfrac{1}{2})\,u_\beta(k+\tfrac{1}{2}) - u_\beta(k)\,v_\varkappa(k) = \epsilon_{\beta\varkappa}. \tag{23d}$$

$$K_{\varkappa\beta}(k) = u_\varkappa(k)\,v_\beta(k) - k\epsilon_{\varkappa\beta} = v_\varkappa(k+\tfrac{1}{2})\,u_\beta(k+\tfrac{1}{2}) + (k+1)\epsilon_{\varkappa\beta}. \tag{23e}$$

The bar denotes Hermitian conjugate. It is easy to show that on account of the Hermitian character of $\mathbf{K}$, $u_\varkappa(k)$ and $v_\beta(k)$ can always be chosen to satisfy (23c). Notice that (23c) is a proper spinor equation, since for spatial rotations a lower spinor index transforms the same way as the complex conjugate of an upper spinor index.

Now define two matrices $U^\varkappa$ and $V^\varkappa$ as follows:

$$(k \mid U^\varkappa \mid k-\tfrac{1}{2}) = u^\varkappa(k), \quad (k-\tfrac{1}{2} \mid V^\varkappa \mid k) = v^\varkappa(k), \tag{24}$$

all the other elements being zero. Then clearly (23) can be written as

$$U_\varkappa V^\varkappa = 2k, \quad V^\varkappa U_\varkappa = 2(k+1), \tag{25a}$$

$$U_\varkappa U^\varkappa = V^\varkappa V_\varkappa = 0, \tag{25b}$$

$$V^\varkappa = \overline{U_\varkappa}, \quad U_\varkappa = \overline{V^\varkappa}, \tag{25c}$$

$$V_\varkappa U_\beta - U_\beta V_\varkappa = \epsilon_{\beta\varkappa}, \tag{25d}$$

$$K_{\varkappa\beta} = U_\varkappa V_\beta - k\epsilon_{\varkappa\beta} = V_\varkappa U_\beta + (k+1)\epsilon_{\varkappa\beta}. \tag{25e}$$

Here k is the diagonal matrix whose component inside $\mathfrak{R}_{k'}$ is simply k' times the unit matrix. Notice that from (24)

$$V^\varkappa k = (k+\tfrac{1}{2})\,V^\alpha, \tag{26a}$$

$$U^\varkappa k = (k-\tfrac{1}{2})\,U^\varkappa. \tag{26b}$$

† Spinor indices are always denoted by Greek letters and tensor indices by Latin letters.

Now construct similarly a spinor $L_{\alpha\beta}$ corresponding to the vector $\mathbf{L}$,

$$L_{12} = L_{21} = -L_3, \quad L_{11} = L^-, \quad L_{22} = -L^+. \tag{27}$$

Then (6) can be written as

$$[K_{\alpha\beta}, K_{\lambda\mu}] = \tfrac{1}{2}\underset{(\alpha\beta)}{S}\,\underset{(\lambda\mu)}{S}\,\epsilon_{\lambda\alpha}K_{\mu\beta}, \tag{28a}$$

$$[K_{\alpha\beta}, L_{\lambda\mu}] = [L_{\alpha\beta}, K_{\lambda\mu}] = \tfrac{1}{2}\underset{(\alpha\beta)}{S}\,\underset{(\lambda\mu)}{S}\,\epsilon_{\lambda\alpha}L_{\mu\beta}, \tag{28b}$$

$$[L_{\alpha\beta}, L_{\lambda\mu}] = -\tfrac{1}{2}\underset{(\alpha\beta)}{S}\,\underset{(\lambda\mu)}{S}\,\epsilon_{\lambda\alpha}K_{\mu\beta}, \tag{28c}$$

where $\underset{(\alpha\beta)}{S}$ denotes that terms arising from the interchange of α and β are to be added.
Also

$$-\tfrac{1}{2}K^{\alpha\beta}K_{\alpha\beta} = \mathbf{K}^2, \quad -\tfrac{1}{2}L^{\alpha\beta}L_{\alpha\beta} = \mathbf{L}^2, \tag{29a}$$

$$-\tfrac{1}{2}K^{\alpha\beta}L_{\alpha\beta} = (\mathbf{KL}) = I. \tag{29b}$$

Now it follows from (10) and (19) that

$$L_{\alpha\beta} = \frac{I}{k(k+1)}K_{\alpha\beta} + i\sqrt{\frac{(k-j)(k+j)}{(2k-1)(2k+1)}}\sqrt{\left(1+\frac{\mu^2}{k^2}\right)}U_\alpha U_\beta$$
$$+ V_\alpha V_\beta i\sqrt{\frac{(k-j)(k+j)}{(2k-1)(2k+1)}}\sqrt{\left(1+\frac{\mu^2}{k^2}\right)}. \tag{30}$$

As before, the matrix square root $\sqrt{\dfrac{(k-j)(k+j)}{(2k-1)(2k+1)}}$ is always taken to be positive while $\sqrt{\left(1+\dfrac{\mu^2}{k^2}\right)}$ is defined arbitrarily. To verify (30) observe that (28b) follows from the spinor character of $L_{\alpha\beta}$ while (28c) is easily obtained on using (30) and (24). Similarly (29b) and (15) are also easily verified. This fixes the j and μ value of the representation uniquely, and therefore as shown in §2, (29) must lead to the same representation, apart from equivalence, as (18) and (19), i.e. if $u_\alpha(k)$ and $v_\alpha(k)$ are suitably chosen the two representations must coincide.

Now put
$$I_{\alpha\beta} = \tfrac{1}{2}(K_{\alpha\beta} - iL_{\alpha\beta}), \tag{31a}$$

and notice that
$$[I_{\alpha\beta}, I_{\lambda\mu}] = \tfrac{1}{2}\underset{(\alpha\beta)}{S}\,\underset{(\lambda\mu)}{S}\,\epsilon_{\lambda\alpha}I_{\mu\beta}, \tag{31b}$$

$$-\tfrac{1}{2}I^{\alpha\beta}I_{\alpha\beta} = \frac{J^2 - 2iI}{4} = \tfrac{1}{4}(j - i\mu)^2 - \tfrac{1}{4}. \tag{31c}$$

Consider the product space $\Re \times \Re_1$, where $\Re_1$ is a two-dimensional space transforming according to the fundamental spinor representation $\mathfrak{D}_{\frac{1}{2},0}$ of the Lorentz group. The components of a vector in this space may be written as ψ_α ($\alpha = 1, 2$), where ψ_1 and ψ_2 are one-column matrices enumerating the components in $\Re$. With respect to the index α, ψ_α transforms as a spinor. Now define

$$\psi^* = U^\alpha \frac{1}{\sqrt{(2k+1)}}\psi_\alpha, \tag{32a}$$

$$\psi = V^\alpha \frac{1}{\sqrt{(2k+1)}}\psi_\alpha. \tag{32b}$$

Then from (25)
$$\psi_\alpha = \frac{1}{\sqrt{(2k+1)}} U_\alpha \psi - \frac{1}{\sqrt{(2k+1)}} V_\alpha \psi^*. \tag{32c}$$

Now put
$$I_\lambda{}^\mu \psi_\mu = \psi'_\lambda. \tag{33}$$

From (31a) and (30)
$$I_{\alpha\beta} = a(k) K_{\alpha\beta} + b(k) U_\alpha U_\beta + V_\alpha V_\beta b(k), \tag{34a}$$

where
$$a(k) = \frac{1}{2}\left(1 - \frac{iI}{k(k+1)}\right), \quad b(k) = \frac{1}{2}\sqrt{\frac{(k-j)(k+j)}{(2k-1)(2k+1)}}\sqrt{\left(1+\frac{\mu^2}{k^2}\right)}. \tag{34b}$$

On using (25), (26), (32) and (34) one finds that

$$\psi' \equiv V^\lambda \cdot \frac{1}{\sqrt{(2k+1)}} \psi'_\lambda = a(k+\tfrac{1}{2})(k+\tfrac{3}{2})\psi + b(k+\tfrac{1}{2})\sqrt{\{2k(2k+2)\}}\psi^*, \tag{35a}$$

and similarly

$$\psi^{*\prime} \equiv U^\lambda \frac{1}{\sqrt{(2k+1)}} \psi'_\lambda = -a(k-\tfrac{1}{2})(k-\tfrac{1}{2})\psi^* - b(k+\tfrac{1}{2})\sqrt{\{2k(2k+2)\}}\psi. \tag{35b}$$

(35) may be looked upon as a linear transformation in the two 'variables' ψ and ψ^*. The eigenvalues of this transformation are easily obtained on using (34b). They are $\frac{1}{2}(1\pm\gamma)$, where γ is a complex number defined, except for an ambiguity in sign, by the equation
$$\gamma^2 = (j-i\mu)^2. \tag{36}$$

Write $\gamma_{k+\frac{1}{2}} = k+\tfrac{1}{2} - \dfrac{iI}{k+\frac{1}{2}}$, and put

$$\sqrt{2}\,\psi_{\gamma-1} = \sqrt{(\gamma_{k+\frac{1}{2}}+\gamma)}\,\psi + \sqrt{(\gamma_{k+\frac{1}{2}}-\gamma)}\,\psi^*, \tag{37a}$$

$$\sqrt{2}\,\psi_{\gamma+1} = \sqrt{(\gamma_{k+\frac{1}{2}}-\gamma)}\,\psi + \sqrt{(\gamma_{k+\frac{1}{2}}+\gamma)}\,\psi^*, \tag{37b}$$

the square roots being so chosen that

$$\sqrt{(\gamma_{k+\frac{1}{2}}+\gamma)}\sqrt{(\gamma_{k+\frac{1}{2}}-\gamma)} = \sqrt{\{(k+\tfrac{1}{2}-j)(k+\tfrac{1}{2}+j)\}}\sqrt{\left(1+\frac{\mu^2}{(k+\frac{1}{2})^2}\right)}. \tag{37c}$$

From (35) and (37)
$$\sqrt{2}\,\psi'_{\gamma-1} \equiv \sqrt{(\gamma_{k+\frac{1}{2}}+\gamma)}\,\psi' + \sqrt{(\gamma_{k+\frac{1}{2}}-\gamma)}\,\psi^{*\prime}$$
$$= \tfrac{1}{2}(1+\gamma)\sqrt{2}\,\psi_{\gamma-1}. \tag{38a}$$

Similarly
$$\sqrt{2}\,\psi'_{\gamma+1} \equiv \sqrt{(\gamma_{k+\frac{1}{2}}-\gamma)}\,\psi' + \sqrt{(\gamma_{k+\frac{1}{2}}+\gamma)}\,\psi^{*\prime}$$
$$= \tfrac{1}{2}(1-\gamma)\sqrt{2}\,\psi_{\gamma+1}. \tag{38b}$$

Thus the transformation (37) brings (35) to the diagonal form. Also it follows from (37) that

$$\sqrt{2}\,\gamma\psi = \sqrt{(\gamma_{k+\frac{1}{2}}+\gamma)}\,\psi_{\gamma-1} - \sqrt{(\gamma_{k+\frac{1}{2}}-\gamma)}\,\psi_{\gamma+1}, \tag{39a}$$

$$\sqrt{2}\,\gamma\psi^* = -\sqrt{(\gamma_{k+\frac{1}{2}}-\gamma)}\,\psi_{\gamma-1} + \sqrt{(\gamma_{k+\frac{1}{2}}+\gamma)}\,\psi_{\gamma+1}. \tag{39b}$$

Now corresponding to (36) define γ^* by the equation

$$\gamma^{*2} = (j+i\mu)^2, \tag{40}$$

and write $\gamma = 2\kappa + 1, \gamma^* = 2\kappa^* + 1$. Then

$$iI = \frac{\gamma^{*2} - \gamma^2}{4} = \kappa^*(\kappa^* + 1) - \kappa(\kappa + 1) = (\kappa^* - \kappa)(\kappa^* + \kappa + 1), \qquad (41)$$

$$\gamma_{k+\frac{1}{2}} = k + \frac{1}{2} - \frac{iI}{k + \frac{1}{2}} = \frac{(k + \frac{1}{2})^2 - (\kappa^* + \frac{1}{2})^2 + (\kappa + \frac{1}{2})^2}{(k + \frac{1}{2})}.$$

Hence $\quad \gamma_{k+\frac{1}{2}} + \gamma = \dfrac{(k + \kappa + 1)^2 - (\kappa^* + \frac{1}{2})^2}{(k + \frac{1}{2})} = \dfrac{(k + \kappa + \kappa^* + \frac{3}{2})(k + \kappa - \kappa^* + \frac{1}{2})}{(k + \frac{1}{2})}, \qquad (42a)$

$$\gamma_{k+\frac{1}{2}} - \gamma = \frac{(k - \kappa)^2 - (\kappa^* + \frac{1}{2})^2}{k + \frac{1}{2}} = \frac{(k - \frac{1}{2} - \kappa - \kappa^*)(k + \frac{1}{2} - \kappa + \kappa^*)}{(k + \frac{1}{2})}. \qquad (42b)$$

Put $\psi_{\gamma-1} \equiv \psi_{\kappa-\frac{1}{2}}, \psi_{\gamma+1} \equiv \psi_{\kappa+\frac{1}{2}}$. Then it follows from (37) and (32) that

$$\psi_{\kappa-\frac{1}{2}} = \mathfrak{B}^\alpha(\kappa, \kappa^*)\,\psi_\alpha, \qquad (43a)$$

$$\psi_{\kappa+\frac{1}{2}} = \mathfrak{U}^\alpha(\kappa + \tfrac{1}{2}, \kappa^*)\,\psi_\alpha, \qquad (43b)$$

where $\mathfrak{B}^\alpha$ and $\mathfrak{U}^\alpha$ are defined by

$$\mathfrak{U}^\alpha(\kappa, \kappa^*) = \tfrac{1}{2} V^\alpha \sqrt{\left|\frac{(k + 1 + \kappa + \kappa^*)(k + \kappa - \kappa^*)}{k(k + \frac{1}{2})}\right|}$$
$$+ \frac{1}{2}\sqrt{\left|\frac{(k + \frac{1}{2} - \kappa + \kappa^*)(k - \frac{1}{2} - \kappa - \kappa^*)}{k(k + \frac{1}{2})}\right|}\, U^\alpha, \qquad (44a)$$

$$\mathfrak{B}^\alpha(\kappa, \kappa^*) = \tfrac{1}{2} V^\alpha \sqrt{\left|\frac{(k + \frac{1}{2} - \kappa + \kappa^*)(k - \frac{1}{2} - \kappa - \kappa^*)}{k(k + \frac{1}{2})}\right|}$$
$$+ \frac{1}{2}\sqrt{\left|\frac{(k + 1 + \kappa + \kappa^*)(k + \kappa - \kappa^*)}{k(k + \frac{1}{2})}\right|}\, U^\alpha. \qquad (44b)$$

Further from (32c), (39) and (44) it is found that

$$(2\kappa + 1)\psi_\alpha = \mathfrak{U}_\alpha(\kappa, \kappa^*)\,\psi_{\kappa-\frac{1}{2}} - \mathfrak{B}_\alpha(\kappa + \tfrac{1}{2}, \kappa^*)\,\psi_{\kappa+\frac{1}{2}}. \qquad (45)$$

When $\gamma = 2\kappa + 1 \neq 0$ the two eigenvalues $\frac{1}{2}(1 \pm \gamma)$ of (35) are distinct and hence the two 'eigenvectors' $\psi_{\kappa-\frac{1}{2}}$ and $\psi_{\kappa+\frac{1}{2}}$ are linearly independent. Now from (43) and (45)

$$(2\kappa + 1)\psi_\alpha = \mathfrak{U}_\alpha(\kappa, \kappa^*)\,\mathfrak{B}^\beta(\kappa, \kappa^*)\,\psi_\beta - \mathfrak{B}_\alpha(\kappa + \tfrac{1}{2}, \kappa^*)\,\mathfrak{U}^\beta(\kappa + \tfrac{1}{2}, \kappa^*)\,\psi_\beta, \qquad (46a)$$

$$(2\kappa + 1)\psi_{\kappa-\frac{1}{2}} = \mathfrak{B}^\alpha(\kappa, \kappa^*)\,\mathfrak{U}_\alpha(\kappa, \kappa^*)\,\psi_{\kappa-\frac{1}{2}} - \mathfrak{B}^\alpha(\kappa, \kappa^*)\,\mathfrak{B}_\alpha(\kappa + \tfrac{1}{2}, \kappa^*)\,\psi_{\kappa+\frac{1}{2}}, \qquad (46b)$$

$$(2\kappa + 1)\psi_{\kappa+\frac{1}{2}} = \mathfrak{U}^\alpha(\kappa + \tfrac{1}{2}, \kappa^*)\,\mathfrak{U}_\alpha(\kappa, \kappa^*)\,\psi_{\kappa-\frac{1}{2}} - \mathfrak{U}^\alpha(\kappa + \tfrac{1}{2}, \kappa^*)\,\mathfrak{B}_\alpha(\kappa + \tfrac{1}{2}, \kappa^*)\,\psi_{\kappa+\frac{1}{2}}. \qquad (46c)$$

Also from (33), (38) and (43)

$$(\kappa + 1)\psi_{\kappa-\frac{1}{2}} = \mathfrak{B}^\alpha(\kappa, \kappa^*)\,I_\alpha{}^\beta\psi_\beta, \qquad (47a)$$

$$-\kappa\psi_{\kappa+\frac{1}{2}} = \mathfrak{U}^\alpha(\kappa + \tfrac{1}{2}, \kappa^*)\,I_\alpha{}^\beta\psi_\beta, \qquad (47b)$$

so that from (45)

$$(2\kappa + 1)I_\alpha{}^\beta\psi_\beta = (\kappa + 1)\mathfrak{U}_\alpha(\kappa, \kappa^*)\,\psi_{\kappa-\frac{1}{2}} + \kappa\mathfrak{B}_\alpha(\kappa + \tfrac{1}{2}, \kappa^*)\,\psi_{\kappa+\frac{1}{2}}$$
$$= (\kappa + 1)\mathfrak{U}_\alpha(\kappa, \kappa^*)\,\mathfrak{B}^\beta(\kappa, \kappa^*)\,\psi_\beta + \kappa\mathfrak{B}_\alpha(\kappa + \tfrac{1}{2}, \kappa^*)\,\mathfrak{U}^\beta(\kappa + \tfrac{1}{2}, \kappa^*)\,\psi_\beta. \qquad (48)$$

Now ψ_{α} is completely arbitrary. Therefore from (46a) and (48)

$$(2\kappa+1)\epsilon_{\alpha}{}^{\beta} = \mathfrak{U}_{\alpha}(\kappa,\kappa^*)\mathfrak{B}^{\beta}(\kappa,\kappa^*) - \mathfrak{B}_{\alpha}(\kappa+\tfrac{1}{2},\kappa^*)\mathfrak{U}^{\beta}(\kappa+\tfrac{1}{2},\kappa^*), \tag{49a}$$

$$(2\kappa+1)I_{\alpha}{}^{\beta} = (\kappa+1)\mathfrak{U}_{\alpha}(\kappa,\kappa^*)\mathfrak{B}^{\beta}(\kappa,\kappa^*) + \kappa\mathfrak{B}_{\alpha}(\kappa+\tfrac{1}{2},\kappa^*)\mathfrak{U}^{\beta}(\kappa+\tfrac{1}{2},\kappa^*). \tag{49b}$$

Now when $\gamma = 2\kappa+1 \neq 0$ the transformation (37) is non-singular since it has an inverse, namely, (39). Hence the correspondence between ψ_{α} and the pair $(\psi_{\kappa-\frac{1}{2}}, \psi_{\kappa+\frac{1}{2}})$ is a 1-1 correspondence. Thus every arbitrarily chosen pair $(\psi_{\kappa-\frac{1}{2}}, \psi_{\kappa+\frac{1}{2}})$ corresponds to a unique ψ_{α}, and therefore (46) and (47) hold for arbitrary $\psi_{\kappa-\frac{1}{2}}$ and $\psi_{\kappa+\frac{1}{2}}$ if $2\kappa+1 \neq 0$. This implies that

$$\mathfrak{B}^{\alpha}(\kappa,\kappa^*)\mathfrak{U}_{\alpha}(\kappa,\kappa^*) = 2\kappa+1, \quad \mathfrak{B}^{\alpha}(\kappa,\kappa^*)\mathfrak{B}_{\alpha}(\kappa+\tfrac{1}{2},\kappa^*) = 0,$$

$$\mathfrak{U}^{\alpha}(\kappa+\tfrac{1}{2},\kappa^*)\mathfrak{U}_{\alpha}(\kappa,\kappa^*) = 0, \quad \mathfrak{U}^{\alpha}(\kappa+\tfrac{1}{2},\kappa^*)\mathfrak{B}_{\alpha}(\kappa+\tfrac{1}{2},\kappa^*) = -2\kappa-1.$$

After a little algebraic manipulation all these relations can be expressed as follows:

$$\mathfrak{U}_{\alpha}(\kappa,\kappa^*)\mathfrak{B}^{\alpha}(\kappa,\kappa^*) = 2\kappa, \quad \mathfrak{B}^{\alpha}(\kappa,\kappa^*)\mathfrak{U}_{\alpha}(\kappa,\kappa^*) = 2\kappa+1, \tag{50a}$$

$$\mathfrak{U}_{\alpha}(\kappa+\tfrac{1}{2},\kappa^*)\mathfrak{U}^{\alpha}(\kappa,\kappa^*) = \mathfrak{B}^{\alpha}(\kappa,\kappa^*)\mathfrak{B}_{\alpha}(\kappa+\tfrac{1}{2},\kappa^*) = 0, \tag{50b}$$

$$\mathfrak{B}_{\alpha}(\kappa+\tfrac{1}{2},\kappa^*)\mathfrak{U}_{\beta}(\kappa+\tfrac{1}{2},\kappa^*) - \mathfrak{U}_{\beta}(\kappa+\tfrac{1}{2},\kappa^*)\mathfrak{B}_{\alpha}(\kappa+\tfrac{1}{2},\kappa^*) = \epsilon_{\beta\alpha}, \tag{50c}$$

$$I_{\alpha\beta} = \mathfrak{U}_{\alpha}(\kappa,\kappa^*)\mathfrak{B}_{\beta}(\kappa,\kappa^*) - \kappa\epsilon_{\alpha\beta} = \mathfrak{B}_{\alpha}(\kappa+\tfrac{1}{2},\kappa^*)\mathfrak{U}_{\beta}(\kappa+\tfrac{1}{2},\kappa^*) + (\kappa+1)\epsilon_{\alpha\beta}. \tag{50d}$$

For the exceptional case $\kappa = -\tfrac{1}{2}$ (50) can be verified directly from (44) and (34). Thus the analogy between (50) and (23) is complete. The following relations are also worth noting:

$$\mathfrak{U}^{\alpha}(\kappa,\kappa^*) = \mathfrak{U}^{\alpha}(\kappa,-\kappa^*-1), \quad \mathfrak{B}^{\alpha}(\kappa,\kappa^*) = \mathfrak{B}^{\alpha}(\kappa,-\kappa^*-1); \tag{51a}$$

$$\mathfrak{B}^{\alpha}(-\kappa-\tfrac{1}{2},\kappa^*) = \mathfrak{U}^{\alpha}(\kappa,\kappa^*), \quad \mathfrak{U}^{\alpha}(-\kappa-\tfrac{1}{2},\kappa^*) = \mathfrak{B}^{\alpha}(\kappa,\kappa^*). \tag{51b}$$

Now put
$$U_{\alpha} = \overline{V^{\dot{\alpha}}} = iU^{\dot{\alpha}}, \quad V^{\alpha} = \overline{U}_{\dot{\alpha}} = iV_{\dot{\alpha}}. \tag{52a}$$

The equations $(25a)$ to $(25d)$ then hold unchanged with dotted indices provided these are raised and lowered by means of the spinors $\epsilon^{\dot{\alpha}\dot{\beta}}$, $\epsilon_{\dot{\alpha}\dot{\beta}}(\epsilon^{\dot{1}\dot{2}} = \epsilon_{\dot{1}\dot{2}} = 1)$.

Also
$$K_{\alpha\beta} = U_{\alpha}V_{\beta} - k\epsilon_{\alpha\beta} = U^{\dot{\alpha}}V^{\dot{\beta}} - k\epsilon^{\dot{\alpha}\dot{\beta}} \equiv K^{\dot{\alpha}\dot{\beta}} \tag{52b}$$

and
$$L_{\alpha\beta} = \frac{I}{k(k+1)}K^{\dot{\alpha}\dot{\beta}} - i\sqrt{\frac{(k-j)(k+j)}{(2k-1)(2k+1)}}\sqrt{\left(1+\frac{\mu^2}{k^2}\right)}U^{\dot{\alpha}}U^{\beta}$$
$$- V^{\dot{\alpha}}V^{\dot{\beta}}i\sqrt{\frac{(k-j)(k+j)}{(2k-1)(2k+1)}}\sqrt{\left(1+\frac{\mu^2}{k^2}\right)}$$
$$\equiv L^{\dot{\alpha}\dot{\beta}}. \tag{52c}$$

Put
$$I^{\dot{\alpha}\dot{\beta}} = \tfrac{1}{2}(K^{\dot{\alpha}\dot{\beta}} + iL^{\dot{\alpha}\dot{\beta}}). \tag{53a}$$

Then
$$[I_{\alpha\beta}, I_{\lambda\mu}] = 0, \tag{53b}$$

$$[I_{\dot{\alpha}\dot{\beta}}, I_{\lambda\dot{\mu}}] = \tfrac{1}{2}\underset{(\dot{\alpha}\dot{\beta})(\lambda\dot{\mu})}{S\ S}\,\epsilon_{\lambda\dot{\alpha}}I_{\dot{\mu}\dot{\beta}}, \tag{53c}$$

$$-\tfrac{1}{2}I^{\dot{\alpha}\dot{\beta}}I_{\dot{\alpha}\dot{\beta}} = \tfrac{1}{4}(J^2 + 2iI) = \tfrac{1}{4}(j+i\mu)^2 - \tfrac{1}{4} = \frac{\gamma^{*2}-1}{4}$$
$$= \kappa^*(\kappa^*+1). \tag{53d}$$

Now consider the product space $\Re \times \Re_{\frac{1}{2}}$ where $\Re_{\frac{1}{2}}$ is a two-dimensional space whose vectors transform as dotted spinors under transformations of the Lorentz group. As before, the components of a vector of the product space may be written as $\psi_{\dot\mu}$. Put

$$\psi^* = U^{\dot\alpha}\cdot\frac{1}{\sqrt{(2k+1)}}\,\psi_{\dot\alpha}, \tag{54a}$$

$$\psi = V^{\dot\alpha}\cdot\frac{1}{\sqrt{(2k+1)}}\,\psi_{\dot\alpha}. \tag{54b}$$

Then
$$\psi_{\dot\alpha} = \frac{1}{\sqrt{(2k+1)}}\,U_{\dot\alpha}\psi - \frac{1}{\sqrt{(2k+1)}}\,V_{\dot\alpha}\psi^*. \tag{54c}$$

As before, consider $\psi'_{\dot\lambda} = I_{\dot\lambda}{}^{\dot\mu}\psi_{\dot\mu}$ and construct ψ', $\psi^{*\prime}$. Write $\gamma^*_{k+\frac{1}{2}} = k+\frac{1}{2}+\dfrac{iI}{k+\frac{1}{2}}$ and put

$$\sqrt{2}\,\psi_{\gamma^*-1} = \sqrt{(\gamma^*_{k+\frac{1}{2}}+\gamma^*)}\,\psi + \sqrt{(\gamma^*_{k+\frac{1}{2}}-\gamma^*)}\,\psi^*, \tag{55a}$$

$$\sqrt{2}\,\psi_{\gamma^*+1} = \sqrt{(\gamma^*_{k+\frac{1}{2}}-\gamma^*)}\,\psi + \sqrt{(\gamma^*_{k+\frac{1}{2}}+\gamma^*)}\,\psi^*, \tag{55b}$$

the square roots being so chosen that

$$\sqrt{(\gamma^*_{k+\frac{1}{2}}+\gamma^*)}\,\sqrt{(\gamma^*_{k+\frac{1}{2}}-\gamma^*)} = \sqrt{\{(k+\tfrac{1}{2}-j)(k+\tfrac{1}{2}+j)\}}\,\sqrt{\left(1+\frac{I^2}{k^2}\right)}.$$

From (55) one finds that

$$\sqrt{2}\,\psi'_{\gamma^*-1} = \tfrac{1}{2}(1+\gamma^*)\sqrt{2}\,\psi_{\gamma^*-1}, \quad \sqrt{2}\,\psi'_{\gamma^*+1} = \tfrac{1}{2}(1-\gamma^*)\sqrt{2}\,\psi_{\gamma^*+1}.$$

Put $\psi_{\gamma^*-1} = \psi_{\kappa^*-\frac{1}{2}}$, $\psi_{\gamma^*+1} = \psi_{\kappa^*+\frac{1}{2}}$, and write

$$\mathfrak{V}^{\dot\alpha}(\kappa,\kappa^*) = \tfrac{1}{2}V^{\dot\alpha}\sqrt{\left(\frac{(k+1+\kappa+\kappa^*)(k+\kappa^*-\kappa)}{k(k+\frac{1}{2})}\right)}$$
$$+\frac{1}{2}\sqrt{\left(\frac{(k+\frac{1}{2}-\kappa^*+\kappa)(k-\frac{1}{2}-\kappa^*-\kappa)}{k(k+\frac{1}{2})}\right)}U^{\prime\dot\alpha}, \tag{56a}$$

$$\mathfrak{U}^{\dot\alpha}(\kappa,\kappa^*) = \tfrac{1}{2}V^{\dot\alpha}\sqrt{\left(\frac{(k+\frac{1}{2}-\kappa^*+\kappa)(k-\frac{1}{2}-\kappa^*-\kappa)}{k(k+\frac{1}{2})}\right)}$$
$$+\frac{1}{2}\sqrt{\left(\frac{(k+1+\kappa+\kappa^*)(k+\kappa^*-\kappa)}{k(k+\frac{1}{2})}\right)}U^{\prime\dot\alpha}, \tag{56b}$$

then
$$\psi_{\kappa^*-\frac{1}{2}} = \mathfrak{V}^{\dot\alpha}(\kappa,\kappa^*)\,\psi_{\dot\alpha}, \tag{57a}$$

$$\psi_{\kappa^*+\frac{1}{2}} = \mathfrak{U}^{\dot\alpha}(\kappa,\kappa^*+\tfrac{1}{2})\,\psi_{\dot\alpha}, \tag{57b}$$

$$(2\kappa^*+1)\,\psi_{\dot\alpha} = \mathfrak{U}_{\dot\alpha}(\kappa,\kappa^*)\,\psi_{\kappa^*-\frac{1}{2}} - \mathfrak{V}_{\dot\alpha}(\kappa,\kappa^*+\tfrac{1}{2})\,\psi_{\kappa^*+\frac{1}{2}}. \tag{57c}$$

Corresponding to (50) one gets

$$\mathfrak{U}_{\dot\alpha}(\kappa,\kappa^*)\,\mathfrak{V}^{\dot\alpha}(\kappa,\kappa^*) = 2\kappa^*, \quad \mathfrak{V}^{\dot\alpha}(\kappa,\kappa^*)\,\mathfrak{U}_{\dot\alpha}(\kappa,\kappa^*) = 2\kappa^*+1, \tag{58a}$$

$$\mathfrak{U}_{\dot\alpha}(\kappa,\kappa^*+\tfrac{1}{2})\,\mathfrak{U}^{\dot\alpha}(\kappa,\kappa^*) = \mathfrak{V}^{\dot\alpha}(\kappa,\kappa^*)\,\mathfrak{V}_{\dot\alpha}(\kappa,\kappa^*+\tfrac{1}{2}) = 0, \tag{58b}$$

$$\mathfrak{V}_{\dot\alpha}(\kappa,\kappa^*+\tfrac{1}{2})\,\mathfrak{U}_{\dot\beta}(\kappa,\kappa^*+\tfrac{1}{2}) - \mathfrak{U}_{\dot\beta}(\kappa,\kappa^*)\,\mathfrak{V}_{\dot\alpha}(\kappa,\kappa^*) = \epsilon_{\dot\beta\dot\alpha}, \tag{58c}$$

$$I_{\dot\alpha\beta} = \mathfrak{U}_{\dot\alpha}(\kappa,\kappa^*)\,\mathfrak{V}_{\beta}(\kappa,\kappa^*) - \kappa^*\epsilon_{\alpha\beta} = \mathfrak{V}_{\dot\alpha}(\kappa,\kappa^*+\tfrac{1}{2})\,\mathfrak{U}_{\beta}(\kappa,\kappa^*+\tfrac{1}{2}) + (\kappa^*+1)\epsilon_{\alpha\beta}, \tag{58d}$$

and $\qquad \mathfrak{U}^{\dot{a}}(\kappa, \kappa^*) = \mathfrak{U}^{\dot{a}}(-\kappa-1, \kappa^*), \quad \mathfrak{B}^{\dot{a}}(\kappa, \kappa^*) = \mathfrak{B}^{\dot{a}}(-\kappa-1, \kappa^*),$ \hfill (59a)

$$\mathfrak{U}^{\dot{a}}(\kappa, -\kappa^*-\tfrac{1}{2}) = \mathfrak{B}^{\dot{a}}(\kappa, \kappa^*), \quad \mathfrak{B}^{\dot{a}}(\kappa, -\kappa^*-\tfrac{1}{2}) = \mathfrak{U}^{\dot{a}}(\kappa, \kappa^*). \tag{59b}$$

Now let the numbers γ and γ^* be given for a particular representation. Then from (36) and (40)

$$j = \frac{\epsilon\gamma + \epsilon^*\gamma^*}{2}, \quad i\mu = \frac{\epsilon^*\gamma^* - \epsilon\gamma}{2}, \quad iI = \frac{\gamma^{*2} - \gamma^2}{4}, \tag{60}$$

where $\epsilon = \pm 1, \epsilon^* = \pm 1$, the two ambiguities being independent. Since $2j$ is integral and $\geqslant 0$ it follows that at least one of the two quantities $\gamma \pm \gamma^*$ must be an integer. In case only one of these quantities is integral the value of j is completely determined from those of γ and γ^*. This determines ϵ and ϵ^* and therefore μ except in the case when $j = 0$. But even in this case I and μ^2 are completely determined. Since only j, I and μ^2 enter in the representation it follows that the representation is completely determined by the pair (γ, γ^*). However, when both $\gamma + \gamma^*$ and $\gamma - \gamma^*$ are integral and distinct there are two possible values of j, say j^+ and j^-, such that $j^+ > j^-$. Therefore in this case there are two distinct representations corresponding to the pair (γ, γ^*). They may be denoted by $\mathfrak{D}^+_{\kappa,\kappa^*}$ and $\mathfrak{D}^-_{\kappa,\kappa^*}$ respectively, corresponding to the j value j^+ and j^-. Thus a given pair (κ, κ^*) completely determines the irreducible representation unless κ, κ^* are both of the form $\tfrac{1}{2}n\,(n \neq -2)$. It is easy to see that $\mathfrak{D}^+_{\kappa,\kappa^*}$ is always an infinite representation. $\mathfrak{D}^-_{\kappa,\kappa^*}$ is finite only if j^- and j^+ differ by an integer which is the case when both κ and κ^* are of the form $\tfrac{1}{2}n\,(n \neq -1)$. Since reversing the sign of γ or γ^* does not alter the representation the four pairs (κ, κ^*), $(-\kappa-1, \kappa^*), (\kappa, -\kappa^*-1), (-\kappa-1, -\kappa^*-1)$ determine the same representation. From (60) it follows that at least one of the two quantities $\kappa - \kappa^*$ and $\kappa + \kappa^* + 1$ is of the form $\tfrac{1}{2}n$. It may therefore be fixed for convenience that κ, κ^* are to be so chosen that $\kappa - \kappa^*$ is an integer or half-odd integer. This is the only condition to be imposed on the choice of κ, κ^*.

It is useful to examine the procedure which led to the matrices $\mathfrak{U}_\alpha, \mathfrak{B}_\alpha$ and $\mathfrak{U}_{\dot{a}}, \mathfrak{B}_{\dot{a}}$ a little more closely. From (15), (36) and (40)

$$\gamma^2 = J^2 + 1 - 2iI, \quad \gamma^{*2} = J^2 + i + 2iI. \tag{61}$$

(61) may be used to define γ and γ^* as operators in terms of $\mathbf{K}$ and $\mathbf{L}$ even for those reducible representations in which J^2 and I are no longer represented by multiples of the unit matrix. In particular in the product space $\mathfrak{R} \times \mathfrak{R}_1$ the infinitesimal transformations $\mathfrak{R}$ and $\mathfrak{L}$ corresponding to $\mathbf{K}$ and $\mathbf{L}$ are given by

$$\mathfrak{R} = \mathbf{K} + \tfrac{1}{2}\boldsymbol{\sigma}, \quad \mathfrak{L} = \mathbf{L} + \tfrac{1}{2}\boldsymbol{\sigma},$$

where $\boldsymbol{\sigma} = (\sigma_1, \sigma_2, \sigma_3)$ is the vector constructed from the three Pauli matrices $(\sigma_1\sigma_2 = i\sigma_3)$. Denote by $\mathcal{I}$ and $\mathcal{J}^2$ the operators in the product space corresponding to I and J^2 so that

$$\mathcal{I} = (\mathfrak{R}\mathfrak{L}) = (\mathbf{K} + \tfrac{1}{2}\boldsymbol{\sigma})(\mathbf{L} + \tfrac{1}{2}i\boldsymbol{\sigma}) = I + \tfrac{3}{4}i + iA, \tag{62a}$$

$$\mathcal{J}^2 = \mathfrak{R}^2 - \mathfrak{L}^2 = (\mathbf{K} + \tfrac{1}{2}\boldsymbol{\sigma})^2 - (\mathbf{L} + \tfrac{1}{2}i\boldsymbol{\sigma})^2 = J^2 + \tfrac{3}{2} + 2A, \tag{62b}$$

where $\qquad\qquad A = \tfrac{1}{2}(\mathbf{K} . \boldsymbol{\sigma}) - \tfrac{1}{2}i(\mathbf{L} . \boldsymbol{\sigma}). \tag{62c}$

Now notice that A is an invariant since

$$A = -\tfrac{1}{2}I^{kl}I_{kl}(\tfrac{1}{2}),$$

where I^{kl} and $I^{kl}(\tfrac{1}{2})$ are the infinitesimal transformations of the spaces $\mathfrak{R}$ and $\mathfrak{R}_{\tfrac{1}{2}}$ respectively. Hence A commutes with $\mathfrak{K}$ and $\mathfrak{L}$. Therefore in an irreducible subspace of $\mathfrak{R} \times \mathfrak{R}_{\tfrac{1}{2}}$ it must be represented by a multiple of the unit matrix. From (62c) one finds that

$$A^2 = \tfrac{1}{4}(J^2 - 2iI) - A,$$

i.e.

$$(A + \tfrac{1}{2})^2 = \tfrac{1}{4}(J^2 - 2iI + 1) = \tfrac{1}{4}\gamma^2.$$

So that the eigenvalues of A are $-\tfrac{1}{2}(1 \pm \gamma)$. If Γ and Γ^* denote the operators of the product space corresponding to γ and γ^* respectively, then

$$\Gamma^{*2} = \mathscr{J}^2 + 2i\mathscr{J} + 1 = J^2 + 1 + 2iI = \gamma^{*2}, \tag{63a}$$

$$\Gamma^2 = \mathscr{J}^2 - 2i\mathscr{J} + 1 = J^2 + 1 - 2iI + 4A + 3 = \gamma^2 + 4A + 3. \tag{63b}$$

Hence for the two irreducible subspaces corresponding to the two eigenvalues of A one gets

$$\Gamma^2 = (\gamma \mp 1)^2.$$

Hence one may choose $\Gamma^* = \gamma^*$, $\Gamma = \gamma \mp 1$, the choice of sign being dependent on the corresponding choice in the eigenvalues $-\tfrac{1}{2}(1 \pm \gamma)$ of A. $-A$ is precisely the transformation $\psi_\mu \rightarrow \psi'_\mu = I_\mu{}^\lambda \psi_\lambda$ (the negative sign coming from the fact that the spinor component $K_{12} = -K_3$). Hence it is clear that $\psi_{\kappa \pm \frac{1}{2}}$ and $\psi_{\kappa^* \pm \frac{1}{2}}$ transform according to the representations $(\kappa \pm \tfrac{1}{2}, \kappa^*)$ and $(\kappa, \kappa^* \pm \tfrac{1}{2})$ respectively. In view of this it is more logical to write $\psi(\kappa \pm \tfrac{1}{2}, \kappa^*)$ and $\psi(\kappa, \kappa^* \pm \tfrac{1}{2})$ respectively. The j values of these representations are given by $|\kappa \pm \tfrac{1}{2} - \kappa^*|$ and $|\kappa - \kappa^* \mp \tfrac{1}{2}|$ respectively, if that of the original representation $\mathfrak{D}_{\kappa, \kappa^*}$ was $|\kappa - \kappa^*|$.

Let κ_0 and κ_0^* be any two complex numbers such that $2(\kappa_0 - \kappa_0^*)$ is an integer. Consider all numbers of the forms $\kappa = \kappa_0 + \tfrac{1}{2}n, \kappa^* = \kappa_0^* + \tfrac{1}{2}n^*$, where n, n^* are integers, and construct the matrices $\mathfrak{U}^\alpha, \mathfrak{B}^\alpha, \mathfrak{U}^\beta, \mathfrak{B}^\beta$ as follows

$$
\left.
\begin{aligned}
(\kappa, \kappa^* \,|\, \mathfrak{U}^\alpha \,|\, \kappa - \tfrac{1}{2}, \kappa^*) &= \mathfrak{U}^\alpha(\kappa, \kappa^*), \\
(\kappa - \tfrac{1}{2}, \kappa^* \,|\, \mathfrak{B}^\alpha \,|\, \kappa, \kappa^*) &= \mathfrak{B}^\alpha(\kappa, \kappa^*), \\
(\kappa, \kappa^* \,|\, \mathfrak{U}^{\dot\alpha} \,|\, \kappa, \kappa^* - \tfrac{1}{2}) &= \mathfrak{U}^{\dot\alpha}(\kappa, \kappa^*), \\
(\kappa, \kappa^* - \tfrac{1}{2} \,|\, \mathfrak{B}^{\dot\alpha} \,|\, \kappa, \kappa^*) &= \mathfrak{B}^{\dot\alpha}(\kappa, \kappa^*),
\end{aligned}
\right\} \tag{64}
$$

all the other matrix elements being zero. Then it is easy to verify directly from (44), (52a) and (56) that $\mathfrak{U}^\alpha, \mathfrak{B}^\alpha$ commute with $\mathfrak{U}^\beta, \mathfrak{B}^\beta$, so that, for example,

$$\mathfrak{U}^\alpha(\kappa, \kappa^*)\mathfrak{B}^\beta(\kappa - \tfrac{1}{2}, \kappa^* + \tfrac{1}{2}) = \mathfrak{B}^\beta(\kappa, \kappa^* + \tfrac{1}{2})\mathfrak{U}^\alpha(\kappa^* + \tfrac{1}{2}, \kappa).$$

Similarly define $J^{\alpha\beta}$ and $J^{\dot\alpha\beta}$ by

$$J^{\alpha\beta} = \mathfrak{U}^\alpha\mathfrak{B}^\beta + \mathfrak{U}^\beta\mathfrak{B}^\alpha, \tag{65a}$$

$$J^{\dot\alpha\beta} = \mathfrak{U}^{\dot\alpha}\mathfrak{B}^\beta + \mathfrak{U}^\beta\mathfrak{B}^{\dot\alpha}. \tag{65b}$$

$(\kappa, \kappa^* \mid J^{\alpha\beta} \mid \kappa, \kappa^*)$ and $(\kappa, \kappa^* \mid J^{\dot\alpha\dot\beta} \mid \kappa, \kappa^*)$ are equal to the $I^{\alpha\beta}$ and $I^{\dot\alpha\dot\beta}$ of the representation $\mathfrak{D}_{\kappa,\kappa^*}$. Now, in order that an operator C_α may transform as an undotted spinor under Lorentz transformation, it is necessary and sufficient that

$$[J_{\alpha\beta}, C_\gamma] = \tfrac{1}{2} \underset{(\alpha\beta)}{S} \epsilon_{\gamma\alpha} C_\beta, \tag{66a}$$

$$[J_{\dot\alpha\dot\beta}, C_\gamma] = 0. \tag{66b}$$

A similar condition holds with the interchange of dotted and undotted indices for a dotted spinor $C_{\dot\gamma}$. The commutation relations of $\mathfrak{U}^\alpha, \mathfrak{V}^\beta$ and $\mathfrak{U}^{\dot\alpha}, \mathfrak{V}^{\dot\beta}$ thus ensure that the former transform as undotted spinors and the latter as dotted spinors.

4. Unitary representations

Since I is real for unitary representations it follows from (41) that either $\kappa^* + \kappa + 1$ is pure imaginary or else $\kappa^* - \kappa = \pm j = 0$. In the first case (20) is satisfied automatically and κ, κ^* are given by (1a). In the second case it follows from (20) that $\kappa = \kappa^*$ is real and

$$1 \geqslant (\kappa + \kappa^* + 1)^2 = (2\kappa + 1)^2,$$

so that $\tfrac{1}{2} \geqslant |\kappa + \tfrac{1}{2}|$, i.e. $0 \geqslant \kappa = \kappa^* \geqslant -1$ in accordance with (1b). In particular, if κ, κ^* are both of the form $\tfrac{1}{2}n(n + -2)$ then $j^+ = |\kappa + \tfrac{1}{2}| + |\kappa^* + \tfrac{1}{2}|$ and

$$\pm j^- = |\kappa + \tfrac{1}{2}| - |\kappa^* + \tfrac{1}{2}|.$$

Therefore $\mathfrak{D}_{\kappa,\kappa^*}^-$ is unitary only if $\tfrac{1}{2} \geqslant |\kappa + \tfrac{1}{2}| = |\kappa^* + \tfrac{1}{2}|$ while $\mathfrak{D}_{\kappa,\kappa^*}^+$ is unitary whenever $|\kappa + \tfrac{1}{2}| = |\kappa^* + \tfrac{1}{2}|$, since in the latter case

$$\pm iI = (|\kappa + \tfrac{1}{2}| - |\kappa^* + \tfrac{1}{2}|)(|\kappa + \tfrac{1}{2}| + |\kappa^* + \tfrac{1}{2}|) = 0$$

and $j = j^+ \neq 0$. By reflexion I goes over into $-I$ so that a representation is reflexion invariant only if I is zero, i.e. $\kappa + \tfrac{1}{2} = \pm(\kappa^* + \tfrac{1}{2})$. Hence the representations corresponding to (1b) are always reflexion invariant while those corresponding to (1a) are so only if $\nu = 0$.

5. Expinors

Consider two complex variables u_1 and u_2 which transform as components of a spinor under Lorentz transformations. Put

$$u_{\dot 1} = \overline{u_1}, \quad u_{\dot 2} = \overline{u_2}, \tag{67a}$$

$$\dot u_1 = -\overline{u_2}, \quad \dot u_2 = \overline{u_1}, \tag{67b}$$

where the bar denotes complex conjugate. Then $u_{\dot 1}, u_{\dot 2}$ transform as components of a dotted spinor and $\dot u_1, \dot u_2$ transform the same way as u_1, u_2 under spatial rotations, i.e. under unitary spinor transformations (see Van der Waerden 1932). Now consider terms of the type

$$\frac{u_1^p u_2^q \dot u_1^r \dot u_2^s}{\varDelta^{n-\lambda}} = \frac{(-1)^s u_1^p u_2^q \dot u_1^s \dot u_2^r}{\varDelta^{n-\lambda}}, \tag{68}$$

where

$$\varDelta = u_1 u_{\dot 1} + u_2 u_{\dot 2} = u_1 \dot u_2 - u_2 \dot u_1 = |u_1|^2 + |u_2|^2 \tag{69}$$

25-2

and p, q, r, s, n are integers $\geqslant 0$ while λ is a fixed complex number. Notice that on account of (69) the various terms for different values of p, q, r, s, n are in general *not* linearly independent. To remedy this defect one proceeds as follows. Put

$$U_M^{K,j} = \sum_{M=m+m'} \frac{2k!}{k-m!\,k+m!} \frac{2k'!}{k'-m'!\,k'+m'!} u_1^{k-m} u_2^{k+m} \dot{u}_1^{k'-m'} \dot{u}_2^{k'+m'}, \qquad (70a)$$

$$K = k+k', \qquad (70b)$$

$$j = k'-k, \qquad (70c)$$

where $2k, 2k'$ are integers $\geqslant 0$. M can take the values $K, K-1, \ldots, -K$ and similarly m and m' the values $k, k-1, \ldots, -k$ and $k', k'-1, \ldots, -k'$ respectively. The summation in (70a) is over all values of m and m' such that $m+m' = M$. It is well known (see Van der Waerden 1932) that every polynomial $P^{k,\,k'}$ which is homogeneous in each of the two sets of variables u and $\dot{u}$ and has the degrees $2k$ and $2k'$ respectively can be expressed *uniquely* in terms of $U_M^{K,j}$ as follows:

$$P^{k,\,k'} = \sum_{0 \leqslant \nu \leqslant 2k,\, 2k'} \sum_{M=-K+\nu}^{K-\nu} A_{M,\nu} \Delta^{\nu} U_M^{K-\nu,\,j}. \qquad (71)$$

ν runs through all integral values in the range $2k, 2k' \geqslant \nu \geqslant 0$ and $A_{M,\nu}$ are constants. Since every polynomial in u and $\dot{u}$ can be split up into a sum of homogeneous polynomials the expansion (71) is always possible and is *unique*. One can therefore express the numerator of (68) in the form (71) and arrange the result according to descending powers of Δ. Now consider a series of terms of the type (68) which is homogeneous and of degree $\lambda-j$ and $\lambda+j$ in u and $\dot{u}$ respectively, where

$$2j = r+s-p-q$$

is an integer. Then it follows from (71) that it can be put in the form

$$\sum_{K \geqslant |j|}^{\infty} \sum_{M=-K}^{K} a_M^{K,j} \frac{U_M^{K,j}}{\Delta^{K-\lambda}}, \qquad (72)$$

where $a_M^{K,j}$ are constants. K runs through the values $|j|, |j|+1, \ldots, \infty$. Now apply a spinor transformation to u_1 and u_2

$$u_\mu = \alpha_\mu{}^\nu u'_\nu, \qquad u_{\dot\mu} = \overline{\alpha_\mu{}^\nu} u'_{\dot\nu}, \qquad (73)$$

and as before put
$$\dot{u}'_1 = -\overline{u'_2}, \qquad \dot{u}'_2 = \overline{u'_1}. \qquad (74)$$

Then $U_M^{K,j} \equiv U_M^{K,j}(u, \dot{u})$ goes over into a polynomial in u' and $\dot{u}'$ of degree $K-j$ and $K+j$ respectively, which from (71) can again be expressed linearly in terms of $\Delta'^{\nu} U_M^{K-\nu,\,j}$, where $\Delta' = |u'_1|^2 + |u'_2|^2$ and $U_M'^{K,j} = U_M^{K,j}(u', \dot{u}')$. Further

$$\Delta = |u_1|^2 + |u_2|^2 = |\alpha_1{}^1 u'_1 + \alpha_1{}^2 u'_2|^2 + |\alpha_2{}^1 u'_1 + \alpha_2{}^2 u'_1|^2$$

$$= \|\alpha\|(|u'_1|^2 + |u'_2|^2) - (|\alpha_1{}^1 \dot{u}'_1 + \alpha_1{}^2 \dot{u}'_1|^2 + |\alpha_2{}^1 \dot{u}'_1 + \alpha_2{}^2 \dot{u}'_2|^2), \qquad (75)$$

where $\|\alpha\| \equiv |\alpha_1{}^1|^2 + |\alpha_1{}^2|^2 + |\alpha_2{}^1|^2 + |\alpha_2{}^2|^2 > 0$. Thus

$$\Delta = \|\alpha\|\Delta'\left\{1 - \frac{D'}{\|\alpha\|\Delta'}\right\}, \tag{76}$$

where

$$D' = |\alpha_1{}^1\dot{u}_1' + \alpha_1{}^2\dot{u}_2'|^2 + |\alpha_2{}^1\dot{u}_1' + \alpha_2{}^2\dot{u}_2'|^2 \tag{77}$$

and

$$0 < \frac{D'}{\|\alpha\|\Delta'} < 1, \tag{78}$$

since $\Delta, \Delta' > 0$, if one excludes the case $u_1 = u_2 = 0$. Therefore one can expand $1/\Delta^{K-\lambda}$ in (72) by the binomial theorem in ascending powers of $D'/\|\alpha\|\Delta'$. Now D' is a polynomial in u' and $\dot{u}'$. Hence the product of its powers with $U'^{K,j}_M$ can again be expressed linearly in the form (71). Ignoring for the moment the question of convergence, which will be examined presently one may collect the coefficients of $U'^{K,j}_M/\Delta'^{K-\lambda}$ together and write

$$\sum_{K,M}^{\infty} a_M^{K,j}\frac{U_M^{K,j}}{\Delta^{K-\lambda}} = \sum_{K,M}^{\infty} a_M'^{K,j}\frac{U_M'^{K,j}}{\Delta'^{K-\lambda}}, \tag{79}$$

where $a_M'^{K,j}$ are the new coefficients. The numbers $\lambda - j$ and $\lambda + j$ being the degrees of the series in the variables u and $\dot{u}$ respectively, are clearly left unaltered by the homogeneous linear transformation (73). Hence only terms of the type $U'^{K,j}_M/\Delta'^{K-\lambda}$ can occur on the right-hand side of (79). The linear transformation $a_M^{K,j} \to a_M'^{K,j}$ of the coefficients thus constitutes a representation of the Lorentz transformation associated with (73).

To examine rigorously the validity of our considerations make the substitution

$$u_1 = \rho\cos\tfrac{1}{2}\theta\, e^{\frac{1}{2}i\chi}e^{\frac{1}{2}i\phi}, \quad u_{\dot{1}} = \rho\cos\tfrac{1}{2}\theta\, e^{-\frac{1}{2}i\chi}e^{-\frac{1}{2}i\phi}, \tag{80a}$$

$$u_2 = \rho\sin\tfrac{1}{2}\theta\, e^{\frac{1}{2}i\chi}e^{-\frac{1}{2}i\phi}, \quad u_{\dot{2}} = \rho\sin\tfrac{1}{2}\theta\, e^{-\frac{1}{2}i\chi}e^{+\frac{1}{2}i\phi}, \tag{80b}$$

where ρ, θ, χ, ϕ are real and $\rho > 0, \pi \geqslant \theta \geqslant 0, 2\pi > \phi \geqslant 0, 4\pi > \chi \geqslant 0$. Put $x = \cos\theta$ and

$$\frac{1}{\sqrt{(2\pi)}}A_M^{K,j} = \sqrt{\left(\frac{K-j!}{K-M!}\frac{K+j!}{K+M!}\frac{2}{2K+1}\right)}a_M^{K,j}. \tag{81a}$$

Then (72) becomes

$$\rho^{2\lambda}e^{-ij\chi}\sum_{K\geqslant|j|}^{\infty}\sum_{M=-K}^{K}A_M^{K,j}\frac{1}{\sqrt{(2\pi)}}P_M^{K,j}(x)\,e^{-iM\phi}, \tag{81b}$$

where

$$P_M^{K,j}(x) = \frac{(-1)^{K+j}}{2^K}\sqrt{\left(\frac{K-M!}{K+M!}\frac{(K+\tfrac{1}{2})}{K+j!\,K-j!}\right)}$$

$$\times (1-x)^{\frac{1}{2}(M-j)}(1+x)^{\frac{1}{2}(M+j)}\frac{d^{K+M}}{dx^{K+M}}\{(1-x)^{K+j}(1+x)^{K-j}\}. \tag{81c}$$

The functions $Q_M^{K,j} \equiv \dfrac{1}{\sqrt{(2\pi)}}P_M^{K,j}(x)\,e^{-iM\phi}$ are orthogonal and normalized, i.e.

$$\int_{\phi=0}^{2\pi}\int_{x=-1}^{1}\overline{Q_{M'}^{K',j}}\,Q_M^{K,j}\,dx\,d\phi = \begin{cases}1 & \text{if } K=K', M=M', \\ 0 & \text{otherwise.}\end{cases} \tag{82}$$

For convenience arrange these orthogonal functions $Q_M^{K,j}$ in a sequence

$$Q_\nu \ (\nu = 1, 2, \ldots, \infty)$$

and denote the interval $-1 \leqslant x \leqslant 1, 0 \leqslant \phi \leqslant 2\pi$ by G. Then (82) can be written as

$$\int_G \overline{Q_\mu} Q_\nu \, dx \, d\phi = \delta_{\mu\nu}. \tag{83}$$

Denote a series $\sum_1^\infty A_\nu Q_\nu$ by A. A_ν will be called the coefficients of A. It is important to note that A is not the sum of the series since this sum may not exist in general. A is merely an abbreviation for the infinite sequence of complex numbers A_ν ($\nu = 1, 2, \ldots, \infty$) which completely characterizes the series. Two series A and B are to be regarded as distinct unless $A_\nu = B_\nu$ for all ν. If c is a constant cA denotes a series with the coefficients cA_ν. Similarly $A + B$ stands for the series with the coefficients $A_\nu + B_\nu$. $A = 0$ is the series such that $A_\nu = 0$ for all ν. The norm $\| A \|$ of a series A is a non-negative number defined by

$$\| A \|^2 = \sum_1^\infty | A |^2. \tag{84}$$

Clearly $\| A \| > 0$ unless $A = 0$ when $\| A \| = 0$. If the series on the right-hand side of (84) is divergent we say the norm is infinite.

Let $f(x,\phi)$ be a continuous function of x and ϕ defined in G. Then by the norm of f is meant the number $\| f \| \geqslant 0$ given by

$$\| f \|^2 = \int_G | f |^2 \, dx \, d\phi. \tag{85}$$

Now denote by A^N the partial sum $\sum_1^N A_\nu Q_\nu$ of the series A. A^N is a continuous function of x and ϕ. Hence from (85) and (83)

$$\| A^N \|^2 = \sum_1^N | A_\nu |^2, \tag{86a}$$

so that
$$\| A \|^2 = \lim_{N \to \infty} \| A^N \|^2. \tag{86b}$$

This rather natural connexion justifies the use of the same term norm for both A and f. Now it is well known (see Courant & Hilbert 1931) that the functions Q_ν form a complete set in the sense that for every continuous function defined in G there exists a series A such that

$$\lim_{N \to \infty} \| f - A^N \| = 0. \tag{87}$$

Clearly this series is unique, since if A and B be two such series then by Schwarz's inequality which clearly holds for norms

$$\| A - B \| = \lim_{N \to \infty} \| A^N - B^N \| \leqslant \lim_{N \to \infty} \{ \| f - A^N \| + \| f - B^N \| \} = 0.$$

Hence $A = B$. It can be proved from (87) and (83) (see Courant & Hilbert 1931) that

$$A_\nu = \int_G \overline{Q}_\nu f \, dx \, d\phi = (Q_\nu, f) \tag{88a}$$

and

$$\|f\| = \|A\|, \tag{88b}$$

so that A has a finite norm. Now consider (81b). Write it as

$$\rho^{2\lambda} e^{-ijx} A. \tag{89}$$

Apply the spinor transformation to the u's and introduce the four real variables $\rho', \chi', \theta', \phi'$ corresponding to u_1', u_2' by equations similar to (80). Put $x' = \cos\theta'$. Then ρ, χ, x, ϕ are functions of ρ', χ', x', ϕ'. Notice that

$$\frac{\rho^2}{\rho'^2} = \frac{|u_1|^2 + |u_2|^2}{|u_1'|^2 + |u_2'|^2}, \quad e^{i(\chi-\chi')} = \frac{u_1 u_2}{u_1' u_2'} \begin{vmatrix} u_1' u_2' \\ u_1 u_2 \end{vmatrix}, \tag{90}$$

and so ρ^2/ρ'^2 and $e^{i(\chi-\chi')}$ are functions of x' and ϕ' alone. Obviously the same is true of x and ϕ. Therefore

$$\rho^{2\lambda} e^{-ijx} A^N = \rho'^{2\lambda} e^{-ijx'} \left\{ \left(\frac{\rho^2}{\rho'^2}\right)^\lambda e^{-ij(\chi-\chi')} \sum_1^N A_\nu Q_\nu(x,\phi) \right\}$$

$$\equiv \rho'^{2\lambda} e^{-ijx'} f_N(x', \phi'). \tag{91}$$

Clearly $f_N(x', \phi')$ is a continuous function of x', ϕ'. Now put

$$A'_{\nu, N} = \int_G \overline{Q_\nu(x', \phi')} f_N(x', \phi') \, dx' \, d\phi', \tag{92}$$

and denote by A'_N the series whose coefficients are $A'_{\nu, N}$. From the completeness of the set Q_ν it follows that

$$\lim_{n \to \infty} \|f_N - A_N''\|^2 = \lim_{n \to \infty} \int_G |f_N(x', \phi') - A_N''(x', \phi')|^2 \, dx' \, d\phi' = 0, \tag{93}$$

where $A_N''(x', \phi') = \sum_{\nu=1}^n A'_{\nu, N} Q_\nu(x', \phi')$. Therefore

$$\|A'_{N+p} - A'_N\| = \lim_{n \to \infty} \|A_{N+p}''' - A_N'''\|$$

$$\leqslant \lim_{n \to \infty} \left\{ \|f_{N+p} - A_{N+p}''\| + \|f_N - A_N''\| + \|f_{N+p} - f_N\| \right\}$$

$$= \|f_{N+p} - f_N\| \tag{94}$$

from (93). But from (91)

$$\|f_{N+p} - f_N\|^2 = \int_G \left| \left(\frac{\rho^2}{\rho'^2}\right)^\lambda \right|^2 \left| \sum_{N+1}^{N+p} A_\nu Q_\nu(x, \phi) \right|^2 dx' \, d\phi'. \tag{95}$$

Now since the determinant of any spinor transformation is unity it follows that

$$\frac{\partial(u_1, u_2, u_1', u_2')}{\partial(\rho', \chi', \phi', \theta')} = \frac{\partial(u_1', u_2', u_1', u_2')}{\partial(\rho', \chi', \phi', \theta')} = \tfrac{1}{2}\rho'^3 \sin\theta'. \tag{96}$$

(Here the usual notation for the Jacobian has been used.) Therefore

$$\rho^3 \sin\theta \, \frac{\partial(\rho,\chi,\phi,\theta)}{\partial(\rho',\chi',\phi',\theta')} = \rho'^3 \sin\theta',$$

or
$$\frac{\partial(\rho,\chi,x,\phi)}{\partial(\rho',\chi',x',\phi')} = \frac{\rho'^3}{\rho^3}. \tag{97}$$

But since ρ/ρ', $e^{i(\chi-\chi')}$, x and ϕ are functions of x' and ϕ' alone

$$\frac{\partial(\rho,\chi,x,\phi)}{\partial(\rho',\chi',x',\phi')} = \frac{\rho}{\rho'}\,\frac{\partial(x,\phi)}{\partial(x',\phi')},$$

so that from (97)
$$\frac{\partial(x',\phi')}{\partial(x,\phi)} = \left(\frac{\rho}{\rho'}\right)^4 = \left(\frac{\Delta}{\Delta'}\right)^2. \tag{98}$$

Therefore, taking x,ϕ as the independent variables in (95), one gets

$$\|f_{N+p} - f_N\|^2 = \int_G \left|\left(\frac{\Delta}{\Delta'}\right)^{\lambda+1}\right|^2 \left|\sum_{N+1}^{N+p} A_\nu Q_\nu(x,\phi)\right|^2 dx\,d\phi.$$

But from (76)
$$\frac{\Delta}{\Delta'} < \|\alpha\|.$$

Similarly by considering the inverse transformation α^{-1} one finds that $\Delta'/\Delta < \|\alpha^{-1}\|$. Hence

$$\frac{1}{\|\alpha^{-1}\|} < \frac{\Delta}{\Delta'} < \|\alpha\|.$$

Therefore $|(\Delta/\Delta')^{\lambda+1}|^2$ is bounded for every fixed value of λ and x, and so

$$\|f_{N+p} - f_N\|^2 \leqslant C(\alpha,\lambda) \int_G \left|\sum_{N+1}^{N+p} A_\nu Q_\nu(x,\phi)\right|^2 dx\,d\phi, \tag{99}$$

where $C(\alpha,\lambda)$ is a real positive constant depending on α and λ. Hence from (94) and (99)

$$\|A'_{N'} - A'_N\| \leqslant C(\alpha,\lambda) \sum_{N_0}^{\infty} |A_\nu|^2 \quad (N, N' \geqslant N_0).$$

So that from the convergence of $\sum_1^{\infty} |A_\nu|^2$ it follows that

$$\lim_{\substack{N' \to \infty \\ N \to \infty}} \|A'_{N'} - A'_N\| = 0. \tag{100}$$

Therefore $A'_{\nu,N}$ tends to a limit A'_ν as $N \to \infty$ in such a way that if A' be the corresponding series then

$$\lim_{N \to \infty} \|A' - A'_N\| = 0. \tag{101}$$

Also clearly

$$\sum_{\nu=1}^{n} |A'_{N,\nu}|^2 \leqslant \int_G |f_N(x',\phi')|^2 dx'\,d\phi' = \int_G \left|\left(\frac{\Delta}{\Delta'}\right)^{\lambda+1}\right|^2 |A^N(x,\phi)|^2 dx\,d\phi$$

$$\leqslant C(\alpha,\lambda) \sum_{\nu=1}^{N} |A_\nu|^2 \leqslant C(\alpha,\lambda)\|A\|^2.$$

Therefore $\| A' \|$ is finite. A' has the property that

$$\lim_{\substack{N \to \infty \\ N' \to \infty}} \int_G |\, \rho^{2\lambda} e^{-ij\chi} A^N(x,\phi) - \rho'^{2\lambda} e^{-ij\chi'} A'^{N'}(x',\phi')\,|^2 \, dx' d\phi' = 0. \tag{102}$$

This follows from the inequality

$$\| f_N - A'^n \| \leqslant \| f_{N_0} - A'^n_{N_0} \| + \| A'^n - A'^n_{N_0} \| + \| f_N - f_{N_0} \|$$
$$\leqslant \| f_{N_0} - A'^n_{N_0} \| + \| A' - A'_{N_0} \| + \| f_N - f_{N_0} \|. \tag{103}$$

By choosing n_0 and N_0 sufficiently large all the three terms on the right can be made arbitrarily small for $n \geqslant n_0$, $N \geqslant N_0$. On going over to the variables x, ϕ in (102) and using (98) one finds that

$$\lim_{\substack{N \to \infty \\ N' \to \infty}} \int_G |\, \rho^{2\lambda} e^{-ij\chi} A^N(x,\phi) - \rho'^{2\lambda} e^{-ij\chi'} A'^{N'}(x',\phi')\,|^2 \, dx d\phi = 0. \tag{104}$$

Thus (102) implies (104) and vice versa. Also it is clear that A' is the only series satisfying (102), since if B' were another such series then

$$\| A'^n - B'^n \| \leqslant \| f_N - A'^n \| + \| f_N - B'^n \|,$$

so that on making N and n tend to infinity

$$\| A' - B' \| = \lim_{n \to \infty} \| A'^n - B'^n \| = 0.$$

Hence $A' = B'$. The transformation $A \to A'$ is thus uniquely determined by the spinor transformation α. Clearly it is a linear transformation in the sense that if $A \to A'$ and $B \to B'$ then $A + B \to A' + B'$ and $cA \to cA'$, where c is a complex number. In fact it is a linear transformation of the Hilbert space consisting of all sequences for which $\sum_1^\infty |\, A_\nu \,|^2$ is convergent. These transformations thus form a representation of the Lorentz group. Also notice that

$$\| A'_N \|^2 = \int_G |\, f_N(x',\phi')\,|^2 \, dx' d\phi' = \int_G \left(\frac{\varDelta}{\varDelta'} \right)^{\lambda+1}{}^2 |\, A^N(x,\phi)\,|^2 \, dx d\phi,$$

so that if $\lambda + 1$ is pure imaginary or zero then

$$\| A'_N \|^2 = \| A^N \|^2.$$

Therefore
$$\| A' \| = \| A \|,$$

since
$$\lim_{N \to \infty} |\, \| A' \| - \| A'_N \| \,| \leqslant \lim_{N \to \infty} \| A' - A'_N \| = 0$$

from (101). Thus the norm of the 'vector' A of the Hilbert space is unchanged by the transformation. The representation is therefore unitary. Besides these unitary representations there are others corresponding to the case $j = 0$, λ real and $|\, \lambda + 1 \,| \leqslant 1$ (cf. (1b)), but they do not leave the norm $\| A \|$ invariant.

To obtain the same matrix representation as that given in § 2 one has to choose

$$i^K \sqrt{\left(\frac{K-\lambda-1!}{K+\lambda+1!}\right)} Q_M^{K,j} e^{-ijx}\rho^{2\lambda} \tag{105a}$$

as the basic vectors. Here $z! = \Gamma(z+1)$ for any complex number z. In the exceptional case where λ is an integer $K-\lambda-1!$ or $K+\lambda+1!$ may become infinite. Then one can make use of the relation

$$\Gamma(z)\,\Gamma(1-z) = \frac{\pi}{\sin z\pi},$$

and ignore constant factors (including $1/\sin\pi = \infty$). Thus one gets the alternative forms

$$\frac{Q_M^{K,j} e^{-ijx}\rho^{2\lambda}}{\sqrt{(\lambda+1+K!\,\lambda-K!)}}, \tag{105b}$$

and

$$\sqrt{(K-\lambda-1!\,-\lambda-2-K!)}\, Q_M^{K,j} e^{-ijx}\rho^{2\lambda}. \tag{105c}$$

$(105a, b, c)$ respectively are applicable to the three cases $|j| \geqslant |\lambda+1|$, $\lambda \geqslant |j|$ and $-\lambda \geqslant |j|+2$, λ being an integer. Then $(105b)$ and $(105c)$ are finite representations. $(105c)$ corresponds to an expansion in ascending powers of $1/\Delta$ and ignoring all powers higher than $1/\Delta^{-(2\lambda+2)}$, i.e. calculating the expansion mod $1/\Delta^{-(2\lambda+2)}$.

The theory developed so far for functions of the two variables x,ϕ can easily be extended to those of the three variables x,ϕ,χ defined in the interval H given by $-1 \leqslant x \leqslant 1, 0 \leqslant \phi \leqslant 2\pi, 0 \leqslant \chi \leqslant 4\pi$. Put

$$S_M^{K,j} = \frac{1}{\sqrt{(4\pi)}} Q_M^{K,j} e^{-ijx}, \tag{106}$$

and denote a series

$$\sum_{\nu=1}^{\infty} \mathfrak{A}_\nu S_\nu \equiv \sum_{j=-\infty}^{\infty} \sum_{K \geqslant |j|}^{\infty} A_M^{K,j} S_M^{K,j}$$

by $\mathfrak{A}$. The summation for j is over all integral and half-integral values. Put

$$\|\mathfrak{A}\|^2 = \sum_{\nu=1}^{\infty} |\mathfrak{A}_\nu|^2.$$

Let $f(x,\phi,\chi)$ be a continuous function of x,ϕ,χ defined in the interval H. Put

$$\|f\|^2 = \int_H |f|^2 \, dx\, d\phi\, d\chi.$$

Now it is well known that the set S_ν is complete and therefore for every continuous f there exists a series $\mathfrak{A}$ such that

$$\lim_{N\to\infty} \|f - \mathfrak{A}^N\| = 0, \tag{107a}$$

where $\mathfrak{A}^N = \sum_{\nu=1}^{N} \mathfrak{A}_\nu S_\nu$. Also

$$\|f\|^2 = \|\mathfrak{A}\|^2 = \lim_{N\to\infty} \|\mathfrak{A}^N\|^2, \tag{107b}$$

$$\mathfrak{A}_\nu = (S_\nu, f), \tag{107c}$$

where the symbol (g,f) is defined for any two continuous functions g and f by the equation

$$(g,f) = (\overline{f,g}) = \int_H \bar{g} f \, dx \, d\phi \, d\chi. \tag{108}$$

Let $\mathfrak{A}$ be a series with finite norm and let f be a continuous function defined on H. Then f is bounded, i.e. there exists a constant C such that $|f| \leqslant C$. Clearly $f\mathfrak{A}^N$ is a continuous function. Hence there exists a series $\mathfrak{B}_N$ satisfying the equation

$$\lim_{n \to \infty} \| f\mathfrak{A}^N - \mathfrak{B}_N^n \| = 0.$$

Now
$$\| \mathfrak{B}_N^n - \mathfrak{B}_{N_\bullet}^n \| \leqslant \| f\mathfrak{A}^{N_\bullet} - \mathfrak{B}_{N_\bullet}^n \| + \| f\mathfrak{A}^N - \mathfrak{B}_N^n \| + \| f\mathfrak{A}^N - f\mathfrak{A}^{N_\bullet} \|,$$

and so
$$\| \mathfrak{B}_N - \mathfrak{B}_{N_\bullet} \| \leqslant \| f\mathfrak{A}^N - f\mathfrak{A}^{N_\bullet} \|.$$

i.e.
$$\| \mathfrak{B}_N - \mathfrak{B}_{N_\bullet} \| \leqslant C^2 \sum_{N_\bullet}^{\infty} | \mathfrak{A}_\nu |^2 \quad (N \geqslant N_0).$$

Therefore $\mathfrak{B}_N$ tends to a limit $\mathfrak{B}$ as $N \to \infty$ and $\| \mathfrak{B} \|$ is finite, since

$$\| \mathfrak{B} \|^2 = \lim_{N \to \infty} \int_H | f\mathfrak{A}^N |^2 \, dx \, d\phi \, d\chi \leqslant C^2 \| \mathfrak{A} \|^2.$$

Hence we conclude that
$$\mathfrak{B}_\nu = \lim_{N \to \infty} (S_\nu, f\mathfrak{A}^N). \tag{109}$$

Now let D be any one of the three operators $\partial/\partial x$, $\partial/\partial \phi$ or $\partial/\partial \chi$ and let $f(x, \phi, \chi)$ be a continuous function whose three partial differential coefficients exist and are continuous everywhere on H. Also assume that

$$f(x, \phi, 0) = f(x, \phi, 4\pi) \quad \text{and} \quad f(x, 0, \chi) = f(x, 2\pi, \chi \pm 2\pi),$$

where the sign in $\chi \pm 2\pi$ is determined from the fact that the resulting value must lie between 0 and 2π. Then it is clear by partial integration that

$$(S_\nu, Df) = -(D^*S_\nu, f), \tag{110a}$$

where D^*S_ν is defined by
$$D^*S_\nu = \overline{D\bar{S}_\nu}. \tag{110b}$$

But it is well known that D^*S_ν can be expressed as a finite linear combination of the S_σ's, i.e.

$$D^*S_\nu = \sum_\sigma S_\sigma(S_\sigma, D^*S_\nu), \tag{111}$$

where (S_σ, D^*S_ν) is zero for all except a finite number of values of σ. Hence from (110a) and (111)

$$(S_\nu, Df) = -\sum_\sigma (\overline{S_\sigma, D^*S_\nu}) (S_\sigma, f)$$

$$= \sum_\sigma (\overline{DS_\sigma, S_\nu}) (S_\sigma, f)$$

$$= \sum_\sigma (S_\nu, DS_\sigma) (S_\sigma, f). \tag{112}$$

Since Df is a continuous function the series $\mathfrak{B}$ with $\mathfrak{B}_\nu = (S_\nu, Df)$ 'represents' Df, i.e.

$$\lim_{n\to\infty} \| Df - \mathfrak{B}^n \| = 0,$$

and therefore

$$\| \mathfrak{B} \|^2 = \int_H | Df |^2 \, dx \, d\phi \, d\chi$$

is finite. Now (112) shows that the coefficients $\mathfrak{B}_\nu$ can be obtained by differentiating the series $\mathfrak{A}$ representing f ($\mathfrak{A}_\sigma = (S_\sigma, f)$) term by term, expressing DS_σ as a linear combination of S_ν and collecting coefficients of S_ν. Thus if $\mathfrak{A}$ 'represents' a continuous function with continuous partial derivatives then the series $D\mathfrak{A}$ obtained by term-by-term differentiation represents Df and therefore $\| D\mathfrak{A} \|$ is finite.

Now from a series $\mathfrak{A}$ one can always go back to the series of the type (72) by making use of the relation

$$S_M^{K,j} = \frac{(-1)^{K+j}}{\sqrt{8\pi}} \sqrt{\left(-\frac{K-M!\,K+M!}{K-j!\,K+j!} \, (K+\tfrac{1}{2}) \right)} \frac{U_M^{K,j}}{\Delta^K}.$$

Formally put

$$\psi(u, \dot{u}) = \sum_{j=-\infty}^{\infty} \sum_{K\geqslant|j|}^{\infty} \sum_M a_M^{K,j} \frac{U_M^K}{\Delta^{K-\lambda}}, \tag{113a}$$

and write

$$\| \psi \|^2 = 8\pi^2 \sum_{j=-\infty}^{\infty} \sum_{K\geqslant|j|}^{\infty} \sum_M \frac{K-j!\,K+j!}{K-M!\,K+M!} \frac{2}{2K+1} | a_M^{K,j} |^2$$

$$= \| \mathfrak{A} \|^2. \tag{113b}$$

For brevity we say that ψ is a continuous function of u and $\dot{u}$ if there exists a continuous function $f(x, \phi, \chi)$ defined on H with

$$f(x, \phi, 0) = f(x, \phi, 4\pi) \quad \text{and} \quad f(x, 0, \chi) = f(x, 2\pi, \chi \pm 2\pi)$$

such that

$$\mathfrak{A}_\nu = (S_\nu, f).$$

Such a function, if it exists, is uniquely determined by $\mathfrak{A}$, since if f and f' were two such functions then

$$\| f - f' \| \leqslant \lim_{N\to\infty} \| f - \mathfrak{A}^N \| + \lim_{N\to\infty} \| f' - \mathfrak{A}^N \| = 0,$$

so that

$$\| f - f' \|^2 = \int_H | f - f' |^2 \, dx \, d\phi \, d\chi = 0.$$

But as f and f' are continuous this implies that $f = f'$ everywhere on H. Similarly, if f has got continuous partial derivatives everywhere on H we say ψ is continuously differentiable. Hence the above result can be expressed by saying that *if ψ is continuously differentiable it can be differentiated term by term.*

Notice that if ψ is continuous in one Lorentz frame, then it is continuous in every other. Since if $\mathfrak{A}$ is the corresponding series and $\mathfrak{A}'$ its transform in any other Lorentz frame then

$$\lim_{\substack{N\to\infty \\ N'\to\infty}} \int_H | \rho^{2\lambda} \mathfrak{A}^N(x, \phi, \chi) - \rho'^{2\lambda} \mathfrak{A}'^{N'}(x', \phi', \chi') |^2 \, dx' \, d\phi' \, d\chi' \tag{114a}$$

and

$$\lim_{N\to\infty} \int_H | \rho^{2\lambda} \mathfrak{A}^N - \rho^{2\lambda} f |^2 \, dx \, d\phi \, d\chi = 0. \tag{114b}$$

where f is continuous. Define

$$f'(x', \phi', \chi') = \left(\frac{\rho}{\rho'}\right)^{2\lambda} f(x, \phi, \chi), \tag{114c}$$

then clearly $f'(x, \phi, \chi)$ is continuous, and as is easily proved

$$\lim_{N \to \infty} \|f' - \mathfrak{A}'^N\| = 0, \tag{114d}$$

Therefore the transform $\psi'(u, \dot{u})$ is continuous. The same is true for the property of continuous differentiability. In particular, if $\psi(u, \dot{u})$ is such that in some particular Lorentz frame it has only a finite number of terms in the series, i.e. $\mathfrak{A}_\nu = 0$ for all $\nu > N$ then clearly $\psi(u, \dot{u})$ is continuously differentiable an arbitrary number of times.

One can now introduce the operators $u_\mu, \partial/\partial u_\mu, u_{\dot\mu}$ and $\partial/\partial u_{\dot\mu}$. $\partial/\partial u_\mu$ is defined by

$$\frac{\partial}{\partial u_\mu} = \frac{\partial \rho}{\partial u_\mu} \frac{\partial}{\partial \rho} + \frac{\partial \chi}{\partial u_\mu} \frac{\partial}{\partial \chi} + \frac{\partial x}{\partial u_\mu} \frac{\partial}{\partial x} + \frac{\partial \phi}{\partial u_\mu} \frac{\partial}{\partial \phi},$$

where $\partial\xi/\partial u_\mu$ is $\left\{\dfrac{\partial(u_1, u_2, u_{\dot1}, u_{\dot2})}{\partial(\rho, \chi, x, \phi)}\right\}^{-1}$ times the minor of $\partial u_\mu/\partial\xi$ in the Jacobian

$$\frac{\partial(u_1, u_2, u_{\dot1}, u_{\dot2})}{\partial(\rho, \chi, x, \phi)} \quad (\xi = \rho, \chi, x \text{ or } \phi).$$

$\partial/\partial u_{\dot\mu}$ is defined similarly. $u_\mu, u_{\dot\mu}, \partial\xi/\partial u_\mu$ and $\partial\xi/\partial u_{\dot\mu}$ are clearly continuous functions of x, ϕ, χ (apart from a possible factor ρ or $1/\rho$) and therefore from (109) and (112) the operators $u_\mu, u_{\dot\mu}, \partial/\partial u_\mu$ and $\partial/\partial u_{\dot\mu}$ can be applied to every continuously differentiable series ψ. In this term by term application $\partial/\partial u_1$ corresponds formally to partial differentiation with respect to u_1 keeping $u_2, u_{\dot1}, u_{\dot2}$ constant. Similarly for $\partial/\partial u_2, \partial/\partial u_{\dot1}$ and $\partial/\partial u_{\dot2}$.

Consider an infinitesimal spinor transformation

$$u_\mu = u'_\mu + \eta_\mu{}^\nu u'_\nu, \tag{115a}$$

$$u_{\dot\mu} = u'_{\dot\mu} + \eta_{\dot\mu}{}^{\dot\nu} u'_{\dot\nu}. \tag{115b}$$

$(\eta_{\dot\mu}{}^{\dot\nu} = \overline{\eta_\mu{}^\nu})$ applied to the expinor space. From the condition that the determinant of the transformation should be 1 so that the fundamental spinor $\epsilon_{\mu\nu}$ remains invariant it follows that

$$\eta_{\mu\nu} = \eta_{\nu\mu}. \tag{115c}$$

Let $\psi(u, \dot{u})$ be a series which is homogeneous and continuously differentiable so that $\psi = \rho^{2\lambda}\mathfrak{A}$. Under (115) ψ goes over to $\psi' = \rho^{2\lambda}\mathfrak{A}'$ where $\mathfrak{A}'$ is defined by (114a). Since ψ is continuously differentiable there exist continuously differentiable functions $f(x, \phi, \chi)$ and $f'(x', \phi', \chi')$ satisfying (114b, c, d). Write $\rho^{2\lambda}f(x, \phi, \chi) = \Psi(u, \dot{u})$ and $\rho'^{2\lambda}f'(x', \phi', \chi') = \Psi'(u', \dot{u}')$. Then

$$\Psi'(u', \dot{u}') = \Psi(u, \dot{u}). \tag{116}$$

From (116), (115) and the definition of $\partial^\mu \equiv \partial/\partial u_\mu$ and $\partial^{\dot\mu} \equiv \partial/\partial u_{\dot\mu}$, it follows that

$$\Psi'(u,\dot u) = \Psi'(u',\dot u') + \eta_\mu{}^\nu u'_\nu \frac{\partial}{\partial u'_\mu}\Psi' + \eta_{\dot\mu}{}^{\dot\nu} u'_{\dot\nu} \frac{\partial}{\partial u'_{\dot\mu}}\Psi'$$

$$= \Psi(u,\dot u) + \eta_\mu{}^\nu u_\nu \partial^\mu\Psi + \eta_{\dot\mu}{}^{\dot\nu} u_{\dot\nu} \partial^{\dot\mu}\Psi$$

$$= \{1 - \eta_{\mu\nu}I^{\mu\nu} - \eta_{\dot\mu\dot\nu}I^{\dot\mu\dot\nu}\}\,\Psi(u,\dot u) \tag{117}$$

up to quantities of the first order. Here

$$I^{\mu\nu} = \tfrac{1}{2}(u^\mu\partial^\nu + u^\nu\partial^\mu), \tag{118a}$$

$$I^{\dot\mu\dot\nu} = \tfrac{1}{2}(u^{\dot\mu}\partial^{\dot\nu} + u^{\dot\nu}\partial^{\dot\mu}). \tag{118b}$$

Since ψ is continuously differentiable the functions $I^{\mu\nu}\Psi(u,\dot u)$ and $I^{\dot\mu\dot\nu}\Psi(u,\dot u)$ are represented by the series $I^{\mu\nu}\psi$ and $I^{\dot\mu\dot\nu}\psi$ respectively.

Now the Lorentz transformation

$$x_k = x'_k + \epsilon_k{}^l x'_l, \quad \epsilon_{kl} = -\epsilon_{lk} \tag{119a}$$

associated to (115) is given by

$$\epsilon^{kl} = -\tfrac{1}{2}\eta_\alpha{}^\beta \sigma^k_{\beta\lambda}\sigma^{l,\,\alpha\lambda} - \tfrac{1}{2}\eta_\lambda{}^{\dot\mu}\sigma^k_{\alpha\dot\mu}\sigma^{l,\,\alpha\lambda}, \tag{119b}$$

where the σ-symbols have their usual meaning (see Van der Waerden 1932). (119b) is obtained immediately from the condition that a spinor transformation applied together with the associated Lorentz transformation leaves $\sigma^k_{\alpha\lambda}$ unaltered. From (3) the transformation operator corresponding to (119a) is $1 - \tfrac{1}{2}\epsilon_{kl}I^{kl}$. Hence the connexion between $I^{\mu\nu}$, $I^{\dot\mu\dot\nu}$ and I_{kl} is given by

$$I_\alpha{}^\beta = \tfrac{1}{4}I_{kl}\sigma^k_{\alpha\lambda}\sigma^{l,\,\lambda\beta}, \quad I_\lambda{}^{\dot\mu} = \tfrac{1}{4}I_{kl}\sigma^k_{\lambda\alpha}\sigma^{l,\,\alpha\dot\mu}. \tag{120}$$

On using the usual representation for $\sigma^k_{\alpha\lambda}$ it is easily proved that $I_{\alpha\beta}$ and $I_{\lambda\dot\mu}$ so obtained are the same as those introduced in §3.

Notice that
$$[\partial^\mu, u^\nu] = \epsilon^{\nu\mu} = [u^\mu, \partial^\nu], \tag{121a}$$

$$[\partial^{\dot\mu}, u^{\dot\nu}] = \epsilon^{\dot\nu\dot\mu} = [u^{\dot\mu}, \partial^{\dot\nu}], \tag{121b}$$

and the dotted operators commute with the undotted ones. Also if ψ is homogeneous in the dotted and undotted pair of variables separately and of degree $2\kappa^* = \lambda + j$ and $2\kappa = \lambda - j$ respectively, then it follows immediately on term-by-term differentiation that

$$u_\mu\partial^\mu\psi = 2\kappa\psi, \tag{122a}$$

$$u_{\dot\mu}\partial^{\dot\mu}\psi = 2\kappa^*\psi. \tag{122b}$$

From (118), (121) and (122) it is clear that the properties of u_μ, ∂_μ are the same as those of $\mathfrak{U}_\mu$, $\mathfrak{B}_\mu$ respectively as defined in (64). In fact the matrix representation of u_μ, ∂_μ is identical with that given for $\mathfrak{U}_\mu$, $\mathfrak{B}_\mu$ in §3 if one chooses either of the three functions (105a), (105b) or (105c) for all values of λ and j as basic vectors. Similarly $u_{\dot\mu} = \mathfrak{U}_{\dot\mu}$ and $\partial_{\dot\mu} = \mathfrak{B}_{\dot\mu}$.

The meaning of (43) and (45) is now obvious in terms of u_α and ∂_α. Let ψ_α be a vector of the space $\Re \times \Re_i$ such that the corresponding pair of series $\psi_\alpha(u, \dot{u})$ are continuously differentiable and are of degree 2κ and $2\kappa^*$ in u and $\dot{u}$ respectively. Then clearly $u^\alpha \psi_\alpha(u, \dot{u})$ and $\partial^\alpha \psi_\alpha(u, \dot{u})$ are two series of degrees $(2\kappa + 1, 2\kappa^*)$ and $(2\kappa - 1, 2\kappa^*)$ respectively and therefore transform according to the representation $\mathfrak{D}_{\kappa+\frac{1}{2}}, \kappa^*$ and $\mathfrak{D}_{\kappa-\frac{1}{2}}, \kappa^*$.

Also notice that the commutation relation $(121a)$ remains unchanged under the formal substitution

$$u_\alpha \to \partial_\alpha, \quad \partial_\alpha \to u_\alpha.$$

By the same substitution $2\kappa = u_\alpha \partial^\alpha$ goes over into $\partial_\alpha u^\alpha = -(2\kappa + 2)$, i.e. κ goes over into $-\kappa - 1$. This corresponds to (59).

6. Physical applications

The irreducible representations of the Lorentz group discussed above may be used to describe the transformation properties of the wave function of a particle. Let ψ_α and ψ_λ be two vectors of the product spaces $\Re \times \Re_i$ and $\Re \times \Re_i$ respectively. The corresponding series will be denoted by $\psi_\alpha(u, \dot{u})$ and $\psi_\lambda(u, \dot{u})$, their coefficients being functions of the space-time coordinates $x_k (k = 0, 1, 2, 3)$. Assume that the functions Ψ_α and Ψ_λ associated to these series (cf. § 5) are continuous and differentiable a sufficient number of times with respect to $u_\mu, u_{\dot\mu}$ and x_k. Then in analogy with the case of finite representations (Fierz 1939) one can set up an equation of motion of the particle as follows:

$$i\partial^{\alpha\lambda}\psi_\alpha = \chi\psi^\lambda, \tag{123a}$$

$$i\partial_{\alpha\lambda}\psi^\lambda = \chi\psi_\alpha, \tag{123b}$$

where

$$\partial_{\alpha\lambda} = \sigma^k_{\alpha\lambda}\partial_k, \quad \partial_k \equiv \frac{\partial}{\partial x^k}$$

and χ is a constant $\neq 0$. The second-order equation

$$\{\partial_k\partial^k + \chi^2\}\,\psi_\alpha = 0 \tag{124}$$

follows immediately. (123) can be put in a more elegant form by making use of the formalism of a recent paper (Harish-Chandra 1947). Introduce the 4×1 matrices T_α and T_λ (see Harish-Chandra 1946) and put

$$\Psi = T_\alpha\psi^\alpha + iT_\lambda\psi^\lambda. \tag{125}$$

Then there exist 1×4 matrices Γ^*_β and $\Gamma^*_{\dot\mu}$ having the following properties:

$$\Gamma^*_{\dot\alpha}T_\beta = 0 = \Gamma^*_\alpha T_{\dot\mu}, \tag{126a}$$

$$\Gamma^*_\alpha T_\beta = i\epsilon_{\alpha\beta}, \quad \Gamma^*_{\dot\lambda}T_{\dot\mu} = i\epsilon_{\lambda\dot\mu}, \tag{126b}$$

$$iT_\alpha\Gamma^{*\alpha} + iT_{\dot\alpha}\Gamma^{*\dot\alpha} = 1, \tag{126c}$$

$$A_{\lambda\dot\mu} = 2(T_\lambda\Gamma_{\dot\mu} + T_{\dot\mu}\Gamma^*_\lambda), \tag{126d}$$

where $A_{\lambda\mu} = \alpha_k \sigma^k_{\lambda\mu}$, α_k being the usual 4×4 Dirac matrices satisfying the commutation rules

$$\alpha_k \alpha_l + \alpha_l \alpha_k = 2g_{kl}. \tag{127}$$

Therefore

$$\psi_\alpha = i\Gamma^*_\alpha \Psi, \quad \psi_\mu = \Gamma^*_\mu \Psi, \tag{128a}$$

and (123) is equivalent to

$$i\alpha^k \partial_k \Psi + \chi \Psi = 0. \tag{128b}$$

Ψ is a 4×1 matrix, each element of which is a series in $u, \dot{u}$ of degree $(2\kappa, 2\kappa^*)$ whose coefficients are functions of x_k.

Now choose a representation with $\lambda + 1 = 0$. Then it is unitary. For two given series $\psi = \rho^{2\lambda} e^{-i\slashed{x}} A$ and $\phi = \rho^{2\lambda} e^{-i\slashed{x}} B$ define the scalar product (ψ, ϕ) by

$$(\psi, \phi) = \lim_{N \to \infty} \int \bar{A}^{\cdot N} B^N dx \, d\phi = \sum_{\nu=1}^{\infty} \bar{A}_\nu B_\nu. \tag{129}$$

From Schwarz's inequality

$$\left| \sum_{\nu=N_0}^{N} \bar{A}_\nu B_\nu \right|^2 \leqslant \sum_{\nu=N_0}^{N} |A_\nu|^2 \sum_{\nu=N_0}^{N} |B_\nu|^2 \leqslant \| A \|^2 \| B \|^2,$$

and therefore (ψ, ϕ) exists and is finite. Also it is easy to show from (98) that (ψ, ϕ) is invariant under Lorentz transformations. Notice that

$$(\psi, \phi) = \overline{(\phi, \psi)} \quad \text{and} \quad (\psi, \psi) = \| \psi \|^2.$$

The charge and current density corresponding to (128b) can be given by

$$s_k = e\Psi^* \alpha_k \Psi, \tag{130}$$

where Ψ^* is defined by

$$\Psi^* = \tilde{\Psi}\Lambda, \tag{131a}$$

Λ being a non-singular Hermitian matrix such that

$$\alpha^\dagger_k = \Lambda \alpha_k \Lambda^{-1}. \tag{131b}$$

($\dagger$ denotes Hermitian conjugate.) $\tilde{\Psi}$ is the transposed of Ψ and the right side of (130) stands for

$$e \sum_{\rho, \sigma=1}^{4} (\alpha_k)_{\rho\sigma} (\Psi^*_\rho, \Psi_\sigma).$$

The charge density is therefore positive definite as in Dirac's theory of the electron. In fact (128b) can be derived from the Lagrangian

$$L = \frac{1}{2i} (\partial_k \Psi^* a^k \Psi - \Psi^* \alpha^k \partial_k \Psi) + \chi \Psi^* \Psi$$

in the usual way. The analogy with the case of the electron is complete. Neither the energy density nor the total energy is positive, and therefore the theory can be quantized only in accordance with the exclusion principle. Write $\rho^{2\lambda} e^{-i\slashed{x}} Q_\nu = V_\nu$ and $\Psi_\rho(u, \dot{u}) = \sum_{V=1}^{\infty} \Psi_{\rho, \nu} V_\nu$ and use a representation for α_k for which $\Lambda = x_0$. Then in the quantized theory the coefficients $\Psi_{\rho, \nu}$ appear as operators satisfying the commutation rules

$$\Psi^\dagger_{\sigma, \mu}(\mathbf{x}, t) \Psi_{\rho, \nu}(\mathbf{x}', t) + \Psi_{\rho, \nu}(\mathbf{x}', t) \Psi^\dagger_{\sigma, \mu}(\mathbf{x}, t) = \delta_{\sigma\rho} \delta_{\nu\mu} \delta(\mathbf{x} - \mathbf{x}'). \tag{132}$$

Here $\mathbf{x}$ and t are the space and time coordinates and $\Psi^{\dagger}_{\sigma,\mu}$ is the Hermitian conjugate of $\Psi_{\sigma,\mu}$.

Notice that no subsidiary conditions are necessary to make the charge-density positive definite (cf. Fierz 1939). Further, this scheme describes particles of integral or half-integral spin according as j is half-integral or integral. This follows immediately on reducing the total representation space with respect to the rotation group. Hence particles with integral spin must be quantized here according to Fermi statistics. Such a state of affairs, however, is not possible for finite representations (Pauli 1940).

Now instead of (128b) one can equally well take the equation of motion of the particle to be

$$i\beta_k \partial^k \Psi + \chi \Psi = 0, \tag{133a}$$

where β_k are the Duffin-Kemmer matrices satisfying the commutation rules

$$\beta_k \beta_l \beta_m + \beta_m \beta_l \beta_k = g_{kl}\beta_m + g_{ml}\beta_k. \tag{133b}$$

In this case the energy density as calculated from the energy momentum tensor

$$\begin{aligned}
T_{kl} &= \Psi^*(g_{kl} - \beta_k\beta_l - \beta_l\beta_k)\,\Psi, \\
\Psi^* &= \Psi\Lambda, \\
\beta_k^{\dagger} &= \Lambda\beta_k\Lambda^{-1}, \quad \Lambda^{\dagger} = \Lambda
\end{aligned} \tag{134}$$

is positive definite. The equation describes particles of integral or half-integral spin according as j is integral or half-integral. As usual it can be quantized according to Bose statistics through the commutation rules

$$[(\Psi^*\beta_0^2)_{\sigma,\mu}, (\beta_0\Psi')_{\rho,\nu}] = -(\beta_0^2)_{\rho\sigma}\delta_{\nu\mu}\delta(\mathbf{x}-\mathbf{x}'), \tag{135}$$

where $\quad (\Psi^*\beta_0^2)_{\sigma,\mu} = \sum_\rho \Psi^*_{\rho,\mu}(\beta_0^2)_{\rho,\sigma}\quad$ and $\quad (\beta_0\Psi')_{\rho,\nu} = \sum_\sigma (\beta_0)_{\rho\sigma}\Psi'_{\sigma,\nu}.$

Ψ' stands for $\Psi(\mathbf{x}',t)$. Electromagnetic interaction can be introduced in (128) and (133) in the usual way. Also it is clear that (128) and (133) are both reflexion invariant since the unitary representations with $\lambda + 1 = 0$ are reflexion invariant.

So far no subsidiary conditions have been imposed on Ψ. As a result the particle thus described does not possess a definite spin in the sense that in the rest system the non-vanishing components of the wave function do not all necessarily transform according to one irreducible representation $\mathfrak{D}_s$ of the rotation group. The subsidiary conditions imposed by Dirac (1936) and Fierz (1939) on $\psi_\varkappa$ and ψ_λ are as follows:

$$\partial^\alpha \psi_\varkappa = \partial^\lambda \psi_\lambda = 0.$$

From (126) and (128b) they may be written as

$$\partial^\mu \Gamma^*_\mu \Psi = \partial^\mu \Gamma^*_{\dot\mu} \Psi = 0, \tag{136a}$$

or otherwise as $\qquad \partial^\lambda \partial^\mu A_{\lambda\dot\mu}\Psi = 0. \tag{136b}$

Then as shown by Fierz (1939) all non-vanishing components of Ψ in the rest system transform according to $\mathfrak{D}_s$ with $s = \kappa + \kappa^*$ (where κ, κ^* are both real, $\geqslant 0$ and of the form $\frac{1}{2}n$). One can, however, consider this question in a more general way at least in the unquantized force-free case. Consider the quantity $\partial^{\alpha\lambda}\partial^{\beta\dot\mu}I_{\alpha\beta}I_{\lambda\dot\mu}\Psi$, where $I_{\alpha\beta}, I_{\lambda\dot\mu}$ refer to the total representation space so that, for example, in the particular case of (128)

$$I_{\alpha\beta} = \tfrac{1}{2}i(T_\alpha \Gamma_\beta^* + T_\beta \Gamma_\alpha^*) + \tfrac{1}{2}(u_\alpha \partial_\beta + u_\beta \partial_\alpha).$$

In the rest system of the particle $\partial_k = 0$ $(k \neq 0)$ and $i\partial_0 = \pm\chi$ so that from (53a), (52), (31a) and (29)

$$
\begin{aligned}
\partial^{\alpha\lambda}\partial^{\beta\dot\mu}I_{\alpha\beta}I_{\lambda\dot\mu}\Psi &= -\chi^2 \sigma_0{}^{\alpha\lambda}\sigma_0{}^{\beta\dot\mu} I_{\alpha\beta}I_{\lambda\dot\mu}\Psi \\
&= -\chi^2\{I_{11}I_{\dot1\dot1} + 2I_{12}I_{\dot1\dot2} + I_{22}I_{\dot2\dot2}\}\Psi \\
&= -\tfrac{1}{4}\chi^2(K_{\alpha\beta} + iL_{\alpha\beta})(K^{\alpha\beta} - iL^{\alpha\beta})\Psi \\
&= \tfrac{1}{2}\chi^2(\mathbf{K}^2 + \mathbf{L}^2)\Psi \\
&= \chi^2(\mathbf{K}^2 - \tfrac{1}{2}J^2)\Psi \\
&= \chi^2\mathbf{K}^2\Psi + \chi^2\{\tfrac{1}{2}I_{\alpha\beta}I^{\alpha\beta} + \tfrac{1}{2}I_{\lambda\dot\mu}I^{\lambda\dot\mu}\}\Psi \\
&= \chi^2\{s(s+1) + \tfrac{1}{2}I_{\alpha\beta}I^{\alpha\beta} + \tfrac{1}{2}I_{\lambda\dot\mu}I^{\lambda\dot\mu}\}\Psi,
\end{aligned}
$$

since in the rest system $\mathbf{K}^2\Psi = s(s+1)\Psi$ if the particle has the 'pure' spin s. Therefore

$$[\partial^{\alpha\lambda}\partial^{\beta\dot\mu}I_{\alpha\beta}I_{\lambda\dot\mu} - \chi^2(s(s+1) + \tfrac{1}{2}I_{\alpha\beta}I^{\alpha\beta} + \tfrac{1}{2}I_{\lambda\dot\mu}I^{\lambda\dot\mu})]\Psi = 0. \tag{137}$$

Now (137) is a spinor equation and therefore since it holds in the rest system it must hold in every system. (137) is therefore the condition which must be satisfied in order that the particle may possess the 'pure' spin s. However, there is no real reason for preferring (137) to the simpler condition

$$[\partial^{\alpha\lambda}\partial^{\beta\dot\mu}I^{(u)}_{\alpha\beta}I^{(u)}_{\lambda\dot\mu} - \chi^2(s(s+1) + \tfrac{1}{2}I^{(u)}_{\alpha\beta}I^{\alpha\beta(u)} + \tfrac{1}{2}I^{(u)}_{\lambda\dot\mu}I^{\lambda\dot\mu(u)})]\Psi = 0, \tag{138}$$

where $\qquad I^{(u)}_{\alpha\beta} = \tfrac{1}{2}(u_\alpha \partial_\beta + u_\beta \partial_\alpha)$ and $\quad I^{(u)}_{\lambda\dot\mu} = \tfrac{1}{2}(u_\lambda \partial_{\dot\mu} + u_{\dot\mu}\partial_\lambda).$

(138) implies that in the rest system each element of Ψ is a series all whose coefficients lying outside the subspace $\mathfrak{R}_\varkappa$ vanish. In the case (128) the elements Ψ_ρ themselves transform according to $\mathfrak{D}_{\frac{1}{2},0}$ and $\mathfrak{D}_{0,\frac{1}{2}}$ with respect to the matrix index and therefore the entire set of components of the wave function in the rest system can be reduced into two subsets transforming according to $\mathfrak{D}_{s+\frac{1}{2}}$ and $\mathfrak{D}_{s-\frac{1}{2}}$ under the rotation group. Clearly a similar argument is applicable to (133). Now (138) can be written as

$$\partial^{\alpha\lambda}\partial^{\beta\dot\mu}u_\alpha u_\lambda \partial_\beta \partial_{\dot\mu}\Psi + \chi^2\{(\kappa + \kappa^* - s)(\kappa + \kappa^* + 1 + s)\}\Psi = 0.$$

Since in our case $\kappa + \kappa^* + 1 = \lambda + 1 = 0$ it follows that (138) is equivalent to

$$\partial^{\alpha\lambda}\partial^{\beta\dot\mu}u_\alpha u_\lambda \partial_\beta \partial_{\dot\mu}\Psi - \chi^2 s(s+1)\Psi = 0. \tag{139a}$$

The following equations are easily obtained from (139a):

$$\partial^{\alpha\lambda}\partial^{\beta\dot\mu}\partial_\alpha \partial_\lambda u_\beta u_{\dot\mu}\Psi - \chi^2 s(s+1)\Psi = 0, \tag{139b}$$

$$\partial^{\alpha\lambda}\partial^{\beta\dot\mu}u_\alpha \partial_\lambda \partial_\beta \partial_{\dot\mu}\Psi - \chi^2(s-j)(s+j+1)\Psi = 0, \tag{139c}$$

$$\partial^{\alpha\lambda}\partial^{\beta\dot\mu}\partial_\alpha u_\lambda u_\beta \partial_{\dot\mu}\Psi - \chi^2(s+j)(s-j+1)\Psi = 0. \tag{139d}$$

These four equations are completely equivalent. It is easily seen that they can be satisfied by imposing any one of the following four subsidiary conditions on Ψ:

$$\partial^{\beta\mu}\partial_\beta\partial_\mu\Psi = 0 \quad (s = j = 0), \tag{140a}$$

$$\partial^{\beta\mu}u_\beta u_\mu\Psi = 0 \quad (s = j = 0), \tag{140b}$$

$$\partial^{\beta\mu}\partial_\beta u_\mu\Psi = 0 \quad (s = j \geqslant 0), \tag{140c}$$

$$\partial^{\beta\mu}u_\beta\partial_\mu\Psi = 0 \quad (s = -j \geqslant 0). \tag{140d}$$

$(140a)$ and $(140b)$ are equivalent since one goes over into the other by the substitution $\kappa \to -\kappa - 1, \kappa^* \to -\kappa^* - 1$. The same is true of $(140c)$ and $(140d)$. The equation proposed by Dirac (1945) for a spinless particle corresponds to (133) and $(140b)$, using the five-row representation for β_k.

Notice that (137) or (138) ensure that Ψ_ρ reduces to a finite series in the rest system. Therefore as proved in § 5 it is continuously differentiable an arbitrary number of times. In an arbitrary system Ψ_ρ is in general an infinite series having components in every subspace $\Re_k$. This, however, in my opinion, does not justify the interpretation given by Dirac (1945), that the particle develops new spins during motion. Even in the case of finite representations Ψ has components lying outside the subspace $\Re_s$ in an arbitrary frame of reference. This is usually not interpreted to mean that the spin of the particle is altered by motion.

Finally, I should like to thank Professor Dirac for suggesting the investigation of these infinite representations.

<h2 style="text-align:center">References</h2>

Courant & Hilbert 1931 *Methoden der Mathematischen Physik*, 1, chapter II. Berlin: Springer.

Dirac 1936 *Proc. Roy. Soc.* A, **155**, 447–459.

Dirac 1945 *Proc. Roy. Soc.* A, **183**, 284–295.

Fierz 1939 *Helv. Phys. Acta*, **12**, 3–37.

Harish-Chandra 1947 *Proc. Camb. Phil. Soc.* (in the Press).

Pauli 1940 *Phys. Rev.* **58**, 716–722.

Pontrjagin 1939 *Topological groups*. Princeton: Princeton University Press.

Van der Waerden 1932 *Die gruppentheoretische methode der Quantummechanik*. Berlin: Springer.

Weyl 1931 *The theory of groups and quantum mechanics*. London: Methuen.

Weyl 1939 *The classical groups*. Princeton: Princeton University Press.

Reprinted from
Proc. Royal Soc. A.
189 (1947), 372–401

26-2

Reprinted without change of pagination from the
Proceedings of the Royal Society, A, *volume* 192, 1948

Relativistic equations for elementary particles‡

BY HARISH-CHANDRA, *Gonville and Caius College, University of Cambridge*§

(*Communicated by D. R. Hartree, F.R.S.—Received* 3 *December* 1946)

From the general principles of quantum mechanics it is deduced that the wave equation of a particle can always be written as a linear differential equation of the first order with matrix coefficients. The principle of relativity and the elementary nature of the particle then impose certain restrictions on these coefficient matrices. A general theory for an elementary particle is set up under certain assumptions regarding these matrices. Besides, two physical assumptions concerning the particle are made, namely, (i) that it satisfies the usual second-order wave equation with a fixed value of the rest mass, and (ii) either the total charge or the total energy for the particle-field is positive definite. It is shown that in consequence of (ii) the theory can be quantized in the interaction free case. On introducing electromagnetic interaction it is found that the particle exhibits a pure magnetic moment in the non-relativistic approximation. The well-known equations for the electron and the meson are included as special cases in the present scheme. As a further illustration of the theory the coefficient matrices corresponding to a new elementary particle are constructed. This particle is shown to have states of spin both $\frac{3}{2}$ and $\frac{1}{2}$. In a certain sense it exhibits an inner structure in addition to the spin. In the non-relativistic approximation the behaviour of this particle in an electromagnetic field is the same as that of the Dirac electron. Finally, the transition from the particle to the wave form of the equations of motion is effected and the field equations are given in terms of tensors and spinors.

‡ A short summary of this paper was read at the Physical Society Conference held in July 1946 at Cambridge.
§ Now at The Institute for Advanced Study, Princeton, N.J., U.S.A.

1. Introduction and general theory

In quantum mechanics every particle is described through a wave function ψ. A complete description of the behaviour of the particle therefore implies the knowledge of the value of ψ at all space-time points. Any set of physical laws governing the particle are to be translated into the corresponding mathematical equations satisfied by ψ. The entire system of these equations is called the equation of motion of the particle. In accordance with the principle of near action one assumes that these equations can always be put in the differential form. The principle of superposition then requires that these differential equations be linear in ψ.

In general it may be necessary to use more than one wave function to describe the particle as, for example, in the case of the electron. Assuming that a finite number of them are sufficient one can write them as $\psi_1, \psi_2, ..., \psi_n$. It is usual to call these n functions the components of one wave function ψ. This terminology, which has been taken over from the theory of vector spaces in mathematics, has the following two physical justifications, except for which it would have been simpler to treat the n functions separately and thus regard the particle as composed of n physically simpler entities. First, these functions are in general not independent, since they are interconnected through the equations of motion. Therefore it may not be possible to put, say, all ψ_r except ψ_1 equal to zero for all time. The second justification depends on the principle of relativity according to which all physical laws should be expressible in the same mathematical form in different co-ordinate systems moving uniformly with respect to each other. In order to be able to preserve the form of the equation of motion under a Lorentz transformation it is, in general, necessary to assume that in the new co-ordinate system the components of the same wave function are given by suitable linear combinations of those in the old one. This is illustrated in the case of the electromagnetic field where a pure electric field gives rise to magnetic components in a transformed system. It is therefore not possible to separate the electric and the magnetic components of the field in a relativistic way.

Returning to the equations of motion one notices that any system of linear differential equations can always be replaced by a suitable system of linear differential equations of the first order merely by defining new functions which are equal to the derivatives of the previous ones (Courant & Hilbert 1937, p. 10). Therefore one can bring the equations to the form

$$\sum_{q=1}^{n} c_{pq}^{k} \frac{\partial}{\partial x^k} \psi_q + \sum_{q=1}^{n} c_{pq} \psi_q = 0 \quad (p = 1, ..., m), \tag{1}$$

where c_{pq}^{k} and c_{pq} are suitable coefficients. Here the summation convention for the tensor index k ($k = 0, 1, 2, 3$) is used, and as usual the metric tensor g_{kl} ($g_{kl} = 0, k \neq l$; $g_{00} = -g_{11} = -g_{22} = -g_{33} = 1$) is employed for raising and lowering the tensor indices. m and n are suitable integers. Considering the interaction free case so that no physical quantity other than ψ can enter into (1), it follows that the c's must be given functions of x_k. It will be assumed that they are constants. (1) can then be written as

$$i\gamma^k\partial_k\psi + \chi\beta\psi = 0 \quad (\partial_k \equiv \partial/\partial x^k), \tag{2}$$

where $\gamma^k = \dfrac{1}{i}\| c_{pq}^k \|$ and $\beta = \dfrac{1}{\chi}\| c_{pq} \|$ are $m \times n$ matrices operating on the $n \times 1$ matrix $\psi = \| \psi_q \|$ and χ is a real numerical constant $\neq 0$. Now in accordance with the principles of relativity and quantum mechanics one requires that in a new Lorentz frame the components of the same wave function be given by suitable linear combinations of those in the old one. Therefore for every Lorentz transformation t given by

$$x_k' = t_k^{\ l} x_l \tag{3a}$$

there exists an $n \times n$ matrix $T(t)$ with constant coefficients such that the new $n \times 1$ matrix $\psi'(x')$ formed from the components in the new system is given by

$$\psi'(x') = T(t)\,\psi(x). \tag{3b}$$

However, in present quantum mechanics, it is sufficient for all physical purposes to determine ψ' to within a constant factor of modulus unity. Therefore, in order that one may be able to calculate ψ' unambiguously from ψ to within this factor, it is necessary and sufficient that for any two Lorentz transformations s and t and their product st the following relation should hold:

$$T(s)\,T(t) = \omega(s,t)\,T(st). \tag{4}$$

Here $\omega(s,t)$ is a unimodular complex number. By multiplying the matrices $T(t)$ with suitable unimodular factors it can then always be arranged that $\omega(s,t) = \pm 1$ (Wigner 1939). (4) then implies that the matrices $T(t)$ constitute a representation of the Lorentz group, the double-valued representations being permitted. Further, in order to ensure the relativistic invariance of the equations of motion (2), one assumes that for every Lorentz transformation t there exists an $m \times m$ matrix $T'(t)$ such that

$$T'(t)\gamma_k T^{-1}(t) = t_k^{\ l}\gamma_l, \tag{5a}$$

$$T'(t)\beta T^{-1}(t) = \beta. \tag{5b}$$

It is now necessary to introduce the concept of decomposability. Let $\Re$ be a vector space and let $\Re_i$ $(i = 1, \ldots, j; j \geqslant 2)$ be subspaces of $\Re$ such that

$$\Re = \Re_1 \oplus \Re_2 \oplus \ldots \oplus \Re_j, \tag{6}$$

where $\oplus$ denotes a direct sum. Then $\Re$ is said to be decomposed into a direct sum of $\Re_i$. Let $\mathfrak{S}$ be another vector space and let T be a linear transformation of $\Re$ into‡ $\mathfrak{S}$. Then one writes $T : \Re \to \mathfrak{S}$. If T transforms $\Re_i$ into a subspace $\mathfrak{S}_i$ of $\mathfrak{S}$ one writes $T \mid \Re_i : \Re_i \to \mathfrak{S}_i$. A simultaneous decomposition

$$\Re = \Re_1 \oplus \ldots \oplus \Re_j, \tag{7a}$$

$$\mathfrak{S} = \mathfrak{S}_1 \oplus \ldots \oplus \mathfrak{S}_j \tag{7b}$$

‡ The preposition 'into' is used in the usual sense. It includes the case when the transformation is 'on to' (see Lefschetz 1942, p. 2).

($j \geqslant 2$) of $\mathfrak{R}$, $\mathfrak{S}$ is said to be allowed under T if $T \mid \mathfrak{R}_i : \mathfrak{R}_i \to \mathfrak{S}_i$ ($i = 1, ..., j$). In particular, if $\mathfrak{S}$ is the same as $\mathfrak{R}$, (7) is an allowed decomposition‡ of $\mathfrak{R}$, $\mathfrak{R}$. On the other hand, the decomposition (6) of $\mathfrak{R}$ is said to be allowed under T if $T : \mathfrak{R} \to \mathfrak{R}$ and $T \mid \mathfrak{R}_i : \mathfrak{R}_i \to \mathfrak{R}_i$ ($i = 1, ..., j$). $\mathfrak{R}_i$ are then said to be invariant under T. Now let Σ be a set of linear transformations of the following three types:

$$(1) \ \ T_1 : \mathfrak{R} \to \mathfrak{R}, \quad (2) \ \ T_2 : \mathfrak{R} \to \mathfrak{S}, \quad (3) \ \ T_3 : \mathfrak{S} \to \mathfrak{S}.$$

Then the simultaneous decomposition (7) is said to be allowed under Σ if (i) the decomposition ($7a$) of $\mathfrak{R}$ is allowed under all T_1, (ii) the simultaneous decomposition (7) of $\mathfrak{R}$, $\mathfrak{S}$ is allowed under all T_2, (iii) the decomposition ($7b$) of $\mathfrak{S}$ is allowed under all T_3. An allowed decomposition of $\mathfrak{R}$, $\mathfrak{S}$ is said to be proper if $\mathfrak{R}_i$, $\mathfrak{S}_i$ are never both zero for any i. If a proper allowed decomposition of $\mathfrak{R}$, $\mathfrak{S}$ exists, then the pair $\mathfrak{R}$, $\mathfrak{S}$ is said to be decomposable under Σ and also Σ itself is said to be decomposable. One easily proves that either the pair $\mathfrak{R}$, $\mathfrak{S}$ is indecomposable or there exists a proper allowed decomposition of $\mathfrak{R}$, $\mathfrak{S}$ such that each pair $\mathfrak{R}_i$, $\mathfrak{S}_i$ ($i = 1, ..., j$) is indecomposable under Σ. Such a decomposition is said to be complete. A complete decomposition of Σ has the same meaning. Moreover, if

$$\mathfrak{R} = \mathfrak{R}_1' \oplus ... \oplus \mathfrak{R}_{j'}', \tag{8a}$$

$$\mathfrak{S} = \mathfrak{S}_1' \oplus ... \oplus \mathfrak{S}_{j'}' \tag{8b}$$

is another complete decomposition of $\mathfrak{R}$, $\mathfrak{S}$, then it can be shown that $j' = j$ and the pair $(\mathfrak{R}_i, \mathfrak{S}_i)$ is isomorphic to $(\mathfrak{R}_i', \mathfrak{S}_i')$ under Σ, provided the suffixes in (8) are suitably rearranged. More explicitly, this means that there exists an isomorphism between $\mathfrak{R}_i$ and $\mathfrak{R}_i'$ and another between $\mathfrak{S}_i$ and $\mathfrak{S}_i'$ such that they are both preserved under all transformations of Σ. Thus a complete decomposition is unique to within isomorphism.

In particular, let Σ be a set of $n \times n$, $m \times n$ and $m \times m$ matrices. Let $\mathfrak{R}$ and $\mathfrak{S}$ be the corresponding n- and m-dimensional representation spaces. Decompose $\mathfrak{R}$, $\mathfrak{S}$ completely under Σ and choose co-ordinate systems in $\mathfrak{R}$ and $\mathfrak{S}$ adapted to this decomposition. Then the linear transformation corresponding to any matrix of Σ is represented by a new matrix when referred to these new co-ordinate systems. This new matrix is then said to be a completely decomposed form of the old matrix. In particular, let $\Sigma = (\gamma^k, \beta, T(t))$, where t runs through the whole Lorentz group. The decomposability of Σ then implies the existence of two non-singular $n \times n$ and $m \times m$ matrices U and U' respectively, such that $U'\gamma^k U$ and $U'\beta U$ have the form

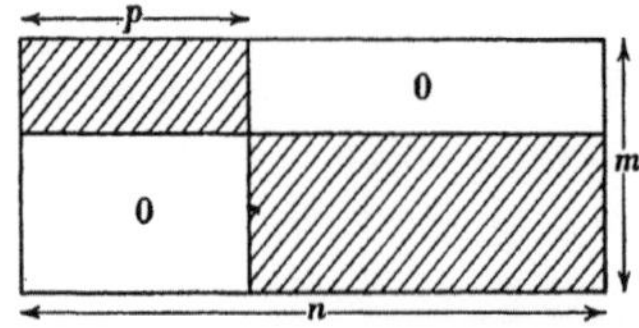

‡ The word 'simultaneous' is often omitted.

(where at least one of the white rectangles consisting of zeroes actually exists) and $U^{-1}T(t)\,U$ has the form

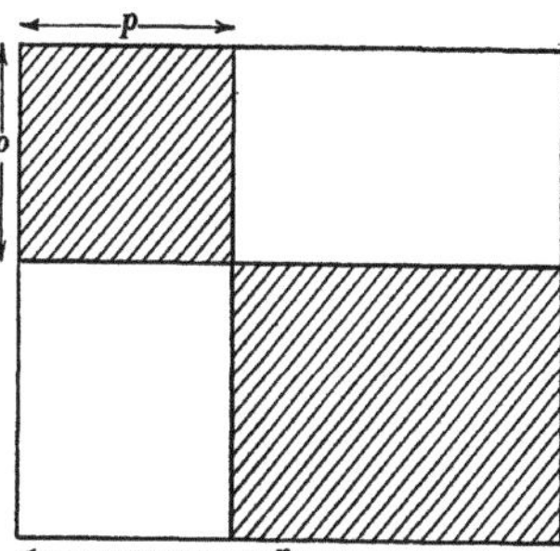

(2) is then equivalent to

$$iU'\gamma^k U\partial_k(U^{-1}\psi) + \chi U'\beta U(U^{-1}\psi) = 0. \tag{9}$$

Clearly (9) then decomposes into two entirely independent sets of equations, each set *separately* being relativistically invariant since

$$U^{-1}\psi' = U^{-1}T(t)\,U(U^{-1}\psi).$$

In such circumstances one says that (2) is decomposable. The complete decomposition of Σ similarly brings about a decomposition of (2) into indecomposable equations. (2) is then said to be completely decomposed. The complete decomposition of (2) is unique to within isomorphism.

Now the physical system described by (2) can be considered elementary only if it cannot be decomposed into two independent subsystems both having relativistically invariant equations of motion. Therefore the equation of motion of an elementary particle must be indecomposable. Also since every equation of the type (2) can be decomposed into two or more independent and indecomposable equations it is sufficient to study only the latter.

Hence let (2) be an indecomposable equation. In the present paper only the case $m = n$ will be considered. Two essentially different cases arise according as β is singular or non-singular. Attention will be confined to the latter case. One can then multiply (2) by β^{-1} on the left. This has the effect of replacing β by 1. Therefore it can be assumed that $\beta = 1$ in (2). Then it follows immediately from (5b) that $T'(t) = T(t)$.

Since $m = n$, one can take $\mathfrak{S} = \mathfrak{R}$. Then the decomposability of $\mathfrak{R}$, $\mathfrak{R}$ in this case is equivalent to that of $\mathfrak{R}$. For, clearly, if $\mathfrak{R}$ is decomposable so is $\mathfrak{R}$, $\mathfrak{R}$. Also if $\mathfrak{R}$, $\mathfrak{R}$ is decomposable there exists an allowed decomposition

$$\mathfrak{R} = \mathfrak{R}_1 \oplus \mathfrak{R}_2, \tag{10a}$$

$$\mathfrak{S} = \mathfrak{S}_1 \oplus \mathfrak{S}_2 \tag{10b}$$

under $\Sigma = (\gamma^k,\ \beta = 1,\ T(t))$. But since $\beta = 1$, it follows$\ddagger$ that $\mathfrak{R}_1 \subset \mathfrak{S}_1$, $\mathfrak{R}_2 \subset \mathfrak{S}_2$. However, as both the sums are direct one gets immediately $\mathfrak{R}_1 = \mathfrak{S}_1$, $\mathfrak{R}_2 = \mathfrak{S}_2$.

$\ddagger$ The usual symbols $\subset$ and $\in$ for 'is contained in' and 'belongs to' respectively are used in this paper.

Hence (10a) is an allowed decomposition of $\Re$ which is therefore decomposable. Thus in order to ensure the elementary character of the particle it is sufficient to assume that $\Re$ is indecomposable. A stronger condition to assume is that $\Re$ is irreducible under $\Sigma = (\gamma^k, T(t))$, i.e. there is no subspace of $\Re$ other than itself and zero, which is invariant under Σ. However, the still stronger assumption that $\Re$ is irreducible under $\Sigma_0 = (\gamma^k)$ will be made. Besides, the following two additional physical assumptions concerning the particle will be made:

(i) That it satisfies the second-order wave equation

$$(\partial_k \partial^k + \chi^2)\,\psi = 0. \tag{11}$$

(ii) Either the total charge or the total energy of the field, associated with the particle is positive definite.

(i) ensures that the particle has a fixed rest mass, while (ii) is necessary if the field is to be quantized. The consequences of (i), together with the relativistic invariance of the equation of motion, have been examined in a recent paper (Harish-Chandra 1946a, referred to as A). It was shown there that in order that every solution of the wave equation

$$i\gamma^k \partial_k \psi + \chi\psi = 0 \tag{12}$$

may satisfy (11), it is necessary and sufficient that the minimum equation of γ_0 be of the form

$$\gamma_0^{n-2}(\gamma_0^2 - 1) = 0, \tag{13}$$

n being some integer $\geqslant 2$. Moreover, it seems essential for constructing an energy-momentum tensor and a current vector, according to the usual principles of quantum mechanics, that the γ-representation be equivalent to its Hermitian conjugate, i.e. there should exist a non-singular matrix Λ such that

$$\gamma_k^\dagger = \Lambda \gamma_k \Lambda^{-1}, \tag{14a}$$

where $\dagger$ denotes Hermitian conjugate. If such a matrix exists it can always be chosen to be Hermitian (cf. § 2) so that

$$\Lambda^\dagger = \Lambda. \tag{14b}$$

(14) enables one to derive (12) by variation of the Lagrangian

$$L = \frac{1}{2i}(\partial_k \psi^* \gamma^k \psi - \psi^* \gamma^k \partial_k \psi) + \chi\psi^*\psi, \tag{15}$$

where $\psi^* = \psi^\dagger \Lambda$. The current vector s_k and the energy-momentum tensor T_{kl}, derived from (15) in the usual way, are

$$s_k = e\psi^* \gamma_k \psi, \tag{16a}$$

$$T_{kl} = \frac{i}{2}(\partial_k \psi^* \gamma_l \psi - \psi^* \gamma_l \partial_k \psi). \tag{16b}$$

Using (16) it will be shown in §3 that (ii) is equivalent to the condition that either $\Lambda\gamma_0^{m+1}$ or $\Lambda\gamma_0^m$ (where m is the greatest even integer $\leqslant n-1$) be a non-negative‡ matrix.

Notice that the present initial assumptions differ from those made by previous authors (Fierz & Pauli 1939; Madhavarao 1942; Bhabha 1945 a,b). The theory given by Fierz & Pauli (1939) fulfils the conditions (i) and (ii), but the corresponding wave equation cannot, in general, be written in the form (12). On the other hand, Bhabha (1945 a, b) starts from an equation of this form (with the γ_k's irreducible), but he assumes that $I_{kl} = I(\gamma_k\gamma_l - \gamma_l\gamma_k)$, where I_{kl} are the matrices corresponding to the infinitesimal Lorentz transformations (cf. equation (20)), and I is a numerical constant. It can be shown that this assumption is incompatible with (11) except in the two cases corresponding to the Dirac and the Duffin-Kemmer matrices. Also, the condition (ii) is, in general, not fulfilled in his theory.

2. The matrix Λ

Assuming that Λ exists one can study its properties more closely. From (14a) it follows that $\Lambda^{\dagger-1}\Lambda$ commutes with γ_k. Therefore it is a multiple of the unit matrix since the γ_k's form an irreducible set. Hence one can arrange that $\Lambda = \Lambda^\dagger$.

Now $t \to T(t)$ is a representation of the Lorentz group. It is uniquely determined by the condition that

$$t_k{}^l\gamma_l = T(t)\,\gamma_k\,T^{-1}(t). \tag{17}$$

For if $t \to T'(t)$ is another representation satisfying (17), then $T^{-1}(t)\,T'(t)$ commutes with γ_k and therefore $T'(t) = c(t)\,T(t)$, where $c(t)$ is a number. Clearly $t \to c(t)$ is a one-dimensional representation of the Lorentz group. But the only possible one-dimensional representation is $t \to \pm 1$, where the $+$ sign is valid at least for all proper Lorentz transformations. Hence $T(t) = T'(t)$ at least for the proper Lorentz group. Now since $t_k{}^l$ is real it follows from (14) and (17) that

$$t_k{}^l\gamma_l = \Lambda^{-1}T^\dagger(t)^{-1}\Lambda\gamma_k\Lambda^{-1}T^\dagger(t)\,\Lambda. \tag{18}$$

Also $$t \to \Lambda^{-1}T^\dagger(t)^{-1}\Lambda$$

is a representation of the Lorentz group. Hence

$$T(t) = \pm\,\Lambda^{-1}T^\dagger(t)^{-1}\Lambda,$$

i.e. $$T^\dagger(t)\,\Lambda T(t) = \pm\,\Lambda. \tag{19}$$

Now consider the infinitesimal Lorentz transformation $t(\epsilon)$ given by

$$t_k{}^l = \delta_k^l + \epsilon_k{}^l \quad (\epsilon_{kl} = -\epsilon_{lk}), \tag{20a}$$

ϵ_{kl} being infinitesimal quantities. The corresponding matrix T may be written as

$$T = 1 + \tfrac{1}{2}\epsilon_{kl}I^{kl}. \tag{20b}$$

‡ A Hermitian matrix is called non-negative if all its eigenvalues, which are necessarily real, are $\geqslant 0$.

I_{kl} satisfy the following commutation rules:

$$[I_{kl}, I_{mn}]_- = -g_{km}I_{ln} + g_{lm}I_{kn} + g_{kn}I_{lm} - g_{ln}I_{km}, \qquad (21a)$$

$$[\gamma_k, I_{lm}]_- = g_{kl}\gamma_m - g_{km}\gamma_l, \qquad (21b)$$

where $[A, B]_- \equiv AB - BA$. From (19) it follows that

$$I_{kl}^\dagger \Lambda + \Lambda I_{kl} = 0. \qquad (22)$$

Let J be the reflexion matrix so that

$$J\gamma_0 J^{-1} = \gamma_0, \quad J\gamma_{k^\circ} J^{-1} = -\gamma_{k^\circ},$$

where the suffix k° runs from 1 to 3 only. Since J^2 commutes with γ_k it can always be arranged that $J^2 = 1$. Now it is well known that by a suitable equivalence transformation one can always go over to a representation in which

$$I_{kl}^\dagger = -JI_{kl}J^{-1}, \qquad (23a)$$

$$J^\dagger = J^{-1} = J. \qquad (23b)$$

For brevity such a representation will be called a U-representation. Consider a non-singular matrix Λ' in a U-representation such that

$$I_{kl}\Lambda' = \Lambda'I_{kl}, \qquad (24a)$$

$$\gamma_k^\dagger = \Lambda'J\gamma_k J^{-1}\Lambda'^{-1}. \qquad (24b)$$

Such a matrix exists since $\Lambda' = \Lambda J$ is a possible choice. From (24a) it follows that Λ' is a scalar or pseudoscalar element of the algebra $\mathfrak{A}(\gamma)$ generated by γ_k (see A). Conversely, if Λ' is a scalar (pseudoscalar) element such that in a U-representation

$$\gamma_0^\dagger = \Lambda'\gamma_0\Lambda'^{-1}, \qquad (25)$$

then Λ' satisfies (24) on account of (21b). Hence $\Lambda = \Lambda'J$ fulfils (14a).

Choose Λ' so that it is Hermitian. Then it can be expressed uniquely in the form

$$\Lambda' = \Lambda^*\eta, \qquad (26)$$

where Λ^* is a positive definite matrix which commutes with the unitary matrix η. Also $\eta = \eta^\dagger$, so that $\eta^2 = 1$ (Murnaghan 1938, p. 26). Let $\sqrt{\Lambda^*}$ denote the positive definite Hermitian square root of Λ^*. Then Λ^* and $\sqrt{\Lambda^*}$ are, respectively, the positive definite Hermitian square root and fourth root of Λ'^2 which commutes with I_{kl}, J and η. Hence Λ^* and $\sqrt{\Lambda^*}$ do the same. Therefore from (24a) η is a scalar or pseudoscalar. Further, the equivalence transformation $\gamma_k \to \sqrt{\Lambda^*}\gamma_k(\sqrt{\Lambda^*})^{-1}$ leaves J and I_{kl} unaltered but transforms Λ' into $(\sqrt{\Lambda^*})^{-1}\Lambda'(\sqrt{\Lambda^*})^{-1} = \eta$. Hence in a suitable U-representation $\Lambda' = \eta$, where η is a scalar or pseudoscalar element such that

$$\eta = \eta^\dagger, \quad \eta^2 = 1. \qquad (27a)$$

Put
$$\Lambda = \epsilon\eta J, \qquad (27b)$$

where $\epsilon = 1$ or i according as η is a scalar or pseudoscalar. Then Λ fulfils (14).

Now in the case of the Dirac and the Duffin-Kemmer matrices it is easily proved on using the results of A (§ 5) that every scalar or pseudoscalar element of $\mathfrak{A}(\gamma)$ satisfies the equation

$$\gamma_0 \eta \gamma_0 \eta = \eta \gamma_0 \eta \gamma_0.$$

Then it follows from (27) that in a suitable U-representation

$$\gamma_0 \gamma_0^\dagger = \gamma_0^\dagger \gamma_0,$$

so that γ_0 is a normal matrix (Murnaghan 1938, p. 26) which can therefore be brought to the diagonal form by a unitary transformation. Since unitary transformations preserve the U-property of the representation it follows that it is possible to have γ_0 diagonal in a U-representation. But clearly in the diagonal form $\gamma_0^\dagger = \gamma_0$, since the eigenvalues of γ_0 are real. Therefore $\Lambda' = 1$ is a possible value of Λ' in (25) and so $\Lambda = J$ in a suitable representation.

Returning to the general case, suppose η is a pseudoscalar. Then

$$J\Lambda\gamma_0^{m+1}J^{-1} = -\Lambda\gamma_0^{m+1}, \quad J\Lambda\gamma_0^m J^{-1} = -\Lambda\gamma_0^m,$$

and therefore neither $\Lambda\gamma_0^{m+1}$ nor $\Lambda\gamma_0^m$ can be non-negative. This contradicts our assumption (ii). Therefore η is a scalar and

$$\Lambda = \eta J. \tag{28}$$

Hence only the $+$ sign is possible in (19). It then follows immediately that $\psi^*\psi$ is a scalar and $\psi^*\gamma_k\psi$ transforms as the components of a vector. The Lagrangian L is then invariant under all Lorentz transformations.

3. Quantization

First one must derive the matrix conditions corresponding to the positive definiteness of the total charge or the total energy. For this purpose decompose the field into Fourier waves. Then it follows from (16) in the usual way that the total charge or the total energy of the field is a sum of terms each of which corresponds to a Fourier wave occurring in this decomposition. As $\chi \neq 0$ every Fourier wave can be brought to rest by a suitable Lorentz transformation without affecting the value of the total charge or the sign of the total energy. Now since for a particle at rest the charge and the energy-densities are independent of the space co-ordinates it is necessary and sufficient for our purpose to ensure that either the charge-density or the energy-density is positive definite for a particle at rest. Due to (11), the wave function ψ of such a particle is given by

$$\psi = \Psi_+ e^{i\chi x_0} + \Psi_- e^{-i\chi x_0}, \tag{29}$$

where Ψ_+ and Ψ_- are independent of x_k. Put

$$\Psi = \Psi_+ + \Psi_-, \tag{30a}$$

$$\psi_+ = \Psi_+ e^{i\chi x_0}, \quad \psi_- = \Psi_- e^{-i\chi x_0}. \tag{30b}$$

Then from (12) one finds that

$$(1 \mp \gamma_0)\,\psi_\pm = 0, \tag{31a}$$

$$(1 - \gamma_0^2)\,\psi = 0, \tag{31b}$$

$$\psi_\pm = \tfrac{1}{2}(1 \pm \gamma_0)\,\psi. \tag{31c}$$

Therefore the charge-density is given by

$$s_0 = e\psi^\dagger \Lambda \gamma_0 \psi = e\Psi^\dagger \Lambda \gamma_0 \Psi = e\Psi^\dagger \Lambda \gamma_0^{m+1} \Psi, \tag{32}$$

where m is the greatest even integer $\leqslant n-1$. Conversely, let Ψ be any arbitrary one-column matrix independent of x_k. Then since $m = n-1$ or $n-2$ it follows from (13) that

$$(1 - \gamma_0^2)\,\gamma_0^m = 0, \tag{33}$$

and therefore $\qquad \psi = \tfrac{1}{2}(1 + \gamma_0)\,\gamma_0^m\,\Psi e^{i\chi x_0} + \tfrac{1}{2}(1 - \gamma_0)\,\gamma_0^m\,\Psi e^{-i\chi x_0} \tag{34}$

is a solution of (12) and hence represents a possible state of the particle at rest. Remembering that m is even and $\gamma_0^{m+2} = \gamma_0^m$, one finds that the charge-density corresponding to (34) is

$$s_0 = e\Psi^\dagger \Lambda \gamma_0^{m+1} \Psi. \tag{35}$$

From (32) and (35) it is clear that the total charge is positive definite if

$$\Psi^\dagger \Lambda \gamma_0^{m+1} \Psi > 0 \tag{36a}$$

for every Ψ for which $\qquad\qquad \gamma_0^m \Psi \neq 0. \tag{36b}$

Now from $\Lambda \gamma_0^{m+1} \Psi = 0$ it follows that

$$\gamma_0 \Lambda^{-1} \Lambda \gamma_0^{m+1} \Psi = \gamma_0^{m+2} \Psi = \gamma_0^m \Psi = 0,$$

and obviously the converse is also true. Hence (36b) can be replaced by the condition

$$\Lambda \gamma_0^{m+1} \Psi \neq 0. \tag{36c}$$

From (14) it follows that $\Lambda \gamma_0^{m+1}$ is Hermitian, and so it can be brought to the diagonal form by a unitary transformation. Then it is obvious from (36a) and (36c) that all eigenvalues of $\Lambda \gamma_0^{m+1}$ must be $\geqslant 0$. $\Lambda \gamma_0^{m+1}$ is therefore a non-negative matrix. Conversely, if $\Lambda \gamma_0^{m+1}$ is non-negative (36) holds. Hence the non-negative character of $\Lambda \gamma_0^{m+1}$ is a necessary and sufficient condition for the positive definiteness of the total charge.

On the other hand, from (16b) one finds that the energy-density corresponding to (29) is given by $\qquad T_{00} = \chi \Psi^\dagger \Lambda \Psi = \chi \Psi^\dagger \Lambda \gamma_0^m \Psi. \tag{37}$

Then by a similar argument one shows that the total energy is positive definite if and only if $\Lambda \gamma_0^m$ is a non-negative matrix.

It remains to be shown that the field can be quantized. Consider the commutation rules

$$[\psi_\rho^*(x'), \psi_\sigma(x)]_\pm = \frac{1}{2}\left[\left(\frac{\gamma^k \partial_k}{i\chi}\right)^m \left(1 + \frac{\gamma^l \partial_l}{i\chi}\right) D(x - x')\right]_{\sigma\rho}, \tag{38}$$

where $[A, B]_\pm = AB \pm BA$ and the $+$ or $-$ sign is to be taken according as $\Lambda\gamma_0^{m+1}$ or $\Lambda\gamma_0^m$ is non-negative. σ and ρ are matrix indices and $D(x-x')$ is the well-known generalization of Pauli & Jordan's Delta-function for the case $\chi \neq 0$ (see Pauli 1940). Clearly these commutation rules are relativistically invariant. Further, they are consistent with the equation of motion (12). For it is shown in A that as a result of (13) and the relativistic invariance of (12)

$$\sum_Q Q(\gamma_{k_1}\gamma_{k_2}\cdots\gamma_{k_n} - g_{k_1 k_2}\gamma_{k_3}\cdots\gamma_{k_n}) = 0, \tag{39}$$

where Q denotes any permutation of the indices $k_1, \ldots, k_n$ and the sum is over all Q. Therefore

$$(i\gamma^k\partial_k + \chi)\left(\frac{\gamma^{k'}\partial_{k'}}{i\chi}\right)^m \left(1 + \frac{\gamma^l\partial_l}{i\chi}\right) D(x-x') = \frac{1}{\chi}\left(\frac{\gamma^k\partial_k}{i\chi}\right)^m (\chi^2 + \partial^l\partial_l) D(x-x') = 0.$$

Now transform (38) to the momentum space by the transformation

$$\psi = \frac{1}{\sqrt{V}} \sum_{\mathbf{k}} a_{\mathbf{k}} e^{i(\mathbf{k}\mathbf{x})}, \tag{39a}$$

where, as usual, V is a large volume, $\mathbf{x} = (x_1, x_2, x_3')$ and $\mathbf{k}$ is the spatial momentum vector. Then since

$$D(x) = \frac{1}{V} \sum_{\mathbf{k}} \frac{\sin k_0 x_0}{k_0} e^{i(\mathbf{k}\mathbf{x})}, \tag{39b}$$

$$k_0 = \sqrt{(\mathbf{k}^2 + \chi^2)} > 0, \tag{39c}$$

one finds on putting $\mathbf{k} = 0$ and $a_0 = a$ that

$$[(a_{\mathbf{k}}^\dagger \Lambda\gamma_0)_\rho, a_\sigma]_\pm = \begin{cases} (\gamma_0^m)_{\sigma\rho} & \text{if } \mathbf{k} = 0, \\ 0 & \text{otherwise.} \end{cases} \tag{40}$$

Now corresponding to (31b) one gets

$$(1 - \gamma_0^2) a = 0. \tag{41}$$

Hence, from (40) and (41),

$$[a_\rho^\dagger, a_\sigma]_\pm = (\gamma_0^{m+1}\Lambda^{-1})_{\sigma\rho}. \tag{42}$$

Now the minimum equations of γ_0^m and γ_0^{m+1} have no repeated factors since

$$\gamma_0^m(\gamma_0^m - 1) = 0, \tag{43a}$$

$$\gamma_0^{m+1}(\gamma_0^{2m+2} - 1) = 0. \tag{43b}$$

Hence one can choose a representation in which γ_0^m and γ_0^{m+1} are diagonal and therefore Hermitian, since their eigenvalues $0, \pm 1$ are real. Therefore from (14) Λ commutes with γ_0^m and γ_0^{m+1} in this representation. Therefore by a unitary transformation Λ can be brought to the diagonal form without disturbing γ_0^m and γ_0^{m+1}. Now consider the case when $\Lambda\gamma_0^{m+1}$ is non-negative. Then $\gamma_0^{m+1}\Lambda^{-1} = \Lambda\gamma_0^{m+1}\Lambda^{-2}$ is also non-negative. The diagonal elements $(\gamma_0^{m+1}\Lambda^{-1})_{\sigma\sigma} = 0$ correspond to $(\gamma_0^m)_{\sigma\sigma} = 0$, since $\gamma_0\gamma_0^{m+1} = \gamma_0^m$. Hence for $\sigma = \rho$ the right side of (42) is > 0 unless $(\gamma_0^m)_{\sigma\sigma} = 0$, in which case $a_\sigma = 0$ from (41). The commutation rules (40) are therefore consistent in the case of Fermi statistics. The validity of (38) then follows by Lorentz transformation. The other case corresponding to Bose statistics does not need any further investigation.

4. Electromagnetic interaction

One can introduce electromagnetic interaction in (12) in the usual way by replacing ∂_k by $\partial_{\bar{k}} = \partial_k - ie\phi_k$, where e is the charge and ϕ_k the electromagnetic potentials. The resulting equation is

$$i\gamma^k \partial_k \psi + e\gamma^k \phi_k \psi + \chi\psi = 0. \tag{44}$$

It is not at all obvious whether this equation is self-consistent, i.e. whether it admits of any solutions other than $\psi = 0$ for a given set of functions ϕ_k. The argument that since (44) can be derived by variation of a Lagrangian it must be self-consistent is invalid because the Lagrangian is not a positive definite from and therefore it need not have a minimum at all. Also the method of counting the number of 'subsidiary conditions' (Fierz & Pauli 1939) satisfied by ψ at a given time is equally unsatisfactory so long as the mutual consistency of these conditions is not demonstrated. In fact, this problem of the existence of a solution of the above system of partial differential equations is not an easy one at all, especially since the γ_k's are in general all singular. I shall therefore have to content myself merely with some remarks which make the existence of such a solution plausible.

First notice that the most general solution of (12) is

$$\psi = D^m(1+D)\,\Psi, \tag{45}$$

where $D \equiv \gamma^k \partial_k / i\chi$ and Ψ is an arbitrary solution of (11). For, clearly, (45) is always a solution of (12) in consequence of (39), and every solution ψ of (12) can be put in the form (45) by taking $\Psi = \tfrac{1}{2}\psi$. Hence in considering the equation

$$i\gamma^k \partial_k \psi + \chi\psi = \psi_0 \tag{46}$$

with ψ_0 given, one need obtain only one particular solution. One such solution is given by

$$\chi\psi = D^m(1+D)\,\Psi_0 + (1+D+D^2+\ldots+D^{m-1})\,\psi_0, \tag{47a}$$

where Ψ_0 is any solution of the equation

$$(\partial_k \partial^k + \chi^2)\,\Psi_0 = \chi^2\psi_0. \tag{47b}$$

Thus (46) always has solutions. Now assume that a solution of (44) can be expanded in powers of e so that

$$\psi = \psi_0 + e\psi_1 + \ldots + e^r\psi_r + \ldots. \tag{48a}$$

Substituting (48a) in (44) and equating coefficients of powers of e one gets

$$i\gamma^k \partial_k \psi_r + \chi\psi_r = -\gamma^k \phi_k \psi_{r-1}. \tag{48b}$$

Using (47) one can determine ψ_r by induction. If the series, so determined, converges for sufficiently small e in some domain it defines a solution of (44).

Put
$$\pi_k = \frac{1}{i}\partial_{\bar{k}} \tag{49a}$$

and
$$\rho_k = \Lambda\gamma_k. \tag{49b}$$

Then multiplying (44) by Λ on the left one gets‡
$$\rho_0\pi_0\psi + \rho^k\pi_{k^\circ}\psi = \chi\Lambda\psi. \tag{50}$$

Since Λ is non-singular (50) is completely equivalent to (44). Notice that due to (14), ρ_k are Hermitian. Choose a representation in which ρ_0 is diagonal. If $n > 2$, γ_0 and therefore ρ_0 is singular, and so some of the eigenvalues of ρ_0 are zero. Let σ_0 and σ denote any two matrix indices such that $(\rho_0)_{\sigma_0\sigma_0} = 0$ and $(\rho_0)_{\sigma\sigma} \neq 0$. Then (50) determines $\pi_0\psi_\sigma$ but not $\pi_0\psi_{\sigma_0}$. Corresponding to every σ_0 one gets a 'subsidiary condition' (Fierz & Pauli 1939) for ψ.

So far it has not been possible to quantize the field, in the general case, in presence of the electromagnetic interaction.

5. NON-RELATIVISTIC APPROXIMATION

Put
$$P_+ = \tfrac{1}{2}\gamma_0^m(1+\gamma_0), \quad P_- = \tfrac{1}{2}\gamma_0^m(1-\gamma_0), \quad P_0 = 1 - \gamma_0^m. \tag{51}$$

Then P_+, P_- and P_0 are three mutually orthogonal idempotents. In (39) put $k_1 = k^\circ$, $k_2 = k_3 = \ldots = k_n = 0$ and multiply by P_+ on both sides. Remembering that $P_+\gamma_0 = P_+$, one gets
$$P_+\gamma_{k^\circ}P_+ = 0. \tag{52a}$$

Similarly
$$P_-\gamma_{k^\circ}P_- = 0. \tag{52b}$$

The equation of motion with the electromagnetic interaction can be written as
$$\pi_k\gamma^k\psi = \chi\psi. \tag{53}$$

Multiply (53) by P_+, P_- and P_0 respectively. Remembering that $P_+ + P_- + P_0 = 1$ and using (52) one finds that
$$\pi_0\psi_+ + \pi_{k^\circ}P_+\gamma^{k^\circ}(\psi_- + \psi_0) = \chi\psi_+, \tag{54a}$$
$$-\pi_0\psi_- + \pi_{k^\circ}P_-\gamma^{k^\circ}(\psi_+ + \psi_0) = \chi\psi_-, \tag{54b}$$
$$\pi_0\gamma_0\psi_0 + \pi_{k^\circ}P_0\gamma^{k^\circ}\psi = \chi\psi_0, \tag{54c}$$

where $\psi_\pm = P_\pm\psi$ and $\psi_0 = P_0\psi$. (54b) can also be written as
$$\psi_- = \frac{1}{2}\left(1 - \frac{\pi_0}{\chi}\right)\psi_- + P_-\gamma^{k^\circ}\frac{\pi_{k^\circ}}{2\chi}\psi. \tag{55}$$

Similarly (54c) is equivalent to
$$(1-\gamma_0)\psi_0 = \left(\frac{\pi_0}{\chi} - 1\right)\gamma_0\psi_0 + P_0\frac{\gamma^{k^\circ}\pi_{k^\circ}}{\chi}\psi. \tag{56}$$

‡ Throughout this paper any index with a $^\circ$ at the top runs from 1 to 3 only.

Multiplying (56) with $1+\gamma_0+\gamma_0^2+\dots+\gamma_0^{m-1}$ and bearing in mind that $\gamma_0^m P_0 = 0$, one gets

$$\psi_0 = \left(\frac{\pi_0}{\chi}-1\right)(\gamma_0+\gamma_0^2+\dots+\gamma_0^m)\,\psi_0 + (1+\gamma_0+\dots+\gamma_0^{m-1})\,P_0\,\frac{\gamma^{k^\circ}\pi_{k^\circ}\psi}{\chi}. \qquad (57)$$

Now consider the non-relativistic case when

$$\left|\left(\frac{\pi_0}{\chi}-1\right)\psi\right| \ll \left|\frac{\pi_{k^\circ}}{\chi}\psi\right| \ll |\psi|, \qquad (58)$$

where $|\psi| = (\psi^\dagger\psi)^{\frac12}$. Then it is clear from (54), (55) and (57) that up to the first approximation

$$\psi_- \approx P_-\gamma^{k^\circ}\frac{\pi_{k^\circ}}{2\chi}\,\psi,$$

$$\psi_0 \approx (1+\gamma_0+\dots+\gamma_0^{m-1})\,P_0\,\frac{\gamma^{k^\circ}\pi_{k^\circ}}{\chi}\,\psi.$$

Thus ψ_- and ψ_0 are both small in comparison to ψ and therefore to ψ_+, since $\psi = \psi_+ + \psi_- + \psi_0$. Thus neglecting ψ_0 and ψ_- in comparison to ψ_+, one gets up to the same approximation

$$\psi_- \approx P_-\frac{\gamma^{k^\circ}\pi_{k^\circ}}{2\chi}\,\psi_+, \qquad (59a)$$

$$\psi_0 \approx (1+\gamma_0+\dots+\gamma_0^{m-1})\,P_0\,\frac{\gamma^{k^\circ}\pi_{k^\circ}}{\chi}\,\psi_+. \qquad (59b)$$

Substituting (59) in (54a) one finds

$$\pi_0\psi_+ + \frac{\pi_{k^\circ}\pi_{l^\circ}}{2\chi}P_+\gamma^{k^\circ}[2(1+\gamma_0+\dots+\gamma_0^{m-1})P_0+P_-]\gamma^{l^\circ}P_+\psi_+ = \chi\psi_+. \qquad (60)$$

Now on putting $k_1 = k^\circ$, $k_2 = k_3 = \dots = k_{n-1} = 0$ and $k_n = l^\circ$ in (39) and multiplying by P_+ on both sides, it follows that

$$P_+\{\gamma_{k^\circ}[\gamma_0^{n-2}+2(1+\gamma_0+\dots+\gamma_0^{n-3})]\gamma_{l^\circ}$$
$$+\gamma_{l^\circ}[\gamma_0^{n-2}+2(1+\gamma_0+\dots+\gamma_0^{n-3})]\gamma_{k^\circ}-2g_{k^\circ l^\circ}\}P_+ = 0. \qquad (61)$$

Now $m = n-2$ or $n-1$ according as n is even or odd. Hence

$$\gamma_0^{n-2}+2(1+\gamma_0+\dots+\gamma_0^{n-3}) = P_-+2(1+\gamma_0+\dots+\gamma_0^{m-1})\pm P_+,$$

where the $+$ or $-$ sign is to be taken according as n is even or odd. Therefore, from (52a)

$$P_+\gamma_{k^\circ}[\gamma_0^{n-2}+2(1+\gamma_0+\dots+\gamma_0^{n-3})] = P_+\gamma_{k^\circ}[2(1+\gamma_0+\dots+\gamma_0^{m-1})+P_-] \qquad (62)$$

in both cases. From (60), (61) and (62) one gets

$$\pi_0\psi_+ + \frac{\pi_{k^\circ}\pi^{k^\circ}}{2\chi}\psi_+ + \frac{ie}{2\chi}F_{k^\circ l^\circ}P_+\gamma^{k^\circ}\left(\frac{1-P_+}{1-\gamma_0}\right)\gamma^{l^\circ}P_+\psi_+ = \chi\psi_+, \qquad (63)$$

where $\quad F_{kl} = \partial_k\phi_l - \partial_l\phi_k \quad$ and $\quad \dfrac{1-P_+}{1-\gamma_0} \equiv 1+\gamma_0+\gamma_0^2+\dots+\gamma_0^{m-1}+\dfrac{\gamma_0^m}{2}.$

The charge-density is given by

$$s_0 = e\psi^*\gamma_0\psi = e\psi^*_+\psi_+ + e\psi^*_0\gamma_0\psi_0 - e\psi^*_-\psi_- \approx e\psi^*_+\psi_+ \tag{64a}$$

up to the first order of approximation. Here $\psi^*_+ = (\psi_+)^{\dagger}\Lambda$ and similarly for the others. Notice that

$$\psi^*_+\psi_+ = \psi^{\dagger}_+\Lambda\gamma_0^m\psi_+ = \psi^{\dagger}_+\Lambda\gamma_0^{m+1}\psi_+ > 0,$$

since either $\Lambda\gamma_0^m$ or $\Lambda\gamma_0^{m+1}$ is non-negative. Also

$$s_{k^\circ} = e\psi^*\gamma_{k^\circ}\psi = e\psi^*_+\gamma_{k^\circ}(\psi_- + \psi_0) + e(\psi^*_- + \psi^*_0)\gamma_{k^\circ}\psi_+.$$

Now from (59) $\qquad P_+\gamma_{k^\circ}(\psi_- + \psi_0) \approx \dfrac{1}{2\chi}P_+\gamma_{k^\circ}\left(\dfrac{1-P_+}{1-\gamma_0}\right)\gamma_{l^\circ}\pi^{l^\circ}\psi_+.$

Hence

$$s_{k^\circ} \approx \frac{e}{4\chi i}\left\{\psi^*_+\gamma_{k^\circ}\left(\frac{1-P_+}{1-\gamma_0}\right)\gamma^{l^\circ}\partial_{\bar{l^\circ}}\psi_+ - \partial^{\dagger}_{\bar{l^\circ}}\psi_+\gamma^{l^\circ}\left(\frac{1-P_+}{1-\gamma_0}\right)\gamma_{k^\circ}\psi_+\right\}$$

$$\approx \frac{e}{2\chi i}\{\psi^*_+\partial_{\bar{k^\circ}}\psi_+ - \partial^{\dagger}_{\bar{k^\circ}}\psi^*_+\cdot\psi\} + \frac{e}{4\chi i}\partial^{l^\circ}\left\{\psi^*_+\left[\gamma_{k^\circ}\left(\frac{1-P_+}{1-\gamma_0}\right)\gamma_{l^\circ} - \gamma_{l^\circ}\left(\frac{1-P_+}{1-\gamma_0}\right)\gamma_{k^\circ}\right]\psi_+\right\},$$

$$\tag{64b}$$

where $\partial^{\dagger}_k \equiv \partial_k + ie\phi_k$. The particle therefore has a magnetic moment **M** given by

$$\mathbf{M} = \frac{e}{\chi}\left[P_+\boldsymbol{\gamma}\cdot\left(\frac{1-P_+}{1-\gamma_0}\right)\boldsymbol{\gamma}P_+\right], \tag{65}$$

where $\boldsymbol{\gamma} = (\gamma_1, \gamma_2, \gamma_3)$ and the bracket denotes the usual vector product.

6. A new γ-representation

In the two well-known cases of the electron and the meson, the equation of motion can be put in the form (12) and the corresponding γ_k satisfy all the conditions imposed on them so far. It is therefore of interest to know whether any other γ-representations, besides these two, exist which fulfil all the above requirements. Corresponding to every such representation one can set up a theory of a new elementary particle based on (12). In this section one such new representation will be constructed. This suffices to show that the basic assumptions given do not preclude the possibility of the existence of new elementary particles.

Define γ_k by the following algebraic relations:

$$\gamma_k\gamma_l + \gamma_l\gamma_k = 2g_{kl} - B_kB_l - B_lB_k, \tag{66a}$$

$$\gamma_kB_l + B_l\gamma_k = i(B_kB_l + B_lB_k), \tag{66b}$$

$$B_kB_lB_m + B_mB_lB_k = g_{kl}B_m + g_{ml}B_k. \tag{66c}$$

Thus B_k satisfy the Duffin-Kemmer commutation rules. Notice that on putting $B_k = 0$ one gets the usual Dirac commutation rules for γ_k. Hence (66) may be looked upon as a generalization of the latter. Let $\mathfrak{A}(\gamma)$ and $\mathfrak{A}(B)$ denote the abstract algebras generated by $(\gamma_k, 1)$ and $(B_k, 1)$ respectively. Then there exists an element R in $\mathfrak{A}(B)$ such that

$$RB_k + B_k R = 0, \tag{67a}$$

$$R^2 = 1. \tag{67b}$$

This is seen as follows. $-B_k$ satisfy the same commutation rules as B_k. Hence $B_k \to -B_k$ is an automorphism of $\mathfrak{A}(B)$ which leaves the elements of the centre invariant (see A, §5). Since $\mathfrak{A}(B)$ is semi-simple, every such automorphism is an inner automorphism. Therefore there exists a unit R' of $\mathfrak{A}(B)$ such that

$$-B_k = R' B_k R'^{-1}.$$

Clearly $C = R'^{-2}$ commutes with B_k and therefore belongs to the centre. By considering the regular representation it is obvious that an element c exists in the centre such that $c^2 = C$. $R = R'c$ then fulfils (67). Moreover, R commutes with $I_{kl} = B_k B_l - B_l B_k$ and the 'reflexion operator' $1 - 2B_0^2$. Hence it is a 'scalar' (cf. A). Therefore as shown in A (§5) it is a polynomial in $B_k B^k$. Now put

$$A_k = (\gamma_k - iB_k)\, R.$$

Then from (66b) and (67) $A_k B_l = B_l A_k,$

so that A_k and B_l commute. Therefore $A_k R = R A_k$ and from (66a)

$$A_k A_l + A_l A_k = 2g_{kl}.$$

Hence the algebra $\mathfrak{A}(A)$ generated by A_k is the usual Dirac algebra. Thus

$$\gamma_k = A_k R + iB_k, \tag{68}$$

where A_k and B_l commute with each other and satisfy the Dirac and the Duffin-Kemmer commutation rules respectively. Now from (66a) it is clear that $R \in \mathfrak{A}(\gamma)$, so that

$$A_k = \tfrac{1}{2}(R\gamma_k + \gamma_k R) \in \mathfrak{A}(\gamma).$$

Hence $$B_k = \frac{1}{i}(\gamma_k - A_k R) \in \mathfrak{A}(\gamma).$$

Thus both $\mathfrak{A}(A)$ and $\mathfrak{A}(B)$ are contained in $\mathfrak{A}(\gamma)$. Therefore $\mathfrak{A}(A) \times \mathfrak{A}(B) \subset \mathfrak{A}(\gamma)$, where $\times$ denotes Kronecker product. But from (68) it is obvious that

$$\mathfrak{A}(\gamma) \subset \mathfrak{A}(A) \times \mathfrak{A}(B).$$

Hence $$\mathfrak{A}(\gamma) = \mathfrak{A}(A) \times \mathfrak{A}(B). \tag{69}$$

Now choose an irreducible representation for B_k in (68) so that the corresponding $\mathfrak{A}(B)$ is simple. Then since $\mathfrak{A}(A)$ is also simple it follows from (69) that $\mathfrak{A}(\gamma)$ is simple (see Artin, Nesbitt & Thrall 1944, p. 62). Hence the γ-representation so obtained is irreducible.

For physical application it is found more convenient to express γ_k in a slightly modified form. Put

$$\alpha_k = A_k R, \tag{70a}$$

$$\omega = i\alpha_0\alpha_1\alpha_2\alpha_3 = iA_0A_1A_2A_3, \tag{70b}$$

$$\beta_k = \omega B_k. \tag{70c}$$

Then α_k and β_l commute with each other and satisfy the Dirac and the Duffin-Kemmer commutation rules respectively. (66) and (68) can now be written as

$$\gamma_k\gamma_l + \gamma_l\gamma_k = 2g_{kl} - \beta_k\beta_l - \beta_l\beta_k, \tag{71a}$$

$$\gamma_k\omega\beta_l + \beta_l\omega\gamma_k = i(\beta_k\beta_l + \beta_l\beta_k), \tag{71b}$$

$$\beta_k\beta_l\beta_m + \beta_m\beta_l\beta_k = g_{kl}\beta_m + g_{ml}\beta_k, \tag{71c}$$

$$\gamma_k = \alpha_k + i\omega\beta_k. \tag{71d}$$

The tensor form of (66) ensures the invariance of $\mathfrak{A}(\gamma)$ under the Lorentz group (cf. A). In fact, the operators I_{kl} are given by

$$I_{kl} = \tfrac{1}{4}(\alpha_k\alpha_l - \alpha_l\alpha_k) + (\beta_k\beta_l - \beta_l\beta_k). \tag{72}$$

The reflexion operator is $\alpha_0(1 - 2\beta_0^2)\,R$. Moreover, from $(71a)$,

$$(1 - \gamma_0^2) = \beta_0^2,$$

so that

$$\gamma_0^2(\gamma_0^2 - 1) = -\beta_0^2(\beta_0^2 - 1) = 0. \tag{73}$$

However, except for the representation in which $\beta_k = 0$,

$$\gamma_0^3 \neq \gamma_0,$$

and therefore the minimum equation of γ_0 is of the fourth degree so that $n = 4$. Further if Λ_α and Λ_β are two non-singular Hermitian matrices in the α- and β-spaces respectively such that

$$\alpha_k^\dagger = \Lambda_\alpha\alpha_k\Lambda_\alpha^{-1}, \tag{74a}$$

$$\beta_k^\dagger = \Lambda_\beta\beta_k\Lambda_\beta^{-1}, \tag{74b}$$

then

$$\Lambda = \Lambda_\alpha \times \Lambda_\beta \tag{74c}$$

satisfies (14) since $\Lambda_\alpha\omega\Lambda_\alpha^{-1} = -\omega^\dagger$. In a suitable representation $\Lambda_\alpha = \alpha_0$ and $\Lambda_\beta = 1 - 2\beta_0^2$. Then

$$\Lambda\gamma_0^3 = \alpha_0(1 - 2\beta_0^2)(1 - \beta_0^2)\alpha_0 = 1 - \beta_0^2 \tag{75}$$

is a non-negative matrix, its eigenvalues being 0 and 1. Hence the total charge is positive definite and quantization in accordance with Fermi statistics is possible.

For the 10-row representation of β_k the corresponding γ_k's are 40×40 matrices. However, on account of $(71d)$ and the rather complete knowledge of the structure of the α- and the β-algebras which we now possess (Harish-Chandra 1946b,c) the wave equation (12) can be handled quite easily (cf. §§ 7, 9). The α-space is the direct sum of two subspaces $\mathfrak{R}_{\frac{1}{2},0}$ and $\mathfrak{R}_{0,\frac{1}{2}}$ transforming respectively according to the irreducible representations‡ $\mathfrak{D}_{\frac{1}{2},0}$ and $\mathfrak{D}_{0,\frac{1}{2}}$ of the Lorentz group. Similarly, the β-space is the direct sum of three subspaces $\mathfrak{R}_{1,0}$, $\mathfrak{R}_{\frac{1}{2},\frac{1}{2}}$, $\mathfrak{R}_{0,1}$ transforming according to $\mathfrak{D}_{1,0}$, $\mathfrak{D}_{\frac{1}{2},\frac{1}{2}}$, $\mathfrak{D}_{0,1}$ respectively. The product space $\mathfrak{R}_\alpha \times \mathfrak{R}_\beta$ can therefore be reduced with respect to the proper Lorentz group in the usual way by reducing each product occurring in the product representation

$$(\mathfrak{D}_{\frac{1}{2},0}\oplus\mathfrak{D}_{0,\frac{1}{2}})\times(\mathfrak{D}_{1,0}\oplus\mathfrak{D}_{\frac{1}{2},\frac{1}{2}}\oplus\mathfrak{D}_{0,1}).$$

Here $\oplus$ denotes a direct sum. As usual one finds on reduction that

$$(\mathfrak{D}_{\frac{1}{2},0}\oplus\mathfrak{D}_{0,\frac{1}{2}})\times(\mathfrak{D}_{1,0}\oplus\mathfrak{D}_{\frac{1}{2},\frac{1}{2}}\oplus\mathfrak{D}_{0,1})$$
$$= \mathfrak{D}_{\frac{3}{2},0}\oplus 2\mathfrak{D}_{1,\frac{1}{2}}\oplus 2\mathfrak{D}_{\frac{1}{2},1}\oplus\mathfrak{D}_{0,\frac{3}{2}}\oplus 2\mathfrak{D}_{\frac{1}{2},0}\oplus 2\mathfrak{D}_{0,\frac{1}{2}}, \tag{76}$$

where the factor 2 denotes that the representation in question occurs twice in the reduction. Similarly for the five-row representation of β_k, the γ_k's are 20×20 matrices and the following scheme of reduction holds:

$$(\mathfrak{D}_{\frac{1}{2},0}\oplus\mathfrak{D}_{0,\frac{1}{2}})\times(\mathfrak{D}_{0,0}\oplus\mathfrak{D}_{\frac{1}{2},\frac{1}{2}}) = \mathfrak{D}_{1,\frac{1}{2}}\oplus\mathfrak{D}_{\frac{1}{2},1}\oplus 2\mathfrak{D}_{\frac{1}{2},0}\oplus 2\mathfrak{D}_{0,\frac{1}{2}}. \tag{77}$$

(76) and (77) would lead one to expect that the corresponding wave equations describe particles which are capable of both spins $\frac{3}{2}$ and $\frac{1}{2}$. This question will be discussed more closely in §8.

7. The equation of motion

As shown in A the second-order equation (11) follows from (12) on account of (73). However, from (12), one finds that

$$\gamma^k\gamma^l\partial_k\partial_l\psi + \chi^2\psi = 0,$$

so that
$$(g^{kl}-\gamma^k\gamma^l)\,\partial_k\partial_l\psi = 0.$$

Or
$$\beta^k\beta^l\partial_k\partial_l\psi = 0$$

from (71a). Multiplying this equation by $\beta^m\partial_m$ and using (71c) one gets

$$\beta^l\partial_l\psi = 0. \tag{78a}$$

Thus from (71d) and (78a), (12) reduces to

$$i\alpha^k\partial_k\psi + \chi\psi = 0. \tag{78b}$$

The equation with interaction may be written as

$$\pi_k\gamma^k\psi = \chi\psi. \tag{79}$$

Therefore
$$\pi_k\pi_l\gamma^k\gamma^l\psi = \chi^2\psi.$$

Or from (71a)
$$\pi_k\pi_l(g^{kl}-\beta^k\beta^l)\,\psi + \tfrac{1}{2}[\pi_k,\pi_l]_-\,(\gamma^k\gamma^l+\beta^k\beta^l)\,\psi = \chi^2\psi,$$

i.e.
$$\pi_k\pi_l(g^{kl}-\beta^k\beta^l)\,\psi + \frac{ie}{2}\,F_{kl}(\gamma^k\gamma^l+\beta^k\beta^l)\,\psi = \chi^2\psi. \tag{80}$$

---

‡ $\mathfrak{D}_{k,l}$ denotes that irreducible representation of the proper Lorentz group which is induced in the space of all symmetric spinors with $2k$ undotted and $2l$ dotted indices.

Multiply (80) by $\pi_m \beta^m$. Then from (71c) one gets‡

$$\chi^2 \pi_m \beta^m \psi = \pi_m \pi_k \pi_l \beta^m (g^{kl} - \beta^k \beta^l)\,\psi + \frac{ie}{2}\pi_m \cdot F_{kl}\beta^m (\gamma^k \gamma^l + \beta^k \beta^l)\,\psi$$

$$= \tfrac{1}{2}[\pi_m \pi_k \pi_l - \pi_l \pi_k \pi_m]\beta^m (g^{kl} - \beta^k \beta^l)\,\psi + \frac{ie}{2}\pi_m \cdot F_{kl}\beta^m (\alpha^k \alpha^l - 2i\omega\beta^k \beta^l)\,\psi. \quad (81)$$

Now
$$\pi_m \pi_k \pi_l - \pi_l \pi_k \pi_m = [\pi_m, \pi_k \pi_l]_- + [\pi_k, \pi_l]_- \pi_m$$
$$= ie[F_{ml}\pi_k + F_{mk}\pi_l + F_{kl}\pi_m] + e(\partial_k F_{ml}).$$

Therefore

$$\tfrac{1}{2}[\pi_m \pi_k \pi_l - \pi_l \pi_k \pi_m]\beta^m (g^{kl} - \beta^k \beta^l)\,\psi$$

$$= \frac{ie}{2}F_{kl}\pi_m\{-\beta^m \beta^k \beta^l + 2\beta^k g^{lm} - \beta^k \beta^m \beta^l - \beta^k \beta^l \beta^m\}\,\psi - \frac{e}{2}\partial_k F_{lm}\cdot\beta^m (g^{kl} - \beta^k \beta^l)\,\psi. \quad (82)$$

Also

$$\frac{ie}{2}\pi_m \cdot F_{kl}\beta^m (\alpha^k \alpha^l - 2i\omega\alpha^k \beta^l)\,\psi$$

$$= \frac{e}{2}\partial_m F_{kl}\cdot\beta^m (\alpha^k \alpha^l - 2i\omega\alpha^k \beta^l)\,\psi + \frac{ie}{2}F_{kl}\alpha^k \alpha^l \pi_m \beta^m \psi + \frac{ie}{2}F_{kl}(-2i\omega\alpha^k \beta^m \beta^l)\pi_m \psi. \quad (83)$$

Therefore from (81), (82) and (83)

$$\pi_m \beta^m \psi = \frac{ie}{2\chi^2}F_{kl}\{3\beta^k g^{lm} - \beta^k \beta^m \beta^l - 2i\omega\alpha^k \beta^m \beta^l\}\pi_m \psi + \frac{ie}{2\chi^2}F_{kl}(\alpha^k \alpha^l - 2\beta^k \beta^l)\pi_m \beta^m \psi$$

$$+ \frac{e}{2\chi^2}\partial_k F_{lm}\cdot(\beta^k \alpha^l \alpha^m - 2i\omega\beta^k \alpha^l \beta^m - \beta^m g^{kl} + \beta^m \beta^k \beta^l)\,\psi. \quad (84)$$

(84) is the correct generalization of (78a), in the presence of the electromagnetic interaction. The second-order equation with interaction now follows immediately. From (80)

$$\pi_k \pi^k \psi + \frac{ie}{2}F_{kl}(\alpha^k \alpha^l - 2i\omega\alpha^k \beta^l)\,\psi - \pi_k \beta^k \pi_l \beta^l \psi = \chi^2 \psi. \quad (85)$$

One can now substitute for $\pi_l \beta^l \psi$ from (84).

The charge-current density s^k is given by

$$s^k = e\psi^* \gamma^k \psi = \frac{e}{2\chi i}\{\psi^* \gamma^k \gamma^l \partial_l^- \psi - \partial_l^+ \psi^* \cdot \gamma^k \gamma^l \psi\}$$

$$= \frac{e}{4\chi i}[\psi^* \partial^{-k}\psi - \partial^{+k}\psi^* \cdot \psi] + \frac{e}{4\chi i}\partial_l[\psi^*(\gamma^k \gamma^l + \beta^k \beta^l - \gamma^l \gamma^k - \beta^l \beta^k)\,\psi]$$

$$- \frac{e}{2\chi i}[\psi^* \beta^k \beta^l \partial_l^- \psi - \partial_l^+ \psi^* \cdot \beta^l \beta^k \psi]$$

from (71a). Write $-\dfrac{1}{i}\partial_l^+ \psi^* = \psi^* \pi_l^\dagger$. Then the above equation may be written as

$$s^k = \frac{e}{2\chi i}[\psi^* \partial^{-k}\psi - \partial^{+k}\psi^* \cdot \psi] + \frac{e}{4\chi i}\partial_l\{\psi^*(\alpha^k \alpha^l - \alpha^l \alpha^k - 2i\omega\alpha^k \beta^l + 2i\omega\alpha^l \beta^k)\,\psi\}$$

$$- \frac{e}{2\chi}\{\psi^* \beta^k \beta^l \pi_l \psi + \psi^* \pi_l^\dagger \beta^l \beta^k \psi\}. \quad (86)$$

‡ For any two operators p, q, $p\cdot q\psi = p(q\psi)$, while $pq\cdot\psi = (pq)\,\psi$.

The first term corresponds to the usual orbital current. The second term may be interpreted as the current due to a dipole density M^{kl} given by

$$M^{kl} = \frac{e}{\chi}\psi^*\left[\frac{1}{4i}(\alpha^k\alpha^l - \alpha^l\alpha^k) - \frac{\omega}{2}(\alpha^k\beta^l - \alpha^l\beta^k)\right]\psi. \tag{87}$$

On account of (84) the third term in (86) is proportional to e^2 and must be interpreted as the polarization current (cf. Kemmer 1939). It vanishes in the absence of the external electromagnetic field.

Now assume that the electromagnetic field does not vary appreciably within a Compton wave-length and the interaction energy is small compared to the rest mass so that

$$\frac{1}{\chi}\,|\,\partial_k F_{lm}\,| \ll |\,F_{lm}\,| \quad \text{and} \quad \frac{e}{\chi}\,|\,F_{lm}\,| \ll \chi.$$

Hence
$$\frac{e}{\chi^3}\,|\,\partial_k F_{lm}\,| \ll \frac{e}{\chi^2}\,|\,F_{lm}\,| \ll 1. \tag{88}$$

Further, exclude the extreme relativistic case so that $\left|\dfrac{\pi_m}{\chi}\psi\right|$ is not large compared to $|\,\psi\,|$. Then it follows from (84) that up to the first approximation

$$\frac{1}{\chi}\pi_m\beta^m\psi \approx \frac{ie}{2\chi^2}F_{kl}\{3\beta^k g^{lm} - \beta^k\beta^m\beta^l - 2i\omega\alpha^k\beta^m\beta^l\}\frac{\pi_m}{\chi}\psi, \tag{89}$$

and therefore to the same approximation

$$\frac{1}{\chi^2}\pi_n\beta^n\pi_m\beta^m\psi \approx \frac{ie}{2\chi^2}F_{kl}\{3\beta^n\beta^k g^{lm} - \beta^n\beta^k\beta^m\beta^l - 2i\omega\alpha^k\beta^n\beta^m\beta^l\}\frac{\pi_m\pi_n}{\chi^2}\psi$$

$$\approx \frac{ie}{2\chi^2}F_{kl}\{3\beta^n\beta^k g^{ml} - g^{nk}\beta^m\beta^l - 2i\omega\alpha^k g^{nm}\beta^l\}\frac{\pi_m\pi_n}{\chi^2}\psi$$

from (71c). Therefore from (85)

$$\frac{1}{\chi^2}\pi_n\beta^n\pi_m\beta^m\psi \approx \frac{ie}{2\chi^2}F_{kl}\left\{4\beta^n\beta^k\frac{\pi^l\pi_n}{\chi^2} - 2i\omega\beta^k\beta^l\right\}\psi. \tag{90}$$

Therefore from (79), (85), (89) and (90)

$$\pi_k\alpha^k\psi - \frac{e}{2\chi}F_{kl}\{3\omega\beta^k g^{lm} - \omega\beta^k\beta^m\beta^l - 2i\alpha^k\beta^m\beta^l\}\frac{\pi_m\psi}{\chi} = \chi\psi, \tag{91a}$$

$$\pi_k\pi^k\psi + \frac{ie}{2}F_{kl}\alpha^k\alpha^l\psi - \frac{ie}{2}F_{kl}4\beta^n\beta^k\frac{\pi^l\pi_n}{\chi^2}\psi = \chi^2\psi. \tag{91b}$$

8. Spin

Define two three-dimensional vectors $\boldsymbol{\sigma}$ and $\mathbf{K}$ as follows:

$$\sigma_3 = \frac{i}{2}(\alpha_1\alpha_2 - \alpha_2\alpha_1), \tag{92a}$$

$$K_3 = i(\beta_1\beta_2 - \beta_2\beta_1). \tag{92b}$$

the other components being obtained by the cyclic permutations of 1, 2, 3. More-over, define a non-negative Hermitian operator k by the relation

$$\mathbf{K}^2 = k(k+1). \tag{93}$$

k has only two eigenvalues 0 and 1. Choose α_0, σ_3 and β_0, K_3, k diagonal in the α- and β-spaces respectively. This corresponds to a reduction of the two spaces separately with respect to the three-dimensional rotation group. Now consider the free particle in the rest system. From (78a) it follows that $\beta_0 \psi = 0$. Hence only the states corresponding to the eigenvalue 0 of β_0 are permitted. First consider the ten-row representation of β_k. Then the eigenvalue 0 of β_0 occurs four times. There are therefore $4 \times 4 = 16$ independent states in the rest system. The energy density is given by

$$\chi \psi^* \psi = \chi \psi^\dagger \Lambda \psi = \chi \psi^\dagger \alpha_0 (1 - 2\beta_0^2) \psi = \chi \psi^\dagger \alpha_0 \psi.$$

Hence the eigenvalues ± 1 of α_0 correspond to states of positive and negative energy respectively. Thus there are eight independent states corresponding to each sign of energy. Now the angular-momentum tensor is given by

$$M_{kl,m} = \frac{1}{2i}\{(x_k \partial_l \psi^* - x_l \partial_k \psi^*)\gamma_m \psi - \psi^* \gamma_m (x_k \partial_l - x_l \partial_k)\psi\} + \frac{i}{2}\psi^*(I_{kl}\gamma_m + \gamma_m I_{kl})\psi, \tag{94}$$

where I_{kl} is the same as in (72). Hence the spin (i.e. the angular momentum in the rest system) is given by

$$M_{k°l°,0} = \frac{i}{2}\psi^*(I_{k°l°}\gamma_0 + \gamma_0 I_{k°l°})\psi = i\psi^\dagger I_{k°l°}\psi. \tag{95}$$

Thus the spin is represented by the operator $\frac{1}{2}\boldsymbol{\sigma} + \mathbf{K}$. Since $\boldsymbol{\sigma}$ and $\mathbf{K}$ commute this can be interpreted as the resultant of two independent spins $\frac{1}{2}\boldsymbol{\sigma}$ and $\mathbf{K}$ respectively. The first of these has two independent states corresponding to the eigenvalues $\sigma_3 = \pm 1$, and the second has four independent states corresponding to $k = 1$, $K_3 = 0, \pm 1$ and $k = 0$, $K_3 = 0$. This accounts for the eight independent states for each sign of energy. Similarly for the five-row representation there are six independent states for each sign of the energy corresponding to the eigenvalues $\sigma_3 = \pm 1$ and $k = 1$, $K_3 = 0, \pm 1$. In both cases the total spin has the eigenvalues $\frac{3}{2}$ and $\frac{1}{2}$. Notice that in order to describe completely the state of the free particle in the rest system in the ten-row representation it is, in general, not sufficient to give the sign of the energy and the direction and magnitude of the total spin. One must, in addition, specify whether the two component spins $\boldsymbol{\sigma}$ and $\mathbf{K}$ are pointed along or opposite to each other. In this sense therefore *the particle has an inner structure beyond the spin.*

From (63) one easily shows that the non-relativistic approximation of (79) is the same as that of the Dirac electron. In this approximation the electromagnetic field acts only on the $\boldsymbol{\sigma}$-spin leaving the $\mathbf{K}$-spin free. Hence each non-relativistic state with a given direction of the magnetic moment (i.e. $\boldsymbol{\sigma}$-spin) is fourfold and threefold degenerate in the ten-row and five-row representations

Harish-Chandra

respectively. Thus the Pauli exclusion principle would be apparently inoperative for these particles in the non-relativistic approximation until all the degenerate levels are filled up. This difference in behaviour from a Dirac particle can therefore be detected in the case of systems consisting of several particles.

9. Field equations

One can go over from the above 'particle formulation' to the 'wave formulation' (cf. Kemmer 1939) by using the Γ-formalism developed in two previous papers (Harish-Chandra 1946 b, c). Greek alphabets will be used to denote spinor indices, Latin letters being reserved for tensor indices. Two one-row matrices Γ_λ^* and $\Gamma_{\dot\lambda}^*$ and two one-column matrices T_μ and $T_{\dot\mu}$ are defined in the α-space so as to satisfy the following relations:

$$\Gamma_\alpha^* T_\beta = \Gamma_{\dot\alpha}^* T_\beta = 0, \tag{96a}$$

$$\Gamma_\alpha^* T_\beta = i\epsilon_{\alpha\beta}, \quad \Gamma_{\dot\alpha}^* T_\beta = i\epsilon_{\dot\alpha\beta}, \tag{96b}$$

$$\alpha_{\lambda\dot\mu} = \alpha_k \sigma_{\lambda\dot\mu}^k = 2(T_\lambda \Gamma_{\dot\mu}^* + T_{\dot\mu} \Gamma_\lambda^*). \tag{96c}$$

Here $\epsilon_{\alpha\beta}$, $\epsilon_{\dot\alpha\beta}$ and $\sigma_{\lambda\dot\mu}^k$ have their usual meaning. It can be shown (Harish-Chandra 1946c) that

$$\omega = iT_\alpha \Gamma^{*\alpha} - iT_{\dot\alpha} \Gamma^{*\dot\alpha}, \tag{97a}$$

$$1 = iT_\alpha \Gamma^{*\alpha} + iT_{\dot\alpha} \Gamma^{*\dot\alpha}. \tag{97b}$$

Then (79) may be written as

$$i\alpha^k \partial_{\bar k}^- \psi - \omega \beta^k \partial_{\bar k}^- \psi + \chi\psi$$
$$= i(T_\lambda \Gamma_{\dot\mu}^* + T_{\dot\mu} \Gamma_\lambda^*) \partial^{-\lambda\dot\mu} \psi - i(T_\alpha \Gamma^{*\alpha} - T_{\dot\alpha} \Gamma^{*\dot\alpha}) \beta^k \partial_{\bar k}^- \psi + \chi\psi = 0, \tag{98}$$

where $\partial^{-\lambda\dot\mu} = \partial^{-k} \sigma_k^{\lambda\dot\mu}$. Now put $\Gamma_\lambda^* \psi = \psi_\lambda$ and $\Gamma_{\dot\mu}^* \psi = \psi_{\dot\mu}$ and multiply (98) by Γ_ν^* and $\Gamma_{\dot\nu}^*$ respectively on the left. Then one finds that

$$\partial_{\bar\lambda}^{-\dot\mu} \psi_{\dot\mu} - \beta^k \partial_{\bar k}^- \psi_\lambda + \chi\psi_\lambda = 0, \tag{99a}$$

$$\partial_{\dot\mu}^{-\lambda} \psi_\lambda + \beta^k \partial_{\bar k}^- \psi_{\dot\mu} + \chi\psi_{\dot\mu} = 0. \tag{99b}$$

Now in the β-space one can similarly introduce one-row and one-column matrices Γ_k^* and T_l respectively with the following properties (Harish-Chandra 1946b):

$$\Gamma_k^* T_l = g_{kl}, \quad \Gamma_k^* \beta_l T_m = 0, \tag{100a}$$

$$\left.\begin{aligned}
\Gamma_k^* \beta_l &= -\Gamma_l^* \beta_k, \quad \beta_k T_l = -\beta_l T_k, \\
\Gamma_k^* \beta_l \beta_m &= \Gamma_k^* g_{lm} - \Gamma_l^* g_{km}, \quad \beta_k \beta_l T_m = g_{kl} T_m - g_{km} T_l,
\end{aligned}\right\} \quad \begin{aligned}&\text{for the ten-row} \\ &\text{representation}\end{aligned} \tag{100b}$$

$$\left.\begin{aligned}
\Gamma_k^* \beta_l &= g_{kl} \tfrac{1}{4} \Gamma_m^* \beta^m, \quad \beta_k T_l = g_{kl} \tfrac{1}{4} \beta_m T^m, \\
\Gamma_k^* \beta_l \beta_m &= g_{kl} \Gamma_m^*, \quad \beta_k \beta_l T_m = T_k g_{lm}.
\end{aligned}\right\} \quad \begin{aligned}&\text{for the five-row} \\ &\text{representation}\end{aligned} \tag{100c}$$

First consider the ten-row representation. Multiply (99) by Γ_m^* and $\Gamma_m^* \beta_n$ respectively and put

$$\left.\begin{aligned}
\Gamma_k^* \psi_\lambda &= U_{k,\lambda}, \quad \Gamma_k^* \beta_l \psi_\lambda = \frac{1}{\chi} G_{kl,\lambda}, \\[2mm]
\Gamma_k^* \psi_{\dot\lambda} &= U_{k,\dot\lambda}, \quad \Gamma_k^* \beta_l \psi_{\dot\lambda} = -\frac{1}{\chi} G_{kl,\dot\lambda}.
\end{aligned}\right\} \tag{101}$$

Then one finds that
$$\partial^{-k}G_{kl,\lambda} + \chi^2 U_{l,\lambda} = \chi\partial^{-}_{\lambda\mu}U_l^{\dot\mu}, \tag{102a}$$

$$\partial^{-}_{k}U_{l,\lambda} - \partial^{-}_{l}U_{k,\lambda} + G_{kl,\lambda} = \frac{1}{\chi}\partial^{-\mu}_{\lambda}G_{kl,\mu}, \tag{102b}$$

$$\partial^{-k}G_{kl,\dot\mu} + \chi^2 U_{l,\dot\mu} = \chi\partial^{-}_{\dot\mu\lambda}U_l^{\lambda}, \tag{102c}$$

$$\partial^{-}_{k}U_{l,\dot\mu} - \partial^{-}_{l}U_{k,\dot\mu} + G_{kl,\dot\mu} = \frac{1}{\chi}\partial^{-\lambda}_{\dot\mu}G_{kl,\lambda}. \tag{102d}$$

From (78) it follows that in the force-free case these equations reduce to

$$\left.\begin{aligned}
\partial_{\lambda\dot\mu}U_k^{\dot\mu} = \chi U_{k,\lambda}, \quad \partial_{\dot\mu\lambda}U_k^{\lambda} = \chi U_{k,\dot\mu}, \\
\partial_k U_{l,\lambda} - \partial_l U_{k,\lambda} = \partial_k U_{l,\dot\mu} - \partial_l U_{k,\dot\mu} = 0,
\end{aligned}\right\} \tag{103a}$$

$$\left.\begin{aligned}
\partial^{\dot\mu}_{\lambda}G_{kl,\dot\mu} = \chi G_{kl,\lambda}, \quad \partial^{\lambda}_{\dot\mu}G_{kl,\lambda} = \chi G_{kl,\dot\mu}, \\
\partial^{k}G_{kl,\dot\mu} = \partial^{k}G_{kl,\lambda} = 0.
\end{aligned}\right\} \tag{103b}$$

For the five-row representation one similarly finds on putting

$$\left.\begin{aligned}
\Gamma_k^{*}\psi_{\lambda} = \frac{1}{\chi}U_{k,\lambda}, \quad \Gamma_k^{*}\beta_l\psi_{\lambda} = -g_{kl}U_{\lambda}, \\
\Gamma_k^{*}\psi_{\dot\lambda} = \frac{1}{\chi}U_{k,\dot\lambda}, \quad \Gamma_k^{*}\beta_l\psi_{\dot\mu} = g_{kl}U_{\dot\mu},
\end{aligned}\right\} \tag{104}$$

that (99) is equivalent to
$$\partial^{-}_{k}U_{\lambda} + U_{k,\lambda} = \frac{1}{\chi}\partial^{-}_{\lambda\dot\mu}U_k^{\dot\mu}, \tag{105a}$$

$$\partial^{-}_{k}U^{k}_{\lambda} + \chi^2 U_{\lambda} = \frac{1}{\chi}\partial^{-\dot\mu}_{\lambda}U_{\dot\mu}, \tag{105b}$$

$$\partial^{-}_{k}U_{\dot\mu} + U_{k,\dot\mu} = \frac{1}{\chi}\partial^{-}_{\lambda\dot\mu}U_k^{\lambda}, \tag{105c}$$

$$\partial^{-}_{k}U^{k}_{\dot\mu} + \chi^2 U_{\dot\mu} = \chi\partial^{-\lambda}_{\dot\mu}U_{\lambda}. \tag{105d}$$

From (78a) one finds that in the force-free case
$$\partial_k U_{.} = 0, \quad \partial^k U_{k,.} = 0,$$

where the dot denotes that one can put either of the two suffixes $\lambda, \dot\mu$. Hence
$$U_{\lambda} = U_{\dot\mu} = 0,$$

and
$$\partial_{\lambda\dot\mu}U_k^{\dot\mu} = \chi U_{k,\lambda}, \quad \partial_{\dot\mu\lambda}U_k^{\lambda} = \chi U_{k,\dot\mu}, \quad \partial^k U_{k,\lambda} = \partial^k U_{k,\dot\mu} = 0. \tag{106}$$

REFERENCES

Artin, Nesbitt & Thrall 1944 *Rings with minimum condition.* University of Michigan Press.
Bhabha 1945a *Rev. Mod. Phys.* **17**, 200–216.
Bhabha 1945b *Proc. Ind. Acad. Sci.* A, **21**, 264–291.
Courant & Hilbert 1937 *Methoden der Mathematischen Physik*, **2**. Berlin: Springer.
Fierz & Pauli 1939 *Proc. Roy. Soc.* A, **173**, 211–232.
Harish-Chandra 1946a *Phys. Rev.* **71**, 793–805 (referred to as A).

Harish-Chandra 1946b *Proc. Roy. Soc.* A, **186**, 502–525.
Harish-Chandra 1946c *Proc. Camb. Phil. Soc.* **43**, 414–421.
Kemmer 1939 *Proc. Roy. Soc.* A, **173**, 91–116.
Lefschetz 1942 *Algebraic topology.* American Mathematical Society.
Madhavarao 1942 *Proc. Ind. Acad. Sci.* A, **15**, 139–147.
Murnaghan 1938 *The theory of group representation.* The John Hopkins Press.
Pauli 1940 *Phys. Rev.* **58**, 716–722.
Wigner 1939 *Ann. Math.* **40**, 149–204.

PRINTED IN GREAT BRITAIN AT THE UNIVERSITY PRESS, CAMBRIDGE
(BROOKE CRUTCHLEY, UNIVERSITY PRINTER)

Motion of an Electron in the Field of a Magnetic Pole

Harish-Chandra

The Institute for Advanced Study, Princeton, New Jersey

(Received June 21, 1948)

The motion of an electron in the field of a magnetic pole is considered. It is shown that in spite of its magnetic moment the electron has no bound states.

RECENTLY Dirac has revived interest in a theory of the electromagnetic field proposed by him some time ago[1] which allows the existence of free magnetic poles. As a consequence of his theory he is able to deduce the fact that the charge of any elementary particle is always an integral multiple of a certain fixed unit. The motion of an electron in the field of a magnetic pole was considered by Dirac[1] and he found that the electron cannot have any bound states. However, Dirac did not take into account the spin of the electron and since the spin gives rise to a magnetic moment, it seemed conceivable that the above conclusion might cease to hold if one used the correct equation of motion. The object of this note is to investigate this point. It turns out that Dirac's result is still valid and so the electron cannot be bound to a magnetic pole.

We put the velocity of light and the Planck's constant equal to 1 and 2π respectively. It is convenient to use tensor notation. Let $(x, y, z) = (x^1, x^2, x^3)$ be the Cartesian coordinates in space and let ϵ^{ijk} be the antisymmetric tensor such that $\epsilon^{123} = 1$. Further, let g_{ik} be the metric tensor corresponding to the quadratic form

$$dx^2 + dy^2 + dz^2 = g_{ik}dx^i dx^k.$$

The equations determining the vector potential A_k can then be written as

$$\epsilon^{ijk}(\partial A_j/\partial x^i) = H^k, \tag{1}$$

where H_k are the components of the magnetic field. Now since we wish to consider the field of a magnetic pole it is convenient to introduce polar coordinates (r, θ, φ) given by

$$x = r\sin\theta\cos\varphi, \quad y = r\sin\theta\sin\varphi, \quad z = r\cos\theta,$$

so that

$$dx^2 + dy^2 + dz^2 = dr^2 + r^2 d\theta^2 + r^2\sin^2\theta d\varphi^2.$$

Therefore if $g_{\alpha\beta}(\alpha, \beta$ running over $r, \theta, \varphi)$ denotes the metric tensor in the polar coordinates, we have

$$g_{rr} = 1, \quad g_{\theta\theta} = r^2, \quad g_{\varphi\varphi} = r^2\sin^2\theta,$$
$$g^{rr} = 1, \quad g^{\theta\theta} = 1/r^2, \quad g^{\varphi\varphi} = 1/r^2\sin^2\theta, \tag{2}$$
$$g_{\alpha\beta} = g^{\alpha\beta} = 0, \quad \alpha \neq \beta.$$

On transforming (1) to the new coordinate system (r, θ, φ) according to the usual rules of tensor calculus we get

$$\epsilon^{\alpha\beta\gamma}(\partial A_\beta/\partial\xi^\alpha) = H^\gamma, \tag{3}$$

where α, β, γ run over the three indices r, θ, φ and $(\xi^r, \xi^\theta, \xi^\varphi) = (r, \theta, \varphi)$. It follows from the transformation laws that

$$\epsilon^{r\theta\varphi} = 1/(g)^{\frac{1}{2}} = 1/r^2\sin\theta,$$

where $g = r^4\sin^2\theta$ is the determinant of the matrix formed by the components $g_{\alpha\beta}$. Let $-e$ be the charge of the electron and n an integer and consider a magnetic pole of strength $\frac{1}{2}n/e$ at the origin. Then $H^\theta = H^\varphi = 0$ and $H^r = \frac{1}{2}n/er^2$. A possible solution of (3) is now obtained by putting $A_r = A_\theta = 0$ and choosing A_φ such that

$$\frac{1}{r^2\sin\theta}\frac{\partial A_\varphi}{\partial\theta} = \frac{n}{2er^2}. \tag{4}$$

This gives $A_\varphi = (n/2e)(1 - \cos\theta)$, the constant of integration having been so chosen as to make $A_\varphi = 0$ for $\theta = 0$ so that the nodal line[1] runs from the origin along the line $\theta = \pi$. This is seen as follows. Consider the integral

$$\int_c A_k dx^k = \int_c A_\alpha d\xi^\alpha = \int_c A_\varphi d\varphi,$$

taken round a closed curve c surrounding the line $\theta = \theta_0$. Clearly if $\theta_0 \neq 0$ or π, $\int_c d\varphi = 0$ and therefore the integral tends to zero as c shrinks to a point on $\theta = \theta_0$. On the other hand if $\theta_0 = 0$ or π, $\int_c d\varphi = 2\pi$ and the value of the integral tends

---

[1] P. A. M. Dirac, Proc. Roy. Soc. **133**, 60 (1931).

"

to $(\pi n/e)(1-\cos\theta_0)$ i.e., 0 or $2\pi n/e$ according as $\theta_0=0$ or π.

Now we come to the equation of motion of the electron. As usual let σ_1, σ_2, σ_3 and ρ_1, ρ_2, ρ_3 be two independent sets of Pauli matrices.[2] Transform σ_1, σ_2, σ_3 as components of a vector to polar coordinates so that

$$\sigma_r=\sigma_k\frac{\partial x^k}{\partial r}=\frac{1}{r}(\boldsymbol{\sigma}\mathbf{x}),$$

$$\sigma_\theta=\sigma_k\frac{\partial x^k}{\partial \theta}=\frac{1}{r\sin\theta}(\mathbf{x}(\boldsymbol{\sigma}\mathbf{x})-(\mathbf{x}\mathbf{x})\boldsymbol{\sigma})_3, \qquad (5)$$

$$\sigma_\varphi=\sigma_k\frac{\partial x^k}{\partial \varphi}=[\mathbf{x}\times\boldsymbol{\sigma}]_3,$$

where $\mathbf{x}=(x_1,\ x_2,\ x_3)$ and $\boldsymbol{\sigma}=(\sigma_1,\ \sigma_2,\ \sigma_3)$ and the usual notations for scalar and vector products are used. The Hamiltonian H for the electron can now be written as

$$H=-\rho_1(\boldsymbol{\sigma}\mathbf{p})-\rho_3\mu-\rho_1 e\sigma^\varphi A_\varphi$$
$$=-\rho_1(\boldsymbol{\sigma}\mathbf{p})-\rho_3\mu-\rho_1\frac{[\mathbf{x}\times\boldsymbol{\sigma}]_3}{r^2\sin^2\theta}\frac{n}{2}(1-\cos\theta), \qquad (6)$$

where $\mathbf{p}=1/i(\partial/\partial x,\ \partial/\partial y,\ \partial/\partial z)$ is the momentum and μ the mass of the electron. Our problem is to find a wave function ψ such that

$$H\psi=E\psi, \qquad (7)$$

where E is some eigenvalue of H. Now notice that

$$[\mathbf{x}\times\mathbf{p}]_3=\frac{1}{i}\left(x\frac{\partial}{\partial y}-y\frac{\partial}{\partial x}\right)=\frac{1}{i}\frac{\partial}{\partial\varphi},$$

and $[\mathbf{x}\times\mathbf{p}]+\tfrac{1}{2}\boldsymbol{\sigma}$ commutes with $H_0=-\rho_1(\boldsymbol{\sigma}\mathbf{p})-\rho_3\mu$ while $[\mathbf{x}\times\mathbf{p}]_3+\tfrac{1}{2}\sigma_3$ commutes with $[\mathbf{x}\times\boldsymbol{\sigma}]_3$. Therefore

$$[\mathbf{x}\times\mathbf{p}]_3+\tfrac{1}{2}\sigma_3=\frac{1}{i}\frac{\partial}{\partial\varphi}+\tfrac{1}{2}\sigma_3,$$

commutes with H. Hence we can choose ψ in such a way that

$$\left(\frac{1}{i}\frac{\partial}{\partial\varphi}+\tfrac{1}{2}\sigma_3\right)\psi=M\psi,$$

where M is half-an-odd integer. Therefore

$$\psi=e^{i(M-\frac{1}{2}\sigma_3)\varphi}\psi', \qquad (8)$$

---

[2] See P. A. M. Dirac, *Principles of Quantum Mechanics* (Clarendon Press, Oxford, 1947), Chap. XI.

where ψ' is independent of φ. Now

$$\sigma^k\frac{\partial}{\partial x^k}=\sigma^\alpha\frac{\partial}{\partial\xi^\alpha}=\sigma_r\frac{\partial}{\partial r}+\frac{\sigma_\theta}{r^2}\frac{\partial}{\partial\theta}+\frac{\sigma_\varphi}{r^2\sin^2\theta}\frac{\partial}{\partial\varphi}. \qquad (9)$$

From (5) we get

$$\sigma_r=\sin\theta(\sigma_1\cos\varphi+\sigma_2\sin\varphi)+\sigma_3\cos\theta$$
$$=\sin\theta\,\sigma_1 e^{i\sigma_3\varphi}+\sigma_3\cos\theta$$
$$=e^{-i\sigma_3\varphi/2}(\sigma_3\cos\theta+\sigma_1\sin\theta)e^{i\sigma_3\varphi/2}$$
$$=e^{-i\sigma_3\varphi/2}e^{-i\sigma_2\theta/2}\sigma_3 e^{i\sigma_2\theta/2}e^{i\sigma_3\varphi/2}. \qquad (10a)$$

Similarly

$$\sigma_\theta=r\cos\theta(\sigma_1\cos\varphi+\sigma_2\sin\varphi)-\sigma_3 r\sin\theta$$
$$=r\cos\theta\,\sigma_1 e^{i\sigma_3\varphi}-\sigma_3 r\sin\theta$$
$$=re^{-i\sigma_3\varphi/2}(\sigma_1\cos\theta-\sigma_3\sin\theta)e^{i\sigma_3\varphi/2}$$
$$=re^{-i\sigma_3\varphi/2}e^{-i\sigma_2\theta/2}\sigma_1 e^{i\sigma_2\theta/2}e^{i\sigma_3\varphi/2}, \qquad (10b)$$

$$\sigma_\varphi=r\sin\theta(\sigma_2\cos\varphi-\sigma_1\sin\varphi)$$
$$=r\sin\theta e^{-i\sigma_3\varphi/2}e^{-i\sigma_2\theta/2}\sigma_2 e^{i\sigma_2\theta/2}e^{i\sigma_3\varphi/2}. \qquad (10c)$$

Also

$$e^{i\sigma_2\theta/2}\frac{\partial}{\partial\theta}=\left(\frac{\partial}{\partial\theta}-i\frac{\sigma_2}{2}\right)e^{i\sigma_2\theta/2},$$

$$e^{i\sigma_2\theta/2}e^{i\sigma_3\varphi/2}\frac{\partial}{\partial\varphi}=e^{i\sigma_2\theta/2}\left(\frac{\partial}{\partial\varphi}-i\frac{\sigma_3}{2}\right)e^{i\sigma_3\varphi/2}$$
$$=\left(\frac{\partial}{\partial\varphi}-\frac{i}{2}e^{i\sigma_2\theta}\sigma_3\right)e^{i\sigma_2\theta/2}e^{i\sigma_3\varphi/2}$$
$$=\left\{\frac{\partial}{\partial\varphi}-\frac{i}{2}(\sigma_3\cos\theta-\sigma_1\sin\theta)\right\}e^{i\sigma_2\theta/2}e^{i\sigma_3\varphi/2}.$$

Hence

$$\sigma^k\partial/\partial x^k=e^{-i\sigma_2\theta/2}e^{-i\sigma_3\varphi/2}\left[\sigma_3\frac{\partial}{\partial r}+\frac{\sigma_1}{r}\left(\frac{\partial}{\partial\theta}-\frac{i\sigma_2}{r}\right)\right.$$
$$\left.+\frac{\sigma_2}{r\sin\theta}\left\{\frac{\partial}{\partial\varphi}-\frac{i}{2}(\sigma_3\cos\theta-\sigma_1\sin\theta)\right\}\right]e^{i\sigma_2\theta/2}e^{i\sigma_3\varphi/2}.$$

Similarly

$$ie\sigma^\varphi A_\varphi=i\frac{\sigma_\varphi}{r^2\sin^2\theta}\frac{n}{2}(1-\cos\theta)$$
$$=ie^{-i\sigma_3\varphi/2}e^{-i\sigma_2\theta/2}\frac{\sigma_2}{r\sin\theta}\frac{n}{2}(1-\cos\theta)$$
$$\times e^{i\sigma_2\theta/2}e^{i\sigma_3\varphi/2}.$$

Hence on substituting (8) in (7) and putting $e^{i\sigma_2\theta/2}\psi'=\psi_0$, we get

$$\left[\tfrac{1}{2}\rho_1\left\{\sigma_3\left(\frac{\partial}{\partial r}+\frac{1}{r}\right)\right.\right.$$
$$+\frac{\sigma_1}{r}\left(\frac{\partial}{\partial\theta}-\frac{\sigma_3}{\sin\theta}\left\{M+\frac{n}{2}(1-\cos\theta)\right.\right.$$
$$\left.\left.\left.-\frac{\sigma_3}{2}\cos\theta\right\}\right)\right\}+\rho_3\mu+E\right]\psi_0=0. \quad (11)$$

It is clear that

$$-K^2=\left\{\sigma_1\left[\frac{\partial}{\partial\theta}\right.\right.$$
$$\left.\left.-\frac{\sigma_3}{\sin\theta}\left(M+\frac{n}{2}(1-\cos\theta)-\frac{\sigma_3\cos\theta}{2}\right)\right]\right\}^2,$$

commutes with the operator acting on ψ_0 in (11). Hence we can choose ψ_0 to be an eigenvector of K^2. But

$$-K^2=\left\{\frac{\partial}{\partial\theta}+\tfrac{1}{2}\cot\theta+\frac{\sigma_3}{\sin\theta}\left(M+\frac{n}{2}(1-\cos\theta)\right)\right\}\left\{\frac{\partial}{\partial\theta}+\tfrac{1}{2}\cot\theta-\frac{\sigma_3}{\sin\theta}\left(M+\frac{n}{2}(1-\cos\theta)\right)\right\}$$

$$=\frac{1}{\sin\theta}\left\{\frac{\partial}{\partial\theta}-\tfrac{1}{2}\cot\theta+\frac{\sigma_3}{\sin\theta}\left(M+\frac{n}{2}(1-\cos\theta)\right)\right\}\left\{\sin\theta\frac{\partial}{\partial\theta}+\tfrac{1}{2}\cos\theta-\sigma_3\left(M+\frac{n}{2}(1-\cos\theta)\right)\right\}$$

$$=\frac{1}{(1-u^2)}\left\{(1-u^2)\frac{d}{du}-\sigma_3\left(M+\frac{n}{2}\right)+\left(\frac{n}{2}\sigma_3+\tfrac{1}{2}\right)u\right\}\left\{(1-u^2)\frac{d}{du}+\sigma_3\left(M+\frac{n}{2}\right)-\left(\frac{n}{2}\sigma_3+\tfrac{1}{2}\right)u\right\}$$

$$=(1-u^2)\frac{d^2}{du^2}-2u\frac{d}{du}-\frac{\left\{\sigma_3\left(M+\frac{n}{2}\right)-\left(\frac{n}{2}\sigma_3+\tfrac{1}{2}\right)u\right\}^2}{1-u^2}-\left(\frac{n}{2}\sigma_3+\tfrac{1}{2}\right)^2$$

$$=(1-u^2)\frac{d^2}{du^2}-2u\frac{d}{du}-\frac{\left\{\left(M+\frac{n}{2}\right)-\left(\frac{n}{2}+\frac{\sigma_3}{2}\right)u\right\}^2}{1-u^2}-\left(\frac{n}{2}+\frac{\sigma_3}{2}\right)^2+\frac{n^2-1}{4}, \quad (12)$$

where $u=\cos\theta$. Now it is known that if m and j are both integral or both half-integral the only eigenfunctions of the operator

$$(1-u^2)\frac{d^2}{du^2}-2u\frac{d}{du}-\frac{(m-ju)^2}{1-u^2}-j^2, \quad (13)$$

corresponding to the interval $-1\leqslant u\leqslant 1$ are the Jacobi polynomials $P^k_{m,j}(u)$ and the corresponding eigenvalues are $-k(k+1)$, where k is to be so chosen that $k\geqslant|m|$, $|j|$, and $k-j$ is an integer. $P^k_{m,j}(\cos\theta)$ is defined by the identity

$$\frac{\left(t_1\cos\frac{\theta}{2}+t_2\sin\frac{\theta}{2}\right)^{k-j}\left(-t_1\sin\frac{\theta}{2}+t_2\cos\frac{\theta}{2}\right)^{k+j}}{(k-j!k+j!)^{\frac{1}{2}}}$$

$$=\sum_m\frac{t_1^{k-m}t_2^{k+m}}{(k-m!k+m!)^{\frac{1}{2}}}P^k_{m,j}(\cos\theta), \quad (14)$$

where m runs through the set of values k, $k-1$,

$\cdots$, $-k$. Write

$$w_1=t_1\cos\theta/2+t_2\sin\theta/2,$$
$$w_2=t_2\cos\theta/2-t_1\sin\theta/2.$$

If we observe that $\partial w_1/\partial\theta=w_2$, $\partial w_2/\partial\theta=-w_1$, we get immediately on differentiating (14) with respect to θ

$$((k-j)(k+j+1))^{\frac{1}{2}}P^k_{m,j+1}(\cos\theta)$$
$$-((k+j)(k-j+1))^{\frac{1}{2}}P^k_{m,j-1}(\cos\theta)$$
$$=2\frac{d}{d\theta}P^k_{m,j}(\cos\theta). \quad (15)$$

Now keep θ fixed and transform from the variables t_1, t_2 to w_1, w_2. Then

$$t_1\frac{\partial}{\partial t_1}-t_2\frac{\partial}{\partial t_2}=\cos\theta\left(w_1\frac{\partial}{\partial w_1}-w_2\frac{\partial}{\partial w_2}\right)$$
$$-\sin\theta\left(w_1\frac{\partial}{\partial w_2}+w_2\frac{\partial}{\partial w_1}\right).$$

On applying this operator to (14) we get

$$-2j\cos\theta P^k_{m,j}(\cos\theta)$$
$$-\sin\theta(((k-j)(k+j+1))^{\frac{1}{2}}P^k_{m,j+1}(\cos\theta)$$
$$+((k+j)(k-j+1))^{\frac{1}{2}}P^k_{m,j-1}(\cos\theta))$$
$$=-2mP^k_{m,j}(\cos\theta). \quad (16)$$

From (15) and (16) we find at once that

$$\left\{(1-u^2)\frac{d}{du}+m-ju\right\}P^k_{m,j}(u)$$
$$=((k+j)(k-j+1))^{\frac{1}{2}}(1-u^2)^{\frac{1}{2}}P^k_{m,j-1}(u), \quad (17a)$$

$$\left\{(1-u^2)\frac{d}{du}-m+ju\right\}P^k_{m,j}(u)$$
$$=((k-j)(k+j+1))^{\frac{1}{2}}(1-u^2)^{\frac{1}{2}}P^k_{m,j+1}(u). \quad (17b)$$

Now for (12) $m=M+n/2$ and

$$j=(n/2)+(\sigma_3/2)=\begin{bmatrix}\dfrac{n+1}{2}\\[2mm]\dfrac{n-1}{2}\end{bmatrix}$$

if we choose a representation in which σ_3 is diagonal. Therefore the eigenfunction ψ_0 can be written as

$$\psi_0=\begin{pmatrix}P^k_{M+(n/2),(n+1)/2}(\cos\theta) & \psi^+\\ P^k_{M+(n/2),(n-1)/2}(\cos\theta) & \psi^-\end{pmatrix},$$

where ψ^+, ψ^- depend only on r and

$$k\geqslant\left|M+\frac{n}{2}\right|,\ \left|\frac{n-1}{2}\right|,\ \left|\frac{n+1}{2}\right|\ \text{ and }\ k+\frac{n+1}{2}$$

is an integer. Making use of (17) we find that

$$\sigma_1\left[\frac{\partial}{\partial\theta}-\frac{\sigma_3}{\sin\theta}\left(M+\frac{n}{2}-\frac{n+\sigma_3}{2}\cos\theta\right)\right]$$
$$\times\begin{pmatrix}P^k_{M+(n/2),(n+1)/2} & \psi^+\\ P^k_{M+(n/2),(n-1)/2} & \psi^-\end{pmatrix}$$
$$=-\left(\left(k+\frac{n}{2}+\frac{1}{2}\right)\left(k-\frac{n}{2}+\frac{1}{2}\right)\right)^{\frac{1}{2}}\sigma_1$$
$$\times\begin{pmatrix}P^k_{M+(n/2),(n-1)/2} & \psi^+\\ P^k_{M+(n/2),(n+1)/2} & \psi^-\end{pmatrix}$$

$$=-\left(\left(k+\frac{n}{2}+\frac{1}{2}\right)\left(k-\frac{n}{2}+\frac{1}{2}\right)\right)^{\frac{1}{2}}$$
$$\times\begin{pmatrix}P^k_{M+(n/2),(n+1)/2} & \psi^-\\ P^k_{M+(n/2),(n-1)/2} & \psi^+\end{pmatrix}$$

Therefore (11) now becomes

$$\left[\frac{1}{i}\rho_1\left\{\sigma_3\left(\frac{\partial}{\partial r}+\frac{1}{r}\right)-\frac{\sigma_1}{r}K\right\}+\rho_3\mu+E\right]\begin{pmatrix}\psi^+\\ \psi^-\end{pmatrix}=0,$$

where

$$K=\left(\left(k+\frac{n+1}{2}\right)\left(k-\frac{n-1}{2}\right)\right)^{\frac{1}{2}}.$$

If we write $\psi=\begin{pmatrix}\psi^+\\ \psi^-\end{pmatrix}$ this equation may be written as

$$\left\{\left(\frac{\partial}{\partial r}+\frac{1}{r}\right)+\frac{i\sigma_2}{r}K\right\}\psi=i\sigma_3(i\rho_2-E\rho_1)\psi. \quad (19)$$

Choose ρ_3 diagonal and split ψ into $\begin{pmatrix}\psi_1\\ \psi_2\end{pmatrix}$ with respect to ρ_3 so that

$$\rho_3\psi=\rho_3\begin{pmatrix}\psi_1\\ \psi_2\end{pmatrix}=\begin{pmatrix}\psi_1\\ -\psi_2\end{pmatrix}$$

Then

$$(i\rho_2\mu-E\rho_2)\psi=\begin{pmatrix}0 & \mu & -E\\ -\mu & -E & 0\end{pmatrix}\begin{pmatrix}\psi_1\\ \psi_2\end{pmatrix}$$

Put

$$\mu+E=\frac{1}{a_1},\quad \mu-E=\frac{1}{a_2}.$$

Then (19) can be written as

$$\left\{\left(\frac{\partial}{\partial r}+\frac{1}{r}\right)+\frac{i\sigma_2}{r}K\right\}\psi_1=\frac{i\sigma_3}{a_2}\psi_2, \quad (20a)$$

$$\left\{\left(\frac{\partial}{\partial r}+\frac{1}{r}\right)+\frac{i\sigma_2}{r}K\right\}\psi_2=-\frac{i\sigma_3}{a_1}\psi_1. \quad (20b)$$

Now choose a representation in which σ_2 is diagonal so that

$$\sigma_2=\begin{pmatrix}1 & 0\\ 0 & -1\end{pmatrix},\quad \sigma_3=\begin{pmatrix}0 & 1\\ 1 & 0\end{pmatrix},\quad \sigma_1=\begin{pmatrix}0 & -i\\ i & 0\end{pmatrix}.$$

Write

$$\psi_1=\begin{pmatrix}\psi_1{}^+\\ \psi_1{}^-\end{pmatrix},\quad \psi_2=\begin{pmatrix}\psi_2{}^+\\ \psi_2{}^-\end{pmatrix}.$$

Then (20) is equivalent to the set of equations

$$\left\{\frac{\partial}{\partial r}+\frac{1}{r}+\frac{iK}{r}\right\}\psi_1{}^+=\frac{i}{a_2}\psi_2{}^-,\qquad (21a)$$

$$\left\{\frac{\partial}{\partial r}+\frac{1}{r}-\frac{iK}{r}\right\}\psi_2{}^-=-\frac{i}{a_2}\psi_1{}^+,\qquad (21b)$$

$$\left\{\frac{\partial}{\partial r}+\frac{1}{r}-\frac{iK}{r}\right\}\psi_1{}^-=\frac{i}{a_2}\psi_2{}^+,\qquad (21c)$$

$$\left\{\frac{\partial}{\partial r}+\frac{1}{r}+\frac{iK}{r}\right\}\psi_2{}^+=-\frac{i}{a_2}\psi_1{}^-.\qquad (21d)$$

Put $a=(a_1a_2)^{\frac12}$ the square root being positive in case $a_1a_2>0$. Then we get

$$\left(\frac{\partial}{\partial r}+\frac{1}{r}-\frac{iK}{r}\right)\left(\frac{\partial}{\partial r}+\frac{1}{r}+\frac{iK}{r}\right)\psi_1{}^+$$

$$=\left(\frac{i}{a_1}\right)\left(\frac{-i}{a_2}\right)\psi_1{}^+,$$

i.e.,

$$\left\{\frac{\partial^2}{\partial r^2}+\frac{2}{r}\frac{\partial}{\partial r}+\frac{K^2-iK}{r^2}-\frac{1}{a^2}\right\}\psi_1{}^+=0.\quad (22a)$$

Similarly

$$\left\{\frac{\partial^2}{\partial r^2}+\frac{2}{r}\frac{\partial}{\partial r}+\frac{K^2+iK}{r^2}-\frac{1}{a^2}\right\}\psi_1{}^-=0,\quad (22b)$$

(22) is completely equivalent to (21) since (21a) and (21c) can be regarded as the definitions of $\psi_2{}^-$ and $\psi_2{}^+$ respectively. Since the operator

$$\frac{\partial^2}{\partial r^2}+\frac{2}{r}\frac{\partial}{\partial r}+\frac{K^2\pm iK}{r^2}$$

is homogeneous in r it is clear that if $\psi(r)$ is one of its eigenfunctions corresponding to the eigenvalue $1/a^2$ and α is any real constant, the function $\psi(\alpha r)$ is another eigenfunction corresponding to the eigenvalue α^2/a^2. Hence the eigenvalues $E=(\mu^2-1/a^2)^{\frac12}$ cannot form a discrete spectrum. We shall now show that for all permissible solutions of (22) $a^2<0$. It is sufficient to consider (22a). Put $\varphi_1{}^+=fe^{-r/a}$. Then

$$\left\{\frac{\partial^2}{\partial r^2}+\left(\frac{2}{r}-\frac{2}{a}\right)\frac{\partial}{\partial r}+\frac{K^2-iK}{r^2}-\frac{2}{ra}\right\}f=0.$$

The point $r=0$ is a singular point of this equation. According to the usual procedure for solving second-order linear differential equations with analytic coefficients, we make the substitution

$$f=\sum_{\nu\geq0}c_\nu r^{\nu+\alpha},$$

where $c_0=1$ and ν runs over all non-negative integers. The indical equation is

$$\alpha(\alpha-1)+2\alpha+K^2-iK=0,$$

i.e.,

$$\left(\alpha+\tfrac12\right)^2+\left(K-\frac{i}{2}\right)^2=0.$$

Hence $\alpha=iK$ or $-iK-1$. However the boundary condition at $r=0$ requires[3] that $r\psi_1{}^+\to0$ as $r\to0$. Hence only $\alpha=iK$ is permissible. On substituting

$$f=\sum_{\nu\geq0}c_\nu r^{\nu+iK},$$

in (23) and equating the coefficients of the various powers of r to zero we get the recurrence relation

$$c_{\nu+1}=\frac{2}{a}\frac{\nu+iK+1}{(\nu+1)(\nu+2iK+2)}.$$

Since K is real it is clear that the series cannot terminate. It converges like $e^{2r/a}$ and therefore for large r, $\psi_1{}^+=fe^{-r/a}$ behaves like $e^{r/a}$. Therefore from the boundary condition at infinity it follows that only those solutions are permissible for which a is pure imaginary, i.e., $a^2<0$, or $E^2>\mu^2$. Thus the electron is never bound to the magnetic pole.

[3] See reference 2, p. 269.

Reprinted from
The Physical Review
74 (1948), 883–887

ANNALS OF MATHEMATICS
Vol. 50, No. 1, January, 1949

FAITHFUL REPRESENTATIONS OF LIE ALGEBRAS

By Harish-Chandra

(Received April 12, 1948)

The object of this paper is to give a new proof of the theorem that every Lie algebra over a field K of characteristic zero, has a faithful representation. The first proof of this result, at least when K is algebraically closed, is due to Ado (1). Later Cartan (2) gave a simpler and entirely different proof for the case when K is the field of either real or complex numbers. Cartan's proof depends on the integration of the Maurer-Cartan equations and therefore is of a non-algebraic character.[1] The present proof is of course algebraic and seems to differ from the earlier ones in approaching the problem quite directly. Also the result established is slightly sharper than the usual one in so far as we assert the existence of a faithful representation in which every element of the maximal nilpotent ideal of the given Lie algebra is mapped on a nilpotent matrix.

I am very much indebted to Professor C. Chevalley for his advice and help in improving the presentation of the proof. Also I should like to thank Dr. G. D. Mostow for many interesting and valuable discussions.

All algebras (whether Lie algebras or associative algebras) and vector spaces appearing in this paper are to be understood over the basic field K. A linear Lie algebra $\mathfrak{L}$ is a Lie algebra whose elements are endomorphisms of some given vector space, the bracket operation in $\mathfrak{L}$ being defined by $[X,Y] = XY - YX$. As far as possible we follow the notation and terminology of Chevalley's book (3) and his papers (4). In particular, if $\mathfrak{L}$ is a Lie algebra and $X \in \mathfrak{L}$ we denote by ad X the derivation of $\mathfrak{L}$ defined by $(\text{ad } X)Y = [X,Y](Y \in \mathfrak{L})$.[2]

The following notion of the semidirect sum of a Lie algebra and its algebra of derivations[3] is important for our purpose.

DEFINITION. *Let $\mathfrak{L}$ be a Lie algebra and $\mathfrak{D}$ the algebra of its derivations. By the semidirect sum of $\mathfrak{L}$ and $\mathfrak{D}$ is meant a Lie algebra $\mathfrak{L} + \mathfrak{D}$ defined as follows. Considered as a vector space $\mathfrak{L} + \mathfrak{D}$ is the direct sum of $\mathfrak{L}$ and $\mathfrak{D}$ so that an element of $\mathfrak{L} + \mathfrak{D}$ is a pair (X, D) with $X \in \mathfrak{L}$ and $D \in \mathfrak{D}$. The bracket operation in $\mathfrak{L} + \mathfrak{D}$ is defined by*

$$(1) \qquad [(X, D), (X', D')] = ([X, X'] + DX' - D'X, [D, D']).$$

It is easy to verify that the Jacobi identities are fulfilled and therefore $\mathfrak{L} + \mathfrak{D}$ is actually a Lie algebra.

The set $\mathfrak{L}_0$ of all elements of $\mathfrak{L} + \mathfrak{D}$ of the form $(X, 0)$ is clearly an ideal which is isomorphic to $\mathfrak{L}$. Similarly the set $\mathfrak{D}_0$ of all elements of the form $(0, D)$ is

---

[1] Professor Chevalley has kindly informed me that Dr. Hochschild has succeeded in constructing an algebraic proof which imitates Cartan's procedure.

[2] Our notation differs from Chevalley's only in two respects. Chevalley puts (ad $X)Y$ [Y,X] and $[X,Y] = YX - XY$ in case X, Y are endomorphisms.

[3] The algebra of derivations is of course a linear Lie algebra.

68

a subalgebra isomorphic to $\mathfrak{D}$. We can therefore identify $\mathfrak{L}_0$ with $\mathfrak{L}$ and $\mathfrak{D}_0$ with $\mathfrak{D}$ and write $(X, D) = (X, 0) + (0, D) = X + D$.

The following proposition is needed in proving our main lemma:

PROPOSITION 1. *Let $\mathfrak{L}$ be a solvable Lie algebra and $\mathfrak{D}$ the algebra of its derivations. Let $\mathfrak{N}$ be the set of all $X \epsilon \mathfrak{L}$ such that ad X is nilpotent. Then $\mathfrak{N}$ is a nilpotent ideal in $\mathfrak{L}$ and every nilpotent ideal of $\mathfrak{L}$ is contained in $\mathfrak{N}$. Also $\mathfrak{D}\mathfrak{L} \subset \mathfrak{N}$ i.e. for every $D \epsilon \mathfrak{D}$ and $X \epsilon \mathfrak{L}$ $DX \epsilon \mathfrak{N}$.*

Since $\mathfrak{L}$ is solvable the derived algebra $\mathfrak{L}'$ is nilpotent. Hence for every $X \epsilon \mathfrak{L}'$ ad X is nilpotent and so $\mathfrak{N} \supset \mathfrak{L}'$. Therefore in order to prove that $\mathfrak{N}$ is an ideal it is sufficient to show that $\mathfrak{N}$ is a linear set. First suppose that K is algebraically closed. Then by Lie's Theorem it is possible to choose a basis $X_1, X_2, \cdots, X_n$ for $\mathfrak{L}$ such that

$$[X_i X_j] = \sum_{k \geq i, j}^{n} c_{ij}{}^{k} X_k \qquad (c_{ij}{}^{k} \epsilon K).$$

Let $X = \sum_i \xi^i X_i (\xi^i \epsilon K)$. Then the Killing polynomial of X is clearly

$$(s - \lambda^1(X))(s - \lambda^2(X)) \cdots (s - \lambda^n(X))$$

where $\lambda^j(X) = \sum_i \xi^i c_{ij}{}^j$. Hence in order that $X \epsilon \mathfrak{N}$ it is necessary and sufficient that $\lambda^j(X) = 0$ for all j. These conditions are clearly linear. Hence $\mathfrak{N}$ is a linear set and therefore an ideal. Since ad X is nilpotent for every $X \epsilon \mathfrak{N}$ it follows that $\mathfrak{N}$ is nilpotent. Also if X belongs to any nilpotent ideal of $\mathfrak{L}$, ad X is nilpotent. Hence $X \epsilon \mathfrak{N}$. Therefore $\mathfrak{N}$ is the maximal nilpotent ideal of $\mathfrak{L}$. Further, since $\mathfrak{L}$ is an ideal in $\mathfrak{L} + \mathfrak{D}$ the Killing polynomial of X in $\mathfrak{L} + \mathfrak{D}$ is

$$s^m \prod_{j=1}^{n} (s - \lambda^j(x))$$

where m is the dimension of $\mathfrak{D}$. Since any factor of the Killing polynomial is an invariant of the adjoint representation it follows that

$$\lambda^j([D, X]) = \lambda^j(DX) = 0$$

for any $D \epsilon \mathfrak{D}$. Since this holds for every j it follows that $DX \epsilon \mathfrak{N}$ for every $D \epsilon \mathfrak{D}$, $X \epsilon \mathfrak{L}$. Thus the proposition is proved when K is algebraically closed.

If K is not algebraically closed, consider its algebraic closure $\bar{K}$. Extend $\mathfrak{L}$ to a Lie algebra $\bar{\mathfrak{L}}$ over $\bar{K}$ in the obvious way. Define $\bar{\mathfrak{N}}$ for $\bar{\mathfrak{L}}$ as above. Then $\bar{\mathfrak{N}}$ is the maximal nilpotent ideal of $\bar{\mathfrak{L}}$ and $\bar{\mathfrak{L}}$ is mapped into $\bar{\mathfrak{N}}$ by every derivation of $\bar{\mathfrak{L}}$. Clearly $\mathfrak{N} = \mathfrak{L} \cap \bar{\mathfrak{N}}$. If D is any derivation of $\mathfrak{L}$ it can obviously be extended to a derivation $\bar{D}$ of $\bar{\mathfrak{L}}$. Then $\bar{D}\bar{\mathfrak{L}} \subset \bar{\mathfrak{N}}$. Hence

$$D\mathfrak{L} \subset \bar{D}\bar{\mathfrak{L}} \cap \mathfrak{L} \subset \bar{\mathfrak{N}} \cap \mathfrak{L} = \mathfrak{N}.$$

The proof of the fact that $\mathfrak{N}$ is the maximal nilpotent ideal of $\mathfrak{L}$ is the same as before. The proposition is therefore proved completely.

Let $\mathfrak{L}$ be any Lie algebra and Γ its maximal solvable ideal. Let $\mathfrak{N}$ be the maximal nilpotent ideal of Γ.   Since every nilpotent ideal is solvable it follows that every nilpotent ideal of $\mathfrak{L}$ is contained in Γ and therefore in $\mathfrak{N}$.   Further $\mathfrak{N}$ is itself an ideal in $\mathfrak{L}$ since $D\Gamma \subset \mathfrak{N}$ for every derivation D of Γ.   Hence $\mathfrak{N}$ is the maximal nilpotent ideal of $\mathfrak{L}$.

The following lemma is the main step of our proof.

LEMMA 1.   *Let $\mathfrak{L}$ be a solvable linear Lie algebra and $\mathfrak{D}$ the algebra of its derivations.   Suppose that all elements of the maximal nilpotent ideal $\mathfrak{N}$ of $\mathfrak{L}$ are nilpotent endomorphisms.   Then there exists a faithful representation θ of $\mathfrak{L} + \mathfrak{D}$ such that if $Y \in \mathfrak{N}, D \in \mathfrak{D}$ and D is nilpotent, then $\theta(Y + D)$ is nilpotent.*

Choose a base $X_1, \cdots, X_n$ in $\mathfrak{L}$.   Consider the ring $\mathfrak{A}$ of non-commutative polynomials in n indeterminates $x_1, \cdots, x_n$ with coefficients in K.   Let $\mathfrak{A}_0$ be the subring of $\mathfrak{A}$ consisting of all polynomials whose constant term is zero. Let π be the linear mapping of $\mathfrak{A}_0$ into the enveloping algebra of $\mathfrak{L}$ defined by

$$\pi(x_{i_1} x_{i_2} \cdots x_{i_s}) = X_{i_1} X_{i_2} \cdots X_{i_s} \qquad 1 \leqq i_1, \cdots, i_s \leqq n, s \geqq 1.$$

Clearly π is a homomorphism.   Let $\mathfrak{X}$ be the kernel of this homomorphism. $\mathfrak{X}$ is an ideal in $\mathfrak{A}_0$ and therefore in $\mathfrak{A}$.

Let ω be the linear mapping of $\mathfrak{L}$ into $\mathfrak{A}$ given by $\omega(X_i) = x_i \ (1 \leqq i \leqq n)$. Then $\pi(\omega(X)) = X$ for $X \in \mathfrak{L}$.   For any $z, w \in \mathfrak{A}$ put $[z, w] = zw - wz$, and let

$$(2) \qquad\qquad u(X, Y) = [\omega(X), \omega(Y)] - \omega([X,Y])$$

for $X, Y \in \mathfrak{L}$.   Clearly $\pi(u(X, Y)) = 0$ and therefore $u(X, Y) \in \mathfrak{X}$.   Let $\mathfrak{U}$ be the ideal in $\mathfrak{A}$ generated by $u(X, Y)$ for all $X, Y \in \mathfrak{L}$ (i.e. $\mathfrak{U}$ is the smallest possible ideal containing $u(X, Y)$ for all $X, Y \in \mathfrak{L}$).   Then $\mathfrak{U} \subset \mathfrak{X}$.

Let D be any derivation of $\mathfrak{L}$.   Corresponding to D we define a linear mapping d_D of $\omega(\mathfrak{L})$ into itself by

$$(3) \qquad\qquad d_D\omega(X) = \omega(DX).$$

d_D can be extended to a linear mapping of $\mathfrak{A}$ into itself by putting

$$d_D 1 = 0$$

$$(4) \quad
\begin{aligned}
d_D(x_{i_1}x_{i_2} \cdots x_{i_s}) &= (d_D x_{i_1})x_{i_2} \cdots x_{i_s} + x_{i_1}(d_D x_{i_2}) \cdots x_{i_s} \\
&\quad + \cdots + x_{i_1} \cdots x_{i_{s-1}}(d_D x_{i_s}) \\
& 1 \leqq i_1, \cdots, i_s \leqq n, s \geqq 1.
\end{aligned}$$

The mapping d_D so defined is clearly a derivation of $\mathfrak{A}$, i.e. for any $z, w \in \mathfrak{A}$

$$(5) \qquad\qquad d_D(zw) = (d_D z)w + z(d_D w).$$

Further it is clear that

$$(6) \qquad\qquad d_{[D_1, D_2]} = [d_{D_1}, d_{D_2}]$$

where $D_1, D_2 \in \mathfrak{D}$ and $[d_{D_1}, d_{D_2}] = d_{D_1}d_{D_2} - d_{D_2}d_{D_1}$.   Let Δ be the (associative) algebra of linear mappings generated by d_D for all $D \in \mathfrak{D}$. Δ is the set of all those linear mappings of $\mathfrak{A}$ into itself which can be expressed as a linear combination of products of the mappings $d_D (D \epsilon \mathfrak{D})$. Let $\mathfrak{X}_0$ be the set of all elements z of $\mathfrak{X}$ such that $\delta z \epsilon \mathfrak{X}$ for every $\delta \epsilon \Delta$. Clearly $\mathfrak{X}_0$ is a linear set. We shall show it is an ideal in $\mathfrak{A}$. Let $a \epsilon \mathfrak{A}$, $z \epsilon \mathfrak{X}_0$ and $\delta \epsilon \Delta$ then $az \epsilon \mathfrak{X}$ and from (4) $\delta(az)$ is a linear combination of elements of the form bz and $b\delta'z$ where $b \epsilon \mathfrak{A}$ and $\delta' \epsilon \Delta$. But since $z \epsilon \mathfrak{X}_0$, $\delta'z \epsilon \mathfrak{X}$ and therefore bz and $b\delta'z \epsilon \mathfrak{X}$ since $\mathfrak{X}$ is an ideal. Hence $\delta(az) \epsilon \mathfrak{X}$ for every $\delta \epsilon \Delta$. Therefore $az \epsilon \mathfrak{X}_0$ and similarly $za \epsilon \mathfrak{X}_0$. Hence $\mathfrak{X}_0$ is an ideal. It is clear from its definition that $\mathfrak{X}_0$ is invariant under Δ, i.e. $\delta \mathfrak{X}_0 \subset \mathfrak{X}_0$ for every $\delta \epsilon \Delta$.

First we shall prove that $\mathfrak{U} \subset \mathfrak{X}_0$. From (2)

$$\begin{aligned}
d_D u(X, Y) &= [d_D \omega(X), \omega(Y)] + [\omega(X), d_D \omega(Y)] - d_D \omega([X, Y]) \\
&= [\omega(DX), \omega(Y)] + \omega(X), \omega(DY)] - \omega(D[X, Y)] \\
&= [\omega(DX), \omega(Y)] + [\omega(X), \omega(DY)] - \omega([DX, Y]) - \omega([X, DY]) \\
&= u(DX, Y) + u(X, DY).
\end{aligned}$$

Hence $d_D u(X, Y) \epsilon \mathfrak{U}$. Therefore on making use of (4) and the fact that $\mathfrak{U}$ is an ideal we get

$$d_D(au(X, Y)b) \epsilon \mathfrak{U}$$

for any $a, b \epsilon \mathfrak{A}$. Since every element of $\mathfrak{U}$ can be written as a linear combination of terms of the form $au(X, Y)b$ it follows that $d_D \mathfrak{U} \subset \mathfrak{U}$. Hence $\delta \mathfrak{U} \subset \mathfrak{U} \subset \mathfrak{X}$ for every $\delta \epsilon \Delta$. Therefore $\mathfrak{U} \subset \mathfrak{X}_0$.

Since $\mathfrak{N}$ is a Lie algebra of nilpotent endomorphisms, its enveloping algebra is nilpotent (see Jacobsen (5) Theorem 5). Hence there exists an r such that $Y_1 Y_2 \cdots Y_r = 0$ for any $Y_1, Y_2 \cdots, Y_r \epsilon \mathfrak{N}$. Hence

$$(7) \qquad \omega(Y_1)\omega(Y_2) \cdots \omega(Y_r) \epsilon \mathfrak{X}.$$

Let $\mathfrak{Y}$ be the ideal generated by $\omega(Y)$ for all $Y \epsilon \mathfrak{N}$. We shall show that $\mathfrak{Y}^r \subset \mathfrak{X}$. Since $\mathfrak{U} \subset \mathfrak{X}$ it is sufficient to prove that every element of $\mathfrak{Y}^r$ belongs to $\mathfrak{X}$ modulo $\mathfrak{U}$. Now from (2)

$$(8) \qquad \omega(X)\omega(Y) \equiv \omega(Y)\omega(X) + \omega([X,Y]) \qquad (\text{mod } \mathfrak{U}).$$

Since $[X,Y] \epsilon \mathfrak{N}$, $\omega([X,Y] \epsilon \mathfrak{Y}$. On making use of (8) we easily prove by induction on s that

$$x_{i_1} x_{i_2} \cdots x_{i_s} \omega(Y) \; 1 \leqq i_1, \cdots, i_s \leqq n, \; Y \epsilon \mathfrak{N}$$

can be expressed mod $\mathfrak{U}$ as a linear combination of elements of the form $\omega(Y')a$ where $Y' \epsilon \mathfrak{N}$ and $a \epsilon \mathfrak{A}$. Now consider a term of the type

$$(9) \qquad a_1\omega(Y_1)a_2\omega(Y_2) \cdots a_s\omega(Y_s)a_{s+1} \qquad (s \geqq 1)$$

where $a_1, a_2, \cdots, a_s, a_{s+1} \epsilon \mathfrak{A}$, $Y_1, Y_2, \cdots, Y_s \epsilon \mathfrak{N}$. For $s = 1$ we already know that it can be expressed mod $\mathfrak{U}$ as a linear combination of elements of the form $\omega(Y_1')\omega(Y_2') \cdots \omega(Y_s')b$ where $Y_1', \cdots, Y_s' \epsilon \mathfrak{N}$ and $b \epsilon \mathfrak{A}$. For $s > 1$ the same statement is proved by an easy induction on s. Since all elements of $\mathfrak{Y}'$ are linear combinations of elements of the type (9) with $s = r$, we have proved that every element of $\mathfrak{Y}'$ can be written mod $\mathfrak{U}$ as a linear combination of elements of the form $\omega(Y_1)\omega(Y_2) \cdots \omega(Y_r)b$ where $Y_1, Y_2, \cdots, Y_r \in \mathfrak{N}$. From (7) it then follows that $\omega(Y_1)\omega(Y_2) \cdots \omega(Y_r)b \in \mathfrak{X}$ and therefore $\mathfrak{Y}' \subset \mathfrak{X}$.

From Proposition 1 we know that for any $X \in \mathfrak{L}$ and $D \in \mathfrak{D}$, $DX \in \mathfrak{N}$. Hence

$$d_D \,\omega(X) = \omega(DX) \in \mathfrak{Y}.$$

Therefore from (4) $d_D\mathfrak{A} \in \mathfrak{Y}$ and so in particular $\delta\mathfrak{Y} \subset \mathfrak{Y}$ for any $\delta \in \Delta$. Hence $\delta\mathfrak{Y}' \subset \mathfrak{Y}' \subset \mathfrak{X}$ for any $\delta \in \Delta$. This shows that $\mathfrak{Y}' \subset \mathfrak{X}_0$.

We shall now prove that $\mathfrak{X}' \subset \mathfrak{X}_0$. Clearly $\mathfrak{X}' \subset \mathfrak{X}$. Further if $z_1, z_2, \cdots,$ $z_r \in \mathfrak{X}$ and $\delta_1, \cdots, \delta_r \in \Delta$ then since $d_D\mathfrak{A} \subset \mathfrak{Y}$, $(\delta_1 z_1)(\delta_2 z_2) \cdots (\delta_r z_r) \in \mathfrak{Y}' \subset \mathfrak{X}$. Now consider $\delta(z_1 z_2 \cdots z_r)$ for any $\delta \in \Delta$. Then from (4) $\delta(z_1 z_2 \cdots z_r)$ is a linear combination of elements of the form $(\delta_1 z_1)(\delta_2 z_2) \cdots (\delta_r z_r)$ where either $\delta_i \in \Delta$ or $\delta_i = I$ $(1 \leqq i \leqq r)$ where I is the identity mapping of $\mathfrak{A}$ onto itself. If all $\delta_i \in \Delta$ we have seen that $(\delta_1 z_1)(\delta_2 z_2) \cdots (\delta_r z_r) \in \mathfrak{X}$. On the other hand, if for some i, $\delta_i = I$ then $\delta_i z_i = z_i \in \mathfrak{X}$, and therefore since $\mathfrak{X}$ is an ideal $(\delta_1 z_1)(\delta_2 z_2) \cdots (\delta_r z_r) \in \mathfrak{X}$. Therefore in every case $\delta(z_1 z_2 \cdots z_r) \in \mathfrak{X}$. This shows that $z_1 z_2 \cdots z_r \in \mathfrak{X}_0$. Hence $\mathfrak{X}' \subset \mathfrak{X}_0$.

Now we prove that the dimension of the factor algebra $\mathfrak{A}^* = \mathfrak{A}/\mathfrak{X}_0$ is finite. We observe that $[x_k, x_{i_1} x_{i_2} \cdots x_{i_s}]$ $(1 \leqq i_1, \cdots, i_s, k \leqq n)$ can be written mod $\mathfrak{U}$ as a linear combination of $x_{j_1} x_{j_2} \cdots x_{j_s}$ $(1 \leqq j_1, j_2 \cdots, j_s \leqq n)$. This is clear from (2) if we notice that

$$[x_k, x_{i_1} x_{i_2} \cdots x_{i_s}] = [x_k x_{i_s}]x_{i_s} \cdots x_{i_s} + x_{i_1}[x_k x_{i_s}] \cdots x_{i_s}$$
$$+ \cdots + x_{i_1} x_{i_2} \cdots x_{i_{s-1}}[x_k x_{i_s}].$$

We assert that any element $x_{i_1} x_{i_2} \cdots x_{i_s}$ $(1 \leqq i_1, \cdots, i_s \leqq n)$ can be written mod $\mathfrak{U}$ as a linear combination of elements of the form $x_1^{m_1} x_2^{m_2} \cdots x_n^{m_n}$ where $m_1, m_2, \cdots, m_n \geqq 0$ and $m_1 + m_2 + \cdots + m_n \leqq s$. (Here $x_i^{m_i}$ is to be understood to mean 1 in case $m_i = 0$.) The assertion is trivially true when $s = 1$. We can therefore use induction on s. Since the induction assumption holds for $s - 1$ it is clearly sufficient to prove that $x_1^{m_1} x_2^{m_2} \cdots x_n^{m_n} x_i(m_1, m_2, \cdots, m_n \geqq 0$, $m_1 + m_2 + \cdots + m_n \leqq s - 1)$ can be expressed mod $\mathfrak{U}$ as a linear combination of $x_1^{p_1} x_2^{p_2} \cdots x_n^{p_n}$ with $p_1, p_2, \cdots, p_n \geqq 0$ and $p_1 + p_2 + \cdots + p_n \leqq s$ If $i = n$ the assertion is obviously true. Hence we can assume $i < n$. Consider

$$\begin{aligned}
(10) \qquad & x_1^{m_1} x_2^{m_2} \cdots x_n^{m_n} x_i - x_1^{m_1} \cdots x_i^{m_i+1} \cdots x_n^{m_n} \\
& \qquad\qquad = -x_1^{m_1} \cdots x_i^{m_i}[x_i, x_{i+1}^{m_{i+1}} \cdots x_n^{m_n}].
\end{aligned}$$

We have seen above that $[x_i, x_{i+1}^{m_{i+1}} \cdots x_n^{m_n}]$ can be written mod $\mathfrak{U}$ as a linear combination of $x_{j_1} x_{j_2} \cdots x_{j_k}(1 \leqq j_1, \cdots, j_k \leqq n, k \leqq m_{i+1} + \cdots + m_n)$. Since $m_1 + \cdots + m_i + k \leqq m_1 + \cdots + m_i + m_{i+1} + \cdots + m_n \leqq s - 1$, it follows from the induction hypothesis that the right side of (10) can be written mod $\mathfrak{U}$ as a linear combination of $x_1^{p_1} x_2^{p_2} \cdots x_n^{p_n}$ $(p_1, p_2, \cdots p_n \geqq 0$, $p_1 + p_2 + \cdots + p_n \leqq s - 1)$. Since

$$m_1 + m_2 + \cdots + (m_2 + 1) + \cdots + m_n \leqq s - 1 + 1 = s,$$

the term $x_1^{m_1} x_2^{m_2} \cdots x_i^{m_i+1} \cdots x_n^{m_n}$ also has the required form and therefore the assertion is proved. Since $\mathfrak{U} \subset \mathfrak{X}_0$ the same statement holds mod $\mathfrak{X}_0$ instead of mod $\mathfrak{U}$. Now there exists a non-zero polynomial $f_i(t)$ without a constant term, such that $f_i(X_i) = 0$. This implies that $f_i(x_i) \in \mathfrak{X}$. Hence $f_i^r(x_i) \in \mathfrak{X}^r \subset \mathfrak{X}_0$. Let q_i be the degree of $f_i(t)$. Then every power $x_i^p (p \geqq 0)$ can be written mod $\mathfrak{X}_0$ as a linear combination of x_i^q with $0 \leqq q \leqq rq_r - 1$. This, together with the result proved above, implies that every element of $\mathfrak{A}$ can be written mod $\mathfrak{X}_0$ as a linear combination of terms of the form $x_1^{p_1} x_2^{p_2} \cdots x_n^{p_n}$ with $0 \leqq p_i \leqq rq_i - 1$. Therefore $\mathfrak{A}^* = \mathfrak{A}/\mathfrak{X}_0$ has a finite basis.

Let $z \to z^*$ denote the natural homomorphism of $\mathfrak{A}$ onto $\mathfrak{A}^*$ so that in particular 1^* is the unit element of $\mathfrak{A}^*$. Since $\mathfrak{X}_0$ is invariant under Δ, d_D for $D \in \mathfrak{D}$ induces a derivation d_D^* of $\mathfrak{A}^*$ given by

$$(11) \qquad (d_D z)^* = d_D^* z^*.$$

For any $X \in \mathfrak{L}$ define a linear mapping A_X of $\mathfrak{A}$ into itself as follows:

$$(12) \qquad A_X z = \omega(X) z \qquad (z \in \mathfrak{A}).$$

Since $\mathfrak{X}_0$ is an ideal, $\mathfrak{X}_0$ is invariant under A_X. Let A_X^* be the mapping induced by A_X in $\mathfrak{A}^*$ so that

$$(13) \qquad A_X^* z^* = (\omega(X))^* z^* \qquad (z^* \in \mathfrak{A}^*).$$

Since $\mathfrak{U} \subset \mathfrak{X}_0$ it follows immediately from (2) that

$$(14) \qquad [A_X^*, A_Y^*] = A_{[X,Y]}^*$$

where $[A^*, B^*] = A^*B^* - B^*A^*$ for any two linear mappings A^*, B^* of $\mathfrak{A}^*$ into itself. Also

$$
\begin{aligned}
d_D^* A_X^* z^* &= d_D^*((\omega(X))^* z^*) \\
&= (d_D^*(\omega(X))^*) z^* + (\omega(X))^* d_D^* z^* \\
&= (d_D \omega(X))^* z^* + A_X^* d_D^* z^* \\
&= (\omega(DX))^* z^* + A_X^* d_D^* z^* \\
&= A_{DX}^* z^* + A_X^* d_D^* z^*.
\end{aligned}
$$

Hence

$$(15) \qquad [d_D^*, A_X^*] = A_{DX}^*.$$

Also from (6)

$$(16) \qquad d_{[D_1,D_2]}^* = [d_{D_1}^*, d_{D_2}^*].$$

Therefore if we put $\theta(X + D) = A_X^* + d_D^*$, θ is a representation of $\mathfrak{L} + \mathfrak{D}$. We must now show that it is a faithful representation. Suppose $A_X^* + d_D^* = 0$. Then

$$0 = (A_X^* + d_D^*) 1^* = A_X^* 1^* = (\omega(X))^*.$$

This implies that $\omega(X) \in \mathfrak{X}_0 \subset \mathfrak{X}$. Hence $X = \pi\omega(X) = 0$. Therefore $A_X^* = 0$ and

$$0 = d_D^*(\omega(Y))^* = (d_D\omega(Y))^* = (\omega(DY))^*$$

for every $Y \in \mathfrak{L}$. Hence $\omega(DY) \in \mathfrak{X}_0 \subset \mathfrak{X}$. Therefore $DY = \pi\omega(DY) = 0$. Since this is true for every Y, $D = 0$. Hence θ is a faithful representation. Also notice that if $Y \in \mathfrak{N}$, $\theta(Y) = A_Y^*$ is nilpotent. For, since $(\omega(Y))^r \in \mathfrak{Y}^r \subset \mathfrak{X}_0$,

$$(A_Y^*)^r z^* = ((A_Y)^r z)^* = ((\omega(Y))^r z)^* = 0 \qquad (z \in \mathfrak{A}).$$

Let $\mathfrak{G}$ be the enveloping algebra of the linear mappings $\theta(X + D)$ $x \in \mathfrak{L}$, $D \in \mathfrak{D}$. Let $\mathfrak{H}$ be the enveloping algebra of the linear mappings $\theta(Y)$, $Y \in \mathfrak{N}$. Then $\mathfrak{H} \subset \mathfrak{G}$ and $\mathfrak{H}$ is nilpotent ((5), Theorem 5). From Proposition 1 it is clear that $\mathfrak{N}$ is an ideal in $\mathfrak{L} + \mathfrak{D}$. Hence $\theta(\mathfrak{N})$ is an ideal in $\theta(\mathfrak{L} + \mathfrak{D})$. By a theorem of Jacobson ((5) Lemma 1, p. 876) it follows that $\mathfrak{G}\mathfrak{H}$ and $\mathfrak{H}\mathfrak{G}$ are nilpotent (associative) ideals in $\mathfrak{G}$. Let $Y \in \mathfrak{N}$ and let D be a nilpotent derivation of $\mathfrak{L}$ so that $D^p = 0$. Let $\mathfrak{R}$ be the radical of $\mathfrak{G}$. Then

$$(17) \qquad \{\theta(D + Y)\}^s = (d_D^* + A_Y^*)^s \equiv (d_D^*)^s \ (\mathrm{mod}\ \mathfrak{R}), \qquad s \geqq 2$$

since $\mathfrak{R} \supset \mathfrak{G}\mathfrak{H} + \mathfrak{H}\mathfrak{G}$. Now we prove by induction that

$$(d_D^*)^{sp}(x_{i_1}^* x_{i_2}^* \cdots x_{i_s}^*) = 0 \qquad 1 \leqq i_1, \cdots, i_s \leqq n, s \geqq 1.$$

This is clear for $s = 1$ since $(d_D^*)^p x_i^* = (d_D^p x_i)^* = (\omega(D^p x_i))^* = 0$. By using the Leibnitz rule for derivations we find that $(d_D^*)^{sp}(x_{i_1}^* \cdots x_{i_s}^*)$ is a linear combination of terms of the form

$$\{(d_D^*)^{sp-q}(x_{i_1}^* \cdots x_{i_{s-1}}^*)\}\{(d_D^*)^q x_{i_s}^*\} \qquad (0 \leqq q \leqq sp)$$

where $(d_D^*)^m$ is to be regarded as the identity mapping in case $m = 0$. By induction hypothesis the first factor is zero if $q \leqq p$ and the second factor vanishes if $q \geqq p$. Hence the result is established. Since $\mathfrak{A}^*$ is a finite dimensional algebra there exists a t such that the elements 1^* and

$$x_{i_1}^* x_{i_2}^* \cdots x_{i_s}^* \qquad (1 \leqq i_1, \cdots, i_s \leqq n, 1 \leqq s \leqq t)$$

span the whole of $\mathfrak{A}^*$. Hence $(d_D^*)^{tp} = 0$. From (17) this implies that

$$(d_D^* + A_Y^*)^{tp} \in \mathfrak{R}.$$

But $\mathfrak{R}$ is nilpotent. Hence $(d_D^* + A_Y^*)^{tp}$ and therefore $\theta(D + Y)$ is nilpotent. Thus the lemma is proved completely.

REMARK. Let $\mathfrak{L}$ be a solvable Lie algebra and $\mathfrak{D}$ the algebra of its derivations. Let $\mathfrak{N}$ be the maximal nilpotent ideal of $\mathfrak{L}$. Suppose $\mathfrak{L}$ has a faithful representation which maps every element of $\mathfrak{N}$ on a nilpotent matrix. Then the above lemma is applicable to the image $\mathfrak{L}_0$ of $\mathfrak{L}$ under this representation. Since $\mathfrak{L}_0$ is isomorphic to $\mathfrak{L}$ it follows immediately that $\mathfrak{L} + \mathfrak{D}$ has a faithful representation θ such that if $Y \in \mathfrak{N}$, $D \in \mathfrak{D}$ and D is nilpotent, then $\theta(Y + D)$ is also nilpotent.

CoROLLARY 1. *Let $\mathfrak{L}$ be a solvable Lie algebra and $\mathfrak{N}$ the maximal nilpotent ideal of $\mathfrak{L}$. Then $\mathfrak{L}$ has a faithful representation which maps every element of $\mathfrak{N}$ on a nilpotent matrix.*

We shall prove it by induction on the dimension of $\mathfrak{L}$. Suppose dim $\mathfrak{L} = 1$. Let X be any non-zero element of $\mathfrak{L}$. Then X forms a base for $\mathfrak{L}$ and the linear mapping which maps X on the matrix $\begin{pmatrix} 0 & 0 \\ 1 & 0 \end{pmatrix}$ is clearly a faithful representation of the required type. Now let dim $\mathfrak{L} = n > 1$. First let $\mathfrak{L} \neq \mathfrak{N}$. Since $\mathfrak{N}$ contains the derived algebra of $\mathfrak{L}$, $\mathfrak{L}/\mathfrak{N}$ is abelian and therefore we can find an ideal $\mathfrak{M}$ in $\mathfrak{L}$ such that $\mathfrak{M} \supset \mathfrak{N}$ and dim $\mathfrak{M} = n - 1$. On the other hand if $\mathfrak{L} = \mathfrak{N}$ choose $\mathfrak{M}$ to be any ideal in $\mathfrak{L}$ such that dim $\mathfrak{M} = n - 1$. Such an ideal exists since $\mathfrak{L}$ is solvable. In either case let X be any element of $\mathfrak{L}$ such that $X \notin \mathfrak{M}$. If ad $X = 0$, $X \in \mathfrak{N}$ and therefore $\mathfrak{L} = \mathfrak{N}$ and $\mathfrak{L}$ is the direct sum of $\mathfrak{M}$ and the one-dimensional Lie algebra KX. By our induction hypothesis the assertion is true for both $\mathfrak{M}$ and KX and therefore also for their direct sum. We have merely to take the direct sum of the respective representations of $\mathfrak{M}$ and KX as the representation for $\mathfrak{L}$. If we observe that the direct sum of two nilpotent matrices is nilpotent, it is clear that this representation, in addition to being faithful, is of the required type. Hence we can now suppose that ad $X \neq 0$. Let D be the derivation of $\mathfrak{M}$ induced by ad X. Then $D \neq 0$. Let $\mathfrak{D}_{\mathfrak{M}}$ be the algebra of derivations of $\mathfrak{M}$. Let $\mathfrak{L}_0$ be the subalgebra of $\mathfrak{M} + \mathfrak{D}_{\mathfrak{M}}$ spanned by $(\mathfrak{M}, D)$. Clearly $\mathfrak{L}$ is isomorphic to $\mathfrak{L}_0$ under the linear mapping which maps $\mathfrak{M}$ identically onto itself and X on D. Hence we can consider $\mathfrak{L}_0$ instead of $\mathfrak{L}$. From our induction hypothesis and the remark made at the end of Lemma 1 it follows that $\mathfrak{M} + D_{\mathfrak{M}}$ has a faithful representation θ. Now first let $\mathfrak{L} = \mathfrak{N}$. Then ad X and therefore D is nilpotent. Also $\mathfrak{M}$ is nilpotent. Therefore from Lemma 1, θ can be so chosen that $\theta(cD + Y)$ is nilpotent for every $Y \in \mathfrak{M}$ and $c \in K$. This means that every element of $\mathfrak{L}_0$ is mapped on a nilpotent matrix. If $\mathfrak{L} \neq \mathfrak{N}$, $\mathfrak{M} \supset \mathfrak{N}$ and so, again from Lemma 1, we can assume that $\theta(Y)$ is nilpotent for every $Y \in \mathfrak{N}$. In either case θ defines a faithful representation of $\mathfrak{L}_0$ which is of the required type.

CoROLLARY 2. *Let $\mathfrak{L}$ be a solvable Lie algebra and $\mathfrak{D}$ the algebra of its derivations. Let $\mathfrak{N}$ be the maximal nilpotent ideal of $\mathfrak{L}$. Then $\mathfrak{L} + \mathfrak{D}$ has a faithful representation θ such that if $Y \in \mathfrak{N}$, $D \in \mathfrak{D}$ and D is nilpotent, then $\theta(Y + D)$ is nilpotent.*

This follows immediately from Corollary 1, and the remark made at the end of Lemma 1.

THEOREM. *Let $\mathfrak{L}$ be a Lie algebra and Γ its maximal solvable ideal. Let $\mathfrak{N}$ be the maximal nilpotent ideal of $\mathfrak{L}$. There exists a faithful representation ϕ of $\mathfrak{L}$ such that $\phi(Y)$ is nilpotent for every $Y \in \mathfrak{N}$.*

By Levi's Theorem $\mathfrak{L}$ can be decomposed into a sum of Γ and a semi-simple Lie algebra $\mathfrak{S}$. Let $\mathfrak{D}$ be the algebra of derivations of Γ. For every $S \in \mathfrak{S}$ we define a derivation D_S of Γ by $D_S X = [S,X](X \in \Gamma)$. The mapping $S \to D_S$ of $\mathfrak{S}$ into $\mathfrak{D}$ is clearly a homomorphism. Hence it follows easily that the maping $X + S \to X + D_S (X \in \Gamma, S \in \mathfrak{S})$ is a homomorphism of $\mathfrak{L}$ into $\Gamma + \mathfrak{D}$.

Let θ be the faithful representation of $\Gamma + \mathfrak{D}$ whose existence is asserted in Corollary 2. Then the mapping ϕ_1 given by

$$\phi_1 (X + S) = \theta(X + D_S) \qquad (X \in \Gamma, S \in \mathfrak{S})$$

is a representation of $\mathfrak{L}$. Let ad S be the derivation of $\mathfrak{S}$ corresponding to S in adjoint representation of $\mathfrak{S}$. Since the mappings $S + X \to S$ and $S \to$ ad S are both homomorphic, the mapping θ_2 defined by

$$\phi_2 (S + X) = \text{ad } S$$

is a representation of $\mathfrak{L}$. Let ϕ be the direct sum of the representations ϕ_1 and ϕ_2. Then ϕ it is a faithful representation of $\mathfrak{L}$. For $\phi(S + X) = 0$ implies $\phi_1 (S + X) = \theta (X + D_S) = 0$ and $\phi_2 (S + X) = $ ad $S = 0$. Since $\mathfrak{S}$ is semi-simple, the adjoint representation is faithful. Hence $S = 0$. Also as θ is a faithful representation of $\Gamma + \mathfrak{D}$, $X = 0$. Further, if $Y \in \mathfrak{N} \subset \Gamma$, $\phi_1 (Y)$ $= \theta(Y)$ is nilpotent and $\phi_2(Y) = 0$. Hence $\phi(Y)$ is nilpotent. The theorem is therefore proved.

Institute for Advanced Study

References

1. I. D. Ado, Kazan. Univ. Fiziko-mat. obshchestvo, Bull. vol. 7 (1934–35), p. 3.
2. E. Cartan, Jour. de Math. pures, sér. 9, vol. 17 (1938), p. 1.
3. C. Chevalley, Theory of Lie Groups, I. Princeton University Press, 1946.
4. C. Chevalley, Ann. of Math., vol. 48 (1947), p. 91.
5. N. Jacobson, Ann. of Math., vol. 36 (1936), p. 875.

ANNALS OF MATHEMATICS
Vol. 50, No. 4, October, 1949

ON REPRESENTATIONS OF LIE ALGEBRAS

By Harish-Chandra

(Received July 14, 1948)

1. Introduction

Let $\mathfrak{A}$ be an associative algebra over a field K, the dimension of $\mathfrak{A}$ over K not being necessarily finite. By a representation of $\mathfrak{A}$ we mean as usual a homomorphic mapping of $\mathfrak{A}$ into the algebra of endomorphisms of some finite dimensional vector space over K. We shall say that $\mathfrak{A}$ has sufficiently many representations if given any element $z \in \mathfrak{A}$, $z \neq 0$ we can find a representation π such that $\pi(z) \neq 0$.

Let $\mathfrak{L}$ be a Lie algebra over K. Let $X_1, X_2, \cdots, X_n$ be a base for $\mathfrak{L}$. Let $\mathfrak{A}$ be the free algebra with n generators i.e. the ring of all non-commutative polynomials in n variables $x_1, x_2, \cdots, x_n$ with coefficients in K. For any $z, w \in \mathfrak{A}$ write $[z, w] = zw - wz$. Let ω be the linear mapping of $\mathfrak{L}$ into $\mathfrak{A}$ defined by

$$(1) \qquad \omega(X_i) = x_i \qquad\qquad 1 \leqq i \leqq n.$$

Put

$$(2) \qquad u(X, Y) = [\omega(X), \omega(Y)] - \omega([X, Y]).$$

where $X, Y \in \mathfrak{L}$ and $[X, Y]$ is their bracket product defined in $\mathfrak{L}$. Let ρ be any representation of $\mathfrak{L}$. Then ρ defines uniquely a representation π of $\mathfrak{A}$ as follows:

$$(3) \qquad \begin{aligned} \pi(\omega(X)) &= \rho(X) \qquad X \in \mathfrak{L} \\ \pi(1) &= I \end{aligned}$$

where I is the identity mapping of the vector space on which $\rho(\mathfrak{L})$ operates. Since $\mathfrak{A}$ is generated by 1 and x_i $(1 \leqq i \leqq n)$ these relations determine π completely. Notice that[1]

$$\begin{aligned} \pi(u(X, Y)) &= [\pi(\omega(X)), \quad \pi(\omega(Y))] - \pi(\omega([X, Y])) \\ &= [\rho(X), \quad \rho(Y)] - \rho([X, Y]) = 0 \end{aligned}$$

since ρ is a representation of $\mathfrak{L}$. Therefore if $\mathfrak{U}$ is the ideal generated by $u(X, Y)$ (i.e. the smallest possible ideal in $\mathfrak{A}$ containing $u(X, Y)$ for all $X, Y \in \mathfrak{L}$) then π maps $\mathfrak{U}$ on zero. Hence ρ defines, in fact, a representation π^* of the factor algebra $\mathfrak{A}^* = \mathfrak{A}/\mathfrak{U}$. Conversely every representation π^* of $\mathfrak{A}^*$ defines a representation ρ of $\mathfrak{L}$ by the relation $\rho(X) = \pi^*((\omega(X))^*)$ where $z \to z^*$ is the natural homomorphism of $\mathfrak{A}$ onto $\mathfrak{A}^*$. In view of this 1-1 correspondence between the representations of $\mathfrak{L}$ and $\mathfrak{A}^*$ it is appropriate to call $\mathfrak{A}^*$ the general enveloping algebra of $\mathfrak{L}$.[2] It is clear that $\mathfrak{A}^*$ is independent of the particular base $X_1, \cdots, X_n$

---

[1] If V is a vector space over K (not necessarily finite dimensional) and A, B are any two endomorphisms of V, we always write $[A, B] = AB - BA$.

[2] This algebra is the same as the one considered by Birkhoff (1) and Witt (7), though their method of construction is different.

900

used in its construction since a different base gives rise to an isomorphic algebra. The main purpose of this paper is to prove the fact that the general enveloping algebra $\mathfrak{A}^*$ of any Lie algebra $\mathfrak{L}$ over a field of characteristic zero has sufficiently many representations. It will appear later that this result, in its complete form, can be regarded as a generalisation of Ado's Theorem. A number of corollaries can be obtained from it. For example it will be deduced that $\mathfrak{A}^*$ is an integral domain and $\mathfrak{L}$ has infinitely many inequivalent irreducible representations. Also we shall prove quite simply that if $\mathfrak{L}$ is semisimple it has only a finite number of inequivalent irreducible representations whose degree is less than a given integer r. This would prove that a semisimple Lie algebra has irreducible representations of arbitrarily high degree. Finally we shall construct an example of an algebra with a finite number of generators, which has no nontrivial representations over any field whatsoever.[3]

2. Definitions and preliminaries

Henceforward the characteristic of the basic field K is always assumed to be zero. All vector spaces and algebras are to be understood over K unless explicitly mentioned otherwise.

Let $\mathfrak{A}$ be the free algebra with n generators $x_1, x_2, \cdots, x_n$. Every element z of $\mathfrak{A}$ can be written uniquely as a linear combination of elements of the form $x_{i_1} x_{i_2} \cdots x_{i_r}$, $1 \leq i_1, i_2, \cdots, i_r \leq n, r \geq 0$. (In case $r = 0$ $x_{i_1} x_{i_2} \cdots x_{i_r}$ is to be interpreted to mean 1.) Let

$$(4) \qquad z = \sum_{s \geq r \geq 0} \sum_{1 \leq i_1, \cdots, i_r \leq n} a^{i_1 \cdots i_r} x_{i_1} \cdots x_{i_r}.$$

z is said to have the degree s provided not all $a^{i_1 \cdots i_s}$ are zero. z is said to be a homogeneous element of degree s if $a^{i_1 \cdots i_r} = 0$ whenever $r \neq s$. By convention 0 is assumed to possess every degree $s \geq 0$. z is called canonical if for $s \geq r \geq 0$, $1 \leq i_1, \cdots, i_r \leq n$ and an arbitrary permutation $j_1, \cdots, j_r$ of $i_1, \cdots, i_r$ $a^{j_1 \cdots j_r} = a^{i_1 \cdots i_r}$. An element is called normal if it is a linear combination of elements of the form $x_1^{m_1} x_2^{m_2} \cdots x_n^{m_n}$, $m_1, m_2, \cdots, m_n \geq 0$ where $x_i^{m_i}$ is to be understood to mean 1 in case $m_i = 0$. An element of degree 1 is called a linear element.

An ideal $\mathfrak{B}$ in $\mathfrak{A}$ is called a Lie ideal if for any two linear elements $x, y \in \mathfrak{A}$ one can always find a linear element z such that $[x, y] \equiv z \pmod{\mathfrak{B}}$.

LEMMA 1. *Let $\mathfrak{A}$ be the free algebra with n generators and let $\mathfrak{B}$ be a Lie ideal in $\mathfrak{A}$. Then every element $z \in \mathfrak{A}$ of degree s is congruent modulo $\mathfrak{B}$ to a canonical (normal) element z_0 of degree $\leq s$.*

Since every linear element is canonical the lemma is true for $s = 1$. Hence we

---

[3] If K is a subfield of K' and $\mathfrak{B}$ is an algebra over K, then by a representation of $\mathfrak{B}$ over K' we mean a homomorphic mapping of $\mathfrak{B}$ into the algebra of endomorphisms of a finite dimensional vector space over K'. Professor Jacobson has pointed out to me that the algebra generated over K by two elements p, q with the single relation $pq - qp = 1$ has no nontrivial representations over any extension of K if the characteristic of K is 0.

can use induction on s and assume $s \geqq 2$. Since the canonical elements clearly form a linear set it is sufficient to consider elements of the form $x_{i_1} x_{i_2} \cdots x_{i_s}$, $1 \leqq i_1, \cdots, i_s \leqq n$. As $\mathfrak{B}$ is a Lie ideal

$$x_{i_1} x_{i_r} \cdots x_{i_r} x_{i_{r+1}} \cdots x_{i_s} - x_{i_1} x_{i_2} \cdots x_{i_{r+1}} x_{i_r} \cdots x_{i_s}$$
$$= x_{i_1} \cdots [x_{i_r}, x_{i_{r+1}}] \cdots x_{i_s}$$

is congruent mod $\mathfrak{B}$ to an element of degree $\leqq s - 1$. Now any permutation of the indices $i_1, \cdots, i_s$ to $j_1, \cdots, j_s$ can be effected by successive transposition of adjacent indices. Hence if we denote by $\sum$ the sum over all such permutations

$$x_{i_1} x_{i_2} \cdots x_{i_s} - \frac{1}{s!} \sum x_{j_1} \cdots x_{j_s}$$

is congruent mod $\mathfrak{B}$ to an element w of degree $\leqq s - 1$. By induction hypothesis w may be assumed to be canonical. Hence

$$x_{i_1} x_{i_2} \cdots x_{i_s} \equiv \frac{1}{s!} \sum x_{j_1} \cdots x_{j_s} + w \pmod{\mathfrak{B}}.$$

But clearly $1/s! \sum x_{j_1} \cdots x_{j_s}$ is canonical and therefore $1/s! \sum x_{j_1} \cdots x_{j_s} + w$ is a canonical element. The proof in the other case when z_0 is required to be normal is entirely similar.

Let $\mathfrak{A}$ be the free algebra with n generators and $\mathfrak{B}$ any ideal in $\mathfrak{A}$. Let $z \to z^*$ denote the natural homomorphism of $\mathfrak{A}$ onto $\mathfrak{A}^* = \mathfrak{A}/\mathfrak{B}$. z^* is then the residue class consisting of all those elements of $\mathfrak{A}$ which are congruent to z mod $\mathfrak{B}$. Any element z of $\mathfrak{A}$ such that $z \epsilon z^*$ will be called an expression for z^* and z^* will be called the value of this expression in $\mathfrak{A}^*$. The degree of any non-zero element $z^* \epsilon \mathfrak{A}^*$ is defined to be the least possible degree of any of its expressions. Lemma 1 asserts that if $\mathfrak{B}$ is a Lie ideal every element of $\mathfrak{A}^*$ has a canonical expression. In order that any element z^* of $\mathfrak{A}^*$ may have at most one canonical expression it is clearly necessary and sufficient that $\mathfrak{B}$ should contain no canonical element of $\mathfrak{A}$ other than zero.

3. The main lemma

Let V be a finite dimensional vector space and E the ring of all its endomorphisms. Choose a basis $X_0, X_1, \cdots, X_m$ for E such that $X_0 = I$ is the identity mapping of V and $\mathrm{sp}(X_i) = 0$ $1 \leqq i \leqq m$. Clearly this is always possible. Let E_0 be the subspace of E spanned by $X_1, \cdots, X_m$. Then E_0 is a Lie algebra under the usual bracket operation. Let V_r $(r \geqq 1)$ be the Kronecker product of V with itself r times and let E_r be the ring of all endomorphisms of V_r. It is well known that, as a vector space, E_r can be considered to be the Kronecker product of E with itself r times. Hence

$$X_{i_1} \mathbf{x} X_{i_2} \mathbf{x} \cdots \mathbf{x} X_{i_r} \qquad 0 \leqq i_1, \cdots, i_r \leqq m$$

form a base for E_r. Here $\mathbf{x}$ denotes Kronecker product. For any $X \in E_0$ we denote by $X^{(r)}$ the element of E_r defined as follows[4]

$$(5) \qquad X^{(r)} = X^{(r,1)} + X^{(r,2)} + \cdots + X^{(r,r)}$$

where

$$X^{(r,j)} = I \mathbf{x} \cdots \mathbf{x} I \mathbf{x} X \mathbf{x} I \mathbf{x} \cdots \mathbf{x} I$$

all the factors in the Kronecker product being I except the jth factor which is X. The identity

$$(6) \qquad [X^{(r)}, Y^{(r)}] = [X, Y]^{(r)}$$

is verified easily.

Let $\mathfrak{B}$ be the free algebra with m generators $x_1, \cdots, x_m$. Let ω be the linear mapping of E into $\mathfrak{B}$ given by

$$(7) \qquad \omega(I) = 1, \qquad \omega(X_i) = x_i \qquad\qquad 1 \leqq i \leqq m.$$

For $X, Y \in E$ put

$$(8) \qquad u(X, Y) = [\omega(X), \omega(Y)] - \omega([X, Y])$$

and let $\mathfrak{B}$ be the ideal in $\mathfrak{B}$ generated by $u(X, Y)$ for all $X, Y \in E$. Clearly $\mathfrak{B}$ is a Lie ideal in $\mathfrak{B}$.

Let π_r be the linear mapping of $\omega(E)$ into E_r defined by

$$\pi_r(\omega(I)) = \pi_r(1) = I \mathbf{x} I \mathbf{x} \cdots \mathbf{x} I$$
$$\pi_r(\omega(X)) = X^{(r)}, \qquad X \in E_0 .$$

Clearly π_r can be extended to a homomorphism of $\mathfrak{B}$ into E_r. Let $\mathfrak{X}_r$ be the kernel of this homomorphism. Put $\mathfrak{X} = \bigcap_{r \geqq 1} \mathfrak{X}_r$. $\mathfrak{X} \supset \mathfrak{B}$ since for every r and $X, Y \in E_0$.

$$\pi_r(u(X, Y)) = [\pi_r(\omega(X)), \pi_r(\omega(Y))] - \pi_r(\omega[X, Y])$$
$$= [X^{(r)}, Y^{(r)}] - [X, Y]^{(r)} = 0$$

and $u(I, X) = 0$ $(X \in E)$. We shall now show that $\mathfrak{X} = \mathfrak{B}$. This result which we state as a lemma is the essential step of our proof.

LEMMA 2. $\mathfrak{X} = \mathfrak{B}$ *and* $\mathfrak{B}$ *contains no canonical element other than zero.*

First we show that zero is the only canonical element contained in $\mathfrak{X}$. For otherwise let $z \neq 0$ be a canonical element of degree s such that $z \in \mathfrak{X}$. Clearly $s \geqq 1$. We can write $z = z_0 + w$ where z_0 and w are both canonical and z_0 is homogeneous and of degrees s while the degree of w is less than s. Let

$$(9) \qquad z_0 = \sum_{1 \leqq i_1, \cdots, i_s \leqq m} a^{i_1 i_2 \cdots i_s} x_{i_1} x_{i_2} \cdots x_{i_s} .$$

---

[4] In Chevalley's terminology $X^{(r)}$ is the Kronecker sum of X with itself r times.

Consider $\pi_s(z)$. Let F_s be the subspace of E_s spanned by all those elements of the form $Y_1 \times Y_2 \times \cdots \times Y_s$ ($Y_j \in E$, $1 \le j \le s$) for which $Y_j = I$ for at least one j. We first prove that

$$(10) \quad \pi_s(z) \equiv s! \sum_{1 \le i_1 \cdots, i_s \le m} a^{i_1 i_2 \cdots i_s} (X_{i_1} \times X_{i_2} \times \cdots \times X_{i_s}) \pmod{F_s}.$$

Notice that $\pi_s(x_{i_1} x_{i_2} \cdots x_{i_r}) = X_{i_1}^{(s)} X_{i_2}^{(s)} \cdots X_{i_r}^{(s)}$ $(1 \le i_1, \cdots i_r \le m)$. From (5)

$$X_{i_1}^{(s)} X_{i_2}^{(s)} \cdots X_{i_r}^{(s)} = \sum_{1 \le j_1, \cdots, j_r \le s} X_{i_1}^{(s,j_1)} \cdots X_{i_r}^{(s,j_r)}$$

If $r < s$ there is at least one integer j in the set $(1, 2, \cdots, s)$ which is not present among $(j_1, \cdots, j_r)$ and therefore

$$X_{i_1}^{(s,j_1)} \cdots X_{i_r}^{(s,j_r)} = Y_1 \times Y_2 \times \cdots \times Y_r$$

with $Y_j = I$. Hence

$$(11) \qquad \pi_s(x_{i_1} x_{i_2} \cdots x_{i_r}) \equiv 0 \pmod{F_s} \qquad\qquad 1 \le i_1, \cdots, i_r \le m, r < s.$$

Further if $r = s$ the same argument is again applicable unless $j_1, j_2, \cdots, j_r$ are all distinct. Hence

$$X_{i_1}^{(s)} X_{i_2}^{(s)} \cdots X_{i_s}^{(s)} \equiv \sum X_{i_1}^{(s,j_1)} \cdots X_{i_s}^{(s,j_s)} \pmod{F_s}$$

where $\sum$ denotes a sum over all possible permutations $j_1, \cdots, j_s$ of $1, 2, \cdots, s$. But the right hand side is clearly equal to

$$\sum (X_{k_1} \times X_{k_2} \times \cdots \times X_{k_s})$$

where now the sum extends over all possible permutations $k_1, k_2, \cdots, k_s$ of $i_1, i_2, \cdots, i_s$. Hence

$$(12) \qquad X_{i_1}^{(s)} X_{i_2}^{(s)} \cdots X_{i_s}^{(s)} \equiv \sum (X_{k_1} \times X_{k_2} \times \cdots \times X_{k_s}) \pmod{F_s}.$$

Making use of (11) and (12) and the fact that $a^{i_1 \cdots i_s}$ are symmetric with respect to all permutations of the indices $i_1, \cdots, i_s$ we get (10). Now by hypothesis $z \in \mathfrak{X} \subset \mathfrak{X}_s$. Therefore $\pi_s(z) = 0$ and hence

$$(13) \qquad \sum_{1 \le i_1, \cdots, i_s \le m} a^{i_1 \cdots i_s} (X_{i_1} \times X_{i_2} \times \cdots \times X_{i_s}) \equiv 0 \pmod{F_s}.$$

But

$$X_{i_1} \times X_{i_2} \times \cdots \times X_{i_s} \qquad\qquad 0 \le i_1, \cdots, i_s \le m$$

constitute a base for E_s. Among them those elements for which $i_j = 0$ for at least one j $(1 \le j \le s)$ obviously form a base for F_s. Hence it is clear that the elements $X_{i_1} \times X_{i_2} \times \cdots \times X_{i_s}$, $1 \le i_1, \cdots, i_s \le m$ are linearly independent mod F_s. Therefore it follows from (13) that

$$a^{i_1 i_2 \cdots i_s} = 0 \qquad\qquad (1 \le i_1, \cdots, i_s \le m)$$

which contradicts the fact that the degree of z is s. Therefore $\mathfrak{X}$ contains no canonical element except zero.

It is now easy to prove that $\mathfrak{X} = \mathfrak{B}$. Let z be any element of $\mathfrak{X}$. Since $\mathfrak{B}$ is a Lie ideal in $\mathfrak{B}$, there exists a canonical element z_0 such that $z \equiv z_0 \pmod{\mathfrak{B}}$. Since $\mathfrak{B} \subset \mathfrak{X}$ it is clear that $z_0 \in \mathfrak{X}$. There $z_0 = 0$. Hence $z \equiv 0 \pmod{\mathfrak{B}}$. The lemma is thus proved completely.

COROLLARY 1. *Every element in $\mathfrak{B}^* = \mathfrak{B}/\mathfrak{B}$ has a unique canonical expression.*

In view of the remarks made at the end of Section 2, this follows immediately from the fact that $\mathfrak{B}$ is a Lie ideal in $\mathfrak{B}$ and it contains no canonical element other than zero.

COROLLARY 2. *$\mathfrak{B}^*$ has sufficiently many representations.*

Let $z^* \neq 0$ be any element of $\mathfrak{B}^*$. From Corollary 1 there exists a unique canonical element $z \in \mathfrak{B}$ such that $z \in z^*$. Since $z^* \neq 0$, $z \neq 0$. Hence $z \notin \mathfrak{X}$. Therefore there exists an s such that $z \notin \mathfrak{X}_s$. Since $\mathfrak{X}_s \supset \mathfrak{B}$, π_s defines a representation π_s^* of $\mathfrak{B}^*$ given by $\pi_s(w) = \pi_s^*(w^*)$, $w \to w^*$ being the natural homomorphism of $\mathfrak{B}$ onto $\mathfrak{B}^*$. Hence $\pi_s^*(z^*) = \pi_s(z) \neq 0$.

4. Proof of the Theorem

We are now in a position to prove our main result.

THEOREM 1. *Let $\mathfrak{L}$ be a Lie algebra over a field K of characteristic zero, and let $\mathfrak{A}^*$ be its general enveloping algebra. Then $\mathfrak{A}^*$ has sufficiently many representations and every element of $\mathfrak{A}^*$ has a unique canonical expression.*

Let dim $\mathfrak{L} = n$. Since $\mathfrak{L}$ has a faithful representation[5] we can assume, without loss of generality that $\mathfrak{L}$ is a linear Lie algebra i.e. its elements are endomorphisms of a certain vector space W. Define V to be the direct sum of W and a one-dimensional vector space Ke spanned by a vector e. Every element $X \in \mathfrak{L}$ is already defined on W. We can extend it to V by defining $Xe = -s(X)e$ where $s(X)$ is the trace of X on W. Now if $X, Y, Z \in \mathfrak{L}$ and $[X, Y] = Z$ on W, it is clear that $s(Z) = 0$ and therefore

$$(XY - YX)e = 0 = -s(Z)e = Ze.$$

This shows that $[X, Y] = Z$ on V. Further sp $X = 0$ on V for every $X \in \mathfrak{L}$. In this way $\mathfrak{L}$ is imbedded isomorphically in the space E_0 of all endomorphisms of V of trace zero. Let $X_1, \cdots, X_n$ be a base for $\mathfrak{L}$ and let E be the space of all endomorphisms of V. It is clear that I, the identity mapping of V is linearly independent of $X_1, \cdots, X_n$ since sp $I \neq 0$. Hence we can choose a base for E such that its first $n + 1$ elements are $X_0 = I, X_1, \cdots, X_n$ and its other elements belong to E_0. We denote this base by $X_0, X_1, \cdots, X_m$ $(m \geq n)$.

Let $\mathfrak{B}$ and $\mathfrak{A}$ be the free algebras with m and n generators $x_1, \cdots, x_m$ and $x_1, \cdots, x_n$ respectively, so that $\mathfrak{A}$ is a subalgebra of $\mathfrak{B}$. Define ω and $u(X, Y)$ again by (7) and (8). Let $\mathfrak{B}$ be the ideal in $\mathfrak{B}$ generated by $u(X, Y)$ for all

---

[5] An algebraic proof of the fact that every Lie algebra over a field of characteristic zero has a faithful representation has recently been given by the author (2).

$X, Y \in E$. It is clear that if $X, Y \in \mathfrak{L}$, $u(X, Y) \in \mathfrak{A}$. Let $\mathfrak{U}$ be the ideal in $\mathfrak{A}$ generated by $u(X, Y)$ for all $X, Y \in \mathfrak{L}$. Then $\mathfrak{U} \subset \mathfrak{B} \cap \mathfrak{A}$. Further every canonical element of $\mathfrak{A}$ is obviously also canonical in $\mathfrak{B}$. Since from Lemma 2, $\mathfrak{B}$ contains no canonical elements except zero and $\mathfrak{U} \subset \mathfrak{B}$, the same is true for $\mathfrak{U}$ in $\mathfrak{A}$. Therefore as $\mathfrak{U}$ is a Lie ideal in $\mathfrak{A}$, it follows from Lemma 1 that every $z \in \mathfrak{A}$ is congruent mod $\mathfrak{U}$ to a uniquely determined canonical element z_0. Since z_0 is also canonical in $\mathfrak{B}$ and congruence mod $\mathfrak{U}$ implies congruence mod $\mathfrak{B}$, z_0 is also the unique canonical element in $\mathfrak{B}$ which is congruent to z mod $\mathfrak{B}$. Therefore if $z \in \mathfrak{B}$, $z_0 = 0$ so that $z \equiv 0$ (mod $\mathfrak{U}$). Hence $z \in \mathfrak{U}$. Therefore $\mathfrak{U} = \mathfrak{B} \cap \mathfrak{A}$.

Let $z \to z^*$ denote the natural homomorphism of $\mathfrak{B}$ onto $\mathfrak{B}^* = \mathfrak{B}/\mathfrak{B}$. Let $\mathfrak{A}^*$ be the image of $\mathfrak{A}$ under this homomorphism. Then $\mathfrak{A}^*$ is isomorphic to $\mathfrak{A}/\mathfrak{B} \cap \mathfrak{A} = \mathfrak{A}/\mathfrak{U}$. Therefore $\mathfrak{A}^*$ is the general enveloping albegra of $\mathfrak{L}$. Since $\mathfrak{A}^*$ is a subalgebra of $\mathfrak{B}^*$ it follows from Corollary 2 that it has sufficiently many representations. Further, since $\mathfrak{U}$ is a Lie ideal in $\mathfrak{A}$ which contains no canonical element except zero, the second assertion of the theorem follows immediately.

COROLLARY 1.1. *The degree of a nonzero element z^* in the general enveloping algebra $\mathfrak{A}^*$ of $\mathfrak{L}$ is equal to the degree of its unique canonical expression.*

Let s be the degree of the canonical expression z of z^*. Then it is clear that the degree of z^* is $\leq s$. Suppose contrary to the assertion it is $<s$. Then there exists an element $z' \in \mathfrak{A}$ of degree $\leq s - 1$ such that $z' \in z^*$. From Lemma 1 we can find a canonical element z_0 of degree $\leq s - 1$ such that $z' \equiv z_0$ (mod $\mathfrak{U}$) and therefore $z_0 - z \equiv 0$ (mod $\mathfrak{U}$). Since $\mathfrak{U}$ contains no canonical element other than zero it follows that $z_0 = z$ which contradicts the fact that the degree of z is s.

COROLLARY 1.2. $\mathfrak{A}^*$ *is an integral domain and the product of any two nonzero elements of degree s and t respectively is of degree $s + t$.*

Let $m_1, \cdots, m_n$ be any n integers ≥ 0. Put $m_1 + \cdots + m_n = s$ and

$$z(0, \cdots, 0) = 1$$

$$z(m_1, \cdots, m_n) = \frac{1}{s!} \sum x_{i_1} \cdots x_{i_s} \ (s \geq 1)$$

where $\sum$ denotes a sum over all permutations $i_1, \cdots, i_s$ of the set of s integers of which m_k are $k(1 \leq k \leq n)$. It is clear that $z(m_1, \cdots, m_n)$ is canonical and its degree is $m_1 + \cdots + m_n = s$. We call the various elements $z(m_1, \cdots, m_n)$ $m_k \geq 0$ $(1 \leq k \leq n)$ the basic canonical elements in $\mathfrak{A}$. It is obvious that they are linearly independent and every canonical element can be written as a linear combination of them. We shall say that $z(m_1, \cdots, m_n)$ is of higher rank than $z(m_1', \cdots, m_n')$ if $m_1 + \cdots + m_n \geq m_1' + \cdots + m_n'$ and in case $m_1 + \cdots + m_n = m_1' + \cdots + m_n'$ if $m_k - m_k' > 0$ where k is the least possible integer such that $m_k - m_k' \neq 0$. Since $\mathfrak{U}$ contains no canonical element except zero it follows that the basic canonical elements are also linearly independent mod $\mathfrak{U}$.

Put $m_1 + \cdots + m_n = s$ and $m_1' + \cdots + m_n' = s'$. Then from the reasoning of Lemma 1 it is clear that

$$z(m_1, \cdots, m_n)z(m_1', \cdots, m_n') - z(m_1 + m_1', \cdots, m_n + m_n')$$

is congruent modulo $\mathfrak{U}$ to a canonical element of degree $\leq s + s' - 1$. Further it is obvious that if the ranks of $z(m_1, \cdots, m_n)$ and $z(m_1', \cdots, m_n')$ are higher than those of $z(p_1, \cdots, p_n)$ and $z(p_1', \cdots, p_n')$ respectively, then the rank of $z(m_1 + m_1', \cdots, m_n + m_n')$ is higher than that of $z(p_1 + p_1', \cdots, p_n + p_n')$.

Now let z and w be any two nonzero canonical elements of $\mathfrak{A}$ of degrees s and t respectively, so that

$$z = \sum_{m_1 + \cdots + m_n \leq s} a(m_1, \cdots, m_n) z(m_1, \cdots, m_n)$$

$$w = \sum_{m_1 + \cdots + m_n \leq t} b(m_1, \cdots, m_n) z(m_1, \cdots, m_n)$$

where $a(m_1, \cdots, m_n), b(m_1, \cdots, m_n) \in K$. Let $z(p_1, \cdots, p_n)$ and $z(q_1, \cdots, q_n)$ be the elements of the highest possible rank such that $a(p_1, \cdots, p_n)$ and $b(q_1, \cdots, q_n)$ respectively are different from zero. Then it is clear from what has been said above that

$$(14) \qquad zw \equiv a(p_1, \cdots, p_n)\, b(q_1, \cdots, q_n)\, z(p_1 + q_1, \cdots, p_n + q_n)$$

$$+ \text{ linear combination of basic canonical elements of}$$
$$\text{lower rank} \qquad\qquad (\mathrm{mod}\ \mathfrak{U}).$$

Since elements of different rank are linearly independent mod $\mathfrak{U}$ it follows that $zw \not\equiv 0 \ (\mathrm{mod}\ \mathfrak{U})$.

Let z^*, w^* be any two nonzero elements of $\mathfrak{A}^*$ of degree s and t respectively and let z and w be their canonical expressions. From Corollary 1 the degrees of z and w are s and t respectively. Therefore from (14) it follows immediately that $z^* w^* \neq 0$ and it has the degree $s + t$.

COROLLARY 1.3. *Every Lie algebra $\mathfrak{L}$ over K has infinitely many inequivalent indecomposable representations.*

Let $\rho_1, \cdots, \rho_k$ be any k inequivalent indecomposable representations of $\mathfrak{L}$. Consider the general enveloping algebra $\mathfrak{A}^*$ of $\mathfrak{L}$ and let $\pi_j^*(1 \leq j \leq k)$ be the representation of $\mathfrak{A}^*$ defined by ρ_j corresponding to (3). Let $f_j(t) \neq 0$ be the polynomial of the least possible degree in a variable t such that

$$\pi_j^*(f_j(x_1^*)) = f_j(\rho_j(X_1)) = 0.$$

Put

$$F(t) = \prod_{j=1}^{k} f_j(t).$$

Clearly $F(t) \neq 0$ and $F(x_1)$ is a canonical expression. Hence $F(x_1^*) \neq 0$. Therefore from Theorem 1 there exists a representation π^* of $\mathfrak{A}^*$ such that $\pi^*(F(x_1^*)) \neq 0$. π^* determines a representation ρ of $\mathfrak{L}$ by the relation $\pi^*((\omega(x))^*) = \rho(X)$. Therefore $F(\rho(X_1)) \neq 0$. This implies that ρ cannot be decomposed into a direct sum of representations each of which is equivalent to some ρ_j $(1 \leq j \leq k)$ for if this were possible we would have $F(\rho(X_1)) = 0$. Hence in the complete decomposition of ρ into indecomposable representations there must occur at least one representation which is not equivalent to any ρ_j $(1 \leqq j \leqq k)$. Since $\rho_1, \cdots, \rho_k$ was any arbitrary finite set of inequivalent indecomposable representations, the corollary is proved.

COROLLARY 1.4. *Every Lie algebra $\mathfrak{L}$ over K has infinitely many inequivalent irreducible representations.*

If $\mathfrak{L}$ is semisimple every representation of $\mathfrak{L}$ is fully reducible (Whitehead (6), Hochschild (3)) and so every indecomposable representation is irreducible. Therefore in this case the assertion follows from Corollary 1.3.

In the general case let Γ be the maximal solvable ideal of $\mathfrak{L}$. Then $\mathfrak{L}/\Gamma$ is semisimple and therefore if it is not zero it has infinitely many inequivalent irreducible representations. Since every representation of $\mathfrak{L}/\Gamma$ defines in the obvious way a representation of $\mathfrak{L}$, the same statement is true for $\mathfrak{L}$. On the other hand if $\mathfrak{L}/\Gamma = \{0\}$, $\mathfrak{L}$ is solvable and we can find an ideal $\mathfrak{M}$ in $\mathfrak{L}$ such that $\mathfrak{L}/\mathfrak{M}$ is one-dimensional. Since every representation of $\mathfrak{L}/\mathfrak{M}$ also defines a representation of $\mathfrak{L}$ is is sufficient to show that a one-dimensional Lie algebra has infinitely many inequivalent irreducible representations. But this is obvious since if $X \neq 0$ is any element of such an algebra the representation ρ_a defined by $\rho_a(X) = a$, $a \, \epsilon \, K$ is irreducible and clearly ρ_a and ρ_b are inequivalent if $a \neq b$. Since K is of characteristic zero it contains infinitely many different elements and therefore the desired result follows.

We now wish to compare Theorem 1 with Ado's Theorem which states that every Lie algebra $\mathfrak{L}$ over a field K of characteristic zero has a faithful representation. We keep to the above notation and first show that Ado's Theorem is equivalent to the following proposition.

PROPOSITION 1. *Let $\mathfrak{A}^*$ be the general enveloping algebra of $\mathfrak{L}$. No nonzero homogeneous linear expression has the value zero in $\mathfrak{A}^*$. Given any nonzero homogeneous linear expression with the value z^* in $\mathfrak{A}^*$ there exists a representation π^* of $\mathfrak{A}^*$ such that $\pi^*(z^*) \neq 0$.*

In the view of the 1-1 correspondence between representations of $\mathfrak{A}^*$ and $\mathfrak{L}$ it is clear that the existence of a faithful representation of $\mathfrak{L}$ implies the above proposition. We have therefore only to prove the converse. Let ρ be any representation of $\mathfrak{L}$ with the kernel $\mathfrak{M}$. Such a representation always exists for we can take the zero representation. Let $\dim \mathfrak{M} = r$ so that $0 \leqq r \leqq n$, where $n = \dim \mathfrak{L}$. If $r = 0$, ρ is faithful and so there is nothing to prove. Hence we can suppose $r \geqq 1$. Choose any $X \, \epsilon \, \mathfrak{M}$, $X \neq 0$. $\omega(X)$ is a nonzero homogeneous linear expression and therefore $(\omega(X))^* \neq 0$. Hence we can find a representation π^* of $\mathfrak{A}^*$ such that $\pi^*((\omega(X))^*) \neq 0$. As mentioned before, π^* defines a representation ρ' of $\mathfrak{L}$ such that

$$\rho'(X) = \pi^*((\omega(X))^*) \neq 0.$$

Let $\mathfrak{M}'$ be the kernel of ρ' and consider the direct sum ρ_1 of the representations ρ and ρ'. Clearly the kernel of ρ_1 is $\mathfrak{M} \cap \mathfrak{M}'$. Since $X \, \epsilon \, \mathfrak{M}'$

$$\dim \mathfrak{M} \cap \mathfrak{M}' \leqq \dim \mathfrak{M} - 1 \leqq r - 1.$$

If ρ_1 is not faithful we can proceed the same way with ρ_1 instead of ρ. Thus finally after at most r steps we arrive at a faithful representation of $\mathfrak{L}$.

Since every linear expression is canonical it is evident that Theorem 1 implies Proposition 1 and therefore also the existence of a faithful representation of $\mathfrak{L}$. Hence Theorem 1 can be regarded as a generalization of Ado's Theorem.

5. Irreducible representations of a Lie algebra

First we give a simple proof of the following theorem.

THEOREM 2. *Any semisimple Lie algebra $\mathfrak{L}$ over K has only a finite number of inequivalent representations whose degree is $\leq r$.*

Let $\mathfrak{A}^*$ be the general enveloping algebra of $\mathfrak{L}$. Clearly the mapping ω^* defined by $\omega^*(X) = (\omega(X))^*$ $(X \in \mathfrak{L})$ maps $\mathfrak{L}$ isomorphically onto $\omega^*(\mathfrak{L})$ which is a Lie algebra under the bracket operation $[z^*, w^*] = z^*w^* - w^*z^*$ defined in $\mathfrak{A}^*$. Hence we can identify $\mathfrak{L}$ with $\omega^*(\mathfrak{L})$ and therefore regard $\mathfrak{L}$ as imbedded in $\mathfrak{A}^*$. Also it is now convenient to drop the star in denoting elements of $\mathfrak{A}^*$.

First suppose K is algebraically closed. Let $\mathfrak{H}$ be a Cartan subalgebra of $\mathfrak{L}$ and let $\alpha, \beta, \gamma, \cdots$ be the roots.[6] Then, as is well known

$$\mathfrak{L} = \mathfrak{H} + \sum_{\alpha} K\, x_\alpha$$

where x_α ($\neq 0$) is an element of $\mathfrak{L}$ belonging to the root α and the sum is over all the roots. If α is a root $-\alpha$ is also a root and the following relations hold

(15a) $$[h, h'] = 0, \qquad [h, x_\alpha] = \alpha(h)x_\alpha$$

(15b) $$[x_\alpha, x_\beta] = \begin{cases} c_{\alpha,\beta} x_{\alpha+\beta} & \text{if} \quad \alpha + \beta \text{ is a root } (c_{\alpha,\beta} \in k) \\[4pt] 0 & \text{if} \quad \alpha + \beta \text{ is neither a root nor zero} \\[4pt] h_\alpha & \text{if} \quad \beta = -\alpha \end{cases}$$

where $h, h', h_\alpha \in \mathfrak{H}$. Since $\mathfrak{L}$ is its own derived algebra it is clear that h_α taken together for all α span the whole of $\mathfrak{H}$.

For any given r, let $\mathfrak{X}$ be the ideal in $\mathfrak{A}^*$ generated by the elements x_α^r, $\Pi_{-r+1 \leq s \leq r-1}(h_\alpha - (s/2)\alpha(h_\alpha))$. Let $\mathfrak{B} = \mathfrak{A}^*/\mathfrak{X}$ and π the natural homomorphism of $\mathfrak{A}^*$ on $\mathfrak{B}$. First we prove that $\mathfrak{B}$ is finite dimensional. Let $\dim \mathfrak{H} = l$. Choose l linearly independent elements $h_1, \cdots, h_l$ among h_α and let $x_1, \cdots, x_n$ be a base for $\mathfrak{L}$ such that its first l elements are $h_1, \cdots, h_l$ and the last $n - l$ elements are x_α for the various roots α. Since $\mathfrak{A}^* = \mathfrak{A}/\mathfrak{U}$ and $\mathfrak{U}$ is a Lie ideal, it follows from Lemma 1 that every element in $\mathfrak{A}$ is congruent to a normal expression mod $\mathfrak{U}$. Hence any element in $\mathfrak{A}^*$ can be written as a linear combination of $x_1^{m_1} x_2^{m_2} \cdots x_n^{m_n}$, $m_1, \cdots, m_n \geq 0$ where $x_i^{m_i}$ is to be interpreted as 1 in case $m_i = 0$. But it is clear from the definition of $\mathfrak{X}$ that $x_i^{m_i}$ is linearly dependent mod $\mathfrak{X}$ on $x_i^{p_i}$ $0 \leq p_i \leq 2r - 1$. Hence every element of $\mathfrak{A}^*$ is linearly dependent mod $\mathfrak{X}$ on $x_1^{p_1}, x_2^{p_2} \cdots x_n^{p_n}$, $0 \leq p_i \leq 2r - 1$. This proves our assertion.

---

[6] The roots are linear functions defined on $\mathfrak{H}$. See for example Weyl (4).

It is obvious that every representation of $\mathfrak{B}$ is also a representation of $\mathfrak{A}^*$ and therefore of $\mathfrak{L}$. Since $\mathfrak{L}$ is semisimple every such representation is completely reducible. Hence $\mathfrak{B}$ is a semisimple associative algebra.

We shall now prove that every representation of $\mathfrak{A}^*$ of degree $\leq r$ is actually a representation of $\mathfrak{B}$. Let π be such a representation. Put $\pi(x_\alpha) = X_\alpha$ and $\pi(h_\alpha) = H_\alpha$ and let $\mathfrak{R}$ denote the representation space. As usual, a linear function $\Lambda(h)$ defined on $\mathfrak{H}$ will be called a weight of the representation π if there exists a nonzero vector $\psi \, \epsilon \, \mathfrak{R}$ such that $\pi(h)\psi = \Lambda(h)\psi$ for all $h \, \epsilon \, \mathfrak{H}$. The vector ψ is then said to belong to the weight Λ. A nonzero vector will be called a weight vector if it belongs to some weight. It is easily seen that weight vectors belonging to different weights are linearly independent. Therefore since dim $\mathfrak{R} \leq r$ there cannot be more than r distinct weights. Further it is known that the weight vectors span the whole space $\mathfrak{R}$. Therefore we can choose a set of linearly independent weight vectors $\psi_1, \cdots, \psi_p$ which form a base for $\mathfrak{R}$. The matrix for $\pi(h)$ referred to this base, is diagonal for every $h \, \epsilon \, \mathfrak{H}$. Further it follows from (15a) that if ψ belongs to the weight Λ, $X_\alpha\psi$ would belong to the weight $\Lambda + \alpha$ unless $X_\alpha\psi = 0$. Suppose none of the vectors $\psi, X_\alpha\psi, X_\alpha^2\psi, \cdots, X_\alpha^k\psi$ is zero. Then they belong respectively to the weights $\Lambda, \Lambda + \alpha, \cdots, \Lambda + k\alpha$ which are all distinct since $\alpha \neq 0$. Hence $k + 1 \leq r$. Therefore $X_\alpha^k\psi = 0$ for some $k \leq r$ and so $X_\alpha^r\psi = 0$. Since we can choose ψ to be any one of the basic vectors $\psi_1, \cdots, \psi_p$ it follows that $X_\alpha^r = 0$. Also it is known (see Weyl (4)) that corresponding to any weight Λ we can find two integers $k, k' \geq 0$ such that $\Lambda + \nu\alpha$ is a weight for all integers $- k \leq \nu \leq k'$ but not for $\nu = k' + 1$ or $\nu = -k - 1$, and $\Lambda(h_\alpha) = \dfrac{k - k'}{2}\, \alpha(h_\alpha)$. Again since $\alpha \neq 0$ the $k + k' + 1$ weights $\Lambda + \nu\alpha, \, -k \leq \nu \leq k'$ are all distinct and therefore $k + k' + 1 \leq r$. Hence $|\,(k - k')/2\,| \leq (k + k')/2 \leq (r - 1)/2$. Hence $\Lambda(h_\alpha) = (s/2)\alpha(h_\alpha)$ where $-(r - 1) \leq s \leq r - 1$. In particular this holds for the weights $\Lambda_1, \cdots, \Lambda_p$ to which $\psi_1, \cdots, \psi_p$ belong. Hence from the diagonal form of the matrix H_α it is clear that

$$\prod_{-1+r \leq s \leq r-1} \left(H_\alpha - \frac{s}{2}\, \alpha(h_\alpha) \right) = 0.$$

This proves that x_α^r and $\Pi_{-1+r \leq s \leq r-1}\,(h_\alpha - (s/2)\alpha(h_\alpha))$ belong to the kernel of π. Therefore the kernel of π contains $\mathfrak{X}$ and therefore π can actually be regarded as a representation of $\mathfrak{B}$.

Since $\mathfrak{B}$ is a finite dimensional semisimple associative algebra it has only a finite number of inequivalent irreducible representations. Since we have shown that every representation of $\mathfrak{A}^*$ of degree $\leq r$ is actually a representation of $\mathfrak{B}$, the required result follows.

If K is not algebraically closed let $\bar{K}$ be its algebraic closure. Extend $\mathfrak{L}$ to a Lie algebra $\bar{\mathfrak{L}}$ over $\bar{K}$. $\bar{\mathfrak{L}}$ is semisimple and so the above discussion holds for $\bar{\mathfrak{L}}$. Clearly every representation ρ of $\mathfrak{L}$ over K can be extended to a representation $\bar{\rho}$ of $\bar{\mathfrak{L}}$ over $\bar{K}$. Also, two representations ρ and ρ' are equivalent if and only if

$\bar{\rho}$ and $\bar{\rho}'$ are equivalent. Therefore since $\bar{\mathfrak{L}}$ has only a finite number of inequivalent representations of degree $\leq r$ it is clear that the same is true of $\mathfrak{L}$. The theorem is thus proved completely.

COROLLARY 2.1. *Any semisimple algebra has only a countable number of inequivalent irreducible representations.*

COROLLARY 2.2. *Every semisimple algebra has irreducible representations of arbitrarily high degree.*

These assertions follow immediately from Theorem 2 and Corollary 1.4.

COROLLARY 2.3. *Let $\mathfrak{L}$ be a Lie algebra over K. If there exists an integer r such that the degree of every irreducible representation of $\mathfrak{L}$ is $\leq r$, then $\mathfrak{L}$ is solvable.*

For otherwise suppose $\mathfrak{L}$ is not solvable so that $\mathfrak{L} \neq \Gamma$ where Γ is the maximal solvable ideal of $\mathfrak{L}$. Then $\mathfrak{L}/\Gamma$ is semisimple and therefore it has irreducible representations of arbitrarily high degree. But every representation of $\mathfrak{L}/\Gamma$ can also be regarded as a representation of $\mathfrak{L}$ and therefore this contradicts the hypothesis that every irreducible representation of $\mathfrak{L}$ is of degree $\leq r$.

Let $\mathfrak{A}^*$ be the general enveloping algebra of a Lie algebra $\mathfrak{L}$ over K. We regard $\mathfrak{L}$ as imbedded in $\mathfrak{A}^*$. Let $x_1, \cdots, x_n$ be a base for $\mathfrak{L}$. It is clear that the center $\mathfrak{C}$ of $\mathfrak{A}^*$ consists precisely of those elements $z \in \mathfrak{A}^*$ which commute with x_i, $1 \leq i \leq n$. An ideal $\mathfrak{X}$ in $\mathfrak{A}^*$ will be called admissible if $\mathfrak{A}^*/\mathfrak{X}$ is a finite dimensional algebra. The intersection of a finite number of admissible ideals is again an admissible ideal. For let $\mathfrak{X}_\alpha, 1 \leq \alpha \leq r$ be a set of admissible ideals and let $\mathfrak{X} = \bigcap_{1 \leq \alpha \leq r} \mathfrak{X}_\alpha$. Since $\mathfrak{X}_\alpha$ is admissible there exists a polynomial $f_{\alpha,i}(t) \neq 0$ such that $f_{\alpha,i}(x_i) \in \mathfrak{X}_\alpha$. Therefore $F_i(x_i) = \Pi_{\alpha=1}^{r} f_{\alpha,i}(x_i) \in \mathfrak{X}$. By the same argument as the one used in the course of the proof of Theorem 2 it is now easy to prove that $\mathfrak{X}$ is admissible.

We now wish to prove the following theorem.

THEOREM 3. *Let $\mathfrak{A}^*$ be the general enveloping algebra of a semisimple algebra $\mathfrak{L}$ over K. Let $\mathfrak{C}$ be the centre of $\mathfrak{A}^*$ and let $\mathfrak{X}$ and $\mathfrak{Y}$ be any two admissible ideals in $\mathfrak{A}^*$. Then $\mathfrak{Y} \supset \mathfrak{X}$ if and only if $\mathfrak{Y} \cap \mathfrak{C} \supset \mathfrak{X} \cap \mathfrak{C}$.*

For any representation π of $\mathfrak{A}^*$ put $s_\pi(z) = \mathrm{sp}\,(\pi(z))$ $(z \in \mathfrak{A}^*)$. First we prove the following lemma.

LEMMA 3. *Let π, π' be two irreducible representations of $\mathfrak{A}^*$. If there exist two elements $m, m' \in K$ not both zero such that*

$$ms_\pi(z) = m's_{\pi'}(z)$$

for all $z \in \mathfrak{A}^*$ then π and π' are equivalent.[7]

Let π_0 be the direct sum of π and π'. Let $\mathfrak{P}_0$, $\mathfrak{P}$ and $\mathfrak{P}'$ be the kernels of π_0, π and π' respectively so that $\mathfrak{P}_0 = \mathfrak{P} \cap \mathfrak{P}'$. Put $\mathfrak{B}_0 = \mathfrak{A}^*/\mathfrak{P}_0$, $\mathfrak{B} = \mathfrak{A}^*/\mathfrak{P}$ and $\mathfrak{B}' = \mathfrak{A}^*/\mathfrak{P}'$. Since π and π' are irreducible representations $\mathfrak{B}$ and $\mathfrak{B}'$ are simple algebras, while $\mathfrak{B}_0$ is semisimple. Clearly $\mathfrak{B}$ and $\mathfrak{B}'$ can be considered as factor algebras of $\mathfrak{B}_0$. Decompose $\mathfrak{B}_0$ into a direct sum of simple algebras. If π and π' are inequivalent $\mathfrak{B}$ and $\mathfrak{B}'$ would respectively be isomorphic to two

---

[7] This lemma follows directly from the Schur-Frobenius generalization of Burnside's Theorem. We give here a proof only for completeness.

distinct simple algebras $\mathfrak{B}_1$ and $\mathfrak{B}_2$ occurring in this decomposition.   Let e_1 and e_2 be the unity quantities of $\mathfrak{B}_1$ and $\mathfrak{B}_2$ respectively, and let z_1, $z_2 \in \mathfrak{A}^*$ be any two elements which are mapped on e_1 and e_2 respectively by the natural homomorphism of $\mathfrak{A}^*$ on $\mathfrak{B}_0$.   Then since $e_1 e_2 = 0$ it follows that $\pi(z_1) = \pi(1)$, $\pi'(z_2) = 0$ and $\pi(z_1) = 0$, $\pi'(z_2) = \pi'(1)$.   Suppose $m \neq 0$ and let r be the degree of the representation π.   Then $s_\pi(z_1) = r$ while $s_{\pi'}(z_1) = 0$.   This contradicts the hypothesis that $m s_\pi(z) = m' s_{\pi'}(z)$ for every $z \in \mathfrak{A}^*$.   Hence π and π' are equivalent.

Now let $B(x, y)$ be the fundamental bilinear form of $\mathfrak{L}$.[8]   Put $g_{ik} = B(x_i, x_k)$. Since $B(x, y)$ is non-degenerate there exists a unique solution g^{ik} of the equations

$$\sum_k g^{ik} g_{kj} = \sum_k g_{jk} g^{ki} = \delta_j^i \qquad\qquad 1 \leqq i, j \leqq n$$

where $\delta_j^i = 1$ or 0 according as $i = j$ or $i \neq j$.   From now on we use the summation convention for contracted indices and raise and lower the indices by means of g^{ij} and g_{ij} respectively.

Define $\epsilon_{ij}{}^k \in K$ by the relations

$$[x_i, x_j] = \epsilon_{ij}{}^k x_k.$$

It is well known that ϵ_{ijk} is antisymmetric with respect to every pair of indices. Consider

$$[x_j, s_\pi(x_{i_1} \cdots x_{i_r}) x^{i_1} x^{i_2} \cdots x^{i_r}] = \sum_{1 \leqq t \leqq r} s_\pi(x_{i_1} \cdots x_{i_r}) x^{i_1} \cdots [x_j, x^{i_t}] \cdots x^{i_r}$$

$$= \sum_{1 \leqq t \leqq r} s_\pi(x_{i_1} \cdots x_{i_r}) x^{i_1} \cdots \epsilon_j{}^{i_t}{}_k x^k \cdots x^{i_r}$$

$$= \sum_{1 \leqq t \leqq r} s_\pi(x_{i_1} \cdots x_k \epsilon_j{}^k{}_{i_t} \cdots x_{i_r}) x^{i_1} \cdots x^{i_t} \cdots x^{i_r}$$

by interchanging the dummy indices k and i_t.   Therefore it is

$$= - \sum_{1 \leqq t \leqq r} s_\pi(x_{i_1} \cdots [x_j, x_{i_t}] \cdots x_{i_r}) x^{i_1} \cdots x^{i_r}$$

$$= - s_\pi([x_j, x_{i_1} \cdots x_{i_r}]) x^{i_1} \cdots x^{i_r} = 0.$$

Hence $s_\pi(x_{i_1} \cdots x_{i_r}) x^{i_1} \cdots x^{i_r} \in \mathfrak{C}$.   In exactly the same way we prove that $s_\pi(x_{i_{\theta_1}}, x_{i_{\theta_2}} \cdots x_{i_{\theta_r}}) x^{i_1} x^{i_2} \cdots x^{i_r} \in \mathfrak{C}$ where $\theta_1, \cdots, \theta_r$ is any permutation of $1, 2, \cdots, r$.

Now we need the following lemma.

LEMMA 4.   *Let K be algebraically closed and let π_α $(1 \leqq \alpha \leqq p)$ be any p inequivalent irreducible representations of the general enveloping algebra $\mathfrak{A}^*$ of a semisimple algebra $\mathfrak{L}$ over K.   Then there exists an element $z \in \mathfrak{C}$ such that $\pi_\alpha(z) = 0$, $\alpha \neq 1$ and $\pi_1(z) = \pi_1(1)$.*

For $p = 1$ the assertion is clearly true since we can take $z = 1$.   Hence we can

---

[8] If adx $(x \in \mathfrak{L})$ be the linear mapping of $\mathfrak{L}$ into itself defined by $(adx)y = [x, y]$, $y \in \mathfrak{L}$, then $B(x, y) = \mathrm{sp}(adx\, ady)$ for $x, y \in L$.

use induction on p. Let m_α be the degree of π_α. We can find a set of indices $i_1, \cdots, i_r$ such that

$$(16) \qquad \frac{1}{m_1} s_{\pi_1}\left(\sum x_{i_1} \cdots x_{i_r}\right) \neq \frac{1}{m_p} s_{\pi_p}\left(\sum x_{i_1} \cdots x_{i_r}\right)$$

where $\sum$ denotes summation over all possible permutations of $i_1, \cdots, i_r$. For otherwise, since every element of $\mathfrak{A}^*$ has a canonical expression, we would have

$$\frac{1}{m_1} s_{\pi_1}(x_{j_1} x_{j_2} \cdots x_{j_r}) = \frac{1}{m_p} s_{\pi_p}(x_{j_1} x_{j_2} \cdots x_{j_r})$$

for all $1 \leq j_1, \cdots, j_r \leq n, r \geq 0$. But since π_1 and π_p are inequivalent this is impossible from Lemma 3. From (16) it is clear that

$$\frac{1}{r!}\left\{\frac{1}{m_1} s_{\pi_1}\left(\sum x_{i_1} \cdots x_{i_r}\right) - \frac{1}{m_p} s_{\pi_p}\left(\sum x_{i_1} \cdots x_{i_r}\right)\right\} \sum x^{i_1} \cdots x^{i_r} \neq 0$$

since the values of different basic canonical expressions are linearly in‑dependent in $\mathfrak{A}^*$. Put $v_1 = (1/m_1)s_{\pi_1}(\sum x_{i_1} \cdots x_{i_r})x^{i_1} \cdots x^{i_r}$ and $v_p = (1/m_p)s_{\pi_p}(\sum x_{i_1} \cdots x_{i_r})x^{i_1} \cdots x^{i_r}$. Since $v_1 - v_p \neq 0$ it follows from Theorem 1 that there exists a representation π such that $\pi(v_1 - v_p) \neq 0$. Since $\mathfrak{L}$ is semisimple every representation is fully reducible and therefore we can choose π so that it is irreducible. Since $v_1, v_p \in \mathfrak{C}$, $\pi(v_1 - v_p)$ commutes with every element in $\pi(\mathfrak{A}^*)$. Hence by Schur's lemma $\pi(v_1 - v_p) = c\pi(1)$ where $c \in K$. Since $\pi(v_1 - v_p) \neq 0$, $c \neq 0$ and therefore $s_\pi(v_1 - v_p) \neq 0$. Put $v = s_\pi(\sum x_{i_1} \cdots x_{i_r})x^{i_1} \cdots x^{i_r}$. Then $s_{\pi_1}(v) = m_1 s_\pi(v_1)$ and $s_{\pi_p}(v) = m_p s_\pi(v_p)$. Since $v \in \mathfrak{C}$ we deduce by a similar argument that $\pi_1(v) = c_1\pi_1(1)$ $(c_1 \in K)$. Therefore

$$s_{\pi_1}(v) = m_1 s_\pi(v_1) = m_1 c_1.$$

Hence $\pi_1(v) = s_\pi(v_1)\pi_1(1)$. Similarly $\pi_p(v) = s_\pi(v_p)\pi_p(1)$. Therefore if we put

$$z_1 = \frac{1}{s_\pi(v_1 - v_p)} (v - s_\pi(v_p))$$

$\pi_1(z_1) = \pi_1(1)$, $\pi_p(z_1) = 0$. On the other hand by induction hypothesis it is possible to find an element $z' \in \mathfrak{C}$ such that $\pi_1(z') = \pi_1(1)$ and $\pi_\alpha(z') = 0$, $z \leq \alpha \leq p - 1$. Put $z = z_1 z'$. Then z has the required properties.

We are now in a position to prove the theorem. Let π be the natural homo-morphism of $\mathfrak{A}$ on $\mathfrak{B} = \mathfrak{A}^*/\mathfrak{X} \cap \mathfrak{Y}$. If $\mathfrak{X}$ is not contained in $\mathfrak{Y}$ there exists an element $z \in \mathfrak{X}$ such that $z \notin \mathfrak{X} \cap \mathfrak{Y}$ so that $\pi(\mathfrak{X}) \neq 0$. Since $\mathfrak{B}$ is semisimple it can be de-composed into a direct sum of simple algebras $\mathfrak{B}_1, \cdots, \mathfrak{B}_p$ and every ideal in $\mathfrak{B}$ is a sum of $\mathfrak{B}_i$ where i runs through some subset of $1, 2, \cdots, p$. Hence some $\mathfrak{B}_i$, say $\mathfrak{B}_1$, is contained in $\pi(\mathfrak{X})$. Let us first suppose that K is algebraically closed. Then, on considering the regular representation of $\mathfrak{B}$ it follows from Lemma 4 that there exists an element $z \in \mathfrak{C}$ such that $\pi(z)$ is the unity quantity of $\mathfrak{B}_1$. Hence $\pi(z) \in \pi(\mathfrak{X})$ and $\pi(z) \neq 0$. Therefore $z \in \mathfrak{C} \cap \mathfrak{X}$ and $z \notin \mathfrak{C} \cap \mathfrak{X} \cap \mathfrak{Y}$.

Therefore $\mathfrak{C} \cap \mathfrak{Y}$ does not contain $\mathfrak{C} \cap \mathfrak{X}$. Since $\mathfrak{Y} \supset \mathfrak{X}$ implies $\mathfrak{C} \cap \mathfrak{Y} \supset \mathfrak{C} \cap \mathfrak{X}$, the theorem is proved in this case.

If K is not algebraically closed let $\bar{K}$ be the algebraic closure of K. Let $\bar{\mathfrak{L}}$ and $\overline{\mathfrak{A}^*}$ be the extensions of $\mathfrak{L}$ and $\mathfrak{A}^*$ respectively over $\bar{K}$. Then $\bar{\mathfrak{L}}$ is semi-simple and $\overline{\mathfrak{A}^*}$ is the general enveloping algebra of $\bar{\mathfrak{L}}$. For any linear set $\mathfrak{B}$ in $\mathfrak{A}^*$ let us denote by $\bar{\mathfrak{B}}$ the smallest possible linear set in $\overline{\mathfrak{A}^*}$ containing $\mathfrak{B}$. It is easily seen that if $\mathfrak{B}_1$, $\mathfrak{B}_2$ are any two linear sets in $\mathfrak{A}^*$ then $\overline{\mathfrak{B}_1 \cap \mathfrak{B}_2} = \bar{\mathfrak{B}}_1 \cap \bar{\mathfrak{B}}_2$. Let $\mathfrak{C}'$ be the center of $\overline{\mathfrak{A}^*}$. We shall first show that $\mathfrak{C}' = \bar{\mathfrak{C}}$. Let $\mathfrak{C}_s$ and $\mathfrak{C}_s'$ be the set of all elements of degree $\leq s$ in $\mathfrak{C}$ and $\mathfrak{C}'$ respectively. Consider

$$ z = \sum_{m_1 + \cdots + m_n \leq s} a(m_1, \cdots, m_n) z(m_1, \cdots, m_n), \qquad a(m_1, \cdots, m_n) \, \epsilon \, \bar{K} $$

where $z(m_1, \cdots, m_n)$ is the value in $\mathfrak{A}^*$ of the corresponding basic canonical expression. $z \, \epsilon \, \mathfrak{C}'$ if and only if $[z, x_i] = 0 \; 1 \leq i \leq n$. But these conditions are equivalent to certain linear homogeneous equations with coefficients in K, to be satisfied by $a(m_1, \cdots, m_n) \; (m_1, \cdots, m_n \geq 0, \, m_1 + \cdots + m_n \leq s)$. Therefore it follows immediately that $\mathfrak{C}_s' = \bar{\mathfrak{C}}_s$. Hence

$$ \mathfrak{C}' = \bigcup_{s \geq 1} \mathfrak{C}_s' = \bigcup_{s \geq 1} \bar{\mathfrak{C}}_s = \overline{\bigcup_{s \geq 1} \mathfrak{C}_s} = \bar{\mathfrak{C}}. $$

Now $\mathfrak{Y} \cap \mathfrak{C} \supset \mathfrak{X} \cap \mathfrak{C}$ implies $\overline{\mathfrak{Y} \cap \mathfrak{C}} \supset \overline{\mathfrak{X} \cap \mathfrak{C}}$ or $\bar{\mathfrak{Y}} \cap \bar{\mathfrak{C}} \supset \bar{\mathfrak{X}} \cap \bar{\mathfrak{C}}$. In view of the proof given above, it then follows that $\bar{\mathfrak{Y}} \supset \bar{\mathfrak{X}}$ and therefore $\bar{\mathfrak{Y}} \cap \mathfrak{A}^* \supset \bar{\mathfrak{X}} \cap \mathfrak{A}^*$ i.e. $\mathfrak{Y} \supset \mathfrak{X}$. Hence $\mathfrak{Y} \cap \mathfrak{C} \supset \mathfrak{X} \cap \mathfrak{C}$ implies $\mathfrak{Y} \supset \mathfrak{X}$ and so the theorem is proved.

6. Example of a finitely generated algebra which has no nontrivial representation

Let $\mathfrak{L}$ be a semisimple algebra over a field K of characteristic zero. Let $\mathfrak{A}^*$ be the general enveloping algebra of $\mathfrak{L}$. Let π be any representation of $\mathfrak{L}$. Then it is known (Whitehead (5) p. 235) that $s_\pi(x_i x^i)$ is a non-negative rational number. Let μ be any element of K which is not a rational number ≥ 0. Consider the smallest possible ideal $\mathfrak{X}$ in $\mathfrak{A}^*$ containing $x_i x^i - \mu$. We assert that $\mathfrak{B} = \mathfrak{A}^*/\mathfrak{X}$ is a finitely generated algebra which has no non-trivial representations. Since $\mathfrak{B}$ is a factor algebra of $\mathfrak{A}^*$ it is obvious that it has a finite number of generators. We first show that $\mathfrak{X} \neq \mathfrak{A}^*$ so that $\mathfrak{B} \neq \{0\}$. Since $g^{ij} = g^{ji}$ it is clear that $x_i x^i = g^{ij} x_i x_j$ is an element of degree 2. Further it is easy to verify that $[x_i x^i, x_j] = 0$ so that $x_i x^i \, \epsilon \, \mathfrak{C}$ and $\mathfrak{X} = (x_i x^i - \mu)\mathfrak{A}^*$. Hence it follows from Corollary 1.2 that the degree of every element of $\mathfrak{X}$ is ≥ 2. Hence $1 \notin \mathfrak{X}$ and therefore $\mathfrak{X} \neq \mathfrak{A}^*$. Now we prove that $\mathfrak{B}$ cannot have any non-trivial representations. For otherwise let π_0 be such a representation. Put $\pi = \pi_0 \rho$ where ρ is the natural homomorphism of $\mathfrak{A}^*$ on $\mathfrak{B}$. Then since $x_i x^i - \mu \, \epsilon \, \mathfrak{X}$, $\pi(x_i x^i) = \mu \pi(1)$. Hence $s_\pi(x_i x^i) = r\mu$ where r is the degree of the representation. But this is impossible since $r\mu$ is not a rational number ≥ 0. The same argument is applicable if we consider representations of $\mathfrak{B}$ over a field K' which is an extension of K. Hence $\mathfrak{B}$ has no non-trivial representation over any field whatsoever.

REFERENCES

1. G. BIRKHOFF, *Representability of Lie algebras and Lie groups by matrices*, Ann. of Math. 38 (1937), 526–532.
2. HARISH-CHANDRA, *Faithful representations of Lie algebras*, Ann. of Math. 50 (1949), 68–76.
3. G. HOCHSCHILD, *Semi-simple algebras and generalized derivations*, Amer. J. Math. 64 (1942), 677–694.
4. H. WEYL, *Theorie der Darstellung Kontinuierlicher halb-einfacher, Gruppen durch lineare Transformationen*, Math. Zeit. 24 (1925), 328–395.
5. J. H. C. WHITEHEAD, *On the decomposition of an infinitesimal group*, Proc. Cambridge Philos. Soc. 32 (1936) 229–237.
6. J. H. C. WHITEHEAD, *Certain equations in the algebra of a semi-simple infinitesimal group*, Quart. J. Math. Oxford Ser. 8 (1937), 220–237.
7. E. WITT, *Treue Darstellung Liescher Ringe*, J. Reine Angew. Math. 177 (1937), 152–160.

INSTITUTE FOR ADVANCED STUDY

ON THE RADICAL OF A LIE ALGEBRA

HARISH-CHANDRA

Let $\mathfrak{L}$ be a Lie algebra over a field K of characteristic zero. For any $X \in \mathfrak{L}$ we denote, as usual, the linear mapping $Y \to [X, Y]$ of $\mathfrak{L}$ into itself by ad X. Let Γ be the radical of $\mathfrak{L}$. Consider the set $\mathfrak{N}$ consisting of all $N \in \Gamma$ such that ad N is nilpotent. It was shown in a recent paper[1] that $\mathfrak{N}$ is the unique maximal nilpotent ideal[2] of $\mathfrak{L}$. Further if D is a derivation of Γ then $D\Gamma \subset \mathfrak{N}$.

For any X, Y, $Z \in \mathfrak{L}$ put $B(X, Y) = sp(\text{ad } X \text{ ad } Y)$ and $T(X, Y, Z) = sp(\text{ad } [X, Y] \text{ ad } Z)$. Then $B(X, Y)$ is a symmetric bilinear form on $\mathfrak{L}$ while $T(X, Y, Z)$ is a skewsymmetric trilinear form. It is easily verified that they are both invariant under all derivations of $\mathfrak{L}$, that is,

$$B(DX, Y) + B(X, DY) = 0,$$
$$T(DX, Y, Z) + T(X, DY, Z) + T(X, Y, DZ) = 0$$

for any derivation D and X, Y, $Z \in \mathfrak{L}$.

An ideal $\mathfrak{M}$ in $\mathfrak{L}$ is called characteristic if $D\mathfrak{M} \subset \mathfrak{M}$ for every derivation D of $\mathfrak{L}$. Our first theorem may now be stated as follows:

THEOREM 1. *An element X of $\mathfrak{L}$ belongs to the radical Γ if and only if $T(X, Y, Z) = 0$ for all Y, $Z \in \mathfrak{L}$.*[3]

As an immediate corollary we get the following:

Received by the editors October 4, 1948 and, in revised form, November 11, 1948.

[1] Ann. of Math. vol. 50 (1949) p. 68.

[2] My attention has been drawn to a paper by Malcev (Bull. Acad. Sci. URSS. vol. 9 (1945) pp. 329–356) where it is shown that $\mathfrak{N}$ is an ideal.

[3] Since $T(X, Y, Z) = -T(Z, Y, X)$ this condition is clearly equivalent to $B(X, Y) = 0$ for all $Y \in \mathfrak{L}' = [\mathfrak{L}, \mathfrak{L}]$. Professor Jacobson has kindly brought it to my notice that this theorem is contained in Cartan's thesis p. 109.

"

COROLLARY 1. *Both Γ and $\mathfrak{N}$ are characteristic ideals in $\mathfrak{L}$.*

For every $N \in \mathfrak{N}$, ad N is a nilpotent derivation of $\mathfrak{L}$. Hence $\sigma_N = \exp$ (ad N) is defined and is an automorphism of $\mathfrak{L}$. Let $\mathfrak{M}$ be any nilpotent ideal in $\mathfrak{L}$. Then $\mathfrak{M} \subset \mathfrak{N}$. By $G_{\mathfrak{M}}$ we denote the group of all automorphisms of $\mathfrak{L}$ of the form $\sigma_{M_1} \sigma_{M_2} \cdots \sigma_{M_r}$ where $M_1, \cdots, M_r \in \mathfrak{M}$ and $r \geq 1$. Clearly every ideal is invariant under $G_{\mathfrak{M}}$. Our second theorem now runs as follows:

THEOREM 2. *Let $\mathfrak{S}$ be a semisimple subalgebra of $\mathfrak{L}$ such that $\mathfrak{L} = \mathfrak{S} + \Gamma$. Then, given any semisimple subalgebra $\mathfrak{M}$ of $\mathfrak{L}$, there exists a $\sigma \in G_{\mathfrak{N}}$ such that $\mathfrak{M} \subset \sigma \mathfrak{S}$.*

The following two corollaries follow immediately from this theorem.

COROLLARY 2. *Any maximal semisimple subalgebra of $\mathfrak{L}$ is isomorphic to $\mathfrak{L}/\Gamma$.*

COROLLARY 3. *Given any two maximal semisimple subalgebras $\mathfrak{S}_1$, $\mathfrak{S}_2$ of $\mathfrak{L}$, there exists a $\tau \in G_{\mathfrak{N}}$ such that $\tau \mathfrak{S}_1 = \mathfrak{S}_2$.*

Corollary 3 is a sharper form of a result due to Malcev.[4]

PROOF OF THEOREM 1. First we shall prove that for any $N \in \mathfrak{N}$, $B(N, Z) = 0$ for all $Z \in \mathfrak{L}$. For any $s \geq 1$ define $\mathfrak{N}_{(s)}$ by induction as follows. $\mathfrak{N}_{(1)} = \mathfrak{N}$, $\mathfrak{N}_{(s+1)} = [\mathfrak{N}, \mathfrak{N}_{(s)}]$. Then $\mathfrak{N}_{(s)}$ is an ideal in $\mathfrak{L}$ and therefore (ad N ad $Z)\mathfrak{L} \subset \mathfrak{N}$ and (ad N ad $Z)\mathfrak{N}_{(s)} \subset \mathfrak{N}_{(s+1)}$. Hence (ad N ad $Z)^{s+1}\mathfrak{L} = \mathfrak{N}_{(s)}$. But $\mathfrak{N}$ is nilpotent and therefore $\mathfrak{N}_{(s)} = \{0\}$ for some s. Hence (ad N ad $Z)^{s+1}\mathfrak{L} = \{0\}$ or (ad N ad $Z)^{s+1} = 0$. Therefore ad N ad Z is nilpotent and sp(ad N ad $Z) = B(N, Z) = 0$.

Let $\mathfrak{M}$ be the set of all $X \in \mathfrak{L}$ such that $T(X, Y, Z) = 0$ for all $Y, Z \in \mathfrak{L}$. Since T is invariant under all derivations of $\mathfrak{L}$, $\mathfrak{M}$ is a characteristic ideal in $\mathfrak{L}$. We have to show that $\mathfrak{M} = \Gamma$. Let $X \in \Gamma$. For any $Y \in \mathfrak{L}$, ad Y is a derivation of $\mathfrak{L}$ and Γ is invariant under ad Y. Hence ad Y induces a derivation of Γ and therefore (ad $Y)\Gamma \subset \mathfrak{N}$. So $[X, Y] = -($ad $Y)X \in \mathfrak{N}$. Hence $B([X, Y], Z) = T(X, Y, Z) = 0$ for all $Z \in \mathfrak{L}$. Since this is true for every Y, $X \in \mathfrak{M}$. Hence $\Gamma \subset \mathfrak{M}$. On the other hand let $\mathfrak{M}' = [\mathfrak{M}, \mathfrak{M}]$. Then it is clear that for any $M \in \mathfrak{M}'$, $B(M, Z) = 0$ for all $Z \in \mathfrak{L}$. Hence by Cartan's criterion for solvability $\mathfrak{M}'$ is solvable. Hence $\mathfrak{M}$ is solvable and therefore $\mathfrak{M} \subset \Gamma$. So the theorem is proved.

Since $\mathfrak{M}$ is a characteristic ideal the same is true of Γ. Hence if D is a derivation of $\mathfrak{L}$, $D\Gamma \subset \Gamma$ and so D induces a derivation of Γ. Therefore $D\Gamma \subset \mathfrak{N}$ and so $D\mathfrak{N} \subset D\Gamma \subset \mathfrak{N}$. Hence $\mathfrak{N}$ is also a characteristic ideal.

---

[4] A. Malcev, C. R. Acad. Sci. URSS. vol. 36 (1942) p. 42.

Proof of Theorem 2. Since $\mathfrak{S}$ is semisimple $\mathfrak{S} \cap \Gamma = \{0\}$. Hence every $P \in \mathfrak{M}$ can be written uniquely as

$$P = S(P) + \nu(P)$$

where $S(P) \in \mathfrak{S}$ and $\nu(P) \in \mathfrak{N}$. Hence

$$[P, Q] = [S(P), S(Q)] + [S(P), \nu(Q)]$$
$$+ [\nu(P), S(Q)] + [\nu(P), \nu(Q)].$$

Therefore

$$S([P, Q]) = [S(P), S(Q)],$$
$$\nu([P, Q]) = [\nu(P), S(Q)] + [S(P), \nu(Q)] + [\nu(P), \nu(Q)].$$

We have already seen that $(\text{ad } X)\Gamma \subset \mathfrak{N}$ for any $X \in \mathfrak{L}$. Hence, $\nu([P, Q]) \in \mathfrak{N}$. Let ν denote the mapping $P \to \nu(P)$. Then $\nu([\mathfrak{M}, \mathfrak{M}]) \subset \mathfrak{N}$. But $\mathfrak{M}$ is semisimple. Therefore $[\mathfrak{M}, \mathfrak{M}] = \mathfrak{M}$ and $\nu(\mathfrak{M}) \subset \mathfrak{N}$. Hence $\mathfrak{M} \subset \mathfrak{S} + \mathfrak{N}$.

Put $\mathfrak{L}_0 = \mathfrak{S} + \mathfrak{N}$. First suppose that $\mathfrak{N}$ is abelian. The mapping $P \to S(P)$ is a homomorphic mapping of $\mathfrak{M}$ into $\mathfrak{S}$. For any $S \in \mathfrak{S}$ let D_S denote the derivation of $\mathfrak{N}$ given by $D_S N = [S, N]$ $(N \in \mathfrak{N})$. Then the mapping ρ defined by $\rho(P) = D_{S(P)}$ is a representation of $\mathfrak{M}$. Also, since $\mathfrak{N}$ is abelian

$$\nu([P, Q]) = [S(P), \nu(Q)] - [S(Q), \nu(P)]$$
$$= \rho(P)\nu(Q) - \rho(Q)\nu(P).$$

Hence ν is a Whitehead mapping of $\mathfrak{M}$ into $\mathfrak{N}$ with respect to the representation ρ. Therefore by the first Whitehead lemma[5] there exists an element $-N \in \mathfrak{N}$ such that $\nu(P) = -\rho(P)N = [N, S(P)]$. Hence

$$P = S(P) + \nu(P) = S(P) + [N, S(P)] = \exp (\text{ad } N)S(P)$$

since $(\text{ad } N)^2 = 0$, $\mathfrak{N}$ being abelian. Therefore $\mathfrak{M} \subset \sigma_N \mathfrak{S}$ and so the theorem is proved in this case.

Now consider the general case. Let $n = \dim \mathfrak{N}$. If $n \leq 1$, $\mathfrak{N}$ is abelian and so the theorem is true. Hence we can assume $n > 1$ and use induction on n. Further we can assume that $\mathfrak{N}$ is not abelian so that $\mathfrak{N}' = [\mathfrak{N}, \mathfrak{N}] \neq \{0\}$. Let $X \to \overline{X}$ denote the natural homomorphism of $\mathfrak{L}_0$ onto $\overline{\mathfrak{L}}_0 = \mathfrak{L}_0/\mathfrak{N}'$. The radical of $\overline{\mathfrak{L}}_0$ is $\overline{\mathfrak{N}} = \mathfrak{N}/\mathfrak{N}'$ which is abelian. Let $\overline{\mathfrak{M}}$ and $\overline{\mathfrak{S}}$ be the images of $\mathfrak{M}$ and $\mathfrak{S}$ respectively in $\overline{\mathfrak{L}}_0$. Then they are both semisimple and

---

[5] See Hochschild, Amer. J. Math. vol. 64 (1942) p. 677.

$$\overline{\mathfrak{M}} \subset \overline{\mathfrak{S}} + \overline{\mathfrak{N}}.$$

Since $\overline{\mathfrak{N}}$ is abelian it follows from the above proof that there exists an $\overline{N} \in \overline{\mathfrak{N}}$ such that

$$\overline{\mathfrak{M}} \subset \sigma_{\overline{N}} \overline{\mathfrak{S}}.$$

Let $N \in \overline{N}$ $(N \in \mathfrak{N})$. The complete inverse image of $\sigma_{\overline{N}} \overline{\mathfrak{S}}$ in $\mathfrak{L}_0$ is $\mathfrak{S}_1 + \mathfrak{N}'$ where $\mathfrak{S}_1 = \sigma_N \mathfrak{S}$. Hence

$$\mathfrak{M} \subset \mathfrak{S}_1 + \mathfrak{N}'.$$

Since $\mathfrak{N}$ is nilpotent, dim $\mathfrak{N}' <$ dim $\mathfrak{N} = n$. Hence the induction hypothesis is applicable to $\mathfrak{L}_1 = \mathfrak{S}_1 + \mathfrak{N}'$ and therefore there exists a $\sigma_1 \in G_{\mathfrak{N}'}$ such that $\mathfrak{M} \subset \sigma_1 \mathfrak{S}_1$. But $G_{\mathfrak{N}'} \subset G_{\mathfrak{N}}$ and therefore $\sigma = \sigma_1 \sigma_N \in G_{\mathfrak{N}}$ and $\mathfrak{M} \subset \sigma \mathfrak{S}$.

Now let $\mathfrak{S}^*$ be any maximal semisimple algebra of $\mathfrak{L}$. It follows from the above theorem that $\mathfrak{S}^* \subset \sigma \mathfrak{S}$ for some $\sigma \in G_{\mathfrak{N}}$. Since σ is an automorphism of $\mathfrak{L}$, $\sigma \mathfrak{S}$ is semisimple. Therefore $\mathfrak{S}^* = \sigma \mathfrak{S} \cong \mathfrak{S} \cong \mathfrak{L}/\Gamma$.

If $\mathfrak{S}_1$ and $\mathfrak{S}_2$ are two maximal semisimple subalgebras of $\mathfrak{L}$, we can find $\tau_1, \tau_2 \in G_{\mathfrak{N}}$ such that $\mathfrak{S}_1 = \tau_1 \mathfrak{S}$, $\mathfrak{S}_2 = \tau_2 \mathfrak{S}$. Then $\tau \mathfrak{S}_1 = \mathfrak{S}_2$ where $\tau = \tau_2 \tau_1^{-1} \in G_{\mathfrak{N}}$.

Summary. Let $\mathfrak{L}$ be a Lie algebra over a field of characteristic zero and let Γ be its radical. It is proved that any $X \in \mathfrak{L}$ belongs to Γ if and only if $sp(\text{ad } [X, Y] \text{ ad } Z) = 0$ for all $Y, Z \in \mathfrak{L}$. Here $X \to \text{ad } X$ is the adjoint representation of $\mathfrak{L}$. Further let $\mathfrak{N}$ be the maximal nilpotent ideal of $\mathfrak{L}$ and let $\mathfrak{S}$ and $\mathfrak{S}^*$ be any two maximal semisimple subalgebras of $\mathfrak{L}$. Then $\mathfrak{S} + \mathfrak{N} = \mathfrak{S}^* + \mathfrak{N}$ and $\mathfrak{S}$ and $\mathfrak{S}^*$ are conjugate in a certain strict sense.

INSTITUTE FOR ADVANCED STUDY

Reprinted from
Proc. Amer. Math. Soc.
1 (1950), 14–17

ON FAITHFUL REPRESENTATIONS OF LIE GROUPS

HARISH-CHANDRA

Let G and H be two connected Lie groups and ϕ a continuous homomorphism of H into the group of automorphisms of G. Then we define a new group $G\times_\phi H$ as follows. The elements of $G\times_\phi H$ are pairs (g, h) $(g\in G,\ h\in H)$ and group multiplication is defined by

$$(g_1, h_1)(g_2, h_2) = (g_1(\phi(h_1)g_2), h_1h_2).$$

Topologically $G\times_\phi H$ is taken to be just the Cartesian product of G and H. It is then easily proved that under this topology $G\times_\phi H$ is a Lie group. It is called the semidirect product of G and H under ϕ. The object of this note is to prove the following theorem.

THEOREM. *Let G be a connected, simply connected solvable Lie group and H a connected Lie group which has a faithful representation. Let ϕ be any continuous homomorphism of H into the group of automorphisms of G. Then $G\times_\phi H$ has a faithful representation.*

The special case of this theorem when H is semisimple is due to Cartan.[1]

Let R be the field of real numbers and K the field of either real or complex numbers. Let G be a connected Lie group with the Lie algebra $\mathfrak{g}$ and θ a representation of G over K of degree d. Then we denote by $d\theta$ the representation of $\mathfrak{g}$ given by[2]

$$d\theta(x) = \lim_{t\to 0} \frac{\theta(\exp tX) - I}{t} \qquad (X \in \mathfrak{g})$$

where $t\in R$ and I is the unit matrix of degree d. Let $GL(K, d)$ denote the group of all nonsingular matrices of degree d with coefficients in K. Any subgroup of $GL(K, d)$ will be called a linear group of degree d. Let θ be the identity representation of a linear Lie group G with the Lie algebra $\mathfrak{g}$ so that $\theta(x)=x$ $(x\in G)$. Then $d\theta$ is a faithful representation of $\mathfrak{g}$ and $\exp d\theta(X)=\theta(\exp X)=\exp X$ for any $x\in\mathfrak{g}$. Hence we may identify $\mathfrak{g}$ with $d\theta(\mathfrak{g})$ under $d\theta$. $\mathfrak{g}$ can therefore be regarded as a linear Lie algebra. In particular the Lie algebra of $GL(K, d)$ then

---

Received by the editors October 4, 1948, and, in revised form, February 11, 1949.

[1] Cartan, J. Math. Pures Appl. vol. 17 (1938) pp. 1–12. See also Malcev, C. R. (Doklady) Acad. Sci. URSS. vol. 40 (1943) pp. 87–89.

[2] For the precise definitions of the terms used in this paper see Chevalley, *Theory of Lie groups*, Princeton University Press, 1946.

205

consists of all matrices of degree $\dot{d}$ with coefficients in K. We denote it by $\mathfrak{gl}(K, \dot{d})$. Given any subalgebra $\mathfrak{h} \subset \mathfrak{gl}(K, \dot{d})$, by the linear Lie group generated by $\mathfrak{h}$ we mean the analytic subgroup of $GL(K, \dot{d})$ corresponding to $\mathfrak{h}$.

Let us call a matrix subtriangular if it has zeros on and below the the diagonal. First we state the following two well known lemmas.

LEMMA 1.[3] *Let $\mathfrak{N}$ be the Lie algebra of all subtriangular $\dot{d} \times \dot{d}$ matrices over K and let G be the linear Lie group generated by $\mathfrak{N}$. Then $X \to \exp X$ is a topological mapping of $\mathfrak{N}$ onto G. Also G consists of all matrices which have zeros below the diagonal and 1 everywhere on the diagonal.*

LEMMA 2.[4] *Let G be a connected, simply connected solvable Lie group. Then every analytic subgroup of G is closed and simply connected.*

From now on I adhere strictly to the notation of my paper, *Faithful representations of Lie algebras*,[5] which will be quoted as FRL.

LEMMA 3. *Let $\mathfrak{L}$, $\mathfrak{N}$, $\mathfrak{D}$ be as in Lemma 1 of FRL. We construct the faithful representation θ of $\mathfrak{L} + \mathfrak{D}$ as described there. Then for any $Y_1, \cdots, Y_s \in \mathfrak{L}$ and $D_1, \cdots, D_s \in \mathfrak{D}$,*

(1) $\exp \theta(Y_1) \cdots \exp \theta(Y_s) \neq I$ *unless* $\exp Y_1 \cdots \exp Y_s = I''$,

(2) $\exp \theta(D_1) \cdots \exp \theta(D_s) = I$

$$\text{if and only if } \exp D_1 \cdots \exp D_s = I',$$

where I, I', and I'' are unit matrices of suitable degrees.

If we use the notation of the proof of Lemma 1 of FRL, (1) follows immediately from the fact that $\mathfrak{A}/\mathfrak{X} \cong \mathfrak{A}^*/\mathfrak{X}^*$ where $\mathfrak{X}^* = \mathfrak{X}/\mathfrak{X}_0$. Now we prove (2). Put $\omega^*(X) = (\omega(X))^*$ for $X \in \mathfrak{L}$. Then it is easily proved by induction on s that for any $D \in \mathfrak{D}$

$$\{\theta(D)\}^s \omega^*(X) = \omega^*(D^s X), \qquad s \geq 1.$$

Hence $(\exp \theta(D)) \omega^*(X) = \omega^*((\exp D)X)$. Also since $\theta(D) = d_D^*$ is a derivation of $\mathfrak{A}^* = \mathfrak{A}/\mathfrak{X}_0$, $\exp \theta(D)$ is an automorphism of $\mathfrak{A}^*$. Put $Y_X = (\exp D_1) \cdots (\exp D_s) X$ $(X \in \mathfrak{L})$. Then if

$$\exp \theta(D_1) \cdots \exp \theta(D_s) = I,$$

$\omega^*(Y_X) - \omega^*(X) = 0$ for every $X \in \mathfrak{L}$. Hence $\omega(Y_X - X) \in \mathfrak{X}_0 \subset \mathfrak{X}$. Therefore $\pi \omega(Y_X - X) = Y_X - X = 0$. Since this is true for every X,

---

[3] Birkhoff, Ann. of Math. vol. 38 (1937) pp. 526–532.

[4] Chevalley, Ann. of Math. vol. 42 (1941) pp. 668–675.

[5] Harish-Chandra, Ann. of Math. vol. 50 (1949) pp. 68–76.

$(\exp D_1) \cdots (\exp D_s) = I'$. The converse is obvious. Hence the lemma is proved.

LEMMA 4. *Let G be a connected, simply connected solvable Lie group with Lie algebra $\mathfrak{g}$. Let $\mathfrak{N}$ be the maximal nilpotent ideal of $\mathfrak{g}$. Then G has a faithful representation ψ such that $d\Psi(X)$ is nilpotent for every $X \in \mathfrak{N}$.*

By Corollary 1 of FRL we can find a faithful representation ρ_0 of $\mathfrak{g}$ such that $\rho_0(X)$ is nilpotent for every $X \in \mathfrak{N}$. We can therefore choose, if necessary, a new base in our representation space such that with respect to this base the matrix representing $\rho_0(X)$ is subtriangular for every $X \in \mathfrak{N}$. Now consider the factor algebra $\mathfrak{g}/\mathfrak{N}$ which is abelian and hence nilpotent. By the same corollary it follows that $\mathfrak{g}/\mathfrak{N}$ has a faithful representation by nilpotent matrices. Hence $\mathfrak{g}$ has a representation ρ_1 such that the kernel of ρ_1 is $\mathfrak{N}$ and $\rho_1(X)$ is nilpotent for all $X \in \mathfrak{g}$. We can again arrange that $\rho_1(X)$ is subtriangular for all X. Put $\rho = \rho_0 + \rho_1$, where $+$ denotes direct sum. Since G is simply connected, there exist representations ψ_0 and ψ_1 of G such that $d\psi_0 = \rho_0$, $d\psi_1 = \rho_1$. Put $\psi = \psi_0 + \psi_1$. Then $d\psi = \rho$. Let N be the analytic subgroup of G corresponding to $\mathfrak{N}$. Then from Lemma 2, N is a closed invariant subgroup. Consider $\psi(N)$. It is clear that $d\psi(Y) = \rho_0(Y) + \rho_1(Y)$ is subtriangular for all $Y \in \mathfrak{N}$. Hence from Lemmas 1 and 2 it follows that the linear Lie group $\psi(N)$ generated by $d\psi(\mathfrak{N})$ is simply connected. Since $d\psi$ is an isomorphism, $\psi(N)$ is locally isomorphic to N. Therefore since $\psi(N)$ is simply connected, ψ maps N isomorphically. Similarly we prove that $\psi_1(G)$ is simply connected. It is clear that the kernel of ψ_1 contains N. Hence ψ_1 defines a representation ψ^* of G/N given by $\psi^*(x^*) = \psi_1(x)$ where $x \to x^*$ is the natural homomorphism of G onto $G/N = G^*$. Since $d\psi^*$ is an isomorphism, the kernel of $d\psi_1 = \rho_1$ being $\mathfrak{N}$, it follows from the simple connectivity of $\psi^*(G^*) = \psi_1(G)$ that ψ^* is an isomorphism. Hence the kernel of ψ_1 is exactly N. Let D be the kernel of ψ. Then D is contained in the kernel of ψ_1 which is N. Also since ψ is faithful on N, $D \cap N = \{e\}$ where e is the unit element of N. Hence $D = \{e\}$ and ψ is a faithful representation. Also $d\psi(X)$ is nilpotent for every $X \in \mathfrak{N}$.

Now we come to the proof of the theorem. Let $\mathfrak{g}$ be the Lie algebra of G and $\mathfrak{N}$ the maximal nilpotent ideal of $\mathfrak{g}$. By Lemma 4, G has a faithful representation. Hence we may assume that G is a linear Lie group such that every element of $\mathfrak{N}$ is nilpotent. We keep to the notation of Lemma 3 except that $\mathfrak{L}$ is replaced by $\mathfrak{g}$. Let $\mathfrak{h}$ be the Lie algebra of H. Define a homomorphism $d\tau$ of $\mathfrak{h}$ into $\mathfrak{D}$ as follows. Let

Aut (G) be the group of automorphisms of G. Then Aut (G) is a Lie group with a Lie algebra $\mathfrak{A}$. It is well known that there exists an isomorphism λ of $\mathfrak{A}$ onto $\mathfrak{D}$ such that

$$(\exp A)\, \exp X = \exp\left((\exp \lambda(A))X\right)$$

for any $A \in \mathfrak{A}$ and $X \in \mathfrak{g}$. We put $d\tau = \lambda \circ d\phi$ where $d\phi$ is the homomorphism of $\mathfrak{h}$ into $\mathfrak{A}$ induced by ϕ. Then for any $P_i \in \mathfrak{h}$, $1 \leq i \leq r$,

$$\phi(\exp P_1 \cdots \exp P_r)\, \exp X = \exp\left((\exp d\tau(P_1) \cdots \exp d\tau(P_r))X\right).$$

Let e and e' denote the identity elements of G and H respectively. Suppose $\exp P_1 \cdots \exp P_r = e'$. Then clearly

$$\begin{aligned}
\exp X &= \phi(\exp P_1 \cdots \exp P_r)\, \exp X \\
&= \exp\left((\exp d\tau(P_1) \cdots \exp d\tau(P_r))X\right).
\end{aligned}$$

Since this is true for every $X \in \mathfrak{g}$,

$$\exp d\tau(P_1) \cdots \exp d\tau(P_r) = I'.$$

Hence we can define a representation τ of H by the rule

$$\tau(\exp P_1 \cdots \exp P_r) = \exp d\tau(P_1) \cdots \exp d\tau(P_r) \quad (P_1, \cdots, P_r \in \mathfrak{h}).$$

Put $d\psi = \theta \circ d\tau$. Then $d\psi$ is a representation of $\mathfrak{h}$. Suppose $\exp P_1 \cdots \exp P_r = e'$ $(P_1, \cdots, P_r \in \mathfrak{h})$. Then

$$\tau(\exp P_1 \cdots \exp P_r) = \exp d\tau(P_1) \cdots \exp d\tau(P_r) = I'$$

and from Lemma 3

$$\exp d\psi(P_1) \cdots \exp d\psi(P_r) = I.$$

Hence we can again define a representation ψ of H by putting

$$\psi(\exp P_1 \cdots \exp P_r) = \exp d\psi(P_1) \cdots \exp d\psi(P_r) \quad (P_1, \cdots, P_r \in \mathfrak{h}).$$

Also since G is simply connected there exists a representation χ of G such that $d\chi(X) = \theta(X)$ for every $X \in \mathfrak{g}$. Hence

$$\chi(\exp Y_1 \cdots \exp Y_r) = \exp \theta(Y_1) \cdots \exp \theta(Y_r) \quad (Y_i \in \mathfrak{G}, 1 \leq i \leq r).$$

From Lemma 3 it follows that χ is faithful.

Consider the mapping μ of $G \times_\phi H$ defined by $\mu(g, h) = \chi(g)\psi(h)$. We claim that μ is a representation. For any $P \in \mathfrak{h}$ and $X \in \mathfrak{g}$ consider

$$\begin{aligned}
\psi(\exp P)\chi(\exp X)(\psi(\exp P))^{-1} &= \exp d\psi(P)\, \exp \theta(X)\, \exp(-d\psi(P)) \\
&= \exp \theta(D)\, \exp \theta(X)\, \exp(-\theta(D))
\end{aligned}$$

where $D = d\tau(P)$. Now for any two elements $A, B \in \mathfrak{gl}(K, d)$,

$$\exp A \, \exp B \, \exp (-A) = \exp ((\exp \operatorname{ad} A)B)$$

where $\operatorname{ad} A$ is defined as in FRL. Since $[\theta(D), \theta(Y)] = \theta([D, Y]) = \theta(DY)$ for any $Y \in \mathfrak{g}$, it follows immediately that

$$\exp \theta(D) \, \exp \theta(X) \, \exp (-\theta(D)) = \exp \theta((\exp D)X)$$
$$= \exp \theta(\tau(\exp P)X)$$
$$= \chi(\exp \tau(\exp P)X)$$
$$= \chi(\phi(\exp P) \exp X).$$

Since any $h \in H$ can be written in the form $\exp P_1 \cdots \exp P_r$, $P_i \in \mathfrak{h}$, $1 \le i \le r$, $r \ge 1$, we get

$$\psi(h)\chi(\exp X)(\psi(h))^{-1} = \chi(\phi(h) \exp X).$$

Similarly since every $g \in G$ can be written as $\exp Y_1 \cdots \exp Y_r$, $Y_i \in \mathfrak{g}$, $1 \le i \le r$, $r \ge 1$, we have

$$\psi(h)\chi(g)(\psi(h))^{-1} = \chi(\phi(h)g).$$

Therefore

$$\mu((g_1, h_1)(g_2, h_2)) = \mu(g_1\phi(h_1)g_2, h_1h_2) = \chi(g_1\phi(h_1)g_2)\psi(h_1h_2)$$
$$= \chi(g_1)\chi(\phi(h_1)g_2)\psi(h_1)\psi(h_2)$$
$$= \chi(g_1)\psi(h_1)\chi(g_2)\psi(h_2) = \mu(g_1, h_1)\mu(g_2, h_2).$$

Since μ is clearly a continuous mapping it is a representation of $G \times_\phi H$. By hypothesis H has a faithful representation ν_0. Define a representation ν of $G \times_\phi H$ by $\nu(g, h) = \nu_0(h)$ and put $\xi = \mu + \nu$. Suppose (g, h) belongs to the kernel of ξ. Then since $\xi(g, h) = \mu(g, h) + \nu_0(h) = \chi(g)\psi(h) + \nu_0(h)$ and since ν_0 is faithful on H, $h = e'$. Hence g belongs to the kernel of χ. But as χ is faithful on G, $g = e$. Therefore $(g, h) = (e, e')$ and ξ is faithful on $G \times_\phi H$.

COROLLARY (MALCEV).[1] *A connected solvable Lie group G has a faithful representation if and only if $G = NA$, where N is a closed, connected, simply connected invariant subgroup and A is a connected, compact abelian subgroup such that $N \cap A = \{e\}$.*

Suppose $G = NA$. For any $a \in A$ let $\phi(a)$ denote the automorphism of N given by $\phi(a)n = ana^{-1}$ ($n \in N$). Then it is easily seen that $(n, a) \to na$ is an isomorphism of $N \times_\phi A$ onto G. Since A is compact, it has a faithful representation. Hence by the above theorem it follows immediately that G has a faithful representation.

In order to establish the converse we make use of the following lemma which follows easily from the results of Chevalley.[4]

LEMMA 5.[6] *If G is a connected solvable Lie group and N a closed, connected invariant subgroup such that G/N is compact, then there exists a compact connected abelian subgroup A of G such that $G = AN$ and $A \cap N$ is finite.*

Returning to the corollary, suppose G is linear. Since $\mathfrak{g}$ is solvable, we deduce in the usual way that every element $X \in [\mathfrak{g}, \mathfrak{g}] = \mathfrak{g}'$ is nilpotent and therefore may be assumed to be subtriangular. Therefore by Lemmas 1 and 2 the group G' generated by $\mathfrak{g}'$ is simply connected. Let $\dot{d}$ be the degree of G and G_0 the group of all matrices in $GL(K, \dot{d})$ which have zero below the diagonal and 1 everywhere on the diagonal. By Lemma 2, G' is closed in G_0. However, since G_0 is clearly closed in $GL(K, \dot{d})$, G' is closed in $GL(K, \dot{d})$ and therefore in G. Let $x \to x^*$ denote the natural homomorphism of G onto $G/G' = G^*$. Since G^* is abelian, $G^* = T^* V^*$ where T^* and V^* are connected subgroups, T^* being compact and V^* simply connected and $T^* \cap V^* = \{e^*\}$. Let N be the complete inverse image of V^* in G. Since $N/G' = V^*$ and G' are both simply connected, N is simply connected and $G/N \cong T^*$ is compact. Therefore by Lemma 5, $G = AN$ where A is compact, connected, and abelian and $A \cap N$ is finite. Let $\sigma \in A \cap N$. Then $\sigma^r = e$ for some $r \geq 1$. Then $\sigma^* \in V^*$ and $(\sigma^*)^r = e^*$. Since V^* is simply connected and abelian, $\sigma^* = e^*$. Hence $\sigma \in A \cap G'$. Since $X \to \exp X$ is a topological mapping of $\mathfrak{g}'$ onto G', it follows that $\sigma \in G'$, $\sigma^r = e$ implies $\sigma = e$. Hence $A \cap N = \{e\}$. The corollary is therefore proved.

Institute for Advanced Study

---

[6] This lemma was pointed out to me by Dr. G. D. Mostow.

Reprinted from
Proc. Amer. Math. Soc.
1 (1950), 205–210

ANNALS OF MATHEMATICS
Vol. 51, No. 2, March, 1950

LIE ALGEBRAS AND THE TANNAKA DUALITY THEOREM

By Harish-Chandra

(Received November 16, 1948, Revised February 4, 1949)

1. Introduction

Let $\mathfrak{L}$ be a Lie algebra over a field K. Let Σ be an extension field of K and d a positive integer. We denote the set of all $d \times d$ matrices with coefficients in Σ by $E(\Sigma, d)$. For any $A, B \in E(\Sigma, d)$ we write $[A, B] = AB - BA$. Then with respect to this bracket operation $E(\Sigma, d)$ becomes a Lie algebra over Σ. By a Σ-representation ρ of $\mathfrak{L}$ we mean a homomorphic mapping of $\mathfrak{L}$ into $E(\Sigma, d)$ for some d. d is called the degree of ρ. For simplicity a K-representation of $\mathfrak{L}$ will be called just a representation. Given any $A \in E(\Sigma, d)$ we write $A^* = -{}^t\! A$ where ${}^t\! A$ is the transposed of the matrix A. Also if $A \in E(\Sigma, d_1)$ and $B \in E(\Sigma, d_2)$ we define their direct sum $A \dotplus B$ and the Kronecker product $A \times B$ in the usual way (see Chevalley (2), Chapter VI) and write $A + B = A \times I_{d_2} + I_{d_1} \times B$ where I_{d_1} and I_{d_2} are the unit matrices of degree d_1 and d_2 respectively. $A + B$ is called the Kronecker sum of A and B. Given any Σ-representations ρ, ρ_1, ρ_2 of $\mathfrak{L}$ we construct the Σ-representations ρ^*, $\rho_1 \dotplus \rho_2$, $\rho_1 + \rho_2$ as follows:

$$\rho^*(X) = (\rho(X))^*, \qquad (\rho_1 \dotplus \rho_2)(X) = \rho_1(X) \dotplus \rho_2(X)$$

$$(\rho_1 + \rho_2)(X) = \rho_1(X) + \rho_2(X) \qquad\qquad (X \in \mathfrak{L}).$$

Also we write $\rho_1 \subset \rho_2$ if $\rho_1 = \rho_2$ or if there exists a Σ-representation ρ_3 such that $\rho_1 = \rho_2 \dotplus \rho_3$.

Let ρ be a Σ-representation of $\mathfrak{L}$ of degree d and let γ be a nonsingular matrix in $E(\Sigma, d)$. We define the Σ-representation $\gamma\rho\gamma^{-1}$ by $(\gamma\rho\gamma^{-1})(X) = \gamma\rho(X)\gamma^{-1}$ $(X \in \mathfrak{L})$. Two Σ-representations ρ_1, ρ_2 are said to be equivalent if they have the same degree d and if there exists a nonsingular matrix $\gamma \in E(\Sigma, d)$ such that $\rho_1 = \gamma\rho_2\gamma^{-1}$. We denote equivalence by writing $\rho_1 \sim \rho_2$. We say that ρ_1 is a component of ρ_2 or is contained in ρ_2 and write $\rho_1 \prec \rho_2$ if there exists a Σ-representation ρ_3 such that $\rho_1 \subset \rho_3$ and $\rho_3 \sim \rho_2$. Also, we denote the one-dimensional zero Σ-representation of $\mathfrak{L}$ by 0_1. It maps every element of $\mathfrak{L}$ into the zero matrix of degree 1.

Definition. A non-empty set $\mathfrak{F}$ of Σ-representations of $\mathfrak{L}$ is called a *ring* if the following conditions are fulfilled:

    1) If ρ_1, ρ_2 are in $\mathfrak{F}$ then $\rho_1 \dotplus \rho_2$ and $\rho_1 + \rho_2$ are also in $\mathfrak{F}$.

    2) If ρ is in $\mathfrak{F}$ and $\rho' \prec \rho$ then ρ' is in $\mathfrak{F}$.

Clearly the set $\mathfrak{R}_\Sigma$ of all Σ-representations of $\mathfrak{L}$ is a ring. Every ring of Σ-representations is therefore a subring of $\mathfrak{R}_\Sigma$.

Definition. *Let F be a subset of $\mathfrak{R}_\Sigma$. The smallest subring of $\mathfrak{R}_\Sigma$ containing F is called the ring of Σ-representations generated by F. Also we write $F^* = \{\rho^* \mid \rho \in F\}$.*

299

We now define a representation ζ of a subring $\mathfrak{F}$ of $\mathfrak{R}_\Sigma$. For any $\rho \, \epsilon \, \mathfrak{R}_\Sigma$ let d_ρ denote the degree of ρ.

DEFINITION. *Let ζ be a mapping which assigns to every $\rho \, \epsilon \, \mathfrak{F}$ a matrix $\zeta(\rho) \, \epsilon \, E(\Sigma, d_\rho)$. Then ζ is called a representation of $\mathfrak{F}$ if for any $\rho, \rho_1, \rho_2 \, \epsilon \, \mathfrak{F}$ the following relations hold.*

$$1) \quad \zeta(\gamma\rho\gamma^{-1}) \quad = \gamma\zeta(\rho)\gamma^{-1}$$

$$2) \quad \zeta(\rho_1 \dotplus \rho_2) = \zeta(\rho_1) \dotplus \zeta(\rho_2)$$

$$3) \quad \zeta(\rho_1 \mathbin{\dot{+}} \rho_2) = \zeta(\rho_1) \mathbin{\dot{+}} \zeta(\rho_2)$$

γ being any nonsingular matrix in $E(\Sigma, d_\rho)$.

For any two representations ζ_1, ζ_2 of $\mathfrak{F}$ and $a_1, a_2 \, \epsilon \, \Sigma$ we define two new representations $a_1\zeta_1 + a_2\zeta_2$ and $[\zeta_1, \zeta_2]$ as follows:

$$(a_1\zeta_1 + a_2\zeta_2)(\rho) = a_1\zeta_1(\rho) + a_2\zeta_2(\rho); \qquad [\zeta_1, \zeta_2](\rho) = [\zeta_1(\rho), \zeta_2(\rho)]$$

for every $\rho \, \epsilon \, \mathfrak{F}$. It is easily verified that under these operations the set $\hat{\mathfrak{L}}(\mathfrak{F})$ of all representations of $\mathfrak{F}$ becomes a Lie algebra (not necessarily finite dimensional) over Σ.

Our first two theorems may now be stated as follows:

THEOREM 1. *Let $\mathfrak{L}$ be a semisimple Lie algebra over a field K of characteristic zero and let $\mathfrak{F}$ be a ring of representations of $\mathfrak{L}$ such that $\mathfrak{F} = \mathfrak{F}^*$. Then there exists a representation $\rho \, \epsilon \, \mathfrak{F}$ such that ρ generates the whole ring $\mathfrak{F}$.*

THEOREM 2. *Let $\mathfrak{L}$ be as in Theorem 1 and let Σ be an algebraically closed extension of K. Let $\hat{\mathfrak{L}}(\mathfrak{F})$ be the Lie algebra of representations of a ring $\mathfrak{F}$ of Σ-representations of $\mathfrak{L}$ such that $\mathfrak{F} = \mathfrak{F}^*$. For each $X \, \epsilon \, \mathfrak{L}$ define ζ_X in $\hat{\mathfrak{L}}(\mathfrak{F})$ by $\zeta_X(\rho) = \rho(X)$ $(\rho \, \epsilon \, \mathfrak{F})$ and put $\mathfrak{L}(\mathfrak{F}) = \{\zeta_X \mid X \, \epsilon \, \mathfrak{L}\}$. Then $X \to \zeta_X$ is a homomorphic mapping of $\mathfrak{L}$ into $\hat{\mathfrak{L}}(\mathfrak{F})$ and $\mathfrak{L}(\mathfrak{F})$ is a Lie algebra over K. Also $\hat{\mathfrak{L}}(\mathfrak{F})$ is the extension of $\mathfrak{L}(\mathfrak{F})$ over Σ.*

Proofs of these theorems are known (see for example Chevalley (2), Chapter VI) at least in case K is the field of real or complex numbers and $\mathfrak{F}$ is the ring of all representations. However these proofs depend, in an essential way, on integration over the group associated with the compact form of $\mathfrak{L}$ and therefore are of an entirely non-algebraic character. Also Cartan (1) has given an algebraic proof of Theorem 1 (in case $\mathfrak{F} = \mathfrak{R}_\Sigma$) by making use of the classification of simple Lie algebras and by explicitly constructing the required representation ρ in each case separately. Our object is to give a general algebraic proof which does not require the consideration of each special case.

Finally we shall apply these results to connected semisimple Lie groups and thus obtain for them an extension of the Tannaka duality theorem.[1] Our method also gives a new proof of the theorem of Weyl according to which the universal covering group of a compact semisimple Lie group is compact.

---

[1] Perhaps it is possible to get the same results also by combining Weyl's 'unitary trick' with the known theory for compact groups.

2. The Casimir operator

We begin with the proof of a known result which will be needed rather frequently.

LEMMA 1. *Let $\mathfrak{M}$ be a three-dimensional Lie algebra over a field K of characteristic zero defined as follows: $\mathfrak{M} = Kx + Ky + Kh$, $[h, x] = x$, $[h, y] = -y$, $[x, y] = h$. Let ρ be any representation of $\mathfrak{M}$. Then $\rho(x)\rho(y)$ and $\rho(y)\rho(x)$ are semisimple[2] matrices and all their eigenvalues are of the form $\frac{1}{2}q$ where q is an integer ≥ 0.*

Define $(X_1, X_2, X_3) = (h, x, y)$ and $(\mathrm{ad}X_i)Y = [X_i, Y]$ $1 \leq i \leq 3$ for any $Y \in \mathfrak{M}$. Then $g_{ij} = \mathrm{sp}(\mathrm{ad}X_i\,\mathrm{ad}X_j)$ $1 \leq i, j \leq 3$ is given by

$$(g_{ij})_{1 \leq i, j \leq 3} = \begin{pmatrix} 2 & 0 & 0 \\ 0 & 0 & 2 \\ 0 & 2 & 0 \end{pmatrix}.$$

Hence $\det (g_{ij}) = -8 \neq 0$. Therefore $\mathfrak{M}$ is semisimple. Hence ρ is completely reducible and therefore it is clearly sufficient to consider the case when ρ is irreducible and $\rho \neq 0$.

Let V be the representation space[3] of ρ. First suppose K is algebraically closed. Let μ be any eigenvalue of $\rho(h)$. Choose k to be the largest integer ≥ 0 such that $\mu + k$ is an eigenvalue of $\rho(h)$. We can find a vector $\psi_0 \neq 0$ in V such that $\rho(h)\psi_0 = (\mu + k)\psi_0$. For $r \geq 1$ define $\psi_r = \{\rho(y)\}^r\psi_0$. First we assert that

$$(1) \qquad \rho(h)\psi_r = (\mu + k - r)\psi_r \qquad\qquad r \geq 0.$$

This is clearly true for $r = 0$ and is easily proved for $r \geq 1$ by induction if we make use of the fact that $\psi_r = \rho(y)\psi_{r-1}$ $(r \geq 1)$ and $\rho(h)\rho(y) = -\rho(y) + \rho(y)\rho(h)$. Since $\rho(h)$ can have only a finite number of distinct eigenvalues it follows that $\psi_r = 0$ for some r. Let J be the least integer ≥ 0 such that $\psi_{J+1} = 0$ and define $\psi_{-1} = 0$. Then we assert that

$$(2) \qquad \rho(x)\psi_r = \{\mu + k - \tfrac{1}{2}(r - 1)\}r\psi_{r-1} \qquad\qquad r \geq 0.$$

This is true for $r = 0$. For,

$$\rho(h)\rho(x)\psi_0 = \rho([h, x])\psi_0 + \rho(x)\rho(h)\psi_0$$

$$= (\mu + k + 1)\rho(x)\psi_0.$$

But $\mu + k + 1$ is not an eigenvalue of $\rho(h)$. Hence $\rho(x)\psi_0 = 0$. The rest follows by induction. If we put $r = J + 1$ in (2) we get

$$0 = \rho(x)\psi_{J+1} = \{\mu + k - \tfrac{1}{2}J\}(J + 1)\psi_J.$$

---

[2] A matrix or an endomorphism is called semisimple if its minimum polynomial has no repeated factors.

[3] Strictly speaking one ought to distinguish between a representation and its abstract form (see Chevalley (2), p. 172). However, we do not do so in order to avoid a cumbersome notation.

But $\psi_J \neq 0$ and $J + 1 \neq 0$. Hence $\mu + k - \frac{1}{2}J = 0$ and

$$\rho(h)\psi_r = (\tfrac{1}{2}J - r)\psi_r$$

$$\rho(x)\psi_r = \tfrac{1}{2}(J - r + 1)r\psi_{r-1}.$$

The space $K\psi_0 + \cdots + K\psi_J$ is therefore clearly invariant under $\rho(\mathfrak{M})$. Since ρ is irreducible it must be the whole space V. But

$$\rho(x)\rho(y)\psi_r = \tfrac{1}{2}(J - r)(r + 1)\psi_r$$

$$\rho(y)\rho(x)\psi_r = \tfrac{1}{2}(J - r + 1)r\psi_r \qquad\qquad 0 \leqq r \leqq J.$$

Hence $\rho(x)\rho(y)$ and $\rho(y)\rho(x)$ are both diagonal and therefore semisimple and their eigenvalues are all of the required form $\frac{1}{2}q$ where q is an integer $\geqq 0$.

If K is not algebraically closed let $\bar{K}$ be its algebraic closure. Extend $\mathfrak{M}$ to $\overline{\mathfrak{M}}$ over $\bar{K}$ and ρ to a representation $\bar{\rho}$ of $\overline{\mathfrak{M}}$. Then it follows from the above proof that $\bar{\rho}(x)\bar{\rho}(y)$ and $\bar{\rho}(y)\bar{\rho}(x)$ are semisimple and all their eigenvalues are of the required form. But $\bar{\rho}(x) = \rho(x)$ and $\bar{\rho}(y) = \rho(y)$. Hence the same holds for $\rho(x)\rho(y)$ and $\rho(y)\rho(x)$.

From now on we use $\mathfrak{L}$ to denote a semisimple Lie algebra over a field K of characteristic zero. If any other additional hypothesis about K or $\mathfrak{L}$ is needed in the proof of some lemma, it will always be stated explicitly. For any representation ρ of $\mathfrak{L}$ we denote the degree of ρ by d_ρ. Also the unit matrix of degree d will be denoted by I_d. We denote the linear mapping $Y \rightarrow [X, Y]$ $(X, Y \in \mathfrak{L})$ of $\mathfrak{L}$ into itself by $\mathrm{ad}X$ and write $B(X, Y) = \mathrm{sp}(\mathrm{ad}X\,\mathrm{ad}Y)$. Since $\mathfrak{L}$ is semisimple $B(X, Y)$ is a nondegenerate bilinear form.

Let $\dim \mathfrak{L} = n$. Choose a base $X_i, 1 \leqq i \leqq n$ for $\mathfrak{L}$. Put $g_{ij} = B(X_i, X_j)$. The matrix $g_{ij}\, 1 \leqq i, j \leqq n$ is then nonsingular. Let $g^{ij}\, 1 \leqq i, j \leqq n$ be its inverse. For any representation ρ of $\mathfrak{L}$ define

$$\Gamma_\rho = \sum_{1 \leqq i,j \leqq n} g^{ij}\rho(X_i)\rho(X_j).$$

It follows immediately from its definition that Γ_ρ is independent of the particular choice of the base $X_i, 1 \leqq i \leqq n$. Γ_ρ is called the Casimir operator of the representation ρ. It is easily verified that Γ_ρ commutes with $\rho(X_i)\, 1 \leqq i \leqq n$ and therefore with $\rho(X)$ for all $X \in \mathfrak{L}$.

LEMMA 2.[4] *If $\rho \neq 0_1$ is an irreducible representation of $\mathfrak{L}$ then $\Gamma_\rho = \omega_\rho I_{d_\rho}$ where ω_ρ is a rational number > 0.*

First suppose K is algebraically closed. Let $\mathfrak{h}$ be a Cartan subalgebra of $\mathfrak{L}$ and let $\dim \mathfrak{h} = l$. Let $\alpha_i\, 1 \leqq i \leqq n - l$ be all the roots $(\neq 0)$ and suppose $\alpha_i\, 1 \leqq i \leqq l$ are linearly independent. (It is known that there always exist exactly l linearly independent roots.) Corresponding to each root α we can find an element X_α in $\mathfrak{L}$ such that $[H, X_\alpha] = \alpha(H)X_\alpha$ for every $H \in \mathfrak{h}$. Further, this choice

---

[4] Lemmas 2, 3 and 4 are of course not new. They are included only for the sake of completeness.

can be so made that $B(X_\alpha, X_{-\alpha}) = 1$. Put $H_\alpha = [X_\alpha, X_{-\alpha}]$ and $H_i = H_{\alpha_i}$ $1 \leq i \leq l$. It is known that for every root α, $\alpha(H_\alpha)$ is a rational number > 0.

Choose a base X_i $1 \leq i \leq n$ for $\mathfrak{L}$ as follows: $X_i = H_i$ for $1 \leq i \leq l$ and $X_{i+l} = X_{\alpha_i}$ for $1 \leq i \leq n - l$. For any two roots α, β and $1 \leq i \leq l$ define $g_{\alpha,\beta} = B(X_\alpha, X_\beta)$ and $g_{\alpha,i} = g_{i,\alpha} = B(X_\alpha, H_i)$. Then it is known that $g_{\alpha,i} = 0$ and $g_{\alpha,\beta} = 0$ if $\alpha + \beta \neq 0$ and $g_{\alpha,-\alpha} = 1$. Put $X^i = \sum_{1 \leq j \leq n} g^{ij}X_j$ $1 \leq i \leq n$, $H^i = X^i$ $1 \leq i \leq l$ and $X^{i+l} = X^{\alpha_i}$ $1 \leq i \leq n - l$. Then for $1 \leq i \leq l$

$$(3) \quad H_i = X_i = \sum_{1 \leq j \leq n} g_{ij}X^j = \sum_{1 \leq j \leq l} g_{ij}H^j + \sum_\alpha g_{i,\alpha}X^\alpha = \sum_{1 \leq j \leq l} g_{ij}H^j$$

since $g_{i,\alpha} = 0$. Here $\sum_\alpha$ denotes summation over all roots α. Similarly

$$(4) \quad X_\alpha = \sum_\beta g_{\alpha,\beta}X^\beta = X^{-\alpha}.$$

It is known that for any $H \in \mathfrak{h}$ and any root α, $B(H_\alpha, H) = \alpha(H)$. Hence in particular $B(H_i, H) = \alpha_i(H)$. Therefore $B(H_i, H) = 0$ for all $1 \leq i \leq l$ implies $H = 0$ since α_i $1 \leq i \leq l$ are linearly independent. Hence the matrix g_{ij} $1 \leq i, j \leq l$ is nonsingular. Let g^{*ij} $1 \leq i, j \leq l$ denote its inverse. Then it follows from (3) that $H^i = \sum_{1 \leq j \leq l} g^{*ij}H_j$. But

$$H^i = \sum_{i \leq j \leq n} g^{ij}X_j = \sum_{1 \leq j \leq l} g^{ij}H_j + \sum_{l+1 \leq j \leq n} g^{ij}X_{\alpha_{j-l}}.$$

Since X_j are linearly independent it follows that $g^{*ij} = g^{ij}$ $1 \leq i, j \leq l$ and $g^{ij} = 0$ $1 \leq i \leq l, j > l$. Therefore, since $g^{ij} = g^{ji}$ we get from (4)

$$\Gamma_\rho = \sum_{1 \leq i,j \leq l} g^{ij}\rho(H_i)\rho(H_j) + \sum_\alpha \rho(X_\alpha)\rho(X^\alpha)$$

$$= \sum_{1 \leq i,j \leq l} g^{ij}\rho(H_i)\rho(H_j) + \sum_\alpha \rho(X_\alpha)\rho(X_{-\alpha}).$$

Now ρ is irreducible and Γ_ρ commutes with $\rho(X)$ $(X \in \mathfrak{L})$. Since K is algebraically closed it follows from Schur's lemma that $\Gamma_\rho = \omega_\rho I_{d_\rho}$ where $\omega_\rho \in K$.

Before proving that ω_ρ is rational it is convenient to introduce some terminology which we shall use constantly throughout this paper. Let V be the representation space of a representation ρ (not necessarily irreducible) of $\mathfrak{L}$. For any linear function λ on $\mathfrak{h}$ we define V_λ to be the set of all $\psi \in V$ such that $\rho(H)\psi = \lambda(H)\psi$ for all $H \in \mathfrak{h}$. Clearly V_λ is a subspace of V. The dimension of V_λ is called the multiplicity of λ in the representation ρ. λ is called a weight of ρ if $V_\lambda \neq \{0\}$. An element $\psi \in V_\lambda$ is called a weight vector belonging to the weight λ if $\psi \neq 0$. It is known that $V = \Sigma_\Lambda V_\Lambda$ where Λ runs over all the weights of the representation ρ and the sum is direct.

Let K_0 be the field of rational numbers. Since the characteristic of K is zero, $K_0 \subset K$. Put $\mathfrak{h}_0 = \sum_\alpha K_0 H_\alpha$. A linear function λ on $\mathfrak{h}$ is called rational if $\lambda = \sum_{1 \leq i \leq l} c_i\alpha_i$ with $c_i \in K_0$. It is known that all weights of every representation and therefore in particular the roots are rational functions. If $\lambda = \sum_{1 \leq i \leq l} c_i\alpha_i$ is rational and $\neq 0$ we say that $\lambda > 0$ or $\lambda < 0$ according as $c_i > 0$ or < 0 where i is the least index such that $c_i \neq 0$. If λ and μ are two rational functions we write $\lambda > \mu$ if $\lambda - \mu > 0$. Thus the rational functions are completely ordered by this relation and therefore every finite set of them has a highest member. For any rational function λ we define $H_\lambda = \sum_{1 \le i \le l} \lambda(H_i) H^i$. Then $H_\lambda \in \mathfrak{h}_0$ and $B(H, H_\lambda) = \lambda(H)$ for all $H \in \mathfrak{h}_0$. This definition is in agreement with that of H_α as is seen from the fact that $B(H, H_\alpha) = \alpha(H)$. Given any two rational functions λ, μ we write $\langle \lambda \mid \mu \rangle = B(H_\lambda, H_\mu) = \lambda(H_\mu) = \mu(H_\lambda)$ and $|\lambda|^2 = \langle \lambda \mid \lambda \rangle$. Then $\langle \lambda \mid \mu \rangle \in K_0$. Further, for any $H \in \mathfrak{h}_0$ we write $|H|^2 = B(H, H)$. It is known that the quadratic form $B(H, H)$ is positive definite on $\mathfrak{h}_0$. Hence it follows that $|\lambda|^2 = |H_\lambda|^2 > 0$ unless $\lambda = 0$ and $|\langle \lambda \mid \mu \rangle| \le |\lambda| \, |\mu|$ where $|\lambda|$ is the real nonnegative square root of $|\lambda|^2$.

Now we return to the irreducible representation ρ on the space $V = \sum_\Lambda V_\Lambda$. We know that $\Gamma_\rho = \omega_\rho I_{d_\rho}$. Therefore in order to show that ω_ρ is rational and > 0 it is sufficient to prove the same for sp Γ_ρ. Let m_Λ denote the multiplicity of the weight Λ. Then for any $H \in \mathfrak{h}$, $\rho(H) = \Lambda(H) I_{m_\Lambda}$ on V_Λ. Hence

$$\sum_{1 \le i,j \le l} g^{ij} \, \mathrm{sp} \, \{\rho(H_i)\rho(H_j)\} = \sum_\Lambda m_\Lambda \sum_{1 \le i,j \le l} g^{ij} \Lambda(H_i)\Lambda(H_j).$$

Now we assert that there exists a weight $\Lambda \ne 0$. For otherwise if $\Lambda = 0$ is the only weight $V = V_0$ and so $\rho(H) = 0$ for all $H \in \mathfrak{h}$. Hence $[\rho(H_\alpha), \rho(X_\alpha)] = \rho([H_\alpha, X_\alpha]) = \alpha(H_\alpha)\rho(X_\alpha) = 0$. But $\alpha(H_\alpha) \ne 0$. Hence $\rho(X_\alpha) = 0$. Since this is true for every α, $\rho = 0$. But this contradicts our hypothesis. Hence $\sum_\Lambda m_\Lambda |\Lambda|^2 > 0$. Now consider $\rho(X_\alpha)\rho(X_{-\alpha})$ for any root α. Let a be some element in K such that $a^2 = \alpha(H_\alpha)$. Then $a \ne 0$. Put $(1/a)X_\alpha = X$, $(1/a)X_{-\alpha} = Y$, $(1/a^2)H_\alpha = H$. (X, Y, H) span a Lie algebra isomorphic to the algebra $\mathfrak{M}$ of Lemma 1 under the mapping $X \to x$, $Y \to y$, $H \to h$. Hence it follows that sp $\{\rho(X)\rho(Y)\} = \tfrac{1}{2} s_\alpha$ where s_α is an integer ≥ 0. Therefore sp $\{\rho(X_\alpha)\rho(X_{-\alpha})\} = \tfrac{1}{2} s_\alpha \alpha(H_\alpha)$. Since $\alpha(H_\alpha)$ is rational and > 0 it follows that

$$\mathrm{sp} \, \Gamma_\rho = \sum_\Lambda m_\Lambda |\Lambda|^2 + \sum_\alpha \tfrac{1}{2} \alpha(H_\alpha) s_\alpha$$

is rational and > 0.

If K is not algebraically closed let $\bar{K}$ be its algebraic closure. Extend $\mathfrak{L}$ to $\bar{\mathfrak{L}}$ over $\bar{K}$ and ρ to a representation $\bar{\rho}$ of $\bar{\mathfrak{L}}$. For any base $X_i, 1 \le i \le n$ of $\mathfrak{L}$, $\bar{\rho}(X_i) = \rho(X_i)$. Hence $\Gamma_{\bar\rho} = \sum_{1 \le i,j \le n} g^{ij} \bar{\rho}(X_i)\bar{\rho}(X_j) = \Gamma_\rho$. Since $\bar{\mathfrak{L}}$ is semisimple, $\bar{\rho}$ is completely reducible. Also $\bar{\rho} \ne 0$. Hence let $\bar{\rho}_1 \ne 0$ be an irreducible component of $\bar{\rho}$. Then it follows from the above proof that $\Gamma_{\bar\rho_1}$ has a rational eigenvalue $\omega > 0$. Since every eigenvalue of $\Gamma_{\bar\rho_1}$ is also an eigenvalue of $\Gamma_{\bar\rho}$ and therefore of Γ_ρ, $\Gamma_\rho - \omega I_{d_\rho}$ is singular. Since ω is rational, $\omega \in K$. Hence by applying Schur's lemma to the irreducible representation ρ of $\mathfrak{L}$ we get $\Gamma_\rho - \omega I_{d_\rho} = 0$. The lemma is therefore proved.

For any representation ρ of $\mathfrak{L}$ define

$$\omega_\rho = \frac{1}{d_\rho} \, \mathrm{sp} \, \Gamma_\rho.$$

By decomposing ρ into a direct sum of irreducible representations it follows immediately that all the eigenvalues of Γ_ρ are rational and ≥ 0. Hence the same is true of ω_ρ. Also it is clear that $\omega_\rho = 0$ implies $\rho = 0$. We shall say that a representation ρ is homogeneous if $\Gamma_\rho = \omega_\rho I_{d_\rho}$.

LEMMA 3. *Let $\mathfrak{L}_s \neq \{0\}$ $1 \leq s \leq r$ be ideals in $\mathfrak{L}$ such that $\mathfrak{L}_s \cap \mathfrak{L}_t = \{0\}$ $s \neq t$ and $\mathfrak{L} = \sum_{1 \leq s \leq r} \mathfrak{L}_s$. For any representation ρ of $\mathfrak{L}$ let $\rho^{(s)}$ denote the representation of $\mathfrak{L}_s$ defined by ρ and let $\Gamma_{\rho(s)}$ be the Casimir operator of $\rho^{(s)}$. Then*

$$\Gamma_\rho = \sum_{1 \leq s \leq r} \Gamma_{\rho(s)}$$

and $\operatorname{sp} \{\rho(X^{(s)})\rho(X^{(t)})\} = 0$ *for any* $X^{(s)} \in \mathfrak{L}_s$, $X^{(t)} \in \mathfrak{L}_t$ $s \neq t$, $1 \leq s, t \leq r$. *Also if ρ is irreducible $\rho^{(s)}$ are all homogeneous.*

Let $n = \dim \mathfrak{L}$ and $n_s = \dim \mathfrak{L}_s$, $1 \leq s \leq r$. Choose a base $X_i^{(s)}$ $1 \leq i \leq n_s$ for $\mathfrak{L}_s$. Define $g_{ij}^{(s,t)} = \operatorname{sp} (\operatorname{ad} X_i^{(s)} \operatorname{ad} X_j^{(t)})$ $1 \leq i \leq n_s$, $1 \leq j \leq n_t$, $1 \leq s, t \leq r$. Since $[\mathfrak{L}_s, \mathfrak{L}_t] = \{0\}$ $(s \neq t)$ it follows that $g_{ij}^{(s,t)} = 0$ for $s \neq t$. Let $X \rightarrow \operatorname{ad}^{(s)}X$ $(X \in \mathfrak{L}_s)$ denote the adjoint representation of $\mathfrak{L}_s$. Then clearly $g_{ij}^{(s,s)} = \operatorname{sp}(\operatorname{ad}^{(s)}X_i^{(s)} \operatorname{ad}^{(s)}X_j^{(s)}) = g_{ij}^{(s)}$ $1 \leq i, j \leq n_s$. Since $\mathfrak{L}_s$ is semisimple, the matrix $g_{ij}^{(s)}$ $1 \leq i, j \leq n_s$ is nonsingular. Let $g^{(s)ij}$ $1 \leq i, j \leq n_s$ denote its inverse. Then it is clear that the matrix $g^{(s,t)ij}$ $1 \leq i \leq n_s$, $1 \leq j \leq n_s$, $1 \leq s, t \leq r$ defined by $g^{(s,t)ij} = 0 (s \neq t)$ and $g^{(s,s)ij} = g^{(s)ij}$ is the inverse of $g_{ij}^{(s,t)}$. Hence it follows immediately that

$$\Gamma_\rho = \sum_{1 \leq s \leq r} \sum_{1 \leq i,j \leq n_s} g^{(s)ij}\rho(X_i^{(s)})\rho(X_j^{(s)}) = \sum_{s=1}^{r} \Gamma_{\rho(s)}.$$

Since $[\mathfrak{L}_s, \mathfrak{L}_t] = \{0\}$ $(s \neq t)$ it is clear that $\Gamma_{\rho(s)}$ commutes with $\rho(X)$ for all $X \in \mathfrak{L}$. We know that the eigenvalues of $\Gamma_{\rho(s)}$ are all rational. Hence we can find an $\omega_s \in K$ such that $\Gamma_{\rho(s)} - \omega_s I_{d_\rho}$ is singular. Therefore if ρ is irreducible it follows from Schur's lemma that $\Gamma_{\rho(s)} = \omega_s I_{d_\rho}$. Hence $\rho^{(s)}$ is homogeneous.

Let $X^{(t)} = [Y^{(t)}, Z^{(t)}]$ where $Y^{(t)}, Z^{(t)} \in \mathfrak{L}_t$. Then for any $X^{(s)} \in \mathfrak{L}_s$, $s \neq t$

$$\operatorname{sp}\{\rho(X^{(s)})\rho(X^{(t)})\} = \operatorname{sp}[\rho(Y^{(t)}), \rho(X^{(s)})\rho(Z^{(t)})] = 0.$$

Therefore it follows by linearity that for any $X^{(t)} \in [\mathfrak{L}_t, \mathfrak{L}_t]$

$$\operatorname{sp}\{\rho(X^{(s)})\rho(X^{(t)})\} = 0.$$

But $\mathfrak{L}_t$ is semisimple and therefore $[\mathfrak{L}_t, \mathfrak{L}_t] = \mathfrak{L}_t$. The lemma is therefore proved.

LEMMA 4. *Let $\mathfrak{L}$ be a simple algebra over K. Suppose $n = \dim \mathfrak{L} > 1$. Then for any representation ρ of $\mathfrak{L}$*

$$\operatorname{sp} \{\rho(X)\rho(Y)\} = \frac{\omega_\rho d_\rho}{n} B(X, Y).$$

Let $\bar{K}$ be the algebraic closure of K. Extend $\mathfrak{L}$ to $\bar{\mathfrak{L}}$ over $\bar{K}$ and ρ to a representation $\bar{\rho}$ of $\bar{\mathfrak{L}}$. We can find a $c \in \bar{K}$ such that the bilinear form

$$C(\bar{X}, \bar{Y}) = \operatorname{sp} \{\bar{\rho}(\bar{X})\bar{\rho}(\bar{Y})\} - cB(\bar{X}, \bar{Y}) \qquad (\bar{X}, \bar{Y} \in \bar{\mathfrak{L}})$$

is degenerate on $\bar{\mathfrak{L}}$. Let $\bar{\mathfrak{L}}_1$ be the set of all $\bar{X} \in \bar{\mathfrak{L}}$ such that $C(\bar{X}, \bar{Y}) = 0$ for all $\bar{Y} \in \bar{\mathfrak{L}}$. Since $C(\bar{X}, \bar{Y})$ is invariant under the adjoint representation of $\bar{\mathfrak{L}}$, $\bar{\mathfrak{L}}_1$ is an ideal in $\bar{\mathfrak{L}}$. Also since $n > 1$, $\mathfrak{L}$ and therefore $\bar{\mathfrak{L}}$ is semisimple. Hence we can find a complementary ideal $\bar{\mathfrak{L}}_2$ such that $\bar{\mathfrak{L}}_1 + \bar{\mathfrak{L}}_2 = \bar{\mathfrak{L}}$, $\bar{\mathfrak{L}}_1 \cap \bar{\mathfrak{L}}_2 = \{0\}$. Then in the notation of Lemma 2,

$$\Gamma_{\bar{\rho}(1)} = \sum_{1 \leq i, j \leq n_1} g^{(1)\,ij} \, \mathrm{sp} \, \{\bar{\rho}(\bar{X}_i^{(1)})\bar{\rho}(\bar{X}_j^{(1)})\}.$$

Also

$$\mathrm{sp} \, \{\bar{\rho}(\bar{X}_i^{(1)})\bar{\rho}(\bar{X}_j^{(i)})\} = c B(\bar{X}_i^{(1)}, \bar{X}_j^{(1)}) = c g_{ij}^{(1)} \qquad 1 \leq i, j \leq n_1$$

since $C(\bar{X}, \bar{Y}) = 0$ for all $\bar{X}, \bar{Y} \in \bar{\mathfrak{L}}_1$. Hence

$$\mathrm{sp} \, \Gamma_{\bar{\rho}(1)} = c n_1.$$

But we know that $\mathrm{sp} \, \Gamma_{\bar{\rho}(1)}$ is rational. Hence $c \in K$. Then it follows that the bilinear form $C(\bar{X}, \bar{Y})$ is actually degenerate on $\mathfrak{L}$. Let $\mathfrak{M}$ be the set of all $X \in \mathfrak{L}$ such that $C(X, Y) = 0$ for all $Y \in \mathfrak{L}$. Then $\mathfrak{M}$ is an ideal in $\mathfrak{L}$ and $\mathfrak{M} \neq \{0\}$. But since $\mathfrak{L}$ is simple it follows that $\mathfrak{M} = \mathfrak{L}$. Hence

$$\mathrm{sp} \, \{\rho(X)\rho(Y)\} = c B(X, Y)$$

for all $X, Y \in \mathfrak{L}$. Therefore if $X_i\, 1 \leq i \leq n$ is any base for $\mathfrak{L}$

$$d_\rho \omega_\rho = \mathrm{sp} \, \Gamma_\rho = \sum_{1 \leq i, j \leq n} g^{ij} \, \mathrm{sp} \, \rho(X_i)\rho(X_j) = cn.$$

Hence $c = (d_\rho \omega_\rho / n)$.

DEFINITION. *Let K be algebraically closed and let $\mathfrak{h}$ be a fixed Cartan subalgebra of $\mathfrak{L}$. Then for any representation ρ of $\mathfrak{L}$ we define the length $|\rho|$ of ρ as follows:*

$$|\rho| = \max_{\alpha, \Lambda} \left| \frac{2\Lambda(H_\alpha)}{\alpha(H_\alpha)} \right|$$

where α runs over all the roots and Λ over all the weights of the representation ρ.

Since for any root α and weight Λ, $(2\Lambda(H_\alpha))/(\alpha(H_\alpha))$ is an integer (see Weyl (6)) it follows that $|\rho|$ is an integer. By the string generated by α and Λ we mean the set St (Λ, α) of all weights of the form $\Lambda + \nu\alpha$ where ν is an integer. The number of elements in this set is called the length of the string. It is known that there exist two integers $k, k' \geq 0$ such that St (Λ, α) is precisely the set $\{\Lambda + \nu\alpha, -k \leq \nu \leq k'\}$. The length of the string is therefore $k + k' + 1$. By a string of the representation ρ we mean a set St (Λ, α) corresponding to some Λ and α. It is easily seen that $|\rho| + 1$ is the length of the longest string of ρ and $|\rho| = 0$ if and only if $\rho = 0$.

LEMMA 5. *Let K be algebraically closed. For any given integer $p \geq 0$ there exist only a finite number of inequivalent irreducible representations ρ of $\mathfrak{L}$ such that $|\rho| \leq p$.*

Let $\mathfrak{A}$ be the universal enveloping algebra of $\mathfrak{L}$ (see Harish-Chandra (4)).[5] We regard $\mathfrak{L}$ imbedded in $\mathfrak{A}$ and use the notation of Lemma 2. Put

$$f_\alpha(H_\alpha) = \prod_{-p \leq s \leq p} (H_\alpha - \frac{s}{2}\alpha(H_\alpha))$$

---

[5] In this paper $\mathfrak{A}$ is called the general enveloping algebra of $\mathfrak{L}$.

and consider the ideal $\mathfrak{X}_p$ in $\mathfrak{A}$ generated by the set of elements X_α^{p+1}, $f_\alpha(H_\alpha)$ where α runs over all the roots. Then, as shown in the paper mentioned above (p. 910), $\mathfrak{B} = \mathfrak{A}/\mathfrak{X}_p$ is a finite dimensional semisimple associative algebra and every representation of $\mathfrak{L}$ of length $\leq p$ can be regarded a representation of $\mathfrak{B}$. Since $\mathfrak{B}$ has only a finite number of inequivalent irreducible representations the result follows immediately.

LEMMA 6. *Let K be algebraically closed. If Λ_0 is the highest weight and Λ any weight of a homogeneous representation ρ of $\mathfrak{L}$ then*

$$\omega_\rho = |\Lambda_0|^2 + \tfrac{1}{2}\sum_\alpha |\Lambda_0(H_\alpha)| \geq |\Lambda|^2 + \tfrac{1}{2}\sum_\alpha |\Lambda(H_\alpha)|.$$

Further, there exist rational numbers κ, $\kappa' > 0$ such that for every homogeneous representation ρ, $\kappa\,|\rho|^2 \leq \omega_\rho \leq \kappa'\,|\rho|^2$.

We keep to the notation of Lemma 2. Then

$$\Gamma_\rho = \sum_{1 \leq i,\,j \leq l} g^{ij}\rho(H_i)\rho(H_j) + \sum_\alpha \rho(X_\alpha)\rho(X_{-\alpha}).$$

Let V be the representation space of ρ. Corresponding to the weight Λ, define V_Λ as before. From $[H, X_\alpha] = \alpha(H)X_\alpha$ it follows that $\rho(H)$ commutes with $\rho(X_\alpha)\,\rho(X_{-\alpha})$. Hence $\rho(X_\alpha)\rho(X_{-\alpha})$ leaves V_Λ invariant. Now

$$\sum_\alpha \rho(X_\alpha)\rho(X_{-\alpha}) = \sum_{\alpha > 0} \{\rho(X_\alpha)\rho(X_{-\alpha}) + \rho(X_{-\alpha})\rho(X_\alpha)\}$$

and

$$\rho(X_\alpha)\rho(X_{-\alpha}) + \rho(X_{-\alpha})\rho(X_\alpha) = 2\rho(X_{-\alpha})\rho(X_\alpha) + \rho(H_\alpha)$$

$$= 2\rho(X_\alpha)\rho(X_{-\alpha}) - \rho(H_\alpha)$$

since $[X_\alpha, X_{-\alpha}] = H_\alpha$. Let P be the set of those positive roots α (i.e. $\alpha > 0$) for which $\Lambda(H_\alpha) \geq 0$ and N the set of the remaining positive roots. Then

$$\sum_\alpha \rho(X_\alpha)\rho(X_{-\alpha}) = \sum_{\alpha \,\epsilon\, P} \{2\rho(X_{-\alpha})\rho(X_\alpha) + \rho(H_\alpha)\}$$

$$+ \sum_{\alpha \,\epsilon\, N} \{2\rho(X_\alpha)\rho(X_{-\alpha}) - \rho(H_\alpha)\}.$$

Now consider

$$\Delta = \Gamma_\rho - \sum_{1 \leq i,\,j \leq l} g^{ij}\rho(H_i)\rho(H_j) - \sum_{\alpha \,\epsilon\, P} \rho(H_\alpha) + \sum_{\alpha \,\epsilon\, N} \rho(H_\alpha)$$

$$= \sum_{\alpha \,\epsilon\, P} 2\rho(X_{-\alpha})\rho(X_\alpha) + \sum_{\alpha \,\epsilon\, N} 2\rho(X_\alpha)\rho(X_{-\alpha}).$$

Clearly Δ leaves V_Λ invariant. For any endomorphism A of V which leaves V_Λ invariant let A_Λ denote the restriction of A to V_Λ. Then it is clear that $\{\rho(H)\}_\Lambda = \Lambda(H)I_\Lambda$ where $I = I_{d_\rho}$. Hence

$$\Delta_\Lambda = \{\omega_\rho - \sum_{1 \leq i,\,j \leq l} g^{ij}\Lambda(H_i)\Lambda(H_j) - \sum_{\alpha \,\epsilon\, P} \Lambda(H_\alpha) + \sum_{\alpha \,\epsilon\, N} \Lambda(H_\alpha)\}I_\Lambda$$

$$= \{\omega_\rho - |\Lambda|^2 - \sum_{\alpha > 0} |\Lambda(H_\alpha)|\}\,I_\Lambda.$$

But

$$\text{sp } \Delta_\Lambda = \sum_{\alpha \, \epsilon \, P} 2 \text{ sp } \{\rho(X_{-\alpha})\rho(X_\alpha)\}_\Lambda + \sum_{\alpha \, \epsilon \, N} 2 \text{ sp } \{\rho(X_\alpha)\rho(X_{-\alpha})\}_\Lambda .$$

By an argument which we have already met in the proof of Lemma 2, we deduce from Lemma 1 that the eigenvalues of $\rho(X_\alpha)\rho(X_{-\alpha})$ and $\rho(X_{-\alpha})\rho(X_\alpha)$ are all of the form $\frac{1}{2}q\alpha(H_\alpha)$ where q is an integer ≥ 0. Hence the same is true of their restrictions to V_Λ. Therefore sp $\{\rho(X_{-\alpha})\rho(X_\alpha)\}_\Lambda$ and sp $\{\rho(X_\alpha)\rho(X_{-\alpha})\}_\Lambda$ are both rational and ≥ 0. Hence sp $\Delta_\Lambda \geq 0$ and therefore

$$(5) \qquad\qquad \omega_\rho \geq |\Lambda|^2 + \tfrac{1}{2} \sum_\alpha |\Lambda(H_\alpha)| .$$

Now let $\psi \neq 0$ be a weight vector belonging to the highest weight Λ_0. Then for any root α,

$$\rho(H)\rho(X_\alpha)\psi = \{\Lambda_0(H) + \alpha(H)\}\rho(X_\alpha)\psi \qquad\qquad (H \, \epsilon \, \mathfrak{h}).$$

If $\alpha > 0$, $\Lambda_0 + \alpha$ is not a weight since Λ_0 is the highest weight. Hence $\rho(X_\alpha)\psi = 0$. But

$$\Gamma_\rho = \sum_{1 \leq i, j \leq l} g^{ij}\rho(H_i)\rho(H_j) + \sum_{\alpha > 0} \{\rho(X_\alpha)\rho(X_{-\alpha}) + \rho(X_{-\alpha})\rho(X_\alpha)\}$$

$$= \sum_{1 \leq i, j \leq l} g^{ij}\rho(H_i)\rho(H_j) + \sum_{\alpha > 0} \{2\rho(X_{-\alpha})\rho(X_\alpha) + \rho(H_\alpha)\}.$$

Hence

$$\Gamma_\rho\psi = \{|\Lambda_0|^2 + \sum_{\alpha > 0} \Lambda_0(H_\alpha)\}\psi.$$

On the other hand $\Gamma_\rho\psi = \omega_\rho\psi$. Since $\psi \neq 0$ it follows that

$$\omega_\rho = |\Lambda_0|^2 + \sum_{\alpha > 0} \Lambda_0(H_\alpha).$$

But from (5) $\omega_\rho \geq |\Lambda_0|^2 + \sum_{\alpha > 0} |\Lambda_0(H_\alpha)|$. Hence $\Lambda_0(H_\alpha) \geq 0$ for all $\alpha > 0$ and $\omega_\rho = |\Lambda_0|^2 + \tfrac{1}{2}\sum_\alpha |\Lambda_0(H_\alpha)|$.

Let $c_i \, 1 \leq i \leq l$ be any l rational numbers such that $\sum_{1 \leq i \leq l} |c_i|^2 \geq 1$. Then we assert that there exists a fixed rational number $\kappa_1 > 0$ such that

$$\left| \sum_{1 \leq i \leq l} c_i H^i \right|^2 \geq \kappa_1 .$$

For otherwise there would exist a sequence $c_i^{(m)} \, 1 \leq i \leq l \, (m = 1, 2, \cdots)$ of rational numbers such that $\sum_{1 \leq i \leq l} |c_i^{(m)}|^2 \geq 1$ and $|H^{(m)}|^2 \to 0$ where $H^{(m)} = \sum_{1 \leq i \leq l} c_i^{(m)} H^i$. But $|H^{(m)}|^2 = $ sp (ad $H^{(m)})^2 = \sum_\alpha \{\alpha(H^{(m)})\}^2$. Since $\alpha(H^{(m)})$ are rational $|H^{(m)}|^2 \to 0$ implies $\alpha(H^{(m)}) \to 0$ for every root α. Hence in particular

$$\alpha_i(H^{(m)}) = \sum_{1 \leq i \leq l} c_j^{(m)}\alpha_i(H^j) = c_i^{(m)} \to 0$$

which contradicts the hypothesis that $\sum_{1 \leq i \leq l} |c_i^{(m)}|^2 \geq 1$. From this it is clear that

$$\left| \sum_{1 \leq i \leq l} c_i H^i \right|^2 \geq \kappa_1 \sum_{i=1}^{l} |c_i|^2$$

for any $c_i \in K_0$. Hence it follows from (5) that

$$\omega_\rho \geq |H_\Lambda|^2 \geq \kappa_1 \sum_{1 \leq i \leq l} |\Lambda(H_i)|^2 \geq \kappa_1 \left|\frac{\Lambda(H_1)}{|\alpha_1|^2}\right|^2 |\alpha_1|^4 \geq \kappa_1 \left|\frac{\Lambda(H_1)}{|\alpha_1|^2}\right|^2 |\alpha|_{\min}^4$$

where $|\alpha|_{\min}^2 = \min_\alpha |\alpha|^2$. Now there are only a finite number of possible ways in which a linearly independent set of l roots can be selected. For each such choice we get a positive rational number κ_1. Let κ_2 be the smallest among these numbers, then

$$\omega_\rho \geq \kappa_2 \left|\frac{\Lambda(H_a)}{|\alpha|^2}\right|^2 |\alpha|_{\min}^4$$

since any root α may be taken to be the first root in some set of l linearly independent roots. Further since this inequality holds for every weight Λ of the representation ρ we get

$$\omega_\rho \geq \kappa_2 \max_{\Lambda,\alpha} \left|\frac{\Lambda(H_a)}{|\alpha|^2}\right|^2 |\alpha|_{\min}^4 = \tfrac{1}{4}\kappa_2 |\alpha|_{\min}^4 |\rho|^2.$$

Hence we can put $\kappa = \tfrac{1}{4}\kappa_2 |\alpha|_{\min}^4$.

Now we shall prove the existence[6] of κ'. Let $\mathfrak{L}_s$, $1 \leq s \leq r$ be the simple components of $\mathfrak{L}$ so that $\mathfrak{L} = \mathfrak{L}_1 + \mathfrak{L}_2 + \cdots + \mathfrak{L}_r$, where the sum is direct. We keep to the notation of Lemma 3. Put $\omega_s = (1/d_\rho)\,\mathrm{sp}\,\Gamma_{\rho(s)}$. Then $\omega_\rho = \sum_{1 \leq s \leq r} \omega_s$. Also from Lemmas 3 and 4,

$$\mathrm{sp}\,\{\rho(X^{(s)})\rho(Y^{(t)})\} = \begin{cases} 0 & \text{if } s \neq t \\ \dfrac{\omega_s d_\rho}{n_s}\,\mathrm{sp}\,(\mathrm{ad}^{(s)}X^{(s)}\,\mathrm{ad}^{(s)}Y^{(s)}) & \text{if } s = t \end{cases}$$

where $X^{(s)} \in \mathfrak{L}_s$, $Y^{(t)} \in \mathfrak{L}_t$ and $1 \leq s, t \leq r$. For any root α write

$H_\alpha = \sum_{1 \leq s \leq r} H_\alpha^{(s)}$ where $H_\alpha^{(s)} \in \mathfrak{L}_s$. Then

$$(6) \qquad \mathrm{sp}\,\{\rho(H_\alpha)\}^2 = \sum_{1 \leq s \leq r} \mathrm{sp}\,\{\rho(H_\alpha^{(s)})\}^2 = \sum_{1 \leq s \leq r} \frac{\omega_s d_\rho}{n_s}\,\mathrm{sp}\,(\mathrm{ad}^{(s)}H_\alpha^{(s)})^2.$$

But clearly $\mathrm{ad}^{(s)}H_\alpha^{(s)}$ is the restricton of $\mathrm{ad}\,H_\alpha$ to $\mathfrak{L}_s$. Hence $\mathrm{ad}^{(s)}H_\alpha^{(s)}$ is a semi-simple matrix and all its eigenvalues are rational. Therefore $\mathrm{sp}\,(\mathrm{ad}^{(s)}H_\alpha^{(s)})^2 > 0$ unless $\mathrm{ad}^{(s)}H_\alpha^{(s)} = 0$. Now consider the sum

$$\sum_\alpha \mathrm{sp}\,(\mathrm{ad}^{(s)}H_\alpha^{(s)})^2$$

taken over all roots α. If it is zero, $\mathrm{ad}^{(s)}H_\alpha^{(s)} = 0$ for every α. Hence for every $X \in \mathfrak{L}_s$, $[H_\alpha, X] = 0$. Since the H_α's span $\mathfrak{h}$ it follows that $[H, X] = 0$ for all $H \in \mathfrak{h}$. But $\mathfrak{h}$ is a maximal abelian subalgebra of $\mathfrak{L}$ and therefore this implies that $X \in \mathfrak{h}$. As this is true for every $X \in \mathfrak{L}_s$, $\mathfrak{L}_s \subset \mathfrak{h}$. Hence $\mathfrak{L}_s$ is abelian which

---

[6] Since $\omega_\rho = |\Lambda_0|^2 + \tfrac{1}{2}\sum_\alpha |\Lambda_0(H_\alpha)|$ it follows immediately that $\omega_\rho \leq \kappa' |\rho|^2$ where $\kappa' = \tfrac{1}{4}\sum_{1 \leq i,j \leq l} |g^{ij}|\,|\alpha_i|^2 |\alpha_j|^2 + \tfrac{1}{2}\sum_\alpha |\alpha|^2$. However in order to be able to deduce the next corollary we give a different proof.

contradicts the fact that it is semisimple. Therefore $\sum_\alpha \mathrm{sp}\ (\mathrm{ad}^{(s)} H_\alpha^{(s)})^2 > 0$. Let

$$\kappa_1 = \min_{1 \leq s \leq r} \sum_\alpha \mathrm{sp}\ (\mathrm{ad}^{(s)} H_\alpha^{(s)})^2 > 0.$$

Since $\omega_s \geqq 0$ it follows from (6) that

$$\sum_\alpha \mathrm{sp}\ \{\rho(H_\alpha)\}^2 \geqq \kappa_1 d_\rho \sum_{1 \leq s \leq r} \frac{\omega_s}{n_s} \geqq \frac{\kappa_1 d_\rho}{n} \sum_{1 \leq s \leq r} \omega_s = \frac{\kappa_1 d_\rho}{n} \omega_\rho$$

Now

$$\frac{1}{d_\rho} \mathrm{sp}\ \{\rho(H_\alpha)\}^2 = \frac{\sum_\Lambda m_\Lambda \mid \Lambda(H_\alpha) \mid^2}{\sum_\Lambda m_\Lambda} \leqq \tfrac{1}{4} \mid \rho \mid^2 \mid \alpha \mid^4$$

where m_Λ is the multiplicity of the weight Λ and the sum is over all the weights of ρ. Hence

$$\tfrac{1}{4} \mid \rho \mid^2 \sum_\alpha \mid \alpha \mid^4 \geqq \frac{1}{d_\rho} \sum_\alpha \mathrm{sp}\ \{\rho(H_\alpha)\}^2 \geqq \frac{\kappa_1}{n} \omega_\rho .$$

Therefore $\kappa' \mid \rho \mid^2 \geqq \omega_\rho$ where $\kappa' = (n/4\kappa_1)\sum_\alpha \mid \alpha \mid^4$.

Let 0 denote the linear function on $\mathfrak{h}$ which maps $\mathfrak{h}$ into zero. Then we get the following interesting corollary immediately.

COROLLARY 1. *Suppose K is algebraically closed. Let m_ρ denote the multiplicity of 0 in any representation ρ of $\mathfrak{L}$. Then there exists a rational number $\delta > 0$ such that*

$$\frac{m_\rho}{d_\rho} \leqq 1 - \delta$$

for every homogeneous representation $\rho \neq 0$.

Notice that

$$\mathrm{sp}\ \{\rho(H_\alpha)\}^2 = \sum_\Lambda m_\Lambda \mid \Lambda(H_\alpha) \mid^2 \leqq \tfrac{1}{4} \mid \rho \mid^2 \mid \alpha \mid^4 (d_\rho - m_\rho).$$

Hence

$$\tfrac{1}{4} \mid \rho \mid^2 \left(1 - \frac{m_\rho}{d_\rho}\right) \sum_\alpha \mid \alpha \mid^4 \geqq \frac{1}{d_\rho} \sum_\alpha \mathrm{sp}\ \{\rho(H_\alpha)\}^2 \geqq \frac{\kappa_1}{n} \omega_\rho \geqq \frac{\kappa_1 \kappa}{n} \mid \rho \mid^2.$$

Since $\rho \neq 0, \mid \rho \mid^2 > 0$ and therefore

$$1 - \frac{m_\rho}{d_\rho} \geqq \delta = \frac{\kappa}{\kappa'} > 0$$

where $\kappa' = (n/4\kappa_1)\sum_\alpha \mid \alpha \mid^4$. Hence $(m_\rho/d_\rho) \leqq 1 - \delta$.

Now we shall prove that given any $\mu \in K_0$ there exist only a finite number of inequivalent irreducible representations ρ of $\mathfrak{L}$ such that $\omega_\rho \leqq \mu$. First we need a few elementary facts about representations.

PROPOSITION 1. *Let ρ and ρ' be any two irreducible representations of $\mathfrak{L}$. Then $\rho' \frown \rho^*$ if and only if $0_1 \prec \rho + \rho'$.*

The proof is well known.

PROPOSITION 2. *Let σ, ρ and ρ_i $1 \leq i \leq r$ be represenations of $\mathfrak{L}$ such that σ is irreducible and $\rho \backsim \rho_1 + \rho_2 + \cdots + \rho_r$. Then if $\sigma < \rho$ there exists an i such that $\sigma < \rho_i$.*

We have only to make use of the fact that every representation of $\mathfrak{L}$ is completely reducible. The proof then follows from standard results (see Chevalley (2), p. 175 proposition 3).

PROPOSITION 3. *Let ρ, ρ' be any two representations of $\mathfrak{L}$. Then $0_1 < \rho + \rho'$ if and only if there exists a representation σ of $\mathfrak{L}$ such that $\sigma < \rho$ and $\sigma^* < \rho'$.*

For let $\rho \backsim \rho_1 + \rho_2 + \cdots + \rho_r$, $\rho' \backsim \rho_1' + \rho_2' + \cdots + \rho_{r'}'$, where ρ_i $1 \leq i \leq r$ and ρ_j' $1 \leq j \leq r'$ are irreducible. Then $\rho + \rho'$ is equivalent to the direct sum of $\rho_i + \rho_j'$ $1 \leq i \leq r$, $1 \leq j \leq r'$. Therefore if $0_1 < \rho + \rho'$, it follows from Proposition 2 that $0_1 < \rho_i + \rho_j'$ for some i, j. But then from proposition 1, $\rho_j' \backsim \rho_i^*$. Hence $\rho_i < \rho$ and $\rho_i^* < \rho'$. The converse is obvious.

PROPOSITION 4. *Let ρ_1, ρ_2 and σ be any three irreducible representations of $\mathfrak{L}$. Then $\rho_1 < \rho_2 + \sigma$ if and only if $\rho_2 < \rho_1 + \sigma^*$.*

If $\rho_1 < \rho_2 + \sigma$, $0_1 < \rho_1 + \rho_1^* < \rho_2 + \sigma + \rho_1^*$. Therefore $0_1 < \rho_2 + (\sigma + \rho_1^*)$. Hence from Proposition 3 there exists a representation ρ such that $\rho < \rho_2$, $\rho^* < \sigma + \rho_1^*$. But since ρ_2 is irreducible $\rho \backsim \rho_2$. Hence $\rho_2^* < \sigma + \rho_1^*$ or $\rho_2 < (\sigma + \rho_1^*)^* \backsim \rho_1 + \sigma^*$. Similarly we prove the converse.

PROPOSITION 5. *Let ρ_1, ρ_2 be two inequivalent irreducible representations of $\mathfrak{L}$ over K. Let $\bar{K}$ be an extension field of K. Extend $\mathfrak{L}$ to $\bar{\mathfrak{L}}$ over $\bar{K}$ and ρ_i $(i = 1, 2)$ to the representations $\bar{\rho}_i$ of $\bar{\mathfrak{L}}$. Then there exists no irreducible representation $\bar{\rho}$ of $\bar{\mathfrak{L}}$ such that $\bar{\rho} < \bar{\rho}_1$, $\bar{\rho} < \bar{\rho}_2$.*

For if such a $\bar{\rho}$ exists it follows from Proposition 2 that $0_1 < \bar{\rho}_1 + \bar{\rho}_2^*$. Choose a base X_j $1 \leq j \leq n$ in $\mathfrak{L}$. Then $\rho_i(X_j) = \bar{\rho}_i(X_j)$ $i = 1, 2$, $1 \leq j \leq n$. Let ξ_{kl}^i $1 \leq k, l \leq d_1$ and η_{kl}^i $1 \leq k, l \leq d_2$ denote the coefficients of the matrices $\rho_1(X_j)$ and $\rho_2^*(X_j)$ respectively. Then $0_1 < \bar{\rho}_1 + \bar{\rho}_2^*$ implies that it is possible to find elements $a_{kl} \in \bar{K}$ $1 \leq k \leq d_1$, $1 \leq l \leq d_2$ not all of them zero such that

$$\sum_{1 \leq p \leq d_1} \xi_{kp}^i a_{pl} + \sum_{1 \leq q \leq d_2} \eta_{lq}^i a_{kq} = 0$$

for all $1 \leq j \leq n$, $1 \leq k \leq d_1$, $1 \leq l \leq d_2$. But since ξ_{kl}^i, $\eta_{kl}^i \in K$ it follows that if these linear homogeneous equations in a_{kl} have a non-trivial solution in $\bar{K}$, they already have such a solution in K. This in turn implies that $0_1 < \rho_1 + \rho_2^*$. But ρ_1, ρ_2 are irreducible. Hence from Proposition 1, $\rho_1 \backsim \rho_2$ which contradicts our hypothesis.

LEMMA 7. *Let μ be any rational number. There exists only a finite number of inequivalent irreducible representations ρ of $\mathfrak{L}$ such that $\omega_\rho \leq \mu$.*

First suppose K is algebraically closed. Then it follows from Lemma 6 that for any such representation $\kappa \mid \rho \mid^2 \leq \omega_\rho \leq \mu$. Hence $\mid \rho \mid \leq p$ where p is some positive integer such that $p^2 \geq (\mu/\kappa)$. Applying Lemma 5 we get the result.

If K is not algebraically closed let $\bar{K}$ be its algebraic closure. Extend $\mathfrak{L}$ to an algebra $\bar{\mathfrak{L}}$ over $\bar{K}$. Let $\bar{\rho}_i$ $1 \leq i \leq r$ be a finite set of inequivalent irreducible representations of $\bar{\mathfrak{L}}$ such that if $\bar{\sigma}$ is any irreducible representation of $\bar{\mathfrak{L}}$ for which $\omega_{\bar{\sigma}} \leq \mu$, then $\bar{\sigma} \backsim \bar{\rho}_i$ for some i. Let $\sigma_1, \cdots, \sigma_k$ be any k inequivalent irreducible representations of $\mathfrak{L}$ such that $\omega_{\sigma_j} \leqq \mu \ 1 \leqq j \leqq k$. Extend σ_j to a representation $\bar{\sigma}_j$ of $\bar{\mathfrak{L}}$. Clearly $\bar{\sigma}_j$ is a homogeneous representation $\bar{\mathfrak{L}}$. Therefore if $\bar{\sigma}$ is any irreducible component of $\bar{\sigma}_j$, $\omega_{\bar{\sigma}} = \omega_{\bar{\sigma}_j} = \omega_{\sigma_j} \leqq \mu$. Hence $\bar{\sigma} \frown \bar{\rho}_i$ for some i $(1 \leqq i \leqq r)$. Therefore for each j $(1 \leqq j \leqq k)$ we can find an i_j $(1 \leqq i_j \leqq r)$ such that $\bar{\rho}_{i_j} < \bar{\sigma}_j$. From Proposition 5 it follows that $i_j \neq i_l$ unless $j = l$. Hence it follows that $k \leqq r$ and therefore $\mathfrak{L}$ cannot have more than r inequivalent irreducible representations σ such that $\omega_\sigma \leqq \mu$.

LEMMA 8. *Let* ρ_1, ρ_2 *be any two homogeneous representations of* $\mathfrak{L}$. *Then for any irreducible component* σ *of* $\rho_1 + \rho_2$

$$\sqrt{\omega_\sigma} \leqq \sqrt{\omega_{\rho_1}} + \sqrt{\omega_{\rho_2}}$$

where $\sqrt{\omega_\rho}$ *is the real nonnegative square root of* ω_ρ.

First suppose K is algebraically closed. It is easily seen that all the weights of the representation $\rho_1 + \rho_2$ are of the form $\lambda + \mu$ where λ, μ are some weights of ρ_1, ρ_2 respectively. Hence the highest weight of σ can also be written in this form. Therefore from Lemma 6,

$$\omega_\sigma = |\lambda + \mu|^2 + \tfrac{1}{2} \sum_\alpha |\lambda(H_\alpha) + \mu(H_\alpha)|$$

$$= |\lambda|^2 + |\mu|^2 + 2\langle\lambda|\mu\rangle + \tfrac{1}{2} \sum_\alpha |\lambda(H_\alpha) + \mu(H_\alpha)|$$

$$\leqq |\lambda|^2 + \tfrac{1}{2} \sum_\alpha |\lambda(H_\alpha)| + |\mu|^2 + \tfrac{1}{2} \sum_\alpha |\mu(H_\alpha)| + 2|\lambda||\mu|$$

$$\leqq \omega_{\rho_1} + \omega_{\rho_2} + 2\sqrt{\omega_{\rho_1}}\sqrt{\omega_{\rho_2}} = (\sqrt{\omega_{\rho_1}} + \sqrt{\omega_{\rho_2}})^2.$$

The assertion is therefore proved in this case.

If K is not algebraically closed extend $\mathfrak{L}$ to $\bar{\mathfrak{L}}$ over the algebraic closure $\bar{K}$ of K. Let $\bar{\rho}_1$, $\bar{\rho}_2$ and $\bar{\sigma}$ denote the extensions of ρ_1, ρ_2 and σ respectively to representations of $\bar{\mathfrak{L}}$. It is clear that $\bar{\rho}_1$, $\bar{\rho}_2$ and $\bar{\sigma}$ are all homogeneous. If $\bar{\sigma}_1$ is any irreducible component of $\bar{\sigma}$ it follows from the above proof that

$$\sqrt{\omega_{\bar{\sigma}}} = \sqrt{\omega_{\bar{\sigma}_1}} \leqq \sqrt{\omega_{\bar{\rho}_1}} + \sqrt{\omega_{\bar{\rho}_2}}.$$

But since $\omega_{\bar{\sigma}}$, $\omega_{\bar{\rho}_1}$, $\omega_{\bar{\rho}_2}$ respectively are equal to ω_σ, ω_{ρ_1}, ω_{ρ_2} we get the result.

LEMMA 9. *Let* $\mathfrak{L}_s$ $1 \leqq s \leqq r$ *be the simple components of* $\mathfrak{L}$ *so that*

$$\mathfrak{L} = \mathfrak{L}_1 + \cdots + \mathfrak{L}_r$$

where the sum is direct. For any representation ρ *of* $\mathfrak{L}$ *let* $\rho^{(s)}$ *denote the representation of* $\mathfrak{L}_s$ *defined by* ρ. *Let* ρ_1, ρ_2 *be any two irreducible representations of* $\mathfrak{L}$. *Then*

$$\Gamma_{\rho_1+\rho_2} = (\Gamma_{\rho_1} + \Gamma_{\rho_2}) + \theta$$

where $\theta = \sum_{1 \leqq s \leqq r} \theta_s$ *and*

$$\theta_s = 2 \sum_{1 \leqq i, j \leqq n_s} g^{(s)ij} \rho_1(X_i^{(s)}) \times \rho_2(X_j^{(s)})$$

in the notation of Lemma 3. Further, sp $\theta = 0$ *so that* $\omega_{\rho_1+\rho_2} = \omega_{\rho_1} + \omega_{\rho_2}$ *and*

$$\text{sp } \theta^2 = 4 \, d_{\rho_1} d_{\rho_2} \sum_{1 \leqq s \leqq r} \frac{1}{n_s} \omega_{\rho_1(s)} \omega_{\rho_2(s)}.$$

Let $g^{(s,t)ij}$ be defined as in the proof of Lemma 3. Then the first statement follows immediately from the fact that $g^{(s,t)ij} = 0$ for $s \neq t$. Also for any $X, Y \in \mathfrak{L}$

$$\mathrm{sp}\ \{\rho_1(X) \times \rho_2(Y)\} = \{\mathrm{sp}\ \rho_1(X)\}\ \{\mathrm{sp}\ \rho_2(Y)\} = 0$$

since $X, Y \in \mathfrak{L} = [\mathfrak{L}, \mathfrak{L}]$. Hence $\mathrm{sp}\ \theta = 0$ and $\omega_{\rho_1+\rho_2} = \omega_{\rho_1} + \omega_{\rho_2}$. Now

$$\theta^2 = 4 \sum_{s,t} \sum_{i,j,k,l} g^{(s)ij} g^{(t)kl} \rho_1(X^{(s)}_i)\rho_1(X^{(t)}_k) \times \rho_2(X^{(s)}_j)\rho_2(X)^{(t)}_l.$$

But from Lemma 3, $\mathrm{sp}\ \{\rho(X^{(s)})\rho(X^{(t)})\} = 0$ for any representation ρ of $\mathfrak{L}$ and $X^{(s)} \in \mathfrak{L}_s$, $X^{(t)} \in \mathfrak{L}_t$, $s \neq t$. Hence

$$\theta^2 = 4 \sum_s \sum_{i,j,k,l} g^{(s)ij} g^{(s)kl}\ \mathrm{sp}\ \{\rho_1(X^{(s)}_i)\rho_1(X^{(s)}_k)\}\ \mathrm{sp}\ \{\rho_2(X^{(s)}_j)\rho_2(X^{(s)}_l)\}$$

$$= 4 \sum_s \sum_{i,j,k,l} g^{(s)ij} g^{(s)kl} g^{(s)}_{ik} \frac{1}{n_s} \omega_{\rho_1(s)}\ d_{\rho_1} g^{(s)}_{jl} \frac{1}{n_s} \omega_{\rho_2(s)}\ d_{\rho_2} \qquad \text{from Lemma 3;}$$

$$= 4\ d_{\rho_1} d_{\rho_2} \sum_s \frac{1}{n_s} \omega_{\rho_1(s)}\omega_{\rho_2(s)}\ .$$

The lemma is therefore proved.

For any semisimple algebra $\mathfrak{L}$ over K put $\omega(\mathfrak{L}) = \min_\rho \omega_\rho$ where ρ runs over all irreducible representations of $\mathfrak{L}$ different from 0_1. From Lemma 7 we know that $\omega(\mathfrak{L}) > 0$.

LEMMA 10. *Let $\mathfrak{F}$ be a ring of representations of $\mathfrak{L}$. For any real number $\mu \geqq 0$ and an irreducible representation ρ of $\mathfrak{L}$ let $\mathfrak{F}(\rho, \mu)$ denote the set of all irreducible components of all representations of the form $\rho + \rho'$ where ρ' is irreducible and belongs to $\mathfrak{F}$ and $\omega_{\rho'} \leqq \mu$. Then there exists a real number μ_0 with the following property. If ρ is any irreducible representation of $\mathfrak{L}$ in $\mathfrak{F}$ such that $\omega_\rho \leqq \omega_\sigma$ for every $\sigma \in \mathfrak{F}(\rho, \mu_0)$ then $\omega_\rho \leqq \mu_0$.*

We use the notation of Lemma 9. By suitably renumbering the simple components $\mathfrak{L}_s$ if necessary, we may assume that $\omega_{\rho(s)} = 0$ for all $\rho \in \mathfrak{F}$ if and only if $r_0 < s \leqq r$. In case $r_0 = 0$, every element in $\mathfrak{F}$ is a zero representation and $\mu_0 = 0$ has the required property.[7] Hence we suppose that $r_0 > 0$. For each $1 \leqq s \leqq r_0$ choose an irreducible representation $\rho_s \in \mathfrak{F}$ such that $\omega_{\rho_s(s)} \neq 0$. Put $\omega_s = \omega_{\rho_s(s)}$, $\Omega = \max_{1 \leqq s \leqq r_0} \omega_{\rho_s}$ and let ω denote the least among the numbers $\omega(\mathfrak{L})$, $\omega_s\ 1 \leqq s \leqq r_0$. Put $\mu_0 = (\tfrac{1}{4} n r_0)^2\ \{2 + \sqrt{\Omega/\omega}\}^2 \Omega^3/\omega^2$. We assert that μ_0 has the required property. For a fixed s consider $\rho + \rho_s$. Let V and V_s be the representation spaces of ρ and ρ_s respectively, so that $V \times V_s$ is the representation space[8] for $\rho + \rho_s$. $V \times V_s$ can be decomposed into a direct sum of irreducible subspaces $W_i\ 1 \leqq i \leqq g$. In each W_i, $\rho + \rho_s$ induces an irreducible representation σ_i. Put $\theta = \Gamma_{\rho+\rho_s} - (\Gamma_\rho + \Gamma_{\rho_s})$. In each W_i, θ is represented by a rational multiple of the unit matrix. Let θ_i and I_i denote the restrictions of θ and $I_{d_\rho, d_{\rho_s}}$ to W_i. Further let p be the number of positive eigenvalues of θ taking their multiplicities into account. Similarly let q be the number of negative eigenvalues. We denote

---

[7] It is perhaps worth mentioning that the condition $\rho \in \mathfrak{F}$ is needed in the proof only in case $r_0 = 0$.

[8] $V \times V_s$ is the Kronecker product of V and V_s (see Chevalley (2), Chapter VI).

these eigenvalues by a_u $1 \leq u \leq p$ and b_v $1 \leq v \leq q$ respectively. Put $a = 1/p \sum_{1 \leq u \leq p} a_u$, $b = 1/q \sum_{1 \leq v \leq q} |b_v|$, $a_m = \max_u a_u$, $b_m = \max_v |b_v|$. Then

$$\text{sp } \theta^2 = \sum_{1 \leq u \leq p} a_u^2 + \sum_{1 \leq v \leq q} b_v^2 \leq paa_m + qbb_m .$$

But since $\text{sp } \theta = 0$, $pa = qb$. Hence

$$\text{sp } \theta^2 \leq (a_m + b_m)qb \leq q(a_m + b_m)b_m \leq d_\rho d_{\rho_s}(a_m + b_m)b_m$$

since $q \leq \dim (V \times V_s) = d_\rho d_{\rho_s}$. Now we assert that $b_m \leq \omega_{\rho_s}$. For otherwise θ would have a negative eigenvalue $b_v < -\omega_{\rho_s}$ and therefore $\theta_i = b_v I_i$ for some i. Then it is clear that

$$\omega_{\sigma_i} = \omega_\rho + \omega_{\rho_s} + b_v < \omega_\rho .$$

But since $\sigma_i < \rho + \rho_s$, ρ_s is irreducible and belongs to $\mathfrak{F}$ and $\omega_{\rho_s} \leq \Omega \leq \mu_0$ it follows that $\sigma_i \epsilon \mathfrak{F}(\rho, \mu_0)$. Therefore this contradicts our hypothesis about ρ. Hence $b_m \leq \omega_{\rho_s}$ and

$$\text{sp } \theta^2 \leq d_\rho d_{\rho_s}(a_m + \omega_{\rho_s})\omega_{\rho_s} .$$

Further, there exists a j $(1 \leq j \leq g)$ such that $\theta_j = a_m I_j$. Hence

$$\omega_{\sigma_j} = \omega_\rho + \omega_{\rho_s} + a_m \leq \omega_\rho + \omega_{\rho_s} + 2\sqrt{\omega_\rho \omega_{\rho_s}}$$

from Lemma 8, since $\sigma_j < \rho + \rho_s$. Therefore $a_m \leq 2\sqrt{\omega_\rho \omega_{\rho_s}}$ and

$$\text{sp } \theta^2 \leq d_\rho d_{\rho_s}(\omega_{\rho_s} + 2\sqrt{\omega_\rho \omega_{\rho_s}})\omega_{\rho_s} .$$

But from Lemma 9 we know that

$$\frac{1}{d_\rho d_{\rho_s}} \text{ sp } \theta^2 = 4 \sum_{1 \leq t \leq r} \frac{1}{n_t} \omega_{\rho(t)} \omega_{\rho_s(t)} \geq \frac{4}{n} \omega_{\rho(s)} \omega_s .$$

Hence

$$\frac{4}{n} \omega_{\rho(s)} \omega_s \leq (\omega_{\rho_s} + 2\sqrt{\omega_\rho \omega_{\rho_s}})\omega_{\rho_s} .$$

Since this is true for all $1 \leq s \leq r_0$ and $\omega_\rho = \sum_{1 \leq s \leq r_0} \omega_{\rho(s)}$, we have

$$\frac{4}{n} \omega_\rho \leq \sum_{1 \leq s \leq r_0} (\omega_{\rho_s} + 2\sqrt{\omega_\rho \omega_{\rho_s}}) \frac{\omega_{\rho_s}}{\omega_s} \leq r_0(\Omega + 2\sqrt{\Omega \omega_\rho}) \frac{\Omega}{\omega} .$$

Now first suppose that $\rho \neq 0$, so that $\omega_o \geq \omega$. Then

$$\frac{4}{n} \leq r_0 \sqrt{\frac{\Omega}{\omega_\rho}} \left(2 + \sqrt{\frac{\Omega}{\omega_\rho}}\right) \frac{\Omega}{\omega_\rho} \leq r_0 \sqrt{\frac{\Omega}{\omega_\rho}} \left(2 + \sqrt{\frac{\Omega}{\omega}}\right) \frac{\Omega}{\omega} .$$

Hence

$$\omega_\rho \leq \left(\frac{nr_0}{4}\right)^2 \left\{2 + \sqrt{\frac{\Omega}{\omega}}\right\}^2 \frac{\Omega^3}{\omega^2} = \mu_0 .$$

On the other hand if $\rho = 0_1$, $\omega_\rho = 0$ and the inequality is true trivially. Therefore the lemma is proved.

Now we are in a position to prove Theorem 1. Using the notation of Lemma 10, we may assume that $r_0 > 0$ since otherwise the proof is trivial. Also we can clearly suppose that $\mu_0 \geqq \Omega = \max_{1 \leqq s \leqq r_0} \omega_{\rho_s}$. From Lemma 7, it is possible to choose a finite number of inequivalent irreducible representations σ_h $1 \leqq h \leqq k$ of $\mathfrak{L}$ in $\mathfrak{F}$ such that $\omega_{\sigma_h} \leqq \mu_0$ and if ρ is any irreducible representation in $\mathfrak{F}$ for which $\omega_\rho \leqq \mu_0$ then $\rho \backsim \sigma_h$ for some h. Let σ_0 be the direct sum of σ_h $1 \leqq h \leqq k$. We claim that σ_0 generates $\mathfrak{F}$. Let $\mathfrak{F}_0$ be the subring of $\mathfrak{F}$ generated by σ_0. Consider the set $\mathfrak{S}$ of all irreducible representations in $\mathfrak{F}$ which are not in $\mathfrak{F}_0$. If $\mathfrak{F} \neq \mathfrak{F}_0$, $\mathfrak{S}$ is not empty. Therefore from Lemma 7, there exists a $\rho \, \epsilon \, \mathfrak{S}$ such that $\omega_\rho = \min_{\sigma \epsilon \mathfrak{S}} \omega_\sigma$. Clearly $\omega_\rho > \mu_0$ for otherwise ρ would belong to $\mathfrak{F}_0$ and therefore not to $\mathfrak{S}$. Hence from Lemma 10, there exist two irreducible representations σ and ν such that $\nu \, \epsilon \, \mathfrak{F}$, $\omega_\nu \leqq \mu_0$, $\sigma < \rho + \nu$ and $\omega_\sigma < \omega_\rho$. From our choice of ρ it is clear that $\sigma \notin \mathfrak{S}$ and therefore $\sigma \, \epsilon \, \mathfrak{F}_0$. Also $\nu^* \, \epsilon \, \mathfrak{F}^* = \mathfrak{F}$ and $\omega_{\nu^*} = \omega_\nu \leqq \mu_0$. Therefore $\nu^* \, \epsilon \, \mathfrak{F}_0$. But from Proposition 4, $\rho < \sigma + \nu^* \, \epsilon \, \mathfrak{F}_0$. Hence $\rho \, \epsilon \, \mathfrak{F}_0$ which gives a contradiction. Therefore $\mathfrak{S}$ is empty and $\mathfrak{F} = \mathfrak{F}_0$.

3. Proof of Theorem 2

First we need two preliminary lemmas.

LEMMA 11. *Let ρ be any representation of $\mathfrak{L}$. Then $0_1 < \rho + \rho + \cdots + \rho = \rho^{(d_\rho)}$ where $\rho^{(d_\rho)}$ is the Kronecker sum of ρ with itself d_ρ times.*

Put $d_\rho = d$ for simplicity and let V be the representation space of ρ. Consider the Kronecker product $V_d = V \times V \times \cdots \times V$ of V with itself d times (see Chevalley (2), Chapter VI). Then ρ induces the representation $\rho^{(d)}$ in V_d. Choose a base e_i $1 \leqq i \leqq d$ for V and let W be the one-dimensional subspace of V_d spanned by the vector

$$f = \sum_{1 \leqq i_1, \cdots, i_d \leqq d} \varepsilon_{i_1 i_2, \cdots, i_d}(e_{i_1} \times e_{i_2} \times \cdots \times e_{i_d})$$

where $\varepsilon_{i_1 i_2 \cdots i_d} = 0$ unless $i_1, \cdots, i_d$ are all distinct and is 1 or -1 according as $i_1, \cdots, i_d$ is an even or odd permutation of $1, 2, \cdots, d$. It is clear that W is invariant under the representation $\rho^{(d)}$. For any $X \, \epsilon \, \mathfrak{L}$ let $\theta(X)$ denote the restriction of $\rho^{(d)}(X)$ to W. Then θ is a one dimensional representation of $\mathfrak{L}$. Therefore, since $\mathfrak{L}$ is semisimple, $\theta = 0$. Hence $0_1 < \rho^{(d)}$.

LEMMA 12. *Let $\mathfrak{M}$ be any Lie algebra (not necessarily semisimple) over K. Let $\mathfrak{A}$ be the universal enveloping algebra of $\mathfrak{M}$ so that $\mathfrak{M} \subset \mathfrak{A}$. For any representation ρ of $\mathfrak{M}$ define π_ρ to be the uniquely determined representation of $\mathfrak{A}$ such that π_ρ coincides with ρ on $\mathfrak{M}$, and let $\mathfrak{X}_\rho$ be the kernel of π_ρ in $\mathfrak{A}$. Also, let $\rho^{(r)}$ denote the Kronecker sum of ρ with itself r times. Suppose ρ is any faithful representation of $\mathfrak{M}$ such that $\mathrm{sp} \, \rho(X) = 0$ for every $X \, \epsilon \, \mathfrak{M}$, then*

$$\bigcap_{r \geqq 1} \mathfrak{X}_{\rho^{(r)}} = \{0\}.$$

The proof of this lemma is essentially contained in Sections 3 and 4 of the paper mentioned above (Harish-Chandra (4)). Since only some very trivial modifications are needed we shall not repeat it here.

Now we come to the proof of Theorem 2. First suppose K is algebraically closed and $\Sigma = K$. Put $\hat{\mathfrak{L}} = \hat{\mathfrak{L}}(\mathfrak{F})$. Let $\rho_0 \,\epsilon\, \mathfrak{F}$ be a representation of $\mathfrak{L}$ which generates the whole ring $\mathfrak{F}$. Consider an element $\zeta \,\epsilon\, \hat{\mathfrak{L}}$ such that $\zeta(\rho_0) = 0$. It is clear that given any $\rho \,\epsilon\, \mathfrak{F}$, $\rho \prec \rho_0^{(r)}$ for some $r \geq 1$. Therefore since ζ is a representation of $\mathfrak{F}$ it follows that $\zeta(\rho) = 0$. Since this is true for every $\rho \,\epsilon\, \mathfrak{F}$, $\zeta = 0$. Now let $\zeta_i\, 1 \leq i \leq r$ be any r elements in $\hat{\mathfrak{L}}$. If $r > d_{\rho_0}^2$ the matrices $\zeta_i(\rho_0)\, 1 \leq i \leq r$ are certainly linearly dependent. Hence we can find $a_i \,\epsilon\, K\, 1 \leq i \leq r$ such that not all a_i are zero and $\sum_{1 \leq i \leq r} a_i \zeta_i(\rho_0) = \eta(\rho_0) = 0$ where $\eta = \sum_{1 \leq i \leq r} a_i \zeta_i$. But we have just seen that this implies $\eta = 0$. Hence ζ_i are linearly dependent. Therefore $\dim \hat{\mathfrak{L}} \leq d_{\rho_0}^2$.

It is easy to verify that the mapping $X \to \zeta_X$ as defined in the statement of Theorem 2 is a homomorphic mapping of $\mathfrak{L}$ into $\hat{\mathfrak{L}}$. Its kernel $\mathfrak{N}$ is the ideal consisting of those $X \,\epsilon\, \mathfrak{L}$ for which $\rho(X) = 0$ for all $\rho \,\epsilon\, \mathfrak{F}$. Let $\mathfrak{M}$ be the ideal in $\mathfrak{L}$ complementary to $\mathfrak{N}$ so that $\mathfrak{L} = \mathfrak{M} + \mathfrak{N}$, $\mathfrak{M} \cap \mathfrak{N} = \{0\}$. Then the mapping $X \to \zeta_X$ maps $\mathfrak{M}$ isomorphically. We may therefore identify X with ζ_X for $X \,\epsilon\, \mathfrak{M}$ and write $X(\rho) = \rho(X)$ for $\rho \,\epsilon\, \mathfrak{F}$ and $X \,\epsilon\, \mathfrak{M}$. In this way $\mathfrak{M}$ becomes a subalgebra of $\hat{\mathfrak{L}}$.

For any $\rho \,\epsilon\, \mathfrak{F}$ the mapping $\dot{\rho}$ given by $\dot{\rho}(\zeta) = \zeta(\rho)$ $(\zeta \,\epsilon\, \hat{\mathfrak{L}})$, is easily seen to be a representation of $\hat{\mathfrak{L}}$. Clearly $\dot{\rho}$ coincides with ρ on $\mathfrak{M}$ and therefore it is obvious that if ρ is irreducible so is also $\dot{\rho}$. Notice that $\dot{\rho}_0$ is a faithful representation of $\hat{\mathfrak{L}}$ since $\dot{\rho}_0(\zeta) = \zeta(\rho_0) = 0$ implies $\zeta = 0$. Without loss of generality we may assume that $\rho_0 = \rho_1 \dotplus \rho_2 \dotplus \cdots \dotplus \rho_r$ where $\rho_i\, 1 \leq i \leq r$ are irreducible. Then

$$\dot{\rho}_0(\zeta) = \zeta(\rho_0) = \zeta(\rho_1) \dotplus \zeta(\rho_2) \dotplus \cdots \dotplus \zeta(\rho_r)$$
$$= \dot{\rho}_1(\zeta) \dotplus \dot{\rho}_2(\zeta) \dotplus \cdots \dotplus \dot{\rho}_r(\zeta).$$

Hence $\dot{\rho}_0 = \dot{\rho}_1 \dotplus \dot{\rho}_2 \dotplus \cdots \dotplus \dot{\rho}_r$, $\dot{\rho}_i\, 1 \leq i \leq r$ being irreducible. It then follows from known results (see Chevalley (3), p. 97, proposition 7) that $\dot{\rho}_0(\hat{\mathfrak{L}})$ is a direct sum of its center and its derived algebra which is semisimple. Since $\dot{\rho}_0$ is a faithful representation of $\hat{\mathfrak{L}}$ the same holds for $\hat{\mathfrak{L}}$. We shall now show that the center of $\hat{\mathfrak{L}}$ is actually zero and so $\hat{\mathfrak{L}}$ is semisimple.

LEMMA 13. *Let ζ be any element in $\hat{\mathfrak{L}}$ such that $[\zeta, X] = 0$ for all $X \,\epsilon\, \mathfrak{M}$. Then $\zeta = 0$.*

Let ρ be any irreducible representation in $\mathfrak{F}$. Then

$$\dot{\rho}([\zeta, X]) = [\zeta(\rho), \rho(X)] = 0.$$

Since K is algebraically closed it follows from Schur's lemma that $\zeta(\rho) = c_\rho I_{d_\rho}$ where $c_\rho \,\epsilon\, K$. Consider a set $\sigma_0, \sigma_1, \cdots, \sigma_h$ of irreducible representations in $\mathfrak{F}$ such that $\sigma_0 \prec \sigma_1 \dotplus \sigma_2 \dotplus \cdots \dotplus \sigma_h$. On making use of the fact that ζ is a representation of $\mathfrak{F}$ it follows that

$$c_{\sigma_0} = c_{\sigma_1} + c_{\sigma_2} + \cdots + c_{\sigma_h} .$$

Put $c_0 = c_\rho$ for $\rho = 0_1$. Then since $0_1 \prec 0_1 + 0_1$ we get $c_0 = c_0 + c_0$. Hence $c_0 = 0$. Further, from Lemma 11, $0_1 \prec \rho^{(d_\rho)}$. Therefore if ρ is irreducible, $0 = d_\rho c_\rho$. Hence $c_\rho = 0$ for every irreducible representation ρ. Then it follows immediately that $\zeta(\rho) = 0$ for all $\rho \, \epsilon \, \mathfrak{F}$ and so $\zeta = 0$.

Thus we have proved that $\mathfrak{L}$ is semisimple. Let $\dot{\mathfrak{M}}$ be the smallest ideal in $\mathfrak{L}$ containing $\mathfrak{M}$. Let $\dot{\mathfrak{N}}$ be the complementary ideal so that $\mathfrak{L} = \dot{\mathfrak{M}} + \dot{\mathfrak{N}}$, $\dot{\mathfrak{M}} \cap \dot{\mathfrak{N}} = \{0\}$. Then $[\dot{\mathfrak{N}}, \mathfrak{M}] \subset [\dot{\mathfrak{N}}, \dot{\mathfrak{M}}] = \{0\}$. Hence from the above lemma $\dot{\mathfrak{N}} = \{0\}$ and $\dot{\mathfrak{M}} = \mathfrak{L}$. Therefore the smallest ideal in $\mathfrak{L}$ which contains $\mathfrak{M}$, is $\mathfrak{L}$ itself.

We have still to prove that $\mathfrak{M} = \mathfrak{L}$. Choose a base $X_i \, 1 \leq i \leq n$ for $\mathfrak{M}$. Define g_{ij} and g^{ij} with respect to this base as before. For any representation θ of $\mathfrak{L}$ put

$$\gamma_\theta = \sum_{1 \leq i, j \leq n} g^{ij} \theta(X_i) \theta(X_j).$$

Then we have the following lemma.

LEMMA 14. *Let θ be any representation of $\mathfrak{L}$. Then for every $\zeta \, \epsilon \, \mathfrak{L}$, $[\gamma_\theta, \theta(\zeta)] = 0$.*

Let $\mathfrak{A}$ be the universal enveloping algebra of $\mathfrak{L}$. For any representation θ of $\mathfrak{L}$ we define the representation π_θ of $\mathfrak{A}$ as in Lemma 12. Let $\mathfrak{X}_\theta$ denote the kernel of π_θ. Consider the faithful representation $\dot{\rho}_0$ of $\mathfrak{L}$. Since $\mathfrak{L}$ is semisimple sp $\dot{\rho}_0(\zeta) = 0$ for every $\zeta \, \epsilon \, \mathfrak{L}$. Hence from Lemma 12,

$$\bigcap_{r \geq 1} \mathfrak{X}_{\dot{\rho}_0(r)} = \{0\}.$$

But it is clear that $(\dot{\rho}_0)^{(r)} = \dot{\sigma}$ where $\sigma = \rho_0^{(r)}$. Therefore

$$\bigcap_{\rho \, \epsilon \, \mathfrak{F}} \mathfrak{X}_{\dot{\rho}} \subset \bigcap_{r \geq 1} \mathfrak{X}_{\dot{\rho}_0(r)} = \{0\}.$$

Now put $\gamma = \sum_{1 \leq i, j \leq n} g^{ij} X_i X_j \, \epsilon \, \mathfrak{A}$. If ρ is any irreducible representation in $\mathfrak{F}$, $\gamma_{\dot{\rho}} = \Gamma_\rho = \omega_\rho I_{d_\rho}$ from Lemma 3. Hence for any $\zeta \, \epsilon \, \mathfrak{L}$. $[\gamma_\rho, \zeta(\rho)] = 0$. Further, given any three representations ρ_1, ρ_2, ρ_3 in $\mathfrak{F}$ it is clear that if $[\gamma_{\dot{\rho}_i}, \zeta(\rho_i)] = 0$ for $i = 1, 2$ then $[\gamma_{\dot{\rho}_3}, \zeta(\rho_3)] = 0$ provided either $\rho_3 = \rho_1 + \rho_2$ or $\rho_3 \frown \rho_1$. Hence it follows immediately that $[\gamma_{\dot{\rho}}, \zeta(\rho)] = 0$ for all $\rho \, \epsilon \, \mathfrak{F}$. Therefore $\gamma\zeta - \zeta\gamma \, \epsilon \bigcap_{\rho \epsilon \mathfrak{F}} \mathfrak{X}_{\dot{\rho}} = \{0\}$. Hence $\gamma\zeta - \zeta\gamma = 0$ in $\mathfrak{A}$ and so for any representation θ of $\mathfrak{L}$

$$[\gamma_\theta, \theta(\zeta)] = \pi_\theta(\gamma\zeta - \zeta\gamma) = 0.$$

LEMMA 15. *Let V be the representation space of a representation θ of $\mathfrak{L}$. Let ψ_0 be an element in V such that $\theta(X)\psi_0 = 0$ for all $X \, \epsilon \, \mathfrak{M}$. Then $\theta(\zeta)\psi_0 = 0$ for all $\zeta \, \epsilon \, \mathfrak{L}$.*

Let U be the subspace of V consisting of all $\psi \, \epsilon \, V$ such that $\gamma_\theta\psi = 0$. Then $\psi_0 \, \epsilon \, U$ and U is invariant under $\theta(\mathfrak{L})$ since $[\gamma_\theta, \theta(\zeta)] = 0$. Let $\varphi(\zeta)$ denote the restriction of $\theta(\zeta)$ to U. Then φ is a representation of $\mathfrak{L}$. Let ρ be the representation of $\mathfrak{M}$ defined by φ. Then $\Gamma_\rho = \gamma_\varphi = 0$. Hence $\rho = 0$ and therefore $\varphi(X) = 0$ for all $X \, \epsilon \, \mathfrak{M}$. So the kernel of φ contains $\mathfrak{M}$ and therefore $\mathfrak{L}$, since the smallest ideal containing $\mathfrak{M}$ is $\mathfrak{L}$. Hence $\varphi = 0$ and $\theta(\zeta)\psi_0 = \varphi(\zeta)\psi_0 = 0$.

LEMMA 16. *Let V be the representation space of a representation θ of $\mathfrak{L}$. Suppose U is a subspace of V which is invariant under $\theta(\mathfrak{M})$. Then U is invariant under $\theta(\mathfrak{L})$.*

Since $\mathfrak{M}$ is semisimple we can find a subspace W of V invariant under $\theta(\mathfrak{M})$ and such that $V = U + W$, $U \cap W = \{0\}$. Choose bases $e_i \, 1 \leq i \leq r$ for U and

$e_i \, r + 1 \leqq i \leqq s$ for W. Clearly we may assume that $U \neq \{0\}$, $W \neq \{0\}$ since if either of them is zero the assertion is true trivially. Let V^* be the space dual to V (see Chevalley (2), p. 140) and $\lambda_i \, 1 \leqq i \leqq s$ a base in V^* dual to the base $e_i \, 1 \leqq i \leqq s$ in V. Then θ induces the representation θ^* in V^*. Consider the representation $\theta + \theta^*$ on $V \times V^*$. Put

$$\psi_0 = \sum_{1 \leqq i \leqq r} e_i \times \lambda_i \, .$$

It is readily verified that

$$(\theta + \theta^*)(X)\psi_0 = 0$$

for all $X \, \epsilon \, \mathfrak{M}$. Hence from the above lemma

$$(\theta + \theta^*)(\zeta)\psi_0 = 0$$

for every $\zeta \, \epsilon \, \mathfrak{L}$. For any given ζ, put

$$\theta(\zeta)e_i = \sum_{1 \leqq j \leqq s} e_j \, a_{ji} \qquad\qquad 1 \leqq i \leqq s$$

where $a_{ji} \, \epsilon \, K$. Then

$$\theta^*(\zeta)\lambda_i = - \sum_{1 \leqq j \leqq s} a_{ij}\lambda_j \qquad\qquad 1 \leqq i \leqq s.$$

Hence

$$0 = (\theta + \theta^*)(\zeta)\psi_0 = \sum_{1 \leqq i \leqq r} \sum_{1 \leqq j \leqq s} (e_j \, a_{ji} \times \lambda_i - e_i \times a_{ij}\lambda_j).$$

Since $e_i \times \lambda_j \, 1 \leqq i, j \leqq s$ form a base for $V \times V^*$ it follows that $a_{ji} = 0$ $r + 1 \leqq j \leqq s, 1 \leqq i \leqq r$. Hence $\theta(\zeta)e_i \, \epsilon \, U$ for $1 \leqq i \leqq r$, and therefore U is invariant under $\theta(\zeta)$.

Now we can complete the proof of the theorem. Consider the adjoint representation $\zeta \to \mathrm{ad} \, \zeta$ of $\mathfrak{L}$. Since $\mathfrak{M}$ is a subalgebra it is clear that $(\mathrm{ad}\ X)\mathfrak{M} \subset \mathfrak{M}$ for every $X \, \epsilon \, \mathfrak{M}$. Hence by the above lemma $(\mathrm{ad}\ \zeta)\mathfrak{M} \subset \mathfrak{M}$ for all $\zeta \, \epsilon \, \mathfrak{L}$. Therefore $\mathfrak{M}$ is an ideal in $\mathfrak{L}$ and so must coincide with $\mathfrak{L}$ and $\mathfrak{L}$ is the smallest ideal containing $\mathfrak{M}$. The theorem is therefore proved in this case.

If $\Sigma \neq K$ let $\bar{\mathfrak{L}}$ be the extension of $\mathfrak{L}$ over K. Clearly every Σ-representation ρ of $\mathfrak{L}$ defines uniquely a representation $\bar{\rho}$ of $\bar{\mathfrak{L}}$ and conversely. We can therefore regard $\mathfrak{F}$ as a ring of representations of $\bar{\mathfrak{L}}$ such that $\mathfrak{F} = \mathfrak{F}^*$ and argue the same way as above. Then it follows that the image of $\bar{\mathfrak{L}}$ under the extension of the mapping $X \to \zeta_X$ to $\bar{\mathfrak{L}}$ is the whole of $\mathfrak{L}$. But clearly this is equivalent to the assertion in the theorem.

4. Application to semisimple Lie groups

Let R and C denote the fields of real and complex numbers respectively. For any $A \, \epsilon \, E(C, d)$ we denote by $A_{ij} \, 1 \leqq i, j \leqq d$ the matrix elements of A and by $\bar{A}$ the complex conjugate of A and write $|A| = (\sum_{1 \leqq i, j \leqq d} |A_{ij}|^2)^{\frac{1}{2}} \geqq 0$. A is called real if $A = \bar{A}$. Let $\mathfrak{g}$ be a Lie algebra over R. For any C-representation ρ of $\mathfrak{g}$ we denote by $\bar{\rho}$ the C-representation $\bar{\rho}(X) = \overline{\rho(X)}$ $(X \, \epsilon \, \mathfrak{g})$. F being any set of C-representations of $\mathfrak{g}$ we put $\bar{F} = \{\bar{\rho} \mid \rho \epsilon F\}$. The group of all nonsingular matrices in $E(C, d)$ with the usual topology will be denoted by $GL(C, d)$.

DEFINITION. *Let $\mathfrak{g}_C$ be a Lie algebra over C. A subset $\mathfrak{g}_R$ of $\mathfrak{g}_C$ is called a real form of $\mathfrak{g}_C$ if the following conditions are fulfilled.*

1) *$X, Y \epsilon \mathfrak{g}_R$ and $a \epsilon R$ imply that aX, $X + Y$ and $[X, Y]$ all belong to $\mathfrak{g}_R$.*

2) *$\mathfrak{g}_R$ spans over C the whole algebra $\mathfrak{g}_C$.*

3) *The dimension of $\mathfrak{g}_R$ over R is the same as the dimension of $\mathfrak{g}_C$ over C.*

It is clear that a real form $\mathfrak{g}_R$ is a Lie algebra over R and every C-representation ρ of $\mathfrak{g}_R$ can be extended uniquely to a C-representation of $\mathfrak{g}_C$.

Let G be a connected Lie group and P a representation[9] of G. We denote the degree of P by $d^\circ(P)$. Given the representations P, P_1, P_2 of G we define $P_1 \dot{+} P_2$, $P_1 \times P_2$, $\gamma P \gamma^{-1}(\gamma \epsilon GL(C, d^\circ(P)))$ as usual (see Chevalley (2), Chapter VI) and P^*, $\bar{P}$ by $P^*(x) = {}^t(P(x^{-1}))$, $\bar{P}(x) = \overline{P(x)}$ $(x \epsilon G)$ respectively. The symbols $P_1 \frown P_2$, $P_1 < P_2$ and $P_1 \subset P_2$ have meanings completely similar to those defined earlier for Lie algebras. P is called real if $P = \bar{P}$. A nonempty set $\mathfrak{F}$ of representations of G is called a ring if 1) whenever P_1, $P_2 \epsilon \mathfrak{F}$, then $P_1 \dot{+} P_2$, $P_1 \times P_2 \epsilon \mathfrak{F}$ and 2) whenever $P \epsilon \mathfrak{F}$ and $P' < P$ then $P' \epsilon \mathfrak{F}$. F being any set of representations of G we put $\bar{F} = (\bar{P} \mid P \epsilon F\}$ and $F^* = \{P^* \mid P \epsilon F\}$. The ring generated by F is the smallest ring containing F.

Let $\mathfrak{E} = \mathfrak{E}^*$ and $\mathfrak{F} = \mathfrak{F}^*$ be two rings of representations of two connected semisimple Lie groups G and H respectively. A univalent mapping θ of $\mathfrak{E}$ into $\mathfrak{F}$ is called an isomorphism if $d^\circ(\theta(P)) = d^\circ(P)$, $\theta(P_1 \times P_2) = \theta(P_1) \times \theta(P_2)$, $\theta(P_1 \dot{+} P_2) = \theta(P_1) \dot{+} \theta(P_2)$, $\theta(\gamma P \gamma^{-1}) = \gamma \theta(P) \gamma^{-1}$, $\theta(P^*) = (\theta(P))^*$ for all P, P_1, $P_2 \epsilon \mathfrak{E}$ and $\gamma \epsilon GL(C, d^\circ(P))$ and if $\theta(P)$ is irreducible whenever P is irreducible $(P \epsilon \mathfrak{E})$. It is easily seen that $\theta(P_1) < \theta(P_2)$ if and only if $P_1 < P_2$ $(P_1, P_2 \epsilon \mathfrak{E})$. Hence it is clear that $\theta(\mathfrak{E})$ is a ring, $\theta(\mathfrak{E}) = (\theta(\mathfrak{E}))^*$ and the inverse mapping θ^{-1} defined on $\theta(\mathfrak{E})$ is also an isomorphism.

Let $\mathfrak{g}$ be the Lie algebra of G. Every representation P of G defines a C-representation dP of $\mathfrak{g}$ by the rule (see Chevalley (2), Chapter IV)

$$dP(X) = \lim_{t \to 0} \frac{P(\exp tX) - I}{t} \qquad (X \epsilon \mathfrak{g})$$

where I is the unit matrix of degree $d^\circ(P)$ and $t \epsilon R$. From the relation $P(\exp tX) = \exp t dP(X)$ it follows at once that $dP^* = (dP)^*$, $d\bar{P} = \overline{dP}$, $d(P_1 \dot{+} P_2) = dP_1 \dot{+} dP_2$, $d(P_1 \times P_2) = dP_1 \dot{+} dP_2$, $d(\gamma P_1 \gamma^{-1}) = \gamma dP_1 \gamma^{-1}$ $(\gamma \epsilon GL(C, d^\circ(P_1)))$. Also P is irreducible if and only if dP is irreducible and the mapping $P \to dP$ is univalent. For any set F of representations of G put $dF = \{dP \mid P \epsilon F\}$. Then dF is a ring if and only if F is a ring and $(dF)^* = dF^*$, $d\bar{F} = \overline{dF}$.

We now give a few definitions which will be required later.

DEFINITION.[10] *Let G be a connected Lie group and $\mathfrak{g}$ its Lie algebra. G will be*

---

[9] By a Lie group we always mean a real Lie group unless some further qualification is made explicitly. A representation of a group is always meant to be over C. For precise definitions of the terms used in this section see Chevalley (2), Chapter IV.

[10] I learn from the referee that a similar treatment of complex Lie groups is contained in Chevalley's unpublished manuscript for the forthcoming Vol. 2 of his book on Lie groups.

*called a complex Lie group if there is given a mapping which assigns to every $c \in C$
and $X \in \mathfrak{g}$ an element $c * X \in \mathfrak{g}$ such that the following conditions are fulfilled:*

1) $a * X = aX$ *for all $a \in R$ and $X \in \mathfrak{g}$.*

2) *Under the mapping $(c, X) \to c * X$ and the bracket operation $[X, Y]$ already
defined in $\mathfrak{g}$, $\mathfrak{g}$ becomes a Lie algebra over C.*

Usually we shall write cX instead of $c * X$ ($c \in C$, $X \in \mathfrak{g}$).

DEFINITION. *Let $\mathfrak{g}$ be the Lie algebra and P a representation of a complex Lie
group G. Then P is called a complex representation of G if dP is a homomorphism of
$\mathfrak{g}$ over C i.e. if $dP(cX) = cdP(X)$ for all $c \in C$ and $X \in \mathfrak{g}$.*

DEFINITION. *Let G be a Lie group. By the kernel $N(G)$ of G we mean the intersection
of the kernels of all representations of G. Clearly $N(G)$ is a closed invariant subgroup
of G. A representation P of G is called quasifaithful if the kernel of P coincides
with $N(G)$.*

From now on G shall denote a connected semisimple Lie group. Any other
extra hypothesis about G will always be mentioned explicitly. First we prove the
following lemma.

LEMMA 17. *G has a quasifaithful representation.*

Let $\mathfrak{F}$ be the ring of all representations of G. Then $d\mathfrak{F}$ is a ring of C-repre-
sentations of the Lie algebra $\mathfrak{g}$ of G and $d\mathfrak{F} = (d\mathfrak{F})^*$. Since $\mathfrak{g}$ is semisimple it
follows from a trivial extension of Theorem 1 that we can find a $P_0 \in \mathfrak{F}$ such
that dP_0 generates $d\mathfrak{F}$. Since the mapping $P \to dP$ is univalent it follows that
P_0 generates $\mathfrak{F}$. Hence for every $P \in \mathfrak{F}$, $P \prec P_0^{(r)}$ where $P_0^{(r)}$ is the Kronecker
product of P_0 with itself r times and $r \geq 1$. Hence it is clear that the kernel of P
contains the kernel of P_0 and therefore P_0 is quasifaithful.

LEMMA 18.[11] *Let P be any representation of G. Then the center of $P(G)$ is finite.*

Since every representation of a connected semisimple group is fully reducible we
may suppose that $P = P_1 \dotplus P_2 \dotplus \cdots \dotplus P_r$ where P_i $1 \leq i \leq r$ are irreducible.
Clearly if $P(x_0)$ ($x_0 \in G$) is in the center of $P(G)$, $P_i(x_0)$ is in the center of $P_i(G)$.
Hence by Schur's lemma $P_i(x_0) = c_i I_i$ $1 \leq i \leq r$ where $c_i \in C$ and I_i is the unit
matrix of degree $d_i = d^\circ(P_i)$. Let $\det P_i(x)$ denote the determinant of $P_i(x)$
($x \in G$). Then the mapping $x \to \det P_i(x)$ is an abelian representation of G. Since
G is connected and semisimple it is its own derived group. Hence $\det P_i(x) = 1$
for all $x \in G$. Therefore $c_i = \exp\left((m_i/d_i)2\pi\sqrt{-1}\right)$ where $0 \leq m_i < d_i$. So there
are only a finite number of possibilities for $P(x_0)$ and therefore the lemma is
proved.

LEMMA 19. *Let H be a compact connected semisimple Lie group. Then the fundamental group of H is finitely generated.*

Let $\tilde{H}$ be the universal covering group of H and φ the natural homomorphism
of $\tilde{H}$ onto H. Let $x \to \operatorname{Ad}(x)$ ($x \in H$) denote the adjoint representation of H.
Let D be the kernel of φ and D_0 the kernel of $\operatorname{Ad} \circ \varphi$. Then $D \subset D_0$. Also since
H is semisimple $\tilde{H}$, H and $\operatorname{Ad}(H)$ are locally isomorphic. Hence the fundamental
groups of H and $\operatorname{Ad}(H)$ are isomorphic to D and D_0 respectively. Since any
subgroup of a finitely generated abelian group is itself finitely generated, it is

---

[11] This Lemma is due to E. Cartan (Journ. de Math. 1929, vol. 94, p. 12.)

sufficient to prove that D_0 is finitely generated. Since H is semisimple, $\mathrm{Ad}(H)$ is the component of identity of an algebraic group (see Chevalley (2), p. 135) namely the group of automorphisms of its Lie algebra. It is easy to prove (Van der Waerden (7), p. 366) that every algebraic variety is triangulable. Therefore since $\mathrm{Ad}(H)$ is compact it is homeomorphic to a finite complex. The fundamental group, being abelian, is isomorphic to the first homology group of this complex and so is finitely generated.

THEOREM 3 (Weyl). *Let $\tilde{G}$ be the universal covering group of a compact connected semisimple Lie group G. Then $\tilde{G}$ is compact.*

Let φ denote the natural homomorphism of $\tilde{G}$ onto G and let D be the kernel of φ. Then D is a discrete group in the center of $\tilde{G}$. From Lemma 17 we can find a quasifaithful representation P_0 of $\tilde{G}$. Let D_0 be the kernel of P_0. Since G is compact it has a faithful representation Q and therefore D is the kernel of $Q_0 = Q \circ \varphi$. Hence $D \supset D_0$ and $G \cong P_0(\tilde{G})/P_0(D)$. But $P_0(D)$ is in the center of $P_0(\tilde{G})$. Hence by Lemma 18, $P_0(D)$ is finite. Since G is compact and $P_0(D)$ is finite, $P_0(\tilde{G})$ is compact. We now claim that $D_0 = \{\tilde{e}\}$ where $\tilde{e}$ is the unit element of $\tilde{G}$. Since D_0 is isomorphic to the fundamental group of the compact connected semisimple Lie group $P_0(\tilde{G})$, it follows from Lemma 19, that it is finitely generated. Hence in case $D_0 \neq \{\tilde{e}\}$ we can find a subgroup $D_1 \subset D_0$ such that $D_1 \neq D_0$ and D_0/D_1 is finite. Then $\tilde{G}/D_0 \cong (\tilde{G}/D_1)/(D_0/D_1)$. Since $\tilde{G}/D_0 \cong P_0(\tilde{G})$ is compact and D_0/D_1 is finite it follows that $\tilde{G}/D_1$ is compact. Hence it has a faithful representation. Therefore there exists a representation P_1 of $\tilde{G}$ whose kernel is exactly D_1. But since P_0 is quasifaithful, $D_1 \supset D_0$. However $D_1 \subset D_0$ and $D_1 \neq D_0$ and so we get a contradiction. Hence $D_0 = \{\tilde{e}\}$ and P_0 is faithful. Therefore $\tilde{G} \cong P_0(\tilde{G})$ is compact.

Let $\mathfrak{F}$ be a ring of representations of G and η a mapping which assigns to every $P \in \mathfrak{F}$ a matrix $\eta(P) \in GL(C, d^\circ(P))$. Then η is called a representation of $\mathfrak{F}$ if for any $P, P_1, P_2 \in \mathfrak{F}$ and $\gamma \in GL(C, d^\circ(P))$, $\eta(\gamma P \gamma^{-1}) = \gamma \eta(P) \gamma^{-1}$, $\eta(P_1 \dotplus P_2) = \eta(P_1) \dotplus \eta(P_2)$, $\eta(P_1 \times P_2) = \eta(P_1) \times \eta(P_2)$. The set $\dot{G}(F)$ of all representations of $\mathfrak{F}$ is turned into a group by defining the product of any two elements η_1, $\eta_2 \in \dot{G}(\mathfrak{F})$ by the rule $\eta_1\eta_2(P) = \eta_1(P)\eta_2(P)$. Also we topologize $\dot{G}(\mathfrak{F})$ as follows. For any $\eta_0 \in \dot{G}(\mathfrak{F})$, $P_i \in \mathfrak{F}$ $1 \leq i \leq r$ and a real number $\varepsilon > 0$, let $U(\eta_0, P_1, \cdots, P_r, \varepsilon)$ denote the set of all $\eta \in \dot{G}(\mathfrak{F})$ such that $|\eta(P_i) - \eta_0(P_i)| < \varepsilon$ $1 \leq i \leq r$. All such sets taken together for arbitrary choices of $\eta_0, \varepsilon, P_1, \cdots, P_r$ and r constitute a neighborhood base in our topology. We shall show that under this topology $\dot{G}(\mathfrak{F})$ is a connected Lie group provided $\mathfrak{F} = \mathfrak{F}^*$.

Henceforth we always suppose that $\mathfrak{F} = \mathfrak{F}^*$. By applying Theorem 1 to $d\mathfrak{F}$ we show as before that there exists a $P_0 \in \mathfrak{F}$ which generates $\mathfrak{F}$. For any $P \in \mathfrak{F}$ let $\dot{P}$ denote the mapping of $\dot{G}(\mathfrak{F})$ into $GL(C, d^\circ(P))$ given by $\dot{P}(\eta) = \eta(P)$ $(\eta \in \dot{G}(\mathfrak{F}))$. Since $\eta_1\eta_2(P) = \eta_1(P)\eta_2(P)$, $\dot{P}$ is a homomorphic mapping. Also since $U(\eta, P, \varepsilon)$ is an open set in $\dot{G}(\mathfrak{F})$, $\dot{P}$ is continuous.

LEMMA 20. *$\dot{P}_0$ is a topological mapping.*

Since P_0 generates $\mathfrak{F}$ it is easily shown that $\eta_1(P_0) = \eta_2(P_0)$ implies $\eta_1(P) = \eta_2(P)$ for all $P \in \mathfrak{F}$ and therefore $\eta_1 = \eta_2$. Hence $\dot{P}$ is univalent. We have therefore only to show that it is open at every point η_0. Consider the neighborhood $U(\eta_0, P_1, \cdots, P_r, \varepsilon)$ defined earlier. For each i $(1 \leq i \leq r)$ we can find an integer $s_i \geq 1$ such that $P_i \subset \gamma_i P_0^{(s_i)} \gamma_i^{-1}$ where $P_0^{(s_i)}$ is the Kronecker product of P_0 with itself s_i times and $\gamma_i \in GL(C, d^\circ(P_0^{(s_i)}))$. It is easily seen that $|\eta(P_0) - \eta_0(P_0)| < \varepsilon_0$ implies

$$|\eta(P_i) - \eta_0(P_i)| \leq |\eta(\gamma_i P_0^{(s_i)} \gamma_i^{-1}) - \eta_0(\gamma_i P_0^{(s_i)} \gamma_i^{-1})|$$

$$\leq |\gamma_i||\gamma_i^{-1}|\,\varepsilon_0 s_i \{|\eta_0(P_0)| + \varepsilon_0\}^{s_i-1} \leq \varepsilon \qquad 1 \leq i \leq r$$

provided ε_0 is sufficiently small. Hence the image of the set $U(\eta_0, P_1, \cdots, P_r, \varepsilon)$ under $\dot{P}_0$ contains the neighborhood $|\eta(P_0) - \eta_0(P_0)| < \varepsilon_0$ of $\eta_0(P_0)$ in $\dot{P}_0(G(\mathfrak{F}))$. Therefore $\dot{P}_0$ is open and hence topological.

For any representation P of G let P_{ij} $1 \leq i, j \leq d^\circ(P)$ denote the functions $P_{ij}(x) = \{P(x)\}_{ij}$ $(x \in G)$ on G to C.

LEMMA 21. *Let P_s $1 \leq s \leq r$ be inequivalent irreducible representations of G. Then the functions P_{sij} $1 \leq i, j \leq d^\circ(P_s)$, $1 \leq s \leq r$ are linearly independent over C.*

The proof follows immediately from the Schur-Frobenius generalization of the Burnside Theorem.

Let $\mathbf{P}(\mathfrak{F})$ denote the ring generated over C by the functions P_{ij} $1 \leq i, j \leq d^\circ(P)$ for all $P \in \mathfrak{F}$. Let $\Omega(\mathfrak{F})$ denote the set of all homomorphisms of $\mathbf{P}(\mathfrak{F})$ into C, and let ξ be a mapping which assigns to every $P \in \mathfrak{F}$ a matrix $\xi(P) \in E(C, d^\circ(P))$. Then, using Lemma 21,[12] it can be shown (see Chevalley (2), p. 194–196 in particular Proposition 1 p. 194) that $\xi \in \dot{G}(\mathfrak{F})$ if and only if the mapping $P_{ij} \to \{\xi(P)\}_{ij}$ $1 \leq i, j \leq d^\circ(P)$ $(P \in \mathfrak{F})$ defines a homomorphism of $\mathbf{P}(\mathfrak{F})$. Here we have to make use of the fact that $\mathfrak{F} = \mathfrak{F}^*$ and therefore $\sum_j P_{ji}^* P_{jk} = \delta_{ik}$ $(1 \leq i, k \leq d^\circ(P))$ where $\delta_{ik} = 1$ if $i = k$ and 0 otherwise. Since P_0 generates $\mathfrak{F}$ it follows that the functions $P_{0,ij}$ $1 \leq i, j \leq d^\circ(P_0)$ constitute a system of generators of $\mathbf{P}(\mathfrak{F})$. Hence there exists (see Chevalley (2), l.c.) a system of polynomial equations with coefficients in C in $\{d^\circ(P_0)\}^2$ variables x_{ij} $1 \leq i, j \leq d^\circ(P_0)$ such that in order that ξ may belong to $\dot{G}(\mathfrak{F})$ it is necessary and sufficient that $x_{ij} = \{\xi(P_0)\}_{ij}$ may be a solution of this system. Therefore $\dot{P}_0(\dot{G}(\mathfrak{F}))$ is an algebraic group (see Chevalley (2), p. 135) and hence it is closed in $GL(C, d^\circ(P_0))$. Therefore it is a Lie group and since $\dot{P}_0$ is a topological and homomorphic mapping the same is true of $\dot{G}(\mathfrak{F})$. Also it is clear that $\dot{P}$ is a continuous representation of $\dot{G}(\mathfrak{F})$ for every $P \in \mathfrak{F}$.

Let $\mathfrak{L}(\mathfrak{F})$ be the Lie algebra of $\dot{G}(\mathfrak{F})$ and $\dot{\mathfrak{g}}(\mathfrak{F})$ the Lie algebra of all representations of $d\mathfrak{F}$.

LEMMA 22. *$\mathfrak{L}(\mathfrak{F})$ is isomorphic (over R) to $\dot{\mathfrak{g}}(\mathfrak{F})$ under the mapping ξ given by $\xi(L)(dP) = d\dot{P}(L)$ $(L \in \mathfrak{L}(\mathfrak{F}), P \in \mathfrak{F})$.*

It is easily verified that $\xi(L)$ so defined is actually a representation of $d\mathfrak{F}$ and that ξ is homomorphic over R. Also $\xi(L) = 0$ implies $d\dot{P}_0(L) = 0$. Therefore

---

[12] The argument in Chevalley (2) (p. 195 lines 15–20) is to be replaced here by Lemma 21.

since $\dot{P}_0$ is a faithful representation of $\dot{G}(\mathfrak{F})$, $L = 0$. Hence the kernel of ξ is $\{0\}$. We shall now show that ξ maps $\mathfrak{L}(\mathfrak{F})$ onto $\dot{\mathfrak{g}}(\mathfrak{F})$. For any $\zeta \in \dot{\mathfrak{g}}(\mathfrak{F})$ and $t \in R$ define a mapping η_t of $\mathfrak{F}$ by the rule $\eta_t(P) = \exp(t\zeta(dP))$ $(P \in \mathfrak{F})$. It is easily verified that η_t is a representation of $\mathfrak{F}$ and $t \to \eta_t$ is a continuous homomorphism of the additive group of real numbers into $\dot{G}(\mathfrak{F})$. Hence there exists an element $L_\zeta \in \mathfrak{L}(\mathfrak{F})$ such that $\eta_t = \exp(tL_\zeta)$ for all $t \in R$. Therefore $\dot{P}(\eta_t) = \exp\{td\dot{P}(L_\zeta)\} = \exp\{t\zeta(dP)\}$. Since this is true for all t, $d\dot{P}(L_\zeta) = \zeta(dP)$ $(P \in \mathfrak{F})$. Hence $\xi(L_\zeta) = \zeta$. The lemma is therefore proved.

In the above lemma the isomorphism ξ was established only over R. We now turn $\mathfrak{L}(\mathfrak{F})$ into a complex Lie algebra by requiring that ξ be an isomorphism over C. Thus for any $c \in C$ and $L \in \mathfrak{L}(\mathfrak{F})$ we define $cL = \xi^{-1}(c\xi(L))$. We may now identify $\dot{\mathfrak{g}}(\mathfrak{F})$ with $\mathfrak{L}(\mathfrak{F})$ under ξ and write $d\dot{P}(\zeta) = \zeta(dP)$ for any $P \in \mathfrak{F}$, $\zeta \in \dot{\mathfrak{g}}(\mathfrak{F})$.

We call a Lie group (or algebra) linear and of degree d if all its elements are matrices of degree d with coefficients in C. Let H be a linear Lie group and $\mathfrak{h}$ its Lie algebra. Let P be the identity representation of H so that $P(x) = x$ $(x \in H)$. Then dP is an isomorphic mapping of $\mathfrak{h}$ and $\exp tX = P(\exp tX) = \exp tdP(X)$ for every $X \in \mathfrak{h}$. Hence we may identify $\mathfrak{h}$ with $dP(\mathfrak{h})$ under the mapping dP and thus regard $\mathfrak{h}$ as a linear Lie algebra. In particular the Lie algebra $\mathfrak{gl}(C, d)$ of $GL(C, d)$ thus consists of all matrices of degree d. Any linear Lie algebra $\mathfrak{h}$ of degree d is therefore a subalgebra of $\mathfrak{gl}(C, d)$. The linear Lie group generated by $\mathfrak{h}$ is defined to be the analytic subgroup of $GL(C, d)$ corresponding to the subalgebra $\mathfrak{h}$ of $\mathfrak{gl}(C, d)$.

DEFINITION. *Let H be a connected linear Lie group of degree d and $\mathfrak{h}$ its Lie algebra. Let $\mathfrak{h}_C$ be the subspace over C of $\mathfrak{gl}(C, d)$ spanned by $\mathfrak{h}$. Clearly $\mathfrak{h}_C$ is a Lie algebra over C. It is called the complexification of $\mathfrak{h}$. The linear Lie group H_C generated by $\mathfrak{h}_C$ is called the complexification of H. Also H is called quasi-real if $\mathfrak{h}$ is a real form of $\mathfrak{h}_C$.*

It is clear that H_C is a complex Lie group and since $\mathfrak{h} \subset \mathfrak{h}_C$, H_C contains H. Let P be any representation of H. Then we claim that there exists at most one complex representation P_C of H_C such that P_C coincides with P on H. For if P_C and Q_C are two such complex representations then dP_C coincides with dQ_C on $\mathfrak{h}$ and therefore on $\mathfrak{h}_C$. Hence $dP_C = dQ_C$ and therefore $P_C = Q_C$. P_C, when it exists, will be called the complexification of P.

Let $\dot{G}_0(\mathfrak{F})$ denote the component of identity of $\dot{G}(\mathfrak{F})$. Then $\dot{G}_0(\mathfrak{F})$ is a complex Lie group. For any $\gamma \in GL(C, d)$ write $\gamma_* = (^t\gamma)^{-1}$ where $^t\gamma$ is the transposed of γ.

LEMMA 23. *For any $P \in \mathfrak{F}$, $\dot{P}(\dot{G}_0(\mathfrak{F})) = \{P(G)\}_C$ and $\dot{G}_0(\mathfrak{F})$ itself is isomorphic to $\{P_0(G)\}_C$. Further $\eta(P^*) = (\eta(P))_*$ for any $\eta \in \dot{G}_0(\mathfrak{F})$.*

For every $x \in G$ define a mapping η_x of $\mathfrak{F}$ by $\eta_x(P) = P(x)$. It is clear that $\eta_x \in \dot{G}(F)$ and $x \to \eta_x$ is a continuous homomorphism. Therefore since G is connected, $\eta_x \in \dot{G}_0(\mathfrak{F})$. Also for every $X \in \mathfrak{g}$ we can find $\zeta \in \dot{\mathfrak{g}}(\mathfrak{F})$ such that $\eta_{\exp tX} = \exp t\zeta$ for all $t \in R$. Therefore

$$\exp(td P(X)) = P(\exp tX) = \eta_{\exp tX}(P) = (\exp t\zeta)(P)$$

$$= \dot{P}(\exp t\zeta) = \exp(td\dot{P}(\zeta)) = \exp(t\zeta(dP)) \qquad (P \in \mathfrak{F}).$$

Therefore $\zeta(dP) = dP(X)$ for all $P \in \mathfrak{F}$. Hence $\zeta = \zeta_X$ where ζ_X is defined as in Theorem 2. Put $\mathfrak{g}(\mathfrak{F}) = \{\zeta_X \mid X \in \mathfrak{g}\}$. Then by Theorem 2, $\mathfrak{g}(\mathfrak{F})$ is a real form of $\dot{\mathfrak{g}}(\mathfrak{F})$. Every element $\eta \in \dot{G}_0(\mathfrak{F})$ can therefore be written in the form[13] $\eta = \exp (c_1\zeta_{X_1}) \cdots \exp (c_r\zeta_{X_r})$ where $c_1, \cdots, c_r \in C$, $X_1, \cdots, X_r \in \mathfrak{g}$ and $r \geqq 1$. Hence for any $P \in \mathfrak{F}$, $\dot{P}(\eta) = \eta(P) = \exp (c_1dP(X_1)) \cdots \exp (c_rdP(X_r))$. Therefore it is clear that $\dot{P}(\dot{G}_0(\mathfrak{F}))$ is the complexification of $P(G)$. In particular since $\dot{P}_0$ is a faithful representation of $\dot{G}_0(\mathfrak{F})$, $\dot{G}_0(\mathfrak{F})$ is isomorphic to $\{P_0(G)\}_c$. Finally

$$\eta(P^*) = \exp (c_1dP^*(X_1)) \cdots \exp (c_rdP^*(X_r)) = (\eta(P))_*.$$

Hence the lemma is proved.

It is clear that for any $P \in \mathfrak{F}$, $\dot{P}$ is a complex representation of $\dot{G}_0(\mathfrak{F})$.

LEMMA 24. *Let $\mathfrak{E}$ and $\mathfrak{F}$ be two rings of representations of G such that $\mathfrak{E} = \mathfrak{E}^*$, $\mathfrak{F} = \mathfrak{F}^*$ and $\mathfrak{E} \subset \mathfrak{F}$. Let ρ be the mapping of $\dot{G}_0(\mathfrak{F})$ into $\dot{G}(\mathfrak{E})$ given by $\rho(\eta)(Q) = \eta(Q)(\eta \in \dot{G}_0(\mathfrak{F}), Q \in \mathfrak{E})$. Then ρ is a continuous homomorphism of $\dot{G}_0(\mathfrak{F})$ onto $\dot{G}_0(\mathfrak{E})$. It is an isomorphism if and only if $\mathfrak{F} = \mathfrak{E}$.*

We shall call ρ the natural mapping of $\dot{G}_0(\mathfrak{F})$ into $\dot{G}(\mathfrak{E})$. Let P_0 and Q_0 be two generating elements of $\mathfrak{F}$ and $\mathfrak{E}$ respectively. The mapping $\eta \to \eta(Q_0) = \rho(\eta)(Q_0)$ is clearly continuous and homomorphic. Since $\xi \to \xi(Q_0)$ $(\xi \in \dot{G}(\mathfrak{E}))$ is a faithful representation of $\dot{G}(\mathfrak{E})$ it follows that ρ is a continuous homomorphism. Therefore since $\dot{G}_0(\mathfrak{F})$ is connected ρ maps it into $\dot{G}_0(\mathfrak{E})$. Now[14] from Lemma 23 $\dot{Q}_0(\dot{G}_0(\mathfrak{E}))$ and $\dot{Q}_0(\dot{G}_0(\mathfrak{F}))$ are both equal to $\{Q_0(G)\}_c$. Therefore for any $\xi \in \dot{G}_0(\mathfrak{E})$ we can find an $\eta \in \dot{G}_0(\mathfrak{F})$ such that $\xi(Q_0) = \eta(Q_0) = \rho(\eta)(Q_0)$. But $\dot{Q}_0$ is faithful on $\dot{G}(\mathfrak{E})$. Hence $\xi = \rho(\eta)$ and therefore ρ maps $\dot{G}_0(\mathfrak{F})$ onto $\dot{G}_0(\mathfrak{E})$.

Now suppose ρ is an isomorphism. Since $\dot{G}_0(\mathfrak{F})$ is semisimple it follows from known results that we can find a compact connected semisimple subgroup G_K of $\dot{G}_0(\mathfrak{F})$ such that its Lie algebra $\mathfrak{g}_K$ is a real form of $\dot{\mathfrak{g}}(\mathfrak{F})$. For any representation P of G in $\mathfrak{F}$ let P_K denote the restriction of $\dot{P}$ to G_K. The mapping $P \to P_K$ is univalent since dP_K determines $d\dot{P}$ and therefore dP by the relation $d\dot{P}(\zeta_X) = dP(X)$ $(X \in \mathfrak{g})$. Also if P_K is reducible then dP_K and therefore $d\dot{P}$, dP and P are reducible. Further, it follows from the relation $\eta(P^*) = (\eta(P))_*$ $(\eta \in \dot{G}_0(\mathfrak{F}), P \in \mathfrak{F})$ that $(P^*)_K = (P_K)^*$ for $P \in \mathfrak{F}$. Therefore the mapping $P \to P_K$ is clearly an isomorphic mapping of $\mathfrak{F}$ onto $\mathfrak{F}_K = \{P_K \mid P \in \mathfrak{F}\}$. Put $\mathfrak{E}_K = \{Q_K \mid Q \in \mathfrak{E}\}$. Then $\mathfrak{F}_K$ and $\mathfrak{E}_K$ are rings and $\mathfrak{F}_K = \mathfrak{F}_K^*$, $\mathfrak{E}_K = \mathfrak{E}_K^*$ and $\mathfrak{F}_K \supset \mathfrak{E}_K$. Also $P_{0,K}$ and $Q_{0,K}$ are generating elements of $\mathfrak{F}_K$ and $\mathfrak{E}_K$ respectively. For any $\eta \in \dot{G}_0(\mathfrak{F})$, $\dot{Q}_0(\eta) = \dot{Q}_0(\rho(\eta))$. Since $\dot{Q}_0$ is faithful on $\dot{G}_0(\mathfrak{E})$ and ρ is an isomorphism, $\dot{Q}_0$ is faithful on $\dot{G}_0(\mathfrak{F})$. Hence $Q_{0,K}$ is a faithful representation of G_K. Therefore since G_K is compact and $Q_{0,K} \in \mathfrak{E}_K = \mathfrak{E}_K^*$ one can show (see Chevalley (2), p. 190) that $\mathfrak{E}_K$ is the ring of all representations of G_K. Hence $\mathfrak{E}_K = \mathfrak{F}_K$. But then

---

[13] Here we make use of the fact that if ζ_i $1 \leqq i \leqq n$ is a base for $\dot{\mathfrak{g}}(\mathfrak{F})$ over R then every element η in a suitable neighborhood of identity can be written in the form $\eta = \exp t_1\zeta_1 \ldots \exp t_n\zeta_n$ $(t_i \in R, 1 \leqq i \leqq n)$.

[14] Strictly speaking we ought to use distinct symbols for the representations $\xi \to \xi(Q)$ $(\xi \in \dot{G}(\mathfrak{E}))$ and $\eta \to \eta(Q)$ $(\eta \in \dot{G}(\mathfrak{F}))$ of $\dot{G}(\mathfrak{E})$ and $\dot{G}(\mathfrak{F})$ respectively defined by any $Q \in \mathfrak{E}$. However it is convenient to denote both of these by $\dot{Q}$.

$\mathfrak{E} = \mathfrak{F}$ since the mapping $P \to P_K$ ($P \in \mathfrak{F}$) is isomorphic. The lemma is therefore proved.

LEMMA 25. *The group $\dot{G}(\mathfrak{F})$ is connected and the mapping $P \to \dot{P}$ ($P \in \mathfrak{F}$) is an isomorphism of $\mathfrak{F}$ onto the ring of all complex representations of $\dot{G}(\mathfrak{F})$.*

We keep to the notation of the above lemma. Let $\dot{G}(\mathfrak{F}_K)$ denote the Lie group of all representations of $\mathfrak{F}_K$. Since $\dot{P}_0$ is a faithful representation of $\dot{G}(\mathfrak{F})$, $P_{0,K}$ is also faithful. Hence by the same argument as before we deduce that $\mathfrak{F}_K$ is the ring of all representations of the compact connected Lie group G_K. Therefore it can be shown (see Chevalley (2), p. 200) that $\dot{G}(\mathfrak{F}_K)$ is connected. For any $\eta \in \dot{G}(\mathfrak{F})$ define η_K to be the mapping of $\mathfrak{F}_K$ given by $\eta_K(P_K) = \eta(P)(P \in \mathfrak{F})$. It is clear that $\eta_K \in \dot{G}(\mathfrak{F}_K)$. Since $\eta \to \eta(P_0) = \eta_K(P_{0,K})$ is an isomorphic mapping of $\dot{G}(\mathfrak{F})$ and since $\dot{P}_{0,K}$ is a faithful representation of $\dot{G}(\mathfrak{F}_K)$ it follows that $\eta \to \eta_K$ is an isomorphic mapping of $\dot{G}(\mathfrak{F})$ into $\dot{G}(\mathfrak{F}_K)$. Now the groups $\dot{G}(\mathfrak{F})$ and $\dot{G}(\mathfrak{F}_K)$ are locally isomorphic since their Lie algebras are both isomorphic to the extension of $\mathfrak{g}_K$ over C. Therefore since $\dot{G}(\mathfrak{F}_K)$ is connected, $\dot{G}(\mathfrak{F})$ is mapped onto $\dot{G}(\mathfrak{F}_K)$ under the mapping $\eta \to \eta_K$. Hence $\dot{G}(\mathfrak{F})$ is isomorphic to $\dot{G}(\mathfrak{F}_K)$ and so is connected.

The mapping $P \to \dot{P}$ ($P \in \mathfrak{F}$) is univalent since $d\dot{P}$ determines dP by the relation $d\dot{P}(\zeta_X) = \zeta_X(dP) = dP(X)$ ($X \in \mathfrak{g}$). Since $\dot{G}(\mathfrak{F}) = \dot{G}_0(\mathfrak{F})$ it follows from Lemma 23 that $(\dot{P})^* = \dot{P}^*$. Further since $\dot{P}(\dot{G}(\mathfrak{F})) = \{P(G)\}_c$ it is obvious that if $\dot{P}$ is reducible, so is also P. Hence $P \to \dot{P}$ ($P \in \mathfrak{F}$) is clearly an isomorphism. Let θ be any complex representation of $\dot{G}(\mathfrak{F})$. Let θ_0 be the restriction of θ to G_K. Since $\mathfrak{F}_K$ is the ring of all representations of G_K there exists a $P \in \mathfrak{F}$ such that $P_K = \theta_0$. $d\dot{P}$ coincides with $d\theta$ on $\mathfrak{g}_K$ and therefore on $\mathfrak{g}(\mathfrak{F})$. Hence $\theta = \dot{P}$ and the lemma is proved.

LEMMA 26. *Let H be a connected semisimple linear Lie group. If ρ is any continuous homomorphism of H onto itself, ρ is an automorphism.*[15]

Let φ be the adjoint representation of H. Since H is semisimple and ρ is onto, $d\varphi$ and $d\rho$ are both isomorphisms. Hence $d(\varphi \circ \rho) = d\varphi \circ d\rho$ is an isomorphism and therefore the kernel of $\varphi \circ \rho$ is contained in the center D of H. Therefore it is clear that $\rho(x) \in D$ if and only if $x \in D$. Hence $\rho(D) = D$. But from Lemma 19, D is a finite group and therefore $\rho(D)$ cannot coincide with D unless the kernel of ρ (which is contained in D) consists of identity alone. Hence ρ is an automorphism.

Our main results can now be summarized in the following theorem.

THEOREM 4. *Let $\mathfrak{F} = \mathfrak{F}^*$ be a ring of representations of G. Then $\dot{G}(\mathfrak{F})$ is a connected Lie group isomorphic to $\{P_0(G)\}_c$ where P_0 is any generating element of $\mathfrak{F}$. Given a subring $\mathfrak{E} = \mathfrak{E}^*$ of $\mathfrak{F}$, $\dot{G}(\mathfrak{E})$ is isomorphic to $\dot{G}(\mathfrak{F})$ if and only if $\mathfrak{E} = \mathfrak{F}$.*

Suppose $\dot{G}(\mathfrak{E})$ is isomorphic to $\dot{G}(\mathfrak{F})$ under a mapping θ. Let ρ be the natural homomorphism of $\dot{G}(\mathfrak{F})$ onto $\dot{G}(\mathfrak{E})$ (cf. Lemma 25). Consider the continuous homomorphism $\theta \circ \rho$ of $\dot{G}(\mathfrak{F})$ onto itself. Since $\dot{G}(\mathfrak{F})$ is semisimple and has a

---

[15] Semisimplicity is essential here, as is seen by considering the mapping $x \to x^2$ in the group of all unimodular complex numbers.

faithful representation $\dot{P}_0$ it follows from Lemma 27 that $\theta \circ \rho$ is an automorphism. Hence ρ is an isomorphism and therefore from Lemma 25, $\mathfrak{E} = \mathfrak{F}$.

Let F be any set of representations of G. For any $P \in F$ let $N(P)$ denote the kernel of P and put $N(F) = \bigcap_{P \in F} N(P)$. If $\mathfrak{E}(F)$ is the ring generated by F it is clear that $N(\mathfrak{E}(F)) = N(F)$. Now suppose $\mathfrak{F} = \mathfrak{F}^* = \bar{\mathfrak{F}}$ is a ring of representations of G. Define an automorphism $\eta \to \eta\dagger$ of $\dot{G}(\mathfrak{F})$ by $\eta\dagger(\rho) = \overline{\eta(\bar{P})}$ $(P \in \mathfrak{F})$. It is obviously continuous and therefore the set $\dot{G}_2(\mathfrak{F})$ of all elements which are left invariant by it is a closed subgroup of $\dot{G}(\mathfrak{F})$. Let $\dot{G}_1(\mathfrak{F})$ denote the component of identity of $\dot{G}_2(\mathfrak{F})$.

LEMMA 27. *Suppose* $\mathfrak{F} = \mathfrak{F}^* = \bar{\mathfrak{F}}$. *Define* $\dot{G}_1(\mathfrak{F})$ *as above. Then* $\dot{G}_1(\mathfrak{F}) \cong G/N(\mathfrak{F})$.

Let ξ denote the mapping $x \to \eta_x$ of G into $\dot{G}(\mathfrak{F})$ defined by $\eta_x(P) = P(x)$ $(x \in G, P \in \mathfrak{F})$. It is clear that ξ is continuous and $(\eta_x)\dagger = \eta_x$. Therefore since G is connected $\eta_x \in \dot{G}_1(\mathfrak{F})$. Since $\dot{P}_0$ is a faithful representation of $\dot{G}(\mathfrak{F})$ and $\dot{P}_0(\eta_x) = \eta_x(P_0) = P_0(x)$, it follows that the kernel of ξ coincides with $N(P_0)$. But P_0 generates $\mathfrak{F}$. Hence $N(P_0) = N(\mathfrak{F})$. It only remains to show that ξ maps G onto $\dot{G}_1(\mathfrak{F})$. Let $\dot{\mathfrak{g}}_1(\mathfrak{F})$ be the Lie algebra of $\dot{G}_1(\mathfrak{F})$. Suppose $\zeta = c_1 \zeta_{x_1} + \cdots + c_r \zeta_{x_r} \in \dot{\mathfrak{g}}_1(F)$ $(X_1, \cdots, X_r \in \mathfrak{g}, c_1, \cdots, c_r \in C)$. Then $(\exp t\zeta)\dagger = \exp t\zeta$ for all $t \in R$. But for any $P \in \mathfrak{F}$,

$$(\exp t\zeta)\dagger(P) = \overline{(\exp t\zeta)(\bar{P})} = \overline{\exp t\zeta(d\bar{P})}$$

$$= \overline{\exp t\{c_1 d\bar{P}(X_1) + \cdots + c_r d\bar{P}(X_r)\}}$$

$$= \exp t\{\bar{c}_1 dP(X_1) + \cdots + \bar{c}_r dP(X_r)\}$$

where $\bar{c}_i$ is the complex conjugate of c_i $1 \leq i \leq r$. Similarly

$$(\exp t\zeta)(P) = \exp t\{c_1 dP(X_1) + \cdots + c_r dP(X_r)\}.$$

Therefore

$$\zeta(dP) = c_1 dP(X_1) + \cdots + c_r dP(X_r) = \bar{c}_1 dP(X_1) + \cdots + \bar{c}_r dP(X_r)$$

$$= a_1 dP(X_1) + \cdots + a_r dP(X_r) = dP(X)$$

where $a_i = \frac{1}{2}(c_i + \bar{c}_i) \in R$ $1 \leq i \leq r$ and $X = a_1 X_1 + \cdots + a_r X_r \in \mathfrak{g}$. In particular

$$d\dot{P}_0(\zeta) = \zeta(dP_0) = dP_0(X) = d\dot{P}_0(\zeta_x).$$

Therefore since $d\dot{P}_0$ is a faithful representation of $\dot{\mathfrak{g}}(\mathfrak{F}), \zeta = \zeta_x$. On the other hand

$$d\xi(X)(dP) = d\dot{P}(d\xi(X)) = d(\dot{P} \circ \xi)(X) = dP(X)$$

since $\dot{P} \circ \xi = P$. Hence $d\xi(X) = \zeta_x$ and therefore $d\xi$ maps $\mathfrak{g}$ onto $\mathfrak{g}_1(F)$. Therefore ξ maps G onto $\dot{G}_1(\mathfrak{F})$ and the lemma is proved.

In particular if we take $\mathfrak{F}$ to be the ring $\mathfrak{R}$ of all representations of G then $N(\mathfrak{R}) = N(G)$. Hence we get the following extension of the Tannaka duality theorem to connected semisimple Lie groups.

THEOREM 5. *Let G be a connected semisimple Lie group and $\mathfrak{R}$ the ring of all its representations. Let $\dot{G}_1(\mathfrak{R})$ be the component of identity of the group $\dot{G}_2(\mathfrak{R})$ of all* *those representations η of $\mathfrak{R}$ for which $\eta\overline{(P)} = \eta(\bar{P})$ $(P \, \epsilon \, \mathfrak{R})$. Then $\dot{G}_1(\mathfrak{R})$ is isomorphic to $G/N(G)$.*

In general $\dot{G}_2(\mathfrak{R})$ is not connected as is shown in the Appendix by a simple example.

We now introduce the notion of a well-imbedded linear Lie group.

Definition. *A connected linear Lie group H is called well-imbedded if every representation of H has a complexification.*

Let H_C denote the complexification of H and $\mathfrak{h}_C$ and $\mathfrak{h}$ the Lie algebras of H_C and H respectively. Let $\tilde{H}_C$ be the universal covering group of H_C and φ the natural homomorphism of $\tilde{H}_C$ onto H_C. Then $d\varphi$ is an isomorphism of the Lie algebra $\tilde{\mathfrak{h}}_C$ of $\tilde{H}_C$, onto $\mathfrak{h}_C$. Put $\tilde{\mathfrak{h}} = d\varphi^{-1}(\mathfrak{h})$ and let $\tilde{H}$ be the analytic subgroup of $\tilde{H}_C$ corresponding to the subalgebra $\tilde{\mathfrak{h}}$ of $\tilde{\mathfrak{h}}_C$. Let $\tilde{N}(\varphi)$ denote the kernel of φ.

Lemma 28.[16] *Suppose H is semisimple. In order that H may be well-imbedded it is necessary and sufficient that H be quasireal and $\tilde{N}(\varphi) \subset \tilde{H}$.*

Suppose H is well-imbedded. Since every complex matrix can be regarded as a real matrix of double the degree, it is clear that H has a faithful real representation P. Let P_C be the complexification of P. Since $dP_C(\mathfrak{h}_C) = (dP(\mathfrak{h}))_C$ and P is real it is evident that $dP_C(\mathfrak{h}) = dP(\mathfrak{h})$ is a real form of $dP_C(\mathfrak{h}_C)$. Since dP_C is an isomorphism over C it follows that $\mathfrak{h}$ is a real form of $\mathfrak{h}_C$ and therefore H is quasi-real. We turn $\tilde{H}_C$ into a complex Lie group by defining $c\tilde{X} = d\varphi^{-1}(cd\varphi(\tilde{X}))$ $(\tilde{X} \, \epsilon \, \tilde{\mathfrak{h}}_C, c \, \epsilon \, C)$. Since $\tilde{N}(\varphi)$ is a discrete group in the center of $\tilde{H}_C$, $\tilde{N}(\varphi) \cap \tilde{H}$ is a closed invariant subgroup of $\tilde{H}_C$. Consider the factor group $L_C = \tilde{H}_C/\tilde{N}(\varphi) \cap \tilde{H}$. Let ψ be the natural homomorphism of $\tilde{H}_C$ into L_C. Denote by $\mathfrak{L}_C$ the Lie algebra of L_C. We turn L_C into a complex semisimple group by requiring that $d\psi$ be an isomorphism over C. It is known[17] that every such group has a faithful complex representation. Let Q be such a representation of L_C. Put $\check{P}_C = Q \circ \psi$. Obviously $\check{P}_C$ is a complex representation of $\tilde{H}_C$ and its kernel is $\tilde{N}(\varphi) \cap \tilde{H}$. Let $\check{P}$ be the restriction of $\check{P}_C$ to $\tilde{H}$. Since the kernel of $\check{P}$ is $\tilde{N}(\varphi) \cap \tilde{H}$ we can define a representation P of H such that $P \circ \varphi_0 = \check{P}$ where φ_0 is the restriction of φ to $\tilde{H}$. By hypothesis P has a complexification Q_C. Put $\check{Q}_C = Q_C \circ \varphi$. Since $d\check{Q}_C$ coincides with dP on $\mathfrak{h}$, $d\check{Q}_C$ coincides with $d\check{P}_C$ on $\tilde{\mathfrak{h}}$ and therefore on $\tilde{\mathfrak{h}}_C$. Hence $\check{P}_C = \check{Q}_C$. But the kernel of $\check{P}_C$ is $\tilde{N}(\varphi) \cap \tilde{H}$ while the kernel of $\check{Q}_C$ contains $\tilde{N}(\varphi)$. Hence $\tilde{N}(\varphi) \subset \tilde{H}$.

Conversely suppose $\tilde{N}(\varphi) \subset \tilde{H}$ and H is quasireal. Let P be any representation of H. Since $\tilde{\mathfrak{h}}$ is a real form of $\tilde{\mathfrak{h}}_C$ we can define a representation $\check{p}$ of the complex Lie algebra $\tilde{\mathfrak{h}}_C$ such that $\check{p}$ coincides with $dP \circ d\varphi_0$ on $\tilde{\mathfrak{h}}$. Since $\tilde{H}_C$ is simply connected we can find a representation $\check{P}_C$ of $\tilde{H}_C$ such that $d\check{P}_C = \check{p}$. Let $\tilde{h} \, \epsilon \, \tilde{H}$. Then $\tilde{h} = \exp a_1\tilde{X}_1 \cdots \exp a_r\tilde{X}_r$, where $\tilde{X}_i \, \epsilon \, \tilde{\mathfrak{h}}$, $a_i \, \epsilon \, R$, $1 \leq i \leq r$. Hence

$$\check{P}_C(\tilde{h}) = \exp a_1 dP(X_1) \cdots \exp a_r dP(X_r) = P(\varphi(\tilde{h}))$$

where $X_i = d\varphi(\tilde{X}_i)$ $1 \leq i \leq r$. Hence if $\tilde{N}(\check{P}_C)$ is the kernel of $\check{P}_C$, $\tilde{N}(\check{P}_C) \supset$

---

[16] Using this lemma we give in the Appendix an example of a connected semisimple linear Lie group which is not well-imbedded.

[17] See for example Malcev (5) p. 87.

$\tilde{N}(\varphi) \cap \tilde{H} = \tilde{N}(\varphi)$. Therefore we can define a representation P_C of H_C such that $P_C \circ \varphi = \tilde{P}_C$. Since $dP_C = d\tilde{P}_C \circ d\varphi$, P_C is clearly a complex representation of H_C. Further as we have seen above $P(\varphi(\tilde{h})) = \tilde{P}_C(\tilde{h}) = P_C(\varphi(\tilde{h}))$ for any $\tilde{h} \in \tilde{H}$. Hence P_C is the complexification of P.

We can now prove the following theorem.

THEOREM 6. *Let $\Re$ be the ring of all representations of G and let P be any element in $\Re$. The set $\{P, P^*\}$ generates $\Re$ if and only if P is quasifaithful and $P(G)$ is well-imbedded.*

Put $P_0 = P + P^*$. Suppose P_0 generates $\Re$. Then $N(P) = N(P^*) = N(P_0) = N(\Re) = N(G)$. Hence P is quasifaithful and from Lemma 25, $\dot{P}_0 = \dot{P} + (\dot{P})^*$. Therefore since $\dot{P}_0$ is a faithful representation of $\dot{G}(\Re)$ the same is true of $\dot{P}$. Hence from Lemma 23 $\dot{G}(\Re)$ is isomorphic to $\{P(G)\}_C$ under the mapping ξ such that $(d\xi)\zeta = \zeta(dP)$ $(\zeta \in \dot{g}(\Re))$. Now let θ be any representation of $P(G)$. Then $P_\theta = \theta \circ P$ is a representation of G and therefore $\dot{P}_\theta(\dot{G}(\Re)) = \{P_\theta(G)\}_C = \{\theta(P(G))\}_C$. Define a representation θ_C of $\{P(G)\}_C$ such that $\theta_C \circ \xi = \dot{P}_\theta$. Then

$$(d\theta_C \circ d\xi)(\zeta) = d\dot{P}_\theta(\zeta) = \zeta(dP_\theta) \ (\zeta \in g(\Re)).$$

Since $\dot{P}_\theta$ is a complex representation and $d\xi$ is an isomorphism over C it follows that θ_C is also a complex representation. Further, for any $X \in g$,

$$(d\theta_C \circ d\xi)(\zeta_X) = d\theta_C(\zeta_X(dP)) = d\theta_C(dP(X))$$

and $\zeta_X(dP_\theta) = dP_\theta(X) = d(\theta \circ P)(X) = d\theta(dP(X))$. Therefore $d\theta_C$ coincides with $d\theta$ on $dP(g)$. Therefore θ_C is a complexification of θ and $P(G)$ is well-imbedded.

Conversely suppose P is quasifaithful and $P(G)$ is well-imbedded. Let $\mathfrak{F}$ be the ring generated by P and P^*. Clearly $\mathfrak{F} = \mathfrak{F}^*$. Let Q_0 be a generating element of $\Re$. P being quasifaithful, $N(\Re) = N(G) = N(P)$. Hence $P(G)$ and $Q_0(G)$ are isomorphic groups. Let θ denote the representation of $P(G)$ such that $Q_0 = \theta \circ P$. Since $P(G)$ is well-imbedded θ has a complexification θ_C. It is clear that $\theta_C(\{P(G)\}_C) = \{Q_0(G)\}_C$. From Lemma 23, $\{P(G)\}_C = \dot{P}(\dot{G}(\Re))$, $\{Q_0(G)\}_C = \dot{Q}_0(\dot{G}(\Re))$. Since $\dot{Q}_0$ is faithful it follows that there exists a homomorphism φ of $\{Q_0(G)\}_C$ onto $\{P(G)\}_C$ such that $\varphi \circ \dot{Q}_0 = \dot{P}$. Hence for any $\zeta \in \dot{g}(\Re)$,

$$(d\varphi \circ d\dot{Q}_0)(\zeta) = d\dot{P}(\zeta)$$

or

$$d\varphi(\zeta(dQ_0)) = \zeta(dP).$$

Therefore $d\varphi$ is a homomorphism over C. Also if we substitute ζ_X for $\zeta(X \in g)$ we get

$$d\varphi(dQ_0(X)) = dP(X).$$

Hence

$$(d\theta_C \circ d\varphi)(dQ_0(X)) = d\theta_C(dP(X)) = d\theta(dP(X)) = d(\theta \circ P)(X)$$
$$= dQ_0(X)$$

since $d\theta_c$ coincides with $d\theta$ on $dP(\mathfrak{g})$ and $\theta \circ P = Q_0$. Therefore $d(\theta_c \circ \varphi)$ is the identity mapping on $dQ_0(\mathfrak{g})$ and hence on $\{dQ_0(\mathfrak{g})\}_c$. Hence $\theta_c \circ \varphi$ is the identity mapping of $\{Q_0(G)\}_c$ and therefore $\dot{Q}_0 = \theta_c \circ \dot{P}$. Hence $\dot{P}$ is faithful. Let ρ be the natural homomorphism of $\dot{G}(\mathfrak{R})$ onto $\dot{G}(\mathfrak{F})$ given by $\rho(\eta)(Q) = \eta(Q)$ ($\eta \in \dot{G}(\mathfrak{R})$, $Q \in \mathfrak{F}$). The kernel of ρ is the same as that of $\dot{P}_0$ or $\dot{P}$. Therefore since $\dot{P}$ is faithful, ρ is an isomorphism. Hence from Lemma 24, $\mathfrak{F} = \mathfrak{R}$. The theorem is therefore proved.

DEFINITION. *Let H be a subgroup of G. We shall say that a representation Q_0 of H can be extended to a representation of G if there exists a representation P of G such that P_0 being the restriction of P to H, $Q_0 < P_0$.*

In case H is compact or is a semisimple analytic subgroup, the set of all representations of H which can be extended to representations of G, clearly from a ring $\mathfrak{F} = \mathfrak{F}^*$. Hence we get from Theorem 6 the following two corollaries immediately.

COROLLARY 1. *Let H be a compact subgroup of G. Every representation of H can be extended to a representation of G if and only if a quasifaithful representation of G is faithful on H.*

COROLLARY 2. *Let H be a semisimple analytic subgroup of G, and P a quasifaithful representation of G such that $P(G)$ is well-imbedded. Then every representation of H can be extended to a representation of G if and only if P is quasifaithful on H and $P(H)$ is well-imbedded.*

5. Appendix

Let H and H_c respectively be the groups of all real and complex 2 x 2 matrices of determinant 1. H is connected and H_c is the complexification of H. Also H_c is simply connected. Let G_1 and G_c respectively be the groups of all real and complex linear transformations of determinant 1, of three variables x_i $1 \leq i \leq 3$ which leave the quadratic form $x_1^2 - x_2^2 - x_3^2$ invariant. For any $t \in G_1$, let t_{ik} $1 \leq i, k \leq 3$ denote the matrix elements of t so that $(tx)_i = \sum_{1 \leq k \leq 3} t_{ik}x_k$. It is known that G_c is connected and G_1 consists of two components corresponding to the two possibilities $t_{11} > 0$ or < 0 ($t \in G_1$). Let G denote the component of identity of G_1. Then G_c is the complexification of G.

Let D be the subgroup of H consisting of I and $-I$ where $I = \begin{pmatrix} 1 & 0 \\ 0 & 1 \end{pmatrix}$. It is known that there exists a homomorphism φ of H_c onto G_c such that D is the kernel of φ and $\varphi(H) = G$. Since G is real and semisimple, H_c is simply connected and $D \subset H$ it follows from Lemma 27 that G is well-imbedded. Let $\mathfrak{R}$ denote the ring of all representations of G and P the identity representation so that $P(t) = t$ ($t \in G$). From Theorem 6, $P + P^*$ generates $\mathfrak{R}$ and therefore $\dot{P}$ is a faithful representation of $\dot{G}(\mathfrak{R})$. Hence from Lemma 23, $\dot{G}(\mathfrak{R})$ is isomorphic to $\{P(G)\}_c = G_c$. Therefore $\eta \in \dot{G}_2(\mathfrak{R})$ if and only if $\eta(P) = \eta^\dagger(P) = \overline{\eta(\bar{P})} = \overline{\eta(P)}$ since P is real. Hence the image of $\dot{G}_2(\mathfrak{R})$ under $\dot{P}$ consists precisely of all the real elements in G_c i.e. of G_1. But G_1 is not connected. Therefore the same holds for $\dot{G}_2(\mathfrak{R})$.

We now give an example of a connected semisimple linear Lie group which is not well-inbedded. Let G be the group of all complex 3 x 3 matrices with determinant 1. Let H be the subgroup consisting of all real elements of G. Let ω denote a primitive cube root of 1. Put $\Omega = \omega I$ where I is the unit matrix of degree 3. Let $N = \{I, \Omega, \Omega^2\}$ be the subgroup of G generated by Ω. Then clearly N is contained in the center of G and $H \cap N = \{I\}$. Let ψ be the natural homomorphism of G onto G/N. Since G is connected and semisimple and $d\psi$ is an isomorphism, G/N can be regarded in a natural way a complex semisimple Lie group. Hence G/N has a faithful complex representation θ. Put $\varphi = \theta \circ \psi$. Clearly the kernel of φ is N. Hence φ maps H isomorphically. Since H is connected and semisimple the same is true of $\varphi(H)$. We shall now show that $\varphi(H)$ is not well-imbedded. Clearly $\varphi(G)$ is the complexification of $\varphi(H)$ and since G is simply connected, G is the universal covering group of $\varphi(G)$. Since $N \not\subset H$ it follows from Lemma 27 that $\varphi(H)$ is not well-imbedded. Let P denote the restriction of φ to H. Then P is a faithful representation of H. However since $P(H)$ is not well-imbedded it follows from Theorem 6 that $\{P, P^*\}$ cannot generate the ring of all representations of H.

REFERENCES

1. E. CARTAN, *Les groupes projectifs qui ne laissent invariante aucune multiplicité plane*, Bull. Soc. Math. France 41 (1913), 53–96.
2. C. CHEVALLEY, Theory of Lie Groups, Princeton University Press, 1946.
3. C. CHEVALLEY, *Algebraic Lie Algebras*, Annals of Math. 48 (1947), 91–100.
4. HARISH-CHANDRA, *On representations of Lie algebras*, Annals of Math. 50 (1949); 900–915.
5. A. MALCEV, *On linear Lie groups*, C. R. (Doklady) Acad. Sci. URSS 40 (1943), 87–89.
6. H. WEYL, *Theorie der Darstellung Kontinuierlicher half-einfacher Gruppen durch lineare Transformationen. II*, Math. Zeit. 24 (1925), 328–376.
7. B. L. VAN DER WAERDEN, *Topologische Bergründung des Kalküls der abzählenden Geometrie*, Math. Ann. 102 (1929), 337–362.

INSTITUTE FOR ADVANCED STUDY.

ON SOME APPLICATIONS OF THE UNIVERSAL ENVELOPING ALGEBRA OF A SEMISIMPLE LIE ALGEBRA

BY

HARISH-CHANDRA[1]

Introduction. The representation theory of semisimple Lie algebras over the field of complex numbers has been developed by Cartan and Weyl. However some of Cartan's proofs (see [2])[2] make explicit use of the classification of semisimple Lie algebras and in fact require a verification of the asserted statement in each case separately. Weyl [12] has given alternative proofs of these results by making use of general arguments depending on the theory of representations of compact groups and in particular on the Peter-Weyl Theorem. His proofs therefore are necessarily of a nonalgebraic nature. In the first part of this paper we propose to give "general" algebraic proofs of some of these theorems. This work happens to overlap considerably with some recent results of Chevalley [3]. In particular the formulation of Theorem 1 and some of the ideas in the proof are due to him. I shall mention them more specifically later in due course.

Recently great interest has arisen in the theory of representations of a Lie group in a Hilbert space. Since every such representation defines a representation of the corresponding Lie algebra (see Gårding [8]) it is natural to study infinite-dimensional representations of a Lie algebra. Part II of this paper contains a theorem (Theorem 4) concerning such representations of complex semisimple Lie algebras. The desirability of proving such a result was pointed out to me by Mautner. Also its significance for unitary representations of complex semisimple Lie groups on a Hilbert space will be brought out by him in a separate paper.

In Part III we define and study the characters of the universal enveloping algebra $\mathfrak{B}$ of a semisimple Lie algebra $\mathfrak{L}$. They are essentially homomorphisms of the center of $\mathfrak{B}$ into the field of complex numbers. We show that every such homomorphism is determined by a linear function on a fixed Cartan subalgebra of $\mathfrak{L}$. Theorems 5 and 6 contain the principal results of Part III.

Part IV is devoted to a brief study of the representations of a complex semisimple Lie group on a Hilbert space. With certain representations of the group (in particular with all those which are irreducible and unitary) we associate in a natural way a character of $\mathfrak{B}$. It follows from Theorem 6 that in order that a character may be associated to some representation of the

---

Received by the editors March 27, 1950.

[1] Frank B. Jewett Fellow.

[2] Numbers in brackets refer to the references at the end of the paper.

group it must satisfy certain conditions. By a method due to Gelfand and Naimark [7] we show in Theorem 7 that these conditions are also sufficient.

I should like to thank Dr, F. I. Mautner for a number of very valuable discussions and also for his help in some questions concerning operator theory which arise in Part IV.

Part I. Representations of semisimple Lie algebras. Let $\mathfrak{L}$ be a semisimple Lie algebra over an algebraically closed field K of characteristic zero. For any $X \in \mathfrak{L}$ let ad X denote the linear mapping $(\text{ad } X)Y = [X, Y]$ $(Y \in \mathfrak{L})$ of $\mathfrak{L}$ into itself. Put $B(X, Y) = \text{sp (ad } X \text{ ad } Y)$. Then $B(X, Y)$ is a nondegenerate bilinear form defined on $\mathfrak{L}$. Let $\mathfrak{h}$ be a Cartan subalgebra of $\mathfrak{L}$ and α a root of $\mathfrak{L}$ with respect to $\mathfrak{h}$. We denote by H_α' the unique element in $\mathfrak{h}$ such that $B(H_\alpha', H) = \alpha(H)$ for all $H \in \mathfrak{h}$. It is known that $\alpha(H_\alpha')$ is a rational number greater than 0. Put $H_\alpha = (2/\alpha(H_\alpha'))H_\alpha'$ so that $\alpha(H_\alpha) = 2$ and let $\mathfrak{F}$ be the set of all linear functions on $\mathfrak{h}$ with values in K. Then $\mathfrak{F}$ is a vector space over K dual to $\mathfrak{h}$. The linear transformation s_α in $\mathfrak{F}$ defined by $s_\alpha \lambda = \lambda - \lambda(H_\alpha)\alpha$ $(\lambda \in \mathfrak{F})$ is called the Weyl reflexion with respect to the root α. It is known that the group W generated by the s_α's for the various roots α is finite, and if α and β are any two roots then $s_\alpha \beta$ is also a root. W is called the Weyl group of $\mathfrak{L}$ (with respect to $\mathfrak{h}$).

Let $\Sigma = \{\alpha_i, 1 \leq i \leq l\}$ $(l = \dim \mathfrak{h})$ be a maximal set of linearly independent roots. We shall say that Σ is a fundamental system of roots if every root α is of the form $\alpha = \sum_{1 \leq i \leq l} d_i \alpha_i$ where d_i are integers which are either all nonnegative or all nonpositive. It is known that fundamental systems always exist. Further if $\Sigma = \{\alpha_i, 1 \leq i \leq l\}$ is a fundamental system then the Weyl reflexions s_{α_i}, $1 \leq i \leq l$, generate the whole group W and every root α can be written in the form $\alpha = \sigma \alpha_i$ $(\sigma \in W, \alpha_i \in \Sigma)$. Put $s_i = s_{\alpha_i}$ and $H_{\alpha_i} = H_i$, $1 \leq i \leq l$. Then $s_i \alpha_j = \alpha_j + a_{ji} \alpha_i$ where $a_{ji} = -\alpha_j(H_i)$. Since $s_i \alpha_j$ is a root, a_{ji} is an integer such that $a_{ii} = -2$, $a_{ij} \geq 0$, $i \neq j$. It is clear that if the integers a_{ij}, $1 \leq i, j \leq l$, are given we can find out which linear combinations of α_i are roots since they are all of the form $s_{i_1} s_{i_2} \cdots s_{i_r}, \alpha_j$ $1 \leq i_1, \cdots, i_r, j \leq l, r \geq 0$. Thus the matrix $A = (a_{ij})$, $1 \leq i, j \leq l$ determines the root diagram of $\mathfrak{L}$ completely. We shall call A the Weyl matrix of $\mathfrak{L}$ (with respect to $\mathfrak{h}$). Notice that A has the following three properties.

(1) $$a_{ii} = -2, \qquad a_{ij} \geq 0, \qquad i \neq j,$$

(2) $$a_{ij} = 0 \quad \text{whenever } a_{ji} = 0,$$

(3) $$\det A \neq 0.$$

The last assertion follows from the fact that α_i, $1 \leq i \leq l$, are linearly independent. A natural question to ask is the following: Given a square matrix A with integral coefficients what are the conditions which A must satisfy in order that A be the Weyl matrix of some semisimple Lie algebra $\mathfrak{L}$? This is one of the two questions considered in Theorem 1.

Now we come to the second question. Let us call an element $\lambda \in \mathfrak{F}$ rational if $\lambda(H_i) \in K_0$, $1 \leq i \leq l$, where K_0 is the prime field of characteristic zero contained in K. It is clear that $\lambda = \sum_{1 \leq i \leq l} c_i \alpha_i$ $(c_i \in K_0)$. We say that $\lambda > 0$ if $\lambda \neq 0$ and $c_i > 0$, i being the least index such that $c_i \neq 0$. Let $\mathfrak{F}_0$ be the set of all rational linear functions on $\mathfrak{h}$. For any λ, $\mu \in \mathfrak{F}_0$ we write $\lambda > \mu$ or $\mu < \lambda$ if $\lambda - \mu > 0$. In this way $\mathfrak{F}_0$ is completely ordered under the relation $>$. This order is called the lexicographic order in $\mathfrak{F}_0$ with respect to the ordered set $\{\alpha_1, \cdots, \alpha_l\}$. We shall call an element $\lambda \in \mathfrak{F}_0$ integral if $\lambda(H_i)$ is an integer for all $1 \leq i \leq l$, and dominant integral if, in addition, $\lambda(H_i) \geq 0$, $1 \leq i \leq l$. Let ρ be a representation of $\mathfrak{L}$ on a finite-dimensional space V. Given any $\lambda \in \mathfrak{F}$ we define V_λ to be the set of all $\psi \in V$ such that $\rho(H)\psi = \lambda(H)\psi$ for all $H \in \mathfrak{h}$. λ is called a weight of ρ if $V_\lambda \neq \{0\}$. It is known that every weight Λ of ρ is an integral linear function on $\mathfrak{h}$ and $V = \sum_\Lambda V_\Lambda$ where the sum is direct and Λ runs over all the weights of the representation, these being only finite in number. Hence ρ has a highest weight Λ_0 and it is known that $\Lambda_0(H_i) \geq 0$, $1 \leq i \leq l$, so that Λ_0 is a dominant integral function. The second question can now be phrased as follows: Given a dominant integral function Λ does there exist a finite-dimensional representation ρ of $\mathfrak{L}$ such that Λ is the highest weight of ρ?

I should like to mention that in my original proof I had considered the second question alone. The idea of dealing with both questions simultaneously is due to Chevalley [3] who obtained independently a proof of the theorem given below. I present here a modified version of my original proof so as to be able to consider the two questions together. But in this modification I have adopted several of Chevalley's ideas. In particular the construction of the algebra $\mathfrak{A}$ and the consideration of its representations on $\mathfrak{G}$ is due to him.

THEOREM 1. *Let a_{ij}, $1 \leq i, j \leq l$, be l^2 integers such that:*
(1) $a_{ii} = -2$, $a_{ij} \geq 0$, $i \neq j$, *and* $a_{ij} = 0$ *whenever* $a_{ji} = 0$, $1 \leq i, j \leq l$.
(2) $\det (a_{ij}) \neq 0$.
(3) *The group W generated by the linear transformations s_i, $1 \leq i \leq l$, given by* $s_i x_j = x_j + a_{ji} x_i$ $(x_i, 1 \leq i \leq l$, *being indeterminates*) *is finite.*
Then there exists a semisimple Lie algebra $\mathfrak{L}$ over K with a Cartan subalgebra $\mathfrak{h}$ such that the following conditions are fulfilled. It is possible to find a set of linear functions α_i, $1 \leq i \leq l$, on $\mathfrak{h}$ such that α_i, $1 \leq i \leq l$, is a fundamental system of roots of $\mathfrak{L}$ with respect to $\mathfrak{h}$ and $\sigma_i \alpha_j = \alpha_j + a_{ji} \alpha_i$ where σ_i is the Weyl reflexion with respect to α_i. Finally if λ_i, $1 \leq i \leq l$, are any given integers greater than or equal to 0 we can find an irreducible finite-dimensional representation ρ of $\mathfrak{L}$ such that the highest weight Λ_0 of ρ is given by[3] $\Lambda_0(H_{\alpha_i}) = \lambda_i$, $1 \leq i \leq l$.

Before proceeding with the proof we make some remarks about terminology. All vector spaces and algebras appearing in our discussion are understood to be over K. A vector space V (or a representation) is not necessarily

---

[3] As before we define H_α for any root α of $\mathfrak{L}$ in such a way that $\alpha(H_\alpha) = 2$.

assumed to be finite-dimensional unless it is explicitly stated to be so. Given any collection $\{U_j; j \in I\}$ of subspaces of V indexed by a set I (finite or infinite) we denote by $\sum_{j \in I} U_j$ the smallest subspace of V containing all U_j. In fact $\sum_j U_j$ consists of all finite sums of the form $\psi_1 + \cdots + \psi_r$, where each ψ_i belongs to some U_j. The sum $\sum_{j \in I} U_j$ is said to be direct if for every finite subset I_0 of I the sum $\sum_{j \in I_0} U_j$ is direct. If Σ is any set of linear mappings of V into itself we say that V is irreducible under Σ if there exists no subspace U of V such that U is invariant under Σ and $U \neq V$, $U \neq \{0\}$.

Given an associative algebra $\mathfrak{A}$, we shall always write $[z, w] = zw - wz$ for any z, $w \in \mathfrak{A}$. We use a similar notation whenever z and w are matrices or linear transformations on a vector space. If $\mathfrak{M}$ is a left ideal in $\mathfrak{A}$ we define the natural representation π of $\mathfrak{A}$ on the factor space $\mathfrak{A}^* = \mathfrak{A}/\mathfrak{M}$ as follows. Let $z \to z^*$ denote the natural mapping of $\mathfrak{A}$ on $\mathfrak{A}^*$. Then $\pi(z)w^* = (zw)^*$, z, $w \in \mathfrak{A}$. From the fact that $\mathfrak{M}$ is a left ideal it is easily verified that π is a representation.

After these preliminary remarks we return to Theorem 1. The proof is rather long but is otherwise not very complicated. It depends on the consideration of the representations of a certain infinite-dimensional associative algebra $\mathfrak{A}$. We shall have to prove a series of lemmas about left ideals in this algebra, some of which are very simple but are nevertheless essential. Let $\mathfrak{A}'$ be the free associative algebra of all noncommutative polynomials in $3l$ independent variables X_i', Y_i', H_i', $1 \leq i \leq l$, with coefficients in K. Let $\mathfrak{h}'$ be the subspace of $\mathfrak{A}'$ spanned by H_i', $1 \leq i \leq l$. We define the linear functions α_i, $1 \leq i \leq l$, on $\mathfrak{h}'$ by $\alpha_i(H_j') = -a_{ij}$, $i \leq i, j \leq l$. Since $\det(a_{ij}) \neq 0$, α_i are linearly independent. Let $\mathfrak{U}'$ be the smallest ideal in $\mathfrak{A}'$ containing the set $\mathfrak{S}$ consisting of the following elements.

$$
(1) \quad
\begin{aligned}
&[H_i', H_j'], \quad [X_i', Y_i'] - H_i', \quad [X_i', Y_j'], &&i \neq j, \\
&[H_j', X_i'] - \alpha_i(H_j')X_i', \quad [H_j', Y_i'] + \alpha_i(H_j')Y_i', &&1 \leq i, j \leq l.
\end{aligned}
$$

Let $\mathfrak{G}$ be the free associative algebra over K with l generators $\eta_1, \cdots, \eta_l$. Given any linear function μ on $\mathfrak{h}'$ we define a representation π_μ' of $\mathfrak{A}'$ on $\mathfrak{G}$ as follows. $\pi_\mu'(1) = I$ where I is the identity mapping of $\mathfrak{G}$ and

$$
\pi_\mu'(Y_i')\eta_{j_1}\eta_{j_2} \cdots \eta_{j_r} = \eta_i \eta_{j_1}\eta_{j_2} \cdots \eta_{j_r},
$$

$$
(2a) \quad
\begin{aligned}
&\pi_\mu'(H')\eta_{j_1}\eta_{j_2} \cdots \eta_{j_r} \\
&\quad = -\{\alpha_{j_1}(H') + \alpha_{j_2}(H') + \cdots + \alpha_{j_r}(H') - \mu(H')\}\eta_{j_1}\eta_{j_2} \cdots \eta_{j_r}
\end{aligned}
$$

where $H' \in \mathfrak{h}'$, $1 \leq i \leq l$, $1 \leq j_1, \cdots, j_r \leq l$, $r \geq 0$ and $\eta_{j_1} \cdots \eta_{j_r} = 1$ if $r = 0$. Finally $\pi_\mu'(X_i')$ is defined by induction on r in the following way.

$$
\pi_\mu'(X_i')1 = 0,
$$

$$
(2b) \quad
\begin{aligned}
\pi_\mu'(X_i')\eta_{j_1} \cdots \eta_{j_r} &= \delta_{ij_1}\pi_\mu'(H_i')\eta_{j_2} \cdots \eta_{j_r} \\
&\quad + \eta_{j_1}(\pi_\mu'(X_i')\eta_{j_2} \cdots \eta_{j_r})
\end{aligned}
$$

where δ_{ij} is the usual Kronecker symbol. Since $\mathfrak{A}'$ is a free algebra and $\eta_{j_1} \cdots \eta_{j_r}$, $1 \leq j_1, \cdots, j_r \leq l$, $r \geq 0$, form a base for $\mathfrak{G}$ it is clear that π'_μ is uniquely defined by the above equations. It is easily verified that the kernel of π'_μ contains the set $\mathfrak{S}$ and therefore $\mathfrak{U}'$. Now suppose $H' \in \mathfrak{h}'$ and $\pi'_\mu (H') = 0$. Then

$$\alpha_{j_1}(H') + \alpha_{j_2}(H') + \cdots + \alpha_{j_r}(H') - \mu(H') = 0$$

for all $1 \leq j_1, \cdots, j_r \leq l$. This implies that $\alpha_j(H') = 0$ for all $1 \leq j \leq l$. Since α_j are linearly independent, $H' = 0$. Hence it is clear that $\mathfrak{h}' \cap \mathfrak{U}' = \{0\}$.

Let $\mathfrak{A}$ be the factor algebra $\mathfrak{A}'/\mathfrak{U}'$ and let X_i, Y_i, H_i respectively be the images of X'_i, Y'_i, H'_i in $\mathfrak{A}$, $1 \leq i \leq l$. Since $\mathfrak{h}' \cap \mathfrak{U}' = \{0\}$, $\mathfrak{h}'$ is mapped isomorphically under the natural mapping of $\mathfrak{A}'$ on $\mathfrak{A}$, on the linear subspace $\mathfrak{h}$ of $\mathfrak{A}$ spanned by H_i, $1 \leq i \leq l$. Hence dim $\mathfrak{h} = $ dim $\mathfrak{h}' = l$ and every linear function λ on $\mathfrak{h}'$ can also be regarded as a linear function on $\mathfrak{h}$ if we put $\lambda(H_i) = \lambda(H'_i)$, $1 \leq i \leq l$. In particular this holds for α_j and therefore $\alpha_j(H_i) = -a_{ji}$. Since the kernel of π'_μ contains $\mathfrak{U}'$, π'_μ defines in the obvious way a representation π_μ of $\mathfrak{A}$ on $\mathfrak{G}$.

Consider the representation π' of $\mathfrak{A}'$ on $\mathfrak{A}$ defined as follows:

$$\pi'(1)a = a, \qquad \pi'(X'_i)a = [X_i, a], \qquad \pi'(Y'_i)a = [Y_i, a],$$
$$\pi'(H'_i)a = [H_i', a], \qquad\qquad 1 \leq i \leq l, a \in \mathfrak{A}.$$

It is easily seen that π' maps every element of the set $\mathfrak{S}$ into zero. Hence the kernel of π' contains $\mathfrak{U}'$. Therefore π' actually determines a representation π of $\mathfrak{A} = \mathfrak{A}'/\mathfrak{U}'$. π is called the adjoint representation of $\mathfrak{A}$ and we shall write ad z instead of $\pi(z)$ for any $z \in \mathfrak{A}$.

The subspace $\mathfrak{h}$ of $\mathfrak{A}$ is an abelian Lie algebra under the bracket operation. Given a representation θ of this Lie algebra on a vector space V and a linear function λ on $\mathfrak{h}$ we denote by V_λ the subspace of V consisting of all elements $\psi \in V$ such that $\theta(H)\psi = \lambda(H)\psi$ $(H \in \mathfrak{h})$.

LEMMA 1. *The sum* $\sum_\lambda V_\lambda$, *where* λ *runs over all linear functions on* $\mathfrak{h}$, *is direct. If U is any subspace of V which is invariant under $\theta(\mathfrak{h})$ then*

$$U \cap \left(\sum_\lambda V_\lambda \right) = \sum_\lambda (U \cap V_\lambda).$$

Let $\psi \in U \cap (\sum_\lambda V_\lambda)$. Then $\psi = \psi_1 + \cdots + \psi_r$ where $\psi_i \in V_{\lambda_i}$ and $\lambda_i \neq \lambda_j$, $1 \leq i, j \leq r$, $i \neq j$. We claim that $\psi_i \in U$, $1 \leq i \leq r$. If $r = 1$ this is true trivially. Hence we may assume $r > 1$ and use induction on r. We can find an $H \in \mathfrak{h}$ such that $\lambda_i(H) \neq \lambda_1(H)$, $2 \leq i \leq r$. Then

$$\theta(H)\psi = \sum_{1 \leq i \leq r} \lambda_i(H)\psi_i \in U.$$

Hence

$$\theta(H)\psi - \lambda_1(H)\psi = \sum_{2 \le i \le r} \{\lambda_i(H) - \lambda_1(H)\}\psi_i \in U.$$

Therefore by induction hypothesis $\{\lambda_i(H) - \lambda_1(H)\}\psi_i \in U$, $2 \le i \le r$. But $\lambda_i(H) - \lambda_1(H) \ne 0$ for $2 \le i \le r$. Hence $\psi_i \in U$, $2 \le i \le r$ and therefore $\psi_i \in U$, $1 \le i \le r$.

If we take $U = \{0\}$ above it follows that the sum $\sum_\lambda V_\lambda$ is direct. Also the above proof shows that $U \cap (\sum_\lambda V_\lambda) = \sum_\lambda (V_\lambda \cap U)$.

Let π be a representation of $\mathfrak{A}$ on a vector space V. λ being any linear function on $\mathfrak{h}$ we define V_λ as above to be the set of all $\psi \in V$ such that $\pi(H)\psi = \lambda(H)\psi$ $(H \in \mathfrak{h})$. ψ is said to be homogeneous of weight λ (or to have weight λ, or to belong to the weight λ) if $\psi \in V_\lambda$ and λ is called a weight of π if $V_\lambda \ne \{0\}$. It is clear that if ψ has the weight λ and λ is not a weight then $\psi = 0$. We shall call the dimension of V_λ the multiplicity of λ in π. Now in particular we may take π to be the adjoint representation of $\mathfrak{A}$ and define the subspaces $\mathfrak{A}_\lambda$. A weight of the adjoint representation will be called a rank and an element z will be said to be homogeneous of rank λ (or to have rank λ) if $z \in \mathfrak{A}_\lambda$.

LEMMA 2. $\mathfrak{A} = \sum_\lambda \mathfrak{A}_\lambda$ and every rank is a linear combination of α_i, $1 \le i \le l$, with integral coefficients.

Let P, Q, M denote any ordered set of integers as follows:

$$
\begin{aligned}
P &= \{i_1, i_2, \cdots, i_p\}, & 1 \le i_1, \cdots, i_p \le l, \; p \ge 0, \\
Q &= \{j_1, j_2, \cdots, j_q\}, & 1 \le j_1, \cdots, j_q \le l, \; q \ge 0, \\
M &= \{m_1, \cdots, m_l\}, & m_i \ge 0, \; 1 \le i \le l,
\end{aligned}
$$

(3)

where P or Q is empty if p or q is zero. Put $|P| = p$, $|Q| = q$, rank $P = \alpha_{i_1} + \alpha_{i_2} + \cdots + \alpha_{i_p}$, rank $Q = \alpha_{j_1} + \alpha_{j_2} + \cdots + \alpha_{j_q}$, the rank being understood to be zero in case the set is empty. We denote by ϕ the empty set and by 0 the set M all of whose elements are zero. Put

$$(4) \qquad z(Q, M, P) = Y_{j_1} Y_{j_2} \cdots Y_{j_q} H_1^{m_1} \cdot H_2^{m_2} \cdots H_l^{m_l} X_{i_1} X_{i_2} \cdots X_{i_p}$$

where $H_i^m = 1$ if $m = 0$. Making use of the relations

$$[H_i, H_j] = 0, \qquad [X_i, Y_i] = H_i, \qquad [X_i, Y_j] = 0, \qquad i \ne j,$$

$$[H, X_i] = \alpha_i(H)X_i, \qquad [H, Y_i] = -\alpha_i(H)Y_i,$$

$1 \le i, j \le l$, $H \in \mathfrak{h}$, which hold in $\mathfrak{A}$ it follows easily that $\mathfrak{A}$ is spanned by the elements $z(Q, M, P)$ taken together for all Q, M, and P. Clearly $z(Q, M, P)$ has the rank rank $P -$ rank Q. Let P denote the set of all linear functions on $\mathfrak{h}$ of the form rank $P -$ rank Q for all P and Q. Then since $z(Q, M, P)$ span $\mathfrak{A}$, $\mathfrak{A} \subset \sum_{\lambda \in P} \mathfrak{A}_\lambda$. Hence $\mathfrak{A} = \sum_{\lambda \in P} \mathfrak{A}_\lambda = \sum_\lambda \mathfrak{A}_\lambda$. But from Lemma 1 the sum $\sum_\lambda \mathfrak{A}_\lambda$ is direct. Hence $\mathfrak{A}_\lambda = \{0\}$ if $\lambda \notin P$. This proves the lemma.

Given $z \in \mathfrak{A}$ and a linear function λ_0 on $\mathfrak{h}$ it follows from Lemmas 1 and 2 that we can find a unique element $z_{\lambda_0} \in \mathfrak{A}_{\lambda_0}$ such that $z - z_{\lambda_0} \in \sum_{\lambda \neq \lambda_0} \mathfrak{A}_\lambda$. z_{λ_0} will be called the homogeneous component of z of rank λ_0. Clearly $z_\lambda = 0$ for all linear functions λ except a finite number and $z = \sum_\lambda z_\lambda$.

Exactly as before we call a linear function λ on $\mathfrak{h}$ rational if $\lambda(H_i) \in K_0$ for all $1 \le i \le l$. We order rational functions lexicographically with respect to the ordered set $\{\alpha_1, \cdots, \alpha_l\}$. Let $\mathfrak{H}$ be the subalgebra of $\mathfrak{A}$ generated by H_i, $1 \le i \le l$, and 1 and let $\mathfrak{P}$ be the left ideal $\sum_{1 \le i \le l} \mathfrak{A} X_i$.

LEMMA 3. *$\mathfrak{P}$ coincides with the subspace spanned by all elements of the form $z(Q, M, P)$, $|P| > 0$. Further $\mathfrak{P} \supset \sum_{\lambda > 0} \mathfrak{A}_\lambda$ where λ runs over all rational functions on $\mathfrak{h}$ which are greater than 0.*

It is clear from the definition of $\mathfrak{P}$ that $z(Q, M, P) \in \mathfrak{P}$ if $|P| > 0$. Conversely let $z \in \mathfrak{P}$. Then $z = \sum_{1 \le i \le l} z_i X_i$ $(z_i \in \mathfrak{A})$. Since z_i is a linear combination of $z(Q, M, P')$ and since $z(Q, M, P') X_i = z(Q, M, P)$ with $|P| > 0$, the first assertion follows.

Let $z \in \mathfrak{A}_\lambda$ where λ is a rational function greater than 0. We know that z is a linear combination of $z(Q, M, P)$. Since $z(Q, M, P)$ has rank rank $P -$ rank Q, we may, in view of Lemma 1, assume that only such elements $z(Q, M, P)$ appear in this linear combination for which rank $P -$ rank $Q = \lambda$. Since $\lambda > 0$ it follows that rank $P > 0$ and therefore $|P| > 0$. Hence $z \in \mathfrak{P}$.

LEMMA 4. $\mathfrak{P} \cap \mathfrak{H} = \{0\}$.

Let $z \in \mathfrak{P} \cap \mathfrak{H}$. Then $z = \sum_{m_1, \ldots, m_l} a(m_1, \cdots, m_l) H_1^{m_1} \cdots H_l^{m_l}$ where $a(m_1, \cdots, m_l) \in K$ and the sum is finite. Let μ be any linear function on $\mathfrak{h}$ and let π_μ be the representation of $\mathfrak{A}$ on $\mathfrak{G}$ corresponding to (2). Then if $\mu_i = \mu(H_i)$, $1 \le i \le l$,

$$\pi_\mu(z)1 = \sum_{m_1, \cdots, m_l} a(m_1, \cdots, m_l) \mu_1^{m_1} \cdots \mu_l^{m_l}.$$

But since $z \in \mathfrak{P}$ and $\pi_\mu(x_i)1 = 0$ it follows that

$$\sum_{m_1, \cdots, m_l} a(m_1, \cdots, m_l) \mu_1^{m_1} \cdots \mu_l^{m_l} = 0.$$

This is true for every μ and therefore for every choice of $\mu_1, \mu_2, \cdots, \mu_l \in K$. Since K is an infinite field it follows that all the coefficients $a(m_1, \cdots, m_l)$ are zero. Hence $z = 0$.

Let Λ be any linear function on $\mathfrak{h}$. Put

$$\mathfrak{Q}_\Lambda = \sum_{1 \le i \le l} \mathfrak{A}(H_i - \Lambda(H_i)), \qquad \mathfrak{H}_\Lambda = \sum_{1 \le i \le l} \mathfrak{H}(H_i - \Lambda(H_i)).$$

LEMMA 5. $\mathfrak{P} + \mathfrak{Q}_\Lambda \neq \mathfrak{A}$.

It is sufficient to show that $1 \notin \mathfrak{P} + \mathfrak{Q}_\Lambda$. Suppose contrary to the assertion

$1 \in \mathfrak{P} + \mathfrak{Q}_\Lambda$. Then we can find z_i, $u_i \in \mathfrak{A}$ such that

$$1 = \sum_{1 \leq i \leq l} z_i X_i + \sum_{1 \leq i \leq l} u_i(H_i - \Lambda(H_i)).$$

Since elements having different ranks are linearly independent we may assume that z_i has rank $-\alpha_i$ and u_i is of rank zero. Now

$$1 \equiv \sum_{1 \leq i \leq l} u_i(H_i - \Lambda(H_i)) \bmod \mathfrak{P}$$

and since u_i is of rank zero, $u_i(H_i - \Lambda(H_i)) = (H_i - \Lambda(H_i))u_i$. Further

$$u_i = \sum a_i(Q, M, P)z(Q, M, P) \qquad (a_i(Q, M, P) \in K)$$

where the sum is only over such (Q, M, P) for which rank Q=rank P since u_i is of rank zero and $z(Q, M, P)$ has the rank rank P−rank Q. Therefore from Lemma 3,

$$u_i \equiv \sum a_i(\phi, M, \phi)z(\phi, M, \phi) \bmod \mathfrak{P}.$$

Since $z(\phi, M, \phi) \in \mathfrak{H}$ it follows that

$$\sum_{1 \leq i \leq l} u_i(H_i - \Lambda(H_i)) = \sum_{1 \leq i \leq l} (H_i - \Lambda(H_i))u_i \in \mathfrak{P} + \mathfrak{H}_\Lambda.$$

Hence $1 \in \mathfrak{P} + \mathfrak{H}_\Lambda$. Therefore $1 = z + h$ where $z \in \mathfrak{P}$, $h \in \mathfrak{H}_\Lambda$. Hence $1 - h = z \in \mathfrak{P} \cap \mathfrak{H} = \{0\}$ from Lemma 4. Therefore $1 = h \in \mathfrak{H}_\Lambda$. Now consider the representation π_Λ of $\mathfrak{A}$ on $\mathfrak{G}$ corresponding to the linear function Λ on $\mathfrak{h}$. Since $h \in \mathfrak{H}_\Lambda$ it follows from the definition of π_Λ that $\pi_\Lambda(h)1 = 0$. But since $1 = h$, $\pi_\Lambda(h) \cdot 1 = \pi_\Lambda(1) \cdot 1 = 1$. Since $1 \neq 0$ in $\mathfrak{G}$ we get a contradiction. The lemma is therefore proved.

LEMMA 6. *Let m be any integer greater than or equal to 0. Then*

$$[X_i, Y_i^m] = mY_i^{m-1}(H_i - m + 1), \qquad 1 \leq i \leq l,$$

where by definition $Y_i^p = 1$ if $p \leq 0$.

The assertion is clearly true for $m = 0, 1$. Hence we may assume $m \geq 2$ and use induction. Then

$$[X_i, Y_i^m] = [X_i, Y_i]Y_i^{m-1} + Y_i[X_i, Y_i^{m-1}]$$
$$= H_i Y_i^{m-1} + (m-1)Y_i^{m-1}(H_i - m + 2)$$

by induction hypothesis. But $[H_i, Y_i] = -\alpha_i(H_i)Y_i = -2Y_i$. Therefore $H_i Y_i = Y_i(H_i - 2)$ and

$$[X_i, Y_i^m] = Y_i^{m-1}(H_i - 2m + 2) + (m-1)Y_i^{m-1}(H_i - m + 2)$$
$$= mY_i^{m-1}(H_i - m + 1).$$

Now put $\theta_{ii}=0$ and

$$\theta_{ij} = (\operatorname{ad} Y_i^{a_{ji}+1})Y_j, \qquad\qquad i \neq j,$$

$1 \leq i, j \leq l$. This is well defined since $a_{ji} \geq 0$ for $i \neq j$.

LEMMA 7 [4]. $[X_k, \theta_{ij}]=0$ *for all* $1 \leq i, j, k \leq l$.

We may assume that $i \neq j$. First suppose $k \neq i$. Then $[X_k, Y_i]=0$ and therefore $[\operatorname{ad} X_k, \operatorname{ad} Y_i]=0$. Hence

$$\begin{aligned}
(\operatorname{ad} X_k)\theta_{ij} &= \operatorname{ad} X_k(\operatorname{ad} Y_i)^{a_{ji}+1}Y_j \\
&= (\operatorname{ad} Y_i)^{a_{ji}+1}([X_k, Y_j]).
\end{aligned}$$

If $k \neq j$, $[X_k, Y_j]=0$ and we get our result. If $k=j$, we get

$$\begin{aligned}
[X_k, \theta_{ij}] &= (\operatorname{ad} Y_i)^{a_{ji}+1}H_j \\
&= (\operatorname{ad} Y_i)^{a_{ji}}[Y_i, H_j] \\
&= \alpha_i(H_j)(\operatorname{ad} Y_i)^{a_{ji}}Y_i.
\end{aligned}$$

If $a_{ji}>0$, $(\operatorname{ad} Y_i)^{a_{ji}}Y_i=0$. On the other hand if $a_{ji}=0$, it follows from assumption (1) of Theorem 1 that $\alpha_i(H_j)=-a_{ij}=0$. Hence in either case $[X_k, \theta_{ij}]=0$.

Finally suppose $k=i$. Then $[X_i, \theta_{ij}]=\operatorname{ad}(X_i Y_i^{a_{ji}+1})Y_j$. From Lemma 6, $[X_i, Y_i^{a_{ji}+1}]=(a_{ji}+1)Y_i^{a_{ji}}(H_i-a_{ji})$. Hence

$$[X_i, \theta_{ij}] = (a_{ji} + 1)\{\operatorname{ad} Y_i^{a_{ji}} \operatorname{ad}(H_i - a_{ji})\}Y_j.$$

But

$$\{\operatorname{ad}(H_i - a_{ji})\}Y_j = [H_i, Y_j] - a_{ji}Y_j = -\alpha_j(H_i)Y_j - a_{ji}Y_j = 0.$$

Hence $[X_i, \theta_{ij}]=0$ and the lemma is proved.

Let λ_i, $1 \leq i \leq l$, be any given set of non-negative integers and let Λ_0 be the linear function on $\mathfrak{h}$ defined by $\Lambda_0(H_i)=\lambda_i$. We consider the left ideal

$$\mathfrak{B}_{\Lambda_0} = \mathfrak{P} + \mathfrak{O}_{\Lambda_0} + \sum_{1 \leq i, j \leq l} \mathfrak{A}\theta_{ij}\mathfrak{A} + \sum_{1 \leq i \leq l} \mathfrak{A}Y_i^{\lambda_i+1}$$

in $\mathfrak{A}$, where $\mathfrak{P}$ and $\mathfrak{O}_{\Lambda_0}$ are defined as in Lemma 5.

LEMMA 8. $\mathfrak{B}_{\Lambda_0} \neq \mathfrak{A}$.

Suppose the assertion is false. Then $1 \in \mathfrak{B}_{\Lambda_0}$. Hence

$$1 \equiv z + w \bmod (\mathfrak{P} + \mathfrak{O}_{\Lambda_0})$$

where $z \in \sum_{1 \leq i \leq l} \mathfrak{A}Y_i^{\lambda_i+1}$ and $w \in \sum_{1 \leq i,j \leq l} \mathfrak{A}\theta_{ij}\mathfrak{A}$. Notice that θ_{ij} has the rank $-(a_{ji}+1)\alpha_i-\alpha_j$. By considering components of different ranks we can

---

[4] This lemma is due to Chevalley.

show, as in the proof of Lemma 5, that z can be chosen to be a linear combination of elements of the form $z(Q, M, P)Y_i^{\lambda_i+1}$ where rank $P-$rank Q $-(\lambda_i+1)\alpha_i=0$. Similarly we may assume that w is a linear combination of elements of the form $z(Q, M, P)\theta_{ij}z(Q', M', P')$ where rank $P+$rank P' $-$rank $Q-$rank $Q'-(a_{ji}+1)\alpha_i-\alpha_j=0$ $(i\neq j)$. Now consider a term $z(Q, M, P)Y_i^{\lambda_i+1}$ such that rank $P-$rank $Q=(\lambda_i+1)\alpha_i$. If rank $Q>0$, rank $P>(\lambda_i+1)\alpha_i$. Hence $z(\phi, M, P)Y_i^{\lambda_i+1}$ has rank $P-(\lambda_i+1)\alpha_i>0$. Therefore by Lemma 3, it is contained in $\mathfrak{P}$. Since $\mathfrak{P}$ is a left ideal, $z(Q, M, P)Y_i^{\lambda_i+1}$ $\in\mathfrak{P}$. Hence the term corresponding to $z(Q, M, P)Y_i^{\lambda_i+1}$ can be dropped from the above congruence. Therefore we may assume that z is a linear combination of elements of the form $z(\phi, M, P)Y_i^{\lambda_i+1}$ with rank $P=(\lambda_i+1)\alpha_i$. But since α_j, $1\leq j\leq l$, are linearly independent rank $P=(\lambda_i+1)\alpha_i$ implies that $z(\phi, 0, P)=X_i^{\lambda_i+1}$. Hence z is a linear combination of elements of the form $H_1^{m_1}\cdots H_l^{m_l}X_i^{\lambda_i+1}Y_i^{\lambda_i+1}$. But from Lemma 6,

$$X_i^{\lambda_i+1}Y_i^{\lambda_i+1} = X_i^{\lambda_i}[X_i, Y_i^{\lambda_i+1}] + X_i^{\lambda_i}Y_i^{\lambda_i+1}X_i$$

$$\equiv (\lambda_i + 1)X_i^{\lambda_i}Y_i^{\lambda_i}(H_i - \lambda_i)\bmod \mathfrak{P}.$$

Since $H_i-\lambda_i\in\mathfrak{Q}_{\Lambda_0}$, $X_i^{\lambda_i+1}Y_i^{\lambda_i+1}\equiv 0\bmod(\mathfrak{P}+\mathfrak{Q}_{\Lambda_0})$. This shows that z can be replaced by zero in our congruence. Hence

$$1 \equiv w\bmod(\mathfrak{P} + \mathfrak{Q}_{\Lambda_0})$$

where w is a linear combination of terms of the form $z(Q, M, P)\theta_{ij}z(Q', M', P')$ with rank $P+$rank $P'-$rank $Q-$rank $Q'=(a_{ji}+1)\alpha_i+\alpha_j$ $(i\neq j)$. Since we are considering a congruence mod $(\mathfrak{P}+\mathfrak{Q}_{\Lambda_0})$ we may clearly assume in addition that $P'=\phi$ and $M'=0$. Hence we have only terms of the form $z(Q, M, P)\theta_{ij}z(Q', 0, \phi)$ with rank $P-$rank $Q-$rank $Q'=(a_{ji}+1)\alpha_i+\alpha_j$. From Lemma 7, X_k commutes with θ_{ij}. Hence

$$z(Q, M, P)\theta_{ij}z(Q', 0, \phi) = z(Q, M, \phi)\theta_{ij}z(\phi, 0, P)z(Q', 0, \phi).$$

But rank $P-$rank $Q'=$rank $Q+(a_{ji}+1)\alpha_i+\alpha_j>0$ $(i\neq j)$ and therefore by Lemma 3, $z(\phi, 0, P)z(Q', 0, \phi)\in\mathfrak{P}$. Therefore $w\in\mathfrak{P}$ and we have $1\in\mathfrak{P}+\mathfrak{Q}_{\Lambda_0}$. But by Lemma 5 this is impossible. Hence $\mathfrak{B}_{\Lambda_0}\neq\mathfrak{A}$.

As usual we call a left ideal $\mathfrak{N}$ in $\mathfrak{A}$ maximal if $\mathfrak{N}\neq\mathfrak{A}$ and if there exists no left ideal $\mathfrak{N}'$ in $\mathfrak{A}$ such that $\mathfrak{N}'\supset\mathfrak{N}$, $\mathfrak{N}'\neq\mathfrak{N}$, and $\mathfrak{N}'\neq\mathfrak{A}$.

LEMMA 9. *Let Λ be any linear function on $\mathfrak{h}$. Then there exists at most one maximal ideal $\mathfrak{N}$ in $\mathfrak{A}$ such that $\mathfrak{N}\supset\mathfrak{P}+\mathfrak{Q}_{\Lambda}$.*

For suppose $\mathfrak{N}_1$, $\mathfrak{N}_2$ are two distinct maximal left ideals containing $\mathfrak{P}+\mathfrak{Q}_{\Lambda}$. Then $\mathfrak{N}_1+\mathfrak{N}_2=\mathfrak{A}$. Hence $1=z_1+z_2$ where $z_i\in\mathfrak{N}_i$, $i=1$, 2. Notice that if $w\in\mathfrak{N}_1$, and $H\in\mathfrak{h}$, then

$$[H, w] = Hw - w(H - \Lambda(H)) - \Lambda(H)w \in \mathfrak{N}_1$$

since $\mathfrak{N}_1 \supset \mathfrak{O}_\Lambda$. Therefore $\mathfrak{N}_1$ is invariant under ad $\mathfrak{h}$ and from Lemmas 1 and 2, $\mathfrak{N}_1 = \sum_\lambda (\mathfrak{N}_1 \cap \mathfrak{A}_\lambda)$. Similarly for $\mathfrak{N}_2$. Hence if $z_{i,0}$ is the homogeneous component of z_i of rank zero $z_{i,0} \in \mathfrak{N}_i$ $(i=1, 2)$ and $1 = z_{1,0} + z_{2,0}$. Therefore we may assume that z_1, z_2 are both of rank zero. But then they can be written as linear combinations of $z(Q, M, P)$ with rank $P -$ rank $Q = 0$. If $|P| > 0$, $z(Q, M, P) \in \mathfrak{P}$. On the other hand if $P = \phi$ then $Q = \phi$ since rank $P =$ rank Q and $z(\phi, M, \phi) \equiv c \bmod \mathfrak{O}_\Lambda$ $(c \in K)$ because $H_i - \Lambda(H_i) \in \mathfrak{O}_\Lambda$, $1 \leq i \leq l$. Hence it follows that $z_i - c_i \in \mathfrak{P} + \mathfrak{O}_\Lambda$ $(i=1, 2)$ for some $c_1, c_2 \in K$. Since $z_1 \in \mathfrak{N}_1$ and $\mathfrak{N}_1 \supset \mathfrak{P} + \mathfrak{O}_\Lambda$, $c_1 \in \mathfrak{N}_1$. But $\mathfrak{N}_1$ is maximal and therefore $1 \notin \mathfrak{N}_1$. Hence $c_1 = 0$. Similarly $c_2 = 0$. Hence $z_1, z_2 \in \mathfrak{P} + \mathfrak{O}_\Lambda$ and therefore $1 \in \mathfrak{P} + \mathfrak{O}_\Lambda$. But, in view of Lemma 5, this is false. Thus the lemma is established.

REMARK. Since $\mathfrak{P} + \mathfrak{O}_\Lambda \neq \mathfrak{A}$ it follows from Zorn's lemma that there exists at least one maximal left ideal in $\mathfrak{A}$ containing $\mathfrak{P} + \mathfrak{O}_\Lambda$. The above lemma then shows that it is unique. However we shall not have to invoke Zorn's lemma for our purpose.

LEMMA 10. *Let π be a representation of $\mathfrak{A}$ on V. If λ and μ are linear functions on $\mathfrak{h}$ and $z \in \mathfrak{A}_\lambda$, $\psi \in V_\mu$ then $\pi(z)\psi \in V_{\lambda+\mu}$.*

Let $H \in \mathfrak{h}$. Then

$$\begin{aligned}
\pi(H)\pi(z)\psi &= \pi([H, z])\psi + \pi(z)\pi(H)\psi \\
&= \lambda(H)\pi(z)\psi + \mu(H)\pi(z)\psi \\
&= (\lambda(H) + \mu(H))\pi(z)\psi.
\end{aligned}$$

Hence $\pi(z)\psi \in V_{\lambda+\mu}$.

We now define a linear transformation σ_i, $1 \leq i \leq l$, in the space of all linear functions on $\mathfrak{h}$ as follows:

$$\sigma_i\lambda = \lambda - \lambda(H_i)\alpha_i.$$

Since σ_i^2 is the identity, σ_i, $1 \leq i \leq l$, generate a group W. Further, we recall that a linear function λ is called integral if $\lambda(H_i)$, $1 \leq i \leq l$, are all integers.

LEMMA 11. *Let π be a representation of $\mathfrak{A}$ on V. Suppose for every $\psi \in V$ we can find an integer $\nu \geq 0$ such that $\pi(X_i^\nu)\psi = \pi(Y_i^\nu)\psi = 0$, $1 \leq i \leq l$. Then every weight of π is an integral function. Also if Λ is a weight of π then for any i, $1 \leq i \leq l$, $\Lambda - k\alpha_i$ is also a weight of π for every integer k such that*

$$\min (0, \Lambda(H_i)) \leq k \leq \max (0, \Lambda(H_i)).$$

In particular $\sigma\Lambda$ is a weight of π for every $\sigma \in W$.

Let Λ be any weight of π. Consider any fixed i and choose an element $\psi \in V_\Lambda$, $\psi \neq 0$. Let k_0 be the least integer greater than or equal to 0 such that $\pi(X_i^{k_0+1})\psi = 0$. Put $\psi_0 = \pi(X_i^{k_0})\psi \neq 0$. Define $\psi_k = \pi(Y_i^k)\psi_0$, $k \geq 1$. Then by hypothesis $\psi_\nu = 0$ for some integer $\nu \geq 0$. Let J be the least integer greater than or equal to 0 such that $\psi_{J+1}=0$. Put $\psi_{-1}=1$. From Lemma 10 it is clear that $\psi_k \in V_{\Lambda+(k_0-k)\alpha_i}$. Using this fact together with the relation $[X_i,\ Y_i]=H_i$ we easily prove by induction on k that

$$\pi(X_i)\psi_k = k[\Lambda(H_i) + (2k_0 - k + 1)]\psi_{k-1}, \qquad k \geq 0.$$

On substituting $k=J+1$ and remembering that $\psi_{J+1}=0$, $\psi_J\neq0$, $J+1\neq0$ we get

$$J = \Lambda(H_i) + 2k_0.$$

This shows that $\Lambda(H_i)$ is an integer. This being true for every i, $1\leq i\leq l$, it follows that Λ is integral. Further $\psi_k \in V_{\Lambda+(k_0-k)\alpha_i}$ and $\psi_k\neq0$ for $0\leq k\leq J$. Hence $\Lambda-k\alpha_i$ is a weight of π for

$$- k_0 \leq k \leq J - k_0 = \Lambda(H_i) + k_0.$$

Similarly let k_0' be the least integer greater than or equal to 0 such that $\pi(Y_i^{k_0'+1})\psi=0$. Put $\phi_0=\pi(Y_i^{k_0'})\psi\neq0$ and $\phi_k=\pi(X_i^k)\phi_0$, $k\geq1$, and $\phi_{-1}=0$. Again we prove by induction on k that

$$\pi(Y_i)\phi_k = k\left\{- \Lambda(H_i) + (2k_0' - k + 1)\right\}\phi_{k-1}, \qquad k \geq 0.$$

Let J' be the least integer greater than or equal to 0 such that $\phi_{J'+1}=0$. Substituting $k=J'$ in the above equation we get, as before,

$$J' = - \Lambda(H_i) + 2k_0'.$$

Now $\phi_k \in V_\Lambda - (k_0' - k)\alpha_i$ and $\phi_k\neq0$ for $0\leq k\leq J'$. Hence $\Lambda-k\alpha_i$ is a weight of π for

$$k_0' - J' = - k_0' + \Lambda(H_i) \leq k \leq k_0'.$$

Combining this with the earlier result we find that $\Lambda-k\alpha_i$ is a weight of π for all integers k such that

$$\min\left\{- k_0, - k_0' + \Lambda(H_i)\right\} \leq k \leq \max\left\{k_0', \Lambda(H_i) + k_0\right\}.$$

Since the integer $k=\Lambda(H_i)$ always lies in this range, $\sigma_i\Lambda=\Lambda-\Lambda(H_i)\alpha_i$ is a weight. This being true for any i and any weight Λ of π it follows immediately that $\sigma\Lambda$ is a weight for every $\sigma\in W$. Finally since $k_0\geq0$, $k_0'\geq0$,

$$\min\left\{- k_0, - k_0' + \Lambda(H_i)\right\} \leq \min(0, \Lambda(H_i)) \leq \max(0, \Lambda(H_i))$$
$$\leq \max\left\{k_0', \Lambda(H_i) + k_0\right\}.$$

Therefore the lemma is proved.

We now return to the left ideal $\mathfrak{B}_{\Lambda_0}$ of Lemma 8.

LEMMA 12. *Let π be the natural representation of $\mathfrak{A}$ on $\mathfrak{A}/\mathfrak{B}_{\Lambda_0}=\mathfrak{A}^*$. Then Λ_0 is a weight of π, and every weight of π is of the form $\Lambda=\Lambda_0-(d_1\alpha_1+\cdots+d_l\alpha_l)$ where d_i are integers greater than or equal to 0. Further $\mathfrak{A}^*=\sum_\Lambda \mathfrak{A}_\Lambda^*$ where the* *sum is direct and is over all weights Λ of π. Finally, given any $z^* \in \mathfrak{A}^*$ we can find an integer $\nu \geq 0$ such that*

$$\pi(X_i^\nu)z^* = \pi(Y_i^\nu)z^* = 0, \qquad\qquad 1 \leq i \leq l.$$

Let $z \to z^*$ denote the natural mapping of $\mathfrak{A}$ on $\mathfrak{A}^*$. Since from Lemma 8, $1 \not\in \mathfrak{B}_{\Lambda_0}$, $1^* \neq 0$. Also $\mathfrak{B}_{\Lambda_0} \supset \mathfrak{Q}_{\Lambda_0}$ and therefore it is clear that $1^* \in \mathfrak{A}^*_{\Lambda_0}$. Hence Λ_0 is a weight of π.

Given any $z \in \mathfrak{A}$ we can write it as a linear combination of $z(Q, M, P)$. But clearly $z(Q, M, P) \in \mathfrak{P}$ if $|P| > 0$ or is congruent to $cz(Q, 0, \phi)$ mod $\mathfrak{Q}_{\Lambda_0}$ $(c \in K)$ if $|P| = 0$. Hence z is congruent mod $\mathfrak{P} + \mathfrak{Q}_{\Lambda_0}$ to a linear combination of $z(Q, 0, \phi)$. Therefore $(z(Q, 0, \phi))^*$ taken together for all Q span $\mathfrak{A}^*$. Since $z(Q, 0, \phi)$ has the rank$-$rank Q and $(z(Q, 0, \phi))^* = \pi(z(Q, 0, \phi))1^*$ it follows from Lemma 10 that $(z(Q, 0, \phi))^* \in \mathfrak{A}^*_{\Lambda_0 - \mathrm{rank}\ Q}$. Hence

$$\mathfrak{A}^* = \sum_Q \mathfrak{A}^*_{\Lambda_0 - \mathrm{rank}\ Q}.$$

We now deduce from Lemma 1 that this sum is direct and every weight Λ of π is of the form $\Lambda_0 - \mathrm{rank}\ Q$. The first part of the lemma is therefore proved.

Since $(z(Q, 0, \phi))^*$ span $\mathfrak{A}^*$ we may, in proving the second part, assume that $z^* = (z(Q, 0, \phi))^*$. Hence $z^* \in \mathfrak{A}^*_{\Lambda_0 - \mathrm{rank}\ Q}$. For a fixed i consider $\pi(X_i^\nu)z^*$. From Lemma 10, $\pi(X_i^\nu)z^* \in \mathfrak{A}^*_{\Lambda_0 - \mathrm{rank}\ Q + \nu\alpha_i}$. Now suppose $\pi(X_i^\nu)z^* \neq 0$. Then $\Lambda_0 - \mathrm{rank}\ Q + \nu\alpha_i$ is a weight and therefore is of the form $\Lambda_0 - \mathrm{rank}\ Q'$ for some Q'. Therefore rank $Q - \nu\alpha_i = $ rank Q'. Let

$$\mathrm{rank}\ Q = d_1\alpha_1 + \cdots + d_l\alpha_l,$$
$$\mathrm{rank}\ Q' = d_1'\alpha_1 + \cdots + d_l'\alpha_l.$$

Then d_j, d_j', $1 \leq j \leq l$, are all integers greater than or equal to 0. Since $\alpha_1, \cdots, \alpha_l$ are linearly independent it follows that $d_i - \nu = d_i'$. Hence $\nu = d_i - d_i' \leq d_i$. Therefore if $\nu > d_i$, $\pi(X_i^\nu)z^* = 0$.

Now we consider $\pi(Y_i^\nu)z^*$. Let R_i, L_i, and D_i denote the linear mappings of $\mathfrak{A}$ defined as follows:

$$R_i w = wY_i, \qquad L_i w = Y_i w, \qquad D_i w = [Y_i, w] \qquad (w \in \mathfrak{A}).$$

Clearly $L_i = D_i + R_i$ and L_i, R_i, D_i all commute with each other. Hence for any integer $m \geq 0$

$$(6) \qquad L_i^m = (D_i + R_i)^m = \sum_{0 \leq p \leq m} \frac{m!}{m - p!\,p!} R_i^{m-p} D_i^p$$

(where $L_i^p = R_i^p = D_i^p$ is the identity mapping if $p = 0$). Notice that $D_i^{a_{ji}+1} Y_j = \theta_{ij}$ if $i \neq j$ and $D_i Y_i = 0$. Hence we can find an integer ν_0 such that if $p > \nu_0$, $D_i^p Y_j \in \mathfrak{B}$, $1 \leq j \leq l$, where $\mathfrak{B}$ is the ideal $\sum_{1 \leq j, k \leq l} \mathfrak{A}\theta_{jk}\mathfrak{A}$. We claim that if

$p > \nu_0 |Q|$, $D_i^p z(Q, 0, \phi) \in \mathfrak{B}$. This is easily proved by induction on $|Q|$ if we make use of the rule

$$D_i^p(uv) = \sum_{0 \le r \le p} \frac{p!}{p - r! r!} D_i^r(u) D_i^{p-r}(v) \qquad (u, v \in \mathfrak{A}).$$

Let $\nu = \nu_0 |Q| + \max_{1 \le j \le l} \lambda_j$ where $\lambda_j = \Lambda_0(H_j)$ and let $m > \nu$. Then from (6),

$$Y_i^m z(Q, 0, \phi) = \sum_{0 \le p \le m} \frac{m!}{m - p! p!} \{D_i^p(z(Q, 0, \phi))\} Y_i^{m-p}$$

Now if $p > \nu_0 |Q|$, $D_i^p(z(Q, 0, \phi)) \in \mathfrak{B}$ and if $p \le \nu_0 |Q|$, $m - p \ge \lambda_i + 1$ and

$$\{D_i^p(z(Q, 0, \phi))\} Y_i^{m-p} \in \mathfrak{A} Y_i^{\lambda_i+1} \subset \mathfrak{B}_{\Lambda_0}.$$

Therefore $Y_i^m z(Q, 0, \phi) \in \mathfrak{B}_{\Lambda_0}$ and

$$\pi(Y_i^m)(z(Q, 0, \phi))^* = 0.$$

Thus the lemma is proved.

Let π be as above. We shall call a weight Λ of π extreme if it is impossible to find an α_i, $1 \le i \le l$, and $\sigma \in W$ such that $\Lambda + \sigma\alpha_i$ and $\Lambda - \sigma\alpha_i$ are both weights of π. Obviously Λ_0 is the highest weight of π and therefore it is also extreme. From Lemmas 11 and 12 it is clear that if Λ is a weight of π then $\sigma\Lambda$ is also a weight of π for all $\sigma \in W$.

LEMMA 13. *Let π be as in Lemma 12 and let Λ be an extreme weight of π. Then for any $\sigma \in W$, $\sigma\Lambda$ is also an extreme weight of π.*

Suppose $\sigma\Lambda$ is not extreme. Then for some α_i and $\sigma_0 \in W$, $\sigma\Lambda + \sigma_0\alpha_i$ and $\sigma\Lambda - \sigma_0\alpha_i$ are both weights of π. But it follows from Lemmas 11 and 12 that $\sigma^{-1}(\sigma\Lambda + \sigma_0\alpha_i) = \Lambda + \sigma^{-1}\sigma_0\alpha_i$ and $\sigma^{-1}(\sigma\Lambda - \sigma_0\alpha_i) = \Lambda - \sigma^{-1}\sigma_0\alpha_i$ are also weights of π. Since $\sigma^{-1}\sigma_0 \in W$ this contradicts the hypothesis that Λ is extreme. Hence $\sigma\Lambda$ must be extreme.

LEMMA 14. *Let π be as above and let Λ be an extreme weight of π such that $\sigma_i\Lambda \ge \Lambda$ for all $1 \le i \le l$. Then $\Lambda - \alpha_i$ is not a weight for any α_i, $1 \le i \le l$.*

Suppose $\Lambda - \alpha_i$ is a weight. Since $\sigma_i\Lambda = \Lambda - \Lambda(H_i)\alpha_i \ge \Lambda$, $\Lambda(H_i) \le 0$. From Lemma 11, $\Lambda - \nu\alpha_i$ is a weight for $\Lambda(H_i) \le \nu \le 0$. Hence if $\Lambda(H_i) < 0$, $\Lambda + \alpha_i$ is a weight. On the other hand if $\Lambda(H_i) = 0$, $\sigma_i(\Lambda - \alpha_i) = \Lambda + \alpha_i$ is again a weight. Therefore in either case both $\Lambda + \alpha_i$ and $\Lambda - \alpha_i$ are weights, thus contradicting our hypothesis that Λ is extreme.

Notice that $\sigma_i\alpha_j = \alpha_j - \alpha_j(H_i)\alpha_i = \alpha_j + a_{ji}\alpha_i$. Since α_i, $1 \le i \le l$, form a base for the space of all linear functions on $\mathfrak{h}$ it follows that the group W generated by σ_i is exactly the same as that appearing in the statement of Theorem 1. So far we have made no use of the hypothesis that W is a finite group. But now it will enter in an essential way in the proof.

LEMMA 15. *If W is a finite group the space $\mathfrak{A}^* = \mathfrak{A}/\mathfrak{B}_{\Lambda_0}$ is finite-dimensional.*

Let π be the natural representation of $\mathfrak{A}$ on $\mathfrak{A}^*$ and let $\mathfrak{S}$ denote the set of all weights of π of the form $\sigma\Lambda_0$ $(\sigma \in W)$. Since W is finite, $\mathfrak{S}$ is a finite set. Also since Λ_0 is extreme it follows from Lemma 13 that every weight in $\mathfrak{S}$ is extreme. Let Λ_1 be the lowest weight in $\mathfrak{S}$. Then $\sigma_i\Lambda_1 \geq \Lambda_1$ for all $1 \leq i \leq l$ and Λ_1 is an extreme weight. Hence from Lemma 14, $\Lambda_1 - \alpha_i$ is not a weight of π for any α_i, $1 \leq i \leq l$. Choose any $z^* \in \mathfrak{A}^*_{\Lambda_1}$ $(z^* \neq 0)$ and put $\mathfrak{N}^* = \pi(\mathfrak{A})z^*$. Then $\mathfrak{N}^*$ is an invariant subspace of $\mathfrak{A}^*$. We claim that $\mathfrak{N}^*$ is finite-dimensional. Corresponding to any P, M, Q put

$$w(P, M, Q) = X_{i_1}X_{i_2} \cdots X_{i_p}H_1^{m_i} \cdots H_l^{m_l}Y_{j_1}Y_{j_2} \cdots Y_{j_q}$$

where $P = \{i_1, \cdots, i_p\}$, $M = \{m_1, \cdots, m_l\}$, and $Q = \{j_1, \cdots, j_q\}$ as in (3). Then exactly as in the case of $z(Q, M, P)$ we prove that $w(P, M, Q)$ taken together for all P, M, Q span $\mathfrak{A}$. From Lemma 10, $\pi(Y_i)z^* \in \mathfrak{A}^*_{\Lambda_1 - \alpha_i}$. But since $\Lambda_1 - \alpha_i$ is not a weight of π, $\pi(Y_i)z^* = 0$. Hence $\pi(w(P, M, Q))z^* = 0$ if $|Q| > 0$. Also $\pi(H)z^* = \Lambda_1(H)z^*$ $(H \in \mathfrak{h})$. Hence it is obvious that $\mathfrak{N}^*$ is spanned by elements of the form $\pi(w(P, 0, \phi))z^*$. But again by Lemma 10, $\pi(w(P, 0, \phi)) \in \mathfrak{A}^*_{\Lambda_1 + \mathrm{rank} P}$. Hence if $\pi(w(P, 0, \phi))z^* \neq 0$, $\Lambda_1 + \mathrm{rank}\ P$ is a weight. Therefore from Lemma 12,

$$\Lambda_1 + \mathrm{rank}\ P = \Lambda_0 - \mathrm{rank}\ Q$$

for some Q. Also $\Lambda_1 = \Lambda_0 - \mathrm{rank}\ Q'$ for a suitable Q'. Therefore

$$\mathrm{rank}\ P = \mathrm{rank}\ Q' - \mathrm{rank}\ Q.$$

Let $\mathrm{rank}\ P = e_1\alpha_1 + \cdots + e_l\alpha_l$, $\mathrm{rank}\ Q = d_1\alpha_1 + \cdots + d_l\alpha_l$, $\mathrm{rank}\ Q' = d_1'\alpha_1 + \cdots + d_l'\alpha_l$. Then e_i, d_i, $d_i' \geq 0$, $1 \leq i \leq l$, and again from the linear independence of α_i we deduce that $e_i = d_i' - d_i$, $1 \leq i \leq l$. Hence $e_i \leq d_i'$. Therefore $|P| = \sum e_i \leq \sum d_i' = |Q'|$. Thus it is clear that there are only a finite number of possibilities for P. Since $\mathfrak{N}^*$ is spanned by $\pi(w(P, 0, \phi))z^*$ with $|P| \leq |Q'|$, it follows that $\dim \mathfrak{N}^*$ is finite. Also since $z^* \in \mathfrak{N}^*$, $z^* \neq 0$, $\mathfrak{N}^* \neq \{0\}$.

Since $\mathfrak{N}^*$ is invariant under π, it follows from Lemmas 1 and 12 that $\mathfrak{N}^* = \sum_\Lambda \mathfrak{N}^*_\Lambda$ where $\mathfrak{N}^*_\Lambda = \mathfrak{A}^*_\Lambda \cap \mathfrak{N}^*$ and Λ runs over all the weights of π. Since the sum $\sum_\Lambda \mathfrak{A}^*_\Lambda$ is direct and $\dim \mathfrak{N}^*$ is finite, $\mathfrak{N}^*_\Lambda \neq \{0\}$ for only a finite number of weights Λ. Let π^* denote the representation of $\mathfrak{A}$ induced on $\mathfrak{N}^*$. Then it is obvious from the above remark that π^* has only a finite number of weights. Let Λ be a weight of π^* and let $w^* \in \mathfrak{N}^*_\Lambda$. Then by Lemma 10, $\pi^*(X_i^\nu)w^* \in \mathfrak{N}^*_{\Lambda + \nu\alpha_i}$. Since $\alpha_i \neq 0$, the linear functions $\Lambda + \nu\alpha_i$, $\nu = 1, 2, \cdots$, are all distinct. Hence they cannot all be weights of π^*. Hence for sufficiently large ν, $\pi^*(X_i^\nu)w^* = 0$. Similarly we prove that $\pi^*(Y_i^\nu)w^* = 0$ for ν sufficiently large. Since $\mathfrak{N}^* = \sum_\Lambda \mathfrak{N}^*_\Lambda$ it follows that for any $w^* \in \mathfrak{N}^*$ we can find an integer $\nu \geq 0$ such that $\pi^*(X_i^\nu)w^* = \pi^*(Y_i^\nu)w^* = 0$, $1 \leq i \leq l$. Hence Lemma 11 is applicable. Since $z^* \in \mathfrak{N}^*_{\Lambda_1}$ $(z^* \neq 0)$ it follows that Λ_1 is a weight of π^*. But

$\Lambda_1 \in \mathfrak{S}$ and therefore $\Lambda_1 = \sigma\Lambda_0 (\sigma \in W)$. Therefore from Lemma 11, $\sigma^{-1}\Lambda_1 = \Lambda_0$ is also a weight of π^*. Now $(z(Q, 0, \phi))^*$ has the weight $\Lambda_0 - \operatorname{rank} Q$. Hence $(z(Q, 0, \phi))^* \in \mathfrak{A}_{\Lambda_0}^*$ and $(z(Q, 0, \phi))^* \neq 0$ implies that rank $Q = 0$, that is, $Q = \phi$. Since the elements $(z(Q, 0, \phi))^*$ span $\mathfrak{A}^*$, it follows from Lemma 1 that $\mathfrak{A}_{\Lambda_0}^* = K \cdot 1^*$. Since $\mathfrak{N}_{\Lambda_0}^* \neq \{0\}$ it follows that $1^* \in \mathfrak{N}^*$. Hence

$$\mathfrak{A}^* = \pi(\mathfrak{A})1^* \subset \mathfrak{N}^*.$$

Therefore $\mathfrak{A}^*$ is finite-dimensional.

Let $\mathfrak{N}^*$ be an invariant subspace of $\mathfrak{A}^*$ of the maximum possible dimension such that $\mathfrak{N}^* \neq \mathfrak{A}^*$. Let $\mathfrak{N}_{\Lambda_0}$ be the complete inverse image of $\mathfrak{N}^*$ in $\mathfrak{A}$. Clearly $\mathfrak{M}_{\Lambda_0}$ is a maximal left ideal containing $\mathfrak{B}_{\Lambda_0}$. Since $\mathfrak{B}_{\Lambda_0} \supset \mathfrak{P} + \mathfrak{Q}_{\Lambda_0}$, it follows from Lemma 9 that $\mathfrak{M}_{\Lambda_0}$ is the unique maximal left ideal containing $\mathfrak{B}_{\Lambda_0}$. The natural representation π_{Λ_0} on $\mathfrak{A}/\mathfrak{M}_{\Lambda_0} \cong \mathfrak{A}^*/\mathfrak{M}^*$ is then irreducible and finite-dimensional. We note for later use that if $\Lambda_0 = 0$ then $\pi_{\Lambda_0}(X_i) = \pi_{\Lambda_0}(Y_i) = \pi_{\Lambda_0}(H_i) = 0$, $1 \leq i \leq l$. This follows from the fact that X_i, Y_i, $H_i \in \mathfrak{B}_{\Lambda_0}$ and therefore $\mathfrak{A}/\mathfrak{B}_{\Lambda_0} = K \cdot 1^*$.

Let π be any representation of $\mathfrak{A}$ on a vector space V. We shall say that λ is the highest weight of π if λ is a weight of π and for any weight μ of π $(\mu \neq \lambda)$, $\lambda - \mu$ is a rational function greater than 0. Given any linear function Λ on $\mathfrak{h}$, let $\mathfrak{M}_\Lambda$ be the unique maximal ideal containing[5] $\mathfrak{P} + \mathfrak{Q}_\Lambda$. We denote by π_Λ the natural representation of $\mathfrak{A}$ on $\mathfrak{A}/\mathfrak{M}_\Lambda$. Since $\mathfrak{M}_\Lambda$ is maximal, π_Λ is irreducible. It is easily seen that Λ is the highest weight of π_Λ.

LEMMA 16. *Let π be an irreducible representation of $\mathfrak{A}$ on V such that Λ is the highest weight of π. Then π is equivalent[6] to π_Λ. Also* dim $V_\Lambda = 1$.

Let $\psi \in V_\Lambda$, $\psi \neq 0$. Since π is irreducible, $\pi(\mathfrak{A})\psi = V$. Let $\mathfrak{M}$ be the left ideal in $\mathfrak{A}$ consisting of all elements z such that $\pi(z)\psi = 0$. Put $\mathfrak{A}^* = \mathfrak{A}/\mathfrak{M}$ and let $z \to z^*$ denote the natural mapping of $\mathfrak{A}$ on $\mathfrak{A}^*$. Let θ be the linear mapping of $\mathfrak{A}^*$ into V defined as follows. For any $z \in \mathfrak{A}$ put $\theta z^* = \pi(z)\psi$. It is easily seen that this mapping is well defined. Since $V = \pi(\mathfrak{A})\psi$, θ maps $\mathfrak{A}^*$ onto V. If $\theta z^* = 0$ then $\pi(z)\psi = 0$. Hence $z \in \mathfrak{M}$ and so $z^* = 0$. Therefore θ is an isomorphism of $\mathfrak{A}^*$ on V. Let π^* be the natural representation of $\mathfrak{A}$ on $\mathfrak{A}^*$. Then if w, $z \in \mathfrak{A}$,

$$\theta\pi^*(w)z^* = \theta(wz)^* = \pi(wz)\psi = \pi(w)\pi(z)\psi$$
$$= \pi(w)\theta z^*.$$

This shows that π and π^* are equivalent. Since π is irreducible the same

---

[5] We have assumed the existence of $\mathfrak{M}_\Lambda$ here and therefore made use of Zorn's lemma. This is done only for convenience. It would be sufficient for our purpose to define π_Λ whenever $\mathfrak{M}_\Lambda$ exists.

[6] $\mathfrak{A}$ being any associative algebra and π, π' two representations of $\mathfrak{A}$ on V, V' respectively we say that π and π' are equivalent if there exists an isomorphism θ of V onto V' such that $\pi'(z)\theta\psi = \theta\pi(z)\psi$ for every $\psi \in V$ and $z \in \mathfrak{A}$.

holds for π^* and therefore $\mathfrak{M}$ is a maximal left ideal. Now it follows from Lemma 10 that

$$\pi(X_i)\psi \in V_{\Lambda+\alpha_i}.$$

Since Λ is the highest weight of π, $\Lambda+\alpha_i$ is not a weight of π. Hence $\pi(X_i)\psi = 0$. Also since $\psi \in V_\Lambda$,

$$\{\pi(H) - \Lambda(H)\}\psi = 0 \qquad (H \in \mathfrak{h}).$$

Since $\mathfrak{M}$ is a left ideal these relations imply that $\mathfrak{M} \supset \mathfrak{P}+\mathfrak{Q}_\Lambda$. Since $\mathfrak{M}$ is maximal it must coincide with $\mathfrak{M}_\Lambda$. Hence $\pi^* = \pi_\Lambda$. Finally, since $\mathfrak{M} \supset \mathfrak{P}+\mathfrak{Q}_\Lambda$ it is clear that $(z(Q, 0, \phi))^*$ span $\mathfrak{A}^*$. Since $z(Q, 0, \phi)$ is of rank rank Q, $(z(Q, 0, \phi))^* \in \mathfrak{A}^*_{\Lambda-\mathrm{rank}\ Q}$. Therefore $\mathfrak{A}^*_\Lambda$ is spanned by 1^*. Since $1 \notin \mathfrak{M}$, dim $\mathfrak{A}^*_\Lambda$ $= 1$. Since π and π^* are equivalent dim $V_\Lambda = $ dim $\mathfrak{A}^*_\Lambda = 1$. The lemma is therefore proved.

We recall that an integral function Λ on $\mathfrak{h}$ is called dominant if $\Lambda(H_i) \geq 0$ for all $1 \leq i \leq l$.

LEMMA 17. *Let π be an irreducible representation of $\mathfrak{A}$ on a finite-dimensional space $V \neq \{0\}$. Then π has a highest weight Λ which is a dominant integral function. Also $V = \sum_\lambda V_\lambda$ where λ runs over all the weights of π.*

Since K is algebraically closed and $\pi(H_i)$, $1 \leq i \leq l$, commute with each other it follows that we can find an element $\psi \in V$ ($\psi \neq 0$) such that ψ is a common eigenvector of $\pi(H_i)$, $1 \leq i \leq l$. Therefore there exists a linear function λ on $\mathfrak{h}$ such that $\pi(H)\psi = \lambda(H)\psi$ ($H \in \mathfrak{h}$). Since V is irreducible, $V = \pi(\mathfrak{A})\psi$. Hence V is spanned by $\pi(z(P, M, Q))\psi$ for all P, M, and Q. But from Lemma 10, $\pi(z(P, M, Q))\psi$ has the weight $\lambda+\mathrm{rank}\ P-\mathrm{rank}\ Q$. Therefore $V = \sum_\mu V_\mu$ where μ runs over all the weights of π. Since dim V is finite it follows from Lemma 1 that π has only a finite number of weights. Now if $\varphi \in V_\mu$ then $\pi(X_i^\nu)\varphi \neq V_{\mu+\nu\alpha_i}$. Since $\mu+\nu\alpha_i$, $\nu = 1, 2, \cdots$, are all distinct linear functions on $\mathfrak{h}$ they cannot all be weights of π. Hence $\pi(X_i^\nu)\varphi = 0$ for some ν. Similarly we show that $\pi(Y_i^\nu)\varphi = 0$ for some ν. Since $V = \sum_\mu V_\mu$ it is clear that the hypotheses of Lemma 11 are fulfilled. Hence every weight of π is an integral function. Since π has only a finite number of weights it has a highest weight Λ. From Lemma 11, $\sigma_i\Lambda$ is also a weight for every i, $1 \leq i \leq l$. Since Λ is highest, $\Lambda \geq \sigma_i\Lambda = \Lambda - \Lambda(H_i)\alpha_i$. Hence $\Lambda(H_i) \geq 0$, and Λ is dominant.

COROLLARY. *Every finite-dimensional irreducible representation of $\mathfrak{A}$ is equivalent to some π_Λ where Λ is a dominant integral function on $\mathfrak{h}$.*

This is an immediate consequence of Lemmas 16 and 17.

Let $\mathfrak{g}$ be the smallest subspace of $\mathfrak{A}$ which contains X_i, Y_i, H_i, $1 \leq i \leq l$, and which is invariant under the adjoint representation of $\mathfrak{A}$. Let $\mathfrak{X}$ be the smallest subspace which contains X_i, $1 \leq i \leq l$, and which is invariant under ad X_i, $1 \leq i \leq l$. Similarly let $\mathfrak{Y}$ be the smallest subspace containing Y_i,

$1 \leqq i \leqq l$, and invariant under ad Y_i, $1 \leqq i \leqq l$.

LEMMA 18. $\mathfrak{g} = \mathfrak{h} + \mathfrak{X} + \mathfrak{Y}$ *and* $(\mathrm{ad}\ z)w = [z,\ w]$ *for any* $z \in \mathfrak{g}$ *and* $w \in \mathfrak{A}$. *Further if* $z,\ w \in \mathfrak{g}$ *then* $[z,\ w] \in \mathfrak{g}$.

It is obvious that $\mathfrak{h} + \mathfrak{X} + \mathfrak{Y} \subset \mathfrak{g}$. Hence in order to prove the equality it is sufficient to show that $\mathfrak{h} + \mathfrak{X} + \mathfrak{Y}$ is invariant under ad $\mathfrak{L}_0$ where $\mathfrak{L}_0$ is the linear space spanned by X_i, Y_i, H_i, $1 \leqq i \leqq l$. It is clear from its definition that $\mathfrak{X}$ is spanned by all elements of the form $\{\mathrm{ad}\ z(\phi, 0, P)\} X_i$. Put $z_P = z(\phi, 0, P)$ and $z = (\mathrm{ad}\ z_P) X_i$ for brevity. Then if $H \in \mathfrak{h}$,

$$\begin{aligned}
(\mathrm{ad}\ H)z &= (\mathrm{ad}\ [H,\ z_P]) X_i + (\mathrm{ad}\ z_P)[H,\ X_i] \\
&= \lambda(H)z
\end{aligned}$$

where $\lambda(H) = \mathrm{rank}\ P + \alpha_i$. This shows that $\mathfrak{X}$ is invariant under ad H. Now consider

$$(\mathrm{ad}\ Y_j)z = (\mathrm{ad}\ [Y_j,\ z_P]) X_i + (\mathrm{ad}\ z_P)[Y_j,\ X_i].$$

Since $[Y_j,\ X_i] = -\delta_{ij} H_i$, it is clear that $(\mathrm{ad}\ z_P)[Y_j,\ X_i] \in \mathfrak{h} + \mathfrak{X}$. We claim further that $(\mathrm{ad}\ [Y_j,\ z_P]) X_i \in \mathfrak{h} + \mathfrak{X}$. If $|P| = 0$ this is true. Hence we may assume $|P| \geqq 1$ and use induction on $|P|$. Then

$$z_P = X_k z(\phi, 0, P')$$

for some k and P' such that $|P'| = |P| - 1$ and $1 \leqq k \leqq l$. Put $z(\phi, 0, P') = z_{P'}$. Then

$$[Y_j,\ z_P] = [Y_j,\ X_k]z_{P'} + X_k[Y_j,\ z_{P'}] = -\delta_{jk} H_k z_{P'} + X_k[Y_j,\ z_{P'}].$$

Therefore

$$(\mathrm{ad}\ [Y_j,\ z_P]) X_i = -\delta_{jk}(\mathrm{ad}\ H_k)(\mathrm{ad}\ z_{P'}) X_i + (\mathrm{ad}\ X_k)(\mathrm{ad}\ [Y_j,\ z_{P'}]) X_i.$$

Clearly $(\mathrm{ad}\ z_{P'}) X_i \in \mathfrak{X}$ and $(\mathrm{ad}\ [Y_j,\ z_{P'}]) X_i \in \mathfrak{h} + \mathfrak{X}$ by induction hypothesis. Since $\mathfrak{h} + \mathfrak{X}$ is invariant under ad H_k and ad X_k the assertion follows. Hence $(\mathrm{ad}\ Y_j)z \in \mathfrak{h} + \mathfrak{X}$. Since $\mathfrak{X}$ is invariant under ad X_j we have shown that $(\mathrm{ad}\ w)z \in \mathfrak{h} + \mathfrak{X}$ for any $w \in \mathfrak{L}_0$ and $z \in \mathfrak{X}$. Similarly we prove that $(\mathrm{ad}\ w)z \in \mathfrak{h} + \mathfrak{Y}$ for any $w \in \mathfrak{L}_0$ and $z \in \mathfrak{Y}$. Finally it is clear that if $w \in \mathfrak{L}_0$ and $H \in \mathfrak{h}$ then $(\mathrm{ad}\ w)H \in \mathfrak{X} + \mathfrak{Y}$. Hence it follows that $\mathfrak{h} + \mathfrak{X} + \mathfrak{Y}$ is invariant under ad $\mathfrak{L}_0$ and therefore $\mathfrak{g} = \mathfrak{h} + \mathfrak{X} + \mathfrak{Y}$.

Keeping to the above notation, let $z = (\mathrm{ad}\ z_P) X_i$ and $w \in \mathfrak{A}$. We claim that $(\mathrm{ad}\ z)w = [z,\ w]$. If $|P| = 0$, $z = X_i$ and this is true. Hence again we may assume $|P| \geqq 1$ and use induction on $|P|$. Then as above $z_P = X_k z_{P'}$ with $|P'| = |P| - 1$. Put $z' = (\mathrm{ad}\ z_{P'}) X_i$. Then

$$z = (\mathrm{ad}\ z_P) X_i = (\mathrm{ad}\ X_k\ \mathrm{ad}\ z_{P'}) X_i = [X_k,\ z'].$$

Hence

$$(\text{ad } z)w = (\text{ad } [X_k, z'])w = [X_k, (\text{ad } z')w] - (\text{ad } z')[X_k, w]$$
$$= [X_k, [z', w]] - [z' [X_k, w]] \quad \text{(by the induction hypothesis)}$$
$$= [[X_k, z'], w] = [z, w].$$

Our assertion is therefore proved. Hence by linearity $(\text{ad } z)w = [z, w]$ for any $z \in \mathfrak{X}$. Similarly we prove that $(\text{ad } z)w = [z, w]$ for any $z \in \mathfrak{Y}$ and $w \in \mathfrak{A}$. Finally if $H \in \mathfrak{h}$, $(\text{ad } H)w = [H, w]$. Hence $(\text{ad } z)w = [z, w]$ for all $z \in \mathfrak{h} + \mathfrak{X} + \mathfrak{Y}$ $= \mathfrak{g}$ and $w \in \mathfrak{A}$. Since $\mathfrak{g}$ is invariant under the adjoint representation it follows that $(\text{ad } z)w \in \mathfrak{g}$ for any $z \in \mathfrak{A}$ and $w \in \mathfrak{g}$. Therefore if $z \in \mathfrak{g}$, $[z, w] = (\text{ad } z)w \in \mathfrak{g}$. This completes the proof of the lemma.

Let Λ_i, $1 \leq i \leq l$, be the dominant integral functions on $\mathfrak{h}$ defined by $\Lambda_i(H_j)$ $= \delta_{ij}$, $1 \leq j \leq l$. Λ_0 being any given dominant integral function on $\mathfrak{h}$, let π denote the direct sum of the finite-dimensional representations π_{Λ_i}, $0 \leq i \leq l$. Then $\pi(\mathfrak{A})$ is a finite-dimensional associative algebra and therefore from Lemma 18, $\pi(\mathfrak{g})$ is a linear[7] Lie algebra.

LEMMA 19. *π maps $\mathfrak{h}$ isomorphically on $\pi(\mathfrak{h})$. Further $\pi(\mathfrak{g})$ is a semisimple Lie algebra and $\pi(\mathfrak{h})$ is a Cartan subalgebra of $\pi(\mathfrak{g})$.*

Let V be the representation space of π and let $H \in \mathfrak{h}$ $(H \neq 0)$. Since Λ_i, $1 \leq i \leq l$, are linearly independent, $\Lambda_i(H) \neq 0$ for some i. Since Λ_i is the highest weight of π_{Λ_i} it is also a weight of π. Hence we can choose $\psi \in V_{\Lambda_i}$ $(\psi \neq 0)$. Then $\pi(H)\psi = \Lambda_i(H_i)\psi \neq 0$. Hence $\pi(H) \neq 0$ and this shows that π maps $\mathfrak{h}$ isomorphically.

Let $\mathfrak{B}$ be the kernel of π and $\mathfrak{C}$ the set of all elements $z \in \mathfrak{g}$ such that $[H, z] \in \mathfrak{B}$ for all $H \in \mathfrak{h}$. Clearly $\mathfrak{C}$ is invariant under ad H $(H \in \mathfrak{h})$. Let $z \in \mathfrak{C}$. From Lemma 1 every homogeneous component of z belongs to $\mathfrak{C}$. Let z_λ be such a component of rank λ. Then $[H, z_\lambda] = \lambda(H)z_\lambda \in \mathfrak{B}$. If $\lambda \neq 0$, $\lambda(H) \neq 0$ for some $H \in \mathfrak{h}$. Hence $z_\lambda \in \mathfrak{B}$ and $\pi(z_\lambda) = 0$. On the other hand let us now suppose that $\lambda = 0$. We have seen above that $\mathfrak{X}$ is spanned by suitable homogeneous elements of rank greater than 0 and $\mathfrak{Y}$ by similar elements of rank less than 0. Since $z_\lambda \in \mathfrak{g} = \mathfrak{h} + \mathfrak{X} + \mathfrak{Y}$ it follows that $z_\lambda \in \mathfrak{h}$. Therefore since $z = \sum_\lambda z_\lambda$, $\pi(z) = \pi(z_0) \in \pi(\mathfrak{h})$ and so $\pi(\mathfrak{C}) \subset \pi(\mathfrak{h})$. This shows that $\pi(\mathfrak{h})$ is a maximal abelian subalgebra of $\pi(\mathfrak{g})$. In particular the center of $\pi(\mathfrak{g})$ is contained in $\pi(\mathfrak{h})$. Let $\pi(H)$ $(H \in \mathfrak{h})$ belong to the center of $\pi(\mathfrak{g})$. Then $[\pi(H), \pi(X_i)] = \pi([H, X_i]) = \alpha_i(H)\pi(X_i) = 0$. But $[\pi(X_i), \pi(Y_i)] = \pi(H_i) \neq 0$ as we saw above. Hence $\pi(X_i) \neq 0$. Therefore $\alpha_i(H) = 0$, $1 \leq i \leq l$. Since α_i are linearly independent this implies that $H = 0$. Hence the center of $\pi(\mathfrak{g})$ is $\{0\}$.

Now for each i, $0 \leq i \leq l$, π_{Λ_i} is an irreducible representation of $\mathfrak{A}$. Since $1, X_j, Y_j, H_j, 1 \leq j \leq l$, generate $\mathfrak{A}$ it follows that $\pi_{\Lambda_i}(\mathfrak{g})$ is an irreducible set of linear transformations, $0 \leq i \leq l$. Since π is the direct sum of π_{Λ_i}, $0 \leq i \leq l$,

---

[7] A Lie algebra consisting of endomorphisms of a finite-dimensional vector space with the usual bracket operation $[A, B] = AB - BA$ is called a linear Lie algebra.

$\pi(\mathfrak{g})$ is a fully reducible set of endomorphisms of V. Since the center of $\pi(\mathfrak{g})$ is $\{0\}$ it follows (see Chevalley [5]) that $\pi(\mathfrak{g})$ is a semisimple Lie algebra.

Since π maps $\mathfrak{h}$ isomorphically we can regard every linear function λ on $\mathfrak{h}$ also as a linear function on $\pi(\mathfrak{h})$ by setting $\lambda(\pi(H)) = \lambda(H)$ $(H \in \mathfrak{h})$. In particular therefore α_i, $1 \leq i \leq l$, are now linear functions also on $\pi(\mathfrak{h})$. We shall now show that $\pi(\mathfrak{h})$ is a Cartan subalgebra of $\pi(\mathfrak{g})$ and α_i, $1 \leq i \leq l$, is a fundamental system of roots of $\pi(\mathfrak{g})$ with respect to $\pi(\mathfrak{h})$.

For a fixed $z \in \mathfrak{A}$ consider the linear mapping $w \to \pi((\mathrm{ad}\ z)w)$ of $\mathfrak{A}$ into $\pi(\mathfrak{A})$. Since the kernel $\mathfrak{B}$ of π is an ideal, it is invariant under the adjoint representation of $\mathfrak{A}$. Hence $\pi(w) = 0$ implies $\pi((\mathrm{ad}\ z)w) = 0$. Therefore we get a linear mapping $\pi(w) \to \pi((\mathrm{ad}\ z)w)$ of $\pi(\mathfrak{A})$ into itself, which we denote by $\rho'(z)$. Now

$$\rho'(z_1 z_2)\pi(w) = \pi(\mathrm{ad}\ (z_1 z_2)w) = \pi((\mathrm{ad}\ z_1)(\mathrm{ad}\ z_2)w)$$
$$= \rho'(z_1)\rho'(z_2)\pi(w) \qquad\qquad (z_1,\ z_2,\ w \in \mathfrak{A}).$$

Hence ρ' is a representation of $\mathfrak{A}$. Since $\mathfrak{g}$ is invariant under the adjoint representation of $\mathfrak{A}$, $(\mathrm{ad}\ z)\ w \in \mathfrak{g}$ if $w \in \mathfrak{g}$ and $z \in \mathfrak{A}$. Hence $\pi(\mathfrak{g})$ is invariant under $\rho'(\mathfrak{A})$. Let ρ be the representation of $\mathfrak{A}$ induced by ρ' on $\pi(\mathfrak{g})$. Then

$$\rho(z)\pi(w) = \pi((\mathrm{ad}\ z)w) = \pi([z,\ w]) = [\pi(z),\ \pi(w)] \qquad (z,\ w \in \mathfrak{g})$$

from Lemma 18. Since $\pi(\mathfrak{g})$ is semisimple its adjoint representation is fully reducible. Hence $\rho(\mathfrak{g})$ is a fully reducible set of endomorphisms of $\pi(\mathfrak{g})$. Since $\rho(\mathfrak{A})$ is generated by $\rho(\mathfrak{g})$ and 1, it follows that ρ is a fully reducible representation. Let $U = \pi(\mathfrak{g})$. Applying Lemma 17 to each irreducible component of ρ we immediately get $U = \sum_\lambda U_\lambda$ where λ runs over all weights of ρ. This shows that $\rho(H)$ is semisimple[8] for each $H \in \mathfrak{h}$. But $\rho(H) = \mathrm{ad}\ \pi(H)$ where $\pi(z) \to \mathrm{ad}\ \pi(z)$ $(z \in \mathfrak{g})$ denotes the adjoint representation of $\pi(\mathfrak{g})$. Hence we have shown that $\pi(\mathfrak{h})$ is a maximal abelian algebra of $\pi(\mathfrak{g})$ every element of which is mapped on a semisimple endomorphism under the adjoint representation of $\pi(\mathfrak{g})$. This proves that $\pi(\mathfrak{h})$ is a Cartan subalgebra of $\pi(\mathfrak{g})$.

LEMMA 20. α_i, $1 \leq i \leq l$, *is a fundamental system of roots of* $\pi(\mathfrak{g})$ *with respect to* $\pi(\mathfrak{h})$. *Also* dim $\pi(\mathfrak{g}) = l + g$ *where g is the number of distinct functions of the form* $\sigma\alpha_i$ $(\sigma \in W,\ 1 \leq i \leq l)$.

We keep to the above notation. We have seen that $U = \sum_\lambda U_\lambda$ where λ runs over all weights of ρ. Notice that a nonzero weight of ρ is exactly the same thing as a root of $\pi(\mathfrak{g})$. λ being a weight of ρ choose a $z \in \mathfrak{g}$ such that $\pi(z) \in U_\lambda$ $(\pi(z) \neq 0)$. Then $\rho(H)\pi(z) = [\pi(H),\ \pi(z)] = \lambda(H)\pi(z)$. Hence $[H,\ z] - \lambda(H)z \in \mathfrak{B}$ for all $H \in \mathfrak{h}$. Let z_μ be the homogeneous component of z of rank μ. Then the corresponding component of $[H,\ z] - \lambda(H)z$ is $[H,\ z_\mu] - \lambda(H)z_\mu$

---

(8) An endomorphism A of a finite-dimensional vector space V is called semisimple if V is fully reducible under A.

$= \{\mu(H) - \lambda(H)\} z_\mu$. Since $\mathfrak{B}$ is invariant under ad $\mathfrak{h}$ it follows from Lemma 1 that $\{\mu(H) - \lambda(H)\} z_\mu \in \mathfrak{B}$ for all $H \in \mathfrak{h}$. If $\mu \neq \lambda$ we can find an $H \in \mathfrak{h}$ such that $\mu(H) \neq \lambda(H)$. Hence $z_\mu \in \mathfrak{B}$. Therefore $\pi(z) = \pi(z_\lambda) \neq 0$. Since $\mathfrak{g}$ is invariant under ad $\mathfrak{h}$ and $z \in \mathfrak{g}$ it follows again by Lemma 1 that $z_\lambda \in \mathfrak{g}$. Since $\mathfrak{g} = \mathfrak{h} + \mathfrak{X} + \mathfrak{Y}$ it is clear that $\mathfrak{g} \cap \mathfrak{A}_\lambda = \{0\}$ unless $\lambda = d_1\alpha_1 + \cdots + d_l\alpha_l$ where d_i are integers which are either all greater than or equal to 0 or all less than or equal to 0. But $z_\lambda \in \mathfrak{g}$ and $z_\lambda \neq 0$. Hence λ must be of the above form. Finally $\pi(X_i) \in U_{\alpha_i}$ and we have already seen in the proof of Lemma 19 that $\pi(X_i) \neq 0$. Hence α_i, $1 \leq i \leq l$, are roots of $\pi(\mathfrak{g})$. This proves that α_i, $1 \leq i \leq l$, is a fundamental system of roots of $\pi(\mathfrak{g})$.

Since $\pi(X_i) \neq 0$ and similarly $\pi(Y_i) \neq 0$, they are the root elements in $\pi(\mathfrak{g})$ corresponding to the roots α_i and $-\alpha_i$ respectively. Since $[\pi(X_i), \pi(Y_i)] = \pi(H_i)$ it follows that if σ_j' is the Weyl reflexion in $\pi(\mathfrak{g})$ with respect to α_j',

$$\sigma_j' \alpha_i = \alpha_i - 2\,\frac{\alpha_i(H_j)}{\alpha_j(H_j)}\,\alpha_j = \sigma_j \alpha_i$$

since $\alpha_j(H_j) = 2$. Hence the Weyl matrix of $\pi(\mathfrak{g})$ with respect to the fundamental system $\{\alpha_1, \cdots, \alpha_l\}$ is (a_{ij}), $1 \leq i, j \leq l$, and the Weyl group of $\pi(\mathfrak{g})$ is W. Therefore $\pi(\mathfrak{g})$ has exactly g distinct roots. Since dim $\pi(\mathfrak{h}) = l$, dim $\pi(\mathfrak{g}) = l + g$.

Now consider the special case when $\Lambda_0 = 0$. As remarked earlier (cf. p. 43), in this case $\pi_{\Lambda_0}(\mathfrak{g}) = \{0\}$. Let $\bar\pi$ be the direct sum of π_{Λ_i}, $1 \leq i \leq l$. Then clearly $\pi(\mathfrak{g})$ is isomorphic to $\bar\pi(\mathfrak{g})$ under the mapping $\pi(z) \leftrightarrow \bar\pi(z)$ $(z \in \mathfrak{g})$. We put $\mathfrak{L} = \bar\pi(\mathfrak{g})$. Again we may regard α_i as linear functions on $\bar\pi(\mathfrak{h})$ in the obvious way. Then $\mathfrak{L}$ is a semisimple Lie algebra with the fundamental system of roots α_i, $1 \leq i \leq l$, with respect to the Cartan subalgebra $\bar\pi(\mathfrak{h})$ and dim $\mathfrak{L}$ = dim $\pi(\mathfrak{g}) = l + g$. Returning to the general case when Λ_0 is arbitrary, it is obvious that $\pi(z) \to \bar\pi(z)$ $(z \in \mathfrak{g})$ is a homomorphism of $\pi(\mathfrak{g})$ onto $\mathfrak{L}$. But since both $\pi(\mathfrak{g})$ and $\mathfrak{L}$ have the same dimension $l + g$, this must be an isomorphism. Also $\pi(z) \to \pi_{\Lambda_0}(z)$ $(z \in \mathfrak{g})$ is evidently a representation of $\pi(\mathfrak{g})$. Therefore $\bar\pi(z) \to \pi_{\Lambda_0}(z)$ $(z \in \mathfrak{g})$ is a representation of $\mathfrak{L}$ whose highest weight is Λ_0. The proof of Theorem 1 is therefore complete.

We now consider the question of the uniqueness of the Lie **algebra** whose existence is asserted in Theorem 1. Let $\bar{\mathfrak{L}}$ be any semisimple Lie algebra with a Cartan subalgebra $\bar{\mathfrak{h}}$ satisfying the requirements of the theorem. Let $\overline{X}_i$, $\overline{Y}_i$ be the root elements of $\bar{\mathfrak{L}}$ corresponding to the roots α_i and $-\alpha_i$, $1 \leq i \leq l$. Put $[\overline{X}_i, \overline{Y}_i] = \overline{H}_i$. Since α_i is a fundamental system of roots, $\alpha_i - \alpha_j$ $(i \neq j)$ is not a root. Hence $[\overline{X}_i, \overline{Y}_j] = 0$. By multiplying $\overline{X}_i$ with an element $c \in K$ $(c \neq 0)$ we may arrange that $\alpha_i(\overline{H}_i) = 2$, $1 \leq i \leq l$. Then it follows that

$$\sigma_i \alpha_j = \alpha_j - \frac{2\alpha_j(\overline{H}_i)}{\alpha_i(\overline{H}_i)}\,\alpha_i = \alpha_j + a_{ji}\alpha_i.$$

Hence $\alpha_j(\overline{H}_i) = \alpha_j(H_i)$, $1 \leq i, j \leq l$. Let ρ' denote the representation of the free algebra $\mathfrak{A}'$ on $\overline{\mathfrak{L}}$ defined uniquely by the equations

$$\rho'(1) = I, \quad \rho'(X_i') = \text{ad } \overline{X}_i, \quad \rho'(Y_i') = \text{ad } \overline{Y}_i, \quad \rho'(H_i') = \text{ad } \overline{H}_i, \quad 1 \leq i \leq l,$$

where I is the identity mapping of $\mathfrak{L}$ and $\overline{X} \to \text{ad } \overline{X}$ $(\overline{X} \in \overline{\mathfrak{L}})$ denotes the adjoint representation of $\overline{\mathfrak{L}}$. It is easily seen that the kernel of ρ' contains the ideal $\mathfrak{U}'$ and so ρ' actually defines a representation $\bar{\rho}$ of $\mathfrak{A}$ on $\overline{\mathfrak{L}}$. Since $\overline{\mathfrak{L}}$ is semi-simple its adjoint representation is fully reducible. Hence $\bar{\rho}$ is also fully reducible. From the corollary to Lemma 17, every finite-dimensional irreducible representation of $\mathfrak{A}$ is equivalent to π_Λ for some dominant integral function Λ. But we have seen above that $\bar{\pi}(z) \to \pi_\Lambda(z)$ $(z \in \mathfrak{g})$ is a representation of $\mathfrak{L}$. Hence it is clear that $\bar{\pi}(z) \to \bar{\rho}(z)$ $(z \neq \mathfrak{g})$ is also a representation of $\mathfrak{L}$. We denote this representation by θ. Then $\theta(\mathfrak{L})$ is a linear Lie algebra and since $\dim \mathfrak{L} = l + g$, $\dim \theta(\mathfrak{L}) \leq l + g$. But clearly $\theta(\overline{\mathfrak{L}}) \supset \text{ad } \overline{\mathfrak{L}}$ and $\dim (\text{ad } \overline{\mathfrak{L}})$ $= \dim \overline{\mathfrak{L}} = l + g$ since g depends only on the group W which is the same for both $\overline{\mathfrak{L}}$ and $\mathfrak{L}$. Hence $\theta(\mathfrak{L}) = \text{ad } \overline{\mathfrak{L}}$ and θ must be an isomorphism. Therefore $\mathfrak{L} \cong \text{ad } \overline{\mathfrak{L}} \cong \overline{\mathfrak{L}}$. This shows that $\mathfrak{L}$ is uniquely determined up to an isomorphism.

Also notice that if θ is any representation of $\mathfrak{L}$ on V then θ defines uniquely a representation φ of $\mathfrak{A}$ by the rule

$$\varphi(1) = I, \qquad \varphi(z) = \theta(\bar{\pi}(z))$$

where I is the identity mapping of V and z is any one of the elements X_i, Y_i, H_i, $1 \leq i \leq l$. We claim that

$$\varphi(z) = \theta(\bar{\pi}(z))$$

for all $z \in \mathfrak{g}$. This is obvious if $z \in \mathfrak{h}$. Now suppose $z \in \mathfrak{X}$. It is sufficient to consider the case when $z = (\text{ad } z_P) X_i$ where $z_P = z(\phi, 0, P)$. Again if $|P| = 0$ the statement is true. Hence we may assume that $|P| \geq 1$ and use induction on $|P|$. Let $z_P = X_k z_{P'}$ where $z_{P'} = z(\phi, 0, P')$, $|P'| = |P| - 1$, $1 \leq k \leq l$. Then

$$z = [X_k, z']$$

where $z' = (\text{ad } z_{P'}) X_i$. Hence

$$\varphi(z) = \varphi([X_k, z']) = [\varphi(X_k), \varphi(z')]$$
$$= [\theta(\bar{\pi}(X_k)), \theta(\bar{\pi}(z'))]$$

by induction hypothesis. Therefore

$$\varphi(z) = \theta([\bar{\pi}(X_k), \bar{\pi}(z')]) = \theta(\bar{\pi}([X_k, z']))$$
$$= \theta(\bar{\pi}(z)).$$

Similarly if $z \in \mathfrak{Y}$. Since $\mathfrak{g} = \mathfrak{h} + \mathfrak{X} + \mathfrak{Y}$ the assertion follows. In particular if θ is irreducible the same is true of φ and the weights of φ are the same as those of θ. Hence if θ is irreducible and finite-dimensional and Λ is the highest weight of θ then it follows from Lemma 16 that φ is equivalent to π_Λ. Hence φ and therefore θ is uniquely determined up to equivalence, and the multiplicity of Λ in θ is 1. Thus we have proved the following theorem.

THEOREM 2. *The Lie algebra $\mathfrak{L}$ of Theorem 1 is unique within isomorphism. Also the irreducible representation ρ of $\mathfrak{L}$ with the highest weight Λ_0 is uniquely determined within equivalence and the multiplicity of Λ_0 in ρ is 1.*

Another way of stating the first part of Theorem 2 is to say that two semisimple Lie algebras with isomorphic root diagrams are isomorphic. In this form this result was first proved by Weyl [12]. The uniqueness of the representation ρ is due to Cartan [2].

Finally we shall prove a result on the degree of the representation ρ. The exact formula for this degree has been obtained by Weyl [12] by using transcendental methods.

THEOREM 3. *Let $\mathfrak{L}$, $\mathfrak{h}$, α_i, $1 \le i \le l$, and ρ be as in Theorems 1 and 2 and let d_ρ be the degree of the representation ρ. Then*

$$\prod_{1 \le i \le l} \left(\frac{2\Lambda_0(H_{\alpha_i})}{\alpha_i(H_{\alpha_i})} + 1 \right) \le d_\rho \le \prod_{\alpha > 0} \left(\frac{2\Lambda_0(H_\alpha)}{\alpha(H_\alpha)} + 1 \right)$$

where α denotes any root of $\mathfrak{L}$ with respect to $\mathfrak{h}$.

Let α_i, $1 \le i \le r$, be the set of all roots of $\mathfrak{L}$ which are greater than 0 and let $x_i \ne 0$ and $y_i \ne 0$ be root elements in $\mathfrak{L}$ corresponding to the roots α_i and $-\alpha_i$ respectively, $1 \le i \le r$. Also let $H_i = [x_i, y_i]$, $1 \le i \le r$. (Notice that H_i corresponds to $\bar\pi(H_i)$, $1 \le i \le l$, in our earlier notation.) Then x_i, y_i, H_j, $1 \le i \le r$, $1 \le j \le l$, form a base for $\mathfrak{L}$. Let $\mathfrak{U}$ be the universal enveloping algebra[9] of $\mathfrak{L}$ and let V be the representation space of ρ. Then ρ defines a representation π of $\mathfrak{U}$ on V by the rule

$$\pi(1) = I, \qquad \pi(x) = \rho(x) \qquad\qquad (x \in \mathfrak{L})$$

where I is the identity mapping of V. Let $\psi \ne 0$ be an element in V which belongs to the weight Λ_0. Let $\mathfrak{M}$ be the left ideal in $\mathfrak{U}$ consisting of all elements $z \in \mathfrak{U}$ such that $\pi(z)\psi = 0$. Since V is irreducible under ρ and therefore under π, $\mathfrak{M}$ is a maximal left ideal. Let π^* denote the natural representation and $z \to z^*$ the natural mapping of $\mathfrak{U}$ on $\mathfrak{U}^* = \mathfrak{U}/\mathfrak{M}$. It is easily verified that π^* is equivalent to π under the isomorphism $z^* \leftrightarrow \pi(z)\psi$ ($z \in \mathfrak{U}$) of $\mathfrak{U}^*$ with V. Hence instead of ρ we may consider the representation π^*. Notice that since Λ_0 is the highest weight of ρ, $\Lambda_0 + \alpha_i$ is not a weight for any i, $1 \le i \le r$. Hence $\pi(x_i)\psi = 0$ and $x_i \in \mathfrak{M}$, $1 \le i \le r$. Also since $\pi(H)\psi = \Lambda_0(H)\psi$, $H - \Lambda_0(H) \in \mathfrak{M}$ for all $H \in \mathfrak{h}$.

---

[9] This concept is due to Birkhoff [1] and Witt [14]. However we follow the definition given in [9]. We shall always assume that the Lie algebra is naturally imbedded in its universal enveloping algebra.

We now introduce the notion of a rank in $\mathfrak{U}$ exactly as we did it in $\mathfrak{A}$. λ being any linear function on $\mathfrak{h}$ we say that an element $z \in \mathfrak{U}$ is of rank λ if $[H, z] = \lambda(H)z$ $(H \in \mathfrak{h})$.

By multiplying x_i by an element $c_i \in K$ $(c_i \neq 0)$ we can arrange that $\alpha_i(H_i) = 2$, $1 \leq i \leq 2$. We shall suppose that this has been done. Now we assert that the elements $y_1^{m_1} y_2^{m_2} \cdots y_l^{m_l}$, $0 \leq m_i \leq \Lambda_0(H_i)$, $1 \leq i \leq l$, are linearly independent[10] mod $\mathfrak{M}$. Notice that $y_1^{m_1} y_2^{m_2} \cdots y_l^{m_l}$ is of rank $-(m_1\alpha_1 + \cdots + m_l\alpha_l)$. Since α_i, $1 \leq i \leq l$, are linearly independent, all these linear functions are distinct. Since $H - \Lambda_0(H) \in \mathfrak{M}$ $(H \in \mathfrak{h})$, $\mathfrak{M}$ is invariant under ad $\mathfrak{h}$. Hence in view of Lemma 1 it is sufficient to prove that $y_1^{m_1} y_2^{m_2} \cdots y_l^{m_l}$ $\not\in \mathfrak{M}$ for $0 \leq m_i \leq \Lambda_0(H_i)$, $1 \leq i \leq l$. Let $M = m_1 + m_2 + \cdots + m_l$. If $M = 0$, $y_1^{m_1} \cdots y_l^{m_l} = 1$ and the assertion is true since $1 \not\in \mathfrak{M}$. Hence we may assume $M \geq 1$ and use induction on M. Let i be the largest index such that $m_i \neq 0$. Put $z = 1$ if $i = 1$ or $z = y_1^{m_1} y_2^{m_2} \cdots y_{i-1}^{m_{i-1}}$ if $i > 1$. Then we have to consider the term $zy_i^{m_i}$. Notice that $[x_i, y_j] = 0$ if $i \neq j$, $1 \leq i, j \leq l$. This follows from the fact that α_i, $1 \leq i \leq l$, is a fundamental system and so $\alpha_i - \alpha_j$ is not a root, $1 \leq i, j \leq l$ $(i \neq j)$. Hence from Lemma 6,

$$
\begin{aligned}
x_i z y_i^{m_i} &= z x_i y_i^{m_i} = z[x_i, y_i^{m_i}] + z y_i^{m_i} x_i \\
&= m_i z y_i^{m_i-1}(H_i - m_i + 1) + z y_i^{m_i} x_i \\
&\equiv m_i z y_i^{m_i-1}(\Lambda_0(H_i) - m_i + 1) \bmod \mathfrak{M}.
\end{aligned}
$$

Now $0 < m_i \leq \Lambda_0(H_i)$. Hence $m_i(\Lambda_0(H_i) - m_i + 1) \neq 0$. Also by induction hypothesis $zy_i^{m_i-1} \not\in \mathfrak{M}$. Hence $x_i z y_i^{m_i} \not\in \mathfrak{M}$. Since $\mathfrak{M}$ is a left ideal, it follows that $zy_i^{m_i} \not\in \mathfrak{M}$. Therefore our assertion is proved. Since the elements $y_1^{m_1} y_2^{m_2} \cdots y_l^{m_l}$, $0 \leq m_i \leq \Lambda_0(H_i)$, $1 \leq i \leq l$, are $\prod_{1 \leq i \leq l}(\Lambda_0(H_i) + 1)$ in number and since they are linearly independent mod $\mathfrak{M}$, it follows that

$$
\dim \mathfrak{U}^* \geq \prod_{1 \leq i \leq l}(\Lambda_0(H_i) + 1).
$$

Hence the first inequality of the theorem is proved.

Now we come to the second inequality. For brevity write

$$
z(q, m, p) = y_1^{q_1} y_2^{q_2} \cdots y_r^{q_r} H_1^{m_1} H_2^{m_2} \cdots H_l^{m_l} x_1^{p_1} \cdots x_r^{p_r}
$$

where q_i, p_i, $m_j \geq 0$, $1 \leq i \leq r$, $1 \leq j \leq l$, and q, m, p denote the corresponding sets of integers. Also put $|q| = q_1 + \cdots + q_r$, $|p| = p_1 + p_2 + \cdots + p_r$, $|m| = m_1 + \cdots + m_l$ and $s = |q| + |m| + |p|$. We shall write 0 for the set p all of whose elements are zero. Similarly for q and m. It is known (see for example [9]) that the elements $z(q, m, p)$ for all $q, m,$ and p span $\mathfrak{U}$. Let U be the subspace of $\mathfrak{U}$ spanned by all elements of the form $y_1^{q_1} y_2^{q_2} \cdots y_r^{q_r}$, $0 \leq q_i \leq \Lambda_0(H_i)$, $1 \leq i \leq r$. We shall prove that $\mathfrak{U} = \mathfrak{M} + U$. First we claim that if $q_j > \Lambda_0(H_j)$ then

---

[10] We use the convention that $z^m = 1$ for any $z \in \mathfrak{U}$ if $m = 0$.

$y_j^{q_j} \in \mathfrak{M}$, $1 \leq j \leq r$. For otherwise $(y_j^{q_j})^* \neq 0$. Since $y_j^{q_j}$ has the rank $-q_j \alpha_j$ it follows that $(y_j^{q_j})^*$ is homogeneous of weight $\Lambda_0 - q_j \alpha_j$. Since $(y_j^{q_i})^* \neq 0$, $\Lambda_0 - q_j \alpha_j$ is a weight of π^* and therefore of ρ. Let σ_j denote the Weyl reflexion with respect to the root α_j. Then[11] $\sigma_j(\Lambda_0 - q_j \alpha_j)$ must also be a weight of ρ. But

$$\sigma_j(\Lambda_0 - q_j \alpha_j) = \Lambda_0 - \Lambda_0(H_j)\alpha_j + q_j \alpha_j > \Lambda_0$$

since $q_j > \Lambda_0(H_j)$. This contradicts our hypothesis that Λ_0 is the highest weight of ρ. Hence $y_j^{q_j} \in \mathfrak{M}$.

In order to prove that $\mathfrak{U} = \mathfrak{M} + U$ it is sufficient to show that $z(q, m, p) \in \mathfrak{M} + U$ for all q, m, p. If $s = |q| + |m| + |p| = 0$ this is true since $1 \in U$. Hence we may assume $s \geq 1$ and use induction on s. Now if $|p| > 0$ it is clear that $z(q, m, p) \in \mathfrak{M}$ since $x_i \in \mathfrak{M}$, $1 \leq i \leq r$. Similarly since $H_i - \Lambda_0(H_i) \in \mathfrak{M}$, it follows that $z(q, m, 0) \equiv cz(q, 0, 0) \bmod \mathfrak{M}$ $(c \in K)$. Hence we need consider only elements of the form $z(q, 0, 0)$ with $|q| = s$. If $q_j \leq \Lambda_0(H_j)$ for all j, it is obvious from the definition of U that $z(q, 0, 0) \in U$. Hence we may assume that $q_j > \Lambda_0(H_j)$ for some j. Choose the greatest such j. If $j = r$ then $z(q, 0, 0) \in \mathfrak{M}$ since $\mathfrak{M}$ is a left ideal and we have seen above that $y_r^{q_r} \in \mathfrak{M}$ if $q_r > \Lambda_0(H_r)$. Hence we may suppose that $j < r$. Put $z_0 = 1$ if $j = 1$ and $z_0 = y_1^{q_1} \cdots y_{j-1}^{q_{j-1}}$ if $j > 1$. Then since $y_j^{q_j} \in \mathfrak{M}$,

$$z(q, 0, 0) \equiv z_0 [y_j^{q_j}, y_{j+1}^{q_{j+1}} \cdots y_r^{q_r}] \bmod \mathfrak{M}.$$

But it is known (see [9]) that $z_0 [y_j^{q_j}, y_{j+1}^{q_{j+1}} \cdots y_r^{q_r}] = z'$ can be written as a linear combination of $z(q', m, p)$ with $|q'| + |m| + |p| < s$. Hence it follows by induction hypothesis that $z' \in \mathfrak{M} + U$. Therefore $z(q, 0, 0) \in \mathfrak{M} + U$ and our assertion is proved.

Since $\mathfrak{M} + U = \mathfrak{U}$ it is clear that $\dim \mathfrak{U}^* \leq \dim U$. Since U is spanned by the $\prod_{1 \leq i \leq r} (\Lambda_0(H_i) + 1)$ elements $y_1^{q_1} y_2^{q_2} \cdots y_r^{q_r}$, $0 \leq q_i \leq \Lambda_0(H_i)$, $1 \leq i \leq r$, it follows that

$$\dim \mathfrak{U}^* \leq \dim U \leq \prod_{1 \leq i \leq r} (\Lambda_0(H_i) + 1).$$

The theorem is now proved completely.

Part II. Infinite-dimensional representations of complex semisimple Lie algebras. Let R and C be the fields of real and complex numbers respectively. Let $\mathfrak{L}_0$ be a Lie algebra over R with a semisimple subalgebra $\mathfrak{L}_{K,0}$ such that there exists a linear mapping Γ of $\mathfrak{L}_{K,0}$ into $\mathfrak{L}_0$ with the following properties:

(1) $\qquad \mathfrak{L}_0 = \mathfrak{L}_{K,0} + \Gamma(\mathfrak{L}_{K,0}), \qquad \mathfrak{L}_{K,0} \cap \Gamma(\mathfrak{L}_{K,0}) = \{0\},$

(2) $\qquad [X_1, \Gamma(X_2)] = \Gamma([X_1, X_2]), \qquad [\Gamma(X_1), \Gamma(X_2)] = -[X_1, X_2]$

for any $X_1, X_2 \in \mathfrak{L}_{K,0}$.

---

[11] This follows either from well known results on finite-dimensional representations of semisimple Lie algebras or from Lemma 11.

Let $\mathfrak{L}$ denote the complexification[12] of $\mathfrak{L}_0$ and let $\mathfrak{L}_K$ be the smallest subspace of $\mathfrak{L}$ containing $\mathfrak{L}_{K,0}$. Then clearly $\mathfrak{L}_K$ is a Lie algebra which is the complexification of $\mathfrak{L}_{K,0}$. We extend Γ on $\mathfrak{L}_K$ by linearity and put[13]

$$\gamma(X) = \tfrac{1}{2}(X + (-1)^{1/2}\Gamma(X)), \qquad \gamma^+(X) = \tfrac{1}{2}(X - (-1)^{1/2}\Gamma(X)) \qquad (X \in \mathfrak{L}_K).$$

Let $\mathfrak{g}=\gamma(\mathfrak{L}_K)$, $\mathfrak{g}^+=\gamma^+(\mathfrak{L}_K)$. Then $\mathfrak{g}$ and $\mathfrak{g}^+$ are ideals in $\mathfrak{L}$ and $\mathfrak{L}=\mathfrak{g}+\mathfrak{g}^+$, $\mathfrak{g}\cap\mathfrak{g}^+=[\mathfrak{g}, \mathfrak{g}^+]=\{0\}$. Also γ and γ^+ are isomorphisms of $\mathfrak{L}_K$ on $\mathfrak{g}$ and $\mathfrak{g}^+$ respectively. Since $\mathfrak{L}_{K,0}$ is semisimple, $\mathfrak{L}_K$, $\mathfrak{g}$, $\mathfrak{g}^+$, and $\mathfrak{L}$ are all semisimple.

Let $\mathfrak{B}$ denote the universal enveloping algebra[9] of $\mathfrak{L}$. Let $\mathfrak{A}$, $\mathfrak{X}$, and $\mathfrak{A}^+$ be the subalgebras of $\mathfrak{B}$ generated by the sets $(1, \mathfrak{g})$, $(1, \mathfrak{L}_K)$, and $(1, \mathfrak{g}^+)$ respectively. We claim that $\mathfrak{A}$, $\mathfrak{X}$, and $\mathfrak{A}^+$ are the universal enveloping algebras of $\mathfrak{g}$, $\mathfrak{L}_K$, and $\mathfrak{g}^+$ respectively. This follows from the following lemma.

LEMMA 21. *Let $\mathfrak{M}$ be a Lie algebra over a field k of characteristic zero and let $\mathfrak{N}$ be a subalgebra of $\mathfrak{M}$. Let $\mathfrak{U}$ be the universal enveloping algebra of $\mathfrak{M}$ and $\mathfrak{B}$ the subalgebra of $\mathfrak{U}$ generated by $\mathfrak{N}$ and 1. Then $\mathfrak{B}$ is isomorphic to the universal enveloping algebra of $\mathfrak{N}$.*

Choose a base x_i, $1 \leq i \leq m$, for $\mathfrak{M}$ such that x_i, $1 \leq i \leq n$, $n \leq m$, is a base for $\mathfrak{N}$. Let $\mathfrak{B}'$ be the universal enveloping algebra of $\mathfrak{N}$. Then clearly there is a homomorphism φ of $\mathfrak{B}'$ onto $\mathfrak{B}$ such that φ leaves every element of $\mathfrak{N}$ fixed and $\varphi(1')=1$, $1'$ being the unit element of $\mathfrak{B}'$. We define as in [9] the basic canonical elements in $\mathfrak{U}$ and $\mathfrak{B}'$ with respect to the bases x_i, $1 \leq i \leq m$, and x_i, $1 \leq i \leq n$, for $\mathfrak{M}$ and $\mathfrak{N}$ respectively. Then it is clear from their definition that if z_j', $1 \leq j \leq r$, are any distinct basic canonical elements in $\mathfrak{B}'$, $\varphi(z_j')$, $1 \leq j \leq r$, are also distinct basic canonical elements in $\mathfrak{U}$. Now let $z' \in \mathfrak{B}'$. Since the basic canonical elements form a base for $\mathfrak{B}'$, $z' = \sum_{1 \leq i \leq r} c_i z_i'$ where $c_i \in k$ and z_i', $1 \leq i \leq r$, are distinct basic canonical elements in $\mathfrak{B}'$. Hence $\varphi(z') = \sum_{1 \leq i \leq r} c_i \varphi(z_i')$. Since distinct basic canonical elements are linearly independent in $\mathfrak{U}$ it follows that if $\varphi(z')=0$, $c_i=0$, $1 \leq i \leq r$, and therefore $z'=0$. Hence φ is an isomorphism.

We make the convention that whenever we speak of a representation of the universal enveloping algebra $\mathfrak{U}$ of a Lie algebra $\mathfrak{M}$ on a vector space V it will always be assumed implicitly that $\pi(1)=I$ where I is the identity mapping of V. Then it is clear that there is a 1-1 correspondence between representations of $\mathfrak{U}$ and those of $\mathfrak{M}$ such that corresponding representations coincide on $\mathfrak{M}$. We shall usually denote these corresponding representations by the same symbol. In particular if $\mathfrak{M}$ is semisimple every finite-dimensional representation of $\mathfrak{M}$ (and therefore of $\mathfrak{U}$) is fully reducible.

Choose a fixed Cartan subalgebra $\mathfrak{h}_K$ of $\mathfrak{L}_K$ and a fundamental system of roots $\{\alpha_1, \cdots, \alpha_l\}$ of $\mathfrak{L}_K$ with respect to $\mathfrak{h}_K$. Let P_K be the set of all dominant

---

[12] This means that $\mathfrak{L}$ is a Lie algebra obtained from $\mathfrak{L}_0$ by extending the ground field from R to C.

[13] We fix once for all an element $(-1)^{1/2}$ in C such that $((-1)^{1/2})^2=-1$.

integral linear functions on $\mathfrak{h}_K$. Then from Theorems 1 and 2 we know that there is a 1-1 correspondence between equivalence classes of finite-dimensional representations of $\mathfrak{L}_K$ (and therefore of $\mathfrak{X}$) and dominant integral functions Λ on $\mathfrak{h}_K$ such that if $\mathfrak{D}_\Lambda$ is the class corresponding to Λ then Λ is the highest weight of any representation in $\mathfrak{D}_\Lambda$. Let $\pi \in \mathfrak{D}_\Lambda$. Then every weight of π will be called a weight of $\mathfrak{D}_\Lambda$. In particular Λ is the highest weight of $\mathfrak{D}_\Lambda$.

Let π be any representation $\mathfrak{X}$ on a vector space V (not necessarily finite-dimensional). For any fixed $\psi \in V$ put $U = \pi(\mathfrak{X})\psi$. Let π' be the representation of $\mathfrak{X}$ induced on U. Given any $\Lambda \in P_K$ we say that ψ transforms under π according to $\mathfrak{D}_\Lambda$ if either $\psi = 0$ or π' is finite-dimensional (and therefore fully reducible) and every irreducible component of π' belongs to $\mathfrak{D}_\Lambda$. Let V_Λ be the set of all elements of V which transform according to $\mathfrak{D}_\Lambda$. It is clear from its definition that V_Λ is an invariant subspace of V.

LEMMA 22. *The sum $\sum_{\Lambda \in P_K} V_\Lambda$ is direct and if U is any invariant subspace of V then*

$$U \cap \left(\sum_{\Lambda \in P_K} V_\Lambda \right) = \sum_{\Lambda \in P_K} (U \cap V_\Lambda).$$

The proof is similar to that of Lemma 1. Let $\psi \in U \cap (\sum_{\Lambda \in P_K} V_\Lambda)$. Then $\psi = \psi_1 + \cdots + \psi_r$ where $\psi_i \in V_{\Lambda_i}$ and $\Lambda_i \neq \Lambda_j$, $1 \leq i$, $j \leq r$ $(i \neq j)$ $(\Lambda_i \in P_K)$. Choose some representation π_{Λ_i} of $\mathfrak{X}$ in $\mathfrak{D}_{\Lambda_i}$, $1 \leq i \leq r$. Then π_{Λ_i} are all inequivalent representations. Hence given any j, $1 \leq j \leq r$, we can, from Lemma 4 of [9], find an element $x \in \mathfrak{X}$ such that $\pi_{\Lambda_j}(x) = \pi_{\Lambda_j}(1)$, $\pi_{\Lambda_i}(x) = 0$ $(i \neq j)$. But then it is clear that $\pi(x)\psi = \psi_j$. Since U is invariant under π, $\psi_j = \pi(x)\psi \in U$. Hence $U \cap (\sum_{\Lambda \in P_K} V_\Lambda) = \sum_{\Lambda \in P_K} (U \cap V_\Lambda)$. Now, if we take $U = \{0\}$ we find that if $\psi = 0$ then $\psi_j = 0$, $1 \leq j \leq r$. Hence the sum $\sum_{\Lambda \in P_K} V_\Lambda$ is direct.

Since V_Λ is invariant under π we can, for any linear function λ on $\mathfrak{h}$, define $V_{\Lambda,\lambda}$ just as in Lemma 1.

LEMMA 23. $V_\Lambda = \sum_\lambda V_{\Lambda,\lambda}$.

Let $\psi \in V_\Lambda$. Put $U = \pi(\mathfrak{X})\psi$. Then by the definition of V_Λ, U is finite-dimensional. But from the theory of finite-dimensional representations it is known that $U = \sum_\lambda U_\lambda$. Since clearly $U_\lambda \subset V_{\Lambda,\lambda}$, it follows that $\psi \in \sum_\lambda V_{\Lambda,\lambda}$. This being true for every $\psi \in V_\Lambda$, $V_\Lambda = \sum_\lambda V_{\Lambda,\lambda}$. Notice that $V_{\Lambda,\lambda} = \{0\}$ if λ is not a weight of $\mathfrak{D}_\Lambda$.

Let π be any representation of $\mathfrak{B}$ on V. Then π defines a representation of $\mathfrak{X}$ on V and for each $\Lambda \in P_K$ we can construct the subspace V_Λ of Lemma 22 with respect to $\pi(\mathfrak{X})$. Given any $\Lambda_0 \in P_K$ we say that $\mathfrak{D}_{\Lambda_0}$ occurs in π if $V_{\Lambda_0} \neq \{0\}$. Also we shall say that $\mathfrak{D}_{\Lambda_0}$ occurs a finite or an infinite number of times according as dim V_{Λ_0} is finite or infinite. Let $\mathfrak{C}$ be the center of $\mathfrak{A}$ and χ a homomorphic mapping of $\mathfrak{C}$ into C. Our object is to prove the following theorem.

THEOREM 4. *Given Λ_0 and χ there exist only a finite number of inequivalent irreducible representations π of $\mathfrak{B}$ such that $\mathfrak{D}_{\Lambda_0}$ occurs in π and[14] $\pi(z-\chi(z))=0$ for all $z\in\mathfrak{C}$. Further if π is such a representation then $\mathfrak{D}_\Lambda$ occurs only a finite number of times in π for every $\Lambda\in P_K$ and $\mathfrak{Z}$ being the center of $\mathfrak{B}$ we can find a homomorphism ξ of $\mathfrak{Z}$ into C such that $\pi(z-\xi(z))=0$ for all $z\in\mathfrak{Z}$.*

First we need some lemmas. Since $\mathfrak{g}$ is isomorphic to $\mathfrak{L}_K$ under the mapping γ every representation of $\mathfrak{L}_K$ determines a representation of $\mathfrak{g}$ and conversely. Hence for any $\Lambda\in P_K$, $\mathfrak{D}_\Lambda$ can also be regarded as a class of representations of $\mathfrak{g}$. If π is any representation of $\mathfrak{A}$ on V we can define exactly as before the subspace V_Λ of V consisting of all elements which transform under $\pi(\mathfrak{A})$ according to $\mathfrak{D}_\Lambda$. In particular let π be the adjoint representation of $\mathfrak{A}$ on itself given by

$$\pi(Z)w = [Z,\, w] \qquad\qquad (Z \in \mathfrak{g},\ w \in \mathfrak{A}).$$

Then for every $\Lambda\in P_K$ we can define $\mathfrak{A}_\Lambda$.

LEMMA 24. $\mathfrak{A} = \sum_{\Lambda\in P_K}\mathfrak{A}_\Lambda$ *and for every Λ, $\mathfrak{A}_\Lambda$ is a finite module over $\mathfrak{C}$.*

We shall derive this lemma from a well known result in the theory of invariants. Let ρ and σ be any two matrix representations of $\mathfrak{g}$ of degree p and q respectively. Let x_i, $1\leq i\leq p$, and y_j, $1\leq j\leq q$, be two sets of independent indeterminates. Let $C[x,\, y]$ denote the (commutative) ring of all polynomials in (x) and (y) with coefficients in C. For every $Z\in\mathfrak{g}$ we define a C-derivation[15] D_Z of $C[x,\, y]$ which is uniquely determined by the relations

$$D_Z x_j = \sum_{1\leq i\leq p} x_i\rho_{ij}(Z),\quad 1\leq j\leq p; \qquad D_Z y_j = \sum_{1\leq i\leq q} y_i\sigma_{ij}(Z),\quad 1\leq j\leq q.$$

The mapping $Z\to D_Z$ $(Z\in\mathfrak{g})$ is easily seen to be a representation of $\mathfrak{g}$ on $C[x,\, y]$. An element $f\in C[x,\, y]$ is said to be an invariant if $D_Z f=0$ for all $Z\in\mathfrak{g}$. Since $\mathfrak{g}$ is semisimple the main theorem of the theory of invariants is applicable in this case (see Weyl [13, p. 274]). It may be stated as follows.

THEOREM. *There exist a finite number of invariants, J_ν, $1\leq\nu\leq N$, such that the ring $C[J_1,\, \cdots,\, J_N]$ contains all invariants.*

We shall now use this theorem to prove Lemma 24. $f(i_1,\, \cdots,\, i_r)$ being any function of r indices (all running from 1 to n) with values in a vector space over C, we denote by $S_{(i_1,\ldots,i_r)}f(i_1,\, \cdots,\, i_r)$ $1/r!$ times the sum of $f(j_1,\, \cdots,\, j_r)$ for all permutations $(j_1,\, \cdots,\, j_r)$ of $(i_1,\, \cdots,\, i_r)$. Such a function will be called symmetric if $f(i_1,\, \cdots,\, i_r)=S_{(i_1,\ldots,i_r)}f(i_1,\, \cdots,\, i_r)$ for all $1\leq i_1,\, \cdots,\, i_r\leq n$.

Choose a base Z_i, $1\leq i\leq n$, for $\mathfrak{g}$. Then $S_{(i_1,\ldots,i_r)}Z_{i_1}Z_{i_2}\cdots Z_{i_r}$ is a basic

---

[14] We shall assume throughout that $\chi(1)=1$. For the only other possibility is $\chi(1)=0$. But in this case $\pi(1)=0$ and so in accordance with our convention π is not a representation.

[15] This means that D_Z is a linear mapping of $C[x,\, y]$ such that $D_Z(fg)=(D_Z f)g+f(D_Z g)$ for any $f,\, g\in C[x,\, y]$.

canonical element in $\mathfrak{A}$ with respect to this base. Let V_r be the subspace of $\mathfrak{A}$ spanned by all such elements for a fixed value of $r \geq 0$. Also put $V_{-1} = \{0\}$. Clearly V_r is invariant under the adjoint representation π of $\mathfrak{A}$. Since V_r is finite-dimensional it is fully reducible under π and therefore $V_r \subset \sum_\Lambda \mathfrak{A}_\Lambda$. But since the basic canonical elements span $\mathfrak{A}$, $\mathfrak{A} = \sum_{r \geq 0} V_r \subset \sum_\Lambda \mathfrak{A}_\Lambda$. Hence $\mathfrak{A} = \sum_\Lambda \mathfrak{A}_\Lambda$. But then from Lemma 22, $V_r = \sum_\Lambda (V_r \cap \mathfrak{A}_\Lambda)$. Hence

$$\sum_\Lambda \mathfrak{A}_\Lambda = \mathfrak{A} = \sum_{r \geq 0} V_r = \sum_{r \geq 0} \sum_\Lambda (V_r \cap \mathfrak{A}_\Lambda) = \sum_\Lambda \sum_{r \geq 0} (V_r \cap \mathfrak{A}_\Lambda) = \sum_\Lambda \mathfrak{A}_\Lambda'$$

where $\mathfrak{A}_\Lambda' = \sum_{r \geq 0} (V_r \cap \mathfrak{A}_\Lambda)$. Since $\mathfrak{A}_\Lambda' \subset \mathfrak{A}_\Lambda$ and since, by Lemma 22, the sum $\sum_\Lambda \mathfrak{A}_\Lambda$ is direct it follows that $\mathfrak{A}_\Lambda' = \mathfrak{A}_\Lambda$. Hence

$$\mathfrak{A}_\Lambda = \sum_{r \geq 0} (V_r \cap \mathfrak{A}_\Lambda).$$

The sum on the right-hand side is direct since distinct basic canonical elements are linearly independent.

Now let ρ be the adjoint representation of $\mathfrak{g}$ and σ^* any irreducible representation of $\mathfrak{g}$ on a vector space U^* such that $\sigma^* \in \mathfrak{D}_\Lambda$. Let U be the space dual to U^* and σ the representation of $\mathfrak{g}$ induced on U. Let e_i, $1 \leq i \leq q$, be a base for U. Then σ can be regarded as a matrix representation with respect to this base, so that

$$\sigma(Z)e_j = \sum_{1 \leq i \leq q} e_i \sigma_{ij}(Z) \qquad (Z \in \mathfrak{g}, \; 1 \leq j \leq q).$$

Similarly

$$\rho(Z)Z_j = [Z, Z_j] = \sum_{1 \leq i \leq n} Z_i \rho_{ij}(Z) \qquad (Z \in \mathfrak{g}, \; 1 \leq j \leq n).$$

Now we apply the above theorem to the pair (ρ, σ). It is clear that if a polynomial $f \in C[x, y]$ is an invariant then all the doubly homogeneous components([16]) of f are also invariants. Hence we may assume that $J_1, \cdots, J_N$ are all themselves doubly homogeneous. Among these let $G_1(x), \cdots, G_m(x)$ be all those which are independent of (y) and $H_\nu(x, y)$, $1 \leq \nu \leq s$, all those which are linear in (y). Then it is evident that any invariant $f(x, y)$ which is homogeneous and linear in (y) must belong to $\sum_{1 \leq \nu \leq s} \Omega H_\nu(x, y)$ where $\Omega = C[G_1, \cdots, G_m]$. Let r_ν be the degree of $H_\nu(x, y)$ in x. Then

$$H_\nu(x, y) = \sum_{1 \leq i_1, \cdots, i_r \leq m; \, 1 \leq j \leq q} h_\nu^{i_1 i_2 \cdots i_{r_\nu}, j} x_{i_1} x_{i_2} \cdots x_{i_{r_\nu}} y_j$$

where $h_\nu^{i_1 i_2 \cdots i_{r_\nu}, j} \in C$ and we may assume that $h_\nu^{i_1 i_2 \cdots i_{r_\nu}, j}$ are symmetric with respect to $i_1, \cdots, i_{r_\nu}$. Put

---

([16]) By doubly homogeneous we mean homogeneous in each of the two sets of variables (x) and (y) separately.

$$H_\nu^j(Z) = \sum_{1 \leq i_1, \cdots, i_{r_\nu} \leq m} h_\nu^{i_1 i_2 \cdots i_{r_\nu}, j} Z_{i_1} Z_{i_2} \cdots Z_{i_{r_\nu}}, \qquad 1 \leq j \leq q.$$

We shall prove that $H_\nu^j(Z)$, $1 \leq \nu \leq s$, $1 \leq j \leq q$, form a $\mathfrak{C}$-basis for $\mathfrak{A}_\Lambda$. Consider the element $\sum_{1 \leq j \leq q} H_\nu^j(Z) \times e_j$. If we extend ρ to a representation of $\mathfrak{g}$ on $\mathfrak{A}$ by setting $\rho(Z)w = [Z, w]$ $(Z \in \mathfrak{g}, w \in \mathfrak{A})$ we get the representation $\rho + \sigma$ of $\mathfrak{g}$ on the Kronecker product $\mathfrak{A} \times U$. Since $H_\nu(x, y)$ is an invariant it is clear that $\sum_{1 \leq j \leq q} H_\nu^j(Z) \times e_j$ is invariant under $\rho + \sigma$. Therefore it follows that the space W spanned by $H_\nu^j(Z)$, $1 \leq j \leq q$, is invariant under ρ and the representation induced on it by ρ must be dual to σ unless $W = \{0\}$. Hence in any case $H_\nu^j(Z) \in \mathfrak{A}_\Lambda$, $1 \leq j \leq q$. Therefore in order to show that

$$\mathfrak{A}_\Lambda = \sum_{j, \nu} \mathfrak{C} H_\nu^j(Z)$$

it is sufficient to prove that

$$\mathfrak{A}_\Lambda \cap V_r \subset \sum_{j, \nu} \mathfrak{C} H_\nu^j(Z), \qquad r \geq -1,$$

since $\mathfrak{A}_\Lambda = \sum_{r \geq 0} (\mathfrak{A}_\Lambda \cap V_r)$. We shall prove this by induction on r. For $r = -1$ this is true trivially. Hence suppose $r \geq 0$. Let $z \in \mathfrak{A}_\Lambda \cap V_r$. We have to prove that $z \in \sum_{j, \nu} \mathfrak{C} H_\nu^j(Z)$. Since $\mathfrak{A}_\Lambda \cap V_r$ is finite-dimensional it is completely reducible under ρ. Hence it is sufficient to consider the case when $z \neq 0$ and $W = \pi(\mathfrak{A})z$ is irreducible under π. (π is the adjoint representation of $\mathfrak{A}$.) Since z transforms according to $\mathfrak{D}_\Lambda$ we can choose a base w_j, $1 \leq j \leq q$, such that $\sum_{1 \leq j \leq q} w_j \times e_j$ is invariant under $\rho + \sigma$. Let

$$w_j = \sum_{1 \leq i_1, \cdots, i_r \leq n} a^{i_1 i_2 \cdots i_r, j} Z_{i_1} Z_{i_2} \cdots Z_{i_r}, \qquad 1 \leq j \leq q,$$

where the coefficients $a^{i_1 i_2 \cdots i_r, j}$ are symmetric in $i_1, \cdots, i_r$. Then it is easily verified that the form

$$f(x, y) = \sum_{1 \leq i_1, \cdots, i_r \leq n, 1 \leq j \leq q} a^{i_1 i_2 \cdots i_r, j} x_{i_1} x_{i_2} \cdots x_{i_r} y_j$$

is invariant with respect to (ρ, σ). Hence

$$f(x, y) = \sum_{1 \leq \nu \leq s} M_\nu(x) H_\nu(x, y)$$

where $M_\nu(x)$ is either zero or an invariant form of degree $r - r_\nu$. We may assume that $M_\nu(x) \neq 0$ for $1 \leq \nu \leq s'$ and $M_\nu(x) = 0$ for $\nu > s'$. Then if $s_\nu = r - r_\nu$,

$$M_\nu(x) = \sum_{1 \leq i_1, \cdots, i_{s_\nu} \leq n} b_\nu^{i_1 i_2 \cdots i_{s_\nu}} x_{i_1} x_{i_2} \cdots x_{i_{s_\nu}}, \qquad 1 \leq \nu \leq s',$$

where the coefficients are again symmetric with respect to $i_1, i_2, \cdots, i_s$. Hence comparing coefficients we have

$$a^{i_1 i_2 \cdots i_r, j} = \sum_{1 \leq \nu \leq s'} \underset{(i_1, \cdots, i_r)}{S} (b_\nu^{i_1 i_2 \cdots i_{s_\nu}} h_\nu^{i_{s_\nu+1} \cdots i_r, j}).$$

Therefore

$$
\begin{aligned}
w_j &= \sum_{1 \leq i_1, \cdots, i_r \leq n} a^{i_1 i_2 \cdots i_r, j} Z_{i_1} Z_{i_2} \cdots Z_{i_r} \\
&= \sum_{1 \leq \nu \leq s'} \sum_{1 \leq i_1, \cdots, i_r \leq n} b_\nu^{i_1 i_2 \cdots i_{s_\nu}} h_\nu^{i_{s_\nu+1} \cdots i_r, j} \underset{(i_1, \cdots, i_r)}{S} Z_{i_1} Z_{i_2} \cdots Z_i \\
&\equiv \sum_{1 \leq \nu \leq s'} \sum_{1 \leq i_1, \cdots, i_r \leq n} b_\nu^{i_1 i_2 \cdots i_{s_\nu}} h_\nu^{i_{s_\nu+1} \cdots i_r, j} Z_{i_1} Z_{i_2} \cdots Z_{i_r} \bmod \sum_{-1 \leq p < r} V_p,
\end{aligned}
$$

since

$$Z_{i_1} Z_{i_2} \cdots Z_{i_r} - \underset{(i_1, \cdots, i_r)}{S} Z_{i_1} Z_{i_2} \cdots Z_{i_r} \in \sum_{-1 \leq p < r} V_r \qquad \text{(see [9, p. 902]).}$$

Also

$$\sum_{1 \leq \nu \leq s'} \sum_{1 \leq i_1, \cdots, i_r \leq n} b_\nu^{i_1 i_2 \cdots i_{s_\nu}} h_\nu^{i_{s_\nu+1} \cdots i_r, j} Z_{i_1} Z_{i_2} \cdots Z_{i_r} = \sum_{1 \leq \nu \leq s'} \omega_\nu H_\nu^j(Z)$$

where

$$\omega_\nu = \sum_{1 \leq i_1, \cdots, i_{s_\nu} \leq n} b_\nu^{i_1 \cdots i_{s_\nu}} Z_{i_1} Z_{i_2} \cdots Z_{i_{s_\nu}} \in \mathfrak{C}$$

since $M_\nu(x)$ is an invariant. Since $w_j,\ H_\nu^j(Z) \in \mathfrak{A}_\Lambda$ and $\omega_\nu \in \mathfrak{C}$ it is clear that $w_j - \sum_{1 \leq \nu \leq s'} \omega_\nu H_\nu^j(Z) \in \mathfrak{A}_\Lambda$. Therefore

$$w_j - \sum_{1 \leq \nu \leq s'} \omega_\nu H_\nu^j(Z) \in \mathfrak{A}_\Lambda \cap \left(\sum_{-1 \leq p < r} V_p \right) = \sum_{-1 \leq p < r} (\mathfrak{A}_\Lambda \cap V_p)$$

since $\mathfrak{A}_\Lambda = \sum_{p \leq 0} (V_p \cap \mathfrak{A}_\Lambda)$. Hence by induction hypothesis

$$w_j - \sum_{1 \leq \nu \leq s'} \omega_\nu H_\nu^j(Z) \in \sum_{1 \leq j \leq q, 1 \leq \nu \leq s} \mathfrak{C} H_\nu^j(Z).$$

Therefore

$$w_j \in \sum_{1 \leq j \leq q, 1 \leq \nu \leq s} \mathfrak{C} H_\nu^j(Z)$$

and the lemma is proved.

For a given $\Lambda_0 \in P_K$ let π be any irreducible representation of $\mathfrak{X}$ on a space V such that $\pi \in \mathfrak{D}_{\Lambda_0}$. Choose an element $\psi \in V$ $(\psi \neq 0)$ which belongs to the highest weight Λ_0. Since from Theorem 2 the multiplicity of Λ_0 in π is 1, ψ

is unique apart from a numerical factor. Let $\mathfrak{Y}_{\Lambda_0}$ be the left ideal in $\mathfrak{X}$ consisting of all elements $x \in \mathfrak{X}$ such that $\pi(x)\psi = 0$. Let $z \to z^*$ denote the natural mapping of $\mathfrak{X}$ on $\mathfrak{X}^* = \mathfrak{X}/\mathfrak{Y}_{\Lambda_0}$ and π^* the natural representation of $\mathfrak{X}$ on $\mathfrak{X}^*$. Since V is irreducible $\mathfrak{Y}_{\Lambda_0}$ is a maximal left ideal and it is easily seen that π is equivalent to π^* under the isomorphism $\pi(z)\psi \leftrightarrow z^*$ $(z \in \mathfrak{X})$ of V with $\mathfrak{X}^*$. Further $\mathfrak{Y}_{\Lambda_0}$ is uniquely determined by Λ_0 since the vector ψ is essentially unique. We now define a representation ρ^* of $\mathfrak{L}$ on $\mathfrak{X}^*$ by

$$(7) \qquad \rho^*(X) = \pi^*(X), \qquad X \in \mathfrak{L}_K; \qquad \rho^*(Z) = 0, \qquad Z \in \mathfrak{g}.$$

It is easily checked that ρ^* is a representation. Let σ denote the representation of $\mathfrak{L}$ on $\mathfrak{A}$ given as follows:

$$(8) \qquad \begin{aligned} \sigma(X)z &= [X, z] & (X \in \mathfrak{L}_K,\, z \in \mathfrak{A}), \\ \sigma(Z)z &= Zz & (Z \in \mathfrak{g},\, z \in \mathfrak{A}). \end{aligned}$$

It is again easy to verify that this is indeed a representation. Let ν be the uniquely determined representation of $\mathfrak{B}$ on the Kronecker product $\mathfrak{A} \times \mathfrak{X}^*$ which coincides with $\sigma + \rho^*$ on $\mathfrak{L}$. For any $\Lambda \in \mathrm{P}_K$ we consider the subspace $(\mathfrak{A} \times \mathfrak{X}^*)_\Lambda$ consisting of all elements of $\mathfrak{A} \times \mathfrak{X}^*$ which transform under $\nu(\mathfrak{X})$ according to $\mathfrak{D}_\Lambda$.

LEMMA 26. $\mathfrak{A} \times \mathfrak{X}^* = \sum_{\Lambda \in \mathrm{P}_K} (\mathfrak{A} \times \mathfrak{X}^*)_\Lambda$ and $(\mathfrak{A} \times \mathfrak{X}^*)_\Lambda$ is a finite $\mathfrak{C}$-module for each $\Lambda \in \mathrm{P}_K$.

Let π denote the adjoint representation of $\mathfrak{A}$, and γ the isomorphism of $\mathfrak{L}_K$ with $\mathfrak{g}$ as defined on p. 53. Since $[X, z] = [\gamma(X), z]$ $(X \in \mathfrak{L}_K,\, z \in \mathfrak{A})$ it follows that $\sigma(X) = \pi(\gamma(X))$ $(X \in \mathfrak{L}_K)$. Hence for any $\Lambda \in \mathrm{P}_K$ the subspace $\mathfrak{A}_\Lambda$ consisting of all elements in $\mathfrak{A}$ which transform under $\sigma(\mathfrak{L}_K)$ according to $\mathfrak{D}_\Lambda$ is the same as $\mathfrak{A}_\Lambda$ of Lemma 24. Therefore we can find $a_1, \cdots, a_r \in \mathfrak{A}_\Lambda$ such that $\mathfrak{A}_\Lambda = \sum_{1 \le i \le r} \mathfrak{C}a_i$. Since $a_i \in \mathfrak{A}_\Lambda$ the space $\sigma(\mathfrak{X})a_i = \pi(\mathfrak{A})a_i$ is finite-dimensional. Since $\mathfrak{X}^*$ is also finite-dimensional, the same holds for the space $\sigma(\mathfrak{X})a_i \times \mathfrak{X}^*$ which is invariant under $\nu(\mathfrak{X})$. Hence it is completely reducible under $\nu(\mathfrak{X})$ and therefore

$$a_i \times \mathfrak{X}^* \subset \sum_{\Lambda' \in \mathrm{P}_K} (\mathfrak{A} \times \mathfrak{X}^*)_{\Lambda'}.$$

Now $\nu(\mathfrak{C})(a_i \times \mathfrak{X}^*) = \mathfrak{C}a_i \times \mathfrak{X}^*$ and clearly each $(\mathfrak{A} \times \mathfrak{X}^*)_{\Lambda'}$ is invariant under $\nu(\mathfrak{C})$. Hence

$$\mathfrak{A}_\Lambda \times \mathfrak{X}^* \subset \sum_{\Lambda' \in \mathrm{P}} (\mathfrak{A} \times \mathfrak{X}^*)_{\Lambda'}.$$

Since $\mathfrak{A} = \sum_\Lambda \mathfrak{A}_\Lambda$ from Lemma 25, we get

$$\mathfrak{A} \times \mathfrak{X}^* = \sum_\Lambda (\mathfrak{A}_\Lambda \times \mathfrak{X}^*) = \sum_{\Lambda \in \mathrm{P}_K} (\mathfrak{A} \times \mathfrak{X}^*)_\Lambda;$$

this proves the first statement in the lemma.

Now we come to the second part. For any fixed $\Lambda \in P_K$ choose an irreducible representation θ^* of $\mathfrak{L}_K$ on a space U^* such that $\theta^* \in \mathfrak{D}_\Lambda$. We extend θ^* to a representation of $\mathfrak{L}$ by defining $\theta^*(Z) = 0$ $(Z \in \mathfrak{g})$. Let U be the space dual to U^*, and θ the representation of $\mathfrak{L}$ induced on U. Let φ denote the representation of $\mathfrak{B}$ on $\mathfrak{A} \times \mathfrak{X} \times U$ which coincides with $\nu + \theta$ on $\mathfrak{L}$. Given any $\Lambda' \in P_K$ choose $a_i \in \mathfrak{A}_{\Lambda'}$, $1 \leq i \leq r$, such that $\mathfrak{A}_{\Lambda'} = \sum_{1 \leq i \leq r} \mathfrak{C} a_i$. Then $\sigma(\mathfrak{X}) a_i \times \mathfrak{X}^* \times U$ is finite-dimensional and invariant under $\varphi(\mathfrak{X})$. Hence

$$\sigma(\mathfrak{X}) a_i \times \mathfrak{X}^* \times U \subset \sum_{\Lambda'' \in P_K} (\sigma(\mathfrak{X}) a_i \times \mathfrak{X}^* \times U)_{\Lambda''}$$

where $(\sigma(\mathfrak{X}) a_i \times \mathfrak{X}^* \times U)_{\Lambda''}$ has the usual meaning. From this it follows as above that

$$\mathfrak{A} \times \mathfrak{X}^* \times U = \sum_{\Lambda' \in P_K} (\mathfrak{A} \times \mathfrak{X}^* \times U)_{\Lambda'}.$$

We now claim that $(\mathfrak{A} \times \mathfrak{X}^* \times U)_0$ is a finite $\mathfrak{C}$-module. Since $\mathfrak{X}^* \times U$ is a finite-dimensional space it is completely reducible under $(\rho^* + \theta)(\mathfrak{L}_K)$. Let $\mathfrak{X}^* \times U = \sum_{1 \leq j \leq N} U_j$ where the sum is direct and the subspaces U_j are invariant and irreducible under $(\rho^* + \theta)(\mathfrak{L}_K)$. Let Λ_j be the highest weight of the representation of $\mathfrak{L}_K$ induced on U_j^* where U_j^* is the space dual to U_j. Then it is well known that $\mathfrak{D}_0$ occurs in the representation of $\mathfrak{L}_K$ induced on $\mathfrak{A}_{\Lambda'} \times U_j$ if and only if $\Lambda' = \Lambda_j$. Hence if $\Lambda' \neq \Lambda_j$, it follows from Lemma 22 that

$$\mathfrak{A}_{\Lambda'} \times U_j \subset \sum_{\Lambda'' \neq 0} (\mathfrak{A} \times \mathfrak{X}^* \times U)_{\Lambda''}.$$

Now

$$\mathfrak{A} \times \mathfrak{X}^* \times U = \sum_{\Lambda, j} (\mathfrak{A}_\Lambda \times U_j) = \sum_j (\mathfrak{A}_{\Lambda_j} \times U_j) + \sum_j \sum_{\Lambda \neq \Lambda_j} (\mathfrak{A}_\Lambda \times U_j)$$
$$\subset \sum_j (\mathfrak{A}_{\Lambda_j} \times U_j) + \sum_{\Lambda \neq 0} (\mathfrak{A} \times \mathfrak{X}^* \times U)_\Lambda.$$

But from Lemma 22,

$$\mathfrak{A}_{\Lambda_j} \times U_j = \sum_\Lambda \left\{ (\mathfrak{A}_{\Lambda_j} \times U_j) \cap (\mathfrak{A} \times \mathfrak{X}^* \times U)_\Lambda \right\}.$$

Hence

$$\mathfrak{A} \times \mathfrak{X}^* \times U \subset \sum_j \left\{ (\mathfrak{A}_{\Lambda_j} \times U_j) \cap (\mathfrak{A} \times \mathfrak{X}^* \times U)_0 \right\}$$
$$+ \sum_{\Lambda \neq 0} (\mathfrak{A} \times \mathfrak{X}^* \times U)_\Lambda.$$

Since the sum $\sum_\Lambda (\mathfrak{A} \times \mathfrak{X}^* \times U)_\Lambda$ is direct it follows that

$$(\mathfrak{A} \times \mathfrak{X}^* \times U)_0 \subset \sum_j \left\{ (\mathfrak{A}_{\Lambda_j} \times U_j) \cap (\mathfrak{A} \times \mathfrak{X}^* \times U)_0 \right\}.$$

Hence

$$(\mathfrak{A} \times \mathfrak{X}^* \times U)_0 = \sum_j (\mathfrak{A}_{\Lambda_j} \times U_j)_0.$$

Thus it is sufficient to prove that for a fixed j, $(\mathfrak{A}_{\Lambda_j} \times U_j)_0$ is a finite $\mathfrak{C}$-module. Choose $a_i \in \mathfrak{A}_{\Lambda_j}$, $1 \leq i \leq r$, such that $\mathfrak{A}_{\Lambda_j} = \sum_{1 \leq i \leq r} \mathfrak{C} a_i$. Since $\sigma(\mathfrak{X}) a_i \times U_j$ is finite-dimensional, it is completely reducible under $\nu(\mathfrak{X})$. Hence

$$\sigma(\mathfrak{X}) a_i \times U_j = \sum_{\Lambda'} (\sigma(\mathfrak{X}) a_i \times U_j)_{\Lambda'}.$$

Clearly $\nu(\mathfrak{C})(\sigma(\mathfrak{X}) a_i \times U_j)_{\Lambda'} \subset (\mathfrak{A}_{\Lambda_j} \times U_j)_{\Lambda'}$. But

$$\mathfrak{A}_{\Lambda_j} \times U_j = \sum_{1 \leq i \leq r} \nu(\mathfrak{C})(\sigma(\mathfrak{X}) a_i \times U_j).$$

Therefore

$$(\mathfrak{A}_{\Lambda_j} \times U_j)_{\Lambda'} = \sum_{1 \leq i \leq r} \nu(\mathfrak{C})(\sigma(\mathfrak{X}) a_i \times U_j)_{\Lambda'}.$$

Since dim $(\sigma(\mathfrak{X}) a_i \times U_j)_{\Lambda'}$ is finite it follows that $(\mathfrak{A}_{\Lambda_j} \times U_j)_{\Lambda'}$ is a finite $\mathfrak{C}$-module. If we take $\Lambda' = 0$ we get the required result.

Let e_j, $1 \leq j \leq p$, be a base for U and let u_i, $1 \leq i \leq r$, be elements in $(\mathfrak{A} \times \mathfrak{X}^* \times U)_0$ such that

$$(\mathfrak{A} \times \mathfrak{X}^* \times U)_0 = \sum_{1 \leq i \leq r} \nu(\mathfrak{C}) u_i.$$

Let $u_i = \sum_{1 \leq j \leq p} a_i^j \times e_j$, $1 \leq i \leq r$, $(a_i^j \in \mathfrak{A} \times \mathfrak{X}^*)$. We claim that

$$(\mathfrak{A} \times \mathfrak{X}^*)_\Lambda = \sum_{1 \leq i \leq r, 1 \leq j \leq p} \nu(\mathfrak{C}) a_i^j.$$

Put $A = \sum_{1 \leq i \leq r, 1 \leq j \leq p} \nu(\mathfrak{C}) a_i^j$. Since u_i transform according to $\mathfrak{D}_0$ under $(\nu + \theta)(\mathfrak{L}_K)$ it is clear that $a_i^j \in (\mathfrak{A} \times \mathfrak{X}^*)_\Lambda$. Hence $A \subset (\mathfrak{A} \times \mathfrak{X}^*)_\Lambda$. Now let $a \in (\mathfrak{A} \times \mathfrak{X}^*)_\Lambda$. We have to show that $a \in A$. We may assume that $a \neq 0$. Then $\nu(\mathfrak{X}) a$ is a finite-dimensional space which is completely reducible under $\nu(\mathfrak{X})$ into a direct sum of invariant irreducible subspaces V_k, $1 \leq k \leq q$. It is sufficient to show that $V_k \subset A$, $1 \leq k \leq q$. For a fixed k write $V = V_k$. Let ψ be the irreducible representation of $\mathfrak{X}$ induced on V. Then $\psi \in \mathfrak{D}_\Lambda$ and therefore we can choose a base v_j, $1 \leq j \leq p$, for V such that

$$\sum_{1 \leq j \leq p} (v_j \times e_j) \in (\mathfrak{A} \times \mathfrak{X}^* \times U)_0.$$

Hence

$$\sum_{1 \leq j \leq p} v_j \times e_j = \sum_{1 \leq i \leq r} \nu(z_i) u_i = \sum_{1 \leq i \leq r, 1 \leq j \leq p} z_i a_i^j \times e_j \qquad (z_i \in \mathfrak{C}).$$

Therefore $v_j = \sum_{1 \leq i \leq r} z_i a_i^j \in A$, $1 \leq j \leq p$. This shows that $V \subset A$ and so the lemma is proved.

LEMMA 27. *Let ν be the representation of $\mathfrak{B}$ on $\mathfrak{A} \times \mathfrak{X}^*$ as defined above. Then*[17] *$\nu(\mathfrak{B})(1 \times 1^*) = \mathfrak{A} \times \mathfrak{X}^*$ and the set of all elements $b \in \mathfrak{B}$ such that $\nu(b)\{1 \times 1^*\} = 0$ coincides with $\mathfrak{A} \mathfrak{Y}_{\Lambda_0}$.*

If $a \in \mathfrak{A}$ and $x \in \mathfrak{X}$, it is easily seen that $\nu(ax)\{1 \times 1^*\} = a \times x^*$. Hence $\nu(\mathfrak{B})(1 \times 1^*) = \mathfrak{A} \times \mathfrak{X}^*$. Also since $[X, a] \in \mathfrak{A}$ for any $X \in \mathfrak{L}_K$ and $a \in \mathfrak{A}$ it is obvious that $\mathfrak{B} = \mathfrak{A} \mathfrak{X}$. Let $\mathfrak{M}$ be the set of all elements $b \in \mathfrak{B}$ such that $\nu(b)(1 \times 1^*) = 0$. Clearly $\mathfrak{M}$ is a left ideal and $\mathfrak{M} \supset \mathfrak{Y}_{\Lambda_0}$. Hence $\mathfrak{M} \supset \mathfrak{A} \mathfrak{Y}_{\Lambda_0}$. On the other hand we can choose elements ω_i, $1 \leq i \leq N$, in $\mathfrak{X}$ such that ω_i^*, $1 \leq i \leq N$, form a base for $\mathfrak{X}^*$. Since $\mathfrak{B} = \mathfrak{A} \mathfrak{X}$ every element $b \in \mathfrak{B}$ can be written as $b \equiv \sum_{1 \leq i \leq N} a_i \omega_i \bmod \mathfrak{A} \mathfrak{Y}_{\Lambda_0}$ $(a_i \in \mathfrak{A})$. Therefore

$$\nu(b)(1 \times 1^*) = \sum_{1 \leq i \leq N} a_i \times \omega_i^*.$$

Hence if $b \in \mathfrak{M}$, $a_i = 0$, $1 \leq i \leq N$, and $b \in \mathfrak{A} \mathfrak{Y}_{\Lambda_0}$. Therefore $\mathfrak{M} = \mathfrak{A} \mathfrak{Y}_{\Lambda_0}$.

From Lemma 26, we can find elements a_i, $1 \leq i \leq r$, in $(\mathfrak{A} \times \mathfrak{X}^*)_{\Lambda_0}$ such that $(\mathfrak{A} \times \mathfrak{X}^*)_{\Lambda_0} = \sum_{1 \leq i \leq r} \nu(\mathfrak{C}) a_i$. Let ν_0 be the representation of $\mathfrak{X}$ induced on $(\mathfrak{A} \times \mathfrak{X}^*)_{\Lambda_0}$ by ν. By Lemma 23, we can write $(\mathfrak{A} \times \mathfrak{X}^*)_{\Lambda_0} = \sum_\lambda (\mathfrak{A} \times \mathfrak{X}^*)_{\Lambda_0, \lambda}$ where λ runs over all the weights of ν_0. Let $\nu_0(\mathfrak{X}) a_i = \sum_\lambda (\nu_0(\mathfrak{X}) a_i)_\lambda$ be the corresponding decomposition of the finite-dimensional space $\nu_0(\mathfrak{X}) a_i$. It is clear that $\nu(\mathfrak{C})(\nu_0(\mathfrak{X}) a_i)_\lambda \subset (\mathfrak{A} \times \mathfrak{X}^*)_{\Lambda_0, \lambda}$. Hence

$$(\mathfrak{A} \times \mathfrak{X}^*)_{\Lambda_0, \lambda} = \sum_{1 \leq i \leq r} \nu(\mathfrak{C})(\nu_0(\mathfrak{X}) a_i)_\lambda.$$

Since $\nu_0(\mathfrak{X}) a_i$ is finite-dimensional this shows that each $(\mathfrak{A} \times \mathfrak{X}^*)_{\Lambda_0, \lambda}$ is a finite $\mathfrak{C}$-module. In particular this holds for $A = (\mathfrak{A} \times \mathfrak{X}^*)_{\Lambda_0, \Lambda_0}$. Now we turn A into an associative algebra as follows. For any $u \in A$ consider the finite-dimensional space $\nu_0(\mathfrak{X}) u$. It follows from the definition of $(\mathfrak{A} \times \mathfrak{X}^*)_{\Lambda_0}$ that the representation of $\mathfrak{X}$ induced on $\nu_0(\mathfrak{X}) u$ is a direct sum of irreducible representations each of which belongs to $\mathfrak{D}_{\Lambda_0}$. Since u belongs to the highest weight Λ_0, it follows that $\nu(\mathfrak{Y}_{\Lambda_0}) u = 0$ and therefore $\nu(\mathfrak{A} \mathfrak{Y}_{\Lambda_0}) u = 0$. Given any $v \in A$ we can, by Lemma 27, find a $b \in \mathfrak{B}$ such that $\nu(b)(1 \times 1^*) = v$. If b' is another such element then by the same lemma, $b - b' \in \mathfrak{A} \mathfrak{Y}_{\Lambda_0}$ and therefore $\nu(b - b') u = 0$. Hence $\nu(b) u$ is uniquely determined by v alone. Now we define multiplication by setting $vu = \nu(b) u$. If b' is any element in $\mathfrak{B}$ such that $\nu(b')(1 \times 1^*) = u$, then $vu = \nu(b) \nu(b')(1 \times 1^*)$. This immediately shows that the multiplication is associative. For any $z \in \mathfrak{C}$, $\nu(z)(1 \times 1^*) = z \times 1^* \in A$. Also if $z \neq 0$ then $z \times 1^* \neq 0$. Therefore the mapping $z \to z \times 1^*$ $(z \in \mathfrak{C})$ is clearly an isomorphism of $\mathfrak{C}$ into A. We may therefore identify $\mathfrak{C}$ with its image under this mapping. Then it is easily seen that $\mathfrak{C}$ lies in the center of A and A is a finite $\mathfrak{C}$-module (under the

---

[17] We denote the natural mapping of $\mathfrak{X}$ on $\mathfrak{X}^*$ by $x \to x^*$ $(x \in \mathfrak{X})$ throughout.

multiplication defined in A). Let $\mathfrak{Z}$ be the center of $\mathfrak{B}$. We note for later use that if $z \in \mathfrak{Z}$, $\nu(z)(1 \times 1^*)$ lies in the center of A.

Now let χ be any homomorphism of $\mathfrak{C}$ into C such that $\chi(1) = 1$. Let $\mathfrak{N}_\chi$ be the left ideal in $\mathfrak{A}$ generated by all elements of the form $z - \chi(z)$ ($z \in \mathfrak{C}$). Then it is obvious that $\mathfrak{N}_\chi$ is actually an ideal and therefore $[X, \mathfrak{N}_\chi] = [\gamma(X), \mathfrak{N}_\chi] \subset \mathfrak{N}_\chi$ ($X \in \mathfrak{L}_K$). Therefore the space $\mathfrak{N}_\chi \times \mathfrak{X}^*$ is invariant under $\nu(\mathfrak{B})$. Let $z \to \bar{z}$ denote the natural mapping of $\mathfrak{A}$ on $\bar{\mathfrak{A}} = \mathfrak{A}/\mathfrak{N}_\chi$ and θ the natural mapping of $\mathfrak{A} \times \mathfrak{X}^*$ on the factor space $(\mathfrak{A} \times \mathfrak{X}^*)/(\mathfrak{N}_\chi \times \mathfrak{X}^*)$. Then it is easily seen that $\bar{\mathfrak{A}} \times \mathfrak{X}^*$ and $(\mathfrak{A} \times \mathfrak{X}^*)/(\mathfrak{N}_\chi \times \mathfrak{X}^*)$ are isomorphic under the mapping $\bar{z} \times x^* \leftrightarrow \theta(z \times x^*)$ ($z \in \mathfrak{A}$, $x \in \mathfrak{X}$). We therefore identify the two spaces under this mapping. Let $\bar{\nu}$ be the representation of $\mathfrak{B}$ induced on $(\bar{\mathfrak{A}} \times \mathfrak{X}^*) = \theta(\mathfrak{A} \times \mathfrak{X}^*)$. Then it is easy to verify that

$$\bar{\nu}(X)(\bar{z} \times \omega^*) = \overline{[X, z]} \times \omega^* + \bar{z} \times (X\omega)^*,$$

$$\bar{\nu}(w)(\bar{z} \times \omega^*) = \overline{wz} \times \omega^* \qquad (X \in \mathfrak{L}_K, z, w \in \mathfrak{A}, \omega \in \mathfrak{X}).$$

Also it is clear that

$$(\bar{\mathfrak{A}} \times \mathfrak{X}^*)_\Lambda \supset \theta((\mathfrak{A} \times \mathfrak{X}^*)_\Lambda) \qquad (\Lambda \in \mathrm{P}_K).$$

Since $\mathfrak{A} \times \mathfrak{X}^* = \sum_\Lambda (\mathfrak{A} \times \mathfrak{X}^*)_\Lambda$, it follows that

$$(9) \qquad \bar{\mathfrak{A}} \times \mathfrak{X} = \sum_\Lambda (\bar{\mathfrak{A}} \times \mathfrak{X})_\Lambda$$

and $(\bar{\mathfrak{A}} \times \mathfrak{X}^*)_\Lambda = \theta((\mathfrak{A} \times \mathfrak{X}^*)_\Lambda)$. We have seen that $(\mathfrak{A} \times \mathfrak{X}^*)_\Lambda$ is a finite module over $\mathfrak{C}$. Since $\bar{\nu}(z - \chi(z)) = 0$ ($z \in \mathfrak{C}$) it follows that dim $(\bar{\mathfrak{A}} \times \mathfrak{X}^*)_\Lambda < \infty$. Now consider $A \cap (\mathfrak{N}_\chi \times \mathfrak{X}^*)$. Since $\mathfrak{N}_\chi \times \mathfrak{X}^*$ is invariant under $\nu(\mathfrak{B})$, $A \cap (\mathfrak{N}_\chi \times \mathfrak{X}^*)$ is a left ideal in A. We shall now prove that it is actually an ideal in A. First notice that $\mathfrak{N}_\chi \mathfrak{X}$ is an ideal in $\mathfrak{B}$ since $[X, \mathfrak{N}_\chi] \subset \mathfrak{N}_\chi$ ($X \in \mathfrak{L}_K$) and $\mathfrak{A}\mathfrak{X} = \mathfrak{X}\mathfrak{A} = \mathfrak{B}$. Let $u \in A \cap (\mathfrak{N}_\chi \times \mathfrak{X}^*)$, $v \in A$. Then we can find $b \in \mathfrak{N}_\chi \mathfrak{X}$, $b' \in \mathfrak{B}$ such that $u = \nu(b)(1 \times 1^*)$, $v = \nu(b')$ $(1 \times 1^*)$. Hence $uv = \nu(bb')(1 \times 1^*)$. Since $\mathfrak{N}_\chi \mathfrak{X}$ is an ideal, $bb' \in \mathfrak{N}_\chi \mathfrak{X}$ and therefore $uv \in (\mathfrak{N}_\chi \times \mathfrak{X}^*) \cap A$. Put $\bar{A} = \theta(A)$. Then we can regard $\bar{A} \cong A/A \cap (\mathfrak{N}_\chi \times \mathfrak{X}^*)$ as a factor algebra of A. Again we verify that if $\bar{\nu}(b_1)(\bar{1} \times 1^*) = \bar{u}_1$, $\bar{\nu}(b_2)(\bar{1} \times 1^*) = \bar{u}_2$ then $\bar{u}_1 \bar{u}_2 = \bar{\nu}(b_1 b_2)(\bar{1} \times 1^*)$ ($b_1, b_2 \in \mathfrak{B}$, $\bar{u}_1, \bar{u}_2 \in A$). Since dim $\bar{A} \leq$ dim $(\bar{\mathfrak{A}} \times \mathfrak{X}^*)_{\Lambda_0} < \infty$, $\bar{A}$ is a finite-dimensional associative algebra. We note that $\bar{1} \times 1^*$ is the unit element of $\bar{A}$.

Put $\mathfrak{M}_{\Lambda_0} = \mathfrak{A}\mathfrak{Y}_{\Lambda_0} + \mathfrak{N}_\chi \mathfrak{X}$. It follows easily from Lemma 27 that $\bar{\nu}(\mathfrak{B})$ $(\bar{1} \times 1^*) = \bar{\mathfrak{A}} \times \mathfrak{X}^*$ and $\nu(b)(\bar{1} \times 1^*) = 0$ ($b \in \mathfrak{B}$) if and only if $b \in \mathfrak{M}_{\Lambda_0}$.

LEMMA 28. *Let $\mathfrak{M}$ be a maximal left ideal in $\mathfrak{B}$ such that $\mathfrak{M} \supset \mathfrak{M}_{\Lambda_0}$. Put $\mathfrak{M}_* = \bar{\nu}(\mathfrak{M})(\bar{1} \times 1^*)$. Then $\mathfrak{M}_* \cap \bar{A}$ is a maximal left ideal in $\bar{A}$.*

Let $z \to z_*$ denote the natural mapping of $\mathfrak{B}$ on $\mathfrak{B}_* = \mathfrak{B}/\mathfrak{M}_{\Lambda_0}$ and let π_* be the natural representation of $\mathfrak{B}$ on $\mathfrak{B}_*$. It is obvious that $\bar{\nu}$ is equivalent to π_* under the isomorphism $\bar{\nu}(b)(\bar{1} \times 1^*) \leftrightarrow b_*$ ($b \in \mathfrak{B}$) of $\bar{\mathfrak{A}} \times \mathfrak{X}^*$ with $\mathfrak{B}_*$. We may therefore identify these two spaces under this isomorphism. Since $\mathfrak{M}$ is a left ideal it is clear that $\mathfrak{M}_*$ is invariant under $\bar{\nu}(\mathfrak{B})$ and therefore $\mathfrak{M}_* \cap \overline{A}$ is a left ideal in $\overline{A}$. Also $\overline{1} \times 1^* \notin \mathfrak{M}_* \cap \overline{A}$ for otherwise we would have $\mathfrak{B}_* = \bar{\nu}(\mathfrak{B})$ $(\overline{1} \times 1^*) \subset \mathfrak{M}_*$. Since $\mathfrak{M} \supset \mathfrak{M}_{\Lambda_0}$ this would imply that $\mathfrak{B}/\mathfrak{M} \cong \mathfrak{B}_*/\mathfrak{M}_* = \{0\}$ thus contradicting the fact that $\mathfrak{M}$ is maximal.

Let ρ be the natural representation of $\mathfrak{B}$ on $\mathfrak{B}/\mathfrak{M}$. Since $\mathfrak{M}$ is maximal ρ is irreducible. Also since $\mathfrak{M} \supset \mathfrak{M}_{\Lambda_0}$, ρ is equivalent to the representation induced by $\bar{\nu}$ on $\mathfrak{B}_*/\mathfrak{M}_*$. Let u, $v \in \overline{A}$, $u \notin \mathfrak{M}_*$. Since ρ is irreducible we can find a $b \in \mathfrak{B}$ such that $\bar{\nu}(b)u \equiv v$ mod $\mathfrak{M}_*$. We have seen above (equation (9), p. 63) that $\overline{\mathfrak{A}} \times \mathfrak{X}^* = \sum_{\Lambda} (\overline{\mathfrak{A}} \times \mathfrak{X}^*)_{\Lambda}$. Hence $\bar{\nu}(b)u = u_0 + u_1 + \cdots + u_r$ where $u_i \in (\overline{\mathfrak{A}} \times \mathfrak{X}^*)_{\Lambda_i}$, $\Lambda_0, \cdots, \Lambda_r$ being distinct. Then

$$\bar{\nu}(b)u - v = (u_0 - v) + u_1 + \cdots + u_r \in \mathfrak{M}_*.$$

From Lemma 4 of [9, p. 912], we can find an $x \in \mathfrak{X}$ such that $\bar{\nu}(x)z = 0$ for all $z \in (\overline{\mathfrak{A}} \times \mathfrak{X}^*)_{\Lambda_i}$, $1 \leq i \leq r$, and $\bar{\nu}(x)z = z$ for all $z \in (\overline{\mathfrak{A}} \times \mathfrak{X}^*)_{\Lambda_0}$. Hence

$$\bar{\nu}(x)\{\bar{\nu}(b)u - v\} = \bar{\nu}(xb)u - v = u_0 - v \in \mathfrak{M}_*.$$

Further if $\lambda_0, \lambda_1, \cdots, \lambda_s$ ($\lambda_0 = \Lambda_0$) are all the distinct weights of $\mathfrak{D}_{\Lambda_0}$, put

$$x' = \frac{\prod_{1 \leq i \leq s} (H - \lambda_i(H))}{\prod_{1 \leq i \leq s} (\lambda_0(H) - \lambda_i(H))}$$

where H is any element in $\mathfrak{h}_K$ such that $\lambda_0(H) \neq \lambda_i(H)$, $1 \leq i \leq s$. Then clearly $\nu(x')z \in \overline{A}$ for any $z \in (\overline{\mathfrak{A}} \times \mathfrak{X}^*)_{\Lambda_0}$ and $\bar{\nu}(x')z = z$ if $z \in \overline{A}$. Hence

$$\bar{\nu}(x'xb)u - v = \nu(x')(u_0 - v) \in \mathfrak{M}_* \cap \overline{A}.$$

Put $b' = x'xb$. Then $\bar{\nu}(b')u - v \in \mathfrak{M}_* \cap \overline{A}$ and since $v \in \overline{A}$, $\bar{\nu}(b')u \in \overline{A}$. Let $\mathfrak{M}_u$ be the left ideal in $\mathfrak{B}$ consisting of all $z \in \mathfrak{B}$ such that $\bar{\nu}(z)u = 0$. It is clear that $\mathfrak{M}_u \supset \mathfrak{M}_{\Lambda_0}$. Let $\mathfrak{M}_{u*} = \bar{\nu}(\mathfrak{M}_u)(\overline{1} \times 1^*)$ and let θ be the natural mapping of $\mathfrak{B}_*$ on $\mathfrak{B}_*/\mathfrak{M}_{u*}$. Also let $\bar{\nu}_0$ be the representation of $\mathfrak{B}$ on $\mathfrak{B}_*/\mathfrak{M}_{u*}$ induced by $\bar{\nu}$. Then it is obvious that

$$\bar{\nu}(\mathfrak{B})u \cong \mathfrak{B}/\mathfrak{M}_u \cong \mathfrak{B}_*/\mathfrak{M}_{u*} = (\overline{\mathfrak{A}} \times \mathfrak{X}^*)/\mathfrak{M}_{u*}$$

and the representation of $\mathfrak{B}$ on $\bar{\nu}(\mathfrak{B})u$ is equivalent to $\bar{\nu}_0$ under the isomorphism $\bar{\nu}(z)u \leftrightarrow \theta(\bar{\nu}(z)(\overline{1} \times 1^*))$ ($z \in \mathfrak{B}$) of $\bar{\nu}(\mathfrak{B})u$ with $\mathfrak{B}_*/\mathfrak{M}_{u*}$. Since $\bar{\nu}(b')u \in \overline{A}$ it follows that

$$\theta(\bar{\nu}(b')(\overline{1} \times 1^*)) \in (\mathfrak{B}_*/\mathfrak{M}_{u*})_{\Lambda_0, \Lambda_0}$$

in the notation of Lemma 23. But it is obvious that

$$(\mathfrak{B}_*/\mathfrak{M}_{u*})_{\Lambda_0, \Lambda_0} = \theta(\overline{A}).$$

Hence $\bar{\nu}(b')(\overline{1} \times 1^*) \in \overline{A} + \mathfrak{M}_{u*}$. Therefore we can find a $z \in \mathfrak{M}_u$ such that

$\bar{v}(b'-z)(\bar{1}\times 1^*) = u' \in \bar{A}$. But then

$$u'u - v = \bar{v}(b' - z)u - v = \bar{v}(b')u - v \in \mathfrak{M}_* \cap \bar{A}$$

since $z \in \mathfrak{M}_u$. Since u, v were any two elements in $\bar{A}$ such that $u \notin \mathfrak{M}_* \cap \bar{A}$, this shows that the natural representation of $\bar{A}$ on $\bar{A}/\mathfrak{M}_* \cap \bar{A}$ is irreducible. Since $\bar{1}\times 1^* \notin \mathfrak{M}_* \cap A$ it follows that $\mathfrak{M}_* \cap \bar{A}$ is a maximal left ideal in $\bar{A}$.

We have seen above that every maximal left ideal $\mathfrak{M}$ in $\mathfrak{B}$ such that $\mathfrak{M} \supset \mathfrak{M}_{A_0}$ defines an irreducible representation of $\bar{A}$ namely the natural representation of $\bar{A}$ on $\bar{A}/\mathfrak{M}_* \cap \bar{A}$. Since $\bar{A}$ is a finite-dimensional associative algebra it has only a finite number of inequivalent *irreducible* representations. Hence there exist a finite number of maximal left ideals $\mathfrak{M}_i$, $1 \leq i \leq r$, in $\mathfrak{B}$ each containing $\mathfrak{M}_{A_0}$ such that if $\mathfrak{M}$ is any maximal left ideal containing $\mathfrak{M}_{A_0}$ the representation of $\bar{A}$ defined by $\mathfrak{M}$ is equivalent to the one defined by $\mathfrak{M}_i$ for some i, $1 \leq i \leq r$. Let ρ_i denote the natural representation of $\mathfrak{B}$ on $\mathfrak{B}/\mathfrak{M}_i$.

LEMMA 29. *Let $\mathfrak{M}$ be a maximal left ideal in $\mathfrak{B}$ such that $\mathfrak{M} \supset \mathfrak{M}_{A_0}$ and let ρ be the natural representation of $\mathfrak{B}$ on $\mathfrak{B}/\mathfrak{M}$. Then ρ is equivalent to ρ_i for some i, $1 \leq i \leq r$. Moreover if $\mathfrak{Z}$ is the center of $\mathfrak{B}$ there exists a homomorphism ξ of $\mathfrak{Z}$ into C such that $z - \xi(z) \in \mathfrak{M}$ for all $z \in \mathfrak{Z}$.*

We keep to the notation of the proof of the preceding lemma. Then $\mathfrak{M}_* = \bar{v}(\mathfrak{M})(\bar{1}\times 1^*)$, $\mathfrak{M}_{i*} = \bar{v}(\mathfrak{M}_i)(\bar{1}\times 1^*)$, $1 \leq i \leq r$, and we can choose an i such that the natural representations of $\bar{A}$ on $\bar{A}/\mathfrak{M}_* \cap \bar{A}$ and $\bar{A}/\mathfrak{M}_{i*} \cap \bar{A}$ are equivalent. Hence we can find an element $v \in \bar{A}$ ($v \notin \mathfrak{M}_* \cap \bar{A}$) such that for any $u \in \bar{A}$, $uv \in \mathfrak{M}_*$ if and only if $u \in \mathfrak{M}_{i*} \cap \bar{A}$. Let $\mathfrak{M}_v$ be the left ideal in $\mathfrak{B}$ consisting of all $b \in \mathfrak{B}$ such that $\bar{v}(b)v \in \mathfrak{M}_*$. Let β be the natural mapping of $\mathfrak{B}$ on $\mathfrak{B}/\mathfrak{M}$ and θ the natural mapping of $\mathfrak{B}_*$ on $\mathfrak{B}_*/\mathfrak{M}_*$. Since $\mathfrak{M} \supset \mathfrak{M}_{A_0}$, ρ is equivalent to the representation of $\mathfrak{B}$ on $\mathfrak{B}_*/\mathfrak{M}_*$ induced by $\bar{v}$, under the natural isomorphism

$$\beta(b) \leftrightarrow \theta(b_*) = \theta(\bar{v}(b)(\bar{1} \times 1^*)) \qquad\qquad (b \in \mathfrak{B})$$

of $\mathfrak{B}/\mathfrak{M}$ with $\mathfrak{B}_*/\mathfrak{M}_*$. Since $v \notin \mathfrak{M}$, $\theta(v) \neq 0$. Hence $\rho(\mathfrak{B})\theta(v) = \mathfrak{B}/\mathfrak{M}$ since ρ is irreducible. Let $\mathfrak{M}'$ be the set of all elements $b \in \mathfrak{B}$ such that $\rho(b)\theta(v) = 0$. Clearly $\mathfrak{M}'$ is a maximal left ideal in $\mathfrak{B}$ and if ρ' is the natural representation of $\mathfrak{B}$ on $\mathfrak{B}/\mathfrak{M}'$, ρ is equivalent to ρ' under the isomorphism $\beta(ba) \leftrightarrow \beta'(b)$ $(b \in \mathfrak{B})$ where β' is the natural mapping of $\mathfrak{B}$ on $\mathfrak{B}/\mathfrak{M}'$ and a is any element in $\mathfrak{B}$ such that $\beta(a) = \theta(v)$. Hence it is sufficient to prove that ρ' is equivalent to ρ_i.

Let $\mathfrak{M}'_* = \bar{v}(\mathfrak{M}')(\bar{1}\times 1^*)$. Then $u \in \mathfrak{M}'_* \cap \bar{A}$ if and only if $u \in \bar{A}$ and $uv \in \mathfrak{M}_*$, that is, if and only if $u \in \mathfrak{M}_{i*} \cap \bar{A}$. Hence $\mathfrak{M}'_* \cap \bar{A} = \mathfrak{M}_{i*} \cap \bar{A}$. We now claim that $\mathfrak{M}' = \mathfrak{M}_i$. For otherwise suppose $\mathfrak{M}' \neq \mathfrak{M}_i$. Since they are both maximal left ideals, $1 \in \mathfrak{M}' + \mathfrak{M}_i$. Hence $\bar{1}\times 1^* = z_* + w_*$ where $z_* \in \mathfrak{M}'_*$, $w_* \in \mathfrak{M}_{i*}$. Since $\mathfrak{M}'_*$ and $\mathfrak{M}_{i*}$ are invariant under $\bar{v}(\mathfrak{X})$, it follows from Lemmas 1, 22, and 23 and the relation $\bar{\mathfrak{A}}\times\mathfrak{X}^* = \sum_A (\bar{\mathfrak{A}}\times\mathfrak{X}^*)_A$, that

$$\mathfrak{M}'_* = \sum_{\Lambda,\mu} \mathfrak{M}'_* \cap \overline{(\mathfrak{A} \times \mathfrak{X}^*)}_{\Lambda,\mu}$$

in the notation of Lemma 23. A similar equation holds for $\mathfrak{M}_{i*}$. Since the sum $\sum_{\Lambda,\mu} (\mathfrak{A} \times \mathfrak{X}^*)_{\Lambda,\mu}$ is direct and $\overline{1} \times 1^* \in \overline{A} = (\mathfrak{A} \times \mathfrak{X}^*)_{\Lambda_0,\Lambda_0}$ it follows that $\overline{1} \times 1^*$ $= z'_* + w'_*$ where z'_* and w'_* respectively are the components of z_* and w_* in $\overline{A}$. From the above remarks it is clear that $z'_* \in \mathfrak{M}'_* \cap \overline{A}$, $w'_* \in \mathfrak{M}_{i*} \cap \overline{A}$. But since $\mathfrak{M}'_* \cap \overline{A} = \mathfrak{M}_{i*} \cap \overline{A}$, we have $\overline{1} \times 1^* \in \mathfrak{M}_{i*} \cap \overline{A}$ which of course is false since $\mathfrak{M}_{i*} \cap \overline{A} \neq \overline{A}$ from Lemma 28. Therefore $\mathfrak{M}_i = \mathfrak{M}'$ and $\rho_i = \rho'$.

Furthermore we know that $\overline{v}(z)(\overline{1} \times 1^*)$ lies in the center of $\overline{A}$ for any $z \in \mathfrak{Z}$. Since the natural representation of $\overline{A}$ on $\overline{A}/\overline{A} \cap \mathfrak{M}_*$ is irreducible it follows from Schur's lemma that we can find $\xi(z) \in C$ such that $\overline{v}(z)(\overline{1} \times 1^*)$ $- \xi(z)(\overline{1} \times 1^*) \in \overline{A} \cap \mathfrak{M}_*$. Hence $z - \xi(z) \in \mathfrak{M}$. Since the mapping $z \rightarrow \xi(z) (z \in \mathfrak{Z})$ is clearly a homomorphism the lemma is proved completely.

It is now easy to deduce Theorem 4. Let V be the representation space of π. Since by hypothesis $V_{\Lambda_0} \neq \{0\}$ we can find an element $\psi \in V_{\Lambda_0}$, $\psi \neq 0$. We may clearly suppose that the space $\pi(\mathfrak{X})\psi$ is irreducible under $\pi(\mathfrak{X})$ and ψ belongs to the highest weight Λ_0. Then it follows that $\pi(\mathfrak{Y}_{\Lambda_0})\psi = 0$. Let $\mathfrak{M}$ be the set of all elements $b \in \mathfrak{B}$ such that $\pi(b)\psi = 0$. Since π is irreducible, $\mathfrak{M}$ is a maximal left ideal and $\mathfrak{M} \supset \mathfrak{A}\mathfrak{Y}_{\Lambda_0}$. Also it is clear that $\mathfrak{M} \supset \mathfrak{N}_x$ and therefore $\mathfrak{M} \supset \mathfrak{X}\mathfrak{N}_x = \mathfrak{N}_x\mathfrak{X}$. Hence $\mathfrak{M} \supset \mathfrak{M}_{\Lambda_0} = \mathfrak{A}\mathfrak{Y}_{\Lambda_0} + \mathfrak{N}_x\mathfrak{X}$. Let ρ be the natural representation of $\mathfrak{B}$ on $\mathfrak{B}/\mathfrak{M}$. Then from Lemma 29, ρ is equivalent to ρ_i for some $1 \leq i \leq r$. Since π is clearly equivalent to ρ the first assertion of the theorem is established.

Now we come to the second part. We have already seen (cf. p. 63) that $\dim (\overline{\mathfrak{A} \times \mathfrak{X}^*})_\Lambda < \infty$ and $\mathfrak{B}/\mathfrak{M} = \theta(\overline{\mathfrak{A} \times \mathfrak{X}^*}) = \sum_\Lambda \theta((\mathfrak{A} \times \mathfrak{X}^*)_\Lambda)$. Let $(\mathfrak{B}/\mathfrak{M})_\Lambda$ denote the set of all elements of $\mathfrak{B}/\mathfrak{M}$ which transform under ρ according to $\mathfrak{D}_\Lambda$. Then it is easily seen that $(\mathfrak{B}/\mathfrak{M})_\Lambda = \theta((\mathfrak{A} \times \mathfrak{X}^*)_\Lambda)$. Hence

$$\dim (\mathfrak{B}/\mathfrak{M})_\Lambda \leq \dim (\overline{\mathfrak{A} \times \mathfrak{X}^*})_\Lambda < \infty.$$

Since the existence of the homomorphism ξ has already been established in Lemma 29, the proof of Theorem 4 is now complete.

Part III. Characters. Let $\mathfrak{L}$ be a semisimple Lie algebra over C and $\mathfrak{B}$ the universal enveloping algebra of $\mathfrak{L}$. Choose a fixed Cartan subalgebra $\mathfrak{h}$ of $\mathfrak{L}$ and a fundamental system of roots $\{\alpha_1, \cdots, \alpha_l\}$ of $\mathfrak{L}$ with respect to $\mathfrak{h}$. Let $\mathfrak{Z}$ denote the center of $\mathfrak{B}$.

DEFINITION. A complex-valued linear function χ on $\mathfrak{B}$ will be called a character if the following conditions are fulfilled:

(1) $\chi(b_1 b_2) = \chi(b_2 b_1)$ for all $b_1, b_2 \in \mathfrak{B}$.

(2) $\chi(1) = 1$ and $\chi(z_1 z_2) = \chi(z_1)\chi(z_2)$ for all $z_1, z_2 \in \mathfrak{Z}$.

Let X_i, $1 \leq i \leq n$, be a base for $\mathfrak{L}$. Put $g_{ij} = \mathrm{sp}(\mathrm{ad}\, X_i\, \mathrm{ad}\, X_j)$. Since $\mathfrak{L}$ is semisimple the matrix $(g_{ij})_{1 \leq i,j \leq n}$ is nonsingular. Let $(g^{ij})_{1 \leq i,j \leq n}$ denote its inverse. Put $X^i = \sum_{1 \leq j \leq n} g^{ij} X_j$.

LEMMA 30. *If χ is a character and $(j_1, \cdots, j_r)$ is any permutation of the set $(1, 2, \cdots, r)$, then*

$$\sum_{1 \leq i_1, \cdots, i_r \leq n} \chi(X_{i_1} X_{i_2} \cdots X_{i_r}) X^{i_{j_1}} X^{i_{j_2}} \cdots X^{i_{j_r}} \in \mathfrak{Z}.$$

The proof of this lemma is exactly parallel to a similar assertion proved in [9, p. 912].

LEMMA 31. *If χ_1 and χ_2 are two distinct characters of $\mathfrak{B}$ we can find a $z \in \mathfrak{Z}$ such that $\chi_1(z) \neq \chi_2(z)$.*

We know that the basic canonical elements[18] $S_{(i_1, \cdots, i_r)} X_{i_1} X_{i_2} \cdots X_{i_r}$, $1 \leq i_1, \cdots, i_r \leq n$, $r \geq 0$, form a base for $\mathfrak{B}$. Since $\chi_1 \neq \chi_2$ we can find an $r \geq 0$ and $1 \leq i_1, \cdots, i_r \leq n$ such that

$$\chi_1\left(\underset{(i_1, \cdots, i_r)}{S} X_{i_1} X_{i_2} \cdots X_{i_r}\right) \neq \chi_2\left(\underset{(i_1, \cdots, i_r)}{S} X_{i_1} X_{i_2} \cdots X_{i_r}\right).$$

Put

$$a_{j_1 j_2 \cdots j_r} = \chi_1\left(\underset{(j_1, \cdots, j_r)}{S} X_{j_1} \cdots X_{j_r}\right) - \chi_2\left(\underset{(j_1, \cdots, j_r)}{S} X_{j_1} \cdots X_{j_r}\right),$$

$$1 \leq j_1, \cdots, j_r \leq n.$$

Then $a_{j_1 j_2 \cdots j_r}$ are symmetric and not all of them are zero. Hence

$$w = \sum_{1 \leq j_1, \cdots, j_r \leq n} a_{j_1 j_2 \cdots j_r} X^{j_1} X^{j_2} \cdots X^{j_r} \neq 0$$

(cf. [9, p. 913]). Also from Lemma 30, $w \in \mathfrak{Z}$. Hence from Theorem 1 of [9, p. 905] we can find a finite-dimensional representation π' of $\mathfrak{B}$ such that $\pi'(w) \neq 0$. Since $\mathfrak{L}$ is semisimple π' is fully reducible. Hence $\pi(w) \neq 0$ for some irreducible component π of π'. Since π is irreducible and $w \in \mathfrak{Z}$, it follows from Schur's lemma that $\pi(w) = c\pi(1)$ $(c \in C)$. Then $c \neq 0$ and therefore sp $\pi(w) \neq 0$. Now put

$$z = \sum_{1 \leq j_1, \cdots, j_r \leq n} \left\{\text{sp } \pi\left(\underset{(j_1, \cdots, j_r)}{S} X_{j_1} X_{j_2} \cdots X_{j_r}\right)\right\} X^{j_1} X^{j_2} \cdots X^{j_r}.$$

If d is the degree of π the function sp $\pi(b)/d$ $(b \in \mathfrak{B})$ is clearly a character. Hence from Lemma 30, $z \in \mathfrak{Z}$. Also

$$\chi_1(z) - \chi_2(z) = \text{sp } \pi(w) \neq 0.$$

Hence the lemma is proved.

The above lemma shows that a character is uniquely determined by its value on $\mathfrak{Z}$. Our object is to obtain all the characters of $\mathfrak{B}$.

Let P be the set of all dominant integral functions on $\mathfrak{h}$. For any $\Lambda \in P$ let π_Λ denote a finite-dimensional irreducible representation of $\mathfrak{B}$ with the

---

[18] The symbol $S_{(i_1, \cdots, i_r)}$ has the same meaning as on p. 55.

highest weight Λ and let d_Λ be the degree of π_Λ. For any root α let $X_\alpha \neq 0$ be the root element corresponding to α. Put $[X_\alpha, X_{-\alpha}] = H_\alpha$, $H_i = H_{\alpha_i}$, $1 \leq i \leq l$. As before we may assume that $\alpha(H_\alpha) = 2$. Let $\mathfrak{H}$ be the subalgebra of $\mathfrak{B}$ generated by H_i, $1 \leq i \leq l$, and 1, and let $C[x]$ be the (commutative) ring of all polynomials in l independent variables $x_1, \cdots, x_l$ with coefficients in C. We know that $H_1^{m_1} H_2^{m_2} \cdots H_l^{m_l}$, $m_1, \cdots, m_l \geq 0$, are linearly independent. Let β denote the isomorphism of $\mathfrak{H}$ onto $C[x]$ defined by $\beta(H_1^{m_1} \cdots H_l^{m_l})$ $= x_1^{m_1} \cdots x_l^{m_l}$, $m_1, \cdots, m_l \geq 0$. Moreover if λ is any linear function on $\mathfrak{h}$ and $f(x) \in C[x]$ we denote by $f(\lambda)$ the value of $f(x)$ at the point $x_i = \lambda(H_i)$, $1 \leq i \leq l$. We shall constantly make use of the following simple lemma which is easily proved by induction on l.

LEMMA 32. *If $f \in C[x]$ $(f \neq 0)$ we can find l integers $\lambda_1, \cdots, \lambda_l$ all greater than or equal to 0 such that $f(\lambda_1, \cdots, \lambda_l) \neq 0$.*

Put $\mathfrak{P} = \sum_{\alpha > 0} \mathfrak{B} X_\alpha$. First we prove a few preliminary lemmas.

LEMMA 33. $\mathfrak{P} \cap \mathfrak{H} = \{0\}$.

Let $h \in \mathfrak{P} \cap \mathfrak{H}$. Suppose $h \neq 0$. Then $\beta(h) = f(x) \neq 0$. Hence from Lemma 32, we can find a $\Lambda \in P$ such that $f(\Lambda) \neq 0$. Let $\psi \neq 0$ be a vector belonging to the highest weight Λ in the representation space of π_Λ. Then clearly $\pi_\Lambda(h)\psi$ $= f(\Lambda)\psi \neq 0$. On the other hand since $h \in \mathfrak{P}$, $\pi_\Lambda(h)\psi = 0$. Thus we get a contradiction and the lemma is proved.

We recall that the adjoint representation ρ of $\mathfrak{B}$ is defined by the relation $\rho(X)b = [X, b]$ $(X \in \mathfrak{L}, b \in \mathfrak{B})$.

LEMMA 34. *The smallest subspace of $\mathfrak{B}$ which contains $\mathfrak{H}$ and which is invariant under the adjoint representation of $\mathfrak{B}$, is $\mathfrak{B}$ itself.*

Let $\mathfrak{M}$ be the smallest subspace satisfying the required conditions. For any α, exp $(t \text{ ad } X_\alpha) = \sigma_\alpha(t)$ $(t \in C)$ is a well defined automorphism of $\mathfrak{L}$. Clearly this can be extended uniquely to an automorphism of $\mathfrak{B}$. Let G be the group generated by $\sigma_\alpha(t)$ for all roots α and all $t \in C$. Then it is known (see Chevalley [4]) that there exists a polynomial f in n variables with coefficients in C such that $f \neq 0$ and if $X = \sum_{1 \leq i \leq n} t_i X_i$ $(t_i \in C)$ and $f(t_1, \cdots, t_n)$ $\neq 0$ we can find a $\sigma \in G$ and an $H \in \mathfrak{h}$ such that $X = \sigma H$. It is clear that $\mathfrak{M}$ is invariant under σ. Therefore since $\mathfrak{H} \subset \mathfrak{M}$ and σ is an automorphism of $\mathfrak{B}$, $X^m = \sigma H^m \in \mathfrak{M}$ $(m \geq 1)$. Let V_m denote the subspace of $\mathfrak{B}$ spanned by $S_{(i_1, \ldots, i_m)} X_{i_1} X_{i_2} \cdots X_{i_m}$, $1 \leq i_1, \cdots, i_m \leq n$; $m \geq 0$. We claim that $V_m \subset \mathfrak{M}$. For $m = 0$ this is immediate since $1 \in \mathfrak{H} \subset \mathfrak{M}$. Hence we may assume $m \geq 1$. It will be sufficient to show that $V_m / V_m \cap \mathfrak{M} = \{0\}$. Suppose this is false. Then we can find a base ω_μ, $1 \leq \mu \leq N$, for $V_m / V_m \cap \mathfrak{M}$. Let θ denote the natural mapping of V_m on $V_m / V_m \cap \mathfrak{M}$. Then

$$\theta(\underset{(i_1, \cdots, i_m)}{S} X_{i_1} X_{i_2} \cdots X_{i_m}) = \sum_{1 \leq \mu \leq N} a_{i_1 i_2 \cdots i_m, \mu} \omega_\mu \qquad (a_{i_1 \cdots i_m, \mu} \in C).$$

Hence

$$\theta(X^m) = \sum_{1 \le i_1, \cdots, i_m \le n} \sum_{1 \le \mu \le N} t_{i_1} t_{i_2} \cdots t_{i_m} a_{i_1 i_2 \cdots i_m, \mu} \omega_\mu.$$

But if $f(t_1, \cdots, t_n) \neq 0$, $X^m = \sigma H^m \in V_m \cap \mathfrak{M}$. Hence

$$\sum_{1 \le i_1, \cdots, i_m \le n} a_{i_1 i_2 \cdots i_m, \mu} t_{i_1} t_{i_2} \cdots t_{i_m} = 0, \qquad\qquad 1 \le \mu \le N,$$

whenever $f(t_1, \cdots, t_n) \neq 0$. Since $a_{i_1 i_2 \cdots i_m, \mu}$ are symmetric in $i_1, \cdots, i_m$ this implies that $a_{i_1 i_2 \cdots i_m, \mu} = 0$. But since $S_{(i_1, \cdots, i_m)} X_{i_1} X_{i_2} \cdots X_{i_m}$ span V_m, we must have $\theta(V_m) = \{0\}$ which contradicts our hypothesis. Hence $V_m \subset \mathfrak{M}$ and this being true for all m, $\mathfrak{B} = \sum_{m \ge 0} V_m \subset \mathfrak{M}$.

Given any linear function λ on $\mathfrak{h}$ we propose to associate with λ a character χ_λ of $\mathfrak{B}$. Put $\rho = (1/2) \sum_{\alpha > 0} \alpha$ and let W be the Weyl group of $\mathfrak{L}$ with respect to $\mathfrak{h}$. Let Δ denote the function $\prod_{\alpha > 0} (\exp ((-1)^{1/2}\alpha/2) - \exp (-(-1)^{1/2}\alpha/2))$ defined on $\mathfrak{h}$. Since every $s \in W$ induces a permutation $\alpha \to s\alpha$ of the roots of $\mathfrak{L}$ it is clear that

$$s\Delta = \prod_{\alpha > 0}\left(\exp \left((-1)^{1/2} \frac{s\alpha}{2} \right) - \exp \left(-(-1)^{1/2} \frac{s\alpha}{2} \right) \right) = \pm \Delta.$$

We say that s is even or odd according as $s\Delta = +\Delta$ or $-\Delta$ and write $\epsilon(s) = 1$ or -1 accordingly. It is known that the Weyl reflexion s_α with respect to a root α is odd (see Weyl [12]).

Let $\mathfrak{F}$ be the space of all linear functions on $\mathfrak{h}$. Then $\mathfrak{F}$ and $\mathfrak{h}$ are dual spaces and every $s \in W$ can be made to act on $\mathfrak{h}$ by duality so that

$$\lambda(sH) = s^{-1}\lambda(H)$$

for all $\lambda \in \mathfrak{F}$ and $H \in \mathfrak{h}$. Let $x_1, \cdots, x_l, t_1, \cdots, t_l$ be $2l$ independent variables. For any $H \in \mathfrak{h}$ let $x(H)$ denote the linear form $\sum_{1 \le i \le l} c_i x_i$, $H = \sum_{1 \le i \le l} c_i H_i$. Also set $sx(H) = x(s^{-1}H)$ and $(sx)_i = sx(H_i)$, $1 \le i \le l$. For any $\lambda \in \mathfrak{F}$ we denote by $\lambda(H_t)$ the linear form $\sum_{1 \le i \le l} \lambda(H_i) t_i$ and write $\lambda(sH_t) = s^{-1}\lambda(H_t)$. Finally we put

$$x(sH_t) = s^{-1}x(H_t) = \sum_{1 \le i \le l} t_i x(sH_i).$$

Consider the power series $\theta(x, t)$ in $x_1, \cdots, x_l, t_1, \cdots, t_l$ with coefficients in C given by[19]

$$\theta(x, t) = \sum_{s \in W} \epsilon(s) \exp ((-1)^{1/2} s x(H_t)).$$

Since $\epsilon(s_\alpha s) = -\epsilon(s)$,

$$2\theta(x, t) = \sum_{s \in W} \epsilon(s) \{ \exp ((-1)^{1/2} s x(H_t)) - \exp ((-1)^{1/2} s_\alpha s x(H_t)) \}.$$

---

[19] As usual $\exp z$ stands for the power series $1 + z + z^2/2! + z^3/3! + \cdots$.

But $s_\alpha s x = s x - (s x(H_\alpha))\alpha$. Hence

$$\exp\left((-1)^{1/2} s x(H_i)\right) - \exp\left((-1)^{1/2} s_\alpha s x(H_i)\right)$$
$$= \exp\left((-1)^{1/2} s x(H_i)\right)\left\{1 - \exp\left(-(-1)^{1/2} s x(H_\alpha)\alpha(H_i)\right)\right\}.$$

Now $1 - \exp\left(-(-1)^{1/2} s x(H_\alpha)\alpha(H_i)\right)$ is divisible by $\alpha(H_i)$ in the ring of power series. Hence $\theta(x, t)$ is divisible by $\alpha(H_i)$. Similarly

$$2\theta(x, t) = \sum_{s \in W} \epsilon(s)\left\{\exp\left((-1)^{1/2} s x(H_i)\right) - \exp\left((-1)^{1/2} s s_\alpha x(H_i)\right)\right\}.$$

Since $s s_\alpha x(H_i) = s_\alpha x(s^{-1} H_i) = x(s^{-1} H_i) - x(H_\alpha)\alpha(s^{-1} H_i)$, we get

$$\exp\left((-1)^{1/2} s x(H_i)\right) - \exp\left((-1)^{1/2} s s_\alpha x(H_i)\right)$$
$$= \exp\left((-1)^{1/2} s x(H_i)\right)\left\{1 - \exp\left(-(-1)^{1/2} x(H_\alpha)\alpha(s^{-1} H_i)\right)\right\}.$$

Since $1 - \exp\left(-(-1)^{1/2} x(H_\alpha)\alpha(s^{-1} H_i)\right)$ is divisible by $x(H_\alpha)$ it follows that $\theta(x, t)$ is divisible also by $x(H_\alpha)$. Now if α, β are roots greater than 0 and $\alpha \neq \beta$, $x(H_\alpha)$, $\alpha(H_i)$, $x(H_\beta)$, $\beta(H_i)$ are all relatively prime. Since a power series ring over C is a unique factorisation domain it follows that $\theta(x, t)$ is divisible by $\prod_{\alpha>0}\{x(H_\alpha)\alpha(H_i)\}$. If r is the number of positive roots of $\mathfrak{L}$, $\prod_{\alpha>0}\{x(H_\alpha)\alpha(H_i)\}$ is of degree r in (x) and (t) each. Since each homogeneous term of $\theta(x, t)$ is clearly of the same degree in (x) and (t) the same must hold for the power series $\theta(x, t)/\prod_{\alpha>0}\{x(H_\alpha)\alpha(H_i)\}$. Now

$$\frac{\exp\left((-1)^{1/2}(\alpha(H_i)/2)\right) - \exp\left(-(-1)^{1/2}(\alpha(H_i)/2)\right)}{\alpha(H_i)}$$

is a unit in the power series ring since its constant term is $(-1)^{1/2} \neq 0$. Consider

$$\frac{\theta(x, t)}{\prod_{\alpha>0}\{x(H_\alpha)\alpha(H_i)\}} \prod_{\alpha>0}\left\{\frac{\exp\left((-1)^{1/2}(\alpha(H_i)/2)\right) - \exp\left(-(-1)^{1/2}(\alpha(H_i)/2)\right)}{\alpha(H_i)}\right\}^{-1}.$$

Since the last factor is a power series in (t) only, it is obvious from the above remark that the product is a power series in (x), (t) such that the coefficient of each power product $t_1^{m_1} t_2^{m_2} \cdots t_l^{m_l}$ is a *polynomial* in (x) whose degree is less than or equal to $m_1 + m_2 + \cdots + m_l$. Put

$$\varphi(x, t) = \prod_{\alpha>0} \rho(H_\alpha) \frac{\theta(x, t)}{\prod_{\alpha>0}\{x(H_\alpha)\alpha(H_i)\}}$$

$$\cdot \prod_{\alpha>0}\left\{\frac{\exp\left((-1)^{1/2}(\alpha(H_i)/2)\right) - \exp\left(-(-1)^{1/2}(\alpha(H_i)/2)\right)}{\alpha(H_i)}\right\}^{-1} \tag{10}$$

$$= \frac{\prod_{\alpha>0}\rho(H_\alpha)}{\prod_{\alpha>0}x(H_\alpha)} \frac{\theta(x, t)}{\Delta(H_i)}$$

where $\Delta(H_t)$ is the power series $\prod_{\alpha>0}\{\exp((-1)^{1/2}(\alpha(H_t)/2)) - \exp(-(-1)^{1/2} \cdot (\alpha(H_t)/2)))\}$. Then

$$(11) \qquad \varphi(x, t) = \sum_{m_1,\cdots,m_l \geq 0} s_{m_1,\cdots,m_l}(x) t_1^{m_1} \cdots t_l^{m_l}$$

where $s_{m_1,\cdots,m_l}(x) \in C[x]$.

Now put $x_i' = x_i + \rho(H_i)$ and consider the power series

$$(12) \qquad \chi(x, t) = \sum_{m_1,\cdots,m_l \geq 0} s_{m_1,\cdots,m_l}(x') t_1^{m_1} t_2^{m_2} \cdots t_l^{m_l}$$

where $s_{m_1,\cdots,m_l}(x')$ is obtained by replacing each x_i in $s_{m_1,\cdots,m_l}(x)$ by x_i', $1 \leq i \leq l$. Then $\chi(x, t)$ is a power series in (t) with coefficients in $C[x]$. Define a linear mapping χ_x of $\mathfrak{H}$ into $C[x]$ by the rule

$$(13) \qquad \chi_x(H_1^{m_1} H_2^{m_2} \cdots H_l^{m_l}) = (1/((-1)^{1/2})^{m_1+\cdots+m_l}) s_{m_1,\cdots,m_l}(x').$$

Since $H_1^{m_1} H_2^{m_2} \cdots H_l^{m_l}$, $m_1, \cdots, m_l \geq 0$, form a base for $\mathfrak{H}$ this defines χ_x completely. Given any $\lambda \in \mathfrak{F}$ and $h \in \mathfrak{H}$ we denote by $\chi_\lambda(h)$ the values of the polynomial $\chi_x(h)$ at the point $x_i = \lambda(H_i)$, $1 \leq i \leq l$. In this way we get a linear mapping χ_λ of $\mathfrak{H}$ into C.

If $\Lambda \in P$ and if π_Λ is the corresponding finite-dimensional irreducible representation of $\mathfrak{B}$ with the highest weight Λ, then it is known (see Weyl [12]) that the degree d_Λ of π_Λ is given by

$$(14) \qquad d_\Lambda = \frac{\prod_{\alpha>0}\Lambda'(H_\alpha)}{\prod_{\alpha>0}\rho(H_\alpha)}$$

where $\Lambda' = \Lambda + \rho$. Moreover

$$(15) \qquad \text{sp}(\exp(-1)^{1/2}\pi_\Lambda(H_t)) = d_\Lambda \sum_{m_1,\cdots,m_l \geq 0} s_{m_1,\cdots,m_l}(\Lambda') t_1^{m_1} t_2^{m_2} \cdots t_l^{m_l}$$

for any $t_1, \cdots, t_l \in C$ such that $|t_1|, \cdots, |t_l| \leq \eta$, η being a suitable real number greater than 0. Here $H_t = \sum_{1 \leq i \leq l} t_i H_i$ and $s_{m_1,\cdots,m_l}(\Lambda')$ is the value of the polynomial $s_{m_1,\cdots,m_l}(x)$ at the point $x_i = \Lambda'(H_i)$, $1 \leq i \leq l$. Also $\exp(-1)^{1/2}\pi_\Lambda(H_t)$ denotes the usual exponential of the matrix $(-1)^{1/2}\pi_\Lambda(H_t)$. Hence by comparing coefficients of the two power series in (15) we get

$$(16) \qquad \frac{1}{d_\Lambda} \text{sp}\, \pi_\Lambda(H_1^{m_1} H_2^{m_2} \cdots H_l^{m_l}) = \frac{1}{((-1)^{1/2})^{m_1+\cdots+m_l}} s_{m_1,\cdots,m_l}(\Lambda')$$

and therefore

$$(17) \qquad \chi_\Lambda(h) = \frac{1}{d_\Lambda} \text{sp}\, \pi_\Lambda(h) \qquad\qquad (h \in \mathfrak{H})$$

in this case.

Let $\mathfrak{B}'$ be the set of all elements in $\mathfrak{B}$ which can be written as linear combination of elements of the form $[b_1, b_2]$ $(b_1, b_2 \in \mathfrak{B})$.

LEMMA 35. *If $h \in \mathfrak{H} \cap \mathfrak{B}'$, $\chi_x(h) = 0$. Furthermore $\chi_x(1) = 1$.*

Let $\chi_x(h) = g(x)$. If $g(x) \neq 0$ we can find a $\Lambda \in P$ such that $g(\Lambda) \neq 0$. Consider the irreducible finite-dimensional representation π_Λ of $\mathfrak{B}$ whose highest weight is Λ. Then $\chi_\Lambda(h) = g(\Lambda) \neq 0$. On the other hand from (17)

$$\chi_\Lambda(h) = \frac{1}{d_\Lambda} \operatorname{sp} \pi_\Lambda(h) = 0$$

since $\in \mathfrak{B}'$. Thus we get a contradiction. Hence $\chi_x(h) = 0$. We prove in exactly the same way that $\chi_x(1) - 1 = 0$.

From Lemma 34 we know that $\mathfrak{B} = \mathfrak{B}' + \mathfrak{H}$. Given any $z \in \mathfrak{B}$ let $z = b + h$ where $b \in \mathfrak{B}'$, $h \in \mathfrak{H}$. If $b' \in \mathfrak{B}'$, $h' \in \mathfrak{H}$ are two other elements such that $z = b' + h'$ then $h - h' \in \mathfrak{H} \cap \mathfrak{B}'$ and therefore by the above lemma $\chi_x(h - h') = 0$. Hence it follows that $\chi_x(h)$ is uniquely determined by z alone. We now extend χ_x to a linear mapping of $\mathfrak{B}$ into $C[x]$ by setting $\chi_x(z) = \chi_x(h)$. For any $\lambda \in \mathfrak{F}$ we define χ_λ on $\mathfrak{B}$ by $\chi_\lambda(z) = (\chi_x(z))_{x_i = \lambda_i}$, $\lambda_i = \lambda(H_i)$, $1 \leq i \leq l$. It is clear that $\chi_x(b) = \chi_\lambda(b) = 0$ for any $b \in \mathfrak{B}'$.

LEMMA 36. *Given any $z \in \mathfrak{Z}$ there exists a unique element $f_z(x) \in C[x]$ such that $z - \beta^{-1}(f_z(x)) \in \mathfrak{P}$. Moreover the mapping $z \to f_z$ $(z \in \mathfrak{Z})$ is an isomorphism of $\mathfrak{Z}$ into $C[x]$. Finally, $f_z = \chi_x(z)$.*

$\alpha_1, \cdots, \alpha_r$ being all the positive roots of $\mathfrak{L}$, put $X_{\alpha_i} = X_i$, $X_{-\alpha_i} = Y_i$, $1 \leq i \leq r$, and

$$z(q, m, p) = Y_1^{q_1} Y_2^{q_2} \cdots Y_r^{q_r} H_1^{m_1} H_2^{m_2} \cdots H_l^{m_l} X_1^{p_1} \cdots X_r^{p_r}$$

as in the proof of Theorem 2. Also we define ranks in $\mathfrak{B}$ exactly as there. Then $z(q, m, p)$ is of rank $\operatorname{rank} p - \operatorname{rank} q$ where $\operatorname{rank} p = \sum_{1 \leq i \leq r} p_i \alpha_i$, $\operatorname{rank} q = \sum_{1 \leq i \leq r} q_i \alpha_i$. It is clear that every $z \in \mathfrak{Z}$ is of zero rank. Hence from Lemma 1, z is a linear combination of $z(q, m, p)$ with $\operatorname{rank} q = \operatorname{rank} p$. But if $\operatorname{rank} p > 0$, $z(q, m, p) \in \mathfrak{P}$. Hence

$$z \equiv \sum_m a(m) z(0, m, 0) \bmod \mathfrak{P} \qquad (a(m) \in C).$$

This shows that there exists an $h \in \mathfrak{H}$ such that $z - h \in \mathfrak{P}$. Put $f(x) = \beta(h)$. Now if $g \in C[x]$ and $z - \beta^{-1}(g) \in \mathfrak{P}$ it follows that

$$\beta^{-1}(g) - h \in \mathfrak{P} \cap \mathfrak{H} = \{0\}$$

from Lemma 32. Hence $g - \beta(h) = g - f = 0$. Therefore f is unique.

Let $z_1, z_2 \in \mathfrak{Z}$. Then if $h_1 = \beta^{-1}(f_{z_1})$, $h_2 = \beta^{-1}(f_{z_2})$

$$z_1z_2 - h_1h_2 = z_2(z_1 - h_1) + h_1(z_2 - h_2) \equiv 0 \bmod \mathfrak{P}.$$

Hence it follows from the uniqueness established above that $f_{z_1z_2} = \beta(h_1h_2)$ $= f_{z_1}f_{z_2}$. Now let $z \in \mathfrak{Z}$, $z \neq 0$. Then we can find a finite-dimensional irreducible representation π_Λ of $\mathfrak{B}$ with the highest weight Λ such that $\pi_\Lambda(z) \neq 0$ (see [9]). Since π_Λ is irreducible it follows from Schur's lemma that $\pi_\Lambda(z) = c\pi_\Lambda(1)$ where $c \in C$. Let $\psi \neq 0$ be a vector in the representation space of π_Λ which belongs to the highest weight Λ. Then since $z - \beta^{-1}(f_z) \in \mathfrak{P}$,

$$\pi_\Lambda(z - \beta^{-1}(f_z))\psi = \{c - f_z(\Lambda)\}\psi = 0.$$

Since $\psi \neq 0$, $f_z(\Lambda) = c \neq 0$. Therefore $f_z \neq 0$. This proves that $z \to f_z$ $(z \in \mathfrak{Z})$ is an isomorphism.

For any $z \in \mathfrak{Z}$ put $g(x) = \chi_x(z) - f_z(x)$. If $g(x) \neq 0$ we can find a $\Lambda \in P$ such that $g(\Lambda) \neq 0$. Then

$$\chi_\Lambda(z) - f_z(\Lambda) = g(\Lambda) \neq 0.$$

Let $\psi \neq 0$ be a vector belonging to the highest weight Λ of π_Λ. Since π_Λ is irreducible and $z \in \mathfrak{Z}$, $\pi_\Lambda(z) = c\pi_\Lambda(1)$ $(c \in C)$ by Schur's lemma. Hence $c = (1/d_\Lambda)\,\mathrm{sp}\,\pi_\Lambda(z)$.

Now

$$\pi_\Lambda(\beta^{-1}(f_z))\psi = f_z(\Lambda)\psi.$$

Since $z - \beta^{-1}(f_z) \in \mathfrak{P}$, $\pi_\Lambda(z - \beta^{-1}(f_z))\psi = 0$. Hence

$$\left\{\frac{1}{d_\Lambda}\,\mathrm{sp}\,\pi_\Lambda(z) - f_z(\Lambda)\right\}\psi = 0.$$

Since $\psi \neq 0$,

$$\frac{1}{d_\Lambda}\,\mathrm{sp}\,\pi_\Lambda(z) - f_z(\Lambda) = \chi_\Lambda(z) - f_z(\Lambda) = g(\Lambda) = 0$$

from (17). Thus we get a contradiction and therefore $\chi_x(z) = f_z(x)$.

We are now in a position to begin the proof of the following theorem.

THEOREM 5. χ_λ *is a character for every linear function* λ *on* $\mathfrak{h}$. *Given any homomorphism* χ *of* $\mathfrak{Z}$ *into* C *such that* $\chi(1) = 1$ *we can find a linear function* λ *on* $\mathfrak{h}$ *such that* $\chi(z) = \chi_\lambda(z)$ *for all* $z \in \mathfrak{Z}$. *If* λ_1, λ_2 *are two linear functions on* $\mathfrak{h}$ *then* $\chi_{\lambda_1} = \chi_{\lambda_2}$ *if and only if* $\lambda_2 + \rho = s(\lambda_1 + \rho)$ *for some* $s \in W$.

COROLLARY. *Every homomorphism* χ *of* $\mathfrak{Z}$ *into* C *such that* $\chi(1) = 1$ *can be extended uniquely to a character of* $\mathfrak{B}$. *Every character of* $\mathfrak{B}$ *is of the form* χ_λ *where* λ *is a linear function on* $\mathfrak{h}$.

This is an immediate consequence of Theorem 5 and Lemma 31.

The first assertion of the theorem follows directly from Lemmas 35 and

36 and the definition of χ_λ. In order to prove the rest we first need some lemmas.

LEMMA 37. *Let $C\{x\}$ be the power series ring in $x_1, \cdots, x_l$ with coefficients in C and let $\mathfrak{N}$ be the ideal in $C\{x\}$ generated by $x_1, \cdots, x_l$. Given any $f(x)$ $\in C[x]$ and an integer $N \geq 1$ we can find a finite number of linear forms $\lambda_j(x)$, $1 \leq j \leq r$, such that*

$$\sum_{1 \leq j \leq r} c_j \exp \lambda_j(x) \equiv f(x) \bmod \mathfrak{N}^N \qquad\qquad (c_j \in C).$$

If $N=1$ we may clearly take $r=1$ and $\lambda_1(x)=0$. Hence we may assume $N \geq 2$ and use induction on N. Notice that

$$x_i - (\exp x_i) + 1 \in \mathfrak{N}^2, \qquad\qquad 1 \leq i \leq l.$$

By induction hypothesis we can find linear forms $\lambda_j(x)$, $1 \leq j \leq r$, and $c_j \in C$ such that

$$\sum_{1 \leq j \leq r} c_j \exp \lambda_j(x) \equiv f(x) \bmod \mathfrak{N}^{N-1}.$$

Hence

$$f(x) \equiv \sum_{1 \leq j \leq r} c_j \exp \lambda_j(x) + g(x) \bmod \mathfrak{N}^N$$

where $g(x)$ is a form of degree $N-1$. Let G denote the power series obtained by replacing each x_i by $\exp (x_i) - 1$ in $g(x)$. Clearly $g(x) \equiv G \bmod \mathfrak{N}^N$. Hence

$$f(x) \equiv \sum_{1 \leq j \leq r} c_j \exp \lambda_j(x) + G \bmod \mathfrak{N}^N.$$

The expression on the right is of the required type since G is clearly a linear combination of exponentials of linear forms. The lemma is therefore proved.

Notice that if $\lambda(x)$ is any given linear form we can find an $H \in \mathfrak{h}$ such that $\lambda(x) = (-1)^{1/2} x(H)$. Hence by the above lemma

$$f(x) \equiv \sum_{1 \leq j \leq r} c_j \exp (-1)^{1/2} x(H_{(j)}) \bmod \mathfrak{N}^N$$

for suitable $c_j \in C$ and $H_{(j)} \in \mathfrak{h}$.

For any $s \in W$ the mapping $x_i \rightarrow (sx)_i = sx(H_i)$, $1 \leq i \leq l$, can be extended uniquely to an automorphism of $C\{x\}$. We denote this extension again by s. Let $f(x)$ be a polynomial such that $sf(x) = f(x)$ for all $s \in W$. Put

$$g(x) = f(x) \prod_{\alpha > 0} x(H_\alpha).$$

It is known (see Weyl [12]) that $\prod_{\alpha > 0} sx(H_\alpha) = \epsilon(s) \prod_{\alpha > 0} x(H_\alpha)$. Hence

$$sg(x) = \epsilon(s) g(x).$$

Now by Lemma 37,

$$g(x) \equiv \sum_{1 \leq j \leq r} c_j \exp\left((-1)^{1/2}x(H_{(j)})\right) \mod \mathfrak{N}^N \qquad (c_j \in C,\ H_{(j)} \in \mathfrak{h}).$$

Since $\mathfrak{N}$ is invariant under W it follows that

$$g(x) = \frac{1}{h} \sum_{s \in W} \epsilon(s)(sg(x))$$

$$\equiv \frac{1}{h} \sum_{1 \leq j \leq r} c_j \sum_{s \in W} \epsilon(s) \exp\left((-1)^{1/2}sx(H_{(j)})\right) \mod \mathfrak{N}^N$$

where h is the order of W. We prove exactly as before that the power series on the right-hand side is divisible by $\prod_{\alpha>0} x(H_\alpha)$. Hence

$$f(x) \equiv \sum_{1 \leq j \leq r} c_j' \left\{ \frac{\sum_{s \in W}\epsilon(s) \exp (-1)^{1/2}sx(H_{(j)})}{\prod_{\alpha>0}x(H_\alpha)} \right\} \mod \mathfrak{N}^M \quad (c_j' \in C)$$

where $N-M$ is the degree of the form $\prod_{\alpha>0} x(H_\alpha)$. Since N could be chosen arbitrarily large the same is true of M. Let us choose M greater than the degree of $f(x)$. Then it follows immediately that $f(x)$ is a finite linear combination of the coefficients of the power series

$$\varphi_0(x,\ t) = \frac{\sum_{s \in W}\epsilon(s) \exp \left((-1)^{1/2}sx(H_i)\right)}{\prod_{\alpha>0}x(H_\alpha)}$$

if we regard it as a power series in $(t_1, \cdots, t_l)$ with coefficients in $C[x]$. Now put $x_i' = x_i + \rho(H_i)$, $1 \leq i \leq l$, as before. From (11) and (12) it is clear that the coefficients of $\chi(x, t)$ are linear combinations of the coefficients of $\varphi_0(x', t)$ and conversely. Hence $f(x')$ is a linear combination of $\chi_x(H_1^{m_1}H_2^{m_2} \cdots H_l^{m_l})$, $m_1, \cdots, m_l \geq 0$. Thus we have proved the following lemma.

LEMMA 38. *If $f(x)$ is a polynomial in $C[x]$ such that $sf(x) = f(x)$ for all $s \in W$ then $f(x') = \chi_x(b)$ for some $b \in \mathfrak{B}$.*

Let $\mathfrak{N}$ be the set of all elements in $C[x]$ of the form $\chi_x(z)$ $(z \in \mathfrak{Z})$. We shall show that $\mathfrak{N}$ coincides with the set of all elements of the form $\chi_x(b)$ $(b \in \mathfrak{B})$.

LEMMA 39. *Let $\mathfrak{B}' = [\mathfrak{B}, \mathfrak{B}]$ as before. Then $\mathfrak{B} = \mathfrak{B}' + \mathfrak{Z}$.*

Let $V = \mathfrak{Z} + \mathfrak{B}'$. Suppose $V \neq \mathfrak{B}$. Then we can find a subspace U of $\mathfrak{B}$ such that $\mathfrak{B} = V + U$, $V \cap U = \{0\}$, $U \neq 0$. Clearly dim U is either finite or countable. Let e_j, $1 \leq j < N$, be a base for U, where N is either a positive integer or ∞. λ being any linear function on $\mathfrak{h}$ we define a linear function χ on $\mathfrak{B}$ as follows:

$$\chi(z) = \chi_\lambda(z),\quad z \in V, \qquad \chi(e_1) = \chi_\lambda(e_1) + 1, \qquad \chi(e_j) = \chi_\lambda(e_j),\quad j \geq 2.$$

Since χ coincides with χ_λ on V it is clear that χ is a character. However

$\chi \neq \chi_\lambda$ since $\chi(e_1) \neq \chi_\lambda(e_1)$. But this contradicts Lemma 31. Hence $V = \mathfrak{B}$.

COROLLARY. $\mathfrak{R}$ *coincides with the set of all polynomials of the form* $\chi_z(b)$ $(b \in \mathfrak{B})$.

For, by the above lemma, $b = z + b'$ when $z \in \mathfrak{Z}$, $b' \in \mathfrak{B}'$. Hence $\chi_z(b) = \chi_z(z) \in \mathfrak{R}$.

LEMMA 40. *Let U be an indeterminate. The coefficients of the polynomial*

$$\prod_{s \in W} (U - sx(H_i) - s\rho(H_i))$$

regarded as a polynomial in U lie in $\mathfrak{R}[t]$.

Consider

$$g(U, x, t) = \prod_{s \in W} (U - sx(H_i)).$$

Clearly this polynomial is invariant under the substitution $x_i \rightarrow (sx)_i$, $1 \leq i \leq l$. Hence it follows from Lemma 38 and the above corollary that the coefficients of $g(U, x', t)$ regarded as a polynomial in U and (t) lie in $\mathfrak{R}$. This proves the lemma.

COROLLARY. $C[x]$ *is integrally dependent on* $\mathfrak{R}$.

On making the substitution $t_j = 0$, $1 \leq j \leq l$, $i \neq j$, $t_i = 1$, we find from the above lemma that x_i' and therefore x_i is integrally dependent on $\mathfrak{R}$. These being true for every i, $1 \leq i \leq l$, the assertion follows.

We can now prove the second part of Theorem 5. χ being any homomorphism of $\mathfrak{Z}$ into C such that $\chi(1) = 1$, put $\chi'(\chi_z(z)) = \chi(z)$ $(z \in \mathfrak{Z})$. Then χ' is a homomorphism of $\mathfrak{R}$ into C and $\chi'(1) = 1$. Since $C[x]$ is integrally dependent on $\mathfrak{R}$ it follows from well known results in algebra that every homomorphism of $\mathfrak{R}$ into C can be extended to a homomorphism of $C[x]$ into C. We denote such an extension of χ' again by χ'. Let λ be the linear function on $\mathfrak{h}$ such that $\chi'(x_i) = \lambda(H_i)$, $1 \leq i \leq l$. Then $\chi(z) = \chi'(\chi_z(z)) = \chi_\lambda(z)$. This proves the second part of the theorem.

For the proof of the last part we proceed as follows. Let $C\{t\}$ and $C[t]$ denote the power series and the polynomial rings respectively in (t) with coefficients in C. We denote by $\partial/\partial t_i$ the uniquely determined C-derivation of $C\{t\}$ such that $\partial t_j/\partial t_i = \delta_{ij}$, $1 \leq i, j \leq l$.

LEMMA 41. *Let $\lambda_1, \cdots, \lambda_r$ be any r distinct linear functions on $\mathfrak{h}$. Then the elements* $\exp(\lambda_i(H_i)) \in C\{t\}$, $1 \leq i \leq r$, *are linearly independent over* $C[t]$, *that is,*

$$\sum_{1 \leq j \leq r} f_j(t) \exp(\lambda_j(H_i)) = 0, \qquad f_j(t) \in C[t]$$

implies $f_j(t) = 0$, $1 \leq j \leq r$.

Suppose the assertion is false. Let m be the least integer such that there exist polynomials $f_j(t)$, $1 \leq j \leq r$, not all zero and all of degree less than or equal to m such that $\sum_{1 \leq j \leq r} f_j(t) \exp (\lambda_j(H_i)) = 0$. Let s be the number of polynomials among these which are not zero and which have degree m. Clearly $s \geq 1$. We choose that particular set of f_j for which s has the least possible value. Then

$$(18) \qquad \sum_{1 \leq j \leq r} f_j(t) \exp (\lambda_j(H_i)) = 0$$

and by differentiation

$$(19) \qquad \sum_{1 \leq j \leq r} \left\{ \lambda_j(H_i) f_j(t) + \frac{\partial f_j(t)}{\partial t_i} \right\} \exp (\lambda_j(H_i)) = 0, \qquad 1 \leq i \leq l.$$

We may assume that $f_1(t) \neq 0$ and its degree is m. Since $\lambda_1, \cdots, \lambda_r$ are all distinct we can find $c_i \in C$, $1 \leq i \leq l$, such that if $H = \sum_{1 \leq i \leq l} c_i H_i$, $\lambda_1(H) \neq \lambda_j(H)$, $2 \leq j \leq r$. Put

$$g_j(t) = \{\lambda_j(H) - \lambda_1(H)\} f_j(t) + \sum_{1 \leq i \leq l} c_i \frac{\partial f_j(t)}{\partial t_i}, \qquad 1 \leq j \leq r.$$

Then from (18) and (19)

$$\sum_{1 \leq j \leq r} g_j(t) \exp \{\lambda_j(H_i)\} = 0.$$

Not all $g_j(t)$ are zero. For otherwise

$$(20) \qquad \{\lambda_j(H) - \lambda_1(H)\} f_j(t) = - \sum_{1 \leq i \leq l} c_i \frac{\partial f_j(t)}{\partial t_i}, \qquad 1 \leq j \leq r.$$

If $j \neq 1$, $\lambda_j(H) \neq \lambda_1(H)$ and (20) is impossible unless $f_j(t) = 0$. But then $f_1(t) \exp (\lambda_1(H_i)) = 0$. Since $C\{t\}$ is an integral domain and since $\exp (\lambda_1(H_i)) \neq 0$, $f_1(t) = 0$. Therefore $f_j(t) = 0$, $1 \leq j \leq r$, which contradicts our hypothesis. Moreover it is clear that the degree of $g_j(t)$ is not greater than that of $f_j(t)$. Also $g_1(t)$ is of degree less than or equal to $m - 1$. Hence at most $s - 1$ polynomials among $g_j(t)$ are of degree m. But this contradicts the definition of s. Hence the lemma is proved.

Using the same notation as before, we consider the power series

$$\varphi_0(x, t) = \frac{\prod_{s \in W} \epsilon(s) \exp ((-1)^{1/2} s x(H_i))}{\prod_{\alpha > 0} x(H_\alpha)}$$

as a power series in (t) with coefficients in $C[x]$. Given a power series $\xi(x, t)$ in (t) with coefficients in $C[x]$ and any $\lambda \in \mathfrak{F}$ we denote by $\xi(\lambda, t)$ the series in $C\{t\}$ obtained by substituting $x_i = \lambda(H_i)$, $i \leq i \leq l$, in the coefficients of $\xi(x, t)$.

LEMMA 42. *Let λ_1, λ_2 be two linear functions on $\mathfrak{h}$ such that $c_1 \varphi_0(\lambda_1, t)$*

$= c_2\varphi_0(\lambda_2, t)$ *where* c_1, c_2 *are complex numbers not both zero. Then* $\lambda_2 = s\lambda_1$ *for some* $s \in W$.

Let $\partial/\partial x_i$ denote differentiation with respect to x_i in $C[x]$ and for any power series $\theta(x, t)$ with coefficients in $C[x]$ let $\partial\theta(x, t)/\partial x_i$ denote the series obtained by differentiating the coefficients of $\theta(x, t)$ with respect to x_i. Then

$$\varphi_0(x, t) \prod_{\alpha>0} x(H_\alpha) = \sum_{s \in W} \epsilon(s) \exp((-1)^{1/2}sx(H_t)).$$

Differentiating r times we get

$$\frac{\partial^r}{\partial x_{i_1}\partial x_{i_2}\cdots\partial x_{i_r}}\left\{\varphi_0(x, t) \prod_{\alpha>0} x(H_\alpha)\right\}$$

$$= \sum_{s \in W} \epsilon(s)((-1)^{1/2})^r \frac{\partial sx(H_t)}{\partial x_{i_1}} \cdots \frac{\partial sx(H_t)}{\partial x_{i_r}} \exp((-1)^{1/2}sx(H_t)).$$

Put

$$f^{(s)}_{i_1i_2\cdots i_r}(t) = ((-1)^{1/2})^r \frac{\partial sx(H_t)}{\partial x_{i_1}} \cdots \frac{\partial sx(H_t)}{\partial x_{i_r}},$$

$$g_{i_1i_2\cdots i_r}(x) = \frac{\partial^r}{\partial x_{i_1}\partial x_{i_2}\cdots\partial x_{i_r}} \prod_{\alpha>0} x(H_\alpha).$$

$f^{(s)}(t) = 1$ and $g(x) = \prod_{\alpha>0} x(H_\alpha)$. Since $g(x) \neq 0$, it is clear that there exists an $r \geq 0$ such that $g_{i_1i_2\cdots i_r}(\lambda_1) \neq 0$ for some i_1, i_2, $\cdots$, i_r. (We define $g_{i_1\cdots i_r}(x) = g(x)$ if $r = 0$. Similarly for $f^{(s)}_{i_1\cdots i_r}(t)$.) Choose the least such r. Then

$$\sum_{s \in W} \epsilon(s)f^{(s)}_{i_1i_2\cdots i_r}(t) \exp((-1)^{1/2}s\lambda_1(H_t))$$

$$= \left[\frac{\partial^r}{\partial x_{i_1}\cdots\partial x_{i_r}}\left\{\varphi_0(x, t) \prod_{\alpha>0} x(H_\alpha)\right\}\right]_{x_i=\lambda_1(H_i)}$$

$$= g_{i_1i_2\cdots i_r}(\lambda_1)\varphi_0(\lambda_1, t).$$

Since $g_{i_1i_2\cdots i_r}(\lambda_1) \neq 0$,

$$\varphi_0(\lambda_1, t) = \frac{\sum_{s \in W}\epsilon(s)f^{(s)}_{i_1i_2\cdots i_r}(t) \exp((-1)^{1/2}s\lambda_1(H_t))}{g_{i_1i_2\cdots i_r}(\lambda_1)}.$$

Similarly

$$\varphi_0(\lambda_2, t) = \frac{\sum_{s \in W}\epsilon(s)f^{(s)}_{j_1\cdots j_{r'}}(t) \exp((-1)^{1/2}s\lambda_2(H_t))}{g_{j_1\cdots j_{r'}}(\lambda_2)}.$$

where $g_{j_1j_2\cdots j_{r'}}(\lambda_2) \neq 0$. It is clear from the definition of $\varphi_0(x, t)$ that

$$\varphi_0(x,\, t) = \varphi(x,\, t)\, \frac{\Delta(H_t)}{\prod_{\alpha>0}\rho(H_\alpha)},$$

where $\varphi(x,\, t)$ is defined by (10). Since the coefficient of 1 in $\varphi(x,\, t)$ is $\chi_z(1)=1$, it follows that $\varphi(\lambda_1,\, t)\neq 0$ and therefore $\varphi_0(\lambda_1,\, t)\neq 0$. Similarly $\varphi_0(\lambda_2,\, t)\neq 0$. On the other hand $c_1\varphi_0(\lambda_1,\, t)-c_2\varphi_0(\lambda_2,\, t)=0$ where $c_1,\, c_2$ are not both zero. Hence from Lemma 41, $s_1\lambda_1=s_2\lambda_2$ for some $s_1,\, s_2\in W$. Therefore $\lambda_2=s\lambda_1$ where $s=s_2^{-1}s_1$.

Now we can complete the proof of Theorem 5. Let $\chi(x,\, t)$ be defined as in (13). Then $\chi_{\lambda_1}=\chi_{\lambda_2}$ implies that $\chi(\lambda_1,\, t)=\chi(\lambda_2,\, t)$, and therefore $\varphi_0(\lambda_1+\rho, t)=\varphi_0(\lambda_2+\rho,\, t)$. Hence from Lemma 42, $\lambda_2+\rho=s(\lambda_1+\rho)$ for some $s\in W$, and Theorem 5 is proved completely.

Notice that in the definition of the series $\chi(x,\, t)$ on p. 71 we made use of the positive roots rather than of the negative roots. We could have equally well considered the series

$$\chi^-(x,\, t) = \frac{\sum_{s\in W}\epsilon(s)\,\exp\,((-1)^{1/2}sx''(H_t))}{\prod_{\alpha<0}x''(H_\alpha)\Delta'(H_t)}\,\prod_{\alpha<0}\rho'(H_\alpha),$$

$x_i''=x_i+\rho'(H_i),\, 1\leq i\leq l,\, \rho'=2^{-1}\sum_{\alpha<0}\alpha=-\rho,\, \Delta'=\prod_{\alpha<0}(e^{(-1)^{1/2}\alpha/2}-e^{-(-1)^{1/2}\alpha/2})$. Then

$$(21)\qquad\qquad \chi^-(-x,\, t) = \overline{\chi(x,\, t)} = \chi(x,\, -t)$$

where $\overline{\chi(x,\, t)}$ is obtained from $\chi(x,\, t)$ by changing the coefficients to their complex conjugates. Let $((-1)^{1/2})^{m_1+\cdots+m_l}\chi_z^-(H_1^{m_1}\cdots H_l^{m_l})$ denote the coefficient of $t_1^{m_1}\cdots t_l^{m_l}$ in $\chi^-(x,\, t)$. We define $\chi_z^-(h)$ $(h\in\mathfrak{H})$ by linearity and extend χ_z^- to a linear mapping of B into $C[x]$ exactly as in the case of χ_z by setting $\chi_z^-(z)=\chi_z^-(h)$ where $z=h+b$ $(h\in\mathfrak{H},\, b\in\mathfrak{B}'=[\mathfrak{B},\,\mathfrak{B}],\, z\in\mathfrak{B})$. Let $\mathfrak{N}=\sum_{\alpha<0}\mathfrak{B}X_\alpha$. Then corresponding to Lemma 36, we prove that $z-\beta^{-1}(\chi_z^-(z))\in\mathfrak{N}$ $(z\in\mathfrak{Z})$. Let φ denote the linear mapping $X\to -X$ $(X\in\mathfrak{L})$ of $\mathfrak{L}$ into itself. Since $\varphi([X,\, X'])=[X',\, X]=[\varphi(X'),\,\varphi(X)]$ $(X,\, X'\in\mathfrak{L})$, φ can be extended uniquely to an anti-automorphism of $\mathfrak{B}$. Clearly $\varphi(\mathfrak{Z})=\mathfrak{Z}$. Now for any $z\in\mathfrak{Z}$,

$$z - \beta^{-1}(\chi_z(z)) \in \mathfrak{P}.$$

Hence

$$z - \beta^{-1}(\chi_z(z)) = \sum_{\alpha>0} z_\alpha X_\alpha$$

where z_α is of rank $-\alpha$. Hence

$$\varphi(z) - \varphi(\beta^{-1}(\chi_z(z))) = \sum_{\alpha>0}\varphi(X_\alpha)\varphi(z_\alpha).$$

But

$$[H,\,\varphi(z_\alpha)] = -[\varphi(H),\,\varphi(z_\alpha)] = \varphi([H,\, z_\alpha]) = -\alpha(H)\varphi(z_\alpha)$$

for any $H \in \mathfrak{h}$. Hence $\varphi(z_\alpha)$ is of rank $-\alpha$ and therefore lies in $\mathfrak{N}$. Hence

$$(22) \qquad \chi_{\bar{x}}^-(\varphi(z)) = \beta(\varphi(\beta^{-1}(\chi_x(z)))) = \chi_{-x}(z)$$

where the polynomial $\chi_{-x}(z)$ is obtained from $\chi_x(z)$ by the substitution $x_i \rightarrow -x_i$, $1 \leq i \leq l$. For any $\lambda \in \mathfrak{F}$, we denote by χ_λ^- the linear function on $\mathfrak{B}$ such that $\chi_\lambda^-(z) = f(\lambda)$ where $f(x) = \chi_x^-(z)$ $(z \in \mathfrak{B})$. We prove exactly as in the case of χ_λ that χ_λ^- is a character. Therefore from Theorem 5, $\chi_\lambda^- = \chi_\mu$ for some $\mu \in \mathfrak{F}$. But then

$$\chi^-(\lambda, t) = \sum_{m_1, \cdots, m_l \geq 0} ((-1)^{1/2})^{m_1 + \cdots + m_l} \chi_\lambda^- (H_1^{m_1} \cdots H_l^{m_l}) t_1^{m_1} \cdots t_l^{m_l}$$

$$= \sum_{m_1, \cdots, m_l \geq 0} ((-1)^{1/2})^{m_1 + \cdots + m_l} \chi_\mu (H_1^{m_1} \cdots H_l^{m_l}) t_1^{m_1} \cdots t_l^{m_l} = \chi(\mu, t).$$

Hence from Lemma 42, $\lambda - \rho = s(\mu + \rho)$ for some $s \in W$. Changing λ to $-\lambda$, we get the result that $\chi_{-\lambda}^- = \chi_\mu$ if and only if

$$(23) \qquad -(\lambda + \rho) = s(\mu + \rho).$$

Let π be a representation of $\mathfrak{B}$ and χ a character of $\mathfrak{B}$. We shall say that π has the character χ if $\pi(z - \chi(z)) = 0$ for all $z \in \mathfrak{Z}$. From the corollary to Theorem 5 it is clear that π has a uniquely determined character provided $\pi(z)$ is a multiple of $\pi(1)$ for every $z \in \mathfrak{Z}$.

We shall now apply the above results to the situation considered in Theorem 4. From now on we adhere strictly to the notation of Part II. In particular $\mathfrak{L}$, $\mathfrak{g}$, $\mathfrak{g}^+$, $\mathfrak{L}_K$, $\mathfrak{B}$, $\mathfrak{A}$, and $\mathfrak{X}$ have the same meaning as there. $\mathfrak{A}^+$ is the algebra generated by $\mathfrak{g}^+$ and 1. The isomorphisms γ and γ^+ of $\mathfrak{L}_K$ with $\mathfrak{g}$ and $\mathfrak{g}^+$ respectively have been defined on p. 53. Then $\gamma(\mathfrak{h}_K)$ is a Cartan subalgebra of $\mathfrak{g}$ and every linear function λ on $\mathfrak{h}_K$ can also be regarded as a linear function on $\gamma(\mathfrak{h}_K)$ and conversely by the rule $\lambda(\gamma(H)) = \lambda(H)$ $(H \in \mathfrak{h}_K)$. Under this correspondence roots of $\mathfrak{L}_K$ with respect to $\mathfrak{h}_K$ are also roots of $\mathfrak{g}$ with respect to $\gamma(\mathfrak{h}_K)$. Hence the Weyl group W of $\mathfrak{L}_K$ is also the Weyl group of $\mathfrak{g}$. Similar remarks hold for $\mathfrak{g}^+$. Since $\mathfrak{L}$ is the direct sum of $\mathfrak{g}$ and $\mathfrak{g}^+$, $\mathfrak{h} = \gamma(\mathfrak{h}_K) + \gamma^+(\mathfrak{h}_K)$ is a Cartan subalgebra of $\mathfrak{L}$. Every linear function ν on $\mathfrak{h}$ can be regarded as a pair (λ, μ) of linear functions on $\mathfrak{h}_K$ by the rule

$$\nu(H) = \lambda(H_1) + \mu(H_2)$$

where $H = \gamma(H_1) + \gamma^+(H_2)$ $(H \in \mathfrak{h}, H_1, H_2 \in \mathfrak{h}_K)$. It is easily seen that the roots of $\mathfrak{L}$ are exactly those pairs which are of the form $(\alpha, 0)$ or $(0, \alpha)$ where α is a root of $\mathfrak{L}_K$. Let $\{\alpha_1, \cdots, \alpha_l\}$ be a fundamental system of roots of $\mathfrak{L}_K$ with respect to $\mathfrak{h}_K$. Then the set $(\alpha_i, 0)$ $(0, \alpha_i)$, $1 \leq i \leq l$, is a fundamental system of roots of $\mathfrak{L}$ with respect to $\mathfrak{h}$ and therefore the Weyl group of $\mathfrak{L}$ is the direct product $W \times W$.

Let (λ, μ) be a pair of linear functions on $\mathfrak{h}_K$. This pair defines a linear function on $\mathfrak{h}$ and therefore a character of $\mathfrak{B}$. We denote this character by $\chi_{\lambda,\mu}$. Let X_α be a root element in $\mathfrak{L}_K$ corresponding to the root α. Put $[X_\alpha, X_{-\alpha}] = H_\alpha$, $H_i = H_{\alpha_i}$, $1 \leq i \leq l$. Let $x_1, \cdots, x_l, y_1, \cdots, y_l, t_1, \cdots, t_l, u_1, \cdots, u_l$ be independent variables. Then corresponding to (12) we have to consider the power series

$$\chi(x, y, t, u)$$
$$(24) = \left\{ \prod_{\alpha>0} \rho(H_\alpha) \right\}^2 \frac{\sum_{s,\sigma \in W} \epsilon(s)\epsilon(\sigma) \exp \left((-1)^{1/2} s x'(H_t) + (-1)^{1/2} \sigma y'(H_u) \right)}{\prod_{\alpha>0} x'(H_\alpha) \prod_{\alpha>0} y'(H_\alpha) \Delta(H_t) \Delta(H_u)}$$
$$= \chi(x, t)\chi(y, t)$$

where the notation is analogous to that of pp. 71–72. $x_i' = x_i + \rho(H_i)$, $y_i' = y_i + \rho(H_i)$, and $\rho = 2^{-1}\sum_{\alpha>0} \alpha$. We define the linear mapping $\chi_{x,y}$ of $\mathfrak{B}$ into $C[x, y]$ exactly as on p. 72. Similarly let χ_x and χ_y denote the corresponding linear mappings of $\mathfrak{X}$ into $C[x]$ and $C[y]$ respectively. Whenever necessary we shall also regard χ_x and χ_y as linear mappings of $\mathfrak{A}$ and $\mathfrak{A}^+$ so that

$$\chi_x(\gamma(\omega)) = \chi_x(\omega), \qquad \chi_y(\gamma^+(\omega)) = \chi_y(\omega) \qquad (\omega \in \mathfrak{X})$$

where the isomorphisms γ and γ^+ have now been extended (uniquely) on $\mathfrak{X}$. It follows from (24) that

$$(25)\ \chi_{x,y}(\gamma(H_1^{m_1} \cdots H_l^{m_l})\gamma^+(H_1^{n_1} \cdots H_l^{n_l})) = \chi_x(H_1^{m_1} \cdots H_l^{m_l})\chi_y(H_1^{n_1} \cdots H_l^{n_l}),$$
$$m_i, n_i \geq 0.$$

Let χ denote the linear mapping of $\mathfrak{B}$ into $C[x, y]$ such that $\chi(aa^+) = \chi_x(a)\chi_y(a^+)$ $(a \in \mathfrak{A}, a^+ \in \mathfrak{A}^+)$. Then clearly $\chi([b_1, b_2]) = 0$ for any $b_1, b_2 \in \mathfrak{B}$. Also χ coincides with $\chi_{x,y}$ on the algebra $\mathfrak{H}$ generated by $\mathfrak{h}$ and 1. Hence from Lemma 34, $\chi = \chi_{x,y}$ and therefore $\chi_{x,y}(aa^+) = \chi_x(a)\chi_y(a^+)$ $(a \in \mathfrak{A}, a^+ \in \mathfrak{A}^+)$. We express this relation symbolically in the form $\chi_{x,y} = \chi_x \times \chi_y$. It is clear that $\chi_{\lambda,\mu}(aa^+) = \chi_\lambda(a)\chi_\mu(a^+)$. Hence we again write $\chi_{\lambda,\mu} = \chi_\lambda \times \chi_\mu$.

Let $\mathfrak{Z}$, $\mathfrak{C}$, and $\mathfrak{C}^+$ be the centers of $\mathfrak{B}$, $\mathfrak{A}$, and $\mathfrak{A}^+$ respectively. Clearly $\chi_{x,y}(z) = \chi_x(z)$ if $z \in \mathfrak{C}$ and $\chi_{x,y}(z^+) = \chi_y(z^+)$ if $z^+ \in \mathfrak{C}^+$. Hence $\chi_{x,y}(\mathfrak{Z})$ contains $\chi_x(\mathfrak{C})$ and $\chi_y(\mathfrak{C})$. Let Ω be the set of all elements $\chi_{x,y}(z)$ $(z \in \mathfrak{Z})$ such that [20] $\nu(z)(1 \times 1^*) = 0$.

LEMMA 43. *Let Λ_i, $0 \leq i \leq N$, be all the distinct weights of $\mathfrak{D}_{\Lambda_0}$. Then the coefficients of the polynomial*

$$F(x, y, t) = \prod_{0 \leq i \leq N} \prod_{s,\sigma \in W} (\sigma y(H_t) + \sigma\rho(H_t) + sx(H_t) + s\rho(H_t) - \Lambda_i(H_t))$$

regarded as a polynomial in (t) are all in Ω.

Exactly as in the proof of Lemma 40, we show that these coefficients lie

---

[20] We recall that ν is the representation of $\mathfrak{B}$ on $\mathfrak{A} \times \mathfrak{X}^*$ (see p. 59). Λ_0 is a fixed dominant integral function on $\mathfrak{h}_K$ and $\mathfrak{X}^* = \mathfrak{X}/\mathfrak{Y}_{\Lambda_0}$.

in the ring generated by $\chi_x(\mathbb{C})$ and $\chi_y(\mathbb{C}^+)$ and therefore in $\chi_{x,y}(\mathfrak{Z})$. Therefore we can write

$$F(x, y, t) = \sum_{m_1, \cdots, m_l \geq 0} \chi_{x,y}(z(m_1, \cdots, m_l)) t_1^{m_1} t_2^{m_2} \cdots t_l^{m_l}, \quad z(m_1, \cdots, m_l) \in \mathfrak{Z}.$$

We have to show that $\nu(z(m_1, \cdots, m_l))(1 \times 1^*) = 0$, $m_1, \cdots, m_l \geq 0$. Suppose this is false. Then for some z among $z(m_1, \cdots, m_l)$

$$\nu(z)(1 \times 1^*) \neq 0.$$

Choose ω_i, $1 \leq i \leq p$, in $\mathfrak{X}$ such that ω_i^* form a base for $\mathfrak{X}^*$. Then

$$\nu(z)(1 \times 1^*) = \sum_{1 \leq i \leq p} a_i \times \omega_i^*, \qquad a_i \in \mathfrak{A}.$$

Since $\nu(z)(1 \times 1^*) \neq 0$ we can find an a_i, say a_1, such that $a_1 \neq 0$. By Theorem 1 of [9] there exists a finite-dimensional representation π of $\mathfrak{A}$ such that $\pi(a_1) \neq 0$. Clearly we may choose π so that it is irreducible.

Let θ denote the linear mapping of $\mathfrak{g}^+$ on $\mathfrak{g}$ given by $\theta(\gamma^+(X)) = -\gamma(X)$ $(X \in \mathfrak{L}_K)$. It is easily verified that

$$\theta([Z_1^+, Z_2^+] = - [\theta(Z_1^+), \theta(Z_2^+)] \qquad (Z_1^+, Z_2^+ \in \mathfrak{g}^+).$$

Hence θ can be extended uniquely to an anti-isomorphism of $\mathfrak{A}^+$ onto $\mathfrak{A}$. Let σ denote the representation of $\mathfrak{B}$ on $\pi(\mathfrak{A})$ defined by

$$\sigma(a')\pi(a) = \pi(a'a), \qquad \sigma(X)\pi(a) = \pi([X, a]) \qquad (X \in \mathfrak{L}_K; a', a \in \mathfrak{A}).$$

Then

$$\sigma(Z^+)\pi(a) = \sigma(Z^+)\sigma(a)\pi(1) = \sigma(a)\sigma(Z^+)\pi(1)$$
$$= \pi(a)\pi(\theta(Z^+)) \qquad (Z^+ \in \mathfrak{g}^+, a \in \mathfrak{A})$$

since $[Z^+, a] = 0$ and $Z^+ = X - \gamma(X)$ if $Z^+ = \gamma^+(X)$ $(X \in \mathfrak{L}_K)$. From this it follows easily that

$$\sigma(a^+)\pi(a) = \pi(a)\pi(\theta(a^+))$$
$$\sigma(a')\pi(a) = \pi(a')\pi(a) \qquad (a^+ \in \mathfrak{A}^+; a, a' \in \mathfrak{A}).$$

Since π is irreducible, $\pi(\mathfrak{A})$ is a simple algebra. Therefore it has no ideals other than $\{0\}$ and itself. Hence σ is an irreducible representation.

Let P_K be the set of all dominant integral functions on $\mathfrak{h}_K$. For any $\Lambda_1, \Lambda_2 \in P_K$ let φ_1, φ_2 be two irreducible representations of $\mathfrak{L}_K$ belonging to $\mathfrak{D}_{\Lambda_1}$ and $\mathfrak{D}_{\Lambda_2}$ respectively. Extend φ_1, φ_2 to representations of $\mathfrak{L}$ by setting $\varphi_1(Z^+) = 0$ for all $Z^+ \in \mathfrak{g}^+$ and $\varphi_2(Z) = 0$ for all $Z \in \mathfrak{g}$. The representation $\varphi_1 + \varphi_2$ of $\mathfrak{L}$ is then known to be irreducible. We denote by $\mathfrak{D}_{\Lambda_1, \Lambda_2}$ the class of all representations of $\mathfrak{L}$ equivalent to $\varphi_1 + \varphi_2$. It is known that $\mathfrak{D}_{\Lambda_1, \Lambda_2} \neq \mathfrak{D}_{\Lambda_1', \Lambda_2'}$ unless $\Lambda_1 = \Lambda_1'$, $\Lambda_2 = \Lambda_2'$, and every irreducible finite-dimensional representation of $\mathfrak{L}$ is contained in some $\mathfrak{D}_{\Lambda_1,\Lambda_2}$. Hence $\sigma \in \mathfrak{D}_{\lambda,\mu}$ for some $\lambda,\ \mu \in P_K$. By Schur's lemma, $\sigma(z) = \chi(z)\sigma(1)$ $(z \in \mathfrak{Z})$ where $z \to \chi(z)$ is a homomorphism of $\mathfrak{Z}$ into C. Hence by Theorem 5, χ can be extended uniquely to a character of $\mathfrak{B}$. From the theory of finite-dimensional representations it is known that this character is $\chi_{\lambda,\mu}$. Finally we note that the zero representation of $\mathfrak{L}_K$ occurs in σ since $\sigma(X)\pi(1) = 0$ $(X \in \mathfrak{L}_K)$ and $\pi(1) \neq 0$. Hence again it follows from known theory that

$$\chi(\mu,\ t) = \overline{\chi(\lambda,\ t)}$$

where $\overline{\chi(\lambda,\ t)}$ is obtained by changing the coefficients of the series $\chi(\lambda,\ t)$ (regarded as a series in (t)) to their complex conjugates. Applying Lemma 42, we get immediately

$$\mu + \rho = -s(\lambda + \rho) \qquad\qquad (s \in W).$$

Now consider the representation π^+ of $\mathfrak{B}$ induced on $\pi(\mathfrak{A}) \times \mathfrak{X}^*$. It is given by

$$\pi^+(X)(\pi(a) \times \omega^*) = \pi([X,\ a]) \times \omega^* + \pi(a) \times (X\omega)^*,$$

$$\pi^+(Z)(\pi(a) \times \omega^*) = \pi(Z)\pi(a) \times \omega^* \qquad (X \in \mathfrak{L}_K,\ \omega \in \mathfrak{X},\ a \in \mathfrak{A},\ Z \in \mathfrak{g}).$$

Let $\mathfrak{N}_\pi$ be the kernel of π in $\mathfrak{A}$. It is easily seen that if ξ is the natural mapping of $\mathfrak{A} \times \mathfrak{X}^*$ on $\mathfrak{A} \times \mathfrak{X}^*/\mathfrak{N}_\pi \times \mathfrak{X}^*$ then π^+ is equivalent to the representation of $\mathfrak{B}$ induced by ν on $\mathfrak{A} \times \mathfrak{X}^*/\mathfrak{N}_\pi \times \mathfrak{X}^*$ under the isomorphism $\pi(a) \times \omega^* \leftrightarrow \xi(a \times \omega^*)$ $(a \in \mathfrak{A},\ \omega^* \in \mathfrak{X}^*)$ of $\pi(\mathfrak{A}) \times \mathfrak{X}^*$ with $\mathfrak{A} \times \mathfrak{X}^*/\mathfrak{N}_\pi \times \mathfrak{X}^*$. We may therefore identify $\pi(\mathfrak{A}) \times \mathfrak{X}^*$ and $\mathfrak{A} \times \mathfrak{X}^*/\mathfrak{N}_\pi \times \mathfrak{X}^*$ under this isomorphism. Since $\pi(a_1) \neq 0$, $\pi^+(z)(\pi(1) \times 1^*) = \sum_{1 \leq i \leq N} \pi(a_i) \times \omega^* \neq 0$. Since $\pi(\mathfrak{A}) \times \mathfrak{X}^*$ is finite-dimensional it is fully reducible under $\pi^+(\mathfrak{B})$. Then it follows easily that we can find a maximal invariant[21]subspace $\mathfrak{M}_*$ of $\pi(\mathfrak{A}) \times \mathfrak{X}^*$ such that $\pi^+(z)(\pi(1) \times 1^*) \notin \mathfrak{M}_*$. Let π' be the irreducible representation of $\mathfrak{B}$ induced on $(\pi(\mathfrak{A}) \times \mathfrak{X}^*)/\mathfrak{M}_*$. Let $\pi' \in \mathfrak{D}_{\lambda',\mu'}$ $(\lambda',\ \mu' \in P_K)$. Since the representation of $\mathfrak{B}$ on $\mathfrak{X}^*$ is of the type $\mathfrak{D}_{0,\Lambda_0}$ it follows from known results that $\lambda' = \lambda$ and $\mu' = \mu + \Lambda_i$ where Λ_i is some weight of $\mathfrak{D}_{\Lambda_0}$. Since $\mu + \rho = -s(\lambda + \rho)$ $(s \in W)$,

$$s\lambda'(H_i) + s\rho(H_i) + \mu'(H_i) + \rho(H_i) - \Lambda_i(H_i) = 0.$$

Therefore $F(\lambda',\ \mu',\ t) = 0$. On the other hand

$$F(x,\ y,\ t) = \sum_{m_1,\cdots,m_l \geq 0} \chi_{x,y}(z(m_1,\ \cdots,\ m_l)) t_1^{m_1} t_2^{m_2} \cdots t_l^{m_l}.$$

Hence

$$F(\lambda',\ \mu',\ t) = \sum_{m_1,\cdots,m_l \geq 0} \chi_{\lambda',\mu'}(z(m_1,\ \cdots,\ m_l)) t_1^{m_1} \cdots t_l^{m_l}.$$

---

[21] $\mathfrak{M}_*$ is maximal independently of the property of not containing $\pi^+(z)(\pi(1) \times 1^*)$.

Since $\pi^+(z)(\pi(1)\times 1^*)\not\in\mathfrak{M}_*$, $\pi'(z)\neq 0$. Therefore as $z\in\mathfrak{Z}$, $\pi'(z)=\chi_{\lambda',\mu'}(z)\pi'(1)$ $\neq 0$. Hence $\chi_{\lambda',\mu'}(z)\neq 0$. Since z is one of the elements $z(m_1,\cdots,m_l)$, it follows that $F(\lambda',\mu',t)\neq 0$. Thus we get a contradiction. This proves the lemma.

Let π be any representation of $\mathfrak{B}$ with the character χ such that $\mathfrak{D}_{\Lambda_0}$ occurs in π. Let $f(x,y)\in\Omega$. Then $f(x,y)=\chi_{x,y}(z)$ where $z\in\mathfrak{Z}$ and $\nu(z)(1\times 1^*)=0$. Let V be the representation space of π. By hypothesis $V_{\Lambda_0}\neq 0$. Choose ψ $\in V_{\Lambda_0,\Lambda_0}$, $\psi\neq 0$, and let $\mathfrak{M}$ be the left ideal in $\mathfrak{B}$ consisting of all $b\in\mathfrak{B}$ such that $\pi(b)\psi=0$. Then $\mathfrak{M}\supset\mathfrak{AY}_{\Lambda_0}$. Since $\nu(z)(1\times 1^*)=0$, $z\in\mathfrak{AY}_{\Lambda_0}$ from Lemma 27. Hence $z\in\mathfrak{M}$ and $\pi(z)=0$. Therefore $f(\lambda,\mu)=\chi_{\lambda,\mu}(z)=0$. Applying this to the polynomial $F(x,y,t)$ of the above lemma we find that

$$s\lambda(H_i)+s\rho(H_i)+\sigma\mu(H_i)+\sigma\rho(H_i)-\Lambda_i(H_i)=0$$

for some $s,\sigma\in W$ and i, $0\leq i\leq N$. Now for every Λ_i and $\tau\in W$, $\tau\Lambda_i$ is also a weight of $\mathfrak{D}_{\Lambda_0}$. Hence

$$\lambda(H_i)+\rho(H_i)+\tau\mu(H_i)+\tau\rho(H_i)-\Lambda_j(H_i)=0$$

for some $\tau\in W$ and some j, $0\leq j\leq N$. Put $\tau(\mu+\rho)=-(\mu'+\rho)$. Then $\lambda-\mu'=\Lambda_j$ and $\chi_\mu=\chi^-_{\mu'}$ from (23). Thus we have proved the following theorem.

THEOREM 6. *Let π be a representation of $\mathfrak{B}$ with the character χ such that $\mathfrak{D}_{\Lambda_0}$ occurs in π ($\Lambda_0\in P_K$). Then $\chi=\chi_\lambda\times\chi^-_\mu$ where $\lambda-\mu$ is an integral function on $\mathfrak{h}_K$ which is a weight of $\mathfrak{D}_{\Lambda_0}$.*

COROLLARY. *Let π be any representation of $\mathfrak{B}$ with the character $\chi=\chi_\lambda$ $\times\chi^-_\mu$. For any $\Lambda\in P_K$, $\mathfrak{D}_\Lambda$ cannot occur in π unless $(\lambda+\rho)-\sigma(\mu+\rho)$ is a weight of $\mathfrak{D}_\Lambda$ for some $\sigma\in W$.*

Let $\mathfrak{L}_0$ and $\mathfrak{L}_{K,0}$ be the Lie algebras over the field R of real numbers as defined on p. 52. Henceforward we suppose that $\mathfrak{L}_{K,0}$ is compact, that is, the quadratic form sp $(\mathrm{ad}\,X)^2$ $(X\in\mathfrak{L}_{K,0})$ is negative definite. Let $\mathfrak{B}_0$ be the universal enveloping algebra of $\mathfrak{L}_0$. Then $\mathfrak{B}_0$ is an algebra over R and $\mathfrak{B}$ can be regarded, in the obvious way, as the extension of $\mathfrak{B}_0$ over C. Then every $b\in\mathfrak{B}$ can be written uniquely in the form $b=b_1+(-1)^{1/2}b_2$ $(b_1,b_2\in\mathfrak{B}_0)$. Let $\mathfrak{Z}_0$ be the center of $\mathfrak{B}_0$. It is easy to show (see [9, p. 914]) that the elements of $\mathfrak{Z}_0$ span over C the center $\mathfrak{Z}$ of $\mathfrak{B}$. Let φ denote the linear mapping of $\mathfrak{L}_0$ into itself given by $\varphi(X)=-X$ $(X\in\mathfrak{L}_0)$. φ can be extended uniquely to an anti-automorphism of $\mathfrak{B}_0$. We now extend φ on $\mathfrak{B}$ as follows. If $b=b_1+(-1)^{1/2}b_2$ $(b_1,b_2\in\mathfrak{B})$ we put $\varphi(b)=\varphi(b_1)-(-1)^{1/2}\varphi(b_2)$. Let χ be any character of $\mathfrak{B}$. We shall denote by χ^* the linear function on $\mathfrak{B}$ defined as follows:

$$(26)\qquad\qquad\qquad\chi^*(b)=\overline{\chi(\varphi(b))}\qquad\qquad\qquad(b\in\mathfrak{B})$$

where the bar denotes complex conjugate. It is easy to verify that χ^* is also a character.

Let $\mathfrak{h}_{K,0}$ be a maximal abelian subalgebra of $\mathfrak{L}_{K,0}$ and $\mathfrak{h}_K$ the complexification of $\mathfrak{h}_{K,0}$. Then $\mathfrak{h}_K$ is a Cartan subalgebra. If α is any root of $\mathfrak{L}_K$ with respect to $\mathfrak{h}_K$ and H_α the corresponding element in $\mathfrak{h}_K$ then it is known that $(-1)^{1/2}H_\alpha \in \mathfrak{h}_{K,0}$. In particular if $\{\alpha_1, \cdots, \alpha_l\}$ is a fundamental set of roots and $H_{\alpha_i} = H_i$, $1 \leq i \leq l$, $H_i' = (-1)^{1/2}H_i \in \mathfrak{h}_{K,0}$. Moreover $\varphi(\gamma(X)) = -\gamma^+(X)$, $\varphi(\gamma^+(X)) = -\gamma(X)$ for any $X \in \mathfrak{L}_{K,0}$. Hence

$$\chi(\varphi(\gamma(H_1'^{m_1} \cdots H_l'^{m_l})\gamma^+(H_1'^{n_1} \cdots H_l'^{n_l})))$$
$$= (-1)^{m_1+\cdots+m_l+n_1+\cdots+n_l}\chi(\gamma^+(H_1'^{m_1} \cdots H_l'^{m_l})\gamma(H_1'^{n_1} \cdots H_l'^{n_l})),$$

since φ is clearly an anti-automorphism of $\mathfrak{B}$ over R. Hence

$$\chi^*(\gamma^+(H_1'^{m_1} \cdots H_l'^{m_l})\gamma(H_1'^{n_1} \cdots H_l'^{n_l}))$$

$$(27) \qquad = (-1)^{m_1+\cdots+m_l+n_1+\cdots+n_l}\ \overline{\chi(\gamma(H_1'^{m_1} \cdots H_l'^{m_l})\gamma^+(H_1'^{n_1} \cdots H_l'^{n_l}))}$$

$$(m_1, \cdots, m_l, n_1, \cdots, n_l \geq 0).$$

For any linear function λ on $\mathfrak{h}_K$ define $\bar\lambda$ by $\bar\lambda(H_i) = \overline{\lambda(H_i)}$, $1 \leq i \leq l$. Further if ξ is any polynomial (or series) we denote by $\bar\xi$ the polynomial (or series) obtained by changing the coefficients of ξ to their complex conjugates. Suppose $\chi = \chi_\lambda \times \chi_\mu$ and $\chi^* = \chi_{\lambda^*} \times \chi_{\mu^*}$ where $\lambda, \mu, \lambda^*, \mu^*$ are linear functions on $\mathfrak{h}_K$. Then from (27) it follows that

$$\overline{\chi(\lambda^*, \mu^*, t, u)} = \chi(\mu, \lambda, -t, -u)$$

in the notation of (24). But, from (21), $\chi(\mu, \lambda, -t, -u) = \overline{\chi(\bar\mu, \bar\lambda, t, u)}$. Hence

$$\chi(\lambda^*, \mu^*, t, u) = \chi(\bar\mu, \overline{\bar\lambda}, t, u).$$

Therefore from Lemma 42, $\chi^* = \chi_{\lambda^*} \times \chi_{\mu^*} = \chi_{\bar\mu} \times \chi_{\bar\lambda}$. Thus we have the following lemma.

LEMMA 44. Let $\chi = \chi_\lambda \times \chi_\mu$ be a character of $\mathfrak{B}$. Then $\chi^* = \chi_{\bar\mu} \times \chi_{\bar\lambda}$.

COROLLARY. $\chi = \chi^*$ if and only if $\chi = \chi_\lambda \times \chi_\lambda$ for some linear function λ on $\mathfrak{h}_K$.

Part IV. Representations of a complex semisimple Lie group in a Hilbert space. So far we have considered only abstract representations of the Lie algebra $\mathfrak{L}_0$. Now we come to the representations of the corresponding group in a Hilbert space. Let G be the simply connected Lie group whose Lie algebra is $\mathfrak{L}_0$, the latter being defined as in Part II. We assume that $\mathfrak{L}_{K,0}$ is compact and $\mathfrak{h}_{K,0}$ is a maximal abelian subalgebra of $\mathfrak{L}_{K,0}$ and $\mathfrak{h}_K = \mathfrak{h}_{K,0} + (-1)^{1/2}\mathfrak{h}_{K,0}$. We shall adhere closely to the notation of Parts II and III.

Let V be a Hilbert space and π a mapping which associates to each $g \in G$ a bounded operator $\pi(g)$ on V such that $\pi(g_1 g_2) = \pi(g_1)\pi(g_2)$ $(g_1, g_2 \in G)$ and $\pi(1) = I$, where 1 is the unit element of G and I is the unit operator on V.

π is called a representation of G on V if for every $\psi \in V$ the mapping $g \to \pi(g)\psi$ $(g \in G)$ is continuous. Moreover it is called an irreducible representation if there exists no closed subspace other than V and $\{0\}$ which is invariant under $\pi(g)$ for all $g \in G$.

Since G is a Lie group it is also an analytic manifold. Let $C_c^\infty(G)$ denote the class of all complex-valued functions on G which are infinitely differentiable everywhere and which vanish outside a compact set. For any $X \in \mathfrak{L}_0$ and $f \in C_c^\infty(G)$ we define $\_Xf \in C_c^\infty(G)$ by the rule

$$(28) \qquad (Xf)(g) = \left\{ \frac{d}{dt} f((\exp(-tX))g) \right\}_{t=0}$$

It is easily verified that if X_1, X_2, $X_3 \in \mathfrak{L}_0$ and $[X_1, X_2] = X_3$ then

$$X_1(X_2 f) \_X_2(X_1 f) = X_3 f.$$

Thus we get a representation of $\mathfrak{L}_0$ (and therefore by linearity a representation of $\mathfrak{L}$) on $C_c^\infty(G)$. We extend this uniquely to a representation of $\mathfrak{B}$. Then for any $z \in \mathfrak{B}$ and $f \in C_c^\infty(G)$ the function $-zf$ is well defined. Let V_1 be the subspace spanned by all elements $\varphi \in V$ of the form

$$(29) \qquad \varphi = \int_G f(g)\pi(g)\psi dg$$

where $\psi \in V$ and $f \in C_c^\infty(G)$ and dg is the left invariant Haar measure on G. It has been shown by Gårding [8] that V_1 is dense in V and

$$(30) \qquad \lim_{t \to 0} \frac{1}{t} \{ \pi(\exp tX)\varphi - \varphi \} = \int_G \_Xf(g)\pi(g)\psi dg \qquad (X \in \mathfrak{L}_0)$$

where φ is given by (29). Let $\pi(X)$ denote the operator on V_1 defined by

$$(31) \qquad \pi(X)\varphi = \lim_{t \to 0} \frac{1}{t} \{ \pi(\exp tX)\varphi - \varphi \} \qquad (X \in \mathfrak{L}_0, \varphi \in V_1, t \in R).$$

Then $\pi(X)$ maps V_1 into itself and it is obvious from (30) that $\pi([X_1, X_2]) = \pi(X_1)\pi(X_2) - \pi(X_2)\pi(X_1)$ $(X_1, X_2 \in \mathfrak{L}_0)$. Hence we get a representation of $\mathfrak{L}_0$ and therefore of $\mathfrak{B}$ on V_1 which we shall denote again by π. We shall call V_1 the Gårding subspace of V.

Let $\mathfrak{Z}$ be the center of $\mathfrak{B}$ and χ a character. We shall say that the representation π of G on V has the character χ if $\pi(z)\varphi = \chi(z)\varphi$ for every $\varphi \in V_1$ and $z \in \mathfrak{Z}$. It is known that if $\pi(G)$ is an irreducible unitary representation then $\pi(z)$ $(z \in \mathfrak{Z})$ is a multiple of the unit operator on V_1. Hence π has a uniquely determined character in this case.

Let G_K be the analytic subgroup of G corresponding to $\mathfrak{L}_{K,0}$. G_K is compact and simply connected. Hence there is a 1-1 correspondence between representations of G_K and those of $\mathfrak{L}_{K,0}$ (and therefore of $\mathfrak{L}_K$). Hence for any $\Lambda \in P_K$, $\mathfrak{D}_\Lambda$ can also be regarded as an equivalence class of representations of G_K. For any $\Lambda \in P_K$ we denote by V_Λ the set of all elements $\psi \in V$ such that either $\psi = 0$ or the subspace U spanned by all vectors of the form $\pi(g)\psi$ ($g \in G_K$) is finite-dimensional and the representation of G_K induced on U is fully reducible into components each of which belongs to $\mathfrak{D}_\Lambda$. Put $V_\Lambda^0 = V_1 \cap V_\Lambda$ and $V^0 = \sum_\Lambda V_\Lambda^0$. It can be shown (Mautner [11]) that V^0 is dense in V.

Choose a fundamental system $\{\alpha_1, \cdots, \alpha_l\}$ of roots of $\mathfrak{L}_K$ with respect to $\mathfrak{h}_K$. For any root α we define the corresponding element $H_\alpha \in \mathfrak{h}_K$ as before. We may again suppose that $\alpha(H_\alpha) = 2$. Put $H_i = H_{\alpha_i}$, $1 \leq i \leq l$. We note that $(-1)^{1/2} H_\alpha \in \mathfrak{h}_{K,0}$ for every root α.

Let π be a representation of G on V with the character χ. Since V^0 is dense in V, $V_{\Lambda_0}^0 \neq \{0\}$ for some $\Lambda_0 \in P_K$. Hence $\mathfrak{D}_{\Lambda_0}$ occurs in $\pi(\mathfrak{B})$. Therefore from Theorem 6, $\chi = \chi_\lambda \times \chi_{-\mu}^-$ where $\lambda - \mu$ is an integral function on $\mathfrak{h}_K$. The following theorem shows that the converse is also true.

THEOREM 7. *Let λ and μ be linear functions on $\mathfrak{h}_K$ such that $\lambda - \mu$ is integral. Then we can find a representation π of G on a Hilbert space such that π has the character $\chi_\lambda \times \chi_{-\mu}^-$. Moreover if $\lambda + \bar{\lambda}$ is integral, we can find a unitary representation of G with the character $\chi_\lambda \times \chi_{\bar{\lambda}}$.*

For the proof of this theorem we follow a method which is due to Gelfand and Naimark [7]. Let P be the set of all positive roots of $\mathfrak{L}_K$. We can choose elements U_α, V_α ($\alpha \in P$) in $\mathfrak{L}_{K,0}$ such that $U_\alpha + (-1)^{1/2} V_\alpha = X_\alpha \neq 0$ and $U_\alpha - (-1)^{1/2} V_\alpha = X_{-\alpha} \neq 0$ are root elements corresponding to the roots α and $-\alpha$ respectively. Then the elements $(-1)^{1/2} H_i = H_i'$, $1 \leq i \leq l$, U_α, V_α ($\alpha \in P$) form a base for $\mathfrak{L}_{K,0}$. We define the linear mappings Γ, γ, and γ^+ as in Part II (pp. 52–53) and put

$$(32) \qquad
\begin{aligned}
W_\alpha &= U_\alpha - \Gamma(V_\alpha), & W_\alpha' &= V_\alpha + \Gamma(U_\alpha), \\
W_{-\alpha} &= U_\alpha + \Gamma(V_\alpha), & W_{-\alpha}' &= V_\alpha - \Gamma(U_\alpha),
\end{aligned}$$

$$(33) \qquad
\begin{aligned}
Z_\alpha &= \gamma(X_\alpha) = (W_\alpha + (-1)^{1/2} W_\alpha')/2, \\
Z_{-\alpha} &= \gamma(X_{-\alpha}) = (W_{-\alpha} - (-1)^{1/2} W_{-\alpha}')/2, \\
Z_\alpha^+ &= \gamma^+(X_\alpha) = (W_{-\alpha} + (-1)^{1/2} W_{-\alpha}')/2, \\
Z_{-\alpha}^+ &= \gamma^+(X_{-\alpha}) = (W_\alpha - (-1)^{1/2} W_\alpha')/2.
\end{aligned}$$

Let $\mathfrak{h} = \gamma(\mathfrak{h}_K) + \gamma^+(\mathfrak{h}_K)$, $\mathfrak{h}_0 = \mathfrak{h}_{K,0} + \Gamma(\mathfrak{h}_{K,0})$. We have already identified (see p. 80) linear functions on $\mathfrak{h}$ with pairs (λ, μ) of linear functions on $\mathfrak{h}_K$. λ being any linear function on $\mathfrak{h}_K$ we denote by λ^+ and λ^- the linear functions on $\mathfrak{h}$ defined by the pairs $(\lambda, 0)$ and $(0, \lambda)$ respectively. Also we write $\lambda(\Gamma(H)) = -(-1)^{1/2} \lambda(H)$ ($H \in \mathfrak{h}_K$).

Let $\mathfrak{N}$ be the subspace of $\mathfrak{L}_0$ spanned by $W_{-\alpha}$, $W_{-\alpha}'$ ($\alpha \in P$). Then $\mathfrak{N}$ is a nilpotent subalgebra of $\mathfrak{L}_0$. Put $\mathfrak{h}_0^* = \Gamma(\mathfrak{h}_{K,0})$ and $\mathfrak{S} = \mathfrak{h}_0^* + \mathfrak{N}$. Then

$$(34) \qquad \mathfrak{L}_0 = \mathfrak{L}_{K,0} + \mathfrak{h}_0^* + \mathfrak{N} = \mathfrak{L}_{K,0} + \mathfrak{S}$$

where all the sums are direct. Let S, A, and N respectively be the analytic subgroups of G corresponding to the subalgebras $\mathfrak{S}$, $\mathfrak{h}^*$, and $\mathfrak{N}$. Then it is known (see for example Iwasawa [10]) that the mapping $\Phi\colon (u, h, n) \to uhn$ ($u \in G_K$, $h \in A$, $n \in N$) is a topological and analytic mapping of $G_K \times A \times N$ onto G. The tangent space of $G_K \times A \times N$ at any point (u, h, n) is the Cartesian product of the tangent spaces of G_K, A, and N at u, h, and n respectively. But for a Lie group the tangent space at any point may be identified under left translation with the Lie algebra, which is the tangent space at the unit element (see Chevalley [6, Chap. IV]). Therefore we can, in a natural way, identify the tangent space of $G_K \times A \times N$ at any point (u, h, n) with $\mathfrak{L}_0$. Let $d\Phi$ be the differential([22]) of Φ. Then for any $(u, h, n) \in G_K \times A \times N$, $(d\Phi)_{u,h,n}$ is a linear mapping of $\mathfrak{L}_0$ into itself. Our object is to evaluate the Haar measure on G in terms of the Haar measures on G_K, A, and N.

Let φ, ψ, χ be left invariant differential forms ($\neq 0$) of degree n, l, and $n-l$ on G_K, A, and N respectively. Here $n = \dim \mathfrak{L}_{K,0}$ and $l = \dim \mathfrak{h}_{K,0}$. Then we define a differential form ξ of degree $2n$ on $G_K \times A \times N$ as follows:

$$
\begin{aligned}
(35) \quad &\xi(U_1, \cdots, U_n; H^{(1)}, \cdots, H^{(l)}; X_1, \cdots, X_{n-l}) \\
&\qquad = \varphi(U_1, \cdots, U_n)\psi(H^{(1)}, \cdots, H^{(l)})\chi(X_1, \cdots, X_{n-l})
\end{aligned}
$$

for any $U_i \in \mathfrak{L}_{K,0}$, $H^{(j)} \in \mathfrak{h}_0^*$, $X_k \in \mathfrak{N}$, $1 \le i \le n$, $1 \le j \le l$, $1 \le k \le n-l$. Let $\theta \neq 0$ denote a left invariant differential form of degree $2n$ on G and let $\delta\Phi$ be the mapping dual to $d\Phi$. Then $\delta\Phi$ maps θ on a differential form θ' (of degree $2n$) on $G_K \times A \times N$. If $\theta'_{u,h,n}$ denotes the value of θ' at any point (u, h, n)

$$
\begin{aligned}
\theta'_{u,h,n}(U_1, \cdots, U_n; H^{(1)}, \cdots, H^{(l)}; X_1, \cdots, X_{n-l}) \\
= \theta(U'_1, \cdots, U'_n, H'^{(1)}, \cdots, H'^{(l)}, X'_1, \cdots, X'_{n-l})
\end{aligned}
$$

where U_i, $H^{(j)}$, and X_k are as in (35) and U'_i, $H'^{(j)}$, X'_k respectively are their transforms under $d\Phi_{u,h,n}$. Therefore we get

$$
(36) \qquad \theta'_{u,h,n} = c(\det d\Phi_{u,h,n})\xi_{u,h,n}
$$

where $\xi_{u,h,n}$ is the value of ξ at (u, h, n) and c is a fixed real number $\neq 0$. Let f be a function on G which is analytic around uhn. Put $F = f \bigcirc \Phi$. Then F is analytic around (u, h, n) on $G_K \times A \times N$. Let $X \in \mathfrak{N}$. Then

$$
\begin{aligned}
(37a) \quad XF(u, h, n) &= \left\{ \frac{d}{dt} F(u, h, n \exp tX) \right\}_{t=0} \\
&= \left\{ \frac{d}{dt} f(uhn \exp tX) \right\}_{t=0} \\
&= Xf(u\,h\,n).
\end{aligned}
$$

([22]) We use here and in the sequel the terminology of [6].

Similarly if $H \in \mathfrak{h}_0^*$,

$$HF(u, h, n) = \left\{ \frac{d}{dt} F(u, h \exp tH, n) \right\}_{t=0}$$

$$(37\text{b}) \qquad = \left\{ \frac{d}{dt} f(uh \exp tHn) \right\}_{t=0}$$

$$= \left\{ \frac{d}{dt} f(unh (\exp t \, \mathrm{Ad} \, (n^{-1})H)) \right\}_{t=0} = (\mathrm{Ad} \, (n^{-1})H)f(uhn)$$

where $g \to \mathrm{Ad} \, (g)$ denotes the adjoint representation of G, so that

$$g(\exp W)g^{-1} = \exp (\mathrm{Ad} \, (g)W)$$

for any $W \in \mathfrak{L}_0$ and $g \in G$. Finally if $U \in \mathfrak{L}_{K,0}$,

$$(37\text{c}) \quad UF(u, h, n) = \left\{ \frac{d}{dt} F(u \exp tU, h, n) \right\}_{t=0} = \left\{ \mathrm{Ad} \, (n^{-1}h^{-1})U \right\}f(uhn).$$

These relations show that

$$(38) \qquad \begin{aligned} d\Phi_{u,h,n}U &= \mathrm{Ad} \, (n^{-1}h^{-1})U & (U \in \mathfrak{L}_{K,0}), \\ d\Phi_{u,h,n}H &= \mathrm{Ad} \, (n^{-1})H & (H \in \mathfrak{h}_0^*), \\ d\Phi_{u,h,n}X &= X & (X \in \mathfrak{N}). \end{aligned}$$

Let $W \to \mathrm{ad} \, W$ denote the adjoint representation of $\mathfrak{L}_0$. Then it is easily seen that ad X is nilpotent for all $X \in \mathfrak{N}$ and therefore det Ad $(n) = 1$ for all $n \in N$. Hence det $(d\Phi_{u,h,n}) = \det D$, where $D = \mathrm{Ad} \, (n) \, d\Phi_{u,h,n}$. Now $\mathfrak{N}$ is invariant under Ad (n). Let $(\mathrm{Ad} \, (n))_\mathfrak{N}$ denote the restriction of Ad (n) on $\mathfrak{N}$. Then again we prove in the same way as above that det $(\mathrm{Ad} \, (n))_\mathfrak{N} = 1$. Also we notice from (38) that $\mathfrak{S} = \mathfrak{h}_0^* + \mathfrak{N}$ is invariant under D. Let D^* be the linear mapping of the factor space $\mathfrak{L}_0/\mathfrak{S}$ induced by D. Since det $(\mathrm{Ad} \, (n))_\mathfrak{N} = 1$ it follows from (38) that det $D = \det D^*$. Clearly $\mathfrak{S}$ is also invariant under Ad (h^{-1}) and the linear mapping of $\mathfrak{L}_0/\mathfrak{S}$ induced by Ad (h^{-1}) coincides with D^*. Now every $U \in \mathfrak{L}_{K,0}$ can be written uniquely in the form

$$U = H' + \sum_{\alpha \in P} 2a_\alpha U_\alpha + \sum_{\alpha \in P} 2b_\alpha V_\alpha$$

$$= H' + \sum_{\alpha \in P} a_\alpha(W_\alpha + W_{-\alpha}) + \sum_{\alpha \in P} b_\alpha(W'_\alpha + W'_{-\alpha})$$

where $H' \in \mathfrak{h}_{K,0}$ and $a_\alpha, b_\alpha \in R$. Now for any $H \in \mathfrak{h}_0^*$ and $\alpha \in P$,

$$[H, W_\alpha] = \alpha(H)W_\alpha, \qquad\qquad [H, W'_\alpha] = \alpha(H)W'_\alpha,$$

$$[H, W_{-\alpha}] = -\alpha(H)W_{-\alpha}, \qquad [H, W'_{-\alpha}] = -\alpha(H)W'_{-\alpha},$$

where $\alpha(H) = -(-1)^{1/2}\alpha(\Gamma^{-1}(H))$ is real. Hence if $h = \exp H$ $(H \in \mathfrak{h}_0^*)$,

$$
\begin{aligned}
\mathrm{Ad}\ (h^{-1})U &= H' + \sum_{\alpha \in P} a_\alpha(e^{-\alpha(H)}W_\alpha + e^{\alpha(H)}W_{-\alpha}) \\
&\quad + \sum_{\alpha \in P} b_\alpha(e^{-\alpha(H)}W'_\alpha + e^{\alpha(H)}W'_{-\alpha}) \\
&\equiv H' + \sum_{\alpha \in P} a_\alpha e^{-\alpha(H)}(W_\alpha + W_{-\alpha}) \\
&\quad + \sum_{\alpha \in P} b_\alpha e^{-\alpha(H)}(W'_\alpha + W'_{-\alpha}) \qquad\qquad \mathrm{mod}\ \mathfrak{S} \\
&\equiv H' + \sum_{\alpha \in P} 2a_\alpha e^{-\alpha(H)}U_\alpha + \sum_{\alpha \in P} 2b_\alpha e^{-\alpha(H)}V_\alpha \qquad \mathrm{mod}\ \mathfrak{S}.
\end{aligned}
$$

This shows that $\det D^* = \exp\left(-2\sum_{\alpha \epsilon P} \alpha(H)\right) = e^{-4\rho(H)}$. Hence

$$
\theta'_{u,h,n} = ce^{-4\rho(H)}\xi_{u,h,n} \qquad\qquad (h = \exp H,\ H \in \mathfrak{h}_0^*)
$$

and it follows that the left invariant Haar measure on G is given by

$$
e^{-4\rho(H)}du\,dh\,dn
$$

where $du,\ dh,\ dn$ respectively are the left invariant Haar measures on G_K, A, and N and $h = \exp H$ $(H \in \mathfrak{h}_0^*)$.

Let $x \in G$ and $u \in G_K$. Then $xu = u_x h(x, u)n(x, u)$ where $u_x \in G_K$, $h(x, u) \in A$, $n(x, u) \in N$. Since A is simply connected there is a unique element $H(x, u) \in \mathfrak{h}_0^*$ such that $h(x, u) = \exp H(x, u)$. Clearly u_x, $H(x, u)$, and $n(x, u)$ are continuous functions of (x, u) and for a fixed x, $u \to u_x$ $(u \in G_K)$ is a topological mapping of G_K onto itself. Let $y = uhn$ $(u \in G_K, h \in A, n \in N)$, and let dy denote the left invariant Haar measure of G. Then for a fixed $x \in G$,

$$
d(xy) = dy = e^{-4\rho(H)}du\,dh\,dn.
$$

On the other hand

$$
\begin{aligned}
xy &= xuhn = u_x h(x, u)n(x, u)hn \\
&= u_x h(x, u)h(h^{-1}n(x, u)hn).
\end{aligned}
$$

Since $\mathfrak{N}$ is an ideal in $\mathfrak{S}$, N is an invariant subgroup of S. Hence $h^{-1}n(x, u)h = n(x, u, h) \in N$. Also for fixed (x, u), $d(h(x, u)h) = dh$ and for fixed (x, u, h), $d(n(x, u, h)n) = dn$ on account of the left invariance of the Haar measures. Hence

$$
e^{-4\rho(H)}du\,dh\,dn = d(xy) = e^{-4\rho(H)-4\rho(H(x,u))}du_x\,dh\,dn
$$

and therefore

$$
(40) \qquad\qquad du_x = e^{4\rho(H(x,u))}du.
$$

Let f be any function on G_K. For any $x \in G$ we define a new function f^x on G_K as follows:

$$f^x(u) = f(u_{x^{-1}}).$$

Let $L_2(G_K)$ be the Hilbert space consisting of all measurable functions f on G_K such that $\int_{G_K}|f(u)|^2 du < \infty$. Then if $f \in L_2(G_K)$,

$$\int_{G_K}|f^x(u)|^2 du = \int_{G_K}|f(u)|^2 du_x = \int_{G_K}|f(u)|^2 e^{4\rho(H(x,u))} du < \infty$$

since the function $e^{4\rho(H(x,u))}$, being continuous, is bounded on the compact set G_K. Notice that if $y \in G$,

$$yxu = yu_x h(x, u)n = (u_x)_y h(y, u_x)h(x, u)n'$$

where $n, n' \in N$. Hence $u_{yx} = (u_x)_y$ and $H(u, yx) = H(y, u_x) + H(x, u)$. ν being any linear function on $\mathfrak{h}_K$ and $x \in G$, $f \in L_2(G_K)$, we define $\pi(x)f = f' \in L_2(G_K)$ as follows

$$f'(u_x) = e^{\nu(H(x,u))}f(u) \qquad\qquad (u \in G_K).$$

Then by using the above relations it is easily verified that π is a representation of G on $L_2(G_K)$. We shall write $\pi(x)f(u)$ for $f'(u)$ $(u \in G_K)$.

For any $x \in G$ and $W \in \mathfrak{L}_0$ put $W^x = \mathrm{Ad}\,(x)W$. Then we have the following lemma.

LEMMA 45. *Let $u \in G_K$ and $\varphi \in L_2(G_K)$. Then if $x \in uNu^{-1}$, $\pi(x)\varphi(u) = \varphi(u)$. Moreover if $x = \exp H^u$ $(H \in \mathfrak{h}_0{}^*)$, $\pi(x)\varphi(u) = e^{\nu(H)}\varphi(u)$.*

Let $x = unu^{-1}$ $(n \in N)$. Then $xu = un$. Hence $u_x = u$ and $H(x, u) = 0$. Therefore $\pi(x)\varphi(u) = \pi(x)\varphi(u_x) = \varphi(u)$. Now let $x = \exp H^u$ $(H \in \mathfrak{h}_0{}^*)$. Then $xu = u \exp H$. Hence $u_x = u$, $H(x, u) = H$. Hence $\pi(x)\varphi(u) = \pi(x)\varphi(u_x) = e^{\nu(H)}\varphi(u)$.

Let A_K be the analytic subgroup of G_K corresponding to the subalgebra $\mathfrak{h}_{K,0}$ of $\mathfrak{L}_{K,0}$.

LEMMA 46. *Let $u \in G_K$ and $h \in A_K$. Then for any $x \in G$,*

$$(uh)_x = u_x h, \qquad H(x, uh) = H(x, u).$$

Notice that elements of A_K and A commute and $\mathfrak{N}$ is invariant under ad H for all $H \in \mathfrak{h}_{K,0}$. Hence $h^{-1}Nh \subset N$. But

$$xuh = u_x h(x, u)nh = u_x h(x, u)h(h^{-1}nh)$$

where $n \in N$. Hence

$$u_x = u_x h, \qquad h(x, uh) = h(x, u).$$

This proves the lemma.

Let Λ be any integral function on $\mathfrak{h}_K$. Then for some $\Lambda_0 \in P_K$, $-\Lambda$ is a weight of $\mathfrak{D}_{\Lambda_0}$. Since every finite-dimensional irreducible representation of G_K occurs in the right regular representation of G_K we can find a continuous function $\psi \neq 0$ on G_K such that ψ transforms according to $\mathfrak{D}_{\Lambda_0}$ under the right regular representation of G_K and belongs to the weight $-\Lambda$. Then $\psi(u \exp H)$ $= e^{-\Lambda(H)}\psi(u)$ for any $u \in G_K$ and $H \in \mathfrak{h}_{K,0}$.

LEMMA 47. *Let $x \in G$, $u \in G_K$, and $H \in \mathfrak{h}_{K,0}$. Then*

$$\pi(\exp H^u)\pi(x)\psi(u) = e^{\Lambda(H)}\pi(x)\psi(u).$$

Now

$$\pi(\exp H^u)\pi(x)\psi(u) = \pi(x)\psi((\exp(-H^u))u) = \pi(x)\psi(uh^{-1})$$

where $h = \exp H$. But from Lemma 46,

$$\begin{aligned}
\pi(x)\psi(uh^{-1}) &= \exp(\nu(H(x, (uh^{-1})_{x^{-1}})))\psi((uh^{-1})_{x^{-1}}) \\
&= \exp(\nu(H(x, u_{x^{-1}}h^{-1})))\psi(u_{x^{-1}})e^{\Lambda(H)} \\
&= \exp(\nu(H(x, u_{x^{-1}})))\psi(u_{x^{-1}})e^{\Lambda(H)} \\
&= e^{\Lambda(H)}\pi(x)\psi(u).
\end{aligned}$$

Hence the result.

LEMMA 48. *Let V_ψ be the set of all $\varphi \in L_2(G)$ which are of the form*

$$\varphi = \int_G f(x)\pi(x)\psi dx \qquad\qquad (f \in C_c^\infty).$$

Then $\pi(z)\varphi = \chi(z)\varphi$ ($z \in \mathfrak{Z}$, $\varphi \in V_\psi$) where $\chi = \chi_\lambda \times \chi_{-\mu}^-$ and $\lambda = (\Lambda+\nu)/2$, $\mu = (\nu-\Lambda)/2$.

It is clear that V_ψ is invariant under $\pi(\mathfrak{B})$ and $\pi(G)$. Also it follows from Lemmas 45 and 47 that

$$\pi(\exp tW_{-\alpha}^u)\varphi(u) = \pi(\exp tW_{-\alpha}'^u)\varphi(u) = \varphi(u) \qquad (\alpha \in P),$$

$$\pi(\exp tH^u)\varphi(u) = e^{t\nu(H)}\varphi(u) \qquad (H \in \mathfrak{h}_0^*),$$

$$\pi(\exp tH^u)\varphi(u) = e^{t\Lambda(H)}\varphi(u) \qquad (H \in \mathfrak{h}_{0,K})$$

for any $\varphi \in V_\psi$ and $u \in G_K$. Since V_ψ is invariant under $\pi(\mathfrak{B})$, the same relations hold true if we replace φ by $\pi(b)\varphi$ ($b \in \mathfrak{B}$). Hence it is clear that

$$\pi(W_{-\alpha}^u b)\varphi(u) = \pi(W_{-\alpha}'^u b)\varphi(u) = 0 \qquad (\alpha \in P),$$

$$\pi(H^u b)\varphi(u) = \nu(H)\pi(b)\varphi(u) \qquad (H \in \mathfrak{h}_0^*),$$

$$\pi(H^u b)\varphi(u) = \Lambda(H)\pi(b)\varphi(u) \qquad (H \in \mathfrak{h}_{0,K}).$$

Now for any $Z = W_1 + (-1)^{1/2}W_2 \in \mathfrak{L}$ ($W_1, W_2 \in \mathfrak{L}_0$) put

$$Z^x = W_1^x + (-1)^{1/2}W_2^x \qquad\qquad (x \in G).$$

Then from (33) it follows that

$$(41) \qquad \begin{aligned} \pi(Z^u_{-\alpha}b)\varphi(u) &= \pi(Z^{+u}_{\alpha}b)\varphi(u) = 0 & (\alpha \in P) \\ \pi(H^u b)\varphi(u) &= (\lambda^+(H) - \mu^-(H))\pi(b)\varphi(u) & (H \in \mathfrak{h}) \end{aligned}$$

where $\lambda = (\Lambda+\nu)/2$, $\mu = (\nu-\Lambda)/2$. For any $x \in G$, the mapping $Z \to Z^x$ ($Z \in \mathfrak{L}$) is an automorphism of $\mathfrak{L}$ which can be extended uniquely to an automorphism of $\mathfrak{B}$ which we denote by $b \to b^x$ ($b \in \mathfrak{B}$). It is clear that if $z \in \mathfrak{Z}$, $z^x = z$. Let $\mathfrak{H}$ be the subalgebra of $\mathfrak{B}$ generated by 1 and $\mathfrak{h} = \gamma(\mathfrak{h}_K) + \gamma^+(\mathfrak{h}_K)$. Let β denote the isomorphism of $\mathfrak{H}$ with the ring $C[x, y]$ of polynomials in the indeterminates $(x_1, \cdots, x_l, y_1, \cdots, y_l)$ with coefficients in C given by

$$\beta(\gamma(H_1^{m_1} \cdots H_l^{m_l})\gamma^+(H_1^{n_1} \cdots H_l^{n_l})) = x_1^{m_1} \cdots x_l^{m_l} y_1^{n_1} \cdots y_l^{n_l}, \qquad m_i, n_i \geqq 0.$$

Then if we take[23] $(\alpha_i, 0)$, $(0, -\alpha_i)$, $1 \leqq i \leqq l$, as a fundamental system of roots of $\mathfrak{L}$ with respect to $\mathfrak{h}$, for every $z \in \mathfrak{Z}$ we can find a unique element $f_z(x, y) \in C[x, y]$ such that

$$z - \beta^{-1}(f_z(x, y) \in \sum_{\alpha \in P} \mathfrak{B} Z_\alpha + \sum_{\alpha \in P} \mathfrak{B} Z^+_{-\alpha}.$$

Hence

$$z - \beta^{-1}(f_z(x, y)) = \sum_{\alpha \in P} a_{-\alpha} Z_\alpha + \sum_{\alpha \in P} b_\alpha Z^+_{-\alpha} \qquad (a_{-\alpha}, b_\alpha \in \mathfrak{B}).$$

Since the left-hand side is of rank zero, we may suppose that $a_{-\alpha}$ is of rank $(-\alpha, 0)$ and b_α is of rank $(0, \alpha)$ with respect to $\mathfrak{h}$. Then we can show exactly as in the proof of Lemma 3, that $a_{-\alpha} \in \sum_{\alpha \in P} Z_{-\alpha}\mathfrak{B}$, $b_\alpha = \sum_{\alpha \in P} Z^+_\alpha \mathfrak{B}$. Hence

$$z - \beta^{-1}(f_z(x, y)) = \sum_{\alpha \in P} Z_{-\alpha} c_\alpha + \sum_{\alpha \in P} Z^+_\alpha d_\alpha \qquad (c_\alpha, d_\alpha \in \mathfrak{B})$$

and therefore

$$z = z^u = (\beta^{-1}(f_z(x, y)))^u + \sum_{\alpha \in P} Z^u_{-\alpha} c^u_\alpha + \sum_{\alpha \in P} (Z^+_\alpha)^u d^u_\alpha \qquad (u \in G_K).$$

But then it follows from (41) that

$$\pi(z)\varphi(u) = f_z(\lambda, -\mu)\varphi(u).$$

Since this is true for all $u \in G_K$,

$$\pi(z)\varphi = f_z(\lambda, -\mu)\varphi.$$

But it is clear that $f_z(\lambda, -\mu) = \chi(z)$ where $\chi = \chi_\lambda \times \chi^-_{-\mu}$. Hence the lemma is proved.

---

[23] As already mentioned earlier, we identify linear functions on $\mathfrak{h}$ with pairs of linear function on $\mathfrak{h}_K$.

The next two lemmas are due to Mautner. We use the usual terminology of Hilbert space. If Q is an operator with a dense domain we denote by Q^* its adjoint. Also if Q_1, Q_2 are two operators we say that $Q_1 \subset Q_2$ if the domain D_1 of Q_1 is contained in the domain of Q_2 and Q_1 and Q_2 coincide on D_1.

LEMMA 49. *Let Q_0 and Q be two operators on a Hilbert space V such that Q_0 and Q^* are densely defined and $Q_0 \subset Q$. Then if Q_0 is bounded so also is Q.*

Since $Q_0 \subset Q$ and Q_0 is bounded, $Q^* \subset Q_0^*$ and Q_0^* is also bounded. Hence Q^* and therefore Q^{**} is bounded. But $Q^{**} \supset Q$. Therefore Q is also bounded.

LEMMA 50. *Let π be a representation of G on a Hilbert space V. Then for any $b \in \mathfrak{B}$, $(\pi(b))^*$ has a dense domain.*

Let V_1^* denote the set of all vectors φ in V of the form

$$\varphi = \int_G f(g)\pi^*(g)\psi\,dg \qquad\qquad (f \in C_c^\infty(G),\ \psi \in V)$$

where $\pi^*(g)$ is the adjoint of $\pi(g)$. Also for any $X \in \mathfrak{L}_0$ and $f \in C_c^\infty(G)$ put

$$(X^*f)(g) = \left\{\frac{d}{dt} f(g \exp tX)\right\}_{t=0}.$$

Now

$$\pi^*(\exp(-tX))\varphi = \int_G f(g)\pi^*(g \exp(-tX))\psi\,dg$$

$$= \int_G f(g \exp tX)\pi^*(g)\psi\,dg$$

since the Haar measure on G is both left and right invariant. Hence

$$\lim_{t \to 0} \frac{1}{t}\left\{\pi^*(\exp tX)\varphi - \varphi\right\} = -\int_G X^*f(g)\pi^*(g)\psi\,dg \in V_1^*.$$

From this it is clear that for any $X \in \mathfrak{L}_0$, $(\pi(X))^*$ is defined on V_1^* and leaves it invariant. Hence it follows immediately that for any $b \in \mathfrak{B}$, $(\pi(b))^*$ is defined on V_1^*. Since V_1^* is dense in V the assertion follows.

Now we return to the notation of Lemma 48. Let U be the closure of V_ψ. Then U is clearly invariant under $\pi(G)$. Let π_0 be the representation of G induced on U. Let U_1 be the Gårding subspace of U.

LEMMA 51. $\pi(z)\varphi = \chi(z)\varphi$ *for any $z \in \mathfrak{Z}$ and $\varphi \in U_1$. Hence π_0 has the character $\chi = \chi_\lambda \times \chi_{-\mu}^-$.*

Clearly $V_\psi \subset U_1$. For any fixed $z \in \mathfrak{Z}$ let Q_0 and Q be the operators in U with the domains V_ψ and U_1 respectively such that $Q_0\varphi = \pi(z)\varphi$ ($\varphi \in V_\psi$)

and $Q\varphi = \pi(z)\varphi$ $(\varphi \in U_1)$. Then $Q_0 \subset Q$ and Q_0 is densely defined. Moreover by applying Lemma 46 to the representation π_0 of G on U we see that Q^* is also densely defined. On the other hand we know from Lemma 48 that $Q_0\varphi = \chi(z)\varphi$ $(\varphi \in V_\psi)$. Hence Q_0 is bounded. Therefore from Lemma 46, Q is also bounded. Hence Q_0 and Q have a unique common extension on U which must be $\chi(z)I$ where I is the unit operator. Therefore $Q\varphi = \chi(z)\varphi$ $(\varphi \in U_1)$ and the lemma is proved.

We are now in a position to prove Theorem 7. Put $\lambda - \mu = \Lambda$, $\lambda + \mu = \nu$. Since Λ is integral by hypothesis, the first assertion of the theorem follows from Lemma 51. Notice that if $\varphi \in L_2(G_K)$ and $x \in G$ we get from (40)

$$\int_{G_K} |\pi(x)\varphi(u)|^2 du = \int_{G_K} |e^{\nu(H(x,u))}\varphi(u)|^2 du_x = \int_{G_K} |e^{\nu'(H(x,u))}\varphi(u)|^2 du$$

where $\nu' = \nu + 2\rho$. Now suppose $\nu' = (-1)^{1/2}\sigma$ where σ is a real linear function on $\mathfrak{h}_K$ (that is, $\sigma(H_i)$, $1 \leq i \leq l$, are all real). Then $\nu'(H)$ is pure imaginary for all $H \in \mathfrak{h}_0{}^*$ and it is clear that the representation π of G on $L_2(G_K)$ is unitary. Moreover in this case $\lambda + \rho = (\Lambda + (-1)^{1/2}\sigma)/2$, $-(\mu + \rho) = (\Lambda - (-1)^{1/2}\sigma)/2$. Hence $-(\mu + \rho) = \bar\lambda + \rho$ and so from equation (23) of Part III, $\chi^-_{-\mu} = \chi_{\bar\lambda}$. Therefore π_0 is a unitary representation with the character $\chi_\lambda \times \chi_{\bar\lambda}$. Since Λ can be any integral function and σ any real function on $\mathfrak{h}_K$ it is obvious that λ is arbitrary apart from the condition that $\lambda + \bar\lambda = \Lambda - 2\rho$ be integral. This completes the proof of Theorem 7.

From the theory of representations of compact groups it follows that the function ψ of Lemma 48 is a linear combination of the matrix elements of some representation of G_K which belongs to $\mathfrak{D}_{\Lambda_0}$. Let $\mathfrak{D}_{\Lambda_0}{}^* = \mathfrak{D}_{\Lambda_1}$ $(\Lambda_1 \in P_K)$ be the equivalence class of representations of G_K which are dual to the representations in $\mathfrak{D}_{\Lambda_0}$. It is easily seen that under the left regular representation of G_K, ψ transforms according to $\mathfrak{D}_{\Lambda_1}$. Hence $\mathfrak{D}_{\Lambda_1}$ occurs in the representation π_0 of Lemma 51. Moreover it is clear that $-\Lambda$ is a weight of $\mathfrak{D}_{\Lambda_0}$ if and only if Λ is a weight of $\mathfrak{D}_{\Lambda_1}$. Therefore we get the following extension of Theorem 7.

THEOREM 8. *Given any $\Lambda_0 \in P_K$ such that $\lambda - \mu$ is a weight of $\mathfrak{D}_{\Lambda_0}$ we can find a representation π of G on a Hilbert space such that π has the character $\chi_\lambda \times \chi^-_{-\mu}$ and $\mathfrak{D}_{\Lambda_0}$ occurs in π. Similarly if $\lambda + \bar\lambda + 2\rho$ is a weight of $\mathfrak{D}_{\Lambda_0}$, we can find a unitary representation π of G with the character $\chi_\lambda \times \chi_{\bar\lambda}$ such that $\mathfrak{D}_{\Lambda_0}$ occurs in π.*

This is a sort of converse of Theorem 6. We have mentioned earlier that every irreducible unitary representation π of G has a character χ. It is not difficult to prove that in this case $\chi = \chi^*$ in the notation of Lemma 44. Taking into account the corollary to Lemma 44, we see that Theorem 7 provides a partial answer to the problem of determining those characters of $\mathfrak{B}$ which correspond to some unitary representation of G (see also Gelfand and Naimark [7]).

References

1. G. Birkhoff, Ann. of Math. vol. 38 (1937) pp. 526–532.
2. E. Cartan, Bull. Soc. Math. France vol. 41 (1913) pp. 53–96.
3. C. Chevalley, C. R. Acad. Sci. Paris vol. 227 (1948) pp. 1136–1138.
4. ———, Amer. J. Math. vol. 63 (1941) pp. 785–793.
5. ———, Ann. of Math. vol. 48 (1947) pp. 91–100.
6. ———, *Theory of Lie groups*, Princeton University Press, 1946.
7. I. M. Gelfand and M. A. Naimark, Doklady Akademii Nauk SSSR (N.S.) vol. 63 (1948) pp. 225–228.
8. L. Gårding, Proc. Nat. Acad. Sci. U.S.A. vol. 33 (1947) pp. 331–332.
9. Harish-Chandra, Ann. of Math. vol. 50 (1949) pp. 900–915.
10. K. Iwasawa, Ann. of Math. vol. 50 (1949) pp. 507–557.
11. F. I. Mautner, Ann. of Math. vol. 52 (1951) pp. 528–556.
12. H. Weyl, Math. Zeit. vol. 24 (1925) pp. 328–395.
13. ———, *The classical groups*, Princeton University Press, 1939.
14. E. Witt, J. Reine Angew. Math. vol. 177 (1937) pp. 152–160.

Harvard University,
    Cambridge, Mass.

Reprinted from
Trans. Amer. Math. Soc.
70 (1951), 28–96

Reprinted from the Proceedings of the National Academy of Sciences,
Vol. 37, No. 3, pp. 170–173. March, 1951

REPRESENTATIONS OF SEMISIMPLE LIE GROUPS ON A BANACH SPACE

By Harish-Chandra

Department of Mathematics, Columbia University

Communicated by O. Zariski, January 26, 1951

Let G be a connected semisimple Lie group and $\mathfrak{H}$ a Banach space. By a representation of G on $\mathfrak{H}$ we mean a mapping π which assigns to every $x \,\epsilon\, G$ a bounded linear operator $\pi(x)$ on $\mathfrak{H}$ such that the following two conditions are fulfilled:

(1) $\pi(xy) = \pi(x)\pi(y)$ and $\pi(1) = I$ (here 1 is the unit element of G and I the identity operator on $\mathfrak{H}$).

(2) The mapping $(x, \psi) \to \pi(x)\psi(x \,\epsilon\, G, \psi \,\epsilon\, \mathfrak{H})$ is a continuous mapping of $G \times \mathfrak{H}$ into $\mathfrak{H}$.

The object of this note is to announce a few theorems on these representations. No attempt is made to give proofs here. A detailed account with complete proofs will appear elsewhere in another paper.

Let R and C, respectively, be the fields of real and complex numbers. Let $\mathfrak{g}_0$ be the Lie algebra of G and $\mathfrak{g}$ the complexification of $\mathfrak{g}_0$. We denote the universal enveloping[1] algebra of $\mathfrak{g}$ by $\mathfrak{B}$. Let $C_c^\infty(G)$ be the set of all complex-valued functions on G which are indefinitely differentiable everywhere and which vanish outside a compact set. Let π be a representation of G on $\mathfrak{H}$ and V the set of all elements in $\mathfrak{H}$ which can be written as finite linear combinations of elements of the form

$$\int_G f(x)\pi(x)\psi \, dx \qquad (\psi \,\epsilon\, \mathfrak{H}, f \,\epsilon\, C_c^\infty(G))$$

where dx is the left invariant Haar measure on G. V is called the Gårding subspace[2] of $\mathfrak{H}$ (with respect to π). For every $X \,\epsilon\, \mathfrak{g}_0$ we can define a linear transformation $\pi_V(X)$ of V into itself such that

$$\pi_V(x)\psi = \lim_{\cdot \to 0} \frac{1}{t} \left\{ \pi(\exp tX) - I \right\}\psi \qquad (\psi \,\epsilon\, V, t \,\epsilon\, R).$$

The mapping $X \to \pi_V(X)$ is a representation of $\mathfrak{g}_0$ and therefore it can be extended uniquely to a representation π_V of $\mathfrak{B}$ on V. Let $\psi \in V$. Consider $\overline{\pi_V(\mathfrak{B})\psi}$ where the bar denotes closure in $\mathfrak{H}$. It turns out that in general $\overline{\pi_V(\mathfrak{B})\psi}$ is not invariant under $\pi(G)$. In order to avoid this unpleasant state of affairs we replace the Gårding subspace V by the space of all well-behaved elements. This is defined as follows. Let U be a subspace of $\mathfrak{H}$ (not necessarily closed). We say that U is well-behaved under π if the following conditions hold.

(1)   There exists a representation π_U of $\mathfrak{g}_0$ on U such that

$$\pi_U(X)\psi = \lim_{t \to 0} \frac{1}{t} \left\{ \pi(\exp tX) - I \right\} \psi \qquad (X \in \mathfrak{g}_0, \psi \in U).$$

(2)   For any continuous linear function f on $\mathfrak{H}$ and $\psi \in U$ the function

$$f(\pi(x)\psi) \qquad (x \in G)$$

is an analytic function on G. It is clear that if U_1, U_2 are two well-behaved subspaces of $\mathfrak{H}$ then $U_1 + U_2$ is also well-behaved. From this it follows that the union W of all well-behaved subspaces in $\mathfrak{H}$ is itself a well-behaved subspace. An element $\psi \in \mathfrak{H}$ will be called well-behaved if $\psi \in W$ and W is called the space of all well-behaved elements. It is clear that the mapping $X \to \pi_W(X)$ $(X \in \mathfrak{g}_0)$ can be extended uniquely to a representation π_W of $\mathfrak{B}$ on W. The following theorem justifies the notion of well-behaved elements.

THEOREM 1.   *Let ψ be a well-behaved element of $\mathfrak{H}$. Then $\overline{\pi_W(\mathfrak{B})\psi}$ is invariant under $\pi(G)$.*

Let G^* be the adjoint group of G and let K^* be a maximal compact subgroup of G^*. Let K be the complete inverse image of K^* in G. Then K is connected though it is not necessarily compact. Let P be the set of all equivalence classes of finite-dimensional irreducible representations of K. Let $\mathfrak{D} \in$ P and $\psi \in \mathfrak{H}$. We say that ψ transforms under $\pi(K)$ according to $\mathfrak{D}$ if the space U spanned by $\pi(u)\psi$ for all $u \in K$ is finite-dimensional and the representation of K induced on U is fully reducible into a direct sum of irreducible components each of which lies in $\mathfrak{D}$. Let $\mathfrak{H}_{\mathfrak{D}}$ be the set of all elements in $\mathfrak{H}$ which transform under $\pi(K)$ according to $\mathfrak{D}$. Put $W_{\mathfrak{D}} = W \cap \mathfrak{H}_{\mathfrak{D}}$. Let Z be the center of G and $\mathfrak{Z}$ the center of $\mathfrak{B}$. Let V be the Gårding subspace of $\mathfrak{H}$ and π_V the representation of $\mathfrak{B}$ on V as defined above. We shall say that π is a quasi-simple representation of G if there exist homomorphisms c and χ of Z and $\mathfrak{Z}$ respectively, into C such that the following two conditions hold:

$$(1) \quad \pi(d) \;\; = c(d)I \qquad (d \in Z).$$

$$(2) \quad \pi_V(z)\psi = \chi(z)\psi \qquad (z \in \mathfrak{Z}, \psi \in V).$$

In case condition (2) is satisfied we say that π has the character χ.

THEOREM 2.   *Let π be a quasi-simple representation of G on $\mathfrak{H}$.   Then $\sum_{\mathfrak{D} \epsilon P} W_{\mathfrak{D}}$ is dense in $\mathfrak{H}$ and $\mathfrak{H}_{\mathfrak{D}} = \overline{W_{\mathfrak{D}}}$ ($\mathfrak{D} \epsilon$ P).*

Here $\sum_{\mathfrak{D} \epsilon P} W_{\mathfrak{D}}$ is the space consisting of all finite linear combinations of elements in $\underset{\mathfrak{D} \epsilon P}{\cup} W_{\mathfrak{D}}$.

THEOREM 3.   *Let π be quasi-simple and let $\psi \epsilon \sum_{\mathfrak{D} \epsilon P} W_{\mathfrak{D}}$.   Put $U = \overline{\pi_W(\mathfrak{B})\psi}$ and $U_{\mathfrak{D}} = U \cap \mathfrak{H}_{\mathfrak{D}} (\mathfrak{D} \epsilon$ P).   Then $\pi_W(\mathfrak{B})\psi = \sum_{\mathfrak{D} \epsilon P} U_{\mathfrak{D}}$ and dim $U_{\mathfrak{D}} < \infty$ for every $\mathfrak{D} \epsilon$ P.*

Let U, V be two subspaces of $\mathfrak{H}$.   We write $U > V$ or $V < U$ if $U \supset V$ and $U \neq V$.   Let U and V be two closed subspaces of $\mathfrak{H}$ invariant under $\pi(G)$.   We shall say that V is maximal in U if $U > V$ and there exists no closed invariant subspace U_1 such that $U > U_1 > V$.

THEOREM 4.   *Let π be a quasi-simple representation of G on a Banach space $\mathfrak{H} \neq \{0\}$.   Then there exist two closed invariant subspaces U and V in $\mathfrak{H}$ such that V is maximal in U.*

Now we shall consider the special case when $\mathfrak{H}$ is a Hilbert space and π is a unitary representation.   From Theorem 4 we deduce the following result.

THEOREM 5.   *Let π be a quasi-simple unitary representation of G on a Hilbert space $\mathfrak{H} \neq \{0\}$.   Then there exists a closed invariant subspace U of $\mathfrak{H}$ such that $\{0\}$ is maximal in U (i.e., $U > \{0\}$ and U is irreducible under $\pi(G)$).*

The above theorem has the following significance in relation to the theory of factors of Murray and von Neumann.[3]   Let π be a unitary representation of G on a Hilbert space $\mathfrak{H}$.   Let $\mathfrak{A}$ be the smallest weakly closed algebra of bounded operators on $\mathfrak{H}$ containing $\pi(G)$.   It is known[4] that if $\mathfrak{A}$ is a factor, i.e., if the center of $\mathfrak{A}$ consists of scalar multiples of I, then π is quasi-simple.   Therefore in this case Theorem 5 is applicable and so it follows from the results of Murray and von Neumann[3] that $\mathfrak{A}$ must be a factor of type I_n or I_∞.   This shows that factors of Type II and Type III cannot arise from a unitary representation of a semisimple Lie group on a Hilbert space.   For the consequences of this result we refer the reader to the above-quoted paper of Mautner.

Let $\mathfrak{D} \epsilon$ P.   We say that $\mathfrak{D}$ occurs in π if $\mathfrak{H}_{\mathfrak{D}} \neq \{0\}$.   It is known that all unitary irreducible representations of G are quasi-simple.

THEOREM 6.   *Let χ be a homomorphism of $\mathfrak{Z}$ into C such that $\chi(1) = 1$ and let $\mathfrak{D} \epsilon$ P.   Let $\mathfrak{E}(\chi, \mathfrak{D})$ denote the set of all representations π of G which have the following properties:*

(1)   *π is an irreducible unitary representation of G on some Hilbert space.*

(2)   *$\mathfrak{D}$ occurs in π.*

(3) π *has the character* χ.

Then $\mathfrak{E}(\chi, \mathfrak{D})$ *contains only a finite number of inequivalent representations.*

Let $d(\mathfrak{D})$ denote the degree of any representation in $\mathfrak{D}$. Suppose $d(\mathfrak{D}) = 1$, $\pi \epsilon \mathfrak{E}(\chi, \mathfrak{D})$ and $\mathfrak{H}$ is the representation space of π. Then it can be shown that dim $\mathfrak{H}_{\mathfrak{D}} = 1$ and the maximum number of inequivalent representations in $\mathfrak{E}(\chi, \mathfrak{D})$ is not greater than the order of the Weyl group of $\mathfrak{g}$ with respect to some Cartan subalgebra. Moreover if G is a complex semi-simple Lie group all representations in $\mathfrak{E}(\chi, \mathfrak{D})$ are equivalent (provided $d(\mathfrak{D}) = 1$).

Let π be an irreducible unitary representation of G on a Hilbert space $\mathfrak{H}$. For any $\mathfrak{D} \epsilon P$ let $E_{\mathfrak{D}}$ denote the orthogonal projection of $\mathfrak{H}$ on $\mathfrak{H}_{\mathfrak{D}}$. Since π is irreducible and unitary it is quasi-simple and therefore it follows from Theorem 4 that dim $\mathfrak{H}_{\mathfrak{D}} < \infty$. Hence the function $\varphi_{\mathfrak{D}}{}^{\pi}(x) = sp(E_{\mathfrak{D}} \pi(x) E_{\mathfrak{D}})(x \epsilon G)$ is well defined. It is an analytic function on G and $\varphi_{\mathfrak{D}}{}^{\pi}(u\, x\, u^{-1}) = \varphi_{\mathfrak{D}}{}^{\pi}(x)$ $(u \epsilon K)$.

THEOREM 7. *Let* π_1 *and* π_2 *be irreducible unitary representations of* G *on two Hilbert spaces. Let* P_{π_1} *be the set of all elements in* P *which occur in* π_1. *Suppose that for some* $\mathfrak{D} \epsilon P_{\pi_1}$ *and* $c \epsilon C$,

$$\varphi_{\mathfrak{D}}{}^{\pi_1} = c\varphi_{\mathfrak{D}}{}^{\pi_2}.$$

Then π_1 *and* π_2 *are equivalent. Conversely if* π_1 *and* π_2 *are equivalent*

$$\varphi_{\mathfrak{D}}{}^{\pi_1} = \varphi_{\mathfrak{D}}{}^{\pi_2}$$

for all $\mathfrak{D} \epsilon P$.

In case $d(\mathfrak{D}) = 1$ the function $\varphi_{\mathfrak{D}}{}^{\pi}$ is the same as the spherical function introduced by Gelfand and Naimark.[5] Now suppose G is a complex semi-simple Lie group and $d(\mathfrak{D}) = 1$. Then we have seen above that the representation π is completely determined within equivalence by its character χ. Actually in this case an explicit formula for $\varphi_{\mathfrak{D}}{}^{\pi}$ in terms of χ can be obtained quite easily. This formula is very similar to the one given by Gelfand and Naimark[5] for representations of the principal series.

[1] See, for example, Harish-Chandra, *Ann. Math.*, **50**, 900–915 (1949).

[2] Garding, L., PROC. NATL. ACAD. SCI., **33**, 331–332 (1947).

[3] Murray, F. J., and von Neumann, J., *Ann. Math.*, **37**, 116–229 (1936).

[4] See Mautner, F. I., *Ann. Math.*, **52**, 528–556 (1950).

[5] Gelfand, I. M., and Naimark, M. A., *Doklady Akad. Nauk SSR (N. S.)*, **63**, 225-228 (1948).

Reprinted from the Proceedings of the NATIONAL ACADEMY OF SCIENCES,
Vol. 37, No. 6, pp. 362–365.   June, 1951

REPRESENTATIONS OF SEMISIMPLE LIE GROUPS. II

BY HARISH-CHANDRA

DEPARTMENT OF MATHEMATICS, COLUMBIA UNIVERSITY

Communicated by P. A. Smith, April 23, 1951

The object of this note is to announce some further results on representations of a connected semisimple Lie group on a Hilbert space.   We shall omit all proofs and assume that the reader is familiar with the contents of an earlier note[1] (quoted henceforward as RI).

Let R and C denote the fields of real and complex numbers, respectively, and let G be a connected, simply connected, semisimple Lie group and $\mathfrak{g}_0$ its Lie algebra over R.   Let $x \to Ad(x)(x \in G)$ denote the adjoint representation of G and let K be the complete inverse image in G of some maximal compact subgroup of $Ad(G)$. Then K is a closed connected subgroup of G. Let $\mathfrak{L}_0$ be the Lie algebra of K and let $X \to adX(X \,\epsilon\, \mathfrak{g}_0)$ denote the adjoint representation of $\mathfrak{g}_0$. Put $B(X, Y) = sp(adXadY)(X, Y \,\epsilon\, \mathfrak{g}_0)$. Let $\mathfrak{P}_0$ be the set of all elements $Y \,\epsilon\, \mathfrak{g}_0$ such that $B(X, Y) = 0$ for all $X \,\epsilon\, \mathfrak{L}_0$. Then $\mathfrak{g}_0 = \mathfrak{L}_0 + \mathfrak{P}_0$, $\mathfrak{L}_0 \cap \mathfrak{P}_0 = \{0\}$ and there exists an automorphism θ of $\mathfrak{g}_0$ over R such that $\theta(X + Y) = X - Y$ for $X \,\epsilon\, \mathfrak{L}_0$, $Y \,\epsilon\, \mathfrak{P}_0$. Let $\mathfrak{h}_{\mathfrak{P}_0}$ be a maximal abelian subspace of $\mathfrak{P}_0$. We extend $\mathfrak{h}_{\mathfrak{P}_0}$ to a maximal abelian subalgebra $\mathfrak{h}_0$ of $\mathfrak{g}_0$. Then $\mathfrak{h}_0 = \mathfrak{h}_{\mathfrak{L}_0} + \mathfrak{h}_{\mathfrak{P}_0}$ where $\mathfrak{h}_{\mathfrak{L}_0} = \mathfrak{h}_0 \cap \mathfrak{L}_0$. Let $\mathfrak{g}$ be the complexification of $\mathfrak{g}_0$ and let $\mathfrak{P}$, $\mathfrak{L}$, $\mathfrak{h}$, $\mathfrak{h}_{\mathfrak{P}}$, $\mathfrak{h}_{\mathfrak{L}}$ be the subspaces of $\mathfrak{g}$ spanned by $\mathfrak{P}_0$, $\mathfrak{L}_0$, $\mathfrak{h}_0$, $\mathfrak{h}_{\mathfrak{P}_0}$, $\mathfrak{h}_{\mathfrak{L}_0}$ respectively, over C. We extend the bilinear form B and the automorphism θ on $\mathfrak{g}$ by linearity over C. Choose bases $(H_1, \ldots, H_p)$ and $(H_{p+1}, \ldots, H_l)$, respectively, for $\mathfrak{h}_{\mathfrak{P}_0}$ and $\sqrt{-1}\, \mathfrak{h}_{\mathfrak{L}_0}$ over R. Let $\mathfrak{F}$ be the space of all linear functions on $\mathfrak{h}$. Given $\lambda \,\epsilon\, \mathfrak{F}$ we can find a unique element $H_\lambda \,\epsilon\, \mathfrak{h}$ such that $\lambda(H) = B(H, H_\lambda)$ for all $H \,\epsilon\, \mathfrak{h}$. We shall say that λ is real if $H_\lambda \,\epsilon\, \mathfrak{h}_{\mathfrak{P}_0} + \sqrt{-1}\, \mathfrak{h}_{\mathfrak{L}_0}$. Moreover if λ is real and $H_\lambda = \sum_{1 \leqslant i \leqslant l} c_i H_i (c_i \,\epsilon\, R)$ we say that $\lambda > 0$ if $\lambda \neq 0$ and $c_j > 0$ where j is the least index $(1 \leqslant j \leqslant l)$ such that $c_j \neq 0$. We know that $\mathfrak{h}$ is a Cartan subalgebra of $\mathfrak{g}$ and $\theta\mathfrak{h} = \mathfrak{h}$. For every root α of $\mathfrak{g}$ with respect to $\mathfrak{h}$ let $X_\alpha \neq 0$ denote an element in $\mathfrak{g}$ such that $[H, X_\alpha] = \alpha(H)X_\alpha(H \,\epsilon\, \mathfrak{h})$. Let P be the set of all roots $\alpha > 0$. For any $\lambda \,\epsilon\, \mathfrak{F}$ let $\theta\lambda$ denote the linear function given by $\theta\lambda(H) = \lambda(\theta H)(H \,\epsilon\, \mathfrak{h})$. Then if α is a root $\theta\alpha$ is also a root. Let P_+ be the set of all $\alpha \,\epsilon\, P$ such that $\theta\alpha < 0$ and P_- the remaining set of positive roots. Iwasawa[2] has shown that $X_\alpha, X_{-\alpha} \,\epsilon\, \mathfrak{L}$ for $\alpha \,\epsilon\, P_-$. Put

$$\mathfrak{N} = \sum_{\alpha\,\epsilon\,P_+} CX_\alpha, \quad \mathfrak{M} = \mathfrak{h}_{\mathfrak{L}} + \sum_{\alpha\,\epsilon\,P_-} CX_\alpha + \sum_{\alpha\,\epsilon\,P_-} CX_{-\alpha}.$$

Then $\mathfrak{N}$ is a nilpotent subalgebra of $\mathfrak{g}$ and $\mathfrak{M}$ is a subalgebra of $\mathfrak{L}$. Put $\mathfrak{N}_0 = \mathfrak{N} \cap \mathfrak{g}_0$, $\mathfrak{M}_0 = \mathfrak{M} \cap \mathfrak{L}_0$. Then $\mathfrak{g}_0 = \mathfrak{L}_0 + \mathfrak{h}_{\mathfrak{P}_0} + \mathfrak{N}_0$ where the sum is direct[2] and $[\mathfrak{M}, \mathfrak{N}] \subset \mathfrak{N}$, $[\mathfrak{M}, \mathfrak{h}_{\mathfrak{P}}] = \{0\}$. Moreover $\mathfrak{L}$ and $\mathfrak{M}$ are reductive algebras, i.e., they are the direct sums of their centers and their derived algebras which are semisimple. Let $\mathfrak{C}$ be the center of $\mathfrak{L}$. Put $\mathfrak{C}_0 = \mathfrak{C} \cap \mathfrak{g}_0$ and $\mathfrak{L}_0' = [\mathfrak{L}_0, \mathfrak{L}_0]$. Let D, K', A_+, M and N, respectively, be the analytic subgroups of G corresponding to $\mathfrak{C}_0$, $\mathfrak{L}_0'$, $\mathfrak{h}_{\mathfrak{P}_0}$, $\mathfrak{M}_0$ and $\mathfrak{N}_0$. Then M is closed and the mappings $(\gamma, u) \to \gamma u(\gamma \,\epsilon\, D, u \,\epsilon\, K')$ and $(v, h, n) \to vhn(v \,\epsilon\, K, h \,\epsilon\, A_+, n \,\epsilon\, N)$ are analytic isomorphisms of $D \times K'$ with K and $K \times A_+ \times N$ with G, respectively.

Let Ω, Ω' and ω, respectively, be the set of all equivalence classes of finite-dimensional irreducible representations of K, K' and M. Then if $\mathfrak{D} \,\epsilon\, \Omega$ and $\sigma \,\epsilon\, \mathfrak{D}$ it is easily seen that $\sigma(K')$ is irreducible and $\sigma(M)$ is fully reducible. We denote by $\mathfrak{D}'$ the equivalence class of the irreducible representation of K' defined by σ. Also if $\delta \,\epsilon\, \omega$ we say that $\delta < \mathfrak{D}$ if δ occurs in the reduction of $\sigma(M)$. Let $\alpha \,\epsilon\, \delta \,\epsilon\, \omega$. Then α defines a representation β of $\mathfrak{M}_0$ and therefore of $\mathfrak{M}$. Let λ and μ be any two weights of β with respect to $\mathfrak{h}_{\Re}$. We extend λ and μ on $\mathfrak{h}$ by defining them to be zero on $\mathfrak{h}_{\mathfrak{P}}$. Then it is easily seen that $\lambda - \mu$ is a real linear function on $\mathfrak{h}$. Hence β has a weight λ such that $\lambda - \mu > 0$ for every other weight μ. Clearly λ depends on δ alone and we shall call it the highest weight of δ.

Let $\mathfrak{B}$ be the universal enveloping algebra of $\mathfrak{g}$ and let $\mathfrak{Z}$ be the center of $\mathfrak{B}$. Let $C[x]$ denote the ring of all commutative polynomials in l independent variables $x_1, \ldots, x_l$ with coefficients in C. We denote by β the isomorphic mapping of $C[x]$ into $\mathfrak{B}$ given by $\beta(x_1^{m_1} x_2^{m_2} \ldots x_l^{m_l}) = H_1^{m_1} \ldots H_l^{m_l}$. For any $z \in \mathfrak{Z}$ there exists a unique element[3] $\chi_z(z) \in C[x]$ such that $z - \beta(\chi_z(z)) \in \sum_{\alpha \in P} \mathfrak{B} X_{\alpha}$. Λ being any linear function on $\mathfrak{h}$ we denote by $\chi_\Lambda(z)$ the value of the polynomial $\chi_z(z)$ at $x_i = \Lambda(H_i) 1 \leqslant i \leqslant l$. Then the mapping $\chi_\Lambda : z \to \chi_\Lambda(z)(z \in \mathfrak{Z})$ is a homomorphism of $\mathfrak{Z}$ into C and conversely every homomorphism of $\mathfrak{Z}$ into C is of the form[3] χ_Λ for some $\Lambda \in \mathfrak{F}$. Let $\rho = {}^1/_2 \sum_{\alpha \in P} \alpha$ and let W be the Weyl group of $\mathfrak{g}$ with respect to $\mathfrak{h}$. Then if $\Lambda_1, \Lambda_2 \in \mathfrak{F}$, $\chi_{\Lambda_1} = \chi_{\Lambda_2}$ if and only if[3] $s(\Lambda_1 + \rho) = \Lambda_2 + \rho$ for some $s \in W$.

We denote by Ω_F the set of all $\mathfrak{D} \in \Omega$ for which there exists a finite-dimensional representation π of G such that $\mathfrak{D}'$ occurs in the reduction of $\pi(K')$.

THEOREM 1. *Let π be a quasisimple[1] representation of G on a Banach space with the infinitesimal character[4] χ. Suppose $\mathfrak{D}$ is an element in Ω_F such that $\mathfrak{D}$ occurs[5] in π. Then there exists a linear function Λ on $\mathfrak{h}$ and a $\delta \in \omega$ such that $\delta < \mathfrak{D}$ and $\chi = \chi_\Lambda$ and $\Lambda(H) = \lambda_\delta(H)(H \in \mathfrak{h}_\Re)$, where λ_δ is the highest weight of δ. Conversely suppose we are given a linear function Λ on $\mathfrak{h}$ and $\mathfrak{D} \in \Omega$ such that Λ coincides on $\mathfrak{h}_\Re$ with the highest weight λ_δ of some $\delta < \mathfrak{D}$ $(\delta \in \omega)$. Then there exists a quasisimple irreducible representation π of G on a Hilbert space with the infinitesimal character χ_Λ such that $\mathfrak{D}$ occurs in π.*

Remarks.—It seems likely that the first part of the above theorem is actually true for all $\mathfrak{D} \in \Omega$ and not merely for $\mathfrak{D} \in \Omega_F$, but so far it has not been possible to prove this. Notice that if G is a complex semisimple group $\Omega = \Omega_F$ and so in this case the theorem holds without any restriction on $\mathfrak{D}$.

Let π_1 and π_2 be two quasisimple representations of G on the Hilbert spaces $\mathfrak{H}_1$ and $\mathfrak{H}_2$, respectively. For any $\mathfrak{D} \in \Omega$ let $\mathfrak{H}_{i, \mathfrak{D}}$ denote the set of all elements in $\mathfrak{H}_i$ which transform[1] under $\pi_i(K)$ according to $\mathfrak{D}(i = 1, 2)$. Put $\mathfrak{H}_i^0 = \sum_{\mathfrak{D} \in \Omega} \mathfrak{H}_{i, \mathfrak{D}}$. Then we get a representation π_i^0 of $\mathfrak{B}$ on $\mathfrak{H}_i^0$ such that

$$\pi_i^0(X)\psi = \lim_{t \to 0} \frac{1}{t} \{\pi_i(\exp t X)\psi - \psi\} \qquad (\psi \in \mathfrak{H}_i^0, X \in \mathfrak{g}_0, t \in R).$$

We say that π_1 and π_2 are infinitesimally equivalent if the representations π_1^0 and π_2^0 are algebraically equivalent, i.e., if there exists an isomorphism α of $\mathfrak{H}_1^0$ onto $\mathfrak{H}_2^0$ such that $\pi_2^0(b)\alpha\psi = \alpha\pi_1^0(b)\psi (b \epsilon \mathfrak{B}, \psi \epsilon \mathfrak{H}_1^0)$. Clearly if π_1 and π_2 are equivalent they are also infinitesimally equivalent. Conversely it can be shown that if π_1 and π_2 are both unitary, their infinitesimal equivalence implies their equivalence in the usual sense.

Let π be a quasisimple irreducible representation of G on a Hilbert space. We know (see Theorem 3 of *RI*) that dim $\mathfrak{H}_\mathfrak{D} < \infty$ for all $\mathfrak{D} \epsilon \Omega$. Moreover, we may assume without loss of generality that the subspaces $\mathfrak{H}_\mathfrak{D}$ are all mutually orthogonal for distinct $\mathfrak{D}$. Let $E_\mathfrak{D}$ denote the orthogonal projection on $\mathfrak{H}$ on $\mathfrak{H}_\mathfrak{D}$. Put

$$\varphi_\mathfrak{D}{}^\pi(x) = sp(E_\mathfrak{D}\pi(x)E_\mathfrak{D}) \qquad (x \epsilon G).$$

Then we can restate Theorem 7 of *RI* in a slightly improved form as follows.

THEOREM 2. *Let π_1, π_2 be irreducible quasisimple representations of G on two Hilbert spaces. Suppose that for some $\mathfrak{D} \epsilon \Omega$ and $c \epsilon C$, $\varphi_\mathfrak{D}{}^{\pi_1} = c\varphi_\mathfrak{D}{}^{\pi_2} \neq 0$. Then π_1 and π_2 are infinitesimally equivalent. Conversely if π_1 and π_2 are infinitesimally equivalent $\varphi_\mathfrak{D}{}^{\pi_1} = \varphi_\mathfrak{D}{}^{\pi_2}$ for all $\mathfrak{D} \epsilon \Omega$.*

We have seen above that every element $x \epsilon G$ can be written uniquely in the form $x = vhn(v \epsilon K, h \epsilon A_+, n \epsilon N)$. For any $v \epsilon K$ and $x \epsilon G$ let v_x and $H(x, v)$ denote the unique elements in K and $\mathfrak{h}_{\mathfrak{P}_0}$, respectively, such that $xv = v_x(\exp H(x, v))n$ for some $n \epsilon N$. Moreover let $\Gamma(v)$ denote the element in $\mathfrak{C}_0$ such that $v = (\exp \Gamma(v))u$ for some $u \epsilon K'$. Put $\Gamma(x, v) = \Gamma(v_x) - \Gamma(v)$. Let Z be the center of G and let $z \to z^*$ denote the adjoint representation of G. Then $K \supset Z$ and K^* is compact. It is easily seen that $(va)_x = v_xa, H(x, va) = H(x, v), \Gamma(x, va) = \Gamma(x, v)(a \epsilon Z)$. Hence we may write $H(x, v) = H(x, v^*), \Gamma(x, v) = \Gamma(x, v^*)$. Let dv^* denote the Haar measure on K^* such that $\int_K{}^* dv^* = 1$. For any $\mathfrak{D} \epsilon \Omega$ let $\mu_\mathfrak{D}$ denote the linear function on $\mathfrak{C}$ such that

$$\sigma(\exp \Gamma) = e^{\mu_\mathfrak{D}(\Gamma)}\sigma(1) \qquad (\Gamma \epsilon \mathfrak{C}_0)$$

for $\sigma \epsilon \mathfrak{D}$. Also let $d(\mathfrak{D})$ denote the degree of σ.

THEOREM 3. *Let π be a quasisimple irreducible representation of G on a Hilbert space $\mathfrak{H}$ and let $\mathfrak{D}$ be an element in Ω such that $d(\mathfrak{D}) = 1$ and $\mathfrak{D}$ occurs[5] in π. Then dim $\mathfrak{H}_\mathfrak{D} = 1$ and there exists a linear function Λ on $\mathfrak{h}$ such that χ_Λ is the infinitesimal[4] character of π and*

$$\varphi_\mathfrak{D}{}^\pi(x) = \int_K{}^* e^{\mu_\mathfrak{D}(\Gamma(x,\, v^*))}e^{\Lambda(H(x,\, v^*))}\, dv^* \qquad (x \epsilon G).$$

[1] Harish-Chandra, PROC. NATL. ACAD. SCI., **37**, 170–173 (1951).

[2] Iwasawa, K., *Ann. Math.*, **50**, 507–558 (1949).

[3] Harish-Chandra, *Trans. Am. Math. Soc.*, **70**, 28–96 (1951).

[4] The infinitesimal character of π was called simply the character of π in *RI*.

[5] This means that $\mathfrak{D}$ occurs in the reduction of $\pi(K)$.

Reprinted from the Proceedings of the NATIONAL ACADEMY OF SCIENCES,
Vol. 37, No. 6, pp. 366–369. June, 1951

REPRESENTATIONS OF SEMISIMPLE LIE GROUPS. III. CHARACTERS

BY HARISH-CHANDRA

DEPARTMENT OF MATHEMATICS, COLUMBIA UNIVERSITY

Communicated by P. A. Smith, April 23, 1951

We shall adhere strictly to the notation of the preceding note.[1] Making use of an unpublished result of Chevalley one can prove the following theorem.

THEOREM 1. *Let π be a quasisimple irreducible representation of G on a Hilbert space $\mathfrak{H}$. Then there exists an integer N such that*

$$dim \ \mathfrak{H}_{\mathfrak{D}} \leq N(d(\mathfrak{D}))^2$$

for any $\mathfrak{D} \in \Omega$.

Moreover if $\mathfrak{H}_{\mathfrak{D}} \neq \{0\}$ for some $\mathfrak{D} \in \Omega_F$ then it can be shown that we may take N equal to the order of the Weyl group W.

Let π be as above and let $C_c^\infty(G)$ denote the class of all complex-valued functions on G which are indefinitely differentiable everywhere and which vanish outside a compact set. Let A be a bounded operator on $\mathfrak{H}$. We say that A has a trace if for every complete orthonormal set[2] ($\psi_1, \psi_2, \ldots, \psi_n, \ldots$) in $\mathfrak{H}$ the series $\sum_{i \geq 1} (\psi_i, A\psi_i)$ converges to a finite number independent of the choice of this orthonormal set. We denote this number by spA. Let A^* be the adjoint of A. We say that A is of the Hilbert-Schmidt class if AA^* has a trace.

THEOREM 2. *Let π be a quasisimple irreducible representation of G on a Hilbert space $\mathfrak{H}$. Then for any $f \in C_c^\infty(G)$ the operator[3] $\int_G f(x)\pi(x) \, dx$ has a trace. Put*

$$T_\pi(f) = sp(\textstyle\int_G f(x)\pi(x) \, dx)$$

and for any $a \in G$ let $_af_a$ denote the function $_af_a(x) = f(a^{-1}xa) \ (x \in G)$. Then T_π is a distribution in the sense of L. Schwartz[4] and

$$T_\pi(_af_a) = T_\pi(f) \qquad (f \in C_c^\infty(G), a \in G).$$

We shall call the distribution T_π the character of the representation π.

THEOREM 3. *Let π_1 and π_2 be quasisimple irreducible representations of G on two Hilbert spaces. If $T_{\pi_1} = cT_{\pi_2}(c \in C)$ then π_1 and π_2 are infinitesimally equivalent. Conversely if π_1 and π_2 are infinitesimally equivalent $T_{\pi_1} = T_{\pi_2}$.*

Since for irreducible unitary representations infinitesimal equivalence is the same as ordinary equivalence, such a representation is completely determined within equivalence by its character.

THEOREM 4. *Let π be a quasisimple irreducible representation of G and let f be any measurable function on G such that f vanishes outside a compact set and $\int_G |f(x)|^2 dx < \infty$. Then the operator $\int_G f(x)\pi(x) dx$ is of the Hilbert-Schmidt class.*

For any $X \in \mathfrak{g}_0$ and $f \in C_c^\infty(G)$ put

$$(*Xf)(x) = \left\{\frac{d}{dt} f(\exp(-tX)x)\right\}_{t=0}.$$

Then the mapping $X \to *X$ is a representation of $\mathfrak{g}_0$ on $C_c^\infty(G)$ which can be extended to a representation $b \to *b$ ($b \in \mathfrak{B}$) of $\mathfrak{B}$. Let φ be the anti-automorphism of $\mathfrak{B}$ such that $\varphi(X) = -X$ ($X \in \mathfrak{g}$). If T is any distribution on G we define bT ($b \in \mathfrak{B}$) as follows:

$$bT(f) = T(*(\varphi(b))f) \qquad (f \in C_c^\infty(G)).$$

Moreover for any function f on G we denote by $_yf$, f_z and $_yf_z$ (y, z, $\in G$) the functions

$$_yf(z) = f(y^{-1}x), \ f_z(x) = f(xz), \ _yf_z(x) = f(y^{-1}xz) \ (z \in G).$$

Let Z denote the center of G and let π be a quasisimple irreducible representation of G on a Hilbert space. Let χ be the infinitesimal character of π and let η be the homomorphism of Z into C such that $\pi(a) = \eta(a)\pi(1)$ ($a \in Z$). Then if T_π is the character of π it is easily seen that

$$zT_\pi = \chi(z)T_\pi \qquad (z \in \mathfrak{Z})$$

$$T_\pi(_yf_y) = T_\pi(f), \ T_\pi(f_a) = \eta(a)T_\pi(f) \ (f \in C_c^\infty(G), \ y \in G, \ a \in Z).$$

Now put[5] $M_1 = MZ$ and let $x \to x^*$ denote the adjoint representation of G. Put $(x)^{y^*} = yxy^{-1}(x, y \in G)$ and let $C_c(G)$ denote the class of all continuous functions on G which vanish outside a compact set. Let α and g be continuous functions[5] on A_+ and M_1, respectively. Put[6]

$$T(f) = \int f((n^{-1}h^{-1}m^{-1})^{u^*})\alpha(h)g(m) \ dm \ dh \ dn \ du^* \qquad (f \in C_c(G))$$

Here dm, dh, dn, du^* are the left invariant Haar measures on M_1, A_+, N, K^*, respectively, and the integral extends over $K^* \times M_1 \times A_+ \times N$. It can be shown that $T(f) = T(_yf_y)(y \in G)$. Let σ be an irreducible representation of M_1 on a finite-dimensional Hilbert space U. Let δ be the equivalence class of the representation σ_0 of M defined by σ. Let $\psi_0 \neq 0$ be an element in U belonging to the highest weight λ_δ of σ_0 and let η be the homomorphism of Z into C such that $\sigma(a) = \eta(a)\sigma(1)$ ($a \in Z$). Put $g(m) = (\psi_0, \sigma(m^{-1})\psi_0) \ (m \in M_1)$ and $\alpha(h) = e^{-(\nu+2\rho)(\log h)}$ where ν is a linear function on $\mathfrak{h}_\mathfrak{B}$ and $\log h$ is the unique element in $\mathfrak{h}_{\mathfrak{R}_0}$ such that $\exp(\log h) = h$. If we regard T as a distribution it is easily seen that

$$zT = \chi_\Lambda(z)T \qquad (z \in \mathfrak{Z})$$

where $\Lambda(H_1 + H_2) = \nu(H_1) + \lambda_\delta(H_2)$ $(H_1 \in \mathfrak{h}_{\mathfrak{P}}, H_2 \in \mathfrak{h}_{\mathfrak{R}})$. Moreover

$$T(f_a) = \eta(a)T(f) \qquad (a \in Z, f \in C_c(G))$$

and it is easy to check that

$$T(f) = \int f((n^{-1}h^{-1}m^{-1})^{u*})e^{-(\nu+2\rho)(\log h)} \xi(m^{-1}) \, dm \, dh \, dn \, du^* \quad (f \in C_c(G))$$

where

$$\xi(m) = \frac{1}{d(\delta)}|\psi_0|^2 \, sp\sigma(m)$$

and $d(\delta)$ is the degree of σ. Let A_- be the analytic subgroup of M corresponding to $\mathfrak{h}_{\mathfrak{R}_0}$. Put $A = A_+ A_-$ and $A_1 = AZ$. Let V be the set of all elements in G which can be written in the form xyx^{-1} with $x \in G$ and $y \in A_1N$. We shall say that an element $y \in V$ is regular if $y = xhx^{-1}$ for some $x \in G$ and $h \in A_1$ and y^* has exactly[5] l eigen-values equal to 1. Let V_0 be the set of all elements in V which are regular. Then V_0 is open in G and V is the closure of V_0. Let W_0 be the subgroup of the Weyl group W consisting of those elements $s \in W$ for which there exists an $x \in G$ such that $sH = x^*H$ for all $H \in \mathfrak{h}$. It is easily seen that every $s \in W_0$ leaves both $\mathfrak{h}_{\mathfrak{R}}$ and $\mathfrak{h}_{\mathfrak{P}}$ invariant. Put

$$\Delta^-(H) = \prod_{\alpha \in P_-} (e^{\frac{1}{2}\alpha(H)} - e^{-\frac{1}{2}\alpha(H)}), \ \Delta^+(H) = \prod_{\alpha \in P_+} (e^{\frac{1}{2}\alpha(H)} - e^{-\frac{1}{2}\alpha(H)}) \qquad (H \in \mathfrak{h})$$

and define $\epsilon(s) = \pm 1$ $(s \in W_0)$ in such a way that

$$\Delta^-(sH) = \epsilon(s)\Delta^-(H)$$

for all $H \in \mathfrak{h}_{\mathfrak{R}}$. In particular if P_- is empty $\epsilon(s) = 1$ for all $s \in W_0$. Consider the function $\Theta_{\Lambda,\eta}$ on V_0 defined as follows:

$$\Theta_{\Lambda,\eta}(y) = \eta(\gamma) \frac{\sum\limits_{s \in W_0} \epsilon(s)e^{s(\Lambda+\rho)(H_1+H_2)}}{\Delta^-(H_1)|\Delta^+(H_1 + H_2)|} \qquad (y \in V_0)$$

where $y = x(\gamma \exp(H_1 + H_2))x^{-1}$ for some $x \in G$, $\gamma \in Z$, $H_1 \in \mathfrak{h}_{\mathfrak{R}_0}$ and $H_2 \in \mathfrak{h}_{\mathfrak{P}_0}$. It can be shown that in spite of the ambiguity in the choice of γ, H_1 and H_2, $\Theta_{\Lambda,\eta}$ is well defined on V_0. We extend $\Theta_{\Lambda,\eta}$ on G by defining it to be zero outside V_0. Then it can be proved that[6]

$$T(f) = \int_G f(x) \Theta_{\Lambda,\eta}(x) \, dx \qquad (f \in C_c(G))$$

provided the Haar measure dx on G is suitably normalized. Let $s_1 = 1$ $s_2, \ldots, s_r$ be a maximal set of distinct elements in the Weyl group W with the following properties:

(I) Let $\Lambda_i = s_i(\Lambda + \rho) - \rho$ $1 \leq i \leq r$. Then $\Lambda_i + \rho \neq s(\Lambda_j + \rho)$ if $i \neq j$ $(1 \leq i, j \leq r)$ and $s \in W_0$.

(II) For each i $(1 \leq i \leq r)$ there exists a $\delta_i \in \omega$ such that Λ_i coincides on

$\mathfrak{h}_2$ with the highest weight[5] λ_{δ_i} of δ_i. Moreover if $\sigma \in \delta_i$ and $\gamma \in M \cap Z$

$$\sigma(\gamma) = \eta(\gamma)\sigma(1).$$

Put $\Theta_i = \Theta_{\Lambda_i, \eta}$, $1 \leq i \leq r$ and $T_i(f) = \int_G f(x)\Theta_i(x)\, dx$ $(f \in C_c(G))$. Then we see that the distributions T_i $1 \leq i \leq r$ are solutions of the equations,

$$zT = \chi_\Lambda(z)T \qquad (z \in \mathfrak{Z})$$

$$T({}_y f_y) = T(f), \quad T(f_a) = \eta(a)T(f) \; (f \in C_c^\infty(G),\; y \in G,\; a \in Z).$$

We have seen above that if π is any quasisimple irreducible representation of G such that χ_Λ is the infinitesimal character of π and $\pi(a) = \eta(a)\pi(1)$ $(a \in Z)$, then its character T_π is also a solution of the above equations. Hence one might hope that in most cases $T\pi$ would be a linear combination of T_i $1 \leq i \leq r$.

In conclusion I should like to thank Professor C. Chevalley for his help and advice on several questions connected with the results of this note.

[1] "Representations of semisimple Lie groups. II," Proc. Natl. Acad. Sci., **37**, 362 (1951), quoted hereafter as *RII*.

[2] Since π is irreducible it follows easily that $\mathfrak{H}$ is separable.

[3] Here dx denotes the left invariant Haar measure on G.

[4] Schwartz, L., *Theorie des distributions*, Hermann, Paris, 1950.

[5] See *RII* for the meaning of the various symbols.

[6] Compare with Gelfand I. M., and Naimark M. A., *Izvestiya Akad. Nauk SSR*, Ser. Mat., 1947, vol. 11, pp. 411–504; *Mat. Sbornik N. S.*, 1947, vol. 21 (63), pp. 405–434; *Doklady Akad. Nauk SSSR (N. S.)*, 1948, vol. 61, pp. 9–11.

Reprinted from the Proceedings of the NATIONAL ACADEMY OF SCIENCES,
Vol. 37, No. 10, pp. 691–694.   October, 1951

REPRESENTATIONS OF SEMISIMPLE LIE GROUPS.   IV

By HARISH-CHANDRA

DEPARTMENT OF MATHEMATICS, COLUMBIA UNIVERSITY

Communicated by Oscar Zariski, August 8, 1951

The present note is meant to be a continuation of the three earlier ones of this series which have appeared in these PROCEEDINGS.[1]   In order to save space we shall adhere closely to the notation of RII and thus agree that all symbols shall have the same meaning as there unless they are explicitly defined anew.[2]

In RII we have defined the notion of the infinitesimal equivalence of two quasi-simple irreducible representations of G on two Hilbert spaces.   This definition may clearly be extended, without any change whatsoever, to the case when the two representation spaces are Banach spaces.   Let π be a representation of G on a Hilbert space $\mathfrak{H}$.   Let $\mathfrak{H}_1$, $\mathfrak{H}_2$ be two closed invariant subspaces of $\mathfrak{H}$ such that $\mathfrak{H}_1 > \mathfrak{H}_2$.   We regard the factor space $\mathfrak{H}_1/\mathfrak{H}_2$ as a Hilbert space in the usual way and consider the representation π' of G induced on it under π.   Any representation π' obtained in this fashion will be said to be *deducible* from π.

Let Z be the center of G.   Put $K^* = K/Z \cap D$.   Then K^* is compact. Let $v \to v^*$ be the natural mapping of K on K^* and dv^* the element of Haar measure on K^* such that $\int_{K^*} dv^* = 1$.   Let $L_2(K^*)$ be the Hilbert space consisting of all measurable and square-integrable functions on K^*.   For any $x \, \epsilon \, G$ and $v \, \epsilon \, K$ we have defined v_x, $H(x, v)$ aud $\Gamma(x, v)$ in RII.   It is easily seen that for x fixed $(v_x)^*$, $H(x, v)$ and $\Gamma(x, v)$ depend only on v^*.

Hence we may write them as v_x^*, $H(x, v^*)$ and $\Gamma(x, v^*)$. Let μ and ν be two linear functions on $\mathfrak{C}$ and $\mathfrak{h}_{\mathfrak{B}}$ respectively. We define a representation $\pi_{\mu, \nu}$ of G on $L_2(K^*)$ as follows:

$$\pi_{\mu, \nu}(x)f(v^*) = e^{-\mu(\Gamma(x^{-1}, v^*))}e^{-\nu(H(x^{-1}, v^*))} f(v^*_{x^{-1}}).$$

Here $x \in G$, $f \in L_2(K^*)$ and $v^* \in K^*$ and $\pi_{\mu, \nu}(x)f(v^*)$ denotes the value of $\pi_{\mu, \nu}(x)f$ at v^*. Let $\mathfrak{E}$ denote the set of all irreducible representations of G each of which is deducible from some $\pi_{\mu,\nu}$. It is easy to verify that every representation in $\mathfrak{E}$ is quasi-simple.

Let $\mathfrak{X}$ be the subalgebra of $\mathfrak{B}$ generated by $(1, \mathfrak{L})$. Since there is a natural 1-1 correspondence between finite-dimensional representations of K and $\mathfrak{L}$ (and therefore also of $\mathfrak{X}$), any $\mathfrak{D} \in \Omega$ may also be regarded as an equivalence class of finite-dimensional irreducible representations of $\mathfrak{L}$ (or $\mathfrak{X}$). Our first theorem may now be stated as follows:

THEOREM 1. *Let $\mathfrak{R}$ be a maximal left ideal in $\mathfrak{B}$ with the following two properties:*

(1)   There exists a homomorphism χ of $\mathfrak{Z}$ into C such that $z - \chi(z) \in \mathfrak{R}$ for all $z \in \mathfrak{Z}$.

(2)   The natural representation of $\mathfrak{X}$ on $\mathfrak{X}/\mathfrak{R} \cap \mathfrak{X}$ is finite-dimensional and the equivalence class of at least one of its irreducible components lies in Ω_F.

Then there exists a quasi-simple irreducible representation π of G on a Hilbert space $\mathfrak{H}$ and a well-behaved element $\psi \in \mathfrak{H}$ such that $\pi(b)\psi = 0$ $(b \in \mathfrak{B})$ if and only if $b \in \mathfrak{R}$. Moreover π may be chosen in $\mathfrak{E}$.

COROLLARY. *Let π be an irreducible quasi-simple representation of G on a Banach space $\mathfrak{H}$ such that some $\mathfrak{D} \in \Omega_F$ occurs in π. Then π is infinitesimally equivalent to some representation in $\mathfrak{E}$.*

Now let π be an irreducible quasi-simple representation of G on a Hilbert $\mathfrak{H}$. For any $\mathfrak{D} \in \Omega$ let $\mathfrak{H}_{\mathfrak{D}}$ denote the set of all elements in $\mathfrak{H}$ which transform under $\pi(K)$ according to $\mathfrak{D}$. We know from Theorem 3 of *RI* that dim $\mathfrak{H}_{\mathfrak{D}} < \infty$ for all $\mathfrak{D} \in \Omega$. Moreover we may suppose without loss of generality that the subspaces $\mathfrak{H}_{\mathfrak{D}}$ are mutually orthogonal for different $\mathfrak{D}$. Let $E_{\mathfrak{D}}$ denote the orthogonal projection of $\mathfrak{H}$ on $\mathfrak{H}_{\mathfrak{D}}$ $(\mathfrak{D} \in \Omega)$. Let μ be a linear function on $\mathfrak{C}$ such that $\pi(\exp \Gamma) = e^{\mu(\Gamma)}\pi(1)$ $(\Gamma \in \mathfrak{C}_0)$ whenever $\exp \Gamma \in D \cap Z$. Define a representation π^* of K^* on $\mathfrak{H}$ as follows:

$$\pi^*((u \exp \Gamma)^*) = e^{-\mu(\Gamma)}\pi(u \exp \Gamma)   (u \in K', \Gamma \in \mathfrak{C}_0).$$

It is clear that for every $\mathfrak{D} \in \Omega$, $\mathfrak{H}_{\mathfrak{D}}$ is invariant under $\pi^*(K^*)$. Let $\mathfrak{D}^*$ denote the equivalence class of any irreducible component of the representation of K^* induced on $\mathfrak{H}_{\mathfrak{D}}$ under π^*. We choose an orthonormal base for each $\mathfrak{H}_{\mathfrak{D}}$. These bases all taken together form an orthonormal base for $\mathfrak{H}$. Let M be the subgroup of K as defined in *RII* and let ω be the set of all equivalence classes of finite-dimensional irreducible representations of M. For any $\delta \in \omega$ we denote by λ_δ the highest weight of δ.

THEOREM 2. *Suppose $\mathfrak{H}_\mathfrak{D} \neq \{0\}$ for some $\mathfrak{D} \in \Omega_F$. Then there exists a $\delta \in \omega$ and a linear function Λ on $\mathfrak{h}$ such that the following conditions are fulfilled:*

(1) χ_Λ *is the infinitesimal character of* π.

(2) Λ *coincides with* λ_δ *on* $\mathfrak{h}_\mathfrak{R}$.

(3)   *For any* $\mathfrak{D} \in \Omega$

$$\dim \mathfrak{H}_\mathfrak{D} \leqq d(\mathfrak{D})n(\mathfrak{D}, \delta)$$

where $d(\mathfrak{D})$ is the degree of any representation $\sigma \in \mathfrak{D}$ and $n(\mathfrak{D}, \delta)$ is the number of times δ occurs in the reduction of $\sigma(M)$.

(4)   *Let $\mathfrak{D}_1, \mathfrak{D}_2$ be two elements in Ω such that*

$$\mathfrak{H}_{\mathfrak{D}_i} \neq \{0\} \quad i = 1, 2.$$

Then the matrix coefficients of $E_{\mathfrak{D}_1}\pi(x)E_{\mathfrak{D}_2}$ $(x \in G)$ with respect to the above base are finite linear combinations with constant coefficients of functions of the form

$$\int_{K^*} g_{\mathfrak{D}_1*}(v_x{}^*)g_{\mathfrak{D}_2*}(v^{*-1})e^{\Lambda(H(x, \, v^*))}e^{\mu(\Gamma(x, \, v^*))} \, dv^*$$

where $g_{\mathfrak{D}_1*}$ and $g_{\mathfrak{D}_2*}$ are some matrix coefficients of representations in $\mathfrak{D}_1{}^*$ and $\mathfrak{D}_2{}^*$ respectively.

In case $\mathfrak{D}_1{}^*$, $\mathfrak{D}_2{}^*$ both correspond to the trivial representation of K^* the above result gives theorem 3 of *RII*.

THEOREM 3.   *Let π be a quasi-simple irreducible representation of G on a Hilbert space $\mathfrak{H}$. Put $V = \sum\limits_{\mathfrak{D} \, \in \, \Omega} \mathfrak{H}_\mathfrak{D}$. Suppose it is possible to define a new scalar product $(\varphi, \psi)'$ $(\varphi, \psi \in V)$ in V such that*

$$(\pi(X)\varphi, \psi)' = -(\varphi, \pi(X)\psi)' \qquad (X \in \mathfrak{g}_0).$$

Let $\mathfrak{H}'$ be the Hilbert space obtained by completing V with respect to the corresponding metric. Then there exists an irreducible unitary representation π' of G on $\mathfrak{H}'$ such that

$$\pi(X)\psi = \operatorname*{Lim}_{t \to 0} \frac{1}{t}\{\pi'(\exp tX)\psi - \psi\} \qquad (t \in R, \text{ limit in } \mathfrak{H}')$$

or all $X \in \mathfrak{g}_0$ and $\psi \in V$. Moreover π' is uniquely determined.

In view of theorems 2 and 3 it is clear that the problem of constructing all irreducible unitary representations of G is now largely reduced to that of determining the irreducible representations of $\mathfrak{B}$ which are "formally unitary." In fact it seems likely that all the above theorems actually remain true if we replace Ω_F by Ω everywhere even though our proofs are then no longer applicable. In case G has a finite-dimensional representation which is faithful on K' it is easily seen that $\Omega = \Omega_F$ and so the above theorems then hold in all generality.   This is so in particular if G is a complex semisimple Lie group.

[1] Proc. Natl. Acad. Sci., **37,** 170–173, 362–365, 366–369 (1951), quoted hereafter as *RI, RII* and *RIII* respectively.

[2] I take this opportunity of acknowledging the fact that one or two minor errors have crept into the latter portion of *RIII*.   Since they are of a rather computational nature and do not, in any way, affect the general line of argument, it seems best to wait until the publication of the full details of the proof.   However, I should like to correct some misprints in *RII*.   On p. 363 in line 22 the last P should be P_-.   Also in lines 23 and 25 P should be replaced by P_- everywhere, so that we now have

$$\mathfrak{M} = \mathfrak{h}_\varrho + \sum_{\alpha \,\epsilon\, P_-} C \, X_\alpha + \sum_{\alpha \,\epsilon\, P_-} C X_{-\alpha}$$

Plancherel Formula for Complex Semisimple Lie Groups

HARISH-CHANDRA

DEPARTMENT OF MATHEMATICS, COLUMBIA UNIVERSITY
Communicated by P. A. Smith, October 5, 1951

Let R and C be the fields of real and complex numbers respectively and let G be a connected (but not necessarily simply connected) complex semisimple Lie group and $\mathfrak{g}_0$ its Lie algebra over R. Let K be a maximal compact subgroup of G and let $\mathfrak{L}_0$ be the corresponding subalgebra of $\mathfrak{g}_0$. Define $\mathfrak{P}_0$, $\mathfrak{h}_{\mathfrak{L}_0}$, $\mathfrak{h}_{\mathfrak{P}_0}$ and $\mathfrak{h}_0$ as in a previous note.[1] Since G is a complex group there exists a 1-1 linear mapping Γ of $\mathfrak{L}_0$ on $\mathfrak{P}_0$ such that $[X, \Gamma(Y)] = \Gamma([X, Y])$ and $[\Gamma(X), \Gamma(Y)] = -[X, Y] (X, Y \in \mathfrak{L})$. We extend Γ to a linear mapping of $\mathfrak{g}_0$ on itself by defining $\Gamma(\Gamma(X)) = -X (X \in \mathfrak{L}_0)$. Let $\sqrt{-1}$ be a fixed square root of -1 in C. For any $c \in C$ and $X \in \mathfrak{g}_0$ put $c * X = aX + b\Gamma(X)$ where $c = a + \sqrt{-1}\, b \ (a, b \in R)$. Under this multiplication $\mathfrak{g}_0$ becomes a Lie algebra over C. We shall denote this complex algebra by $\mathfrak{g}^*$. Similarly the algebra $\mathfrak{h}_0$ regarded as a (complex) subalgebra of $\mathfrak{g}^*$ will be denoted by $\mathfrak{h}^*$. Then $\mathfrak{h}^*$ is a Cartan subalgebra of $\mathfrak{g}^*$. Let $X \to ad\, X (X \in \mathfrak{g}^*)$ be the adjoint representation of $\mathfrak{g}^*$ and let $B(X, Y) = sp(ad\, X\, ad\, Y)(X, Y \in \mathfrak{g}^*)$. Given any linear function λ on $\mathfrak{h}^*$ we denote by H_λ the unique element in $\mathfrak{h}^*$ such that $\lambda(H) = B(H, H_\lambda)$ for all $H \in \mathfrak{h}^*$. Let $H_1, \ldots, H_l$ be a base for $\mathfrak{h}_{\mathfrak{P}_0}$ over R. Then it is also a base for $\mathfrak{h}^*$ over C. We shall say that λ is real if $H_\lambda = \Sigma_{1 \le i \le l} c_i H_i (c_i \in R)$ and furthermore that $\lambda > 0$ if $\lambda \ne 0$ and $c_j > 0$ where j is the least index $(1 \le j \le l)$ such that $c_j \ne 0$. For every root α of $\mathfrak{g}^*$ (with respect to $\mathfrak{h}^*$) we choose an element $X_\alpha \ne 0$ in $\mathfrak{g}^*$ such that $[H, X_\alpha] = \alpha(H) * X_\alpha (H \in \mathfrak{h}^*)$. We can do this in such a way that $B(X_\alpha, X_{-\alpha}) = 1$ and $X_\alpha + X_{-\alpha}$, $\sqrt{-1} * (X_\alpha - X_{-\alpha})$ are both in $\mathfrak{L}_0$. Put $H_\alpha = \Sigma_{1 \le i \le l} \alpha^i H_i (\alpha^i \in R)$ and let $\mathfrak{N}^* = \Sigma_{\alpha \in P} C * X_\alpha$ where P is the set of all positive roots. Then $\mathfrak{N}^*$ is a nilpotent subalgebra of $\mathfrak{g}^*$ to which there corresponds an analytic subgroup N of G.

Let $C_c^\circ(G)$ be the class of complex-valued functions on G which are everywhere defined and indefinitely differentiable and which vanish outside a compact set. For any complex number c we denote by $\bar{c}$ its complex conjugate. Moreover if $z = x + \sqrt{-1}\, y \ (x, y \in R)$ is a complex variable and f a complex-valued differentiable function of x and y, we write

$$\frac{\partial}{\partial z}f = \frac{1}{2}\left(\frac{\partial}{\partial x} - \sqrt{-1}\,\frac{\partial}{\partial y}\right)f, \qquad \frac{\partial}{\partial \bar{z}}f = \frac{1}{2}\left(\frac{\partial}{\partial x} + \sqrt{-1}\,\frac{\partial}{\partial y}\right)f.$$

We shall now first prove a formula which has been obtained by Gelfand and Naimark[2] in the case when G is the $n \times n$ complex unimodular group. Let $X \to \exp X(X \in \mathfrak{g}_0)$ denote the exponential mapping of $\mathfrak{g}_0$ into G. Put $\rho = \frac{1}{2}\Sigma_{\alpha \in P}\alpha$ and let du and dn denote the elements of the invariant Haar measures on K and N respectively. We assume that $\int_K du = 1$.

Theorem 1. *Put* $H_a = \Sigma_{1 \le i \le l}a_i * H_i \,(a_i \in C)$ *and*

$$D_\alpha = \sum_{1 \le i \le l} \alpha^i \frac{\partial}{\partial a_i}, \qquad \bar{D}_\alpha = \sum_{1 \le i \le} \alpha^i \frac{\partial}{\partial \bar{a}_i} \qquad (\alpha \in P).$$

Then with a suitable normalization of dn we have

$$f(1) = \lim_{H_a \to 0} \prod_{\alpha \in P} D_\alpha \bar{D}_\alpha \left\{ e^{\rho(H_a) + \overline{\rho(H_a)}} \int_{K \times N} f\left[u(\exp H_a)nu^{-1}\right] du\, dn \right\}$$

for any[3] $f \in C_c^\infty(G)$.

We give below a rapid sketch of the various steps leading to the proof of this theorem. Let θ denote the automorphism of $\mathfrak{g}_0$ over R given by $\theta(X + Y) = X - Y(X \in \mathfrak{L}_0, Y \in \mathfrak{P}_0)$. Let $\mathfrak{N}_0$ denote the set $\mathfrak{N}^*$ regarded as a vector space over R and let $dX(X \in \mathfrak{N}_0)$ be the element of the usual Euclidean measure on $\mathfrak{N}_0$.

Lemma 1. *There exists a real constant $c > 0$ such that*

$$\lim_{t \to 0} \frac{d}{dt}\left\{ e^{t\rho(H - \theta H)} \int_N f\left[(\exp tH)n\right] dn \right\} =$$

$$c \int_{\mathfrak{N}_0}\left\{ \frac{d}{dt}f\left[\exp(X + tH)\right] \right\}_{t=0} dX \qquad (t \in R)$$

for any $f \in C_c^\infty(G)$ *and* $H \in \mathfrak{h}_0$.

Since $\mathfrak{g}_0$ is a real Euclidean space it may be regarded as an analytic manifold. Let $C_c^\infty(\mathfrak{g}_0)$ be the class of all complex-valued functions on $\mathfrak{g}_0$ which are everywhere indefinitely differentiable and which vanish outside a compact set. Put

$$X = \sum_{1 \le i \le l} a_i * H_i + \sum_{\alpha \in P} z_\alpha * X_\alpha + \sum_{\alpha \in P} z_{-\alpha} * X_{-\alpha}$$

where $a_i, z_\alpha, z_{-\alpha}(1 \le i \le l, \alpha \in P)$ are independent complex variables. For any complex variable $z = x + \sqrt{-1}\,y(x, y \in R)$ let $d\mu(z)$ denote the element $dx\, dy$ of Euclidean measure on the corresponding complex plane. Let $x \to Ad(x)\,(x \in G)$ be the adjoint representation of G. Consider a function $F \in C_c^\infty(\mathfrak{g}_0)$ such that $F(Ad(u)X) = F(X)(u \in K)$. Put

$$g(Y) = \frac{1}{(2\pi)^n}\int_{\mathfrak{g}_0} \exp\left(\tfrac{1}{2}\sqrt{-1}\left[B(X, Y) + \overline{B(X, Y)}\right]\right)F(X)\, dX \qquad (Y \in \mathfrak{g}_0). \quad (1)$$

Here $n = \frac{1}{2}\dim_R \mathfrak{g}_0$ and $dX = \Pi_{1 \le i \le l} d\mu(a_i) \Pi_{\alpha \in P} d\mu(z_\alpha) d\mu(z_{-\alpha})$. Then if we assume, as we may, that $B(H_i, H_j) = \delta_{ij} (1 \le i, j \le l)$ it follows that

$$F(0) = \frac{1}{(2\pi)^n} \int_{\mathfrak{g}_0} g(X)\, dX \qquad (2)$$

and $g(Ad(u)X) = g(X)(u \in K, X \in \mathfrak{g}_0)$. Now it is known that $\cup_{u \in K} Ad(u)(\mathfrak{h}_0 + \mathfrak{N}_0) = \mathfrak{g}_0$ and from this we can deduce the following lemma.

Lemma 2. *Let $g(X)$ be a measurable function on $\mathfrak{g}_0$ such that*

$$g[Ad(u)X] = g(X)(u \in K, X \in \mathfrak{g}_0) \text{ and } \int_{\mathfrak{g}_0} |g(X)|\, dX < \infty.$$

Then

$$\int_{\mathfrak{g}_0} g(X)\, dX = \int_{\substack{Z \in \mathfrak{N}_0 \\ H \in \mathfrak{h}_0}} \prod_{\alpha \in P} |\alpha(H)|^2 g(Z + H)\, dZ\, dH$$

where dZ and dH are the elements of the (suitably normalized) Euclidean measures on $\mathfrak{N}_0$ and $\mathfrak{h}_0$ respectively.

Applying this lemma to equation (2) we get

$$F(0) = \lim_{H_\alpha \to 0} \left\{ c \int_{Z \in \mathfrak{N}_0} \prod_{\alpha \in P} D_\alpha \bar{D}_\alpha F(Z + H_\alpha)\, dZ \right\}$$

where c is a positive real constant depending only on the normalization of dZ. The assertion of the theorem now follows without much difficulty if we take into account lemma 1.

Now we assume that the base $H_1, \ldots, H_l$ is so chosen that $\exp H_a = 1$ if and only if $\sqrt{-1}(a_i/2\pi)1 \le i \le l$ are all rational integers. Let A, A_+ and A_- be the analytic subgroups of G corresponding to $\mathfrak{h}_0$, $\mathfrak{h}_{\mathfrak{P}_0}$ and $\mathfrak{h}_{\mathfrak{Q}_0}$ respectively. Then A_- is compact while A_+ is simply connected. For any $h \in A$ we denote by h_+ and h_- the unique elements in A_+ and A_- respectively such that $h = h_+ h_-$. Also let $\log h_+$ denote the unique element $H \in \mathfrak{h}_{\mathfrak{P}_0}$ such that $h_+ = \exp H$. Let $\mathfrak{F}_+$ be the set of all linear functions ν on $\mathfrak{h}^*$ such that $\nu(H)$ is real for all $H \in \mathfrak{h}_{\mathfrak{P}_0}$. Moreover let $\mathfrak{F}_-$ denote the set of all linear functions Λ on $\mathfrak{h}^*$ such that $\Lambda(H_i)1 \le i \le l$ are all integers. Given any $\nu \in \mathfrak{F}_+$ and $\Lambda \in \mathfrak{F}_-$ put

$$\xi_{\nu,\Lambda}(h) = e^{\sqrt{-1}\,\nu(\log h_+)} e^{\Lambda(\log h_-)} \qquad (h \epsilon A)$$

$\log h_-$ being any element in $\mathfrak{h}_{\mathfrak{Q}_0}$ such that $\exp(\log h_-) = h_-$. It is known that the mapping $(u, h, n) \to uhn(u \in K, h \in A_+, n \in N)$ is a topological mapping of $K \times A_+ \times N$ on G. We may normalize the Haar measure on G in such a way that

$$dx = e^{4\rho(\log h)}\, du\, dh\, dn \qquad (x = uhn, u \in K, h \in A_+, n \in N)$$

dh being the Haar measure on A_+. Moreover we assume that the normalization of the various Haar measures is such that Theorem 1 holds and

$$\int_K du = 1, \int_{A_-} dh_- = 1, dh = dh_+\, dh_- \qquad (h \in A).$$

For any $x \in G$ and $u \in K$ define $u_x \in K$ and $H(x, u) \in \mathfrak{h}_{\mathfrak{P}_0}$ by the relation

$$xu = u_x [\exp H(x, u)] n \qquad (n \in N).$$

Given any $\Lambda \in \mathfrak{F}_-$ let $\mathfrak{H}_\Lambda'$ denote the set of all continuous functions ψ on K such that

$$\psi(u \exp H) = e^{-\Lambda(H)} \psi(u) \qquad (u \in K, H \in \mathfrak{h}_{\mathfrak{Q}_0}).$$

Let $L_2(K)$ be the Hilbert space consisting of all measurable and square-integrable functions on K, taken with the usual norm. Then the closure $\mathfrak{H}_\Lambda$ of $\mathfrak{H}_\Lambda'$ in $L_2(K)$ is a Hilbert space. For any $\nu \in \mathfrak{F}_+$ we define a unitary representation $\pi_{\nu, \Lambda}$ of G on $\mathfrak{H}_\Lambda$ as follows. If $\psi \in \mathfrak{H}_\Lambda$ its transform $\pi_{\nu, \Lambda}(x)\psi = \varphi$ is given by

$$\varphi(u) = e^{-\sqrt{-1}\,\nu(H(x^{-1}, u))} e^{-2\rho(H(x^{-1}, u))} \psi(u_{x^{-1}}) \qquad (u \in K).$$

It is easily proved that if $f \in C_c^\infty(G)$, the operator

$$\int_G f(x) \pi_{\nu, \Lambda}(x)\, dx$$

has a trace $T_{\nu, \Lambda}(f)$ which is given by

$$T_{\nu, \Lambda}(f) = \int f(uhnu^{-1}) \xi_{\nu, \Lambda}(h) e^{2\rho(\log h_+)}\, du\, dh\, dn$$

where the integral extends over all $u \in K, h \in A, n \in N$. Now $\mathfrak{F}_+$ is clearly a vector space over R of finite dimension. Let $d\nu$ denote the element of Euclidean measure in $\mathfrak{F}_+$. Then the following result is easily obtained from Theorem 1.

Theorem 2. *Put*

$$m(\nu, \Lambda) = \prod_{\alpha \in P} |\sqrt{-1}\,\nu(H_\alpha) + \Lambda(H_\alpha)|^2 \qquad (\nu \in \mathfrak{F}_+, \Lambda \in \mathfrak{F}_-).$$

Then if $d\nu$ is suitably normalized we have the formula

$$f(1) = \sum_{\Lambda \in \mathfrak{F}_-} \int_{\mathfrak{F}_+} m(\nu, \Lambda) T_{\nu, \Lambda}(f)\, d\nu \qquad [f \in C_c^\infty(G)]$$

the series being absolutely convergent.

Now suppose $f \in C_c^\infty(G)$ and

$$F(x) = \int_G \overline{f(y)} f(yx)\, dy.$$

Then $F \in C_c^\infty(G)$ and the operator $\int_G F(x) \pi_{\nu, \Lambda}(x)\, dx$ is self-adjoint and positive semidefinite. Hence $T_{\nu, \Lambda}(F)$ is real and non-negative. In fact

$$T_{\nu, \Lambda}(F) = \int_{v, u \in K} |f_{\nu, \Lambda}(v, u)|^2\, dv\, du$$

where

$$f_{\nu, \Lambda}(v, u) = \int_{AN} f(vhnu^{-1}) \xi_{\nu, \Lambda}(h) e^{2\rho(\log h_+)}\, dh\, dn.$$

Therefore

$$\int_G |f(x)|^2 dx = F(1) = \sum_{\Lambda \in \mathfrak{F}_-} \int_{\mathfrak{F}_+} m(\nu, \Lambda)\, d\nu \int_{K \times K} |f_{\nu, \Lambda}(v, u)|\, dv\, du$$

from Theorem 2. Since $m(\nu, \Lambda)$ is real and non-negative the following analogue of the Plancherel theorem is now easily obtained.

Theorem 3. *Let f be a measurable function on G such that*

$$\int_G |f(x)|^2 dx < \infty \quad \text{and} \quad \int_G |f(x)|\, dx < \infty.$$

Then

$$\int_G |f(x)|^2 dx =$$

$$\sum_{\Lambda \in \mathfrak{F}_-} \int_{\mathfrak{F}_+} m(\nu, \Lambda)\, d\nu \int_{K \times K} dv\, du \left| \int_{AN} f(vhnu^{-1}) \xi_{\nu, \Lambda}(h) e^{2\rho(\log h_+)}\, dh\, dn \right|^2.$$

[1] Harish-Chandra, PROC. NATL. ACAD. SCI., **37**, 362–365 (1951).
[2] Gelfand and Naimark, *Trudi Mat. Inst. Steklova*, **36**, 198 (1950).
[3] We denote the unit element of G by 1.

Reprinted from the Proceedings of the NATIONAL ACADEMY OF SCIENCES,
Vol. 38, No. 4, pp. 337–342.   April, 1952

PLANCHEREL FORMULA FOR THE 2 × 2 REAL UNIMODULAR GROUP

BY HARISH-CHANDRA

DEPARTMENT OF MATHEMATICS, COLUMBIA UNIVERSITY.

Communicated by P. A. Smith, February 27, 1952

In an earlier note[1] in these PROCEEDINGS we have obtained an *explicit* Plancherel formula for a connected complex semisimple Lie group.   The corresponding problem for a real group appears to be much more difficult due to the circumstance that two Cartan subgroups in it are in general not conjugate.   In fact it seems likely that there is a close connection between classes of conjugate Cartan subgroups and the various "series" of unitary representations which occur in the Plancherel formula.   The object of this note is to make a beginning in the study of this question by deriving an *explicit* Plancherel formula in the simple case of 2 × 2 real unimodular group.[2]

Let R be the field of real numbers and G the group of all 2 × 2 real matrices with determinant 1.   Then its Lie algebra $\mathfrak{g}_0$ consists of all matrices with trace zero.   Put

$$H = \begin{pmatrix} 1/2 & 0 \\ 0 & -1/2 \end{pmatrix}, \quad X = \begin{pmatrix} 0 & 1 \\ 0 & 0 \end{pmatrix}, \quad Y = \begin{pmatrix} 0 & 0 \\ 1 & 0 \end{pmatrix}, \quad U = \begin{pmatrix} 0 & 1/2 \\ -1/2 & 0 \end{pmatrix}$$

and

$$h_t = \exp tH, \quad n_s = \exp sX, \quad u_\theta = \exp \theta U \qquad (t, s, \theta \,\epsilon\, R).$$

Let K and A be the one parameter subgroups G corresponding to U and

H, respectively. Then the mapping $(u, t, s) \rightarrow uh_t n_s (u \in K, t, s \in R)$ is a topological mapping of $K \times R \times R$ onto G. Let G^* denote the adjoint group of G and $x \rightarrow x^*$ $(x \in G)$ the natural mapping of G on G^*. We denote by η_j the character of K given by

$$\eta_j(u_\theta) = e^{ij\theta} \qquad (j = 0, \tfrac{1}{2}).$$

(i is a fixed square-root of -1). Let du^* denote the element of Haar measure on K^* such that $\int_{K^*} du^* = 1$ and let $\mathfrak{H}$ be the Hilbert space of all square-integrable functions on K^*. Then for any complex number ν we define a representation $\pi_{\nu, j}$ $(j = 0, \tfrac{1}{2})$ of G on $\mathfrak{H}$ as follows. If $x \in G$ and $u \in K$ we write

$$xu = u_x \exp (t(x, u)H) \exp (s(x, u)X)$$

where $u_x \in K$ and $t(x, u)$, $s(x, u) \in R$. It is easy to verify that for fixed x, $(u_x)^*$, $u^{-1}u_x$ and $t(x, u)$ depend only on u^* and so we may denote them by u_x^*, $v(x, u^*)$ and $t(x, u^*)$, respectively. Now for any $f \in \mathfrak{H}$ we define $\pi_{\nu, j}(x)f = g$ by the rule

$$g(u^*) = \eta_j(v(x^{-1}, u^*))e^{-\nu t(x^{-1},\, u^*)}f(u_{x^{-1}}^*).$$

It can be checked without difficulty that $\pi_{\nu, j}$ is actually a representation and if $\nu + \tfrac{1}{2}$ is purely imaginary $\pi_{\nu, j}$ is unitary. Put $\tau_\lambda^+ = \pi_{\nu, 0}$ and $\tau_\lambda^- = \pi_{\nu, 1/2}$ where $\nu = i\lambda - \tfrac{1}{2}(\lambda \in R)$. Then τ_λ^+ and τ_λ^- are unitary representations of G on $\mathfrak{H}$.

Now consider the representation $\pi_{n+j, j}$ where n is an integer ≥ 0. For any integer m let f_m denote the function $f_m(u_\theta^*) = e^{im\theta}$. Then these functions form an orthonormal base for $\mathfrak{H}$. Let $V_{n, j}$ be the subspace of $\mathfrak{H}$ spanned by the set $\{f_m; \ |m + j| \leq n + j\}$ and let $W_{n, j}$ be the orthogonal complement of $V_{n, j}$ in $\mathfrak{H}$. Then $V_{n, j}$, $W_{n, j}$ are invariant subspaces and we denote by $\rho_{n, j}$, $\sigma_{n, j}$ the corresponding representations of G induced on them. It is known that there exists a unitary representation $\sigma_{j, n}$ which is infinitesimally[3] equivalent to $\sigma'_{j, n}$. Let $\chi_{n, j}$ be the character of $\rho_{n, j}$. Then

$$\left. \begin{array}{l} \chi_{n, j}(h_t) = \dfrac{e^{(n+j+1/2)t} - e^{-(n+j+1/2)t}}{e^{1/2t} - e^{-1/2t}} \\[2em] \chi_{n, j}(u_\theta) = \dfrac{\sin (n + j + 1/2)\theta}{\sin \dfrac{\theta}{2}} \end{array} \right\} \qquad (1)$$

Let $\gamma = \exp (2\pi U)$. For any function f on G put $f_+(x) = f(x) + f(\gamma x)$ and $f_-(x) = f(x) - f(\gamma x)$. Let $C_c^\infty(G)$ denote the set of all functions on G which vanish outside a compact set and which are indefinitely differentiable. Put

$$T_\lambda^{\pm}(f) = \int\!\!\int f_{\pm}((h_t n_s)^{u^*})e^{(i\lambda+1/2)t} \, dt \, ds \, du^* \qquad (2)$$

and

$$S_k{}^{\pm}(f) = \int f_{\pm}((h_t n_s)^{u^*})e^{(k+1)t}\, dt\, ds\, du^* - \int_G f_{\pm}(x)\chi_k(x)\, dx \quad (3)$$

Here $y^{x^*} = xyx^{-1}$ $(x, y \in G)$ and $dx = e^t\, dt\, ds\, du$ $(x = uh_t n_s)$, du being so normalized that $\int_K du = 1$. $2k$ is an integer ≥ 0 and $\chi_k = \chi_{k, 0}$ or $\chi_{k-1/2, 1/2}$ according as $2k$ is even or odd and we have to take the plus or minus sign in (3) accordingly. It can be seen without difficulty[4] that $T_\lambda{}^{\pm}(f)$, $S_k{}^{\pm}(f)$, respectively, are the traces of the operators $\int_G f(x)\tau_\lambda{}^{\pm}(x)\, dx$ and $\int_G f(x)\sigma_k{}^{\pm}(x)\, dx$ where $\sigma_k{}^+ = \sigma_{k, 0}$ or $\sigma_k{}^- = \sigma_{k-1/2, 1/2}$ according as $2k$ is even or odd. Moreover since the conjugates of K and A taken together cover G except for a set of measure zero it follows easily from (1) that

$$S_k{}^{\pm}(f) = \int f_{\pm}((h_t n_s)^{u^*})e^{1/2 t}e^{-(k+1/2)|t|}\, dt\, ds\, du^* -$$

$$\frac{2}{\pi} \int_{-\pi}^{\pi} d\theta \sin\left(k + \frac{1}{2}\right)\theta \sin\frac{\theta}{2} \int_{G^*} f_{\pm}((u_\theta)^{x^*})\, dx^* \quad (4)$$

where $dx^* = e^t\, dt\, ds\, du^*$ $(x^* = u^*(h_t n_s)^*)$.

THEOREM. *Let $f \in C_c^\infty(G)$. Then[5]*

$$8\pi f(1) = \frac{1}{2} \int_{-\infty}^{\infty} \lambda \tanh \pi\lambda \; T_\lambda{}^+(f)\, d\lambda + \frac{1}{2} \int_{-\infty}^{\infty} \lambda \coth \pi\lambda \; T_\lambda{}^-(f)\, d\lambda +$$

$$\sum_{n \geq 0} \left(n + \frac{1}{2}\right) S_n{}^+(f) + \sum_{n \geq 0} (n + 1)S_{n+1/2}^{-}(f)$$

the two series being absolutely convergent.

The proof is based on the following lemma.

LEMMA. *Let $F(x_1, x_2)$ be an indefinitely differentiable function of two real variables which vanishes outside a compact set. Put*

$$g(\theta) = \int_0^\infty (e^t - e^{-t})F(\theta e^t, \theta e^{-t})\, dt \qquad (\theta > 0)$$

Then

$$\operatorname*{Lim}_{\theta \to 0} \theta g(\theta) = \int_0^\infty F(s, 0)\, ds$$

$$\operatorname*{Lim}_{\theta \to 0} \frac{d}{d\theta}(\theta g(\theta)) = -2F(0, 0)$$

and there exist positive constants a and b such that

$$\left|\frac{1}{\theta}\left(\frac{d}{d\theta}(\theta g(\theta)) + 2F(0, 0)\right)\right| \leq a + b|\log \theta|$$

for $0 < \theta \leq 1$.

In proving the theorem we may assume that $f(uxu^{-1}) = f(x) = f(x^{-1})$ $(x \epsilon G, u \epsilon K)$. Put

$$F_{\pm}(\theta) = \int_{G^*} f_{\pm}((u_\theta)^{x*}) \, dx^*.$$

Then making use of the above lemma we find that

$$\underset{m \to \infty}{\mathrm{Lim}} \left\{ (m+1) \int_{-\infty}^{\infty} f_+(n_s) \, ds - \sum_{0 \le n \le m} \left(n + \frac{1}{2} \right)\frac{1}{\pi} \times \right.$$
$$\left. \int_{-\pi}^{\pi} \sin \left(n + \frac{1}{2} \right)\theta \sin \frac{\theta}{2} F_+(\theta) \, d\theta \right\} = - \underset{m \to \infty}{\mathrm{Lim}} \sum_{0 \le n \le m} \frac{1}{\pi} \times$$
$$\int_{-\pi}^{\pi} \cos \left(n + \frac{1}{2} \right) \theta \frac{d}{d\theta} \left(\sin \frac{\theta}{2} F_+(\theta) \right) d\theta = 2\pi f_+(1).$$

Now put $\varphi_{\pm}(t) = e^{t/2} \int_{-\infty}^{\infty} f_{\pm}((h_t n_s) \, ds$. Then

$$\sum_{0 \le n \le m} \left(n + \frac{1}{2} \right) \int_0^{\infty} \varphi_+(t) e^{-(n+1/2)t} \, dt =$$
$$(m+1)\varphi_+(0) + \sum_{0 \le n \le m} \int_0^{\infty} \frac{d\varphi_+(t)}{dt} e^{-(n+1/2)t} \, dt.$$

Hence

$$4\pi f_+(1) = \underset{m \to \infty}{\mathrm{Lim}} \left\{ \sum_{0 \le n \le m} \left(n + \frac{1}{2} \right) S_n{}^+(f) - \right.$$
$$\left. \sum_{0 \le n \le m} 2 \int_0^{\infty} \frac{d\varphi_+(t)}{dt} e^{-(n+1/2)t} \, dt \right\}.$$

Since $\varphi_+(t) = \varphi_+(-t)$, $\frac{1}{t} \frac{d\varphi_+(t)}{dt}$ is a differentiable function of t and therefore

$$\underset{m \to \infty}{\mathrm{Lim}} \sum_{0 \le n \le m} \int_0^{\infty} \frac{d\varphi_+(t)}{dt} e^{-(n+1/2)t} \, dt =$$
$$\int_0^{\infty} \frac{1}{t} \frac{d\varphi_+(t)}{dt} \frac{t}{(e^{t/2} - e^{-t/2})} \, dt = -\frac{1}{4} \int_{-\infty}^{\infty} \lambda \tanh \pi\lambda \, T_\lambda{}^+(f) \, d\lambda$$

by Fourier transformation since $T_\lambda{}^+(f) = \int_{-\infty}^{\infty} \varphi_+(t) e^{i\lambda t} \, dt$. Hence

$$4\pi f_+(1) = \frac{1}{2} \int_{-\infty}^{\infty} \lambda \tanh \pi\lambda \, T_\lambda{}^+(f) \, d\lambda + \sum_{0 \le n < \infty} \left(n + \frac{1}{2} \right) S_n^+(f). \quad (5)$$

Similarly we prove that

$$\sum_{n \geq 0} (n + 1) S_{n+\frac{1}{2}}^{-} (f) = 4\pi f_{-}(1) + \int_{0}^{\infty} \frac{d\varphi_{-}(t)}{dt} \coth \frac{t}{2} dt =$$

$$4\pi f_{-}(1) - \frac{1}{2} \int_{-\infty}^{\infty} \lambda \coth \pi\lambda \, T_{\lambda}^{-}(f) \, d\lambda. \quad (6)$$

Adding (5) and (6) we get the result. The Plancherel formula for G can now be obtained in the usual way.

Notice that if $f(uxv) = f(x)$ $(u, v \in K)$ then $T_{\lambda}^{-}(f) = S_n^{+}(f) = S_{n+\frac{1}{2}}^{-}(f) = 0$ and therefore

$$8\pi f(1) = \frac{1}{2} \int_{-\infty}^{\infty} \lambda \tanh \pi\lambda \, T_{\lambda}^{+}(f) \, d\lambda. \quad (7)$$

If we assume, as we may without essential loss of generality, that $f(x) = f(x^{-1})$ $(x \in G)$ and put

$$\varphi(t) = e^{t/2} \int_{-\infty}^{\infty} f(h_t n_s) \, ds$$

then (7) is equivalent to

$$2\pi f(1) = - \int_{-\infty}^{\infty} \frac{d\varphi(t)}{dt} \, (e^{t/2} - e^{-(t/2)})^{-1} dt.$$

This formula is quite easy to verify directly if we take into account the fact that the function $F(t, s) = f(h_t n_s)$ depends only on $(1 + s^2)e^t + e^{-t} = (e^{t/2} - e^{-(t/2)})^2 + s^2 e^t - 2$ which is the trace of the product of the matrix $h_t n_s$ with its transposed. Put $x = (e^{t/2} - e^{-(t/2)})$, $y = s e^{t/2}$ and $\Phi(r) = F(t, s)$ where $r = \sqrt{x^2 + y^2} \geq 0$. Then

$$\varphi(t) = \int_{-\infty}^{\infty} \Phi(r) \, dy$$

and therefore

$$\frac{d\varphi(t)}{dt} = \frac{dx}{dt} \int_{-\infty}^{\infty} \frac{x}{r} \frac{d\Phi(r)}{dr} \, dy.$$

Hence

$$\int_{-\infty}^{\infty} \frac{d\varphi(t)}{dt} \, (e^{t/2} - e^{-(t/2)})^{-1} dt = \int_{-\infty}^{\infty} \int_{-\infty}^{\infty} \frac{1}{r} \frac{d\Phi(r)}{dr} \, dx \, dy =$$

$$2\pi \int_{0}^{\infty} \frac{d\Phi(r)}{dr} \, dr = -2\pi\Phi(0) = -2\pi f(1).$$

Formula (7) has also been obtained independently by R. Godement.

[1] PROC. NATL. ACAD. SCI., **37**, 813–818 (1951).

[2] The representations of this group have been studied by Bargmann, *Ann. Math.*, **48**, 568–640 (1947).

[3] For the definition of infinitesimal equivalence see an earlier note (Proc. Natl. Acad. Sci., **37**, 362–365 (1951)). $\sigma_{j, n}$ is the direct sum of two irreducible representations which in Bargmann's notation are $D^+_{n\ +1\ +j}$ and $D^-_{n\ +1\ +j}$.

[4] See Proc. Natl. Acad. Sci., **37**, 366–369 (1951).

[5] Compare this result with formulas (11.7), (12.12) and (12.24) of Bargmann's paper.

REPRESENTATIONS OF A SEMISIMPLE LIE GROUP ON A BANACH SPACE. I.

BY

HARISH-CHANDRA

1. Introduction. During the past few years several authors have studied unitary representations of a Lie group or, more generally, of a locally compact group on a Hilbert space. In the present paper we approach this problem from a somewhat different point of view and do not require that the representation be necessarily unitary. In fact, following a suggestion of Chevalley, we allow the representation space to be a Banach space. But on the other hand we restrict ourselves to the case when the given group is a connected semisimple Lie group and attack the problem by means of a closer study of the corresponding representations of the Lie algebra. Roughly speaking, our plan of investigation can be divided into two separate parts. The first consists in a careful study of infinite-dimensional representations of a semisimple Lie algebra from a purely algebraic point of view. The second step is then to establish a sufficiently close relationship between the representations of the group and those of the algebra so that the information obtained about the latter can be put to use.

This paper is divided into three parts. Part I contains a purely algebraic theorem (Theorem 1) about infinite-dimensional representations of a semisimple Lie algebra. In the special case of a complex semisimple algebra this result had been obtained in a previous paper [8(e)]. Part II is concerned with the second step of the program mentioned above. It has been shown by Gårding [5] that every representation of a Lie group G gives rise in a natural way to a representation of its Lie algebra. However, it turns out that this correspondence between the representations of the group and of its algebra, as it stands, is not very satisfactory for our purpose. This is due to the fact that the Taylor series of an indefinitely differentiable function on G does not necessarily converge to this function. In order to remedy this defect we replace such functions by suitable analytic functions in Gårding's construction and thus obtain what we call well-behaved vectors in the representation space. The remainder of Part II is then devoted to the problem of approximating an arbitrary vector by well-behaved vectors. Theorem 3 deals with this question in a special case which is of particular importance for us.

In Part III we combine the results of Part I with those of Part II and this enables us to derive information about representations of a semisimple group G. We introduce the notion of infinitesimal equivalence of two representations and show that in the case of irreducible unitary representations

---

Received by the editors June 19, 1951.

185

this coincides with the usual notion of equivalence. It follows from Theorem 7 that any factor arising from a unitary representation of G is necessarily of type I in the terminology of Murray and von Neumann [14]. (This fact had been conjectured by Mautner [12].) Theorem 9 gives a purely infinitesimal criterion for the existence of a unitary representation which is infinitesimally equivalent to a given irreducible representation. In the last section we give a method for obtaining a large class of representations of G.

All the results of this paper, except Theorem 9, were obtained before the end of 1950. Due to various circumstances their full publication has been delayed although these results were announced, along with some others, in a short note [8(c)]. In the mean time Godement has obtained alternative proofs for some of them. However the present paper contains only a small proportion of the results which have appeared in a series of notes in the Proceedings of the National Academy [8(d)]. Their proofs will be published in subsequent papers.

I am greatly indebted to Professor Chevalley for reading the original draft of this paper and suggesting a large number of valuable improvements. In fact Part I was almost entirely rewritten by him and in Part II the whole treatment of well-behaved vectors has been very much simplified due to his suggestions. In particular the method of constructing well-behaved functions on a "quasi-nilpotent" group, as given in this paper, is due to Chevalley. My original construction was more complicated.

PART I. REPRESENTATIONS OF A SEMISIMPLE LIE ALGEBRA

2. Preliminary lemmas. Let k be a field of characteristic zero and $\mathfrak{l}$ a Lie algebra over k. Let $\mathfrak{k}$ be a subalgebra of $\mathfrak{l}$ and σ the representation of $\mathfrak{k}$ on $\mathfrak{l}$ given by $\sigma(X)Y = [X, Y]$ $(X \in \mathfrak{k}, Y \in \mathfrak{l})$. Following Koszul [11] we say that $\mathfrak{k}$ is reductive in $\mathfrak{l}$ if σ is a semisimple([1]) representation. Moreover $\mathfrak{l}$ is called reductive if it is reductive in itself. It is obvious that if $\mathfrak{k}$ is reductive in $\mathfrak{l}$, then $\mathfrak{k}$ is reductive.

LEMMA 1. *Suppose $\mathfrak{l}$ is reductive. Then $\mathfrak{l} = \mathfrak{c} + \mathfrak{l}'$, $\mathfrak{c} \cap \mathfrak{l}' = \{0\}$ where $\mathfrak{c}$ is the center of $\mathfrak{l}$ and $\mathfrak{l}' = [\mathfrak{l}, \mathfrak{l}]$ the derived algebra. Moreover $\mathfrak{l}'$ is semisimple. A finite-dimensional representation ρ of $\mathfrak{l}$ is semisimple if and only if $\rho(\Gamma)$ is semisimple for every $\Gamma \in \mathfrak{c}$.*

Since the adjoint representation of $\mathfrak{l}$ is semisimple, $\mathfrak{l}$ can be written as a direct sum $\sum_{1 \le j \le m} \mathfrak{l}_j$ of simple ideals $\mathfrak{l}_j$. Then $\mathfrak{l}_j$ is either semisimple or abelian. Suppose $\mathfrak{l}_j$ is abelian if $1 \le j \le m_1$, and semisimple if $j > m_1$ $(0 \le m_1 \le m)$. Put $\mathfrak{l}' = \sum_{j > m_1} \mathfrak{l}_j$ and $\mathfrak{c} = \sum_{j \le m_1} \mathfrak{l}_j$. Then $\mathfrak{l}'$ is semisimple and it is clear that $\mathfrak{c}$ is the kernel of the adjoint representation. Hence $\mathfrak{c}$ is the center and

---

([1]) We use here the terminology of Chevalley [4, Chap. VI]. A linear transformation of a vector space V (of finite dimension) is called semisimple if V is fully reducible under A.

$$[\mathfrak{l}, \mathfrak{l}] = \sum_{1 \le j \le m} [\mathfrak{l}_j, \mathfrak{l}_j] = \sum_{j > m_1} \mathfrak{l}_j = \mathfrak{l}'.$$

This proves the first part. Now $\rho(\mathfrak{l}) = \rho(\mathfrak{c}) + \rho(\mathfrak{l}')$. Since $\mathfrak{l}'$ is semisimple the same holds for the linear Lie algebra $\rho(\mathfrak{l}')$. So the center of $\rho(\mathfrak{l}')$ is $\{0\}$ and therefore $\rho(\mathfrak{c})$ is the center of $\rho(\mathfrak{l})$. Our second assertion now follows from a result of Jacobson [10].

LEMMA 2. *Suppose $\mathfrak{k}$ is reductive in $\mathfrak{l}$ and ρ is a finite-dimensional semisimple representation of $\mathfrak{l}$. Then ρ induces a semisimple representation of $\mathfrak{k}$.*

Let $\mathfrak{D}$ be the center of $\mathfrak{k}$. Since $\mathfrak{k}$ is reductive, $\rho(\mathfrak{D})$ is the center of $\rho(\mathfrak{k})$. Put $\mathfrak{l}_1 = \rho(\mathfrak{l})$, $\mathfrak{k}_1 = \rho(\mathfrak{k})$, and $\mathfrak{D}_1 = \rho(\mathfrak{D})$. Since $\mathfrak{k}$ is reductive in $\mathfrak{l}$, it is clear that $\mathfrak{k}_1$ is reductive in $\mathfrak{l}_1$. Moreover since ρ is semisimple, $\mathfrak{l}_1$ is reductive (see Jacobson [10]). Let $\mathfrak{c}_1$ be the center and $\mathfrak{l}_1'$ the derived algebra of $\mathfrak{l}_1$. Let σ denote the adjoint representation of $\mathfrak{l}_1$. Then we know from the above lemma that $\sigma(D)$ is semisimple for any $D \in \mathfrak{D}_1$. Choose an abelian subalgebra $\mathfrak{h}_1$ of $\mathfrak{l}_1$ such that $\mathfrak{h}_1 \supset \mathfrak{D}_1$, $\sigma(H)$ is semisimple for all $H \in \mathfrak{h}_1$, and $\mathfrak{h}_1$ is not contained in a larger such algebra. Then $\mathfrak{h}_1 \supset \mathfrak{c}_1$, and it is clear that $\mathfrak{l}_1' \cap \mathfrak{h}_1$ is a Cartan subalgebra of the *semisimple* algebra $\mathfrak{l}_1'$. Now if ϕ is a semisimple representation of $\mathfrak{l}_1$ it follows from the above lemma that $\phi(\Gamma)$ is semisimple for all $\Gamma \in \mathfrak{c}_1$. Moreover $\phi(H)$ is semisimple for every $H \in \mathfrak{l}_1' \cap \mathfrak{h}_1$, since $\mathfrak{l}_1' \cap \mathfrak{h}_1$ is a Cartan subalgebra of the *semisimple* algebra $\mathfrak{l}_1'$. Hence $\phi(H)$ is semisimple for all $H \in \mathfrak{h}_1$ and therefore in particular for $H \in \mathfrak{D}_1$. Therefore, again by the above lemma, the representation of $\mathfrak{l}_1$ defined by ϕ is semisimple. In particular if we take for ϕ the representation $\phi: \rho(X) \to \rho(X)$ $(X \in \mathfrak{l})$, we get the required result.

Let R and C be the fields of real and complex numbers respectively and $(-1)^{1/2}$ a fixed square root of -1 in C. Let $\mathfrak{g}_0$ be a semisimple Lie algebra over R and $\mathfrak{g}$ its complexification. Let $X \to \operatorname{ad} X$ $(X \in \mathfrak{g})$ denote the adjoint representation of $\mathfrak{g}$. Put $B(X, Y) = \operatorname{sp}(\operatorname{ad} X \operatorname{ad} Y)$. A real form $\mathfrak{g}_k$ of $\mathfrak{g}$ is called compact if $B(X, X) < 0$ for all $X \in \mathfrak{g}_k$ $(X \ne 0)$. It is well known (see Cartan [3] and also Mostow [13]) that there exists a compact real form $\mathfrak{g}_k$ and an automorphism θ of order 2 of $\mathfrak{g}$ with the following properties: $\theta \mathfrak{g}_0 \subset \mathfrak{g}_0$, $\theta \mathfrak{g}_k \subset \mathfrak{g}_k$ and

$$\mathfrak{g}_0 = \mathfrak{k}_0 + \mathfrak{p}_0, \qquad \mathfrak{g}_k = \mathfrak{k}_0 + (-1)^{1/2}\mathfrak{p}_0$$

where $\mathfrak{k}_0$ is the set of all $X \in \mathfrak{g}_0$ such that $\theta X = X$ and $\mathfrak{p}_0$ the set of all $Y \in \mathfrak{g}_0$ such that $\theta Y = -Y$. Let $\mathfrak{k}$ and $\mathfrak{p}$ be the subspaces of $\mathfrak{g}$ spanned by $\mathfrak{k}_0$ and $\mathfrak{p}_0$ respectively over C. Since θ is an automorphism, $\theta([X, Y]) = [\theta X, \theta Y]$ $(X, Y \in \mathfrak{g})$ and from this it follows immediately that

$$[\mathfrak{k}, \mathfrak{k}] \subset \mathfrak{k}, \qquad [\mathfrak{k}, \mathfrak{p}] \subset \mathfrak{p}, \qquad [\mathfrak{p}, \mathfrak{p}] \subset \mathfrak{k}.$$

Hence in particular $\mathfrak{k}$ is a subalgebra of $\mathfrak{g}$.

LEMMA 3. *$\mathfrak{k}$ is reductive in $\mathfrak{g}$.*

Let G be the component of identity of the adjoint group of $\mathfrak{g}_0$ and K the analytic subgroup of G corresponding to $\mathfrak{k}_0$. Then it is known (see Mostow [13]) that K is compact. From this it follows immediately that $\mathfrak{k}_0$ is reductive in $\mathfrak{g}_0$ and therefore $\mathfrak{k}$ in $\mathfrak{g}$.

Let $\mathfrak{h}_{\mathfrak{p}_0}$ be an abelian subalgebra of $\mathfrak{g}_0$ contained in $\mathfrak{p}_0$ and having the maximal possible dimension. Extend $\mathfrak{h}_{\mathfrak{p}_0}$ to a maximal abelian subalgebra $\mathfrak{h}_0$ of $\mathfrak{g}_0$. Let $X \in \mathfrak{h}_0$. Then if $Y \in \mathfrak{h}_{\mathfrak{p}_0}$, $[\theta X, Y] = - [\theta X, \theta Y] = -\theta([X, Y]) = 0$. Hence $X - \theta X$ commutes with all elements in $\mathfrak{h}_{\mathfrak{p}_0}$. But $X - \theta X \in \mathfrak{p}_0$ and therefore in view of our choice of $\mathfrak{h}_{\mathfrak{p}_0}$, $X - \theta X \in \mathfrak{h}_{\mathfrak{p}_0}$. This proves that $\theta \mathfrak{h}_0 = \mathfrak{h}_0$ and therefore $\mathfrak{h}_0 = \mathfrak{h}_{\mathfrak{k}_0} + \mathfrak{h}_{\mathfrak{p}_0}$ where $\mathfrak{h}_{\mathfrak{k}_0} = \mathfrak{h}_0 \cap \mathfrak{k}_0$. Since $\mathfrak{g}_k = \mathfrak{k}_0 + (-1)^{1/2} \mathfrak{p}_0$ is compact, $\operatorname{ad} X$ is semisimple for every $X \in \mathfrak{k}_0 \cup \mathfrak{p}_0$. Therefore if $\mathfrak{h}$ is the subspace of $\mathfrak{g}$ spanned by $\mathfrak{h}_0$, $\mathfrak{h}$ is a Cartan subalgebra of $\mathfrak{g}$. Put $\mathfrak{h}_{\mathfrak{p}} = \mathfrak{p} \cap \mathfrak{h}$, $\mathfrak{h}_{\mathfrak{k}} = \mathfrak{k} \cap \mathfrak{h}$. Let $H_1, \cdots, H_l$ be a base for $\mathfrak{h}_0$ (over R) such that $H_1, \cdots, H_m$ is a base for $\mathfrak{h}_{\mathfrak{p}_0}$ and $H_{m+1}, \cdots, H_l$ is a base for $\mathfrak{h}_{\mathfrak{k}_0}$. Put $H_i^* = H_i$ or $(-1)^{1/2} H_i$ according as $i \leq m$ or $i > m$ and let $\mathfrak{h}^* = \mathfrak{h}_{\mathfrak{p}_0} + (-1)^{1/2} \mathfrak{h}_{\mathfrak{k}_0}$. For any linear function λ on $\mathfrak{h}$ let H_λ denote the unique element in $\mathfrak{h}$ such that

$$B(H_\lambda, H) = \lambda(H) \qquad (H \in \mathfrak{h}).$$

We say that λ is real if $H_\lambda \in \mathfrak{h}^*$. Now let $H = \sum_{i=1}^l c_i H_i^*$ $(c_i \in R)$ be any element in $\mathfrak{h}^*$. We say that $H > 0$ if $H \neq 0$ and $c_j > 0$ where j is the least index such that $c_j \neq 0$. If λ and μ are two linear functions on $\mathfrak{h}$ such that $\lambda - \mu$ is real, then we write $\lambda > \mu$ or $\mu < \lambda$ if $H_{\lambda - \mu} = H_\lambda - H_\mu > 0$. Moreover we denote by $\theta \lambda$ the linear function $H \to \lambda(\theta H)$ $(H \in \mathfrak{h})$.

It is known that every root α of $\mathfrak{g}$ with respect to $\mathfrak{h}$ is a real linear function and if α is a root $\theta \alpha$ is also a root. For every root α choose an element $X_\alpha \in \mathfrak{g}$ such that $X_\alpha \neq 0$ and $[H, X_\alpha] = \alpha(H) X_\alpha$ for all $H \in \mathfrak{h}$. Let P be the set of all roots $\alpha > 0$. Define the subsets P_+ and P_- of P as follows. A root $\alpha \in P$ belongs to P_+ or P_- according as $\theta \alpha \neq \alpha$ or $\theta \alpha = \alpha$. It can be shown (see Iwasawa [9]) that $\theta \alpha < 0$ for $\alpha \in P_+$ and $H_\beta \in \mathfrak{h}_{\mathfrak{k}}$ and $X_\beta, X_{-\beta} \in \mathfrak{k}$ for $\beta \in P_-$. Moreover $\alpha > \beta$ for any $\alpha \in P_+$ and $\beta \in P_-$. Let $\mathfrak{n}$ be the subspace of $\mathfrak{g}$ spanned by X_α $(\alpha \in P_+)$ and $\mathfrak{m}$ the subspace spanned by $\mathfrak{h}_{\mathfrak{k}}$, X_β and $X_{-\beta}$ $(\beta \in P_-)$. Then $\mathfrak{n}$ is a nilpotent subalgebra of $\mathfrak{g}$. Put $\mathfrak{n}_0 = \mathfrak{g}_0 \cap \mathfrak{n}$. Iwasawa [9] has proved that $\mathfrak{g}_0$ is the direct sum of $\mathfrak{k}_0$, $\mathfrak{h}_{\mathfrak{p}_0}$, and $\mathfrak{n}_0$.

LEMMA 4. *The centraliser of $\mathfrak{h}_{\mathfrak{p}}$ in $\mathfrak{g}$ is $\mathfrak{h}_{\mathfrak{p}} + \mathfrak{m}$ and so $\mathfrak{m}$ is the centraliser of $\mathfrak{h}_{\mathfrak{p}}$ in $\mathfrak{k}$. Moreover there exists an element $H \in \mathfrak{h}_{\mathfrak{p}}$ whose centraliser (in $\mathfrak{g}$) is exactly $\mathfrak{h}_{\mathfrak{p}} + \mathfrak{m}$. Finally $\mathfrak{m}$ is a subalgebra of $\mathfrak{k}$ and*

$$\dim \mathfrak{k} - \dim \mathfrak{m} = \dim \mathfrak{p} - \dim \mathfrak{h}_{\mathfrak{p}} = \dim \mathfrak{n}.$$

Let $X = H_0 + \sum_{\alpha \in P} (c_\alpha X_\alpha + c_{-\alpha} X_{-\alpha})$ $(H_0 \in \mathfrak{h}; c_\alpha, c_{-\alpha} \in C)$. Then

$$[H, X] = \sum_{\alpha \in P} (c_\alpha \alpha(H) X_\alpha - c_{-\alpha} \alpha(H) X_{-\alpha}) \qquad (H \in \mathfrak{h}).$$

Therefore if X commutes with $\mathfrak{h}_{\mathfrak{p}}$, $c_\alpha = c_{-\alpha} = 0$ unless α is identically zero on

$\mathfrak{h}_\mathfrak{p}$. But α is zero on $\mathfrak{h}_\mathfrak{p}$ if and only if $\alpha = \theta\alpha$. This proves that $\mathfrak{h}_\mathfrak{p} + \mathfrak{m}$ is the centraliser of $\mathfrak{h}_\mathfrak{p}$ in $\mathfrak{g}$. Choose $H \in \mathfrak{h}_\mathfrak{p}$ such that $\alpha(H) \neq 0$ for $\alpha \in P_+$. Then it is clear from the above argument that the centraliser of H is $\mathfrak{h}_\mathfrak{p} + \mathfrak{m}$.

If $\alpha \in P_+$, $\theta\alpha < 0$ and so $-\theta\alpha > 0$. Hence $-\theta\alpha \neq -\alpha$ and this shows that $-\theta\alpha \in P_+$. The mapping $\alpha \to -\theta\alpha$ $(\alpha \in P_+)$ therefore defines a permutation of order 2 on P_+. Thus when α runs through P_+, $\theta\alpha$ runs through the set $\{-\alpha; \alpha \in P_+\}$. Since θX_α and $X_{\theta\alpha}$ differ only by a nonzero constant factor, it follows that X_α, θX_α $(\alpha \in P_+)$ are linearly independent modulo $\mathfrak{h}_\mathfrak{p} + \mathfrak{m}$. The same therefore holds for $X_\alpha - \theta X_\alpha$, $X_\alpha + \theta X_\alpha$ $(\alpha \in P_+)$. Let $\mathfrak{q} = \sum_{\alpha \in P_+} C(X_\alpha - \theta X_\alpha)$ and $\mathfrak{l} = \sum_{\alpha \in P_+} C(X_\alpha + \theta X_\alpha)$. Then $\mathfrak{q} \subset \mathfrak{p}$ and $\mathfrak{l} \subset \mathfrak{k}$ and $\mathfrak{q} + \mathfrak{l} + \mathfrak{h}_\mathfrak{p} + \mathfrak{m} = \mathfrak{g} = \mathfrak{p} + \mathfrak{k}$. Hence $\mathfrak{q} + \mathfrak{h}_\mathfrak{p} = \mathfrak{p}$, $\mathfrak{l} + \mathfrak{m} = \mathfrak{k}$, and if q is the number of roots in P_+, it is clear that dim $\mathfrak{n} = q = $ dim $\mathfrak{q} = $ dim $\mathfrak{p} - $ dim $\mathfrak{h}_\mathfrak{p}$. Similarly $q = $ dim $\mathfrak{l} = $ dim $\mathfrak{k} - $ dim $\mathfrak{m}$.

Since $\mathfrak{m}$ is the centraliser of $\mathfrak{h}_\mathfrak{p}$ in $\mathfrak{k}$, $\mathfrak{m}$ is a subalgebra of $\mathfrak{k}$.

LEMMA 5. $\mathfrak{m}$ *is reductive both in* $\mathfrak{k}$ *and in* $\mathfrak{g}$ *and* $[\mathfrak{m}, \mathfrak{n}] \subset \mathfrak{n}$. *Furthermore if* σ *is the adjoint representation of the algebra* $\mathfrak{h}_\mathfrak{p} + \mathfrak{n}$, *then* sp $\sigma(H) = \sum_{\alpha \in P} \alpha(H)$ $(H \in \mathfrak{h}_\mathfrak{p})$.

Put $\mathfrak{m}_0 = \mathfrak{m} \cap \mathfrak{g}_0$ and define G and K as in the proof of Lemma 3. Let M be the set of all $x \in K$ such that $xH = H$ $(H \in \mathfrak{h}_{\mathfrak{p}_0})$. Clearly M is a closed subgroup of K and so it is compact. Moreover from Lemma 4 $\mathfrak{m}_0$ is the subalgebra of $\mathfrak{k}_0$ corresponding to the component of identity of M. Hence $\mathfrak{m}_0$ is reductive in $\mathfrak{k}_0$ and $\mathfrak{g}_0$ and therefore the same is true of $\mathfrak{m}$ in $\mathfrak{k}$ and $\mathfrak{g}$.

Let $\alpha \in P_+$ and $\beta \in P_-$. Then $\alpha + \beta$ and $\alpha - \beta$ are both greater than 0. Moreover $\theta(\alpha + \beta) = \theta\alpha + \beta \neq \alpha + \beta$ since $\alpha \neq \theta\alpha$. Similarly $\theta(\alpha - \beta) \neq \alpha - \beta$. This shows that $[X_\alpha, X_\beta]$ and $[X_\alpha, X_{-\beta}]$ are both in $\mathfrak{n}$. Since $[\mathfrak{h}, \mathfrak{n}] \subset \mathfrak{n}$, we conclude that $[\mathfrak{m}, \mathfrak{n}] \subset \mathfrak{n}$.

It is clear that sp $\sigma(H) = \sum_{\alpha \in P_+} \alpha(H)$ $(H \in \mathfrak{h}_\mathfrak{p})$. Since α is zero on $\mathfrak{h}_\mathfrak{p}$ if $\alpha \in P_-$, it follows that sp $\sigma(H) = \sum_{\alpha \in P} \alpha(H)$.

Let ρ be a representation of $\mathfrak{k}$ on a vector space V whose dimension need not be finite. A subspace W of V will be called ρ-stable if $\rho(X)W \subset W$ for all $X \in \mathfrak{k}$. Let ρ_W denote the representation of $\mathfrak{k}$ induced on a ρ-stable subspace W. We shall call ρ_W simple (semisimple) if dim $W < \infty$ and ρ_W is simple (semisimple) in the usual sense (see Chevalley [4, Chap VI]). Also we then say that W is a simple (semisimple) subspace of V.

Let Ω denote the set of all (equivalence) classes of simple representations of $\mathfrak{k}$. For any $\mathfrak{D} \in \Omega$ we denote by $V_\mathfrak{D}$ the sum of all ρ-stable subspaces W of V such that $\rho_W \in \mathfrak{D}$. $(V_\mathfrak{D} = \{0\}$ if no such W exists.)

LEMMA 6. *If* W *is any ρ-stable subspace of* V, *then*

$$W \cap \left(\sum_{\mathfrak{D} \in \Omega} V_\mathfrak{D} \right) = \sum_{\mathfrak{D} \in \Omega} W \cap V_\mathfrak{D}.$$

Moreover the sum in $\sum_{\mathfrak{D} \in \Omega} V_\mathfrak{D}$ *is direct.*

For any $x \in \sum_{\mathfrak{D} \in \Omega} V_{\mathfrak{D}}$ let V_x be the smallest ρ-stable subspace containing x. It is clear that V_x is contained in the sum of a finite number of simple spaces. Hence V_x is semisimple (see Chevalley [4, Chap. VI]) and therefore $V_x = \sum_{\mathfrak{D}} V_x \cap V_{\mathfrak{D}}$. Applying this result to the elements of $W \cap (\sum_{\mathfrak{D} \in \Omega} V_{\mathfrak{D}})$ we get the first statement of the lemma. Let Ω_1 be a subset of Ω. Then for any $x \in \sum_{\mathfrak{D} \in \Omega_1} V_{\mathfrak{D}}$, V_x is contained in the sum of a finite number of simple spaces W' such that the class of $\rho_{W'}$ lies in Ω_1. Hence it follows (see Chevalley loc. cit.) that every simple component of ρ_{V_x} belongs to a class in Ω_1 and therefore $V_x \cap V_{\mathfrak{D}} = \{0\}$ if $\mathfrak{D} \in \Omega_1$. This proves the directness of the sum $\sum_{\mathfrak{D} \in \Omega} V_{\mathfrak{D}}$.

An element $x \in V$ will be said to transform (under ρ) according to $\mathfrak{D}$ if $x \in V_{\mathfrak{D}}$ and we shall say that the representation ρ is quasi-semisimple if $V = \sum_{\mathfrak{D} \in \Omega} V_{\mathfrak{D}}$. In order that this be the case it is necessary and sufficient that every element $x \in V$ should belong to some semisimple subspace W of V. The following lemma is well known (see Godement [7(b), p. 102]).

LEMMA 7. *Assume that ρ is quasi-semisimple. Let V_0 be the space of all $x \in V$ such that $\rho(X)x = 0$ for all $X \in \mathfrak{k}$ and let V_1 be the space spanned by all elements $\rho(X)x$ ($X \in \mathfrak{k}$, $x \in V$). Then V is the direct sum of V_0 and V_1.*

Let $\mathfrak{D}_0$ be the class of the simple zero representation of degree 1 of $\mathfrak{k}$. Then $V_0 = V_{\mathfrak{D}_0}$ and it is clear that $V_1 \subset \sum_{\mathfrak{D} \neq \mathfrak{D}_0} V_{\mathfrak{D}}$ which shows that the sum $V_0 + V_1$ is direct. In order to prove that $V_0 + V_1 = V$ it is sufficient to show that if W is a ρ-stable subspace of V such that $\rho_W \in \mathfrak{D}$ ($\mathfrak{D} \neq \mathfrak{D}_0$), then $W \subset V_1$. The subspace W' of W spanned by the elements $\rho(X)x$ ($X \in \mathfrak{k}$, $x \in W$) is clearly ρ-stable and not equal to $\{0\}$ since ρ_W is simple and not in $\mathfrak{D}_0$. Hence $W = W'$, which proves our assertion.

Let V and V' be the spaces of the representations ρ and ρ' of $\mathfrak{k}$. Then the tensor product $V \times V'$ is the space of a representation $\rho + \rho'$ (the tensor sum of ρ and ρ') which is defined as follows: if $X \in \mathfrak{k}$, $x \in V$, and $x' \in V'$, then $((\rho + \rho')(X))(x \times x') = (\rho(X)x) \times x' + x \times (\rho(X')x')$.

LEMMA 8. *Assume that the representations ρ and ρ' are quasi-semisimple. Then $\rho + \rho'$ is also quasi-semisimple.*

Every element of $V \times V'$ is contained in a space of the form $W \times W'$ where W and W' are semisimple subspaces of V and V' respectively. Let $\mathfrak{c}$ be the center of $\mathfrak{k}$. Since ρ_W is semisimple, it follows from Lemma 1 that we can choose a base $(x_1, \cdots, x_m)$ in W such that the matrix representing $\rho_W(\Gamma)$ ($\Gamma \in \mathfrak{c}$) relative to this base is diagonal. A similar base $(x_1', \cdots, x_{m'}')$ can be found for W' with respect to $\rho_{W'}'$. But then it is clear that $(\rho + \rho')_{W + W'}(\Gamma)$ ($\Gamma \in \mathfrak{c}$) is represented by a diagonal matrix relative to the base $x_i \times x_j'$, $1 \leq i \leq m$; $1 \leq j \leq m'$, for $W \times W'$. In view of Lemma 1 this shows that $W \times W'$ is semisimple under $\rho + \rho'$. Hence every element of $V \times V'$ is contained in a semisimple subspace and this proves the lemma.

Let $\mathfrak{B}$ be the universal enveloping[2] algebra of $\mathfrak{g}$ and $\mathfrak{X}$ the subalgebra of $\mathfrak{B}$ generated by $(1, \mathfrak{k})$. Then $\mathfrak{X}$ is the universal enveloping algebra of $\mathfrak{k}$ (see [8(e)]) and there is a natural 1-1 correspondence between representations of $\mathfrak{k}$ and their unique extensions[3] on $\mathfrak{X}$. It is convenient to identify the representations of $\mathfrak{k}$ and $\mathfrak{X}$ under this correspondence. Let π be a representation of $\mathfrak{B}$ on V. Then the restriction of π on $\mathfrak{k}$ is a representation of $\mathfrak{k}$. For any $\mathfrak{D} \in \Omega$ we define $V_{\mathfrak{D}}$ as the set of all elements of V which transform under $\pi(\mathfrak{k})$ according to $\mathfrak{D}$.

LEMMA 9. *The subspace $\sum_{\mathfrak{D} \in \Omega} V_{\mathfrak{D}}$ is invariant under $\pi(\mathfrak{B})$.*

Let $V_0 = \sum_{\mathfrak{D} \in \Omega} V_{\mathfrak{D}}$. Since $(1, \mathfrak{g})$ generates $\mathfrak{B}$ it would be sufficient to show that $\pi(\mathfrak{g}) V_0 \subset V_0$. But $\mathfrak{g} = \mathfrak{k} + \mathfrak{p}$ and V_0 is obviously invariant under $\pi(\mathfrak{k})$. Hence we have only to show that $\pi(\mathfrak{p}) V_0 \subset V_0$. Let σ be the representation of $\mathfrak{k}$ on $\mathfrak{p}$ defined by $\sigma(X) Y = [X, Y]$ $(X \in \mathfrak{k}, Y \in \mathfrak{p})$ and let ρ be the representation of $\mathfrak{k}$ on V_0. Consider $\mathfrak{p} \times V_0$ and the representation $\sigma + \rho$ induced on it. Since $\mathfrak{k}$ is reductive in $\mathfrak{g}$, σ is semisimple. Moreover for any given $x \in V_0$ we can find a ρ-stable semisimple subspace W of V_0 such that $x \in W$. Then $\mathfrak{p} \times W$ is semisimple under the representation ν induced on it under $\sigma + \rho$ (Lemma 8). Let λ be the linear mapping of $\mathfrak{p} \times W$ into V defined by $\lambda(Y \times w) = \pi(Y)w$ $(Y \in \mathfrak{p}, w \in W)$. Then it is easy to verify that $\pi(X)\lambda = \lambda\nu(X)$. This means that λ is a homomorphism of the $\mathfrak{k}$-module $\mathfrak{p} \times W$ into the $\mathfrak{k}$-module V. Since $\mathfrak{p} \times W$ is semisimple the same holds for $\lambda(\mathfrak{p} \times W)$ (see Chevalley [4, Chap. VI]). Since $\pi(\mathfrak{p})x \subset \lambda(\mathfrak{p} \times W)$ it follows that $\pi(\mathfrak{p})x$ is contained in a semisimple $\mathfrak{k}$-submodule of V and therefore $\pi(\mathfrak{p})x \subset V_0$. This proves the lemma.

Let V be a vector space over C of finite dimension. Then V may be considered as an abelian Lie algebra and the universal enveloping algebra of this Lie algebra is called the *symmetric algebra over* V. We shall denote it by $S(V)$. If $\{x_1, \cdots, x_n\}$ is a base for V, the monomials, $x_1^{e_1} x_2^{e_2} \cdots x_n^{e_n}$ $(e_i \geq 0; 1 \leq i \leq n)$ form a base for $S(V)$ and so $S(V)$ may be regarded as the algebra of polynomials in $x_1, \cdots, x_n$. For any integer $d \geq 0$ we denote by $S_d(V)$ the subspace of $S(V)$ consisting of all forms in $(x_1, \cdots, x_n)$ of degree d. It is clear that $S_d(V)$ is independent of the particular choice of the base $(x_1, \cdots, x_n)$. We call an element $F \in S(V)$ homogeneous of degree d if $F \in S_d(V)$.

Let T be an endomorphism of V. Then it follows from the structure of $S(V)$ as an algebra of polynomials that T may be extended uniquely to a derivation d_T of $S(V)$. If $f = P(x_1, \cdots, x_n)$ where P is a polynomial, then $d_T f = \sum_{1 \leq i \leq n} (\partial P(x_1, \cdots, x_n)/\partial x_i)(Tx_i)$. Hence d_T maps $S_d(V)$ into itself. Moreover the mapping $T \to d_T$ is linear and[4] $d_{[T_1, T_2]} = [d_{T_1}, d_{T_2}]$.

----

[2] For a definition of the universal enveloping algebra see [8(a)].

[3] Let $\mathfrak{A}$ be an associative algebra with a unit element 1 and π a representation of $\mathfrak{A}$. We shall always assume that $\pi(1)$ is the identity mapping of the representation space and this condition will be included in the definition of a representation.

[4] Throughout this paper we write $[a, b] = ab - ba$ whenever a, b lie in an associative algebra or are endomorphisms of a vector space.

Now consider in particular the algebra $S(\mathfrak{g})$. The product of elements of $\mathfrak{g}$ do not have the same meaning in $S(\mathfrak{g})$ as in $\mathfrak{B}$. In order to avoid confusion we shall represent by $\overline{X}$ an element $X \in \mathfrak{g}$ when we want to consider it as an element of $S(\mathfrak{g})$ (this is less cumbersome than introducing different notations for multiplications in $S(\mathfrak{g})$ and in $\mathfrak{B}$). Let $(X_1, \cdots, X_n)$ be a base for $\mathfrak{g}$. Define a linear mapping λ of $S(\mathfrak{g})$ into $\mathfrak{B}$ as follows:

$$\lambda(\overline{X}_1^{e_1} \cdots \overline{X}_n^{e_n}) = (1/m!) \sum X_{i_1} X_{i_2} \cdots X_{i_m} \qquad (0 \leqq e_i < \infty, 1 \leqq i \leqq n)$$

where $m = e_1 + \cdots + e_n$ and the summation is over all sequences $(i_1, \cdots, i_m)$ which have exactly e_j terms equal to j $(1 \leqq j \leqq n)$. It is clear that if $Y_1, \cdots, Y_k$ are any elements of $\mathfrak{g}$, then

$$\lambda(\overline{Y}_1 \overline{Y}_2 \cdots \overline{Y}_k) = (1/k!) \sum_{\omega} Y_{\omega(1)} \cdots Y_{\omega(k)}$$

where the summation is over all permutations ω of $\{1, 2, \cdots, k\}$. This shows that the definition of λ is independent of the choice of the base $(X_1, \cdots, X_m)$. We know from theorems of $[8(a)]$ that λ is a 1-1 linear mapping of $S(\mathfrak{g})$ onto $\mathfrak{B}$. We shall call it the *canonical mapping of $S(\mathfrak{g})$ onto $\mathfrak{B}$*.

Let $\sigma(X)$ $(X \in \mathfrak{g})$ be the derivation of $S(\mathfrak{g})$ which coincides with ad X on $\mathfrak{g}$. Then $\sigma \colon X \to \sigma(X)$ is a representation of $\mathfrak{g}$ on $S(\mathfrak{g})$. Let $\sigma_{\mathfrak{k}}$ be the restriction of σ to $\mathfrak{k}$.

LEMMA 10. *$\sigma_{\mathfrak{k}}$ is a quasi-semisimple representation of $\mathfrak{k}$.*

Let $\sigma_d(X)$ $(X \in \mathfrak{g})$ denote the restriction of $\sigma(X)$ on $S_d(\mathfrak{g})$. Then σ_d is a finite-dimensional representation of $\mathfrak{g}$. Since $\mathfrak{g}$ is semisimple and $\mathfrak{k}$ is reductive in $\mathfrak{g}$, σ_d induces a semisimple representation of $\mathfrak{k}$ on $S_d(\mathfrak{g})$. But $S(\mathfrak{g}) = \sum_{d \geqq 0} S_d(\mathfrak{g})$ and so the lemma follows.

LEMMA 11. *Let σ be the representation of $\mathfrak{g}$ on $S(\mathfrak{g})$ as defined above. Then if $F \in S(\mathfrak{g})$ and $X \in \mathfrak{g}$,*

$$\lambda(\sigma(X)F) = [X, \lambda(F)] = X\lambda(F) - \lambda(F)X.$$

It is clearly sufficient to prove this formula when F is of the form $\overline{Y}_1 \overline{Y}_2 \cdots \overline{Y}_k$, $Y_1, \cdots, Y_k$ being in $\mathfrak{g}$. Set $Y_i' = [X, Y_i]$ $(1 \leqq i \leqq k)$; since $\sigma(X)$ is a derivation, $\sigma(X)F$ is the sum of the k products obtained from $\overline{Y}_1 \cdots \overline{Y}_k$ by replacing successively each one of the factors $\overline{Y}_i$ by $\overline{Y}_i'$. On the other hand the mapping $b \to [X, b]$ $(b \in \mathfrak{B})$ is a derivation of $\mathfrak{B}$. It follows that for any permutation ω of $(1, 2, \cdots, k)$, $[X, Y_{\omega(1)} Y_{\omega(2)} \cdots Y_{\omega(k)}]$ is the sum of the products obtained from $Y_{\omega(1)} Y_{\omega(2)} \cdots Y_{\omega(k)}$ by replacing successively each one of the factors $Y_{\omega(i)}$ by its transform $[X, Y_{\omega(i)}] = Y_{\omega(i)}'$. The formula $\lambda(\sigma(X)F) = [X, \lambda(F)]$ now follows immediately.

COROLLARY. *The center of the algebra $\mathfrak{B}$ is the image under λ of the set of elements F in $S(\mathfrak{g})$ such that $\sigma(X)F = 0$ for all $X \in \mathfrak{g}$.*

This is an immediate consequence of Lemma 11.

If a vector space V (of finite dimension) is the direct sum of two subspaces V' and V'', we may obviously identify $S(V')$ and $S(V'')$ with subalgebras of $S(V)$ and the bilinear mapping $(F', F'') \to F'F''$ defines an isomorphism of the tensor product $S(V') \times S(V'')$ with $S(V)$ as is easily seen by making use of a base of V which is composed of a base of V' and a base of V''.

LEMMA 12. *Let $\mathfrak{g}$ be the direct sum of two subspaces $\mathfrak{g}'$ and $\mathfrak{g}''$. The bilinear mapping $(F', F'') \to \lambda(F')\lambda(F'')$ $(F' \in S(\mathfrak{g}'), F'' \in S(\mathfrak{g}''))$ defines a linear isomorphism of $S(\mathfrak{g}') \times S(\mathfrak{g}'')$ with $\mathfrak{B}$. If d is an integer ≥ 0, denote by $S_d(\mathfrak{g})$, $S_d(\mathfrak{g}')$, and $S_d(\mathfrak{g}'')$ the spaces of homogeneous elements of degree d in $S(\mathfrak{g})$, $S(\mathfrak{g}')$, and $S(\mathfrak{g}'')$ respectively. Then*

$$\sum_{e \leq d} \lambda(S_e(\mathfrak{g})) = \sum_{d'+d'' \leq d} \lambda(S_{d'}(\mathfrak{g}'))\lambda(S_{d''}(\mathfrak{g}'')).$$

In this lemma, as in the rest of this paper, we make the following convention: if A and B are linear subspaces of an associative algebra, we denote by AB the vector space spanned by all elements ab $(a \in A, b \in B)$.

Consider a base $\{X_1, \cdots, X_n\}$ of $\mathfrak{g}$ which is composed of a base $\{X_1, \cdots, X_{n'}\}$ of $\mathfrak{g}'$ and a base $\{X_{n'+1}, \cdots, X_{n'+n''}\}$ of $\mathfrak{g}''$ $(n = n'+n'')$. We denote by M the set of all monomials in $\overline{X}_1, \cdots, \overline{X}_n$, by M' the set of monomials in $\overline{X}_1, \cdots, \overline{X}_{n'}$, by M'' the set of monomials in $\overline{X}_{n'+1}, \cdots, \overline{X}_{n'+n''}$, by M_d the set of monomials of degree d in M, by M_d' the set $M' \cap M_d$, and by M_d'' the set $M'' \cap M_d$. The space $\sum_{e \leq d} \lambda(S_e(\mathfrak{g}))$ is spanned by the elements $\lambda(\mu)$, $\mu \in \cup_{e \leq d} M_e$. We shall prove by induction on d that it is also spanned by $\lambda(\mu')\lambda(\mu'')$ where $\mu' \in M_{d'}'$, $\mu'' \in M_{d''}''$, and $d'+d'' \leq d$. This is obviously true for $d = 0$. Assume that it is true for d. It is proved in [8(a)] that if $F \in S_k(\mathfrak{g})$, $F' \in S_{k'}(\mathfrak{g})$ then

$$(1) \qquad \lambda(F)\lambda(F') \equiv \lambda(FF') \bmod \sum_{m < k+k'} \lambda(S_m(\mathfrak{g})).$$

It follows immediately that the condition $d'+d'' \leq d+1$ implies $\lambda(S_{d'}(\mathfrak{g}'))\lambda(S_{d''}(\mathfrak{g}'')) \subset \sum_{e \leq d+1} \lambda(S_e(\mathfrak{g}))$. On the other hand every $\mu \in M_{d+1}$ may be written in the form $\mu'\mu''$ where $\mu' \in M_{d'}'$, $\mu'' \in M_{d''}''$, and $d'+d'' = d+1$. Hence it follows from the above formula and our inductive assumption that $\lambda(\mu) \in \sum_{d'+d'' \leq d+1} \lambda(S_{d'}(\mathfrak{g}'))\lambda(S_{d''}(\mathfrak{g}''))$ and this proves our assertion for $d+1$. Since every element $\mu \in M$ can be written in exactly one way in the form $\mu'\mu''$ $(\mu' \in M', \mu'' \in M'')$, we see that for every d the elements $\lambda(\mu')\lambda(\mu'')$ $(\mu' \in M_{d'}', \mu'' \in M_{d''}'', d'+d'' \leq d)$ form a base of $\sum_{e \leq d} \lambda(S_e(\mathfrak{g}))$. This being true for every d, the lemma follows.

It is clear that $\lambda(S(\mathfrak{k})) = \mathfrak{X}$. *We shall denote by $\mathfrak{P}$ the set $\lambda(S(\mathfrak{p}))$ and by $\mathfrak{P}_d$ the set $\lambda(S_d(\mathfrak{p}))$.*

LEMMA 13. *We have $\lambda(S_d(\mathfrak{p})S(\mathfrak{k})) \subset \sum_{e \leq d} \mathfrak{P}_d\mathfrak{X}$ and $\lambda(S_d(\mathfrak{p})S(\mathfrak{k}))\lambda(S_{d'}(\mathfrak{p})S(\mathfrak{k}))$ $\subset \sum_{e \leq d+d'} \mathfrak{P}_e\mathfrak{X}.$*

The first formula follows immediately from Lemma 12. In order to prove the second formula, it will be sufficient to show that

$$(\mathfrak{P}_d\mathfrak{X})(\mathfrak{P}_{d'}\mathfrak{X}) \subset \sum_{e \leq d+d'} \mathfrak{P}_e\mathfrak{X}.$$

Let $\mathfrak{p}^d$ denote the space spanned by the products of d elements in $\mathfrak{p}$. First we prove that $\mathfrak{X}\mathfrak{p}^d \subset \mathfrak{p}^d\mathfrak{X}$. This is true for $d=0$. Let $\mathfrak{X}'$ be the set of all $u \in \mathfrak{B}$ such that $u\mathfrak{p} \subset \mathfrak{p}\mathfrak{X}$. Then $\mathfrak{X}'$ is a subalgebra of $\mathfrak{B}$ and $1 \in \mathfrak{X}'$. Moreover if $X \in \mathfrak{k}$ and $Y \in \mathfrak{p}$, then $XY = YX + [X, Y] \in \mathfrak{p}\mathfrak{X}$ since $[X, Y] \in \mathfrak{p}$. Hence $\mathfrak{k}$ and therefore $\mathfrak{X}$ is contained in $\mathfrak{X}'$ which shows that $\mathfrak{X}\mathfrak{p} \subset \mathfrak{p}\mathfrak{X}$. Now assume that our assumption is true for some d. Then $\mathfrak{X}\mathfrak{p}^{d+1} = (\mathfrak{X}\mathfrak{p}^d)\mathfrak{p} \subset \mathfrak{p}^d\mathfrak{X}\mathfrak{p} \subset \mathfrak{p}^{d+1}\mathfrak{X}$. This proves it for $d+1$. But it is clear that $\mathfrak{P}_d \subset \mathfrak{p}^d$ and therefore

$$(\mathfrak{P}_d\mathfrak{X})(\mathfrak{P}_{d'}\mathfrak{X}) \subset \mathfrak{p}^d\mathfrak{X}\mathfrak{p}^{d'}\mathfrak{X} \subset \mathfrak{p}^{d+d'}\mathfrak{X}.$$

Hence it would be sufficient to prove that $\mathfrak{p}^d\mathfrak{X} \subset \sum_{e \leq d} \mathfrak{P}_e\mathfrak{X}$. This is true for $d=0, 1$. Assume that it is true for d. Then

$$\mathfrak{p}^{d+1}\mathfrak{X} \subset \sum_{e \leq d} \mathfrak{p}\mathfrak{P}_e\mathfrak{X}.$$

But $\mathfrak{p}\mathfrak{P}_e \subset \lambda(S_1(\mathfrak{g}))\lambda(S_e(\mathfrak{g}))$ and therefore from formula (1) above $\mathfrak{p}\mathfrak{P}_e \subset \sum_{e' \leq e+1} \lambda(S_{e'}(\mathfrak{g}))$ and this is contained in $\sum_{e' \leq e+1} \mathfrak{P}_{e'}\mathfrak{X}$ in virtue of Lemma 12. Since this proves our formula for $d+1$ the lemma follows.

Let $\mathfrak{g}^*$ be the space dual to $\mathfrak{g}$, i.e., the space of linear functions on $\mathfrak{g}$. Then the fundamental bilinear form $B(X, Y)$ of $\mathfrak{g}$ defines an isomorphism I of $\mathfrak{g}$ with $\mathfrak{g}^*$, which assigns to every $X \in \mathfrak{g}$ the linear function $Y \to B(X, Y)$ on $\mathfrak{g}$. (This is an isomorphism since B is nondegenerate.) I may be extended uniquely to an isomorphism of $S(\mathfrak{g})$ with the symmetric algebra $S(\mathfrak{g}^*)$ on $\mathfrak{g}^*$. We denote this extension again by I. Let $\{x_1^*, \cdots, x_n^*\}$ be a base for $\mathfrak{g}^*$ and F any element in $S(\mathfrak{g}^*)$. Then $F = P(x_1^*, \cdots, x_n^*)$ where P is a polynomial. Since $x_1^*, \cdots, x_n^*$ are linear functions on $\mathfrak{g}$, $P(x_1^*, \cdots, x_n^*)$ represents a function on $\mathfrak{g}$, namely the function F': $X \to P(x_1^*(X), \cdots, x_n^*(X))$ $(X \in \mathfrak{g})$. It is clear that the mapping $F \to F'$ is isomorphic and does not depend on the choice of the base $\{x_1^*, \cdots, x_n^*\}$. The functions F' are called the *polynomial functions* on $\mathfrak{g}$. *Henceforth we shall identify $S(\mathfrak{g})$ with $S(\mathfrak{g}^*)$ and the algebra of polynomial functions on $\mathfrak{g}$ under the mappings $F \to I(F)$ $\to (I(F))'$ $(F \in S(\mathfrak{g}))$.* Since $B(X, Y) = 0$, $X \in \mathfrak{k}$, and $Y \in \mathfrak{p}$, $\mathfrak{p}$ is identified with the space of linear functions on $\mathfrak{g}$ which are zero on $\mathfrak{k}$. We may therefore identify $S(\mathfrak{p})$ with the algebra of polynomial functions on the subspace $\mathfrak{p}$ of $\mathfrak{g}$. Let $\{x_1, \cdots, x_p, x_{p+1}, \cdots, x_n\}$ be a base for $\mathfrak{g}$ such that $\{x_1, \cdots, x_p\}$ and $\{x_{p+1}, \cdots, x_n\}$ are bases for $\mathfrak{p}$ and $\mathfrak{k}$ respectively. Then if P is a polynomial in n variables, the restriction of the polynomial function $F = P(x_1, \cdots, x_n)$ on $\mathfrak{p}$ is $P(x_1, \cdots, x_p, 0, \cdots, 0)$.

3. **Proof of Theorem 1.** Let $\mathfrak{A}$ be an associative algebra with a unit element 1 and $\mathfrak{M}$ a left ideal in $\mathfrak{A}$. Then the space $\mathfrak{A}^* = \mathfrak{A}/\mathfrak{M}$ is an $\mathfrak{A}$-module and if

we assign to every $x \in \mathfrak{A}$ the mapping $\pi(x): a^* \to xa^*$ of $\mathfrak{A}^*$ into itself we obtain a representation π of $\mathfrak{A}$, which we call the *natural representation* of $\mathfrak{A}$ on $\mathfrak{A}/\mathfrak{M}$. Let $\mathfrak{l}$ be a finite-dimensional subspace of $\mathfrak{A}$ which is a Lie algebra under the bracket operation $[X, Y] = XY - YX$ $(X, Y \in \mathfrak{l})$. Then π defines a representation of this Lie algebra which will also be called the natural representation of $\mathfrak{l}$ on $\mathfrak{A}/\mathfrak{M}$.

THEOREM 1. *Let $\mathfrak{Y}$ be a left ideal in $\mathfrak{X}$. Assume that $\mathfrak{X}/\mathfrak{Y}$ is finite-dimensional and the natural representation of $\mathfrak{l}$ on $\mathfrak{X}/\mathfrak{Y}$ is semisimple. Then the natural representation of $\mathfrak{l}$ on $\mathfrak{B}/\mathfrak{B}\mathfrak{Y}$ is quasi-semisimple. Let $\mathfrak{D}$ be any equivalence class of finite-dimensional simple representations of $\mathfrak{l}$ and let $\mathfrak{B}^*_\mathfrak{D}$ be the set of all elements of $\mathfrak{B}^* = \mathfrak{B}/\mathfrak{B}\mathfrak{Y}$ which transform according to $\mathfrak{D}$ under the natural representation π of $\mathfrak{l}$ on $\mathfrak{B}^*$. Then $\mathfrak{B}^*_\mathfrak{D}$ is a finite module over the center $\mathfrak{Z}$ of $\mathfrak{B}$.*

In order to clarify the last assertion we first observe that $\mathfrak{B}^*_\mathfrak{D}$ is actually a module over $\mathfrak{Z}$. Let x be an element of $\mathfrak{B}^*_\mathfrak{D}$ and let W be the smallest π-stable subspace of $\mathfrak{B}^*$ containing x. Let $\pi_W(X)$ $(X \in \mathfrak{l})$ denote the restriction of $\pi(X)$ on W. Then π_W is semisimple and every simple component of π_W is contained in $\mathfrak{D}$. Let $z \in \mathfrak{Z}$; then the mapping $y \to zy$ $(y \in W)$ maps W upon a space zW. If $X \in \mathfrak{l}$, $Xzy = zXy \in zW$; thus zW is π-stable and $\pi(X)zy = z(\pi(X)y)$. It follows immediately (see Chevalley [4, Chap. VI]) that the representation of $\mathfrak{l}$ induced on zW is semisimple and all its simple components are in $\mathfrak{D}$. Hence $zx \in \mathfrak{B}^*_\mathfrak{D}$ and this shows that $\mathfrak{B}^*_\mathfrak{D}$ is a $\mathfrak{Z}$-module.

The proof of Theorem 1 will be divided into several parts.

1. Proof of the first assertion. We have defined above the canonical mapping λ of $S(\mathfrak{g})$ onto $\mathfrak{B}$. We consider now the tensor product $S(\mathfrak{p}) \times (\mathfrak{X}/\mathfrak{Y})$. Since $\mathfrak{Y} \subset \mathfrak{B}\mathfrak{Y}$, there is a natural mapping $f \to f^*$ $(f \in \mathfrak{X}/\mathfrak{Y})$ of $\mathfrak{X}/\mathfrak{Y}$ into $\mathfrak{B}^* = \mathfrak{B}/\mathfrak{B}\mathfrak{Y}$. Define a linear mapping Γ of $S(\mathfrak{p}) \times (\mathfrak{X}/\mathfrak{Y})$ into $\mathfrak{B}^*$ as follows:

$$\Gamma(F \times f) = \lambda(F)f^* \qquad (F \in S(\mathfrak{p}), f \in \mathfrak{X}/\mathfrak{Y}).$$

(We recall that $\mathfrak{B}^*$ is a $\mathfrak{B}$-module and so $\lambda(F)$ operates on $\mathfrak{B}^*$.) We shall prove that Γ is a linear isomorphism of $S(\mathfrak{p}) \times (\mathfrak{X}/\mathfrak{Y})$ with $\mathfrak{B}^*$. Select a base $(s_i)_{i \in I}$ for $S(\mathfrak{p})$ and a base $(t_j)_{j \in J}$ for $S(\mathfrak{l})$. We may suppose that one of the elements s_i, say s_0, is equal to 1. The mapping λ induces an isomorphism of $S(\mathfrak{l})$ with $\mathfrak{X}$. Hence we may assume that the base $(t_j)_{j \in J}$ is so chosen that for a suitable subset J' of J the elements $(\lambda(t_j))_{j \in J'}$ form a base for $\mathfrak{Y}$. From Lemma 12 the elements $\lambda(s_i)\lambda(t_j)$ $(i \in I, j \in J)$ form a base for $\mathfrak{B}$ and, if $j \in J'$, $\lambda(s_i)\lambda(t_j) \in \mathfrak{B}\mathfrak{Y}$. Conversely every element of $\mathfrak{B}\mathfrak{Y}$ is in the space spanned by the elements $\lambda(s_i)\lambda(t_j)\lambda(t_{j'})$ $(i \in I, j \in J, j' \in J')$. But since $\mathfrak{Y}$ is a left ideal in $\mathfrak{X}$, $\lambda(t_j)\lambda(t_{j'}) \in \mathfrak{Y}$ and so $\lambda(t_j)\lambda(t_{j'})$ lies in the space spanned by $\lambda(t_{j''})$ $(j'' \in J')$. This proves that the elements $\lambda(s_i)\lambda(t_{j'})$ $(i \in I, j' \in J')$ form a base for $\mathfrak{B}\mathfrak{Y}$. Since $\mathfrak{X}$ is spanned by the elements $\lambda(s_0)\lambda(t_j)$ $(j \in J)$, it follows that $\mathfrak{B}\mathfrak{Y} \cap \mathfrak{X} = \mathfrak{Y}$. We may therefore identify $\mathfrak{X}/\mathfrak{Y}$ with its image in $\mathfrak{B}^*$. Let J'' be the complement of J' in J. Let $\lambda^*(t_j)$ $(j \in J'')$ denote the residue class of $\lambda(t_j)$ modulo

$\mathfrak{Y}$ (or $\mathfrak{BY}$). Then the elements $\lambda(s_i)\,\lambda^*(t_j)$ $(i\in I,\ j\in J'')$ form a base for $\mathfrak{B}^*$. On the other hand the elements $\lambda^*(t_j)$ $(j\in J'')$ form a base of $\mathfrak{X}/\mathfrak{Y}$ and $\Gamma(s_i\times\lambda^*(t_j))=\lambda(s_i)\lambda^*(t_j)$ $(i\in I,\ j\in J'')$. This proves that Γ is a linear isomorphism of $S(\mathfrak{p})\times(\mathfrak{X}/\mathfrak{Y})$ with $\mathfrak{B}^*$.

For any $X\in\mathfrak{k}$ let $\sigma(X)$ denote the derivation of $S(\mathfrak{g})$ which coincides with ad X on $\mathfrak{g}$. Then we have seen in Lemma 10 that σ is a quasi-semisimple representation of $\mathfrak{k}$. On the other hand ad X maps $\mathfrak{p}$ into itself and therefore $\sigma(X)$ maps $S(\mathfrak{p})$ into itself. *Let $\rho(X)$ denote the restriction of $\sigma(X)$ on $S(\mathfrak{p})$.* The representation ρ is quasi-semisimple (Lemma 6). Let μ be the natural representation of $\mathfrak{k}$ on $\mathfrak{X}/\mathfrak{Y}$. Then μ is of finite degree and semisimple. From Lemma 8, the representation $\rho+\mu$ on $S(\mathfrak{p})\times(\mathfrak{X}/\mathfrak{Y})$ is quasi-semisimple. The first assertion of Theorem 1 will therefore be proved if we show that π is equivalent to $\rho+\mu$. Now if $F\in S(\mathfrak{p})$, $f\in\mathfrak{X}/\mathfrak{Y}$, and $X\in\mathfrak{k}$,

$$((\rho+\mu)(X))(F\times f) = \rho(X)F\times f + F\times Xf.$$

But from Lemma 11,

$$\lambda(\rho(X)F) = [X,\lambda(F)].$$

Hence

$$\Gamma(\rho(X)F\times f) + \Gamma(F\times Xf) = ([X,\lambda(F)])f + \lambda(F)Xf$$
$$= X\lambda(F)f = \pi(X)(\Gamma(F\times f)).$$

Since Γ is an isomorphism, π is equivalent to $\rho+\mu$.

2.1. Proof of the second assertion. First reduction. Let ν be a simple representation of $\mathfrak{k}$ which is contragredient to the representations of the class $\mathfrak{D}$ and let V be the space of the representation ν. Consider the representation $\pi+\nu$ of $\mathfrak{k}$ on the tensor product $\mathfrak{B}^*\times V$. An element $\phi\in\mathfrak{B}^*\times V$ will be called an *invariant* if $((\pi+\nu)(X))\phi=0$ for all $X\in\mathfrak{k}$. Let $\mathfrak{I}$ be the set of these invariants. Suppose $\{v_1,\cdots,v_m\}$ is a base for V and $\phi=\sum_{i=1}^m b_i^*\times v_i$ $(b_i^*\in\mathfrak{B}^*)$ is in $\mathfrak{I}$. Then the elements b_i^* belong to $\mathfrak{B}_{\mathfrak{D}}^*$. For, if $X\in\mathfrak{k}$, $\sum_{i=1}^m \pi(X)b_i^*\times v_i$ $=-\sum_{i=1}^m b_i^*\times\nu(X)v_i$ from which it follows immediately that the elements $\pi(X)b_i^*$ lie in the space V^* spanned by $b_1^*,\cdots,b_m^*$ and therefore V^* is π-stable. Let $\pi_{V^*}(X)$ denote the restriction of $\pi(X)$ on V^*. Now $\phi\in V^*\times V$ and $((\pi_{V^*}+\nu)(X))\phi=0$ for all $X\in\mathfrak{k}$. If $\phi\neq0$ this means that π_{V^*} contains a simple representation contragredient to ν. Since dim $V^*\leq m$, it follows that π_{V^*} itself is contragredient to ν and so lies in $\mathfrak{D}$. Therefore $b_i^*\in\mathfrak{B}_{\mathfrak{D}}^*$ $(1\leq i\leq m)$. On the other hand if $\phi=0$, $b_i^*=0$ $(1\leq i\leq m)$ and so our assertion is true trivially. Conversely let U be a π-stable subspace of $\mathfrak{B}^*$ such that the representation π_U induced on U lies in $\mathfrak{D}$. Then π_U is contragredient to ν and so there exists a base $\{b_1^*,\cdots,b_m^*\}$ for U such that $\sum_{i=1}^m b_i^*\times v_i$ is an invariant of $\pi+\nu$.

Since $\mathfrak{B}^*$ is a $\mathfrak{B}$-module we may also regard $\mathfrak{B}^*\times V$ as a $\mathfrak{B}$-module as follows:

$$a(b^* \times v) = (ab^*) \times v \quad (a \in \mathfrak{B}, b^* \in \mathfrak{B}^*, v \in V).$$

If $z \in \mathfrak{Z}$, the operator z on $\mathfrak{B}^* \times V$ commutes with $(\pi + \nu)(X)$ $(X \in \mathfrak{k})$. Hence it follows immediately that $\mathfrak{Y}$ is a $\mathfrak{Z}$-module. If this $\mathfrak{Z}$-module is finite, then $\mathfrak{B}_{\mathfrak{D}}^*$ is a finite $\mathfrak{Z}$-module. For let ϕ_k $(1 \le k \le q)$ be a finite set of elements in $\mathfrak{Y}$ such that $\mathfrak{Y} = \sum_{k=1}^{q} \mathfrak{Z}\phi_k$. Let $\phi_k = \sum_{i=1}^{m} b_{ki}^* \times v_i$ $(b_{ki}^* \in \mathfrak{B}^*)$. Then it follows immediately from what we said above that $\mathfrak{B}_{\mathfrak{D}}^* = \sum_{k=1}^{q} \sum_{i=1}^{m} \mathfrak{Z}b_{ki}^*$.

We have defined above an isomorphism Γ of $S(\mathfrak{p}) \times (\mathfrak{X}/\mathfrak{Y})$ with $\mathfrak{B}^*$. Now consider the linear isomorphism

$$F \times f \times v \rightarrow \Gamma(F \times f) \times v \qquad (F \in S(\mathfrak{p}), f \in \mathfrak{X}/\mathfrak{Y}, v \in V)$$

of $S(\mathfrak{p}) \times (\mathfrak{X}/\mathfrak{Y}) \times V$ with $\mathfrak{B}^* \times V$. This we shall denote again by Γ. Put $W = (\mathfrak{X}/\mathfrak{Y}) \times V$ and $\xi = \rho + \mu + \nu$, $\mu' = \mu + \nu$, $\pi' = \pi + \nu$. Then if $X \in \mathfrak{k}$ and $A \in S(\mathfrak{p}) \times W$, $\Gamma(\xi(X)A) = \pi'(X)\Gamma(A)$. Let $\mathfrak{Y}'$ be the set of elements $A \in S(\mathfrak{p}) \times W$ such that $\xi(X)A = 0$ for all $X \in \mathfrak{k}$. Thus Γ induces an isomorphism of $\mathfrak{Y}'$ with $\mathfrak{Y}$.

We now regard $S(\mathfrak{p}) \times W$ as a module over $S(\mathfrak{p})$ by the rule $F_1(F_2 \times w) = (F_1 F_2) \times w$ $(F_1, F_2 \in S(\mathfrak{p}), w \in W)$. If $X \in \mathfrak{k}$, $F \in S(\mathfrak{p})$, $A \in S(\mathfrak{p}) \times W$, we have $\xi(X)FA = (\rho(X)F)A + F(\xi(X)A)$. *Let Ω denote the set of elements $F \in S(\mathfrak{p})$ such that $\rho(X)F = 0$ for all $X \in \mathfrak{k}$.* Then $\mathfrak{Y}'$ is clearly an Ω-module.

On the other hand for any $Z \in \mathfrak{g}$ the operation ad Z may be extended to a derivation $\tau(Z)$ of the algebra $S(\mathfrak{g})$. If $Z \in \mathfrak{k}$, then the operation $\sigma(Z)$ considered above is identical with $\tau(Z)$. We shall denote by $\mathfrak{J}$ the set of elements $J \in S(\mathfrak{g})$ such that $\tau(Z)J = 0$ for all $Z \in \mathfrak{g}$; this is a subalgebra of $S(\mathfrak{g})$. We now regard the elements of $S(\mathfrak{g})$ as polynomial functions on $\mathfrak{g}$ in the manner explained above and denote by $\mathfrak{J}_{\mathfrak{p}}$ the set of restrictions to $\mathfrak{p}$ of all functions J in $\mathfrak{J}$. Then $\mathfrak{J}_{\mathfrak{p}} \subset \Omega$. For, let $J \in \mathfrak{J}$ and let J' be its restriction to $\mathfrak{p}$. Then J' is the unique element of $S(\mathfrak{p})$ such that $J - J'$ belongs to the ideal $\mathfrak{K}$ generated by $\mathfrak{k}$ in $S(\mathfrak{g})$. Let X be an element of $\mathfrak{k}$. Then $\sigma(X)J = 0$. On the other hand since ad X maps the spaces $\mathfrak{k}$ and $\mathfrak{p}$ into themselves, $\sigma(X)$ maps $\mathfrak{K}$ and $S(\mathfrak{p})$ into themselves and so $\rho(X)J' = \tau(X)J' \in \mathfrak{K} \cap S(\mathfrak{p}) = \{0\}$. Hence $J' \in \Omega$. It is clear that $\mathfrak{J}_{\mathfrak{p}}$ is a algebra of Ω. We shall see that *in order to prove that $\mathfrak{Y}$ is a finite $\mathfrak{Z}$-module it is sufficient to show that $\mathfrak{Y}'$ is a finite $\mathfrak{J}_{\mathfrak{p}}$-module.*

Assume that $\mathfrak{Y}'$ is a finite $\mathfrak{J}_{\mathfrak{p}}$-module. For any integer $d \ge 0$ let $S_d(\mathfrak{p})$ denote the space of homogeneous elements of degree d of $S(\mathfrak{p})$. Then $S_d(\mathfrak{p}) \times W$ is mapped into itself by the operations of $\xi(\mathfrak{k})$. Therefore we can find a finite number of elements $A_i \in \mathfrak{Y}'$ $(1 \le i \le r)$ each of which lies in some $S_d(\mathfrak{p}) \times W$, say $S_{d_i}(\mathfrak{p}) \times W$, such that $\mathfrak{Y}' = \sum_{i=1}^{r} \mathfrak{J}_{\mathfrak{p}} A_i$. The space $\mathfrak{Y}_\infty = \sum_{i=1}^{r} \mathfrak{Z}\Gamma(A_i)$ is contained in $\mathfrak{Y}$. We now intend to prove that $\mathfrak{Y}_\infty = \mathfrak{Y}$. Put

$$\mathfrak{Y}'_d = \mathfrak{Y}' \cap \left(\sum_{0 \le e \le d} S_e(\mathfrak{p}) \times W \right), \quad \mathfrak{Y}_d = \Gamma(\mathfrak{Y}'_d).$$

The set of those A_i for which $d_i = 0$ obviously generates the space $\mathfrak{Y}' \cap W$

(W being identified in the usual manner to the subspace $1 \times W$ of $S(\mathfrak{p}) \times W$).
It follows that $\mathfrak{J}_0 \subset \mathfrak{J}_\infty$. It will therefore be sufficient to prove that for every
$d \geq 0$, $\mathfrak{J}_{d+1} \subset \mathfrak{J}_d + \mathfrak{J}_\infty$.

Let $A \in \mathfrak{J}' \cap (S_{d+1}(\mathfrak{p}) \times W)$ and let $A = \sum_{i=1}^r F_i A_i$ with $F_i \in \mathfrak{J}_\mathfrak{p}$. It is clear
that the homogeneous components of any element of $\mathfrak{J}_\mathfrak{p}$ also belong to $\mathfrak{J}_\mathfrak{p}$.
We may therefore assume that $F_i \in S_{d+1-d_i}(\mathfrak{p})$ $(S_e(\mathfrak{p}) = \{0\}$ if $e < 0)$. Let
$\{w_1, \cdots, w_a\}$ be a base for W and let $A_i = \sum_{j=1}^a A_{ij} \times w_j$ with $A_{ij} \in S_{d_i}(\mathfrak{p})$.
Then $\Gamma(A) = \sum_{i=1}^r \sum_{j=1}^a \lambda(F_i A_{ij}) w_j$. Each F_i is the restriction to $\mathfrak{p}$ of an element J_i of $\mathfrak{J}$ and we may assume that J_i is homogeneous of degree $d+1-d_i$
which is the degree of F_i. It follows that $F_i - J_i \in \sum_{e < d+1-d_i} S_e(\mathfrak{p}) S(\mathfrak{k})$.
Hence

$$F_i A_{ij} - J_i A_{ij} \in \sum_{e \leq d} S_e(\mathfrak{p}) S(\mathfrak{k}).$$

Therefore from Lemma 13,

$$\Gamma(A) - \sum_{i=1}^r \sum_{j=1}^a \lambda(J_i A_{ij}) w_j \in \sum_{e \leq d} (\mathfrak{P}_e \mathfrak{X}) W$$

where $\mathfrak{P}_e = \lambda(S_e(\mathfrak{p}))$. On the other hand we know (see [8(a)]) that

$$\lambda(J_i A_{ij}) \equiv \lambda(J_i) \lambda(A_{ij}) \bmod \sum_{e \leq d} \lambda(S_e(\mathfrak{g}))$$

where $S_e(\mathfrak{g})$ is the space of homogeneous elements of degree e in $S(\mathfrak{g})$. Since
$\lambda(S_e(\mathfrak{g})) \subset \sum_{e' \leq e} \mathfrak{P}_{e'} \mathfrak{X}$ from Lemma 12, it follows that

$$\Gamma(A) - \sum_{i=1}^r \sum_{j=1}^a \lambda(J_i) \lambda(A_{ij}) w_j \in \sum_{e \leq d} (\mathfrak{P}_e \mathfrak{X}) W.$$

Since W is the subspace $((\mathfrak{X} + \mathfrak{B}\mathfrak{Y})/\mathfrak{B}\mathfrak{Y}) \times V$ of $\mathfrak{B}^* \times V$, it is clear that $\mathfrak{X} W \subset W$
and therefore

$$\sum_{e \leq d} (\mathfrak{P}_e \mathfrak{X}) W \subset \Gamma\left(\left(\sum_{e \leq d} S_e(\mathfrak{p}) \right) \times W \right).$$

On the other hand

$$\sum_{i=1}^r \sum_{j=1}^a \lambda(J_i) \lambda(A_{ij}) w_j = \sum_{i=1}^r \lambda(J_i) \Gamma(A_i).$$

The elements $\lambda(J_i)$ are in $\mathfrak{Z}$ in virtue of the corollary to Lemma 11. Hence
$\sum_{i=1}^r \lambda(J_i) \Gamma(A_i) \in \mathfrak{J}_\infty$ and therefore

$$\Gamma(A) - \sum_{i=1}^r \lambda(J_i) \Gamma(A_i) \in \mathfrak{J} \cap \Gamma\left(\sum_{e \leq d} S_e(\mathfrak{p}) \times W \right) = \mathfrak{J}_d.$$

This proves that $\Gamma(A) \in \mathfrak{J}_d + \mathfrak{J}_\infty$ and therefore $\mathfrak{J}_{d+1} \subset \mathfrak{J}_d + \mathfrak{J}_\infty$. Hence $\mathfrak{J} = \mathfrak{J}_\infty$.

2.2. Second reduction. We shall now establish that the ring Ω introduced above is a finite module over its subring $\mathfrak{J}_\mathfrak{p}$. This will reduce the problem of proving that $\mathfrak{J}'$ is a finite $\mathfrak{J}_\mathfrak{p}$-module to that of proving that it is a finite Ω-module.

Let $\mathfrak{h}$ be the Cartan subalgebra of $\mathfrak{g}$ as defined in Lemma 4. Let $p = \dim \mathfrak{p}$ and $h = \dim (\mathfrak{h} \cap \mathfrak{p})$. We shall first show that Ω cannot contain more than h algebraically independent elements. Let $(Y_1, \cdots, Y_p)$ be a base of $\mathfrak{p}$. We have identified every element Y of $\mathfrak{g}$ to the linear function $Z \to B(Y, Z)$ on $\mathfrak{g}$. Let $y_1, \cdots, y_p$ be the elements $Y_1, \cdots, Y_p$ regarded as linear functions on $\mathfrak{g}$. These functions are zero on $\mathfrak{k}$. An element of Ω may be written as a polynomial $F(y_1, \cdots, y_p)$ in $y_1, \cdots, y_p$ and $\sum_{i=1}^{p} \partial F(y_1, \cdots, y_p)/\partial y_i \, (\rho(X)y_i) = 0$ for all $X \in \mathfrak{k}$. The function $\rho(X)y_i$ is the function $Z \to B([X, Y_i], Z) = -B(Y_i, [X, Z])$ on $\mathfrak{g}$. Let Z be any element of $\mathfrak{p}$. Put $\zeta_i = y_i(Z)$ $(1 \leq i \leq p)$. Then

$$\sum_{i=1}^{p} \frac{\partial F}{\partial y_i}(\zeta_1, \cdots, \zeta_p) y_i([X, Z]) = 0.$$

Let $\{X_1, \cdots, X_k\}$ be a base for $\mathfrak{k}$. Then the rank $r(Z)$ of the matrix $(y_i([X_j, Z]))_{1 \leq i \leq p, 1 \leq j \leq k}$ is the dimension of the space spanned by the elements $[X, Z]$ for all $X \in \mathfrak{k}$. Let $r = \max_{Z \in \mathfrak{p}} r(Z)$. Then there exists a polynomial function $G \neq 0$ on $\mathfrak{p}$ such that $r(Z) = r$ if $G(Z) \neq 0$ $(Z \in \mathfrak{p})$. It follows from the above relation that if $F_1, \cdots, F_q$ are in Ω, the rank of the Jacobian matrix $(\partial F_i/\partial y_j)_{1 \leq i \leq q, 1 \leq j \leq p}$ for $y_i = \zeta_i$ is $\leq p - r(Z)$. Hence there cannot be more than $p - r$ algebraically independent elements among $F_1, \cdots, F_q$. Now it is clear that $r(Z)$ is equal to the difference between the dimension of $\mathfrak{k}$ and that of the space of elements $X \in \mathfrak{k}$ which commute with Z. Making use of Lemma 4, we see that there exists an element $Z \in \mathfrak{p}$ such that $r(Z) = p - h$. Hence $r \geq p - h$ and therefore there cannot be more than h algebraically independent elements among $F_1, \cdots, F_q$. This proves our assertion.

The bilinear form B is nondegenerate on $\mathfrak{h}$. Since $\mathfrak{h} = \mathfrak{h}_\mathfrak{p} + \mathfrak{h}_\mathfrak{k}$ and $\mathfrak{k}$ and $\mathfrak{p}$ are mutually orthogonal under B, it follows that B is nondegenerate on $\mathfrak{h}_\mathfrak{p}$. We may therefore assume that the base $\{Y_1, \cdots, Y_p\}$ for $\mathfrak{p}$ is so chosen that $Y_1, \cdots, Y_h$ form a base for $\mathfrak{h}_\mathfrak{p} = \mathfrak{h} \cap \mathfrak{p}$ and $B(Y_i, Y_j) = 0$ if $i \leq h < j$. The restrictions of $y_1, \cdots, y_h$ to $\mathfrak{h}_\mathfrak{p}$ are then linearly independent and the ring generated by $y_1, \cdots, y_h$ may be identified with the algebra $S(\mathfrak{h}_\mathfrak{p})$ of polynomial functions on $\mathfrak{h}_\mathfrak{p}$ (i.e., those functions on $\mathfrak{h}_\mathfrak{p}$ which may be written as polynomials in linear functions). The restriction of an element $F(y_1, \cdots, y_p)$ of $S(\mathfrak{p})$ to $\mathfrak{h}_\mathfrak{p}$ is $F(y_1, \cdots, y_h, 0, \cdots, 0)$. Let $\mathfrak{J}_{\mathfrak{h}_\mathfrak{p}}$ be the algebra of restrictions to $\mathfrak{h}_\mathfrak{p}$ of elements of $\mathfrak{J}_\mathfrak{p}$. We shall prove $\mathfrak{J}_{\mathfrak{h}_\mathfrak{p}}$ contains h algebraically independent elements and $S(\mathfrak{h}_\mathfrak{p})$ is a finite module over $\mathfrak{J}_{\mathfrak{h}_\mathfrak{p}}$. Let $\mathfrak{J}_\mathfrak{h}$ be the ring of restrictions to $\mathfrak{h}$ of elements of $\mathfrak{J}$. Then $\mathfrak{J}_{\mathfrak{h}_\mathfrak{p}}$ coincides with ring of restrictions to $\mathfrak{h}_\mathfrak{p}$ of elements of $\mathfrak{J}_\mathfrak{h}$. Let $\mathfrak{w}$ be the Weyl group of $\mathfrak{g}$ with respect to $\mathfrak{h}$ and $S(\mathfrak{h})$

the algebra of polynomial functions on $\mathfrak{h}$. Then Chevalley[5] has proved that every element of $S(\mathfrak{h})$ which is invariant under $\mathfrak{w}$ lies in $\mathfrak{I}_\mathfrak{h}$. It follows that every element $\phi \in S(\mathfrak{h})$ is integral over $\mathfrak{I}_\mathfrak{h}$ since it is a root of the polynomial $\prod_{s \in \mathfrak{w}} (T - s\phi)$ whose coefficients are in $\mathfrak{I}_\mathfrak{h}$. Moreover since $\mathfrak{w}$ is a finite group it follows from the theory of invariants that the ring of invariants of $\mathfrak{w}$ (in $S(\mathfrak{h})$) is finitely generated. Hence $\mathfrak{I}_\mathfrak{h}$ is a Noetherian ring. Now the operation of restriction from $\mathfrak{h}$ to $\mathfrak{h}_\mathfrak{p}$ is a homomorphism of $S(\mathfrak{h})$ onto $S(\mathfrak{h}_\mathfrak{p})$ which maps $\mathfrak{I}_\mathfrak{h}$ onto $\mathfrak{I}_{\mathfrak{h}_\mathfrak{p}}$. Hence $\mathfrak{I}_{\mathfrak{h}_\mathfrak{p}}$ is a Noetherian ring and every element of $S(\mathfrak{h}_\mathfrak{p})$ is integral over $\mathfrak{I}_{\mathfrak{h}_\mathfrak{p}}$. This shows that $\mathfrak{I}_{\mathfrak{h}_\mathfrak{p}}$ contains h algebraically independent elements. Furthermore since $S(\mathfrak{h}_\mathfrak{p})$ may be obtained from $\mathfrak{I}_{\mathfrak{h}_\mathfrak{p}}$ by the adjunction of a finite number of elements which are integral over it, $S(\mathfrak{h}_\mathfrak{p})$ is a finite module over $\mathfrak{I}_{\mathfrak{h}_\mathfrak{p}}$. Let $\Omega_{\mathfrak{h}_\mathfrak{p}}$ be the set of restrictions of elements of Ω to $\mathfrak{h}_\mathfrak{p}$. Then this restriction is a homomorphism of Ω onto $\Omega_{\mathfrak{h}_\mathfrak{p}}$. The ring Ω does not contain more than h algebraically independent elements and contains $\mathfrak{I}_{\mathfrak{h}_\mathfrak{p}}$ which does contain h algebraically independent elements. Since Ω and $\Omega_{\mathfrak{h}_\mathfrak{p}}$ are integral domains. this implies that the operation of restriction to $\mathfrak{h}_\mathfrak{p}$ induces an *isomorphism* of Ω onto $\Omega_{\mathfrak{h}_\mathfrak{p}}$. But since $\mathfrak{I}_{\mathfrak{h}_\mathfrak{p}} \subset \Omega_{\mathfrak{h}_\mathfrak{p}} \subset S(\mathfrak{h}_\mathfrak{p})$ and since $\mathfrak{I}_{\mathfrak{h}_\mathfrak{p}}$ is Noetherian and $S(\mathfrak{h}_\mathfrak{p})$ is a finite $\mathfrak{I}_{\mathfrak{h}_\mathfrak{p}}$-module, it follows (see van der Waerden [16, vol. II, §99]) that $\Omega_{\mathfrak{h}_\mathfrak{p}}$ is also a finite $\mathfrak{I}_{\mathfrak{h}_\mathfrak{p}}$-module. Therefore in view of the above isomorphism Ω is a finite $\mathfrak{I}_\mathfrak{p}$-module.

2.3. The last step. We have now to prove that $\mathfrak{J}'$ is a finite Ω-module. For this we shall use the classical argument of Hilbert. Let $\mathfrak{A}$ be the $S(\mathfrak{p})$-module generated by $\mathfrak{J}'$ (i.e. the set of linear combinations of elements of $\mathfrak{J}'$ with coefficients in $S(\mathfrak{p})$). Since the ring $S(\mathfrak{p})$ is Noetherian and $S(\mathfrak{p}) \times W$ is a finite module over $S(\mathfrak{p})$, $\mathfrak{A}$ is a finite $S(\mathfrak{p})$-module and therefore $\mathfrak{A} = \sum_{i=1}^{r} S(\mathfrak{p}) A_i$ $(A_i \in \mathfrak{J}')$. We shall prove that $\mathfrak{J}' = \sum_{i=1}^{r} \Omega A_i$. From Lemma 7, $S(\mathfrak{p}) \times W$ is the direct sum of $\mathfrak{J}'$ and the space $\mathfrak{M}$ spanned by the elements $\xi(X) A$ $(X \in \mathfrak{k}$ and $A \in S(\mathfrak{p}) \times W)$. Similarly $S(\mathfrak{p})$ may be written on the direct sum of Ω and the space $\mathfrak{N}$ spanned by the elements $\rho(X) F (X \in \mathfrak{k}, F \in S(\mathfrak{p}))$. Let $A = \sum_{i=1}^{r} F_i A_i$ $(F_i \in S(\mathfrak{p}))$ be an element of $\mathfrak{J}'$. Then $F_i = F_i' + N_i$ when $F_i' \in \Omega$ and $N_i \in \mathfrak{N}$. If $G \in S(\mathfrak{p})$, $\xi(X)(G A_i) = (\rho(X) G) A_i$ $(X \in \mathfrak{k})$ since $\xi(X) A_i = 0$. Hence $(\rho(X) G) A_i \in \mathfrak{M}$ and therefore $N_i A_i \in \mathfrak{M}$ $(1 \leq i \leq r)$. Now $F_i' A_i \in \mathfrak{J}'$ and $A = \sum_{i=1}^{r} F_i' A_i + \sum_{i=1}^{r} N_i A_i$. The sum $\mathfrak{J}' + \mathfrak{M}$ being direct, it follows that $A = \sum_{i=1}^{r} F_i' A_i$ and this proves our assertion. Theorem 1 is now completely proved.

PART II. WELL-BEHAVED FUNCTIONS ON A LIE GROUP

4. Preliminary remarks. Let G be a connected Lie group and $\mathfrak{H}$ a (complex) Banach space. For any bounded linear operator A on $\mathfrak{H}$ we write $|A| = \sup_{|\psi| \leq 1} |A\psi|$ $(\psi \in \mathfrak{H})$. Let I denote the unit operator on $\mathfrak{H}$. Then by a representation of G on $\mathfrak{H}$ we mean a mapping π which assigns to every

---

(5) Chevalley's results are not yet published. I am thankful to Professor Chevalley for being good enough to let me use them.

$x \in G$ a bounded linear operator $\pi(x)$ on $\mathfrak{H}$ such that the following two conditions hold:

(1) $\pi(xy) = \pi(x)\pi(y)$ $(x, y \in G)$ and $\pi(1) = I$ where 1 is the unit element of G.

(2) The mapping[6] $(x, \psi) \rightarrow \pi(x)\psi$ $(x \in G, \psi \in \mathfrak{H})$ is a continuous mapping of $G \times \mathfrak{H}$ into $\mathfrak{H}$.

It follows from the second condition that $|\pi(x)|$ is bounded on every compact set in G. Conversely assuming that (1) is fulfilled, it is easy to prove that if the mapping $x \rightarrow \pi(x)\psi$ $(x \in G)$ is continuous at $x = 1$ for every $\psi \in \mathfrak{H}$ and $|\pi(x)|$ is bounded on some neighbourhood of 1 in G, then π is a representation.

Let $C_c^\infty(G)$ denote the set of all (complex-valued) functions on G which are indefinitely differentiable everywhere and which vanish outside a compact set. Let V be the subspace of $\mathfrak{H}$ spanned by all elements ϕ of the form

$$\phi = \int_G f(x)\pi(x)\psi \, dx \qquad (f \in C_c^\infty(G), \ \psi \in \mathfrak{H})$$

where dx is the element of the left invariant Haar measure on G. Let $\mathfrak{g}_0$ be the Lie algebra of G and $\mathfrak{g}$ its complexification. We denote by $\mathfrak{B}$ the universal enveloping algebra of $\mathfrak{g}$. Gårding [5] has shown that for any $X \in \mathfrak{g}_0$ and $\phi \in V$ the limit

$$\lim_{t \to 0} \frac{1}{t} \{\pi(\exp tX)\phi - \phi\} \qquad (t \in R)$$

exists and lies in V and if we denote this limit by $\pi_V(X)\phi$ we get a representation π_V of $\mathfrak{g}$ (and therefore of $\mathfrak{B}$) on V. We shall call V the Gårding subspace of $\mathfrak{H}$ and π_V the Gårding representation of $\mathfrak{g}$ (or $\mathfrak{B}$). Unfortunately this representation has one serious shortcoming. If U is a π_V-stable linear subspace of V, then its closure $\overline{U}$ is not necessarily invariant under $\pi(G)$. Thus one of the main links which connect representations of G with those of $\mathfrak{g}_0$ in the finite-dimensional case is absent in the relationship between π and π_V. Our principal objective now will be to restore this link by replacing the Gårding subspace by the space of all "well-behaved" elements in $\mathfrak{H}$.

5. **Power series in a Banach space.** Let $\{\psi_\alpha\}_{\alpha \in J}$ be an indexed set of elements in $\mathfrak{H}$. For any finite subset F of J let s_F denote the sum $\sum_{\alpha \in F} \psi_\alpha$. We say that the series $\sum_{\alpha \in J} \psi_\alpha$ converges if there exists an element $\phi \in \mathfrak{H}$ such that for any $\epsilon > 0$ we can find a finite subset F_0 of J with the property that $|s_F - \phi| \leq \epsilon$ whenever F is a finite subset of J containing F_0. ϕ is then called the sum of the series and we write $\phi = \sum_{\alpha \in J} \psi_\alpha$. (It is clear that ϕ if it exists is unique.) Moreover we say that the series converges absolutely if

---

[6] It can be shown that condition (2) can be replaced by the apparently weaker requirement that the mapping $x \rightarrow \pi(x)\psi$ $(x \in G)$ be continuous for every $\psi \in \mathfrak{H}$.

$\sum_{\alpha \in J} |\psi_\alpha| < \infty$. It is obvious that an absolutely convergent series is convergent.

Let $\sum_{e_1, \cdots, e_n \geq 0} \psi(e_1, \cdots, e_n) t_1^{e_1} \cdots t_n^{e_n}$ $(\psi(e_1, \cdots, e_n) \in \mathfrak{H})$ be a power series which converges for $t_1 = a_1, \cdots, t_n = a_n$ $(a_i \in C)$. Then it follows from the usual arguments that it converges absolutely for all $t_1, \cdots, t_n$ such that $|t_i| < |a_i|$. We shall say that the series[7] $\sum_{(e)} \psi(e_1, \cdots, e_n) t_1^{e_1} \cdots t_n^{e_n}$ converges near the origin if it converges when $|t_1|, \cdots, |t_n|$ are all sufficiently small. It is clear that convergence near the origin implies absolute convergence for sufficiently small values of $|t_1|, \cdots, |t_n|$. Moreover a power series converges to zero near the origin if and only if all its coefficients are zero. Since the field of complex numbers is a Banach space, the above terminology is also applicable to ordinary power series with complex coefficients.

LEMMA 14. *Let $\sum_{e_1, \cdots, e_n \geq 0} \psi(e_1, \cdots, e_n) t_1^{e_1} \cdots t_n^{e_n}$ be a power series with coefficients in $\mathfrak{H}$ which converges near the origin. Let $g_1, \cdots, g_n$ be n power series in m variables $u_1, \cdots, u_m$ with complex coefficients which are all convergent near the origin and which take the values 0 at the origin (i.e. at $u_1 = u_2 = \cdots = u_m = 0$). Let*

$$\sum_{(d)} a(e_1, \cdots, e_n, d_1, \cdots, d_m) u_1^{d_1} \cdots u_m^{d_m} \qquad (a(e, d) \in C)$$

be the power series expansion of $g_1^{e_1} \cdots g_n^{e_n}$. Then the series

$$\sum_{(e),(d)} \psi(e_1, \cdots, e_n) a(e_1, \cdots, e_n, d_1, \cdots, d_n) u_1^{d_1} \cdots u_m^{d_m}$$

converges absolutely near the origin.

Choose $\epsilon > 0$ such that the series $\sum_{(e)} \psi(e_1, \cdots, e_n) t_1^{e_1} \cdots t_n^{e_n}$ and the series $g_1, \cdots, g_n$ all converge absolutely if $|t_i| \leq \epsilon, |u_j| \leq \epsilon$ $(1 \leq i \leq n, 1 \leq j \leq m)$. Let $g_i = \sum_{(d)} b_i(d_1, \cdots, d_m) u_1^{d_1} \cdots u_m^{d_m}$ $(b_i(d) \in C, 1 \leq i \leq n)$. Since the constant term of g_i is zero, it is possible to choose $\delta > 0$ $(\delta \leq \epsilon)$ such that

$$\sum_{(d)} \left| b_i(d_1, \cdots, d_m) u_1^{d_1} \cdots u_m^{d_m} \right| \leq \epsilon$$

if $|u_j| \leq \delta$ $(1 \leq j \leq m)$. Then it is clear that

$$\sum_{(e),(d)} \left| \psi(e_1, \cdots, e_n) a(e_1, \cdots, e_n, d_1, \cdots, d_m) u_1^{d_1} \cdots u_m^{d_m} \right|$$

$$\leq \sum_{(e)} \left| \psi(e_1, \cdots, e_n) \right| \epsilon^{e_1 + \cdots + e_n} < \infty$$

provided $|u_j| \leq \delta$. This proves the lemma.

---

[7] We shall often abbreviate $(e_1, \cdots, e_n)$ to (e) and $\psi(e_1, \cdots, e_n)$ to $\psi(e)$. Similarly for other symbols.

LEMMA 15. *Let $\sum_{e\geq 0} \psi(e)t^e$ $(\psi(e)\in\mathfrak{H})$ be a power series which converges to $f(t)$ if $|t| < r$ $(r>0)$. Then the power series $\sum_{e\geq 1}\psi(e)et^{e-1}$ is convergent for $|t| < r$ and its sum is equal to the limit*

$$\lim_{h\to 0}\frac{f(t+h)-f(t)}{h} \qquad (h\in C,\; |t| < r).$$

It is clear that $\sum_{e\geq 0}|\psi(e)t^e|$ converges if $|t| < r$ and from this it follows that the same is true of the series $\sum_{e\geq 1}|\psi(e)et^{e-1}|$. Therefore if $g(t)$ is the sum of the series $\sum_{e\geq 1}\psi(e)et^{e-1}$ $(|t| < r)$, we have

$$\left|\frac{f(t+h)-f(t)}{h}-g(t)\right| \leq \sum_{e\geq 0}|\psi(e)|\left|\frac{(t+h)^e-t^e}{h}-et^{e-1}\right|.$$

Choose δ so small $(\delta>0)$ that $|t|+2\delta<r$. Then if $|h|<\delta/2$,

$$\left|\frac{(t+h)^e-t^e}{h}-et^{e-1}\right| = \left|h\oint\frac{z^e}{(z-t)^2(z-t-h)}dz\right| \leq |h|\frac{4\pi}{\delta^2}(r-\delta)^e$$

where $\oint$ denotes complex integration on the circle $|z-t|=\delta$. Hence

$$\left|\frac{f(t+h)-f(t)}{h}-g(t)\right| \leq \frac{4\pi}{\delta^2}|h|\sum_{e\geq 0}|\psi(e)|(r-\delta)^e.$$

Since $\sum_{e\geq 0}|\psi(e)|(r-\delta)^e<\infty$ it follows that

$$g(t) = \lim_{h\to 0}\frac{f(t+h)-f(t)}{h}.$$

Put

$$\frac{d}{dt}f(t) = \lim_{h\to 0}\frac{f(t+h)-f(t)}{h}.$$

Then

$$\frac{df(t)}{dt} = \sum_{e\geq 1}e\psi(e)t^{e-1} \qquad (|t| < r).$$

COROLLARY. *Let $\sum_{(e)}\psi(e_1,\cdots,e_n)\,t_1^{e_1}\cdots t_n^{e_n}$ $(\psi(e_1,\cdots,e_n)\in\mathfrak{H})$ be a power series which converges to $f(t_1,\cdots,t_n)=f(t)$ if $|t_i|<r_i$ $(r_i>0,\; 1\leq i\leq n)$. Then the series $\sum_{(e)}\psi(e_1,\cdots,e_n)\,\partial(t_1^{e_1}\cdots t_n^{e_n})/\partial t_i$ converges in the same region to*

$$\frac{\partial f(t)}{\partial t_i} = \lim_{h\to 0}\frac{f(t_1,\cdots,t_i+h,\cdots,t_n)-f(t_1,\cdots,t_n)}{h}.$$

We know that $\sum_{(e)}\psi(e_1,\cdots,e_n)\,t_1^{e_1}\cdots t_n^{e_n}$ converges absolutely for $|t_j|<r_j$. Hence the series obtained by collecting together the coefficients of

$t_i^{e_i}$ is also absolutely convergent. Let $\psi'(e_i)$ denote its sum. Then

$$f(t) = \sum_{e_i \geq 0} \psi'(e_i)t_i^{e_i}$$

and our assertion now follows from the above lemma.

Let M be a real (complex) analytic manifold. A mapping f of M into a Banach space $\mathfrak{H}$ is called analytic (holomorphic) at a point $x_0 \in M$ if for every coordinate system $(t_1, \cdots, t_n)$ at x_0 such that $t_i(x_0) = 0$, there exists a neighbourhood V of x_0 in M (on which our coordinates are valid) and a power series $\sum_{e_1, \ldots, e_n \geq 0} \psi(e_1, \cdots, e_n)\tau_1^{e_1} \cdots \tau_n^{e_n}$ $(\psi(e_1, \cdots, e_n) \in \mathfrak{H})$ in n variables $\tau_1, \cdots, \tau_n$ such that this series converges to $f(x)$ at $\tau_i = t_i(x)$, $1 \leq i \leq n$ $(x \in V)$. It follows easily from Lemma 14 that if the above condition is fulfilled for one coordinate system $(t_1, \cdots, t_n)$, then it necessarily holds for all coordinate systems $(u_1, \cdots, u_n)$ at x_0 such that $u_i(x_0) = 0$, $1 \leq i \leq n$.

LEMMA 16. *Let M and M' be real analytic manifolds and g a mapping of M' into M which is analytic at a point $x_0' \in M'$. Let $x_0 = g(x_0')$ and let f be a mapping of M into $\mathfrak{H}$ which is analytic at x_0. Then the mapping $f \circ g$ of M' into $\mathfrak{H}$ is analytic at x_0'.*

This again is an immediate consequence of Lemma 14.

LEMMA 17. *Let f be a mapping of M into $\mathfrak{H}$ analytic at x_0. Let u be a (complex-valued) function on M which is analytic at x_0. Then the mapping uf is analytic at x_0.*

Choose a coordinate system $(t_1, \cdots, t_n)$ at x_0 such that $t_i(x_0) = 0$, $1 \leq i \leq n$. Then if x lies in a suitable neighbourhood V of x_0,

$$f(x) = \sum_{(e)} \psi(e_1, \cdots, e_n)t_1^{e_1}(x) \cdots t_n^{e_n}(x) \qquad (\psi(e) \in \mathfrak{H}),$$

$$u(x) = \sum_{(e)} a(e_1, \cdots, e_n)t_1^{e_1}(x) \cdots t_n^{e_n}(x) \qquad (a(e) \in C)$$

both series being absolutely convergent. Hence

$$\sum_{(e),(d)} \psi(e_1, \cdots, e_n)a(d_1, \cdots, d_n)t_1^{e_1+d_1}(x) \cdots t_n^{e_n+d_n}(x)$$

converges absolutely to $u(x)f(x)$ for $x \in V$. From this the lemma follows.

LEMMA 18. *Let f be a mapping of M into $\mathfrak{H}$ which is analytic at x_0 and let A be a continuous linear mapping of $\mathfrak{H}$ into a Banach space $\mathfrak{H}'$. Then the mapping $A \circ f$ of M into $\mathfrak{H}'$ is analytic at x_0.*

Let ϕ be the sum of a convergent series $\sum_{\alpha \in J} \psi_\alpha$ in $\mathfrak{H}$. Since A is linear and continuous, the series $\sum_{\alpha \in J} A\psi_\alpha$ is convergent in $\mathfrak{H}'$ and its sum is $A\phi$. The assertion in the lemma is an immediate consequence of this fact.

Let f be a mapping of M into $\mathfrak{H}$ which is analytic at x_0 and let X_0 be a vector[8] in the tangent space of M at x_0. Let $(t_1, \cdots, t_n)$ be a coordinate system at x_0 such that $t_i(x_0) = 0$ and let

$$f(x) = \sum_{(e)} \psi(e_1, \cdots, e_n) t_1^{e_1}(x) \cdots t_n^{e_n}(x) \qquad (\psi(e_1, \cdots, e_n) \in \mathfrak{H})$$

for x sufficiently near x_0. Then we put

$$X_0 f = \sum_{1 \le i \le n} \psi_i (X_0 t_i)$$

where ψ_i is the value of $\psi(e_1, \cdots, e_n)$ for $e_i = 1$ and $e_j = 0$, $j \ne i$. It is easy to verify that the value of $X_0 f$ does not depend on the particular choice of the coordinate system $(t_1, \cdots, t_n)$ used in its definition.

LEMMA 19. *Let f and X_0 be as above and ϕ a continuous linear function on $\mathfrak{H}$. Then the function $\phi(f(x))$ is analytic at x_0 and $X_0(\phi(f)) = \phi(X_0 f)$.*

ϕ is a continuous linear mapping of $\mathfrak{H}$ into C. Hence from Lemma 18, $\phi(f(x))$ is analytic at x_0. Moreover $\phi(f(x)) = \sum_{(e)} \phi(\psi(e_1, \cdots, e_n)) t_1^{e_1}(x) \cdots t_n^{e_n}(x)$ if x is sufficiently near x_0. Hence

$$X_0(\phi(f)) = \sum_{1 \le i \le n} \phi(\psi_i)(X_0 t_i) = \phi(X_0 f).$$

LEMMA 20. *Suppose M' and M are real analytic manifolds and g is a mapping of M' into M which is analytic at $x_0' \in M'$. Put $x_0 = g(x_0')$ and let f be a mapping of M into $\mathfrak{H}$ which is analytic at x_0. Then if X_0' is a vector in the tangent space of M' at x_0',*

$$X_0'(f \circ g) = (dg\, X_0')f$$

where dg is the differential of g.

This follows without difficult from Lemmas 16 and 14.

We regard the field R of real numbers as a real analytic manifold in the usual way. For a given $t_0 \in R$ let T_0 be the vector in the tangent space of R at t_0 such that $T_0 g = \{dg(t)/dt\}_{t = t_0}$ for any function g which is analytic at t_0.

LEMMA 21. *Let f be a mapping of R into $\mathfrak{H}$ which is analytic at t_0. Then*

$$T_0 f = \lim_{h \to 0} \frac{1}{h} \{f(t_0 + h) - f(t_0)\} = \left(\frac{df}{dt}\right)_{t_0}.$$

Put $h(t) = t - t_0$ $(t \in R)$. Then $h(t_0) = 0$ and h can be chosen as a coordinate on R. Since f is analytic at t_0 we can write

$$f(t) = \sum_{e \ge 0} \psi(e) h^e(t) \qquad (\psi(e) \in \mathfrak{H})$$

---

for t sufficiently near t_0. Hence

$$\frac{f(t) - f(t_0)}{h(t)} = \sum_{\bullet \geq 1} \psi(e) h^{\bullet - 1}(t)$$

and therefore

$$\lim_{t \to 0} \frac{f(t) - f(t_0)}{h(t)} = \psi(1) = T_0 f.$$

This proves the lemma.

COROLLARY. *Let g be a mapping of R in M which is analytic at t_0 and f a mapping of M into $\mathfrak{H}$ which is analytic at $x_0 = g(t_0)$. Then if $X_0 = dgT_0$,*

$$X_0 f = \left(\frac{d}{dt} f(g(t))\right)_{t = t_0} = \lim_{h \to 0} \frac{f(g(t_0 + h)) - f(g(t_0))}{h}.$$

This follows immediately from Lemmas 20 and 21.

Let f be a mapping of a real analytic manifold M into $\mathfrak{H}$. We say f is analytic if it is analytic at every point of M.

LEMMA 22. *Let f be an analytic mapping of M into $\mathfrak{H}$ and X an analytic infinitesimal transformation on M. Then the mapping Xf (i.e. $x \to X(x)f$, $x \in M$) is also analytic. Moreover if Y is another infinitesimal transformation and $Z = [X, Y] = XY - YX$, then*

$$Zf = X \cdot Yf - Y \cdot Xf.$$

Let $x_0 \in M$ and let $(t_1, \cdots, t_n)$ be a coordinate system at x_0 such that $t_i(x_0) = 0$, $1 \leq i \leq n$. Then there exists an open neighbourhood U of x_0 (on which the system $(t_1, \cdots, t_n)$ is valid) and functions $u_1, \cdots, u_n$ which are analytic on U such that $X = \sum_{1 \leq i \leq n} u_i \partial/\partial t_i$ on U. Moreover we may choose U so small that we have the expansion

$$f(x) = \sum_{(e)} \psi(e_1, \cdots, e_n) t_1^{e_1}(x) \cdots t_n^{e_n}(x) \qquad (\psi(e) \in \mathfrak{H}, \, x \in U).$$

It follows from the above corollary that

$$\left(\frac{\partial f}{\partial t_i}\right)(x) = \lim_{h \to 0} \frac{1}{h} \left\{ \sum_{(e)} \psi(e_1, \cdots, e_n) t_1^{e_1}(x) \cdots (t_i(x) + h)^{e_i} \cdots t_n^{e_n}(x) \right.$$

$$\left. - \sum_{(e)} \psi(e_1, \cdots, e_n) t_1^{e_1}(x) \cdots t_i^{e_i}(x) \cdots t_n^{e_n}(x) \right\}$$

$$= \sum_{(e)} \psi(e_1, \cdots, e_n) \left\{ \frac{\partial}{\partial t_i} (t_1^{e_1} \cdots t_n^{e_n}) \right\}_{t_j = t_j(x)}$$

in view of the corollary to Lemma 15. This proves that $\partial f/\partial t_i$ is analytic at x_0. Since the functions u_i are analytic at x_0 we conclude from Lemma 17 that Xf is analytic at x_0.

Now we come to the second part. Let ϕ be a continuous linear function on $\mathfrak{H}$ and g an analytic mapping of M into $\mathfrak{H}$. Then from Lemma 19 the function $\phi_g\colon x \to \phi(g(x))$ is an analytic function on M and $X\phi_g = \phi_{Xg}$ for any analytic linear transformation X on M. From this it follows that $\phi_{Zf} = \phi_{XYf} - \phi_{YXf}$. Now put $g = Zf - (XYf - YXf)$. Then $\phi(g(x)) = 0$. Since this is true for every ϕ it follows from the Hahn-Banach Theorem that $Zf = XYf - YXf$.

6. Banach spaces of functions. Let L be a closed finite interval on the real line and $t \to \psi_t$ a continuous mapping of L into a Banach space $\mathfrak{H}$. Then $|\psi_t|$ is a continuous function on L and $\int_L |\psi_t| \, dt < \infty$. Therefore the integral $\int_L \psi_t \, dt$ is defined and for any bounded linear function α on $\mathfrak{H}$, $\alpha(\int_L \psi_t \, dt) = \int_L \alpha(\psi_t) \, dt$. Let D be a domain (i.e. an open connected set) in the complex plane and $z \to \psi_z$ a continuous mapping of D into $\mathfrak{H}$. Let Γ be a rectifiable curve in D. Then it is clear that the complex integral $\int_\Gamma \psi_z \, dz$ exists and $\alpha(\int_\Gamma \psi_z \, dz) = \int_\Gamma \alpha(\psi_z) \, dz$.

Let E be a locally compact Hausdorff space with a (regular) positive measure μ given on it. Let $\mathfrak{H}$ denote the Banach space of all μ-summable functions on E.

LEMMA 23. *Let D be a domain in C^n and $(z_1, \cdots, z_n, x) \to f(z_1, \cdots, z_n; x)$ a continuous function on $D \times E$ which satisfies the following two conditions:*

(a) *there exists a μ-summable function g on E such that $|f(z_1, \cdots, z_n; x)| \leqq |g(x)|$ for all $(z_1, \cdots, z_n, x) \in D \times E$;*

(b) *for each $x \in E$ the function $(z_1, \cdots, z_n) \to f(z_1, \cdots, z_n; x)$ is holomorphic on D.*

Let $\psi(z_1, \cdots, z_n)$ be the element of $\mathfrak{H}$ represented by the function $x \to f(z_1, \cdots, z_n; x)$. Then the mapping $(z_1, \cdots, z_n) \to \psi(z_1, \cdots, z_n)$ is a holomorphic mapping of D into $\mathfrak{H}$.

Let $(z_1, \cdots, z_n) \in D$. Given $\epsilon > 0$ and a compact set K in E we can find $\delta > 0$ such that if $\max_i |h_i| < \delta$, then $(z_1 + h_1, \cdots, z_n + h_n) \in D$ and

$$\left| f(z_1 + h_1, \cdots, z_n + h_n; x) - f(z_1, \cdots, z_n; x) \right| \leqq \epsilon$$

for x in K. Since $|g|$ is μ-summable we can choose K such that $\int_{E-K} |g| \, d\mu \leqq \epsilon$. Hence

$$\left| \psi(z_1 + h_1, \cdots, z_n + h_n) - \psi(z_1, \cdots, z_n) \right|$$

$$= \int \left| f(z_1 + h_1, \cdots, z_n + h_n; x) - f(z_1, \cdots, z_n; x) \right| d\mu$$

$$\leqq \epsilon \int_K d\mu + 2 \int_{E-K} |g| \, d\mu \leqq (\mu(K) + 2)\epsilon$$

if $|h_i| < \delta$. This shows that the mapping $(z_1, \cdots, z_n) \to \psi(z_1, \cdots, z_n)$ is a continuous mapping of D into $\mathfrak{H}$.

Now let $(a_1, \cdots, a_n)$ be any point of D. Choose positive real numbers $r_1, \cdots, r_n$ such that the polycylinder P defined by the conditions $|z_i - a_i| \leq r_i$ $(1 \leq i \leq n)$ lies in D. Let $(z_1, \cdots, z_n)$ be any interior point of P. Put

$$\phi(z_1, \cdots, z_n) = \left(\frac{1}{2\pi(-1)^{1/2}}\right)^n \oint_1 \cdots \oint_n \frac{\psi(\zeta_1, \cdots, \zeta_n)}{(\zeta_1 - z_1) \cdots (\zeta_n - z_n)} \, d\zeta_1 \cdots d\zeta_n$$

where $\oint_k$ denotes complex integration on the circle $|\zeta_k - a_k| = r_k$. For any measurable set X in E and a μ-summable function F on E put

$$\alpha_X(F) = \int_X F d\mu.$$

Then α_X may clearly be regarded as a bounded linear function on $\mathfrak{H}$. Moreover $\alpha_X(F) = 0$ for all X implies $F = 0$ μ-almost everywhere. Therefore if $\alpha_X(\psi) = 0$ $(\psi \in \mathfrak{H})$ for all X then $\psi = 0$. Now

$$\alpha_X(\phi(z_1, \cdots, z_n))$$

$$= \left(\frac{1}{2\pi(-1)^{1/2}}\right)^n \oint_1 \cdots \oint_n \frac{\alpha_X(\psi(\zeta_1, \cdots, \zeta_n))}{(\zeta_1 - z_1) \cdots (\zeta_n - z_n)} \, d\zeta_1 \cdots d\zeta_n.$$

But $\alpha_X(\psi(\zeta_1, \cdots, \zeta_n)) = \int_X f(\zeta_1, \cdots, \zeta_n; x) d\mu$ and therefore

$$\left| \alpha_X(\psi(\zeta_1, \cdots, \zeta_n)) \right| \leq \int |g| \, d\mu.$$

Hence it is clear that

$$\alpha_X(\phi(z_1, \cdots, z_n))$$

$$= \left(\frac{1}{2\pi(-1)^{1/2}}\right)^n \int_X d\mu \oint_1 \cdots \oint_n \frac{f(\zeta_1, \cdots, \zeta_n; x)}{(\zeta_1 - z_1) \cdots (\zeta_n - z_n)} \, d\zeta_1 \cdots d\zeta_n$$

$$= \int_X f(z_1, \cdots, z_n; x) d\mu$$

by Cauchy's Theorem. Therefore

$$\alpha_X(\phi(z_1, \cdots, z_n)) = \alpha_X(\psi(z_1, \cdots, z_n)).$$

This being true for all X we conclude that

$$\psi(z_1, \cdots, z_n) = \phi(z_1, \cdots, z_n)$$

$$= (2\pi(-1)^{1/2})^{-n} \oint_1 \cdots \oint_n \frac{\psi(\zeta_1, \cdots, \zeta_n)}{(\zeta_1 - z_1) \cdots (\zeta_n - z_n)} \, d\zeta_1 \cdots d\zeta_n.$$

It follows immediately from this formula, by the classical argument, that there exists a power series in $(z_1 - a_1), \cdots, (z_n - a_n)$ with coefficients in $\mathfrak{H}$ which converges in the interior of P to $\psi(z_1, \cdots, z_n)$. Therefore the mapping $(z_1, \cdots, z_n) \to \psi(z_1, \cdots, z_n)$ is holomorphic at $(a_1, \cdots, a_n)$.

7. Well-behaved vectors in a representation space. Let π be a representation of a connected Lie group G on a Banach space $\mathfrak{H}$. We shall say that an element $\psi \in \mathfrak{H}$ is well-behaved (under π) if the mapping $f: x \to \pi(x)\psi$ is an analytic mapping of G into $\mathfrak{H}$. For any $y \in G$ let ϕ_y denote the mapping $x \to y^{-1}x$ $(x \in G)$. Then it is easy to verify that $\pi(y) \circ f \circ \phi_y = f$. Since ϕ_y is an analytic isomorphism of G with itself it follows from Lemmas 16 and 18 that if f is analytic at $x = 1$ then $f = \pi(y) \circ f \circ \phi_y$ is analytic at y. Therefore in order that f be analytic it is sufficient that it should be analytic at 1.

Let W be the set of all elements in $\mathfrak{H}$ which are well-behaved under π. Clearly W is a linear subspace of $\mathfrak{H}$. Moreover since for a fixed $y \in G$ the mapping $x \to xy$ is an analytic mapping of G, it follows that if $\psi \in W$ then $\pi(y)\psi \in W$. Let $\mathfrak{g}_0$ be the Lie algebra of G. For any $\psi \in W$ and $X \in \mathfrak{g}_0$ consider the mapping $t \to \pi(\exp tX)\psi$ $(t \in R)$. Clearly this mapping is analytic and therefore from Lemma 21 the limit

$$\lim_{t \to 0} \frac{1}{t} \left\{ \pi(\exp tX)\psi - \psi \right\}$$

exists. We denote this limit by $\pi_W(X)\psi$. Let f be the mapping $x \to \pi(x)\psi$ and g the mapping $x \to \pi(x)\pi_W(X)\psi$. Then it follows from Lemmas 20 and 21 that $g = Xf$ and so g is analytic (Lemma 22). Hence $\pi_W(X)\psi \in W$. Thus we get a linear transformation $\pi_W(X)$ of W into itself. Moreover if $[X, Y] = Z$ $(X, Y, Z \in \mathfrak{g}_0)$ we know from Lemma 22 that $Zf = X \cdot Yf - Y \cdot Xf$ and therefore $\pi_W([X, Y]) = \pi_W(X)\pi_W(Y) - \pi_W(Y)\pi_W(X)$. This shows that the mapping $X \to \pi_W(X)$ is a representation of $\mathfrak{g}_0$ on W. Let $\mathfrak{g}_0$ be the complexification of $\mathfrak{g}$ and $\mathfrak{B}$ the universal enveloping algebra of $\mathfrak{g}$. We denote by π_W the representation of $\mathfrak{B}$ on W which coincides on $\mathfrak{g}_0$ with this representation.

Since $\mathfrak{g}_0$ is a vector space of finite dimension over R we may regard it as a real analytic manifold in the obvious way. Then the most important property of well-behaved elements may be expressed as follows.

THEOREM 2. *Let ψ be an element in W. Then there exists a neighbourhood V of zero in $\mathfrak{g}_0$ such that the series $\sum_{m \geq 0} (1/m!)\pi_W(X^m)\psi$ converges to $\pi(\exp X)\psi$ for $X \in V$.*

Let $(t_1, \cdots, t_n)$ be the Cartesian coordinate system in $\mathfrak{g}_0$ corresponding to a base $(X_1, \cdots, X_n)$ so that $X = \sum_{i=1}^{n} t_i(X)X_i$ $(X \in \mathfrak{g}_0)$. Since the mapping $X \to \pi(\exp X)\psi$ is an analytic mapping of $\mathfrak{g}_0$ into $\mathfrak{H}$, we can find a neighbourhood V of zero in $\mathfrak{g}_0$ such that

$$\pi(\exp X)\psi = \sum_{(e)} \psi(e_1, \cdots, e_n)t_1^{e_1}(X) \cdots t_n^{e_n}(X)$$

$(\psi(e_1, \cdots, e_n) \in \mathfrak{H}, X \in V)$, the series being absolutely convergent. We shall now prove that

$$\sum_{e_1+\cdots+e_n=m} \psi(e_1, \cdots, e_n)t_1^{e_1}(X) \cdots t_n^{e_n}(X) = \frac{1}{m!}\pi_W(X^m)\psi \quad (X \in \mathfrak{g}_0).$$

For a fixed $X \in \mathfrak{g}_0$ choose $r > 0$ so small that $uX \in V$ whenever $|u| \leqq r$ $(u \in R)$. Then

$$\pi(\exp uX)\psi = \sum_{m \geqq 0} \psi_m u^m \qquad (|u| < r)$$

where $\psi_m = \sum_{e_1+\ldots+e_n=m} \psi(e_1, \cdots, e_n)t_1^{e_1}(X) \cdots t_n^{e_n}(X)$. We shall now prove by induction on m that

$$\pi(\exp uX)\pi_W(X^m)\psi = \sum_{\mu \geqq m} \psi_\mu\left(\frac{d^m}{du^m}u^\mu\right) \qquad (|u| < r).$$

This is true for $m = 0$. Now

$$\pi_W(X^{m+1})\psi = \pi_W(X)\pi_W(X^m)\psi$$

$$= \lim_{s \to 0} \frac{1}{s}\left\{\pi(\exp sX)\pi_W(X^m)\psi - \pi_W(X^m)\psi\right\}.$$

Since $\pi(\exp uX)$ is a bounded operator,

$$\pi(\exp uX)\pi_W(X^{m+1})\psi = \lim_{s \to 0} \frac{1}{s}\left[\pi(\exp (s+u)X) - \pi(\exp uX)\right]\psi'$$

where $\psi' = \pi_W(X^m)\psi$. But

$$\pi(\exp uX)\psi' = \sum_{\mu \geqq m} \psi_\mu\left(\frac{d^m}{du^m}u^\mu\right) \qquad (|u| < r)$$

by our induction hypothesis. Hence

$$\pi(\exp uX)\pi_W(X^{m+1})\psi = \sum_{\mu \geqq m+1} \psi_\mu \frac{d^{m+1}}{du^{m+1}}u^\mu \qquad (|u| < r)$$

from Lemma 21 and so our assertion is proved. Now if we put $u = 0$ we get

$$\pi_W(X^m)\psi = m!\psi_m = m! \sum_{e_1+\cdots+e_n=m} \psi(e_1, \cdots, e_n)t_1^{e_1}(X) \cdots t_n^{e_n}(X).$$

Therefore if $X \in V$

$$\pi(\exp X)\psi = \sum_{m \geqq 0} \frac{1}{m!}\pi_W(X^m)\psi.$$

COROLLARY. *Let ψ be a well-behaved element and let* Cl $(\pi_W(\mathfrak{B})\psi)$ *denote the* *closure of $\pi_W(\mathfrak{B})\psi$ in $\mathfrak{H}$. Then* $\mathrm{Cl}\,(\pi_W(\mathfrak{B})\psi)$ *is invariant under* $\pi(G)$.

Let ψ_0 be any element in $U = \pi_W(\mathfrak{B})\psi$ and consider any continuous linear function ϕ on $\mathfrak{H}$ which vanishes on U. Since $\psi_0 \in W$ the function $x \to \phi(\pi(x)\psi_0)$ $(x \in G)$ is an analytic function on G. On the other hand in view of the above theorem there exists a neighbourhood V of zero in $\mathfrak{g}_0$ such that

$$\pi(\exp X)\psi_0 = \sum_{m \geq 0} \frac{1}{m!}\, \pi_W(X^m)\psi_0). \qquad (X \in V).$$

Therefore

$$\phi(\pi(\exp X)\psi_0) = \sum_{m \geq 0} \frac{1}{m!}\, \phi(\pi_W(X^m)\psi_0).$$

But $\pi_W(X^m)\psi_0 \in U$ and therefore $\phi(\pi_W(X^m)\psi_0) = 0$. Hence $\phi(\pi(\exp X)\psi_0) = 0$ for $X \in V$ and therefore $\phi(\pi(x)\psi_0)$ vanishes on a neighbourhood of 1 in G. But since it is an analytic function on G it follows that $\phi(\pi(x)\psi_0) = 0$ for all $x \in G$. Keeping x fixed and varying ϕ we deduce from the Hahn-Banach theorem that $\pi(x)\psi_0 \in \mathrm{Cl}\,(U)$. This proves that $\pi(x) U \subset \mathrm{Cl}\,(U)$ $(x \in G)$ and therefore by continuity $\pi(x)\,\mathrm{Cl}\,(U) \subset \mathrm{Cl}\,(U)$.

8. **Well-behaved functions.** Let π be a representation of G on a Banach space $\mathfrak{H}$. Then $|\pi(x)| = \sup_{|\psi| \leq 1} |\pi(x)\psi|$ $(\psi \in \mathfrak{H})$ is a semi-continuous function on G and therefore it is measurable (with respect to the Haar measure). Let $L(G, \pi)$ denote the space of all (complex-valued) measurable functions f on G such that

$$\|f\| = \int_G |f(x)|\,|\pi(x)|\,dx < \infty$$

where dx is the element of the left-invariant Haar measure on G. Then with respect to the norm $\|\ \|$, $L(G, \pi)$ is a Banach space and we get a representation λ of G on $L(G, \pi)$ if we define $\lambda(y)f$ to be the function whose value at x is $f(y^{-1}x)$ $(x, y \in G, f \in L(G, \pi))$. In fact,

$$\|\lambda(y)f\| = \int |f(y^{-1}x)|\,|\pi(x)|\,dx = \int |f(x)|\,|\pi(yx)|\,dx \leq |\pi(y)|\,\|f\|$$

and therefore $\lambda(y)$ is a bounded operator and

$$\|\lambda(y)\| = \sup_{\|f\|=1} \|\lambda(y)f\| \leq |\pi(y)|.$$

This shows that $\|\lambda(y)\|$ remains bounded on a compact set. Therefore, in order to prove that λ is a representation, it is sufficient to show that $\lim_{x \to 1} \|\lambda(x)f_0 - f_0\| = 0$ for any $f_0 \in L(G, \pi)$. Let V be a compact neighbourhood of 1 in G and m an upper bound for $\|\lambda(x)\|$ on V. Given any $\epsilon > 0$ we can choose a continuous function g which vanishes outside a compact set such that $\|f_0 - g\| \leqq \epsilon$. Since $|\pi(x)|$ remains bounded on a compact set, it is clear that we can find a neighbourhood U of 1 $(U \subset V)$ such that

$$\|\lambda(x)g - g\| = \int |g(x^{-1}y) - g(y)||\pi(y)| dy \leqq \epsilon.$$

Then

$$\|\lambda(x)f_0 - f_0\| = \|\lambda(x)(f_0 - g) - (f_0 - g) + (\lambda(x)g - g)\|$$
$$\leqq \|\lambda(x)\|\epsilon + \epsilon + \epsilon \leqq (m + 2)\epsilon \qquad (x \in U).$$

This proves that $\lim_{x \to 1} \|\lambda(x)f_0 - f_0\| = 0$ and therefore λ is indeed a representation.

Let $\mathfrak{O}$ be the Banach space of all bounded linear operators on $\mathfrak{H}$ with the norm $|A| = \sup_{|\psi| \leq 1} |A\psi|$ $(A \in \mathfrak{O}, \psi \in \mathfrak{H})$. If $f \in L(G, \pi)$, the integral $\int_G f(x)\pi(x)dx$ has a well-defined meaning in $\mathfrak{O}$ since $\int_G |f(x)\pi(x)| dx = \|f\| < \infty$. Put $T_f = \int f(x)\pi(x)dx$. Then the mapping $T: f \to T_f$ is a continuous linear mapping of $L(G, \pi)$ into $\mathfrak{O}$ since $|T_f| \leqq \|f\|$. Now let $\mathfrak{W}_\pi$ denote the set of all elements in $L(G, \pi)$ which are well-behaved under λ. Then for any $\psi \in \mathfrak{H}$,

$$\pi(x)T_f\psi = \int_G f(y)\pi(xy)\psi dy = \int_G f(x^{-1}y)\pi(y)\psi dy = T_{\lambda(x)f}\psi.$$

But the mapping $x \to \lambda(x)f$ is analytic and T is linear and continuous. Similarly $A \to A\psi$ $(A \in \mathfrak{O})$ is a continuous linear mapping of $\mathfrak{O}$ into $\mathfrak{H}$ since $|A\psi| \leqq |A||\psi|$. Hence $x \to T_{\lambda(x)f}\psi = \pi(x)T_f\psi$ is an analytic mapping of G into $\mathfrak{H}$ (Lemma 18). Thus we have the following result.

LEMMA 24. *Let $f \in \mathfrak{W}_\pi$ and $\psi \in \mathfrak{H}$. Then $\int_G f(x)\pi(x)\psi dx$ is a well-behaved element of $\mathfrak{H}$ (with respect to π).*

We shall now investigate the question of approximating arbitrary elements in $\mathfrak{H}$ by well-behaved elements. Let $L_1(G)$ denote the space of all functions on G which are summable with respect to the Haar measure.

DEFINITIONS. *Let $\mu(x)$ be any measurable function on G which is real and non-negative and let $\{f_n(x)\}$ be a sequence of functions in $L_1(G)$. We shall call it a Dirac μ-sequence if the following conditions hold:*

(1) $\int f_n(x)dx = 1$ *and* $\lim_{n \to \infty} \int_G |f_n(x)| dx = 1$.

(2) *For any measurable neighbourhood V of 1 in G,*

$$\lim_{n \to \infty} \int_{G-V} |f_n(x)| (1 + \mu(x))dx = 0.$$

If $\mu = 1$ we call it just a Dirac sequence.

LEMMA 25. *Suppose there exists a Dirac $|\pi(x)|$-sequence $\{f_n(x)\}$ on G such that $f_n \in \mathfrak{W}_\pi$. Then the space W of all well-behaved elements is dense in $\mathfrak{H}$.*

Let ψ be an element in $\mathfrak{H}$. Given any $\epsilon > 0$ we can find, due to the continuity of the representation π, an open neighbourhood V of 1 in G such that $|\pi(x)\psi - \psi| \leq \epsilon|\psi|$ for $x \in V$. Then if $\psi_n = T_{f_n}\psi$,

$$\psi_n - \psi = \int_G f_n(x)(\pi(x)\psi - \psi)dx$$

$$= \int_V f_n(x)(\pi(x)\psi - \psi)dx + \int_{G-V} f_n(x)(\pi(x)\psi - \psi)dx.$$

Therefore

$$|\psi_n - \psi| \leq \epsilon|\psi| \int_G |f_n(x)| \, dx + |\psi| \int_{G-V} |f_n(x)| (1 + |\pi(x)|)dx \leq 2\epsilon|\psi|$$

if n is sufficiently large. Hence $\psi = \lim_{n\to\infty} \psi_n$. But since $\psi_n \in W$ (Lemma 24), it follows that W is dense in $\mathfrak{H}$.

We shall now give a method for constructing such a sequence under suitable assumptions regarding G.

Suppose $\mathfrak{g}_0$ is the direct sum of two subalgebras $\mathfrak{k}_0$ and $\mathfrak{s}_0$ and K and S are the analytic subgroups of G corresponding to $\mathfrak{k}_0$ and $\mathfrak{s}_0$ respectively. Consider the mapping $\Phi \colon (u, s) \to us$ ($u \in K$, $s \in S$) of $K \times S$ into G. Clearly it is analytic.

LEMMA 26. Φ is everywhere regular[8] on $K \times S$.

Since the tangent space of a Lie group at any point may be identified under left translation by its Lie algebra (which is the tangent space at 1), it is clear that the tangent space of $K \times S$ at any point is the direct sum of $\mathfrak{k}_0$ and $\mathfrak{s}_0$ and therefore it may be identified with $\mathfrak{g}_0$. Let $d\Phi$ be the differential of Φ and let $u_0 \in K$ and $s_0 \in S$ be two fixed elements. Then we see without difficulty that $(d\Phi)_{u_0,s_0} U = \mathrm{Ad}\ (s_0^{-1}) U$, $(d\Phi)_{u_0,s_0} Y = Y$ for $Y \in \mathfrak{k}_0$ and $U \in \mathfrak{s}_0$. Here $x \to \mathrm{Ad}\ (x)$ is the adjoint representation of G. Let $s \to \mathrm{Ad}\ (s)$ ($s \in S$) denote the adjoint representation of S. Then it follows that

$$\det ((d\Phi)_{u_0,s_0}) = \frac{\det \mathrm{Ad}_S (s_0)}{\det \mathrm{Ad} (s_0)} \neq 0$$

and so Φ is regular at (u_0, s_0).

COROLLARY. *If Φ is a 1-1 mapping of $K \times S$ onto G, it is an analytic isomorphism of $K \times S$ with G.*

For since Φ is regular, the inverse mapping is analytic.

LEMMA 27. *Suppose Φ is a 1-1 mapping of $K \times S$ onto G. Let du, ds, and dx denote the left-invariant Haar measure on K, S, and G respectively. Then under suitable normalisation of these measures we have*

$$dx = \frac{\det \operatorname{Ad}_{\mathcal{S}}(s)}{\det \operatorname{Ad}(s)} \, du\,ds \qquad (x = us,\; u \in K,\; s \in S).$$

This follows immediately from the fact that

$$\det\; ((d\Phi)_{u,s}) = \det \operatorname{Ad}_{\mathcal{S}}(s)/\det \operatorname{Ad}(s).$$

DEFINITION. *Let $\mathcal{S}$ be a Lie algebra over a field of characteristic zero. We say that $\mathcal{S}$ is quasi-nilpotent if it can be written as the direct sum of an abelian subalgebra $\mathfrak{a}$ and a nilpotent ideal $\mathfrak{n}$. A connected Lie group S is called quasi-nilpotent if its Lie algebra is quasi-nilpotent.*

From now on let S denote a simply-connected quasi-nilpotent Lie group and $\mathcal{S}_0$ its Lie algebra. Let $\mathcal{S}_0 = \mathfrak{a}_0 + \mathfrak{n}_0$ where $\mathfrak{a}_0$ is an abelian subalgebra and $\mathfrak{n}_0$ a nilpotent ideal in $\mathcal{S}_0$ and the sum is direct. It is known that there exists a faithful matrix representation of S (see for example [8 (b)]). Therefore we may consider S a linear group and $\mathcal{S}_0$ a linear Lie algebra. Since every matrix with complex coefficients may also be regarded as a real matrix of twice the degree, we may further assume that all matrices in S and $\mathcal{S}_0$ are real. Let $\mathcal{S}$ be the Lie algebra over C spanned by $\mathcal{S}_0$. Then $\mathcal{S}$ is the complexification of $\mathcal{S}_0$. Let S_c be the analytic group of matrices corresponding to $\mathcal{S}$. Then $S_c \supset S$ and, since $\mathcal{S}$ is a complex algebra, S_c is a complex analytic group. Thus we have imbedded S in a complex analytic group S_c whose Lie algebra $\mathcal{S}$ is the complexification of $\mathcal{S}_0$. We may assume that S_c is simply-connected, for otherwise we could replace it by its universal covering group and thus obtain an imbedding of S into a simply-connected complex group. Let $\mathfrak{a}$ and $\mathfrak{n}$ be the subalgebra of $\mathcal{S}$ spanned by $\mathfrak{a}_0$ and $\mathfrak{n}_0$ respectively, and A_c, N_c, A, and N the analytic subgroups corresponding to $\mathfrak{a}$, $\mathfrak{n}$, $\mathfrak{a}_0$, and $\mathfrak{n}_0$ respectively. All these groups are simply-connected and $(n, a) \to na$ $(n \in N_c,\; a \in A_c)$ is a topological mapping of $N_c \times A_c$ onto S_c. Moreover, $na \in S$ if and only if $n \in N$ and $a \in A$. Since $\mathfrak{a}$ and $\mathfrak{n}$ are complex Euclidean spaces we may regard them as complex analytic manifolds in the obvious way. Then the mappings $H \to \exp H$ and $X \to \exp X$ $(H \in \mathfrak{a},\; X \in \mathfrak{n})$ define holomorphic isomorphisms of $\mathfrak{a}$ with A_c and $\mathfrak{n}$ with N_c respectively.

For any $x \in S_c$ and $n \in N_c$ write $xn = \tau_x(n)a$ where $\tau_x(n) \in N_c$ and $a \in A_c$. Then τ_x is a topological mapping of N_c onto itself and $\tau_{x^{-1}y} = (\tau_x)^{-1}\tau_y$ $(x, y \in S_c)$. Let θ denote the conjugation of $\mathcal{S}$ with respect to its real form $\mathcal{S}_0$, so that $\theta(X + (-1)^{1/2}Y) = X - (-1)^{1/2}Y$ $(X, Y \in \mathcal{S}_0)$. We extend θ to an automorphism $x \to \theta x$ of S_c.

Let f be a complex-valued function on N_c. We shall say that f is linear (polynomial) if the function $X \to f(\exp X)$ $(X \in \mathfrak{n})$ is a linear (polynomial) function on $\mathfrak{n}$. Similarly, we call a function f on $N_c \times N_c$ a polynomial function if $f(n, n') = \sum_{1 \leq i \leq k} f_i(n) f_i'(n')$ $(n, n' \in N_c)$ where f_i, f_i' $(1 \leq i \leq k)$ are polynomial functions on N_c. Let $(l_1, \cdots, l_r)$ be a base for linear functions on N_c. Then every polynomial function f on N_c is uniquely expressible as a polynomial $P(l_1, \cdots, l_r)$ in $l_1, \cdots, l_r$. We define the degree of f to be the degree of P. It is then clear that the polynomial functions of degree $\leq \mu$ form a vector space of finite dimension.

For any function f on N_c we denote by f_x $(x \in S_c)$ the function $n \to f(\tau_{x^{-1}}(n))$ $(n \in N_c)$. If f is a polynomial function, then so is f_x. It is clearly sufficient to prove this when f is linear. It is then known that for any $n_0 \in N_c$ the function $n \to f(n_0 n)$ is a polynomial function (see Birkhoff [2]). Now if $x^{-1} = n_0 a_0$ $(n_0 \in N_c, a_0 \in A_c)$, $f(\tau_{x^{-1}}(n)) = f(n_0 a_0 n a_0^{-1}) = f_1(a_0 n a_0^{-1})$ where f_1 is a polynomial function. On the other hand

$$a_0(\exp X)a_0^{-1} = \exp (\mathrm{Ad}(a_0)X) \qquad\qquad (X \in \mathfrak{n})$$

where $y \to \mathrm{Ad}\ (y)$ is the adjoint representation of S_c. Hence the functions $n \to l_i(a_0 n a_0^{-1})$ $(1 \leq i \leq r)$ are all linear and this proves that the function $f_x\colon n \to f_1(a_0 n a_0^{-1})$ is a polynomial function. Moreover if l is a linear function on N_c, the function $(n_0, n) \to l(n_0 n)$ is a polynomial function on $N_c \times N_c$ (see Birkhoff [2]) and so it follows that there exists an integer μ independent of n_0 such that every function of the form $n \to l(n_0 n)$ (l linear) is of degree μ. This means that the vector space $\mathfrak{Q}$ spanned by the functions $(l_i)_x$ $(1 \leq i \leq r, x \in S_c)$ is finite-dimensional. Since $(f_x)_y = f_{yx}$ for any function f on N_c $(x, y \in G)$, it follows that $Q_x \in \mathfrak{Q}$ for every $Q \in \mathfrak{Q}$.

We may choose the functions l_i, $1 \leq i \leq r$, in such a way that they are real on N. Then $l_i(\theta n) = \overline{l_i(n)}$ $(n \in N_c)$ where the bar denotes complex-conjugate. From this it follows that $(l_i)_{\theta x}$ coincides with $\overline{(l_i)_x}$ on N. Hence for every $Q \in \mathfrak{Q}$ there is a function $Q' \in \mathfrak{Q}$ which coincides with $\overline{Q}$ on N. This shows that $\mathfrak{Q}$ is spanned by functions which are real on N. Moreover $l_i \in \mathfrak{Q}$, $1 \leq i \leq r$. Let $(Q_1, \cdots, Q_m)$ be a base for $\mathfrak{Q}$ such that the functions Q_j are real on N and the functions l_i, $1 \leq i \leq r$, are included among $(Q_1, \cdots, Q_m)$. Then

$$(Q_i)_x = \sum_{j=1}^{m} a_{ij}(x)Q_j \qquad\qquad (x \in S_c)$$

and the mapping $x \to (a_{ij}(x)) = \alpha(x)$ is a matrix representation of S_c. If l is a linear function on N_c, $(n, x) \to l_x(n)$ is clearly a holomorphic function on $N_c \times S_c$. Therefore the same holds for the functions $(n, x) \to f_x(n)$ where f is a polynomial function on N_c. Therefore the functions a_{ij} are holomorphic on S_c. Let $\mathfrak{R}c$ denote the real part of a complex number c.

LEMMA 28. *Let c_j, $1 \leq j \leq m$, be given complex numbers and let $x_i' + c_i = \sum_{j=1}^{m} b_{ij}(x_j + c_j)$ $(b_{ij} \in C, x_j \in R)$. There exists a constant $\epsilon > 0$ such that if* $\max_{i,j} |b_{ij} - \delta_{ij}| \leq \epsilon$, *then*

$$\mathfrak{R}\left(\sum_{i=1}^{m} x_i'^2 \right) \geq \frac{1}{2} \sum_{i=1}^{m} x_i^2$$

for all systems of real numbers $x_1, \cdots, x_m$ such that $\sum_{i=1}^{m} x_i^2 \geq 1$.

Put $c_{ij} = b_{ij} - \delta_{ij}$, $M = \max_{i,j} |c_{ij}|$, and $c = \max_j |c_j|$. Then

$$x_i' = x_i + \sum_j c_{ij}(x_j + c_j)$$

and therefore

$$\sum_i x_i'^2 = \sum_i x_i^2 + 2\sum_{i,j} c_{ij} x_i (x_j + c_j) + \sum_i \left(\sum_j c_{ij}(x_j + c_j) \right)^2.$$

Hence

$$\left| \sum_i (x_i'^2 - x_i^2) \right| \leq m(2M + 2Mc + M^2(1 + c)^2) \sum_i x_i^2$$

provided $\sum_i x_i^2 \geq 1$. Hence if M is sufficiently small

$$\left| \sum_i (x_i'^2 - x_i^2) \right| \leq \frac{1}{2} \sum_i x_i^2$$

and therefore

$$\Re\left(\sum_{i=1}^{m} x_i'^2 \right) \geq \frac{1}{2} \sum_{i=1}^{m} x_i'^2.$$

Since the functions $a_{ij}(x)$ are continuous and $a_{ij}(1) = \delta_{ij}$ it follows from the above lemma that there exists a neighbourhood V' of 1 in S_c such that if

$$u(na) = \sum_{i=1}^{n} (Q_i(n) - Q_i(1))^2 \qquad (n \in N_c, \, a \in A_c),$$

then

$$\Re(u(x_0 na)) \geq \frac{1}{2} u(n) \geq \frac{1}{2} \sum_{i=1}^{r} l_i^2(n)$$

for $n \in N$, $a \in A$, and $x_0 \in V'$ provided $\sum_{i=1}^{r} l_i^2(n) \geq 1$.

Let us call a function λ on A_c linear if the function $H \to \lambda(\exp H)$ ($H \in \mathfrak{a}$) is linear on $\mathfrak{a}$. Choose a base $(\lambda_1, \cdots, \lambda_s)$ for the space of linear functions on A_c such that λ_i are real on A and let λ_0 denote the constant function 1. Then if $a, a' \in A_c$, $\lambda_i(aa') = \lambda_i(a') + \lambda_i(a)\lambda_0(a')$, $1 \leq i \leq s$. Put

$$v(na) = \sum_{j=1}^{s} \lambda_j^2(a) \qquad (n \in N_c, \, a \in A_c).$$

Then making use of an argument similar to that of Lemma 28, we see that there exists a compact neighbourhood V of 1 in S_c ($V \subset V'$) such that

$$\mathfrak{R}(v(x_0 na)) \geqq \frac{1}{2} v(na) \geqq \frac{1}{2} \sum_{j=1}^{s} \lambda_j^2(a)$$

for $x_0 \in V$, $n \in N$, and $a \in A$. Now put $F_\nu(x) = e^{-\nu\{u(x)+v(x)\}}$ ($x \in S_c$) where ν is a positive real number. Then F_ν is holomorphic on S_c and

$$\left| F_\nu(x_0 x) \right| \leqq M_\nu \exp \left\{ -\frac{\nu}{2} \left(\sum_{i=1}^{r} l_i^2(n) + \sum_{j=1}^{s} \lambda_j^2(a) \right) \right\}$$

for $x_0 \in V$ and $x \in S$ ($x = na$, $n \in N$, $a \in A$). Here M_ν is the supremum of $e^{\nu/2} \left| e^{-\nu u(x_0 n)} \right|$ for $x_0 \in V$, and $\sum l_i^2(n) \leqq 1$ ($n \in N$).

Now let π be a representation of S on a Banach space $\mathfrak{H}$. We shall prove that $F_\nu \in \mathfrak{W}_\pi$. Let us say that a function f on S is of at most exponential growth if there exist real constants M_1, $M_2 \geqq 0$ such that

$$\left| f(na) \right| \leqq M_1 \exp \left(M_2 \max_{i,j} \left(\left| l_i(n) \right|, \left| \lambda_j(a) \right| \right) \right) \qquad (n \in N, \, a \in A).$$

Clearly the product of two such functions is again of the same type. Now consider the function $\left| \pi(x) \right|$. Let M denote an upper bound for $\left| \pi(x) \right|$ on the compact set consisting of all points $x = na$ in S ($n \in N$, $a \in A$) such that $\max_{i,j} \left\{ \left| l_i(n) \right|, \left| \lambda_j(a) \right| \right\} \leqq 1$. For any $x \in S$ let q denote the least integer such that $q \geqq \max_{i,j} \left\{ \left| l_i(n) \right|, \left| \lambda_j(a) \right| \right\}$ ($x = na$). Then we can choose elements $n' \in N$ and $a' \in A$ such that $n = n'^q$ and $a = a'^q$ and

$$\left| \pi(x) \right| \leqq \left| \pi(n) \right| \left| \pi(a) \right| \leqq \left| \pi(n') \right|^q \left| \pi(a') \right|^q \leqq M^{2q}$$

since $\left| l_i(n') \right| = (1/q) \left| l_i(n) \right| \leqq 1$ and $\left| \lambda_j(a') \right| = (1/q) \left| \lambda_j(a) \right| \leqq 1$. Since $q \leqq 1 + \max_{i,j} \left\{ \left| l_i(n) \right|, \lambda_j(a) \right\}$, we see that $\left| \pi(x) \right|$ is at most of exponential growth. Now consider the Haar measure dx on S. Making use of Lemma 27 we find that

$$dx = \det (\mathrm{Ad}_S (a^{-1})) dn da \qquad (x = na)$$

when Ad_S is the adjoint representation of S and dn, da are the Haar measures of N and A respectively. Since N is nilpotent and A is abelian,

$$dn da = dl_1 \cdots dl_r d\lambda_1 \cdots d\lambda_s.$$

Moreover, since the function $na \to \det \mathrm{Ad}_S (a^{-1})$ is clearly of at most exponential growth, it follows that

$$\left| \pi(x) \right| dx = \mu(x) dl_1 \cdots dl_r d\lambda_1 \cdots d\lambda_s,$$

where $\mu(x)$ is a function of at most exponential growth. Therefore it is clear that

$$\int g_\nu(x) \left| \pi(x) \right| dx < \infty,$$

where

$$g_\nu(na) = \exp\left\{-\frac{\nu}{2}\left(\sum_{i=1}^{r} l_i^2(n) + \sum_{i=1}^{s} \lambda_i^2(a)\right)\right\} \qquad (n \in N,\ a \in A).$$

We have seen above that we can choose an open connected neighbourhood W of 0 in $\mathfrak{s}$ such that

$$\left| F_\nu(\exp(-Z)x) \right| \leq M_\nu g_\nu(x) \qquad (Z \in W,\ x \in S)$$

where $Z \to \exp Z$ is the exponential mapping of $\mathfrak{s}$ into S_c. Therefore the function $x \to F_\nu(\exp(-Z)x)$ is summable with respect to the measure $|\pi(x)|\,dx$ on S. Let $L(S, \pi)$ denote the Banach space of functions corresponding to the representation π, which was introduced in §7, and let $\psi_\nu(Z)$ be the element in $L(S, \pi)$ which is represented by the function $x \to F_\nu(\exp(-Z)x)$ $(Z \in W)$. Then all the conditions of Lemma 23 are fulfilled and therefore $Z \to \psi_\nu(Z)$ is a holomorphic mapping of W into $L(S, \pi)$. Let σ denote the representation of S on $L(S, \pi)$ as it was defined in §7. Then $\sigma(\exp X)\psi_\nu = \psi_\nu(X)(X \in \mathfrak{s}_0)$ where ψ_ν is the element in $L(S, \pi)$ corresponding to F_ν. Therefore the mapping $X \to \sigma(\exp X)\psi_\nu$ $(X \in \mathfrak{s}_0)$ is analytic at $X = 0$ and so $F_\nu \in \mathfrak{W}_\pi$.

Now consider $J_\nu = \int F_\nu(x)dx$. If $n \in N$ and $a \in A$, let n_ν and a_ν denote two elements in N and A respectively such that $l_i(n_\nu) = \nu^{1/2}l_i(n)$, $1 \leq i \leq r$, and $\lambda_j(a_\nu) = \nu^{1/2}\lambda_j(a)$, $1 \leq j \leq s$. Moreover if

$$P = \sum_{(p)} c(p_1, \cdots, p_r)l_1^{p_1} \cdots l_r^{p_r} \qquad (c(p_1, \cdots, p_r) \in R)$$

is any polynomial function on N_c, we put $P^*(n) = \sum_{(p)} \left| c(p_1, \cdots, p_r) \right|$ $\left| l_1(n) \right|^{p_1} \cdots \left| l_r(n) \right|^{p_r}$. Then it is clear that $\nu^{1/2}\left| P(n) \right| \leq \nu^{1/2}P^*(n) \leq P^*(n_\nu)$ if $P(1) = 0$. Hence if $Q_i'(n) = Q_i(n) - Q_i(1)$,

$$F_\nu(na) = \exp\left\{-\nu \sum_{i=1}^{m} Q_i'^2(n) - \nu\sum_{j=1}^{s} \lambda_j^2(a)\right\}$$

$$\geq \exp\left\{-\sum_{i=1}^{m} Q_i'^{*2}(n_\nu) - \sum_{j=1}^{s} \lambda_j^2(a_\nu)\right\}.$$

On the other hand

$$J_\nu = \int F_\nu(na)\,\det(\exp \mathrm{Ad}_S(a^{-1}))dl_1 \cdots dl_r d\lambda_1 \cdots d\lambda_s$$

and we can find real numbers $M_1,\ M_2 > 0$ such that

$$\det(\exp \mathrm{Ad}_S(a)) \leq M_1 \exp\left(M_2\sum_{i=1}^{s} |\lambda_i(a)|\right).$$

Hence

$$\det \left(\exp \mathrm{Ad}_S \left(a^{-1}\right)\right) \geq M_1^{-1} \exp \left(-M_2 \sum_{i=1}^{s} |\lambda_i(a)|\right)$$

and

$$J_\nu \geq M_1^{-1} \int \exp \left\{- \sum_{i=1}^{m} Q_i'^{*2}(n_\nu) - \sum_{j=1}^{s} \lambda_j^2(a_\nu)\right.$$

$$\left. - M_2 \sum_{j=1}^{s} |\lambda_j(a_\nu)|\right\} dl_1 \cdots dl_r d\lambda_1 \cdots d\lambda_s = \frac{c}{\nu^{(r+s)/2}} > 0,$$

where c is the value of the right-hand side when $\nu = 1$. On the other hand, let V_δ be the neighbourhood of 1 in S consisting of all points $na \in S$ such that $|l_i(n)| < \delta$, $|\lambda_i(a)| < \delta$ $(\delta > 0)$. Then let us consider the integral

$$J_\nu(\delta) = \int_{S - V_\delta} F_\nu(x)(1 + |\pi(x)|)dx.$$

We can find positive constants M_3 and M_4 such that

$$(1 + |\pi(na)|)\det\left(\exp \mathrm{Ad}_S\left(a^{-1}\right)\right) \leq M_3 \exp M_4\left(\sum_{i=1}^{r} |l_i(n)| + \sum_{j=1}^{s} |\lambda_j(a)|\right).$$

Put

$$P_1(n, a) = \sum_{i=1}^{r} l_i^2(n) + \sum_{j=1}^{s} \lambda_j^2(a) \quad \text{and} \quad P_2(n, a) = \sum_{i=1}^{r} |l_i(n)| + \sum_{j=1}^{s} |\lambda_j(a)|.$$

Then $\nu P_1(n, a) = P_1(n_\nu, a_\nu)$ and $\nu^{1/2} P_2(n, a) = P_2(n_\nu, a_\nu)$ and therefore

$$J_\nu(\delta) \leq M_3 \int_{S - V_\delta} \exp \left\{-\nu P_1(n, a) + \nu^{1/2} M_4 P_2(n, a)\right\} dl_1 \cdots dl_r d\lambda_1 \cdots d\lambda_s$$

$$\leq M_3 e^{-\nu(r+s)\delta^2/2} \int_{S - V_\delta} \exp \left\{- \frac{\nu}{2} P_1(n, a)\right.$$

$$\left. + \nu^{1/2} M_4 P_2(n, a)\right\} dl_1 \cdots dl_r d\lambda_1 \cdots d\lambda_s$$

$$\leq e^{-\nu(r+s)\delta^2/2} \frac{c'}{\nu^{(r+s)/2}}$$

where

$$c' = M_3 \int_{S - V_\delta} \exp \left\{- \frac{1}{2} P_1(n, a) + M_4 P_2(n, a)\right\} dl_1 \cdots dl_r d\lambda_1 \cdots d\lambda_s < \infty.$$

Therefore

$$\frac{J_\nu(\delta)}{J_\nu} \leqq \frac{c'}{c} e^{-\nu(r+s)\delta^2/2} \cdot \nu^{r+s}$$

and so

$$\lim_{\nu \to \infty} \frac{J_\nu(\delta)}{J_\nu} = 0.$$

Hence if we put $f_\nu = F_\nu/J_\nu$, the sequence $\{f_1, f_2, \cdots\}$ is a $|\pi(x)|$-Dirac-sequence of functions in $\mathfrak{W}_r$. Therefore from Lemma 25 the space of well-behaved elements is dense in $\mathfrak{H}$. We can now prove the following theorem:

THEOREM 3. *Suppose a connected Lie group G has two analytic subgroups K and S such that the following two conditions are fulfilled*: (a) K *is compact and S is quasi-nilpotent*; (b) *every element in G can be written uniquely in the form us ($u \in K$, $s \in S$). Let π be a representation of G on a Banach space $\mathfrak{H}$ and let $\mathfrak{H}_0$ denote the set of all well-behaved elements ψ in $\mathfrak{H}$ such that the linear space spanned by the elements $\pi(u)\psi$ ($u \in K$) is of finite dimension. Then $\mathfrak{H}_0$ is dense in $\mathfrak{H}$.*

The restriction of π on S defines a representation ρ of S on $\mathfrak{H}$. Since ρ may also be regarded as a representation of the simply-connected covering group of S, it follows from the result proved above that the set of elements which are well-behaved under ρ is dense in $\mathfrak{H}$. Hence given any $\psi_0 \in \mathfrak{H}$ and $\epsilon > 0$ we can find an element ψ which is well-behaved under ρ and such that $|\psi_0 - \psi| \leqq \epsilon$. Choose a neighborhood U of 1 in K such that $|\pi(u)\psi - \psi| \leqq \epsilon$ for $u \in U$ and let du denote the element of Haar measure on K such that $\int_K du = 1$. Select a continuous, real, non-negative function f on K such that $f = 0$ outside U and $\int_K f du = 1$. Then if $\psi_1 = \int_K f(u)\pi(u)\psi du$,

$$|\psi_1 - \psi| \leqq \int_K f(u) |\pi(u)\psi - \psi| du \leqq \epsilon.$$

Let $\mathfrak{R}$ be the set of all finite linear combinations (with complex coefficients) of the coefficients of finite-dimensional matrix representations of K. Then for any $\eta > 0$ we can find a function $\omega \in \mathfrak{R}$ such that $|f(u) - \omega(u)| \leqq \eta (u \in K)$. Hence

$$\left| \int_K (f(u) - \omega(u))\pi(u)\psi du \right| \leqq M\eta |\psi| \leqq M\eta(|\psi_0| + \epsilon)$$

where M is an upper bound for $|\pi(u)|$ on the compact set K. Therefore if

$$\psi_2 = \int_K \omega(u)\pi(u)\psi du, \quad |\psi_0 - \psi_2| \leqq 2\epsilon + M\eta(|\psi_0| + \epsilon) \leqq 3\epsilon$$

if η is sufficiently small. Since $\omega \in \mathfrak{R}$ we see easily that the elements

$\pi(u)\psi_2$ $(u \in K)$ span a finite-dimensional subspace of $\mathfrak{H}$. Therefore in order to prove the theorem it is sufficient to show that the element ψ_2 is well-behaved under $\pi(G)$. This follows from the lemma given below.

LEMMA 29. *Let ω be an analytic function on K and ψ an element in $\mathfrak{H}$ which is well-behaved under ρ. Then $\int_K \omega(u)\pi(u)\psi du$ is well-behaved under $\pi(G)$.*

For any $x \in G$ and $u \in K$ put $xu = u_x s(x, u)$ where $u_x \in K$ and $s(x, u) \in S$. Then we know (see the corollary to Lemma 26) that $(x, u) \to u_x$ and $(x, u) \to s(x, u)$ are analytic mappings of $G \times K$ into K and S respectively. Now put $\psi_0 = \int_K \omega(u)\pi(u)\psi du$. Then

$$\pi(x)\psi_0 = \int_K \omega(u)\pi(xu)\psi du.$$

But for x fixed, the mapping $u \to u_x$ $(u \in K)$ is an analytic isomorphism of the manifold structure of K with itself. Since G is connected, this isomorphism is orientation preserving. From this we conclude that

$$du_x = D(x, u)du$$

where $D(x, u)$ is an analytic function of $G \times K$ which is everywhere positive. Hence

$$\pi(x)\psi_0 = \int_K \omega(u_x^{-1})\pi(xu_x^{-1})\psi du_x^{-1}$$

$$= \int_K \omega(u_x^{-1})D(x^{-1}, u)\pi(u)\pi((s(x^{-1}, u))^{-1})\psi du.$$

On the other hand, since ψ is well-behaved under ρ, it follows from Lemmas 16 and 17 that the mapping is an analytic mapping of $G \times K$ into $\mathfrak{H}$. Let $t_1, \cdots, t_n$ be a coordinate system in G at 1 such that $t_j(1) = 0$ $(1 \leq j \leq n)$. Then for each $u_0 \in K$ we can find an open connected neighbourhood V_{u_0} of u_0 in K and a coordinate system $q_1, \cdots, q_k$ in K valid on V_{u_0} with the following properties: (1) $q_i(u_0) = 0$, $1 \leq i \leq k$; (2) there exists a power series $P_{u_0}(q, t)$ in $(q_1, \cdots, q_k)$ and $(t_1, \cdots, t_n)$ with coefficients in $\mathfrak{H}$ and an open neighbourhood W_{u_0} of 1 in G such that the coordinates $(t_1, \cdots, t_n)$ are valid on W_{u_0} and the power series converges to $\psi(x, u)$ on $W_{u_0} \times V_{u_0}$. Suppose

$$P_{u_0}(q, t) = \sum_{(a),(e)} \psi_{u_0}(a, e)q_1^{a_1} \cdots q_k^{a_k} t_1^{e_1} \cdots t_n^{e_n} \qquad (\psi_{u_0}(a, e) \in \mathfrak{H})$$

and put

$$\psi_{u_0}(a, e) = \sum_{(a)} \psi_{u_0}(a, e)q_1^{a_1}(u) \cdots q_k^{a_k}(u) \qquad (u \in V_{u_0}).$$

Then the mapping $u \to \psi_{u_0}(u, e)$ is an analytic mapping of V_{u_0} into $\mathfrak{H}$. Choose

a number $r(u_0) > 0$ such that the series $P_{u_0}(q, t)$ converges if $\max_{i,j} (|q_j|, |t_i|)$
$\leq r(u_0)$. Let U_{u_0} be a neighbourhood of u_0 (in K) such that $U_{u_0} \subset V_{u_0}$ and $\max_j$
$|q_j(u)| \leq r(u_0)$ $(u \in U_{u_0})$. Similarly we may suppose that $\max_i |t_i(x)| \leq r(u_0)$
for $x \in W_{u_0}$. Since K is compact, there exists a finite set of points $u_1, \cdots, u_p$
such that $K = U_{u_1} \cup U_{u_2} \cup \cdots \cup U_{u_p}$. Put $W = W_{u_1} \cap W_{u_2} \cap \cdots \cap W_{u_p}$ and
$r = \min \{r(u_1), \cdots, r(u_p)\}$. Then if $u \in U_{u_i} \cap U_{u_j}$ $(1 \leq i, j \leq p)$ and $x \in W$,

$$\psi(x, u) = \sum_{(e)} \psi_{u_i}(u, e) t_1^{e_1}(x) \cdots t_n^{e_n}(x) = \sum_{(e)} \psi_{u_j}(u, e) t_1^{e_1}(x) \cdots t_n^{e_n}(x)$$

and therefore $\psi_{u_i}(u, e) = \psi_{u_j}(u, e)$. Thus for each (e) there exists a mapping
$u \to \psi(u, e)$ of K into $\mathfrak{H}$ which coincides on U_{u_i} with $u \to \psi_{u_i}(u, e)$. This mapping is clearly analytic and

$$\psi(x, u) = \sum_{(e)} \psi(u, e) t_1^{e_1}(x) \cdots t_n^{e_n}(x) \qquad (u \in K, x \in W).$$

Now if $u \in U_{u_i}$ and $x \in W$,

$$\sum_{(e)} | \psi(u, e) t_1^{e_1}(x) \cdots t_n^{e_n}(x) | \leq \sum_{(e)} | \psi_{u_i}(u, e) | (r(u_i))^{e_1 + \cdots + e_n} < \infty.$$

Moreover since K is compact, we can find a real number M such that $|\pi(u)|$
$\leq M$ for all $u \in K$. Then

$$\sum_{(e)} | \pi(u) \psi(u, e) t_1^{e_1}(x) \cdots t_n^{e_n}(x) | \leq M \sum_{(e)} | \psi(u, e) t_1^{e_1}(x) \cdots t_n^{e_n}(x) |$$

and therefore the series

$$\sum_{(e)} \pi(u) \psi(u, e) t_1^{e_1}(x) \cdots t_n^{e_n}(x)$$

converges uniformly to $\pi(u) \psi(x, u)$ on $K \times W$. Therefore

$$\pi(x) \psi_0 = \int_K \pi(u) \psi(x, u) du$$

$$= \sum_{(e)} \left(\int_K \pi(u) \psi(u, e) du \right) t_1^{e_1}(x) \cdots t_n^{e_n}(x) \qquad (x \in W).$$

This proves that the mapping $x \to \pi(x) \psi_0$ is analytic at 1 and so ψ_0 is well-
behaved under π.

<h3 style="text-align:center">PART III. REPRESENTATIONS OF A SEMISIMPLE LIE GROUP</h3>

9. Permissible representations. We shall now apply the results of Part II
to representations of a connected semisimple Lie group G on a Banach space
$\mathfrak{H}$. Since every such representation may also be regarded as a representation
of the simply-connected covering group of G, we may assume that G itself

is simply-connected. Let $\mathfrak{g}_0$ be its Lie algebra. We define $\mathfrak{k}_0$, $\mathfrak{h}_{p_0}$, and $\mathfrak{n}_0$ as in Part I, §2. Let K, A_+, and N be the analytic subgroups of G corresponding to $\mathfrak{g}_0$, $\mathfrak{h}_{p_0}$, and $\mathfrak{n}_0$ respectively. They are all simply-connected. Moreover, $S = A_+N$ is a quasi-nilpotent group, and Iwasawa [9] has shown that the mapping $(u, s) \to us$ $(u \in K, s \in S)$ is a topological mapping of $K \times S$ onto G. However, in general K is not compact and therefore Theorem 3 is not applicable immediately. Therefore we shall now give a method for extending the results of this theorem to the present case.

We know that $\mathfrak{k}_0$ is reductive (Lemma 3). Let $\mathfrak{c}_0$ be the center of $\mathfrak{k}_0$ and D the analytic subgroup of G corresponding to $\mathfrak{c}_0$. Then K is the direct product of its commutator group K' and D. Moreover if Z is the center of G, then $K/D \cap Z$ is compact[9]. We shall say that a *representation* π of G on $\mathfrak{H}$ is *permissible if $\pi(z)$ is a scalar multiple of the unit operator for $z \in D \cap Z$.* Assuming that π is permissible, we can choose a complex-valued linear function μ on $\mathfrak{c}_0$ such that $\pi(\exp \Gamma) = e^{\mu(\Gamma)}\pi(1)$ whenever $\exp \Gamma \in D \cap Z$ $(\Gamma \in \mathfrak{c}_0)$. Let $u \to u^*$ denote the natural mapping of K on $K^* = K/D \cap Z$. Then it is easy to verify that $e^{-\mu(\Gamma)}\pi(u \exp \Gamma)$ $(u \in K', \Gamma \in \mathfrak{c}_0)$ depends only on $(u \exp \Gamma)^*$ and so, if we denote it by $\pi^*((u \exp \Gamma)^*)$, the mapping $u^* \to \pi^*(u^*)$ is a representation of K^* on $\mathfrak{H}$. Let ψ_0 be a given element in $\mathfrak{H}$ and ϵ a positive number. Since S is quasi-nilpotent, we can find an element $\psi \in \mathfrak{H}$ which is well-behaved under $\pi(S)$ and such that $|\psi - \psi_0| \leq \epsilon$. Moreover, since K^* is compact, we can choose a finite linear combination ω of the coefficients of a finite-dimensional matrix representation of K^* such that

$$\left| \int_{K^*} \omega(u^*)\pi^*(u^*)\psi du^* - \psi \right| \leq \epsilon$$

and therefore

$$\left| \int_{K^*} \omega(u^*)\pi^*(u^*)\psi du^* - \psi_0 \right| \leq 2\epsilon.$$

(Here du^* is the element of the Haar measure on K^* normalised in such a way that $\int_{K^*} du^* = 1$.) We now claim that $\phi = \int_{K^*}\omega(u^*)\pi^*(u^*)\psi du^*$ is well-behaved under $\pi(G)$. Consider the element

$$\pi(x)\phi = \int_{K^*} \omega(u^*)\pi(x)\pi^*(u^*)\psi du^* \qquad (x \in G).$$

For any $u \in K$ let $\Gamma(u)$ denote the unique element in $\mathfrak{c}_0$ such that $u = (\exp \Gamma(u))v$ $(v \in K')$. Moreover let $xu = u_x s(x, u)$ $(x \in G, u_x \in K, s(x, u) \in S)$. Then it is easy to verify that for a fixed x the elements $(u_x)^*$, $\Gamma(u_x) - \Gamma(u)$, and $s(x, u)$ depend only on u^*. Hence we may write them as u_x^*, $\Gamma(x, u^*)$, and

---

[9] This follows from the fact that the image of K into the adjoint group of G is compact (see for example Mostow [13]).

$s(x, u^*)$ respectively. It is clear that the mappings $(x, u^*) \rightarrow u_x^*$, $(x, u^*) \rightarrow \Gamma(x, u^*)$, $(x, u^*) \rightarrow s(x, u)$ are all analytic and

$$e^{-\mu(\Gamma(u))}\pi(x)\pi(u) = e^{-\mu(\Gamma(u))}\pi(u_x)\pi(s(x, u))$$

$$= e^{\mu(\Gamma(x, u^*))}\pi^*(u_x^*)\pi(s(x, u^*)).$$

Therefore

$$\pi(x)\phi = \int_{K^*} \omega(u^*)e^{\mu(\Gamma(x, u^*))}\pi^*(u_x^*)\pi(s(x, u^*))\psi du^*$$

and we conclude exactly as in the proof of Lemma 29 that

$$\pi(x)\phi = \int_{K^*} \pi^*(u^*)\psi(x, u^*)du^*$$

where $(x, u^*) \rightarrow \psi(x, u^*)$ is an analytic mapping of $G \times K^*$ into $\mathfrak{H}$. Since K^* is compact we can now use the same argument as in the proof of Lemma 29 and show that the mapping $x \rightarrow \pi(x)\phi$ is analytic at $x = 1$. Therefore ϕ is well-behaved under π.

Let Ω be the set of all equivalence classes of finite-dimensional irreducible representations of K. For any $\mathfrak{D} \in \Omega$ we denote by $\mathfrak{H}_\mathfrak{D}$ the set of all elements $\psi \in \mathfrak{H}$ with the following property: there exists a finite-dimensional linear space U containing ψ which is invariant and semisimple under $\pi(K)$ and is such that the representation of K induced on every simple subspace of U lies in $\mathfrak{D}$. We shall say that an element $\psi \in \mathfrak{H}$ transforms under K (or $\pi(K)$) according to $\mathfrak{D}$ if $\psi \in \mathfrak{H}_\mathfrak{D}$. Let W be the space of all elements in $\mathfrak{H}$ which are well-behaved under $\pi(G)$. We have seen above that $W \cap (\sum_{\mathfrak{D} \in \Omega} \mathfrak{H}_\mathfrak{D})$ is dense in $\mathfrak{H}$. But since W is stable under $\pi(G)$, it follows that $W \cap (\sum_{\mathfrak{D} \in \Omega} \mathfrak{H}_\mathfrak{D})$ $= \sum_{\mathfrak{D} \in \Omega} W_\mathfrak{D}$ where $W_\mathfrak{D} = W \cap \mathfrak{H}_\mathfrak{D}$ (see Lemma 6). Thus we have the following theorem:

THEOREM 4. *Let π be a permissible representation of G on $\mathfrak{H}$ and W the set of all well-behaved elements in $\mathfrak{H}$. Then if $W_\mathfrak{D} = W \cap \mathfrak{H}_\mathfrak{D}$ ($\mathfrak{D} \in \Omega$), the space $\sum_{\mathfrak{D} \in \Omega} W_\mathfrak{D}$ is dense in $\mathfrak{H}$.*

For any linear subspace V of $\mathfrak{H}$ which is invariant under $\pi(K)$ put $V_\mathfrak{D} = V \cap \mathfrak{H}_\mathfrak{D}$. Then we shall prove the following lemma:

LEMMA 30. *Let π be a permissible representation of G on $\mathfrak{H}$ and V a subspace of $\mathfrak{H}$ invariant under $\pi(K)$. Then if $\sum_{\mathfrak{D} \in \Omega} V_\mathfrak{D}$ is dense in $\mathfrak{H}$, $\mathfrak{H}_\mathfrak{D} = \mathrm{Cl}\,(V_\mathfrak{D})$[10].*

We keep to the notation introduced above. For any $\mathfrak{D} \in \Omega$ let $\chi_\mathfrak{D}$ denote the character of the class $\mathfrak{D}$. Suppose $\mathfrak{H}_\mathfrak{D} \neq \{0\}$. Then it is easily seen that $\chi_\mathfrak{D}(u)e^{-\mu(\Gamma(u))}$ ($u \in K$) depends only on u^*. We may therefore denote it by

---

[10] Cl denotes closure in $\mathfrak{H}$.

$\chi_{\mathfrak{D}}^*(u^*)$. Let $E_{\mathfrak{D}} = d(\mathfrak{D}) \int_K \overline{\chi_{\mathfrak{D}}^*(u^*)} \pi^*(u^*) du^*$ where $d(\mathfrak{D})$ is the degree in $\mathfrak{D}$ and the bar denotes complex conjugate. $E_{\mathfrak{D}}$ is a bounded linear operator such that $E_{\mathfrak{D}}^2 = E_{\mathfrak{D}}$ and $\mathfrak{H}_{\mathfrak{D}} = E_{\mathfrak{D}}\mathfrak{H}$. Hence $\mathfrak{H}_{\mathfrak{D}}$ as closed in $\mathfrak{H}$. Moreover $V_0 = \sum_{\mathfrak{D} \in \Omega} V_{\mathfrak{D}}$ is stable under $\pi(K)$ and the linear space spanned by $\pi(u)\psi$ ($u \in K$) for any $\psi \in V_0$ is of finite dimension. Hence V_0 is stable under $E_{\mathfrak{D}}$. Now let $\psi \in \mathfrak{H}_{\mathfrak{D}}$. We can choose a sequence $\psi_n \in V_0$ such that $\psi = \lim_{n \to \infty} \psi_n$. Then $\psi = E_{\mathfrak{D}}\psi = \lim_{n \to \infty} E_{\mathfrak{D}}\psi_n$. But $E_{\mathfrak{D}}\psi_n \in V_0 \cap E_{\mathfrak{D}}\mathfrak{H} \subset V_{\mathfrak{D}}$. Therefore $V_{\mathfrak{D}}$ is dense in $\mathfrak{H}_{\mathfrak{D}}$ and so $\mathfrak{H}_{\mathfrak{D}} = \mathrm{Cl}\,(V_{\mathfrak{D}})$.

Notice that the operator $E_{\mathfrak{D}}$ is bounded and $E_{\mathfrak{D}}\psi = \psi$ or 0 according as $\psi \in \mathfrak{H}_{\mathfrak{D}}$ or $\psi \in \mathfrak{H}_{\mathfrak{D}'}$ ($\mathfrak{D}' \neq \mathfrak{D}$). (We define $E_{\mathfrak{D}}$ to be zero in case $\mathfrak{H}_{\mathfrak{D}} = \{0\}$.) Since $\sum_{\mathfrak{D}' \in \Omega} \mathfrak{H}_{\mathfrak{D}'}$ is dense in $\mathfrak{H}$, $E_{\mathfrak{D}}$ is uniquely characterised by these properties. We shall call $E_{\mathfrak{D}}$ the canonical projection of $\mathfrak{H}$ on $\mathfrak{H}_{\mathfrak{D}}$. It is clear that $E_{\mathfrak{D}}E_{\mathfrak{D}'} = E_{\mathfrak{D}}$ or 0 according as $\mathfrak{D}' = \mathfrak{D}$ or $\mathfrak{D}' \neq \mathfrak{D}$ ($\mathfrak{D}, \mathfrak{D}' \in \Omega$).

Let π be a representation of G on $\mathfrak{H}$ and U a linear subspace of $\mathfrak{H}$. We shall say that U is *differentiable* if there exists a representation π_U of $\mathfrak{g}_0$ on U such that $\pi_U(X)\psi = \lim_{t \to 0} (1/t)\,(\pi(\exp tX)\psi - \psi)(X \in \mathfrak{g}_0, \psi \in U)$. It is clear that the sum of two differentiable spaces is again differentiable. Hence there exists a largest differentiable space, namely the union of all differentiable subspaces of $\mathfrak{H}$. We say that an element $\psi \in \mathfrak{H}$ is differentiable (under π) if it lies in this union. It is obvious that the Gårding subspace and the space of well-behaved elements are both differentiable. We now state without proof the following lemma.

LEMMA 31. *Let π be a permissible representation of G on $\mathfrak{H}$ and let $E_{\mathfrak{D}}$ denote the projection of $\mathfrak{H}$ on $\mathfrak{H}_{\mathfrak{D}}$. Then if ψ is a differentiable element in $\mathfrak{H}$, the series $\sum_{\mathfrak{D} \in \Omega} E_{\mathfrak{D}}\psi$ converges to ψ.*

We shall not make use of this result anywhere in this paper except in the proof of Theorem 9. A proof of this lemma will be given in a subsequent paper.

10. Quasi-simple representations. Let π be a representation of G on $\mathfrak{H}$ and V the Gårding subspace of $\mathfrak{H}$. We denote by $\mathfrak{g}$ the complexification of $\mathfrak{g}_0$, by $\mathfrak{B}$ the universal enveloping algebra of $\mathfrak{g}$, by $\mathfrak{Z}$ the center of $\mathfrak{B}$, and by π_V the Gårding representation of $\mathfrak{B}$ on V.

DEFINITION. *The representation π is called quasi-simple if* (1) *π is permissible and* (2) *there exists a homomorphism χ of $\mathfrak{Z}$ into C such that $\pi_V(z)\psi = \chi(z)\psi$ for all $z \in \mathfrak{Z}$ and $\psi \in V$. χ is then called the infinitesimal character of π.*

Let U be the space of all elements in $\mathfrak{H}$ which are differentiable under a representation π. We denote by π_U the representation of $\mathfrak{B}$ on U such that

$$\pi_U(X)\psi = \lim_{t \to 0} \frac{1}{t} \{\pi(\exp tX)\psi - \psi\} \qquad (X \in \mathfrak{g}_0, \psi \in U).$$

LEMMA 32. *Let U_0 be a linear subspace of U and χ a homomorphism of $\mathfrak{Z}$ into C such that $\pi_U(z)\psi = \chi(z)\psi$ for all $z \in \mathfrak{Z}$ and $\psi \in U_0$. Then if U_0 is dense in*

$\mathfrak{H}$, $\pi_U(z)\psi = \chi(z)\psi$ *for all* $z \in \mathfrak{Z}$ *and* $\psi \in U$.

If we put $U_0 = V$ in the above lemma we get the following corollary.

COROLLARY. *Let* π *be a quasi-simple representation of* G *on* $\mathfrak{H}$ *and* π_W *the corresponding representation of* $\mathfrak{B}$ *on the space* W *of all well-behaved elements. Then* $\pi_W(z)\psi = \chi(z)\psi$ $(z \in \mathfrak{Z}, \psi \in W)$ *where* χ *is the infinitesimal character of* π.

In order to prove the above lemma, we first establish a simple result which was pointed out to me by Mautner. Let $\tilde{\mathfrak{H}}$ be the space of all bounded linear functions α on $\mathfrak{H}$ taken with the weak topology. If $\tilde{\psi} \in \tilde{\mathfrak{H}}$ and $\psi \in \mathfrak{H}$ we denote by $(\tilde{\psi}, \psi)$ the value of the linear function $\tilde{\psi}$ at ψ. Let A be a bounded linear operator on $\mathfrak{H}$. Then the mapping $\tilde{A}: \tilde{\psi} \to \tilde{A}\tilde{\psi}$ $(\tilde{\psi} \in \tilde{\mathfrak{H}})$ defined by the condition $(\tilde{A}\tilde{\psi}, \psi) = (\tilde{\psi}, A\psi)$ $(\psi \in \mathfrak{H})$ is a continuous linear transformation on $\tilde{\mathfrak{H}}$. $\tilde{A}$ is called the adjoint of A. Let M be a linear subspace of $\mathfrak{H}$ and $\mathfrak{B}$ a linear transformation of M into $\mathfrak{H}$. We put $|B|_M = \sup_{|\psi| \leq 1} |B\psi|$ $(\psi \in M)$.

LEMMA 33. *Let* M *and* N *be linear subspaces of* $\mathfrak{H}$ *such that* $M \supset N$ *and* N *is dense in* $\mathfrak{H}$. *Let* A *and* B *be linear transformations in* M *and* N *respectively such that* A *coincides with* B *on* N. *Let* $\tilde{M}$ *be a dense linear subspace of* $\tilde{\mathfrak{H}}$. *Suppose there exists a linear transformation* $\tilde{A}$ *in* $\tilde{M}$ *such that* $(\tilde{A}\tilde{\phi}, \psi)$ $= (\tilde{\phi}, A\psi)$ *for all* $\tilde{\phi} \in \tilde{M}$ *and* $\psi \in M$. *Then* $|A|_M = |B|_N$.

It is sufficient to prove that $|A|_M \leq |B|_N$ and so we may assume that $|B_N| < \infty$. Now

$$| (\tilde{A}\tilde{\phi}, \psi) | = | (\tilde{\phi}, A\psi) | = | (\tilde{\phi}, B\psi) | \leq | B |_N | \tilde{\phi} | | \psi |$$

if $\tilde{\phi} \in \tilde{M}, \psi \in N$. Since N is dense in $\mathfrak{H}$ we get by continuity

$$| (\tilde{A}\tilde{\phi}, \psi) | \leq | B |_N | \tilde{\phi} | | \psi | \qquad\qquad (\tilde{\phi} \in \tilde{M}).$$

So in particular if $\psi \in M$

$$| (\tilde{\phi}, A\psi) | = | (\tilde{A}\tilde{\phi}, \psi) | \leq | B |_N | \tilde{\phi} | | \psi | \qquad\qquad (\tilde{\phi} \in \tilde{M}).$$

Since $\tilde{M}$ is dense in $\tilde{\mathfrak{H}}$ we conclude that

$$| (\tilde{\phi}, A\psi) | \leq | B |_N | \tilde{\phi} | | \psi | \qquad\qquad (\tilde{\phi} \in \tilde{\mathfrak{H}}, \psi \in M).$$

But from a theorem of Banach [1, p. 55], we can choose $\tilde{\phi}$ so that $(\tilde{\phi}, A\psi)$ $= |A\psi|$ and $|\tilde{\phi}| = 1$. Hence $|A\psi| \leq |B|_N|\psi|$. This proves that $|A|_M \leq |B|_N$.

Now we come to the proof of Lemma 32. For any $f \in C_c^\infty(G)$ let A_f denote the operator $\int f(x^{-1})\pi(x)dx$ and $\tilde{A}_f$ its adjoint. Let $\tilde{V}$ be the subspace of $\tilde{\mathfrak{H}}$ spanned by the elements of the form $\tilde{A}_f\tilde{\psi}$ $(f \in C_c^\infty(G), \tilde{\psi} \in \tilde{\mathfrak{H}})$. It is well known that there exists a sequence $\{f_n\}$ of functions in $C_c^\infty(G)$ with the following properties: (1) $\int f_n dx = 1$ and $\lim_{n \to \infty} \int |f_n| dx = 1$; (2) all but a finite number of the functions f_n are zero outside any given neighbourhood of 1. From this it follows immediately that $\lim_{n \to \infty} A_{f_n}\psi = \psi$ $(\psi \in \mathfrak{H})$ and therefore $\lim_{n \to \infty} \tilde{A}_{f_n}\tilde{\psi}$ $= \tilde{\psi}$ $(\tilde{\psi} \in \tilde{\mathfrak{H}})$. Hence $\tilde{V}$ is dense in $\tilde{\mathfrak{H}}$. If $X \in \mathfrak{g}_0$ and $f \in C_c^\infty(G)$ we denote by Xf

the function $x \rightarrow \{df(x \exp tX)/dt\}_{t=0}$. This defines a representation of $\mathfrak{g}_0$ on $C_c^\infty(G)$ which may be extended uniquely to a representation of $\mathfrak{B}$. Let $\sigma(x)$ denote the adjoint of $\pi(x^{-1})$. Then if $\psi \in \mathfrak{H}$ and $\tilde{\psi} \in \tilde{\mathfrak{H}}$ we find that

$$\lim_{t \to 0} \frac{1}{t}(\sigma(\exp tX)\tilde{A}_f\tilde{\psi} - \tilde{A}_f\tilde{\psi}, \psi) = \lim_{t \to 0} \frac{1}{t}(\tilde{\psi}, \pi(\exp -tX)A_f\psi - A_f\psi)$$

$$= (\tilde{\psi}, A_{-X_f}\psi) = (\tilde{A}_{-X_f}\tilde{\psi}, \psi).$$

This shows that

$$\lim_{t \to 0} \frac{1}{t}(\sigma(\exp tX)\tilde{A}_f\tilde{\psi} - \tilde{A}_f\tilde{\psi}) = \tilde{A}_{-X_f}\tilde{\psi}$$

and therefore we get a representation $\sigma\tilde{v}$ of $\mathfrak{B}$ on $\tilde{V}$ such that

$$\sigma\tilde{v}(X)\tilde{\psi} = \lim_{t \to 0} \frac{1}{t}(\sigma(\exp tX)\tilde{\psi} - \tilde{\psi}) \qquad (\tilde{\psi} \in \tilde{V}, \, X \in \mathfrak{g}_0).$$

Hence

$$(\sigma\tilde{v}(X)\tilde{\psi}, \psi) = \lim_{t \to 0} \frac{1}{t}(\sigma(\exp tX)\tilde{\psi} - \tilde{\psi}, \psi)$$

$$= \lim_{t \to 0} \frac{1}{t}(\tilde{\psi}, \pi(\exp -tX)\psi - \psi) = (\tilde{\psi}, \pi_V(-X)\psi)$$

if $\psi \in U$, $\tilde{\psi} \in \tilde{V}$ and $X \in \mathfrak{g}_0$. Let ϕ denote the antiautomorphism of $\mathfrak{B}$ (over C) such that $\phi(X) = -X$ $(X \in \mathfrak{g})$. Then $(\tilde{\psi}, \pi_U(b)\psi) = (\sigma\tilde{v}(\phi(b))\tilde{\psi}, \psi)$ $(\psi \in U, \tilde{\psi} \in \tilde{V}, b \in \mathfrak{B})$. Now let z be any element in $\mathfrak{Z}$ and B the restriction of the operator $\pi_U(z)$ on U_0. Then B is $\chi(z)$ times the unit operator and so is bounded. On the other hand, $\tilde{V}$ is dense in $\tilde{\mathfrak{H}}$ and $(\tilde{\psi}, \pi_U(z)\psi) = (\sigma\tilde{v}(\phi(z)\tilde{\psi}, \psi)$ $(\tilde{\psi} \in \tilde{V}, \psi \in U)$. Since U_0 is dense in $\mathfrak{H}$ we conclude from Lemma 33 that $\pi_U(z)$ is also bounded. Let A denote the unique bounded extension of the operator $\pi_U(z)$ on $\mathfrak{H}$. Then $A - \chi(z)I$ is bounded and since it is zero on U_0 it must be zero. Therefore $\pi_U(z)\psi = A\psi = \chi(z)\psi$ for $\psi \in U$. This proves the lemma.

Let π be a representation of G on $\mathfrak{H}$ and U any linear subspace of $\mathfrak{H}$ which is stable under $\pi(K)$. Then for any $\mathfrak{D} \in \Omega$, we denote by $U_\mathfrak{D}$ the set of all elements in U which transform under $\pi(K)$ according to $\mathfrak{D}$.

LEMMA 33. *Let π be a quasi-simple representation of G on $\mathfrak{H}$ and π_W the corresponding representation of $\mathfrak{B}$ on the space W of all well-behaved elements. Suppose $\psi_0 \in \sum_{\mathfrak{D} \in \Omega} W_\mathfrak{D}$ and $U = \mathrm{Cl}\,(\pi_W(\mathfrak{B})\psi_0)$ (where Cl denotes closure in $\mathfrak{H}$). Then U is invariant under $\pi(G)$, $\pi_W(\mathfrak{B})\psi_0 = \sum_{\mathfrak{D} \in \Omega} U_\mathfrak{D}$, and $\dim U_\mathfrak{D} < \infty$ $(\mathfrak{D} \in \Omega)$.*

We use the notation of Part I and denote by $\mathfrak{k}$ the subspace of $\mathfrak{g}$ spanned by $\mathfrak{k}_0$ over C. Let $\mathfrak{X}$ be the subalgebra of $\mathfrak{B}$ generated by $(1, \mathfrak{k})$. Since K is simply-connected there is a 1-1 correspondence between finite-dimensional irreducible representations of $\mathfrak{k}_0$ (and therefore of $\mathfrak{k}$ or $\mathfrak{X}$) and those of K. Hence any $\mathfrak{D} \in \Omega$ may be looked upon as an equivalence class of finite-dimensional irreducible representations of $\mathfrak{k}$. We shall regard it in this way whenever it is convenient to do so.

We already know from the corollary to Theorem 2 that U is stable under $\pi(G)$. Put $U_0 = \pi_W(\mathfrak{B})\psi_0$. It follows from Lemma 9 that $U_0 \subset \sum_{\mathfrak{D} \in \Omega} W_{\mathfrak{D}}$ and therefore $U_0 = \sum_{\mathfrak{D} \in \Omega} V_0 \cap W_{\mathfrak{D}}$ (Lemma 6). Let $\mathfrak{Y}$ be the set of all elements $x \in \mathfrak{X}$ such that $\pi_W(x)\psi_0 = 0$. Then $\mathfrak{Y}$ is a left ideal in $\mathfrak{X}$ which satisfies the conditions of Theorem 1. Let $a \to a^*$ $(a \in \mathfrak{B})$ denote the natural mapping and π^* the natural representation of $\mathfrak{B}$ on $\mathfrak{B}^* = \mathfrak{B}/\mathfrak{B}\mathfrak{Y}$. Put $\alpha(b^*) = \pi_W(b)\psi_0$ $(b \in \mathfrak{B})$. Then $b^* \to \alpha(b^*)$ is a well-defined mapping of $\mathfrak{B}^*$ into U_0 and $\alpha(\pi^*(a)b^*) = \pi_W(a)\alpha(b^*)$ $(a, b \in \mathfrak{B})$. Now $\mathfrak{B}^* = \sum_{\mathfrak{D} \in \Omega} \mathfrak{B}_{\mathfrak{D}}^*$ in the notation of Theorem 1 and therefore $\alpha(\mathfrak{B}_{\mathfrak{D}}^*) = U_0 \cap W_{\mathfrak{D}}$ $(\mathfrak{D} \in \Omega)$. Moreover for every $\mathfrak{D} \in \Omega$ we can choose a finite set of elements $b_1^*, \cdots, b_k^* \in \mathfrak{B}_{\mathfrak{D}}^*$ such that $\mathfrak{B}_{\mathfrak{D}}^* = \sum_{i=1}^k \pi^*(\mathfrak{Z})b_i^*$. Hence $U_0 \cap W_{\mathfrak{D}} = \sum_{i=1}^k \pi_W(\mathfrak{Z})\alpha(b_i^*)$. But since π is quasi-simple, it follows from the corollary to Lemma 32 that $\pi_W(z) = \chi(z)\pi_W(1)$ $(z \in \mathfrak{Z})$ where χ is the infinitesimal character of π. Therefore $\alpha(b_i^*)$, $1 \leq i \leq k$, span $U_0 \cap W_{\mathfrak{D}}$ and so dim $(U_0 \cap W_{\mathfrak{D}}) < \infty$. Now $U_0 \cap W_{\mathfrak{D}} = U_0 \cap \mathfrak{H}_{\mathfrak{D}} = U_0 \cap U_{\mathfrak{D}}$. Since $U_0 = \sum_{\mathfrak{D} \in \Omega} U_0 \cap U_{\mathfrak{D}}$ is dense in U, it follows from Lemma 30 that $U_{\mathfrak{D}} = U_0 \cap U_{\mathfrak{D}}$. But the dimension of $U_0 \cap U_{\mathfrak{D}}$ is finite and therefore $U_{\mathfrak{D}} = U_0 \cap U_{\mathfrak{D}}$. Hence $U_0 = \sum_{\mathfrak{D} \in \Omega} U_{\mathfrak{D}}$ and dim $U_{\mathfrak{D}} < \infty$.

LEMMA 34. *Let π be a permissible representation of G on $\mathfrak{H}$ such that* dim $\mathfrak{H}_{\mathfrak{D}} < \infty$ *for every* $\mathfrak{D} \in \Omega$. *Then every element ψ_0 in $\sum_{\mathfrak{D} \in \Omega} \mathfrak{H}_{\mathfrak{D}}$ is well-behaved and $\pi_W(\mathfrak{B})\psi_0 = \sum_{\mathfrak{D} \in \Omega} U_{\mathfrak{D}}$ where $U = \mathrm{Cl}\,(\pi_W(\mathfrak{B})\psi_0)$.*

We know from Theorem 4 and Lemma 31 that $\mathfrak{H}_{\mathfrak{D}} = \overline{W}_{\mathfrak{D}}$. But since dim $\mathfrak{H}_{\mathfrak{D}} < \infty$, $\mathfrak{H}_{\mathfrak{D}} = W_{\mathfrak{D}}$ and therefore ψ_0 is well-behaved. Put $V = \pi_W(\mathfrak{B})\psi_0$ and $V_{\mathfrak{D}} = V \cap U_{\mathfrak{D}}$. Then $\psi_0 \in \sum_{\mathfrak{D} \in \Omega} V_{\mathfrak{D}}$ and therefore, from Lemma 6, $V = \sum_{\mathfrak{D} \in \Omega} V_{\mathfrak{D}}$. Since V is dense in U we conclude from Lemma 31 that $V_{\mathfrak{D}}$ is dense in $U_{\mathfrak{D}}$. But dim $V_{\mathfrak{D}} \leq$ dim $\mathfrak{H}_{\mathfrak{D}} < \infty$. Hence $U_{\mathfrak{D}} = V_{\mathfrak{D}}$ and therefore $V = \sum_{\mathfrak{D} \in \Omega} U_{\mathfrak{D}}$.

The following theorem is of decisive significance for our purpose.

THEOREM 5. *Let π be a permissible representation of G on $\mathfrak{H}$ such that* dim $\mathfrak{H}_{\mathfrak{D}} < \infty$ *for all* $\mathfrak{D} \in \Omega$. *For any $\psi_0 \in \sum_{\mathfrak{D} \in \Omega} \mathfrak{H}_{\mathfrak{D}}$ put $U' = \pi_W(\mathfrak{B})\psi_0$ and $U = \mathrm{Cl}\,(U')$. Let M be the set of all closed linear subspaces of U which are invariant under $\pi(G)$. Similarly let M' be the set of all linear subspaces of U' which are invariant under $\pi_W(\mathfrak{B})$. Then there exists a 1-1 mapping $V \to V'$ of M onto M' such that $V' = V \cap U'$, $V = \mathrm{Cl}(V')$, and $V' = \sum_{\mathfrak{D} \in \Omega} V_{\mathfrak{D}}$.*

Notice that since dim $\mathfrak{H}_{\mathfrak{D}} < \infty$ it follows from Lemma 34 that ψ_0 is well-behaved, U is invariant under $\pi(G)$, and $U' = \sum_{\mathfrak{D} \in \Omega} U_{\mathfrak{D}}$. For any $V \in M$ put $V' = V \cap U'$. Then if $\psi \in V'$ and $X \in \mathfrak{g}_0$,

$$\pi_W(X)\psi = \lim_{t \to 0} \frac{1}{t} \{\pi(\exp tX)\psi - \psi\} \in V \cap U' = V'$$

since V is closed. Hence V is stable under $\pi_W(\mathfrak{B})$ and therefore, from Lemma 6, $V' = \sum_{\mathfrak{D} \in \Omega} V' \cap U_{\mathfrak{D}} = \sum_{\mathfrak{D} \in \Omega} V_{\mathfrak{D}}$. Moreover $\sum_{\mathfrak{D} \in \Omega} V_{\mathfrak{D}}$ is dense in V (Theorem 4). Hence $V = \mathrm{Cl}(V')$.

Conversely let $V' \in M'$. Then, again from Lemma 6, $V' = \sum_{\mathfrak{D} \in \Omega} V' \cap U_{\mathfrak{D}}$. Since every element in V' is well-behaved $V = \mathrm{Cl}\ (V')$ is invariant under $\pi(G)$. But then, $V_{\mathfrak{D}} = \mathrm{Cl}\ (V' \cap U_{\mathfrak{D}})$ (Lemma 30). Since dim $(V' \cap U_{\mathfrak{D}})$ $\leqq \dim \mathfrak{H}_{\mathfrak{D}} < \infty$, $V_{\mathfrak{D}} = V' \cap U_{\mathfrak{D}}$ and so $V' = \sum_{\mathfrak{D} \in \Omega} V_{\mathfrak{D}}$. Moreover since $V \in M$ we know from the earlier part of the proof that $V \cap U' = \sum_{\mathfrak{D} \in \Omega} V_{\mathfrak{D}}$. Therefore $V \cap U' = V'$ and the theorem is proved.

Let $\mathfrak{H}_1$, $\mathfrak{H}_2$ be two closed subspaces of $\mathfrak{H}$ both invariant under $\pi(G)$. We say that $\mathfrak{H}_2$ is maximal in $\mathfrak{H}_1$ (with respect to π) if[11] $\mathfrak{H}_1 > \mathfrak{H}_2$ and there exists no closed subspace $\mathfrak{H}_3$ invariant under $\pi(G)$ such that $\mathfrak{H}_1 > \mathfrak{H}_3 > \mathfrak{H}_2$. Then we get the following corollary from Theorem 5.

COROLLARY 1. *If $U \neq \{0\}$ there exists an element $V \in M$ such that V is maximal in U.*

Let $\mathfrak{M}_0$ be the left ideal in $\mathfrak{B}$ consisting of all elements $b \in \mathfrak{B}$ such that $\pi_W(b)\psi_0 = 0$. Since $U \neq \{0\}$, $\psi_0 \neq 0$ and therefore $1 \notin \mathfrak{M}_0$. Therefore by Zorn's lemma there exists a maximal left ideal $\mathfrak{M}$ in $\mathfrak{B}$ containing $\mathfrak{M}_0$. Put $V' = \pi_W(\mathfrak{M})\psi_0$. Then it is clear that V' is not properly contained in any element of M' other than U'. Put $V = \mathrm{Cl}\ (V')$. We claim V is maximal in U. For let $U \supset S \supset V$ $(S \in M)$. Then $U' \supset S' \supset V'$ and therefore $S' = U'$ or V'. Since the mapping from M to M' is 1-1 this proves that $S = U$ or V and $U \neq V$ and this establishes our assertion.

COROLLARY 2. *Let π be a quasi-simple representation of G on $\mathfrak{H} \neq \{0\}$. Then it is possible to find two closed invariant subspaces U and V in $\mathfrak{H}$ such that V is maximal in U.*

As before let W be the space of well-behaved elements in $\mathfrak{H}$. Then $\sum_{\mathfrak{D} \in \Omega} W_{\mathfrak{D}}$ is dense in $\mathfrak{H}$ (Theorem 4). Choose $\psi_0 \in \sum_{\mathfrak{D} \in \Omega} W_{\mathfrak{D}}$, $\psi_0 \neq 0$, and put $U = \mathrm{Cl}\ (\pi_W(\mathfrak{B})\psi_0)$. From Lemma 33, dim $U_{\mathfrak{D}} < \infty$ $(\mathfrak{D} \in \Omega)$ and so our result follows immediately from Corollary 1 above.

Let π_1, π_2 be two representations of G on the Banach spaces $\mathfrak{H}_1$ and $\mathfrak{H}_2$ respectively. We say that they are *equivalent* if there exists a linear mapping S which maps $\mathfrak{H}_1$ topologically onto $\mathfrak{H}_2$ such that $\pi_2(x)S = S\pi_1(x)$ for all $x \in G$. However in case π_1 and π_2 are both permissible we introduce the concept of a new kind of equivalence as follows. Let W_i be the space of all well-behaved elements in $\mathfrak{H}_i$, and let π_{W_i} be the representation of $\mathfrak{B}$ on W_i $(i = 1, 2)$. Put

---

[11] A and B being two sets we write $A > B$ or $A < B$ if $B \supset A$ and $A \neq B$.

$W_{i,\mathfrak{D}} = W_i \cap (\mathfrak{H}_i)_{\mathfrak{D}}$ and $\mathfrak{H}_i^0 = \sum_{\mathfrak{D} \in \Omega} W_{i,\mathfrak{D}}$. Then we know from Theorem 4 and Lemma 9 that $\mathfrak{H}_i^0$ is dense in $\mathfrak{H}_i$, and it is stable under $\pi_{W_i}(\mathfrak{B})$. Let π_i^0 denote the representation of $\mathfrak{B}$ induced on $\mathfrak{H}_i^0$. We shall say that the representations π_1, π_2 are *infinitesimally equivalent* if there exists a 1-1 linear mapping α of $\mathfrak{H}_1^0$ *onto* $\mathfrak{H}_2^0$ such that $\pi_2^0(b)\alpha(\psi) = \alpha(\pi_1^0(b)\psi)$ for all $b \in \mathfrak{B}$ and $\psi \in \mathfrak{H}_1^0$. It is clear that if π_1, π_2 are equivalent they are also infinitesimally equivalent, but the converse is not true in general. However we shall see later (§11) that for irreducible unitary representations on Hilbert spaces these two concepts of equivalence actually coincide.

11. Unitary representations on a Hilbert space. In this section $\mathfrak{H}$ shall stand for a Hilbert space. A representation π of G on $\mathfrak{H}$ is called unitary if the operator $\pi(x)$ is unitary for every $x \in G$. For any two elements ϕ, ψ in $\mathfrak{H}$ we denote by (ϕ, ψ) their scalar product.

THEOREM 6. *Let π be an irreducible unitary representation of G on a Hilbert space $\mathfrak{H}$. Then every element in $\sum_{\mathfrak{D} \in \Omega} \mathfrak{H}_{\mathfrak{D}}$ is well-behaved and* dim $\mathfrak{H}_{\mathfrak{D}} < \infty$ *$(\mathfrak{D} \in \Omega)$. Moreover $\sum_{\mathfrak{D} \in \Omega} \mathfrak{H}_{\mathfrak{D}}$ is dense in $\mathfrak{H}$.*

It is known that an irreducible unitary representation of G on $\mathfrak{H}$ is quasi-simple (see Mautner [12] and Segal [15]). Let W be the space of all well-behaved elements in $\mathfrak{H}$. Then $\sum_{\mathfrak{D} \in \Omega} W_{\mathfrak{D}}$ is dense in $\mathfrak{H}$ (Theorem 4). Choose $\psi_0 \in \sum_{\mathfrak{D} \in \Omega} W_{\mathfrak{D}}, \psi_0 \neq 0$. Since π is irreducible it follows from Lemma 33 that $\mathfrak{H} = \mathrm{Cl}\ (\pi_W(\mathfrak{B})\psi_0)$, $\pi_W(\mathfrak{B})\psi_0 = \sum_{\mathfrak{D} \in \Omega} \mathfrak{H}_{\mathfrak{D}}$, and dim $\mathfrak{H}_{\mathfrak{D}} < \infty$. This proves the theorem.

THEOREM 7. *Let π be a quasi-simple unitary representation of G on $\mathfrak{H} \neq \{0\}$. Then there exists a minimal[12] closed invariant subspace in $\mathfrak{H}$.*

From Corollary 2 of Theorem 5 we can choose two closed invariant subspaces U and S such that S is maximal in U. Let V be the orthogonal complement of S in U. Since π is unitary it is clear that V is invariant and minimal.

The above theorem has the following significance in relation to the theory of Murray and von Neumann [14]. Let π be a unitary representation of G on $\mathfrak{H}$ and $\mathfrak{A}$ the smallest weakly closed algebra of bounded operators on $\mathfrak{H}$ which contains $\pi(G)$. Suppose $\mathfrak{A}$ is a factor, i.e. the center of $\mathfrak{A}$ consists of scalar multiples of the unit operator. Then it can be shown that π must be quasi-simple and we conclude from Theorem 7 that there exists a closed subspace $V \neq \{0\}$ which is invariant and irreducible under $\pi(G)$ and therefore under $\mathfrak{A}$. But then it follows from the results of Murray and von Neumann (Lemmas 5.3.1, 5.3.8, 8.6.1) that $\mathfrak{A}$ is of type I_n or I_∞. Thus we have proved that any factor arising from a unitary representation of semisimple Lie group in a Hilbert space is necessarily of type I.

The following theorem shows that for irreducible unitary representations infinitesimal equivalence is the same as ordinary equivalence.

---

[12] A closed invariant subspace V is called minimal if $\{0\}$ is maximal in V.

THEOREM 8. *Let π_1, π_2 be two irreducible unitary representations of G on the Hilbert spaces $\mathfrak{H}_1$ and $\mathfrak{H}_2$ respectively. Then they are infinitesimally equivalent if and only if they are equivalent.*

It follows from Theorem 6 that dim $\mathfrak{H}_{i,\mathfrak{D}} < \infty$ $(\mathfrak{D} \in \Omega)$, every element in $\mathfrak{H}_i^0 = \sum_{\mathfrak{D} \in \Omega} \mathfrak{H}_{i,\mathfrak{D}}$ is well-behaved, and $\mathfrak{H}_i^0$ is dense in $\mathfrak{H}_i$ $(i=1,\ 2)$. Let π_i^0 denote the representation of $\mathfrak{B}$ on $\mathfrak{H}_i^0$. Then if π_1, π_2 are infinitesimally equivalent there exists a 1-1 linear mapping α of $\mathfrak{H}_1^0$ onto $\mathfrak{H}_2^0$ such that $\alpha(\pi_1^0(b)\psi) = \pi_2^0(b)\alpha\psi$ for all $\psi \in \mathfrak{H}_1^0$ and $b \in \mathfrak{B}$. It is clear that $\alpha(\mathfrak{H}_{1,\mathfrak{D}}) = \mathfrak{H}_{2,\mathfrak{D}}$ $(\mathfrak{D} \in \Omega)$ and the representations of K induced on $\mathfrak{H}_{1,\mathfrak{D}}$ and $\mathfrak{H}_{2,\mathfrak{D}}$ are unitary and equivalent to each other. Therefore there exists a 1-1 linear mapping $\beta_{\mathfrak{D}}$ of $\mathfrak{H}_{1,\mathfrak{D}}$ onto $\mathfrak{H}_{2,\mathfrak{D}}$ such that if $\phi_1,\ \psi_1 \in \mathfrak{H}_{1,\mathfrak{D}}$ and $\phi_2 = \beta_{\mathfrak{D}}\phi_1$, $\psi_2 = \beta_{\mathfrak{D}}\psi_1$ then $(\phi_1, \pi_1(u)\psi_1) = (\phi_2, \pi_2(u)\psi_2)$ for all $u \in K$. Let β be the 1-1 linear mapping of $\mathfrak{H}_1^0$ onto $\mathfrak{H}_2^0$ such that β coincides with $\beta_{\mathfrak{D}}$ on $\mathfrak{H}_{1,\mathfrak{D}}$. Now it is easily seen that for two distinct elements $\mathfrak{D}, \mathfrak{D}' \in \Omega$ the spaces $\mathfrak{H}_{i,\mathfrak{D}}$ and $\mathfrak{H}_{i,\mathfrak{D}'}$ are mutually orthogonal and therefore $(\phi, \pi_1(u)\psi) = (\beta\phi, \pi_2(u)\beta\psi)$ for all $\phi, \psi \in \mathfrak{H}_1^0$ and $u \in K$. In particular $|\psi| = |\beta\psi|$ $(\psi \in \mathfrak{H}_1^0)$ and since $\mathfrak{H}_i^0$ is dense in $\mathfrak{H}_i$ it follows that β can be extended uniquely to an isometric mapping of $\mathfrak{H}_1$ onto $\mathfrak{H}_2$. Let S denote the linear transformation of $\mathfrak{H}_1^0$ onto itself given by $S\psi = \beta^{-1}\alpha\psi$ $(\psi \in \mathfrak{H}_1^0)$. Since α and β are isomorphisms of $\mathfrak{H}_1^0$ onto $\mathfrak{H}_2^0$ it follows that the inverse transformation $S^{-1} = \alpha^{-1}\beta$ exists. Moreover it is obvious that S and S^{-1} leave $\mathfrak{H}_{1,\mathfrak{D}}$ invariant $(\mathfrak{D} \in \Omega)$. Since dim $\mathfrak{H}_{1,\mathfrak{D}} < \infty$ it follows immediately that there exists a linear transformation S^* of $\mathfrak{H}_1^0$ into itself such that

$$(S^*\phi, \psi) = (\phi, S\psi) \qquad\qquad (\phi, \psi \in \mathfrak{H}_1^0).$$

Furthermore the inverse transformation S^{*-1} exists and S^*, S^{*-1} leave $\mathfrak{H}_{1,\mathfrak{D}}$ invariant for all $\mathfrak{D} \in \Omega$. Put $A = S^*S$. Then we claim that

$$A\pi_1^0(b)A^{-1} = \pi_1^0(b) \qquad\qquad (b \in \mathfrak{B}).$$

First notice that if $X \in \mathfrak{g}_0$ and $\phi_i, \psi_i \in \mathfrak{H}_i^0$ $(i=1,\ 2)$ then

$$(\phi_i, \pi_i^0(X)\psi_i) = \lim_{t \to 0} \frac{1}{t}(\phi_i, \pi_i(\exp tX)\psi_i - \psi_i)$$

$$= \lim_{t \to 0} \frac{1}{t}(\pi_i(\exp -tX)\phi_i - \phi_i, \psi_i) = -(\pi_i^0(X)\phi_i, \psi_i)$$

since π_i is a unitary representation. Hence if $\phi, \psi \in \mathfrak{H}_1^0$,

$$(\phi, A\pi_1^0(X)A^{-1}\psi) = (\phi, S^*S\pi_1^0(X)S^{-1}S^{*-1}\psi)$$

$$= (S\phi, S\pi_1^0(X)S^{-1}S^{*-1}\psi)$$

$$= (\beta S\phi, \pi_2^0(X)\beta S^{*-1}\psi)$$

since $\beta S\pi_1^0(X)S^{-1} = \alpha\pi_1^0(X)\alpha^{-1}\beta = \pi_2^0(X)\beta$. Therefore

$$(\phi,\, A\overset{0}{\pi}_1(X)A^{-1}\psi) = (\alpha\phi,\, \overset{0}{\pi}_2(X)\beta S^{*^{-1}}\psi) = -\,(\overset{0}{\pi}_2(X)\alpha\phi,\, \beta S^{*^{-1}}\psi)$$

$$= -\,(\alpha\overset{0}{\pi}_1(X)\phi,\, \beta S^{*^{-1}}\psi)$$

$$= -\,(\beta^{-1}\alpha\overset{0}{\pi}_1(X)\phi,\, S^{*^{-1}}\psi)$$

$$= -\,(S\overset{0}{\pi}_1(X)\phi,\, S^{*^{-1}}\psi)$$

$$= -\,(\overset{0}{\pi}_1(X)\phi,\, \psi) = (\phi,\, \overset{0}{\pi}_1(X)\psi).$$

Since this is true for all $\phi\in\mathfrak{H}_1^0$ and since $\mathfrak{H}_1^0$ is dense in $\mathfrak{H}_1$ we conclude that

$$A\overset{0}{\pi}_1(X)A^{-1}\psi = \overset{0}{\pi}_1(X)\psi \qquad\qquad (\psi\in\mathfrak{H}_1^0,\, X\in\mathfrak{g}_0)$$

and from this our assertion follows immediately.

Now choose $\mathfrak{D}_0\in\Omega$ such that $\mathfrak{H}_{1,\mathfrak{D}_0}\neq\{0\}$. Since A leaves $\mathfrak{H}_{1,\mathfrak{D}_0}$ invariant and since dim $\mathfrak{H}_{1,\mathfrak{D}_0}<\infty$ we can find an element $\psi_0\in\mathfrak{H}_{1,\mathfrak{D}_0}$ $(\psi_0\neq0)$ and a complex number c such that $A\psi_0=c\psi_0$. But $c|\psi_0|^2=(\psi_0,\, S^*S\psi_0)=|S\psi_0|^2$ and $S\psi_0\neq0$. Therefore c is real and positive. Moreover $A\pi_1^0(b)\psi_0=\pi_2^0(b)A\psi_0$ $=c\pi_1^0(b)\psi_0$ $(b\in\mathfrak{B})$. Let $V=\pi_1^0(\mathfrak{B})\psi_0$. Since π_1 is irreducible it follows from Lemma 33 that $\overline{V}=\mathfrak{H}_1$ and $V=\sum_{\mathfrak{D}\in\Omega}\mathfrak{H}_{1,\mathfrak{D}}=\mathfrak{H}_1^0$. Therefore $A\psi=c\psi$ for all $\psi\in\mathfrak{H}_1^0$. Now $|\alpha\psi|^2=|S\psi|^2=(\psi,\, A\psi)=c|\psi|^2$ $(\psi\in\mathfrak{H}_1^0)$ and so α is continuous. Hence it may be extended uniquely to a linear mapping of $\mathfrak{H}_1$ into $\mathfrak{H}_2$ such that $|\alpha\psi|=c^{1/2}|\psi|$ $(\psi\in\mathfrak{H}_1)$. Since $c>0$, α maps $\mathfrak{H}_1$ topologically into $\mathfrak{H}_2$ and therefore $\alpha\mathfrak{H}_1$, being complete, is closed in $\mathfrak{H}_2$. But $\alpha\mathfrak{H}_1\supset\mathfrak{H}_2^0$ and $\mathfrak{H}_2^0$ is dense in $\mathfrak{H}_2$. Hence $\alpha\mathfrak{H}_1=\mathfrak{H}_2$.

Now let $\psi\in\mathfrak{H}_1^0$. Then by Theorem 2 we can choose a neighbourhood V of zero in $\mathfrak{g}_0$ such that if $X\in V$, $\pi_1(\exp X)\psi=\sum_{n\geq0}(1/n!)\pi_1^0(X^n)\psi$ and $\pi_2(\exp X)\alpha\psi=\sum_{n\geq0}(1/n!)\pi_2^0(X^n)\alpha\psi$. But since α is a continuous linear mapping we find that

$$\alpha\pi_1(\exp X)\psi = \sum_{n\geq0}\frac{1}{n!}\alpha\overset{0}{\pi}_1(X^n)\psi = \sum_{n\geq0}\frac{1}{n!}\overset{0}{\pi}_2(X^n)\alpha\psi = \pi_2(\exp X)\alpha\psi$$

$$(X\in V).$$

Put $\psi(X)=\alpha\pi_1(\exp X)\psi-\pi_2(\exp X)\alpha\psi$ $(X\in\mathfrak{g}_0)$. It is obvious that the mapping $X\to\psi(X)$ is analytic. Hence if ϕ is an element in $\mathfrak{H}_2$, $(\phi,\, \psi(X))$ is an analytic function on $\mathfrak{g}_0$ which is zero on V. This however implies that $(\phi,\, \psi(X))=0$ for all $X\in\mathfrak{g}_0$. This being true for every $\phi\in\mathfrak{H}_2$ it follows that $\psi(X)=0$ $(X\in\mathfrak{g}_0)$. Hence $\alpha\pi_1(\exp X)\psi=\pi_2(\exp X)\alpha\psi$ $(X\in\mathfrak{g}_0,\, \psi\in\mathfrak{H}_1^0)$. But $\mathfrak{H}_1^0$ is dense in $\mathfrak{H}_1$, and therefore this relation is actually true for all $\psi\in\mathfrak{H}_1$. Moreover since G is generated by the elements $\exp X$ $(X\in\mathfrak{g}_0)$, it follows that $\alpha\pi_1(x)\psi=\pi_2(x)\alpha\psi$ $(x\in G,\, \psi\in\mathfrak{H}_1)$ and therefore π_1, π_2 are equivalent. Conversely, if π_1, π_2 are equivalent, it is obvious that they are infinitesimally equivalent. Hence the theorem.

Let π be a permissible representation of G on a Banach space $\mathfrak{H}$ and let W be the space of all well-behaved elements in $\mathfrak{H}$. Put $\mathfrak{H}_0=\sum_{\mathfrak{D}\in\Omega}W_{\mathfrak{D}}$. We know that $\mathfrak{H}_0$ is stable under the representation of $\mathfrak{B}$ on W (Lemma 9). Let π_0 denote the representation of $\mathfrak{B}$ on $\mathfrak{H}_0$. We shall say that π is *infinitesimally unitary* if it is possible to define a scalar product (ψ, ϕ) $(\psi, \phi \in \mathfrak{H}_0)$ such that under this product $\mathfrak{H}_0$ becomes a (possibly incomplete) Hilbert space and $(\psi, \pi_0(X)\phi) = -(\pi_0(X)\psi, \phi)$ for all $\psi, \phi \in \mathfrak{H}_0$ and $X \in \mathfrak{g}_0$.

THEOREM 9. *Let π be a quasi-simple irreducible representation of G on a Banach space. Then if π is infinitesimally unitary, it is infinitesimally equivalent to a unitary representation σ of G on a Hilbert space. Moreover, σ is unique, apart from equivalence, and it is irreducible.*

We keep to the above notation. Then it follows from Lemma 33 that $\mathfrak{H}_0 = \sum_{\mathfrak{D} \in \Omega} \mathfrak{H}_\mathfrak{D}$. Let (ϕ, ψ) be the scalar product in $\mathfrak{H}_0$ satisfying the above-mentioned properties with respect to π_0. We can then complete $\mathfrak{H}_0$ with respect to the norm $\|\psi\| = (\psi, \psi)^{1/2}$ $(\psi \in \mathfrak{H}_0)$ and thus obtain a (complete) Hilbert space V. We shall now define a representation σ of G on V.

Let ψ_0 be a fixed element in $\mathfrak{H}_0$ $(\psi_0 \neq 0)$. Consider the mapping

$$T: (x, y, z) \to \pi(x)\pi(y)\pi(z)\psi_0 = \pi(xyz)\psi_0 \qquad (x, y, z \in G)$$

of $G \times G \times G$ into $\mathfrak{H}$. Since ψ_0 is well-behaved since $(x, y, z) \to xyz$ is an analytic mapping of $G \times G \times G$ into G, T is analytic (Lemma 16). Hence we can find an open convex neighbourhood U of zero in $\mathfrak{g}_0$ such that $U = -U$ and

$$\pi(\exp rX)\pi(\exp sY)\pi(\exp tZ)\psi_0 = \sum_{m,n,p \geq 0} \psi_{m,n,p}(X, Y, Z)r^m s^n t^p$$

$(\psi_{m,n,p}(X, Y, Z) \in \mathfrak{H})$ for $X, Y, Z \in U$ provided $|r|, |s|, |t| \leq 1$. Now from the corollary to Lemma 15 we find that

$$\left\{ \frac{\partial^m}{\partial r^m} \frac{\partial^n}{\partial s^n} \frac{\partial^p}{\partial t^p} \left(\pi(\exp rX)\pi(\exp sY)\pi(\exp tZ)\psi_0 \right) \right\}_{r=s=t=0}$$
$$= m!n!p!\psi_{m,n,p}(X, Y, Z).$$

Moreover we know from Theorem 2 that

$$\pi(\exp tZ)\psi_0 = \sum_{p \geq 0} \frac{t^p}{p!} \pi_0(Z^p)\psi_0$$

if $|t|$ is sufficiently small. Since $\pi(\exp rX)\pi(\exp sY)$ is a bounded operator,

$$\pi(\exp rX)\pi(\exp sY)\pi(\exp tZ)\psi_0 = \sum_{p} \frac{t^p}{p!} \pi(\exp rX)\pi(\exp sY)\pi_0(Z^p)\psi_0.$$

Hence

$$\left\{ \frac{\partial^p}{\partial t^p} \left(\pi(\exp rX)\pi(\exp sY)\pi(\exp tZ)\psi_0 \right) \right\}_{t=0} = \pi(\exp rX)\pi(\exp sY)\pi_0(Z^p)\psi_0.$$

Again since $\pi_0(Z^p)\psi_0$ is well-behaved,

$$\pi(\exp sY)\pi_0(Z^p)\psi_0 = \sum \frac{s^n}{n!}\pi_0(Y^n)\pi_0(Z^p)\psi_0$$

provided $|s|$ is sufficiently small and so we find in the same way as above that

$$\left\{\frac{\partial^n}{\partial s^n}(\pi(\exp rX)\pi(\exp sY)\pi_0(Z^p)\psi_0)\right\}_{s=0} = \pi(\exp rX)\pi_0(Y^n)\pi_0(Z^p)\psi_0.$$

Repeating the same argument once more we get finally

$$\left\{\frac{\partial^{m+n+p}}{\partial r^m\partial s^n\partial t^p}(\pi(\exp rX)\pi(\exp sY)\pi(\exp tZ)\psi_0)\right\}_{r=s=t=0}$$
$$= \pi_0(X^m)\pi_0(Y^n)\pi_0(Z^p)\psi_0.$$

Hence

$$\psi_{m,n,p}(X,\ Y,\ Z) = \frac{1}{m!n!p!}\pi_0(X^mY^nZ^p)\psi_0$$

and

$$\pi(\exp rX)\pi(\exp sY)\pi(\exp tZ)\psi_0 = \sum_{m,n,p\geq 0}\frac{r^ms^nt^p}{m!n!p!}\pi_0(X^mY^nZ^p)\psi_0$$

provided $|r|,\ |s|,\ |t|\leq 1$. Let $Z_1,\ \cdots,\ Z_q$ be a base for $\mathfrak{g}_0$ over R. We can choose $\epsilon>0$ such that if $|t_1|,\ \cdots,\ |t_q|\leq\epsilon$, $Z_t=t_1Z_1+\cdots+t_qZ_q$ lies in U $(t_i\in R,\ 1\leq i\leq q)$. Hence

$$\pi(\exp rX)\pi(\exp sY)\pi(\exp Z_t)\psi_0 = \sum_{m,n,p\geq 0}\frac{r^ms^n}{m!n!p!}\pi_0(X^mY^nZ_t^p)\psi_0.$$

But

$$Z_t^p = p!\sum\frac{t_1^{p_1}t_2^{p_2}\cdots t_q^{p_q}}{p_1!p_2!\cdots p_q!}Z(p_1,\ p_2,\ \cdots,\ p_q)$$

where the sum is over all integers $p_1,\ \cdots,\ p_q\geq 0$ such that $p_1+p_2+\cdots+p_q=p$ and

$$Z(p_1,\ p_2,\ \cdots,\ p_q) = \frac{1}{p!}\sum_{\omega}Z_{i_{\omega(1)}}Z_{i_{\omega(2)}}\cdots Z_{i_{\omega(p)}}.$$

Here $\{i_1,\ i_2,\ \cdots,\ i_p\}$ is a set of p integers such that exactly p_j of them are equal to j $(1\leq j\leq q)$ and the sum is over all permutations ω of the p integers $(1,\ 2,\ \cdots,\ p)$. Therefore

$$\left\{\frac{\partial^{p_1+\cdots+p_q}}{\partial t_1^{p_1}\cdots\partial t_q^{p_q}}\left(\pi(\exp rX)\pi(\exp sY)\pi(\exp Z_t)\psi_0\right)\right\}_{t_1=t_2=\cdots=t_q=0}$$

$$=\sum_{m,n\geq 0}\frac{r^m s^n}{m!n!}\pi_0(X^m Y^n)\pi_0(Z(p_1,\cdots,p_q))\psi_0.$$

Since the elements $Z(p_1, p_2, \cdots, p_q)$ all taken together span $\mathfrak{B}$ (see [8(a)]) it follows that

$$\pi(\exp rX)\pi(\exp sY)\pi_0(b)\psi_0 = \sum_{m,n}\frac{r^m s^n}{m!n!}\pi_0(X^m Y^n)\pi_0(b)\psi_0 \qquad (X, Y \in U)$$

for any $b\in\mathfrak{B}$ and $|r|$, $|s| \leq 1$. However, since π is quasi-simple and irreducible, it follows from Lemma 33 that $\pi_0(\mathfrak{B})\psi_0=\mathfrak{H}_0$. Therefore if we let $r=s=1$ we get

$$\pi(\exp X)\pi(\exp Y)\psi = \sum_{m,n\geq 0}\frac{1}{m!n!}\pi_0(X^m Y^n)\psi$$

for all $\psi\in\mathfrak{H}_0$ and X, $Y\in U$. Now put $\exp_m Z=1+Z+Z^2/2!+\cdots+Z^m/m!$ $\in\mathfrak{B}$ for any $Z\in\mathfrak{g}_0$. Then the above equation may be written as follows:

$$\pi(\exp X)\pi(\exp Y)\psi = \lim_{m\to\infty,n\to\infty}(\mathfrak{H})\pi_0(\exp_m X)\pi_0(\exp_n Y)\psi \quad (X, Y \in U, \psi \in \mathfrak{H}_0).$$

Here $\lim(\mathfrak{H})$ means limit in $\mathfrak{H}$.

Let $E_\mathfrak{D}$ denote the projection of $\mathfrak{H}$ on $\mathfrak{H}_\mathfrak{D}$ (see §8). Since $\mathfrak{H}_\mathfrak{D}$ is a finite-dimensional subspace of V it may itself be regarded a Hilbert space. Moreover $\mathfrak{H}_\mathfrak{D}$ is invariant under $\pi(K)$ and since $(\phi_1, \pi_0(X)\phi_2) = -(\pi_0(X)\phi_1, \phi_2)$ $(X\in\mathfrak{k}_0,$ $\phi_1, \phi_2\in\mathfrak{H}_\mathfrak{D})$ it is clear that the representation of K induced on $\mathfrak{H}_\mathfrak{D}$ is unitary. From this it follows immediately that if $\mathfrak{D}_1\neq\mathfrak{D}_2$, then $\mathfrak{H}_{\mathfrak{D}_1}$ and $\mathfrak{H}_{\mathfrak{D}_2}$ are mutually orthogonal subspaces of $\mathfrak{H}_0$ $(\mathfrak{D}_1, \mathfrak{D}_2\in\Omega)$. Hence $(E_\mathfrak{D}\phi_1, \phi_2) = (\phi_1, E_\mathfrak{D}\phi_2)$ $(\phi_1, \phi_2\in\mathfrak{H}_0)$. For any finite subset F of Ω put $E_F = \sum_{\mathfrak{D}\in F} E_\mathfrak{D}$. Then E_F is a bounded linear operator on $\mathfrak{H}$ and therefore if $\psi\in\mathfrak{H}_0$,

$$\lim_{m\to\infty,n\to\infty}(\mathfrak{H})E_F\pi_0(\exp_m X)\pi_0(\exp_n Y)\psi = E_F\pi(\exp X)\pi(\exp Y)\psi \quad (X, Y \in U).$$

But $E_F\mathfrak{H} = \sum_{\mathfrak{D}\in F}\mathfrak{H}_\mathfrak{D}$ is a finite-dimensional subspace contained in $\mathfrak{H}_0$. Hence the two topologies induced in $E_F\mathfrak{H}$ by $\mathfrak{H}$ and V are the same. Therefore

$$\lim_{m\to\infty,n\to\infty}(V)E_F\pi_0(\exp_m X)\pi_0(\exp_n Y)\psi = E_F\pi(\exp X)\pi(\exp Y)\psi \quad (X, Y \in U)$$

where $\lim(V)$ denotes limit in V. Now

$$\|\pi_0(\exp_m X)\psi\|^2 = (\pi_0(\exp_m X)\psi, \pi_0(\exp_m X)\psi)$$

$$= (\psi, \pi_0(\exp_m (-X) \exp_m X)\psi) \qquad (X \in \mathfrak{g}_0)$$

since $(\pi_0(Z)\phi_1, \phi_2) = -(\phi_1, \pi_0(Z)\phi_2)$ for all $\phi_1, \phi_2 \in \mathfrak{H}_0$ and $Z \in \mathfrak{g}_0$. Choose F such that $\psi \in E_F \mathfrak{H}$. Then if $X \in U$ we have

$$\lim_{m \to \infty} \|\pi_0(\exp_m X)\psi\|^2 = \lim_{m \to \infty} (\psi, \pi_0(\exp_m (-X))\pi_0(\exp_m X)\psi)$$

$$= \lim_{m \to \infty} (\psi, E_F \pi_0(\exp_m (-X))\pi_0(\exp_m X)\psi)$$

$$= (\psi, \lim_{m \to \infty} E_F \pi_0(\exp_m (-X))\pi_0(\exp_m X)\psi)$$

$$= (\psi, E_F \pi(\exp -X)\pi(\exp X)\psi) = (\psi, E_F \psi) = \|\psi\|^2.$$

This proves that the sequence $\pi_0(\exp_m X)\psi$ is bounded in V. Moreover, for any $\phi \in \mathfrak{H}_0$ we can choose a finite subset F of Ω such that $\phi \in E_F \mathfrak{H}$ and therefore

$$\lim_{m \to \infty} (\phi, \pi_0(\exp_m X)\psi) = \lim_{m \to \infty} (\phi, E_F \pi_0(\exp_m X)\psi) = (\phi, E_F \pi(\exp X)\psi).$$

Since $\mathfrak{H}_0$ is dense in V, it follows that the sequence $\pi_0 (\exp_m X)\psi$ is weakly convergent in V. Let $A(X)\psi$ denote the weak limit of this sequence in V. Then it is obvious that $A(X)$ is a linear operator with domain $\mathfrak{H}_0$. We shall now show that $A(X)$ is bounded. Let $\phi \in \mathfrak{H}_0$. Then

$$(\phi, A(X)\psi) = \lim_{n \to \infty} (\phi, \pi_0(\exp_n X)\psi).$$

Hence $|(\phi, A(X)\psi)| \leq \|\phi\| \lim_{n \to \infty} \|\pi_0(\exp_n X)\psi\| = \|\phi\|\|\psi\|$ as we have seen above. Since $\mathfrak{H}_0$ is dense in V it follows that

$$|(\phi, A(X)\psi)| \leq \|\phi\| \|\psi\|$$

for all $\phi \in V$ and therefore $\|A(X)\psi\| \leq \|\psi\|$. As this is true for all $\psi \in \mathfrak{H}_0$, $A(X)$ is bounded and therefore it may be extended in a unique manner to a bounded linear operator on V. We shall denote this extension again by $A(X)$. Then it is clear that

$$\|A(X)\| = \sup_{\|\psi\| \leq 1} \|A(X)\psi\| \leq 1 \qquad\qquad (\psi \in V).$$

We now claim that $A(X)$ is a unitary operator. For if $X, Y \in U$ and $\phi, \psi \in \mathfrak{H}_0$,

$$(\phi, A(X)A(Y)\psi) = (A^*(X)\phi, A(Y)\psi)$$

where $A^*(X)$ is the adjoint of $A(X)$. Therefore

$$(\phi, A(X)A(Y)\psi) = \lim_{n \to \infty} (A^*(X)\phi, \pi_0(\exp_n Y)\psi)$$

$$= \lim_{n \to \infty} (\phi, A(X)\pi_0(\exp_n Y)\psi)$$

$$= \lim_{n \to \infty} (\lim_{m \to \infty} (\phi, \pi_0(\exp_m X)\pi_0(\exp_n Y)\psi)$$

since $\pi_0(\exp_n Y) \in \mathfrak{H}_0$. Now choose F such that $\phi \in E_F \mathfrak{H}$. Then we have seen above that

$$(\phi, E_F \pi(\exp X)\pi(\exp Y)\psi) = \lim_{m \to \infty, n \to \infty} (\phi, \pi_0(\exp_m X)\pi_0(\exp_n Y)\psi).$$

Therefore

$$(\phi, A(X)A(Y)\psi) = (\phi, E_F \pi(\exp X)\pi(\exp Y)\psi).$$

In particular if we put $Y = -X$ we get

$$(\phi, A(X)A(-X)\psi) = (\phi, E_F\psi) = (\phi, \psi).$$

Since $A(X)A(-X)$ is a bounded operator and since $\mathfrak{H}_0$ is dense in $\mathfrak{H}$, we conclude that $A(X)A(-X) = I$ where I is the unit operator. Replacing X by $-X$ we obtain $A(-X)A(X) = I$. This shows that $A(X)$ is regular (i.e. it has a two-sided bounded inverse). Moreover,

$$\|\psi\| = \|A(-X)A(X)\psi\| \leq \|A(X)\psi\| \leq \|\psi\|$$

since $\|A(-X)\|$ and $\|A(X)\|$ are both ≤ 1. Therefore $\|A(X)\psi\| = \|\psi\|$ and this proves that $A(X)$ is unitary.

For any $\mathfrak{D} \in \Omega$ let $E'_\mathfrak{D}$ denote the orthogonal projection of V on $\mathfrak{H}_\mathfrak{D}$. Since $\mathfrak{H}_0$ is dense in V, it is clear that for any $\psi \in V$ the series $\sum_{\mathfrak{D} \in \Omega} E'_\mathfrak{D}\psi$ converges to ψ in V. For any finite subset of F of Ω put $E'_F = \sum_{\mathfrak{D} \in F} E'_\mathfrak{D}$. It is obvious that the operators E'_F and E_F coincide on $\mathfrak{H}_0$. We shall now prove that if $X_1, \cdots, X_r \in U$ and $\psi \in \mathfrak{H}_0$ then

$$E'_F A(X_1)A(X_2) \cdots A(X_r)\psi = E_F \pi(\exp X_1)\pi(\exp X_2) \cdots \pi(\exp X_r)\psi$$

for any finite subset F of Ω. First notice that since strong and weak convergence in $E'_F V$ are the same,

$$E'_F A(X_1)\psi = \lim_{n \to \infty} E_F \pi_0(\exp_n X_1)\psi = E_F \pi(\exp X_1)\psi$$

and therefore our statement is true if $r = 1$. So we may assume $r \geq 2$ and use induction on r. Put $\phi = \pi(\exp X_2) \cdots \pi(\exp X_r)\psi$ and $\phi' = A(X_2) \cdots A(X_r)\psi$. Then if F_1 is any finite subset of Ω it follows from our induction hypothesis that

$$E'_{F_1}\phi' = E_{F_1}\phi = \phi_{F_1} \text{ (say)}.$$

Moreover $\phi_{F_1} \in E_{F_1}\mathfrak{H} \subset \mathfrak{H}_0$, and therefore

$$E'_F A(X_1)\phi_{F_1} = E_F \pi(\exp X_1)\phi_F..$$

Since ψ is well-behaved in $\mathfrak{H}$, the same holds for ϕ. Therefore from Lemma 31 the series $\sum_{\mathfrak{D} \in \Omega} E_\mathfrak{D}\phi$ converges to ϕ in $\mathfrak{H}$. On the other hand, we have seen above that $\sum_{\mathfrak{D}\in\mathfrak{a}} E'_{\mathfrak{D}}\phi'$ converges to ϕ' in V. Therefore, since $E'_F A(X_1)$ and $E_F\pi(\exp X_1)$ are bounded operators in V and $\mathfrak{H}$ respectively and since $\phi_{F_1} = \sum_{\mathfrak{D}\in F_1} E_{\mathfrak{D}}\phi = \sum_{\mathfrak{D}\in F_1} E'_{\mathfrak{D}}\phi'$, it follows from the above result that

$$E'_F A(X_1)\phi' = E_F\pi(\exp X_1)\phi$$

and this proves our assertion.

Now suppose $X_1, \cdots, X_r$ is a finite set of elements in U such that $\exp X_1 \cdots \exp X_r = 1$. Then for any $\phi, \psi \in \mathfrak{H}_0$,

$$\begin{aligned}
(\phi, A(X_1)A(X_2) \cdots A(X_r)\psi) &= (\phi, E'_F A(X_1) \cdots A(X_r)\psi) \\
&= (\phi, E_F\pi(\exp X_1) \cdots \pi(\exp X_r)\psi) \\
&= (\phi, E_F\psi) = (\phi, \psi)
\end{aligned}$$

where F is so chosen that $\phi \in E_F\mathfrak{H}$. Since $\mathfrak{H}_0$ is dense in V and since $A(X_1)A(X_2) \cdots A(X_r)$ is a bounded operator in V, this proves that $A(X_1)A(X_2) \cdots A(X_r) = I$.

We know that every $x \in G$ can be written in the form $x = \exp X_1 \exp X_2 \cdots \exp X_r$ $(X_i \in U, 1 \leq i \leq r)$. Then if we put $\sigma(x) = A(X_1) \cdots A(X_r)$ it is clear from what we have proved above that $\sigma(x)$ depends only on x and not on the choice of $X_1, \cdots, X_r$ and $\sigma(xy) = \sigma(x)\sigma(y)$. Moreover $\sigma(x)$ is obviously a unitary operator and $\sigma(1) = I$. Hence it follows from well known arguments (see Godement $[7(a)]$) that in order to show that σ is a representation of G on V it is sufficient to prove that $\lim_{x\to1} (\phi, \sigma(x)\psi) = (\phi, \psi)$ for any two elements $\phi, \psi \in V$. First suppose $\phi, \psi \in \mathfrak{H}_0$. Choose F so that $\phi \in E_F\mathfrak{H}$. Then

$$E'_F A(X)\psi = E_F\pi(\exp X)\psi \qquad\qquad (X \in U)$$

and therefore

$$(\phi, \sigma(\exp X)\psi) = (\phi, E_F\pi(\exp X)\psi).$$

Now as $X \to 0$, $\pi(\exp X)\psi \to \psi$ and since E_F is a bounded operator on $\mathfrak{H}$, $E_F\pi(\exp X)\psi \to E_F\psi$. Hence

$$\lim_{X\to0} (\phi, \sigma(\exp X)\psi) = (\phi, E_F\psi) = (\phi, \psi).$$

Now if we take into account the fact that $\mathfrak{H}_0$ is dense in V and $\sigma(x)$ is unitary, it follows immediately that $\lim_{x\to1} (\phi, \sigma(x)\psi) = (\phi, \psi)$ for all $\phi, \psi \in V$. Therefore σ is a unitary representation of G on V. Moreover, it is obvious that for any $u \in K$ the operators $\sigma(u)$ and $\pi(u)$ coincide on $\mathfrak{H}_0$ and therefore since π is a permissible representation and since $\mathfrak{H}_0$ is dense in V, σ is also permissible. Hence it follows from Lemmas 30 and 34 that $V_{\mathfrak{D}} = \mathfrak{H}_{\mathfrak{D}}$ and every element in $\mathfrak{H}_0$ is well-behaved under σ. Let $X \in U$ and $\phi, \psi \in \mathfrak{H}_0$. Then

$$(\phi, \sigma(\exp X)\psi) = (\phi, E'_F\sigma(\exp X)\psi) = (\phi, E_F\pi(\exp X)\psi)$$

where F is so chosen that $\phi \in E_F \mathfrak{H}$. Therefore

$$\lim_{t \to 0} \frac{1}{t} (\phi, \sigma(\exp tX)\psi - \psi) = \lim_{t \to 0} \frac{1}{t} (\phi, E_F(\pi(\exp tX)\psi - \psi))$$

$$= (\phi, E_F \pi_0(X)\psi) = (\phi, \pi_0(X)\psi).$$

Since $\mathfrak{H}_0$ is dense in V and since ψ is well-behaved under σ this shows that

$$\lim_{t \to 0} \frac{1}{t} (\sigma(\exp tX)\psi - \psi) = \pi_0(X)\psi.$$

Therefore the representation of $\mathfrak{B}$ on $\mathfrak{H}_0$ corresponding to σ coincides with π_0, and this proves that π and σ are infinitesimally equivalent. Since π is irreducible, it follows from Theorem 5 that $\mathfrak{H}_0$ is irreducible under π_0 and therefore σ is also irreducible.

Finally, if τ is any permissible unitary representation of G on a Hilbert space $\mathfrak{R}$ which is infinitesimally equivalent to π it is clear that τ is quasi-simple and dim $\mathfrak{R}_\mathfrak{D} = \dim \mathfrak{H}_\mathfrak{D} < \infty$. Therefore, from Theorem 5, τ is irreducible. Since σ and τ are irreducible unitary representations which are infinitesimally equivalent, it follows from Theorem 8 that they are equivalent.

12. Explicit construction of some representations. We shall now give a method for constructing a certain class[13] of representations of G. This is a generalisation of a method given by Gelfand and Naimark [6(a), (b)] in the case when G is a complex semi-simple Lie group (see also [8(e), Part IV]).

First we compute the Haar measure of G in terms of the Haar measures[14] of K, A_+, and N, respectively. The group A_+ being simply-connected, we denote by $\log h$ $(h \in A_+)$ the unique element $H \in \mathfrak{h}_{\mathfrak{p}_0}$ such that $h = \exp H$. Moreover set $\rho = 2^{-1} \sum_{\alpha \in P} \alpha$ where the notation is that of Lemma 5.

LEMMA 35. *Let dx, du, dh, and dn denote the elements of Haar measures on G, K, A_+, and N respectively. Then*

$$dx = e^{2\rho(\log h)} du\, dh\, dn \qquad (x = uhn,\ u \in K,\ h \in A_+,\ n \in N).$$

Let $S = A_+ N$. Then in the notation of Lemma 27 $dx = \mu(s) du\, ds$ $(x = us,$ $u \in K$, $s \in S$ where $\mu(s) = \det \mathrm{Ad}_S (s)/\det \mathrm{Ad} (s)$ and ds is the element of the left-invariant measure on S). But since G is semi-simple, $\det \mathrm{Ad} (x) = 1$ for all $x \in G$. Hence $\mu(s) = \det \mathrm{Ad}_S (s)$. Moreover, by applying the same lemma to $S = A_+ N$ we get

$$ds = \frac{\det \mathrm{Ad}_N (n)}{\det \mathrm{Ad}_S (n)} dh\, dn \quad (s = hn,\ h \in A_+,\ n \in N)$$

---

[13] It seems very likely that this method gives all irreducible quasi-simple representations within infinitesimal equivalence (see [8(d)]).

[14] For the meaning of the various symbols see the beginning of Part III (§8).

where $n \to \mathrm{Ad}_N (n)$ is the adjoint representation of N. Since N is nilpotent, $\det \mathrm{Ad}_N (n) = 1$ and therefore

$$dx = \det \mathrm{Ad}_S (h) du\,dh\,dn = e^{\mathrm{sp}(\sigma(\log h))} du\,dh\,dn = e^{2\rho(\log h)} du\,dh\,dn$$

in the notation of Lemma 5.

Let $u \to u^*$ denote the natural mapping of K on $K^* = K/D \cap Z$ and let du^* denote the element of Haar measure of K^* so normalised that $\int_{K^*} du^* = 1$. Let $L_2(K^*)$ denote the Hilbert space of all square integrable functions f on K^*. We shall now define a class of representations of G on $L_2(K^*)$. For any $x \in G$ and $u \in K$ write $xu = u_x h(x, u) n(x, u)$ $(u_x \in K, h(x, u) \in A_+, n(x, u) \in N)$. Moreover, let $\Gamma(u)$ denote the unique element in $\mathfrak{c}_0$ such that $u \exp (-\Gamma(u))$ lies in the commutator group K' of K. Then it is easy to verify that $(u_x)^*$, $\Gamma(u_x) - \Gamma(u)$, and $h(x, u)$ depend only on x and u^*. Hence we may write them as u_x^*, $\Gamma(x, u^*)$, and $h(x, u^*)$ respectively. It is obvious that these elements depend continuously on (x, u^*). Put $H(x, u^*) = \log h(x, u^*)$.

LEMMA 36. *If $x, y \in G$, then $u_{yx}^* = (u_x^*)_y$ and*

$$H(yx, u^*) = H(y, u_x^*) + H(x, u^*), \qquad \Gamma(yx, u^*) = \Gamma(y, u_x^*) + \Gamma(x, u^*).$$

For we have $xu = u_x h(x, u) n_1$ $(n_1 \in N)$ and therefore

$$yxu = yu_x h(x, u) n_1 = (u_x)_y h(y, u_x) n_2 h(x, u) n_1 \qquad (n_2 \in N)$$
$$= (u_x)_y h(y, u_x) h(x, u) n_2' n_1$$

where $n_2' = (h(x, u))^{-1} n_2 h(x, u) \in N$. Therefore

$$(u)_{yx} = (u_x)_y, \qquad h(yx, u) = h(y, u_x) h(x, u)$$

and

$$\Gamma(yx, u) = \Gamma(u_{yx}) - \Gamma(u) = \Gamma((u_x)_y) - \Gamma(u_x) + \Gamma(u_x) - \Gamma(u)$$
$$= \Gamma(y, u_x) + \Gamma(x, u).$$

The statements of the lemma now follow immediately.

Let ν and μ be (complex-valued) linear functions on $\mathfrak{h}_{\mathfrak{p}_0}$ and $\mathfrak{c}_0$ respectively. Then we define a representation $\pi_{\mu,\nu}$ of G on $L_2(K^*)$ as follows. For any $f \in L_2(K^*)$, $\pi_{\mu,\nu}(x) f$ $(x \in G)$ is the function g given by

$$g(u^*) = e^{-\mu(\Gamma(x^{-1} u^*))} e^{-\nu'(H(x^{-1}, u^*))} f(u_{x^{-1}}^*)$$

where $\nu' = \nu + 2\rho$. First of all, we have to verify that $g \in L_2(K^*)$. Notice that if $y = uhn$ $(u \in K, h \in A_+, n \in N)$ then

$$xy = u_x h(x, u) h(h^{-1} n(x, u) hn).$$

Hence if x is fixed,

$$e^{2\rho(\log h)} du\,dh\,dn = dy = d(xy) = e^{2\rho(\log (h(x,u)h))} du_x\,dh\,dn$$

since $d(h(x, u)h) = dh$ for fixed x and u and $d((h^{-1}n(x, u)h)n) = dn$ for fixed x, u, and h. This shows that $du = e^{2\rho(H(x,u))}du_x$ and from this it follows immediately that

$$du^* = e^{2\rho(H(x,u^*))}du_x^*$$

since K and K^* are locally isomorphic. Hence

$$\int_{K^*} |g(u^*)|^2 du^* = \int_{K^*} |g(u_x^*)|^2 du_x^* = \int_{K^*} |g(u_x^*)e^{-\rho(H(x,u^*))}|^2 du^*.$$

Now $\Gamma(x^{-1}, u_x^*) = -\Gamma(x, u^*)$ and $H(x^{-1}, u_x^*) = -H(x, u^*)$ from Lemma 36. Therefore

$$g(u_x^*)e^{-\rho(H(x,u^*))} = e^{\mu(\Gamma(x,u^*))}e^{(\nu+\rho)(H(x,u^*))}f(u^*).$$

The function $\left|e^{\mu(\Gamma(x,u^*))}e^{(\nu+\rho)(H(x,u^*))}\right|$ being continuous is bounded on the compact set K^* and therefore if M_x is a bound for it

$$\int_{K^*} |g(u^*)|^2 du^* \leq M_x^2 \int |f(u^*)|^2 du^* < \infty.$$

Hence $g \in L_2(K^*)$ and $|\pi_{\mu,\nu}(x)| \leq M_x$, so that $\pi_{\mu,\nu}(x)$ is bounded. Moreover, it follows easily from Lemma 36 that $\pi_{\mu,\nu}(y)\pi_{\mu,\nu}(x) = \pi_{\mu,\nu}(yx)$ $(x, y \in G)$. Therefore, it only remains to verify the continuity of $\pi_{\mu,\nu}$. Let U be a compact neighbourhood of 1 in G. Then

$$\sup_{x \in U, u^* \in K^*} \left|e^{\mu(\Gamma(x,u^*))}e^{(\nu+\rho)(H(x,u^*))}\right| = M_U < \infty$$

since $e^{\mu(\Gamma(x,u^*))}e^{(\nu+\rho)(H(x,u^*))}$ is a continuous function on $G \times K^*$. Hence $|\pi_{\mu,\nu}(x)| \leq M_U$ for $x \in U$. Therefore in order to show that $\pi_{\mu,\nu}$ is a representation, it is sufficient to prove that $\lim_{x\to 1}\|\pi_{\mu,\nu}f - f\| = 0$ $(f \in L_2(K^*))$ where $\|\cdot\|$ denotes the usual norm in $L_2(K^*)$. Given $\epsilon > 0$, choose a continuous function g on K^* such that $\|f - g\| \leq \epsilon$. Then it is clear that we can find a neighbourhood V of 1 in G $(V \subset U)$ such that $|\pi_{\mu,\nu}(x)g - g| \leq \epsilon$ $(x \in V)$. Hence

$$\|\pi_{\mu,\nu}(x)g - g\|^2 = \int_{K^*} |\pi_{\mu,\nu}(x)g - g|^2 du^* \leq \epsilon^2 \qquad (x \in V)$$

and

$$\|\pi_{\mu,\nu}(x)f - f\| = \|\pi_{\mu,\nu}(x)(f - g) - (f - g) + (\pi_{\mu,\nu}(x)g - g)\|$$
$$\leq |\pi_{\mu,\nu}(x)|\epsilon + \epsilon + \epsilon \leq (M_? + 2)\epsilon \qquad (x \in V).$$

This proves that π is a representation of G.

Let λ be the left regular representation of K^* on $L_2(K^*)$ so that $\lambda(v^*)f = g$ $(f \in L_2(K^*), v^* \in K^*)$ where $g(u^*) = f(v^{*-1}u^*)$. Then we check easily that

$$\lambda(u^*) = e^{-\mu(\Gamma(u))}\pi_{\mu,\nu}(u) \qquad\qquad (u \in K).$$

Let $\mathfrak{H} = L_2(K^*)$ and let $\mathfrak{H}_\mathfrak{D}$ ($\mathfrak{D} \in \Omega$) denote the set of all elements in $\mathfrak{H}$ which transform under $\pi_{\mu,\nu}(K)$ according to $\mathfrak{D}$. Then if $\mathfrak{H}_\mathfrak{D} \neq \{0\}$, it is clear that there exists an equivalence class $\mathfrak{D}^*$ of finite-dimensional irreducible representations of K^* such that every element of $\mathfrak{H}_\mathfrak{D}$ transforms under λ according to $\mathfrak{D}^*$. The degrees of the representations in $\mathfrak{D}$ and $\mathfrak{D}^*$ are obviously the same. Hence if $d(\mathfrak{D})$ denotes this degree, it follows from well known results on compact groups that dim $\mathfrak{H}_\mathfrak{D} \leq (d(\mathfrak{D}))^2$. Moreover $\pi_{\mu,\nu}$ is permissible since $\pi_{\mu,\nu}(x) = e^{\mu(\Gamma(x))}\pi_{\mu,\nu}(1)$ if $x \in D \cap Z$. Therefore from Lemma 34 every element in $\sum_{\mathfrak{D} \in \Omega} \mathfrak{H}_\mathfrak{D}$ is well-behaved under $\pi_{\mu,\nu}$. We have seen above that if $f \in L_2(K^*)$

$$\int_{K^*} \left| \pi_{\mu,\nu}(x)f \right|^2 du^* = \int_{K^*} \left| e^{\mu(\Gamma(x,u^*))}e^{(\nu+\rho)(H(x,u^*))}f(u^*) \right|^2 du^*.$$

Therefore if the linear functions μ and $\nu+\rho$ take purely imaginary values on $\mathfrak{c}_0$ and $\mathfrak{h}_{\mathfrak{p}_0}$, the representation is unitary

Let M be the analytic subgroup of G corresponding to the algebra[15] $\mathfrak{m}_0$ $= \mathfrak{m} \cap \mathfrak{g}_0$ and let $x \to x^*$ denote the natural mapping of G on $G/D \cap Z = G^*$. Since $D \subset K$, the mapping $(u^*, a, n) \to u^*a^*n^*$ ($u^* \in K^*$, $a \in A_+$, $n \in N$) is a homeomorphism of $K^* \times A_+ \times N$ onto G^*. Therefore A_+^* is closed in G^*. Since $\mathfrak{m}_0$ is the centraliser of $\mathfrak{h}_{\mathfrak{p}_0}$ in $\mathfrak{k}_0$ (Lemma 4), M^* is the component of identity of the centraliser of A_+^* in K^*. Therefore M^* is closed and hence compact.

Let τ denote the right regular representation of K^* so that

$$(\tau(u^*)f)(v^*) = f(v^*u^*) \quad (f \in L_2(K^*); u^*, v^* \in K^*).$$

LEMMA 37. *Let $m \in M$ and $x \in G$. Then $\tau(m^*)$ commutes with $\pi_{\mu,\nu}(x)$.*

Let $u \in K$. Then $xu = u_x h(x, u)n$ ($n \in N$). From Lemmas 4 and 5, $n' = m^{-1}nm \in N$ and $h(x, u)m = mh(x, u)$. Hence $xum = u_x mh(x, u)n'$ and therefore $(um)_x = u_x m$, $h(x, um) = h(x, u)$ and $\Gamma(x, um) = \Gamma(x, u)$. Our assertion follows immediately from these facts.

Let σ be an irreducible representation of M on a finite-dimensional space V. It is clear that we can choose μ such that σ (exp Γ) $= e^{\mu(\Gamma)} \sigma(1)$ if exp Γ $\in M \cap Z$ ($\Gamma \in \mathfrak{c}_0$). Put $\sigma^*(m^*) = e^{-\mu(\Gamma(m))} \sigma(m)$ ($m \in M$). Then σ^* is an irreducible representation of M^* on V. Now it follows from well known results (see [17]) on compact groups that we can choose a continuous function ϕ on K^* such that the linear space spanned by the functions $\tau(m^*)\phi$ ($m^* \in M^*$) is of finite dimension and the representation of M^* induced on it under τ is irreducible and dual to σ^*. Let $\mathfrak{H}_\phi$ be the smallest closed subspace of $L_2(K^*)$ containing ϕ which is invariant under $\pi_{\mu,\nu}$. Let $\pi'_{\mu,\nu}$ denote the representation

---

[15] See §2 (Lemma 4) for the definition of $\mathfrak{m}$.

of G induced on $\mathfrak{H}_\phi$ under $\pi_{\mu,\nu}$. Then it can be shown that $\pi'_{\mu,\nu}$ is quasi-simple. Let δ be the equivalence class of the representation σ. Since $\mathfrak{m}$ is reductive in $\mathfrak{k}$ (Lemma 5), every class $\mathfrak{D} \in \Omega$ is fully reducible with respect to M. Let $(D:\delta)$ denote the number of times δ occurs in this reduction. Then it follows from the Frobenius reciprocity relation (see [17, p. 83]) that $\dim (\mathfrak{H}_\phi)_\mathfrak{D} \leq d(\mathfrak{D}) (\mathfrak{D}:\delta)$.

We shall return to a more detailed study of these representations of G in another paper.

REFERENCES

1. S. Banach, *Leçons sur les éspaces linéares*, Warsaw, 1932.
2. G. Birkhoff, Ann. of Math. vol. 38 (1937) pp. 526–532.
3. E. Cartan, J. Math. Pures Appl. vol. 8 (1929) pp. 1–33.
4. C. Chevalley, *Theory of Lie groups*, Princeton University Press, 1946.
5. L. Gårding, Proc. Nat. Acad. Sci. U.S.A. vol. 33 (1947) pp. 331–332.
6. I. M. Gelfand and M. A. Naimark, (a) Doklady Akademii Nauk SSSR. N.S. vol. 63 (1948) pp. 225–228.
——, (b) Trudi Mat. Inst. Steklova vol. 36 (1950).
7. R. Godement, (a) Trans. Amer. Math. Soc. vol. 63 (1948) pp. 1–84.
——, (b) J. Math. Pures Appl. vol. 30 (1951) pp. 1–110.
——, (c) *On spherical functions*, Trans. Amer. Math. Soc. vol. 73 (1952) pp. 496–566.
8. Harish-Chandra, (a) Ann. of Math. vol. 50 (1949) pp. 900–915.
——, (b) Proc. Amer. Math. Soc. vol. 1 (1950) pp. 205–210.
——, (c) Proc. Nat. Acad. Sci. U.S.A. vol. 37 (1951) pp. 170–173.
——, (d) Proc. Nat. Acad. Sci. U.S.A. vol. 37 (1951) pp. 362–365, 366–369, 691–694.
——, (e) Trans. Amer. Math. Soc. vol. 70 (1951) pp. 28–96.
9. K. Iwasawa, Ann. of Math. vol. 50 (1949) pp. 507–557.
10. N. Jacobson, Ann. of Math. vol. 36 (1935) pp. 875–881.
11. L. Koszul, Bull. Soc. Math. France vol. 78 (1950) pp. 65–127.
12. F. I. Mautner, Ann. of Math. vol. 52 (1950) pp. 528–556.
13. G. D. Mostow, Bull. Amer. Math. Soc. vol. 55 (1949) pp. 969–980.
14. F. J. Murray and J. von Neumann, Ann. of Math. vol. 37 (1936) pp. 116–229.
15. I. E. Segal, Proc. Amer. Math. Soc. vol. 3 (1952) pp. 13–15.
16. B. L. van der Waerden, *Moderne Algebra*, Berlin, Springer, 1937.
17. A. Weil, *L'intégration dans les groupes topologiques et ses applications*, Paris, Hermann, 1940.

COLUMBIA UNIVERSITY,
    NEW YORK, N Y.

Reprinted from
Trans. Amer. Math. Soc.
75 (1953), 185–243

REPRESENTATIONS OF SEMISIMPLE LIE GROUPS. II

BY

HARISH-CHANDRA

In an earlier paper [5] we have established a close relationship between an irreducible representation of a semisimple Lie group on a Banach space and the corresponding representation of its Lie algebra. The object of the present paper is to make a deeper study of the representations of the algebra. We shall show that every irreducible representation of the group (at least in case it is complex) is infinitesimally equivalent to one constructed in a certain standard way (Theorem 4). This is the principal result of this paper. Although the proof is rather long the idea behind it is quite simple and has been employed before (see [3]). It can be explained as follows. Let $\mathfrak{B}$ be the universal enveloping algebra of the Lie algebra $\mathfrak{g}$ of our group. The finite-dimensional representations of $\mathfrak{B}$ are very well known and their knowledge enables us to deduce certain algebraic identities in $\mathfrak{B}$. Since these identities must be preserved in every representaton, they provide us with useful information about infinite-dimensional representations as well. Roughly speaking one can say that the representations of finite degree are characterised by a certain set of parameters which take integral values. If we keep to the same algebraic pattern but give arbitrary complex values to these parameters we obtain the infinite-dimensional representations.

The results of this paper were obtained in the summer of 1951 and they have been announced in a short note [4(b)].

1. **Proof of Theorem 1.** Let $\mathfrak{g}_0$ be a semisimple Lie algebra over the field R of real numbers. We denote by $\mathfrak{g}$ its complexification. Define $\mathfrak{f}$, $\mathfrak{p}$, $\mathfrak{f}_0$, and $\mathfrak{p}_0$ as in [5, Part I]. Let $\mathfrak{c}$ be the center of $\mathfrak{f}$ and $\mathfrak{f}'$ the derived algebra of $\mathfrak{f}$. Then $\mathfrak{f}'$ is semisimple and $\mathfrak{f}$ is the direct sum of $\mathfrak{c}$ and $\mathfrak{f}'$. Moreover $\mathfrak{f}$ is reductive in $\mathfrak{g}$ (see [5, Lemma 3]). Let Ω' be the set of all equivalence classes of finite-dimensional simple representations of $\mathfrak{f}'$. Let Ω'_F: denote the subset consisting of all $\mathfrak{D}' \in \Omega'$ which occur in the reduction of some finite-dimensional representation of $\mathfrak{g}$ with respect to $\mathfrak{f}'$. Let $\mathfrak{B}$ be the universal enveloping algebra of $\mathfrak{g}$ and $\mathfrak{X}$, $\mathfrak{X}'$ the subalgebras of $\mathfrak{B}$ generated by $(1, \mathfrak{f})$ and $(1, \mathfrak{f}')$ respectively. Our main object at present is to prove the following theorem which provides the basis for most of our subsequent arguments.

THEOREM 1. *Let $\mathfrak{Y}'$ be a maximal left ideal in $\mathfrak{X}'$ such that the natural representation*[1] *of $\mathfrak{f}'$ on $\mathfrak{X}'/\mathfrak{Y}'$ lies in some class in Ω'_F. Let $F(\mathfrak{Y}')$ be the set of all maximal left ideals $\mathfrak{M}$ in $\mathfrak{B}$ with the following two properties: (a) $\mathfrak{M} \supset \mathfrak{Y}'$; (b) the space $\mathfrak{B}/\mathfrak{M}$ is of finite dimension. Then $\mathfrak{B}\mathfrak{Y}' = \bigcap_{\mathfrak{M} \in F(\mathfrak{Y}')} \mathfrak{M}$.*

Received by the editors October 30, 1952.

[1] See [5], beginning of §3.

26

"

We first introduce some terminology. Let C be the field of complex numbers and V a vector space over C. Let $y_1, \cdots, y_r$ be a finite set of independent commutative indeterminates and $C[y]$ the ring of polynomials in (y) with coefficient in C. Consider the tensor product $V \times C[y]$ which we regard as a two-sided $C[y]$-module so that if f, $g \in C[y]$ and $v \in V$ we have $f(v \times g) = (v \times g)f = v \times fg$. We may identify V with a subspace of this tensor product under the mapping $v \to v \times 1$. Then every element in $V \times C[y]$ can be written as a sum of elements of the form vf where $v \in V$ and $f \in C[y]$. We shall often refer to the elements of $V \times C[y]$ as polynomials in (y) with coefficients in V. In accordance with this interpretation we shall write $V[y]$ to denote $V \times C[y]$. The *value* of such a polynomial at the point $y_j = \mu_j$, $1 \leq j \leq r$ $(\mu_j \in C)$, is the element in V obtained by replacing each y_j by μ_j. If V is an algebra we define multiplication in $V[y]$ according to the rule $(v_1 f_1)(v_2 f_2) = (v_1 v_2)(f_1 f_2)$ $(v_1, v_2 \in V; f_1, f_2 \in C[y])$. This makes $V[y]$ into an algebra.

Now consider the symmetric algebra $S(\mathfrak{g})$ over $\mathfrak{g}$. We identify $S(\mathfrak{g})$ with the algebra of polynomial functions on $\mathfrak{g}$ as explained in [5]. Moreover if $\mathfrak{l}$ is a linear subspace of $\mathfrak{g}$ the symmetric algebra $S(\mathfrak{l})$ may be regarded as a subalgebra of $S(\mathfrak{g})$. Let λ denote the canonical mapping of $S(\mathfrak{g})$ into $\mathfrak{B}$ which was defined in [5, §2]. Since $\mathfrak{g}$ is the direct sum of $\mathfrak{p}$, $\mathfrak{k}'$, and $\mathfrak{c}$, it follows from Lemma 10 of [5] that the mapping $(f, g, h) \to \lambda(f)\lambda(g)\lambda(h)$ is a linear isomorphism of the tensor product $S(\mathfrak{p}) \times S(\mathfrak{k}') \times S(\mathfrak{c})$ with $\mathfrak{B}$. Put $\mathfrak{P} = \lambda(S(\mathfrak{p}))$, $\mathfrak{C} = \lambda(S(\mathfrak{c}))$. If $(w_i)_{i \in I}$, $(x_j)_{j \in I'}$, $(\gamma_k)_{k \in I''}$ are bases for $\mathfrak{P}$, $\mathfrak{X}'$, and $\mathfrak{C}$ respectively, then $w_i x_j \gamma_k$ and $x_j \gamma_k$ respectively are bases for $\mathfrak{B}$ and $\mathfrak{X}$. This fact should be constantly borne in mind during the following discussion.

We know that the simply-connected analytic group G corresponding to $\mathfrak{g}_k = \mathfrak{k}_0 + (-1)^{1/2} \mathfrak{p}_0$ is compact. Let K and K' be the analytic subgroups of G corresponding to $\mathfrak{k}_0$ and $\mathfrak{k}_0' = \mathfrak{k}' \cap \mathfrak{k}_0$. We can choose a base $\Gamma_1', \cdots, \Gamma_r'$ for $\mathfrak{c}_0 = \mathfrak{c} \cap \mathfrak{k}_0$ over R such that $\exp(t_1 \Gamma_1' + \cdots + t_r \Gamma_r') \in K'$ $(t_1, \cdots, t_r \in R)$ if and only if $t_j / 2\pi$, $1 \leq j \leq r$, are all integers. Put $\Gamma_j = (-1)^{1/2} \Gamma_j'$. Then $\Gamma_1, \cdots, \Gamma_r$ is a base for $\mathfrak{c}$ over C. Let M be a linear function on $\mathfrak{c}$. We shall say that M is integral if $m_j = M(\Gamma_j)$ are all integers ≥ 0 and in that case we write $\Gamma^M = \Gamma_1^{m_1} \Gamma_2^{m_2} \cdots \Gamma_r^{m_r}$. It is clear that every element $b \in \mathfrak{B}$ can be written uniquely in the form $b = \sum_M w_M \Gamma^M$ where $w_M \in \mathfrak{P}\mathfrak{X}'$ and M runs over all integral functions. Let $y_1, \cdots, y_r$ be r independent commuting indeterminates. Put $y^M = y_1^{m_1} \cdots y_r^{m_r}$ where $m_j = M(\Gamma_j)$ and M is integral. Then if we put $\alpha_y(b) = \sum_M w_M y^M$, the mapping α_y is a linear isomorphism of $\mathfrak{B}$ with $\mathfrak{P}\mathfrak{X}'[y]$. If μ is a linear function on $\mathfrak{c}$ we shall denote by $\alpha_\mu(b)$ the value of the polynomial $\alpha_y(b)$ at the point $y_j = \mu(\Gamma_j)$, $1 \leq j \leq r$.

Now we come to the proof of Theorem 1 which would consist of three stages.

First reduction. Let b_0 be an element in $\mathfrak{B}$ which is not in $\mathfrak{B}\mathfrak{Y}'$. Then we have to find an $\mathfrak{M} \in F(\mathfrak{Y}')$ such that $b_0 \notin \mathfrak{M}$. We can write b_0 in the form $b_0 = \sum_M w_M \Gamma^M$ $(w_M \in \mathfrak{P}\mathfrak{X}')$. Since $b_0 \notin \mathfrak{B}\mathfrak{Y}'$, the elements w_M cannot all lie in $\mathfrak{P}\mathfrak{Y}'$. Hence $\alpha_\nu(b_0) = \sum_M w_M y^M \overline{\in} \mathfrak{P}\mathfrak{Y}'[y]$. Let $w_1, \cdots, w_j$ be a maximal set of elements among w_M which are linearly independent mod $\mathfrak{P}\mathfrak{Y}'$. Then $\alpha_\nu(b_0) \equiv \sum_{i=1}^s w_i f_i(y) \mod \mathfrak{P}\mathfrak{Y}'[y]$ where $f_i(y) \in C[y]$ and not all $f_i(y)$ are zero. So let us assume $f_1(y) \neq 0$. Then we can find an integral linear function μ on $\mathfrak{c}$ such that $f_1(\mu) \neq 0$ ($f_1(\mu)$ is the value of the polynomial $f_1(y)$ at $y_j = \mu(\Gamma_j)$, $1 \leq j \leq r$). Put $\mathfrak{Y} = \mathfrak{X}\mathfrak{Y}' + \sum_{j=1}^r \mathfrak{X}(\Gamma_j - \mu(\Gamma_j))$. Then $\mathfrak{Y}$ is a left ideal in $\mathfrak{X}$. It would be sufficient to find a maximal left ideal $\mathfrak{M}$ in $\mathfrak{B}$ such that $b_0 \overline{\in} \mathfrak{M}$, $\mathfrak{M} \supset \mathfrak{Y}$, and dim $\mathfrak{B}/\mathfrak{M} < \infty$. First of all we claim that $b_0 \overline{\in} \mathfrak{B}\mathfrak{Y}$. For since $\Gamma_j \equiv \mu(\Gamma_j) \mod \mathfrak{Y}$ $(1 \leq j \leq r)$, $b_0 \equiv \alpha_\mu(b_0) \mod \mathfrak{B}\mathfrak{Y}$. Therefore if $b_0 \in \mathfrak{B}\mathfrak{Y}$, $\alpha_\mu(b_0) \in \mathfrak{B}\mathfrak{Y} \cap \mathfrak{P}\mathfrak{X}'$. But $\mathfrak{B}\mathfrak{Y} = \mathfrak{P}\mathfrak{Y}$ and therefore $\mathfrak{P}\mathfrak{Y} \cap \mathfrak{P}\mathfrak{X}' = \mathfrak{P}(\mathfrak{Y} \cap \mathfrak{X}') = \mathfrak{P}\mathfrak{Y}'$. (In order to prove these statements we have to make use of our earlier remarks about bases.) Hence $\alpha_\mu(b_0) \in \mathfrak{P}\mathfrak{Y}'$. But $\alpha_\mu(b_0) \equiv \sum_{i=1}^s f_i(\mu) w_i \mod \mathfrak{P}\mathfrak{Y}'$ and w_i $(1 \leq i \leq s)$ are linearly independent mod $\mathfrak{P}\mathfrak{Y}'$ and $f_1(\mu) \neq 0$. Therefore $\alpha_\mu(b_0) \overline{\in} \mathfrak{P}\mathfrak{Y}'$ and we get a contradiction. This shows that $b_0 \overline{\in} \mathfrak{B}\mathfrak{Y}$. Moreover since $\mathfrak{Y}' = \mathfrak{X}' \cap \mathfrak{Y}$ and $\Gamma \equiv \mu(\Gamma) \mod \mathfrak{Y}$ $(\Gamma \in \mathfrak{c})$ we may identify $\mathfrak{X}/\mathfrak{Y}$ and $\mathfrak{X}'/\mathfrak{Y}'$ in a natural way. Let ρ be the natural representation of $\mathfrak{k}$ on $\mathfrak{X}/\mathfrak{Y}$. Then ρ coincides on $\mathfrak{k}'$ with its natural representation ρ' on $\mathfrak{X}'/\mathfrak{Y}'$. Since the equivalence class of ρ' lies in $\Omega_{\mathfrak{k}'}$ it is clear that ρ' can be extended to a representation of K' on $\mathfrak{X}'/\mathfrak{Y}'$. Hence if we bear in mind the fact that the linear function μ is integral, we see easily that ρ can also be extended to a representation of K on $\mathfrak{X}/\mathfrak{Y}$. Now K is a closed[2] subgroup of the compact group G and therefore every finite-dimensional simple representation of K occurs in the reduction of some representation of G of the same type. This shows that the class $\mathfrak{D}$ of ρ occurs in the reduction of some finite-dimensional representation of $\mathfrak{g}$. Hence it is sufficient to prove the following lemma.

LEMMA 1. *Let $\mathfrak{Y}$ be a maximal left ideal in $\mathfrak{X}$ such that the natural representation of $\mathfrak{k}$ on $\mathfrak{X}/\mathfrak{Y}$ is finite-dimensional and its equivalence class occurs in the reduction of some finite-dimensional representation of $\mathfrak{g}$ with respect to $\mathfrak{k}$. Then*

$$\mathfrak{B}\mathfrak{Y} = \bigcap_{\mathfrak{M} \in F(\mathfrak{Y})} \mathfrak{M}$$

where $F(\mathfrak{Y})$ is the set of all left ideals $\mathfrak{M}$ in $\mathfrak{B}$ containing $\mathfrak{Y}$ for which dim $\mathfrak{B}/\mathfrak{M}$ $< \infty$.

The second step. We shall first prove this lemma in the special case when $\mathfrak{Y} = \mathfrak{X}\mathfrak{k}$. Let Σ denote the set of all finite-dimensional simple representations π of $\mathfrak{g}$ such that the zero representation of $\mathfrak{k}$ occurs in π. Let $\mathfrak{n}_\pi$ denote the kernel of π in $\mathfrak{g}$.

---

[2] This is seen as follows. We know that $\mathfrak{k}_0$ is the subalgebra of $\mathfrak{g}_k$ which consists of all points which are left fixed by an automorphism of $\mathfrak{g}_k$ of order two. K is therefore the component of identity of the subgroup of G consisting of the fixed points of the corresponding automorphism of G. Hence K is closed.

LEMMA 2. $\bigcap_{\pi \in \Sigma} \mathfrak{n}_\pi \subset \mathfrak{k}$.

Consider the compact groups G and K defined above and let V be the factor space G/K consisting of all cosets of the form xK $(x \in G)$. Then G operates on V in the usual fashion. Let $C(V)$ be the space of all continuous functions on V. For any $f \in C(V)$ and $x \in G$ define $\rho(x)f$ as the function $v \to f(x^{-1}v)$ $(v \in V)$ on V. Then the mapping $\rho : x \to \rho(x)$ is a representation of G on $C(V)$. Let $C_0(V)$ be the set of those elements $f \in C(V)$ whose transforms $\rho(x)f$ $(x \in G)$ span a finite-dimensional subspace. Then it is well known that every continuous function on V is the uniform limit of functions in $C_0(V)$. Let us denote by v_0 the coset K regarded as a point of V. Then if $x \notin K$, $x^{-1}v_0 \neq v_0$. Hence we can find a continuous function f on V such that $f(x^{-1}v_0) \neq f(v_0)$. Since f can be approximated uniformly by elements in $C_0(V)$ we may assume that f lies in $C_0(V)$. Let U be the linear space spanned by $\rho(y)f$ $(y \in G)$ and let ρ_U be the representation of G induced on U. Then ρ_U is a finite-dimensional continuous representation of G and since $\rho(x)f \neq f$, x does not lie in the kernel of ρ_U. G being compact, ρ_U is fully reducible and so we can find a simple component σ_x of ρ_U such that x is not in the kernel of σ_x. On the other hand it follows from the Frobenius reciprocity relation (see Weil [7, p. 83]) that the trivial representation of K occurs in σ_x. Now let $X \in \mathfrak{g}_k$, $X \notin \mathfrak{k}_0$. Then we can find a real number t such that $\exp t X \notin K$. Consider the representation σ_x corresponding to $x = \exp t X$. Then if π is the representation of $\mathfrak{g}_k$ (and therefore of $\mathfrak{g}$) corresponding to σ_x it is clear that $\pi \in \Sigma$ and $X \notin \mathfrak{n}_\pi$. This shows $\mathfrak{n} \cap \mathfrak{g}_k \subset \mathfrak{k}_0$ where $\mathfrak{n} = \bigcap_{\pi \in \Sigma} \mathfrak{n}_\pi$. Let η be the conjugation of $\mathfrak{g}$ with respect to $\mathfrak{g}_k$ so that $\eta(X + (-1)^{1/2} Y) = X - (-1)^{1/2} Y$ $(X, Y \in \mathfrak{g}_k)$. Let Σ_0 be the set of all matrix representations in Σ. Then it is clear that $\mathfrak{n} = \bigcap_{\pi \in \Sigma_0} \mathfrak{n}_\pi$. Now corresponding to any $\pi \in \Sigma_0$ define a representation $\bar\pi$ by the rule $\bar\pi(X) = \overline{\pi(\eta(X))}$ $(X \in \mathfrak{g})$ where $\overline{\pi(\eta(X))}$ is the complex-conjugate of the matrix $\pi(\eta(X))$. It is obvious that $\bar\pi \in \Sigma_0$, and $\mathfrak{n}_{\bar\pi} = \eta(\mathfrak{n}_\pi)$ and therefore $\eta(\mathfrak{n}) = \mathfrak{n}$. Hence if $X \in \mathfrak{n}$, $X + \eta(X)$ and $(-1)^{1/2}(X - \eta(X))$ are both in $\mathfrak{n} \cap \mathfrak{g}_k \subset \mathfrak{k}_0$. Therefore $X \in \mathfrak{k}$ and this proves the lemma.

Now let $\mathfrak{A}$ be the intersection of all left ideals in $F(\mathfrak{X}\mathfrak{k})$. We have to prove that $\mathfrak{A} = \mathfrak{B}\mathfrak{k}$. First we shall show that $\mathfrak{A} \cap \mathfrak{g} = \mathfrak{k}$. Since $\mathfrak{A} \supset \mathfrak{k}$ it is enough to prove that $\mathfrak{A} \cap \mathfrak{p} = \{0\}$. In order to do this we make use of an argument due to E. Cartan [1]. Let $\mathfrak{p}_1 = \mathfrak{A} \cap \mathfrak{p}$ and let $\mathfrak{p}_2$ be the set of all $Y \in \mathfrak{p}$ such that $B(X, Y) = \mathrm{sp}(\mathrm{ad}\, X\, \mathrm{ad}\, Y) = 0$ for all $X \in \mathfrak{p}_1$. Since the bilinear form B is nondegenerate on $\mathfrak{g}$ and since $\mathfrak{k}$ and $\mathfrak{p}$ are orthogonal under B it follows that B is nondegenerate on $\mathfrak{p}$ and therefore $\mathfrak{p} = \mathfrak{p}_1 + \mathfrak{p}_2$, $\mathfrak{p}_1 \cap \mathfrak{p}_2 = \{0\}$. Let $X \in \mathfrak{p}_1$, $Y \in \mathfrak{p}_2$, and $Z \in \mathfrak{k}$. Then $B([X, Y], Z) = -B(Y, [X, Z])$. Since $\mathfrak{A}$ is a left ideal containing $\mathfrak{k}$ and $[\mathfrak{k}, \mathfrak{p}] \subset \mathfrak{p}$ it is clear that $[\mathfrak{k}, \mathfrak{p}_1] \subset \mathfrak{p}_1$. Hence $[X, Z] \in \mathfrak{p}_1$ and therefore $B(Y, [X, Z]) = 0$. This means that $B([X, Y], Z) = 0$ for all $Z \in \mathfrak{k}$. On the other hand $[X, Y] \in [\mathfrak{p}, \mathfrak{p}] \subset \mathfrak{k}$ and since B is nondegenerate on $\mathfrak{k}$ it follows that $[X, Y] = 0$. Hence $[\mathfrak{p}_1, \mathfrak{p}_2] = \{0\}$. Moreover $[\mathfrak{k}, \mathfrak{p}_1] \subset \mathfrak{p}_1$ and therefore

$[\mathfrak{k}, \mathfrak{p}_2] \subset \mathfrak{p}_2$. Let I be the centraliser of $\mathfrak{p}_2$ in $\mathfrak{k}$. Then we claim that $\mathfrak{p}_1 + I$ is an ideal in $\mathfrak{g}$. Since $[\mathfrak{p}_1 + I, \mathfrak{p}_2] = \{0\}$ it follows that

$$[\mathfrak{p}_1 + I, \mathfrak{p}] = [\mathfrak{p}_1 + I, \mathfrak{p}_1] = [\mathfrak{p}_1, \mathfrak{p}_1] + [I, \mathfrak{p}_1].$$

But $[\mathfrak{p}_1, \mathfrak{p}_1] \subset \mathfrak{k}$ and $[\mathfrak{p}_2, [\mathfrak{p}_1, \mathfrak{p}_1]] = \{0\}$. Hence $[\mathfrak{p}_1, \mathfrak{p}_1] \subset I$. Furthermore we know that $[\mathfrak{k}, \mathfrak{p}_1] \subset \mathfrak{p}_1$. Therefore

$$[\mathfrak{p}_1 + I, \mathfrak{p}] \subset I + \mathfrak{p}_1.$$

On the other hand since $[\mathfrak{k}, \mathfrak{p}_2] \subset \mathfrak{p}_2$ it is obvious that $[\mathfrak{k}, I] \subset I$. Hence $[\mathfrak{p}_1 + I, \mathfrak{k}] \subset \mathfrak{p}_1 + I$. This proves that $\mathfrak{p}_1 + I$ is an ideal in $\mathfrak{g}$ and it is clear that $\mathfrak{A} \supset \mathfrak{p}_1 + I$.

Now suppose $\mathfrak{p}_1 \neq \{0\}$ and let $X \neq 0$ be an element in $\mathfrak{p}_1$. Then from Lemma 2 we can find a finite-dimensional simple representation π of $\mathfrak{g}$ such that $\pi(X) \neq 0$ and the zero representation of $\mathfrak{k}$ occurs in π. Choose a vector $\psi \neq 0$ in the representation space such that $\pi(Z)\psi = 0$ for all $Z \in \mathfrak{k}$. We extend π to a representation of $\mathfrak{B}$ and consider the set $\mathfrak{M}$ of all elements $b \in \mathfrak{B}$ such that $\pi(b)\psi = 0$. Then it is obvious that $\mathfrak{M} \in F(\mathfrak{X}\mathfrak{k})$ and therefore $\mathfrak{M} \supset \mathfrak{A} \supset \mathfrak{p}_1 + I$. Since π is irreducible and $\mathfrak{p}_1 + I$ is an ideal in $\mathfrak{g}$ and since $\pi(\mathfrak{p}_1 + I)\psi \subset \pi(\mathfrak{M})\psi = \{0\}$, it follows that $\pi(Y) = 0$ for all $Y \in \mathfrak{p}_1 + I$. Therefore in particular $\pi(X) = 0$. Since this contradicts our choice of π, $\mathfrak{p}_1 = \{0\}$.

Let $\mathfrak{p}'$ be the linear subspace of $\mathfrak{B}$ spanned by the set $(\mathfrak{p}, 1)$. We claim that $\mathfrak{p}' \cap \mathfrak{A} = \{0\}$. For suppose $c + Y \in \mathfrak{A}$ $(c \in C, Y \in \mathfrak{p})$. Then it follows from the above result that if $Y \neq 0$ we can find an irreducible representation π of $\mathfrak{g}$ on a finite-dimensional space V and a vector $\psi \neq 0$ in V such that $\pi(X)\psi = 0$ for all $X \in \mathfrak{k}$ and $\pi(Y)\psi \neq 0$. Consider the representation π_2 of $\mathfrak{g}$ (and therefore of $\mathfrak{B}$) induced on $V \times V$. Let $\mathfrak{M}_\pi$ and $\mathfrak{M}_{\pi_2}$ respectively be the set of those elements $b \in \mathfrak{B}$ for which $\pi(b)\psi = 0$ and $\pi_2(b)(\psi \times \psi) = 0$. Then it is obvious that $\mathfrak{M}_\pi$ and $\mathfrak{M}_{\pi_2}$ are both in $F(\mathfrak{X}\mathfrak{k})$ and therefore $\mathfrak{A} \subset \mathfrak{M}_\pi \cap \mathfrak{M}_{\pi_2}$. Now let $Y' = c + Y$. Then

$$0 = \pi_2(Y')(\psi \times \psi) = \pi(Y')\psi \times \psi + \psi \times \pi(Y')\psi - c(\psi \times \psi)$$
$$= - c(\psi \times \psi)$$

since $Y' \in \mathfrak{A}$. Hence $c = 0$ and therefore $Y \in \mathfrak{A} \cap \mathfrak{p} = \{0\}$. This proves that $\mathfrak{A} \cap \mathfrak{p}' = \{0\}$. Since dim $\mathfrak{p}'$ is finite it follows that we can find a finite set $\mathfrak{M}_1, \cdots, \mathfrak{M}_s$ of left ideals in $F(\mathfrak{X}\mathfrak{k})$ such that $\mathfrak{M}_1 \cap \cdots \cap \mathfrak{M}_s \cap \mathfrak{p}' = \{0\}$. Put $\mathfrak{M}_0 = \mathfrak{M}_1 \cap \cdots \cap \mathfrak{M}_s$. Then $\mathfrak{M}_0 \supset \mathfrak{X}\mathfrak{k}$ and dim $\mathfrak{B}/\mathfrak{M}_0 \leq \sum_{i=1}^{s}$ dim $\mathfrak{B}/\mathfrak{M}_i < \infty$. Hence $\mathfrak{M}_0 \in F(\mathfrak{X}\mathfrak{k})$.

We are now in a position to prove Lemma 1 in the case $\mathfrak{Y} = \mathfrak{X}\mathfrak{k}$. Choose a base $Y_1, \cdots, Y_p$ for $\mathfrak{p}$ and let Q be any monomial in $S(\mathfrak{p})$ constructed from the elements of this base. Then the elements $\lambda(Q)$ taken together for all Q form a base for $\mathfrak{P}$. Since $\mathfrak{B} = \mathfrak{P}\mathfrak{X}$ it follows that $\mathfrak{B} = \mathfrak{P} + \mathfrak{B}\mathfrak{k}$. Now let b_0 be an element of $\mathfrak{B}$ which is not in $\mathfrak{B}\mathfrak{k}$. Then $b_0 \equiv w \bmod \mathfrak{B}\mathfrak{k}$ where $w \in \mathfrak{P}$ and $w \neq 0$. Since w is a linear combination of $\lambda(Q)$ it could be written in the form

$$w = \sum_{0 \leq r \leq M} \sum_{1 \leq i_1, \cdots, i_r \leq p} a^{i_1 i_2 \cdots i_r} Y_{i_1} Y_{i_2} \cdots Y_{i_r} \qquad (a^{i_1 i_2 \cdots i_r} \in C)$$

where the a's are symmetric, i.e. $a^{i_1 i_2 \cdots i_r} = a^{i_{j_1} i_{j_2} \cdots i_{j_r}}$ for any permutation $(j_1, \cdots, j_r)$ of $(1, 2, \cdots, r)$. We choose M in such a way that some $a^{i_1 i_2 \cdots i_M} \neq 0$. Let $b \to b^*$ denote the natural mapping and π the natural representation of $\mathfrak{B}$ on $\mathfrak{B}/\mathfrak{M}_0 = \mathfrak{B}^*$. We know that $\mathfrak{M}_0 \supset \mathfrak{k}$ and dim $\mathfrak{B}^* < \infty$. Let $\mathfrak{B}_M^*$ denote the Kronecker product of $\mathfrak{B}^*$ with itself M times. Since $\mathfrak{M}_0 \cap \mathfrak{p}' = \{0\}$ it follows that $1^*, Y_1^*, \cdots, Y_p^*$ are linearly independent in $\mathfrak{B}^*$. Choose a base z_j^*, $0 \leq j \leq h$, for $\mathfrak{B}^*$ such that $z_0^* = 1^*$ and $z_i^* = Y_i^*$, $1 \leq i \leq p$, and let E_M^* be the subspace of $\mathfrak{B}_M^*$ spanned by elements of the form $z_{i_1}^* \times z_{i_2}^* \times \cdots \times z_{i_M}^*$ $(0 \leq i_1, \cdots, i_M \leq h)$ where $i_r = 0$ for at least one r $(1 \leq r \leq M)$. Let π_M denote the representation of $\mathfrak{g}$ (and therefore of $\mathfrak{B}$) induced on the Kronecker product $\mathfrak{B}_M^*$. Then it is easy to verify (see [2, p. 904]) that

$$\pi_M(w)(1^* \times 1^* \times \cdots \times 1^*)$$

$$\equiv M! \sum_{1 \leq i_1, \cdots, i_M \leq p} a^{i_1 i_2 \cdots i_M}(Y_{i_1}^* \times \cdots \times Y_{i_M}^*) \bmod E_M^*.$$

Since $z_{j_1}^* \times z_{j_2}^* \times \cdots \times z_{j_M}^*$ $(0 \leq j_1, \cdots, j_M \leq h)$ is a base for $\mathfrak{B}_M^*$, the elements $Y_{i_1}^* \times Y_{i_2}^* \times \cdots \times Y_{i_M}^*$ $(1 \leq i_1, \cdots, i_M \leq p)$ are linearly independent mod E_M^*. Since some $a^{i_1 i_2 \cdots i_M} \neq 0$ it follows that the right-hand side is not congruent to zero mod E_M^* and therefore $\pi_M(w)$ $(1^* \times 1^* \times \cdots \times 1^*) \neq 0$. On the other hand since $\mathfrak{M}_0 \supset \mathfrak{k}$ it is clear that $\pi_M(X)$ $(1^* \times 1^* \times \cdots \times 1^*) = 0$ for all $X \in \mathfrak{k}$. Therefore $\pi_M(b_0)$ $(1^* \times 1^* \times \cdots \times 1^*) = \pi_M(w)$ $(1^* \times \cdots \times 1^*) \neq 0$. Let $\mathfrak{M}_M$ be the set of all elements $b \in \mathfrak{B}$ such that $\pi_M(b)$ $(1^* \times \cdots \times 1^*) = 0$. Then $b_0 \notin \mathfrak{M}_M$ and $\mathfrak{M}_M \in F(\mathfrak{X}\mathfrak{k})$ since dim $\mathfrak{B}/\mathfrak{M}_M \leq$ dim $\mathfrak{B}_M^* < \infty$. This shows that $b_0 \notin \mathfrak{A}$ and therefore we conclude that $\mathfrak{A} = \mathfrak{B}\mathfrak{k}$.

The final step. Now we come to the general case of Lemma 1. We recall that $\mathfrak{P}$ is the image of $S(\mathfrak{p})$ under λ. First we prove the following lemma.

LEMMA 3. *Let $w_1, \cdots, w_r$ be a finite set of elements in $\mathfrak{P}$ which are linearly independent. Then there exists a left ideal $\mathfrak{M} \in F(\mathfrak{X}\mathfrak{k})$ such that $w_1, \cdots, w_r$ are linearly independent* modulo $\mathfrak{M}$.

Let W be the space spanned by $w_1, \cdots, w_r$. Then $W \cap \mathfrak{B}\mathfrak{k} \subset \mathfrak{P} \cap \mathfrak{B}\mathfrak{k} = \mathfrak{P} \cap \mathfrak{P}\mathfrak{X}\mathfrak{k}$. Since λ is a linear isomorphism of $S(\mathfrak{p}) \times S(\mathfrak{k})$ onto $\mathfrak{B}$ it follows that $\mathfrak{P} \cap \mathfrak{P}\mathfrak{X}\mathfrak{k} = \{0\}$ and therefore $W \cap \mathfrak{B}\mathfrak{k} = \{0\}$. Since dim $W < \infty$ it follows from what we have proved above that we can find a finite set of elements $\mathfrak{M}_1, \cdots, \mathfrak{M}_s$ in $F(\mathfrak{X}\mathfrak{k})$ such that $W \cap \mathfrak{M}_1 \cap \cdots \cap \mathfrak{M}_s = \{0\}$. Put $\mathfrak{M} = \mathfrak{M}_1 \cap \cdots \cap \mathfrak{M}_s$. Then $\mathfrak{M} \supset \mathfrak{k}$ and dim $\mathfrak{B}/\mathfrak{M} \leq \sum_{1 \leq i \leq s}$ dim $\mathfrak{B}/\mathfrak{M}_i < \infty$ and therefore $\mathfrak{M} \in F(\mathfrak{X}\mathfrak{k})$. Since $\mathfrak{M} \cap W = \{0\}$ our assertion is proved.

Let $\mathfrak{D}_0$ be the (equivalence) class of the natural representation of $\mathfrak{k}$ on $\mathfrak{X}/\mathfrak{Y}$ in Lemma 1. Then we can find an irreducible representation σ of $\mathfrak{g}$ (and therefore of $\mathfrak{B}$) on a finite-dimensional space V and a vector $\psi \neq 0$ in V such that $\sigma(\mathfrak{Y})\psi = \{0\}$. Let $\mathfrak{B}_1$ be the set of all elements $b \in \mathfrak{B}$ such that $\sigma(b)\psi = 0$. Then $\mathfrak{B}_1 \supset \mathfrak{Y}$ and we may regard σ as the natural representation of $\mathfrak{B}$ on $\mathfrak{B}/\mathfrak{B}_1$.

Let a be an element in $\mathfrak{B}$ which is not in $\mathfrak{B}\mathfrak{Y} = \mathfrak{P}\mathfrak{Y}$. Since $\lambda(Q)$ form a base for $\mathfrak{P}$ we can write a in the form $a = \sum_Q \lambda(Q)x_Q$ where $x_Q \in \mathfrak{X}$ and not all x_Q are in $\mathfrak{Y}$. Let $|Q|$ denote the degree of the monomial Q and let M be the highest integer for which there exists a monomial Q_0 with degree M such that $x_{Q_0} \notin \mathfrak{Y}$. From Lemma 3 there exists a left ideal $\mathfrak{B} \in F(\mathfrak{X}\mathfrak{k})$ such that the elements $\lambda(Q)$ $(|Q| \leq M)$ are linearly independent mod $\mathfrak{B}$. Let $b \to b^*$ denote the natural mapping and π the natural representation of $\mathfrak{B}$ on $\mathfrak{B}^* = \mathfrak{B}/\mathfrak{B}$. Similarly let $b \to \bar{b}$ denote the natural mapping of $\mathfrak{B}$ on $\overline{\mathfrak{B}} = \mathfrak{B}/\mathfrak{B}_1$. Consider the representation ν of $\mathfrak{g}$ induced on $\mathfrak{B}^* \times \overline{\mathfrak{B}}$. Then it is obvious that (see [5, Lemma 13])

$$\nu(a)(1^* \times \bar{1}) = \sum_{|Q|=M} (\lambda(Q))^* \times \bar{x}_Q + \sum_{|Q|<M} (\lambda(Q))^* \times \bar{b}_Q \quad (b_Q \in \mathfrak{B}).$$

Since $(\lambda(Q))^*$ $(|Q| \leq M)$ are linearly independent and since not all $\bar{x}_Q$ $(|Q| = M)$ are zero it follows that $\nu(a)(1^* \times \bar{1}) \neq 0$. Let $\mathfrak{M}$ be the set of all $b \in \mathfrak{B}$ for which $\nu(b)(1^* \times \bar{1}) = 0$. Since $\pi(X)1^* = 0$ $(X \in \mathfrak{k})$ it follows that $\mathfrak{M} \supset \mathfrak{Y}$. Moreover dim $\mathfrak{B}/\mathfrak{M} \leq$ dim $(\mathfrak{B}^* \times \overline{\mathfrak{B}}) < \infty$ and therefore $\mathfrak{M} \in F(\mathfrak{Y})$. Since $a \notin \mathfrak{M}$, Lemma 1 and therefore Theorem 1 is proved.

2. **Proof of Theorem 2.** Let $\mathfrak{A}$ be an associative algebra with 1 and let $\mathfrak{M}_1$, $\mathfrak{M}_2$ be two left ideals in $\mathfrak{A}$. Let π_1, π_2 be the natural representations of $\mathfrak{A}$ on $\mathfrak{A}/\mathfrak{M}_1$ and $\mathfrak{A}/\mathfrak{M}_2$ respectively. We shall say that $\mathfrak{M}_1$, $\mathfrak{M}_2$ are equivalent if there exists a linear *isomorphism* α of $\mathfrak{A}/\mathfrak{M}_1$ onto $\mathfrak{A}/\mathfrak{M}_2$ such that $\pi_2(a)\alpha = \alpha\pi_1(a)$ $(a \in \mathfrak{A})$.

Let $\mathfrak{Q}$ be the centraliser of $\mathfrak{k}$ in $\mathfrak{B}$ and $\mathfrak{Z}$ the center of $\mathfrak{B}$. If $\chi \neq 0$ is a homomorphism of $\mathfrak{Z}$ into C we denote by $\mathfrak{Z}_\chi$ the kernel of χ in $\mathfrak{Z}$. Similarly for any $\mathfrak{D}_0 \in \Omega$ we denote by $\mathfrak{N}_{\mathfrak{D}_0}$ the kernel (in $\mathfrak{X}$) of any representation of $\mathfrak{X}$ which lies in $\mathfrak{D}_0$.

THEOREM 2. *Let $M(\mathfrak{D}_0, \chi)$ be the set of all maximal left ideals in $\mathfrak{B}$ which contain $\mathfrak{N}_{\mathfrak{D}_0} + \mathfrak{Z}_\chi$. Then for any $\mathfrak{M} \in M(\mathfrak{D}_0, \chi)$, $\mathfrak{M} \cap (\mathfrak{Q}\mathfrak{X})$ is a maximal left ideal in $\mathfrak{Q}\mathfrak{X}$. Two elements $\mathfrak{M}_1$, $\mathfrak{M}_2$ in $M(\mathfrak{D}_0, \chi)$ are equivalent in $\mathfrak{B}$ if and only if $\mathfrak{M}_1 \cap \mathfrak{Q}\mathfrak{X}$ and $\mathfrak{M}_2 \cap \mathfrak{Q}\mathfrak{X}$ are equivalent in $\mathfrak{Q}\mathfrak{X}$.*

First we shall prove the following simple lemma.

LEMMA 4. *Let π be a quasi semisimple representation of $\mathfrak{X}$ on a space V. Suppose $\mathfrak{D}_1, \cdots, \mathfrak{D}_N$ are all the distinct classes in Ω which occur in π and let $V_{\mathfrak{D}_j}$ be the set of those elements in V which transform under π according to $\mathfrak{D}_j$. Then there exist elements $e_1, \cdots, e_N$ in $\mathfrak{X}$ such that $\pi(e_i)\psi = \psi$ or 0 according as $\psi \in V_{\mathfrak{D}_i}$ or $\psi \in V_{\mathfrak{D}_j}, j \neq i$ $(1 \leq i \leq N)$.*

Let π_j be the representation of $\mathfrak{X}$ induced on $V_{\mathfrak{D}_j}$ $(1 \leq j \leq N)$ and let $\mathfrak{N}_j$ be the kernel of π_j in $\mathfrak{X}$. Since the class $\mathfrak{D}_j$ is irreducible, $\mathfrak{N}_j$ is a maximal two-sided ideal in $\mathfrak{X}$ and $\mathfrak{N}_i \neq \mathfrak{N}_j$ if $i \neq j$. Put $\mathfrak{N} = \cap_{j=1}^{N} \mathfrak{N}_j$. Then $\mathfrak{N}$ is the kernel of π and since $\mathfrak{D}_1, \cdots, \mathfrak{D}_N$ are all the distinct classes which occur in π it is clear that $\mathfrak{N}_1, \cdots, \mathfrak{N}_N$ are the only maximal ideals in $\mathfrak{X}$ which contain $\mathfrak{N}$. It is obvious that the algebra $\mathfrak{X}^* = \mathfrak{X}/\mathfrak{N}$ is finite-dimensional and semisimple and $\mathfrak{Y}_j^* = \mathfrak{N}_j/\mathfrak{N}$, $1 \le j \le N$, are all the distinct maximal ideals in $\mathfrak{X}^*$. Let $\mathfrak{X}_j^*$ be the ideal complementary to $\mathfrak{Y}_j^*$. Then it follows that $\mathfrak{X}_1^*, \cdots, \mathfrak{X}_N^*$ is a complete set of distinct minimal ideals in $\mathfrak{X}^*$. Now since $\mathfrak{N}$ is the kernel of π, we may regard π as a faithful representation of $\mathfrak{X}^*$ on V. Since $\mathfrak{Y}_j^*\mathfrak{X}_j^* = \{0\}$ it follows that $\pi(\mathfrak{N}_j)\pi(\mathfrak{X}_j^*)V = \{0\}$ and therefore $\pi(\mathfrak{X}_j^*)V \subset V_{\mathfrak{D}_j}$. Moreover $V = \pi(\mathfrak{X}^*)V = \sum_{j=1}^{N}\pi(\mathfrak{X}_j^*)V$. Since the sum $\sum_j V_{\mathfrak{D}_j}$ is direct we conclude that $\pi(\mathfrak{X}_j^*)V = V_{\mathfrak{D}_j}$. Let e_j^* denote the unit element of the simple algebra $\mathfrak{X}_j^*$. Then $e_i^*e_j^* = 0$ if $i \ne j$. Select an element $e_j \in \mathfrak{X}$ such that its residue class mod $\mathfrak{N}$ is e_j^*. Then $\pi(e_j)V = \pi(e_j^*)V = V_{\mathfrak{D}_j}$. Therefore if $\psi \in V_{\mathfrak{D}_j}$, $\psi = \pi(e_j)\phi$ ($\phi \in V$). Hence $\pi(e_i)\psi = \pi(e_i^*e_j^*)\phi = \delta_{ij}\pi(e_j^*)\phi = \delta_{ij}\psi$. This proves the lemma.

Now we return to Theorem 2. Put $M = M(\mathfrak{D}_0, \chi)$ and $\mathfrak{N} = \mathfrak{N}_{\mathfrak{D}_0}$ for brevity. Let $b \to b^*$ denote the natural mapping and π the natural representation of $\mathfrak{B}$ on $\mathfrak{B}^* = \mathfrak{B}/(\mathfrak{B}\mathfrak{N} + \mathfrak{B}\mathfrak{Z}_\chi)$. Let A be the set of all $a \in \mathfrak{B}$ such that $\mathfrak{N}a \subset \mathfrak{B}\mathfrak{N} + \mathfrak{B}\mathfrak{Z}_\chi$. Then A is a subalgebra of $\mathfrak{B}$ and $A \supset \mathfrak{Q}\mathfrak{X} + \mathfrak{B}\mathfrak{N} + \mathfrak{B}\mathfrak{Z}_\chi$. Let $\mathfrak{B}_{\mathfrak{D}}^*$ denote the set of all elements $b^* \in \mathfrak{B}^*$ which transform under $\pi(\mathfrak{k})$ according to $\mathfrak{D}$ ($\mathfrak{D} \in \Omega$). Then it is obvious that $\mathfrak{B}_{\mathfrak{D}_0}^* = A^* = A/\mathfrak{B}\mathfrak{N} + \mathfrak{B}\mathfrak{Z}_\chi$.

Let $\mathfrak{M} \in M$. Denote by $b \to \bar{b}$ the natural mapping and by $\bar{\pi}$ the natural representation of $\mathfrak{B}$ on $\overline{\mathfrak{B}} = \mathfrak{B}/\mathfrak{M}$. Since $\mathfrak{M} \supset \mathfrak{B}\mathfrak{N} + \mathfrak{B}\mathfrak{Z}_\chi$ we may identify $\overline{\mathfrak{B}}$ with $\mathfrak{B}^*/\mathfrak{M}^*$ in a natural way ($\mathfrak{M}^* = \mathfrak{M}/\mathfrak{B}\mathfrak{N} + \mathfrak{B}\mathfrak{Z}_\chi$). Let $b^* \to \bar{b}^*$ denote the natural mapping of $\mathfrak{B}^*$ on $\overline{\mathfrak{B}} = \mathfrak{B}^*/\mathfrak{M}^*$. Then if $\overline{\mathfrak{B}}_{\mathfrak{D}}$ is the set of all elements in $\overline{\mathfrak{B}}$ which transform under $\bar{\pi}(\mathfrak{k})$ according to $\mathfrak{D}$ ($\mathfrak{D} \in \Omega$), it is obvious that $\overline{\mathfrak{B}}_{\mathfrak{D}} = \overline{\mathfrak{B}_{\mathfrak{D}}^*}$. Hence $\overline{\mathfrak{B}}_{\mathfrak{D}_0} = \overline{A^*} = \overline{A}$. *We shall now show that $\overline{A}$ is irreducible under* $\bar{\pi}(A)$. Let a_0, a_1 be two elements in A such that $\bar{a}_0 \ne 0$. Since $\mathfrak{M}$ is maximal, $\overline{\mathfrak{B}}$ is irreducible under $\bar{\pi}(\mathfrak{B})$ and therefore there exists an element $b_0 \in \mathfrak{B}$ such that $\bar{\pi}(b_0)\bar{a}_0 = \bar{a}_1$. Since $\mathfrak{B}^* = \sum_{\mathfrak{D} \in \Omega} \mathfrak{B}_{\mathfrak{D}}^*$ and $\dim \mathfrak{B}_{\mathfrak{D}}^* < \infty$ (see Theorem 1 of [5]), it follows that there exists a finite number of distinct elements $\mathfrak{D}_1, \cdots, \mathfrak{D}_N \in \Omega$ ($\mathfrak{D}_j \ne \mathfrak{D}_0$, $1 \le j \le N$) such that $\pi(b_0)\mathfrak{B}_{\mathfrak{D}_0}^* \subset \sum_{0 \le j \le N} \mathfrak{B}_{\mathfrak{D}_j}^*$. From Lemma 4 we can find an element $e \in \mathfrak{X}$ such that $\pi(e)a^* = a^*$ or 0 according as $a^* \in \mathfrak{B}_{\mathfrak{D}_0}^*$ or $a^* \in \mathfrak{B}_{\mathfrak{D}_j}^*$, $1 \le j \le N$. Then if we put $b = eb_0$, $\pi(b)\mathfrak{B}_{\mathfrak{D}_0}^* \subset \mathfrak{B}_{\mathfrak{D}_0}^*$ and $\bar{\pi}(b)\bar{a}_0 = \bar{\pi}(e)\bar{a}_1 = \bar{a}_1$. Since $\pi(b)\mathfrak{B}_{\mathfrak{D}_0}^* \subset \mathfrak{B}_{\mathfrak{D}_0}^*$, $b \in A$ and hence $\bar{a}_1 \in \bar{\pi}(A)\bar{a}_0$. This shows that $\overline{A}$ is irreducible under $\bar{\pi}(A)$. Since $1 \notin \mathfrak{M}$, $\mathfrak{M} \cap A \ne A$ and therefore $A \cap \mathfrak{M}$ is maximal in A.

Let M_A be the set of all maximal ideals in A which contain $\mathfrak{B}\mathfrak{N} + \mathfrak{B}\mathfrak{Z}_\chi$. We have just seen that if $\mathfrak{M} \in M$ then $\mathfrak{M} \cap A \in M_A$. *We shall now show that the mapping* $\mathfrak{M} \to \mathfrak{M} \cap A$ ($\mathfrak{M} \in M$) *is a 1-1 mapping of M onto M_A.* (This is actually more than we need for Theorem 2.) Let $\mathfrak{M}$ and $\mathfrak{M}'$ be two distinct elements in M. Since they are both maximal, $\mathfrak{M} + \mathfrak{M}' = \mathfrak{B}$ and therefore we can select $b \in \mathfrak{M}$ and $b' \in \mathfrak{M}'$ such that $b + b' = 1$. Choose a finite set of distinct classes $\mathfrak{D}_1, \cdots, \mathfrak{D}_N$ in Ω ($\mathfrak{D}_0 \ne \mathfrak{D}_j$, $1 \le j \le N$) such that $\pi(b)\mathfrak{B}_{\mathfrak{D}_0}^*$ and $\pi(b')\mathfrak{B}_{\mathfrak{D}}^*$ are both contained in $\sum_{j=0}^{N} \mathfrak{B}_{\mathfrak{D}_j}^*$. In accordance with Lemma 4 we can pick an element $e \in \mathfrak{X}$ such that $\pi(e)\mathfrak{B}_{\mathfrak{D}_j}^* = \{0\}$, $1 \leq j \leq N$, and $\pi(e)a^* = a^*$ if $a^* \in \mathfrak{B}_{\mathfrak{D}_0}^*$. Then if $a = eb$ and $a' = eb'$, $a^* + (a')^* = \pi(e)1^* = 1^*$ since $1^* \in \mathfrak{B}_{\mathfrak{D}_0}^*$. It is clear that $a \in \mathfrak{M} \cap A$ and $a' \in \mathfrak{M}' \cap A$. Since $\mathfrak{M} \cap \mathfrak{M}' \cap A \supset \mathfrak{B}\mathfrak{N} + \mathfrak{B}\mathfrak{Z}_x$ it follows that $\mathfrak{M} \cap A + \mathfrak{M}' \cap A = A$ and therefore $\mathfrak{M} \cap A \neq \mathfrak{M}' \cap A$. Now let B be any element in M_A. Put $\mathfrak{M}_0 = \mathfrak{B}B$. We claim $\mathfrak{M}_0 \neq \mathfrak{B}$. For otherwise $\mathfrak{B}^* = \pi(\mathfrak{B})B^*$ where B^* is the image of B in $\mathfrak{B}^*$. Hence $1^* = \sum_{i=1}^r \pi(b_i)u_i^*$ ($b_i \in \mathfrak{B}$, $u_i^* \in B^*$). Again we may choose a finite set of classes $\mathfrak{D}_1, \cdots, \mathfrak{D}_N$ in Ω ($\mathfrak{D}_j \neq \mathfrak{D}_0$, $1 \leq j \leq N$) such that $\pi(b_i)\mathfrak{B}_{\mathfrak{D}_0}^* \subset \sum_{0 \leq j \leq N} \mathfrak{B}_{\mathfrak{D}_j}^*$ ($1 \leq i \leq r$). Now select $e \in \mathfrak{X}$ such that $\pi(e)\mathfrak{B}_{\mathfrak{D}_j}^* = \{0\}$, $1 \leq j \leq N$, and $\pi(e)a^* = a^*$ if $a^* \in \mathfrak{B}_{\mathfrak{D}_0}^*$. Then

$$1^* = \pi(e)1^* = \sum_{i=1}^r \pi(eb_i)u_i^*.$$

Put $eb_i = a_i$ and let u_i be an element in B such that its image in B^* is u_i^*. Then since $\pi(a_i)\mathfrak{B}_{\mathfrak{D}_0}^* \subset \mathfrak{B}_{\mathfrak{D}_0}^*$, $a_i \in A$, and therefore $\sum_{i=1}^r a_i u_i \in B$. On the other hand

$$1^* = \left(\sum_{i=1}^r a_i u_i \right)^*$$

and $B \supset \mathfrak{B}\mathfrak{N} + \mathfrak{B}\mathfrak{Z}_x$. Therefore $B = A$ and we get a contradiction with the fact that $B \in M_A$. Hence $\mathfrak{M}_0 \neq \mathfrak{B}$ and therefore by Zorn's lemma we can find a maximal left ideal $\mathfrak{M}$ in $\mathfrak{B}$ which contains $\mathfrak{M}_0$. Then $\mathfrak{M} \in M$ and $\mathfrak{M} \cap A$ is a maximal left ideal in A containing B. Since B is maximal, $\mathfrak{M} \cap A = B$. This proves that the mapping $\mathfrak{M} \to \mathfrak{M} \cap A$ ($\mathfrak{M} \in M$) is a 1-1 mapping of M onto M_A.

Now suppose $\mathfrak{M}_1$, $\mathfrak{M}_2$ are two elements in M. Put $B_i = \mathfrak{M}_i \cap A$ ($i = 1, 2$). *We shall prove that $\mathfrak{M}_1$ and $\mathfrak{M}_2$ are equivalent in $\mathfrak{B}$ if and only if B_1, B_2 are equivalent in A.* Suppose $\mathfrak{M}_1$ and $\mathfrak{M}_2$ are equivalent. Then we can pick an element $b_0 \in \mathfrak{B}$ such that $bb_0 \in \mathfrak{M}_2$ ($b \in \mathfrak{B}$) if and only if $b \in \mathfrak{M}_1$. Choose $\mathfrak{D}_1, \cdots, \mathfrak{D}_N \in \Omega$ ($\mathfrak{D}_j \neq \mathfrak{D}_0$, $1 \leq j \leq N$) such that $(b_0)^* \in \sum_{0 \leq j \leq N} \mathfrak{B}_{\mathfrak{D}_j}^*$ and let e be an element in $\mathfrak{X}$ such that $\pi(e-1)\mathfrak{B}_{\mathfrak{D}_0}^* = \{0\}$ and $\pi(e)\mathfrak{B}_{\mathfrak{D}_j}^* = \{0\}$ ($1 \leq j \leq N$). Then if we put $a = eb_0$, $a \in A$. Moreover it is clear that $1 - e \in \mathfrak{N} \subset \mathfrak{M}_1$ and therefore $b_0 - a = (1-e)b_0 \in \mathfrak{M}_2$. Therefore $ba \in \mathfrak{M}_2$ ($b \in \mathfrak{B}$) if and only if $b \in \mathfrak{M}_1$. Hence $ba \in B_2$ ($b \in A$) if and only if $b \in B_1$. This proves that B_1 and B_2 are equivalent in A. Conversely let B_1, B_2 be two elements in M_A which are equivalent in A. Then we can select an $a \in A$ such that $ba \in B_2$ ($b \in A$) if and only if $b \in B_1$. It is obvious that $a \notin B_2$. Let $\mathfrak{M}_1$, $\mathfrak{M}_2$ be the left ideals in M such that $B_i = \mathfrak{M}_i \cap A$ ($i = 1, 2$) and let $\mathfrak{M}_1'$ be the set of all elements b in $\mathfrak{B}$ for which $ba \in \mathfrak{M}_2$. Since $\mathfrak{M}_2$ is maximal the natural representation of $\mathfrak{B}$ on $\mathfrak{B}/\mathfrak{M}_2$ is irreducible and therefore since $a \notin \mathfrak{M}_2$ it is clear that $\mathfrak{M}_1'$ is a maximal left ideal equivalent to $\mathfrak{M}_2$. Moreover since $a \in A$, $a^* \in \mathfrak{B}_{\mathfrak{D}_0}^*$ and therefore $\mathfrak{M}_1' \supset \mathfrak{B}\mathfrak{N} + \mathfrak{B}\mathfrak{Z}_x$. Hence $\mathfrak{M}_1' \in M$ and $\mathfrak{M}_1' \cap A \supset B_1$. Since B_1 is maximal in A, $B_1 = \mathfrak{M}_1' \cap A$ and therefore $\mathfrak{M}_1' = \mathfrak{M}_1$ in view of the result above. Hence $\mathfrak{M}_1$

and $\mathfrak{M}_2$ are equivalent and our assertion is proved.

Now let $\mathfrak{A} = \mathfrak{Q}\mathfrak{X}$. Then $A \supset \mathfrak{A}$. We shall prove that if $\mathfrak{M} \in M$, $\mathfrak{M} \cap \mathfrak{A}$ is maximal in $\mathfrak{A}$. Let $a \to \bar{a}$ be the natural mapping and ν the natural representation of A on $A/\mathfrak{M} \cap A$. We know that dim $\bar{A} \leqq \dim \mathfrak{B}_{\mathfrak{D}_0}^* < \infty$ and ν is irreducible. In order to prove that $\mathfrak{M} \cap \mathfrak{A}$ is maximal in $\mathfrak{A}$ it is sufficient to prove that $\bar{A}$ is irreducible under $\nu(\mathfrak{A})$. Since $\mathfrak{A} \supset \mathfrak{X}$ and $\bar{A}$ is semisimple under $\nu(\mathfrak{X})$ it would be enough to show that if U and V are any two subspaces of $\mathfrak{A}$ which are stable and simple under $\nu(\mathfrak{X})$, then $V \subset \nu(\mathfrak{A}) U$. It is obvious that the representations of $\mathfrak{X}$ induced on V and U lie in $\mathfrak{D}_0$. Hence there exists a linear isomorphism α of U with V such that $\nu(z)\alpha = \alpha\nu(z)$ $(z \in \mathfrak{X})$. Since $\nu(A)$ is irreducible it follows from Burnside's Theorem that we can find an element $a \in A$ such that $\nu(a)u = \alpha u$ for all $u \in U$. Then $\nu(za)u = \nu(z)\alpha u = \alpha\nu(z)u = \nu(az)u$ $(u \in U, z \in \mathfrak{X})$. For any $X \in \mathfrak{k}$ let $\rho(X)$ denote the mapping $b \to [X, b]$ $(b \in A)$ of A into itself. It is obvious that ρ is a quasi semisimple representation of $\mathfrak{k}$ (and therefore of $\mathfrak{X}$) on A. Let B be the set of all $b \in A$ such that $\nu(b)u = 0$ for all $u \in U$. Then $\rho(X)a \in B$ $(X \in \mathfrak{k})$ and it is obvious that B is a left ideal in A and $B\mathfrak{X} \subset B$. Hence B is stable under $\rho(\mathfrak{X})$ and therefore $\rho(\mathfrak{X}\mathfrak{k})a = \rho(\mathfrak{k}\mathfrak{X})a \subset B$. Put $W = \rho(\mathfrak{X})a$. Then $\mathfrak{Q} \cap W$ is the set of all elements $w \in W$ such that $\rho(X)w = 0$ $(X \in \mathfrak{k})$. Hence, from Lemma 7 of [5], $W = \mathfrak{Q} \cap W + \rho(\mathfrak{k})W$. In particular $a \equiv q \bmod (\rho(\mathfrak{k})W)$ where $q \in \mathfrak{Q} \cap W$. Since $\rho(\mathfrak{k})W \subset B$, $\nu(a)u = \nu(q)u$ for all $u \in U$. Hence $V = \nu(a)U = \nu(q)U \subset \nu(\mathfrak{A})U$ and our assertion is proved.

REMARK. We note for later use the fact that $\bar{A} = \nu(\mathfrak{A})\bar{1}$ which follows from the irreducibility of $\bar{A}$ under $\nu(\mathfrak{A})$. If $\bar{\mathfrak{A}}$ is the image of $\mathfrak{A}$ in $\bar{\mathfrak{B}} = \mathfrak{B}/\mathfrak{M}$ under the natural mapping of $\mathfrak{B}$ on $\bar{\mathfrak{B}}$, then $\bar{\mathfrak{A}} = \bar{\mathfrak{B}}_{\mathfrak{D}_0} = \bar{A}$.

Let $\mathfrak{M}'$ be another ideal in M and let ν' denote the natural representation of A on $A/\mathfrak{M}' \cap A$. If $\mathfrak{M}$ and $\mathfrak{M}'$ are equivalent in $\mathfrak{B}$, the representations ν and ν' of A are equivalent. Therefore the representations of $\mathfrak{A}$ defined by ν and ν' are also equivalent and this proves that $\mathfrak{M} \cap \mathfrak{A}$ and $\mathfrak{M}' \cap \mathfrak{A}$ are equivalent in $\mathfrak{A}$. Conversely suppose $\mathfrak{M} \cap \mathfrak{A}$ and $\mathfrak{M}' \cap \mathfrak{A}$ are equivalent in $\mathfrak{A}$. Then $\mathrm{sp}\,\nu(q) = \mathrm{sp}\,\nu'(q)$ for all $q \in \mathfrak{A}$. Let a be any element in A. Define $W = \rho(\mathfrak{X})a$ as above. Then $W = W \cap \mathfrak{Q} + \rho(\mathfrak{k})W$ and therefore $a \equiv q \bmod \rho(\mathfrak{k})W$ where $q \in \mathfrak{Q}$. Now $\mathrm{sp}(\nu([X, b])) = \mathrm{sp}(\nu(Xb - bX)) = 0$ $(X \in \mathfrak{k}, b \in A)$. Hence $\mathrm{sp}\,\nu(a) = \mathrm{sp}\,\nu(q)$. Similarly $\mathrm{sp}\,\nu'(a) = \mathrm{sp}\,\nu'(q)$. Therefore $\mathrm{sp}\,\nu(a) = \mathrm{sp}\,\nu'(a)$ for all $a \in A$ and then it follows immediately from the theory of semisimple representations of an associative algebra (see Appendix, Lemma 16) that ν and ν' are equivalent. Hence $\mathfrak{M} \cap A$ and $\mathfrak{M}' \cap A$ are equivalent in A. In view of the result proved above this implies that $\mathfrak{M}$ and $\mathfrak{M}'$ are equivalent in $\mathfrak{B}$. Hence Theorem 2 is established.

We have established above a 1-1 equivalence-preserving correspondence between elements of M and the maximal left ideals in $A^* = A/\mathfrak{B}\mathfrak{M} + \mathfrak{B}\mathfrak{Z}_x$. (We observe that $\mathfrak{B}\mathfrak{M} + \mathfrak{B}\mathfrak{Z}_x$ is a two-sided ideal in A and therefore A^* is an algebra.) Since dim $A^* < \infty$, A^* has only a finite number of inequivalent irreducible representations. Therefore we get the following corollary.

COROLLARY 1. $M(\mathfrak{D}_0, \chi)$ *contains only a finite number of inequivalent left ideals.*

This is the generalisation of a result which has been proved in an earlier paper (Theorem 4 of [3]).

Let us call a representation π of $\mathfrak{B}$ on V quasisimple if it maps every element of $\mathfrak{Z}$ into a scalar multiple of the unit operator and if the representation $X \to \pi(X)$ $(X \in \mathfrak{k})$ is quasi semisimple (see [5, §2]). We denote by $V_{\mathfrak{D}}$ $(\mathfrak{D} \in \Omega)$ the set of those elements of V which transform under $\pi(\mathfrak{k})$ according to $\mathfrak{D}$.

COROLLARY 2. *Let π_1, π_2 be two quasisimple irreducible representations of $\mathfrak{B}$ on the vector spaces V_1, V_2 respectively. Choose a class $\mathfrak{D}_i \in \Omega$ which occurs in π_i and let ν_i denote the representation of $\mathfrak{A} = \mathfrak{Q}\mathfrak{X}$ on $V_{i,\mathfrak{D}_i}$ induced under π_i $(i=1, 2)$. Then ν_1, ν_2 are both irreducible. Moreover, if they are equivalent the same holds for π_1, π_2.*

Let χ_i be the homomorphism of $\mathfrak{Z}$ into C such that $\pi_i(z - \chi_i(z)) = 0$ $(z \in \mathfrak{Z})$. We have seen above that dim $V_{i,\mathfrak{D}_i} < \infty$ and ν_i is irreducible. Now if ν_1, ν_2 are equivalent, $\chi_1 = \chi_2$ and $\mathfrak{D}_1 = \mathfrak{D}_2$. Therefore Theorem 2 is applicable and we conclude that π_1 is equivalent to π_2.

3. **Some auxiliary results.** In this section we obtain two results whose full significance will appear in a subsequent paper. We keep to the above notation.

THEOREM 3. *Let $\mathfrak{M}$ be a left ideal in $M(\mathfrak{D}_0, \chi)$ and π the natural representation of $\mathfrak{B}$ on $\mathfrak{B}^* = \mathfrak{B}/\mathfrak{M}$. For any $\mathfrak{D} \in \Omega$ let $\mathfrak{B}_{\mathfrak{D}}^*$ denote the set of elements in $\mathfrak{B}^*$ which transform under $\pi(\mathfrak{X})$ according to $\mathfrak{D}$. Then there exists an integer N such that*

$$\dim \mathfrak{B}_{\mathfrak{D}}^* \leq Nd(\mathfrak{D})^2 \qquad (\mathfrak{D} \in \Omega)$$

where $d(\mathfrak{D})$ is the degree of any representation in $\mathfrak{D}$.

The proof of this theorem is based on an unpublished result of Chevalley. Consider the symmetric algebra $S(\mathfrak{p})$ over $\mathfrak{p}$. For any $X \in \mathfrak{k}$ define a derivation d_X of $S(\mathfrak{p})$ by the rule $d_X Y = [X, Y]$ $(Y \in \mathfrak{p})$. Let $\mathfrak{J}$ be the set of those $f \in S(\mathfrak{p})$ for which $d_X f = 0$ for all $X \in \mathfrak{k}$. Let σ be a simple representation of $\mathfrak{k}$ on a space U of dimension d. Consider the Kronecker product $S(\mathfrak{p}) \times U$. Let $S_m(\mathfrak{p})$ denote the set of homogeneous elements in $S(\mathfrak{p})$ of degree m. We shall say that an element in $S(\mathfrak{p}) \times U$ is homogeneous of degree m if it lies in $S_m(\mathfrak{p}) \times U$. We turn $S(\mathfrak{p}) \times U$ into an $S(\mathfrak{p})$-module by defining $f(g \times u) = fg \times u$ $(f, g \in S(\mathfrak{p}), u \in U)$. Let ν be the representation of $\mathfrak{k}$ on $S(\mathfrak{p}) \times U$ defined as follows:

$$\nu(X)(f \times u) = (d_X f) \times u + f \times \sigma(X)u \qquad (X \in \mathfrak{k}, f \in S(\mathfrak{p}), u \in U).$$

An element $e \in S(\mathfrak{p}) \times U$ is called an invariant if $\nu(X)e = 0$ for all $X \in \mathfrak{k}$. Let

$\mathfrak{R}$ be the set of all invariants. Then it is obvious that $\mathfrak{R}$ is a module over $\mathfrak{J}$.

CHEVALLEY'S THEOREM. *We can choose 2d homogeneous elements e_1, e_2, $\cdots$, e_{2d} in $\mathfrak{R}$ such that $\mathfrak{R} = \sum_{1 \leq i \leq 2d} \mathfrak{J} e_i$.*

COROLLARY. *The above result holds also if σ instead of being simple is semi-simple.*

Let $U = \sum_{1 \leq k \leq s} U_k$ where each U_k is stable and simple under σ and the sum is direct. Then $\mathfrak{R} = \sum_{1 \leq k \leq s} \mathfrak{R}_k$ where $\mathfrak{R}_k = \mathfrak{R} \cap (S(\mathfrak{p}) \times U_k)$. If we apply the above theorem to each $\mathfrak{R}_k$ we get the result.

We shall now make use of Chevalley's theorem to prove Theorem 3. Our method is the same as that used in §2 of [5]. Let $S(\mathfrak{g})$ be the symmetric algebra over $\mathfrak{g}$. For any $X \in \mathfrak{g}$ define a derivation D_X of $S(\mathfrak{g})$ by the rule $D_X Y = [X, Y]$ $(Y \in \mathfrak{g})$. Let J be the set of $F \in S(\mathfrak{g})$ for which $D_X F = 0$ for all $X \in \mathfrak{g}$. For any $F \in J$ we denote by $F_{\mathfrak{p}}$ the restriction of F on $\mathfrak{p}$ (see [5, end of §2]). Let $J_{\mathfrak{p}}$ be the image of J under this mapping. Then we know from [5, §3] that $\mathfrak{J}$ is a finite $J_{\mathfrak{p}}$-module and therefore we may select a finite set of homogeneous elements ω_β, $1 \leq \beta \leq \alpha$, in $\mathfrak{J}$ such that $\mathfrak{J} = \sum_{1 \leq \beta \leq \alpha} J_{\mathfrak{p}} \omega_\beta$. Since $\mathfrak{M} \in M(\mathfrak{D}_0, \chi)$ it is clear that $\mathfrak{B}^*_{\mathfrak{D}_0} \neq \{0\}$. Choose an element $\psi_0 \in \mathfrak{B}^*_{\mathfrak{D}_0}$ $(\psi_0 \neq 0)$ such that $U = \pi(\mathfrak{X})\psi_0$ is irreducible under $\pi(\mathfrak{X})$. Let ψ_i, $1 \leq i \leq d_0$, be a base for U. Let $\mathfrak{D}$ be any class in Ω. Put $d = d(\mathfrak{D})$ and choose a representation σ of $\mathfrak{f}$ on a vector space V of dimension d such that σ is dual (or contragredient) to any representation in $\mathfrak{D}$. We consider the space $S(\mathfrak{p}) \times U \times V$ and the set $\mathfrak{R}$ of all invariants in it. Then in view of the above corollary we can find $2dd_0$ homogeneous elements e_k $(1 \leq k \leq 2dd_0)$ in $\mathfrak{R}$ such that

$$\mathfrak{R} = \sum_{1 \leq k \leq 2dd_0} \mathfrak{J} e'_k = \sum_{1 \leq k \leq 2dd_0} \sum_{1 \leq \beta \leq \alpha} J_{\mathfrak{p}} \omega_\beta e'_k.$$

Put $N = 2\alpha d_0$ and the elements $\omega_\beta e'_k$ $(1 \leq \beta \leq \alpha, 1 \leq k \leq 2dd_0)$ in a sequence e_β $(1 \leq \beta \leq Nd)$. Then e_β are homogeneous elements in $\mathfrak{R}$ and $\mathfrak{R} = \sum_{1 \leq \beta \leq Nd} J_{\mathfrak{p}} e_\beta$. Let v_j, $1 \leq j \leq d$, be a base for V and let

$$e_\beta = \sum_{1 \leq j \leq d} e_{\beta,j} \times v_j \qquad (1 \leq \beta \leq Nd)$$

where $e_{\beta,j} \in S(\mathfrak{p}) \times U$. Let λ denote the canonical mapping of $S(\mathfrak{g})$ into $\mathfrak{B}$ (see [5, §2]). We define a linear mapping Γ of $S(\mathfrak{p}) \times U$ into $\mathfrak{B}^*$ as follows:

$$\Gamma(f \times u) = \pi(\lambda(f))u \qquad (f \in S(\mathfrak{p}), u \in U).$$

We intend to show that the Nd^2 elements $\Gamma(e_{\beta,j})$ $(1 \leq \beta \leq Nd, 1 \leq j \leq d)$ span $\mathfrak{B}^*_{\mathfrak{D}}$. Since N is independent of $\mathfrak{D}$ this would prove Theorem 3.

First of all we claim that $\Gamma(e_{\beta,j}) \in \mathfrak{B}^*_{\mathfrak{D}}$. Let ν be the representation of $\mathfrak{f}$ on $S(\mathfrak{p}) \times U$ defined as follows:

$$\nu(X)(f \times u) = (d_X f) \times u + f \times \pi(X)u \qquad (f \in S(\mathfrak{p}), u \in U, X \in \mathfrak{f}).$$

Since $\lambda(d_X f) = [X, \lambda(f)]$ (see Lemma 11 of [5]) it follows that $\Gamma(\nu(X)(f \times u)) = \pi(X)\Gamma(f \times u)$. Now consider the element e_β in $S(\mathfrak{p}) \times U \times V$. If $e_\beta \neq 0$ the representation of $\mathfrak{k}$ induced under ν on the space W_β spanned by $e_{\beta,j}$ ($1 \leq j \leq d$) is clearly contragredient to σ. On the other hand if $e_\beta = 0$, $W_\beta = \{0\}$ and therefore in either case $\Gamma(W_\beta) \subset \mathfrak{B}_{\mathfrak{D}}^*$. Hence $\Gamma(e_{\beta,j}) \in \mathfrak{B}_{\mathfrak{D}}^*$, $1 \leq \beta \leq Nd$, $1 \leq j \leq d$. Now put $\mathfrak{B}_m^* = \Gamma(S_m(\mathfrak{p}) \times U)$ $(m \geq 0)$. Since $\mathfrak{B} = \mathfrak{P} \mathfrak{X}$ where $\mathfrak{P} = \lambda(S(\mathfrak{p}))$, it is obvious that $\mathfrak{B}^* = \pi(\mathfrak{P})U = \Gamma(S(\mathfrak{p}) \times U)$. Hence $\mathfrak{B}^* = \sum_{m \geq 0} \mathfrak{B}_m^*$. Moreover since $S_m(\mathfrak{p}) \times U$ is stable under $\nu(\mathfrak{k})$ the same holds for $\mathfrak{B}_m^*$ under $\pi(\mathfrak{k})$. Also we know that the natural representation of $\mathfrak{k}$ on $\mathfrak{B}^*$ is quasi semisimple (see Theorem 1 of [5]) and so it follows (see Lemma 6 of [5]) that $\mathfrak{B}_m^* = \sum_{\mathfrak{D}' \in \Omega} \mathfrak{B}_m^* \cap \mathfrak{B}_{\mathfrak{D}'}^*$. Therefore

$$\sum_{\mathfrak{D}' \in \Omega} \mathfrak{B}_{\mathfrak{D}'}^* = \mathfrak{B}^* = \sum_{m \geq 0} \mathfrak{B}_m^* = \sum_{m \geq 0} \sum_{\mathfrak{D}' \in \Omega} \mathfrak{B}_m^* \cap \mathfrak{B}_{\mathfrak{D}'}^* = \sum_{\mathfrak{D}' \in \Omega} \sum_{m \geq 0} \mathfrak{B}_m^* \cap \mathfrak{B}_{\mathfrak{D}'}^*.$$

Since the sum on the left-hand side is direct we conclude that

$$\mathfrak{B}_{\mathfrak{D}}^* = \sum_{m \geq 0} (\mathfrak{B}_m^* \cap \mathfrak{B}_{\mathfrak{D}}^*).$$

Hence if $A = \sum_{\beta,j} C\Gamma(e_{\beta,j})$ it would be sufficient to prove that

$$\mathfrak{B}_m^* \cap \mathfrak{B}_{\mathfrak{D}}^* \subset A \qquad\qquad (m \geq 0).$$

Put $\mathfrak{B}_{-1}^* = \{0\}$. Then the above statement is true for $m = -1$. Hence we may use induction on m and assume that $m \geq 0$. Let z^* be an element in $\mathfrak{B}_m^* \cap \mathfrak{B}_{\mathfrak{D}}^*$. We have to show that $z^* \in A$. The representation of $\mathfrak{X}$ induced on the space $\pi(\mathfrak{X})z^*$ is semisimple. Hence it would be enough to prove that if W is any simple subspace of $\pi(\mathfrak{X})z^*$ then $W \subset A$. It is obvious that the representation of $\mathfrak{k}$ induced on W lies in $\mathfrak{D}$. Hence we can choose a base $w_1^*, \cdots, w_d^*$ for W such that

$$\xi(X)\left(\sum_{1 \leq j \leq d} w_j^* \times v_j\right) = 0 \qquad\qquad (X \in \mathfrak{k})$$

where ξ is the representation of $\mathfrak{k}$ induced on the Kronecker product $W \times V$. Since $W \subset \mathfrak{B}_m^* \cap \mathfrak{B}_{\mathfrak{D}}^*$ and since $S_m(\mathfrak{p}) \times U$ is stable and semisimple under $\nu(\mathfrak{X})$, it is easily seen that we can select elements $w_j \in S_m(\mathfrak{p}) \times U$ such that $\Gamma(w_j) = w_j^*$ and the space spanned by w_j, $i \leq j \leq d$, is invariant and irreducible under $\nu(\mathfrak{X})$. Then it is obvious that $\sum_{1 \leq j \leq d} w_j \times v_j$ belongs to $\mathfrak{R}$ and therefore

$$\sum_j w_j \times v_j = \sum_{1 \leq \beta \leq Nd} f_\beta e_\beta \qquad\qquad (f_\beta \in J_\mathfrak{p}).$$

Since $w_j^* \neq 0$ not all e_β can be zero. Suppose $e_\beta \neq 0$ for $1 \leq \beta \leq \beta_0$ and $e_\beta = 0$ for $\beta_0 < \beta \leq Nd$ $(1 \leq \beta_0 \leq Nd)$. Let δ_β be the degree of e_β $(1 \leq \beta \leq \beta_0)$. It is clear that the homogeneous components of any element in $J_\mathfrak{p}$ or J also belong to $J_\mathfrak{p}$ or

J respectively. Hence we may assume that f_β is either zero or homogeneous of degree $m - \delta_\beta$. We may assume that $f_\beta \neq 0$ if $1 \leq \beta \leq \beta_1$ and $f_\beta = 0$ if $\beta_1 < \beta \leq \beta_0$ $(1 \leq \beta_1 \leq \beta_0)$. Choose an element $F_\beta \in J$ such that $f_\beta = (F_\beta)_\mathfrak{p}$. Clearly we may assume that F_β is homogeneous of degree $m_\beta = m - \delta_\beta$ $(1 \leq \beta \leq \beta_1)$. But then $F_\beta - f_\beta \in \sum_{m_\beta > \mu \geq 0} S_\mu(\mathfrak{p}) S(\mathfrak{k})$ where $S(\mathfrak{k})$ is the symmetric algebra over $\mathfrak{k}$.

Now put

$$w_j^{*\prime} = \sum_{1 \leq \beta \leq \beta_1} \pi(\lambda(F_\beta))\Gamma(e_{\beta,j}).$$

Since $F_\beta \in J$, $\lambda(F_\beta) \in \mathfrak{Z}$ (see the corollary to Lemma 11 of [5]) and therefore

$$w_j^{*\prime} = \sum_{1 \leq \beta \leq \beta_1} \chi(\lambda(F_\beta))\Gamma(e_{\beta,j}) \in A.$$

Hence $w_j^* - w_j^{*\prime} \in \mathfrak{B}_\mathfrak{D}^*$. Moreover

$$w_j^* - w_j^{*\prime} = \sum_{1 \leq \beta \leq \beta_1} \left\{ \Gamma(f_\beta e_{\beta,j}) - \pi(\lambda(F_\beta))\Gamma(e_{\beta,j}) \right\},$$

$$\Gamma(f_\beta e_{\beta,j}) - \pi(\lambda(F_\beta))\Gamma(e_{\beta,j})$$

$$= \left\{ \Gamma(f_\beta e_{\beta,j}) - \pi(\lambda(f_\beta))\Gamma(e_{\beta,j}) \right\} - \pi(\lambda(F_\beta - f_\beta))\Gamma(e_{\beta,j}).$$

Since $F_\beta - f_\beta \in \sum_{m_\beta > \mu \geq 0} S_\mu(\mathfrak{p}) S(\mathfrak{k})$ $(1 \leq \beta \leq \beta_1)$ and $\Gamma(e_{\beta,j}) \in \mathfrak{B}_{\delta_\beta}^*$ it follows that

$$\pi(\lambda(F_\beta - f_\beta))\Gamma(e_{\beta,j}) \in \sum_{m_\beta > \mu \geq 0} \pi(\lambda(S_\mu(\mathfrak{p}))\mathfrak{B}_{\delta_\beta}^*$$

$$= \sum_{m_\beta > \mu \geq 0} \pi(\lambda(S_\mu(\mathfrak{p}))\lambda(S_{\delta_\beta}(\mathfrak{p})) \, U \subset \sum_{m > \mu \geq 0} \mathfrak{B}_\mu^*$$

since

$$\lambda(S_\mu(\mathfrak{p}))\lambda(S_{\delta_\beta}(\mathfrak{p})) \subset \sum_{\mu + \delta_\beta \geq \mu_1 \geq 0} \lambda(S_{\mu_1}(\mathfrak{p}))\mathfrak{X}$$

from Lemma 13 of [5]. Moreover

$$\Gamma(f_\beta e_{\beta,j}) - \pi(\lambda(f_\beta))\Gamma(e_{\beta,j}) \in \sum_{0 \leq \mu < m} \pi(\lambda(S_\mu(\mathfrak{g}))U$$

(see the proof of Lemma 1 of [2]) where $S_\mu(\mathfrak{g})$ is the set of all homogeneous elements of $S(\mathfrak{g})$ of degree μ. Since $\lambda(S_\mu(\mathfrak{g})) \subset \sum_{0 \leq \mu_1 \leq \mu} \lambda(S_{\mu_1}(\mathfrak{p}))\mathfrak{X}$ it follows that

$$\Gamma(f_\beta e_{\beta,j}) - \pi(\lambda(f_\beta))\Gamma(e_{\beta,j}) \in \sum_{m > \mu \geq 0} \mathfrak{B}_\mu^*.$$

Hence

$$w_j^* - w_j^{*\prime} \in \mathfrak{B}_\mathfrak{D}^* \cap \left(\sum_{m > \mu \geq 0} \mathfrak{B}_\mu^* \right) = \sum_{m > \mu \geq 0} (\mathfrak{B}_\mu^* \cap \mathfrak{B}_\mathfrak{D}^*) \subset A$$

by induction hypothesis. Since $w_j^{*\prime} \in A$ it follows that $w_j^* \in A$ and therefore $W \subset A$. This proves our assertion and so the theorem is established.

We shall now prove a lemma which will be of use later.

LEMMA 4. *Let $\mathfrak{l}$ be a semisimple algebra over C and $\mathfrak{U}$ the universal enveloping algebra of $\mathfrak{l}$. Then there exists an element z in the center of $\mathfrak{U}$ such that if π is any irreducible finite-dimensional representation of $\mathfrak{U}$ of degree d, then $\pi(z) = d^2\pi(1)$.*

Let Γ be a Cartan subalgebra of $\mathfrak{l}$ and W the Weyl group of $\mathfrak{l}$ with respect to Γ. Let $l = \dim \Gamma$. We choose a fundamental system of roots $\{\alpha_1, \cdots, \alpha_l\}$ of $\mathfrak{l}$ (with respect to Γ) and introduce a lexicographic ordering in the set of all weights of finite-dimensional representations of $\mathfrak{l}$ with respect to this set (see [3, Part I]). Let $B(X, Y)$ denote the bilinear form $\mathrm{sp}(\mathrm{ad}\ X\ \mathrm{ad}\ Y)$ on $\mathfrak{l}$ where $X \to \mathrm{ad}\ X$ is the adjoint representation of $\mathfrak{l}$. For every root α there exists a unique element $H_\alpha \in \Gamma$ such that $B(H, H_\alpha) = \alpha(H)$ for all $H \in \Gamma$. Let 2ρ be the sum of all positive roots of $\mathfrak{l}$ and let π be an irreducible finite-dimensional representation of $\mathfrak{l}$ (or $\mathfrak{U}$) whose highest weight is Λ. Then we know that the degree of π is

$$d_\Lambda = \frac{\displaystyle\prod_{\alpha>0} \Lambda'(H_\alpha)}{\displaystyle\prod_{\alpha>0} \rho(H_\alpha)}$$

where $\Lambda' = \Lambda + \rho$ (see Weyl [8]). Now we use the notation of Part III of [3]. Consider the polynomial

$$f(x) = \left(\frac{\displaystyle\prod_{\alpha>0} x(H_\alpha)}{\displaystyle\prod_{\alpha>0} \rho(H_\alpha)} \right)^2$$

in $x_1, \cdots, x_l$. It is clear that $sf = f$ for all $s \in W$. Hence it follows from Lemmas 38 and 39 of [3] that $f(x') = \chi_z(z)$ where z is some element in the center of $\mathfrak{U}$. Hence $d_\Lambda^2 = \chi_\Lambda(z)$. But we know that $\pi(z) = \chi_\Lambda(z)\pi(1)$ (see proof of Lemma 36 of [3]). Therefore $\pi(z) = d_\Lambda^2\pi(1)$ and this proves our assertion.

4. **Proof of the main theorem.** Let G be the simply connected Lie group with the Lie algebra $\mathfrak{g}_0$. Let K and D be the analytic subgroups of G corresponding to the subalgebras $\mathfrak{k}_0$ and $\mathfrak{c}_0$ ($\mathfrak{c}_0$ is the center of $\mathfrak{k}_0$). Let Z be the center of G and let $u \to u^*$ denote the natural mapping of K on $K^* = K/D \cap Z$. We know that K is simply connected and K^* is compact (see for example Mostow [6]). Since there is a natural 1-1 correspondence between the finite-dimensional simple representations of K and those of $\mathfrak{k}$ we may regard any $\mathfrak{D} \in \Omega$ as a class of representations of K. Let Ω^* be the set of all equivalence classes of finite-dimensional simple representations of K^*. Since every representation σ of K^* can also be regarded as a representation $u \to \sigma(u^*)$ ($u \in K$) we may consider Ω^* as a subset of Ω.

Let V be a vector space over C and $y_1, \cdots, y_r$ a finite set of inde-terminates. In §1 we have defined the space $V[y]$. Let W be another vector space and β a linear mapping of $V[y]$ into $W[y]$. We shall say that β is (y)-linear if $\beta(fv) = f\beta(v)$ for $f \in C[y]$ and $v \in V[y]$. It is clear that any linear mapping of V into W can be extended uniquely to a (y)-linear mapping of $V[y]$ into $W[y]$. Similarly if V is an algebra and π is a representation of $V[y]$ on $W[y]$, then π will be called (y)-linear if $\pi(f)w = fw$ for $f \in C[y]$ and $w \in W[y]$.

Let $\Gamma_1, \cdots, \Gamma_r$ be the base of $\mathfrak{c}$ introduced in §1 and let $y_1, \cdots, y_r$ be r independent indeterminates. If M is any integral linear function on $\mathfrak{c}$ (see §1) we put as before $\Gamma^M = \Gamma_1^{m_1} \Gamma_2^{m_2} \cdots \Gamma_r^{m_r}$ and $y^M = y_1^{m_1} \cdots y_r^{m_r}$ where $m_i = M(\Gamma_i)$, $1 \leq i \leq r$. We shall now define two (y)-linear mappings β_y and β_{-y} of $\mathfrak{B}[y]$ into itself. It is sufficient to define $\beta_y(b)$ and $\beta_{-y}(b)$ for $b \in \mathfrak{B}$. We have seen in §1 that b can be written uniquely in the form $b = \sum_M w_M \Gamma^M$ where $w_M \in \mathfrak{P}\mathfrak{X}'$. (Here $\mathfrak{P}$ and $\mathfrak{X}'$ have the same meaning as in §1.) Now we write $\beta_y(b) = \sum_M w_M \Gamma_y^M$ and $\beta_{-y}(b) = \sum_M w_M \Gamma_{-y}^M$ where

$$\Gamma_y^M = (\Gamma_1 + y_1)^{m_1} (\Gamma_2 + y_2)^{m_2} \cdots (\Gamma_r + y_r)^{m_r},$$
$$\Gamma_{-y}^M = (\Gamma_1 - y_1)^{m_1} (\Gamma_2 - y_2)^{m_2} \cdots (\Gamma_r - y_r)^{m_r} \quad (m_j = M(\Gamma_j)).$$

It is obvious that $\beta_y(\beta_{-y}(b)) = \beta_{-y}(\beta_y(b)) = b$ for all $b \in \mathfrak{B}[y]$. Moreover since Γ^M belongs to the center of $\mathfrak{X}$ it is clear that $\beta_y(bz) = \beta_y(b)\beta_y(z)$ for $b \in \mathfrak{B}$ and $z \in \mathfrak{X}$.

For any $x \in G$ and $v^* \in K^*$ we define v_x^*, $H(x, v^*)$, $\Gamma(x, v^*)$ as in §11 of [5]. Choose a base $Z_1, \cdots, Z_n$ for $\mathfrak{g}_0$ over R and put $Z(t) = t_1 Z_1 + \cdots + t_n Z_n$ $(t_j \in R)$, $x_t = \exp Z(t)$. Let $H_1, \cdots, H_p$ be the base for $\mathfrak{h}_{\mathfrak{p}_0}$ over R which was selected in §2 of [5]. Put

$$H(x_t^{-1}, v^*) = \sum_{1 \leq i \leq p} H_i f_i(t, v^*),$$
$$\Gamma(x_t^{-1}, v^*) = \sum_{1 \leq j \leq r} \Gamma_j g_j(t, v^*),$$
$$\chi_{\mathfrak{D}^*}((v_{x_t^{-1}}^*)^{-1}) = \chi(\mathfrak{D}^*, t, v^*)$$

where $\chi_{\mathfrak{D}^*}$ is the character of K^* corresponding to the class $\mathfrak{D}^* \in \Omega^*$. It is obvious (see the corollary to Lemma 26 of [5]) that $f_i(t, v^*)$, $g_j(t, v^*)$, $\chi(\mathfrak{D}^*, t, v^*)$ are all analytic functions on $R^n \times K^*$. Put $|t| = \max_j |t_j|$. We may choose $\epsilon > 0$ so small that each of these functions can be expanded as a power series in $(t_1, \cdots, t_n)$ (with coefficients which are analytic functions on K^*) which is uniformly convergent for all $v^* \in K^*$ and $|t| \leq \epsilon$ (see the proof of Lemma 29 of [5]). Let $(u_1, \cdots, u_p)$ be p indeterminates independent of $(y_1, \cdots, y_r)$. If $H = \sum_{1 \leq i \leq p} a_i H_i$, $\Gamma = \sum_{1 \leq j \leq r} b_j \Gamma_j$ $(a_i, b_j \in C)$ put $u(H) = \sum_{1 \leq i \leq p} a_i u_i$, $y(\Gamma) = \sum_{1 \leq j \leq r} b_j y_j$. For any linear form T in (u, y) we define e^T as the element $1 + T + T^2/2! + T^3/3! + \cdots$ in the ring of power series in (u, y) with coefficients in C. Then it is clear that the coefficients of the series

$e^{-y(\Gamma(x_i^{-1},v^*))}e^{-u(H(x_i^{-1},v^*))}$ are polynomials in $f_i(t,v^*)$ and $g_j(t,v^*)$. Moreover if Λ and μ are any linear functions on $\mathfrak{h}_\mathfrak{p}$ and $\mathfrak{c}$ respectively, then the series resulting under the substitution $y\to\mu$ and $u\to\Lambda$ (that is, $y_j\to\mu(\Gamma_j)$), $1\leq j\leq r$, and $u_i \to\Lambda(H_i)$ $(1\leq i\leq p)$ converges absolutely to $e^{-\Lambda(H(x_i^{-1},v^*))}e^{-\mu(\Gamma(x_i^{-1},v^*))}$. Let 2ρ denote the sum of all positive roots of $\mathfrak{g}$ (with respect to $\mathfrak{h}$) under the ordering introduced in §2 of [5]. Consider linear functions P, M, N on $\mathfrak{g}$, $\mathfrak{c}$, and $\mathfrak{h}_\mathfrak{p}$ respectively such that $P(Z_i)=p_i$, $M(\Gamma_j)=m_j$, $N(H_k)=n_k$ $(1\leq i\leq n$, $1\leq j\leq r$, $1\leq k \leq p)$ are all non-negative integers. Put $P!=p_1!p_2!\cdots p_n!$ and

$$t^P = t_1^{p_1}t_2^{p_2}\cdots t_n^{p_n}, \qquad y^M = y_1^{m_1}\cdots y_r^{m_r}, \qquad u^N = u_1^{n_1}\cdots u_p^{n_p}.$$

Then it follows from the above remarks that

$$d(\mathfrak{D}^*)\chi(\mathfrak{D}^*,\,t,\,v^*)e^{-y(\Gamma(x_i^{-1},v^*))}e^{-u(H(x_i^{-1},v^*))}e^{-2\rho(H(x_i^{-1},v^*))}$$

$$= \sum_{M,N,P} \zeta(M,\,N,\,P,\,v^*)y^M u^N \frac{t^P}{P!} \qquad\qquad (\,|t|\leq\epsilon)$$

where $\zeta(M,N,P,v^*)$ are analytic functions on K^*. $(d(\mathfrak{D}^*)$ is the degree of any representation in $\mathfrak{D}^*$.) The above equality is to be understood in the following sense. Firstly the coefficient of $y^M u^N$ on the right-hand side is a power series in (t) which converges uniformly for $|t|\leq\epsilon$ and $v^*\in K^*$ to the coefficient of $y^M u^N$ on the left. Secondly if Λ and μ are linear functions on $\mathfrak{h}_\mathfrak{p}$ and $\mathfrak{c}$ respectively, then the series obtained after the substitution $y\to\mu$, $u\to\Lambda$ converges uniformly to

$$d(\mathfrak{D}^*)\chi(\mathfrak{D}^*,\,t,\,v^*)e^{-\mu(\Gamma(x_i^{-1},v^*))}e^{-\Lambda(H(x_i^{-1},v^*))}e^{-2\rho(H(x_i^{-1},v^*))}$$

for all $|t|\leq\epsilon$ and $v^*\in K^*$. Notice that $H(x^{-1},v^*)=\Gamma(x^{-1},v^*)=0$ if $x=1$. Hence the constant terms in the power series for $f_i(t,v^*)$ and $g_j(t,v^*)$ are all zero. From this it follows immediately that for a fixed P there are only a finite number of values of M and N such that $\zeta(M,N,P,v^*)\neq 0$. Therefore the expression

$$\zeta_{u,y}(P,\,v^*) = \sum_{M,N} \zeta(M,\,N,\,P,\,v^*)y^M u^N$$

is a polynomial in (u,y).

Let $\mathfrak{N}(\mathfrak{D}^*)$ denote the kernel (in $\mathfrak{X}$) of any representation of $\mathfrak{X}$ lying in $\mathfrak{D}^*$. Let $z\to\bar{z}$ denote the natural mapping and σ the natural representation of $\mathfrak{X}$ on $\mathfrak{X}/\mathfrak{N}(\mathfrak{D}^*)=\bar{\mathfrak{X}}$. Since $\mathfrak{D}^*\in\Omega^*$ and $\dim\bar{\mathfrak{X}}<\infty$ we can "extend" σ to a representation of K^* on $\bar{\mathfrak{X}}$. Put

$$\eta_{u,y}^{\mathfrak{D}^*}(P) = \int_{K^*} \zeta_{u,y}(P,\,v^*)\sigma(v^*)\bar{1}_{dv^*}$$

$$= \sum_{M,N} y^M u^N \int_{K^*} \zeta(M,\,N,\,P,\,v^*)\sigma(v^*)\bar{1}dv^*$$

where dv^* is the Haar measure on K^* normalised so as to make the total measure of K^* equal to 1. Then $\eta_{u,v}^{\mathfrak{D}^*}(P) \in \bar{\mathfrak{X}}[u, y]$. Let W_i denote the element $Z_i \in \mathfrak{g}$ regarded as an element of $S(\mathfrak{g})$ $(1 \leq i \leq n)$. Consider the monomial $W(P) = W_1^{p_1} W_2^{p_2} \cdots W_n^{p_n}$ $(p_i = P(Z_i), 1 \leq i \leq n)$ in $S(\mathfrak{g})$. Put $Z(P) = \lambda(W(P))$ where λ is the canonical mapping of $S(\mathfrak{g})$ onto $\mathfrak{B}$ (see §2 of [5]). We know that the elements $Z(P)$ form a base for $\mathfrak{B}$. Hence we can define a (u, y)-linear mapping $\eta_{u,v}^{\mathfrak{D}^*}$ of $\mathfrak{B}[u, y]$ into $\bar{\mathfrak{X}}[u, y]$ by setting $\eta_{u,v}^{\mathfrak{D}^*}(Z(P)) = \eta_{u,v}^{\mathfrak{D}^*}(P)$. If Λ and μ are linear functions on $\mathfrak{h}_\mathfrak{p}$ and $\mathfrak{c}$ respectively and b is an element of $\mathfrak{B}$, we shall denote by $\eta_{\Lambda,\mu}^{\mathfrak{D}^*}(b)$ the value of $\eta_{u,v}^{\mathfrak{D}^*}(b)$ obtained under the substitution $u \to \Lambda$, $y \to \mu$. Obviously $\eta_{\Lambda,\mu}^{\mathfrak{D}^*}$ is a linear mapping of $\mathfrak{B}$ into $\bar{\mathfrak{X}}$. We shall use a similar notation in other cases as well, without further comment.

In the next few lemmas we study in detail the properties of the mapping $b \to \eta_{u,v}^{\mathfrak{D}^*}(b)$ $(b \in \mathfrak{B})$.

Put $\mathfrak{N}_y(\mathfrak{D}^*) = \beta_{-y}(\mathfrak{N}(\mathfrak{D}^*)[y])$ and let $A_y(\mathfrak{D}^*)$ denote the set of all elements $a \in \mathfrak{B}[y]$ such that $\mathfrak{N}_y(\mathfrak{D}^*)a \subset \mathfrak{B}\mathfrak{N}_y(\mathfrak{D}^*)$. Clearly $A_y(\mathfrak{D}^*)$ is a subalgebra of $\mathfrak{B}[y]$ and $A_y(\mathfrak{D}^*) \supset \mathfrak{Q}\mathfrak{X}$.

LEMMA 5. *Let $\mathfrak{N}_2$ be the set of all elements $b \in A_y(\mathfrak{D}^*)[u]$ such that $\eta_{u,v}^{\mathfrak{D}^*}(b) = 0$. Then $\mathfrak{N}_2$ is a left ideal in $A_y(\mathfrak{D}^*)[u]$.*

Put $A_2 = A_y(\mathfrak{D}^*)[u]$. Clearly $\mathfrak{N}_2$ is a linear subspace of A_2. We have to show that if $a(u, y) \in A_2$ and $b(u, y) \in \mathfrak{N}_2$ then $a(u, y)b(u, y) \in \mathfrak{N}_2$ (we write $a(u, y)$ and $b(u, y)$ to emphasise the fact that they are polynomials in (u) and (y) with coefficients in $\mathfrak{B}$). Suppose this is false. Then $\eta_{u,v}(a(u, y)b(u, y)) \neq 0$. (We write $\eta_{u,v}$ instead of $\eta_{u,v}^{\mathfrak{D}^*}$ for convenience.) Hence we can choose linear functions ν and μ on $\mathfrak{h}_\mathfrak{p}$ and $\mathfrak{c}$ respectively such that the value of $\eta_{u,v}(a(u, y)b(u, y))$ obtained by the substitution $u \to \nu$ and $y \to \mu$ is not zero. Let a_0, b_0 denote the values of $a(u, y)$ and $b(u, y)$ under this substitution. Then $\eta_{\nu,\mu}(a_0 b_0) \neq 0$. Since $\bar{\mathfrak{X}}$ is a simple algebra, there exists a maximal left ideal $\bar{\mathfrak{Y}}$ in $\bar{\mathfrak{X}}$ such that $\eta_{\nu,\mu}(a_0 b_0) \notin \bar{\mathfrak{Y}}$. Let $\mathfrak{Y}$ be the complete inverse image of $\bar{\mathfrak{Y}}$ in $\mathfrak{X}$. Then it is clear that the natural representation of $\mathfrak{X}$ on $\mathfrak{X}/\mathfrak{Y} \cong \bar{\mathfrak{X}}/\bar{\mathfrak{Y}}$ lies in $\mathfrak{D}^*$.

Let $L_2(K^*) = \mathfrak{H}$ be the Hilbert space of all square-integrable functions on K^*. We define (see [5, §12]) a representation $\pi_{\nu,\mu}$ of G on $\mathfrak{H}$ as follows:

$$(\pi_{\nu,\mu}(x)f)(v^*) = e^{-\mu(\Gamma(x^{-1}, v^*))} e^{-(\nu+2\rho)(H(x^{-1}, v^*))} f(v_{x^{-1}}^*)$$

where $x \in G$, $v^* \in K^*$, and $f \in \mathfrak{H}$. Put

$$\psi(v^*) = d(\mathfrak{D}^*)\chi_{\mathfrak{D}^*}(v^{*-1}) \qquad (v^* \in K^*)$$

and let τ be the left-regular representation of K^* on $L_2(K^*)$. Put

$$E = d(\mathfrak{D}^*)\int_{K^*} \chi_{\mathfrak{D}^*}(v^{*-1})\tau(v^*)dv^* = \int_{K^*} \psi(v^*)\tau(v^*)dv^*.$$

Then E is the orthogonal projection of $\mathfrak{H}$ on the space $\mathfrak{H}_{\mathfrak{D}^*}$ consisting of all those elements in $\mathfrak{H}$ which transform under τ according to $\mathfrak{D}^*$. Put $\pi = \pi_{\nu,\mu}$ and consider $E\pi(x)\psi$. T being any linear operator on $\mathfrak{H}$ we denote by $Tf(v^*)$ the value of Tf at v^* $(v^* \in K^*)$ whenever f lies in the domain of T. Then

$$E\pi(x)\psi(v^*) = \int_{K^*} \psi(u^*)\pi(x)\psi(u^{*-1}v^*)du^*$$

$$= \int_{K^*} \psi(v^*u^{*-1})\pi(x)\psi(u^*)du^* = \int_{K^*} \psi(u^{*-1}v^*)\pi(x)\psi(u^*)du^*$$

since $\psi(v^*u^{*-1}) = \psi(u^{*-1}v^*)$. Hence

$$E\pi(x)\psi(v^*) = \int_{K^*} e^{-\mu(\Gamma(x^{-1},u^*))}e^{-(\nu+2\rho)(H(x^{-1},u^*))}\psi(u^*{}_{x}{}^{-1})\tau(u^*)\psi(v^*)du^*$$

or

$$E\pi(x)\psi = \int_{K^*} e^{-\mu(\Gamma(x^{-1},v^*))}e^{-(\nu+2\rho)(H(x^{-1},v^*))}\psi(\overset{*}{v}{}_{x}{}^{-1})\tau(v^*)\psi dv^*.$$

Let U be the linear space spanned by $\tau(v^*)\psi$ for all $v^* \in K^*$. Then $U = \mathfrak{H}_{\mathfrak{D}^*}$ and dim $U = (d(\mathfrak{D}^*))^2$. We get a representation of K^* on U which we again denote by τ. τ may therefore also be regarded as a representation of $\mathfrak{k}$ (or $\mathfrak{X}$) on U. Now it is clear that $\tau(z)f = 0$ if $z \in \mathfrak{N}(\mathfrak{D}^*)$ and $f \in U$. Hence we get a representation $\bar{\tau}$ of $\bar{\mathfrak{X}}$ on U if we put $\bar{\tau}(\bar{z})f = \tau(z)f$ $(z \in \mathfrak{X}, f \in U)$, $z \to \bar{z}$ being the natural mapping of $\mathfrak{X}$ on $\bar{\mathfrak{X}}$. Let τ' be the right regular representation of G on $\mathfrak{H}$. Since $\psi(u^{*-1}v^*) = \psi(v^*u^{*-1})$ it follows that $\tau(u^*)\psi = \tau'(u^{*-1})\psi$. Hence

$$\bar{\tau}(\bar{z})\tau(u^*)\psi = \bar{\tau}(\bar{z})\tau'(u^{*-1})\psi = \tau'(u^{*-1})\bar{\tau}(\bar{z})\psi \qquad\qquad (\bar{z} \in \bar{\mathfrak{X}},\, u^* \in K^*)$$

since right translations commute with the left translations. Now suppose $\bar{\tau}(\bar{z})\psi = 0$. Then it follows that $\bar{\tau}(\bar{z})\tau(u^*)\psi = 0$ for all $u^* \in K^*$. Therefore $\bar{\tau}(\bar{z}) = 0$ which implies that $\bar{z} = 0$. Thus $\bar{\tau}(\bar{z})\psi = 0$ if and only if $\bar{z} = 0$.

Let $\zeta_{\nu,\mu}(P, v^*)$ denote the value of the polynomial $\zeta_{u,y}(P, v^*)$ under the substitution $u \to \nu$, $y \to \mu$. Then we have seen that the series $\sum_P \zeta_{\nu,\mu}(P, v^*)t^P/P!$ converges uniformly to

$$\psi(\overset{*}{v}{}_{x_i}{}^{-1})e^{-\mu(\Gamma(x_i^{-1},v^*))}e^{-(\nu+2\rho)(H(x_i^{-1},v^*))} \qquad\qquad \text{if } |t| \leq \epsilon \text{ and } v^* \in K^*.$$

Hence

$$E\pi(x_i)\psi = \sum_P \frac{t^P}{P!}\int_{K^*} \zeta_{\nu,\mu}(P, u^*)\tau(u^*)\psi du^* \qquad\qquad (|t| \leq \epsilon).$$

But it is clear from the definition of $\eta_{\nu,\mu}(Z(P))$ that

$$\bar{\tau}(\eta_{\nu,\mu}(Z(P)))\psi = \int_{K^*} \zeta_{\nu,\mu}(P, u^*)\tau(u^*)\psi du^*.$$

Hence

$$E\pi(x_t)\psi = \sum_P \frac{t^P}{P!} \bar{\tau}(\eta_{\nu,\mu}(Z(P)))\psi \qquad (|t| \leq \epsilon).$$

But we know that ψ is well-behaved under $\pi(G)$ (see Lemma 34 of [5]). Hence if $|t|$ is sufficiently small

$$\pi(x_t)\psi = \sum_P \frac{t^P}{P!} \pi(Z(P))\psi$$

where $w \rightarrow \pi(w)$ ($w \in \mathfrak{B}$) denotes the representation of $\mathfrak{B}$ induced on the space of all well-behaved elements in $\pi(G)$. The series above is convergent in $\mathfrak{H}$ (see [5, §4]). Since E is a bounded linear operator on $\mathfrak{H}$ we get

$$E\pi(x_t)\psi = \sum_P \frac{t^P}{P!} \times E\pi(Z(P))\psi.$$

Therefore

$$\sum_P \frac{t^P}{P!} E\pi(Z(P))\psi = \sum_P \frac{t^P}{P!} \bar{\tau}(\eta_{\nu,\mu}(Z(P)))\psi$$

if $|t|$ is sufficiently small. Hence comparing coefficients of t^P on both sides we conclude that

$$\bar{\tau}(\eta_{\nu,\mu}(Z(P)))\psi = E\pi(Z(P))\psi$$

or

$$\bar{\tau}(\eta_{\nu,\mu}(a))\psi = E\pi(a)\psi \qquad (a \in \mathfrak{B}).$$

Now $b(u, y) \in \mathfrak{R}_2$ and therefore $\eta_{u,y}(b(u, y)) = 0$. Hence $\eta_{\nu,\mu}(b_0) = 0$. Therefore $E\pi(b_0)\psi = \bar{\tau}(\eta_{\nu,\mu}(b_0))\psi = 0$. On the other hand

$$E\pi(a_0 b_0)\psi = \bar{\tau}(\eta_{\nu,\mu}(a_0 b_0))\psi \neq 0$$

since $\eta_{\nu,\mu}(a_0 b_0) \neq 0$.

Now $b(u, y) \in A_2$. Hence $\beta_{-\nu}(\mathfrak{R}(\mathfrak{D}^*))b(u, y) \subset \mathfrak{B}_2 \beta_{-\nu}(\mathfrak{R}(\mathfrak{D}^*))$ where $\mathfrak{B}_2 = \mathfrak{B}[u, y]$. But it is easy to verify that $\pi(u) = e^{\mu(\Gamma(u))}\tau(u^*)$ ($u \in K$) where $\Gamma(u)$ is the unique element in $\mathfrak{c}_0$ such that $u \exp(-\Gamma(u))$ belongs to the analytic subgroup K' of G corresponding to $\mathfrak{k}_0' = [\mathfrak{k}_0, \mathfrak{k}_0]$. From this it follows immediately that there exists a class $\mathfrak{D} \in \Omega$ such that $\beta_{-\mu}(\mathfrak{R}(\mathfrak{D}^*)) = \mathfrak{R}(\mathfrak{D})$ ($\mathfrak{R}(\mathfrak{D})$ being the kernel in $\mathfrak{X}$ of any representation of $\mathfrak{X}$ which lies in $\mathfrak{D}$), and U is exactly the set of all elements in $\mathfrak{H}$ which transform under $\pi(K)$ according to $\mathfrak{D}$. Then $\mathfrak{R}(\mathfrak{D})b_0 \subset \mathfrak{B}\mathfrak{R}(\mathfrak{D})$. Hence $\pi(b_0)\psi \in U$ and therefore $\pi(b_0)\psi = E\pi(b_0)\psi = 0$. Hence

$$E\pi(a_0 b_0)\psi = E\pi(a_0)\pi(b_0)\psi = 0.$$

As this contradicts the result above the lemma is proved.

We extend the natural mapping $z \to \bar{z}$ of $\mathfrak{X}$ on $\bar{\mathfrak{X}}$ to a (u, y)-linear mapping of $\mathfrak{X}[u, y]$ on $\bar{\mathfrak{X}}[u, y]$. Since $\bar{\mathfrak{X}}$ is an algebra the same is true of $\bar{\mathfrak{X}}[u, y]$.

LEMMA 6. *If* $z \in \mathfrak{X}$ *and* $b \in \mathfrak{B}$ *then* $\eta_{u,y}^{\mathfrak{D}^*}(zb) = \overline{\beta_y(z)}\eta_{u,y}^{\mathfrak{D}^*}(b)$. *Moreover* $\eta_{u,y}^{\mathfrak{D}^*}(1) = 1$.

Again we write $\eta_{u,y}$ instead of $\eta_{u,y}^{\mathfrak{D}^*}$. Suppose $\eta_{u,y}(zb) - \overline{\beta_y(z)}\eta_{u,y}(b) \neq 0$. Then we choose ν, μ such that $\eta_{\nu,\mu}(zb) - \overline{\beta_\mu(z)}\eta_{\nu,\mu}(b) \neq 0$. Hence in the above notation we have

$$E\pi(zb)\psi = \bar{\tau}(\eta_{\nu,\mu}(zb))\psi.$$

It is obvious that E commutes with $\pi(u)$ $(u \in K)$ and therefore

$$E\pi(zb)\psi = \pi(z)E\pi(b)\psi = \pi(z)\bar{\tau}(\eta_{\nu,\mu}(b))\psi.$$

Moreover since $\pi(u) = e^{\mu(\Gamma(u))}\tau(u^*)$ $(u \in K)$ it follows that $\pi(X)\phi = \tau(X)\phi$ $(X \in \mathfrak{k}')$ and $\pi(\Gamma)\phi = \mu(\Gamma)\phi + \tau(\Gamma)\phi$ $(\Gamma \in \mathfrak{c})$ for any $\phi \in U$. Hence $\pi(z)\phi = \tau(\beta_\mu(z))\phi$ $(z \in \mathfrak{X}, \phi \in U)$. Therefore

$$E\pi(zb)\psi = \tau(\beta_\mu(z))\bar{\tau}(\eta_{\nu,\mu}(b))\psi = \overline{\tau(\overline{\beta_\mu(z)}\eta_{\nu,\mu}(b))}\psi.$$

This shows that

$$\bar{\tau}(\eta_{\nu,\mu}(zb) - \overline{\beta_\mu(z)}\eta_{\nu,\mu}(b))\psi = 0$$

and therefore $\eta_{\nu,\mu}(zb) - \overline{\beta_\mu(z)}\eta_{\nu,\mu}(b) = 0$ which contradicts our choice of ν, μ. Hence $\eta_{u,y}(zb) = \beta_y(z)\eta_{u,y}(b)$ and similarly we prove that $\eta_{u,y}(1) = \bar{1}$.

We use the notation of §2 of [5] and define $\mathfrak{m} = \mathfrak{h}_I + \sum_{\alpha \in P_-} CX_\alpha + \sum_{\alpha \in P_-} CX_{-\alpha}$. Then $\mathfrak{m}$ is the centraliser of $\mathfrak{h}_\mathfrak{p}$ in $\mathfrak{k}$.

LEMMA 7. *Let* $\mathfrak{U}$ *be the subalgebra of* $\mathfrak{B}$ *generated by* $(1, \mathfrak{m})$. *Then if* $b \in \mathfrak{B}$ *and* $z \in \mathfrak{U}$,

$$\eta_{u,y}^{\mathfrak{D}^*}(bz) = \eta_{u,y}^{\mathfrak{D}^*}(b)\overline{\beta_y(z)}.$$

For otherwise again choose ν, μ such that $\eta_{\nu,\mu}(bz) - \eta_{\nu,\mu}(b)\overline{\beta_\mu(z)} \neq 0$. Then we have seen above that

$$E\pi(x)\psi = \int_{K^*} e^{-\mu(\Gamma(x^{-1}, u^*))} e^{-(\nu+2\rho)(H(x^{-1}, u^*))} \psi(u_{x^{-1}}^*)\tau(u^*)\psi du^*$$

in the notation used above. Let M be the analytic subgroup of K corresponding to $\mathfrak{m}_0 = \mathfrak{m} \cap \mathfrak{k}_0$. Then if $m \in K$ we know (see Lemma 37 of [5]) that $\pi(x)$ commutes with $\tau'(m^*)$ where τ' is the right regular representation of K^* on $L_2(K^*)$. Hence

$$E\pi(x)\tau'(m^{*-1})\psi = \tau'(m^{*-1})E\pi(x)\psi$$

$$= \int_{K^*} e^{-\mu(\Gamma(x^{-1}, v^*))} e^{-(\nu+2\rho)(H(x^{-1}, v^*))} \psi(v_{x^{-1}}^*)\tau'(m^{*-1})\tau(v^*)\psi dv^*.$$

But we have seen above that $\tau'(m^{*-1})\psi = \tau(m^*)\psi$ and therefore $\tau'(m^{*-1})$ $\cdot \tau(v^*)\psi = \tau(v^*m^*)\psi$. Therefore

$$E\pi(x)\tau(m^*)\psi = \int_{K^*} e^{-\mu(\Gamma(x^{-1},v^*))} e^{-(\nu+2\rho)(H(x^{-1},v^*))} \psi(\overset{*}{v}_{x^{-1}}) \tau(v^*m^*)\psi \, dv^*.$$

From this it follows immediately that

$$E\pi(b')\tau(z')\psi = \bar{\tau}(\eta_{\nu,\mu}(b'))\tau(z')\psi \qquad (b' \in \mathfrak{B}, \, z' \in \mathfrak{U}).$$

But we have seen above that $\pi(z')\psi = \tau(\beta_\mu(z'))\psi$ $(z' \in \mathfrak{X})$. Therefore $E\pi(bz)\psi$ $= E\pi(b)\tau(\beta_\mu(z))\psi$. Let $\mathfrak{U}'$ be the set of all elements $w \in \mathfrak{X}$ such that $\beta_\mu(w) \in \mathfrak{U}$. Since $w \to \beta_\mu(w)$ $(w \in \mathfrak{X})$ is a homomorphism it follows that $\mathfrak{U}'$ is a subalgebra of $\mathfrak{X}$. Moreover $\mathfrak{U}' \supset \mathfrak{m}$ and therefore $\mathfrak{U}' \supset \mathfrak{U}$. Hence $\beta_\mu(z) \in \mathfrak{U}$ and therefore

$$E\pi(bz)\psi = \bar{\tau}(\eta_{\nu,\mu}(b))\tau(\beta_\mu(z))\psi$$
$$= \bar{\tau}(\eta_{\nu,\mu}(b)\overline{\beta_\mu(z)})\psi.$$

On the other hand $E\pi(bz)\psi = \bar{\tau}(\eta_{\nu,\mu}(bz))\psi$ and so we conclude that

$$\bar{\tau}(\eta_{\nu,\mu}(bz) - \eta_{\nu,\mu}(b)\overline{\beta_\mu(z)})\psi = 0$$

which contradicts the fact that $\eta_{\nu,\mu}(bz) - \eta_{\nu,\mu}(b)\overline{\beta_u(z)} \neq 0$. Thus the lemma is proved.

We define the automorphism θ of $\mathfrak{g}$ and the sets P, P_+, and P_- of positive roots of $\mathfrak{g}$ (with respect to $\mathfrak{h}$) as in §2 of [5]. Let $\alpha_1 < \alpha_2 < \cdots < \alpha_r$ be all the roots in P. We recall that if $\alpha \in P_-$ and $\alpha' \in P_+$ then $\alpha' > \alpha$. Suppose then that $\alpha_1, \cdots, \alpha_s \in P_-$ and $\alpha_{s+1}, \cdots, \alpha_r \in P_+$ $(0 \leq s \leq r)$. First we make the following observation.

LEMMA 8. *Let p_i, q_j be non-negative integers such that $p_1\alpha_1 + \cdots + p_r\alpha_r$* $= q_1\alpha_1 + \cdots + q_s\alpha_s$. *Then $p_{s+1} = p_{s+2} = \cdots = p_r = 0$.*

For $p_{s+1}\alpha_{s+1} + \cdots + p_r\alpha_r = (q_1 - p_1)\alpha_1 + \cdots + (q_s - p_s)\alpha_s$. Applying θ to both sides we get

$$p_{s+1}\theta\alpha_{s+1} + \cdots + p_r\theta\alpha_r = (q_1 - p_1)\alpha_1 + \cdots + (q_s - p_s)\alpha_s$$

since $\theta\alpha = \alpha$ for $\alpha \in P_-$. Therefore

$$p_{s+1}(\alpha_{s+1} - \theta\alpha_{s+1}) + p_{s+2}(\alpha_{s+2} - \theta\alpha_{s+2}) + \cdots + p_r(\alpha_r - \theta\alpha_r) = 0.$$

But $\theta\alpha < 0$ for $\alpha \in P_+$ and therefore $\alpha_t - \theta\alpha_t > 0$ for $t > s$. Hence the above relation is possible only if $p_{s+1} = \cdots = p_r = 0$.

Let Λ be a linear function on $\mathfrak{h}$. We say that an element $b \in \mathfrak{B}$ is of rank Λ if $[H, b] = \Lambda(H)b$ for all $H \in \mathfrak{h}$.

LEMMA 9. *Let $\mathfrak{B}$ be the subalgebra of $\mathfrak{B}$ generated by $(1, \mathfrak{m} + \mathfrak{h}_\mathfrak{p})$. Then for any $z \in \mathfrak{z}$ there exists an element $z_0 \in \mathfrak{B}$ such that*

$$z - z_0 \in \sum_{\alpha,\beta \in P_+} X_{-\alpha} \mathfrak{B} X_\beta$$

and z_0 is of the rank zero.

We use the notation of [5, §2] and the lemma above. Choose a base $H_{p+1}, \cdots, H_l$ for $\mathfrak{h}_{\mathfrak{k}}$. Then $X_{-\alpha_r}, \cdots, X_{-\alpha_1}, H_1, \cdots, H_l, X_{\alpha_1}, \cdots, X_{\alpha_r}$ is a base for $\mathfrak{g}$. Hence z can be written uniquely (see [2]) in the form

$$z = \sum_{(p),(m),(q)} a((p),(m),(q)) X_{-\alpha_r}^{p_r} \cdots X_{-\alpha_1}^{p_1} H_1^{m_1} \cdots H_l^{m_l} X_{\alpha_1}^{q_1} \cdots X_{\alpha_r}^{q_r}$$

where $a((p),(m),(q)) \in C$. Since $z \in \mathfrak{Z}$, z is of rank zero and therefore it follows (see [3, proof of Lemma 36]) that $a((p),(m),(q)) = 0$ unless $p_1\alpha_1 + \cdots + p_r\alpha_r = q_1\alpha_1 + \cdots + q_r\alpha_r$. Now

$$X_{-\alpha_r}^{p_r} \cdots X_{-\alpha_1}^{p_1} H_1^{m_1} \cdots H_l^{m_l} X_{\alpha_1}^{q_1} \cdots X_{\alpha_r}^{q_r} \in \sum_{\alpha,\beta \in P_+} X_{-\alpha} \mathfrak{B} X_\beta$$

unless either $p_{s+1} = \cdots = p_r = 0$ or $q_{s+1} = \cdots = q_r = 0$. But if $a((p),(m),(q)) \neq 0$ we conclude from Lemma 8 that $p_{s+1} = \cdots = p_r = 0$ if and only if $q_{s+1} = \cdots = q_r = 0$. Hence $z \equiv z_0 \bmod \sum_{\alpha,\beta \in P_+} X_{-\alpha} \mathfrak{B} X_\beta$ where

$$z_0 = \sum a((p),(m),(q)) X_{-\alpha_s}^{p_s} \cdots X_{-\alpha_1}^{p_1} H_1^{m_1} \cdots H_l^{m_l} X_{\alpha_1}^{q_1} \cdots X_{\alpha_s}^{q_s} \in \mathfrak{B},$$

the sum extending only to those terms for which $p_j = q_j = 0$ $(s < j \le r)$. Since z_0 is of rank zero the lemma is proved.

Since $\mathfrak{g}$ is the direct sum of $\mathfrak{k}$, $\mathfrak{h}_{\mathfrak{p}}$, and $\mathfrak{n} = \sum_{\alpha \in P_+} C X_\alpha$ it follows (see [2]) that for any $b \in \mathfrak{B}$ there exist unique elements $x(m_1, \cdots, m_p)$ in $\mathfrak{X}$ such that

$$b \equiv \sum_{(m)} H_1^{m_1} \cdots H_p^{m_p} x(m_1, \cdots, m_p) \bmod \sum_{\alpha \in P_+} X_\alpha \mathfrak{B}.$$

We now define a linear mapping ξ_u of $\mathfrak{B}$ into $\mathfrak{X}[u]$ by setting

$$\xi_u(b) = \sum_{(m)} u_1'^{m_1} \cdots u_p'^{m_p} x(m_1, \cdots, m_p)$$

where $u_i' = u_i + 2\rho(H_i)$, $1 \le i \le p$. As before let $\mathfrak{Q}$ denote the centraliser of $\mathfrak{X}$ in $\mathfrak{B}$.

LEMMA 10. *If $a \in \mathfrak{Q}$ and $b \in \mathfrak{B}$, then $\xi_u(ba) = \xi_u(a)\xi_u(b)$.*

For in the above notation

$$ba \equiv \sum_{(m)} H_1^{m_1} \cdots H_p^{m_p} a x(m_1, \cdots, m_p) \bmod \sum_{\alpha \in P_+} X_\alpha \mathfrak{B},$$

since $ax(m_1, \cdots, m_p) = x(m_1, \cdots, m_p)a$. Hence if we observe that

$$H_1^{m_1} \cdots H_p^{m_p} X_\alpha \in X_\alpha \mathfrak{B} \qquad\qquad (\alpha \in P_+)$$

our assertion follows.

The significance of the mapping ξ_u derives from the following result.

LEMMA 11. *If* $a \in \mathfrak{Q}$, *then* $\eta_{u,v}^{\mathfrak{D}^*}(a) = \overline{\beta_v(\xi_u(a))}$.

For otherwise we could choose ν, μ such that $\eta_{\nu,\mu}(a) - \overline{\beta_\mu(\xi_\nu(a))} \neq 0$. Let N be the analytic subgroup of G corresponding to the subalgebra $\mathfrak{n}_0 = \mathfrak{n} \cap \mathfrak{g}_0$ of $\mathfrak{g}_0$. For any $x \in G$ and $v \in K$ put $x^v = vxv^{-1}$ and let $x \to \mathrm{Ad}(x)$ denote the adjoint representation of G. Extend $\mathrm{Ad}\,(x)$ to an automorphism of $\mathfrak{B}$ over C and put $b^v = \mathrm{Ad}\,(v)b$ $(b \in \mathfrak{B})$. Then if $f \in L_2(K^*)$ and $\pi = \pi_{\nu,\mu}$ (in the notation of the proof of Lemma 5) we find that

$$\pi(n^v)f(v^*) = f(v^*) \qquad\qquad (n \in N),$$

$$\pi(u^v)f(v^*) = e^{\mu(\Gamma(u))}f(v^*u^{*-1}) \qquad\qquad (u \in K),$$

$$\pi(h^v)f(v^*) = e^{(\nu+2\rho)(\log h)}f(v^*) \qquad\qquad (h \in A_+).$$

Here A_+ is the analytic subgroup of G corresponding to $\mathfrak{h}_{\mathfrak{p}_0}$ and $\log h$ $(h \in A_+)$ denotes the unique element $H \in \mathfrak{h}_{\mathfrak{p}_0}$ such that $\exp H = h$. From this it follows that if f is an indefinitely differentiable function on K^* which is well-behaved under π then

$$\pi(X^v)f(v^*) = 0 \qquad\qquad (X \in \mathfrak{n}),$$

$$\pi(H^v)f(v^*) = (\nu(H) + 2\rho(H))f(v^*) \qquad\qquad (H \in \mathfrak{h}_p).$$

Now let us consider $\pi(a)\psi$ where ψ is the function introduced in the proof of Lemma 5. Let v be a fixed element in K. Since $a \in \mathfrak{Q}$, $a = a^v$. Let $a \equiv a_0$ mod $\sum_{\alpha \in P_+} X_\alpha \mathfrak{B}$ where

$$a_0 = \sum_{(m)} H_1^{m_1} \cdots H_p^{m_p} z(m_1, \cdots, m_p) \qquad\qquad (z(m_1, \cdots, m_p) \in \mathfrak{X}).$$

Then $a^v - a_0^v \in \sum_{\alpha \in P_+} X_\alpha^v \mathfrak{B}$ and therefore

$$\pi(a)\psi(v^*) = \pi(a^v)\psi(v^*) = \pi(a_0^v)\psi(v^*) + \sum_{\alpha \in P_+} \pi(X_\alpha^v)\pi(b_\alpha)\psi(v^*),$$

where $b_\alpha \in \mathfrak{B}$. But if we put $f_\alpha = \pi(b_\alpha)\psi$ $(\alpha \in P_+)$ we know from our discussion above that $\pi(X_\alpha^v)f_\alpha(v^*) = 0$. Hence

$$\pi(a)\psi(v^*) = \pi(a_0^v)\psi(v^*).$$

We find similarly that

$$\pi(a_0^v)\psi(v^*) = \sum_{(m)} \nu_1'^{m_1} \cdots \nu_p'^{m_p} \pi((z(m_1, \cdots, m_p))^v)\psi(v^*) = \pi((\xi_\nu(a))^v)\psi(v^*)$$

where $\nu_i' = \nu(H_i) + 2\rho(H_i)$, $1 \leq i \leq p$. But $\psi(v^*u^{*-1}) = \psi(u^{*-1}v^*)$ $(u^* \in K^*)$ and therefore

$$\pi(u^v)\psi(v^*) = e^{\mu(\Gamma(u))}\psi(u^{*-1}v^*) = \pi(u)\psi(v^*).$$

Hence we conclude that

$$\pi(z^v)\psi(v^*) = \pi(z)\psi(v^*) \qquad\qquad (z \in \mathfrak{X})$$

and therefore

$$\pi(a)\psi(v^*) = \pi((\xi_\nu(a))^v)\psi(v^*) = \pi(\xi_\nu(a))\psi(v^*).$$

Since this is true for every $v^* \in K^*$, $\pi(a)\psi = \pi(\xi_\nu(a))\psi$. Now define the projection E and the representation $\bar{\tau}$ of $\bar{\mathfrak{X}}$ as in the proof of Lemma 5. Since $a \in \mathfrak{Q}$, $\pi(a)\psi = E\pi(a)\psi = \bar{\tau}(\eta_{\nu,\mu}(a))\psi$, and

$$\pi(\xi_\nu(a))\psi = \bar{\tau}(\overline{\beta_\mu(\xi_\nu(a))})\psi$$

from Lemma 6. Therefore $\bar{\tau}(\eta_{\nu,\mu}(a) - \overline{\beta_\mu(\xi_\nu(a))})\psi = 0$ which implies that $\eta_{\nu,\mu}(a) - \overline{\beta_\mu(\xi_\nu(a))} = 0$. Since this contradicts our choice of ν, μ the lemma follows.

In view of Lemmas 6 and 11 the mapping $\eta_{u,v}^{\mathfrak{D}^*}$ is now completely determined on $\mathfrak{A} = \mathfrak{Q}\mathfrak{X} = \mathfrak{X}\mathfrak{Q}$. If $a = \sum_{j=1}^s x_j a_j$ $(x_j \in \mathfrak{X},\ a_j \in \mathfrak{Q})$ then

$$\eta_{u,v}^{\mathfrak{D}^*}(a) = \sum_j \overline{\beta_v(x_j)}\eta_{u,v}^{\mathfrak{D}^*}(a_j) = \sum_j \overline{\beta_v(x_j\xi_u(a_j))}.$$

Notice that if $a = 0$, $\overline{\beta_v(\sum_j x_j\xi_u(a_j))} = 0$ and therefore

$$\beta_v\left(\sum_j x_j\xi_u(a_j)\right) \in \mathfrak{N}(\mathfrak{D}^*)[u, y].$$

Since this is true for every $\mathfrak{D}^* \in \Omega^*$ and since we know that $\bigcap_{\mathfrak{D}^* \in \Omega^*}\mathfrak{N}(\mathfrak{D}^*) = \{0\}$ (see [2]) it follows that $\beta_v(\sum_j x_j\xi_u(a_j)) = 0$. Hence the mapping $\xi_{u,v}$: $\sum_j x_j a_j \rightarrow \beta_v(\sum x_j\xi_u(a_j))$ $(x_j \in \mathfrak{X},\ a_j \in \mathfrak{Q})$ is a well-defined linear mapping of $\mathfrak{A}$ into $\mathfrak{X}[u, y]$ and $\eta_{u,v}^{\mathfrak{D}^*}(a) = \overline{\xi_{u,v}(a)}$ $(a \in \mathfrak{A})$.

Let $\mathfrak{D}_0^*$ be a class in Ω^*. As before we denote by $A_v(\mathfrak{D}_0^*)$ the set of all $a \in \mathfrak{B}[y]$ such that $\beta_{-v}(\mathfrak{N}(\mathfrak{D}_0^*))a \subset \mathfrak{B}_1\beta_{-v}(\mathfrak{N}(\mathfrak{D}_0^*))$ where $\mathfrak{B}_1 = \mathfrak{B}[y]$. Let μ be a linear function on $\mathfrak{C}$. We define a 1-1 mapping $\mathfrak{D} \rightarrow \mathfrak{D}_\mu$ of Ω onto itself such that $\mathfrak{N}(\mathfrak{D}_\mu) = \beta_{-\mu}(\mathfrak{N}(\mathfrak{D}))$ $(\mathfrak{N}(\mathfrak{D})$ is the kernel in $\mathfrak{X}$ of any representation in $\mathfrak{D})$. In particular put $\mathfrak{D}_0 = (\mathfrak{D}_0^*)_\mu$. For any $\mathfrak{D} \in \Omega$ let $A(\mathfrak{D})$ denote the set of all elements $a \in \mathfrak{B}$ such that $\mathfrak{N}(\mathfrak{D})a \subset \mathfrak{B}\mathfrak{N}(\mathfrak{D})$. If $a(y) \in \mathfrak{B}[y]$ we shall write $a(\mu)$ to denote the value of the polynomial under the substitution $y \rightarrow \mu$. It is obvious that if $a(y) \in A_v(\mathfrak{D}_0^*)$, then $a(\mu) \in A(\mathfrak{D}_0)$.

LEMMA 12. *For any $a_0 \in A(\mathfrak{D}_0)$ we can find an element $a(y) \in A_v(\mathfrak{D}_0^*)$ such that $a_0 = a(\mu)$.*

Before proving this lemma we make some general observations. Put

$\mathfrak{B}_1 = \mathfrak{B}[y]$ and $\mathfrak{U} = \mathfrak{B}_1\beta_{-\nu}(\mathfrak{N}(\mathfrak{D}_0^*))$. Let π denote the natural representation and $b \rightarrow b^*$ the natural mapping of $\mathfrak{B}_1$ on $\mathfrak{B}_1^* = \mathfrak{B}_1/\mathfrak{U}$. For any $\mathfrak{D} \in \Omega$ we denote by $\mathfrak{B}_1^*(\mathfrak{D})$ the set of all $b^* \in \mathfrak{B}_1^*$ such that $\pi(\beta_{-\nu}(\mathfrak{N}(\mathfrak{D})))b^* = 0$. Put $\pi^*(z) = \pi(\beta_{-\nu}(z))$ $(z \in \mathfrak{X})$. Then π^* is a representation of $\mathfrak{X}$ on $\mathfrak{B}_1^*$ and $\mathfrak{B}_1^*(\mathfrak{D})$ is exactly the set of those elements in $\mathfrak{B}_1^*$ which transform under π^* according to $\mathfrak{D}$. We shall show that $\mathfrak{B}_1^* = \sum_{\mathfrak{D} \in \Omega} \mathfrak{B}_1^*(\mathfrak{D})$. Since $1^* \in \mathfrak{B}_1^*(\mathfrak{D}_0^*)$ it is sufficient to prove that $V^* = \sum_{\mathfrak{D} \in \Omega} \mathfrak{B}_1^*(\mathfrak{D})$ is invariant under $\pi(\mathfrak{g}[y])$. But it is obvious that V^* is invariant under $\pi(\mathfrak{k}[y])$. Therefore it would be sufficient to show that $\pi(Y)b^* \in V^*$ for $b^* \in \mathfrak{B}_1^*(\mathfrak{D})$ $(\mathfrak{D} \in \Omega)$ and $Y \in \mathfrak{p}$. Let $U = \pi^*(\mathfrak{X})b^*$. Then $\dim U < \infty$ and the representation π_U^* of $\mathfrak{X}$ induced on U under π^* is semisimple. Moreover if $X \in \mathfrak{k}$ and $Y \in \mathfrak{p}$, $XY = [X, Y] + YX$ and $[X, Y] \in \mathfrak{p}$. Hence

$$\pi^*(X)\pi(Y)a^* = \pi(\sigma(X)Y)a^* + \pi(Y)\pi^*(X)a^* \qquad (a^* \in U)$$

where σ is the representation of $\mathfrak{k}$ on $\mathfrak{p}$ given by $\sigma(X)Y = [X, Y]$. Now we regard $\mathfrak{p} \times U$ as a $\mathfrak{k}$-module under the representation $\sigma + \pi_U^*$ of $\mathfrak{k}$. Consider the mapping ϕ: $Y \times a^* \rightarrow \pi(Y)a^*$ $(Y \in \mathfrak{p}, a^* \in U)$ of $\mathfrak{p} \times U$ onto $\pi(\mathfrak{p})U$. Then if $X \in \mathfrak{k}$ we may write the above relation as follows:

$$\phi(\sigma(X)Y \times a^* + Y \times \pi^*(X)a^*) = \pi(X)\phi(Y \times a^*).$$

This proves that the $\mathfrak{k}$-module $\pi(\mathfrak{p})U$ is a homomorphic image of the $\mathfrak{k}$-module $\mathfrak{p} \times U$. Since σ and π_U^* are semisimple representations, $\mathfrak{p} \times U$ is a semisimple $\mathfrak{k}$-module (see Lemma 8 of [5]). The same therefore holds for $\pi(\mathfrak{p})U$. Hence $\pi(\mathfrak{p})U \subset V^*$ and so our assertion is proved. Thus the representation π^* is quasi semisimple and the sum $\sum_{\mathfrak{D} \in \Omega} \mathfrak{B}_1^*(\mathfrak{D})$ is direct (see Lemma 6 of [5]).

Now we come to the lemma. Put $\mathfrak{B}_\mu = \sum_{1 \leq i \leq r} \mathfrak{B}_1(y_i - \mu(\Gamma_i))$ where $\Gamma_1, \cdots, \Gamma_r$ is the base for $\mathfrak{c}$ which we chose in §2. Let $\bar\pi$ be the natural representation and $b \rightarrow \bar{b}$ $(b \in \mathfrak{B}_1)$ the natural mapping of $\mathfrak{B}_1$ on $\mathfrak{B}_1/(\mathfrak{B}_\mu + \mathfrak{U})$. We may clearly identify $\overline{\mathfrak{B}}_1$ with $\mathfrak{B}_1^*/\mathfrak{B}_\mu^*$ in the natural way. Let $\overline{\mathfrak{B}(\mathfrak{D})}$ denote the image of $\mathfrak{B}_1^*(\mathfrak{D})$ in $\mathfrak{B}_1^*/\mathfrak{B}_\mu^* = \overline{\mathfrak{B}}_1$. Then $\overline{\mathfrak{B}}_1 = \sum_{\mathfrak{D} \in \Omega} \overline{\mathfrak{B}(\mathfrak{D})}$. Moreover if $a^* \in \mathfrak{B}_1^*(\mathfrak{D})$, $\pi(\beta_{-\nu}(\mathfrak{N}(\mathfrak{D}))a^* = \{0\}$. Hence if $\overline{a^*}$ is the image of a^* in $\overline{\mathfrak{B}(\mathfrak{D})}$, $\bar\pi(\beta_{-\mu}(\mathfrak{N}(\mathfrak{D})))\overline{a^*} = \{0\}$ and therefore $\overline{a^*}$ transforms under $\bar\pi(\mathfrak{k})$ according to $\mathfrak{D}_\mu$. This shows that the representation $\bar\pi(\mathfrak{k})$ is quasi semisimple. Now it is obvious that if $a \in \mathfrak{B}[y]$ then $a^* \in \mathfrak{B}_1^*(\mathfrak{D}_0^*)$ if and only if $a \in A_\nu(\mathfrak{D}_0^*)$. Similarly an element a of $\mathfrak{B}$ lies in $A(\mathfrak{D}_0)$ if and only if $\bar{a} \in \overline{\mathfrak{B}(\mathfrak{D}_0^*)}$. Now let $a_0 \in A(\mathfrak{D}_0)$. Then it follows from what we have said above that we can find $a_1(y) \in A_\nu(\mathfrak{D}_0^*)$ such that $\bar{a}_0 = \overline{a_1(y)}$. Let $a_1(\mu)$ be the value of $a_1(y)$ under the substitution $y \rightarrow \mu$. It is obvious that $\overline{a_1(y)} = \overline{a_1(\mu)}$. Hence $\overline{a_0 - a_1(\mu)} = 0$. Now it is easy to see that if $b \in \mathfrak{B}$ then $\bar{b} = 0$ if and only if $b \in \mathfrak{B}\mathfrak{N}(\mathfrak{D}_0)$. Hence $a_0 - a(\mu) \in \mathfrak{B}\mathfrak{N}(\mathfrak{D}_0)$ and therefore $\beta_\mu(a_0 - a_1(\mu)) \in \mathfrak{B}\mathfrak{N}(\mathfrak{D}_0^*)$ and $\beta_{-\nu}(\beta_\mu(a_0 - a_1(\mu))) \in \mathfrak{B}_1\beta_{-\nu}(\mathfrak{N}(\mathfrak{D}_0^*)) \subset A_\nu(\mathfrak{D}_0^*)$. So $a_2(y) = \beta_{-\nu}(\beta_\mu(a_0 - a_1(\mu))$ is in $A_\nu(\mathfrak{D}_0^*)$ and $a_2(\mu) = \beta_{-\mu}(\beta_\mu(a_0 - a_1(\mu)) = a_0 - a_1(\mu)$. Therefore if we put $a(y) = a_1(y) + a_2(y)$, $a(\mu) = a_0$ and this proves our result.

We shall keep $\mathfrak{D}_0^*$ fixed in the rest of this discussion. Put $A_2 = A_y(\mathfrak{D}_0^*)[u]$ and let $\mathfrak{R}_2$ be the set of all $a \in A_2$ such that $\eta_{u,y}(a) = 0$ $(\eta_{u,y} = \eta_{u,y}^{\mathfrak{D}_0^*})$. Extend the natural mapping of $\mathfrak{X}$ on $\bar{\mathfrak{X}} = \mathfrak{X}/\mathfrak{N}(\mathfrak{D}_0^*)$ to a (u, y)-linear mapping $z \to \bar{z}$ of $\mathfrak{X}[u, y]$ on $\bar{\mathfrak{X}}[u, y]$. Then if $a \in A_2$ we can choose $w \in \mathfrak{X}[u, y]$ such that $\eta_{u,y}(a) = \bar{w}$. Hence if $z = \beta_{-y}(w)$, $\eta_{u,y}(a - z) = 0$ from Lemma 6. This shows that if $z_1, \cdots, z_d$ is a base for $\mathfrak{X}$ mod $\mathfrak{N}(\mathfrak{D}_0^*)$ then $A_2 = \mathfrak{R}_2 + \sum_{1 \le i \le d} C[u, y]z_i'$ where $z_i' = \beta_{-y}(z_i)$, $1 \le i \le d$. Therefore if $b \in A_y(\mathfrak{D}_0^*)$ we can choose $c_{ij}(u, y) \in C[u, y]$ such that

$$bz_i' - \sum_{1 \le j \le d} c_{ij}(u, y)z_j' \in \mathfrak{R}_2 \qquad (1 \le i \le r).$$

Put

$$f(T, u, y) = \det (T\delta_{ij} - c_{ij}(u, y))_{1 \le i,\, j \le d},$$

where T is an indeterminate. Then $f(T, u, y)$ is a polynomial of degree d in T with coefficients in $C[u, y]$. Let $x_1, \cdots, x_l$ be l independent indeterminates $(l = \dim \mathfrak{h})$. Put $x(H_i) = x_i$, $1 \le i \le l$, and define $x(H)$ $(H \in \mathfrak{h})$ by linearity. Let W be the Weyl group of $\mathfrak{g}$ with respect to $\mathfrak{h}$. We denote by $f^{(\sigma)}(T, x, y)$ $(\sigma \in W)$ the polynomial in T (with coefficients in $C[x, y]$) obtained from $f(T, u, y)$ under the substitution $u_i \to x(\sigma^{-1}H_i) + \rho(\sigma^{-1}H_i) - \rho(H_i)$ $(1 \le i \le p)$. Put

$$F(T, x, y) = \prod_{\sigma \in W} f^{(\sigma)}(T, x, y).$$

Then if h is the order of the group W, F is of degree dh in T. Let χ_x denote the mapping of $\mathfrak{Z}$ into $C[x]$ defined as follows. If $z \equiv \sum_{(m)} a(m_1, \cdots, m_l)H_1^{m_1}H_2^{m_2} \cdots H_l^{m_l}$ mod $\sum_{\alpha \in P} \mathfrak{B}X_\alpha$ $(a(m_1, \cdots, m_l) \in C,\ z \in \mathfrak{Z})$ then $\chi_x(z) = \sum_{(m)} a(m_1, \cdots, m_l)x_1^{m_1} \cdots x_l^{m_l}$. We have seen in [3, Lemma 36] that χ_x is an isomorphism of $\mathfrak{Z}$ into $C[x]$. Extend χ_x to a (y)-linear mapping of $\mathfrak{Z}[y]$ into $C[x, y]$. It follows from Lemma 38 of [3] that every coefficient of F (regarded as a polynomial in T) is of the form $\chi_x(z(y))$ where $z(y)$ is some element in $\mathfrak{Z}[y]$. Therefore

$$F(T, x, y) = \sum_{0 \le q \le hd} T^q \chi_x(z_q(y)) \qquad (z_q(y) \in \mathfrak{Z}[y]).$$

For any $\mathfrak{D} \in \Omega^*$ let $\mathfrak{D}'$ denote the equivalence class of the representation σ' of $\mathfrak{k}' = [\mathfrak{k}, \mathfrak{k}]$ defined by any $\sigma \in \mathfrak{D}$. We denote by Ω_F the set of those $\mathfrak{D} \in \Omega$ for which $\mathfrak{D}' \in \Omega_F'$ (see §1 for the definition of Ω_F'). Then we have the following result which is one of the main steps of our argument.

LEMMA 13. *Suppose* $\mathfrak{D}_0^* \in \Omega_F^* = \Omega_F \cap \Omega^*$. *Then*

$$\beta_y\left(\sum_{0 \le q \le hd} b^q z_q(y)\right) \in \mathfrak{B}\mathfrak{N}(\mathfrak{D}_0^*)[y].$$

First we observe that if $\{\mathfrak{Y}\}$ is any collection of left ideals in $\mathfrak{X}$ then

$\cap(\mathfrak{B}\mathfrak{Y}) = \mathfrak{B}(\cap\mathfrak{Y})$. For let λ be the canonical mapping of the symmetric algebra $S(\mathfrak{g})$ onto $\mathfrak{B}$. Then if $\{Q\}$ is a base for $S(\mathfrak{p})$ we know that every element a in $\mathfrak{B}$ can be written uniquely in the form $a = \sum_Q \lambda(Q)z_Q$ ($z_Q \in \mathfrak{X}$) (see Lemma 12 of [5]). Since $\mathfrak{B} = \mathfrak{P}\mathfrak{X}$ ($\mathfrak{P} = \lambda(S(\mathfrak{p}))$), $\mathfrak{B}\mathfrak{Y} = \mathfrak{P}\mathfrak{Y}$ and it follows that $a \in \mathfrak{B}\mathfrak{Y}$ if and only if $z_Q \in \mathfrak{Y}$ for all Q. Our assertion is an immediate consequence of this fact.

Since $\bar{\mathfrak{X}} = \mathfrak{X}/\mathfrak{N}(\mathfrak{D}_0{}^*)$ is a finite-dimensional simple algebra, $\mathfrak{N}(\mathfrak{D}_0{}^*) = \cap\mathfrak{Y}$ where $\mathfrak{Y}$ runs over all maximal left ideals in $\mathfrak{X}$ containing $\mathfrak{N}(\mathfrak{D}_0{}^*)$. Therefore $\mathfrak{B}\mathfrak{N}(\mathfrak{D}_0{}^*) = \cap(\mathfrak{B}\mathfrak{Y})$ and so it would be sufficient to prove that

$$\beta_y\left(\sum_{0 \leq q \leq hd} b^q z_q(y)\right) \in \mathfrak{B}\mathfrak{Y}[y]$$

for every maximal left ideal $\mathfrak{Y}$ (in $\mathfrak{X}$) containing $\mathfrak{N}(\mathfrak{D}_0{}^*)$. Suppose then that the above statement is false for some $\mathfrak{Y}$. We shall often write $b(y)$ instead of b to put in evidence the fact that, being an element of $A_y(\mathfrak{D}_0{}^*)$, it is a polynomial in (y). Since $\mathfrak{Y}$ is maximal it is clear that we can find $m_i \in C$ ($1 \leq i \leq r$) such that $\Gamma_i - m_i \in \mathfrak{Y}$ ($1 \leq i \leq r$). Therefore

$$\beta_y\left(\sum_{0 \leq q \leq hd} b^q z_q(y)\right) \equiv \sum_{1 \leq j \leq e} a_j(y)w_j \bmod \mathfrak{B}\mathfrak{Y}[y]$$

where $w_1, \cdots, w_e \in \mathfrak{P}\mathfrak{X}'$ and $a_j(y) \in C[y]$. (Here the notation is the same as in §1.) We may evidently assume that $w_1, \cdots, w_e$ are linearly independent mod $\mathfrak{B}\mathfrak{Y}$ and $a_j(y) \neq 0$, $1 \leq j \leq e$. Choose a linear function μ on $\mathfrak{c}$ such that $a_1(\mu) \neq 0$ and $\mu(\Gamma_i) + m_i$, $1 \leq i \leq r$, are all integers. It is clear that this is possible. Then

$$\beta_\mu\left(\sum_{0 \leq q \leq hd} b^q(\mu)z_q(\mu) - \sum_{1 \leq j \leq e} a_j(\mu)w_j\right) \in \mathfrak{B}\mathfrak{Y}$$

since $w_j \in \mathfrak{P}\mathfrak{X}'$ and therefore $\beta_\mu(w_j) = w_j$. Here $b(\mu)$, $z_q(\mu)$, and $a_j(\mu)$ are the values of $b(y)$, $z_q(y)$, and $a_j(y)$ respectively under the substitution $y \to \mu$. Put $\mathfrak{Y}_\mu = \beta_{-\mu}(\mathfrak{Y})$. Then since $\mathfrak{B}\mathfrak{Y} = \mathfrak{P}\mathfrak{Y}$ it is clear that

$$\sum_{0 \leq q \leq hd} b^q(\mu)z_q(\mu) - \sum_{1 \leq j \leq e} a_j(\mu)w_j \in \mathfrak{B}\mathfrak{Y}_\mu.$$

But $\sum_{1 \leq j \leq e} a_j(\mu)w_j \notin \mathfrak{B}\mathfrak{Y}$ because $a_1(\mu) \neq 0$. Hence

$$\sum_{1 \leq j \leq e} a_j(\mu)w_j = \beta_{-\mu}\left(\sum_{1 \leq j \leq e} a_j(\mu)w_j\right) \notin \mathfrak{B}\mathfrak{Y}_\mu$$

and so

$$\sum_{0 \leq q \leq hd} b^q(\mu)z_q(\mu) \notin \mathfrak{B}\mathfrak{Y}_\mu.$$

Let $\mathfrak{D}_0'^*$ be the equivalence class of the representation of $\mathfrak{k}'$ defined by any representation of $\mathfrak{k}$ in $\mathfrak{D}_0{}^*$. It is clear that $\mathfrak{D}_0'^*$ is an irreducible class and since $\mathfrak{D}_0{}^* \in \Omega_F^*$, $\mathfrak{D}_0'^* \in \Omega_F'$. (See §1 for notation.) Moreover $\Gamma_i - (\mu_i + m_i) \in \mathfrak{Y}_\mu$ $(\mu_i = \mu(\Gamma_i), \ 1 \leq i \leq r)$ and $\mu_i + m_i$ are all integers. Let $\mathfrak{D}_0$ be the class of the natural representation of $\mathfrak{k}$ on $\mathfrak{X}/\mathfrak{Y}_\mu$. In view of our choice of the base $(\Gamma_1, \cdots, \Gamma_r)$ it follows from the arguments of §1 that $\mathfrak{D}_0$ occurs in some finite-dimensional representation of $\mathfrak{g}$. Therefore from Lemma 1 there exists a left ideal $\mathfrak{M}$ in $\mathfrak{B}$ such that $\mathfrak{M} \supset \mathfrak{Y}_\mu$, $\dim \mathfrak{B}/\mathfrak{M} < \infty$, and $\sum_q b^q(\mu) z_q(\mu) \notin \mathfrak{M}$. Since $\mathfrak{g}$ is semisimple, we may assume that $\mathfrak{M}$ is a maximal left ideal in $\mathfrak{B}$. Let $a \to \bar{a}$ denote the natural mapping and ν the natural representation of $\mathfrak{B}$ on $\mathfrak{B}/\mathfrak{M} = \bar{\mathfrak{B}}$. Then $\nu(\sum_q b^q(\mu) z_q(\mu)) \bar{1} \neq 0$. Let $\psi \neq 0$ be an element in $\bar{\mathfrak{B}}$ belonging to the highest weight Λ of ν. We "extend" ν to a representation of the group G on $\bar{\mathfrak{B}}$. Also we define a representation ν^* of K^* by setting $\nu^*(v^*) = e^{-(\mu\Gamma(v))}\nu(v)$ $(v \in K)$. It is obvious that $\bar{1}$ transforms according to $\mathfrak{D}_0{}^*$ under ν^*. Put

$$E = d(\mathfrak{D}_0^*) \int_{K^*} \chi_{\mathfrak{D}_0{}^*}(v^{*-1}) \nu^*(v^*) dv^*$$

where $\chi_{\mathfrak{D}_0{}^*}$ is the character of K^* corresponding to the class $\mathfrak{D}_0{}^*$. Then

$$\nu(x) E\psi = d(\mathfrak{D}_0^*) \int_{K^*} \chi_{\mathfrak{D}_0{}^*}(v^{*-1}) \nu(x) \nu^*(v^*) \psi \, dv^*.$$

But $xv = v_x \exp H(x, v^*) \cdot n$ where $n \in N$ $(x \in G, v \in K)$ (the notation is the same as in [5, §12]). Since ψ belongs to the highest weight Λ, $\nu(n)\psi = \psi$ and therefore $\nu(xv)\psi = e^{\Lambda(H(x,v^*))} \nu(v_x)\psi$. So we get

$$\nu(x) E\psi = d(\mathfrak{D}_0^*) \int_{K^*} \chi_{\mathfrak{D}_0{}^*}(v^{*-1}) e^{\mu(\Gamma(x,v^*))} e^{\Lambda(H(x,v^*))} \nu^*(v_x^*) \psi \, dv^*.$$

Now $E^2 = E$ and $E\nu^*(v^*) = \nu^*(v^*)E$. Therefore

$$E\nu(x) E\psi = d(\mathfrak{D}_0^*) \int_{K^*} \chi_{\mathfrak{D}_0{}^*}((v_{x-1}^*)^{-1}) e^{-\mu(\Gamma(x^{-1},v^*))} e^{-(\Lambda+2\rho)(H(x^{-1},v^*))} \nu^*(v^*) \psi \, dv^*$$

if we take into account the relation $dv_{x-1}^* = e^{-2\rho(H(x^{-1},v^*))} \, dv^*$ (see [5, §11]). Now if we recall the definition of $\eta_{u,\psi}(a) = \eta_{u,\psi}^{\mathfrak{D}_0^*}(a)$ $(a \in \mathfrak{B})$ it follows immediately that

$$E\nu(a) E\psi = \gamma(\eta_{\Lambda_\mathfrak{p},\mu}(a)) \qquad\qquad (a \in \mathfrak{B}).$$

Here $\Lambda_\mathfrak{p}$ is the restriction of Λ on $\mathfrak{h}_\mathfrak{p}$ and γ is the linear mapping of $\bar{\mathfrak{X}}$ into $E\bar{\mathfrak{B}}$ defined by $\gamma(\bar{z}) = \nu^*(z) E\psi$ $(z \in \mathfrak{X})$. (It is convenient to denote the corresponding representations of K^*, $\mathfrak{k}$, and $\mathfrak{X}$ by the same symbol ν^*.)

Extend ν to a representation of $\mathfrak{B}[u, y]$ by setting $\nu(y_i) = \mu(\Gamma_i)\nu(1)$ $(1 \leq i \leq r)$ and $\nu(u_i) = \Lambda(H_i)\nu(1)$ $(1 \leq i \leq p)$. Now if $a(y) \in A_y(\mathfrak{D}_0^*)$, $\beta_{-\mu}(\mathfrak{N}(\mathfrak{D}_0^*)) a(\mu) \subset \mathfrak{B}\beta_{-\mu}(\mathfrak{N}(\mathfrak{D}_0^*))$. But it is clear that $\beta_{-\mu}(\mathfrak{N}(\mathfrak{D}_0^*)) = \mathfrak{N}(\mathfrak{D}_0)$ and

$E\overline{\mathfrak{B}}$ is exactly the set of those elements in $\overline{\mathfrak{B}}$ which transform under $\nu(\mathfrak{k})$ according to $\mathfrak{D}_0$. From this it follows that $E\nu(a(y))E = \nu(a(y))E$. Hence

$$\nu\left(b(y)\beta_{-\nu}(z_i) - \sum_{j=1}^{r} c_{ij}(u, y)\beta_{-\nu}(z_j)\right)E\psi$$

$$= E\nu\left(b(\mu)\beta_{-\mu}(z_i) - \sum_{j=1}^{r} c_{ij}(\Lambda_{\mathfrak{p}}, \mu)\beta_{-\mu}(z_j)\right)E\psi$$

$$= \gamma\left(\eta_{\Lambda_{\mathfrak{p}},\mu}\left(b(\mu)\beta_{-\mu}(z_i) - \sum_{j=1}^{r} c_{ij}(\Lambda_{\mathfrak{p}},\mu)\beta_{-\mu}(z_j)\right)\right).$$

But we know that

$$\eta_{u,y}\left(b(y)\beta_{-\nu}(z_i) - \sum_j c_{ij}(u, y)\beta_{-\nu}(z_j)\right) = 0.$$

Hence

$$\nu\left(b(y)\beta_{-\nu}(z_i) - \sum_j c_{ij}(u, y)\beta_{-\nu}(z_j)\right)E\psi = 0$$

or

$$\nu(b(\mu))\psi_i - \sum_j c_{ij}(\Lambda_{\mathfrak{p},\mu})\psi_j = 0 \qquad\qquad (1 \leq i \leq d)$$

where $\psi_i = \nu(\beta_{-\mu}(z_i))E\psi = \nu^*(z_i)E\psi$. Put $g(T) = \det\,(\delta_{ij}T - c_{ij}(\Lambda_{\mathfrak{p},\mu}))_{1\leq i,\,j\leq d}$. Then $g(T)$ is a polynomial in T with coefficients in C and it is obvious that $\nu(g(b(\mu)))\psi_i = 0$, $1\leq i\leq d$. On the other hand $\mathfrak{g} = \mathfrak{k} + \mathfrak{h}_{\mathfrak{p}} + \mathfrak{n}$ $(\mathfrak{n} = \sum_{\alpha\in P_+} CX_\alpha)$ and $\nu(\mathfrak{n})\psi = \{0\}$ while $\nu(\mathfrak{h}_{\mathfrak{p}})\psi \subset C\psi$. Moreover every element in $\mathfrak{B}$ can be written as a linear combination of elements of the form zan where $z\in\mathfrak{X}$ and a and n are products of elements in $\mathfrak{h}_{\mathfrak{p}}$ and $\mathfrak{n}$ respectively (see Lemma 12 of [5]). Hence

$$E\overline{\mathfrak{B}} = E\nu(\mathfrak{B})\psi = E\nu(\mathfrak{X})\psi = \nu(\mathfrak{X})E\psi = \nu^*(\mathfrak{X})E\psi.$$

Since $z_1, \cdots, z_d$ is a base for $\mathfrak{X}$ mod $\mathfrak{N}(\mathfrak{D}_0^*)$ it follows that $E\overline{\mathfrak{B}}$ is spanned by $\psi_i = \nu^*(z_i)E\psi, 1\leq i\leq d$. Moreover as $\overline{1}$ transforms under ν^* according to $\mathfrak{D}_0^*$, $\overline{1}\in E\overline{\mathfrak{B}}$ and so $\nu(g(b(\mu))\overline{1} = 0$. But it is obvious that $g(T) = f(T, \Lambda_{\mathfrak{p}}, \mu)$ where $f(T, \Lambda_{\mathfrak{p}}, \mu)$ is the value of $f(T, u, y)$ under the substitution $u\to\Lambda_{\mathfrak{p}}$, $y\to\mu$. Let $F(T, \Lambda, \mu)$ denote the value of $F(T, x, y)$ under the substitution $x\to\Lambda$ (i.e. $x_i\to\Lambda(H_i)$, $1\leq i\leq l$) and $y\to\mu$. Since $f^{(1)}(T, x, y)$ is one of the factors of $F(T, x, y)$, $g(T)$ divides $F(T, \Lambda, \mu)$. Hence $\nu(F(b(\mu), \Lambda, \mu))\overline{1} = 0$. On the other hand

$$F(T, \Lambda, \mu) = \sum_{0\leq q\leq hd} T^q \chi_\Lambda(z_q(\mu))$$

where $\chi_\Lambda(z_q(\mu))$ is the value of $\chi_x(z_q(\mu))$ under the substitution $x\to\Lambda$. Since ν

is an irreducible representation with the highest weight Λ we know that (see the proof of Lemma 36 of [3])

$$\nu(z_q(\mu)) = \chi_\Lambda(z_q(\mu))\nu(1).$$

Hence

$$\nu(F(b(\mu), \Lambda, \mu)) = \sum_{0 \leq q \leq hd} \nu(b^q(\mu))\chi_\Lambda(z_q(\mu)) = \nu\left(\sum_{0 \leq q \leq hd} b^q(\mu)z_q(\mu)\right).$$

Therefore

$$\nu\left(\sum_{0 \leq q \leq hd} b^q(\mu)z_q(\mu)\right)\bar{1} = 0.$$

Since this contradicts our earlier result the lemma is proved.

In order to derive the consequences of the above lemma we need the following simple result.

LEMMA 14. *Let $\mathcal{A}$ be an associative algebra and $\sigma_1, \cdots, \sigma_r, \pi$ a finite set of finite-dimensional representations of $\mathcal{A}$. Let σ be the direct sum of σ_i, $1 \leq i \leq r$. Suppose π is irreducible and $\pi(a) = 0$ whenever $\sigma(a) = 0$ $(a \in \mathcal{A})$. Then there exists an index i such that $\pi(a) = 0$ whenever $\sigma_i(a) = 0$.*

Obviously $\sigma(\mathcal{A})$ is a finite-dimensional associative algebra. Hence $\sigma(\mathcal{A}) = S_0 + N_0$ where N_0 is the radical and S_0 a semisimple subalgebra of $\sigma(\mathcal{A})$. Let S and N be the complete inverse images of S_0 and N_0 respectively under σ. Since $\sigma(a) \to \pi(a)$ $(a \in \mathcal{A})$ is an irreducible representation of $\sigma(\mathcal{A})$ it is obvious that $\pi(N) = \{0\}$. Now consider the representation $\phi: \sigma(s) \to \pi(s)$ $(s \in S)$ of $\sigma(S)$. Since $\sigma(S)$ is semisimple and π is irreducible, ϕ must be equivalent to some irreducible component of the identical representation $\sigma(s) \to \sigma(s)$. Therefore since σ is the direct sum of σ_i, ϕ is equivalent to an irreducible component of one of the representations $\sigma(s) \to \sigma_i(s)$. Hence we can choose an index i such that $\sigma_i(s) = 0$ implies $\pi(s) = 0$. Now suppose $\sigma_i(a) = 0$ $(a \in \mathcal{A})$. Let $a = s + n$ $(s \in S, n \in N)$. Then $\sigma_i(s) = -\sigma_i(n) \in \sigma_i(N)$. However $\sigma_i(S)$ is a semisimple algebra and $\sigma_i(N)$ a nilpotent ideal in $\sigma_i(\mathcal{A})$. Therefore $\sigma_i(S) \cap \sigma_i(N) = \{0\}$. Hence $\sigma_i(s) = 0$ and therefore $\pi(a) = \pi(s+n) = \pi(s) = 0$. Thus $\sigma_i(a) = 0$ implies $\pi(a) = 0$ and so our lemma is proved.

We shall now apply Lemma 13 to the theory of representations of $\mathfrak{B}$. Let π be a representation of $\mathfrak{B}$ on a vector space V. For any $\mathfrak{D} \in \Omega$ let $V_\mathfrak{D}$ denote the subspace consisting of all those elements in V which transform under $\pi(\mathfrak{k})$ according to $\mathfrak{D}$. Let $\mathfrak{N}(\mathfrak{D})$ denote the kernel (in $\mathfrak{X}$) of any representation of $\mathfrak{X}$ in $\mathfrak{D}$. As before, we shall say that π is *quasi-simple* if $V = \sum_{\mathfrak{D} \in \Omega} V_\mathfrak{D}$ and there exists a homomorphism χ of $\mathfrak{Z}$ into C such that $\pi(z)\phi = \chi(z)\phi$ for all $z \in \mathfrak{Z}$ and $\phi \in V$. χ is then called the *infinitesimal character* of π. Consider the mapping $z \to \chi_x(z)$ $(z \in \mathfrak{Z})$ of $\mathfrak{Z}$ into $C[x]$ which we have already discussed above. We have seen in [3, Part III] that there exists a linear function Λ on $\mathfrak{h}$ such that $\chi(z) = \chi_\Lambda(z)$ $(z \in \mathcal{Z})$ where $\chi_\Lambda(z)$ is the value of $\chi_x(z)$ under the substitution $x \to \Lambda$. As usual we say that π is irreducible if $\pi(\mathfrak{B})\psi = V$ for every nonzero ψ in V.

Let π be a quasi-simple irreducible representation of $\mathfrak{B}$ on a vector space V such that $V_{\mathfrak{D}_0} \neq \{0\}$ for some $\mathfrak{D}_0 \in \Omega_F$. Choose a linear function μ on $\mathfrak{c}$ such that $\Gamma - \mu(\Gamma) \in \mathfrak{N}(\mathfrak{D}_0)$ for all $\Gamma \in \mathfrak{c}$. It is clear that there exists a class $\mathfrak{D}_0^* \in \Omega^*$ such that $\mathfrak{N}(\mathfrak{D}_0^*) = \beta_\mu(\mathfrak{N}(\mathfrak{D}_0))$. Let $A(\mathfrak{D}_0)$ be the set of all $a \in \mathfrak{B}$ for which $\mathfrak{N}(\mathfrak{D}_0)a \subset \mathfrak{B}\mathfrak{N}(\mathfrak{D}_0)$.

LEMMA 15. *There exists a linear function Λ on $\mathfrak{h}$ with the following two properties*:

(1) χ_Λ *is the infinitesimal character of π.*

(2) *Let $\Lambda_\mathfrak{p}$ denote the restriction of Λ on $\mathfrak{h}_\mathfrak{p}$. Suppose a is an element in $A(\mathfrak{D}_0)$ such that $\eta_{\Lambda_{\mathfrak{p},\mu}}^{\mathfrak{D}_0^*}(az) = 0$ for all $z \in \mathfrak{X}$. Then $\pi(a)\psi = 0$ for every $\psi \in V_{\mathfrak{D}_0}$.*

Let A_1 be the set of all $a \in \mathfrak{B}_1 = \mathfrak{B}[y]$ such that $\beta_{-y}(\mathfrak{N}(\mathfrak{D}_0^*))a \subset \mathfrak{B}_1 \beta_{-y}(\mathfrak{N}(\mathfrak{D}_0^*))$ $(A_1 = A_y(\mathfrak{D}_0^*))$ in the notation of Lemma 12). Let b be any element in A_1. Since $\mathfrak{D}_0 \in \Omega_F$ and since $\mathfrak{c} \subset \mathfrak{N}(\mathfrak{D}_0^*)$ it follows from the argument given in §1 that $\mathfrak{D}_0^* \in \Omega_F^*$ and therefore

$$\beta_y\left(\sum_{0 \leq q \leq hd} b^q z_q(y)\right) \in \mathfrak{B}\mathfrak{N}(\mathfrak{D}_0^*)[y]$$

in the notation of Lemma 13. Hence

$$\sum_{0 \leq q \leq hd} b^q z_q(y) \in \mathfrak{B}_1 \beta_{-y}(\mathfrak{N}(\mathfrak{D}_0^*)).$$

Extend π to a representation of $\mathfrak{B}_1$ by putting $\pi(y_i) = \mu(\Gamma_i)\pi(1)$ $(1 \leq i \leq r)$. Then if $\psi \in V_{\mathfrak{D}_0}$

$$\pi(\beta_{-y}(\mathfrak{N}(\mathfrak{D}_0^*)))\psi = \pi(\beta_{-\mu}(\mathfrak{N}(\mathfrak{D}_0^*)))\psi = \pi(\mathfrak{N}(\mathfrak{D}_0))\psi = \{0\}.$$

Hence

$$\pi\left(\sum_{0 \leq q \leq hd} b^q z_q(y)\right)\psi = 0.$$

It is obvious that $V_{\mathfrak{D}_0}$ is invariant under $\pi(A_1)$. Hence we get a representation ν of A_1 on $V_{\mathfrak{D}_0}$. Now $A_1 \supset \mathfrak{Q}\mathfrak{X}$ and we know from Corollary 2 to Theorem 2 that $V_{\mathfrak{D}_0}$ is irreducible under $\pi(\mathfrak{Q}\mathfrak{X})$. Therefore ν is an irreducible representation of A_1 on $V_{\mathfrak{D}_0}$.

Let Λ be a linear function on $\mathfrak{h}$ such that χ_Λ is the infinitesimal character of π. We use the notation of Lemma 13 and consider the polynomials $f(T, u, y)$, $f^{(\sigma)}(T, x, y)$ $(\sigma \in W)$ corresponding to the element $b \in A_1$. Let $f^{(\sigma)}(T, \Lambda, \mu)$ denote the value of $f^{(\sigma)}(T, x, y)$ under the substitution $x \to \Lambda$, $y \to \mu$. Put $\Lambda^\sigma(H) = \Lambda(\sigma^{-1}H) + \rho(\sigma^{-1}H) - \rho(H)$ $(H \in \mathfrak{h})$ and let $\Lambda_\mathfrak{p}^\sigma$ denote the restriction of Λ^σ on $\mathfrak{h}_\mathfrak{p}$. Then it follows from the definition of $f^{(\sigma)}(T, x, y)$ that $f^{(\sigma)}(T, \Lambda, \mu) = f(T, \Lambda_\mathfrak{p}^\sigma, \mu)$ where $f(T, \Lambda_\mathfrak{p}^\sigma, \mu)$ is the value of $f(T, u, y)$ under the substitution $u \to \Lambda_\mathfrak{p}^\sigma$, $y \to \mu$. Therefore

$$0 = \pi\left(\sum_q b^q z_q(y)\right)\psi = \sum_q \chi_\Lambda(z_q(\mu))\pi(b^q)\psi$$

$$= \nu\left(\prod_{\sigma \in W} f^{(\sigma)}(b, \Lambda, \mu)\right)\psi \quad (\psi \in V_{\mathfrak{D}_0}).$$

Hence

$$\nu\left(\prod_{\sigma \in W} f^{(\sigma)}(b, \Lambda, \mu)\right) = \nu\left(\prod_{\sigma \in W} f(b, \Lambda_\mathfrak{p}^\sigma, \mu)\right) = 0.$$

We have the base $z_1, \cdots, z_d$ for $\mathfrak{X}$ mod $\mathfrak{N}(\mathfrak{D}_0^*)$. Put $z_i' = \beta_{-y}(z_i)$, $1 \leq i \leq d$. Let $\mathfrak{N}_2$ denote the set of all elements $a \in A_1[u] = A_2$ such that $\eta_{u,y}(a) = 0$ ($\eta_{u,y} = \eta_{u,y}^{\mathfrak{D}_0^*}$). We know from Lemma 5 that $\mathfrak{N}_2$ is a left ideal in A_2. Put

$$\mathfrak{N}'_{\Lambda_\mathfrak{p},\mu}^\sigma = \mathfrak{N}_2 + \sum_{1 \leq i \leq p} A_2(u_i - \Lambda^\sigma(H_i)) + \sum_{1 \leq i \leq r} A_2(y_i - \mu(\Gamma_i)) \quad (\sigma \in W).$$

For any $a \in A_2$ let $\eta'^\sigma_{\Lambda_\mathfrak{p},\mu}(a)$ denote the value of $\eta_{u,y}(a)$ under the substitution $u \to \Lambda_\mathfrak{p}^\sigma$, $y \to \mu$. It is clear that $\eta'^\sigma_{\Lambda_\mathfrak{p},\mu}(a) = 0$ for $a \in \mathfrak{N}'^\sigma_{\Lambda_\mathfrak{p},\mu}$. Moreover we have seen (see p. 52) that $A_2 = \mathfrak{N}_2 + \sum_{i=1}^d C[u, y]z_i'$. Hence

$$A_2 = \mathfrak{N}'^\sigma_{\Lambda_\mathfrak{p},\mu} + \sum_{1 \leq i \leq d} C z_i'.$$

Therefore for any $a \in A_2$ we can choose $c_i \in C$ such that

$$a \equiv \sum_{1 \leq i \leq d} c_i z_i' \bmod \mathfrak{N}'^\sigma_{\Lambda_\mathfrak{p},\mu}$$

and so

$$\eta'^\sigma_{\Lambda_\mathfrak{p},\mu}(a) = \sum_{1 \leq i \leq d} c_i \eta'^\sigma_{\Lambda_\mathfrak{p},\mu}(z_i').$$

But, from Lemma 6, $\eta_{u,y}(z_i') = \bar{z}_i$ where $z \to \bar{z}$ is the natural mapping of $\mathfrak{X}$ on $\bar{\mathfrak{X}} = \mathfrak{X}/\mathfrak{N}(\mathfrak{D}_0^*)$. Hence $\eta'^\sigma_{\Lambda_\mathfrak{p},\mu}(z_i') = \bar{z}_i$ and therefore

$$\eta'^\sigma_{\Lambda_\mathfrak{p},\mu}(a) = \sum_{1 \leq i \leq d} c_i \bar{z}_i.$$

So in particular if $\eta'^\sigma_{\Lambda_\mathfrak{p},\mu}(a) = 0$, $c_i = 0$, $1 \leq i \leq d$, and therefore $a \in \mathfrak{N}'^\sigma_{\Lambda_\mathfrak{p},\mu}$. This proves that an element $a \in A_2$ belongs to $\mathfrak{N}'_{\Lambda_\mathfrak{p},\mu}$ if and only if $\eta'^\sigma_{\Lambda_\mathfrak{p},\mu}(a) = 0$. Since $\eta'^\sigma_{\Lambda_\mathfrak{p},\mu}(\sum_i c_i z_i) = \sum_i c_i \bar{z}_i'$ ($c_i \in C$) we conclude that $z_1', \cdots, z_r'$ are linearly independent over C modulo $\mathfrak{N}'^\sigma_{\Lambda_\mathfrak{p},\mu}$.

Put $\mathfrak{R}' = \bigcap_{\sigma \in W} \mathfrak{R}'^{\sigma}_{\Lambda_{\mathfrak{p},\mu}}$. We have seen above that $A_2 = \mathfrak{R}'^{\sigma}_{\Lambda_{\mathfrak{p},\mu}} + \sum_{i=1}^{d} C_i z_i'$. Hence $\dim A_2/\mathfrak{R}'^{\sigma}_{\Lambda_{\mathfrak{p},\mu}} \leq d$ and therefore $\dim A_2/\mathfrak{R}' \leq hd$ where h is the order of the Weyl group W. Let τ be the natural representation of A_2 on $A_2/\mathfrak{R}'$. Suppose $\tau(b) = 0$ for some $b \in A_1$. Then $bz_i' \in \mathfrak{R}' \subset \mathfrak{R}'^{\sigma}_{\Lambda_{\mathfrak{p},\mu}}$ $(1 \leq i \leq d)$. But in the notation of Lemma 13

$$bz_i' - \sum_j c_{ij}(u, y)z_j' \in \mathfrak{R}_2 \subset \mathfrak{R}'^{\sigma}_{\Lambda_{\mathfrak{p},\mu}}.$$

Since $bz_i' \in \mathfrak{R}$ and $c_{ij}(u, y)z_j' - c_{ij}(\Lambda^{\sigma}_{\mathfrak{p},\mu})z_j' \in \mathfrak{R}'^{\sigma}_{\Lambda_{\mathfrak{p},\mu}}$ it follows that

$$\sum_{j=1}^{d} c_{ij}(\Lambda^{\sigma}_{\mathfrak{p}}, \mu)z_j' \in \mathfrak{R}'^{\sigma}_{\Lambda_{\mathfrak{p},\mu}}.$$

Therefore in view of the linear independence of z_j' over C mod $\mathfrak{R}'^{\sigma}_{\Lambda_{\mathfrak{p},\mu}}$ we conclude that $c_{ij}(\Lambda^{\sigma}_{\mathfrak{p},\mu}) = 0$, $1 \leq i, j \leq d$. But on the other hand

$$f(T, u, y) = \det (T\delta_{ij} - c_{ij}(u, y)), \qquad 1 \leq i, j \leq d.$$

Therefore $f(T, \Lambda^{\sigma}_{\mathfrak{p}}, \mu) = T^d$ and

$$0 = \nu\left(\prod_{\sigma \in W} f(b, \Lambda^{\sigma}_{\mathfrak{p}}, \mu)\right) = \nu(b^{dh}) = 0.$$

Hence $\tau(b) = 0$ $(b \in A_1)$ implies that $\nu(b)$ is nilpotent. Let B_1 be the set of all elements $a \in A_1$ such that $\tau(a) = 0$. Then it follows that $\nu(B_1)$ is a two-sided ideal in $\nu(A_1)$ all of whose elements are nilpotent. But since ν is irreducible the algebra $\nu(A_1)$ is simple. Hence $\nu(B_1) = \{0\}$. This shows that $\tau(a) = 0$ $(a \in A_1)$ implies $\nu(a) = 0$.

Let $\tau^{(\sigma)}$ denote the natural representation of A_1 on $A_1/\mathfrak{R}''^{\sigma}_{\Lambda_{\mathfrak{p},\mu}} \cong A_2/\mathfrak{R}'^{\sigma}_{\Lambda_{\mathfrak{p},\mu}}$ where $\mathfrak{R}''^{\sigma}_{\Lambda_{\mathfrak{p},\mu}} = A_1 \cap \mathfrak{R}'^{\sigma}_{\Lambda_{\mathfrak{p},\mu}}$. Let τ_0 be the direct sum of $\tau^{(\sigma)}$ $(\sigma \in W)$. It is clear that $\tau_0(a) = 0$ $(a \in A_1)$ if and only if $\tau(a) = 0$. Therefore it is possible (from Lemma 14) to choose a $\sigma \in W$ such that $\tau^{(\sigma)}(a) = 0$ $(a \in A_1)$ implies $\nu(a) = 0$. Since $\chi_\Lambda = \chi_{\Lambda^{\sigma}}$ (see [3, Theorem 5]) we may assume without loss of generality that $\sigma = 1$.

For any $a \in A_1$ let $\phi(a)$ denote the value of a under the substitution $y \to \mu$. We have seen in Lemma 12 that ϕ is a homomorphism of A_1 *onto* $A(\mathfrak{D}_0)$. Let $\mathfrak{R}_{\Lambda_{\mathfrak{p}},\mu}$ be the set of all $b \in A(\mathfrak{D}_0)$ such that $\eta_{\Lambda_{\mathfrak{p}},\mu}(b) = 0$. Let σ be the natural representation of $A(\mathfrak{D}_0)$ on $A(\mathfrak{D}_0)/\mathfrak{R}_{\Lambda_{\mathfrak{p}},\mu}$. Suppose $\sigma(a) = 0$ $(a \in A(\mathfrak{D}_0))$. Then $a\beta_{-\mu}(z_i) \in \mathfrak{R}_{\Lambda_{\mathfrak{p}},\mu}$, $1 \leq i \leq d$. Choose $b \in A_1$ such that $a = \phi(b)$. Then $\phi(bz_i') = a\beta_{-\mu}(z_i)$ and therefore $\eta_{\Lambda_{\mathfrak{p}},\mu}(\phi(bz_i')) = 0$, $1 \leq i \leq d$. As we have seen above this implies that $bz_i' \in \mathfrak{R}'_{\Lambda_{\mathfrak{p}},\mu}$, $1 \leq i \leq d$. Since $A_2 = \sum_{i=1}^{d} Cz_i' + \mathfrak{R}'_{\Lambda_{\mathfrak{p}},\mu}$ it follows that $bA_2 \subset \mathfrak{R}'_{\Lambda_{\mathfrak{p}},\mu}$ and so $\tau^{(1)}(b) = 0$ where $\tau^{(1)}$ is the natural representation of A_1 on $A_1/\mathfrak{R}''_{\Lambda_{\mathfrak{p}},\mu}$. This in its turn implies that $\nu(b) = 0$. But $\nu(b) = \nu(\phi(b)) = \nu(a)$. Hence this proves $\nu(a) = 0$ whenever $\sigma(a) = 0$. Since it is obvious that $A(\mathfrak{D}_0) = \mathfrak{R}_{\Lambda_{\mathfrak{p}},\mu} + \mathfrak{X}$ the lemma follows.

Now consider the representation $\pi_{\Lambda\mathfrak{p},\mu}$ of G on $L_2(K^*)$ defined by

$$\pi_{\Lambda\mathfrak{p},\mu}(x)f(v^*) \;=\; e^{-\mu(\Gamma(x^{-1},v^*))}e^{-(\Lambda\mathfrak{p}+2\rho)(H(x^{-1},v^*))}f(v^*_{x^{-1}}) \qquad (x\in G,\; v^*\in K^*).$$

Put $f_0(v^*)=d(\mathfrak{D}_0^*)\chi_{\mathfrak{D}_0^*}(v^{*-1})$ where $\chi_{\mathfrak{D}_0^*}$ is the character of the class $\mathfrak{D}_0^*$. For any $\mathfrak{D}\in\Omega$ let $\mathfrak{H}_{\mathfrak{D}}$ denote the set of those elements in $\mathfrak{H}=L_2(K^*)$ which transform under $\pi_{\Lambda\mathfrak{p},\mu}(K)$ according to $\mathfrak{D}$. Then we know that every element in $\sum_{\mathfrak{D}\in\Omega}\mathfrak{H}_{\mathfrak{D}}$ is well-behaved (see Lemma 34 of [5]). Let $b\to\pi_{\Lambda\mathfrak{p},\mu}(b)$ $(b\in\mathfrak{B})$ denote the corresponding representation of $\mathfrak{B}$ defined on $\sum_{\mathfrak{D}\in\Omega}\mathfrak{H}_{\mathfrak{D}}$ (see Lemma 9 of [5]). It is obvious that $f_0\in\mathfrak{H}_{\mathfrak{D}_0}$ and $\pi_{\Lambda\mathfrak{p},\mu}(\mathfrak{X})f_0=\mathfrak{H}_{\mathfrak{D}_0}$. Let τ be the left regular representation of K^* on $\mathfrak{H}$. Put $\bar{\mathfrak{X}}=\mathfrak{X}/\mathfrak{N}(\mathfrak{D}_0^*)$ and let $\bar{z}\to\bar{\tau}(\bar{z})$ $(\bar{z}\in\bar{\mathfrak{X}})$ denote the corresponding representation of $\bar{\mathfrak{X}}$ on $\mathfrak{H}_{\mathfrak{D}_0}$. Then we have seen in the proof of Lemma 5 that

$$\pi_{\Lambda\mathfrak{p},\mu}(a)f_0 \;=\; \bar{\tau}(\eta_{\Lambda\mathfrak{p},\mu}(a))f_0 \qquad (a\in A(\mathfrak{D}_0)).$$

Moreover $\bar{\tau}(\bar{z})f_0\neq0$ $(\bar{z}\in\bar{\mathfrak{X}})$ unless $\bar{z}=0$. Hence $\pi_{\Lambda\mathfrak{p},\mu}(a)f_0=0$ if and only if $\eta_{\Lambda\mathfrak{p},\mu}(a)=0$ $(a\in A(\mathfrak{D}_0))$. Thus the natural representation of $A(\mathfrak{D}_0)$ on $A(\mathfrak{D}_0)/\mathfrak{N}_{\Lambda\mathfrak{p},\mu}$ is equivalent to its representation on $\mathfrak{H}_{\mathfrak{D}_0}$ induced under $\pi_{\Lambda\mathfrak{p},\mu}$.

Let M^* be the analytic subgroup of K^* corresponding to the subalgebra $\mathfrak{m}_0=\mathfrak{k}_0\cap(\mathfrak{h}_t+\sum_{\alpha\in P_-}(CX_\alpha+CX_{-\alpha}))$. We know that M^* is compact (see [5, §12]). Let τ' be the right regular representation of K^* on $\mathfrak{H}$ so that $\tau'(u^*)f(v^*)=f(v^*u^*)$ $(u^*,v^*\in K^*;\, f\in\mathfrak{H})$. Let $\mathfrak{D}_0'^*$ be the class dual to $\mathfrak{D}_0^*$. It is obvious that every element in $\mathfrak{H}_{\mathfrak{D}_0}$ transforms under $\tau(K^*)$ according to $\mathfrak{D}_0^*$ and therefore under $\tau'(K^*)$ according to $\mathfrak{D}_0'^*$. Let ω^* be the set of all equivalence classes of finite-dimensional simple representations of M^*. Denote by $\delta_1'^*,\cdots,\delta_s'^*$ all the distinct elements in ω^* which occur in the reduction of $\mathfrak{D}_0^{*'}$ with respect to M^*. Let ξ_j' be the character of M^* corresponding to $\delta_j'^*$. Put

$$E_j' \;=\; d_j\int_{M^*}\xi_j'(m^{*-1})\tau'(m^*)dm^*$$

where d_j is the degree of any representation in $\delta_j'^*$ and the Haar measure dm^* on M^* is so normalised that $\int_{M^*}dm^*=1$. It follows from the theory of compact groups that $E_i'E_j'=\delta_{ij}E_j'$ $(1\leq i,\,j\leq s)$ and $f=E_1'f+\cdots+E_s'f$ $(f\in\mathfrak{H}_{\mathfrak{D}_0})$. We know moreover (see Lemma 37 of [5]) that $\tau'(m^*)$ commutes with $\pi_{\Lambda\mathfrak{p},\mu}(x)$ $(m^*\in M^*,\,x\in G)$. Therefore the subspaces $\mathfrak{H}_j=E_j'\mathfrak{H}$, $1\leq j\leq s$, are closed invariant subspaces of $\mathfrak{H}$ and $\mathfrak{H}_{\mathfrak{D}_0}\cap\mathfrak{H}_j=E_j'\mathfrak{H}_{\mathfrak{D}_0}$, $\mathfrak{H}_{\mathfrak{D}_0}=\sum_{j=1}^s E_j'\mathfrak{H}_{\mathfrak{D}_0}$. Let σ_j denote the representation of $A(\mathfrak{D}_0)$ induced on $E_j'\mathfrak{H}_{\mathfrak{D}_0}$ under $\pi_{\Lambda\mathfrak{p},\mu}$. From Lemma 14 we conclude that there exists an index j (say $j=1$) such that [3] $\nu(a)=0$ whenever $\sigma_1(a)=0$ $(a\in A(\mathfrak{D}_0))$. We now drop the index 1 and write E', δ'^*, and σ instead of E_1', $\delta_1'^*$, and σ_1 respec-

---

[3] We recall that ν is the representation of $A(\mathfrak{D}_0)$ on $V_{\mathfrak{D}_0}$.

tively. Let ξ_{ij}, $1 \leq i$, $j \leq d$, be the matrix elements of an irreducible unitary representation of M^* in δ'^*. Put $E'_{ji} = d \int_{M^*} \xi_{ij}(m^{*-1}) \tau'(m^*) dm^*$. Then $E'_{ij} E'_{kl} = E'_{il} \delta_{jk}$ and $E' = \sum_{i=1}^{d} E'_{ii}$. Put $\mathfrak{H}_i = E'_{ii} \mathfrak{H}_{\mathfrak{D}_0} = E'_{ii} E' \mathfrak{H}_{\mathfrak{D}_0}$. Then again since E'_{ii} commutes with $\pi_{\Lambda\mathfrak{p},\mu}(x)$ ($x \in G$), $\mathfrak{H}_i$ is stable under $\sigma(A(\mathfrak{D}_0))$. Let σ_i denote the representation of $A(\mathfrak{D}_0)$ induced on $\mathfrak{H}_i$. We conclude once more that there exists an index i (say $i = 1$) such that $\nu(a) = 0$ whenever $\sigma_1(a) = 0$ ($a \in A(\mathfrak{D}_0)$). Since ν is irreducible we can find two subspaces U_1, U_2 of $\mathfrak{H}_1$ ($U_1 \supset U_2$) which are both stable under $\pi_{\Lambda\mathfrak{p},\mu}(A(\mathfrak{D}_0))$ and such that the representation of $A(\mathfrak{D}_0)$ induced on U_1/U_2 is equivalent to ν. Choose an element f_{00} in U_1 which is not in U_2. Since $f_{00} \in E'_{11} \mathfrak{H}_{\mathfrak{D}_0}$ the space spanned by $\tau'(m^*) f_{00}$ ($m^* \in M^*$) is irreducible and transforms according to δ'^* under $\tau'(M^*)$. Moreover there exists an element $\psi_{00} \in V_{\mathfrak{D}_0}$ ($\psi_{00} \neq 0$) with the property that $\nu(a) \psi_{00} = 0$ ($a \in A(\mathfrak{D}_0)$) if and only if $\pi_{\Lambda\mathfrak{p},\mu}(a) f_{00} \in U_2$. Let B be the set of those $a \in A(\mathfrak{D}_0)$ for which $\nu(a) \psi_{00} = 0$. Then B is a maximal left ideal in $A(\mathfrak{D}_0)$. We shall now prove that $\pi_{\Lambda\mathfrak{p},\mu}(\mathfrak{B}B) f_{00} \neq \pi_{\Lambda\mathfrak{p},\mu}(\mathfrak{B}) f_{00}$. For otherwise $f_{00} = \sum_{i=1}^{t} \pi_{\Lambda\mathfrak{p},\mu}(a_i b_i) f_{00}$ where $a_i \in \mathfrak{B}$, $b_i \in B$. Consider the natural mapping $b \to b^*$ and the natural representation π^* of $\mathfrak{B}$ on $\mathfrak{B}^* = \mathfrak{B}/\mathfrak{B}\mathfrak{N}(\mathfrak{D}_0)$. We know (see Theorem 1 of [5]) that $\mathfrak{B}^* = \sum_{\mathfrak{D} \in \Omega} \mathfrak{B}^*_{\mathfrak{D}}$ where $\mathfrak{B}^*_{\mathfrak{D}}$ is the set of those elements in $\mathfrak{B}^*$ which transform according to $\mathfrak{D}$ under $\pi^*(\mathfrak{k})$. Pick a set of classes $\mathfrak{D}_1, \cdots, \mathfrak{D}_N \in \Omega$ ($\mathfrak{D}_j \neq \mathfrak{D}_0$, $1 \leq j \leq N$) such that a_i^*, $1 \leq i \leq t$, are all contained in $\mathfrak{B}^*_{\mathfrak{D}_0} + \mathfrak{B}^*_{\mathfrak{D}_1} + \cdots + \mathfrak{B}^*_{\mathfrak{D}_N}$. From Lemma 4 we can choose $e \in \mathfrak{X}$ such that $\pi^*(e) b^* = b^*$ if $b^* \in \mathfrak{B}^*_{\mathfrak{D}_0}$ and $\pi^*(e) b^* = 0$ if $b^* \in \mathfrak{B}^*_{\mathfrak{D}_j}$, $1 \leq j \leq N$. From this it follows immediately that $f_{00} = \pi_{\Lambda\mathfrak{p},\mu}(e) f_{00} = \sum_{i=1}^{t} \pi_{\Lambda\mathfrak{p},\mu}(e a_i b_i) f_{00}$. Moreover $(e a_i)^* \in \mathfrak{B}^*_{\mathfrak{D}_0}$ and therefore $e a_i \in A(\mathfrak{D}_0)$. Since B is a left ideal in $A(\mathfrak{D}_0)$, $e a_i b_i \in B$. Hence $f_{00} \in \pi_{\Lambda\mathfrak{p},\mu}(B) f_{00} \subset U_2$. Since this contradicts our choice of f_{00} we conclude that $\pi_{\Lambda\mathfrak{p},\mu}(\mathfrak{B}B) f_{00} \neq \pi_{\Lambda\mathfrak{p},\mu}(\mathfrak{B}) f_{00}$.

Let $\mathfrak{M}_1$ be the left ideal consisting of all $b \in \mathfrak{B}$ for which $\pi_{\Lambda\mathfrak{p},\mu}(b) f_{00} = 0$. Then the above result shows that $\mathfrak{B}B + \mathfrak{M}_1 \neq \mathfrak{B}$. By Zorn's lemma we can find a maximal left ideal $\mathfrak{M}$ in $\mathfrak{B}$ containing $\mathfrak{B}B + \mathfrak{M}_1$. Then $\pi_{\Lambda\mathfrak{p},\mu}(\mathfrak{M}) f_{00} \neq \pi_{\Lambda\mathfrak{p},\mu}(\mathfrak{B}) f_{00}$. Let $\mathfrak{H}'$ and $\mathfrak{H}''$ respectively be the closures in $\mathfrak{H}$ of the linear spaces $\pi_{\Lambda\mathfrak{p},\mu}(\mathfrak{B}) f_{00}$ and $\pi_{\Lambda\mathfrak{p},\mu}(\mathfrak{M}) f_{00}$. It follows from Theorem 5 of [5] that $\mathfrak{H}'$, $\mathfrak{H}''$ are both invariant under $\pi_{\Lambda\mathfrak{p},\mu}(G)$ and $\mathfrak{H}''$ is maximal in $\mathfrak{H}'$. We regard $U = \mathfrak{H}'/\mathfrak{H}''$ as a Hilbert space in the usual way and consider the representation γ of G induced on U under $\pi_{\Lambda\mathfrak{p},\mu}$. Then γ is irreducible. Let f_{00}^* denote the image of f_{00} in U under the natural mapping of $\mathfrak{H}'$ on U. Then f_{00}^* transforms under $\gamma(K)$ according to $\mathfrak{D}_0$ and therefore it is well-behaved under γ (see Lemma 34 of [5]). Moreover $\gamma(\mathfrak{M}) f_{00}^* = \{0\}$. Let $\mathfrak{Z}_\chi$ be the set of all elements of the form $z - \chi(z)$ ($z \in \mathfrak{Z}$). Then $B \supset \mathfrak{Z}_\chi$ and therefore $\mathfrak{M} \supset \mathfrak{B}\mathfrak{N}(\mathfrak{D}_0) + \mathfrak{B}\mathfrak{Z}_\chi$. Moreover since B is a maximal left ideal in $A(\mathfrak{D}_0)$, $\mathfrak{M} \cap A(\mathfrak{D}_0) = B$. Hence $\mathfrak{M} \cap \Omega\mathfrak{X} = B \cap \Omega\mathfrak{X}$. Similarly if $\mathfrak{M}'$ is the set of all $b \in \mathfrak{B}$ such that $\pi(b') \psi_{00} = 0$ then $\mathfrak{M}' \supset \mathfrak{B}\mathfrak{N}(\mathfrak{D}_0) + \mathfrak{B}\mathfrak{Z}_\chi$ and $\mathfrak{M}' \cap \Omega\mathfrak{X} = B \cap \Omega\mathfrak{X}$. Therefore from Theorem 2, $\mathfrak{M}$ and $\mathfrak{M}'$ are equivalent and so the representations γ and π are infinitesimally equivalent (see [5, §10]).

We keep to the above notation. For any $\mathfrak{D}^* \in \Omega^*$ let $\mathfrak{H}'_{\mathfrak{D}^*}$ denote the set of those elements in $\mathfrak{H}'$ which transform under $\tau(K^*)$ according to $\mathfrak{D}^*$ (τ is the left regular representation of K^* on $\mathfrak{H} = L_2(K^*)$). Then it follows from the Frobenius reciprocity relation (Weil [7, p. 83]) and our construction of the element f_{00} that dim $\mathfrak{H}'_{\mathfrak{D}^*} \leqq (\mathfrak{D}^*:\delta^*)d(\mathfrak{D}^*)$ where δ^* is the class in ω^* dual to δ'^* and $(\mathfrak{D}^*:\delta^*)$ is the number of times δ^* occurs in the reduction of $\mathfrak{D}^*$ with respect to M^*. Since γ and π are infinitesimally equivalent we get dim $V_{\mathfrak{D}} \leqq (\mathfrak{D}:\delta)d(\mathfrak{D})$ $(\mathfrak{D} \in \Omega)$. Here δ is an equivalence class of finite-dimensional simple representations[4] of M defined as follows. If σ^* is a representation in δ^* we define a representation σ of class δ by $\sigma(m) = e^{-\mu(\Gamma(m))}\sigma^*(m^*)$ $(m \in M)$. $(\mathfrak{D}:\delta)$ is the number of times δ occurs in the reduction of $\mathfrak{D}$ with respect to M (see Lemma 5 of [5]).

In particular since dim $V_{\mathfrak{D}_0} > 0$, $(\mathfrak{D}:\delta) > 0$ and so there exists an element $\psi_0 \in V_{\mathfrak{D}_0}$ such that $\pi(X_\alpha)\psi_0 = 0$ $(\alpha \in P_-)$ and $\pi(H)\psi_0 = \lambda_\delta(H)\psi_0$ $(H \in \mathfrak{h}_\mathfrak{k})$. Here λ_δ is the highest[5] weight of δ. Let z be an element in the center $\mathfrak{Z}$ of $\mathfrak{B}$. We know (Lemma 36 of [3]) that there exists a unique element h in the subalgebra of $\mathfrak{B}$ generated by $(1, \mathfrak{h})$ such that $z - h \in \sum_{\alpha \in P} \mathfrak{B}X_\alpha$. Since z and h are both of rank zero it follows easily that $z - h \in \sum_{\alpha,\beta \in P_+} X_{-\alpha}\mathfrak{B}X_\beta + \sum_{\alpha \in P_-} \mathfrak{B}X_\alpha$ (see proof of Lemma 9). Therefore if we extend the involution θ of $\mathfrak{g}$ (see §2 of [5]) to an automorphism of $\mathfrak{B}$ we get

$$\theta(z) - \theta(h) \in \sum_{\alpha,\beta \in P_+} X_\alpha \mathfrak{B}X_{-\beta} + \sum_{\alpha \in P_-} \mathfrak{B}X_\alpha.$$

Hence in the notation of Lemma 11,

$$\eta_{u,v}^{\mathfrak{D}_0^*}(\theta(z)) = \overline{\beta_y(\xi_u(\theta(z)))} = \overline{\beta_y(\xi_u(\theta(h) + w))}$$

where $w \in \sum_{\alpha \in P_-} \mathfrak{B}X_\alpha$. Therefore

$$\pi(\theta(z))\psi_0 = \pi(\xi_{\Lambda_\mathfrak{p}}(\theta(h))\psi_0,$$

since $\pi(X_\alpha)\psi_0 = 0$ $(\alpha \in P_-)$. Let Λ'' be the linear function on $\mathfrak{h}$ which coincides with $-(\Lambda_\mathfrak{p} + 2\rho)$ on $\mathfrak{h}_\mathfrak{p}$ and with λ_δ on $\mathfrak{h}_\mathfrak{k}$. Similarly let Λ' be the linear function which is equal to $\Lambda_\mathfrak{p}$ on $\mathfrak{h}_\mathfrak{p}$ and to λ_δ on $\mathfrak{h}_\mathfrak{k}$. Then $\theta(\Lambda'' + \rho) = \Lambda' + \rho$ and one proves easily[6] that $\chi_{\Lambda''}(z) = \chi_{\Lambda'}(\theta(z))$. Now it is clear that

$$\pi(\xi_{\Lambda_\mathfrak{p}}(\theta(h)))\psi_0 = \chi_{\Lambda''}(z)\psi_0 = \chi_{\Lambda'}(\theta(z))\psi_0.$$

Hence $\pi(\theta(z))\psi_0 = \chi_{\Lambda'}(\theta(z))\psi_0$. Since Λ' is equal to Λ on $\mathfrak{h}_\mathfrak{p}$ we can replace Λ by Λ' in Lemma 15. Moreover Λ' coincides on $\mathfrak{h}_\mathfrak{k}$ with the highest weight λ_δ of an irreducible class δ of M which has the property that dim $V_{\mathfrak{D}} \leqq d(\mathfrak{D})$ $(\mathfrak{D}:\delta)$ for any $\mathfrak{D} \in \Omega$. Thus if we denote by ω the set of all equivalence

---

[4] M is the analytic subgroup of K corresponding to $\mathfrak{m}_0 = \mathfrak{k}_0 \cap (\mathfrak{h}_\mathfrak{k} + \sum_{\alpha \in P_-} (CX_\alpha + CX_{-\alpha}))$.

[5] This means that if λ is any other weight of δ $(\lambda \neq \lambda_\delta)$ then $\lambda_\delta - \lambda > 0$ (see §2 of [5]).

[6] The proof depends on a calculation very similar to the one made in [3] on p. 79.

classes of irreducible finite-dimensional representations of M we get the following theorem.

THEOREM 4. *Let π be an irreducible quasi-simple representation of $\mathfrak{B}$ on a vector space V such that some $\mathfrak{D}_0 \in \Omega_F$ occurs in π. Then there exists a linear function Λ on $\mathfrak{h}$ with the following properties:*

(1) χ_Λ is the infinitesimal character of π,

(2) Λ coincides on $\mathfrak{h}_t$ with the highest weight λ_δ of some $\delta \in \omega$,

(3) $\dim V_\mathfrak{D} \leqq d(\mathfrak{D})(\mathfrak{D}:\delta)$ for any $\mathfrak{D} \in \Omega$.

(4) π is infinitesimally equivalent to a representation γ of G which is deducible[7] from $\pi_{\Lambda\mathfrak{p},\mu}$ where μ is a suitable linear function on $\mathfrak{c}$.

The above theorem contains some of the results announced in an earlier note (Theorems 1 and 2 [4(b)]). The fourth assertion of Theorem 2 follows immediately if we take into account the definition of $\pi_{\Lambda\mathfrak{p},\mu}$. For let $\mathfrak{H} = L_2(K^*)$ be the Hilbert space introduced above and let $\mathfrak{D}_1$, $\mathfrak{D}_2$ be two classes in Ω which occur in $\pi_{\Lambda\mathfrak{p},\mu}$. Let $g_i \in \mathfrak{H}_{\mathfrak{D}_i}$ $(i=1, 2)$ and consider the scalar product $(g_1, \pi_{\Lambda\mathfrak{p},\mu}(x)g_2)$ in $\mathfrak{H}$. Then if a bar denotes complex conjugate,

$$(g_1, \pi_{\Lambda\mathfrak{p},\mu}(x)g_2) = \int_{K^*} \overline{g_1(u^*)} e^{-\mu(\Gamma(x^{-1},u^*))} e^{-(\Lambda\mathfrak{p}+2\rho)(H(x^{-1},u^*))} g_2(u^*_{x^{-1}}) du^*$$

$$= \int_{K^*} \overline{g_1(u_x^*)} e^{\mu(\Gamma(x,u^*))} e^{\Lambda\mathfrak{p}(H(x,u^*))} g_2(u^*) du^*$$

if we recall that $du^* = e^{2\rho(H(x,u^*))} du_x^*$ (see [5, §11]). Put

$$\sigma_i^*(u^*) = e^{-\mu(\Gamma(u))} \sigma_i(u) \qquad (u \in K)\ (i = 1, 2)$$

where σ_i is some representation $\mathfrak{D}_i$. Then σ_i^* is a representation of K^* which we may assume to be unitary. The complex conjugates of the matrix coefficients of σ_i^* span the space $\mathfrak{H}_{\mathfrak{D}_i}$. Hence g_i is a linear combination of the complex conjugates of these functions. Let $E_{\mathfrak{D}_i}$ be the orthogonal projection of $\mathfrak{H}$ on $\mathfrak{H}_{\mathfrak{D}_i}$. Then it follows that the matrix coefficients of $E_{\mathfrak{D}_1}\pi_{\Lambda\mathfrak{p},\mu}(x)E_{\mathfrak{D}_2}$ are finite linear combinations of functions of the form

$$\int_{K^*} g_1(u_x^*) e^{\mu(\Gamma(x,u^*))} e^{\Lambda\mathfrak{p}(H(x,u^*))} g_2(u^{*-1}) du^*$$

where g_i is a matrix coefficient of some representation in $\mathfrak{D}_i^*$. Since γ is deducible from $\pi_{\Lambda\mathfrak{p},\mu}$ the same holds for the matrix coefficients for $E'_{\mathfrak{D}_1}\gamma(x)E'_{\mathfrak{D}_2}$ where $E'_{\mathfrak{D}_i}$ is the orthogonal projection of the representation space $\mathfrak{H}'$ of γ on the subspace $\mathfrak{H}'_{\mathfrak{D}_i}$ corresponding to $\mathfrak{D}_i$ $(i=1, 2)$. These matrix coefficients

<hr>

[7] Let σ be a representation of G on a Hilbert space $\mathfrak{H}$. Let $\mathfrak{H}_1$, $\mathfrak{H}_2$ be two closed invariant subspaces of $\mathfrak{H}$ such that $\mathfrak{H}_1 > \mathfrak{H}_2$. We regard the factor space $\mathfrak{H}_1/\mathfrak{H}_2$ as a Hilbert space in the usual way and consider the representation σ' of G induced on it under σ. Any representation σ' obtained in this fashion is said to be *deducible* from σ (see [4(b)]).

are analytic functions (see $[5, \S6]$) on G and so they are completely determined as soon as we know all their partial derivatives at $x=1$. Since γ is infinitesimally equivalent to π, it follows that the values of these derivatives at $x=1$ coincide with the values of the corresponding derivatives of the coefficients of $\pi(x)$ computed with respect to suitable bases in $V_{\mathfrak{D}_i}$ ($i=1, 2$). Hence the two sets of matrix coefficients for $\gamma(x)$ and $\pi(x)$ are identical as analytic functions on G. In view of the above result for γ the fourth assertion of Theorem 2 of $[4(b)]$ now follows for π.

In particular if $\sigma_1{}^*, \sigma_2{}^*$ are both trivial representations of K^*, the corresponding matrix coefficient of $\pi(x)$ is a numerical multiple of $\int_{K^*} e^{\mu(\Gamma(x,u^*))} e^{\Lambda\mathfrak{p}(H(x,u^*))} du^*$. However since the value of this coefficient for $x=1$ is 1 the multiplicative constant involved must be 1.

Theorem(8) 5. *Let π be a quasi-simple irreducible representation of G on a Banach space $\mathfrak{H}$. Let $\mathfrak{D}_0$ be a class in Ω which occurs in π and such that $d(\mathfrak{D}_0) = 1$. Then* $\dim \mathfrak{H}_{\mathfrak{D}_0} = 1$. *Choose $\psi \in \mathfrak{H}_{\mathfrak{D}_0}$ ($\psi \neq 0$) and let μ be the linear function on $\mathfrak{k}$ such that $\pi(X)\psi = \mu(X)\psi$ ($X \in \mathfrak{k}$). Then there exists a linear function Λ on $\mathfrak{h}$ such that the following conditions are fulfilled:*

(1) Λ *coincides with μ on $\mathfrak{h}_\mathfrak{k}$ and χ_Λ is the infinitesimal character of π.*

(2) *If $E_{\mathfrak{D}_0}$ is the canonical projection of $\mathfrak{H}$ on $\mathfrak{H}_{\mathfrak{D}_0}$ (see $[5, \S8]$) then*

$$E_{\mathfrak{D}_0}\pi(x)\psi = \int_{K^*} e^{\mu(\Gamma(x,u^*))} e^{\Lambda(H(x,u^*))} du^*\psi \qquad (x \in G).$$

Since $d(\mathfrak{D}_0) = 1$ every representation $\sigma \in \mathfrak{D}_0$ is abelian and therefore $\sigma(\mathfrak{k}') = \{0\}$. Hence $\mathfrak{D}_0 \in \Omega_F$ and Theorem 3 is applicable. Choose Λ in accordance with Theorem 3. Then $\dim \mathfrak{H}_{\mathfrak{D}_0} \leq d(\mathfrak{D}_0)(\mathfrak{D}_0{:}\delta) \leq d(\mathfrak{D}_0)^2 = 1$. Since $\mathfrak{D}_0$ occurs in π, it follows that $\dim \mathfrak{H}_{\mathfrak{D}_0} = 1$. The remaining statements follow from Theorem 3 and what we have said above.

5. **Appendix.** The following lemma is frequently useful.

Lemma 16. *Let $\mathcal{A}$ be an associative algebra with 1 over a field K of characteristic zero and let $\pi_1, \cdots, \pi_r$ ($\pi_j \neq 0$, $1 \leq j \leq r$) be a finite number of finite-dimensional irreducible representations of $\mathcal{A}$ no two of which are equivalent. Put $s_j(a) = \mathrm{sp}\ \pi_j(a)$, $1 \leq j \leq r$ ($a \in \mathcal{A}$). Then the linear functions $s_1, \cdots, s_r$ on $\mathcal{A}$ are linearly independent over K.*

Let π be the direct sum of the representations $\pi_1, \cdots, \pi_r$. Then $\mathcal{A}' = \pi(\mathcal{A})$ is a finite-dimensional semisimple algebra. Let $\mathfrak{N}_j$ be the kernel of π_j. Since $\pi(a) \to \pi_j(a)(a \in \mathcal{A})$ is an irreducible representation of $\mathcal{A}'$, $\mathfrak{N}_j' = \pi(\mathfrak{N}_j)$ is a maximal two-sided ideal in $\mathcal{A}'$. Since $\mathcal{A}'$ is semisimple there exists a two-sided ideal $\mathcal{A}_j'$ in $\mathcal{A}'$ which is complementary to $\mathfrak{N}_j'$. It is clear that $\mathcal{A}_j'$ is a simple ideal of $\mathcal{A}'$. Now if $k \neq j$, π_k is not equivalent to π_j and therefore $\mathfrak{N}_k' \neq \mathfrak{N}_j'$. Hence $\mathcal{A}_j' \neq \mathcal{A}_k'$ and therefore $\mathcal{A}_1', \cdots, \mathcal{A}_r'$ are distinct

---

(8) This result was announced in $[4(a)]$.

minimal ideals in $\mathcal{A}'$. But $\mathcal{A}'$ is the direct sum of its simple ideals and therefore $\mathfrak{N}_j' \supset \mathcal{A}_k'$ $(k \neq j)$. Choose $e_j \in \mathcal{A}$ such that $\pi(e_j)$ is the unit element of the simple algebra $\mathcal{A}_j'$. Then since $\pi(e_j) \in \mathfrak{N}_k'$ $(k \neq j)$, $\pi_k(e_j) = 0$. On the other hand $\pi(1 - e_j) \in \mathfrak{N}_j'$ and so $\pi_j(1 - e_j) = 0$. This shows that $s_k(e_j) = 0$ if $k \neq j$ and $s_j(e_j) = d_j$ where d_j is the degree of π_j. The assertion of the lemma is now obvious.

REFERENCES

1. E. Cartan, Ann. École Norm. vol. 44 (1927).

2. Harish-Chandra, Ann. of Math. vol. 50 (1949) pp. 900–915.

3. ———, Trans. Amer. Math. Soc. vol. 70 (1951) pp. 28–96.

4. ———, (a) Proc. Nat. Acad. Sci. U.S.A. vol. 37 (1951) pp. 362–365. (b) Proc. Nat. Acad. Sci. U.S.A. vol. 37 (1951) pp. 691–694.

5. ———, *Representation of a semisimple Lie group on a Banach space.* I, Trans. Amer. Math. Soc. vol. 75 (1953) pp. 185–243.

6. G. D. Mostow, Bull. Amer. Math. Soc. vol. 55 (1949) pp. 969–980.

7. A. Weil, *L'intégration dans les groupes topologiques et ses applications*, Paris, Hermann, 1940.

8. H. Weyl, Math. Zeit. vol. 24 (1925) pp. 328–395.

COLUMBIA UNIVERSITY,
   NEW YORK, N. Y.

TATA INSTITUTE OF FUNDAMENTAL RESEARCH,
   BOMBAY, INDIA.

Reprinted from
Trans. Amer. Math. Soc.
76 (1954), 26–65

REPRESENTATIONS OF SEMISIMPLE LIE GROUPS. III

BY

HARISH-CHANDRA

The main object of this paper is to define the character of an irreducible quasi-simple[1] representation π of a connected semisimple Lie group G on a Hilbert space $\mathfrak{H}$. This will be done as follows. Let $C_c^\infty(G)$ be the class of all functions on G which are indefinitely differentiable and which vanish outside a compact set. For any $f \in C_c^\infty(G)$ we consider the operator $\int f(x)\pi(x)dx$ where dx is the Haar measure on G. It turns out that this operator has a trace (which we denote by $T_\pi(f)$) and the mapping $T_\pi: f \to T_\pi(f)$ $(f \in C_c^\infty(G))$ is a *distribution* in the sense of L. Schwartz [9] such that $T_\pi(f_a) = T_\pi(f)$ where $f_a(x) = f(axa^{-1})$ $(a, x \in G)$. This distribution is defined to be the character of π. We shall see that two such representations π_1, π_2 are infinitesimally equivalent (see [6, §9]) if and only if they have the same character. Therefore in particular a unitary irreducible representation is determined within unitary equivalence by its character (cf. Theorem 8 of [6]).

In the last section we give a simple proof of a formula for "spherical functions" on a *complex* semisimple group. This formula was obtained by Gelfand and Naimark [1; 2] in some special cases by direct computation.

1. Some preliminary results. We keep to the notation of our two earlier papers [6, 7] on the same subject. G is a connected, simply connected, semisimple Lie group and $\mathfrak{g}_0$ is its Lie algebra over the field R of real numbers. $\mathfrak{g}$ is the complexification of $\mathfrak{g}_0$ and $\mathfrak{k}$, $\mathfrak{p}$, $\mathfrak{k}_0$, $\mathfrak{p}_0$, $\mathfrak{c}$, $\mathfrak{k}'$, and $\mathfrak{m}$ are defined as in [6, §2] and [7, §2]. K, K', and D are the analytic subgroups of G corresponding to $\mathfrak{k}_0$, $\mathfrak{k}_0' = \mathfrak{k}' \cap \mathfrak{g}_0$ and $\mathfrak{c}_0 = \mathfrak{c} \cap \mathfrak{g}_0$ respectively. Let Z be the center of G and $\mathfrak{Z}$ the center of the enveloping algebra $\mathfrak{B}$ of $\mathfrak{g}$. If π is a representation of G on a Banach space we shall say that π is quasi-simple if it maps the elements of $D \cap Z$ and $\mathfrak{Z}$ into scalar multiples of the unit operator (see [6, §10]).

Let π be a quasi-simple irreducible representation of G on a Banach space $\mathfrak{H}$. We denote by Ω the set of all equivalence classes of finite-dimensional simple representations of K. Let $\mathfrak{H}_\mathfrak{D}$ $(\mathfrak{D} \in \Omega)$ denote the set of all elements in $\mathfrak{H}$ which transform under $\pi(K)$ according to $\mathfrak{D}$. We know (see [6, Lemma 33]) that dim $\mathfrak{H}_\mathfrak{D} < \infty$. Let $E_\mathfrak{D}$ denote the canonical projection of $\mathfrak{H}$ on $\mathfrak{H}_\mathfrak{D}$ (see [6, §9]). For any $x \in G$ consider the operator $E_\mathfrak{D}\pi(x)E_\mathfrak{D}$. It maps $\mathfrak{H}_\mathfrak{D}$ into itself and $\mathfrak{H}_{\mathfrak{D}'}$ into $\{0\}$ $(\mathfrak{D}' \neq \mathfrak{D})$. Let sp $(E_\mathfrak{D}\pi(x)E_\mathfrak{D})$ denote the trace of the restriction of $E_\mathfrak{D}\pi(x)E_\mathfrak{D}$ on $\mathfrak{H}_\mathfrak{D}$. Since dim $\mathfrak{H}_\mathfrak{D} < \infty$ this trace is well defined. Now any given linear function α on $\mathfrak{H}_\mathfrak{D}$ may be extended to a continuous linear function on $\mathfrak{H}$ by setting $\alpha(\psi) = \alpha(E_\mathfrak{D}\psi)$ $(\psi \in \mathfrak{H})$. In particular if ψ_i,

---

Received by the editors January 9, 1953.

[1] See [6, §10] for the definition of a quasi-simple representation.

234

$1 \leqq i \leqq r$, is a base for $\mathfrak{H}_{\mathfrak{D}}$ and $\tilde{\psi}_i$ is the linear function on $\mathfrak{H}_{\mathfrak{D}}$ which takes the value 1 at ψ_i and zero at ψ_j ($j \neq i$, $1 \leqq i, j \leqq r$) we may extend $\tilde{\psi}_i$ on $\mathfrak{H}$ in the above fashion. Then it is clear that

$$\phi_{\mathfrak{D}}^{\pi}(x) = \mathrm{sp}\, E_{\mathfrak{D}}\pi(x)E_{\mathfrak{D}} = \sum_{i=1}^{r} (\tilde{\psi}_i,\, \pi(x)\psi_i)$$

in the notation of [6, §10]. Hence it follows from Lemmas 19 and 34 of [6] that $\phi_{\mathfrak{D}}^{\pi}$ is an analytic function on G. $X_1, \cdots, X_n$ being a base for $\mathfrak{g}_0$ over R, set $X(t) = t_1 X_1 + \cdots + t_n X_n$ ($t_j \in R$). Then we know (cf. Theorem 2 and Lemma 34 of [6]) that if $|t| = \max_j |t_j|$ is sufficiently small we get the convergent expansions

$$\pi(\exp X(t))\psi_i = \sum_{m \geqq 0} \frac{1}{m!} \pi((X(t))^m)\psi_i, \qquad 1 \leqq i \leqq r.$$

From this it follows immediately that if z is any element in $\mathfrak{B}$ the value of $\sum_{1 \leqq i \leqq r} (\tilde{\psi}_i,\, \pi(z)\psi_i)$ can be obtained in terms of the various partial derivatives of $\phi_{\mathfrak{D}}^{\pi}$ ($\exp X(t)$) with respect to (t) at $t_1 = t_2 = \cdots = t_n = 0$. Let σ be the representation of $\mathfrak{A} = \mathfrak{Q}\mathfrak{X}$ (see [7, §2] for notation) on $\mathfrak{H}_{\mathfrak{D}}$ defined under π. Then the knowledge of the function $\phi_{\mathfrak{D}}^{\pi}$ determines in particular $\mathrm{sp}\, \sigma(z)$ for any $z \in \mathfrak{A}$. Now we know (see Theorem 5 of [6]) that π defines a quasi-simple[2] irreducible representation of $\mathfrak{B}$ on $\mathfrak{H}^{(0)} = \sum_{\mathfrak{D}' \in \mathfrak{Q}} \mathfrak{H}_{\mathfrak{D}'}$ and therefore σ is irreducible (see Corollary 2 to Theorem 2 of [7]). On the other hand a finite-dimensional simple representation of an associative algebra is completely determined within equivalence by its trace (see Lemma 16 of [7]). Hence in view of Theorem 2 of [7] we can conclude that the function $\phi_{\mathfrak{D}}^{\pi}$ determines the representation of $\mathfrak{B}$ on $\mathfrak{H}^{(0)}$ up to equivalence and therefore the representation π of G up to infinitesimal equivalence. This result may be stated in a slightly more general form as follows.

THEOREM[3] 1. *Let $\pi_1, \cdots, \pi_r$ be a finite set of quasi-simple irreducible representations of G on Banach spaces. Suppose no two of them are infinitesimally equivalent. Then all the nonzero functions in the set $\phi_{\mathfrak{D}_1}^{\pi_1}, \cdots, \phi_{\mathfrak{D}_r}^{\pi_r}$ ($\mathfrak{D}_i \in \Omega$, $1 \leqq i \leqq r$) are linearly independent.*

Let C be the field of complex numbers. If our assertion is false we may suppose that $c_1 \phi_{\mathfrak{D}_1}^{\pi_1} + \cdots + c_s \phi_{\mathfrak{D}_s}^{\pi_s} = 0$ where $c_j \phi_{\mathfrak{D}_j}^{\pi_j} \neq 0$, $1 \leqq j \leqq s$ ($c_j \in C$). Let $\mathfrak{H}_i$ be the representation space of π_i. Consider the representation σ_i of $\mathfrak{A}$ on $\mathfrak{H}_{i, \mathfrak{D}_i}$ ($1 \leqq i \leqq s$) induced under π_i. Then $c_1 \mathrm{sp}\, \sigma_1(a) + \cdots + c_s \mathrm{sp}\, \sigma_s(a) = 0$ for all $a \in \mathfrak{A}$. Since $c_j \phi_{\mathfrak{D}_j}^{\pi_j} \neq 0$, $\mathfrak{D}_j$ occurs in π_j. Moreover π_j, π_k ($j \neq k$) are not infinitesimally equivalent ($1 \leqq j, k \leqq s$). Hence it follows from Corollary 2 to Theorem 2 of [7] that the representations $\sigma_1, \cdots, \sigma_s$ are irreducible and no

---

[2] See [7, end of §2] for the definition of quasi-simplicity in this case.

[3] Cf. Theorem 7 of [8(a)] and Theorem 2 of [8(b)].

two of them are equivalent. This however gives a contradiction with Lemma 16 of [7]. So the theorem is proved.

We recall that for two irreducible *unitary* representations on Hilbert spaces the notions of infinitesimal equivalence and ordinary equivalence are the same (see Theorem 8 of [6]). Hence if π_1, π_2 are two such representations which are not equivalent, the corresponding functions $\phi_{\mathfrak{D}_1}^{\pi_1}$, $\phi_{\mathfrak{D}_2}^{\pi_2}$ are always distinct unless they are both zero.

Theorem 5 of [7] can now be rephrased in terms of the function $\phi_{\mathfrak{D}}^{\pi}$ as follows:

THEOREM([4]) 2. *Let π be a quasi-simple irreducible representation of G on a Banach space $\mathfrak{H}$. Suppose $\mathfrak{D}_0$ is a class in Ω occurring in π such that $d(\mathfrak{D}_0) = 1$. Then $\dim \mathfrak{H}_{\mathfrak{D}_0} = 1$ and it is possible to choose linear functions Λ and μ on $\mathfrak{h}$ and $\mathfrak{c}$ respectively such that*

$$\phi_{\mathfrak{D}_0}^{\pi}(x) = \int_{K^*} e^{\mu(\Gamma(x, u^*))} e^{\Lambda(H(x, u^*))} du^* \qquad (x \in G)$$

and the infinitesimal character of π is χ_Λ.

Some properties of the function $\phi_{\mathfrak{D}}^{\pi}$ have been studied by R. Godement [3] (see also [1; 2]).

We shall now state a few immediate consequences of the results proved in [6].

THEOREM([5]) 3. *Let χ be a homomorphism of $\mathfrak{Z}$ into C and $\mathfrak{D}_0$ a class in Ω. Then apart from infinitesimal equivalence there exist only a finite number of irreducible quasi-simple representations π of G which have the infinitesimal character χ and such that $\mathfrak{D}_0$ occurs in π.*

This follows from Theorem 2 of [7]. Similarly the following result is obtained from Theorem 3 of [8].

THEOREM 4. *Let π be a quasi-simple irreducible representation of G on a Banach space $\mathfrak{H}$. Then there exists an integer N such that*

$$\dim \mathfrak{H}_{\mathfrak{D}} \leq N(d(\mathfrak{D}))^2$$

for all $\mathfrak{D} \in \Omega$.

2. **Trace of an operator.** Let $\{c_\alpha\}_{\alpha \in J}$ be an indexed set of complex numbers. We define the convergence of the series $\sum_{\alpha \in J} c_\alpha$ and its sum in the usual manner (see §5 of [6]). Let A be a bounded operator on a Hilbert space $\mathfrak{H}$ and let $\{\psi_\alpha\}_{\alpha \in J}$ be an orthonormal base for $\mathfrak{H}$. We say that A has a trace

---

([4]) Cf. Theorem 3 of [8(b)]. Our notation is the same as that of Theorem 5 of [7].

([5]) Cf. Theorem 6 of [8(a)].

(or A is of the trace class) if for every such base the series[6] $\sum_{\alpha \in J} (\psi_\alpha, A\psi_\alpha)$ converges to a sum which is independent of the choice of the base. The value of this sum is called the trace of A and we shall denote it by sp A.

LEMMA 1. *Let $\{\psi_\alpha\}_{\alpha \in J}$ be an orthonormal base for a Hilbert space $\mathfrak{H}$ and T a bounded operator such that $\sum_{\alpha,\beta \in J} |t_{\alpha\beta}| < \infty$ where $t_{\alpha\beta} = (\psi_\alpha, T\psi_\beta)$. Then if A and B are any bounded operators on $\mathfrak{H}$, ATB, BAT, TBA are all of the trace class and*

$$\mathrm{sp}\ (ATB) = \mathrm{sp}\ (BAT) = \mathrm{sp}\ (TBA).$$

Put $a_{\alpha\beta} = (\psi_\alpha, A\psi_\beta)$, $b_{\alpha\beta} = (\psi_\alpha, B\psi_\beta)$ $(\alpha, \beta \in J)$ and consider the series $\sum_{\alpha,\beta,\gamma \in J} |a_{\alpha\beta} t_{\beta\gamma} b_{\gamma\alpha}|$. Then[7]

$$\sum_\alpha |a_{\alpha\beta} t_{\beta\gamma} b_{\gamma\alpha}| = |t_{\beta\gamma}| \sum_\alpha |a_{\alpha\beta} b_{\gamma\alpha}|$$

$$\leq |t_{\beta\gamma}| \left(\sum_\alpha |a_{\alpha\beta}|^2 \right)^{1/2} \left(\sum_\alpha |b_{\gamma\alpha}|^2 \right)^{1/2} \leq |t_{\beta\gamma}| \, |A| \, |B|$$

since

$$\sum_\alpha |a_{\alpha\beta}|^2 = \sum_\alpha |(\psi_\alpha, A\psi_\beta)|^2 = |A\psi_\beta|^2 \leq |A|^2$$

and similarly for B. Hence

$$\sum_{\alpha,\beta,\gamma} |a_{\alpha\beta} t_{\beta\gamma} b_{\gamma\alpha}| \leq |A| \, |B| \sum_{\beta,\gamma} |t_{\beta\gamma}| < \infty.$$

This proves that the series $\sum_{\alpha,\beta,\gamma} a_{\alpha\beta} t_{\beta\gamma} b_{\gamma\alpha}$ is absolutely convergent and so it follows in the usual way that

$$\sum_\alpha (\psi_\alpha, ATB\psi_\alpha) = \sum_\alpha (\psi_\alpha, TBA\psi_\alpha) = \sum_\alpha (\psi_\alpha, BAT\psi_\alpha).$$

Now let U be a unitary transformation on $\mathfrak{H}$. Consider $U^{-1}ATBU$. Since $U^{-1}A$ and BU are bounded operators, we can conclude from the above result that

$$\sum_\alpha (U\psi_\alpha, ATBU\psi_\alpha) = \sum_\alpha (\psi_\alpha, U^{-1}ATBU\psi_\alpha)$$

$$= \sum_\alpha (\psi_\alpha, TBUU^{-1}A\psi_\alpha) = \sum_\alpha (\psi_\alpha, TBA\psi_\alpha)$$

$$= \sum_\alpha (\psi_\alpha, ATB\psi_\alpha).$$

Since every orthonormal base in $\mathfrak{H}$ is related to the base $\{\psi_\alpha\}_{\alpha \in J}$ by a unitary transformation, this proves that ATB is of the trace class. Since BA is a

---

[6] As usual we denote by (ϕ, ψ) the scalar product of ϕ and ψ in $\mathfrak{H}$.

[7] For any bounded operator Q we put $|Q| = \sup_{|\psi| \leq 1} |Q\psi|$.

bounded operator it follows from this result that BAT and TBA are also of the trace class. Hence in view of the above equalities we conclude that sp $ATB =$ sp $BAT =$ sp TBA.

COROLLARY. *If T satisfies the conditions of the above lemma and if A is a regular operator, then T and ATA^{-1} are both of the trace class and* sp ATA^{-1} $=$ sp T.

3. **An auxiliary lemma.** In order to prove that certain given operators are of the trace class we shall frequently need the following result.

LEMMA 2. *Let $\mathfrak{l}$ be a semisimple Lie algebra over C of rank l. Then the series $\sum_{\mathfrak{D}} d(\mathfrak{D})^{-(l+1)}$ is convergent. Here $\mathfrak{D}$ runs over all equivalence classes of finite-dimensional simple representations of $\mathfrak{l}$ and $d(\mathfrak{D})$ is the degree of any representation in $\mathfrak{D}$.*

Let $\mathfrak{h}$ be a Cartan subalgebra of $\mathfrak{l}$. Choose a fundamental system of roots and let $\Lambda_1, \cdots, \Lambda_l$ be a fundamental set of dominant integral functions on $\mathfrak{h}$ with respect to this system (see [5, Part I]). Then every such function can be written as $m_1\Lambda_1 + \cdots + m_l\Lambda_l$ where m_i are all nonnegative integers. Let $H_1, \cdots, H_l$ be a base for $\mathfrak{h}$. Extend this to a base $X_1, \cdots, X_n$ $(n \geq l)$ for $\mathfrak{l}$ so that $X_i = H_i$, $1 \leq i \leq l$. Put $g_{ij} =$ sp (ad X_i ad X_j), $1 \leq i, j \leq n$, where $X \rightarrow$ ad X is the adjoint representation of $\mathfrak{l}$. Since $\mathfrak{l}$ is semisimple the matrix $(g_{ij})_{1 \leq i,j \leq n}$ is nonsingular. Let $(g^{ij})_{1 \leq i,j \leq n}$ denote its inverse. Let $\mathfrak{U}$ be the enveloping algebra of $\mathfrak{l}$. Put $\omega = \sum_{1 \leq i,j \leq n} g^{ij}X_iX_j \in \mathfrak{U}$. ω is called the Casimir operator of $\mathfrak{l}$ and it is well known that ω lies in the center of $\mathfrak{U}$. For any dominant integral function Λ on $\mathfrak{h}$ put

$$|\Lambda|^2 = \sum_{1 \leq i,j \leq l} g^{ij}\Lambda(H_i)\Lambda(H_j).$$

Then it is known (see for example [4]) that $|\Lambda|^2$ is a positive real number unless $\Lambda = 0$. Now let σ be an irreducible finite-dimensional representation of $\mathfrak{U}$ and let $\mathfrak{D}$ be the class of σ. We denote by $\Lambda_{\mathfrak{D}}$ the highest weight of σ and by $\omega_{\mathfrak{D}}$ the number such that $\sigma(\omega) = \omega_{\mathfrak{D}}\sigma(1)$. Then it follows from Lemma 6 of [4] that $\omega_{\mathfrak{D}}$ is real, $\omega_{\mathfrak{D}} \geq |\Lambda_{\mathfrak{D}}|^2$, and there exists a real number κ such that $\kappa d(\mathfrak{D})^2 \geq \omega_{\mathfrak{D}}$ for every irreducible class $\mathfrak{D}$. Hence

$$d(\mathfrak{D})^{-1} \leq \kappa^{1/2} |\Lambda_{\mathfrak{D}}|^{-1}.$$

Now the base $H_1, \cdots, H_l$ can be so chosen that every root of $\mathfrak{l}$ takes real values at $H_1, \cdots, H_l$. For such a base the quadratic form $\sum_{i,j=1}^{l} g^{ij}x_i \cdot x_j$ $(x_i \in R)$ is real and positive definite. We can therefore select $H_1, \cdots, H_l$ in such a way that this form reduces to $x_1^2 + \cdots + x_l^2$. For any dominant integral function Λ let e_Λ denote the vector in the l-dimensional real Euclidean space with the components $\Lambda(H_i)$. Then the set of all points e_Λ form one "octant" of a lattice whose generators are $e_i = e_{\Lambda_i}$, $1 \leq i \leq l$. Now

$$\sum_{\mathfrak{D}}{}' d(\mathfrak{D})^{-(l+1)} \leqq \kappa^{(l+1)/2} \sum_{\mathfrak{D}}{}' \left| \Lambda_{\mathfrak{D}} \right|^{-(l+1)}$$

where $\sum_{\mathfrak{D}}'$ denotes the sum over all irreducible classes $\mathfrak{D}$ except the one corresponding to the zero representation of degree 1. Since each class is completely determined by its highest weight, it follows that

$$\sum_{\mathfrak{D}}{}' \left| \Lambda_{\mathfrak{D}} \right|^{-(l+1)} \leqq \sum_{(m) \geqq 0}{}' \left| m_1 e_1 + \cdots + m_l e_l \right|^{-(l+1)}$$

where $\left| e \right|$ is the Euclidean length of the vector e and $\sum_{(m) \geqq 0}'$ denotes summation over all sets of nonnegative integers $(m_1, \cdots, m_l)$ such that $m_1 + \cdots + m_l > 0$. Since the series on the right is well known to be convergent, the lemma follows.

4. A result on convergence. We use the terminology of [6, §9]. Let π be a permissible representation of G on a Banach space $\mathfrak{H}$ and $E_{\mathfrak{D}}$ the canonical projection of $\mathfrak{H}$ on the space $\mathfrak{H}_{\mathfrak{D}}$ consisting of all elements in $\mathfrak{H}$ which transform under $\pi(K)$ according to $\mathfrak{D}$ ($\mathfrak{D} \in \Omega$). We shall now prove the following lemma[8].

LEMMA 3. *There exists an element $z \in \mathfrak{X}$ such that*

$$\sum_{\mathfrak{D} \in \Omega} \left| E_{\mathfrak{D}} \psi \right| \leqq \left| \pi(z) \psi \right|$$

for any differentiable element ψ in $\mathfrak{H}$. Moreover the series

$$\sum_{\mathfrak{D} \in p} E_{\mathfrak{D}} \psi$$

converges to ψ.

Let $u \to u^*$ ($u \in K$) denote the natural mapping of K on $K^* = K/D \cap Z$. For any $u \in K$ we denote by $\Gamma(u)$ the unique element in $\mathfrak{c}_0$ such that $u \exp(-\Gamma(u)) \in K'$. Choose a base $\Gamma_1, \cdots, \Gamma_r$ for $\mathfrak{c}_0$ over R such that $\exp \Gamma_i$, $1 \leqq i \leqq r$, is a set of generators for $D \cap Z$. Let μ be a linear function on $\mathfrak{c}$ such that $\pi(\exp \Gamma_i) = e^{\mu(\Gamma_i)} \pi(1)$, $1 \leqq i \leqq r$. Let Ω_π be the set of all classes in Ω which occur in π. Then it is clear that if $\mathfrak{D} \in \Omega_\pi$ and σ is any representation in $\mathfrak{D}$, we must have

$$\sigma(\Gamma_i) = (2\pi(-1)^{1/2} n_i + \mu(\Gamma_i)) \sigma(1)$$

where n_i, $1 \leqq i \leqq r$, are all integers. Define a linear function $n_{\mathfrak{D}}$ on $\mathfrak{c}$ by setting $n_{\mathfrak{D}}(\Gamma_i) = n_i$, $1 \leqq i \leqq r$, and put $\left| n_{\mathfrak{D}} \right| = (1 + n_1^2 + \cdots + n_r^2)^{1/2}$. Then if $w = 1 - (1/4\pi^2) \sum_{i=1}^r (\Gamma_i - \mu(\Gamma_i))^2 \in \mathfrak{X}$, $\sigma(w) = \left| n_{\mathfrak{D}} \right|^2 \sigma(1)$. We note that w lies in the center of $\mathfrak{X}$. Let $\mathfrak{X}'$ be the subalgebra of $\mathfrak{B}$ generated by $(1, \mathfrak{k}')$. Since $\mathfrak{k}'$ is semisimple we can find (see Lemma 4 of [7]) an element z_0 in the center of $\mathfrak{X}'$ such that $\sigma(z_0) = d_\sigma^2 \sigma(1)$ for any simple representation σ of $\mathfrak{X}$ of degree d_σ.

Put $\pi^*(u^*) = e^{-\mu(\Gamma(u))} \pi(u)$ ($u \in K$). Then π^* is a representation of K^* on

---

[8] Cf. Lemma 31 of [6] which was stated without proof.

$\mathfrak{H}$ and if $\mathfrak{D}\in\Omega_\pi$

$$E_\mathfrak{D} = d(\mathfrak{D})\int_{K^*} \operatorname{conj}\,(\xi_\mathfrak{D}(u^*))\pi^*(u^*)du^*\,(^9)$$

where $\xi_\mathfrak{D}$ is the character (on K^*) of the class according to which every element in $\mathfrak{H}_\mathfrak{D}$ transforms under $\pi^*(K^*)$. Let M be an upper bound for $|\pi^*(u^*)|$ on the compact set K^*. Then it is clear that

$$|E_\mathfrak{D}| \leqq d(\mathfrak{D})^2 M.$$

Let q and s be two integers $\geqq 0$. Then

$$|E_\mathfrak{D}\pi(z_0^{q+1}w^*)\psi| \leqq Md(\mathfrak{D})^2|\pi(z_0^{q+1}w^*)\psi|.$$

But if $X\in\mathfrak{k}_0$,

$$\operatorname*{Lim}_{t\to 0}\frac{1}{t}(\pi(\exp tX)-1)E_\mathfrak{D}\psi = \operatorname*{Lim}_{t\to 0}E_\mathfrak{D}\frac{1}{t}(\pi(\exp tX)-1)\psi = E_\mathfrak{D}\pi(X)\psi$$

since $E_\mathfrak{D}$ commutes with $\pi(u)$ $(u\in K)$. Hence it follows that $E_\mathfrak{D}\psi$ is differentiable under $\pi(K)$ and $\pi(x)E_\mathfrak{D}\psi = E_\mathfrak{D}\pi(x)\psi$ $(x\in\mathfrak{X})$. Therefore

$$E_\mathfrak{D}\pi(z_0^{q+1}w^*)\psi = \pi(z_0^{q+1}w^*)E_\mathfrak{D}\psi = d(\mathfrak{D})^{2q+2}|n_\mathfrak{D}|^{2s}E_\mathfrak{D}\psi$$

since $E_\mathfrak{D}\psi$ transforms under $\pi(K)$ according to $\mathfrak{D}$. Hence

$$d(\mathfrak{D})^{2q}|n_\mathfrak{D}|^{2s}|E_\mathfrak{D}\psi| \leqq M|\pi(z_0^{q+1}w^*)\psi| \qquad (\mathfrak{D}\in\Omega_\pi),$$

and therefore

$$\sum_{\mathfrak{D}\in\Omega_\pi}|E_\mathfrak{D}\psi| \leqq \left(\sum_{\mathfrak{D}\in\Omega_\pi}d(\mathfrak{D})^{-2q}|n_\mathfrak{D}|^{-2s}\right)M|\pi(z_0^{q+1}w^*)\psi|.$$

For any $\mathfrak{D}\in\Omega_\pi$ let $\mathfrak{D}'$ denote the class of representations of $\mathfrak{k}'$ defined as follows. If $\sigma\in\mathfrak{D}$, $\mathfrak{D}'$ is the class of the restriction of σ on $\mathfrak{k}'$. Clearly $\mathfrak{D}'$ is irreducible and $d(\mathfrak{D}')=d(\mathfrak{D})$. Moreover $\mathfrak{D}$ is completely determined by $\mathfrak{D}'$ and $n_\mathfrak{D}$. Hence

$$\sum_{\mathfrak{D}\in\Omega_\pi}d(\mathfrak{D})^{-2q}|n_\mathfrak{D}|^{-2s} \leqq \sum_{\mathfrak{D}'}d(\mathfrak{D}')^{-2q}\sum_{n_1,\cdots,n_r}(1+n_1^2+\cdots+n_r^2)^{-s}$$

where $\mathfrak{D}'$ runs over all irreducible classes of finite-dimensional representations of $\mathfrak{k}'$. But if $2q$ exceeds the rank of $\mathfrak{k}'$ it follows from Lemma 2 that $\sum'_\mathfrak{D} d(\mathfrak{D}')^{-2q}<\infty$. Similarly if $2s>r$

$$\sum_{(n)}(1+n_1^2+\cdots+n_r^2)^{-s} \leqq \sum_{(n)}(1+n_1^2+\cdots+n_r^2)^{-(r+1)/2} < \infty.$$

$(^9)$ Conj (x) means conjugate of x.

Therefore if we choose q and s sufficiently large and put

$$z = N z_0^{q+1} w^s$$

where

$$N = M \sum_{\mathfrak{D} \in \Omega_\tau} d(\mathfrak{D})^{-2q} \, | \, n_\mathfrak{D} \, |^{-2s},$$

$$\sum_{\mathfrak{D} \in \Omega} | \, E_\mathfrak{D} \psi \, | \leq | \, \pi(z)\psi \, |.$$

This proves the first assertion of the lemma. Now we come to the second part. Since $\sum_{\mathfrak{D} \in \Omega} | E_\mathfrak{D} \psi | < \infty$ the series $\sum_{\mathfrak{D} \in \Omega} E_\mathfrak{D} \psi$ is convergent. Let ϕ denote its sum. We have to show that $\phi = \psi$. Put $\psi' = \psi - \phi$. Since $E_\mathfrak{D} \phi = E_\mathfrak{D} \psi$, $E_\mathfrak{D} \psi' = 0$. From this we shall deduce that $\psi' = 0$.

Suppose $\psi' \neq 0$. Then given any real $\epsilon > 0$ choose a continuous real nonnegative function f on K^* such that $f(u^*) = 0$ if $\left| \pi^*(u^*)\psi' - \psi' \right| > \epsilon |\psi'|$ $(u^* \in K^*)$ and $\int_{K^*} f(u^*) du^* = 1$. Moreover choose a finite linear combination ω of the matrix coefficients of finite-dimensional simple representations of K^* such that $\left| f(u^*) - \omega(u^*) \right| \leq \epsilon$ $(u^* \in K^*)$. Then if

$$\psi'' = \int \omega(u^*) \pi^*(u^*) \psi' du^*,$$

$\psi'' \in \sum_{\mathfrak{D} \in \Omega} \mathfrak{H}_\mathfrak{D}$ and therefore $\psi'' = \sum_{\mathfrak{D} \in \Omega} E_\mathfrak{D} \psi''$. But

$$E_\mathfrak{D} \psi'' = \int \omega(u^*) \pi^*(u^*) E_\mathfrak{D} \psi' du^* = 0$$

since $E_\mathfrak{D} \psi' = 0$. Hence $\psi'' = 0$. On the other hand

$$| \psi'' - \psi' | \leq \int | \omega(u^*) - f(u^*) | \, | \pi^*(u^*)\psi' | \, du^*$$

$$+ \int f(u^*) | \pi^*(u^*)\psi' - \psi' | \, du^*$$

$$\leq M\epsilon | \psi' | + \epsilon | \psi' |$$

where $M = \sup_{u^* \in K^*} | \pi^*(u^*) |$. Therefore if ϵ is sufficiently small

$$| \psi' | = | \psi'' - \psi' | \leq | \psi' | / 2$$

which contradicts our assumption that $\psi' \neq 0$. Therefore $\psi' = 0$ and so $\sum_{\mathfrak{D} \in \Omega} E_\mathfrak{D} \psi$ converges to ψ.

5. Characters. Let $C_c^\infty(G)$ denote the class of all complex-valued functions on G which are indefinitely differentiable everywhere and which vanish outside a compact set. Let π be a quasi-simple irreducible representation of G on a Hilbert space $\mathfrak{H}$. For any $f \in C_c^\infty(G)$ consider the operator

$$T_f = \int f(x)\pi(x)dx$$

where dx is the Haar measure on G. We intend to show that T_f is of the trace class.

Let $\mathfrak{O}'$ be the Banach space of all bounded linear operators A on $\mathfrak{H}$ with the usual norm $|A| = \sup_{|\psi|\leqq 1}|A\psi|$ $(\psi\in\mathfrak{H})$. Let $\mathfrak{O}_0$ be the subspace of $\mathfrak{O}'$ consisting of all operators of the form T_f $(f\in C_c^\infty(G))$. We denote by $\mathfrak{O}$ the closure of $\mathfrak{O}_0$ in $\mathfrak{O}'$. Now if $y\in G$,

$$\pi(y)T_f = \int f(y^{-1}x)\pi(x)dx, \qquad T_f\pi(y^{-1}) = \int f(xy)\pi(x)dx.$$

Hence it follows that if $A\in\mathfrak{O}$ then $\pi(y)A$ and $A\pi(y^{-1})$ are also in $\mathfrak{O}$. We now define two representations l and r of G on $\mathfrak{O}$ as follows:

$$l(x)A = \pi(x)A, \qquad r(x)A = A\pi(x^{-1}) \qquad (x\in G, A\in\mathfrak{O}).$$

In order to verify the conditions for continuity it is sufficient to prove that $\lim_{x\to1,y\to1}|\pi(x)A\pi(y^{-1})-A| = 0$ $(A\in\mathfrak{O})$. This is done as follows. Given $\epsilon>0$, choose $f\in C_c^\infty(G)$ such that $|A-T_f|\leqq\epsilon$. Let $U=U^{-1}$ be a compact neighbourhood of 1 in G and M an upper bound for $|\pi(z)|$ for $z\in U$. Then

$$\big|\pi(x)A\pi(y^{-1}) - \pi(x)T_f\pi(y^{-1})\big| \leqq M^2\epsilon \qquad (x, y\in U)$$

and therefore

$$\big|\pi(x)A\pi(y^{-1}) - A\big| \leqq (M^2+1)\epsilon + \big|\pi(x)T_f\pi(y^{-1}) - T_f\big|$$

$$\leqq (M^2+1)\epsilon + \int\big|f(x^{-1}zy) - f(z)\big|\,\big|\pi(z)\big|\,dz.$$

Let C be a compact set outside which f is zero. We can choose a neighbourhood V of 1 in G $(V\subset U)$ such that $|f(x^{-1}zy)-f(z)|\leqq\epsilon$ if $x, y\in V$. Let F be a real nonnegative continuous function on G which is equal to 1 on C and which vanishes outside some compact set. Then if $N_0 = \sup_{z\in UCU}|\pi(z)|$,

$$\big|\pi(x)A\pi(y^{-1}) - A\big| \leqq (M^2+1)\epsilon + N_0\epsilon\int F(z)dz$$

provided $x, y\in V$. This proves that $\lim_{x\to1,y\to1}|\pi(x)A\pi(y^{-1})-A| = 0$.

Since $l(x)T_f = \int f(x^{-1}z)\pi(z)dz$, it follows easily that T_f is differentiable under l. Similarly we show that it is differentiable under r. It is clear that the representations l and r are permissible. For any $\mathfrak{D}\in\Omega$ let $E_{\mathfrak{D}}$, $P_{\mathfrak{D}}$, and $Q_{\mathfrak{D}}$ denote the canonical projections (see §9 of [6]) corresponding to $\mathfrak{D}$ under π, l, and r respectively. Then it is clear $P_{\mathfrak{D}}A = E_{\mathfrak{D}}A$; $Q_{\mathfrak{D}}A = AE_{\mathfrak{D}'}$ $(\mathfrak{D}\in\Omega)$ where $\mathfrak{D}'$ is the class contragredient to $\mathfrak{D}$. Let λ and ρ denote the left and right regular representations of G. Then every element in $C_c^\infty(G)$ is differentiable under both λ and ρ and $C_c^\infty(G)$ is invariant under $\lambda(\mathfrak{B})$ and $\rho(\mathfrak{B})$. Moreover since $l(x)r(y)T_f = T_{\lambda(x)\rho(y)f}$ $(x,\ y\in G)$ it follows easily that $l(a)r(b)T_f = T_{\lambda(a)\rho(b)f}$ $(a,\ b\in\mathfrak{B})$.

Now define a representation ϕ of the group $G\times G$ on $\mathfrak{O}$ as follows. $\phi(x,y)A = l(x)r(y)A = \pi(x)A\pi(y^{-1})$ $(x,\ y\in G)$. Then ϕ is a permissible representation of the semisimple group $G\times G$ and any element of $\mathfrak{O}_0$ is differentiable under ϕ. Moreover the canonical projections for the representation ϕ (with respect to the subgroup $K\times K$) are exactly the operators $P_{\mathfrak{D}_1}Q_{\mathfrak{D}_2}$ $(\mathfrak{D}_1,\ \mathfrak{D}_2\in\Omega)$. Let z_0 be the element of $\mathfrak{X}$ which was introduced in the proof of Lemma 3. Then if we apply Lemma 3 to the representation ϕ and the differentiable element $T_{\lambda(z_0)\rho(z_0)f}$ we find that

$$\sum_{\mathfrak{D}_1,\mathfrak{D}_2\in\Omega} | P_{\mathfrak{D}_1}Q_{\mathfrak{D}_2}T_{\lambda(z_0)\rho(z_0)f} | < \infty.$$

But

$$P_{\mathfrak{D}_1}Q_{\mathfrak{D}_2}T_{\lambda(z_0)\rho(z_0)f} = P_{\mathfrak{D}_1}Q_{\mathfrak{D}_2}l(z_0)r(z_0)T_f = d(\mathfrak{D}_1)^2 d(\mathfrak{D}_2)^2 P_{\mathfrak{D}_1}Q_{\mathfrak{D}_2}T_f.$$

Hence

$$\sum_{\mathfrak{D}_1,\mathfrak{D}_2\in\Omega} d(\mathfrak{D}_1)^2 d(\mathfrak{D}_2)^2 | E_{\mathfrak{D}_1}T_f E_{\mathfrak{D}_2} | < \infty.$$

Now let us first suppose that the subspaces $\mathfrak{H}_{\mathfrak{D}} = E_{\mathfrak{D}}\mathfrak{H}$ $(\mathfrak{D}\in\Omega)$ are mutually orthogonal. Choose an orthonormal base for each $\mathfrak{H}_{\mathfrak{D}}$. All these put together form an orthonormal base $\{\psi_\alpha\}_{\alpha\in J}$ for $\mathfrak{H}$. In accordance with Theorem 4 we choose an integer N such that dim $\mathfrak{H}_{\mathfrak{D}}\leq Nd(\mathfrak{D})^2$ $(\mathfrak{D}\in\Omega)$. Then

$$\sum_{\alpha,\beta\in J} | (\psi_\alpha,\ T_f\psi_\beta) | = \sum_{\mathfrak{D}_1,\mathfrak{D}_2\in\Omega} \sum_{\alpha\in J_{\mathfrak{D}_1}} \sum_{\beta\in J_{\mathfrak{D}_2}} | (\psi_\alpha,\ T_f\psi_\beta) |$$

where $J_{\mathfrak{D}}$ is the subset of J such that $\{\psi_\alpha\}_{\alpha\in J_{\mathfrak{D}}}$ is a base for $\mathfrak{H}_{\mathfrak{D}}$. But it is clear that

$$\sum_{\alpha\in J_{\mathfrak{D}_1}} \sum_{\beta\in J_{\mathfrak{D}_2}} | (\psi_\alpha,\ T_f\psi_\beta) | \leq \text{dim } \mathfrak{H}_{\mathfrak{D}_1} \text{dim } \mathfrak{H}_{\mathfrak{D}_2} | E_{\mathfrak{D}_1}T_f E_{\mathfrak{D}_2} |$$

$$\leq N^2 d(\mathfrak{D}_1)^2 d(\mathfrak{D}_2)^2 | E_{\mathfrak{D}_1}T_f E_{\mathfrak{D}_2} |.$$

Hence

$$\sum_{\alpha,\beta\in J} | \psi_\alpha,\ T_f\psi_\beta | \leq N^2 \sum_{\mathfrak{D}_1,\mathfrak{D}_2\in\Omega} d(\mathfrak{D}_1)^2 d(\mathfrak{D}_2)^2 | E_{\mathfrak{D}_1}T_f E_{\mathfrak{D}_2} | < \infty$$

and therefore, from Lemma 1, T_f is of the trace class.

Now we discard the assumption about the mutual orthogonality of the spaces $\mathfrak{H}_{\mathfrak{D}}$. Let $x\to x^*$ denote the natural mapping of G on $G^* = G/D\cap Z$ Define a representation π^* of K^* on $\mathfrak{H}$ as in the proof of Lemma 3. Since K^* is a compact group, π^* is equivalent to a unitary representation. Hence there exists a regular operator B on $\mathfrak{H}$ such that the representation π'^*: $u \to B\pi^*(u^*)B^{-1}$ $(u^*\in K^*)$ is unitary. Now put $\pi'(x) = B\pi(x)B^{-1}$. Then π' is a representation of G on $\mathfrak{H}$. Let $\mathfrak{H}'_{\mathfrak{D}}$ be the subspace of $\mathfrak{H}$ consisting of all elements which transform under $\pi'(K)$ according to $\mathfrak{D}$ ($\mathfrak{D} \in \Omega$). Then since π'^{*} is unitary the spaces $\mathfrak{H}'_{\mathfrak{D}}$ are mutually orthogonal. Therefore the above proof is applicable to

$$T_{f}' = \int f(x)\pi'(x)dx = BT_{f}B^{-1}.$$

Hence T_{f}' fulfills the condition of Lemma 1 and therefore $T_{f} = B^{-1}T_{f}'B$ is of the trace class. Moreover if P and Q are two bounded operators on $\mathfrak{H}$, then PB^{-1} and BQ are also bounded and $PT_{f}Q = (PB^{-1})T_{f}'(BQ)$. Therefore, from Lemma 1, $PT_{f}Q$, $T_{f}QP$, QPT_{f} are all of the trace class and their traces are equal. Therefore in particular sp $(AT_{f}A^{-1}) = $ sp T_{f} if A is a regular operator.

Put $T_{\pi}(f) = \mathrm{sp}(T_{f})$ for any $f \in C_{c}^{\infty}(G)$. Then T_{π} is a linear function on the vector space $C_{c}^{\infty}(G)$. Furthermore if a is a fixed element in G and g is the function $g(x) = f(axa^{-1})$ ($x \in G$) then

$$T_{g} = \pi(a^{-1})T_{f}\pi(a)$$

and therefore $T_{\pi}(g) = T_{\pi}(f)$. Hence we may say that T_{π} is invariant under the inner automorphisms of G. We prove similarly that if π and π' are equivalent representations, $T_{\pi} = T_{\pi'}$.

Our next object is to show that T_{π} is actually a distribution in the sense of L. Schwartz [9]. By going over to an equivalent representation, if necessary, we may assume that the spaces $\mathfrak{H}_{\mathfrak{D}}$ ($\mathfrak{D} \in \Omega$) are mutually orthogonal. Then it is clear that

$$\mathrm{sp}\ T_{f} = \sum_{\mathfrak{D} \in \Omega} \mathrm{sp}\ (E_{\mathfrak{D}}T_{f}E_{\mathfrak{D}})$$

and

$$\mathrm{sp}\ (E_{\mathfrak{D}}T_{f}E_{\mathfrak{D}}) = \sum_{i=1}^{d} (\psi_{i},\ E_{\mathfrak{D}}T_{f}\psi_{i})$$

where $(\psi_{1}, \cdots, \psi_{d})$ is an orthonormal base of $\mathfrak{H}_{\mathfrak{D}}$. Therefore

$$\left|\ \mathrm{sp}\ (E_{\mathfrak{D}}T_{f}E_{\mathfrak{D}})\ \right| \leq \dim \mathfrak{H}_{\mathfrak{D}} \left|\ E_{\mathfrak{D}}T_{f}\ \right| \leq Nd(\mathfrak{D})^{2} \left|\ E_{\mathfrak{D}}T_{f}\ \right|$$

and

$$\left|\ T_{\pi}(f)\ \right| \leq N \sum_{\mathfrak{D} \in \Omega} d(\mathfrak{D})^{2} \left|\ E_{\mathfrak{D}}T_{f}\ \right| = N \sum_{\mathfrak{D} \in \Omega} \left|\ E_{\mathfrak{D}}T_{\lambda(z_{0})f}\ \right|$$

where z_{0} has the same meaning as above. Now by applying Lemma 3 to the representation l of G on $\mathfrak{H}$ and the differentiable element $T_{\lambda(z_{0})f}$ we conclude that

$$\sum_{\mathfrak{D} \in \Omega} \left|\ E_{\mathfrak{D}}T_{\lambda(z_{0})f}\ \right| \leq \left|\ l(z)T_{\lambda(z_{0})f}\ \right| = \left|\ T_{\lambda(zz_{0})f}\ \right|$$

where z is an element of $\mathfrak{X}$ (which does not depend on f). Hence

$$\left| T_\pi(f) \right| \leq N \left| T_{\lambda(zz_0)f} \right|.$$

Now suppose C is a compact set in G and f_n is a sequence of functions in $C_c^\infty(G)$ such that f_n vanishes outside C and for any $b \in \mathfrak{B}$, $\lambda(b)f_n \to 0$ uniformly on C. Then

$$\left| T_{\lambda(b)f_n} \right| \leq \int \left| (\lambda(b)f_n)(x) \right| \left| \pi(x) \right| dx \to 0$$

and therefore in particular $\left| T_\pi(f_n) \right| \leq N \left| T_{\lambda(zz_0)f_n} \right| \to 0$. This proves that T_π is a distribution.

6. **Operators of the Hilbert-Schmidt class.** Let B be a bounded operator on the Hilbert space $\mathfrak{H}$ and let B^* be the adjoint of B. We say that B is of the Hilbert-Schmidt (H.S.) class if B^*B has a trace. Let $\{\psi_\alpha\}_{\alpha \in J}$ be an orthonormal base for $\mathfrak{H}$. Then it is well known that $\|B\|^2 = \sum_{\alpha \in J} |B\psi_\alpha|^2$ is independent of the choice of this base and B is of the H.S. class if and only if $\|B\| < \infty$. Moreover sp $BB^* = \|B\|^2 = \|B^*\|^2$ if $\|B\| < \infty$ and $\|A_1 B A_2\| \leq |A_1| \|B\| |A_2|$ for any two bounded operators A_1, A_2.

Let π be a quasi-simple irreducible representation of G on $\mathfrak{H}$. Let f be a complex-valued measurable function on G which vanishes outside a compact set and such that $\int |f(x)|^2 dx < \infty$. It follows from the Schwartz inequality that $\int |f(x)| dx < \infty$ and therefore the operator $\int f(x)\pi(x)dx$ is a well-defined bounded operator. We intend to prove that this operator is of the H.S. class. Let S be a regular operator on $\mathfrak{H}$. Put $\pi'(x) = S\pi(x)S^{-1}$ $(x \in G)$. Then

$$\left\| \int f(x)\pi(x)dx \right\| = \left\| S^{-1} \int f(x)\pi'(x)dxS \right\| \leq |S^{-1}| \left\| \int f(x)\pi(x)dx \right\| |S|.$$

Therefore it is enough to show that the corresponding operator for an equivalent representation is of the H.S. class.

Let $x \to x^*$ denote the natural mapping of G on $G^* = G/D \cap Z$. For any $x \in G$ we define $\Gamma(x)$ to be the unique element in c_0 such that $x = u(\exp \Gamma(x))s$ where $u \in K'$ and s lies in the solvable subgroup of G corresponding to the subalgebra $\mathfrak{g}_0 \cap (\mathfrak{h}_\mathfrak{p} + \mathfrak{n})$ of $\mathfrak{g}_0$ (see [6, §9]). Then if μ is the linear function on c which was introduced in the proof of Lemma 3, it is clear that $\pi(x)e^{-\mu(\Gamma(x))}$ depends only on x^*. Put $\pi^*(x^*) = \pi(x)e^{-\mu(\Gamma(x))}$. Then we verify immediately that $\pi^*(u^*x^*) = \pi^*(u^*)\pi^*(x^*)$ $(u^* \in K^*, x^* \in G^*)$ and therefore $u^* \to \pi^*(u^*)$ is a representation of K^*. In view of the preceding remarks we may assume without loss of generality that this representation is unitary.

Let $x^* \in G^*$ and $y \in G$. We say that y lies above x^* and write $y > x^*$ if $(y)^* = x^*$. Put

$$f^*(x^*) = \sum_{x > x^*} e^{\mu(\Gamma(x))}f(x) \qquad (x^* \in G^*).$$

Let A be a compact set outside which f is zero. Since $D \cap Z$ is discrete, $(D \cap Z) \cap A^{-1}A$ is a finite set. Let N_0 be the number of elements in it. Then it is clear that not more than N_0 distinct elements in A can lie above the same element in G^*. Hence at most N_0 terms in the above sum are different from zero and therefore the function f^* is well-defined. Moreover if A^* is the image of A in G^*, f^* is zero outside A^*. Now let $x^* \in A^*$. Then

$$\left| f^*(x^*) \right| = \left| \sum_{x > x^*} e^{\mu(\Gamma(x))} f(x) \right| \leq \left(\sum_{x > x^*} |f(x)|^2 \right)^{1/2} \left(\sum_{y > x^*} |e^{\mu(\Gamma(y))}|^2 \right)^{1/2}$$

$$\leq M_0 N_0^{1/2} \left(\sum_{x > x^*} |f(x)|^2 \right)^{1/2}$$

where $M_0 = \sup_{y \in A} |e^{\mu(\Gamma(y))}|$. Hence if the Haar measure dx^* on G^* is suitably normalised it follows that

$$\int f(x)\pi(x)dx = \int f^*(x^*)\pi^*(x^*)dx^*$$

and

$$\int |f^*(x^*)|^2 dx^* \leq M_0^2 N \int |f(x)|^2 dx < \infty.$$

Let B^* be a compact neighbourhood of A^*. Choose a real-valued nonnegative function F on G^* such that $F = 1$ on K^*B^* and $F = 0$ outside some compact set. Let g be any continuous function on G^* which vanishes outside B^*. Consider the operator

$$\int g(x^*)\pi(x^*)dx^* = \int du^* \int g(u^*x^*)\pi^*(u^*x^*)dx^*.$$

(Here du^* is the normalised Haar measure on K^* so that $\int du^* = 1$.) Then

$$\left\| \int g(x^*)\pi^*(x^*)dx^* \right\| \leq \int dx^* \left\| \int g(u^*x^*)\pi^*(u^*x^*)du^* \right\|.$$

But

$$\left\| \int g(u^*x^*)\pi^*(u^*x^*)du^* \right\| \leq \left\| \int g(u^*x^*)\pi^*(u^*)du^* \right\| |\pi^*(x)|$$

and from Theorem 4 we can find an integer N such that $\dim \mathfrak{H}_{\mathfrak{D}} \leq N d(\mathfrak{D})^2$ for any $\mathfrak{D} \in \Omega$. Let Ω^* be the set of all classes of irreducible finite-dimensional representations of K^*. Then no $\mathfrak{D}^* \in \Omega^*$ occurs more than $N d(\mathfrak{D}^*)$ times in the reduction of $\pi^*(K^*)$. Since every $\mathfrak{D}^* \in \Omega^*$ occurs exactly $d(\mathfrak{D}^*)$ times in the left regular representation λ of K^* (on the Hilbert space $L_2(K^*)$ of all square integrable functions on K^*) and since the representation $u^*\to\pi^*(u^*)$ $(u^*\in K^*)$ is unitary, we may conclude that

$$\left\|\int g(u^*x^*)\pi^*(u^*)du^*\right\|^2 \leq N\left\|\int g(u^*x^*)\lambda(u^*)du^*\right\|^2 .$$

But from the Peter-Weyl theorem we know that

$$\left\|\int g(u^*x^*)\lambda(u^*)du^*\right\|^2 = \int |g(u^*x^*)|^2 du^*.$$

Hence

$$\left\|\int g(u^*x^*)\pi^*(u^*)du^*\right\| \leq N^{1/2}\left(\int |g(u^*x^*)|^2 du^*\right)^{1/2}.$$

Now it is easy to see that $|\pi^*(x^*)|$ is bounded on the compact set K^*B^*. Let $M=\sup_{x^*\in K^*B^*}|\pi^*(x^*)|$. Then

$$\left\|\int g(u^*x^*)\pi^*(u^*x^*)du^*\right\| \leq MN^{1/2}\left(\int |g(u^*x^*)|^2 du^*\right)^{1/2}$$

and therefore

$$\left\|\int g(x^*)\pi^*(x^*)dx^*\right\| \leq MN^{1/2}\int dx^*\left(\int |g(u^*x^*)|^2 du^*\right)^{1/2}$$

$$= MN^{1/2}\int F(x^*)dx^*\left(\int |g(u^*x^*)|^2 du^*\right)^{1/2}$$

$$\leq M_1\left(\int |g(x^*)|^2 dx^*\right)^{1/2}.$$

where $M_1 = MN^{1/2}(\int |F(x^*)|^2 dx^*)^{1/2}$. Now choose a sequence g_n of continuous functions on G^* which vanish outside B^* and such that $\int |f^*(x^*)-g_n(x^*)|^2 dx^* \to 0$. Then since

$$\int |f^*(x^*) - g_n(x^*)|\, dx^*$$

$$\leq \left(\int |F(x^*)|^2 dx^*\right)^{1/2}\left(\int |f^*(x^*) - g_n(x^*)|^2 dx^*\right)^{1/2}$$

it follows that

$$\left|\int (f^*(x^*) - g_n(x^*))\pi^*(x^*)dx^*\right| \to 0.$$

Moreover we have seen above that

$$\left\| \int (g_m(x^*) - g_n(x^*))\pi^*(x^*)dx^* \right\| \leq M_1 \left(\int |g_m(x^*) - g_n(x^*)|^2 dx^* \right)^{1/2}$$

and therefore the sequence of operators $T_n = \int g_n(x^*)\pi^*(x^*)dx^*$ is a Cauchy sequence with respect to the Hilbert-Schmidt norm $\| \ \|$. Since the space of operators of the H.S. class is complete with respect to this norm, there exists an operator T of this class such that $\|T_n - T\| \to 0$. But $|T_n - T| \leq \|T_n - T\|$. Hence $|T_n - T| \to 0$. However we have seen already that

$$\left| T_n - \int f^*(x^*)\pi^*(x^*)dx^* \right| \to 0$$

and therefore $T = \int f^*(x^*)\pi^*(x^*)dx^*$. This proves that $T = \int f(x)\pi(x)dx$ is of the H.S. class. Thus we have the following theorem[10].

THEOREM 5. *Let π be a quasi-simple irreducible representation of G on a Hilbert space $\mathfrak{H}$ and let f be a measurable and square integrable function on G which vanishes outside a compact set. Then the operator $\int f(x)\pi(x)dx$ is of the Hilbert-Schmidt class.*

7. **Linear independence of characters.** Let T_π be the character of a quasi-simple irreducible representation π of G on a Hilbert space $\mathfrak{H}$. Let $E_\mathfrak{D}$ denote the canonical projection of $\mathfrak{H}$ on $\mathfrak{H}_\mathfrak{D}$ ($\mathfrak{D} \in \Omega$). Then it follows easily from its definition (see §5) that

$$T_\pi(f) = \mathrm{sp}\left(\int f(x)\pi(x)dx \right) = \sum_{\mathfrak{D} \in \Omega} \mathrm{sp}\left(E_\mathfrak{D} \int f(x)\pi(x)dx \cdot E_\mathfrak{D} \right)$$

$$= \sum_{\mathfrak{D} \in \Omega} \int f(x)\phi_\mathfrak{D}^\pi(x)dx$$

in the notation of §1. Now if π_1, π_2 are two infinitesimally equivalent representations, we have seen in §1 that $\phi_\mathfrak{D}^{\pi_1} = \phi_\mathfrak{D}^{\pi_2}$ ($\mathfrak{D} \in \Omega$) and therefore $T_{\pi_1} = T_{\pi_2}$. Hence two infinitesimally equivalent quasi-simple irreducible representations (on Hilbert spaces) have the same character. Conversely we shall show that two such representations having the same character are infinitesimally equivalent[11].

THEOREM 6. *Let $\pi_1, \cdots, \pi_q$ be a finite set of quasi-simple irreducible representations of G on the Hilbert spaces $\mathfrak{H}_1, \cdots, \mathfrak{H}_q$ respectively. Suppose no two of them are infinitesimally equivalent. Then their characters $T_{\pi_1}, \cdots, T_{\pi_q}$ are linearly independent.*

For otherwise suppose $c_1 T_{\pi_1} + \cdots + c_q T_{\pi_q} = 0$ ($c_i \in C$) where, say, $c_1 \neq 0$.

---

(10) Cf. Theorem 4 of [8(c)].
(11) Cf. Theorem 3 of [8(c)].

Let η_i be the homomorphism of $D \cap Z$ into C such that $\pi_i(\gamma) = \eta_i(\gamma)\pi_i(1)$ $(\gamma \in D \cap Z)$. Then if $f \in C_c^\infty(G)$ and $\gamma \in D \cap Z$, the function $f_\gamma \colon x \to f(\gamma^{-1}x)$ $(x \in G)$ is also in $C_c^\infty(G)$ and it is obvious that $T_{\pi_i}(f_\gamma) = \eta_i(\gamma)T_{\pi_i}(f)$. Therefore $\sum_{i=1}^q c_i T_{\pi_i}(f_\gamma) = \sum_{i=1}^q c_i \eta_i(\gamma) T_{\pi_i}(f) = 0$. This proves that

$$\sum_{i=1}^q c_i \eta_i(\gamma) T_{\pi_i} = 0$$

for all $\gamma \in D \cap Z$. Now if we recall that $D \cap Z$ is a free abelian group with r generators $(r = \dim_R c_0)$ we can conclude that $c_1 T_{\pi_1} + \cdots + c_s T_{\pi_s} = 0$ assuming that $\eta_j = \eta_1$ $(1 \leq j \leq s)$ and $\eta_j \neq \eta_i$ for $s < j \leq q$.

Choose a base $\Gamma_1, \cdots, \Gamma_r$ for c_0 over R such that $\exp \Gamma_i$, $1 \leq i \leq r$, is a set of generators for $D \cap Z$. Select a linear function μ on c such that $\eta_1 (\exp \Gamma_i) = e^{\mu(\Gamma_i)}$, $1 \leq i \leq r$. Put $\pi_j^*(x^*) = e^{-\mu(\Gamma(x))}\pi_j(x)$ $(x \in G, 1 \leq j \leq s)$ in the notation of §6. Let $\mathfrak{D}$ be a class in Ω which occurs in π_1. We denote by $E_{\mathfrak{D}}^i$ the canonical projection of $\mathfrak{H}_i$ on $\mathfrak{H}_{i,\mathfrak{D}}$. Then

$$E_{\mathfrak{D}}^i = d(\mathfrak{D}) \int \mathrm{conj}\,(\xi_{\mathfrak{D}^*}(u^*))\pi_i^*(u^*)du^*$$

where $\mathfrak{D}^*$ is the irreducible class according to which every element in $\mathfrak{H}_{1,\mathfrak{D}}$ transforms under $\pi_1^*(K^*)$ and $\xi_{\mathfrak{D}^*}$ is the character of $\mathfrak{D}^*$. Let K_0 be the set of all elements in K of the form $(\exp \Gamma)v$ where $\Gamma = t_1\Gamma_1 + \cdots + t_r\Gamma_r$ $(t_j \in R, |t_j| \leq 1/2)$ and $v \in K'$. Then K_0 is compact and if we put

$$\xi_{\mathfrak{D}}(u) = e^{-\mathrm{conj}\,(\mu(\Gamma(u)))}\xi_{\mathfrak{D}^*}(u^*) \qquad (u \in K),$$

we get

$$E_{\mathfrak{D}}^i = d(\mathfrak{D}) \int_{K_0} \mathrm{conj}\,(\xi_{\mathfrak{D}}(u))\pi_i(u)du$$

where the Haar measure du on K is so normalised that $\int_{K_0} du = 1$.

Now we use the notation of Theorem 1. Put

$$\phi = c_1\phi_{\mathfrak{D}}^{\pi_1} + \cdots + c_s\phi_{\mathfrak{D}}^{\pi_1}.$$

It follows from Theorem 1 that $\phi \neq 0$. Since ϕ is continuous we can find a function $f \in C_c^\infty(G)$ such that $\int f(x)\phi(x)dx \neq 0$. Now

$$E_{\mathfrak{D}}^i \int f(x)\pi_i(x)dx = \int f_{\mathfrak{D}}(x)\pi_i(x)dx \qquad (1 \leq i \leq s)$$

where

$$f_{\mathfrak{D}}(x) = d(\mathfrak{D}) \int_{K_0} \mathrm{conj}\,(\xi_{\mathfrak{D}}(u))f(u^{-1}x)du.$$

Since K_0 is compact it is clear that $f_{\mathfrak{D}} \in C_c^\infty(G)$. On the other hand

$$\mathrm{sp}\left(\int f_{\mathfrak{D}}(x)\pi_i(x)dx\right) = \mathrm{sp}\left(E_{\mathfrak{D}}^{i}\int f(x)\pi_i(x)dx\right)$$

$$= \mathrm{sp}\left(E_{\mathfrak{D}}^{i}\int f(x)\pi_i(x)dx \cdot E_{\mathfrak{D}}^{i}\right) = \int f(x)\phi_{\mathfrak{D}}^{\pi_i}(x)dx.$$

Therefore $c_1 T_{\pi_1}(f_{\mathfrak{D}}) + \cdots + c_s T_{\pi_s}(f_{\mathfrak{D}}) = \int f(x)\phi(x)dx \neq 0$. This however implies that $c_1 T_{\pi_i} + \cdots + c_s T_{\pi_s} \neq 0$ and so we get a contradiction.

COROLLARY. *Two irreducible unitary representations are equivalent if and only if their characters are the same.*

First of all every irreducible unitary representation is quasi-simple (see for example Segal [10]). Moreover infinitesimal equivalence is the same as ordinary equivalence for two such representations (see Theorem 8 of [6]). Hence the corollary is an immediate consequence of the theorem.

8. **Complex semisimple groups.** Suppose the group G is complex. Then K is semisimple and there exists a 1-1 linear mapping i of $\mathfrak{k}_0$ onto $\mathfrak{p}_0$ such that

$$[X, i(Y)] = i([X, Y]), \qquad [i(X), i(Y)] = -[X, Y] \qquad (X, Y \in \mathfrak{k}_0).$$

Let $\mathfrak{h}_{\mathfrak{k}_0}$ be a maximal abelian subalgebra of $\mathfrak{k}_0$. Then $i(\mathfrak{h}_{\mathfrak{k}_0})$ is clearly a maximal abelian subspace of $\mathfrak{p}_0$. Hence we may take $\mathfrak{h}_{\mathfrak{p}_0} = i(\mathfrak{h}_{\mathfrak{k}_0})$. Then $\mathfrak{h}_{\mathfrak{k}_0} + \mathfrak{h}_{\mathfrak{p}_0}$ is a maximal abelian subalgebra of $\mathfrak{g}_0$. Let $\mathfrak{h}_{\mathfrak{k}}$ and $\mathfrak{h}_{\mathfrak{p}}$ be the subspaces of $\mathfrak{g}$ spanned by $\mathfrak{h}_{\mathfrak{k}_0}$ and $\mathfrak{h}_{\mathfrak{p}_0}$ over C. Then $\mathfrak{h} = \mathfrak{h}_{\mathfrak{k}} + \mathfrak{h}_{\mathfrak{p}}$ is a Cartan subalgebra of $\mathfrak{g}$. We extend i to a mapping of $\mathfrak{k}$ into $\mathfrak{p}$ by linearity.

Let $\alpha_1, \cdots, \alpha_p$ be a maximal set of linearly independent roots of $\mathfrak{k}$ (with respect to $\mathfrak{h}_{\mathfrak{k}}$). We order all roots α of $\mathfrak{k}$ lexicographically with respect to this set (see [5, Part I]). For every root α let H_α be the element in $\mathfrak{h}_{\mathfrak{k}}$ such that sp $(\mathrm{ad}'\, H\, \mathrm{ad}'\, H_\alpha) = \alpha(H)$ $(H \in \mathfrak{h}_{\mathfrak{k}})$ where $X \to \mathrm{ad}'\, X$ $(X \in \mathfrak{k})$ is the adjoint representation of $\mathfrak{k}$. We denote by W the Weyl group of $\mathfrak{k}$ and by 2σ the sum of all positive roots of $\mathfrak{k}$. Let λ be a linear function on $\mathfrak{h}_{\mathfrak{k}}$. We put $\lambda' = \lambda + \sigma$ and use the notation of [5, Part III]. We know (see [5, p. 70]) that the power series $\sum_{s \in W} \epsilon(s)e^{\lambda'(sH)}$ is divisible by $\prod_{\alpha>0} \lambda'(H_\alpha) \prod_{\alpha>0} \alpha(H)$ $(H \in \mathfrak{h}_{\mathfrak{k}})$ and therefore the quotient

$$\frac{\displaystyle\prod_{s \in W} \epsilon(s)e^{\lambda'(sH)}}{\displaystyle\prod_{\alpha>0} \lambda'(H_\alpha) \prod_{\alpha>0} \alpha(H)}$$

is an analytic function on $\mathfrak{h}_{\mathfrak{k}}$. Similarly

$$\frac{\displaystyle\prod_{\alpha>0} \alpha(H)}{\displaystyle\prod_{\alpha>0} (e^{\alpha(H)/2} - e^{-\alpha(H)/2})}$$

is a meromorphic function on $\mathfrak{h}_\mathfrak{k}$ all whose singularities lie on hyperplanes of the form $\alpha(H) = 2\pi(-1)^{1/2}n$ where α is a root and n is some nonzero integer. Hence the function

$$\Phi^*(\lambda, H) = \prod_{\alpha>0} \sigma(H_\alpha) \frac{\sum_{s \in W} \epsilon(s) e^{\lambda'(sH)}}{\prod_{\alpha>0} \lambda'(H_\alpha) \prod_{\alpha>0} \alpha(H)} \frac{\prod_{\alpha>0} \alpha(H)}{\prod_{\alpha>0} (e^{\alpha(H)/2} - e^{-\alpha(H)/2})}$$

is a meromorphic function on $\mathfrak{h}_\mathfrak{k}$ and it is analytic everywhere on $(-1)^{1/2}\mathfrak{h}_{\mathfrak{k}_0}$ and also on a suitable neighbourhood of zero in $\mathfrak{h}_\mathfrak{k}$. We know from [5, Part III, p. 71] that if $H \in \mathfrak{h}_\mathfrak{k}$ and $|t|$ is sufficiently small ($t \in C$) then

$$\Phi^*(\lambda, tH) = \sum_{m \geq 0} \frac{t^m}{m!} \xi_\lambda(H^m)$$

where ξ_λ is the (infinitesimal) character of the algebra $\mathfrak{X}$ corresponding to the linear function λ on $\mathfrak{h}_\mathfrak{k}$.

Λ being any linear function on $\mathfrak{h}_\mathfrak{p}$, consider the integral $\int_K e^{\Lambda(H(x,\ u))} du$ which occurs in Theorem 2. (Notice that $K = K^*$ in our case and therefore $\int_K du = 1$.) We shall now express this integral in terms of the function Φ^*.

Consider the representation π_Λ of G on $L_2(K)$ given by

$$\pi_\Lambda(x)f(u) = e^{-(\Lambda+2\rho)(H(x^{-1},u))}f(u_{x^{-1}}) \qquad (x \in G, u \in K, f \in L_2(K)$$

in the notation of [6, §12]. Let $\mathfrak{H}$ be the smallest closed subspace of $L_2(K)$ which is invariant under $\pi_\Lambda(G)$ and which contains the constant function 1. Then we have seen in [5, Part IV] that the representation π of G induced on $\mathfrak{H}$ is quasi-simple and its infinitesimal character[12] is χ_Λ where Λ is to be extended to a linear function on $\mathfrak{h}$ by putting it equal to zero on $\mathfrak{h}_\mathfrak{k}$. Let ψ_0 denote the vector in $\mathfrak{H}$ corresponding to the constant function 1. Then if we denote the scalar product of two elements in the usual way we get

$$(\psi_0, \pi(x)\psi_0) = \int_K e^{\Lambda(H(x,u))} du.$$

On the other hand suppose $x = \exp tH_0$ where $H_0 \in \mathfrak{h}_{\mathfrak{p}_0}$ and $t \in R$. Then since ψ_0 is well-behaved under π (see Lemma 34 of [6]),

$$(\psi_0, \pi(\exp tH_0)\psi_0) = \sum_{m \geq 0} \frac{t^m}{m!} (\psi_0, \pi(H_0^m)\psi_0)$$

provided $|t|$ is sufficiently small. Now put $i_+(X) = (X + (-1)^{1/2}i(X))/2$, $i_-(X) = (X - (-1)^{1/2}i(X))/2$ ($X \in \mathfrak{k}$). Then $\mathfrak{k}_+ = i_+(\mathfrak{k})$, $\mathfrak{k}_- = i_-(\mathfrak{k})$ are ideals in $\mathfrak{g}$ and $\mathfrak{g}$ is their direct sum. Let $\mathfrak{X}_+$ and $\mathfrak{X}_-$ be the subalgebras of $\mathfrak{B}$ generated by $(\mathfrak{k}_+, 1)$ and $(\mathfrak{k}_-, 1)$ respectively. Then if $a \in \mathfrak{X}_+$ and $b \in \mathfrak{X}$, $ab - ba = 0$.

---

[12] This is easily seen by the argument used in the proof of Lemma 48 of [5].

Choose $H_0^* \in \mathfrak{h}_{\mathfrak{k}_0}$ such that $H_0 = i(H_0^*)$. Then

$$H_0 = i(H_0^*) = 2(-1)^{1/2}[i_-(H_0^*) - H_0^*].$$

Moreover H_0^* and $i_-(H_0^*)$ commute and $\pi(\mathfrak{k})\psi_0 = \{0\}$. Hence

$$\pi(H_0^m)\phi_0 = (2(-1)^{1/2})^m \pi(H_-^m)\psi_0$$

where $H_- = i_-(H_0^*)$. On the other hand if X, $Y \in \mathfrak{k}$,

$$[X, i_-(Y)] = [i_-(X), i_-(Y)].$$

Hence $[X, z] = [i_-(X), z]$ $(X \in \mathfrak{k}, z \in \mathfrak{X}_-)$. Therefore we get a representation ν of $\mathfrak{k}$ on $\mathfrak{X}_-$ such that $\nu(X)z = [X, z]$ $(X \in \mathfrak{k}, z \in \mathfrak{X}_-)$. Obviously ν is quasi semisimple (see Lemma 10 of [6]). Therefore if z is any element in $\mathfrak{X}_-$, it follows from Lemma 7 of [6] that $z \equiv z_0 \bmod \nu(\mathfrak{k})\mathfrak{X}_-$ where z_0 is some element of $\mathfrak{X}_-$ which commutes with $\mathfrak{k}$. But then

$$[i_-(X), z_0] = [X, z_0] = 0 \qquad\qquad (X \in \mathfrak{k})$$

and moreover z_0 commutes with $\mathfrak{k}_+$ since it lies in $\mathfrak{X}_-$. Therefore z_0 is in the center of $\mathfrak{B}$. Furthermore the representation of K induced under π is unitary and therefore if $X \in \mathfrak{k}_0$ and $a \in \mathfrak{B}$,

$$(\psi_0, \pi(Xa - aX)\psi_0) = (-\pi(X)\psi_0, \pi(a)\psi_0) = 0$$

since $\pi(\mathfrak{k})\psi_0 = \{0\}$. This shows that

$$(\psi_0, \pi(z)\psi_0) = (\psi_0, \pi(z_0)\psi_0) = \chi_\Lambda(z_0).$$

Now if we extend χ_Λ to a linear function on $\mathfrak{B}$ such that $\chi_\Lambda(ab) = \chi_\Lambda(ba)$ $(a, b \in \mathfrak{B})$ (see Part III of [5]) it follows that $\chi_\Lambda(z_0) = \chi_\Lambda(z)$. Hence

$$(\psi_0, \pi(z)\psi_0) = \chi_\Lambda(z) \qquad\qquad (z \in \mathfrak{X}_-).$$

This proves that

$$(\psi_0, \pi(\exp tH)\psi_0) = \sum_{m \geq 0} \frac{t^m}{m!} (2(-1)^{1/2})^m \chi_\Lambda(H_-^m).$$

We extend i_- to an isomorphism of $\mathfrak{X}$ with $\mathfrak{X}_-$. Then the mapping $z \to \chi_\Lambda(i_-(z))$ $(z \in \mathfrak{X})$ is clearly a character of the algebra $\mathfrak{X}$. Hence from Theorem 5 of [5] there exists a linear function λ on $\mathfrak{h}_{\mathfrak{k}}$ such that $\chi_\Lambda(i_-(z)) = \xi_\lambda(z)$ $(z \in \mathfrak{X})$. Therefore

$$(\psi_0, \pi(\exp tH_0)\psi_0) = \sum_{m \geq 0} \frac{t^m}{m!} (2(-1)^{1/2})^m \xi_\lambda(H_0^{*m}) = \Phi^*(\lambda, 2(-1)^{1/2} t H_0^*)$$

if $|t|$ is sufficiently small. In view of equation (25) (p. 81) of [5], λ may be chosen in such a way that $\lambda(H) = \Lambda(i_-(H))$ $(H \in \mathfrak{h}_{\mathfrak{k}})$. Since Λ vanishes on $\mathfrak{h}_{\mathfrak{k}}$ we conclude that $\lambda(H) = -((-1)^{1/2}/2)\Lambda(i(H))$ and therefore $2(-1)^{1/2}\lambda(H_0^*) = \Lambda(H_0)$. Put $si(H) = i(sH)$ $(s \in W)$ and $\alpha(i(H)) = (-1)^{1/2}\alpha(H)$, $\sigma(i(H))$

$= (-1)^{1/2}\sigma(H)$ $(H \in \mathfrak{h}_{\mathfrak{k}})$. Then $\lambda(H_\alpha) = -((-1)^{1/2}/2)\Lambda(i(H_\alpha)) = \Lambda(H_\alpha')$ where $H_\alpha' = -((-1)^{1/2}/2)i(H_\alpha)$ and $\sigma(H_\alpha') = \sigma(H_\alpha)/2$. Then if we put

$$\Phi(\Lambda, H) = \frac{\displaystyle\prod_{\alpha>0} 2\sigma(H_\alpha')}{\displaystyle\prod_{\alpha>0}(\Lambda + 2\sigma)(H_\alpha')} \frac{\displaystyle\sum_{s \in W} \epsilon(s) e^{(\Lambda+2\sigma)(sH)}}{\displaystyle\prod_{\alpha>0}(e^{\alpha(H)} - e^{-\alpha(H)})} \qquad (H \in \mathfrak{h}_{\mathfrak{p}_0})$$

it is clear that $\Phi(\Lambda, H)$ is an analytic function on $\mathfrak{h}_{\mathfrak{p}_0}$ and

$$\Phi^*(\lambda, 2(-1)^{1/2}tH_0^*) = \Phi(\Lambda, tH_0).$$

Hence $(\psi_0, \pi(\exp tH_0)\psi_0) = \Phi(\Lambda, tH_0)$ for all sufficiently small values of $|t|$. Since both sides are analytic functions of t, the equality must hold for all values of t. Thus we have the following result.

THEOREM 7. *Let Λ be a linear function on $\mathfrak{h}_{\mathfrak{p}}$. Then if $x = \exp H$ $(H \in \mathfrak{h}_{\mathfrak{p}_0})$ we have the formula*

$$\int_K e^{\Lambda(H(x,u))}du = \frac{\displaystyle\prod_{\alpha>0} 2\sigma(H_\alpha')}{\displaystyle\prod_{\alpha>0}(\Lambda + 2\sigma)(H_\alpha')} \frac{\displaystyle\sum_{s \in W} \epsilon(s) e^{(\Lambda+2\sigma)(sH)}}{\displaystyle\prod_{\alpha>0}(e^{\alpha(H)} - e^{-\alpha(H)})}.$$

Put $\phi(x) = \int_K e^{\Lambda(H(x,u))}du = (\psi_0, \pi(x)\psi_0)$ $(x \in G)$. Then it is clear that $\phi(uxv) = \phi(x)$ $(u, v \in K)$. Since every element in G can be written in the form $u(\exp H)v$ $(H \in \mathfrak{h}_{\mathfrak{p}_0}; u, v \in K)$, the above formula determines ϕ completely.

The particular case of this formula for the complex unimodular group has been obtained by Gelfand and Naimark [2, p. 77] by means of a lengthy computation (see also [1]).

REFERENCES

1. Gelfand and Naimark, Doklady Nauk SSR (N.S.) vol. 63 (1948) pp. 225–228.
2. ———, Trudi Mat. Inst. Steklova vol. 36 (1950).
3. R. Godement, Trans. Amer. Math. Soc. vol. 73 (1952) pp. 496–556.
4. Harish-Chandra, Ann. of Math. vol. 51 (1950) pp. 299–330.
5. ———, Trans. Amer. Math. Soc. vol. 70 (1951) pp. 28–96.
6. ———, Trans. Amer. Math. Soc. vol. 74 (1953) pp. 185–243.
7. ———, Trans. Amer. Math. Soc. vol. 76 (1954) pp. 26–65.
8. ———, Proc. Nat. Acad. Sci. U.S.A. vol. 37 (1951) (a) pp. 170–173; (b) pp. 362–365; (c) pp. 366–369; (d) pp. 691–694.
9. L. Schwartz, *Théorie des distributions*, Paris, Hermann, 1950.
10. I. E. Segal, Proc. Amer. Math. Soc. vol. 3 (1952) pp. 13–15.

TATA INSTITUTE OF FUNDAMENTAL RESEARCH,
    BOMBAY, INDIA.
COLUMBIA UNIVERSITY,
    NEW YORK, N. Y.

Reprinted from
Trans. Amer. Math. Soc.
76 (1954), 234–253

THE PLANCHEREL FORMULA FOR COMPLEX SEMISIMPLE LIE GROUPS

BY

HARISH-CHANDRA

1. Introduction. Let G be a connected semisimple Lie group and π an irreducible unitary representation of G on a Hilbert space. Let $C_c^\infty(G)$ denote the class of all (complex-valued) functions on G which vanish outside a compact set and which are indefinitely differentiable everywhere. Then we have seen in [8] that for any $f \in C_c^\infty(G)$ the operator

$$\int f(x)\pi(x)dx$$

(dx is the Haar measure on G) has a trace which we shall denote by $T_\pi(f)$. The mapping $T_\pi: f \to T_\pi(f)$ is then a distribution which depends only on the equivalence class of π. Hence if $\mathcal{E}$ is the set of all equivalence classes of irreducible unitary representations of G, we have a distribution T_ω defined for each $\omega \in \mathcal{E}$. Our object is to find a (positive) measure $d\omega$ on $\mathcal{E}$ such that

$$f(1) = \int_\mathcal{E} T_\omega(f)d\omega \qquad (f \in C_c^\infty(G))$$

at least in case G is a complex group. Let f' be the function[1] $x \to \operatorname{conj}(f(x^{-1}))$ ($x \in G$) and let $F = f' * f$ where $*$ denotes group convolution. Then

$$F(x) = \int f'(y)f(y^{-1}x)dy = \int \operatorname{conj}(f(y))f(yx)dy$$

and therefore $\int |f(x)|^2 dx = F(1) = \int_\mathcal{E} T_\omega(f' * f)d\omega$. But

$$T_\pi(f' * f) = \left\| \int f(x)\pi(x)dx \right\|^2 \qquad (\pi \in \omega)$$

where $\|\cdot\|$ denotes the Hilbert-Schmidt norm. Hence

$$\int |f(x)|^2 dx = \int_\mathcal{E} N_\omega(f)d\omega$$

where $N_\omega(f) = \|\int f(x)\pi(x)dx\|^2$ for any $\pi \in \omega$. This formula may be regarded as the analogue of the Plancherel formula for abelian groups or of the Peter-Weyl completeness relation for compact groups (see Gelfand and Naimark [3, p. 198]).

Received by the editors May 11, 1953.

[1] For any complex number c we denote the conjugate of c by $\operatorname{conj} c$.

485

Although the final formula of this paper is applicable only when G is complex, the complex structure of G plays no essential role in the earlier stages of the computation. Hence, in the hope that the present method could perhaps be extended to arbitrary semisimple Lie groups, we shall avoid making the assumption about the complexity of G until it becomes absolutely necessary.

2. **Some preliminary results.** Let $\mathfrak{g}_0$ be the Lie algebra of G over the field R of real numbers. We define $\mathfrak{k}_0$, $\mathfrak{h}_{\mathfrak{p}_0}$, and $\mathfrak{n}_0$ as in [6, §2]. Let K, A_+, N be the analytic subgroups of G corresponding to $\mathfrak{k}_0$, $\mathfrak{h}_{\mathfrak{p}_0}$, and $\mathfrak{n}_0$ respectively. Then K is closed and it contains the center Z of G. Let $\mathfrak{k}_0' = [\mathfrak{k}_0, \mathfrak{k}_0]$ be the derived algebra and $\mathfrak{c}_0$ the center of $\mathfrak{k}_0$. We denote by K' and D the analytic subgroups of K corresponding to $\mathfrak{k}_0'$ and $\mathfrak{c}_0$ respectively. K' is semisimple and compact and D, being the connected component of the centralizer of K' in K, is closed. Put $G^* = G/D \cap Z$ and let $x \to x^*$ denote the natural mapping of G on G^*. Then K^* is compact. We shall say that a representation π of G on a Banach space is permissible if $\pi(z)$ is a scalar multiple of the unit operator for all $z \in Z \cap D$.

Let Ω be the set of all equivalence classes of finite-dimensional simple representations of K.

LEMMA 1. *Let π be a representation of G on a Hilbert space $\mathfrak{H}$. For any $\mathfrak{D} \in \Omega$ let $\mathfrak{H}_{\mathfrak{D}}$ denote the subspace consisting of all those elements in $\mathfrak{H}$ which transform under $\pi(K)$ according to $\mathfrak{D}$. Suppose the following two conditions are fulfilled.*

(i) *π is permissible.*

(ii) *There exists an integer N such that* dim $\mathfrak{H}_{\mathfrak{D}} \leqq N d(\mathfrak{D})^2$ *for all $\mathfrak{D} \in \Omega$. (Here $d(\mathfrak{D})$ is the degree of any representation in $\mathfrak{D}$.)*

Then if $f \in C_c^\infty(G)$ the operator $\int f(x)\pi(x)dx$ fulfills the conditions of Lemma 1 of [8].

The proof is exactly the same as that given in §5 of [8]. We can therefore conclude from Lemma 1 of [8] that $\int f(x)\pi(x)dx$ has a trace. We denote this trace by $T_\pi(f)$ and prove exactly as in §5 of [8] that the mapping $T_\pi : f \to T_\pi(f)$ $(f \in C_c^\infty(G))$ is a distribution which depends only on the equivalence class of π. We shall call T_π the character of π.

LEMMA 2. *Let π be a permissible unitary representation of G on a Hilbert space $\mathfrak{H}$. Suppose* dim $\mathfrak{H}_{\mathfrak{D}} < \infty$ *for every $\mathfrak{D} \in \Omega$. Then $\mathfrak{H}$ can be written as a sum[2] of a countable number of mutually orthogonal closed subspaces each of which is invariant and irreducible under $\pi(G)$.*

By going over to the simply connected covering group of G it follows that

---

[2] The sum here is understood in the sense of Hilbert space theory. It denotes the closure of the algebraic sum.

for any homomorphism ξ of Z into the field C of complex numbers we can find a homomorphism η of K into C such that $\eta(z) = \xi(z)$ for $z \in D \cap Z$ (see §9 of [6]). Therefore in particular we can choose η such that $\pi(z) = \eta(z)\pi(1)$ $(z \in D \cap Z)$. Then $\eta(u^{-1})\pi(u)$ $(u \in K)$ depends only on u^* and if we denote it by $\pi^*(u^*)$ the mapping $\pi^*\colon u^* \to \pi^*(u^*)$ is a representation of K^* on $\mathfrak{H}$. Let Ω^* be the set of all equivalence classes of finite-dimensional simple representations of K^*. We denote by $\mathfrak{H}_{\mathfrak{D}}^*$ $(\mathfrak{D} \in \Omega^*)$ the subspace of those elements in $\mathfrak{H}$ which transform under $\pi^*(K^*)$ according to $\mathfrak{D}$. Then it is clear that dim $\mathfrak{H}_{\mathfrak{D}}^* < \infty$. Since K^* is a compact Lie group, Ω^* is a countable set. Hence we can arrange its elements in a sequence $\mathfrak{D}_i$ $(i \geq 1)$. We shall now define a sequence of closed subspaces $\mathfrak{H}_j$ $(j \geq 0)$ with the following properties:

(i) $\mathfrak{H}_j$ is invariant under $\pi(G)$.

(ii) $\mathfrak{H}_j \supset \sum_{i=1}^{j} \mathfrak{H}_{\mathfrak{D}_i}^*$.

(iii) $\mathfrak{H}_{j+1} \supset \mathfrak{H}_j$ and the orthogonal complement of $\mathfrak{H}_j$ in $\mathfrak{H}_{j+1}$ is the sum of a finite number of mutually orthogonal closed spaces each of which is invariant and irreducible under $\pi(G)$.

We proceed by induction on j. Put $\mathfrak{H}_0 = \{0\}$. Now suppose $\mathfrak{H}_j$ has been defined. Let V_j be the orthogonal complement of $\mathfrak{H}_j$ in $\mathfrak{H}$. Since π is unitary, V_j is invariant under $\pi(G)$. Let $V_{j,\mathfrak{D}} = V_j \cap \mathfrak{H}_{\mathfrak{D}}^*$ $(\mathfrak{D} \in \Omega^*)$. It is clear that $\mathfrak{H}_{\mathfrak{D}}^* = V_{j,\mathfrak{D}} + \mathfrak{H}_j \cap \mathfrak{H}_{\mathfrak{D}}^*$ $(\mathfrak{D} \in \Omega^*)$. If $V_{j,\mathfrak{D}_{j+1}} = \{0\}$ we put $\mathfrak{H}_{j+1} = \mathfrak{H}_j$ and all the three conditions are verified. Now suppose $V_{j,\mathfrak{D}_{j+1}} \neq \{0\}$. Then

$$0 < \dim V_{j,\mathfrak{D}_{j+1}} \leq \dim \mathfrak{H}_{\mathfrak{D}_{j+1}}^* < \infty.$$

We shall now define a sequence of mutually orthogonal closed subspaces W_r $(r \geq 0)$ which are invariant and irreducible under $\pi(G)$. Put $W_0 = \{0\}$ and suppose W_i $(0 \leq i \leq r)$ have already been defined. Let U_r be the orthogonal complement of $W_1 + \cdots + W_r$ in V_j. If $U_r \cap \mathfrak{H}_{\mathfrak{D}_{j+1}}^* = \{0\}$ put $W_{r+1} = \{0\}$. So now let us suppose dim $(U_r \cap \mathfrak{H}_{\mathfrak{D}_{j+1}}^*) > 0$. Let Σ be the collection of all closed subspaces U of U_r which are invariant under $\pi(G)$ and such that $U \cap \mathfrak{H}_{\mathfrak{D}_{j+1}}^* \neq \{0\}$. Choose $U \in \Sigma$ such that $s = \dim U \cap \mathfrak{H}_{\mathfrak{D}_{j+1}}^*$ has the least possible value and define W_{r+1} to be the smallest subspace in Σ which contains $U \cap \mathfrak{H}_{\mathfrak{D}_{j+1}}^*$. We claim W_{r+1} is irreducible. For let $W_{r+1} = W' + W''$ where W', W'' are two mutually orthogonal closed subspaces of W_{r+1} which are both invariant under $\pi(G)$. Since dim $W_{r+1} \cap \mathfrak{H}_{\mathfrak{D}_{j+1}}^* \geq s > 0$, at least one of the spaces $W' \cap \mathfrak{H}_{\mathfrak{D}_{j+1}}^*$, $W'' \cap \mathfrak{H}_{\mathfrak{D}_{j+1}}^*$ is not zero. Suppose $W' \cap \mathfrak{H}_{\mathfrak{D}_{j+1}}^* \neq \{0\}$. Then $W' \in \Sigma$ and in view of the definition of s, $s \leq \dim W' \cap \mathfrak{H}_{\mathfrak{D}_{j+1}}^*$. But $W' \subset W_{r+1} \subset U$ and therefore

$$s \leq \dim (W' \cap \mathfrak{H}_{\mathfrak{D}_{j+1}}^*) \leq \dim (W_{r+1} \cap \mathfrak{H}_{\mathfrak{D}_{j+1}}^*) \leq \dim (U \cap \mathfrak{H}_{\mathfrak{D}_{j+1}}^*) = s.$$

Hence $W' \cap \mathfrak{H}_{\mathfrak{D}_{j+1}}^* = W_{r+1} \cap \mathfrak{H}_{\mathfrak{D}_{j+1}}^* = U \cap \mathfrak{H}_{\mathfrak{D}_{j+1}}^*$ and so it follows from the definition of W_{r+1} that $W_{r+1} \subset W'$. This proves that $W' = W_{r+1}$ and so W_{r+1} is irreducible.

Notice that $W_{r+1} = \{0\}$ if and only if $U_r \cap \mathfrak{H}^*_{\mathfrak{D}_{j+1}} = \{0\}$ and $W_{r+1} \neq \{0\}$ implies $W_{r+1} \cap \mathfrak{H}^*_{\mathfrak{D}_{j+1}} \neq \{0\}$. Since

$$\dim (U_r \cap \mathfrak{H}^*_{\mathfrak{D}_{j+1}}) > \dim (U_{r+1} \cap \mathfrak{H}^*_{\mathfrak{D}_{j+1}})$$

unless $U_r \cap \mathfrak{H}^*_{\mathfrak{D}_{j+1}} = \{0\}$ and since $\dim \mathfrak{H}^*_{\mathfrak{D}_{j+1}} < \infty$, it follows that $W_r = \{0\}$ for r sufficiently large. Let r be the least integer ≥ 0 such that $W_{r+1} = \{0\}$. Then $U_r \cap \mathfrak{H}^*_{\mathfrak{D}_{j+1}} = \{0\}$ and therefore $V_{j, \mathfrak{D}_{j+1}} \subset W_1 + \cdots + W_r$. Now put $\mathfrak{H}_{j+1} = \mathfrak{H}_j + W_1 + \cdots + W_r$. Then all the three conditions are fulfilled and the induction is therefore complete.

After this preparation we now come to the proof of the lemma. Let $\mathfrak{H}_{j+1} = \mathfrak{H}_j + \sum_{1 \leq r \leq s_j} W_r^{(j)}$ where $W_r^{(j)}$ are closed subspaces which are invariant and irreducible under $\pi(G)$ and which are orthogonal to $\mathfrak{H}_j$ and to each other. Then it is obvious that

$$\mathfrak{H}_{j+1} = \sum_{0 \leq i \leq j} \sum_{1 \leq r \leq s_i} W_r^{(i)} \qquad (j \geq 0).$$

Let $\mathfrak{H}'$ be the closure of $\sum_{j \geq 0} \mathfrak{H}_{j+1}$ in $\mathfrak{H}$. Then $\mathfrak{H}' \supset \sum_{\mathfrak{D} \in \Omega^*} \mathfrak{H}^*_{\mathfrak{D}} = \sum_{\mathfrak{D} \in \Omega} \mathfrak{H}_{\mathfrak{D}}$. Since $\sum_{\mathfrak{D} \in \Omega} \mathfrak{H}_{\mathfrak{D}}$ is dense in $\mathfrak{H}$ (see Theorem 4 of [6, Part III, §9]) it follows that $\mathfrak{H} = \mathfrak{H}'$. Therefore $\mathfrak{H}$ is the closure of $\sum_{j \geq 0} \sum_{1 \leq r \leq s_j} W_r^{(j)}$ and the lemma is proved.

LEMMA 3. *Let π_1, π_2 be two unitary representations of G both satisfying the conditions of Lemma 1. Then if they have the same character they are equivalent.*

Let $\mathfrak{H}_i$ be the representation space and T_{π_i} the character of π_i $(i = 1, 2)$. Suppose $T_{\pi_1} = T_{\pi_2} = T$ (say). Then if $f \in C_c^\infty(G)$ and

$$F(x) = \int \operatorname{conj} (f(y)) f(yx) dy,$$

it is clear that $F \in C_c^\infty(G)$ and

$$T(F) = \left\| \int f(x) \pi_1(x) dx \right\|^2 = \left\| \int f(x) \pi_2(x) dx \right\|^2.$$

Hence $T \neq 0$ unless $\pi_i(x) = 0$ $(i = 1, 2)$ for all $x \in G$. But since $\pi_i(1)$ is the unit operator on $\mathfrak{H}_i$, this is possible only if $\mathfrak{H}_1 = \mathfrak{H}_2 = \{0\}$. Since the lemma is true in this trivial case, we may assume that $T \neq 0$. Choose $f \in C_c^\infty(G)$ such that $T(f) \neq 0$. For any $z \in Z$ put $f_z(x) = f(z^{-1}x)$ $(x \in G)$. Then

$$T_\pi(f_z) = T_{\pi_i}(f_z) = \xi_i(z) T_{\pi_i}(f) = \xi_i(z) T(f) \qquad (i = 1, 2)$$

where $\pi_i(z) = \xi_i(z)\pi_i(1)$. Since $T(f) \neq 0$ it follows that $\xi_1(z) = \xi_2(z)$. Hence we can find a homomorphism η of K into C such that $\eta(z) = \xi_1(z) = \xi_2(z)$ if $z \in Z \cap D$. Now define as above a representation π_i^* of K^* by putting $\pi_i^*(u^*) = \eta(u^{-1})\pi_i(u)$ $(u \in K)$ and let $\mathfrak{H}^*_{i,\mathfrak{D}}$ denote the subspace of those elements in

$\mathfrak{H}_i$ which transform under $\pi_i^*(K^*)$ according to $\mathfrak{D}$ ($\mathfrak{D}\in\Omega^*$, $i=1,2$). Again we arrange the elements of Ω^* in a sequence $\mathfrak{D}_j$ ($j\geq 1$). In view of Lemma 2, $\mathfrak{H}_i$ can be written as a sum[2] of mutually orthogonal subspaces $W_{i,k}\neq\{0\}$ each of which is invariant and irreducible under π_i. Here k runs over some subset N_i of integers ($i=1,2$). We shall now define a 1-1 mapping α of N_1 onto N_2 such that the representations of G induced on $W_{1,k}$ and $W_{2,\alpha(k)}$ under π_1 and π_2 respectively are equivalent. This would prove the lemma.

For any j let $N_i(\mathfrak{D}_j)$ denote the subset of N_i consisting of those $k\in N_i$ for which $W_{i,k}\cap\mathfrak{H}_{i,\mathfrak{D}_j}^*\neq\{0\}$. Since $\sum_{j\geq 1}(W_{i,k}\cap\mathfrak{H}_{i,\mathfrak{D}_j}^*)$ is dense in $W_{i,k}$ (Theorem 4 of [6, §9]) it follows that $\cup_{j\geq 1} N_i(\mathfrak{D}_j)=N_i$. Put $M_{i,j}=\cup_{1\leq r\leq j} N_i(\mathfrak{D}_r)$ ($j\geq 1$) and let $M_{i,0}$ denote the empty set. We shall now define a 1-1 mapping α of N_1 onto N_2 with the following properties:

(i) $\alpha(N_1(\mathfrak{D}_j))=N_2(\mathfrak{D}_j)$ ($j\geq 1$).

(ii) The representations induced on $W_{1,k}$ and $W_{2,\alpha(k)}$ ($k\in N_1$) under π_1 and π_2 respectively are equivalent.

We proceed by induction on j. Suppose α has been defined as a 1-1 mapping $M_{1,r-1}$ onto $M_{2,r-1}$ ($r\geq 1$) satisfying the above two requirements for $j\leq r-1$ and $k\in M_{1,r-1}$. We shall now extend it on $M_{1,r}$. We may clearly assume that at least one of the above two sets $N_1(\mathfrak{D}_r)$, $N_2(\mathfrak{D}_r)$ is not empty since otherwise $M_{i,r}=M_{i,r-1}$ ($i=1,2$) and no extension is needed. Let $P_{i,k}$ and $E_{i,\mathfrak{D}}$ denote the orthogonal projections of $\mathfrak{H}_i$ on $W_{i,k}$ and $\mathfrak{H}_{i,\mathfrak{D}}^*$ respectively ($\mathfrak{D}\in\Omega^*$, $i=1,2$). Put

$$\phi_{i,k}(x) = \operatorname{sp}\,(E_{i,\mathfrak{D}_r}P_{i,k}\pi_i(x)E_{i,\mathfrak{D}_r}),$$
$$\phi_i(x) = \operatorname{sp}\,(E_{i,\mathfrak{D}_r}\pi_i(x)E_{i,\mathfrak{D}_r}) \qquad (k\in N_i,\ x\in G).$$

Then

$$\phi_i = \sum_{k\in N_i(\mathfrak{D}_r)} \phi_{i,k} \qquad\qquad (i=1,2).$$

We claim $\phi_1=\phi_2$. For otherwise $\phi=\phi_1-\phi_2\neq 0$ and we can find a function $f\in C_c^\infty(G)$ such that $\int\phi(x)f(x)dx\neq 0$. Then as we have seen in the proof of Theorem 6 of [8], there exists a function $f'\in C_c^\infty(G)$ such that

$$E_{i,\mathfrak{D}_r}\int f(x)\pi_i(x)dx = \int f'(x)\pi_i(x)dx.$$

Hence

$$T_{\pi_1}(f') - T_{\pi_2}(f') = \int f(x)\phi(x)dx \neq 0$$

which contradicts our hypothesis that $T_{\pi_1}=T_{\pi_2}$. Therefore

$$\sum_{k\in N_1(\mathfrak{D}_r)} \phi_{1,k} = \sum_{k\in N_2(\mathfrak{D}_r)} \phi_{2,k}.$$

Since $W_{i,k} \cap \mathfrak{H}^*_{i,\mathfrak{D}_r} \neq \{0\}$ $(k \in N_i(\mathfrak{D}_r))$ it follows that $\phi_{i,k}(1) \neq 0$ for $k \in N_i(\mathfrak{D}_r)$. Now suppose $k \in N_1(\mathfrak{D}_r) \cap M_{1,r-1}$ so that $\alpha(k)$ has already been defined. Then by induction hypothesis, the representations induced on $W_{1,k}$ and $W_{2,\alpha(k)}$ are equivalent and therefore $\phi_{1,k} = \phi_{2,\alpha(k)}$. Since $\phi_{1,k}(1) \neq 0$, $\phi_{2,\alpha(k)}(1) \neq 0$ and therefore $W_{2,\alpha(k)} \cap \mathfrak{H}^*_{2,\mathfrak{D}_r} \neq \{0\}$. Hence $\alpha(k) \in N_2(\mathfrak{D}_r) \cap M_{2,r-1}$. Conversely suppose $l \in N_2(\mathfrak{D}_r) \cap M_{2,r-1}$. Since α is a 1-1 mapping of $M_{1,r-1}$ onto $M_{2,r-1}$ there is exactly one $k \in M_{1,r-1}$ such that $l = \alpha(k)$. Moreover the representations induced on $W_{1,k}$ and $W_{2,\alpha(k)} = W_{2,l}$ are equivalent. Therefore since $W_{2,l} \cap \mathfrak{H}^*_{2,\mathfrak{D}_r} \neq \{0\}$, it follows that $W_{1,k} \cap \mathfrak{H}^*_{1,\mathfrak{D}_r} \neq \{0\}$. Hence $k \in N_1(\mathfrak{D}_r) \cap M_{1,r-1}$. This proves that α maps $N_1(\mathfrak{D}_r) \cap M_{1,r-1}$ onto $N_2(\mathfrak{D}_r) \cap M_{2,r-1}$. Let $N_i'(\mathfrak{D}_r)$ be the complement of $N_i(\mathfrak{D}_r) \cap M_{i,r-1}$ in $N_i(\mathfrak{D}_r)$ $(i = 1, 2)$. Then it is clear from what we have said that

$$\sum_{k \in N_1'(\mathfrak{D}_r)} \phi_{1,k} = \sum_{k \in N_2'(\mathfrak{D}_r)} \phi_{2,k}.$$

Let $\psi_1, \cdots, \psi_s$ be all the distinct functions among $\phi_{i,k}$ $(k \in N_i'(\mathfrak{D}_r), i = 1, 2)$. We know that none of these are zero and therefore from Theorem 1 of [8] it follows that each ψ_t $(1 \leq t \leq s)$ appears the same number of times in the two sums on either side of the above equation. This means that we can find a 1-1 mapping $k \to \alpha(k)$ of $N_1'(\mathfrak{D}_r)$ onto $N_2'(\mathfrak{D}_r)$ such that $\phi_{1,k} = \phi_{2,\alpha(k)} \neq 0$. In view of Theorem 1 of [8] and Theorem 8 of [6, §11], we can conclude that the representations induced on $W_{1,k}$ and $W_{2,\alpha(k)}$ $(k \in N_1'(\mathfrak{D}_r))$ are equivalent. Thus α is now defined on M_r and satisfies all the requirements. Therefore our induction is complete and the theorem follows.

Notice that if $k \in N_i(\mathfrak{D})$ $(\mathfrak{D} \in \Omega^*)$, the representations induced on $W_{i,k}$ and $W_{i,l}$ cannot be equivalent unless $l \in N_i(\mathfrak{D})$. Hence if σ is any unitary irreducible representation of G, there are only a finite number of values of k such that the representation induced on $W_{i,k}$ is equivalent to σ. Let $n_i(\sigma)$ $(i = 1, 2)$ be this number. Then the above proof shows that $n_1(\sigma) = n_2(\sigma)$. In particular if $\pi_1 = \pi_2 = \pi$ (say), this number, which we now denote by $n(\sigma)$, is independent of the particular decomposition of $\mathfrak{H}$ into mutually orthogonal invariant irreducible subspaces. If ω is the equivalence class of σ, we call $n(\sigma)$ the multiplicity of ω in π. It is clear from Lemma 1 that for any $f \in C_c^\infty(G)$,

$$T_\pi(f) = \sum_{\omega \in \mathcal{E}} n(\omega) T_\omega(f)$$

where T_ω is the character and $n(\omega)$ the multiplicity of ω in π and the series is absolutely convergent. We may therefore write

$$T_\pi = \sum_{\omega \in \mathcal{E}} n(\omega) T_\omega.$$

Let T be a distribution on G. We shall say that T is a character of G if there exists a representation π satisfying the conditions of Lemma 1 such that

T is the character of π. T is said to be unitary or irreducible if π may be chosen to be unitary or irreducible.

3. **Computation of some characters.** We know that the mapping (u, h, n) $\rightarrow uhn$ $(u \in K, h \in A_+, n \in N)$ is a homeomorphism of $K \times A_+ \times N$ onto G. Since $Z \subset K$, A_+ and N are mapped isomorphically under the mapping $x \rightarrow x^*$ of G on $G^* = G/D \cap Z$. Therefore we may identify A_+N with its image under this mapping. Then $(u^*, h, n) \rightarrow u^*hn$ $(u^* \in K^*, h \in A_+, n \in N)$ is a topological mapping of $K^* \times A_+ \times N$ onto G. For any $x \in G$ put

$$x^*u^* = u_x^* h(x, u^*)n$$

where $u_x^* \in K^*$, $h(x, u^*) \in A_+$, and $n \in N$. Then u_x^* and $h(x, u^*)$ are continuous functions of (x, u^*) on $G \times K^*$. Since A_+ is simply connected, the mapping $H \rightarrow \exp H$ $(H \in \mathfrak{h}_{\mathfrak{p}_0})$ maps $\mathfrak{h}_{\mathfrak{p}_0}$ topologically onto A_+. We denote its inverse by $h \rightarrow \log h$ $(h \in A_+)$. Put $H(x, u^*) = \log h(x, u^*)$. Finally let $\gamma(x, u^*)$ denote the unique element in K such that

$$u^{-1}xu \in \gamma(x, u^*)A_+N \qquad (x \in G, u^* \in K^*).$$

Here u is any element in K lying[3] above u^*.

Normalise the Haar measure du^* on K^* so that the total measure of K^* is 1. Let η be a homomorphism of K into C and Λ a (complex-valued) linear function on $\mathfrak{h}_{\mathfrak{p}_0}$. We regard the space $\mathfrak{H} = L_2(K^*)$ of all square-integrable functions on K^* as a Hilbert space in the usual way and define a representation π of G on $\mathfrak{H}$ as follows. If $f \in \mathfrak{H}$ and $x \in G$,

$$\pi(x)f(u^*) = \eta(\gamma(x^{-1}, u^*)) \exp \{ - (\Lambda + 2\rho)(H(x^{-1}, u^*)) \} f(u_x^{*-1}) \qquad (u^* \in K^*).$$

Here $\pi(x)f(u^*)$ denotes the value of the function $\pi(x)f$ at u^* and ρ has the same meaning as in $[6, \S12]$. It is easy to verify (see $[6, \S12]$) that π is in fact a representation.

Let $\mathfrak{m}_0$, $\mathfrak{k}_{t_0}$, and $\mathfrak{h}_0$ be the subalgebras of $\mathfrak{g}_0$ as defined in $[6, \S2]$ and let M_0, A_-^0, and A^0 be the corresponding analytic subgroups of G. Then A^0 is a maximal connected abelian subgroup of G and therefore it is closed. Let M and A_- respectively be the centralizers of A_+ and A^0 in K. Then they are both closed subgroups of K. Since $\mathfrak{m}_0$ and $\mathfrak{k}_{t_0}$ respectively are the centralizers of $\mathfrak{h}_{\mathfrak{p}_0}$ and $\mathfrak{h}_0$ in $\mathfrak{k}_0$ (see Lemma 4, $\S2$ of $[6]$), M_0 and A_-^0 are the components of identity of M and A_- respectively. Put $A = A_+A_-$. We shall see later that A is exactly the centralizer of A^0 in G.

LEMMA 4. *Let m be an element in M. Then $mNm^{-1} = N$.*

Let $\mathfrak{g}$ be the complexification of $\mathfrak{g}_0$ and $\mathfrak{h}$, $\mathfrak{h}_\mathfrak{p}$, $\mathfrak{h}_t$, $\mathfrak{m}$ the subalgebras of $\mathfrak{g}$ spanned by $\mathfrak{h}_0$, $\mathfrak{h}_{\mathfrak{p}_0}$, $\mathfrak{h}_{t_0}$, $\mathfrak{m}_0$ respectively over C. We define positive roots of $\mathfrak{g}$

---

[3] Let V be the space of all cosets xB $(x \in G)$ with respect to a closed subgroup B of G Then we say that x lies above v $(x \in G, v \in V)$ if x lies in the coset v.

(with respect to $\mathfrak{h}$) and divide them into two disjoint classes P_+ and P_- as described in $[6, \S 2]$. For every root α select an element $X_\alpha \neq 0$ in $\mathfrak{g}$ such that $[H, X_\alpha] = \alpha(H)X_\alpha$ $(H \in \mathfrak{h})$. Then if $\mathfrak{n} = \sum_{\alpha \in P_+} C X_\alpha$ and $\mathfrak{n}^- = \sum_{\alpha \in P_+} C X_{-\alpha}$ we have $\mathfrak{g} = \mathfrak{h}_\mathfrak{p} + \mathfrak{m} + \mathfrak{n} + \mathfrak{n}^-$. Let $x \to \mathrm{Ad}\ (x)$ denote the adjoint representation of G on $\mathfrak{g}$. Then if $m \in M$, $\mathrm{Ad}\ (m)H = H$ and therefore $[H, \mathrm{Ad}(m)X_\alpha] = \alpha(H)\mathrm{Ad}\ (m)X_\alpha$ for all $H \in \mathfrak{h}_\mathfrak{p}$ and $\alpha \in P_+$. Since $[\mathfrak{h}_\mathfrak{p}, \mathfrak{m}] = \{0\}$ it follows from the above decomposition of $\mathfrak{g}$ that $\mathrm{Ad}\ (m)X_\alpha \in \mathfrak{n}$ $(\alpha \in P_+)$. Hence $\mathrm{Ad}\ (m)\mathfrak{n} = \mathfrak{n}$ and therefore $mNm^{-1} = N$.

Let M^* denote the image of M in K^*. Then M^* is the centralizer of A_+ in K^*, and therefore it is closed and hence compact. Let M_1 be any subgroup of M containing $M_0 Z$. Let M_1^* and M_0^* denote the images of M_1 and M_0 respectively in K^*. Then M_0^* is the connected component of M^* and therefore M^*/M_0^* is both compact and discrete and hence finite. From this it follows that M_1^* is compact. We normalise the Haar measure dm^* on M_1^* so as to make the total measure of M_1^* equal to 1.

Let τ denote the right regular representation of K^* on $\mathfrak{H}$ so that $\tau(v^*)f(u^*) = f(u^*v^*)$ $(u^*, v^* \in K^*, f \in \mathfrak{H})$. Then it follows easily from Lemma 4 that $\tau(m^*)$ commutes with $\pi(x)$ if $x \in G$ and $m^* \in M^*$. Put

$$T = \int f(x)\pi(x)dx, \qquad S = \int_{M_1^*} g(m^*)\tau(m^*)dm^*$$

where $f \in C_c^\infty(G)$ and g is a continuous function on M_1^*. Now if $v \in K$,

$$\pi(v)\phi(u^*) = \eta(v^{-1})\phi(v^{*-1}u^*) \qquad (\phi \in \mathfrak{H}, u^* \in K^*).$$

Therefore it follows from the Peter-Weyl Theorem for K^* that no irreducible representation of K occurs more often in the reduction of $\pi(K)$ than its degree. Hence Lemma 1 is applicable. Since S is a bounded operator it follows (see Lemma 1 of $[8]$) that TS is of the trace class. We propose to compute $\mathrm{Sp}\ (TS)$.

Let ϕ be a continuous function on K^*. For a fixed m^* in M_1^* put $\phi' = \tau(m^*)\phi$. Then

$$T\phi'(u^*) = \int f(x)\pi(x)\phi'(u^*)dx = \int f(ux^{-1})\pi(ux^{-1})\phi'(u^*)dx$$

where u is some element in K lying above u^*. Let dv, dh, dn denote the Haar measures on K, A_+, and N respectively. We normalise dv in such a way that for any continuous function ψ on K which vanishes outside a compact set,

$$\int_K \psi(v)dv = \int_{K^*} \psi^*(v^*)dv^*$$

where $\psi^*(v^*) = \sum_{\gamma \in D \cap Z} \psi(v\gamma)$ $(v \in K)$. Moreover we assume that $dx, dh,$ and dn are so normalised that

$$dx = \exp\{2\rho(\log h)\}dvdhdn \qquad (x = vhn; v \in K, h \in A_+, n \in N)$$

(see Lemma 35 of [6, §12]). Then

$$T\phi'(u^*) = \int f(u(vhn)^{-1})\eta(u^{-1}v)\phi'(v^*) \exp\{\Lambda(\log h)\}dvdhdn$$

$$= \int f(u(hn)^{-1}v^{-1})\eta(u^{-1}v)\phi(v^*m^*) \exp\{\Lambda(\log h)\}dvdhdn.$$

Now since A_+ is abelian and N is nilpotent, they are both unimodular. Hence

$$\int f(u(hn)^{-1}v^{-1})dhdn = \int f(unhv^{-1})dhdn.$$

But $nh = h(h^{-1}nh)$ and for a fixed h, $d(h^{-1}nh) = \exp\{-2\rho(\log h)\}dn$ as follows easily from Lemma 5 of [6, §2]. Therefore

$$\int f(u(hn)^{-1}v^{-1})dhdn = \int f(uhnv^{-1}) \exp\{2\rho(\log h)\}dhdn$$

and

$$TS\phi(u^*) = \int f(uhnv^{-1})\eta(u^{-1}v)\phi(v^*m^*)g(m^*) \exp\{(\Lambda + 2\rho)(\log h)\}dvdhdndm^*.$$

Put

$$F(u, v) = \int f(uhnv^{-1})\eta(u^{-1}v) \exp\{(\Lambda + 2\rho)(\log h)\}dhdn \qquad (u, v \in K).$$

Since f vanishes outside a compact set it is clear that for a fixed v, $F(u, v)$ vanishes outside a compact set on K. Since $F(u, v\gamma) = F(u\gamma^{-1}, v)$ $(\gamma \in Z)$ it follows that the sum $\sum_{\gamma \in Z \cap D} F(u, v\gamma)$ is defined and depends only on (u^*, v^*). Put

$$F(u^*, v^*) = \sum_{\gamma \in Z \cap D} F(u, v\gamma).$$

Then it is seen without difficulty that F^* is an indefinitely differentiable function on $K^* \times K^*$ and

$$TS\phi(u^*) = \int F^*(u^*, v^*)\phi(v^*m^*)g(m^*)dv^*dm^*$$

$$= \int F^*(u^*, v^*m^{*-1})\phi(v^*)g(m^*)dv^*dm^*$$

$$= \int \Phi(u^*, v^*)\phi(v^*)dv^*$$

where

$$\Phi(u^*, v^*) = \int_{M_1^*} F^*(u^*, v^*m^{*-1})g(m^*)dm^*.$$

Thus TS is represented here as an integral operator with the kernel Φ. It is clear that Φ is also indefinitely differentiable on $K^* \times K^*$. In order to compute $\mathrm{Sp}\, TS$ we make use of the following lemma.

LEMMA 5. *Let* $\lambda(u^*, v^*)$ *be an indefinitely differentiable function on* $K^* \times K^*$ *and let* L *be the bounded linear operator on* $L_2(K^*)$ *defined by*

$$L\phi(u^*) = \int \lambda(u^*, v^*)\phi(v^*)dv^* \qquad (\phi \in L_2(K^*)).$$

Then if[4] L *is of the trace class*

$$\mathrm{sp}\, L = \int_{K^*} \lambda(u^*, u^*)du^*.$$

As above let Ω^* denote the set of all equivalence classes of simple finite-dimensional representations of K^*. For every $\mathfrak{D} \in \Omega^*$ choose a unitary matrix representation $\sigma^{\mathfrak{D}}$ in $\mathfrak{D}$. Let $d(\mathfrak{D})$ denote the degree and $\sigma_{ij}^{\mathfrak{D}}$, $1 \leq i,j \leq d(\mathfrak{D})$, the matrix coefficients of $\sigma^{\mathfrak{D}}$. Then the functions $d(\mathfrak{D})^{1/2}\sigma_{ij}^{\mathfrak{D}}$, $1 \leq i,j \leq d(\mathfrak{D})$ $(\mathfrak{D} \in \Omega^*)$, form a complete orthonormal set in $L_2(K^*)$. Hence if L is of the trace class,

$$\mathrm{sp}\, L = \sum_{\mathfrak{D} \in \Omega^*} d(\mathfrak{D})(\sigma_{ij}^{\mathfrak{D}}, L\sigma_{ij}^{\mathfrak{D}})$$

$$= \sum_{\mathfrak{D} \in \Omega^*} d(\mathfrak{D}) \int_{K^*} \chi_{\mathfrak{D}}(u^{*-1}v^*)\lambda(u^*, v^*)du^*dv^*$$

where $\chi_{\mathfrak{D}}$ is the character of the class $\mathfrak{D}$ and (ψ, ϕ) denotes the usual scalar product in $L_2(K^*)$ $(\psi, \phi \in L_2(K^*))$. Put

$$\lambda_0(v^*) = \int \lambda(u^*, u^*v^*)du^*.$$

Then $\lambda_0(v^*)$ is an idefinitely differentiable function on K^* and

$$\mathrm{sp}\, L = \sum_{\mathfrak{D} \in \Omega^*} d(\mathfrak{D}) \int \chi_{\mathfrak{D}}(v^*)\lambda_0(v^*)dv^*.$$

Let $\mathfrak{H}$ be the Banach space of continuous functions ϕ on K^* with the norm

---

[4] It is not difficult to show by the method of §5 of [8] that L is in fact of the trace class. However we do not need this fact here.

$|\phi| = \sup_{K^*} |\phi(v^*)|$. For $u^* \in K^*$ let $l(u^*)$ denote the bounded linear operator on $\mathfrak{H}$ given by

$$l(u^*)\phi(v^*) = \phi(u^{*-1}v^*) \qquad (\phi \in \mathfrak{H},\ v^* \in K^*).$$

Then $u^* \to l(u^*)$ is a representation of K^* on $\mathfrak{H}$. Since λ_0 is of class C^∞ it is differentiable under l (see [6, §9]). Therefore if we use the arguments of the proof of Lemma 3 of [8] we see that the series $\sum_{\mathfrak{D} \in \Omega^*} d(\mathfrak{D}) \int \chi_{\mathfrak{D}}(v^{*-1}) l(v^*)\lambda_0 dv^*$ converges absolutely in $\mathfrak{H}$ to λ_0. Hence

$$\sum_{\mathfrak{D} \in \Omega^*} d(\mathfrak{D}) \int \chi_{\mathfrak{D}}(v^*)\lambda_0(v^*u^*)dv^*$$

converges uniformly to $\lambda_0(u^*)$ on K^*. Putting $u^* = 1^*$ we get

$$\mathrm{sp}\ L = \lambda_0(1^*) = \int \lambda(u^*,\ u^*)du^*.$$

If we apply the above lemma to the operator TS we get

$$\mathrm{sp}\ TS = \int \Phi(u^*,\ u^*)du^* = \int F^*(u^*,\ u^*m^{*-1})g(m^*)du^*dm^*.$$

Now M_1 is the complete inverse image of M_1^* in K. Therefore it is closed. We normalise the Haar measure dm on M_1 in such a way that for any continuous function α on M_1 which vanishes outside a compact set,

$$\int_{M_1} \alpha(m)dm = \int_{M_1^*} \alpha^*(m^*)dm^*$$

where $\alpha^*(m^*) = \sum_{\gamma \in Z \cap D} \alpha(m\gamma)$ $(m \in M_1)$. Then if we recall the definition of F^* we find that

$$\int F^*(u^*,\ u^*m^{*-1})g(m^*)dm^*$$

$$= \int F(u,\ um^{-1})g(m^*)dm$$

$$= \int f(uhnmu^{-1})\eta(m^{-1})g(m^*) \exp \left\{ (\Lambda + 2\rho)(\log h) \right\} dh\,dn\,dm$$

where u is any element in K lying above u^*. Now $hnm = mh(m^{-1}nm)$ and for fixed m, $d(m^{-1}nm) = dn$. Hence

$$\Phi(u^*,\ u^*) = \int f((mhn)^{u^*})\eta(m^{-1})g(m^*) \exp \left\{ (\Lambda + 2\rho)(\log h) \right\} dh\,dn\,dm$$

(where $x^{u^*} = uxu^{-1}$) and therefore

$$\text{sp } TS = \int f((mhn)^{u^*})\eta(m^{-1})g(m^*) \exp\left\{(\Lambda + 2\rho)(\log h)\right\} du^* dh dn dm.$$

Now let $m_1 \in M_1$. Then

$$(m_1 m m_1^{-1} hn)^{u^*} = (mhn')^{u^* m_1^*} \qquad (m \in M_1,\ h \in A_+,\ n \in N)$$

where $n' = m_1^{-1} n m_1 \in N$. From this it follows that

$$\text{sp } TS = \text{sp } TS'$$

where $S' = \int_{M_1^*} g'(m^*)\tau(m^*) dm^*$ and $g'(m^*) = g((m_1 m m_1^{-1})^*)$ $(m \in M_1)$. Hence it is clear that

$$\text{sp } TS = \int f((mhn)^{u^*})\eta(m^{-1})\xi^*(m^*) \exp\left\{(\Lambda + 2\rho)(\log h)\right\} du^* dh dn dm$$

where $\xi^*(m^*) = \int_{M_1} g(m_1^* m^* m_1^{*-1}) dm_1^*$.

Now let σ be an irreducible unitary matrix representation of M_1^* of degree d and let σ_{ij}, $1 \leq i, j \leq d$, denote its matrix coefficients. Put

$$E_{ij} = d \int_{M_1^*} \sigma_{ij}(m^{*-1})\tau(m^*) dm^*, \qquad 1 \leq i, j \leq d,$$

and $E_i = E_{ii}$ $(1 \leq i \leq d)$. Then E_i are mutually orthogonal projections which commute with $\pi(x)$ $(x \in G)$. Hence if $\mathfrak{H} = L_2(K^*)$, $\mathfrak{H}_i = E_i \mathfrak{H}$, $1 \leq i \leq d$, are closed subspaces which are invariant under $\pi(G)$. Let π_i be the representation of G induced on $\mathfrak{H}_i$ under π. Then it is clear that the operator $\int f(x)\pi_i(x) dx$ is of the trace class and

$$\text{sp } \left(\int f(x)\pi_i(x) dx \right) = \text{sp } TE_i.$$

But

$$d \int_{M_1} \sigma_{ii}(m_1^* m^* m_1^{*-1}) dm_1^* = \xi^*(m^*) \qquad (m^* \in M_1^*)$$

where ξ^* is the character of σ. Hence

$$\text{sp } \left(\int f(x)\pi_i(x) dx \right) = \int f((mhn)^{u^*})\eta(m^{-1})\xi^*(m^{*-1})$$

$$\cdot \exp\left\{(\Lambda + 2\rho)(\log h)\right\} du^* dh dn dm.$$

Let σ' be the representation of M_1^* contragredient to σ. Define a representation σ'' of M_1 as follows:

$$\sigma''(m) = \eta(m^{-1})\sigma'(m^*) \qquad (m \in M_1).$$

Let δ be the equivalence class and ξ_δ the character of σ''. Then

$$\xi_\delta(m) = \eta(m^{-1})\xi^*(m^{*-1}) \qquad (m \in M_1).$$

Hence[5]

$$\mathrm{sp}\left(\int f(x)\pi_i(x)dx\right) = T_{\Lambda,\delta}(f)$$

where

$$T_{\Lambda,\delta}(f) = \int f((mhn)^{u^*})\xi_\delta(m) \exp\left\{(\Lambda + 2\rho)(\log h)\right\}du^*dhdndm.$$

Let ω_{M_1} be the set of all equivalence classes of finite-dimensional simple representations of M_1. Then we have proved the following theorem.

THEOREM 1. *Let Λ be a linear function on $\mathfrak{h}_{\mathfrak{p}_0}$ and δ a class in ω_{M_1}. Let ξ_δ denote the character of δ. Let $T_{\Lambda,\delta}$ denote the distribution given by*

$$T_{\Lambda,\delta}(f) = \int_{M_1} \xi_\delta(m)dm \int f((mhn)^{u^*}) \exp\left\{(\Lambda + 2\rho)(\log h)\right\}du^*dhdn$$

$(f\in C_c^\infty(G))$. *Then $T_{\Lambda,\delta}$ is a character of G.*

The above formula shows that $T_{\Lambda,\delta}$ is not only a distribution but actually a measure (see [10]). Therefore it may be regarded as a continuous linear functional on the space $C_c(G)$ of all continuous functions on G which vanish outside a compact set. Moreover we know (see [6, §12]) that if $\Lambda+\rho$ takes pure imaginary values on $\mathfrak{h}_{\mathfrak{p}_0}$ and $|\eta(x)| = 1$ $(x\in G)$, π (and therefore π_i) is a unitary representation of G. Hence $T_{\Lambda,\delta}$ is a unitary character if $\Lambda+\rho$ is pure imaginary on $\mathfrak{h}_{\mathfrak{p}_0}$ and δ is a unitary class (i.e. the class of a unitary representation of M_1). For any $\mathfrak{D}\in\Omega$ let $(\mathfrak{D}:\delta)$ denote the number of times δ occurs in the reduction of $\mathfrak{D}$ with respect to M_1. Then we know (see A. Weil [12, p. 83] and [6, §12]) that $\mathfrak{D}$ occurs exactly $(\mathfrak{D}:\delta)$ times in the reduction of π_i with respect to K.

We shall now derive another expression for the character $T_{\Lambda,\delta}$ when $M_1 = M_0 Z$. Since $Z\subset K$, every element $h\in A^0 Z$ can be written uniquely as $h = h_+ h_-$ where $h_+\in A_+$ and $h_-\in A^0_- Z$. Let A^{0*}_- be the image of A^0_- in K^*. Then A^{0*}_- is a maximal abelian subgroup of the compact Lie group M_0^*. Hence every element in M_0^* is conjugate (with respect to M_0^*) to some element in A^{0*}_-. Therefore it follows from known theory (see Weyl [13]) that if the Haar measure dh_- on $A^0_- Z$ is suitably normalised

$$\int_{M_0 Z} \gamma(m)dm = \int_{A^0_- Z} \gamma(h_-)\Delta_-(h_-)^2 dh_-$$

---

for any continuous class function γ on $M_0 Z$ which vanishes outside a compact set. Here

$$\Delta_-(h_-) = \left| \prod_{\alpha \in P_-} (e^{\alpha(H)/2} - e^{-\alpha(H)/2}) \right|$$

where H is any element in $\mathfrak{h}_{\mathfrak{l}_0}$ such that $h_-^{-1} \exp H \in Z$. If we put

$$\gamma(m) = \xi_\delta(m) \int f((mhn)^{u^*}) \exp\{(\Lambda + 2\rho)(\log h)\} du^* dh dn$$

we get the following result.

LEMMA 6. *It is possible to normalize the Haar measures dh_- and dh_+ on $A_-^0 Z$ and A_+ in such a way that*

$$T_{\Lambda,\delta}(f) = \int_{A_-^0 Z} \xi_\delta(h_-) \Delta_-(h_-)^2 dh_- \int f((h_- h_+ n)^{u^*})$$
$$\cdot \exp\{(\Lambda + 2\rho)(\log h_+)\} du^* dh_+ dn$$

for every linear function Λ on $\mathfrak{h}_{\mathfrak{p}_0}$, $\delta \in \omega_{M_0 Z}$ and $f \in C_c^\infty(G)$.

4. Transformation of certain integrals. We keep to the notation of §2. Let $X \to \mathrm{ad}\, X$ $(X \in \mathfrak{g}_0)$ denote the adjoint representation of $\mathfrak{g}_0$. We say that an element $X \in \mathfrak{g}_0$ is singular if the characteristic polynomial of ad X in the indeterminate λ is divisible by λ^{l+1} $(l = \dim \mathfrak{h}_0)$. The coefficient of λ^l is clearly a polynomial function[6] $F(X)$ on $\mathfrak{g}_0$ which is not identically zero. Since X is singular if and only if $F(X) = 0$, it follows that the set of singular elements is closed and nowhere dense in the Euclidean space $\mathfrak{g}_0$. We call an element regular if it is not singular. Let $x \to \mathrm{Ad}\,(x)$ $(x \in G)$ denote the adjoint representation of G on $\mathfrak{g}_0$. It is clear that Ad $(x)X$ $(x \in G)$ is regular if and only if X is regular.

LEMMA 7. *Let H be a regular element in $\mathfrak{h}_0$. Suppose x is an element in G such that Ad $(x)H \in \mathfrak{h}_0$. Then $x \in A_-' A_+$ where A_-' is the normalizer of A^0 in K.*

Since H is regular, $\mathfrak{h}_0$ is the centralizer of H in $\mathfrak{g}_0$. Hence Ad $(x)\mathfrak{h}_0$ is the centralizer of Ad $(x)H$. But since Ad $(x)H \in \mathfrak{h}_0$, $\mathfrak{h}_0$ is contained in this centralizer and therefore $\mathfrak{h}_0 \subset \mathrm{Ad}\,(x)\mathfrak{h}_0$. Since $\mathfrak{h}_0$ and Ad $(x)\mathfrak{h}_0$ have the same dimension, $\mathfrak{h}_0 = \mathrm{Ad}\,(x)\mathfrak{h}_0$. Let $x = uhn'$ $(u \in K, h \in A_+, n' \in N)$. Since $hn' = nh$ where $n = hn'h^{-1} \in N$ we get

$$\mathrm{Ad}\,(u^{-1})\mathfrak{h}_0 = \mathrm{Ad}\,(nh)\mathfrak{h}_0 = \mathrm{Ad}\,(n)\mathfrak{h}_0 \subset \mathfrak{h}_0 + \mathfrak{n}_0.$$

On the other hand Ad $(u^{-1})\mathfrak{h}_{\mathfrak{p}_0} \subset \mathfrak{p}_0$ and $\mathfrak{p}_0 \cap (\mathfrak{h}_0 + \mathfrak{n}_0) = \mathfrak{h}_{\mathfrak{p}_0}$ where $\mathfrak{p}_0$ is defined

---

[6] Let V be a vector space over R or C. A complex-valued function f on V is called a polynomial function if it can be written as a polynomial (with complex coefficients) in linear functions on V.

as in $[6, \,\S 2]$. Similarly Ad $(u^{-1})\mathfrak{h}_{\mathfrak{k}_0}\subset\mathfrak{k}_0$ and $\mathfrak{k}_0\cap(\mathfrak{h}_0+\mathfrak{n}_0)=\mathfrak{h}_{\mathfrak{k}_0}$. Therefore

$$\text{Ad } (u^{-1})\mathfrak{h}_{\mathfrak{p}_0} \subset \mathfrak{h}_{\mathfrak{p}_0}, \qquad \text{Ad } (u^{-1})\mathfrak{h}_{\mathfrak{k}_0} \subset \mathfrak{h}_{\mathfrak{k}_0}.$$

Hence Ad $(u^{-1})\mathfrak{h}_0\subset\mathfrak{h}_0$ and so $u\in A'_-$. Therefore in order to prove our assertion we have only to show that $n=1$. This follows from the lemma below.

LEMMA 8. *Let H be a regular element in $\mathfrak{h}_0$. Then $n\to\text{Ad } (n)H$ $(n\in N)$ is a 1-1 mapping of N onto the set of all elements of the form $H+Z$ $(Z\in\mathfrak{n}_0)$.*

Define X_α and $\mathfrak{n}$ as in the proof of Lemma 4. Then $\mathfrak{n}_0=\mathfrak{n}\cap\mathfrak{g}_0$; since N is nilpotent every $n\in N$ can be written in the form $n=\exp X$ $(X\in\mathfrak{n}_0)$. Therefore Ad $(n)H-H=\exp (\text{ad } X)H-H\in\mathfrak{n}_0$ since $[\mathfrak{n}_0, \mathfrak{h}_0]\subset\mathfrak{n}_0$. This proves that Ad $(n)H=H+Z$ where $Z\in\mathfrak{n}_0$. Now suppose Ad $(n)H=H$. Then $\exp (\text{ad } X)H-H=0$ and this implies that $X=0$. For otherwise suppose $X\neq0$. Then $X=\sum_{\alpha\in P_+} a_\alpha X_\alpha$ $(a_\alpha\in C)$ and not all a_α are zero. Let β be the lowest root in P_+ such that $a_\beta\neq0$. Then

$$\exp (\text{ad } X)H - H \equiv (\text{ad } X)H \equiv - a_\beta\beta(H)X_\beta \text{ mod } \sum_{\alpha>\beta} CX_\alpha.$$

Since H is regular, $\beta(H)\neq0$ and therefore it follows from the linear independence of X_α $(\alpha\in P_+)$ over C that $\exp (\text{ad } X)H-H\neq0$ in contradiction with our hypothesis. Now if Ad $(n_1)H=\text{Ad } (n_2)H$ $(n_1, n_2\in N)$, Ad $(n_2^{-1}n_1)H$ $=H$ and therefore $n_2^{-1}n_1=1$ so that $n_1=n_2$.

Finally we claim that every element of the form $H+Z$ $(Z\in\mathfrak{n}_0)$ can be written as Ad $(n)H$ for some $n\in N$. For otherwise choose Z such that such a representation is impossible. Then clearly $Z\neq0$. Let

$$Z = a_\alpha X_\alpha + \sum_{\beta>\alpha} a_\beta X_\beta \qquad\qquad (a_\alpha, a_\beta \in C)$$

where α is a root in P_+ and $a_\alpha\neq0$. We choose Z in such a way that α has the highest possible value. Since H is regular the mapping $X\to[H, X]$ $(X\in\mathfrak{n}_0)$ is a nonsingular linear mapping of $\mathfrak{n}_0$ into itself. Hence there exists a $Y\in\mathfrak{n}_0$ such that $[H, Y]=Z$. Put $n_1=\exp Y$. Then it is clear that

$$\text{Ad } (n_1)(H + Z) - H \equiv [Y, H] + Z \text{ mod } \sum_{\beta>\alpha} CX_\beta$$

$$\equiv 0 \text{ mod } \sum_{\beta>\alpha} CX_\beta.$$

Hence Ad (n_1) $(H+Z)=H+Z'$ where $Z'\in\mathfrak{n}_0\cap(\sum_{\beta>\alpha} CX_\beta)$. In view of our choice of Z it follows that $H+Z'=\text{Ad } (n_2)H$ for some $n_2\in N$. Therefore $H+Z=\text{Ad } (n)H$ where $n=n_1^{-1}n_2$. Since this contradicts our hypothesis the assertion is proved.

COROLLARY. *$A = A_+A_-$ is exactly the centralizer of $A^0=A_+A_-^0$ in G.*

Let x be an element of G which commutes with all elements in A^0. Then it follows from Lemma 7 that $x = uh$ where $u \in A'_-$ and $h \in A_+$. Since x commutes with A^0 the same is true of u. Hence u lies in the centralizer A_- of A^0 in K.

Since $\mathfrak{g}_0$ is a vector space over R of finite dimension we can regard it as an analytic manifold[7] and identify in the usual way the tangent space at each point $X \in \mathfrak{g}_0$ with $\mathfrak{g}_0$ itself. Then if f is a function on $\mathfrak{g}_0$ which is differentiable at X,

$$Yf(X) = \left\{ \frac{d}{dt} f(X + tY) \right\}_{t=0} \qquad (Y \in \mathfrak{g}_0).$$

Similar remarks hold also for any linear subspace of $\mathfrak{g}_0$ which may also be regarded as an analytic manifold.

Consider the subgroup A^0Z in G. It is clear that Ad (A^0) is a maximal connected abelian subgroup of Ad (G) and therefore it is closed. Hence A^0Z is closed in G. Let $\mathcal{E}$ denote the factor space G/A^0Z consisting of all cosets of the form xA^0Z $(x \in G)$. We regard $\mathcal{E}$ as an analytic manifold in the usual way (see [1]) and denote by $x \to \bar{x}$ the natural mapping of G on $\mathcal{E}$. Then for any fixed $H \in \mathfrak{h}_0$, Ad $(x)H$ depends only on $\bar{x}$. We put $\bar{x}H = $ Ad $(x)H$ $(x \in G)$. It is evident that the mapping ϕ: $(\bar{x}, H) \to \bar{x}H$ $(\bar{x} \in \mathcal{E}, H \in \mathfrak{h}_0)$ is an analytic mapping of $\mathcal{E} \times \mathfrak{h}_0$ into $\mathfrak{g}_0$. We consider the differential[7] of ϕ. Let $X_1, \cdots, X_r$ be a base for $\mathfrak{g}_0$ mod $\mathfrak{h}_0$ and let $(d\pi)_x$ denote the differential of the natural mapping of G on $\mathcal{E}$ at x. Then $(d\pi)_x X_i$, $1 \le i \le r$, forms a base for the tangent space of $\mathcal{E}$ at $\bar{x}$ (see [1, p. 110]). Hence if we regard the tangent space of $\mathcal{E} \times \mathfrak{h}_0$ at $(\bar{x}, H)$ as the direct sum of the tangent spaces of $\mathcal{E}$ and $\mathfrak{h}_0$ it is easily seen that

$$d\phi \circ (d\pi)_x X_i = - \text{ Ad } (x)[H, X_i], \qquad 1 \le i \le r,$$
$$d\phi H_0 = \text{ Ad } (x)H_0 \qquad (H_0 \in \mathfrak{h}_0).$$

Let D be the linear mapping of $\mathfrak{g}_0$ into itself defined by $DX_i = -[H, X_i]$, $1 \le i \le r$, and $DH_0 = H_0$ $(H_0 \in \mathfrak{h}_0)$. Then D defines a linear mapping $\overline{D}$ in the factor space $\mathfrak{g}_0/\mathfrak{h}_0$ which is the same as that induced by $-\text{ad}H$. Since G is semisimple det Ad $(x) = 1$. Hence

$$\left| \det \text{ Ad } (x)D \right| = \left| \det D \right| = \left| \det \overline{D} \right| = \prod_{\alpha > 0} \left| \alpha(H) \right|^2$$

where α runs over all positive roots. This shows that if H is regular det Ad $(x)D \ne 0$ and therefore $d\phi$ is regular[7] at $(\bar{x}, H)$. Let $\mathfrak{h}_1$ be the set of all regular elements in $\mathfrak{h}_0$. Then ϕ defines a continuous open mapping of $\mathcal{E} \times \mathfrak{h}_1$ into $\mathfrak{g}_0$. The image of $\mathcal{E} \times \mathfrak{h}_1$ in $\mathfrak{g}_0$ is obviously the set $\mathfrak{g}_1$ of all regular elements in $\mathfrak{g}_0$ which are conjugate (under G) to some element in $\mathfrak{h}_0$. Since $\mathfrak{h}_1$ is open in $\mathfrak{h}_0$, $\mathfrak{g}_1$ is open in $\mathfrak{g}_0$.

------

(7) We shall follow the terminology of Chevalley [1] in the rest of this paper.

Define A'_- as in Lemma 7 and put $A' = A'_- A_+$. Since Ad (A'_-) is the normalizer of Ad (A^0) in Ad (K), it is closed and therefore compact. Since $\mathfrak{h}_{\mathfrak{k}_0}$ is the normalizer of $\mathfrak{h}_0$ in $\mathfrak{k}_0$ it follows that A^0_- is the connected component of A'_- and therefore $A'/A^0 Z \cong A'_-/A^0_- Z \cong$ Ad $(A'_-)/$Ad (A^0_-) is finite. We denote by W the finite group $A'/A^0 Z$. For any $\bar{x} \in \mathcal{E}$ and $s \in W$ we define $\bar{x}s$ as follows. Choose $x \in G$ and $a \in A'$ lying above $\bar{x}$ and s respectively. Then the coset $xaA^0 Z$ depends only on $\bar{x}$ and s and we define $\bar{x}s$ to be this coset. It is clear that $\bar{x}s \neq \bar{x}$ unless $s = 1$. Since $W \subset \mathcal{E}$, sH $(s \in W, H \in \mathfrak{h}_0)$ is defined and lies in $\mathfrak{h}_0$. Now suppose $\bar{x}_1 H_1 = \bar{x}_2 H_2$ $(\bar{x}_1, \bar{x}_2 \in \mathcal{E}; H_1, H_2 \in \mathfrak{h}_1)$. Then if x_i lies in G above $\bar{x}_i$ $(i = 1, 2)$, Ad $(x_1)H_1 =$ Ad $(x_2)H_2$ and therefore, from Lemma 7, $x_1^{-1} x_2 \in A'$ and $H_1 =$ Ad $(x_1^{-1} x_2)H_2$. Hence there exists an $s \in W$ such that $\bar{x}_2 = \bar{x}_1 s$ and $H_2 = s^{-1} H_1$. Moreover $\bar{x}_2 \neq \bar{x}_1$ unless $s = 1$. Therefore if w is the order of the group W, there are exactly w distinct points in $\mathcal{E} \times \mathfrak{h}_1$ which have the same image in $\mathfrak{g}_1$.

Since $A^0 Z$ is abelian, it is unimodular and therefore (see Weil [11, p. 42]) there exists a measure $d\bar{x}$ (which is unique apart from a constant factor) on $\mathcal{E}$ such that it is invariant under the translations induced on $\mathcal{E}$ by G. Let $n = \dim \mathfrak{g}_0$ and let $(H_1, \cdots, H_l)$ be a base for $\mathfrak{h}_0$ so that $r = n - l$ (in the notation used above). As before let $X_1, \cdots, X_r$ be a base for $\mathfrak{g}_0 \bmod \mathfrak{h}_0$. Let ω be a left invariant differential form of degree n on G such that $\omega(X_1, \cdots, X_r, H_1, \cdots, H_l) = 1$. Then if $\bar{\omega}$ is the differential form of degree r on $\mathcal{E}$ corresponding to the (suitably normalized) invariant measure we have[8]

$$\bar{\omega}((d\pi)_x Y_1, \cdots, (d\pi)_x Y_r) = \omega(Y_1, \cdots, Y_r, H_1, \cdots, H_l)$$

for any $Y_1, \cdots, Y_r \in \mathfrak{g}_0$. Let dX denote the Euclidean measure on $\mathfrak{g}_0$ and η the corresponding differential form of degree n on $\mathfrak{g}_0$. We may assume that $\eta(X_1, \cdots, X_r, H_1, \cdots, H_l) = 1$. Consider the image of η under the dual[7] $\delta\phi$ of the mapping $d\phi$. Then $(\delta\phi\eta)_{\bar{x},H}$ is the form defined at $(\bar{x}, H)$ on $\mathcal{E} \times \mathfrak{h}_0$ by the rule

$$(\delta\phi\eta)_{\bar{x},H}((d\pi)_x X_1, \cdots, (d\pi)_x X_r, H_1, \cdots, H_l)$$
$$= \eta_{\bar{x}H}(d\phi o(d\pi)_x X_1, \cdots, d\phi o(d\pi)_x X_r, (d\phi)H_1, \cdots, (d\phi)H_l)$$
$$= \pm \prod_{\alpha > 0} |\alpha(H)|^2$$

as we saw above. Let ξ be the differential form on $\mathfrak{h}_0$ corresponding to the Euclidean measure dH. We assume that $\xi(H_1, \cdots, H_l) = 1$. Then

$$(\delta\phi\eta)_{\bar{x}H}((d\pi)_x X_1, \cdots, (d\pi)_x X_r, H_1, \cdots, H_l)$$
$$= \pm \prod_{\alpha > 0} |\alpha(H)|^2 \bar{\omega}((d\pi)_x X_1, \cdots, (d\pi)_x X_r)\xi(H_1, \cdots, H_l).$$

---

[8] We give here once in some detail the computation of the measure on $\mathfrak{g}_1$ in terms of the measure on $\mathcal{E}$ and $\mathfrak{h}_1$. All subsequent computations of a similar sort will be sketched only very briefly.

This proves that

$$(\delta\phi\eta)_{\tilde{x},H} = \pm \prod_{\alpha>0} \lvert \alpha(H) \rvert^2 \zeta$$

where ζ is the differential form on $\mathcal{E}\times\mathfrak{h}_0$ corresponding to the product measure $d\tilde{x}dH$. Therefore taking into account the fact that every point in $\mathfrak{g}_1$ has exactly w distinct pre-images in $\mathcal{E}\times\mathfrak{h}_1$, we can conclude that

$$w\int_{\mathfrak{g}_1} f(X)dX = \int_{\mathcal{E}\times\mathfrak{h}_1} \prod_{\alpha>0} \lvert \alpha(H) \rvert^2 f(\tilde{x}H)d\tilde{x}dH$$

for any measurable function f on $\mathfrak{g}_1$, whenever at least one of the two sides of this equation remains finite on replacing f by $\lvert f \rvert$.

Now let $G_* = G/Z$ and let $x \to x_*$ denote the natural mapping of G on G_*. Since $Z \subset K$, A_+ and N are mapped isomorphically and so we may identify them with their images under this mapping. Then $G_* = K_* A_+ N$. Now an $= ana^{-1} \cdot a$ $(a \in A_+,\ n \in N)$. Hence $G_* = K_* N A_+$ and $K_* N \cap A_+ = \{1\}$. Therefore G_*/A_+ is homeomorphic to $K_* N$ under the mapping $u_* n \to u_* n A_+$ $(u_* \in K_*,\ n \in N)$. If we identify G_*/A_+ with $K_* N$ under this mapping it is easy to verify that $du_* dn$ is the invariant measure on G_*/A_+. (Here du_* and dn are the Haar measures on K_* and N respectively.) Now $\mathcal{E} = G/A^0 Z$ $= G_*/A_*^0$ where $A_*^0 = A_+ A_{-*}^0$ and A_{-*}^0 is the analytic subgroup of K_* corresponding to $\mathfrak{h}_{\mathfrak{k}_0}$. Let dx_*, dh, and dh_* be the Haar measures on G_*, A_+, and A_{-*}^0 respectively. Then $dx_* = du_* dn dh$ $(x_* = u_* nh)$ if dn is suitably normalized. We shall assume that $\int dh_* = 1$. Let $d\tilde{x}$ be the invariant measure on $\mathcal{E}$. Then if $d\tilde{x}$ is suitably normalized, $dx_* = d\tilde{x}dh_* dh$ in the sense of $[11, \text{p. }42]$. Let $F(\tilde{x})$ be a continuous function on $\mathcal{E}$ which vanishes outside a compact set. Then there exists (see $[11, \text{p. }43]$) a continuous function F_1 on G_* vanishing outside a compact set such that

$$F(\tilde{x}_*) = \int F_1(x_* h h_*) dh_* dh$$

where $x_* \to \tilde{x}_*$ is the natural mapping of G_* on $\mathcal{E} = G_*/A_*$. Put

$$F_2(x_*) = \int F_1(x_* h_*) dh^*.$$

Then

$$\int F_2(x_*)dx_* = \int F(\tilde{x})d\tilde{x}.$$

But

$$\int F_2(x_*)dx_* = \int F_2(u_* nh)du_* dn dh = \int F(\overline{u_* n})du_* dn$$

since

$$\int F_2(u_*nh)dh = \int F_1(u_*nhh_*)dhdh_* = F(\overline{u_*n}).$$

Therefore

$$\int F(\bar{x})d\bar{x} = \int F(\overline{u_*n})du_*dn$$

and from this the same relation follows for any measurable function F on $\mathcal{E}$ provided either one of the two sides of the above equation remains finite on replacing F by $|F|$. Therefore in particular

$$\int_{\mathcal{E}} f(\bar{x}H)d\bar{x} = \int_{K_*N} f(\mathrm{Ad}\ (u_*n)H)du_*dn$$

where $x_* \rightarrow \mathrm{Ad}\ (x_*)$ is the adjoint representation of G_*. Hence

$$w\int_{\mathfrak{g}_1} f(X)dX = \int_{K_*} du_* \int_{N \times \mathfrak{h}_1} \prod_{\alpha>0} |\ \alpha(H)\ |^2 f(\mathrm{Ad}\ (u_*n)H)dndH.$$

We have seen (Lemma 8) that the mapping $(n, H) \rightarrow \mathrm{Ad}\ (n)H$ is a 1-1 mapping of $N \times \mathfrak{h}_1$ onto $\mathfrak{h}_1 + \mathfrak{n}_0$ which is obviously analytic. Now

$$\lim_{t \to 0} \frac{1}{t} [\mathrm{Ad}\ (n \exp tX)H - \mathrm{Ad}\ (n)H] = - \mathrm{Ad}\ (n)[H, X]\quad (X \in \mathfrak{n}_0),$$

$$\lim_{t \to 0} \frac{1}{t} [\mathrm{Ad}\ (n)(H + tH_0) - \mathrm{Ad}\ (n)H] = \mathrm{Ad}\ (n)H_0\qquad (H_0 \in \mathfrak{h}_0).$$

Hence if D is the linear mapping of $\mathfrak{h}_0 + \mathfrak{n}_0$ into itself given by $DX = -(\mathrm{ad}H)X$ $(X \in \mathfrak{n}_0)$ and $DH_0 = H_0$ $(H_0 \in \mathfrak{h}_0)$ it follows that

$$|\det (\mathrm{Ad}\ (n)D)| = |\det D| = \prod_{\alpha \in P_+} |\ \alpha(H)\ |$$

as the determinant of the restriction of $\mathrm{Ad}\ (n)$ on $\mathfrak{h}_0 + \mathfrak{n}_0$ is obviously 1. Since $H \in \mathfrak{h}_1$, $\prod_{\alpha \in P_+} |\ \alpha(H)\ | \neq 0$ and our mapping is regular at (n, H). Therefore in view of Lemma 8, it defines a topological mapping of $\mathfrak{h}_1 \times N$ onto $\mathfrak{h}_1 + \mathfrak{n}_0$ and $\int_{N \times \mathfrak{h}_1} f(\mathrm{Ad}\ (u_*n)H)dndH = \int_{\mathfrak{h}_1 + \mathfrak{n}_0} \prod_{\alpha \in P_+} |\ \alpha(H)\ |^{-1} f(\mathrm{Ad}\ (u_*)(H + Z))dHdZ$ where dZ is the suitably normalised Euclidean measure on $\mathfrak{n}_0$. Therefore

$$w\int_{\mathfrak{g}_1} f(X)dX$$

$$= \int_{\mathfrak{h}_1 + \mathfrak{n}_0} \prod_{\alpha \in P_+} |\ \alpha(H)\ | \prod_{\alpha \in P_-} |\ \alpha(H)\ |^2 dHdZ \int_{K_*} f(\mathrm{Ad}\ (u_*)(H + Z))du_*.$$

Thus we have proved the following result.

LEMMA 9. *It is possible to normalise the Euclidean measures dX, dH, and dZ on $\mathfrak{g}_0$, $\mathfrak{h}_0$, and $\mathfrak{n}_0$ respectively in such a way that for any measurable function $f(X)$ on $\mathfrak{g}_1$,*

$$w \int_{\mathfrak{g}_1} f(X)dX$$

$$= \int_{\mathfrak{h}_1+\mathfrak{n}_0} \prod_{\alpha \in P_+} |\alpha(H)| \prod_{\alpha \in P_-} |\alpha(H)|^2 dHdZ \int_{K_*} f(\mathrm{Ad}\,(u_*)(H+Z))du_*$$

provided at least one of these two integrals remains finite on replacing f by $|f|$.

We shall call an element $x \in G$ singular if the characteristic polynomial of Ad (x) in the indeterminate λ is divisible by $(1-\lambda)^{l+1}$ ($l=\dim \mathfrak{h}_0$). If x is not singular we say it is regular. Let $H \in \mathfrak{h}_0$. Then $h=\exp H$ is singular if and only if $\alpha(H) = 2\pi m(-1)^{1/2}$ for some root α and some integer m. Let $n \in N$. Then it is easily seen that Ad (h), Ad (hn) have the same characteristic polynomials. Hence hn is singular if and only if h is singular.

LEMMA 10. *Let H be an element in $\mathfrak{h}_0$ such that $h=\exp H$ is regular. Then $X \to h^{-1} \exp (H+X)$ ($X \in \mathfrak{n}_0$) is a 1-1 analytic mapping of $\mathfrak{n}_0$ onto N which is everywhere regular.*

We denote by $(1-e^{-\lambda})/\lambda$ the power series $\sum_{m \geq 0} (-1)^m \lambda^m/(m+1)!$ which is convergent for all values of λ. If B is a matrix or an endomorphism of a finite-dimensional (real or complex) vector space we put

$$\frac{1-e^{-B}}{B} = \sum_{m \geq 0} (-1)^m \frac{B^m}{(m+1)!}\,.$$

In the proof of the above lemma we shall make use of the following known result (see Chevalley [1, p. 157]). If $Y \in \mathfrak{g}_0$ and f is a function on G differentiable at $y=\exp Y$, then

$$\left\{\frac{d}{dt} f(\exp (Y + tZ))\right\}_{t=0} = (Z'f)(y) \qquad (Z \in \mathfrak{g}_0)$$

where

$$Z' = \left(\frac{1-e^{-\mathrm{ad}\,Y}}{\mathrm{ad}\,Y}\right)Z.$$

Now let $\phi(X)=h^{-1} \exp (H+X)$ ($X \in \mathfrak{n}_0$). Then

$$(d\phi)_X Y = \frac{1 - \exp\{-\mathrm{ad}\,(H+X)\}}{\mathrm{ad}\,(H+X)} Y \qquad (Y \in \mathfrak{n}_0).$$

Clearly $\mathfrak{n}_0$ is invariant under ad $(H+X)$. Let D denote the restriction of ad $(H+X)$ on $\mathfrak{n}_0$. We extend D on $\mathfrak{n}$ by linearity. Then the matrix of D with respect to the base X_α ($\alpha \in P_+$) is in triangular form and therefore it is clear that

$$\det \left(\frac{1 - e^{-D}}{D} \right) = \prod_{\alpha \in P_+} \frac{1 - e^{-\alpha(H)}}{\alpha(H)} \neq 0$$

since h is regular. This proves that ϕ is everywhere regular.

We claim moreover that $h^{-1} \exp (H+X) \neq 1$ ($X \in \mathfrak{n}_0$) unless $X = 0$. For suppose $X \neq 0$. Then $X = a_\alpha X_\alpha + \sum_{\beta > \alpha} a_\beta X_\beta$ where $\alpha \in P_+$, a_α, $a_\beta \in C$, and $a_\alpha \neq 0$. Hence for any $H_1 \in \mathfrak{h}_0$,

$$\text{Ad } (\exp (H + X))H_1 = \exp (\text{ad } (H + X))H_1$$
$$\equiv H_1 - a_\alpha \alpha(H_1) X_\alpha \bmod \sum_{\beta > \alpha} C X_\beta.$$

Therefore if $\alpha(H_1) \neq 0$, Ad $(\exp (H+X))H_1 \neq H_1$. This proves that $\exp (H+X) \neq h$.

Now suppose $\phi(X_1) = \phi(X_2)$ (X_1, $X_2 \in \mathfrak{n}_0$). Then $\exp (H+X_1) = \exp (H+X_2)$. Since H is regular in $\mathfrak{h}_0$, it follows from Lemma 8 that $H+X_i = \text{Ad } (n_i)H$ ($n_i \in N$, $i = 1, 2$). Hence if $n_1^{-1}n_2 = n$ and $\text{Ad}(n)H = H+X$ ($X \in \mathfrak{n}_0$), we get $\exp (H+X) = h$. Therefore in view of the above result $X = 0$ and so from Lemma 8, $n = 1$. This proves that $n_1 = n_2$ and $X_1 = X_2$. Therefore ϕ is univalent.

Let $V = \phi(\mathfrak{n}_0)$. Since ϕ is everywhere regular, V is an open subset of N. In order to prove the lemma it only remains to show that $V = N$. Let $n \in N$. Since N is nilpotent, $n = \exp X$ for some $X \in \mathfrak{n}_0$. Consider the one-parameter group $\exp tX$ ($t \in R$). Let T be the subset of R consisting of all t such that $\exp tX \in V$. Since V is open, T is an open set containing zero. Let T_0 be the connected component of zero in T. Put $Z(t) = \phi^{-1}(\exp tX)$ ($t \in T_0$). Since ϕ defines an analytic isomorphism of $\mathfrak{n}_0$ with V, $t \to Z(t)$ is an analytic mapping of T_0 into $\mathfrak{n}_0$. Moreover $\exp (H+Z(t)) = \exp H \exp tX$ ($t \in T_0$). From this it follows immediately that

$$\frac{1 - \exp (-\text{ad } (H + Z(t)))}{\text{ad } (H + Z(t))} \dot{Z}(t) = X$$

where $\dot{Z}(t) = \lim_{\epsilon \to 0} (1/\epsilon) (Z(t+\epsilon) - Z(t))$ ($\epsilon \in R$). Hence

$$(1 - \exp (-\text{ad } (H + Z(t)))\dot{Z}(t) = \text{ad } (H + Z(t))X.$$

Let θ be the automorphism of $\mathfrak{g}_0$ such that $\theta(Y_1 + Y_2) = Y_1 - Y_2$ ($Y_1 \in \mathfrak{k}_0$, $Y_2 \in \mathfrak{p}_0$). Then (see [6, §2])

$$Q(Y) = - \text{sp} (\text{ad } (\theta(Y)) \text{ ad } Y) \qquad (Y \in \mathfrak{g}_0)$$

is a positive definite quadratic form on $\mathfrak{g}_0$. Put $|Y|^2 = Q(Y)$ and let $D(t)$ denote the restriction of $1 - \mathrm{Ad}\ ((\exp H \exp tX)^{-1})$ on $\mathfrak{n}_0$. Then $\det D(t)$ $= \prod_{\alpha \in P_+} (1 - e^{-\alpha(H)}) \neq 0$. Hence $D(t)$ is nonsingular and

$$\dot{Z}(t) = D(t)^{-1}\ \mathrm{ad}\ (H + Z(t))X.$$

For any endomorphism A of $\mathfrak{n}_0$ let $\|A\|^2$ denote the sum of the squares of the matrix coefficients of A relative to any base of $\mathfrak{n}_0$ which is orthonormal with respect to the quadratic form Q. Then it is clear that

$$|\dot{Z}(t)| \leq \|D(t)^{-1}\|\ |\ \mathrm{ad}\ (H + Z(t))X|$$

and

$$|\ \mathrm{ad}\ (H + Z(t))X| \leq |\ [H, X]| + |\ [Z(t), X]| \leq p + q|Z(t)|$$

where p and q are some positive numbers independent of t. Now it is evident that $D(t)$, and therefore $D(t)^{-1}$, depends continuously on t. Hence $\|D(t)^{-1}\|$ is bounded on every bounded subset of R. Therefore given any positive number t_0, there exists a constant M such that $\|D(t)^{-1}\| \leq M$ if $|t| \leq t_0$. Now suppose $t \in T_0$ and $|t| \leq t_0$. Then, if we denote the corresponding bilinear form also by Q,

$$\frac{1}{2}\frac{d}{dt}|Z(t)|^2 = Q(Z(t), \dot{Z}(t)) \leq |Z(t)|\,|\dot{Z}(t)|.$$

Hence, if $t \neq 0$,

$$\frac{d}{dt}|Z(t)| \leq |\dot{Z}(t)| \leq M(p + q|Z(t)|).$$

From this it follows by integration that

$$1 + \frac{q}{p}|Z(t)| \leq e^{Mq|t|}$$

provided $|t| \leq t_0$ and $t \in T_0$. This proves that $|Z(t)|$ remains bounded so long as t remains bounded in T_0.

We shall now show that T_0 is closed in R. Let t_k be a sequence in T_0 which converges to $t \in R$. Then t_k remains bounded and therefore $|Z(t_k)|$ also remains bounded. Since every bounded closed subset of $\mathfrak{n}_0$ is compact, we can choose a subsequence t_{k_i} such that $Z(t_{k_i})$ converges to a limit Z in $\mathfrak{n}_0$. Then

$$\phi(Z) = \lim_{i \to \infty} \phi(Z(t_{k_i})) = \lim_{i \to \infty} \exp t_{k_i}X = \exp tX.$$

This proves that $t \in T$. But T_0, being a component of T, is closed in T. Therefore $t \in T_0$. Hence T_0 is both open and closed in R. Since R is connected and $0 \in T_0$, $T_0 = R$. Therefore $n = \exp X \in V$. This proves that $V = N$.

COROLLARY. *Let h be a regular element in A^0. Then the mapping $n \to h^{-1}nhn^{-1}$ ($n \in N$) is a topological mapping of N onto itself.*

Choose $H \in \mathfrak{h}_0$ such that $h = \exp H$. Then $nhn^{-1} = \exp (\operatorname{Ad} (n)H) = \exp (H + X(n))$ where $X(n) = \operatorname{Ad} (n)H - H \in \mathfrak{n}_0$. But if $Y \in \mathfrak{n}_0$,

$$\lim_{t \to 0} \frac{1}{t} \{ X(n \exp tY) - X(n) \} = - \operatorname{Ad} (n)[H, Y].$$

Therefore if D is the restriction of $\operatorname{ad}H$ on $\mathfrak{n}_0$,

$$| \det \operatorname{Ad} (n)D | = | \det D | = \prod_{\alpha \in P_+} | \alpha(H) | \neq 0$$

since H is regular. Therefore the mapping $n \to X(n)$ is everywhere regular. Hence it is a topological mapping of N onto $\mathfrak{n}_0$ from Lemma 8. The corollary now follows immediately from Lemma 10.

We shall also prove the following lemma which will be useful later.

LEMMA 11. *There exists a neighbourhood U of zero in $\mathfrak{h}_0$ such that the exponential mapping is univalent and regular on $U + \mathfrak{n}_0$ and $\exp H$ is regular in G for every $H \neq 0$ in U.*

It is obvious that there exists a neighbourhood U of zero in $\mathfrak{h}_0$ such that $\exp H$ is regular for all $H \neq 0$ in U and the mapping $H \to \exp H$ is univalent on U. Now we know (see Chevalley [1, p. 157]) that the exponential mapping is regular at a point $X \in \mathfrak{g}_0$ if and only if $\det ((1 - \exp (-\operatorname{ad}X))/\operatorname{ad}X) \neq 0$. But if $H \in U$ and $X \in \mathfrak{n}_0$, $\det ((1 - \exp (-\operatorname{ad}(H+X)))/\operatorname{ad}(H+X)) = \det ((1 - e^{-\operatorname{ad}H})/\operatorname{ad}H) = \prod_{\alpha > 0} ((1 - e^{-\alpha(H)})/\alpha(H)) \prod_{\alpha > 0} (e^{\alpha(H)} - 1)/\alpha(H) \neq 0$. Hence the exponential mapping is regular on $U + \mathfrak{n}_0$. Now suppose $\exp (H_1 + X_1) = \exp (H_2 + X_2)$ ($H_1, H_2 \in U$; $X_1, X_2 \in \mathfrak{n}_0$). Let $h_i = \exp H_i$ ($i = 1, 2$). Then it is clear that $h_2 \in h_1 N$ and therefore $h_1^{-1}h_2 \in N \cap A^0 = \{1\}$. Since the exponential mapping is univalent on U it follows that $H_1 = H_2$. Put $H = H_1 = H_2$. Then if $H = 0$, $\exp X_1 = \exp X_2$. Since the exponential mapping is well known to be univalent on $\mathfrak{n}_0$, $X_1 = X_2$. On the other hand suppose $H \neq 0$. Then $\exp H$ is regular and therefore from Lemma 10, $X_1 = X_2$. This proves the lemma.

COROLLARY. *The exponential mapping maps $U + \mathfrak{n}_0$ topologically into G. Moreover if U is compact, $\exp (U + \mathfrak{n}_0)$ is closed in G.*

The first assertion is obvious from Lemma 11. Moreover it follows from Lemma 10 that $\exp (U + \mathfrak{n}_0) = (\exp U)N$. Therefore if U is compact, $\exp U$ is also compact and therefore $(\exp U)N$ is closed.

Let G_1 be the set of all regular elements of G which are conjugate to some element in $A^0 Z$. Let A_1 be the set of all regular elements in A^0. Define the factor space $\mathcal{E} = G/A^0 Z$ as before. For any $\bar{x} \in \mathcal{E}$ and $h \in A^0 Z$ define $h^{\bar{x}} = xhx^{-1}$ where x is any element of the coset $\bar{x}$. Then $\phi: (\bar{x}, h) \to h^{\bar{x}}$ is a continuous mapping of $\mathcal{E} \times A^0 Z$ into G and $\phi(\mathcal{E} \times A_1 Z) = G_1$. We shall prove that ϕ is regular on $\mathcal{E} \times A_1 Z$. Since $A_1 Z$ is obviously open in $A^0 Z$, it would follow that G_1 is open in G. Let $X_1, \cdots, X_r$ be a base for $\mathfrak{g}_0$ mod $\mathfrak{h}_0$ and let π denote the natural mapping of G on $\mathcal{E}$. Then if $x \in \bar{x}$ we know that $(d\pi)_x X_i$, $1 \leq i \leq r$, is a base for the tangent space of $\mathcal{E}$ at $\bar{x}$. Now if $X \in \mathfrak{g}_0$ and $H \in \mathfrak{h}_0$,

$$x \exp tX\, h(x \exp tX)^{-1} = xhx^{-1} \exp t \operatorname{Ad}(xh^{-1})X \exp(-t \operatorname{Ad}(x)X),$$

and

$$xh \exp tH\, x^{-1} = xhx^{-1} \exp t \operatorname{Ad}(x)H.$$

Therefore

$$d\phi \circ (d\pi)_x X_i = \operatorname{Ad}(x)[\operatorname{Ad}(h^{-1}) - 1]X_i,$$
$$d\phi H = \operatorname{Ad}(x)H.$$

Then if D is the endomorphism of $\mathfrak{g}_0$ such that

$$DX_i = (\operatorname{Ad}(h^{-1}) - 1)X_i, \qquad\qquad 1 \leq i \leq r,$$
$$DH = H \qquad\qquad (H \in \mathfrak{h}_0),$$

it is clear that

$$\left| \det \operatorname{Ad}(x)D \right| = \left| \det D \right| = \left| \prod_{\alpha > 0} (e^{-\alpha(H)} - 1) \prod_{\alpha > 0} (e^{\alpha(H)} - 1) \right|$$

where H is any element in $\mathfrak{h}_0$ such that $h^{-1} \exp H \in Z$. Put

$$\Delta(h) = \left| \prod_{\alpha > 0} (e^{\alpha(H)/2} - e^{-\alpha(H)/2}) \right|.$$

Then if h is regular, $\left| \det(\operatorname{Ad}(x)D \right| = \Delta^2(h) \neq 0$ and this shows that ϕ is regular on $\mathcal{E} \times A_1 Z$. Now suppose x and h are two elements in G and $A_1 Z$ respectively such that $xhx^{-1} \in A^0 Z$. The set of all points in $\mathfrak{g}_0$ which are left fixed by $\operatorname{Ad}(h)$ is exactly $\mathfrak{h}_0$. Hence the corresponding set of fixed points for $\operatorname{Ad}(xhx^{-1})$ is $\operatorname{Ad}(x)\mathfrak{h}_0$. But since $xhx^{-1} \in A^0 Z$ it follows that $\mathfrak{h}_0 \subset \operatorname{Ad}(x)\mathfrak{h}_0$. Therefore $\mathfrak{h}_0 = \operatorname{Ad}(x)\mathfrak{h}_0$ (because $\dim \operatorname{Ad}(x)\mathfrak{h}_0 = \dim \mathfrak{h}_0$) and $x \in A'$ (Lemma 7). From this we deduce easily that the complete inverse image under ϕ of any point in G_1 consists of exactly w distinct points (w is the order of $W = A'/A^0 Z$). Therefore the method used in the proof of Lemma 9 permits us to conclude that

$$\int_{G_1} f(x)dx = \int_{K_*} du_* \int_{N \times A_1 Z} f((nhn^{-1})^{u_*})\Delta^2(h)\,dn\,dh$$

for any measurable function f on G_1 for which at least one of the two integrals above remains finite on replacing f by $|f|$. Here dx, dn, dh are the suitably normalised Haar measures on G, N, and $A^0 Z$ respectively.

Now consider the mapping ψ: $(h, n) \to nhn^{-1}$ of $A^0 Z \times N$ into $A^0 ZN$. Then

$$nh \exp tH\, n^{-1} = nhn^{-1} \exp t \operatorname{Ad}(n)H \qquad (H \in \mathfrak{h}_0),$$

$(n \exp tX)h(n \exp tX)^{-1}$

$$= nhn^{-1} \exp t \operatorname{Ad}(nh^{-1})X \exp(-t \operatorname{Ad}(n)X) \qquad (X \in \mathfrak{n}_0).$$

This shows that

$$(d\psi)H = \operatorname{Ad}(n)H \qquad (H \in \mathfrak{h}_0),$$
$$(d\psi)X = \operatorname{Ad}(n)[\operatorname{Ad}(h^{-1}) - 1]X \qquad (X \in \mathfrak{n}_0).$$

Hence

$$\left| \det (d\psi) \right| = \left| \prod_{\alpha \in P_+} (e^{-\alpha(H)} - 1) \right| = \left| e^{-\rho_+(H)} \right| \left| \prod_{\alpha \in P_+} (e^{\alpha(H)/2} - e^{-\alpha(H)/2}) \right|$$

where $\rho_+ = 2^{-1} \sum_{\alpha \in P_+} \alpha$ and H is any element in $\mathfrak{h}_0$ such that $h^{-1} \exp H \in Z$. Suppose $H = H_+ + H_-$ where $H_+ \in \mathfrak{h}_{\mathfrak{p}_0}$ and $H_- \in \mathfrak{h}_{\mathfrak{t}_0}$. Then $\rho_+(H_-)$ is purely imaginary while $\rho_+(H_+)$ is real and equal to $\rho(H_+)$. Hence

$$\left| \det (d\psi) \right| = e^{-\rho(H_+)} \left| \prod_{\alpha \in P_+} (e^{\alpha(H)/2} - e^{-\alpha(H)/2}) \right|.$$

Since $Z \cap A_+ = \{1\}$, H_+ is uniquely determined by h. Therefore if we put

$$\Delta_+(h) = \left| \prod_{\alpha \in P_+} (e^{\alpha(H)/2} - e^{-\alpha(H)/2}) \right|,$$

Δ_+ is a well-defined function on $A^0 Z$. Now if we take into account the fact that the Haar measure on $A^0 ZN$ is $ds = dhdn$ ($s = hn$; $h \in A^0 Z$, $n \in N$) we find from the corollary to Lemma 11 that

$$\int_{N \times A_1 Z} f(nhn^{-1}) \exp\{-\rho(\log h_+)\} \Delta_+(h)\, dhdn = \int_{A_1 Z N} f(hn)\, dhdn.$$

Here h_+ denotes the unique element in A_+ such that $h^{-1} h_+ \in A^0_- Z$. Comparing this with our earlier result we get

$$\int_{G_1} f(x)\, dx = \int_{K_*} du_* \int_{A_1 Z N} f((hn)^{u_*}) e^{\rho(\log h_+)} \Delta_+(h) \Delta(h)^2\, dhdn.$$

Let Λ be a linear function on $\mathfrak{h}_{\mathfrak{p}_0}$ and let $\delta \in \omega_{M_0 Z}$ (see §3 for notation). Put

$$\gamma_1(h) = \{\Delta_+(h)\}^{-1} \exp\{(\Lambda + \rho)(\log h_+)\} \xi_\delta(h_-) \qquad (h \in A_1 Z).$$

Here h_+ and h_- are the unique elements in A_+ and $A^0_- Z$ respectively such that $h = h_+ h_-$. Moreover let

$$\gamma(h) = \sum_{s \in W} \gamma_1(h^s)$$

and consider the mapping $(\bar{x}, h) \to h^{\bar{x}}$ of $\mathcal{E} \times A_1 Z$ onto G_1. It is everywhere open and continuous and the complete inverse image of $h^{\bar{x}}$ under this mapping is the set $(\bar{x}s^{-1}, h^s)$ $(s \in W)$. Therefore if we put

$$\Theta_{\Lambda,\delta}(h^{\bar{x}}) = \gamma(h) \qquad\qquad (\bar{x} \in \mathcal{E}, \, h \in A_1 Z)$$

we clearly get a continuous function on G_1. Then

$$\int_{G_1} \left| f(x)\Theta_{\Lambda,\delta}(x) \right| dx$$

$$= \int_{\mathcal{E} \times A_1 Z} \left| f(h^{\bar{x}})\gamma(h) \right| \Delta^2(h) d\bar{x}\,dh$$

$$\leq w \int_{K_*} du_* \int_{N \times A_1 Z} \left| f((nhn^{-1})^{u_*})\gamma_1(h) \right| \Delta^2(h) d\bar{x}\,dh$$

$$= w \int_{K_*} du_* \int_{A_1 Z N} \left| f((hn)^{u_*})\gamma_1(h) \right| \exp\{\rho(\log h_+)\}\Delta_+(h)\Delta_-^2(h)\,dn\,dh$$

$$= w \int_{K_*} du_* \int_{A_1 Z N} \left| f((hn)^{u_*}) \exp\{(\Lambda + 2\rho)(\log h_+)\}\xi_\delta(h_-) \right| \Delta_-^2(h)\,dn\,dh$$

where $\Delta_-(h) = \Delta(h)/\Delta_+(h) = \left| \prod_{\alpha \in P_-} (e^{\alpha(H)/2} - e^{-\alpha(H)/2}) \right|$, H being any element in $\mathfrak{h}_0$ such that $h^{-1}\exp H \in Z$. Now suppose f is a continuous function on G which vanishes outside a compact set. Then the right-hand side is clearly finite. Hence $f(x)\Theta_{\Lambda,\delta}(x)$ is integrable on G_1. Hence we may apply the above argument to $f(x)\Theta_{\Lambda,\delta}(x)$ instead of $\left| f(x)\Theta_{\Lambda,\delta}(x) \right|$ and conclude from Lemma 6 that

$$\int_{G_1} f(x)\Theta_{\Lambda,\delta}(x)\,dx = cT_{\Lambda,\delta}(f)$$

since $\Delta_-(h) = \Delta_-(h_-)$. Here c is a positive constant which is independent of Λ, δ, or f.

Let s be any element in W. Then there exists an element $u \in K$ such that Ad $(u)H = sH$ for all $H \in \mathfrak{h}_0$. Since Ad (u) leaves $\mathfrak{p}_0$ and $\mathfrak{k}_0$ invariant it follows that Ad $(u)\mathfrak{h}_{\mathfrak{p}_0} \subset \mathfrak{p}_0 \cap \mathfrak{h}_0 = \mathfrak{h}_{\mathfrak{p}_0}$ and Ad $(u)\mathfrak{h}_{\mathfrak{k}_0} \subset \mathfrak{h}_{\mathfrak{k}_0}$. This shows that s leaves $\mathfrak{h}_{\mathfrak{k}_0}$ and $\mathfrak{h}_{\mathfrak{p}_0}$ separately invariant. Hence if ν is any linear function on $\mathfrak{h}_{\mathfrak{p}_0}$ (or $\mathfrak{h}_{\mathfrak{k}_0}$ or $\mathfrak{h}_0$) we can define another such function $s\nu$ by the rule $s\nu(H) = \nu(s^{-1}H)$. Similarly if $\delta \in \omega_{M_0 Z}$ we can define another class $s^{-1}\delta \in \omega_{M_0 Z}$ by the condition $\xi_{s^{-1}\delta}(h) = \xi_\delta(h^s)$ $(h \in A_-^0 Z)$. That such a class actually exists and is unique is seen as follows. Since $\mathfrak{m}_0$ is the centralizer of $\mathfrak{h}_{\mathfrak{p}_0}$ in $\mathfrak{k}_0$, Ad $(u)\mathfrak{m}_0 = \mathfrak{m}_0$ and $uM_0u^{-1} = M_0$. Let σ be any representation of $M_0 Z$ in δ. Define a new representation σ' by the rule

$$\sigma'(m) = \sigma(umu^{-1}) \qquad\qquad (m \in M_0 Z).$$

Then σ' is irreducible and if δ' is its class $\xi_{\delta'}(h) = \xi_\delta(uhu^{-1}) = \xi_\delta(h^s)$ $(h \in A_-^0 Z)$. Since every element in M_0 is conjugate (with respect to M_0) to some element in A_-^0, every class in $\omega_{M_0 Z}$ is completely determined by the restriction of its character on $A_-^0 Z$. Hence $s^{-1}\delta$ is uniquely defined.

Notice that if α is a root, $s\alpha$ is also a root and $s\alpha$ is zero on $\mathfrak{h}_{\mathfrak{p}_0}$ if and only if the same holds for α. From this it follows immediately that

$$\Delta_+(h^s) = \Delta_+(h), \qquad \Delta_-(h^s) = \Delta_-(h) \qquad\qquad (h \in A^0 Z).$$

Therefore

$$\Theta_{\Lambda,\delta}(h) = [\Delta_+(h)]^{-1} \sum_{s \in W} \exp\left\{ s(\Lambda + \rho)(\log h_+) \right\} \xi_{s\delta}(h_-)$$

and we have the following theorem.

THEOREM 2. *There exists a positive real constant c with the following property. If Λ is any linear function on $\mathfrak{h}_{\mathfrak{p}_0}$ and δ a class in $\omega_{M_0 z}$, then*

$$T_{\Lambda,\delta}(f) = c \int_{G_1} f(x)\Theta_{\Lambda,\delta}(x)dx$$

for any $f \in C_c(G)$. Here $\Theta_{\Lambda,\delta}$ is a continuous function on G_1 defined uniquely by the following two properties:

(i) $\qquad\qquad \Theta_{\Lambda,\delta}(yxy^{-1}) = \Theta_{\Lambda,\delta}(x) \qquad\qquad\qquad (x \in G_1,\ y \in G),$

(ii) $\qquad\qquad \Theta_{\Lambda,\delta}(h) = [\Delta_+(h)]^{-1} \sum_{s \in W} \exp\left\{ s(\Lambda + \rho)(\log h_+) \right\} \xi_{s\delta}(h_-).$

Let $\Lambda^s = s(\Lambda + \rho) - \rho$. Then the above theorem shows that $T_{\Lambda^s, s\delta} = T_{\Lambda,\delta}$ $(\delta \in \omega_{M_0 z})$. Conversely suppose Λ_1, Λ_2 are two linear functions on $\mathfrak{h}_0$ and δ_1, δ_2 two classes in $\omega_{M_0 z}$ such that $T_{\Lambda_1,\delta_1} = T_{\Lambda_2,\delta_2}$. Then it follows from the above theorem that $\Theta_{\Lambda_1,\delta_1} = \Theta_{\Lambda_2,\delta_2}$ on G_1. Therefore

$$\sum_{s \in W} \exp\left\{ s(\Lambda_1 + \rho)(\log h_+) \right\} \xi_{s\delta_1}(h_-) = \sum_{s \in W} \exp\left\{ s(\Lambda_2 + \rho)(\log h_+) \right\} \xi_{s\delta_2}(h_-)$$

for all regular $h \in A^0 Z$. But since both sides are continuous functions on $A^0 Z$ and the set of regular elements is dense in $A^0 Z$, they are equal everywhere. Now the exponentials of distinct linear functions are well known to be linearly independent (see for example [8, Lemma 41]). Hence it follows on putting $h_- = 1$ that $\Lambda_2 = \Lambda_1^{s_0}$ for $s_0 \in W$. Let W_0 be the subgroup consisting of all $t \in W$ such that $\Lambda_2^t = \Lambda_2$. Therefore if we compare coefficients of $\exp\left\{ (\Lambda_2 + \rho)(\log h_+) \right\}$ on the two sides we get

$$\sum_{t \in W_0} \xi_{ts_0\delta_1}(h_-) = \sum_{t \in W_0} \xi_{t\delta_2}(h_-) \qquad\qquad (h_- \in A^0 Z).$$

But every element in M_0 is conjugate to some element in A_-^0 and therefore

$$\sum_{t \in W_0} \xi_{ts_0 \delta_1}(m) = \sum_{t \in W_0} \xi_{t \delta_2}(m) \qquad (m \in M_0 Z).$$

On the other hand it is well known that the characters corresponding to distinct irreducible classes are linearly independent. Hence $\delta_2 = t s_0 \delta_1$ for some $t \in W_0$. Put $s = t s_0$. Then $\Lambda_2 = \Lambda_1^s$ and $\delta_2 = s \delta_1$. Thus we have the following lemma.

LEMMA 12. *Let Λ_1, Λ_2 be two linear functions on $\mathfrak{h}_{\mathfrak{p}_0}$ and δ_1, δ_2 two classes in $\omega_{M_0 Z}$. Then $T_{\Lambda_1, \delta_1} = T_{\Lambda_2, \delta_2}$ if and only if there exists an element $s \in W$ such that $\Lambda_2 = \Lambda_1^s$ and $\delta_2 = s \delta_1$.*

5. Plancherel formula for complex semisimple Lie groups. We shall now assume that G is a complex semisimple group. We keep to the notation of §2 of [6]. Since G is complex, there exists a linear mapping Γ of $\mathfrak{k}_0$ on $\mathfrak{p}_0$ such that $[X, \Gamma(Y)] = \Gamma([X, Y])$ and $[\Gamma(X), \Gamma(Y)] = -[X, Y]$ $(X, Y \in \mathfrak{k}_0)$. We extend Γ to a linear mapping of $\mathfrak{g}_0$ onto itself by defining $\Gamma(\Gamma(X)) = -X$ $(X \in \mathfrak{k}_0)$. Let $(-1)^{1/2}$ be a fixed square root of -1 in C. Then if $c = a + (-1)^{1/2} b$ $(a, b \in R)$ we put $c * X = aX + b\Gamma(X)$ $(X \in \mathfrak{g}_0)$. Under this multiplication $\mathfrak{g}_0$ becomes a Lie algebra over C. We shall denote this complex algebra by $\mathfrak{g}^*$. Similarly the algebra $\mathfrak{h}_0 = \mathfrak{h}_{\mathfrak{k}_0} + \mathfrak{h}_{\mathfrak{p}_0}$, regarded as a (complex) subalgebra of $\mathfrak{g}^*$, will be denoted by $\mathfrak{h}^*$. Then $\mathfrak{h}^*$ is a Cartan subalgebra of $\mathfrak{g}^*$. Let $X \to \mathrm{ad}\, X$ $(X \in \mathfrak{g}^*)$ denote the adjoint representation of $\mathfrak{g}^*$ and let $B(X, Y) = \mathrm{sp}\,(\mathrm{ad}\, X\, \mathrm{ad}\, Y)$ $(X, Y \in \mathfrak{g}^*)$. Given any linear function λ on $\mathfrak{h}^*$, we denote by H_λ the unique element in $\mathfrak{h}^*$ such that $\lambda(H) = B(H, H_\lambda)$ for all $H \in \mathfrak{h}^*$. Let $H_1, \cdots, H_l$ be a base for $\mathfrak{h}_{\mathfrak{p}_0}$ over R. Then it is also a base for $\mathfrak{h}^*$ over C. We shall say that λ is real if $H_\lambda = \sum_{1 \leq i \leq l} c_i H_i$ $(c_i \in R)$, and furthermore that $\lambda > 0$ if $\lambda \neq 0$ and $c_j > 0$ where j is the least index $(1 \leq j \leq l)$ such that $c_j \neq 0$. For every root α of $\mathfrak{g}^*$ (with respect to $\mathfrak{h}^*$) we choose an element $X_\alpha \neq 0$ in $\mathfrak{g}^*$ such that $[H, X_\alpha] = \alpha(H) * X_\alpha$ $(H \in \mathfrak{h}^*)$. We can do this in such a way that $B(X_\alpha, X_{-\alpha}) = 1$ and $X_\alpha - X_{-\alpha}$, $(-1)^{1/2} * (X_\alpha + X_{-\alpha})$ are both in $\mathfrak{k}_0$ (The corresponding statement on p. 814 of my earlier note (Proc. Nat. Acad. Sci. U. S. A. vol. 37(1951) pp. 813–818) has wrong signs.) Since every root α is real, $H_\alpha = \sum_{i=1}^l \alpha^i H_i$ $(\alpha^i \in R)$. Let $\mathfrak{n}^* = \sum_{\alpha \in Q} C * X_\alpha$ where Q is the set of all roots $\alpha > 0$. Then $\mathfrak{n}^*$ is a nilpotent sugalgebra of $\mathfrak{g}^*$ to which there corresponds an analytic subgroup N of G. We shall denote by $\mathfrak{n}_0$ the space $\mathfrak{n}^*$ regarded as a real vector-subspace of $\mathfrak{g}_0$.

Let $\mathfrak{g}$ be the complexification of the real algebra $\mathfrak{g}_0$. Let $\mathfrak{k}$, $\mathfrak{p}$, $\mathfrak{h}$, $\mathfrak{h}_{\mathfrak{k}}$, and $\mathfrak{h}_{\mathfrak{p}}$ respectively denote the subspaces of $\mathfrak{g}$ spanned by $\mathfrak{k}_0$, $\mathfrak{p}_0$, $\mathfrak{h}_0$, $\mathfrak{h}_{\mathfrak{k}_0}$, and $\mathfrak{h}_{\mathfrak{p}_0}$ over C. We denote by γ the isomorphism between $\mathfrak{g}^*$ and $\mathfrak{k}$ given by $\gamma(c * X) = cX$ $(c \in C, X \in \mathfrak{k}_0)$. Put

$$\gamma_+(X) = (X - (-1)^{1/2}\Gamma(X))/2,$$
$$\gamma_-(X) = (X + (-1)^{1/2}\Gamma(X))/2 \qquad (X \in \mathfrak{k})$$

and let $\mathfrak{k}_+ = \gamma_+(\mathfrak{k})$, $\mathfrak{k}_- = \gamma_-(\mathfrak{k})$. (Here we have extended Γ on $\mathfrak{g}$ by linearity.)

Then $\mathfrak{l}_+$ and $\mathfrak{l}_-$ are ideals in $\mathfrak{g}$ and $\mathfrak{g}$ is their direct sum. Moreover γ_+ and γ_- are isomorphisms of $\mathfrak{l}$ on $\mathfrak{l}_+$ and $\mathfrak{l}_-$ respectively. Now $\mathfrak{h} = \gamma_+(\mathfrak{h}_\mathfrak{l}) + \gamma_-(\mathfrak{h}_\mathfrak{l})$. Let λ, μ be two linear functions on $\mathfrak{h}^*$. Then we denote by (λ, μ) the linear function ν on $\mathfrak{h}$ defined as follows:

$$\nu(\gamma_+(\gamma(H))) = \lambda(H),$$
$$\nu(\gamma_-(\gamma(H))) = \mu(H) \qquad\qquad (H \in \mathfrak{h}^*).$$

It is easy to verify that

$$\nu(H) = \lambda(H) + \mu(H) \qquad\qquad \text{if } H \in \mathfrak{h}_{\mathfrak{l}_0},$$
$$\nu(H) = \lambda(H) - \mu(H) \qquad\qquad \text{if } H \in \mathfrak{h}_{\mathfrak{p}_0}.$$

Put $\lambda_+ = (\lambda, 0)$ and $\lambda_- = (0, -\lambda)$. Then if λ is real (in the sense described above) $\lambda_+(H) = \lambda(H)$ and $\lambda_-(H) = \operatorname{conj} \lambda(H)$ for $H \in \mathfrak{h}_0$. Therefore, in particular, for every root α (of $\mathfrak{g}^*$ with respect to $\mathfrak{h}^*$) we get two linear functions α_+, α_- on $\mathfrak{h}$ and if we put $X_\alpha^+ = \gamma_+(\gamma(X_\alpha))$, $X_\alpha^- = \gamma_-(\gamma(X_\alpha))$, we have

$$[H, X_\alpha^+] = \alpha_+(H)X_\alpha^+, \qquad [H, X_\alpha^-] = \alpha_-(H)X_\alpha^- \qquad (H \in \mathfrak{h}).$$

Notice that $\alpha_+ \neq \alpha_-$ since α_+ vanishes on $\gamma_-(\mathfrak{h}_\mathfrak{l})$ while α_- vanishes on $\gamma_+(\mathfrak{h}_\mathfrak{l})$ and neither of them is zero. Hence for each root α of $\mathfrak{g}^*$ we get two distinct roots α_+ and α_- of $\mathfrak{g}$. Moreover if we take $(H_1, \cdots, H_l, (-1)^{1/2}\Gamma(H_1), \cdots, (-1)^{1/2}\Gamma(H_l))$ as an ordered base for $\mathfrak{h}_{\mathfrak{p}_0} + (-1)^{1/2}\mathfrak{h}_{\mathfrak{l}_0}$ over R and define the sets P, P_+, and P_- of positive roots of $\mathfrak{g}$ with respect to this base (see [6, §2]), we find that if $\alpha \in Q$, α_+ and α_- are both in P. In view of the isomorphisms γ_+ and γ_- it is clear that every root of $\mathfrak{g}$ is of the form $\pm\alpha_\pm$ ($\alpha \in Q$). Hence every root in P is the form $\alpha_\pm$ for some $\alpha \in Q$. Moreover since α is complex-linear, it cannot vanish on $\mathfrak{h}_{\mathfrak{p}_0}$. Therefore the set P_- is empty and $P = P_+$. Now let $X_\alpha = X_\alpha' + \Gamma(X_\alpha'') \ (X_\alpha', X_\alpha'' \in \mathfrak{l}_0)$. Then $\gamma(X_\alpha) = X_\alpha' + (-1)^{1/2}X_\alpha''$ and it is easily verified that

$$X_\alpha^+ = (X_\alpha - (-1)^{1/2}\Gamma(X_\alpha))/2,$$
$$X_\alpha^- = (X_\alpha + (-1)^{1/2}\Gamma(X_\alpha))/2.$$

Hence $X_\alpha = X_\alpha^+ + X_\alpha^-$ and therefore $\mathfrak{n}^* = \mathfrak{n}_0 \subset \mathfrak{g}_0 \cap \{\sum_{\alpha \in Q} (CX_\alpha^+ + CX_\alpha^-)\}$. Since $\dim_R \mathfrak{n}_0 = 2 \dim_C \mathfrak{n}^*$, $\dim_R \mathfrak{n}_0$ is equal to the number of roots in $P_+ = P$. Hence $\mathfrak{n}_0 = \mathfrak{g}_0 \cap \{\sum_{\alpha \in Q} (CX_\alpha^+ + CX_\alpha^-)$.

If f is a complex-valued differentiable function of two real variables x, y and $z = x + (-1)^{1/2}y$ we write

$$\frac{\partial}{\partial z}f = \frac{1}{2}\left(\frac{\partial}{\partial x} - (-1)^{1/2}\frac{\partial}{\partial y}\right)f, \qquad \frac{\partial}{\partial \bar{z}}f = \frac{1}{2}\left(\frac{\partial}{\partial x} + (-1)^{1/2}\frac{\partial}{\partial y}\right)f$$

where the bar denotes complex conjugate. Put $\rho = (1/2)\sum_{\alpha \in Q} \alpha$ and let K and A_+ be the analytic subgroups of G corresponding to $\mathfrak{l}_0$ and $\mathfrak{h}_{\mathfrak{p}_0}$ respectively. Then K is compact. Moreover $\mathfrak{c}_0 = \{0\}$ in the present case and therefore $D = \{1\}$. Let du and dn denote the Haar measures on K and N respectively. We assume $\int_K du = 1$. The following theorem is the principal step in the proof of the Plancherel formula for G (see Gelfand and Naimark [3, p. 198]).

THEOREM 3. *Put* $H_a = \sum_{1 \leq i \leq l} a_i * H_i \; (a_i \in C)$ *and*

$$D_\alpha = \sum_{1 \leq i \leq l} \alpha^i \frac{\partial}{\partial a_i}, \qquad \overline{D}_\alpha = \sum_{1 \leq i \leq l} \alpha^i \frac{\partial}{\partial \bar{a}_i} \qquad (\alpha \in Q).$$

Then if dn is suitably normalised we have

$$f(1) = \lim_{H_a \to 0} \prod_{\alpha \in Q} D_\alpha \overline{D}_\alpha \left\{ \exp \{ \rho(H_a) + \overline{\rho(H_a)} \} \int_{K \times N} f(u(\exp H_a)nu^{-1}) du\, dn \right\}$$

for every $f \in C_c^\infty(G)$.

The proof depends on the theory of Fourier transforms for functions on $\mathfrak{g}_0$. Let $C_c^\infty(\mathfrak{g}_0)$ be the class of all complex-valued functions on $\mathfrak{g}_0$ which are everywhere indefinitely differentiable and which vanish outside a compact set. Put

$$X = \sum_{1 \leq i \leq l} a_i * H_i + \sum_{\alpha \in Q} (z_\alpha * X_\alpha + \bar{z_\alpha} * X_{-\alpha})$$

where $a_i, z_\alpha, z_\alpha^- \; (1 \leq i \leq l, \alpha \in Q)$ are independent complex variables. For any complex variable $z = x + (-1)^{1/2} y \; (x, y \in R)$ let $d\mu(z)$ denote the Euclidean measure $dx\, dy$ on the corresponding complex plane. Let F be a function in $C_c^\infty(\mathfrak{g}_0)$. Put

$$g(Y) = (2\pi)^{-n} \int_{\mathfrak{g}_0} \exp \{ (-1)^{1/2} \mathfrak{R}(B(X, Y)) \} F(X) dX$$

where $n = (1/2)\dim_R \mathfrak{g}_0$, $dX = \prod_{1 \leq i \leq l} d\mu(a_i) \prod_{\alpha \in Q} d\mu(z_\alpha) d\mu(z_\alpha^-)$ and $\mathfrak{R}c \; (c \in C)$ denotes the real part of c. Then if we assume, as we may, that $B(H_i, H_j) = \delta_{ij}, \; 1 \leq i, j \leq l \; (\delta_{ij}$ is the usual Kronecker symbol), it follows from the theory of Fourier transforms that $\int_{\mathfrak{g}_0} |g(X)| dX < \infty$ and

$$F(0) = (2\pi)^{-n} \int_{\mathfrak{g}_0} g(X) dX.$$

Now suppose $F(\mathrm{Ad}\,(u)X) = F(X)$ for all $u \in K$ and $X \in \mathfrak{g}_0$. We know that the bilinear form $B(X, Y)$ is invariant under the adjoint representation of G and $d(\mathrm{Ad}\,(x)X) = dX \; (x \in G)$ since $\det \mathrm{Ad}\,(x) = 1$. We can now transform the integral $\int_{\mathfrak{g}_0} g(X) dX$ in another form. Let dH and dZ denote the Euclidean measures on $\mathfrak{h}_0$ and $\mathfrak{n}_0$ respectively. Then we have the following lemma.

LEMMA 13. *It is possible to normalise the measures dZ and dH in such a way that*

$$\int_{\mathfrak{g}_0} g(X)dX = \int_{\mathfrak{h}_0+\mathfrak{n}_0} \prod_{\alpha\in Q} |\alpha(H)|^2 g(Z + H)dZdH$$

for any measurable function $g(X)$ on $\mathfrak{g}_0$ such that $g(\mathrm{Ad}\ (u)X) = g(X)$ ($u\in K$, $X\in\mathfrak{g}_0$) and $\int_{\mathfrak{g}_0}|g(X)|dX < \infty$.

Let $\mathfrak{g}_1$ be the set of all regular elements in $\mathfrak{g}_0$. Then we know (see Chevalley [2]) that every $X\in\mathfrak{g}_1$ is conjugate under G to some $H\in\mathfrak{h}_0$. Since the set of singular elements in $\mathfrak{g}_0$ is of measure zero,

$$\int_{\mathfrak{g}_0} g(X)dX = \int_{\mathfrak{g}_1} g(X)dX$$

and Lemma 9 is applicable. Now P_- is empty and it follows from our earlier remarks that

$$\prod_{\beta\in P_+} |\beta(H)| = \prod_{\alpha\in Q} |\alpha(H)|^2 \qquad (H \in \mathfrak{h}_0).$$

Moreover the set of singular elements in $\mathfrak{h}_0$ is also of measure zero (with respect to the Euclidean measure dH on $\mathfrak{h}_0$). The above lemma is therefore an immediate consequence of Lemma 9.

F and g being as above, consider the function $F' = \prod_{\alpha\in Q} D_\alpha \bar{D}_\alpha F$. Its Fourier transform g' is given by

$$g'(Y) = (2\pi)^{-n} \int_{\mathfrak{g}_0} \exp\{(-1)^{1/2}\Re(B(X, Y))\}F'(X)dX \qquad (Y \in \mathfrak{g}_0).$$

Let $X = \sum_{i=1}^l a_i * H_i + \sum_{\alpha\in Q} (z_\alpha * X_\alpha + z_\alpha^- * X_{-\alpha})$ and $Y = \sum_{i=1}^l b_i * H_i + \sum_{\alpha\in Q} (w_\alpha * X_\alpha + w_\alpha^- * X_{-\alpha})$ where (a, z, z^-, b, w, w^-) are all independent complex variables. Put $F(X) = F(a, z, z^-)$, $g(Y) = g(b, w, w^-)$, and $g'(Y) = g'(b, w, w^-)$. Then

$$g'(b, w, w^-) = (2\pi)^{-n} \int \exp\left\{(-1)^{1/2}\Re\left(\sum_{i=1}^l a_ib_i + \sum_{\alpha\in Q} (z_\alpha w_\alpha^- + z_\alpha^- w_\alpha))\right)\right\}$$
$$\cdot \prod_{\alpha\in Q} D_\alpha \bar{D}_\alpha F(a, z, z^-)d\mu(a)d\mu(z)d\mu(z^-)$$

where

$$d\mu(a) = \prod_{i=1}^l d\mu(a_i), \qquad d\mu(z) = \prod_{\alpha\in Q} d\mu(z_\alpha), \qquad d\mu(z^-) = \prod_{\alpha\in Q} d\mu(z_\alpha^-).$$

Hence by partial integration

$$g'(b, w, w^-) = \prod_{\alpha\in Q} |\alpha(H_b)|^2 b(g, w, w^-)$$

where $H_b = \sum_{i=1}^{l} b_i * H_i$. Therefore

$$\int \prod_{\alpha \in Q} |\alpha(H_a)|^2 g(a, z, 0) d\mu(a) d\mu(z)$$

$$= \int g'(a, z, 0) d\mu(a) d\mu(z)$$

$$= (2\pi)^{-n} \int d\mu(a) d\mu(z) \int \exp\left\{(-1)^{1/2}\Re\left(\sum_{i=1}^{l} a_i b_i + \sum_{\alpha \in Q} z_\alpha \bar{w_\alpha}\right)\right\}$$

$$\cdot F'(b, w, w^-) d\mu(b) d\mu(w) d\mu(w^-)$$

$$= (2\pi)^{-n+r} \int F'(0, w, 0) d\mu(w)$$

from the theory of Fourier transforms. Here r is the complex dimension of $\mathfrak{h}^* + \mathfrak{n}^* = \mathfrak{h}_0 + \mathfrak{n}_0$. This shows that

$$F(0) = (2\pi)^{-n} \int_{\mathfrak{g}_0} g(X) dX$$

$$= c(2\pi)^{-n} \int \prod_{\alpha \in Q} |\alpha(H_a)|^2 g(a, z, 0) d\mu(a) d\mu(z) \qquad \text{(from Lemma 13)}$$

$$= c(2\pi)^{-n+r} \int F'(0, w, 0) d\mu(w)$$

$$= c(2\pi)^{-n+r} \int \lim_{H_a \to 0} \left\{\prod_{\alpha \in Q} D_\alpha \overline{D}_\alpha F\left(H_a + \sum_{\alpha \in Q} z_\alpha * X_\alpha\right)\right\} d\mu(z)$$

where c is a positive real constant independent of F. Since F vanishes outside a compact set it follows easily that

$$\int \lim_{H_a \to 0} \left\{\prod_{\alpha \in Q} D_\alpha \overline{D}_\alpha F\left(H_a + \sum_{\alpha \in Q} z_\alpha * X_\alpha\right)\right\} d\mu(z)$$

$$= \lim_{H_a \to 0} \prod_{\alpha \in Q} D_\alpha \overline{D}_\alpha \left\{\int F\left(H_a + \sum_{\alpha \in Q} z_\alpha * X_\alpha\right) d\mu(z)\right\}.$$

Thus we have the following result.

LEMMA 14. *There exists a positive real constant c with the following property: For any $F \in C_c^\infty(\mathfrak{g}_0)$ such that $F(\text{Ad }(u)X) = F(X)$ $(u \in K, X \in \mathfrak{g}_0)$,*

$$F(0) = \lim_{H_a \to 0} c \prod_{\alpha \in Q} D_\alpha \overline{D}_\alpha \int F\left(H_a + \sum_{\alpha \in Q} z_\alpha * X_\alpha\right) d\mu(z).$$

Now we come to the proof of Theorem 3. Let f be a function in $C_c^\infty(G)$.

Put

$$f_1(x) = \int_K f(uxu^{-1})du \qquad\qquad (x \in G).$$

Then f_1 is also in $C_c^\infty(G)$. Let $F_1(X) = f_1(\exp X)$ $(X \in \mathfrak{g}_0)$. Choose a compact neighbourhood U of zero in $\mathfrak{h}_0$ corresponding to Lemma 11. Let γ be the carrier of F_1 (i.e. the smallest closed set outside which F_1 is zero). Then $\gamma_1 = \gamma \cap (U + \mathfrak{n}_0)$ is the complete inverse image in $U + \mathfrak{n}_0$ (under the exponential mapping) of the intersection of the carrier of f_1 with the closed set $\exp(U + \mathfrak{n}_0)$ (see corollary to Lemma 11). Hence γ_1 is compact. Then $E = \bigcup_{u \in K} \mathrm{Ad}\,(u)\gamma_1$ is also compact. Moreover the exponential mapping is regular on γ_1 and therefore on E. Since E is compact, it is clear that there exists a compact neighbourhood V_1 of E in $\mathfrak{g}_0$ such that the exponential mapping is everywhere regular on V_1. Put $V = \bigcup_{u \in K} \mathrm{Ad}\,(u)V_1$. Then V is still compact and the exponential mapping is regular on V. Let V' be an open neighbourhood of V such that the exponential mapping is still regular on V'. We may assume that the closure of V' is compact. Select a function $\phi \in C_c^\infty(\mathfrak{g}_0)$ such that $\phi = 1$ on V and $\phi = 0$ outside V'.

For any $X \in \mathfrak{g}_0$ consider the endomorphism $(1 - e^{-\mathrm{ad}\,X})/\mathrm{ad}\,X$ of $\mathfrak{g}_0$. (Here $X \to \mathrm{ad}\,X$ is the adjoint representation of the real algebra $\mathfrak{g}_0$.) Since the exponential mapping is regular on V', $\det((1 - e^{-\mathrm{ad}\,X})/\mathrm{ad}\,X) \neq 0$ on V'. Therefore the function $|\det((1 - e^{-\mathrm{ad}\,X})/\mathrm{ad}\,X)|^{1/2}$ is indefinitely differentiable on V'. Put

$$F_2(X) = F_1(X)\phi(X)\,|\det((1 - e^{-\mathrm{ad}\,X})/\mathrm{ad}\,X)|^{1/} \qquad (X \in \mathfrak{g}_0).$$

Since ϕ is zero outside V' it follows that $F_2 \in C_c^\infty(\mathfrak{g}_0)$. Now let

$$F(X) = \int_K F_2(\mathrm{Ad}\,(u)X)du \qquad\qquad (X \in \mathfrak{g}_0).$$

If $X \in V$,

$$F_2(\mathrm{Ad}\,(u)X) = F_1'(\mathrm{Ad}\,(u)X) = F_1'(X)$$

where

$$F_1'(X) = F_1(X)\left|\det\left(\frac{1 - e^{-\mathrm{ad}\,X}}{\mathrm{ad}\,X}\right)\right|^{1/2}.$$

Hence $F(X) = F_1'(X)$ if $X \in V$. Now suppose $X \in U + \mathfrak{n}_0$ but $X \notin V$. Then $X \notin \gamma$ and since $\gamma = \mathrm{Ad}\,(u)\gamma$ $(u \in K)$, $\mathrm{Ad}\,(u)X \notin \gamma$ for any $u \in K$. Hence

$$F(X) = \int_K F_2(\mathrm{Ad}\,(u)X)du = 0 = F_1'(X).$$

This proves that $F = F_1'$ on $U + \mathfrak{n}_0$.

Since $F_2 \in C_c^\infty(\mathfrak{g}_0)$ the same holds for F. Moreover $F(\mathrm{Ad}\ (u)X) = F(X)$ $(u \in K,\ X \in \mathfrak{g}_0)$. Therefore from Lemma 14

$$F(0) = c \lim_{a \to 0} \prod_{\alpha \in Q} \dot{D}_\alpha \overline{D}_\alpha \int F\left(H_a + \sum_{\alpha \in Q} z_\alpha * X_\alpha\right) d\mu(z)$$

$$= c \lim_{a \to 0} \prod_{\alpha \in Q} D_\alpha \overline{D}_\alpha \int_{\mathfrak{n}_0} F_1'(H_a + Z) dZ$$

where $dZ = d\mu(z)$ $(Z = \sum_{\alpha \in Q} z_\alpha * X_\alpha)$ is the Euclidean measure on $\mathfrak{n}_0$. But

$$F_1'(H + Z) = F_1(H + Z)\left| \det\left(\frac{1 - e^{-\mathrm{ad}\ (H+Z)}}{\mathrm{ad}\ (H + Z)}\right)\right|^{1/2}$$

$$= F_1(H + Z)\left| \det\left(\frac{1 - e^{-\mathrm{ad}\ H}}{\mathrm{ad}\ H}\right)\right|^{1/2} \qquad (H \in \mathfrak{h}_0, Z \in \mathfrak{n}_0)$$

and

$$\det\left(\frac{1 - e^{-\mathrm{ad}\ H}}{\mathrm{ad}\ H}\right) = \prod_{\beta \in P_+} \frac{1 - e^{-\beta(H)}}{\beta(H)} \prod_{\beta \in P_+} \frac{1 - e^{\beta(H)}}{\beta(H)}$$

$$= \left\{\prod_{\beta \in P_+} \frac{e^{\beta(H)/2} - e^{-\beta(H)/2}}{\beta(H)}\right\}^2.$$

Therefore

$$F_1'(H + Z) = F_1(H + Z)\Delta_+(H)$$

where

$$\Delta_+(H) = \left|\prod_{\beta \in P_+} \frac{e^{\beta(H)/2} - e^{-\beta(H)/2}}{\beta(H)}\right| \qquad (H \in \mathfrak{h}_0).$$

Now let $H \in U$. Then the mapping $\phi: Z \to h^{-1} \exp\ (H+Z)$ $(h = \exp H)$ is a topological and regular mapping of $\mathfrak{n}_0$ onto N (see Lemmas 10 and 11) and

$$\det\ (d\phi)_Z = \det\left(\frac{1 - e^{-D}}{D}\right)$$

where D is the restriction of $\mathrm{ad}\ (H+Z)$ on $\mathfrak{n}_0$. Hence

$$\left|\det\ (d\phi)_Z\right| = \exp\ \left\{-\rho(H) - \overline{\rho(H)}\right\}\Delta_+(H).$$

Therefore

$$\Delta_+(H)\int_{\mathfrak{n}_0} F_1(H + Z) dZ = \exp\ \left\{\rho(H) + \overline{\rho(H)}\right\}\int_N f_1((\exp H)n) dn$$

where $dn = dZ$ ($n = \exp Z$) is the Haar measure on N. This proves that

$$\int_{\mathfrak{n}_0} F_1'(H + Z)dZ = \exp\{\rho(H) + \overline{\rho(H)}\}\int_N f_1((\exp H)n)dn \qquad (H \in U)$$

and therefore

$$F(0) = c \lim_{H_a \to 0} \prod_{\alpha \in Q} D_\alpha \overline{D}_\alpha \left\{\exp\{\rho(H_a) + \overline{\rho(H_a)}\}\int_N f_1((\exp H_a)n)dn\right\}.$$

But

$$F(0) = F_1'(0) = f_1(1) = f(1)$$

since $0 \in V$. Hence the theorem.

We shall now obtain the Plancherel formula from Theorem 3. In the present case $\mathfrak{h}_{\mathfrak{k}_0} = \mathfrak{m}_0$ (in the notation of §§3 and 4) and $\mathfrak{h}_{\mathfrak{k}_0}$ is a maximal abelian subalgebra of $\mathfrak{k}_0$. Hence A_-^0 is a maximal abelian subgroup of K and it is its own centralizer in K (see A. Weil [12]). Now the adjoint representation can be regarded as a *complex* representation of G on $\mathfrak{g}^*$. Let u be an element in the centralizer of A_+ in K. Then Ad (u) leaves every point in $\mathfrak{h}_{\mathfrak{p}_0}$ fixed. But since Ad (u) is an endomorphism of $\mathfrak{g}^*$ over C, it leaves every point in $\mathfrak{h}_{\mathfrak{k}_0}$ also fixed. This proves that $M = A_-^0 = A_- = M_0 Z$ in the present case. Let A_-' be the normalizer of A_- and K. Then the above argument shows that A_-' is also the normalizer of $A^0 = A_+ A_-^0$ in K and therefore $W = A'/A = A_-'/A_-$ in the notation of §4. Moreover for any $s \in W$, the mapping $H \to sH$ ($H \in \mathfrak{h}_0$) is an endomorphism of $\mathfrak{h}^*$ over C.

Since A_- is a torus we can choose the base $H_1, \cdots, H_l$ of $\mathfrak{h}_{\mathfrak{p}_0}$ over R in such a way that $\exp H_a = 1$ if and only if $(-1)^{1/2}a_i/2\pi$, $1 \leq i \leq l$, are all rational integers. For any $h \in A = A_+ A_-$ we denote by h_+ and h_- the unique elements in A_+ and A_- respectively such that $h = h_+ h_-$. Let $\mathfrak{F}_+$ be the set of all linear functions ν on $\mathfrak{h}^*$ which take real values on $\mathfrak{h}_{\mathfrak{p}_0}$. Moreover let $\mathfrak{F}_-$ be the set of all linear functions λ on $\mathfrak{h}^*$ such that $\lambda(H_i)$, $1 \leq i \leq l$, are all rational integers. For every $\lambda \in \mathfrak{F}_-$ we can define a character ξ_λ of A_- by the rule $\xi_\lambda(\exp H) = e^{\lambda(H)}$ ($H \in \mathfrak{h}_{\mathfrak{k}_0}$) and conversely every character of A_- can be obtained from some $\lambda \in \mathfrak{F}_-$ in this way. Put

$$\xi_{\nu,\lambda}(h) = \exp\{(-1)^{1/2}\nu(\log h_+)\}\xi_\lambda(h_-) \qquad (\nu \in \mathfrak{F}_+, \lambda \in \mathfrak{F}_-, h \in A)$$

and

$$S_{\nu,\lambda}(f) = \int_{K \times A \times N} f(uhnu^{-1})\xi_{\nu,\lambda}(h)\exp\{2\rho(\log h_+)\}dudhdn \qquad (f \in C_c^\infty(G))$$

where dh is the Haar measure on A. It is easy to verify that if $h = \exp H$ ($H \in \mathfrak{h}_0$) then $\rho(H) + \overline{\rho(H)} = 2\rho(\log h_+)$ and

$$\lim_{H_a \to 0} \prod_{\alpha \in Q} D_\alpha \overline{D}_\alpha \xi_{\nu,\lambda}(\exp H_a) = \prod_{\alpha \in Q} |(-1)^{1/2}\nu(H_\alpha) + \lambda(H_\alpha)|^2 = m(\nu, \lambda) \qquad \text{(say)}.$$

Now $\mathfrak{F}_+$ and $\mathfrak{F}_-$ are groups under ordinary addition of functions and it is clear that they may be considered as character groups of A_+ and A_- respectively. Let $d\nu$ denote the Haar measure on $\mathfrak{F}_+$. Then it is the Euclidean measure on the real vector space $\mathfrak{F}_+$. A simple application of the theory of Fourier transforms to the abelian Lie group A now shows that

$$\lim_{H_a \to 0} \prod_{\alpha \in Q} D_\alpha \bar{D}_\alpha \left\{ \exp\left\{ \rho(H_a) + \overline{\rho(H_a)} \right\} \int f(u \exp H_a \, nu^{-1}) du dn \right\}$$

$$= \sum_{\lambda \in \mathfrak{F}_-} \int_{\mathfrak{F}_+} m(\nu, \lambda) S_{\nu, \lambda}(f) d\nu \qquad (f \in C_c^\infty(G))$$

provided $d\nu$ is suitably normalised. Here the series on the right is absolutely convergent since the function

$$g(h) = \exp\left\{ 2\rho(\log h_+) \right\} \int f(u \, h \cdot nu^{-1}) du dn \qquad (h \in A)$$

is everywhere indefinitely differentiable on A and vanishes outside a compact set and $S_{\nu,\lambda}(f)$ is the Fourier transform of g. Now $\rho'(H) = 2^{-1} \sum_{\beta \in P} \beta(H)$ $= 2\rho(H)$ $(H \in \mathfrak{h}_{\mathfrak{p}_0})$. Hence it follows from Theorem 1 that $S_{\nu,\lambda}(f) = T_{\Lambda,\delta}(f)$ where $\Lambda(H) + 2\rho(H) = (-1)^{1/2}\nu(H)$ $(H \in \mathfrak{h}_{\mathfrak{p}_0})$ and δ is the class of the one-dimensional representation $h \to \xi_\lambda(h)$ $(h \in A_-)$ of $A_- = M = M_1$. Moreover since $\Lambda + \rho'$ takes pure imaginary values on $\mathfrak{h}_{\mathfrak{p}_0}$, $S_{\nu,\lambda}$ is a unitary character of G. Finally, in view of Lemma 12, $S_{\nu_1,\lambda_1} = S_{\nu_2,\lambda_2}$ $(\nu_1, \nu_2 \in \mathfrak{F}_+; \lambda_1, \lambda_2 \in \mathfrak{F}_-)$ if and only if $\nu_2 = s\nu_1$, $\lambda_2 = s\lambda_1$ for some $s \in W$. Now we need the following theorem.

THEOREM 4. *Given any $\lambda \in \mathfrak{F}_-$ we can find a subset V_λ of $\mathfrak{F}_+$ of measure zero such that the character $S_{\nu,\lambda}$ is irreducible for all ν in $\mathfrak{F}_+$ outside V_λ.*

If we assume this theorem for a moment, we can derive the Plancherel formula as follows. Let $V = \bigcup_{\lambda \in F_-}$. Since $\mathfrak{F}_-$ is a countable set, V is still a set of measure zero. Let V' be the set of all $\nu \in \mathfrak{F}_+$ such that $\nu = s\nu$ for some $s \neq 1$ in W. Then V' is a closed nowhere dense subset of $\mathfrak{F}_+$ and its measure is zero. Let $\mathfrak{F}_+'$ be a connected component of the complement of V' in $\mathfrak{F}_+$. Then $\mathfrak{F}_+'$ is an open subset of $\mathfrak{F}_+$ and it is known (see Weyl [13]) that for every $\nu \in \mathfrak{F}_+$ which is not in V' there exists a unique $s \in W$ such that $s\nu \in \mathfrak{F}_+'$. Since $S_{s\nu,s\lambda}(f) = S_{\nu,\lambda}(f)$ it is clear that

$$\sum_{\lambda \in \mathfrak{F}_-} \int_{\mathfrak{F}_+} m(\nu, \lambda) S_{\nu,\lambda}(f) d\nu = \sum_{\lambda \in \mathfrak{F}_-} w \int_{\mathfrak{F}_+^0} m(\nu, \lambda) S_{\nu,\lambda}(f) d\nu \qquad (f \in C_c^\infty(G))$$

where $\mathfrak{F}_+^0$ is the complement of V in $\mathfrak{F}_+'$ and w is the order of W. Therefore we get

$$f(1) = \sum_{\lambda \in \mathfrak{F}} w \int_{\mathfrak{F}_+^0} m(\nu, \lambda) S_{\nu,\lambda}(f) d\nu \qquad (f \in C_c^\infty(G))$$

from Theorem 3. Now the characters $S_{\nu,\lambda}$ $(\nu \in \mathfrak{F}_+^0, \lambda \in \mathfrak{F}_-)$ are all distinct and they are unitary and irreducible. Hence if $\mathcal{E}$ is the set of all equivalence-classes of irreducible unitary irreducible representations of G, we get a 1-1 mapping of $\mathfrak{F}_+^0 \times \mathfrak{F}_-$ into $\mathcal{E}$ if we assign to each (ν, λ) the unique class $\omega(\nu, \lambda)$ in $\mathcal{E}$ corresponding to the character $S_{\nu,\lambda}$. We may therefore identify $\mathfrak{F}_+^0 \times \mathfrak{F}_-$ with its image under this mapping and thus regard $\mathfrak{F}_+^0 \times \mathfrak{F}_-$ as a subset of $\mathcal{E}$. Let $d\mu$ be the discrete measure on $\mathfrak{F}_-$ which assigns to every point in $\mathfrak{F}_-$ the mass w. Then we can define a (positive) measure $d\omega$ on $\mathcal{E}$ as follows. Let F be a subset of $\mathcal{E}$. We say that F is measurable if $F_0 = F \cap (\mathfrak{F}_+^0 \times \mathfrak{F}_-)$ is measurable in $\mathfrak{F}_+^0 \times \mathfrak{F}_-$ and in case this is so we put

$$\int_F d\omega = \int_{F_0} d\nu d\mu.$$

Then it is clear that

$$f(1) = \int_{\mathcal{E}} T_\omega(f) d\omega \qquad (f \in C_c^\infty(G))$$

where T_ω is the character of the class ω. Let

$$g(x) = \int \mathrm{conj}\ (f(y))f(yx)dy.$$

Then g is also in $C_c^\infty(G)$ and therefore applying the above formula to g we get

$$g(1) = \int |f(x)|^2 dx = \int_{\mathcal{E}} N_\omega(f) d\omega \qquad (f \in C_c^\infty(G))$$

where $N_\omega(f)$ is defined as in §1. This gives the Plancherel formula for functions of class $C_c^\infty(G)$. Since such functions are dense in the Hilbert space $L_2(G)$ of all square-integrable functions on G, the corresponding formula for functions in $L_2(G)$ follows in the usual way by completion.

Now it remains to prove Theorem 4[9]. We shall say that a function $\lambda \in \mathfrak{F}_-$ is dominant if $s\lambda \geqq \lambda$ for all $s \in W$ (with respect to the lexicographic ordering defined in the beginning of §5). Let $\mathfrak{F}_-^0$ be the set of all dominant functions in $\mathfrak{F}_-$. Let $\mathfrak{F}^*$ be the space of linear functions on $\mathfrak{h}^*$. Then $\mathfrak{F}_+$ and $\mathfrak{F}_-$ may be regarded as subsets of $\mathfrak{F}^*$. Let Ω be the set of all equivalence classes of finite-dimensional simple representations of K. A linear function $\mu \in \mathfrak{F}^*$ is called a weight of a class $\mathfrak{D} \in \Omega$ if there exists a vector $\psi \neq 0$ in the representation space of any representation $\sigma \in \mathfrak{D}$ such that $\sigma(\exp H)\psi = e^{\mu(H)}\psi$ $(H \in \mathfrak{h}_{t_0})$. We

---

[9] (Added in proof.) A result considerably stronger than Theorem 4 has recently been obtained by F. Bruhat for the classical groups. Since it is possible to give a direct proof of his fundamental lemma (C.R. Acad. Sci. Paris vol. 238 (1954) p. 437) for all connected semisimple Lie groups, his results hold also for the exceptional groups.

say that μ is the highest (or lowest) weight of $\mathfrak{D}$ if $\mu+\alpha$ (or $\mu-\alpha$) is not a weight of $\mathfrak{D}$ for any $\alpha\in Q$. It is known that every weight lies in $\mathfrak{F}_-$ and every highest weight in $\mathfrak{F}_-^0$ (see for example [5, Part I]). Moreover there is a 1-1 correspondence $\Lambda\leftrightarrow\mathfrak{D}_\Lambda$ between $\mathfrak{F}_-^0$ and Ω such that Λ is the highest weight of $\mathfrak{D}_\Lambda$ [5, Part I]. For any $\mu\in\mathfrak{F}_-$ let δ_μ denote the equivalence class of the one-dimensional representation $h\to\xi_\mu(h)$ $(h\in A_-)$. Then we denote by $(\Lambda:\mu)$ $(\Lambda\in\mathfrak{F}_-^0)$ the number of times δ_μ occurs in the reduction of $\mathfrak{D}_\Lambda$ with respect to A_-. It is known that $(\Lambda:\Lambda)=1$ and $(\Lambda:\mu)=(\Lambda:s\mu)$ for any $s\in W$ (see Weyl [13]).

Put $\mathfrak{H}=L_2(K)$ and define $\mathfrak{H}_\mathfrak{D}$ $(\mathfrak{D}\in\Omega)$ to be the set of all elements in $\mathfrak{H}$ which transform according to $\mathfrak{D}$ under the left regular representation of K on $\mathfrak{H}$. As usual we normalise the Haar measures du and dh on K and A_- in such a way that $\int_K du=\int_{A_-} dh=1$. Let λ be any function in $\mathfrak{F}_-^0$. Put

$$E_\lambda = \int_{A_-} \xi_\lambda(h)\tau(h)dh$$

where τ is the right regular representation of K on $\mathfrak{H}$. Then E_λ is the orthogonal projection of $\mathfrak{H}$ on $\mathfrak{H}_\lambda = E_\lambda\mathfrak{H}$. Since $(\lambda:\lambda)=1$, it follows from the Frobenius reciprocity relation (see A. Weil [11, p. 83]) that $\dim(\mathfrak{H}_\lambda\cap\mathfrak{H}_\mathfrak{D})=d(\mathfrak{D}_\lambda)$. Therefore apart from a constant factor there is exactly one function $f\neq0$ in $\mathfrak{H}_\lambda\cap\mathfrak{H}_\mathfrak{D}$ such that $f(hu)=\xi_\lambda(h)f(u)$ $(u\in K, h\in A_-)$. For any $\mu\in\mathfrak{F}^*$ define a representation π_μ' of G on $\mathfrak{H}$ by the rule

$$\pi_\mu'(x)\phi(u) = \exp\left[-\left\{((-1)^{1/2}\mu + 2\rho)(H(x^{-1}, u))\right\}\right]\phi(u_{x^{-1}})$$

$(u\in K, \phi\in\mathfrak{H}, x\in G)$ in the notation of §3. (We recall that $K^*=K$ in the present case.) Let $\mathfrak{H}_{\mu,\lambda}$ be the smallest closed subspace of $\mathfrak{H}$ containing f which is invariant under $\pi_\mu'(G)$. We denote by $\pi_{\mu,\lambda}$ the representation of G induced on $\mathfrak{H}_{\mu,\lambda}$. Since $\tau(h)$ commutes with $\pi_\mu'(x)$ $(h\in A_-, x\in G)$ it is clear that $\mathfrak{H}_{\mu,\lambda}\subset\mathfrak{H}_\lambda$.

LEMMA 15. *$\pi_{\nu,\lambda}$ is an irreducible unitary representation of G if $\nu\in\mathfrak{F}_+$. Moreover there exists a set V_λ in $\mathfrak{F}_+$ of measure zero such that if ν lies in the complement of V_λ in $\mathfrak{F}_+$, $\mathfrak{H}_{\nu,\lambda}=\mathfrak{H}_\lambda$.*

Let ν be a function in $\mathfrak{F}_+$. Then we have seen that π_ν', and therefore $\pi_{\nu,\lambda}$, is unitary. We shall now show that $\pi_{\nu,\lambda}$ is irreducible. Suppose $\mathfrak{H}_{\nu,\lambda}=\mathfrak{H}_1=\mathfrak{H}_2$, where $\mathfrak{H}_1$, $\mathfrak{H}_2$ are two mutually orthogonal closed subspaces which are both invariant under $\pi_{\nu,\lambda}(G)$. Then $f=f_1+f_2$ $(f_i\in\mathfrak{H}_i, i=1, 2)$ and since $f\neq0$ we may assume $f_1\neq0$. Then it is clear that $f_1\in\mathfrak{H}_1\cap\mathfrak{H}_\mathfrak{D}$ and therefore $\dim(\mathfrak{H}_1\cap\mathfrak{H}_\mathfrak{D})\geq d(\mathfrak{D}_\lambda)$. But we have seen above that $\dim(\mathfrak{H}_\lambda\cap\mathfrak{H}_\mathfrak{D})=d(\mathfrak{D}_\lambda)$. Therefore $\mathfrak{H}_1\cap\mathfrak{H}_\mathfrak{D}=\mathfrak{H}_\lambda\cap\mathfrak{H}_\mathfrak{D}$ and hence $f\in\mathfrak{H}_1$. But then it follows from the definition of $\mathfrak{H}_{\nu,\lambda}$ that $\mathfrak{H}_{\nu,\lambda}\subset\mathfrak{H}_1$. Therefore $\mathfrak{H}_{\nu,\lambda}$ is irreducible.

In order to prove the second part we need some lemmas. Put $\mathfrak{H}_0$

$= \sum_{\mathfrak{D} \in \mathfrak{n}} \mathfrak{H}_\mathfrak{D}$. Then every element in $\mathfrak{H}_0$ is well-behaved (see Theorem 4 and Lemma 30 of [6]) under π_μ' ($\mu \in \mathfrak{F}^*$). Let $\mathfrak{B}$ denote the universal enveloping algebra of $\mathfrak{g}$. We shall also denote by π_μ' the representation of $\mathfrak{B}$ induced on $\mathfrak{H}_0$ (see [6]). Let Λ_0 be a function in $\mathfrak{F}_-^0$ such that $\mathfrak{H}_\lambda \cap \mathfrak{H}_{\mathfrak{D}\Lambda_0} \neq \{0\}$. We denote by (ϕ, ψ) the scalar product of ϕ, ψ in $\mathfrak{H}$.

LEMMA 16. *Let $\phi_1, \cdots, \phi_m$ be a base for $\mathfrak{H}_\lambda \cap \mathfrak{H}_{\mathfrak{D}\Lambda_0}$. Suppose F_i, $1 \leq i \leq m$, are polynomial functions on $\mathfrak{F}^*$ such that*

$$\sum_{i=1}^{m} F_i(\mu)(\phi_i, \pi_\mu'(b)f) = 0$$

for all $b \in \mathfrak{B}$ and $\mu \in \mathfrak{F}^*$. Then $F_i = 0$, $1 \leq i \leq m$.

Notice that

$$(\phi_i, \pi_\mu'(x)f) = \int_K \text{conj } (\phi_i(u)) \exp \left\{ -[(-1)^{1/2}\mu + 2\rho](H(x^{-1}, u)) \right\} f(u_{x^{-1}})du.$$

Let $X_1, \cdots, X_n$ be a base for $\mathfrak{g}_0$ over R. Put $X_t = t_1 X_1 + \cdots + t_n X_n$ ($t_j \in R$) and $|t| = \max_j |t_j|$. Let p denote any ordered set $(p_1, \cdots, p_n)$ of n non-negative integers. Put $t^p = t_1^{p_1} t_2^{p_2} \cdots t_n^{p_n}$, $p! = p_1! p_2! \cdots p_n!$, and

$$X(p) = \frac{1}{s!} \sum X_{k_{i_1}} X_{k_{i_2}} \cdots X_{k_{i_s}} \in \mathfrak{B}$$

where $s = p_1 + p_2 + \cdots + p_n$, $(k_1, \cdots, k_s)$ is a sequence of indices in which j occurs exactly p_j times ($1 \leq j \leq n$) and the sum is over all permutations $(i_i, \cdots, i_s)$ of $(1, 2, \cdots, s)$. Then $X(p)$ taken together for all p form a base for $\mathfrak{B}$ (see [4]) and

$$(\phi_i, \pi_\mu'(\exp X_t)f) = \sum_p (\phi_i, \pi_\mu'(X(p))f) \frac{t^p}{p!}$$

provided $|t|$ is sufficiently small (see [6, Theorem 2]). Since ϕ_i and f are *analytic* functions on K, it follows from the arguments given in the beginning of §4 of [7] that for each p there exists a polynomial function $\zeta_{i,p}$ on $\mathfrak{F}^*$ such that

$$(\phi_i, \pi_\mu'(\exp X_t)f) = \sum_p \zeta_{i,p}(\mu) \frac{t^p}{p!}$$

provided $|t|$ is sufficiently small. Therefore by comparing coefficients

$$(\phi_i, \pi_\mu'(X(p))f) = \zeta_{i,p}(\mu).$$

Since $X(p)$ form a base for $\mathfrak{B}$, it follows that for each fixed $b \in \mathfrak{B}$ the mapping

$$\mu \rightarrow (\phi_i, \pi_\mu'(b)f) \qquad\qquad (\mu \in \mathfrak{F}^*)$$

is a polynomial function on $\mathfrak{F}^*$.

Now let Λ be any function in $\mathfrak{F}_-^0$. Then $\Lambda+\lambda\in\mathfrak{F}_-^0$. Let $\mathfrak{D}_\Lambda^*$ denote the class in Ω which is contragredient to $\mathfrak{D}_\Lambda$. Then $-\Lambda$ is the lowest weight of $\mathfrak{D}_\Lambda^*$. Choose two representations σ_1, σ_2 in $\mathfrak{D}_{\Lambda+\lambda}$ and $\mathfrak{D}_\Lambda^*$ respectively. We denote the corresponding representations of $\mathfrak{k}_0$ (and therefore of $\mathfrak{k}$) also by the same symbols. Let U_1, U_2 be the representation space of σ_1, σ_2 respectively. Define a representation π of $\mathfrak{g}$ on $U_1 \times U_2$ by the rule[10]

$$\pi(\gamma_+(X) + \gamma_-(Y)) = \sigma_1(X) + \sigma_2(Y) \qquad (X, Y \in \mathfrak{k}).$$

Then it is clear that π is irreducible. Let G' be the simply-connected covering group of G, Z' the kernel of the natural homomorphism of G' on G and K' the complete inverse image of K in G'. Then K' is connected and $Z'\subset K'$ (see Mostow [9]). Now π defines a representation of G' which we shall also denote by π. Note that $X=\gamma_+(X)+\gamma_-(X)$ $(X\in\mathfrak{k})$. Hence

$$\pi(X) = \sigma_1(X) + \sigma_2(X) \qquad (X \in \mathfrak{k}).$$

From this it follows that $\pi(u')=\sigma_1(u)\times\sigma_2(u)$ $(u'\in K')$ where u is the image of u' in K. In particular if $u'\in Z'$, $u=1$ and therefore Z' is contained in the kernel of π. Hence π may also be regarded as a representation of G.

Let $\chi(\Lambda'; u)$ $(u\in K)$ denote the character of the class $\mathfrak{D}_{\Lambda'}$ $(\Lambda'\in\mathfrak{F}_-^0)$. Then it is evident that

$$\mathrm{sp}\ \pi(u) = \chi(\Lambda + \lambda; u)\ \mathrm{conj}\ \chi(\Lambda; u) \qquad (u \in K).$$

For any $\mathfrak{D}\in\Omega$ let $(\pi:\mathfrak{D})$ denote the number of times $\mathfrak{D}$ occurs in the reduction of $\pi(K)$. Then it follows from the Schur orthogonality relations for the characters that

$$(\pi:\mathfrak{D}_{\Lambda_0}) = \int_K \mathrm{sp}\ \pi(u)\ \mathrm{conj}\ \chi(\Lambda_0; u)du$$

$$= \int_K \chi(\Lambda + \lambda; u)\ \mathrm{conj}\ \{\chi(\Lambda_0; u)\chi(\Lambda; u)\}du.$$

Now put

$$\Delta(H) = \prod_{\alpha\in Q} (e^{\alpha(H)/2} - e^{-\alpha(H)/2}) \qquad (H \in \mathfrak{h}_{\mathfrak{t}_0})$$

and

$$\Delta(\exp H) = |\Delta(H)| \qquad (H \in \mathfrak{h}_{\mathfrak{t}_0}).$$

Then we know from the theory of compact Lie groups (see Weyl [13]) that

$$(\pi:\mathfrak{D}_{\Lambda_0}) = \frac{1}{w}\int_{A_-} \chi(\Lambda + \lambda; h)\ \mathrm{conj}\ \{\chi(\Lambda; h)\chi(\Lambda_0; h)\}\Delta^2(h)dh$$

---

[10] The operations $\times$ and $+$ have the same meaning as in [6].

where

$$w = \int_{A-} \Delta^2(h)dh = \text{order of } W.$$

But it is well known (see Weyl [13]) that

$$\chi(\Lambda'; \exp H) = \{\Delta(H)\}^{-1} \sum_{s \in W} \epsilon(s) \exp \{s(\Lambda' + \rho)(H)\} \qquad (H \in \mathfrak{h}_{\mathfrak{t}_0})$$

for any $\Lambda' \in \mathfrak{F}_-^0$. Here $\epsilon(s) = \pm 1$ and is determined by the rule

$$\Delta(sH) = \epsilon(s)\Delta(H) \qquad (s \in W, H \in \mathfrak{h}_{\mathfrak{t}_0}).$$

Therefore

$$\chi(\Lambda + \lambda; \exp H) \text{ conj } \{\chi(\Lambda; \exp H)\chi(\Lambda_0; \exp H)\}\{\Delta(H)\}^2$$

$$= \left[\sum_{s,s' \in W} \epsilon(s)\epsilon(s') \exp \{s(\Lambda + \lambda + \rho)(H) - s'(\Lambda + \rho)(H)\} \right]$$

$$\cdot \sum_{j=0}^{r} (\Lambda_0 : \Lambda_j) \exp \{-\Lambda_j(H)\} \qquad (H \in \mathfrak{h}_{\mathfrak{t}_0})$$

where Λ_j, $0 \leq j \leq r$, are all the distinct weights of $\mathfrak{D}_{\Lambda_0}$.

Now for any function $\mu \in \mathfrak{F}^*$ put $\mu(H_t) = t_1\mu(H_1) + \cdots + t_l\mu(H_l)$ where $t_1, \cdots, t_l$ are independent indeterminates. Consider the polynomial $S(\mu, t)$ in (t) given by

$$S(\mu, t) = \prod_{j=0}^{r} \prod_{s \in W, s \neq 1} \{(\mu + \lambda + \rho)(H_t) - s(\mu + \rho)(H_t) - \Lambda_j(H_t)\}.$$

Then $S(\mu, t) = 0$ if and only if

$$s(\mu + \rho) - (\mu + \rho) = \lambda - \Lambda_j$$

for some $s \neq 1$ in W and some j. Since it is obviously possible to choose μ in such a way that none of these conditions is fulfilled, it follows that $S(\mu, t)$ is not identically zero in μ.

Now suppose the assertion of the lemma is false. Then we may suppose that $F_1 \neq 0$. Let F_1' denote the polynomial function on $\mathfrak{F}^*$ such that $F_1'(\mu) = F_1(\mu')$ where $2\mu + \lambda = -[(-1)^{1/2}\mu' + 2\rho]$. Since $F_1 \neq 0$, it is evident that $F_1' \neq 0$. Hence $F_1'(\mu)S(\mu, t)$ is not identically zero in μ. Then the argument of Lemma 32 of [5] is applicable and we can choose $\Lambda \in \mathfrak{F}_-^0$ such that $F_1'(\Lambda)S(\Lambda, t) \neq 0$. Then

$$s(\Lambda + \lambda + \rho) - s'(\Lambda + \rho) \neq \Lambda_j \qquad (s, s' \in W)$$

for any j unless $s = s'$, in which case

$$s(\Lambda + \lambda + \rho) - s(\Lambda + \rho) = s\lambda.$$

Hence it follows from the orthogonality of the characters of A_- that

$$(\pi : \mathfrak{D}_{\Lambda_0)}) = (\Lambda_0 : \lambda).$$

Let U be the representation space of π. We may regard U as a finite-dimensional Hilbert space and assume that $\pi(u)$ is unitary for $u \in K$. Put $\pi'(x) = (\pi(x^{-1}))^*$ where the star denotes adjoint. Then π' is also an irreducible representation of G on U and $\pi'(u) = \pi(u)$ $(u \in K)$. Since $\pi(\gamma_+(X) + \gamma_-(Y)) = \sigma_1(X) + \sigma_2(Y)$, $(X, Y \in \mathfrak{k})$ it is clear that the weights[11] of π (with respect to $\mathfrak{h}$) are exactly the functions (μ_1, μ_2) where μ_1 and μ_2 run independently through all weights of σ_1 and σ_2 respectively. Moreover (μ_1, μ_2) is the highest weight of π if and only if $(\mu_1 + \alpha, \mu_2)$ and $(\mu_1, \mu_2 - \alpha)$ are not weights of π for any $\alpha \in Q$. Hence μ_1 must be the highest weight of σ_1 and μ_2 the lowest weight of σ_2. Therefore $\mu_1 = \Lambda + \lambda$, $\mu_2 = -\Lambda$. Hence the highest weight of π is $(\Lambda + \lambda, -\Lambda)$ and so it coincides with $2\Lambda + \lambda$ on $\mathfrak{h}_{\mathfrak{p}_0}$ and λ on $\mathfrak{h}_{\mathfrak{t}_0}$. Let $\psi_0 \neq 0$ be a vector in U belonging to the highest weight. Then $\pi(n)\psi_0 = \psi_0$, $\pi(h)\psi_0 = \xi_\lambda(h)\psi_0$, and $\pi(\exp H)\psi_0 = \exp\{(2\Lambda + \lambda)(H)\}\psi_0$ $(n \in N, h \in A_-, H \in \mathfrak{h}_{\mathfrak{p}_0})$. Let μ be the function in $\mathfrak{F}^*$ such that

$$(-1)^{1/2}\mu(H) + 2\rho(H) = -\operatorname{conj}(2\Lambda(H) + \lambda(H)) \qquad (H \in \mathfrak{h}_{\mathfrak{p}_0}).$$

For any $\phi \in U$ put

$$F_\phi(u) = (\pi(u)\psi_0, \phi)$$

where the bracket denotes scalar product in U. Then

$$F_{\pi'(x)\phi}(u) = (\pi(u)\psi_0, \pi'(x)\phi) = (\pi(x^{-1}u)\psi_0, \phi) = \pi_\mu'(x)F_\phi(u).$$

Moreover if $F_\phi = 0$, $(\pi(u)\psi_0, \phi) = 0$ for all $u \in K$. But then if $x = uhn$ $(u \in K, h \in A_+, n \in N)$,

$$(\pi(x)\psi_0, \phi) = (\pi(u)\psi_0, \phi) \exp(-[(-1)^{1/2}\mu + 2\rho](\log h)) = 0.$$

Since π is an irreducible representation, U is spanned by the transforms $\pi(x)\psi_0$ of ψ_0 and therefore $\phi = 0$. Finally

$$F_\phi(uh) = (\pi(uh)\psi_0, \phi) = \operatorname{conj}(\xi_\lambda(h))F_\phi(u) \qquad (h \in A_-).$$

Therefore $\phi \to F_\phi$ is a 1-1 linear mapping of U into $\mathfrak{H}_\lambda$ and $F_{\pi'(x)\phi} = \pi_\mu'(x)F_\phi$. Now it is obvious that $\mathfrak{D}_{\Lambda+\lambda}$ occurs in the reduction of $\mathfrak{D}_\Lambda \times \mathfrak{D}_\lambda$. Therefore $\mathfrak{D}_\lambda$ also occurs in the reduction of $\mathfrak{D}_{\Lambda+\lambda} \times \mathfrak{D}_\Lambda^*$. Since $\pi'(u) = \pi(u)$ $(u \in K)$, it follows that there exists a vector $\phi_0 \neq 0$ in U which transforms according to $\mathfrak{D}_\lambda$ under $\pi'(K)$ and which is such that $\pi'(h)\psi_0 = \xi_\lambda(h)\phi_0$ $(h \in A_-)$. Then F_{ϕ_0} is a nonzero function in $\mathfrak{H}_\lambda \cap \mathfrak{H}_{\mathfrak{D}_\lambda}$ and

$$\pi_\mu'(h)F_{\phi_0} = \xi_\lambda(h)F_{\phi_0} \qquad (h \in A_-)^{\cdot}$$

---

[11] The weights of π are defined as usual (see [5]). They are ordered by the lexicographic ordering introduced by the base $(H_1, \cdots, H_l, (-1)^{1/2}\Gamma(H_1), \cdots, (-1)^{1/2}\Gamma(H_l))$ for $\mathfrak{h}_\mathfrak{p} + (-1)^{1/2}\mathfrak{h}_{\mathfrak{t}_0}$ over R.

Since apart from a constant factor, f is the only function in $\mathfrak{H}_\lambda \cap \mathfrak{H}_{\mathfrak{D}_\lambda}$ satisfying this condition $F_{\phi_0} = cf$ where $c \in C$ and $c \neq 0$. Then it is evident that the representation $\pi_{\mu,\lambda}$ on $\mathfrak{H}_{\mu,\lambda}$ is equivalent to π' under the mapping $\phi \to F_\phi$ ($\phi \in U$). Therefore

$$\dim (\mathfrak{H}_{\mu,\lambda} \cap \mathfrak{H}_{\mathfrak{D}_{\Lambda_0}}) = d(\mathfrak{D}_{\Lambda_0})(\pi : \mathfrak{D}_{\Lambda_0}) = d(\mathfrak{D}_{\Lambda_0})(\Lambda_0 : \lambda)$$

since π' coincides with π on K. But we know from the Frobenius reciprocity relation (A. Weil [11, p. 83]) that

$$m = \dim (\mathfrak{H}_\lambda \cap \mathfrak{H}_{\mathfrak{D}_{\Lambda_0}}) = d(\mathfrak{D}_{\Lambda_0})(\Lambda_0 : \lambda).$$

Therefore

$$\mathfrak{H}_{\mu,\lambda} \cap \mathfrak{H}_{\mathfrak{D}_{\Lambda_0}} = \mathfrak{H}_\lambda \cap \mathfrak{H}_{\mathfrak{D}_{\Lambda_0}}.$$

Since π' is irreducible and finite-dimensional the same holds for $\pi_{\mu,\lambda}$. Therefore we can choose $b \in \mathfrak{B}$ such that $\sum_{i=1}^m F_i(\mu)\phi_i = \pi_\mu'(b)f$. Then

$$(\pi_\mu'(b)f, \pi_\mu'(b)f) = \sum_{i=1}^m F_i(\mu)(\phi_i, \pi_\mu'(b)f) = 0$$

and therefore $\pi_\mu'(b)f = 0$. Since ϕ_i, $1 \leq i \leq m$, are linearly independent over C, $F_1(\mu) = F_1'(\Lambda) = 0$. But this contradicts our choice of Λ and so Lemma 16 is proved.

Now choose the base $(\phi_1, \cdots, \phi_m)$ so that it is orthonormal. For any $b \in \mathfrak{B}$ let $\eta_{i,b}$ denote the polynomial function $\mu \to (\phi_i, \pi_\mu'(b)f)$ on $\mathfrak{F}^*$. Let J_0 be the ring of all polynomial functions on $\mathfrak{F}^*$ and let J be the quotient field of J_0. Select elements $b_1, \cdots, b_q$ in $\mathfrak{B}$ such that the matrix $B = (\eta_{i,b_j})_{1 \leq i \leq m, 1 \leq j \leq q}$ (with coefficients in J_0) has the maximum possible rank (over J). We claim this rank is m. For otherwise we can find $F_i \in J_0$, $1 \leq i \leq m$, not all zero such that

$$\sum_{i=1}^m F_i \eta_{i,b_j} = 0, \qquad\qquad 1 \leq j \leq q.$$

But in view of the above lemma we can choose $b \in \mathfrak{B}$ such that

$$\sum_{i=1}^m F_i \eta_{i,b} \neq 0.$$

Hence if we put $b_{q+1} = b$, the matrix $(\eta_{i,b_j})_{1 \leq i \leq m, 1 \leq j \leq q+1}$ has a larger rank than B, which contradicts the definition of B. Hence B has rank m. Therefore $q \geq m$ and we may assume without loss of generality that $F = \det(\eta_{i,b_j})_{1 \leq i,j \leq m} \neq 0$. Now put

$$\phi_j(\nu) = \sum_{i=1}^m (\phi_i, \pi_\nu'(b_j)f)\phi_i = \sum_{i=1}^m \eta_{i,b_j}(\nu)\phi_i \qquad (\nu \in \mathfrak{F}_+).$$

Since $F \neq 0$, the set $V(\Lambda_0, \lambda)$ of all $\nu \in \mathfrak{F}_+$ such that $F(\nu) = 0$ is clearly of measure zero. So if $\nu \notin V(\Lambda_0, \lambda)$, $F(\nu) \neq 0$ and therefore $\phi_j(\nu)$, $1 \leq j \leq m$, span $\mathfrak{H}_\lambda \cap \mathfrak{H}_{\mathfrak{D}_{\Lambda_0}}$. It is obvious that $\phi_j(\nu)$ is the orthogonal component of $\pi'_\nu(b_j)f$ in $\mathfrak{H}_{\mathfrak{D}_{\Lambda_0}}$ and therefore it lies in $\mathfrak{H}_{\nu,\lambda} \cap \mathfrak{H}_{\mathfrak{D}_{\Lambda_0}}$. Hence if $\nu \notin V(\Lambda_0, \lambda)$, $\mathfrak{H}_\lambda \cap \mathfrak{H}_{\mathfrak{D}_{\Lambda_0}} = \mathfrak{H}_{\nu,\lambda} \cap \mathfrak{H}_{\mathfrak{D}_{\Lambda_0}}$.

Now put $V_\lambda = \bigcup_{\Lambda_0 \in \mathfrak{F}^0_-} V(\Lambda_0, \lambda)$ where $V(\Lambda_0, \lambda)$ is defined to be the empty set in case $\mathfrak{H}_\lambda \cap \mathfrak{H}_{\mathfrak{D}_{\Lambda_0}} = \{0\}$. Then V_λ is a set of measure zero and if $\nu \notin V_\lambda$, $\mathfrak{H}_\lambda \cap \mathfrak{H}_{\mathfrak{D}_{\Lambda_0}} = \mathfrak{H}_{\nu,\lambda} \cap \mathfrak{H}_{\mathfrak{D}_{\Lambda_0}}$ for all $\Lambda_0 \in \mathfrak{F}^0_-$. Hence the orthogonal complement of $\mathfrak{H}_{\nu,\lambda}$ in $\mathfrak{H}_\lambda$ is zero and therefore $\mathfrak{H}_{\nu,\lambda} = \mathfrak{H}_\lambda$. This completes the proof of Lemma 15.

It is clear from the work §3 that $S_{\nu,\lambda}$ is the character of the representation of G induced on $\mathfrak{H}_\lambda$ under π'_ν ($\nu \in \mathfrak{F}_+$). Therefore it follows that $S_{\nu,\lambda}$ is an irreducible character if $\nu \in V_\lambda$.

So far we have assumed that $\lambda \in \mathfrak{F}^0_-$. Now let λ be any element in $\mathfrak{F}_-$. Choose $s \in W$ such that $\lambda_1 = s\lambda$ is dominant. Then $S_{\nu,\lambda} = S_{s\nu,\lambda_1}$ for any $\nu \in \mathfrak{F}_+$. We have seen above that there exists a set V_{λ_1} in $\mathfrak{F}_+$ of measure zero such that if $s\nu \notin V_{\lambda_1}$, $S_{\nu,\lambda} = S_{s\nu,\lambda_1}$ is an irreducible character. Put $V_\lambda = s^{-1}V_{\lambda_1}$. Then V_λ is also of measure zero and if $\nu \notin V_\lambda$, $S_{\nu,\lambda}$ is irreducible. Therefore Theorem 4 is now established.

<h3 style="text-align:center">REFERENCES</h3>

1. C. Chevalley, *Theory of Lie groups*, Princeton University Press, 1946.

2. ———, Amer. J. Math. vol. 63 (1941) pp. 785–793.

3. I. M. Gelfand and M. A. Naimark, Trudi Mat. Inst. Steklova vol. 36 (1950).

4. Harish-Chandra, Ann. of Math. vol. 50 (1949) pp. 900–915.

5. ———, Trans. Amer. Math. Soc. vol. 70 (1951) pp. 28–96.

6. ———, *Representations of a semisimple Lie group on a Banach space*. I, Trans. Amer. Math. Soc. vol. 75 (1953) pp. 185–243.

7. ———, *Representation of semisimple Lie groups*. II, Trans. Amer. Math. Soc. vol. 76 (1954) pp. 26–65.

8. ———, *Representations of semisimple Lie groups*. III, Trans. Amer. Math. Soc. vol. 76 (1954) pp. 234–253.

9. G. D. Mostow, Bull. Amer. Math. Soc. vol. 55 (1949) pp. 969–980.

10. L. Schwartz, *Théorie des distributions*, Paris, Hermann, 1950.

11. A. Weil, *L'Intégration dans les groupes topologiques et ses applications*, Paris, Hermann, 1940.

12. ———, C. R. Acad. Sci. Paris vol. 200 (1935) p. 518.

13. H. Weyl, *The structure and representation of continuous groups*, Princeton, The Institute for Advanced Study, 1935.

TATA INSTITUTE OF FUNDAMENTAL RESEARCH,
   BOMBAY, INDIA.
COLUMBIA UNIVERSITY,
   NEW YORK, N. Y.

Reprinted from
Trans. Amer. Math. Soc.
76 (1954), 485–528

Reprinted from the Proceedings of the NATIONAL ACADEMY OF SCIENCES,
Vol. 40, No. 3, pp. 200–204.   March, 1954

556

ON THE PLANCHEREL FORMULA FOR THE RIGHT K-INVARIANT FUNCTIONS ON A SEMISIMPLE LIE GROUP

BY HARISH-CHANDRA

DEPARTMENT OF MATHEMATICS, COLUMBIA UNIVERSITY

Communicated by Paul A. Smith, January 21, 1954

Let G be a connected semisimple Lie group and K the complete inverse image in G of a maximal compact subgroup of its adjoint group.   Let μ denote the invariant measure on the space G/K of all cosets $xK(x \,\epsilon\, G)$.   We construct the space $L_2(G/K)$ with respect to μ and consider the problem of the complete reduction of the natural representation σ of G on $L_2(G/K)$.   Let D be any subgroup contained in the center of G. Then $D \subset K$ and so if we replace G by G/D our problem is not affected in any way. Hence we may assume that the center of G is finite and K is compact. We can now regard $L_2(G/K)$ as a closed subspace of $L_2(G)$ consisting of those functions f for which $f(x) = f(xu)$ $(x \in G, u \in K)$. Then σ is the restriction of the left-regular representation of G to $L_2(G/K)$. Let dx denote the Haar measure on G. Put

$$(f, g) = \int_G \overline{f(x)} g(x) \, dx \qquad (f, g \in L_2(G))$$

where the bar denotes complex conjugate.

Let $\mathfrak{g}$ and $\mathfrak{k}_0$ be the Lie algebras of G and K respectively. In order to save space we shall adhere strictly to the notation of an earlier paper.[1] Thus any symbol should automatically be given the same meaning as there unless it is explicitly defined anew. Let η be the anti-automorphism of $\mathfrak{B}$ over R such that $\eta(X + (-1)^{1/2}Y) = -X + (-1)^{1/2}Y$ $(X, Y \in \mathfrak{g}_0)$. We denote by $\mathfrak{B}_H$ the set of those elements in $\mathfrak{B}$ which are left fixed by η. Let $\mathfrak{Q}$ be the centralizer of $\mathfrak{k}$ in $\mathfrak{B}$. Put $\mathfrak{Q}_0 = \mathfrak{Q} \cap \mathfrak{B}_0$, $\mathfrak{Q}_H = \mathfrak{Q} \cap \mathfrak{B}_H$, and $\mathfrak{Q}_{0, H} = \mathfrak{Q}_0 \cap \mathfrak{Q}_H$. (Here $\mathfrak{B}_0$ is the algebra generated over R by $(1, \mathfrak{g}_0)$.) Let $\mathfrak{P}_d$ $(d \geqslant 0)$ denote the subspace of $\mathfrak{B}$ spanned over C by the basic canonical elements[2] of degree d formed out of a base for $\mathfrak{p}$. Let $\mathfrak{P} = \sum_{d \geq 0} \mathfrak{P}_d$. An element in $\mathfrak{P}$ is called homogeneous of degree d if it lies in $\mathfrak{P}_d$. Let $l_+ = \dim_R \mathfrak{h}_{\mathfrak{p}_0} = \dim A_+$.

LEMMA 1. *The algebra* $\mathfrak{Q}^* = (\mathfrak{Q} + \mathfrak{B}\mathfrak{k})/\mathfrak{B}\mathfrak{k}$ *is abelian and there exist* l_+ *elements* z_i $1 \leq i \leq l_+$ *in* $\mathfrak{Q}_{0, H}$ *such that their residue-classes mod* $\mathfrak{B}\mathfrak{k}$ *generate* $\mathfrak{Q}^*$. z_i *may be chosen to be homogeneous elements in* $\mathfrak{P} \cap \mathfrak{Q}_{0, H}$.

Let $C_c(G)$ denote the space of all complex-valued continuous functions f on G which vanish outside a compact set and $C_c^\infty(G)$ the subspace consisting of those f which are indefinitely differentiable everywhere. Moreover let $C_c(G/K)$ and $I_c(G)$, respectively, be the subsets of those $f \in C_c(G)$ for which $f(xu) = f(x)$ and $f(uxu^{-1}) = f(x) (x \in G, u \in K)$. Put $I_c(G/K) = I_c(G) \cap C_c(G/K)$ and denote the closures of $I_c(G)$ and $I_c(G/K)$ in $L_1(G)$ and $L_2(G/K)$ by $I_1(G)$ and $I_2(G/K)$ respectively. For any f and g in $L_1(G)$ we denote by f_*g their convolution and by $\sigma(f)$ the operator $\int f(x)\sigma(x) \, dx$. Then $f \to \sigma(f)$ is a continuous representation of $L_1(G)$ and it is clear that if $f \in I_1(G)$, $I_2(G/K)$ is invariant under $\sigma(f)$. Let $\tau(f)$ denote its restriction on $I_2(G/K)$. Then $f \to \tau(f)$ is a representation of the sub-algebra $I_1(G)$ on $I_2(G/K)$.

LEMMA 2. τ *is an abelian representation of* $I_1(G)$.

Let V denote the Garding subspace[3] of $L_2(G/K)$ with respect to σ and $b \to \sigma(b)(b \in \mathfrak{B})$ the corresponding representation of $\mathfrak{B}$ on V. It is obvious that for any $z \in \mathfrak{Q}$, $\sigma(z)$ leaves $V_I = V \cap I_2(G/K)$ invariant. We denote by $\tau(z)$ the restriction of $\sigma(z)$ on $V_I(z \in \mathfrak{Q})$.

LEMMA 3. *If* $z \in \mathfrak{Q}_H$, $\tau(z)$ *is an essentially hypermaximal operator on* $I_2(G/K)$.

For any $z \in \mathfrak{Q}_H$ let $U(z)$ denote the Cayley transform of $\tau(z)$. Then $U(z)$ is a unitary operator on $I_2(G/K)$ and

$$U(z)(\sigma(z) + (-1)^{1/2}I)g = (\sigma(z) - (-1)^{1/2}I)g \qquad (g \in V_I)$$

where I is the unit operator on $I_2(G/K)$. Put $U_i = U(z_i)(1 \leq i \leq l_+)$ where z_i are the elements defined in Lemma 1. Let $\mathfrak{A}$ be the smallest C^*-algebra[4] of operators on $I_2(G/K)$ containing $\tau(f)(f \in I_1(G))$ and $U_i\ 1 \leq i \leq l_+$. One proves without difficulty that $\mathfrak{A}$ is abelian. Let Γ be the spectrum of $\mathfrak{A}$. Then Γ is a compact Hausdorff space. For any $\omega \in \Gamma$ we denote by ξ_ω the corresponding continuous character of $\mathfrak{A}$. Let λ and ρ, respectively, denote the left- and the right-regular representations of G on $L_1(G)$. Put

$$E = \int_K \lambda(u)\rho(u)\,du$$

where du is the normalized Haar measure on K so that $\int_K du = 1$. Then E is a projection from $L_1(G)$ to $I_1(G)$ and $||Ef||_1 \leq ||f||_1\ (f \in L_1(G))$ where $||\ \ ||_1$ denotes the norm in $L_1(G)$. Now set

$$\xi_\omega(f) = \xi_\omega(\tau(Ef)) \qquad (f \in L_1(G)).$$

Then if $\tilde{f}(x) = \overline{f(x^{-1})}(f \in L_1(G),\ x \in G)$ it is easy to verify that $|\xi_\omega(f)| \leq ||f||_1$ and $\xi_\omega(\tilde{f}*f)$ is real and non-negative $(f \in L_1(G))$. Let Γ_0 denote the set of all $\omega \in \Gamma$ such that ξ_ω vanishes identically on $L_1(G)$. Then it follows from well-known results[5] that if $\omega \notin \Gamma_0$ there exists a unitary representation π of G on a Hilbert space $\mathfrak{H}$ and a vector $\psi \in \mathfrak{H}$ such that

(i) $\quad |\psi| = 1$ and $\xi_\omega(f) = \int_G f(x)(\psi,\ \pi(x)\psi)\,dx\ (f \in L_1(G))$.

(ii) $\quad$ The space spanned by $\pi(x)\psi(x \in G)$ is dense in $\mathfrak{H}$.

Here $(\psi_1,\ \psi_2)$ denotes the scalar product and $|\psi|$ the norm in $\mathfrak{H}$. It is easy to verify that $\xi_\omega(\lambda(u)f) = \xi_\omega(\rho(u)f) = \xi_\omega(f)(f \in L_1(G),\ u \in K)$. From this one can prove that $\pi(u)\psi = \psi(u \in K)$ and π is irreducible. Moreover the choice of ψ is unique, apart from a unimodular factor, and ψ is well-behaved[6] under π. Put $\xi_\omega(b) = (\psi,\ \pi(b)\psi)(b \in \mathfrak{B})$. Then it is not difficult to show that

$$\xi_\omega(U_i) = (\xi_\omega(z_i) - (-1)^{1/2})(\xi_\omega(z_i) + (-1)^{1/2})^{-1} \qquad 1 \leq i \leq l_+.$$

Now let $\omega_1,\ \omega_2$ be two points in $\Gamma - \Gamma_0$ and $\pi_1,\ \pi_2$ the corresponding irreducible representations. Then if $\xi_{\omega_1}(U_i) = \xi_{\omega_2}(U_i),\ 1 \leq i \leq l_+$, it follows from Lemma 1 that $\xi_{\omega_1}(z) = \xi_{\omega_2}(z)(z \in \mathfrak{Q})$ and therefore[7] π_1 and π_2 are equivalent. Hence $\xi_{\omega_1}(f) = \xi_{\omega_2}(f)$ for all $f \in L_1(G)$ and so $\omega_1 = \omega_2$. Let $\mathfrak{E}$ denote the set of all equivalence classes of irreducible unitary representations π of G such that the trivial representation of K occurs in the reduction of π. Then if $\omega \in \Gamma - \Gamma_0$ and $\alpha(\omega)$ is the class of the representation π defined above for ω, $\omega \rightarrow \alpha(\omega)$ is a 1-1 mapping of $\Gamma - \Gamma_0$ onto a subset $\mathfrak{E}_0$ of $\mathfrak{E}$. We identify $\Gamma - \Gamma_0$ with $\mathfrak{E}_0$ under this mapping.

On the other hand it is well known that there exists a unique projection-valued (regular) measure dP_ω on the class of all Borel subsets of Γ such that

$$A = \int_\Gamma \xi_\omega(A)dP_\omega$$

for every operator A in $\mathfrak{A}$. Since Γ_0 is clearly compact, it is Borel measurable and if P_{Γ_0} is the corresponding projection it is obvious that $\tau(f)P_{\Gamma_0} = 0$ for all $f \in I_1(G)$. Moreover since it is possible to choose a sequence $f_n \in I_1(G)$ such that $\tau(f_n)g \rightarrow g$ for every $g \in I_2(G/K)$, it follows that $P_{\Gamma_0} = 0$. From this we can conclude that Γ_0 contains no interior point and therefore if $\omega_0 \in \Gamma_0$ there exists a sequence $\omega_n \in \Gamma - \Gamma_0$ such that $\omega_n \rightarrow \omega_0$ in Γ. Furthermore

$$A = \int_{\mathfrak{E}_0} \xi_\omega(A)dP_\omega \qquad (A \in \mathfrak{A}).$$

For any $\omega \in \mathfrak{E}_0$ let $Q(\omega)$ denote the point $(\xi_\omega(z_1), \ldots, \xi_\omega(z_{l_+}))$ in the real Euclidean space of dimension l_+.

LEMMA 4. *Let ω_n be a sequence in $\mathfrak{E}_0$ which converges in Γ to a point $\omega_0 \in \Gamma_0$. Then the sequence $Q(\omega_n)$ diverges to infinity.*

In view of the above-mentioned relationship between $\xi_\omega(U_i)$ and $\xi_\omega(z_i)$ for ω in $\mathfrak{E}_0$, it follows that $\xi_{\omega_n}(U_i) \to 1$ for at least one value of i ($1 \le i \le l_+$). This proves that $\xi_{\omega_0}(U_i) = 1$ for some i if and only if $\omega_0 \in \Gamma_0$.

Now let ω_1, ω_2 be two points in Γ such that $\xi_{\omega_1}(U_i) = \xi_{\omega_2}(U_i)$, $i = 1, \ldots, l_+$. Suppose $\omega_1 \in \Gamma_0$. Then $\xi_{\omega_1}(U_i) = 1$ for some i. Therefore since $\xi_{\omega_2}(U_i) = \xi_{\omega_1}(U_i) = 1$ it follows that $\omega_2 \in \Gamma_0$. Hence ω_1, ω_2 are both in Γ_0 and so $\xi_{\omega_1}, \xi_{\omega_2}$ are identically zero on $L_1(G)$. It is then clear that $\omega_1 = \omega_2$. On the other hand if $\omega_1 \notin \Gamma_0$, we conclude in the same way that $\omega_2 \notin \Gamma_0$ and therefore again in view of our earlier remarks $\omega_1 = \omega_2$. This shows that $\xi_\omega(U_i)$ $1 \le i \le l_+$ is a separating set of continuous functions on Γ. Therefore from the Stone-Wierstrass theorem we can conclude that $\mathfrak{A}$ is the smallest C^*-algebra containing the operators U_i, $1 \le i \le l_+$. Then it follows from Lemma 4 that $\omega \to Q(\omega)$ is a topological mapping of $\mathfrak{E}_0$ onto a subset $Q(\mathfrak{E}_0)$ of the Euclidean space. Let τ_i denote the unique hypermaximal extension of $\tau(z_i)$, $1 \le i \le l_+$, and let dP_λ denote the projection-valued Borel measure on $Q(\mathfrak{E}_0)$ corresponding to dP_ω on $\mathfrak{E}_0$. Then it is clear from the above remarks that $Q(\mathfrak{E}_0)$ is the spectrum of the abelian set τ_i, $1 \le i \le l_+$, of hypermaximal operators on $I_2(G/K)$ and

$$\tau_i = \int_{Q(\mathfrak{E}_0)} \lambda_i \, dP_\lambda \qquad 1 \le i \le l_+$$

where $\lambda_i(Q(\omega)) = \xi_\omega(z_i)$ ($\omega \in \mathfrak{E}_0$).

On the other hand it is not difficult to show that there exists a uniquely determined (regular) Borel measure $d\omega$ on $\mathfrak{E}_0 = \Gamma - \Gamma_0$ such that

$$d(f, P_\omega g) = \overline{\xi_\omega(f)}\xi_\omega(g) \, d\omega \qquad (f, g \in I_c(G/K), \omega \in \mathfrak{E}_0).$$

From this one can prove that

$$\int |f(x)|^2 dx = \int_{\mathfrak{E}_0} \xi_\omega(f^*f) \, d\omega \qquad (f \in C_c(G/K).$$

Let π be any unitary representation in the class $\omega \in \mathfrak{E}_0$. Then if $N_\omega(f)$ denotes the square of the Hilbert-Schmidt norm of the operator $\int f(x)\pi(x) \, dx$ ($f \in C_c(G)$) it is easy to verify that $\xi_\omega(\tilde{f}^*f) = N_\omega(f)$ ($f \in C_c(G/K)$). Hence we have the Plancherel formula

$$\int |f(x)|^2 \, dx = \int_{\mathfrak{E}_0} N_\omega(f) \, d\omega \qquad (f \in C_c(G/K)).$$

Our results may therefore be summarized in the following theorem.

THEOREM. *Let F be the spectrum of the abelian family of hypermaximal operators τ_i, $1 \le i \le l_+$, regarded as a subset of the real Euclidean space E of dimension l_+. Let P_λ denote the corresponding resolution of the identity so that*

$$\tau_i = \int_F \lambda_i \, dP_\lambda \qquad (1 \le i \le l_+)$$

where λ_i is the ith coordinate of any point $\lambda \in E$. Then there exists a 1-1 mapping $\lambda \to \omega(\lambda)$ of F on a subset $\mathfrak{E}_0$ of $\mathfrak{E}$ with the following property.

If π is any representation in $\omega(\lambda)$ and ψ is the (essentially unique) vector in its representation space such that $|\psi| = 1$ and $\pi(u)\psi = \psi(u \in K)$, then

$$\lambda_i = (\psi, \pi(z_i)\psi) \qquad 1 \leq i \leq l_+.$$

Put $\varphi_\lambda(x) = (\psi, \pi(x)\psi)(x \in G)$. *Then there exists a uniquely determined (regular) Borel measure $d\lambda$ on F such that*

$$d(f, P_\lambda f) = \left| \int f(x)\varphi_\lambda(x) \, dx \right|^2 d\lambda \qquad (f \in I_c(G/K)).$$

Moreover we have the following Plancherel formula

$$\int |f(x)|^2 \, dx = \int_F N_{\omega(\lambda)}(f) \, d\lambda \qquad (f \in C_c(G/K)).$$

Let W denote the group of all those linear mappings s of $\mathfrak{h}_{\mathfrak{p}_0}$ into itself for which we can find an element $u \in K$ such that $sH = Ad(u)H(H \in \mathfrak{h}_{\mathfrak{p}_0})$. Then W is a finite group. Let $I_c(\mathfrak{h}_{\mathfrak{p}_0})$ denote the set of all continuous functions f on $\mathfrak{h}_{\mathfrak{p}_0}$ vanishing outside a compact set such that $f(H) = f(sH)$ $(s \in W, H \in \mathfrak{h}_{\mathfrak{p}_0})$. For any $f \in I_c(G/K)$ put $f_0(H) = f(\exp H)$ $(H \in \mathfrak{h}_{\mathfrak{p}_0})$. Since $G = KA_+K$ it is easy to show that $f \to f_0$ is a 1-1 mapping of $I_c(G/K)$ onto $I_c(\mathfrak{h}_{\mathfrak{p}_0})$ and

$$\int |f(x)|^2 \, dx = \int_{\mathfrak{h}_{\mathfrak{p}_0}} |f_0(H)|^2 \Delta_+(H) \, dH$$

where $\Delta_+(H) = \prod_{\alpha \in P\lambda} |e^{\alpha(H)} - e^{-\alpha(H)}|$ and dH is the Euclidean measure on $\mathfrak{h}_{\mathfrak{p}_0}$. Let $L_2(\mathfrak{h}_{\mathfrak{p}_0})$ denote the Hilbert space of all functions on $\mathfrak{h}_{\mathfrak{p}_0}$ which are square-integrable with respect to the measure $\Delta_+(H) \, dH$. Let $I_2(\mathfrak{h}_{\mathfrak{p}_0})$ denote the closure of $I_c(\mathfrak{h}_{\mathfrak{p}_0})$ in $L_2(\mathfrak{h}_{\mathfrak{p}_0})$. Then the above mapping $f \to f_0$ can be extended uniquely to a unitary isomorphism of $I_2(G/K)$ with $I_2(\mathfrak{h}_{\mathfrak{p}_0})$. Under this isomorphism the operators $\tau(z_i)$ $1 \leq i \leq l_+$ go over into certain differential operators D_i whose coefficients are meromorphic functions on $\mathfrak{h}_{\mathfrak{p}_0}$. F and dP_λ of our theorem may now, respectively, be interpreted as the spectrum and the spectral measure of the family D_i $1 \leq i \leq l_+$ and the above Plancherel formula then becomes the "Parseval formula" corresponding to the problem of expanding a given function in $I_2(\mathfrak{h}_{\mathfrak{p}_0})$, in terms of eigen-functions of the family D_i $1 \leq i \leq l_+$ of differential operators.

[1] *Trans. Am. Math. Soc.*, **75**, 185–243 (1953). See especially §2 and §12 for notation. We shall refer to this paper as *RI*.

[2] See *RI*, p. 193.

[3] See §4 of *RI* for the definition of the Garding subspace.

[4] A C*-algebra is an algebra $\mathfrak{A}$ of bounded operators on a Hilbert space such that (i) the unit operator is in $\mathfrak{A}$ (ii) the adjoint of any operator in $\mathfrak{A}$ is also in $\mathfrak{A}$ (iii) $\mathfrak{A}$ is closed under uniform topology.

[5] I. Gelfand and D. Raikov, *Recueil Math.* (*Moscow*), **13**, 316 (1943).

[6] See Theorem 6, §11 of *RI*.

[7] See Proc. Natl. Acad. Sci., **37**, 170–173 (1951); Theorem 7.

Representations of Semisimple Lie Groups. V

HARISH-CHANDRA

DEPARTMENT OF MATHEMATICS, COLUMBIA UNIVERSITY
Communicated by O. Zariski, July 27, 1954

Let G be a connected semisimple Lie group and $\mathfrak{g}_0$ its Lie algebra over the field R of real numbers. Let $x \to Ad(x)$ denote the adjoint representation of G. A maximal connected abelian subgroup A of G is called a Cartan subgroup if $\mathfrak{g}_0$ is fully reducible under $Ad(A)$. The corresponding subalgebra of $\mathfrak{g}_0$ is called a Cartan subalgebra. Two Cartan subgroups A_1, A_2 are called conjugate (in G) if $A_2 = xA_1x^{-1}$ for some $x \in G$. Experience shows[1] that there is a close connection between the classes of conjugate Cartan subgroups and the various series of irreducible unitary representations of G which appear in the reduction of its regular representation. The object of this note is to give a general method of constructing irreducible unitary representations of G from each such class.

Let $X \to adX$ denote the adjoint representation of $\mathfrak{g}_0$. Put $B(X, Y) = \mathrm{sp}(adXadY)$. $(X, Y \in \mathfrak{g}_0)$. A subalgebra $\mathfrak{l}_0$ of $\mathfrak{g}_0$ is called compact if the quadratic form $B(X, X)$ is negative definite on $\mathfrak{l}_0$. Let $\mathfrak{k}_0$ be a maximal compact subalgebra of $\mathfrak{g}_0$. We denote by $\mathfrak{p}_0$ the subspace of $\mathfrak{g}_0$ orthogonal to $\mathfrak{k}_0$ under the bilinear form $B(X, Y)$. Then $\mathfrak{g}_0$ is the direct sum of $\mathfrak{k}_0$ and $\mathfrak{p}_0$. Let $\mathfrak{h}_0$ be a Cartan subalgebra of $\mathfrak{g}_0$. It is possible to choose $\mathfrak{k}_0$ in such a way that $\mathfrak{h}_0 = \mathfrak{h}_{\mathfrak{p}_0} + \mathfrak{h}_{\mathfrak{k}_0}$, where $\mathfrak{h}_{\mathfrak{p}_0} = \mathfrak{h}_0 \cap \mathfrak{h}_{\mathfrak{p}_0}$, $\mathfrak{h}_{\mathfrak{k}_0} = \mathfrak{h}_0 \cap \mathfrak{k}_0$. Let $H_1, \ldots, H_p$ be a base for $\mathfrak{h}_{\mathfrak{p}_0}$ over R. We order real linear functions on $\mathfrak{h}_{\mathfrak{p}_0}$ lexicographically with respect to this base. For any such function λ let $\mathfrak{g}_{0,\lambda}$ denote the set of all $X \in \mathfrak{g}_0$ such that $[H, X] = \lambda(H)X$ $(H \in \mathfrak{h}_{\mathfrak{p}_0})$. Then $\mathfrak{n}_0 = \Sigma_{\lambda > 0}\mathfrak{g}_{0,\lambda}$ is a nilpotent subalgebra of $\mathfrak{g}_0$. Let $\mathfrak{m}_0$ denote the set of all $X \in \mathfrak{g}_0$ such that $[X, H] = 0$ and $B(X, H) = 0$ for all $H \in \mathfrak{h}_{\mathfrak{p}_0}$. Then $\mathfrak{m}_0$ is a reductive[2] subalgebra of $\mathfrak{g}_0$, and $\mathfrak{h}_{\mathfrak{k}_0}$ is a maximal abelian subalgebra of $\mathfrak{m}_0$.

Let K, A_+, M_0, N be the analytic subgroups of G corresponding to $\mathfrak{k}_0, \mathfrak{h}_{\mathfrak{p}_0}, \mathfrak{m}_0$, and $\mathfrak{n}_0$. Let M_k be the centralizer and M'_k the normalizer of A_+ in K. Then $W = M'_k/M_k$ is a finite group. Put $M = M_k M_0$ and $S = MA_+N$. Then M and S are closed subgroups of G, and we have the following generalization of a lemma of F. Bruhat:[3]

Lemma 1. *There exist only a finite number of distinct double cosets SxS ($x \in G$).*

In case $\mathfrak{h}_{\mathfrak{p}_0}$ is a maximal abelian subspace of $\mathfrak{p}_0$, these double cosets are in a natural 1-1 correspondence with the elements of W.

Let λ_0 be an irreducible unitary representation of M on a Hilbert space U and ξ a (unitary) character of A_+. If we put $\lambda(man) = \lambda_0(m)\xi(a)$ ($m \in M, a \in A_+, n \in N$), λ defines a representation of S. Let π be the induced[4] representation of G corresponding to λ, and let $\mathfrak{H}$ denote the representation space of π. If Z is the center of G, $Z \subset M$, and therefore both λ and π map Z into scalars. Let Ω denote the set of all equivalence classes of finite-dimensional irreducible representations of K, and let $\mathfrak{H}_{\mathfrak{D}}(\mathfrak{D} \in \Omega)$ be the set of all elements of $\mathfrak{H}$ which transform under $\pi(K)$ according to $\mathfrak{D}$.

Theorem 1. *π is a unitary quasi-simple[5] representation of G, and $\dim \mathfrak{H}_{\mathfrak{D}} < \infty$ for every $\mathfrak{D} \in \Omega$.*

It follows from this theorem that π decomposes into a (discrete) direct sum of at most a countable number of irreducible unitary representations. However, in view of Lemma 1 and Bruhat's results,[6] it seems likely that π is irreducible at least "in general." Since $M/M_0 Z$ is finite, the above procedure will be applicable as soon as we have a method of constructing irreducible unitary representations of M_0. Such a method is given in the following note.[7]

Let $x \to x^*$ denote the natural mapping of G onto $G^* = G/Z$, and let dx^* denote the Haar measure on G^*. Let π be a unitary irreducible representation of G on a Hilbert space $\mathfrak{H}$. Then for any fixed ψ in $\mathfrak{H}$, $|(\psi, \pi(x)\psi)|$ depends only on x^*. We say that π is square-integrable if $\int_{G^*} |(\psi, \pi(x)\psi)|^2 \, dx^* < \infty$ for every $\psi \in \mathfrak{H}$. One can prove that this is indeed the case if there exist a finite number of elements ϕ_i, ψ_i ($1 \leqslant i \leqslant r$) in $\mathfrak{H}$ such that the function $f(x) = \sum_{i=1}^r (\phi_i, \pi(x)\psi_i)$ is not identically zero and $\int_{G^*} |f(x)|^2 \, dx^* < \infty$. The following theorem gives the analogue of the Schur orthogonality relations for square-integrable representations.[8]

Theorem 2. *Let π and π' be two irreducible unitary representations of G on the Hilbert space $\mathfrak{H}$ and $\mathfrak{H}'$, respectively. Suppose that they are both square-integrable and that they both define the same character of Z. Then*

$$\int_{G^*} \overline{(\phi, \pi(x)\psi)}(\phi', \pi'(x)\psi') \, dx^* = 0 \qquad (\phi, \psi \in \mathfrak{H}, \phi', \psi' \in \mathfrak{H}'),$$

unless π and π' are equivalent. On the other hand, if U is a unitary mapping of $\mathfrak{H}$ on $\mathfrak{H}'$ under which π and π' are equivalent, then

$$\int_{G^*} \overline{(\phi, \pi(x)\psi)}(\phi', \pi'(x)\psi') \, dx^* = c(\phi', U\phi)(U\psi, \psi'),$$

where c is a positive real number independent of ϕ, ψ, ϕ', ψ'.

We shall give a general method of constructing square-integrable representations in the following note.[7]

[1] See V. Bargmann, *Ann. Math.*, **48**, 568–640, 1947; Harish-Chandra, these PROCEEDINGS, **38**, 337–342, 1952; and I. M. Gelfand and M. I. Graev, *Doklady Akad. Nauk S.S.S.R.*, N.S., **92**, 221–224, 1953.

[2] See L. Koszul, *Bull. Soc. Math. France*, **78**, 22, 1950.

[3] F. Bruhat, *Compt. rend. Acad. sci.* (Paris), **238**, 550–553, 1954.

[4] G. W. Mackey, *Ann. Math.*, **55**, 106, 1952.

[5] See Harish-Chandra, these PROCEEDINGS, **37**, 170–173, 1951.

[6] F. Bruhat, *Compt. rend. Acad. sci.* (Paris), **238**, 38–40, 1954.

[7] These PROCEEDINGS, **40**, 1078–1080, 1954.

[8] See Bargmann, *op. cit.*, p. 634. R. Godement (*Compt. rend. Acad. sci.* (Paris), **225**, 657–659, 1947) has proved a similar result for any locally compact unimodular group. However, in case the center of G is infinite, his theorem does not seem to include ours.

Representations of Semisimple Lie Groups. VI

HARISH-CHANDRA

DEPARTMENT OF MATHEMATICS, COLUMBIA UNIVERSITY
Communicated by O. Zariski, July 27, 1954

We keep to the notation of the preceding note.[1] Since $\mathfrak{m}_0$ is reductive, there will be no essential loss of generality from the point of view of irreducible unitary representations of M_0 if we assume that $\mathfrak{m}_0$ is semisimple. Then $\mathfrak{m}_0 \cap \mathfrak{k}_0$ is a maximal compact subalgebra of $\mathfrak{m}_0$, and $\mathfrak{h}_{\mathfrak{k}_0}$ is a Cartan subalgebra of $\mathfrak{m}_0$. Moreover, $\mathfrak{m}_0 = \mathfrak{m}_0 \cap \mathfrak{k}_0 + \mathfrak{m}_0 \cap \mathfrak{p}_0$. Our problem of constructing irreducible unitary representations of M_0 is therefore the same as that for G, under the additional assumption that a maximal abelian subalgebra of $\mathfrak{k}_0$ is also maximal abelian in $\mathfrak{g}_0$. Hence we shall now assume that $\mathfrak{h}_0 \subset \mathfrak{k}_0$. Let C be the field of complex numbers and $\mathfrak{g}$ the complexification of $\mathfrak{g}_0$. We denote by $\mathfrak{h}$, $\mathfrak{k}$, and $\mathfrak{p}$ the subspaces of $\mathfrak{g}$ spanned over C by $\mathfrak{h}_0$, $\mathfrak{k}_0$, and $\mathfrak{p}_0$, respectively. For any root α of $\mathfrak{g}$ (with respect to $\mathfrak{h}$), choose an element $X_\alpha \neq 0$ in $\mathfrak{g}$ such that $[H, X_\alpha] = \alpha(H) X_\alpha$ ($H \in \mathfrak{h}$). Then X_α is unique apart from a factor in C, and it lies either in $\mathfrak{k}$ or in $\mathfrak{p}$. We call α a compact root if $X_\alpha \in \mathfrak{k}$. Either α and $-\alpha$ are both compact, or they are both noncompact.

Let $\mathfrak{B}$ be the universal enveloping algebra of $\mathfrak{g}$ and $\mathfrak{X}$ the subalgebra of $\mathfrak{B}$ generated by $(1, \mathfrak{k})$. Let π be a representation of $\mathfrak{B}$ on a vector space V. We shall say that π has an extreme vector if there exists an element $\psi \neq 0$ in V and a linear function Λ on $\mathfrak{h}$ such that the following conditions hold:

1. For every root α at least one of the two vectors $\pi(X_\alpha)\psi$, $\pi(X_{-\alpha})\psi$ is zero.
2. $\pi(H)\psi = \Lambda(H)\psi$ for all $H \in \mathfrak{h}$, and $\dim \pi(\mathfrak{X})\psi$ is finite.

ψ is then called an extreme vector belonging to the weight Λ. Now, if ψ is an extreme vector, it is possible to define a lexicographic order (with respect to a suitable base on $(-1)^{1/2}\mathfrak{h}_0$ over R) in the space of all linear functions on $\mathfrak{h}$ which take real values on $(-1)^{1/2}\mathfrak{h}_0$, such that[2] $\pi(X_\alpha)\psi = 0$ for every root $\alpha > 0$. From now on we shall keep this order fixed and say that an extreme vector ψ is positive if $\pi(X_\alpha)\psi = 0$ for all $\alpha > 0$. Put $H_\alpha = [X_\alpha, X_{-\alpha}]$ and $\epsilon_\alpha = 1$ or -1 according as α is compact or not. Then it is possible to select X_α in such a way that $\alpha(H_\alpha) = 2$ and

$(-1)^{1/2}(X_\alpha + \epsilon_\alpha X_{-\alpha})$, $(X_\alpha - \epsilon_\alpha X_{-\alpha})$ are both in $\mathfrak{g}_0$ for every root α. Let $\alpha_1, \ldots, \alpha_r$ be all the positive compact roots. We say that a root β is totally positive if every root of the form $\beta + m_1\alpha_1 + \cdots + m_r\alpha_r$ (where $m_1, \ldots, m_r$ are integers) is positive. Clearly, if β is totally positive, it is positive and noncompact. Let Q be the set of all totally positive roots and Q' the remaining set of positive roots.

Theorem 1. *Let π be an irreducible representation of $\mathfrak{B}$ on V with a positive extreme vector ψ belonging to the weight Λ. Then, if $\alpha \in Q'$, $\lambda_\alpha = \Lambda(H_\alpha)$ is a nonnegative integer and $\pi(X_{-\alpha}^{\lambda_\alpha + 1})\psi = 0$. Moreover, if Q is empty, $\dim V$ is finite.*

Conversely, suppose that we are given a linear function Λ on $\mathfrak{h}$ such that $\Lambda(H_\alpha)$ is a nonnegative integer for every $\alpha \in Q'$. Then it can be shown that there exists an irreducible representation π of $\mathfrak{B}$ on a vector space V with a positive extreme vector ψ belonging to the weight Λ, and π is unique up to equivalence. Futhermore, assuming that G is simply connected, we can find an irreducible quasisimple[3] representation σ of G on a Hilbert space $\mathfrak{H}$ and an element $\phi \in \mathfrak{H}$ which is well behaved[3] under σ such that the representation of $\mathfrak{B}$ defined on $\mathfrak{H}_0 = \sigma(\mathfrak{B})\phi$ is equivalent to π under the mapping $\sigma(b)\phi \leftrightarrow \pi(b)\psi$ $(b \in \mathfrak{B})$. In order that π be infinitesimally unitary,[4] it is necessary that $\Lambda(H_\beta)$ be real and $\leqslant 0$ for all positive noncompact roots β.

If $\mathfrak{g}_0$ is simple, only the two following cases are possible:

1. There are no totally positive roots.
2. Every positive noncompact root is totally positive.

The second case can occur only if the center of $\mathfrak{k}_0$ is $\neq \{0\}$. Moreover, if $\mathfrak{g}_0$ is not compact, the representation π of Theorem 1 can never be infinitesimally unitary unless we are in case 2. So now let us assume that $\mathfrak{g}_0$ is not compact and we are in case 2. Then we can select a fundamental system of positive roots $\alpha_0, \alpha_1, \ldots, \alpha_l$ such that α_0 is noncompact, while α_i $(1 \leqslant i \leqslant l)$ are compact. A linear function Λ on $\mathfrak{h}$ fulfills the condition of Theorem 1 if and only if $\Lambda(H_{\alpha_i})$ $(1 \leqslant i \leqslant l)$ are nonnegative integers. We denote by π_Λ the corresponding representation of $\mathfrak{B}$. We have seen that π_Λ cannot be infinitesimally unitary unless $\Lambda(H_{\alpha_0})$ is real and $\leqslant 0$. On the other hand, it can be shown that, if λ_i $(1 \leqslant i \leqslant l)$ are given nonnegative integers, there exists a real number $\lambda_0 \leqslant 0$ such that π_Λ is infinitesimally unitary whenever $\Lambda(H_{\alpha_i}) = \lambda_i$ $(1 \leqslant i \leqslant l)$ and $\Lambda(H_{\alpha_0}) \leqslant \lambda_0$.

Let us now suppose that G (which is again semisimple) has a faithful finite-dimensional representation.[5] Then there exists a complex analytic group G_c with the Lie algebra $\mathfrak{g}$, which contains G as the (real) analytic subgroup corresponding to $\mathfrak{g}_0$. Let $\mathfrak{n} = \sum_{\alpha > 0} CX_\alpha$, and let A_c and N_c be the complex analytic subgroups of G_c corresponding to $\mathfrak{h}$ and $\mathfrak{n}$, respectively. Then $G_c^0 = GA_cN_c$ is an open connected subset, and therefore a complex submanifold, of G_c. Let ξ be a complex analytic character of A_c. We consider the space $\mathfrak{H}_\xi$ of all holomorphic functions ϕ on G_c^0 such that $\phi(xan) = \phi(x)\xi(a)$ $(x \in G_c^0, a \in A_c, n \in N_c)$, and

$$\|\phi\|^2 = \int_G |\phi(x)|^2 \, dx < \infty.$$

(Here dx is the Haar measure of G.) It is not difficult to see[6] that $\mathfrak{H}_\xi$ is complete with respect to the above norm, and so it is a Hilbert space. If $\mathfrak{H}_\xi \neq \{0\}$, we can define a unitary representation π_ξ of G on it by the rule $(\pi_\xi(x)\phi)(y) = \phi(x^{-1}y)$ $(x \in G, y \in G_c^{\,0})$. It can be proved that this representation is irreducible and square-integrable.[7] Now assume again that G is simple and not compact. Then $\mathfrak{H}_\xi = \{0\}$ unless we are in case 2. So now let us suppose that we are in case 2. Given two complex analytic characters ξ and ξ' of A_c, we shall say that $\xi' \leqslant \xi$ if $\xi'(\exp H_{\alpha_i}) = \xi(\exp H_{\alpha_i})$ $(1 \leqslant i \leqslant l)$ and $|\xi'(\exp H_{\alpha_0})| \leqslant |\xi(\exp H_{\alpha_0})|$. Again $\mathfrak{H}_\xi = \{0\}$, unless $|\xi(\exp H_{\alpha_i})| \geqslant 1$ $(1 \leqslant i \leqslant l)$. On the other hand, if ξ_0 is any (complex analytic) character of A_c such that $|\xi_0(\exp H_{\alpha_i})| \geqslant 1$ $(1 \leqslant i \leqslant l)$, we can find a character $\xi_1 \leqslant \xi_0$ such that $\mathfrak{H}_\xi \neq \{0\}$ if $\xi \leqslant \xi_1$. If Λ is a linear function on $\mathfrak{h}$ such that $\xi(\exp H) = e^{\Lambda(H)}$ $(H \in \mathfrak{h})$ and $\mathfrak{H}_\xi \neq \{0\}$, the corresponding representation of $\mathfrak{B}$ defined under π_ξ is equivalent to π_Λ.

[1] These PROCEEDINGS, **40**, 1076–1077, 1954.

[2] This result has also been obtained independently by A. Borel.

[3] See these PROCEEDINGS, **37**, 170–173, 1951.

[4] See *Trans. Am. Math. Soc.*, **75**, 233, 1953.

[5] It is possible to avoid this assumption, which is made here only for simplicity.

[6] See S. Bochner and W. T. Martin, *Several Complex Variables* (Princeton, N.J.: Princeton University Press, 1948), p. 117.

[7] Our procedure for constructing π_ξ is a generalization of the method used by Bargmann (*Ann. Math.*, **48**, 620, 1947) and Gelfand and Graev (*Izvest. Akad. Nauk, S.S.S.R.*, **17**, 189–248, 1953) in certain special cases.